Properties, Definitions, and Formulas

Properties of Real Numbers

$a + b = b + a$, $ab = ba$	**Commutative Properties**
$(a + b) + c = a + (b + c)$, $(ab)c = a(bc)$	**Associative Properties**
$a + 0 = a$, $1 \cdot a = a$	**Identity Properties**
$a + (-a) = 0$, $a \cdot \frac{1}{a} = 1$	**Inverse Properties**
$a \cdot 0 = 0$	**Multiplication Property of 0**
$a(b + c) = ab + ac$	**Distributive Property**
$-(a - b) = b - a$	**Opposite of Difference**
$a(-1) = -1a = -a$	**Multiplication Property of −1**
$-\frac{a}{b} = \frac{-a}{b} = \frac{a}{-b}$	**Signs in Quotients**

Order of Operations

1. Operations within grouping symbols
2. Exponents
3. Multiplication and division
4. Addition and subtraction

Absolute Value

$$|a| = \begin{cases} a, & \text{if } a \geq 0 \\ -a, & \text{if } a < 0 \end{cases}$$

Distance and Midpoint Formulas

The distance d between $P(x_1, y_1)$ and $Q(x_2, y_2)$ is $d = \sqrt{(x_2 - x_1)^2 + (y_2 - y_1)^2}$.

The coordinates (x_m, y_m) of the midpoint of $\overline{PQ}$ are $x_m = \frac{x_1 + x_2}{2}$ and $y_m = \frac{y_1 + y_2}{2}$.

Slope Formula

$$m = \frac{y_2 - y_1}{x_2 - x_1}$$

Equation of Line

Slope-Intercept Form: $y = mx + b$

Point-Slope Form: $y - y_1 = m(x - x_1)$

Exponents

$b^m \cdot b^n = b^{m+n}$ $\qquad (b^m)^n = b^{mn}$

$(ab)^n = a^n b^n$ $\qquad b^0 = 1$

$b^{-n} = \frac{1}{b^n}$ $\qquad \frac{b^m}{b^n} = b^{m-n}$

$\left(\frac{a}{b}\right)^n = \frac{a^n}{b^n}$ $\qquad \left(\frac{a}{b}\right)^{-n} = \frac{b^n}{a^n}$

Radicals

$\sqrt[n]{a}\sqrt[n]{b} = \sqrt[n]{ab}$ $\qquad \frac{\sqrt[n]{a}}{\sqrt[n]{b}} = \sqrt[n]{\frac{a}{b}}$

$b^{m/n} = \sqrt[n]{b^m}$

Factoring Patterns

$A^2 - B^2 = (A + B)(A - B)$

$A^2 + 2AB + B^2 = (A + B)^2$

$A^2 - 2AB + B^2 = (A - B)^2$

$A^3 + B^3 = (A + B)(A^2 - AB + B^2)$

$A^3 - B^3 = (A - B)(A^2 + AB + B^2)$

Zero Factor Property

If $ab = 0$ then $a = 0$ or $b = 0$.

Even Root Property

If $x^n = c$ and n is even, then $x = \pm\sqrt[n]{c}$.

Odd Root Property

If $x^n = c$ and n is odd, then $x = \sqrt[n]{c}$.

Quadratic Formula

$$x = \frac{-b \pm \sqrt{b^2 - 4ac}}{2a}$$

Fractions and Rational Expressions

$\frac{a}{c} + \frac{b}{c} = \frac{a + b}{c}$, $\frac{a}{c} - \frac{b}{c} = \frac{a - b}{c}$

$\frac{ac}{bc} = \frac{a}{b}$

$\frac{a}{b} \cdot \frac{c}{d} = \frac{ac}{bd}$, $\frac{a}{b} \div \frac{c}{d} = \frac{a}{b} \cdot \frac{d}{c}$

$\frac{a}{b} = \frac{c}{d}$ if and only if $ad = bc$

Properties of Logarithms

$\log_b b = 1$ $\qquad \log_b 1 = 0$

$\log_b b^x = x$ $\qquad b^{\log_b x} = x$

$\log_b MN = \log_b M + \log_b N$

$\log_b \frac{M}{N} = \log_b M - \log_b N$

$\log_b M^c = c(\log_b M)$

$\log_b x = \frac{\log_a x}{\log_a b}$

Instructor's Annotated Edition

Elementary and Intermediate Algebra

Instructor's Annotated Edition

Elementary and Intermediate Algebra

Second Edition

Elaine Hubbard

Kennesaw State University

Ronald D. Robinson

Houghton Mifflin Company Boston • New York

Editor-in-Chief: Jack Shira
Sponsoring Editor: Paul Murphy
Editorial Associate: Marika Hoe
Senior Project Editor: Nancy Blodget
Editorial Assistant: Kristin Penta
Senior Production/Design Coordinator: Jodi O'Rourke
Senior Manufacturing Coordinator: Marie Barnes
Marketing Manager: Ben Rivera

Cover photo by Ed Bock/The Stock Market

Photo credits: p. 1: © 2001 Corbis. All rights reserved; p. 12: © The Purcell Team/Corbis; p. 29: AP Photo/Nell Redmond; p. 57: AP Photo/CP, Tom Hanson; p. 80: Official U.S. Navy photo by PH2 Dante DeAngelis; p. 100: © Dick Hemingway; p. 107: © Steve Chenn/Corbis; p. 149: © Raymond Gehman/ Corbis; p. 151: © Gary W. Carter/Corbis; p. 187: Digital Imagery © 2001 PhotoDisc, Inc.; p. 226: AP Photo/David Longstreath; p. 230: © Joseph Sohm/ChromoSohm, Inc./Corbis; p. 249: © 2001 Corbis. All rights reserved; p. 307: © Ed Young/Corbis; p. 309: © Raymond Gehman/Corbis; p. 325: NASA; p. 344: © Science Pictures Limited/Corbis; p. 365: Digital Imagery © 2001 PhotoDisc, Inc.; p. 388: © Philip Gould/Corbis; p. 411: AP Photo/Donald Edgar; p. 425: © Jennie Woodcock, Reflections Photo Library/Corbis; p. 459: © Jim Sugar Photography/Corbis; p. 469: © Gail Mooney/Corbis; p. 493: © Joseph Sohm/ChromoSohm, Inc./Corbis; p. 510: © Judy Griesedieck/Corbis; p. 520: © The Purcell Team/Corbis; p. 535: © Duomo/Corbis; p. 555: AP Photo/Gazette Telegraph, Mark Reis; p. 566: AP Photo/Enric F. Marti; p. 599: © 2001 Corbis. All rights reserved; p. 636: AP Photo/Natasha Lane, File; p. 651: AP Photo/Salt Lake City Tribune.

Printed in the U.S.A.

Library of Congress Control Number: 2001093601

ISBN
Student Text: 0-618-16235-6
Instructor's Annotated Edition: 0-618-16236-4

123456789-DOW- 05 04 03 02 01

Contents

Preface

The General Approach

An increasing number of colleges and universities are combining the traditional elementary and intermediate algebra courses into a single course that is offered over one or two terms. *Elementary and Intermediate Algebra: A Discovery Approach,* Second Edition, was written for such courses. The organization of the content allows the student to move through the conventional topics of elementary and intermediate algebra seamlessly and without repetition.

As in our previous edition, we have placed considerable emphasis on multiple approaches as an instructional strategy. Throughout the text, we begin the development of mathematical concepts numerically and graphically. This gives the student a logical basis for definitions and permits the student to make conjectures about basic rules and principles. From there, we use symbolic and verbal approaches to formalize and articulate concepts.

Approaches to Concept Development

Whenever appropriate, we have introduced concepts under the heading "Exploring the Concept." In this feature, the student is given certain information—numerical, symbolic, or verbal—and is guided through conceptual processes to a desired outcome. Often the student is asked to form a conjecture based on the experimental results. The evidence obtained in the exploration is summarized in formal rules that are then illustrated with numerous examples. Used as a classroom activity, Exploring the Concept can promote lively debate and an active student involvement in the development and understanding of basic concepts.

Numerical, symbolic, verbal (and possibly graphical) approaches are most effective when they are used in mutually reinforcing ways with all other techniques. Whenever appropriate, we begin with an intuitive (numerical) foundation and progress to general (symbolic) conclusions. Through such verbal features as Speaking the Language, Writing Exercises, Data Analysis, and Chapter Projects, we synthesize all the approaches into a meaningful whole. In our own classroom experiences, we have found that collaboration is a natural outcome.

Modeling with Real Data and Real-Life Applications

Nearly every section of the book contains a group of exercises that involve modeling with real data. Through the organization, analysis, and interpretation of sourced data and information, students are able to make connections to a wide variety of practical experiences.

A large number of real-life applications appear throughout the text. Some sections are devoted exclusively to such exercises. These problems are written in a student's experiential context to encourage the student to think about applicability and connections to other disciplines rather than only about how to solve the problem.

What's New in the Second Edition?

Content We have paid special attention to the advice of current users and reviewers on the matter of topics to be added or omitted. Accordingly, the following changes have been made in this edition.

1. Some of the introductory material has been moved to point-of-use. For example, concepts and notation originally introduced in Chapter 1 but not needed until later chapters are now introduced where and when they are needed. As a result, Chapter 1, especially Section 1.1, has been streamlined somewhat.
2. The section entitled "Higher Degree and Rational Inequalities" (formerly Section 10.8) has been omitted.
3. An appendix has been added to present the basics of conic sections (circle, parabola, ellipse, and hyperbola).
4. Although the use of a graphing calculator remains optional, the first-edition feature "Links to Technology" has been eliminated. This is intended to bring greater clarity to the difference between this text and our matching text, *Elementary and Intermediate Algebra: Discovery and Visualization,* Third Edition, which is a graphing calculator-based text.

Examples and Exercises Examples and exercises have been carefully constructed in order to provide a pool of varied activities from which instructors can draw. Instructors can create a syllabus that reflects the goals of their curricula.

1. Exercises labeled as "Modeling with Real Data" have been updated or replaced as needed in order to provide the most current data available.
2. Exercises formerly labeled as "Group Projects" are now entitled "Data Analysis" to reflect their actual intent. A calculator icon is used with the "Data Analysis" groups of exercises to indicate that the use of a scientific or graphing calculator is appropriate for such exercises. These exercises have also been updated or replaced as needed.
3. The "thread" of a topic, which begins with the Chapter Opener, is revisited in the Data Analysis group in one of the sections, and is expanded in the Chapter Project, has been retained. However, the entire thread, including the Chapter Opener, has been updated or replaced as needed.

Art Current users will recognize that much of the art has been revised.

1. Graphs that were formerly labeled with "bubbles" are now in color with corresponding color-coded labels.
2. Graphs of inequalities, including systems of inequalities, have been improved to help the student see the solution regions.
3. Additional new art has been included to provide more visual assistance to the student.

The goals of this book are to provide an effective instructional framework for the classroom and to engage the students in a better understanding of the nature of mathematics in their personal and career lives. We earnestly hope that our work will promote achievement and success for all.

Supplements for Students

HM³ CD-ROM Tutorial The text is supported by this state-of-the-art tutorial software featuring algorithmically generated exercises and quizzes, as well as animated solution steps, selected video clips, and lessons presented in a colorful, lively manner.

Student Solutions Manual Contains worked-out solutions for odd-numbered problems from the student text.

Videotapes These comprehensive lecture videos presented by Dana Mosely explain and reinforce the algebraic concepts for each chapter section in the textbook.

Math Study Skills Workbook by Paul D. Nolting This workbook is designed to help reinforce skills and minimize frustration in any math class, lab, or study skills course. It offers a wealth of proven study tips and sound advice on note taking, time management, and reducing math anxiety. In addition, numerous self-assessment opportunities enable you to track your own progress.

SMARTHINKING™ On-line Tutoring Houghton Mifflin has partnered with SMARTHINKING to provide an easy-to-use and effective on-line tutorial service. **Whiteboard Simulations** and **Practice Area** promote real-time visual interaction. SMARTHINKING offers three levels of service:

- **Prescheduled Text-specific Tutoring** provides real-time, one-on-one instruction with a specially qualified e-structor.
- **Questions Any Time** allows students to submit questions to the tutor outside the scheduled hours and receive a reply within 24 hours.
- **Independent Study Resources** connect students with around-the-clock access to additional educational services, ranging from interactive web sites and on-line textbooks to diagnostic tests and Frequently Asked Questions posed to SMARTHINKING e-structors.

Supplements for Instructors

Instructor's Annotated Edition This special version of the text contains teaching tips in the text margins and provides answers to the odd- and even-numbered exercises for the end-of-section activities, review problems, Chapter Tests, Cumulative Tests, and Selected Features including Think About It, Speaking the Language, Looking Back, and Chapter Projects.

Instructor's Solutions Manual Contains worked-out solutions to every problem in the text.

HM^3 CD-ROM Tutorial (Instructor Network Version) In addition to the student version features, this software offers instructors a classroom management system to track student performance.

Printed Test Bank Provides a variety of test items and sample tests for each chapter of the text.

HM Testing This new computerized test generator database contains more than 2500 individual test items, many of which are algorithmic. HM Testing is designed to produce an unlimited number of tests for each chapter of the text, including cumulative tests and final exams. On-line testing and gradebook features are also included in this software. HM Testing is packaged as a hybrid CD-ROM and works on both Microsoft Windows and Macintosh platforms.

Web Site and PowerPoint Slides A book-specific web site features a number of resources, including PowerPoint slides for easy classroom presentations. Go to http://college.hmco.com and choose Go to Your Discipline.

Acknowledgments

We would like to thank the following colleagues who reviewed the second edition manuscript and made many helpful suggestions:

Debra Bryant—*Tennessee Technological University*
Scott Garten—*Northwest Missouri State University*
Stanley Max—*University of Southern Maine*
Bill Naegele—*South Suburban College*
Frank Pecchioni—*Jefferson Community College*
Kimberly Rahm—*California Polytechnic University, Pomona*
Gina Reed—*Gainesville College*
Ralph Selensky—*Eastern Arizona College*
Ellyn Webb—*Midlands Technical College*

We are grateful to Helen Medley, who took great care to ensure accuracy of the text and answers.

Finally, we thank the staff at Houghton Mifflin for their assistance and contributions to the project. Particular thanks go to Jack Shira, Editor-in-Chief; Paul Murphy, Sponsoring Editor; Marika Hoe, Editorial Associate; Nancy Blodget, Senior Project Editor; Kristin Penta, Editorial Assistant; Jodi O'Rourke, Production/Design Coordinator; Marie Barnes, Senior Manufacturing Coordinator; and Ben Rivera, Marketing Manager.

ELAINE HUBBARD

RONALD D. ROBINSON

Opening Features

Chapter Opener

Each chapter begins with a short introduction to a real-data application. The opening application topic is highlighted again in the Data Analysis feature, and then expanded in the Chapter Project found at the end of the chapter. The Chapter Opener also includes a list of the section titles and a helpful summary of the topics that will be covered in the chapter.

Chapter 2

Expressions and Functions

As more and more states have enacted helmet laws for bicycle riders, the use of helmets has increased dramatically. The bar graph shows the percent of riders who wore helmets for selected years in the period 1992–1998.

*We can write an **algebraic expression** to model the trend in helmet usage during this time. By applying this expression, we can estimate the usage in 1997, for example, or predict the usage in 2002. (For more on this topic, see Exercises 91–94 at the end of Section 2.1, and see the Chapter Project.)*

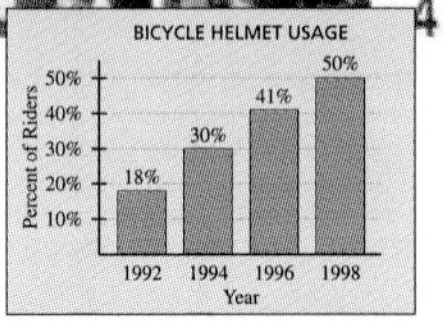

(Source: U.S. Bicycle Federation.)

In this chapter, we consider expressions and how to evaluate and simplify them. Equations and formulas are presented as methods for modeling conditions and relationships. We then introduce the coordinate plane as a model for representing ordered pairs of numbers. Finally, we define a relation and a function, we learn how to recognize and graph a function, and we use a special notation for evaluating functions.

57

58 CHAPTER 2 Expressions and Functions

2.1 Algebraic Expressions and Formulas

Evaluating Algebraic Expressions • Equations and Formulas • Modeling with Real Data

Evaluating Algebraic Expressions

In the previous chapter, we described a **numerical expression** as any combination of numbers and operations. We also described a **variable** as a symbol (usually a letter) used to represent an unknown number.

We can use variables to represent numbers in expressions. An **algebraic expression** is any combination of numbers, variables, grouping symbols, and operation symbols. The following are examples of algebraic expressions.

$$x^2 + 3x - 4 \qquad \frac{a^2+4}{6} \qquad 5(x + 6y) - 4 \qquad r - \sqrt{2t}$$

When we replace the variables in an algebraic expression with specific values, the expression becomes a numerical expression. Performing all indicated operations to determine the value of the expression is called **evaluating the expression.**

EXAMPLE 1 Evaluating Algebraic Expressions

Evaluate each expression for the given value(s) of the variable(s).

(a) $x^2 + 3x - 4,\quad x = -2$

(b) $5x + 2(1 - x),\quad x = -3$

(c) $3x - |y - 2x|,\quad x = 3, y = 1$

(d) $\dfrac{a + 2b}{2a^2b},\quad a = -1, b = 3$

Remind students to use the correct Order of Operations.

Solution

(a) Replace each occurrence of x with -2. Then evaluate the resulting numerical expression.

$$x^2 + 3x - 4 = (-2)^2 + 3(-2) - 4 = 4 - 6 - 4 = -6$$

(b) Replace each occurrence of x with -3.

$$\begin{aligned} 5x + 2(1 - x) &= 5(-3) + 2[1 - (-3)] \\ &= -15 + 2[4] \\ &= -7 \end{aligned}$$

(c) Replace x with 3 and y with 1.

$$3x - |y - 2x| = 3(3) - |1 - 2(3)| = 9 - |-5| = 9 - 5 = 4$$

(d) $\dfrac{a + 2b}{2a^2b} = \dfrac{-1 + 2 \cdot 3}{2(-1)^2 \cdot 3} = \dfrac{5}{6}$ Replace a with -1 and b with 3.

Equations and Formulas

An **equation** is a statement that two expressions have the same value. If both expressions are numerical, then the truth of the equation can be determined by simply evaluating the expressions.

Section Opener

Each section begins with a list of subsection titles that briefly outlines the material found within that section.

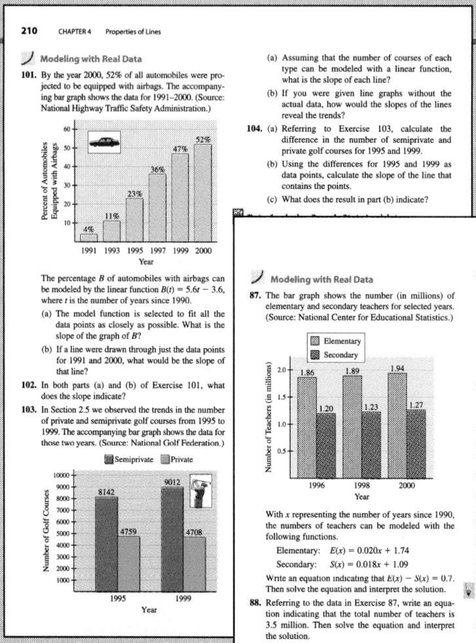

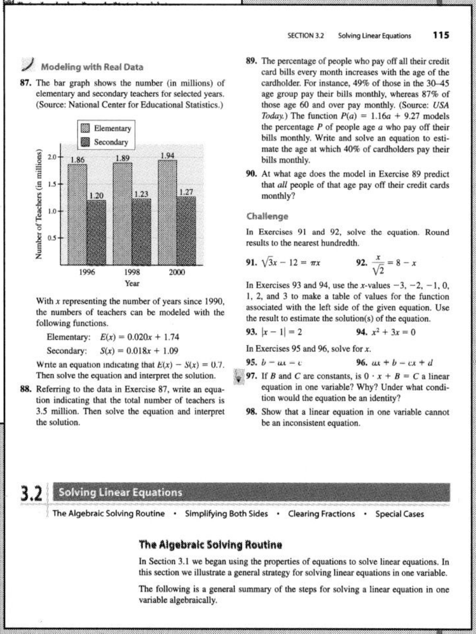

Modeling with Real Data

In accordance with the current standards of professional organizations such as AMATYC and NCTM, there is an abundance of examples and exercises that teach students how to organize and interpret data, how to model it with mathematical functions, and how to use such models to extrapolate and predict outcomes. Sourced data from a variety of subjects are used in numerous examples, groups of modeling exercises, Data Analysis exercises, and Chapter Projects.

Real Data and Real Life

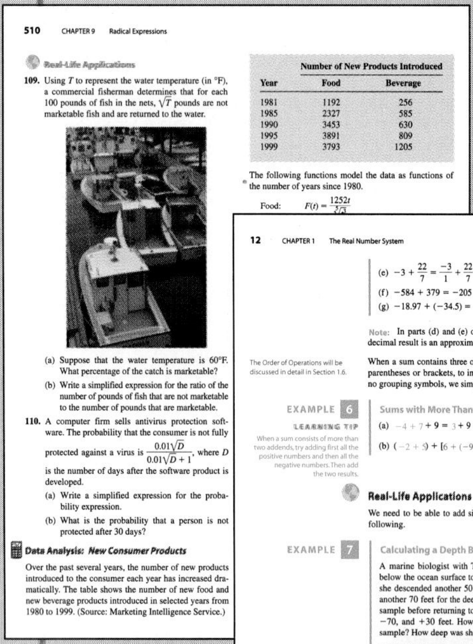

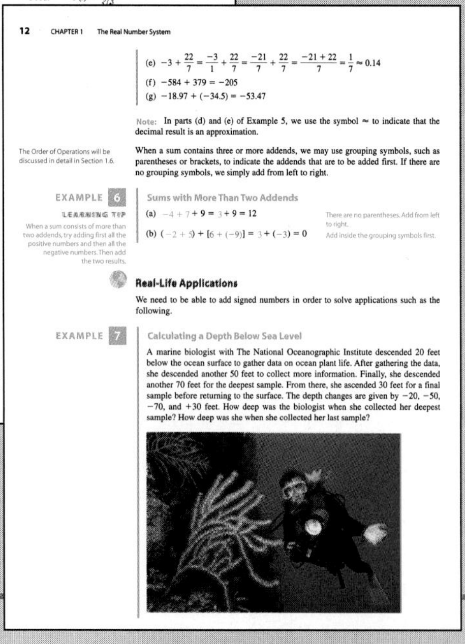

Real-Life Applications

The majority of application problems are written in the context of real-life knowledge and experience. Some sections are devoted exclusively to examples and exercises involving real-life applications, and most other sections contain dedicated blocks of such problems. According to national teaching standards, students gain a better understanding of the practical nature of mathematics when the concepts and topics are connected to real-world experiences and situations.

Systems of Equations

We write equations to describe the conditions and relationships given in applied problems. Up to this point, applications have led to a single equation in one or two variables. However, applications often lead to more than one equation.

Exploring the Concept

Systems of Equations

Suppose that we want to describe two numbers whose sum is 2. If x and y represent the two numbers, then an equation is $x + y = 2$. To sketch the graph, we solve for y and write $y = -x + 2$ and make a table of solutions. Note that there are infinitely many pairs of numbers whose sum is 2. The graph of this equation is shown in Figure 5.1.

Figure 5.1

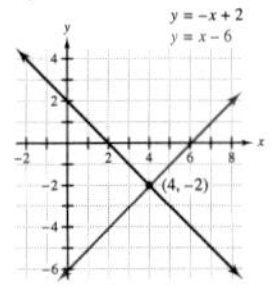

x	−1	0	1	2	3	4	5	6
$-x + 2$	3	2	1	0	−1	−2	−3	−4

Next, suppose that we want to describe two numbers whose difference is 6. If x is the larger number, then an equation is $x - y = 6$. We write $y = x - 6$ and complete a table of solutions. Again, there are infinitely many pairs of numbers whose difference is 6. The graph of this equation is also shown in Figure 5.1.

x	−1	0	1	2	3	4	5	6
$x - 6$	−7	−6	−5	−4	−3	−2	−1	0

Observe that (4, −2) is the *only* pair that satisfies both conditions.

Now suppose that we want to find a pair of numbers whose sum is 2 *and* whose difference is 6. Because the point of intersection in Figure 5.1 belongs to both graphs, its coordinates (4, −2) satisfy both equations. The two tables confirm that (4, −2) satisfies both of the given conditions. Thus the required numbers are 4 and −2.

Exploring the Concept

This effective feature guides students from concrete experiences to generalizations and formal rules. Students are able to anticipate outcomes and arrive at a deeper understanding of important concepts.

Features for Discovery, Visualization, and Support

Examples

All sections contain numerous, titled Examples, many with multiple parts graded by difficulty. These Examples illustrate concepts, procedures, and techniques, and they reinforce the reasoning and critical thinking needed for problem solving. Detailed solutions include helpful comments that justify the steps taken and explain their purpose.

20 CHAPTER 1 The Real Number System

EXAMPLE 5 **Distance Between Points Above and Below Sea Level**

The altitude gauge of a C-130 indicates that the plane is 760 feet above sea level. The crew of a training mission is to drop equipment to a location in Death Valley that is 42 feet below sea level. When the plane is directly above the drop site, what is the distance between the plane and the drop site?

Solution

Consider a vertical number line whose origin is at sea level. The plane is at a point whose coordinate is +760, and the drop site is at a point whose coordinate is −42. (See Fig. 1.11.)

Figure 1.11

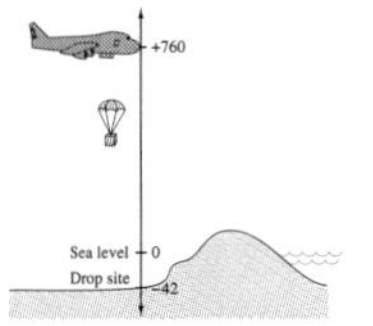

We calculate the distance d between the points as follows.

$$d = |760 - (-42)| = |760 + 42| = |802| = 802$$

The distance from the plane to the drop site is 802 feet.

Quick Reference 1.3

Definition

- In the expression $a - b$, a is the **minuend** and b is the **subtrahend.** Both the expression and the value of the expression are called a **difference.**
- For any real numbers a and b, $a - b$ is defined as $a + (-b)$.

Performing Subtraction

- The definition of subtraction gives us a way to perform the operation by applying the rules for addition. We convert a difference to a sum as follows.
  1. Change the minus sign to a plus sign.
  2. Change the subtrahend to its opposite.
- If we reverse the order of the numbers in a difference, the value of the result is the opposite of the original value.
- Property of the Opposite of a Difference
  For any real numbers a and b, $-(a - b) = b - a$.

Figure 5.15 shows the two half-planes in the same coordinate system. Their intersection is shown in Figure 5.16.

Figure 5.15

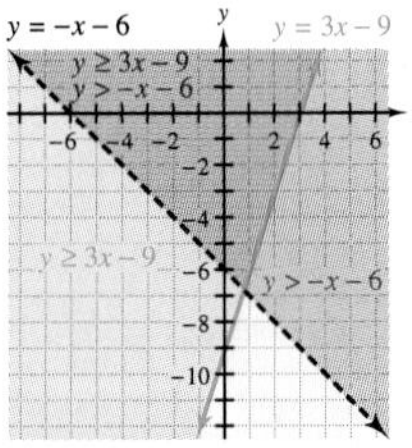

Figure 5.16

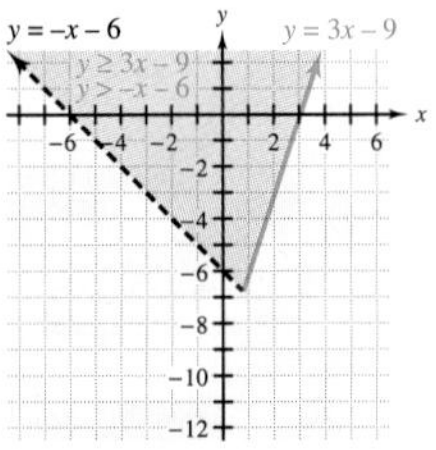

Graphs

Traditional graphs are used throughout the exposition and exercises to assist students in visualizing concepts.

Graphing Calculators

Use of the graphing calculator is completely optional with this text. A graphing calculator icon appears next to the Data Analysis feature and indicates that although the graphing calculator is not required for these exercises, one may be used if so desired.

Notes

Special remarks and cautionary notes that offer additional insight appear throughout the text.

Rate of Change

This topic is optional and is not needed for any future material in this text.

An important concept in mathematics involves the rate at which a function changes. Specifically, for the function $y = mx + b$, the **rate of change** of y with respect to x is the amount by which y changes for a unit change in x.

Note: A *unit change in x* means a 1-unit increase or derease in the value of x. The corresponding change in y can be positive or negative.

Exploring the Concept

The technique in this illustration works well in class in relating rate of change and slope. You may want to repeat it with $y = -\frac{3}{2}x + 10$ or some other line with a negative slope.

Rate of Change and Slope

To see the rate of change for the function $y = 2x - 4$, consider the following table of values.

x	-1	0	1	2	3	4	5
y	-6	-4	-2	0	2	4	6

In order to transform a matrix in this way, we use **row operations.** Row operations correspond to the algebraic operations we perform on the equations of a system. Just as the properties of equations allow us to produce equivalent systems, row operations can be used to produce **row-equivalent matrices.**

The following is a summary of permissible row operations. Included in this summary are examples of the notation we use to indicate each operation.

The Elementary Row Operations on Matrices

1. Two rows may be interchanged: $R_2 \leftrightarrow R_1$.
2. A row may be multiplied or divided by a nonzero number n and the original row may be replaced by the result: $n \cdot R_1 \to R_1$.
3. A row may be replaced by the sum of itself and another row: $R_1 + R_2 \to R_1$.
4. A row may be replaced by the sum of itself and a multiple of another row: $R_1 + n \cdot R_2 \to R_1$.

Although row operation 4 is a combination of row operations 2 and 3, it is convenient to treat it as a separate property.

Definitions, Properties, and Procedures

Important definitions, properties, and procedures are shaded and titled for easy reference.

More Features for Discovery, Visualization, and Support

Think About It

Located in the margin of each section, Think About It is a question or series of questions that requires critical thinking and reasoning in order to broaden and extend concepts. These questions are designed to spark students' imagination and interest. Answers appear in the Instructor's Annotated Edition.

Learning Tip

Every section has at least one Learning Tip in the margin that offers students helpful strategies and alternative ways of thinking about concepts.

32 CHAPTER 1 The Real Number System

EXAMPLE 2 Performing Division

(a) $18 \div (-12 \div 4) = 18 \div (-3) = -6$ — Perform operations within grouping symbols first.

(b) $-4 \cdot 6 \div (-3) = -24 \div (-3) = 8$ — Operate from left to right.

(c) $-\frac{1}{3} \div 5 = -\frac{1}{3} \cdot \frac{1}{5} = -\frac{1}{15}$ — The reciprocal of a positive number is positive.

(d) $\frac{5}{7} \div \left(-\frac{3}{4}\right) = \frac{5}{7} \cdot \left(-\frac{4}{3}\right) = -\frac{20}{21}$ — The reciprocal of a negative number is negative.

Think About It

Explain why you obtain the same result when you divide a number by 2 and when you multiply the number by $\frac{1}{2}$, but you do not obtain the same result when you divide the number by $\frac{1}{2}$. What is an equivalent way to write a number divided by $\frac{1}{2}$?

LEARNING TIP

Knowing that $-(-a) = a$ and $\frac{-a}{-b} = \frac{a}{b}$ allows you to remove pairs of negative (or opposite) signs. Eliminating signs, when possible, simplifies the appearance of the expression and reduces sign errors.

Because the sign rules for division are the same as the sign rules for multiplication, the following generalization can be stated for signs in quotients.

Property of Signs in Quotients

For real numbers a and b ($b \neq 0$), each of the following statements is true.

$$-\frac{a}{b} = \frac{-a}{b} = \frac{a}{-b} \qquad \frac{-a}{-b} = \frac{a}{b} \qquad -\frac{-a}{-b} = \frac{-a}{b} = \frac{a}{-b} = -\frac{a}{b}$$

In words, if a fraction has a single opposite sign, the sign can be placed in the numerator, in the denominator, or in front of the fraction. If a fraction has two opposite signs, they can both be removed. If a fraction has three opposite signs, any two of them can be removed.

End-of-Section Features

A typical section ends with two features—Quick Reference and Speaking the Language.

256 CHAPTER 5 Systems of Linear Equations

Quick Reference 5.1

Systems of Equations

- Two or more equations considered simultaneously form a **system of equations.**
- A **solution** of a system of two linear equations in two variables is an ordered pair (a, b) that satisfies both equations of the system.

Estimating Solutions by Graphing

- To estimate the solution of a system of two linear equations in two variables, use the following procedure.
  1. Sketch the graphs of both equations in the same coordinate system.
  2. Locate the point of intersection of the two graphs.
  3. This point represents the estimated solution of the system of equations.

Substitution Method

- The substitution method is an algebraic method for solving a system of equations.
- In the following summary of the method, we begin by solving one equation for y. However, either x or y may be chosen.
  1. Solve one equation for y to obtain $y =$ expression.
  2. Replace y in the other equation with the expression. This results in an equation in one variable x.
  3. Solve that equation.
  4. Substitute the solution for x in either original equation and solve for y.
  5. Check the solution.

Special Cases

- Two special cases may arise when we are solving a system of two linear equations in two variables.
  1. The system of equations is **inconsistent.** The graphs of the equations are parallel, and the system has no solution.
  2. The equations of the system are **dependent.** The graphs of the equations coincide, and the solution set of the system contains the infinitely many solutions of either equation.

Speaking the Language 5.1

1. If the graphs of a system of equations intersect at a single point, that point represents a(n) ______ of the system.

2. If a system of equations has no solution, then the system is ______.

3. If the graphs of a system of equations coincide, then the equations of the system are called ______.

4. The ______ method is an algebraic method for solving a system of equations.

Quick Reference

Quick Reference appears at the end of all sections except those dealing exclusively with applications. This handy reference and review tool provides detailed summaries of the important rules, definitions, properties, and procedures grouped by subsection.

Speaking the Language

Speaking the Language appears at the end of all sections except those dealing exclusively with applications. This feature helps students to think and communicate in the language of mathematics by reinforcing vocabulary and contextual meanings. Answers are listed in the Instructor's Annotated Edition.

Exercises

A typical exercise set in each section includes exercises from each of the following groups: Concepts and Skills (including Writing and Concept Extension CE), Real-Life Applications and Modeling with Real Data, Data Analysis, and Challenge.

228 CHAPTER 4 Properties of Lines

Exercises 4.4

Concepts and Skills

1. What information do we need to write an equation of a line with the slope–intercept form?
2. Why can the slope–intercept form be used to write an equation of a horizontal line but not an equation of a vertical line?

In Exercises 3–6, write the equation of the line.

3. $m = -4, b = -3$
4. $m = 3, b = 4$
5. $m = \frac{1}{2}, b = 3$
6. $m = -\frac{2}{3}, b = -2$

In Exercises 7–12, write an equation of the line that has the given slope and contains the given y-intercept.

7. $m = -5, (0, 1)$
8. $m = 6, (0, -6)$
9. $m = -\frac{3}{4}, (0, -5)$
10. $m = \frac{5}{8}, (0, 8)$
11. $m = 7, \left(0, -\frac{2}{3}\right)$
12. $m = -4, \left(0, \frac{1}{2}\right)$

In Exercises 13–18, write an equation for the line whose graph is shown in the figure. Remind students to look for a convenient slope triangle.

13.

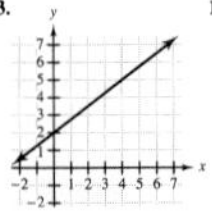

14.

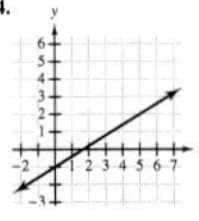

15.

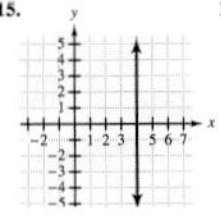

16.

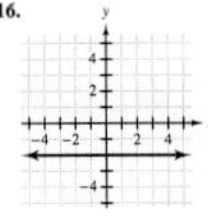

17.

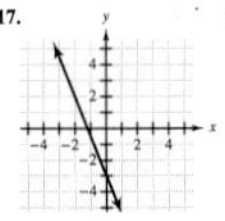

18.

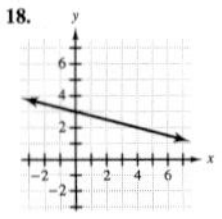

19. Why can the point–slope form be used to write an equation of a horizontal line but not an equation of a vertical line?
20. What information do we need to write an equation of a line with the point–slope form?

In Exercises 21–30, write an equation of the line that has the given slope and contains the given point. Express the equation in the form $y = mx + b$ if possible.

21. $m = 2, (1, 3)$
22. $m = -3, (2, 5)$
23. $m = \frac{2}{5}, (10, -4)$
24. $m = -\frac{1}{3}, (-6, 3)$
25. $m = -\frac{2}{3}, (-2, -5)$
26. $m = \frac{4}{5}, (-1, -2)$
27. $m = 0, (2, -5)$
28. $m = 0, (-3, -4)$
29. m is undefined, $(-3, 4)$
30. m is undefined, $(-4, 5)$

In Exercises 31–36, write an equation of the line that has the given slope and contains the given point. Express the equation in the form $Ax + By = C$.

31. $m = 1, (7, 4)$
32. $m = 2, \left(2, \frac{1}{2}\right)$
33. $m = \frac{5}{3}, (2, 0)$
34. $m = -2, \left(-1, \frac{2}{3}\right)$
35. $m = -\frac{1}{4}, \left(\frac{1}{2}, -1\right)$
36. $m = -\frac{5}{6}, (-3, 3)$
37. If we are given two points of a line but not the slope of the line, how can we use the point–slope or slope–intercept model to write an equation of the line?
38. If a line contains $A(2, -3)$ and $B(2, 5)$, why is the point–slope form not an appropriate model for writing an equation of the line?

Exercises 3.2

Concepts and Skills

1. The following are possible first steps in solving the equation $\frac{2}{3}x = 10$. Explain why each approach is correct.
   (i) Multiply both sides by $\frac{3}{2}$.
   (ii) Divide both sides by $\frac{2}{3}$.
   (iii) Multiply both sides by 3 and then divide both sides by 2.
2. What is an important advantage of the algebraic method over the table method?

In Exercises 3–14, solve the equation algebraically and verify the solution. For special cases, write the solution set.

3. $3(3 - 2x) = 33 - 2x$
4. $2(3y + 8) = -2(2 - y)$
5. $7w - 5 = 11w - 5 - 4w$
6. $9w + 7 = 16w + 7(1 - w)$
7. $9x + 5 = 5x + 3(x - 1)$
8. $11x + 2(4 - 3x) = 3(3 - x) + 15$
9. $16 + 7(6 - x) = 15 - 4(x + 2)$
10. $7 - 4(t - 3) = 6t - 3(t + 3)$
11. $2(3a + 2) = 2(a + 1) + 4a$
12. $8t - 7 - 3t = 5t + 2$
13. $-2(x + 5) = 5(1 - x) + 3(7 - x)$
14. $2y - 3(2y - 3) = 2y - 5(3y - 4)$
15. Consider solving the equation $\frac{1}{3} - 2x = \frac{2}{5}$. To clear fractions, we multiply both sides by 15. Why is it necessary to multiply $2x$ by 15?
16. What is the solution set of $3x = 2x$? Now divide both sides of $3x = 2x$ by x. What is the solution set of the resulting equation? Why are the equations not equivalent?

In Exercises 17–26, solve the given equation.

17. $2x - \frac{3}{4} = -\frac{5}{6}$
18. $\frac{2}{15}y + \frac{3}{5} = 2 - \frac{2}{3}y$
19. $\frac{y}{2} + \frac{y}{4} = \frac{7}{8} - \frac{y}{8}$
20. $\frac{5 + 3x}{2} + 7 = 2x$
21. $\frac{5}{4}(x + 2) = \frac{x}{2}$
22. $\frac{1}{4}(x - 4) = \frac{1}{3}(x + 6)$
23. $x + \frac{3x - 1}{9} = 4 + \frac{3x + 1}{3}$
24. $\frac{2x}{5} - \frac{2x - 1}{2} = \frac{x}{5}$
25. $\frac{1}{4} + \frac{1}{6}(4a + 5) = 2(a - 3) + \frac{5}{12}$
26. $\frac{5}{6}(1 - t) - \frac{t}{2} = -\frac{1}{3}(1 - 3t)$

In Exercises 27–30, solve the given equation.

27. $2.6 = 0.4z + 1$
28. $1 - 0.3c = -0.5$
29. $0.08x + 0.15(x + 200) = 7$
30. $0.6(2x + 1) - 0.4(x - 2) = 1$

CE In Exercises 31–36, determine whether the given equations are equivalent equations.

31. $x + 4 = 7$ and $3x - 5 = 4$
32. $\frac{1}{3}x = 2$ and $x + 3 = 9$
33. $x + 3 = 4 + x$ and $2x - 3 = 4 + 2x$
34. $x + 3 = 3 + x$ and $2x - 5 = -(5 - 2x)$
35. $\frac{1}{4}x = -2$ and $8 - x = 0$
36. $3x = 0$ and $x - 3 = 0$

Concepts and Skills

Most exercise sets begin with the basic skills and concepts discussed within the text. These include Writing Exercises, designed to help students gain confidence in their ability to communicate, and Concept Extension exercises, which go slightly beyond the text examples.

Data Analysis

The Data Analysis exercises appear in many of the end-of-section exercise sets. Students may work independently or in groups to solve these problems by analyzing and interpreting real-life data. Although it is not required, a graphing calculator may be used for these exercises if so desired.

Data Analysis: *College Textbook Sales*

The following table shows the number (in millions) of college textbooks sold during the period 1991–1997. Also shown are the average wholesale prices (in dollars) for college textbooks during the same period. (Source: Book Industry Study Group, Inc.)

Year	Number Sold (millions)	Average Price (dollars)
1991	137	17.33
1993	136	18.37
1995	142	19.07
1997	154	20.19

99. Use the concept of rate of change to explain whether you think a linear model would be appropriate for the number of books sold.
100. Letting x represent the number of years since 1990 and using the data for 1991 and 1997, write a linear function P to model the average wholesale price of textbooks. (Round coefficients to the nearest hundredth.)

Challenge

In Exercises 103–106, multiply.

103. $(x^n + 2)(x^n + 4)$
104. $(x^n - 2)^2$
105. $(x^n + 4)(x^n - 4)$
106. $(x + 1)(x^{2n} - x^n + 1)$
107. In the accompanying figure, the total area of the largest rectangle is the sum of the areas of the four inner rectangles. Use this fact to show the following.

$$(a + b)^2 = a^2 + 2ab + b^2$$

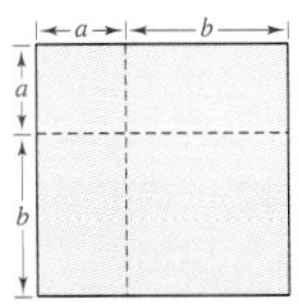

108. In the accompanying figure, the total area of the largest rectangle is the sum of the areas of the three inner rectangles. Use this fact to show the following.

$$(a - b)(a + b) = a^2 - b^2$$

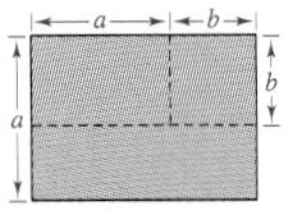

Challenge

These problems appear at the end of most exercise sets and offer more challenging work than the standard and Concept Extension problems.

End-of-Chapter Features

At the end of each chapter, these features appear in the following order: Chapter Project, Chapter Review Exercises, Looking Back (except for Chapter 11), Chapter Test, Cumulative Test (at the end of selected chapters).

Chapter Project

Each Chapter Project is the culmination of the real-data application introduced in the Chapter Opener and continued in the Data Analysis feature. These more in-depth projects emphasize the organization, modeling, and interpretation of data, and they provide opportunities for students to report their findings. Topics are listed in the Table of Contents. Answers appear in the Instructor's Annotated Edition.

Chapter Review Exercises

Each chapter ends with a set of review exercises. These exercises include helpful section references that direct students to the appropriate sections for review. The answers to the odd-numbered review exercises are included at the back of the text.

Looking Back

This short list of review exercises focuses on previously discussed skills and concepts that will be needed in the upcoming chapter.

Chapter Test

A Chapter Test follows each chapter review. The answers to all the test questions, with the appropriate section references, are included at the back of the text.

Cumulative Test

A Cumulative Test appears at the end of Chapters 2, 5, 8, and 11. The answers to all the test questions, with section references, are included at the back of the text.

Chapter 1

The Real Number System

The mean elevation of the continental United States is 2500 feet. The table shows the highest points in six selected states.

State	Highest Point	Altitude (feet)
Colorado	Mount Elbert	14,431
Florida	Iron Mountain	325
Illinois	Charles Mound	1,241
Rhode Island	Durfee Hill	805
Tennessee	Clingmans Dome	6,642
Washington	Mount Rainier	14,408

(Source: U.S. Geological Survey.)

The difference between the altitude of Mount Rainier and the mean elevation is the ***positive number*** *$+11{,}908$, which indicates that Mount Rainier is considerably higher than the mean elevation. The difference between the altitude of Iron Mountain and the mean elevation is the* ***negative number*** *-2175, which indicates that Iron Mountain is lower than the mean elevation. (For more on this topic, see Exercises 97–100 at the end of Section 1.3, and see the Chapter Project.)*

Chapter 1 begins with a discussion of the structure and order of the real numbers. We develop rules for the four basic operations with real numbers and for working with exponents and square roots. We discuss the order in which all these operations are performed and conclude with some of the basic properties of the real number system.

1.1 The Real Numbers

Structure of the Real Numbers • The Number Line • Order of the Real Numbers • Opposites and Absolute Value

Structure of the Real Numbers

It is convenient to organize the numbers of mathematics into sets. A **set** is *any* collection of objects, but our focus will be on sets of numbers.

We use braces to enclose the **members** or **elements** of a set, and we usually assign a letter name to the set. For example, the set N of **natural numbers**, or **counting numbers**, is written as follows.

$$N = \{1, 2, 3, 4, \ldots\}$$

The dots (. . .) mean that the list continues without end.

When the number 0 is included with the natural numbers, the collection of numbers is called the set W of **whole numbers**.

$$W = \{0, 1, 2, 3, \ldots\}$$

When the temperature falls 7° below zero, we use a negative number to describe the temperature: $-7°$. The whole numbers and their opposites form the set J of **integers**.

$$J = \{\ldots, -3, -2, -1, 0, 1, 2, 3, \ldots\}$$

There are infinitely many numbers that are not integers. For example, fractions such as $\frac{1}{2}$, $\frac{2}{3}$, and $-\frac{3}{8}$ are called **rational numbers**.

One way to describe a rational number is with a **variable,** which is a symbol, usually a letter, that represents some unknown number. In the following definition, we use the variables p and q to represent the numerator and the denominator of a fraction.

> A **rational number** is a number of the form $\frac{p}{q}$, where p and q represent integers, and q is not 0.

Note that this definition includes all the integers n because they can be written in the form $\frac{n}{1}$. For example, $2 = \frac{2}{1}$.

Think About It

Do you think that the decimal number 0.121121112...has a pattern? Is it a rational number?

A rational number also has a decimal name that can be determined simply by dividing the numerator by the denominator. The decimal name may be a *terminating decimal,* such as $\frac{3}{4} = 0.75$, or a *repeating decimal,* such as $\frac{3}{11} = 0.27272727\ldots$, in which a block of one or more digits repeats forever. We often write repeating decimals with a bar to indicate the repeating block: $\frac{3}{11} = 0.\overline{27}$.

The set I of decimal numbers that neither terminate nor repeat is called the set of **irrational numbers**. Examples of irrational numbers are π and most square roots, such as $\sqrt{6}$, $\sqrt{15}$, and $-\sqrt{7}$. Note that $\sqrt{25} = 5$, so $\sqrt{25}$ is a rational number.

The rational numbers and the irrational numbers are two distinct sets of numbers. Taken together, the two sets form the set **R** of **real numbers**. We can think of the real numbers as all numbers with decimal names. Figure 1.1 shows the relationships among the various sets of numbers we have discussed.

Figure 1.1 The Real Numbers

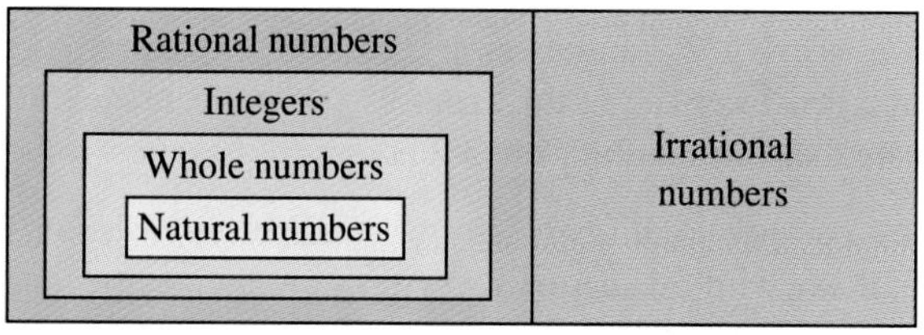

EXAMPLE 1

Classifying the Real Numbers

Indicate every set to which each of the following real numbers belongs.

$$3.2, \quad 0, \quad -4, \quad 3\frac{5}{6}, \quad -\frac{3}{4}, \quad \sqrt{9}, \quad \sqrt{6}$$

Solution

	3.2	**0**	**−4**	$3\frac{5}{6}$	$-\frac{3}{4}$	$\sqrt{9}$	$\sqrt{6}$
Irrational							√
Rational	√	√	√	√	√	√	
Integer		√	√			√	
Whole		√				√	
Natural						√	

The Number Line

We use a **number line** to visualize real numbers and their relationship to each other.

To construct a number line, we choose a point corresponding to the number 0. Points at equally spaced intervals are then associated with the integers. The positive integers are to the right of 0, and the negative integers are to the left of 0. All other real numbers are associated with intervening points.

Because we usually label only the integers when we draw a number line, students sometimes think that integers are the only numbers that can be graphed.

The number associated with a point is called the **coordinate** of the point. The point associated with zero is called the **origin**.

Using a number line to highlight points corresponding to the numbers in a set is called **plotting** the points. The resulting set of plotted points is called the **graph** of the set of numbers.

EXAMPLE 2

Graphing a Set of Real Numbers

Graph the set $\{-4, -1.5, 2, \pi, 4\frac{1}{2}, \sqrt{36}\}$.

Using a calculator to write numbers in decimal form makes it easy to determine the order of the numbers.

Solution

Figure 1.2

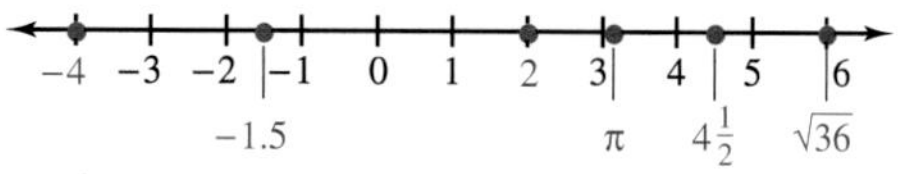

Order of the Real Numbers

LEARNING TIP

Visualizing or drawing a number line is especially helpful when you work with inequalities. For example, by observing that -4.5 is to the left of -4 on the number line, you can easily conclude that $-4.5 < -4$.

We say that two numbers are equal if they are associated with the same point of the number line. We use the symbol $=$ to indicate equality. For example, $\frac{1}{2} = 0.5$. To indicate that two numbers are not equal, we use the symbol $\neq$. For example, $-3 \neq 3$.

We can use a number line to visualize the order of the real numbers, which we can indicate with the inequality symbols $<$ and $>$. Because -2 is to the right of -5 on the line, we say that -2 *is greater than* -5 and write $-2 > -5$. Equivalently, -5 is to the left of -2. Thus we say that -5 *is less than* -2 and write $-5 < -2$.

EXAMPLE 3

Writing Inequalities

Insert an inequality symbol to make the statement true.

(a) 0 ___ -5 (b) $\frac{1}{3}$ ___ $\frac{1}{4}$ (c) -7 ___ -6

Solution

(a) $0 > -5$ (b) $\frac{1}{3} > \frac{1}{4}$ (c) $-7 < -6$

Inequality symbols can be combined with the equality symbol. The symbol $\leq$ means "is less than or equal to," and $\geq$ means "is greater than or equal to."

Opposites and Absolute Value

On the number line, two different numbers that are on opposite sides of zero and that are the same distance from zero are called **opposites** (or **additive inverses**). The numbers 4 and -4 are examples of opposites. (See Fig. 1.3.)

Figure 1.3

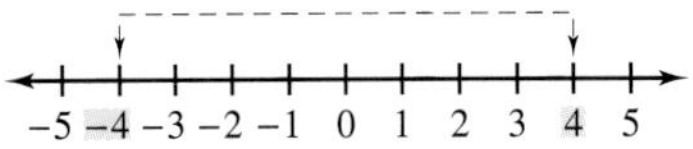

So that every real number has an opposite, we shall agree that the opposite of 0 is 0. For any real number x, we use the symbol $-x$ to represent "the opposite of x."

Note: It is incorrect to read $-x$ as "minus x" because minus refers to subtraction. It is also incorrect to read $-x$ as "negative x" because $-x$ does not necessarily represent a negative number. The symbol $-x$ refers to the *opposite* of the number represented by x.

minus	negative	opposite of
↓	↓	↓
$8 - 5$	-7	$-x$

Figure 1.3 also shows that the opposite of -4 is 4. We write this fact in the following way.

$$-(-4) = 4$$

We read the statement, "The opposite of negative 4 is 4."

The Opposite of the Opposite of a Number

For any real number a, $-(-a) = a$.

On the number line, the distance (or number of units) between any number and 0 is called the **absolute value** of the number. For a real number a, we denote its absolute value by $|a|$. We can also state an algebraic definition of absolute value.

Algebraic Definition of Absolute Value

For any real number a, $|a| = \begin{cases} a & \text{if } a \geq 0 \\ -a & \text{if } a < 0 \end{cases}$

Note: In the definition of absolute value, if $a < 0$, then $-a$ represents a *positive* number. The absolute value of a number is *nonnegative,* which means "not negative."

A **numerical expression** is any combination of numbers and arithmetic operations. When we **evaluate** a numerical expression, we determine the value of the expression by performing the indicated operations.

EXAMPLE 4

Evaluating Absolute Value Expressions

(a) $|10| = 10$

(b) $|-5| = 5$

(c) $-|8| = -8$

(d) $-|-7| = -7$

(e) $|8| + |-7| = 8 + 7 = 15$

(f) $|-12| - |-9| = 12 - 9 = 3$

Quick Reference 1.1

Structure of the Real Numbers

- A set is any collection of objects, such as numbers. The objects are called **members** or **elements** of the set.
- The set $N = \{1, 2, 3, \ldots\}$ is called the set of **natural numbers**, or **counting numbers**.
- The set W of **whole numbers** is $W = \{0, 1, 2, 3, \ldots\}$.
- The set J of **integers** is $J = \{\ldots, -3, -2, -1, 0, 1, 2, 3, \ldots\}$.
- A **variable** is a symbol, usually a letter, that represents some unknown number.
- A **rational number** is a number of the form $\frac{p}{q}$, where p and q represent integers, and q is not 0.
- The decimal names of rational numbers are either *terminating* or *repeating* decimals.
- The set I of decimal numbers that neither terminate nor repeat is called the set of **irrational numbers**.
- The set **R** of **real numbers** is the set of rational numbers together with the set of irrational numbers.

The Number Line

- A **number line** is a set of points each of which is associated with a real number called the **coordinate** of the point. The point whose coordinate is 0 is the **origin**.
- Using a number line to highlight points corresponding to the numbers in a set is called **plotting** the points. The resulting set of plotted points is called the **graph** of the set of numbers.

Order of the Real Numbers

- Two numbers are **equal** if they are associated with the same point of the number line. The symbol $=$ means *is equal to*; the symbol $\neq$ means *is not equal to*.
- The inequality symbols $<$ (*is less than*) and $>$ (*is greater than*) are used to indicate the order of the real numbers. We also use the symbols $\leq$ (*is less than or equal to*) and $\geq$ (*is greater than or equal to*).

Opposites and Absolute Value

- On a number line, two different numbers that are on opposite sides of 0 and the same distance from 0 are called **opposites** (or **additive inverses**).
- The opposite of any real number x is denoted by $-x$.
- For any real number a, $-(-a) = a$.
- On the number line, the distance between any number and 0 is called the **absolute value** of the number. We denote the absolute value of a real number a by $|a|$.
- For any real number a, $|a| = \begin{cases} a & \text{if } a \geq 0 \\ -a & \text{if } a < 0 \end{cases}$
- A **numerical expression** is any combination of numbers and arithmetic operations. When we **evaluate** a numerical expression, we determine the value of the expression by performing the indicated operations.

Speaking the Language 1.1

1. A symbol that represents an unknown number is called a ______.
2. The result of plotting all points associated with the numbers in a set is called a ______.
3. The symbol -6 is read "negative 6," but the symbol $-x$ is read "______."
4. On a number line, the distance between a number and the origin is called the ______ of the number.

For many exercises in this section, it will be helpful to use a calculator to give the decimal names of numbers.

Exercises 1.1

Concepts and Skills

1. Write a sentence to describe the decimal name of a rational number.
2. Write a sentence to describe the decimal name of an irrational number.

In Exercises 3–6, state whether the given number has a terminating or a repeating decimal name.

3. $\frac{3}{8}$ 4. $-2\frac{23}{100}$ 5. $\frac{35}{99}$ 6. $\frac{5}{11}$

In Exercises 7–10, describe, in words, numbers that satisfy the given conditions.

7. Integers that are not whole numbers
8. Rational numbers that are not integers
9. Rational numbers that are integers
10. Integers that are neither negative nor natural numbers

In Exercises 11–16, determine whether the statement is true or false.

11. Every integer is also a whole number.

12. Every integer is a rational number.

13. Every whole number is also an integer.

14. Every natural number is also a whole number.

15. Every nonterminating decimal repeats.

16. The number 0 is a whole number but not an integer.

17. Explain the difference between 1.75 and $1.\overline{75}$.

18. What is the difference between the meanings of the symbols $<$ and $\le$?

In Exercises 19–26, write the given number in decimal form. What does the decimal name suggest about whether the number is rational or irrational?

19. $\frac{35}{37}$ **20.** $\frac{8}{11}$

21. $-\frac{\pi}{3}$ **22.** $\frac{\pi}{4}$

23. $\frac{\sqrt{2}}{\sqrt{8}}$ **24.** $-\frac{\sqrt{3}}{\sqrt{27}}$

25. $-\frac{\sqrt{5}}{2}$ **26.** $\frac{\sqrt{7}}{5}$

In Exercises 27 and 28, list all the numbers in the given set that are members of each of the following sets.

(a) Natural numbers

(b) Whole numbers

(c) Integers

(d) Rational numbers

(e) Irrational numbers

27. $\left\{-4, 0, \frac{3}{5}, \sqrt{7}, 0.25, -17, 0.\overline{63}, \pi, \sqrt{16}\right\}$

28. $\left\{-24, -\sqrt{5}, -4\frac{7}{8}, 0.\overline{36}, \frac{\sqrt{3}}{2}, \frac{\pi}{7}, 0.2863\right\}$

In Exercises 29–34, graph the given set.

29. $\{1, 2, 3, -2, -4\}$ **30.** $\{-1, -3, 5, 3\}$

31. $\left\{\frac{1}{2}, 0.75, -\frac{4}{3}, 0, -1.2\right\}$

32. $\left\{-2\frac{2}{3}, -\frac{3}{2}, -1, 3\frac{4}{5}\right\}$

33. $\left\{-3, -\frac{1}{2}, 3, -\pi, 4.3, \sqrt{4}, -\sqrt{4}, -2\frac{1}{3}\right\}$

34. $\left\{5, -4, 2.5, \frac{\pi}{2}, -\sqrt{9}, -\sqrt{6}, 3\frac{2}{3}, \frac{7}{2}\right\}$

In Exercises 35–44, insert $<$ or $>$ to make the statement true.

35. $-12 \quad\quad -5$ **36.** $11 \quad\quad 5$

37. $-\frac{15}{7} \quad\quad -\frac{7}{15}$ **38.** $\frac{22}{23} \quad\quad \frac{23}{24}$

39. $\frac{22}{7} \quad\quad \pi$ **40.** $3.14 \quad\quad \pi$

41. $\sqrt{7} \quad\quad 2.\overline{645}$

42. $-\sqrt{30} \quad\quad -5.4\overline{7}$

43. $|-4| \quad\quad -4$

44. $-|-3| \quad\quad -(-3)$

In Exercises 45–48, write the opposite of each number.

45. -6

46. 13

47. 0

48. $|-4|$

49. Using the word *distance*, explain why the statement $|7| = |-7|$ is true.

50. Give an example to show that $|-a| = a$ is not necessarily true.

In Exercises 51–54, a number n is described in reference to a number line. State all possible values of n.

51. The number n is 5 units from 0.

52. The number n is 3 units from -2.

53. When increased by 2, n is 4 units from 0.

54. The number n has an absolute value of 8.

In Exercises 55–64, evaluate the given quantity.

55. $|-6|$ **56.** $|-4|$

57. $-|7|$ **58.** $-|-10|$

59. $|0|$ **60.** $-|12|$

61. $|-2| - |2|$ **62.** $|-4| + |5|$

63. $|-6| - |-5|$ **64.** $|-12| - |-10|$

Real-Life Applications

65. Before your trip to England, you converted \$1440 to pounds at an exchange rate of 1 pound per \$1.60. While in England, you spent 600 pounds. When you returned and converted pounds to dollars, the exchange rate had fallen to \$1.50 per pound. How many dollars did you receive?

66. A total of \$9500 is to be divided among four people. Two people are to receive 30% of the total, and one person is to receive \$2000. How much will the fourth person receive?

67. As a produce manager, you have purchased 50 pounds of grapes at a wholesale price of \$75. If you want to make a profit of 40 cents per pound, at what retail price per pound should you sell the grapes?

68. As a store owner, you must collect a 5% state sales tax on all goods sold. If your total receipts (including tax) for the month were \$21,000, how much tax must you remit to your state?

Modeling with Real Data

69. The Jefferson County (Kentucky) school system had 5601 teachers, two-thirds of whom were over 40 years of age. How many teachers were 40 or younger? (Source: U.S. Department of Education.)

70. A person living in Portland, Oregon, and having a median income for that city could afford to buy only 37% of the homes that were being marketed. The accompanying graph shows the five cities with a population of at least 1 million with the lowest percentages of homes that median-income households could afford to buy. (Soure: National Homebuilders Association.)

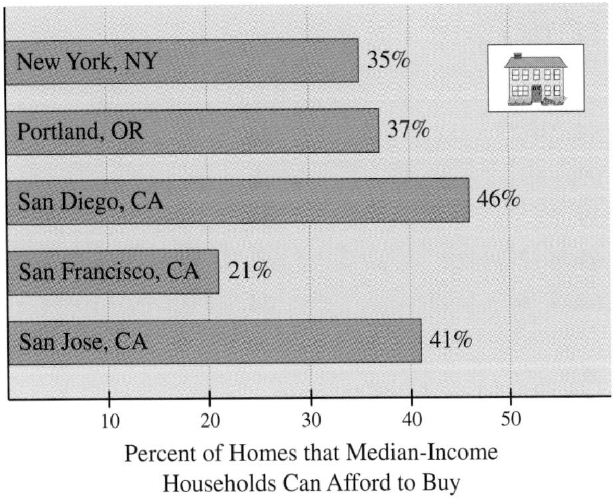

(a) What percent of the homes being marketed in New York City could not be afforded even by median-income households?

(b) Compare the length of the bar for San Francisco to the length of the bar for San Jose. What conclusion can you draw from this comparison?

Data Analysis: *Frozen Desserts Consumption*

The Data Analysis groups that appear throughout this book often involve exercises for which the use of a graphing calculator or a scientific calculator would be appropriate.

In recent years, Americans have consumed an average of about 5 gallons of frozen desserts per person per year. The accompanying figure indicates the popularity of the four basic categories. (Source: U.S. Department of Agriculture.)

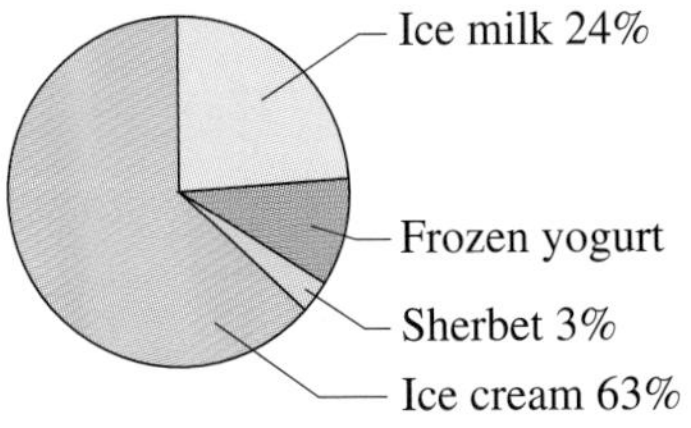

71. On the average, how many gallons of ice cream does each American consume annually?

72. What percent of the total consumption is frozen yogurt?

73. On the average, how many ounces of sherbet does each American consume annually? (There are 128 ounces in a gallon.)

74. According to the graph, how many gallons of ice cream are consumed for every gallon of sherbet?

Challenge

In Exercises 75–78, fill in the blanks with $<$, $>$, or $=$.

75. If $a < 0$, then $-a$ ____ 0.

76. If $a > 0$, then $|a|$ ____ $-a$.

77. If $a < 0$, then $|a|$ ____ a.

78. If $a < 0$, then $-(-a)$ ____ 0.

In Exercises 79–82, write the given number as the simplified quotient of two integers.

79. 2.7

80. 0.352

81. $0.\overline{3}$

82. $1.\overline{6}$

1.2 Addition

Addends with Like Signs • Addends with Unlike Signs • Real-Life Applications

Addends with Like Signs

The basic operations of arithmetic are addition, subtraction, multiplication, and division. In this section we consider addition.

In an expression such as $7 + 5$, the numbers 7 and 5 are called **addends.** The expression, $7 + 5$, and the result of performing the addition, 12, are both called a **sum.**

Exploring the Concept

Sums on a Number Line

We can use a number line to visualize the sum of two real numbers. To do this, we begin at the first addend and agree that adding a positive number implies moving to the right (see Fig. 1.4), whereas adding a negative number implies moving to the left (see Fig. 1.5).

You may want to prepare students for Example 1 in advance. The idea is to compare the absolute value of the sums in parts (a) through (c) with the sum of the absolute values of the addends.

Figure 1.4

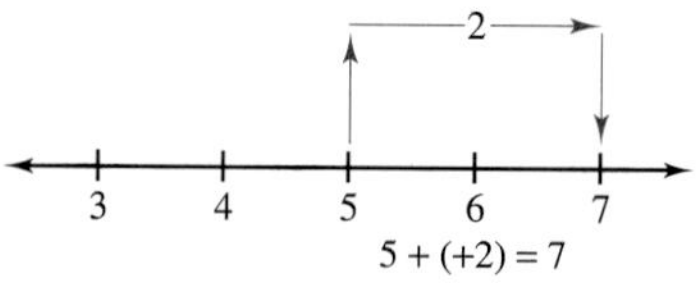

Figure 1.5

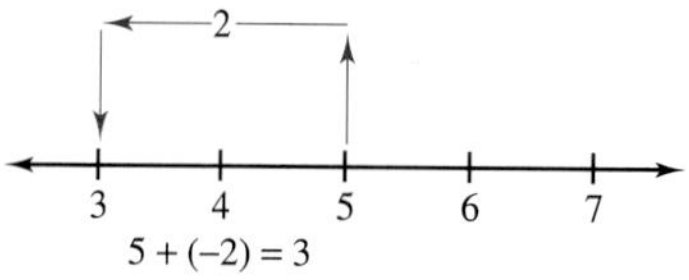

Because we know how to add positive numbers, we will consider only sums of negative numbers in Example 1. As you work through this example, try to formulate rules for adding negative numbers.

EXAMPLE 1

Sums of Negative Numbers

Use a number line to determine the following sums.

(a) $-1 + (-6)$ (b) $-2 + (-1)$ (c) $-4 + (-3)$

Solution

(a)

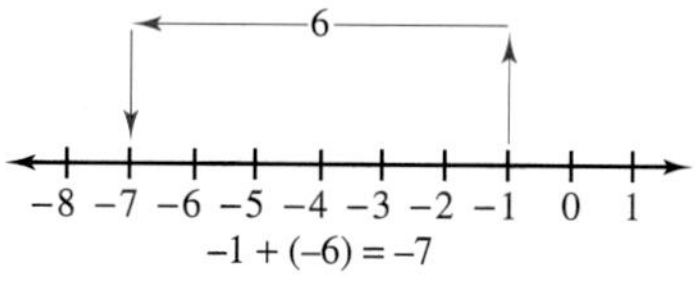

(b)

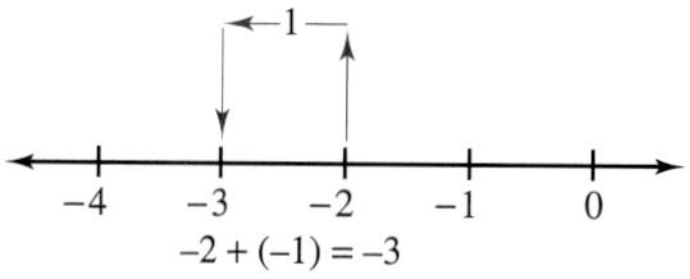

(c)

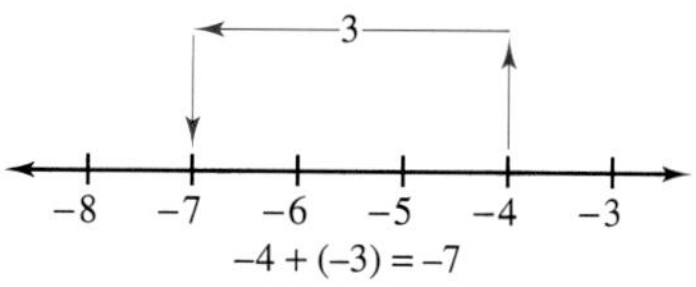

Observe that in each part of Example 1, the result of adding two negative numbers is negative. For the first sum, note that $|-1| + |-6| = 1 + 6 = 7$, which is the absolute value of the sum. It appears that we can determine the absolute value of the sum by adding the absolute values of the addends.

The addition rules suggested by these observations can be proved to be true, but we accept them as given.

> **Rules for Adding Addends with Like Signs**
>
> If the addends of a sum have like signs, then
>
> 1. the sign of the sum is the same as the sign of the addends, and
> 2. the absolute value of the sum is the sum of the absolute values of the addends.

The rules for adding numbers with like signs are summarized less formally in the following procedure.

> **Adding Numbers with Like Signs**
>
> To add two numbers with like signs, follow these steps.
>
> 1. Add their absolute values.
> 2. If the numbers are positive, then the sum is positive; if the numbers are negative, then the sum is negative.

EXAMPLE 2

Adding Addends with Like Signs

(a) $5 + 12 = 17$

(b) $-9 + (-7) = -16$

(c) $-3.7 + (-2.5) = -6.2$

(d) $-\frac{1}{2} + \left(-\frac{3}{4}\right) = -\frac{2}{4} + \left(-\frac{3}{4}\right) = -\frac{5}{4}$

In this first chapter, you may find it necessary to review operations with fractions.

Addends with Unlike Signs

We can use the same number line techniques to investigate methods for adding numbers with unlike signs. As you work Example 3, pay particular attention to the addend that seems to dictate the sign of the result, and try to relate the absolute value of the result to the absolute values of the addends.

EXAMPLE 3

Sums of Numbers with Unlike Signs

Use a number line to determine the following sums.

(a) $6 + (-2)$ (b) $3 + (-6)$

(c) $-3 + 7$ (d) $-8 + 3$

Solution

(a)

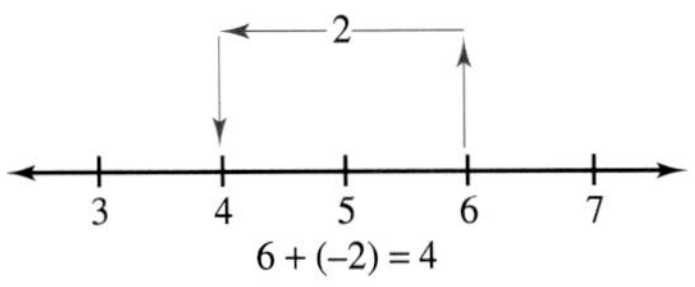

(b)

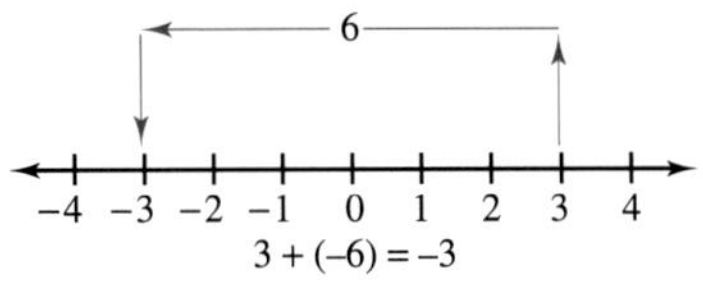

(c)

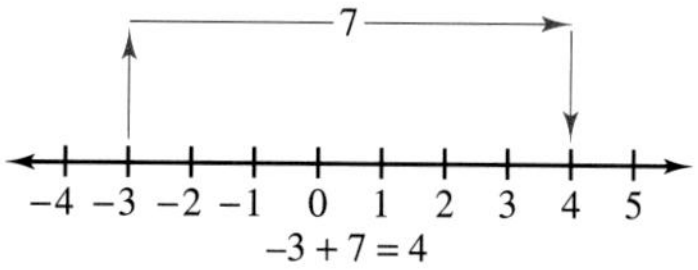

(d)

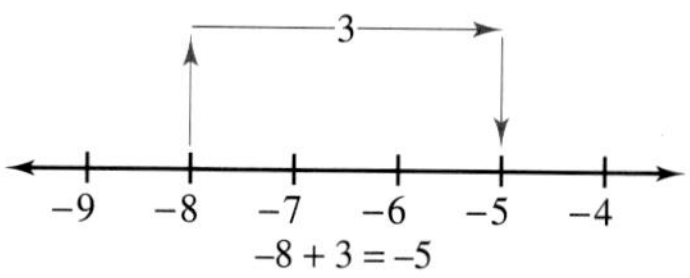

In each part of Example 3, the sign of the result is the same as the sign of the addend with the larger absolute value.

For part (a) of Example 3, note that $|6| - |-2| = 6 - 2 = 4$, which is the absolute value of the sum. In all parts, it appears that we can determine the absolute value of the sum by subtracting the absolute values of the addends (larger minus smaller).

Again, the rules suggested by Example 3 can be proved, but we will simply accept them as given and summarize them as follows.

Rules for Adding Addends with Unlike Signs

If the addends of a sum have unlike signs, then

1. the sign of the sum is the same as the sign of the addend with the larger absolute value, and
2. to determine the absolute value of the sum, take the absolute values of the addends and subtract the smaller absolute value from the larger absolute value.

The rules for adding numbers with unlike signs are summarized less formally in the following procedure.

Adding Numbers with Unlike Signs

To add two numbers with unlike signs, follow these steps.

1. Subtract their absolute values (the smaller from the larger).
2. The sign of the sum is the same as the sign of the number with the larger absolute value.

EXAMPLE 4

Think About It

To determine the sign of the sum of two fractions with unlike signs, you must identify the fraction with the larger absolute value. Why is this easier after writing the fractions with a common denominator?

Adding Addends with Unlike Signs

(a) $15 + (-5) = 10$

(b) $-20 + 14 = -6$

(c) $4.3 + (-7.5) = -3.2$

(d) $-\frac{2}{3} + \frac{1}{2} = -\frac{4}{6} + \frac{3}{6} = -\frac{1}{6}$

EXAMPLE 5

Using the Addition Rules to Calculate Sums

(a) $-8 + 5 = -3$

(b) $-3 + (-12) = -15$

(c) $15 + (-6) = 9$

(d) $\frac{5}{6} + \left(-\frac{4}{9}\right) = \frac{15}{18} + \left(-\frac{8}{18}\right) = \frac{15 + (-8)}{18} = \frac{7}{18} \approx 0.39$

(e) $-3 + \frac{22}{7} = \frac{-3}{1} + \frac{22}{7} = \frac{-21}{7} + \frac{22}{7} = \frac{-21 + 22}{7} = \frac{1}{7} \approx 0.14$

(f) $-584 + 379 = -205$

(g) $-18.97 + (-34.5) = -53.47$

Note: In parts (d) and (e) of Example 5, we use the symbol $\approx$ to indicate that the decimal result is an approximation.

The Order of Operations will be discussed in detail in Section 1.6.

When a sum contains three or more addends, we may use grouping symbols, such as parentheses or brackets, to indicate the addends that are to be added first. If there are no grouping symbols, we simply add from left to right.

EXAMPLE 6

LEARNING TIP

When a sum consists of more than two addends, try adding first all the positive numbers and then all the negative numbers. Then add the two results.

Sums with More Than Two Addends

(a) $-4 + 7 + 9 = 3 + 9 = 12$ — There are no parentheses. Add from left to right.

(b) $(-2 + 5) + [6 + (-9)] = 3 + (-3) = 0$ — Add inside the grouping symbols first.

Real-Life Applications

We need to be able to add signed numbers in order to solve applications such as the following.

EXAMPLE 7

Calculating a Depth Below Sea Level

A marine biologist with The National Oceanographic Institute descended 20 feet below the ocean surface to gather data on ocean plant life. After gathering the data, she descended another 50 feet to collect more information. Finally, she descended another 70 feet for the deepest sample. From there, she ascended 30 feet for a final sample before returning to the surface. The depth changes are given by -20, -50, -70, and $+30$ feet. How deep was the biologist when she collected her deepest sample? How deep was she when she collected her last sample?

Solution

The deepest point is the sum of the depths at which she collected the first three samples.

$$-20 + (-50) + (-70) = -140$$

Students will be inclined just to add ascents and subtract descents. Show how descents are represented by negative numbers, and work the problem only with addition.

We interpret the result as 140 feet below the ocean surface. The location of the final sample is the sum of all of the biologist's depth changes.

$$-20 + (-50) + (-70) + 30 = -110$$

We interpret the result as 110 feet below the ocean surface.

Quick Reference 1.2

Addends with Like Signs

- In the expression $a + b$, a and b are the **addends.** Both the expression and the value of the expression are called a **sum.**
- If a and b are real numbers with like signs, then $a + b$ is determined as follows:
  1. Add the absolute values of the addends.
  2. The sign of the sum is the same as the sign of the addends.

Addends with Unlike Signs

- If a and b are real numbers with unlike signs, then $a + b$ is determined as follows:
  1. Take the absolute values of the addends and subtract the smaller absolute value from the larger absolute value.
  2. The sign of the sum is the same as the sign of the addend with the larger absolute value.
- When a sum consists of three or more addends, we may use grouping symbols, such as parentheses or brackets, to indicate the operations that are to be performed first. Otherwise, add from left to right.

Speaking the Language 1.2

1. In the expression $-8 + 17$, we call -8 and 17 ________.
2. The sign of the sum $x + y$ is the same as the signs of x and y when x and y have ________.
3. The symbol $\approx$ means ________.
4. In the expression $3 + (-2 + 5)$, the parentheses are called ________.

Exercises 1.2

Concepts and Skills

1. Explain how to determine the sum of two numbers that have like signs.
2. Explain how to determine the sum of two numbers that have unlike signs.

In Exercises 3–32, determine the sum.

3. $8 + (-5)$
4. $9 + (-4)$
5. $-6 + 2$
6. $-7 + 4$
7. $-3 + (-7)$
8. $-5 + (-10)$
9. $-6 + (-2)$
10. $-9 + (-5)$

11. $-16 + 16$

12. $-14 + 14$

13. $-3 + 5$

14. $-7 + 5$

15. $12 + (-4)$

16. $13 + (-7)$

17. $-6 + 14$

18. $-14 + 8$

19. $-7 + 0$

20. $-9 + 0$

21. $0 + (-55)$

22. $0 + (-22)$

23. $-9 + 17$

24. $14 + (-20)$

25. $-12 + 12$

26. $17 + (-10)$

27. $-11 + (-11)$

28. $-13 + (-8)$

29. $100 + (-101)$

30. $-50 + (-5)$

31. $-14 + 3$

32. $36 + (-9)$

In Exercises 33–42, determine the sum.

33. $-14 + (-2) + (-1)$

34. $5 + (-8) + (-12)$

35. $-3 + [6 + (-9)]$

36. $-16 + [3 + (-7)]$

37. $6 + (-7) + 2$

38. $-20 + 6 + (-1)$

39. $-12 + (-2) + (-5) + 10$

40. $15 + (-1) + 2 + (-6)$

41. $-6 + 8 + [(-5) + 3]$

42. $(-2) + [(-5) + (-3)] + 9$

In Exercises 43–50, find the sum.

43. $-2.3 + (-4.5)$

44. $2.8 + (-4.6)$

45. $-6.7 + 9.5$

46. $-8.6 + 7.4$

47. $-45.84 + (-5.73)$

48. $-0.35 + 78.3$

49. $24.47 + (-34.78)$

50. $-975.3 + (-537.37)$

CE In Exercises 51–56, write a numerical expression and then evaluate it.

51. What is 11 more than -6?

52. What is -9 increased by 5?

53. If 26, -73, and 40 are the addends, what is the sum?

54. What is the sum of -7, 3, and -10?

55. If 12 is added to the sum of -5 and 2, what is the result?

56. What is the sum of 7 and -2 increased by 3?

In Exercises 57–64, perform the indicated operations.

57. $-\frac{1}{3} + \frac{5}{3}$

58. $-\frac{2}{5} + \left(-\frac{4}{5}\right)$

59. $-\frac{2}{3} + \frac{3}{5}$

60. $\frac{4}{7} + \left(-\frac{1}{3}\right)$

61. $\frac{5}{16} + \left(-\frac{3}{8}\right)$

62. $-\frac{6}{7} + \frac{5}{14}$

63. $-5 + \left(-\frac{2}{3}\right)$

64. $-7 + \left(-\frac{3}{4}\right)$

CE In Exercises 65–68, determine the unknown number.

65. The sum of what number and 2 is -5?

66. The sum of -3 and what number is 7?

67. Two of three addends are -1 and 4. If the sum of the three numbers is -9, what is the third addend?

68. If the sum of three numbers is zero, and two of the numbers are 2 and -2, what is the third number?

In Exercises 69–72, fill in the blank with $<$ or $>$.

69. $8 + (-5)$ ____ $8 + (-5.1)$

70. $-3 + 3$ ____ $-3 + 3.1$

71. $6 + |-6|$ ____ $-6 + |6|$

72. $-4 + (-9)$ ____ $-4 + (-9.5)$

In Exercises 73–92, perform the indicated operations.

Exercises 73–92 are a mixed group.

73. $17 + (-8)$

74. $-36 + (-9)$

75. $6 + (-8) + (-2)$

76. $-5 + 10 + (-7)$

77. $-6 + |-8|$

78. $|4| + |-3|$

79. $187.24 + (-187.24)$

80. $72.72 + (-73.73)$

81. $-6 + \frac{3}{5}$

82. $-\frac{35}{6} + \left(-\frac{23}{4}\right)$

83. $-50 + 5$

84. $-8 + (-9)$

85. $0 + (-18.3)$

86. $-0.5 + 0.3$

87. $-12 + (-3) + 10 + (-2)$

88. $3 + (-8) + (-6) + 11$

89. $|-7| + (-7)$

90. $|-5| + 2$

91. $-\frac{3}{4} + \left(-\frac{21}{8}\right)$

92. $-\frac{2}{3} + \left(-\frac{5}{8}\right)$

Real-Life Applications

93. Suppose that you write an electric bill check for \$76.27, and you notice that your new checkbook balance is $-$\$26.84. You decide not to mail the check to the electric company until you make a deposit of \$35.25. What is your checkbook balance after you have made the deposit and paid your electric bill?

94. On Monday morning, a computer store had 9 color printers in stock, but 3 were sold that day. On Tuesday, 2 more printers were sold, but the store received a delivery of 10 printers. If 6 printers were sold on Wednesday, what was the ending inventory at the close of business that day?

95. A wrestler's weight was 216 pounds when he went on a diet. His weekly change in weight is shown in the Weight column of the table. Complete the table and determine the wrestler's weight at the end of the seven weeks.

Week	Weight (pounds)	Weight at the End of Each Week
1	5 (loss)	
2	1 (gain)	
3	2 (loss)	
4	1 (loss)	
5	4 (gain)	
6	3 (loss)	
7	2 (gain)	

96. The **average,** or **mean,** of n numbers is found by adding the n numbers and dividing the sum by n. During a 1-week period in International Falls, Minnesota, the daily high temperatures (in degrees Fahrenheit) were as follows: $-10°$, $17°$, $5°$, $-9°$, $0°$, $6°$, and $4°$. What was the average high temperature for the week?

Modeling with Real Data

97. The table lists five major league baseball stadiums.

Fenway Park	−38
Wrigley Field	−36
Yankee Stadium	−27
Dodger Stadium	+12
Turner Field	+47

You can learn the year in which a stadium opened by adding the number in the right column to 1950. Determine the age of each stadium in the year 2000. (Source: Major League Baseball.)

98. Southern Company provides electricity to many parts of the Southeast. The graph shows the sources of Southern Company's total capacity (in megawatts). (Source: Southern Company.)

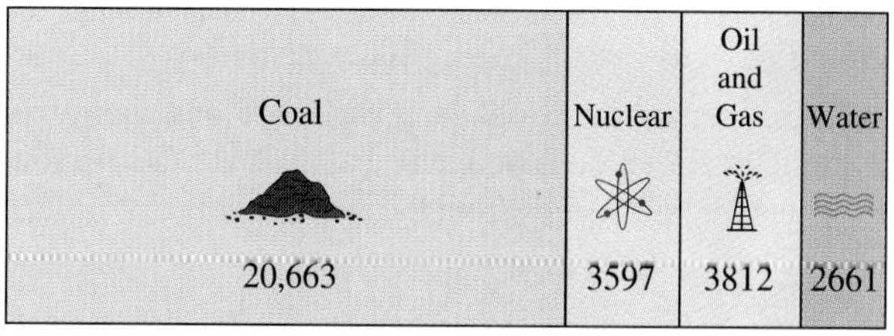

Capacity (in Megawatts) of Southern Company

(a) How many megawatts of capacity are not nuclear?

(b) What is Southern Company's total capacity?

(c) What percent of the total capacity is derived from coal?

Data Analysis: *ATM Cards*

Research indicates that about 60% of American adults have automatic teller machine (ATM) cards. The pie chart shows the percentages of these cardholders who use their cards a certain number of times per month. (Source: Research Partnership.)

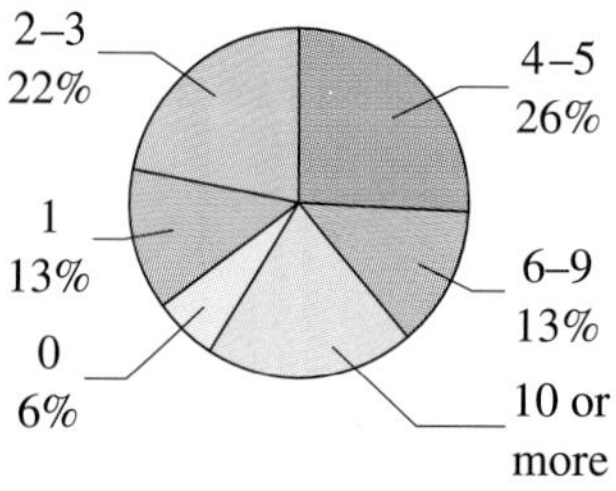

99. What percent of the cardholders use their cards fewer than 4 times per month?

100. What percent of the cardholders use their cards at least 10 times per month?

101. The graph shows that 22% of the cardholders use their cards 2–3 times per month. Can we conclude that 11% use their cards 2 times per month? Why?

102. Why do the given data not enable you to determine the actual number of cardholders in any of the categories?

Challenge

In Exercises 103–106, let a represent a negative number and b a positive number. Determine whether the given sum is positive or negative.

103. $|a| + b$

104. $(-a) + b$

105. $a + (-b)$

106. $-b + b + a$

107. If a person takes one step backward and then two steps forward, how many steps must the person take to be 27 steps from the starting point?

108. (a) Calculate $1 + 2 + (-3) + (-4)$.

(b) Calculate the following sum.

$1 + 2 + (-3) + (-4) + 5 + 6 + (-7) + (-8)$

(c) Consider a sum that follows the same pattern as in part (b) with the last two addends being -47 and -48. What is your conjecture about the value of the expression?

1.3 Subtraction

Definition • Performing Subtraction • Distance Between Points • Real-Life Applications

Definition

In this section we consider subtraction of signed numbers.

In the expression $8 - 5$, the number 8 is called the **minuend** and the number 5 is called the **subtrahend.** Both the expression, $8 - 5$, and the value of the expression, 3, are called a **difference.**

Exploring the Concept

Differences on a Number Line

We can extend our use of the number line to visualize the difference of two numbers.

Because we move to the *right* when we *add* a positive number (see Fig. 1.6), we will move to the *left* when we *subtract* a positive number (see Fig. 1.7).

Figure 1.6

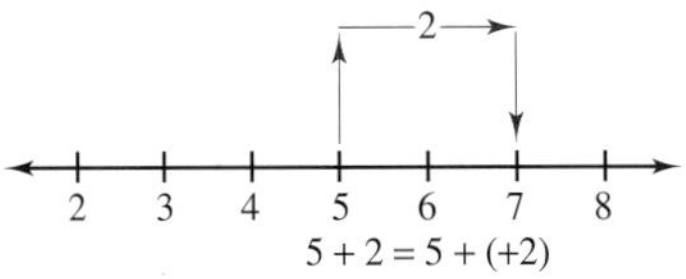

Figure 1.7

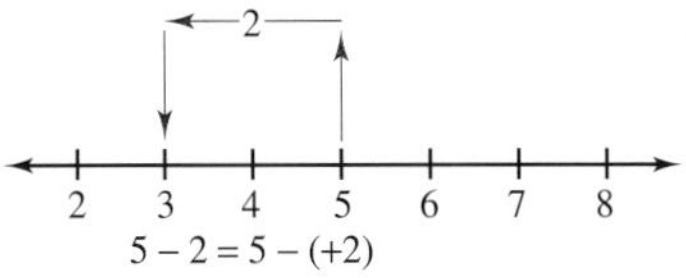

Because we move to the *left* when we *add* a negative number (see Fig. 1.8), we will move to the *right* when we *subtract* a negative number (see Fig. 1.9).

Figure 1.8

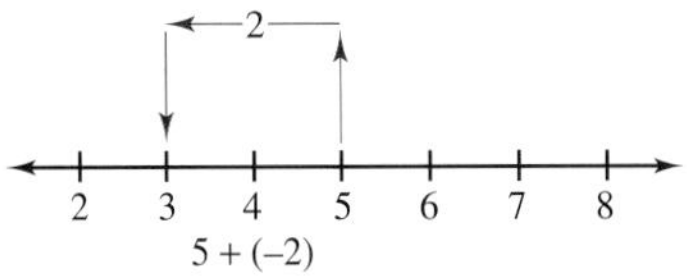

Figure 1.9

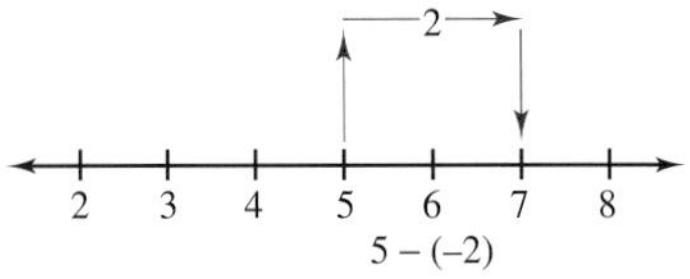

Suppose that we use this number line method to calculate the following differences:

(i) $8 - 5$ (ii) $-2 - 1$ (iii) $5 - (-3)$ (iv) $-1 - (-6)$

In the following summary, you can verify the results in column A. For each part, a related sum is shown in column B.

Column A	*Column B*
$8 - 5 = 3$	$8 + (-5) = 3$
$-2 - 1 = -3$	$-2 + (-1) = -3$
$5 - (-3) = 8$	$5 + 3 = 8$
$-1 - (-6) = 5$	$-1 + 6 = 5$

In each pair, the results are the same.

The important thing to observe is how the difference and the corresponding sum differ.

4 changed to -4

$9 - 4 = 5$ $\qquad$ $9 + (-4) = 5$

minus changed to plus

Subtracting a value yields the same result as adding its opposite.

These observations suggest the following definition of subtraction.

Definition of Subtraction

For any real numbers a and b, $a - b = a + (-b)$.

According to this definition, we can write any difference as an equivalent sum.

Performing Subtraction with Real Numbers

To subtract any two real numbers,

1. change the minus sign to a plus sign, and
2. change the subtrahend to its opposite.

Then evaluate the sum.

After enough practice, students should be able to perform subtraction mentally without writing the addition step. However, the time it takes for students to become comfortable with this will vary.

An obvious benefit of converting a difference into a sum is that we already know the rules of addition. Therefore, no further rules are needed for subtraction.

Performing Subtraction

For simple subtraction problems, you should be able to apply the definition of subtraction mentally. In other cases, use the definition to write the given difference as a sum.

EXAMPLE 1

Performing Subtraction with Real Numbers

Perform the indicated operations.

(a) $5 - 12 = 5 + (-12) = -7$

(b) $-20 - 4 = -20 + (-4) = -24$

(c) $16 - (-5) = 16 + 5 = 21$

(d) $-7 - (-3) = -7 + 3 = -4$

(e) $\frac{8}{3} - \frac{5}{2} = \frac{16}{6} - \frac{15}{6} = \frac{16 - 15}{6} = \frac{16 + (-15)}{6} = \frac{1}{6} \approx 0.167$

(f) $-2.47 - (-6.91) = 4.44$

(g) $-538 - 297 = -835$.

LEARNING TIP

Try not to rush subtraction. You will make fewer sign errors if you actually write the conversion step: $-2 - 9 = -2 + (-9)$.

We now investigate the effect of reversing the order of the numbers in a difference.

Exploring the Concept

Order of Subtraction

Consider the following pairs of differences.

(a) $9 - 5 = 4$ $\qquad 5 - 9 = -4$

(b) $-3 - (-8) = 5$ $\qquad -8 - (-3) = -5$

(c) $-7 - 10 = -17$ $\qquad 10 - (-7) = 17$

Compare with addition, in which the order of the addends does not affect the sum.

In each pair, the results are opposites. This suggests that if a and b are any real numbers, then $a - b$ and $b - a$ are opposites.

Symbolically, $-(a - b) = b - a$ and $-(b - a) = a - b$.

Less formally, we can say that an opposite sign in front of a difference has the effect of turning the difference around.

We summarize these observations with the following rule.

Property of the Opposite of a Difference

For any real numbers a and b, $-(a - b) = b - a$.

As we will see in our later work, this is an important rule. We will use it to simplify expressions and to reverse the order of subtraction when we want to do so.

In our next example, we use the fact that operations inside grouping symbols are to be performed first. Otherwise, we add and subtract from left to right.

EXAMPLE 2

Performing Mixed Operations

(a) $12 - (3 - 7) = 12 - [3 + (-7)] = 12 - (-4) = 12 + 4 = 16$

(b) $4 - (5 + 9) - 7 = 4 - 14 - 7 = 4 + (-14) + (-7) = -17$

(c) $(6 - 9) - (-8 + 2) = -3 - (-6) = -3 + 6 = 3$

(d) $5 - [4 - (1 - 2)] = 5 - [4 - (-1)] = 5 - [4 + 1] = 5 - 5 = 0$

EXAMPLE 3

Translating Phrases into Numerical Expressions

Translate each of the following phrases into a numerical expression and evaluate the expression.

These are some common words that indicate subtraction. They continue to appear in applications.

(a) Four less than -3 (b) Four decreased by -3

(c) The difference between 4 and -3

Solution

(a) $-3 - 4 = -7$

(b) $4 - (-3) = 7$

(c) $4 - (-3) = 7$

The first given number is the minuend, and the second given number is the subtrahend.

Distance Between Points

Figure 1.10

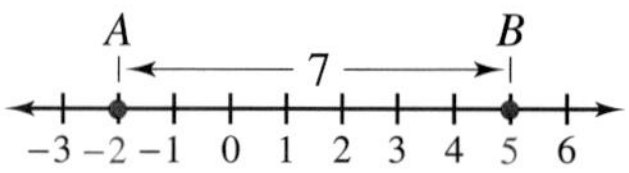

Note that the distance is 7 units whether we count from -2 to 5 or from 5 to -2.

The distance between the two points A and B of the number line in Figure 1.10 can be found by counting the units from -2 to 5. The distance is 7 units.

We can also calculate the distance between points A and B by subtracting the coordinates of the points: $5 - (-2) = 7$.

If we subtract the coordinates in reverse order, the result is $-2 - 5 = -7$. This result is not valid because distance is not negative. However, we can subtract in either order if we agree to take the absolute value of the result.

Think About It

Suppose that you use the number line as a ruler to measure the distance between two points. Does the positioning of the ruler affect the result? Why?

Distance Between Points of the Number Line

If the coordinates of points A and B are a and b, respectively, then the distance d between the points is determined as follows.

$$d = |a - b| = |b - a|$$

EXAMPLE 4

Finding the Distance Between Points of a Number Line

The coordinates of points P and Q are given. Determine the distance d between the points.

(a) -26 and -5 (b) 7 and $-\frac{5}{3}$ (c) -4 and 3.9

Solution

(a) $d = |-26 - (-5)| = |-26 + 5| = |-21| = 21$

(b) $d = \left|7 - \left(-\frac{5}{3}\right)\right| = \left|7 + \frac{5}{3}\right| = \left|\frac{26}{3}\right| = \frac{26}{3}$

(c) $d = |-4 - 3.9| = |-7.9| = 7.9$

Real-Life Applications

Application problems, particularly those involving a calculation of distance, often require subtraction of signed numbers.

EXAMPLE 5 Distance Between Points Above and Below Sea Level

The altitude gauge of a C-130 indicates that the plane is 760 feet above sea level. The crew of a training mission is to drop equipment to a location in Death Valley that is 42 feet below sea level. When the plane is directly above the drop site, what is the distance between the plane and the drop site?

Solution

Consider a vertical number line whose origin is at sea level. The plane is at a point whose coordinate is $+760$, and the drop site is at a point whose coordinate is -42. (See Fig. 1.11.)

Figure 1.11

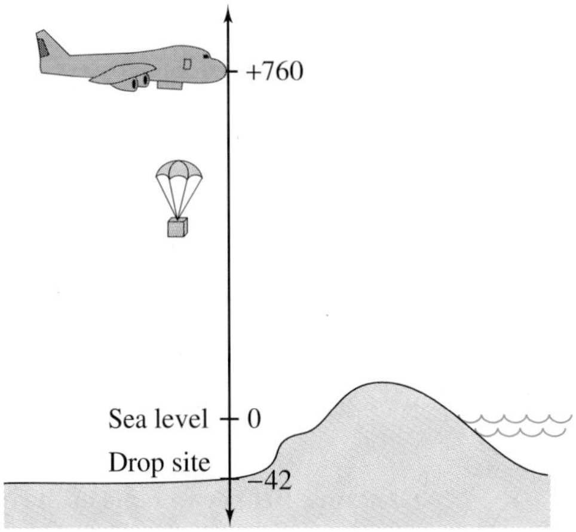

We calculate the distance d between the points as follows.

$$d = |760 - (-42)| = |760 + 42| = |802| = 802$$

The distance from the plane to the drop site is 802 feet.

Quick Reference 1.3

Definition

- In the expression $a - b$, a is the **minuend** and b is the **subtrahend.** Both the expression and the value of the expression are called a **difference.**
- For any real numbers a and b, $a - b$ is defined as $a + (-b)$.

Performing Subtraction

- The definition of subtraction gives us a way to perform the operation by applying the rules for addition. We convert a difference to a sum as follows.
  1. Change the minus sign to a plus sign.
  2. Change the subtrahend to its opposite.
- If we reverse the order of the numbers in a difference, the value of the result is the opposite of the original value.
- Property of the Opposite of a Difference

  For any real numbers a and b, $-(a - b) = b - a$.

Distance Between Points

- We determine the distance between two points of a number line by subtracting the coordinates of the points and taking the absolute value of the result.

Speaking the Language 1.3

1. The expression $4 - (-3)$ is called a ________.
2. To subtract two numbers, we change the minus sign to a plus sign and change the ________ to its opposite.
3. If x and y are the coordinates of two points of a number line, then $|x - y|$ represents the ________.
4. The phrase "6 less than x" translates into a difference, but the statement "6 is less than x" translates into a(n) ________.

Exercises 1.3

Concepts and Skills

1. What two changes must be made to $-5 - 8$ to convert the expression into a sum with the same value?
2. In the expression $7 + 5$, the 7 and the 5 are both called addends. But in the expression $7 - 5$, the 7 and the 5 have two different names. What are they? Why is it necessary for 7 and 5 to have different names in the expression $7 - 5$ but not in the expression $7 + 5$?

In Exercises 3–26, determine the difference.

3. $4 - 15$
4. $3 - 17$
5. $-7 - 3$
6. $-6 - 5$
7. $9 - (-4)$
8. $8 - (-7)$
9. $-2 - (-8)$
10. $-4 - (-12)$
11. $-5 - 8$
12. $-12 - 9$
13. $0 - (-52)$
14. $0 - 25$
15. $13 - (-5)$
16. $15 - (-8)$
17. $-12 - 12$
18. $12 - (-12)$
19. $-6 - (-3)$
20. $-10 - (-5)$
21. $24 - (-24)$
22. $5 - (-24)$
23. $-4 - 16$
24. $-17 - 9$
25. $-8 - (-5)$
26. $-17 - 24$

In Exercises 27–38, perform the indicated operations.

27. $15 - 20 - 5$
28. $-3 - 5 - 7$
29. $-20 - (-14) - (-1)$
30. $6 - (-5) - 8$
31. $10 - (6 - 9)$
32. $-5 - (-2 + 12)$
33. $-8 - 10 - (-3) + 7$
34. $18 - (-2) - 25 + 8$
35. $-3 + (-2) - (-1 + 10)$
36. $6 + 2 - (5 - 12)$
37. $16 - [(-3) + 10 - (-9)]$
38. $-7 + [3 - (-11) - 13]$

CE In Exercises 39–42, determine the unknown number.

39. What is the difference between -17 and -18?
40. What is -3 decreased by 4?
41. What is 9 less than -1?
42. Subtract -3 from 10. What is the result?

In Exercises 43–50, perform the indicated operations.

43. $-\frac{1}{3} - \frac{4}{3}$
44. $-\frac{1}{5} - \left(-\frac{3}{5}\right)$
45. $-\frac{3}{5} - \left(-\frac{1}{2}\right)$
46. $-\frac{3}{4} - \left(-\frac{2}{3}\right)$
47. $-5 - \left(-\frac{1}{3}\right)$
48. $-8 - \left(-\frac{3}{4}\right)$
49. $\frac{7}{16} - \left(-\frac{5}{8}\right)$
50. $-\frac{5}{7} - \frac{9}{14}$

51. In your own words, interpret $-(a - b) = b - a$.

52. Suppose that the value of $x - y$ is -4. What is the value of $y - x$? Why?

In Exercises 53–56, perform the indicated operations.

53. $-22 - 14 - (-13) - 4 + (-7)$

54. $-35 - (-14) + (-23) - (-13) + 5 - (-8)$

55. $13.84 - (-26.19) - (-27.5) - 25 - (-31.7)$

56. $-17.2 - (-82) - 45.23 + (-23.75) - (-54.1)$

CE In Exercises 57–60, determine the unknown number.

57. What number must be subtracted from 3 to obtain -2?

58. The difference between what number and -6 is 4?

59. If decreasing a number by 7 results in -1, what is the number?

60. If 9 less than a number is -2, what is the number?

In Exercises 61 and 62, calculate the distances *AB*, *BC*, and *AC*. Then show that $AB + BC = AC$.

61. 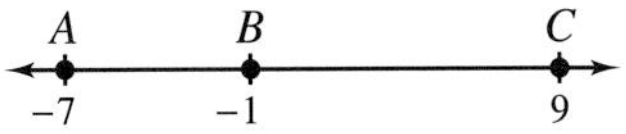

62. 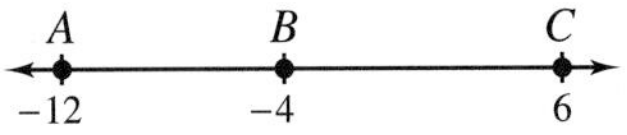

In Exercises 61 and 62, it may be necessary to explain that the symbol AB represents the distance between points A and B.

Exercises 63–92 are a mixed group.

In Exercises 63–92, perform the indicated operations.

63. $-7 - (-2)$

64. $-12 - (-12)$

65. $-47.87 - (-6.74)$

66. $-0.46 - 76.5$

67. $7 - (-36)$

68. $21 - (-35)$

69. $|-8| - |-3|$

70. $|-10| - (-6)$

71. $-\frac{2}{3} - \frac{4}{5}$

72. $-\frac{31}{6} - \left(-\frac{25}{4}\right)$

73. $-6 - 10 + (-5)$

74. $12 - 8 - (-2)$

75. $-16 - 16$

76. $-22 - 37$

77. $-\frac{3}{4} - 4$

78. $-\frac{7}{8} - 6$

79. $0 - (-32)$

80. $-72 - (-72)$

81. $23.89 - (-35.87)$

82. $-974.3 - (-678.39)$

83. $-45 - 0$

84. $23 - (-23)$

85. $4 - 9 - (-5)$

86. $14 - 17 - 7$

87. $|4| - |-12|$

88. $|-3| - 9$

89. $-24 - 56$

90. $35 - 47$

91. $9 - 27 + 2 - (-1)$

92. $-12 + 3 - (-8) - (-5)$

Real-Life Applications

93. In the state of California, the highest elevation is Mount Whitney with an altitude of 14,495 feet. The lowest altitude is Death Valley, at 280 feet below sea level. What is the difference between the two altitudes?

94. Mauna Loa on the island of Hawaii has an elevation of 4169 feet. This peak is 32,024 feet above the ocean floor near Hawaii. How deep is the ocean floor?

Modeling with Real Data

95. The number of Americans with money problems rose in the 1990s. The bar graph for 1995 and 1996 shows how often people worried that their income would not pay their bills. (Source: Marist College Institute for Public Opinion.)

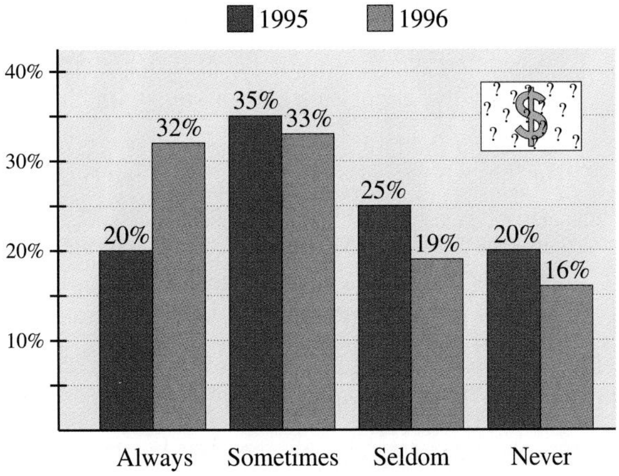

(a) Make a table and use signed numbers to calculate the difference between the 1996 and 1995 percentages for each category.

(b) What is the sum of the differences in part (a)? Why is this so?

96. If you lived in Kansas City and had a median income for that city, you would be able to afford to buy 83% of the homes for sale. Were you to move to San Francisco and have a median income for that city, you would be able to afford only 21% of the homes for sale. (Source: National Home-builders Association.) For each 1000 homes in each city, how many more affordable homes would you find in Kansas City than in San Francisco?

Data Analysis: *U.S. Elevations*

The mean elevation of the continental United States is 2500 feet. The following table shows the highest points in six selected states. (Source: U.S. Geological Survey.)

State	Highest Point	Altitude (feet)	Difference
Colorado	Mount Elbert	14,431	
Florida	Iron Mountain	325	
Illinois	Charles Mound	1,241	
Rhode Island	Durfee Hill	805	
Tennessee	Clingmans Dome	6,642	
Washington	Mount Rainier	14,408	

97. For each state in the accompanying table, record the difference between the altitude of the highest point and the mean elevation of the continental United States. Use positive numbers for altitudes above the mean and negative numbers for altitudes below the mean.

98. Calculate the mean of the differences for the six entries in the table. Why is the mean difference so large?

99. How many miles high is Mount Rainier?

100. As the words are conventionally used, we would rank *mound, hill*, and *mountain* from least to greatest in height. Look at the entries for Illinois, Rhode Island, and Florida, and note the altitudes. What do you observe?

Challenge

In Exercises 101–104, let a represent a negative number and b a positive number. Determine the sign of the expression.

101. $a - b$

102. $b - a$

103. $a - |a|$

104. $|a| - a$

105. What is your opinion of the statement

$$|a - b| = |b - a|$$

for all real numbers a and b? Explain your reasoning.

106. The value of the following expression is 5.

$$1 - 2 + 3 - 4 + 5 - 6 + 7 - 8 + 9$$

Insert parentheses (as many as are needed) so that the value of the expression is 7.

107. The value of the following expression is 45.

$$1 + 2 + 3 + 4 + 5 + 6 + 7 + 8 + 9$$

Is it possible to replace some of the addition symbols with subtraction symbols so that the value of the resulting expression is 0? Why?

1.4 Multiplication

Sign Rules • Real-Life Applications

Sign Rules

Our third basic operation with real numbers is multiplication. In algebra, multiplication is usually indicated by a dot or by parentheses. For example, the product of 3 and 7 could be written in any of the following ways.

$$3 \cdot 7 \qquad 3(7) \qquad (3)(7) \qquad (3)7$$

In the expression $3 \cdot 7$, the numbers 3 and 7 are called **factors.** Both the expression itself, $3 \cdot 7$, and the result of performing the multiplication, 21, are called a **product.**

From arithmetic, we know that the product of any number and 0 is 0. We call this fact the Multiplication Property of 0.

Multiplication Property of 0

For any real number a, $a \cdot 0 = 0 \cdot a = 0$.

We now investigate the sign rules for multiplying two nonzero numbers when at least one factor is negative.

Exploring the Concept

Sign Rules for Multiplication

Consider the following two patterns, in which the first factor of the products is decreasing by 1.

Pattern A	*Pattern B*
$3 \cdot 5 = 15$	$3 \cdot (-5) = -15$
$2 \cdot 5 = 10$	$2 \cdot (-5) = -10$
$1 \cdot 5 = 5$	$1 \cdot (-5) = -5$
$0 \cdot 5 = 0$	$0 \cdot (-5) = 0$
$-1 \cdot 5 = -5$	$-1 \cdot (-5) = 5$
$-2 \cdot 5 = -10$	$-2 \cdot (-5) = 10$

The last two results in Pattern A are based on the observation that the products are decreasing by 5. This pattern suggests that when the factors have unlike signs, the product is negative.

This sign rule is carried over to the start of Pattern B. The last two results of Pattern B follow from the fact that the products are increasing by 5. This pattern suggests that when the factors are both negative, the product is positive.

In both patterns, the absolute value of the product can be found by multiplying the absolute values of the factors.

The multiplication rules suggested by these patterns can be proved to be true.

Rules for Multiplying Two Nonzero Real Numbers

1. If the numbers have like signs, the product is positive; if the numbers have unlike signs, the product is negative.
2. The absolute value of the product is the product of the absolute values of the numbers.

LEARNING TIP

The sign rules for multiplication are quite different from the sign rules for addition. As you do the multiplication exercises, go back and rework some addition exercises to make sure the different rules are clear in your mind.

Note: As is true for addition, the sign rules for multiplication can be proved to be true. It is important to understand that a calculator does not *determine* the sign rules. Rather, a calulator is designed to *obey* the sign rules.

The rules for multiplying signed numbers are summarized less formally in the following procedure.

Multiplying Signed Numbers

To multiply two signed numbers, follow these steps.

1. Multiply their absolute values.
2. If the numbers have like signs, then the product is positive; if the numbers have unlike signs, then the product is negative.

EXAMPLE 1

Evaluating Products

(a) $(-7)(-3) = 21$

(b) $8(-4) = -32$

(c) $-8(4) = -32$

(d) $\frac{2}{3} \cdot \left(-\frac{5}{7}\right) = -\frac{10}{21}$

(e) $8 \cdot \left(-\frac{1}{2}\right) = -4$

(f) $-274(-19) = 5206$

Students may need to be reminded that a common denominator is not needed for multiplication.

Think About It

Which of the following products is easier to estimate? Why?
(i) 9.987(−2.01) (ii) 2.5(−4.6)

Consider the following products in which one factor is -1.

$-1(3) = -3 \qquad -1(-5) = 5 \qquad 4(-1) = -4 \qquad -17(-1) = 17$

In each case, the result of multiplying a number by -1 is the opposite of the number. This fact is generalized in the following property.

Multiplication Property of −1

For any real number a, $a(-1) = -1a = -a$.

Note: The Multiplication Property of -1 entitles us to write the symbol $-x$ as $-1x$ if we want to. As we will see, this property will be useful in simplifying expressions.

We can expand the sign rules for multiplication to include products with more than two factors. If there is an even number of negative factors, the product is positive; if there is an odd number of negative factors, the product is negative. This fact is useful for determining the sign of a product in advance.

If a product consists of three or more factors, grouping symbols may be used to indicate the factors that are to be multiplied first. If a product does not contain grouping symbols, we multiply from left to right.

EXAMPLE 2

Products of Three or More Factors

(a) $-2(-1)(-3)(-2)(-1) = -12$ — Odd number of negative factors

(b) $(-1)(-5)(-2)(-3) = 30$ — Even number of negative factors

(c) $\left(-\frac{3}{2}\right)\left(-\frac{4}{3}\right)\left(-\frac{5}{4}\right)\left(-\frac{6}{5}\right) = 3$ — Divide out common factors.

(d) $(-0.08)(126)(-0.47) = 4.7376$

(e) $(-2)(-2)(-2)(-2)(-2)(-2)(-2) = -128$ — Odd number of negative factors

(f) $(-5)(-5)(-5)(-5) = 625$ — Even number of negative factors

Real-Life Applications

EXAMPLE 3

Opinion Survey

For an opinion survey, the five possible responses were assigned values as shown in the "Code" column.

Response	*Code*	*Tally*
Strongly Agree	$+2$	3
Agree	$+1$	7
No opinion	0	10
Disagree	-1	8
Strongly Disagree	-2	12

The statement "Most Americans have adequate health care insurance" received the numbers of responses shown in the "Tally" column.

The score for this survey item is found by multiplying the code by the tally for each response and then adding the five products. What was the score for this item?

Solution

Code	*Tally*	*Product*
$+2$	3	$+6$
$+1$	7	$+7$
0	10	0
-1	8	-8
-2	12	-24
	Score =	-19

This negative score suggests some level of disagreement with the survey statement.

Quick Reference 1.4

Sign Rules

- In the expression $a \cdot b$, a and b are called **factors.** Both the expression and the value of the expression are called a **product.**
- The following are the rules for multiplying two real numbers.
  1. If at least one factor is 0, then the product is 0.
  2. If neither factor is 0, the product is found as follows.
     (a) If the factors have like signs, the product is positive; if the factors have unlike signs, the product is negative.
     (b) The absolute value of the product is the product of the absolute values of the factors.
- If a product has an even number of negative factors, then the product is positive; if a product has an odd number of negative factors, then the product is negative.

Speaking the Language 1.4

1. The numbers of a product are called ______.
2. The product of two numbers is positive if the numbers ______.
3. A product is negative if a(n) ______ number of the factors are negative.
4. The product of two fractions can be found by multiplying the ______.

Exercises 1.4

Concepts and Skills

1. Explain how to determine the product of two numbers with like signs.
2. Explain how to determine the product of two numbers with unlike signs.

In Exercises 3–20, determine the product.

3. $7 \cdot (-4)$
4. $8 \cdot (-5)$
5. $-5 \cdot 2$
6. $-7 \cdot 6$
7. $-3 \cdot (-4)$
8. $-6 \cdot (-10)$
9. $-6 \cdot (-2)$
10. $-8 \cdot (-5)$
11. $-10 \cdot 10$
12. $-9 \cdot 9$
13. $11 \cdot (-4)$
14. $8 \cdot (-7)$
15. $-7 \cdot 1$
16. $-1 \cdot 0$
17. $4 \cdot (-6)$
18. $5 \cdot (-1)$
19. $0 \cdot (-45)$
20. $-1 \cdot (-92)$

In Exercises 21–26, find the product.

21. $-2.3 \cdot (-2.7)$
22. $-4.5 \cdot (-3.4)$
23. $2.1 \cdot (-2.3)$
24. $0.5 \cdot (-4.3)$
25. $-26.4 \cdot (-57.8)$
26. $-45.1 \cdot 32.3$

CE In Exercises 27–32, write a numerical expression that models the problem and then evaluate it.

27. If the two factors are -1.8 and 1.8, what is the product?
28. What is the product of -8 and -3 times -2?
29. What is $-\frac{3}{4}$ of 32?
30. What is 20% of -19?

Some students are not aware that the word *of* means *multiply.*

31. What is twice the product of 5 and -3?

32. Triple the product of -7 and 2. What is the result?

In Exercises 33–40, multiply the fractions.

33. $-\frac{1}{3} \cdot \frac{3}{5}$

34. $-\frac{4}{5} \cdot \left(-\frac{1}{2}\right)$

35. $-\frac{2}{5} \cdot \left(-\frac{5}{8}\right)$

36. $-\frac{3}{4} \cdot \left(-\frac{8}{9}\right)$

37. $-5 \cdot \left(-\frac{2}{3}\right)$

38. $-\frac{3}{4} \cdot 8$

39. $-\frac{34}{5} \cdot \left(-\frac{25}{4}\right)$

40. $-\frac{21}{4} \cdot \left(-\frac{16}{3}\right)$

41. Suppose a product consists of nine nonzero factors and six of the factors are negative. Explain how to determine the sign of the product.

42. Suppose a product consists of 17 factors. Eight factors are positive, eight factors are negative, and one factor is zero. Explain how to evaluate the product.

In Exercises 43–56, determine the product.

43. $(-2)(-3)(-4)$

44. $-4(-2)(5)$

45. $-5(7)(-2)$

46. $3(-6)(2)$

47. $(-2)(-1)(3)(2)$

48. $(-1)(-4)(3)(-2)$

49. $6(-1)(4)(-2)$

50. $-2(-2)(-3)(10)$

51. $(-1)(-2)(3)(1)(-6)$

52. $(-1)(-1)(2)(-3)(-5)$

53. $(-2)(-2)(-2)(-2)(-2)$

54. $(-5)(-5)$

55. $(-3)(-3)(-3)(-3)$

56. $(-4)(-4)(-4)$

In Exercises 57–60, without evaluating, determine the order of the two given products, and fill in the blank with $<$ or $>$. Then verify by calculating the products.

57. $2(-3)$ ____ $-2(-3)$

58. $(-1)(-1)$ ____ $0 \cdot (-1)$

59. $(-1)(-1)(-1)$ ____ $(-1)(-1)(-1)(-1)$

60. $8965(-213)$ ____ $1 \cdot 2$

CE In Exercises 61–64, determine the unknown number.

61. The product of what number and -8 is 24?

62. Five times what number is -20?

63. What number multiplied by -3 is -18?

64. If the product of three factors is 0 and two of the factors are 120 and -813, what is the third factor?

In Exercises 65–84, perform the indicated operation.

65. $\frac{11}{4} \cdot \left(-\frac{4}{3}\right)$

66. $-\frac{6}{11} \cdot \frac{11}{3}$

67. $-5.76 \cdot (-4.95)$

68. $-0.34 \cdot 25.3$

69. $-5|-4|$

70. $|-9| \cdot |-10|$

71. $5(-2)(4)$

72. $2(-2)(-3)$

73. $36.47 \cdot (-4.47)$

74. $-95.3 \cdot (-57.37)$

75. $(-2)(-1)(-2)(-3)$

76. $(-1)(-1)(-4)(-1)(-3)$

Exercises 65–84 are a mixed group.

77. $|9| \cdot |-8|$

78. $|-7| \cdot (-6)$

79. $\frac{3}{5} \cdot \frac{5}{7} \cdot \left(-\frac{7}{9}\right)$

80. $-\frac{2}{7} \cdot \frac{4}{5} \cdot \left(-\frac{3}{8}\right)$

81. $-4 \cdot (-24)$

82. $-7 \cdot (-9)$

83. $-14 \cdot 10$

84. $-4 \cdot 80$

CE In Exercises 85–92, find two numbers whose sum is the first number and whose product is the second number.

85. 5, 6

86. -6, 8

87. 1, -12

88. 2, -15

89. -8, 12

90. -10, 24

91. -7, -8

92. -4, -21

93. The product $5 \cdot 3$ can be interpreted as repeated addition.

$$5 \cdot 3 = 3 + 3 + 3 + 3 + 3$$

Interpret $7(-2)$ in a similar way and verify that the product and the sum have the same value.

94. Which of the following products is easier to calculate without a calculator? Why?

(i) $(-1)(-1)(-1)(-1)(-1)(-1)(-1)(-1)$

(ii) $(1.37)(6.42)$

Real-Life Applications

95. Instead of recording the actual number of strokes, a golfer used the following to record her scores on each hole.

Eagle: -2 (2 below par)

Birdie: -1 (1 below par)

Par: 0 (standard score)

Bogey: $+1$ (1 above par)

Double bogey: $+2$ (2 above par)

At the end of the round, the golfer found that her performance on the 18 holes was as follows.

Eagle: 1

Birdie: 3

Par: 9

Bogey: 2

Double bogey: 3

If par for the course was 72, what was the golfer's score?

96. To discourage guessing, a test-grading scheme is as follows.

Correct answer	+1
Answer left blank	0
Incorrect answer	−1

On a 100-question test, a student left 13 answers blank and answered 50 questions correctly. What was the student's test score?

Modeling with Real Data

97. The accompanying table shows the population (in millions) of selected countries along with the annual per-person consumption of 8-ounce servings of Coca-Cola products in those countries. (Source: Coca-Cola Company.)

Country	Population (millions)	Annual Per-Person Consumption (8-ounce servings)
Brazil	162	122
China	1221	4
Germany	82	201
Mexico	94	322
Russia	147	6
United States	263	343
Zimbabwe	11	60

(a) Add a column to the table with the heading Total Annual Consumption (billions of gallons) and fill in this column. (There are 128 ounces in a gallon.)

(b) On the average, in which country does every person consume nearly one serving of a Coca-Cola product per day?

98. The pie chart shows that the majority of the assets of the Warburg Pincus Balanced Fund are in common stocks and U.S. Treasury obligations. (Source: Warburg Pincus Balanced Fund.)

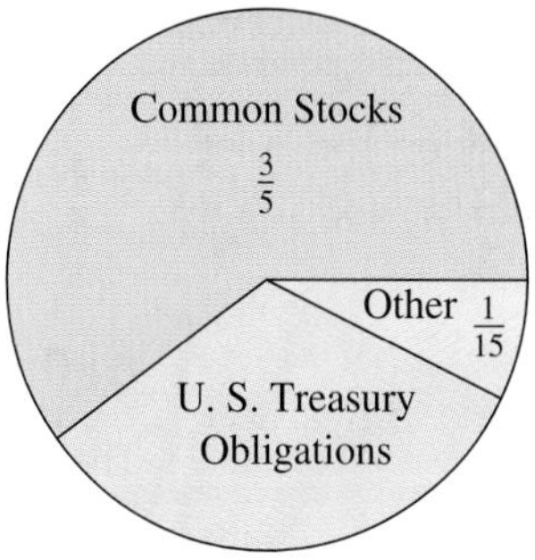

(a) If a person invested $6000 in this fund, how much would be allocated to U.S. Treasury obligations?

(b) What percent of the fund's assets is in common stocks?

Data Analysis: *National Debt*

The federal government has not been free of debt for over 130 years. Assuming a population of 274 million in 1999, the cost of paying off the national debt in 1999 would have been approximately $20,700 for every American citizen. (Source: U.S. Treasury Department.)

99. Using −20,700 as the average cost per citizen, what negative number represents the national debt in 1999? (Round your answer to the nearest tenth of a trillion dollars.)

100. Assuming an average interest rate of 7%, what negative number represents the interest paid on the national debt in 1999? (Round your answer to the nearest billion dollars.)

101. What negative number represents the average cost per citizen just to pay the interest on the national debt in 1999? (Round your answer to the nearest dollar.)

102. Even if the national debt had been cut in half in 1999, what negative number would represent the average cost for a family of four to pay off the national debt?

Challenge

In Exercises 103–108, let a and b represent negative numbers and let c represent a positive number. What is the sign of the given expression?

103. ab

104. bc

105. $a \cdot |b| \cdot c$

106. abc

107. $a \cdot a$

108. $a \cdot b \cdot b$

1.5 Division

Definition and Rules • Division Involving 0 • Multiple Operations and Fractions • Real-Life Applications

Definition and Rules

The last of our four basic operations is division of real numbers. Division is written with the division symbol ÷ or with a fraction bar. For example, we write 19 divided by 7 as $19 \div 7$ or $\frac{19}{7}$. The number 19 is called the **dividend,** and the number 7 is called the **divisor.** Both the expression and the result of the division are called a **quotient.**

Division is defined in terms of multiplication.

The alternative definition

$$a \div b = a \cdot \frac{1}{b}$$

is given in Section 1.7.

> **Definition of Division**
>
> For real numbers a, b, and c, with $b \neq 0$,
>
> $$a \div b = c \quad \text{if} \quad c \cdot b = a$$

In the following exploration, we use this definition to discover the sign rules for division.

Exploring the Concept

Sign Rules for Division

Here are some examples of determining quotients by using the definition of division.

	We Know That	*Because*
(a)	$\frac{18}{9} = 2$	$2 \cdot 9 = 18$
(b)	$\frac{-8}{-2} = 4$	$4 \cdot (-2) = -8$
(c)	$\frac{56}{-7} = -8$	$-8 \cdot (-7) = 56$
(d)	$\frac{-66}{6} = -11$	$-11 \cdot 6 = -66$

In parts (a) and (b), the dividend and the divisor have like signs and the quotient is positive; in parts (c) and (d), the dividend and the divisor have unlike signs and the quotient is negative. In other words, the sign rules for division appear to be exactly the same as the sign rules for multiplication.

We also note that the absolute value of the quotient can be found by dividing the absolute value of the dividend by the absolute value of the divisor.

These observations can be summarized as follows.

Performing Division with Real Numbers

For any real numbers a and b, $b \neq 0$, the quotient $\frac{a}{b}$ is

1. positive if a and b have like signs.
2. negative if a and b have unlike signs.

The absolute value of the quotient $\frac{a}{b}$ equals the quotient of the absolute values of a and b.

Division Involving 0

In our definition of division, we require the divisor to be a *nonzero* number. We can see why this is so if, for example, we try to evaluate $5 \div 0$. The definition states that there must be a number c such that $c \cdot 0 = 5$, but this is impossible. If we try to evaluate $0 \div 0$, then there must be a number c such that $c \cdot 0 = 0$. But this is true for any number c. (We say that $0 \div 0$ is *indeterminate*.)

Students sometimes confuse 0 and *undefined.* It is important that they understand the difference.

In short, dividing by 0 is not defined.

EXAMPLE 1

Evaluating Quotients

Determine the quotient, if possible.

(a) $36 \div (-4) = -9$ — $-9 \cdot (-4) = 36$

(b) $-13 \div (-9) = \frac{13}{9}$ — Like signs, positive quotient

(c) $\frac{0}{-3.2} = 0$ — If $a \neq 0$, then $\frac{0}{a} = 0$.

(d) $\frac{2}{3} \div 0$ is undefined — Division by 0 is undefined.

Multiple Operations and Fractions

Operations in grouping symbols are performed first. If there are no grouping symbols present, we multiply and divide from left to right.

Also, when either the dividend or the divisor is a fraction, we multiply the dividend by the reciprocal of the divisor.

EXAMPLE 2

Performing Division

(a) $18 \div (-12 \div 4) = 18 \div (-3) = -6$ — Perform operations within grouping symbols first.

(b) $-4 \cdot 6 \div (-3) = -24 \div (-3) = 8$ — Operate from left to right.

(c) $-\frac{1}{3} \div 5 = -\frac{1}{3} \cdot \frac{1}{5} = -\frac{1}{15}$ — The reciprocal of a positive number is positive.

(d) $\frac{5}{7} \div \left(-\frac{3}{4}\right) = \frac{5}{7} \cdot \left(-\frac{4}{3}\right) = -\frac{20}{21}$ — The reciprocal of a negative number is negative.

Think About It

Explain why you obtain the same result when you divide a number by 2 and when you multiply the number by $\frac{1}{2}$, but you do not obtain the same result when you divide the number by $\frac{1}{2}$. What is an equivalent way to write a number divided by $\frac{1}{2}$?

LEARNING TIP

Knowing that $-(-a) = a$ and $\frac{-a}{-b} = \frac{a}{b}$ allows you to remove pairs of negative (or opposite) signs. Eliminating signs, when possible, simplifies the appearance of the expression and reduces sign errors.

Because the sign rules for division are the same as the sign rules for multiplication, the following generalization can be stated for signs in quotients.

Property of Signs in Quotients

For real numbers a and b ($b \neq 0$), each of the following statements is true.

$$-\frac{a}{b} = \frac{-a}{b} = \frac{a}{-b} \qquad \frac{-a}{-b} = \frac{a}{b} \qquad -\frac{-a}{-b} = \frac{-a}{b} = \frac{a}{-b} = -\frac{a}{b}$$

In words, if a fraction has a single opposite sign, the sign can be placed in the numerator, in the denominator, or in front of the fraction. If a fraction has two opposite signs, they can both be removed. If a fraction has three opposite signs, any two of them can be removed.

Real-Life Applications

If a quantity has decreased over a period of time, calculating the average decrease may require dividing a negative number by a positive number.

EXAMPLE 3

Average Change in Temperature

On December 24, 1924, the temperature in Fairfield, Montana, dropped from 63°F at noon to −21°F at midnight. (Source: National Oceanic and Atmospheric Administration.) What was the average hourly change in temperature?

Solution

The change in temperature was −84° in 12 hours. Thus the average change was $\frac{-84}{12}$ or −7°F per hour. This means that the temperature dropped an average of 7°F per hour.

Quick Reference 1.5

Definition and Rules

- In the expression $a \div b$, a is called the **dividend** and b is called the **divisor.** Both the expression and the value of the expression are called a **quotient.**
- For real numbers a, b, and c, with $b \neq 0$, $a \div b = c$ if $c \cdot b = a$.

- The sign rules for division are the same as the sign rules for multiplication.
- The absolute value of a quotient is the quotient of the absolute values of the dividend and divisor.

Division Involving 0

- If a is any nonzero real number, then $0 \div a = 0$.
- Division by 0 is undefined.

Multiple Operations and Fractions

- Operations in grouping symbols are performed first. If there are no grouping symbols present, we multiply and divide from left to right.
- When the dividend or divisor is a fraction, multiply the dividend by the reciprocal of the divisor.
- Property of Signs in Quotients

  For real numbers a and b, $b \neq 0$, the following statements are true.

$$-\frac{a}{b} = \frac{-a}{b} = \frac{a}{-b} \qquad \frac{-a}{-b} = \frac{a}{b} \qquad -\frac{-a}{-b} = \frac{-a}{b} = \frac{a}{-b} = -\frac{a}{b}$$

Speaking the Language 1.5

1. In the quotient $y \div x$, the number x is called the ______.
2. The value of $6 \div 0$ is ______, whereas the value of $0 \div 6$ is ______.
3. To divide a number by a fraction, we multiply the number by the ______ of the fraction.
4. A quotient is -1 if the dividend and divisor are ______.

Exercises 1.5

Concepts and Skills

1. Explain how to determine the quotient of two numbers with like signs.
2. Explain how to determine the quotient of two numbers with unlike signs.

In Exercises 3–22, divide, if possible.

3. $8 \div (-4)$ **4.** $-14 \div 2$

5. $\dfrac{15}{-5}$ **6.** $\dfrac{-24}{6}$

7. $-40 \div (-5)$ **8.** $-60 \div (-10)$

9. $-16 \div (-2)$ **10.** $-10 \div 10$

11. $\dfrac{-80}{-5}$ **12.** $\dfrac{-8}{8}$

13. $-21 \div 7$ **14.** $55 \div (-55)$

15. $\dfrac{-20}{5}$ **16.** $\dfrac{56}{-7}$

17. $-7 \div 1$ **18.** $-4 \div 0$

19. $\dfrac{-2}{0}$ **20.** $\dfrac{0}{-1}$

21. $0 \div (-35)$

22. $-92 \div (-92)$

In Exercises 23–30, perform the indicated operations.

23. $-6 \div 3 \div (-4)$ **24.** $-30 \div (-5) \div (-2)$

25. $24 \div [-6 \div (-2)]$ **26.** $-12 \div [-6 \div (-3)]$

27. $-16 \cdot 2 \div 8$ **28.** $20 \div 5 \cdot 3$

29. $18 \div (-6 \cdot 3)$ **30.** $4 \cdot [-8 \div (-16)]$

31. In your own words, interpret $\dfrac{n}{n} = 1, n \neq 0$.

32. In your own words, interpret $\frac{n}{1} = n$.

In Exercises 33–36, determine the quotient.

33. $-5.9 \div (-2.7)$

34. $22.1 \div (-2.3)$

35. $-4.7 \div 9.3$

36. $42.4 \div (-13.5)$

CE In Exercises 37–42, write a numerical expression that models the problem and then evaluate it.

37. What is the quotient of -16.4 and -4.1?

38. What is 500 divided by -1000?

39. What is 0 divided by -10?

40. What is 0 divided into 7?

41. What is -20 divided into -10?

42. What is 12 divided by $\frac{1}{2}$?

In Exercises 43–50, perform the indicated division.

43. $-\frac{1}{3} \div \frac{5}{9}$

44. $-\frac{3}{4} \div \frac{3}{16}$

45. $-\frac{2}{5} \div \left(-\frac{3}{10}\right)$

46. $\frac{5}{16} \div \left(-\frac{3}{8}\right)$

47. $-5 \div \left(-\frac{2}{3}\right)$

48. $-7 \div \left(-\frac{3}{4}\right)$

49. $\left(-\frac{3}{4}\right) \div 8$

50. $\left(-\frac{7}{8}\right) \div 16$

In Exercises 51–56, determine

(a) the opposite of the given number and

(b) the reciprocal of the given number.

51. 5

52. -8

53. $-\frac{3}{4}$

54. $\frac{2}{5}$

55. 0

56. $-\frac{4}{5}$

In Exercises 57–60, estimate the order of the two given quotients, and fill in the blank with < or >.

57. $\frac{3}{7}$ ____ $\frac{3}{8}$

58. $-200 \div 27.68$ ____ $0 \div (-17)$

59. $-8 \div 9$ ____ $-\frac{7}{9}$

60. $(-24 \div 8) \div (-3)$ ____ $(24 \div 8) \div (-3)$

CE In Exercises 61–64, determine the unknown number.

61. The quotient of what number and -7 is -2?

62. Ten divided by what number is -5?

63. If the quotient is 10 and the divisor is -4, what is the dividend?

64. What is the divisor if the dividend is -8 and the quotient is 8?

Exercises 65–88 are a mixed group.

In Exercises 65–88, perform the indicated operations.

65. $-26.4 \div (-57.8)$

66. $-5.76 \div (-4.95)$

67. $-\frac{2}{3} \div \frac{8}{27}$

68. $\frac{6}{7} \div \left(-\frac{3}{28}\right)$

69. $-|-5| \div |-5|$

70. $|-63| \div (-|9|)$

71. $-56 \div 4 \cdot 2$

72. $-8 \cdot 3 \div 12$

73. $\frac{9}{-3}$

74. $\frac{-48}{-6}$

75. $18 \div (-2) \div (-3)$

76. $-32 \div 8 \div (-2)$

77. $12 \div (-2 \cdot 3)$

78. $-50 \div [5 \cdot (-2)]$

79. $0 \div (-1.274)$

80. $-43.56 \div 43.56$

81. $-20 \div (-10) \div (-6)$

82. $-7 \div 7 \div (-4)$

83. $27 \div (-3)$

84. $-27 \div (-9)$

85. $72 \div (-6 \div 3)$

86. $-60 \div (12 \div 4)$

87. $\frac{0}{0}$

88. $\frac{-64}{8}$

89. A nonzero number divided by itself always results in 1. For example, $5 \div 5 = 1$ and $-8 \div (-8) = 1$. Explain, then, why $0 \div 0$ does not equal 1.

90. The definition of division states that $a \div b = c$ if $c \cdot b = a$. Using the definition, we might claim that $0 \div 0 = 0$ because $0 \cdot 0 = 0$. What is your opinion of this?

In Exercises 91–96, rewrite the given fraction with fewer opposite signs.

91. $\frac{-5}{-8}$

92. $-\frac{-6}{-11}$

93. $-\frac{-1}{-6}$

94. $-\frac{3}{-5}$

95. $-\frac{-4}{7}$

96. $\frac{-7}{-9}$

Real-Life Applications

97. An instructor grades a set of 20-answer quizzes by indicating the total number of incorrect answers. Here are the scores for six of the students: -5, -1, -7, -4, -5, -2.

(a) What was the average number of incorrect answers?

(b) What was the average percentage of correct answers?

98. In golf a birdie counts as -1, a par counts as 0, and a bogey counts as $+1$. At the 18th hole of the Masters Tournament, 10 golfers had a birdie, 18 golfers had a par, and the rest of the field of 40 players had a bogey. What decimal number was the average score of all the players?

Modeling with Real Data

99. The number of senior citizens in the United States is projected to increase dramatically over the next half-century. The accompanying bar graph shows the projected numbers (in millions) of people in three age groups for the years 1990, 2020, and 2050. (Source: U.S. Census Bureau.)

(a) Estimate the population of each age group in 2020.

(b) Which age group is projected to almost double in population from 1990 to 2050? to almost triple?

(c) What age group is projected to have the greatest population increase from 2020 to 2050?

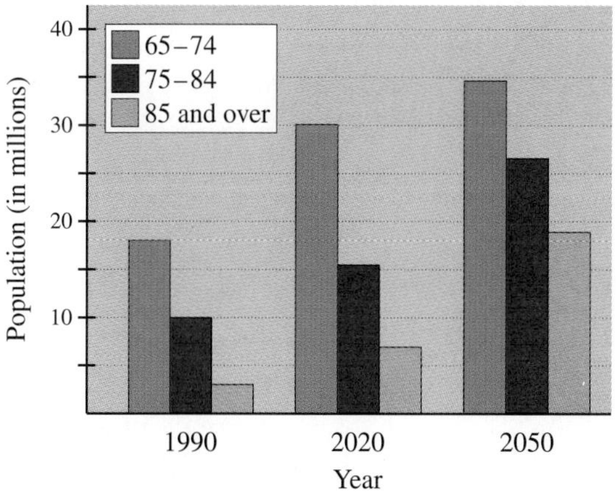

100. Some ecologists estimate that half of the 10 million species of plants and animals on earth in the year 2000 will be extinct by the year 2167. (Source: *Time* magazine.) If these estimates are true, what negative whole number represents the annual rate of extinction?

Data Analysis: *Temperature and Altitude*

Temperature decreases with altitude. Using a reference temperature of 60°F at sea level, the temperature is approximately -67°F at an altitude of 36,000 feet. (Source: American Meteorological Society.)

101. Using the given reference point, what is the total temperature change from sea level to an altitude of 36,000 feet?

102. Assuming a constant decrease in temperature, what negative number represents the temperature decrease for each increase in altitude of 1000 feet? (Round your answer to the nearest tenth of a degree.)

103. Using the assumption and the result in Exercise 102, what temperature would you expect to find at an altitude of 22,000 feet?

104. Using the assumption and the result in Exercise 102, at what altitude would you be flying if the outside temperature were at the freezing point?

Challenge

In Exercises 105–108, let a represent a negative number and b represent a positive number. Determine the sign of the expression.

105. $\frac{a}{b}$

106. $\frac{a}{-b}$

107. $\frac{-a}{b}$

108. $\frac{|a|}{-b}$

In Exercises 109 and 110, evaluate the given quotient.

109. $\dfrac{\frac{14}{3}}{\frac{-7}{6}}$

110. $\dfrac{\frac{-20}{9}}{\frac{10}{-3}}$

1.6 Exponents, Square Roots, Order of Operations

Exponents • Square Roots • Order of Operations

Exponents

To indicate repeated multiplication, we use an exponent.

$$3^4 = 3 \cdot 3 \cdot 3 \cdot 3 = 81$$

Note that the exponent indicates the number of times the base is used as a factor.

The number 3 is called the **base,** and 4 is called the **exponent.** The expression 3^4 is read "three to the fourth power." For exponents of 2 and 3, the words *squared* and *cubed* are sometimes used rather than "to the second power" and "to the third power."

The expression 3^4 is in **exponential form,** and the expression $3 \cdot 3 \cdot 3 \cdot 3$ is in **product** (or **factored**) **form.** When the exponent is a natural number, the base can be any real number.

EXAMPLE 1

Exponential and Product Forms

Write each product form in exponential form and each exponential form in product form.

(a) $5 \cdot 5 \cdot 5 \cdot 5 \cdot 5 \cdot 5 \cdot 5 = 5^7$

(b) $(-3) \cdot (-3) \cdot (-3) \cdot (-3) = (-3)^4$

(c) $2^4 = 2 \cdot 2 \cdot 2 \cdot 2$

(d) $(-4)^5 = (-4) \cdot (-4) \cdot (-4) \cdot (-4) \cdot (-4)$

EXAMPLE 2

Evaluating Exponential Expressions

Evaluate each exponential expression.

(a) 3^2 (b) 2^3 (c) $(-4)^2$ (d) $(-4)^3$

(e) 0^{157} (f) 1^{157} (g) $(-1)^{157}$ (h) $\left(-\frac{2}{3}\right)^3$

Solution

(a) $3^2 = 3 \cdot 3 = 9$

(b) $2^3 = 2 \cdot 2 \cdot 2 = 8$

(c) $(-4)^2 = (-4)(-4) = 16$ — Negative base, even exponent

(d) $(-4)^3 = (-4)(-4)(-4) = -64$ — Negative base, odd exponent

(e) $0^{157} = 0$ — If any factor is 0, the product is 0.

(f) $1^{157} = 1$ — If all factors are 1, the product is 1.

(g) $(-1)^{157} = -1$ — Negative base, odd exponent

(h) $\left(-\frac{2}{3}\right)^3 = \left(-\frac{2}{3}\right)\left(-\frac{2}{3}\right)\left(-\frac{2}{3}\right) = -\frac{8}{27}$ — Negative base, odd exponent

Parts (f) and (g) can be generalized. You might want to explain why $(-1)^n = 1$ for even n and $(-1)^n = -1$ for odd n.

Think About It

In statistics, we often work with squares of differences—that is, quantities of the form $(a - b)^2$. Why are such quantities always nonnegative?

As shown in Example 2(c) and 2(d), if the base b is a negative number, then b^n is

1. positive if n is an even natural number.
2. negative if n is an odd natural number.

Square Roots

Because $4^2 = 16$, 4 is called a **square root** of 16. Similarly, $(-4)^2 = 16$, so -4 is also a square root of 16.

Definition of Square Root

If $b^2 = a$, then b is a **square root** of a.

Note that in the system of real numbers, b^2 cannot be negative. Therefore, a negative number cannot have a square root. Also, because 0 is the only number that can be squared to obtain 0, 0 has only one square root. All positive numbers have two square roots, one positive and one negative.

Sometimes the distinction between the square roots of a number and the principal square root of a number is difficult for students to grasp.

The **principal** (nonnegative) square root of a number is indicated by a **radical** symbol $\sqrt{\ }$. If $b^2 = a$, where $b \geq 0$, then $\sqrt{a} = b$. The number or expression under the radical symbol is called the **radicand**.

EXAMPLE 3 **Determining Square Roots**

Determine each of the following.

(a) The square roots of 81 (b) $\sqrt{81}$

(c) $-\sqrt{81}$ (d) $\sqrt{-81}$ (e) $\sqrt{0}$

Solution

(a) Because $9^2 = 81$ and $(-9)^2 = 81$, the square roots of 81 are 9 and -9.

(b) $\sqrt{81} = 9$ The symbol $\sqrt{81}$ refers to the *principal* square root of 81.

(c) $-\sqrt{81} = -1 \cdot \sqrt{81} = -1 \cdot 9 = -9$ Multiplication Property of -1

(d) Because no real number can be squared to obtain -81, $\sqrt{-81}$ is not a real number.

(e) $\sqrt{0} = 0$

Evaluating square roots such as $\sqrt{9}$ and $\sqrt{100}$ without a calculator is easy because the radicands are familiar perfect squares. However, we can make good use of a calculator for evaluating a square root such as $\sqrt{43}$.

EXAMPLE 4 **Evaluating Square Roots**

(a) $\sqrt{49} = 7$ $7 \cdot 7 = 49$

(b) $\sqrt{\dfrac{9}{25}} = \dfrac{3}{5}$ $\dfrac{3}{5} \cdot \dfrac{3}{5} = \dfrac{9}{25}$

(c) $\sqrt{43} \approx 6.56$

Because $2^3 = 8$, 2 is called a **cube root** of 8. Symbolically, $\sqrt[3]{8} = 2$. Because $3^4 = 81$, 3 is a fourth root of 81. Symbolically, $\sqrt[4]{81} = 3$. The concept can be extended to higher and higher roots. We will discuss roots in greater detail in Chapter 9.

Order of Operations

When more than one operation is required in an expression, the order in which the operations are performed can produce different results.

Consider the *numerical expression* $3 + 2 \cdot 8$. If we add first, the result is

$$3 + 2 \cdot 8 = 5 \cdot 8 = 40$$

If we multiply first, the result is

$$3 + 2 \cdot 8 = 3 + 16 = 19$$

To avoid such ambiguity, we must agree on the order in which operations are to be performed. In particular, we will rank multiplication ahead of addition. Thus the correct result is 19.

A numerical expression may also include **grouping symbols** to indicate which operations are to be performed first. For example, $(12 - 8) - 1 = 4 - 1 = 3$, whereas $12 - (8 - 1) = 12 - 7 = 5$. The most common grouping symbols are parentheses, but brackets [ ] and braces { } are also used. Square root symbols, fraction bars, and absolute value symbols also serve as grouping symbols.

The following is the order in which operations in an expression are performed.

The Order of Operations

1. If grouping symbols are present, perform all operations inside of them according to the following order. For grouping symbols within grouping symbols, start with the innermost group and work outward.
2. Evaluate exponents and roots.
3. Perform multiplication and division from left to right.
4. Perform addition and subtraction from left to right.

By grouping symbols, we usually mean parentheses, brackets, or braces, as in the expressions $2(5 + 4)$ and $3 + \{2[5 - 2(1 - 8)]\}$. However, fraction bars and radical symbols also serve as grouping symbols. For example, $\frac{5 + 3}{7}$ means $\frac{(5 + 3)}{7}$ and $\sqrt{16 - 9}$ means $\sqrt{(16 - 9)}$.

EXAMPLE 5

LEARNING TIP

For each problem, work through the Order of Operations in a disciplined, step-by-step manner. Trying to do too much in each step is a major contributor to errors.

The Order of Operations

Perform the indicated operations.

(a) $5 + 4 \cdot 2$ (b) $5 - 5(3^2 - 7)$ (c) $2 - [4 - (9 - 12)]$

(d) $24 \div 6 \cdot 3 \div 2$ (e) $(-4)^2$ (f) -4^2

(g) $\dfrac{-3 + 5 \cdot 2}{14}$ (h) $7 + 2 \div 2 - 8 \cdot 3^2$ (i) $7 - \sqrt{34 - 6 \cdot 3}$

Solution

(a) $5 + 4 \cdot 2 = 5 + 8 = 13$ Multiplication ranks higher than addition.

(b) $5 - 5(3^2 - 7) = 5 - 5(9 - 7)$ Grouping symbols have the highest priority. Within the parentheses, the exponent ranks first.

$= 5 - 5(2)$

$= 5 - 10$ Multiplication ranks higher than subtraction.

$= -5$

(c) $2 - [4 - (9 - 12)] = 2 - [4 - (-3)]$ Start with the innermost grouping symbols.

$= 2 - [7]$

$= -5$

(d) $24 \div 6 \cdot 3 \div 2 = 4 \cdot 3 \div 2$ Multiplication and division rank equally, so proceed from left to right.

$= 12 \div 2$

$= 6$

(e) $(-4)^2 = (-4)(-4) = 16$

(f) $-4^2 = -1 \cdot 4^2 = -1 \cdot 16 = -16$ Note that -4^2 does not mean $(-4)(-4)$.

(g) The fraction bar acts as a grouping symbol. We evaluate the numerator before dividing by 14.

$$\frac{-3 + 5 \cdot 2}{14} = \frac{-3 + 10}{14} = \frac{7}{14} = \frac{1}{2}$$

(h) $7 + 2 \div 2 - 8 \cdot 3^2 = 7 + 2 \div 2 - 8 \cdot 9$ Exponent ranks first.

$= 7 + 1 - 72$ Multiplication and division rank next.

$= -64$ Addition and subtraction rank last.

(i) The radical symbol serves as a grouping symbol, so the radicand is evaluated first.

$$7 - \sqrt{34 - 6 \cdot 3} = 7 - \sqrt{34 - 18} = 7 - \sqrt{16} = 7 - 4 = 3$$

Quick Reference 1.6

Exponents

- The expression b^n is in **exponential form**. The number b is called the **base**, and the number n is called the **exponent**.
- If the base b is a negative number, then b^n is
  1. positive if n is an even natural number, and
  2. negative if n is an odd natural number.

Square Roots

- If $b^2 = a$, then b is a **square root** of a.
- A negative number cannot have a square root. All positive numbers have two square roots, one positive and one negative. The square root of 0 is 0.
- The **principal** (nonnegative) square root of a number is indicated by a **radical** symbol $\sqrt{}$. The number or expression under the radical symbol is called the **radicand**.

Order of Operations

- The following is the order in which operations are performed.
  1. If grouping symbols are present, perform all operations inside of them according to the following order. For grouping symbols within grouping symbols, start with the innermost group and work outward.
  2. Evaluate exponents and roots.
  3. Perform multiplication and division from left to right.
  4. Perform addition and subtraction from left to right.
- In addition to parentheses, brackets, and braces, fraction bars and radical symbols serve as grouping symbols.

Speaking the Language 1.6

1. A(n) ________ is used to indicate repeated multiplication.

2. In the expression b^n, we call b the ________.

3. In the expression $\sqrt{23}$, the number 23 is called the ________.

4. The conventional sequence of steps for evaluating a numerical expression is called the ________.

Exercises 1.6

Concepts and Skills

1. Use examples to illustrate the meanings of the words *base* and *radicand*.

2. Name five kinds of symbols that are used as grouping symbols. For each kind, give an example of an expression in which the grouping symbols are used.

In Exercises 3–10, write the given product in exponential form.

3. $7 \cdot 7 \cdot 7 \cdot 7 \cdot 7 \cdot 7$

4. $(-11) \cdot (-11) \cdot (-11) \cdot (-11)$

5. $-[(-5) \cdot (-5) \cdot (-5)]$

6. $-(2 \cdot 2 \cdot 2 \cdot 2 \cdot 2)$

7. $a \cdot a \cdot a$

8. $x \cdot x \cdot x \cdot y \cdot y$

9. $(2b)(2b)(2b)(2b)$

10. $(xy)(xy)(xy)$

In Exercises 11–16, write the given exponential form in factored form.

11. $\left(\frac{7}{9}\right)^4$

12. 11^5

13. $2x^2$

14. $-x^4$

15. 5^2y^3

16. $2x^3y^2$

In Exercises 17–28, evaluate the given exponential expression.

17. 5^2

18. 3^4

19. $(-2)^3$

20. $(-4)^2$

21. -7^2

22. -6^2

23. 2^5

24. 10^2

25. $\left(\frac{2}{3}\right)^2$

26. $\left(-\frac{3}{4}\right)^3$

27. $(-1)^6$

28. $(-1)^9$

In Exercises 29–34, use your calculator to perform the indicated operations. Round decimal answers to the nearest hundredth.

29. 12^5

30. 11^4

31. $(-3)^{11}$

32. $(-4)^9$

33. $(3.47)^7$

34. $(2.58)^8$

35. Explain why 36 has two square roots but it is *incorrect* to say $\sqrt{36} = 6$ or -6.

36. Explain how to verify that $\sqrt{2809} = 53$.

In Exercises 37–44, determine the square roots of the given number, if possible.

You may want to emphasize the difference between Exercises 37–44 and Exercises 45–50.

37. 4 **38.** 16

39. 36 **40.** 100

41. −25 **42.** 0

43. 529 **44.** 1764

In Exercises 45–50, evaluate each square root.

45. $\sqrt{4}$ **46.** $\sqrt{16}$ **47.** $-\sqrt{36}$

48. $-\sqrt{100}$ **49.** $\sqrt{\frac{4}{9}}$ **50.** $-\sqrt{\frac{25}{36}}$

In Exercises 51–58, determine the square root. Round results to the nearest hundredth.

51. $\sqrt{676}$ **52.** $\sqrt{1849}$

53. $\sqrt{8.41}$ **54.** $\sqrt{9.61}$

55. $\sqrt{534}$ **56.** $\sqrt{747}$

57. $\sqrt{63{,}876}$ **58.** $\sqrt{0.0076}$

59. Describe the order in which operations are performed for $(-6)^2$ and -6^2. Show that the two expressions have different values.

60. For the following two expressions, describe the difference in the order in which the operations are to be performed.

(i) $5 - 3^2$ (ii) $(5 - 3)^2$

In Exercises 61–78, perform the indicated operations.

61. $6 + 4(-3)$ **62.** $-3 - 2^3$

63. $\frac{240}{6 \cdot 4}$ **64.** $30 \div 5 \cdot 6$

65. $3 \cdot 6 - 4 \cdot 2$ **66.** $-3^2 + 6(-1)^5$

67. $-2|-4| + |5|$ **68.** $-2\sqrt{16} - \sqrt{6 + 3}$

69. $6 - 4 \cdot 2 + 20 \div 5$ **70.** $6 - 5(5 - 3)^2$

71. $\frac{5^2 + 7}{11 - 3^2}$ **72.** $\frac{4 + 6 \cdot 3}{-5 - 3 \cdot 2}$

73. $\frac{9 - 3}{3 - 9}$

74. $\frac{5 \cdot 6 - 2(-3)}{6 - (-6)}$

75. $12 - (5 + 3) \div 2 + 6$

76. $1 - (2 + 7) \div (5 + 4)$

77. $5 - [3(2 - 1) - 4(3 - 5)]$

78. $6 - 4[6 - 4(6 - 4)]$

CE In Exercises 79–82, translate the verbal description into a numerical expression. Then evaluate the expression.

79. Four less than the product of −2 and 7

80. Six more than the square of −8, with the result divided by 10

81. Add −3 to 7 and divide the sum by 2. Then subtract −4 from the quotient.

82. The square of the difference of 3 and −2

CE In Exercises 83 and 84, insert grouping symbols so that the numerical expression has the values given in parts (a), (b), and (c).

Exercises 83 and 84 can lead to some interesting class discussions.

83. $-3 \cdot 4 - 5$

(a) 3

(b) −17

(c) −7

84. $7 - 5^2 + 1$

(a) 5

(b) −19

(c) −17

In Exercises 85–96, perform the operations. Round decimal answers to the nearest hundredth.

Remind students that fractions and radicals have implied grouping symbols.

85. $-4 \cdot 3^2$

86. $-2^4 + 5^1 \cdot 2^3$

87. $-(-2)^4 + 3\sqrt{16}$

88. $2 \cdot 3^2 - 4 \cdot 3 + 1$

89. $5 - 4\sqrt{3 \cdot 7 - 5}$

90. $\frac{5^2 - 3^2}{\sqrt{5^2 - 3^2}}$

91. $(5 - 2)^2 + (-3 - 1)^2$

92. $\sqrt{(-7 - 1)^2 + (2 - 5)^2}$

93. $(-3)^2 - 4(2)(-1)$

94. $\sqrt{6^2 - 4(3)(1)}$

95. $\frac{5 - \sqrt{5^2 - 4(2)(1)}}{2(2)}$

96. $\frac{-3 + \sqrt{(-3)^2 - 4(1)(2)}}{2(1)}$

CE **97.** Complete the table by entering the values of $\sqrt{x}$ and x^2 for the given values of x. Cross out any entries in which the value is not a real number.

x	−4	−1	0	1	4	9	16	25
$\sqrt{x}$								
x^2								

CE **98.** Complete the table by entering the values of $\sqrt{x}$ and $\sqrt{-x}$ for the given values of x. Cross out any entries in which the value is not a real number.

x	-4	-1	0	1	4
$\sqrt{x}$					
$\sqrt{-x}$					

Real-Life Applications

99. Computer scientists often use binary (base-2) notation to represent numbers. The following is an example.

$$\begin{array}{ccccc} 2^4 & 2^3 & 2^2 & 2^1 & 1 \\ \downarrow & \downarrow & \downarrow & \downarrow & \downarrow \\ 1 & 0 & 1 & 1 & 0 \end{array} = 1 \cdot 2^4 + 0 \cdot 2^3 + 1 \cdot 2^2 + 1 \cdot 2^1 + 0 \cdot 1$$

$$= 16 + 0 + 4 + 2 + 0$$

$$= 22$$

If you were a systems engineer at Hewlett-Packard, how would you write 1101101 in decimal form?

100. Suppose that your checking account pays $\frac{1}{4}\%$ monthly interest on your balance and that your credit card company charges 1.5% monthly interest on the amount you owe. How much money will you gain or lose if your checking account balance is \$900 and you owe \$200 on your credit card?

101. A car rental agency charges you \$26 per day plus 12 cents per mile. Write a numerical expression that represents the total cost of a 160-mile trip. Which operation in your expression would you perform first?

102. As a social worker, you must tally your client base each month. You find that you are working with 40 families of which 30% are single-parent households and the rest are two-parent families. The average number of children for all the families is 2.6. What is the total number of men, women, and children in your client base?

Modeling with Real Data

103. Of the 12.8 million people who received Aid to Families with Dependent Children, 8.8 million were 18 years of age or younger. The average age was 7.4 years. (Source: Department of Health and Human Services.) The accompanying bar graph shows the percentages of those who received aid by age group. How many children under 6 years old received assistance?

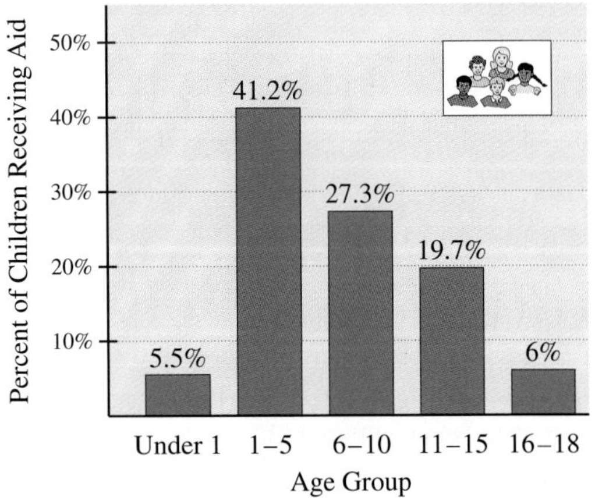

104. In 1999, Ken Griffey, Jr., averaged 1 home run for every 12.625 at-bats, whereas Mark McGwire averaged 1 home run for every 8.015 at-bats. Griffey had 606 at-bats and McGwire had 521 at-bats. (Source: CNNSI.) How many home runs did each hit?

Challenge

In Exercises 105 and 106, for what value(s) of x is the expression

(a) positive? (b) 0? (c) negative?

105. $-x^2$ **106.** $(-x)^2$

107. For the following expressions, $a > 0$, $b > 0$, and $c < 0$. Determine whether the given expression represents a positive number or a negative number or either.

(a) $a - c$ (b) $a + c$

(c) ab (d) ac

(e) $bc - a$ (f) a^2c

(g) ac^2 (h) $\dfrac{a+b}{c}$

108. The value of the following expression is 362,880.

$$1 \cdot 2 \cdot 3 \cdot 4 \cdot 5 \cdot 6 \cdot 7 \cdot 8 \cdot 9$$

Replace some of the multiplication symbols with $+$, $-$, and/or $\div$ so that the value of the resulting expression is 100.

109. One way to write 3 with three 2's is $2 \div 2 + 2$. Write 30 with three 3's.

110. Using four 4's, we can write 1 as $\frac{44}{44}$ and 2 as $\frac{4}{4} + \frac{4}{4}$. Write the numbers 3 through 10 with four 4's.

1.7 Properties of the Real Numbers

Commutative Properties • Associative Properties • Identity Properties • Inverse Properties • Special Multiplication Properties • Distributive Property

In previous sections, we informally used certain rules to perform operations with real numbers. In this section, we summarize and formally state these rules and other important properties of real numbers.

Commutative Properties

In addition, the order of the addends does not affect the sum. In multiplication, the order of the factors does not affect the product.

$$5 + (-8) = -8 + 5 \qquad -7 \cdot 4 = 4(-7)$$

Commutative Properties of Addition and Multiplication

The following statements are true for all real numbers a and b.

$a + b = b + a$ Commutative Property of Addition

$ab = ba$ Commutative Property of Multiplication

Note that subtraction and division are *not* commutative operations.

$$5 - 3 \neq 3 - 5$$
$$12 \div 6 \neq 6 \div 12$$

Associative Properties

In addition, the grouping of three or more addends does not affect the sum. In multiplication, the grouping of three or more factors does not affect the product.

$$-6 + [2 + (-5)] = [-6 + 2] + (-5)$$
$$2 \cdot [(-3)(-7)] = [2(-3)] \cdot (-7)$$

Associative Properties of Addition and Multiplication

The following statements are true for all real numbers, a, b, and c.

$(a + b) + c = a + (b + c)$ Associative Property of Addition

$(ab)c = a(bc)$ Associative Property of Multiplication

LEARNING TIP

Your instructor may or may not ask you to know the names of these properties. Try to use everyday words to assist you. For example, *commute* might suggest travel back and forth, as in $a + b = b + a$, and *associate* might refer to joining or grouping, as in $a(bc) = (ab)c$.

Subtraction and division are *not* associative operations.

$(12 - 8) - 2 = 2,$ but $12 - (8 - 2) = 6$

$(24 \div 4) \div 2 = 3,$ but $24 \div (4 \div 2) = 12$

The associative properties are used to *regroup* addends and factors, whereas the commutative properties are used to *reorder* addends and factors.

Although subtraction is neither commutative nor associative, we frequently wish to reorder and regroup an expression involving subtraction.

EXAMPLE 1

Reordering and Regrouping a Numerical Expression

Simplify $5 - 3 + (-7) - (-8) + 2$.

Solution

By first converting subtractions into additions, we can take advantage of the commutative and associative properties of addition to group the positive numbers and the negative numbers.

Discuss ways to order and group numbers to make mental arithmetic easier.

$$5 - 3 + (-7) - (-8) + 2$$
$$= 5 + (-3) + (-7) + 8 + 2 \quad \text{Definition of Subtraction}$$
$$= 5 + 8 + 2 + (-3) + (-7) \quad \text{Commutative Property of Addition}$$
$$= [5 + 8 + 2] + [(-3) + (-7)] \quad \text{Associative Property of Addition}$$
$$= 15 + (-10)$$
$$= 5$$

Identity Properties

The sum of a number and 0 is the original number, and the product of a number and 1 is the original number. We call 0 the **additive identity**, and we call 1 the **multiplicative identity**.

Identity Properties

The following statements are true for any real number a.

$a + 0 = 0 + a = a$ Additive Identity Property

$a \cdot 1 = 1 \cdot a = a$ Multiplicative Identity Property

Inverse Properties

The sum of a number and its opposite is 0, which is the additive identity. The formal statement of this property usually uses *additive inverse* rather than *opposite*.

Property of Additive Inverses

For every real number a, there is a unique number $-a$ such that

$$a + (-a) = -a + a = 0$$

A corresponding rule exists for multiplication. Each *nonzero* real number a has a **multiplicative inverse**, or **reciprocal**, represented by the symbol $\frac{1}{a}$. The product of a nonzero number and its reciprocal is 1, which is the multiplicative identity.

Property of Multiplicative Inverses

For every nonzero real number a, there is a unique number $\frac{1}{a}$ such that

$$a \cdot \frac{1}{a} = \frac{1}{a} \cdot a = 1$$

From the sign rules for multiplication, we can see that a number and its reciprocal always have the same sign.

For a nonzero rational number $\frac{p}{q}$, the reciprocal is $\frac{q}{p}$. For example, because $\frac{3}{4} \cdot \frac{4}{3} = 1$, the reciprocal of $\frac{3}{4}$ is $\frac{4}{3}$.

EXAMPLE 2

Inverses of Real Numbers

Write the additive inverse (opposite) and the multiplicative inverse (reciprocal) for each number.

(a) -12 (b) $\frac{9}{5}$

Solution

	Additive Inverse	*Multiplicative Inverse*	
(a) -12	12	$-\frac{1}{12}$	$\frac{1}{-12} = -\frac{1}{12}$
(b) $\frac{9}{5}$	$-\frac{9}{5}$	$\frac{5}{9}$	

In Section 1.5 we defined the quotient $a \div b$ $(b \neq 0)$ as the number c such that $c \cdot b = a$. Division is sometimes defined as follows.

Alternative Definition of Division

For real numbers a and b, $b \neq 0$,

$$a \div b = a \cdot \frac{1}{b}$$

In words, to divide a by b, multiply a by the reciprocal of b.

Special Multiplication Properties

In Section 1.4 we stated two multiplication properties that involve 0 and -1.

Multiplication Properties of 0 and −1

The following statements are true for any real number a.

$a \cdot 0 = 0 \cdot a = 0$ — Multiplication Property of 0

$a \cdot (-1) = -1 \cdot a = -a$ — Multiplication Property of −1

Distributive Property

All the properties summarized so far have involved either addition or multiplication. The Distributive Property involves both operations.

Distributive Property

For all real numbers a, b, and c,

$$a(b + c) = ab + ac$$

More precisely, the foregoing property is the Distributive Property of Multiplication Over Addition. There is also a Distributive Property of Multiplication Over Subtraction.

$$a(b - c) = ab - ac$$

We will refer to both versions simply as the Distributive Property.

The Distributive Property can be used to change a product of the form $a(b + c)$ into a sum of the form $ab + ac$ or to change a product of the form $a(b - c)$ into a difference of the form $ab - ac$. We say that we are using the Distributive Property to *multiply*. Note that in doing so, we are removing the parentheses.

EXAMPLE 3 **Using the Distributive Property to Multiply**

Use the Distributive Property to rewrite each product as a sum or difference.

(a) $4(x + 5)$ (b) $a(x - 2)$

(c) $-3(y + 2)$ (d) $-7(x - 3)$

Solution

(a) $4(x + 5) = 4 \cdot x + 4 \cdot 5 = 4x + 20$

(b) $a(x - 2) = a \cdot x - a \cdot 2 = ax - 2a$

(c) $-3(y + 2) = -3 \cdot y + (-3) \cdot 2$
$= -3y + (-6) = -3y - 6$

(d) $-7(x - 3) = -7 \cdot x - (-7) \cdot 3$
$= -7x - (-21) = -7x + 21$

Think About It

Consider the process of evaluating $3(7 + 2)$ and simplifying $3(x + 2)$. Although the Distributive Property can be applied in both expressions, why is it essential in one case but not in the other?

The Distributive Property can also be written in reverse.

$$ab + ac = a(b + c) \qquad ab - ac = a(b - c)$$

With this version, we are changing a sum of the form $ab + ac$ into a product of the form $a(b + c)$ or changing a difference of the form $ab - ac$ into a product of the form $a(b - c)$. In this case, we are using the Distributive Property to *factor*.

EXAMPLE 4

Using the Distributive Property to Factor

Use the Distributive Property to rewrite each sum or difference as a product.

(a) $3x + 3y$ (b) $ac - bc$

(c) $5x - 10$ (d) $-4a - 4z$

Solution

(a) $3x + 3y = 3 \cdot x + 3 \cdot y = 3(x + y)$

(b) $ac - bc = a \cdot c - b \cdot c = c(a - b)$

(c) $5x - 10 = 5 \cdot x - 5 \cdot 2 = 5(x - 2)$

(d) $-4a - 4z = -4a + (-4z)$

$= -4 \cdot a + (-4) \cdot z$

$= -4(a + z)$

In Section 1.3, we stated the following property.

Property of the Opposite of a Difference

For any real numbers a and b, $-(a - b) = b - a$.

This is actually a special case of the Distributive Property, along with the fact that $-x = -1x$. Consider, for example, the expression $-(x - 8)$.

$-(x - 8)$	$= -1(x - 8)$	Multiplication Property of -1
	$= -1 \cdot x - (-1) \cdot 8$	Distributive Property
	$= -1x - (-8)$	
	$= -x + 8$	
	$= 8 + (-x)$	Commutative Property of Addition
	$= 8 - x$	

The Property of the Opposite of a Difference allows us to write $-(x - 8)$ as $8 - x$ without all those steps.

EXAMPLE 5

Identifying the Properties of Real Numbers

Identify the property that justifies the given statement.

(a) $x + (2 + x) = x + (x + 2)$	Commutative Property of Addition
(b) $1 \cdot (y - 2) = y - 2$	Multiplicative Identity Property
(c) $(x + 5) - 1 = x + (5 - 1)$	Associative Property of Addition

(d) $-3(2 - y) = -6 + 3y$ — Distributive Property

(e) $\frac{1}{5} \cdot 5 = 1$ — Property of Multiplicative Inverses

(f) $0 \cdot (5x) = 0$ — Multiplication Property of 0

(g) $-(3b) + (3b) = 0$ — Property of Additive Inverses

(h) $-(3 - w) = w - 3$ — Property of the Opposite of a Difference

EXAMPLE 6

Using the Properties to Rewrite Expressions

Use the given property to rewrite the expression.

(a) Commutative Property of Multiplication: $x \cdot 3 =$ ______

(b) Distributive Property: $2a + 3a =$ ______

(c) Associative Property of Multiplication: $\frac{1}{2}(6x) =$ ______

(d) Additive Identity Property: $0 + y =$ ______

Solution

(a) $x \cdot 3 = 3x$

(b) $2a + 3a = (2 + 3)a$

(c) $\frac{1}{2}(6x) = \left(\frac{1}{2} \cdot 6\right) \cdot x$

(d) $0 + y = y$

Quick Reference 1.7

The following is a summary of the properties of real numbers, where *a*, *b*, and *c* represent any real number.

Commutative Properties

- $a + b = b + a$ — Commutative Property of Addition
- $ab = ba$ — Commutative Property of Multiplication
- Subtraction and division are not commutative.
- The commutative properties are used to reorder addends and factors.

Associative Properties

- $(a + b) + c = a + (b + c)$ — Associative Property of Addition
- $(ab)c = a(bc)$ — Associative Property of Multiplication
- Subtraction and division are not associative.
- The associative properties are used to regroup addends and factors.

Identity Properties

- $a + 0 = 0 + a = a$ — Additive Identity Property
- $a \cdot 1 = 1 \cdot a = a$ — Multiplicative Identity Property

Inverse Properties

- $a + (-a) = -a + a = 0$ — Property of Additive Inverses
- $a \cdot \frac{1}{a} = \frac{1}{a} \cdot a = 1 \quad (a \neq 0)$ — Property of Multiplicative Inverses
- For any real number a, $-a$ is unique; for any nonzero real number a, $\frac{1}{a}$ is unique.
- An alternative definition of division is $a \div b = a \cdot \frac{1}{b}$, where $b \neq 0$.

Special Multiplication Properties

- $a \cdot 0 - 0 \cdot a - 0$ — Multiplication Property of 0
- $a \cdot (-1) = -1 \cdot a = -a$ — Multiplication Property of -1

Distributive Property

- $a(b + c) = ab + ac$ and $a(b - c) = ab - ac$
- We use $a(b + c) = ab + ac$ and $a(b - c) = ab - ac$ to *multiply;* we use $ab + ac = a(b + c)$ and $ab - ac = a(b - c)$ to *factor.*

Speaking the Language 1.7

1. The ______ properties allow us to group addends and factors.
2. The product of a number and its ______ is the multiplicative identity.
3. We can use the ______ Property to remove the parentheses in the expression $4(c - 3)$.
4. Writing an indicated sum as an indicated product is called ______ the expression.

Exercises 1.7

Concepts and Skills

1. Although subtraction is not commutative, it is correct to rewrite $-9 - 4x + x^2$ as $x^2 - 4x - 9$. Why?
2. (a) Name a number that is its own opposite. Show how your answer satisfies the Property of Additive Inverses.

   (b) Name two numbers that are their own reciprocals. Show how your answers satisfy the Property of Multiplicative Inverses.

In Exercises 3–16, name the property that is illustrated.

3. $3 + (2 + b) = (3 + 2) + b$
4. $-x + 2 = 2 - x$
5. $4x + 0 = 4x$
6. $-6x + 6x = 0$
7. $4(x - 5) = 4x - 20$
8. $(x^2 + 1) \cdot \frac{1}{x^2 + 1} = 1$
9. $\frac{2}{3}x - \frac{2}{3} = \frac{2}{3}(x - 1)$
10. $10 \cdot \left(\frac{1}{2}x\right) = \left(10 \cdot \frac{1}{2}\right)x$
11. $5 \cdot 1 = 5$
12. $(2x - 7) \cdot 0 = 0$
13. $-1 \cdot (2y) = -2y$
14. $-3 - x = -1 \cdot (3 + x)$
15. $5x - 3x = (5 - 3)x$
16. $(y + 5)(y - 3) = (y - 3)(y + 5)$

In Exercises 17–24, match the expression in column A with the corresponding expression in column B.

Column A	*Column B*
17. $x \cdot (-7)$	(a) $y - x$
18. $-x + y$	(b) $(y + 2) + 9$
19. $y + (2 + 9)$	(c) $2x - y$
20. $3(x + 7)$	(d) 1
21. $1(2x - y)$	(e) 0
22. $-(x - 2)$	(f) $(2 \cdot 3)x$
23. $2(3x)$	(g) $3x + 21$
24. $2x \cdot \frac{1}{2x}\ (x \neq 0)$	(h) $-1(x - 2)$
	(i) $-7x$

In Exercises 25–34, use the given property to write the expression in an alternative form.

25. Associative Property of Addition:

$(x + 4) + 3 = \underline{\qquad\qquad}$

26. Commutative Property of Multiplication:

$(x + 6) \cdot 3 = \underline{\qquad\qquad}$

27. Associative Property of Multiplication:

$6(2x) = \underline{\qquad\qquad}$

28. Commutative Property of Addition:

$8 - x = \underline{\qquad\qquad}$

29. Additive Identity Property:

$(x + 0) + 6 = \underline{\qquad\qquad}$

30. Property of Additive Inverses:

$-(y + 3) + (y + 3) = \underline{\qquad\qquad}$

31. Property of Multiplicative Inverses:

$\left(3 \cdot \frac{1}{3}\right)x = \underline{\qquad\qquad}$

32. Multiplicative Identity Property:

$1(n - 4) = \underline{\qquad\qquad}$

33. Distributive Property: $12z + 15 = \underline{\qquad\qquad}$

34. Distributive Property:

$-3[a + (-6)] = \underline{\qquad\qquad}$

35. Explain why 0 does not have a reciprocal.

36. Use the definition of reciprocal to show why the reciprocal of $\frac{p}{q}$ is $\frac{q}{p}$, where $p \neq 0$, $q \neq 0$.

CE In Exercises, 37–42, fill in the blank and state the property that justifies your answer.

37. $5 \cdot \underline{\qquad} = 1$

38. $-7 + \underline{\qquad} = 0$

39. $-\frac{2}{3} \cdot \underline{\qquad} = -\frac{2}{3}$

40. $8 + (-x + x) = 8 + \underline{\qquad}$

41. $\underline{\qquad}(2x - 5) = -2x + 5$

42. $-x + 5 = -1 \cdot \underline{\qquad}$

In Exercises 43–46, name the property or definition that justifies each step.

43. $(y + 0) + (-y)$

$= (0 + y) + (-y)$ (a) $\underline{\qquad\qquad}$

$= 0 + [y + (-y)]$ (b) $\underline{\qquad\qquad}$

$= 0 + 0$ (c) $\underline{\qquad\qquad}$

$= 0$ (d) $\underline{\qquad\qquad}$

44. $-(2x - 3)$

$= -1(2x - 3)$ (a) $\underline{\qquad\qquad}$

$= -2x + 3$ (b) $\underline{\qquad\qquad}$

$= 3 - 2x$ (c) $\underline{\qquad\qquad}$

45. $(a - b) + b$

$= [a + (-b)] + b$ (a) $\underline{\qquad\qquad}$

$= a + [(-b) + b]$ (b) $\underline{\qquad\qquad}$

$= a + 0$ (c) $\underline{\qquad\qquad}$

$= a$ (d) $\underline{\qquad\qquad}$

46. $5\left(\frac{1}{5}x\right)$

$= \left(5 \cdot \frac{1}{5}\right)x$ (a) $\underline{\qquad\qquad}$

$= 1x$ (b) $\underline{\qquad\qquad}$

$= x$ (c) $\underline{\qquad\qquad}$

CE In Exercises 47–54, use the associative properties to rewrite the given expression in simpler or exponential form.

47. $(x + 3) + 2$

48. $-4 + (8 + y)$

49. $(2z + 7) - 2$

50. $(3x - 5) + 7$

51. $-5(3x)$

52. $(2y)y$

53. $-\frac{5}{6} \cdot \left(-\frac{9}{10}x\right)$

54. $\frac{3}{4}(2xy)$

In Exercises 55–62, use the Distributive Property to write the product as a sum or difference.

55. $5(4 + 3b)$
56. $-2(3a - 7)$
57. $-\frac{3}{5}(5x + 20)$
58. $\frac{1}{3}(12x - 3)$
59. $-3(5 - 2x)$
60. $-4(-3x + 5)$
61. $x(y + z)$
62. $s(t + 3)$

In Exercises 63–70, use the Distributive Property to write the sum or difference as a product.

63. $5x + 5y$
64. $11a - 11b$
65. $3x + 12$
66. $7y - 14$
67. $5y + 5$
68. $2 + 2b$
69. $6 - 3x$
70. $2x - 2$

In Exercises 71–74, write the given expression without grouping symbols.

71. $-(x - 4)$
72. $-(3 + 5x)$
73. $-(2x + 5y - 3)$
74. $-(-4x - y + 7)$

CE In Exercises 75–82, use the commutative and associative properties to rewrite the given expression in simpler or exponential form.

75. $(3x) \cdot 4$
76. $y(3y)$
77. $(-6x)(2x)$
78. $3x \cdot 4y$
79. $(5 + x) + 3$
80. $x + [7 + (-x)]$
81. $-5x + (3 + 5x)$
82. $-7 + (2x - 3)$

CE In Exercises 83–86, use the commutative and associative properties to write the given numerical expression so that the operations can be easily performed. Then evaluate the expression.

83. $\frac{2}{9} + \frac{3}{7} + \frac{5}{9} + \frac{4}{7} + \frac{2}{9}$

84. $77 + 40 + 23 + 30 + 70 + 60$

85. $2 \cdot (-87) \cdot 5 \cdot (-1)$

86. $5 \cdot \frac{1}{3} \cdot \frac{1}{5} \cdot 3$

The Distributive Property can be useful in performing multiplication mentally. The following is an example.

$$\begin{aligned} 25 \cdot 103 &= 25 \cdot (100 + 3) \\ &= 25 \cdot 100 + 25 \cdot 3 \\ &= 2500 + 75 \\ &= 2575 \end{aligned}$$

CE In Exercises 87–90, show how you would use the preceding method to determine the given product mentally.

87. $7 \cdot 108$
88. $30 \cdot 109$
89. $15 \cdot 98$
90. $4 \cdot 49$

91. As a baseball statistician, you find that you must frequently multiply a number by 9. Use the Distributive Property to explain why multiplying the number by 10 and then subtracting the number produces the desired result.

92. Explain why the opposite of a nonzero number can never be the same as the reciprocal of the number.

In Exercises 93–100, give the additive inverse (opposite) and the multiplicative inverse (reciprocal), if any, of the given number or expression. Assume that all variable expressions represent positive numbers.

93. $\frac{5}{7}$
94. -8
95. $\frac{2}{3}x$
96. $\frac{2}{y}$
97. $x + 3$
98. $y - 5$
99. $4 - 3x$
100. $2y$

Real-Life Applications

101. Suppose that to pay your bills during the month of June you wrote checks totaling the exact amount of your monthly paycheck, which you deposited on June 1. Which property of real numbers guarantees that the sum of your monthly income and expenses is 0?

102. For a theater-in-the-round production, the stage designer decides to use a triangular stage. To purchase construction material, the designer must calculate the stage area, which is found by multiplying 0.5 times the base of the triangle times the height of the triangle. Which property of real numbers guarantees that the calculated area will be the same no matter which two of the three numbers are multiplied first?

103. A tax auditor adds a column of numbers from top to bottom. She then checks her work by adding the column from bottom to top. Which property of real numbers guarantees that the sum is the same either way?

104. At the Two-Dollar Shop, all merchandise is sold for \$2. During one month, 20 key chains and 35 birthday balloons were sold. Write two numerical expressions, one a product and the other a sum, that represent the total income from these items. Then identify the property that guarantees that the expressions have the same value.

Challenge

In Exercises 105–110, for nonzero numbers a and b, indicate whether the given statement is always true, always false, or sometimes true. If sometimes, give examples of a and b for which the statement is true and for which the statement is false.

105. $a - b = b - a$

106. $\frac{a}{b} = \frac{b}{a}$

107. $|a + b| = |a| + |b|$

108. $|a - b| = |a| - |b|$

109. $(a + b)^2 = a^2 + b^2$

110. $\sqrt{a + b} = \sqrt{a} + \sqrt{b}$

111. Show how the Distributive Property can be used to prove that

$$\frac{a+b}{3} = \frac{a}{3} + \frac{b}{3}$$

1 • Chapter Project

Air Pressure and Altitude

The air in an ordinary drinking glass weighs about the same as an aspirin tablet, but the weight of the entire atmosphere is equivalent to the weight we would feel if we were 34 feet under water. Although air pressure varies with weather conditions, the average air pressure at sea level is 14.7 pounds per square inch. The accompanying table shows how air pressure decreases with altitude.

Altitude (feet)	Air Pressure (pounds per square inch)
Sea level	14.7
10,000	10.2
20,000	6.4
30,000	4.3
40,000	2.7
50,000	1.6

1. Produce a bar graph that visually represents the information in the table.
2. Connect the tops of the bars in your bar graph with a smooth curve.
3. Referring to Section 1.3, Exercises 97–100, use your bar graph and curve to estimate the air pressure at the six highest points listed.
4. Determine the highest point in your state, and estimate the air pressure at that point.
5. Suppose you were standing atop Mount Everest and your altimeter showed an air pressure of 4.5 pounds per square inch. Estimate the altitude of Mount Everest.
6. The highest point in the United States is Mount McKinley, Alaska, at 20,320 feet. The lowest point is Death Valley, California, at 282 feet below sea level. For these two points, determine the difference in their elevations and estimate the difference in the air pressure.

Chapter 1 Review Exercises

Section 1.1

In Exercises 1–7, determine whether the statement is true or false.

1. The number -8 is a natural number.
2. The number $-\pi$ is a rational number.
3. Every whole number is also an integer.
4. Because the decimal form of $\sqrt{11}$ is nonterminating and nonrepeating, $\sqrt{11}$ is an irrational number.
5. The value of the expression $\frac{\sqrt{4.8}}{\sqrt{2.7}}$ is an irrational number.
6. The statement $n > 0$ means that n represents a positive number.
7. For any real number n, $-n$ represents a negative real number.
8. Let $A = \left\{-2, \frac{5}{9}, \frac{\pi}{2}, -\sqrt{10}, -\sqrt{9}, 2.34, 5\right\}$. List all the members of set A that are members of the following sets.
   (a) Integers
   (b) Rational Numbers
   (c) Irrational Numbers
9. Explain why the statement $0.5 < \frac{1}{2}$ is false, whereas the statement $0.5 \leq \frac{1}{2}$ is true.
10. The absolute value of a number is nonnegative, but there are numbers n such that $|n| = -n$. Why?
11. Determine the value(s) of n under each of the following conditions.
    (a) The number n is 3 units from 1 on the number line.
    (b) The absolute value of n is 7.
12. Insert $<$ or $>$ to make the statement true.
    (a) $-\frac{28}{29}$ ___ $-\frac{27}{28}$
    (b) 3.1416 ___ π
    (c) $|-3|$ ___ -3
    (d) $-\sqrt{3}$ ___ $-\frac{\pi}{2}$

Section 1.2

13. Explain the meaning of the word *addends*. Give an example.
14. Under what circumstances is subtraction used to determine a sum?

In Exercises 15–20, determine the sum.

15. $8 + (-11)$
16. $(-12) + (-8)$
17. $-\frac{2}{3} + \frac{3}{4}$
18. $-9 + \left(-\frac{4}{5}\right)$
19. $-|-2| + |-5|$
20. $-10 + (-11) + 5 + (-7) + 9$
21. The sum of what number and -5 is 6?
22. What is 3 increased by -5?
23. On a certain day, the high temperature was 26°F and the low was -4°F. What was the average temperature for the day?
24. After writing a check for \$325.50, a person discovers that his bank account is now overdrawn by \$6.38. What was the checking account balance before the check was written?

Section 1.3

25. What are the meanings of the words *minuend* and *subtrahend*? Give an example.
26. Compare $a - b$ with $b - a$ for any two real numbers a and b.

In Exercises 27–32, perform the indicated operations.

27. $7 - (-13)$
28. $(-15) - (-6)$
29. $-\frac{2}{5} - \frac{3}{4}$
30. $-7 - \left(-\frac{2}{3}\right)$
31. $-|-3| - |-7|$
32. $-10 - (-11) - 5 - (-7) - 9$
33. What number must be subtracted from -5 to obtain -9?
34. On the number line, what is the distance from point A at -5 to point B at 8?
35. The mean (average) elevation of the continental United States is 2500 feet. If Montana's highest point is Granite Peak at 12,850 feet, what is the difference between the highest point in Montana and the mean elevation?
36. The temperature range for any day is the difference between the high and low temperatures. What was the temperature range for a day when the high was 19°F and the low was -8°F?

Section 1.4

37. What do we mean by the *factors* of a product?

38. Suppose a product consists of 11 nonzero factors. Describe a quick way to determine the sign of the product.

In Exercises 39–46, multiply.

39. $5 \cdot (-4)$

40. $-7 \cdot (-5)$

41. $-6 \cdot \left(-\frac{5}{12}\right)$

42. $\frac{7}{8} \cdot \left(-\frac{4}{3}\right)$

43. $-\frac{20}{33} \cdot \left(-\frac{22}{15}\right) \cdot \left(-\frac{3}{16}\right)$

44. $(-1)(-2)(-1)(-3)(-3)$

45. $(-1)(-3)(0)(4)$

46. $-|-4| \cdot (-|-5|)$

47. What number must be multiplied by -2 to obtain 9?

48. The product of three factors is 0. If two of the factors are 5 and -5, what is the third factor?

49. A water tank is being drained at a rate of 100 gallons per minute. At the same time, the tank is being replenished at a rate of 60 gallons per minute. At 1:15 P.M., how much water is in the tank compared to the amount that was in the tank at noon?

50. Each of 30 students is given either $+1$ or -1. If the sum of their numbers is $+2$, what is the product of their numbers?

Section 1.5

51. Use an example to illustrate the meanings of the words *divisor* and *dividend*.

52. Using the definition of division, explain why division by 0 is undefined.

In Exercises 53–58, divide, if possible.

53. $\frac{10}{-2}$

54. $\frac{-45}{-9}$

55. $0 \div (-9)$

56. $-\frac{3}{5} \div \frac{5}{6}$

57. $-\frac{4}{9} \div (-9)$

58. $\frac{3}{16} \div \left(-\frac{6}{7}\right)$

59. If the quotient is -12 and the dividend is 6, what is the divisor?

60. If 9.6 inches of snow fell during the previous night, and if it then melts at an average rate of 0.4 inch per hour, how long will it take for the snow to melt completely?

Section 1.6

61. Explain the order of operations used in evaluating -5^2 and $(-5)^2$. Are the values the same?

62. Is the statement $\sqrt{-25} = -\sqrt{25}$ true? Explain your answer.

63. Write $(-3)(-3)(-3)(-3)(-3)$ in exponential form. Then evaluate the exponential expression.

64. Translate the following verbal description into a numerical expression. Then evaluate the expression.

    Multiply the sum of -7 and the square of 3 by -5. Then add -21 to the product.

In Exercises 65–72, perform the indicated operations.

65. 2^4

66. 4^2

67. 0^{33}

68. $(-1)^{33}$

69. $(-1)^{36}$

70. $\left(-\frac{2}{3}\right)^3$

71. $\sqrt{49}$

72. $-\sqrt{49}$

In Exercises 73–78, perform the indicated operations.

73. $5 + 7 \cdot 2 - 6 \div 3 + 8$

74. $8 + 2[8 + 2(8 + 2)]$

75. $-6^2 - 4(5 - 2^3)$

76. $18 - 7 + 2 - 8 \div 2$

77. $18 - (7 + 2) - 8 \div 2$

78. $\frac{2 \cdot (-7) - 4}{9 + 6 \cdot (-3)}$

Section 1.7

79. Explain why we can write $-x$ as $-1x$.

80. Consider the task of putting on your shoes and socks. Is this task commutative? Explain your answer.

In Exercises 81–86, name the property that is illustrated.

81. $3 + (5 + c) = 3 + (c + 5)$

82. $1 \cdot 6 = 6$

83. $4 \cdot (3 \cdot x) = (4 \cdot 3) \cdot x$

84. $7(c + d) = 7c + 7d$

85. $6 + 0 = 6$

86. $8 \cdot (-1) = -8$

In Exercises 87–92, use the stated property to write the expression in an alternative form.

87. Commutative Property of Multiplication:

    $(c + 5) \cdot 7 =$ ______

88. Associative Property of Addition:

    $x + (-2 + 2) =$ ______

89. Distributive Property:

$5 \cdot x + 5 \cdot 8 =$ ______

90. Property of Additive Inverses:

$x + (-2 + 2) =$ ______

91. Multiplicative Identity Property:

$-5\left(-\frac{1}{5}\right) \cdot 1 =$ ______

92. Multiplication Property of -1:

$-1 \cdot 7 =$ ______

Looking Back

The following exercises review concepts and skills that you will need in Chapter 2.

In Exercises 1–4, evaluate the given expression.

1. $-3 - 2(-5 + 1)$

2. $\dfrac{(-1)^2 - 3}{-1 - 4}$

3. $\sqrt{3(7) + 4}$

4. $|7 - 2(15)|$

5. What is the interest on \$4500 at 7% simple interest?

6. At an average speed of 54 mph, the driving time from one city to another is 2 hours and 20 minutes. What is the distance between the cities?

In Exercises 7–10, use the Distributive Property to rewrite the given expression.

7. $3(2x - 7)$

8. $-(3 - y)$

9. $ax + 3x$

10. $-4a - 4b$

11. Determine the distance between the points -7 and 3.

12. Explain why $(5 - 8)^2$ and $|5 - 8|^2$ have same value.

Chapter 1 Test

In Questions 1–3, state which elements of the following set belong to the given sets.

$$\left\{-3.7, -2, -0.\overline{14}, 0, \frac{2}{3}, \pi, \sqrt{7}, 6, \sqrt{9}\right\}$$

1. Integers

2. Irrational numbers

3. Rational numbers

4. Evaluate the following.

(a) $-(-5)$ (b) $|-5|$ (c) $-|-5|$

5. On the number line, what is the distance between the points whose coordinates are -8 and 15?

6. Graph the given set on a number line.

$$\left\{-5, -\sqrt{2}, \frac{1}{2}, 2.7\right\}$$

In Questions 7–16, evaluate the given expression.

7. $-7 - (-3)$

8. $\left(-\dfrac{3}{10}\right)\left(\dfrac{2}{9}\right)$

9. $\dfrac{2 - (-3)}{3(7) - 2(-8)}$

10. $3 - 2[4 - (2 - 3)]$

11. $\dfrac{-3 + \sqrt{3^2 - 4(2)}}{2}$

12. $|6 - 2(5)|$

13. $-\dfrac{3}{4} + \left(-\dfrac{1}{2}\right)$

14. $-2.5 - 6.7$

15. $(-2)(5)(-4)(0)(6)$

16. $-32 \div (-4)$

17. If possible, perform the indicated operation. If the operation is not defined, explain why.

(a) $\dfrac{0}{-3}$ (b) $\dfrac{-3}{0}$

In Questions 18–20, use the indicated property to complete the statement.

18. Distributive Property: $3(x - 4) =$ ______

19. Commutative Property of Addition:

$4 - x =$ ______

20. Associative Property of Addition:

$$(3 + t) - 3r = \underline{\qquad\qquad}$$

21. What is the opposite of $-|-4|$?

22. What is the reciprocal of -2?

23. Evaluate the following expression.

$$\frac{-3.7 + (-4.13)}{4 - (3.1)^2}$$

24. The sum of the numbers 4 and -9 is multiplied by 2. The result is squared and then divided by 3 less than -17. Write the numerical expression and evaluate it.

25. A certain pharmaceutical is manufactured under very carefully controlled temperature conditions. If the temperature in the lab is -2.45°C and is decreased by 1.04°C, what is the new temperature?

Chapter 2

Expressions and Functions

As more and more states have enacted helmet laws for bicycle riders, the use of helmets has increased dramatically. The bar graph shows the percent of riders who wore helmets for selected years in the period 1992–1998.

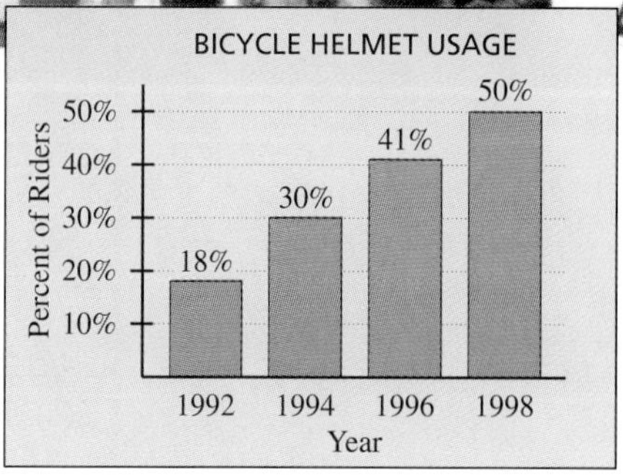

(Source: U.S. Bicycle Federation.)

We can write an ***algebraic expression*** *to model the trend in helmet usage during this time. By applying this expression, we can estimate the usage in 1997, for example, or predict the usage in 2002. (For more on this topic, see Exercises 91–94 at the end of Section 2.1, and see the Chapter Project.)*

In this chapter, we consider expressions and how to evaluate and simplify them. Equations and formulas are presented as methods for modeling conditions and relationships. We then introduce the coordinate plane as a model for representing ordered pairs of numbers. Finally, we define a relation and a function, we learn how to recognize and graph a function, and we use a special notation for evaluating functions.

2.1 Algebraic Expressions and Formulas

Evaluating Algebraic Expressions • Equations and Formulas • Modeling with Real Data

Evaluating Algebraic Expressions

In the previous chapter, we described a **numerical expression** as any combination of numbers and operations. We also described a **variable** as a symbol (usually a letter) used to represent an unknown number.

We can use variables to represent numbers in expressions. An **algebraic expression** is any combination of numbers, variables, grouping symbols, and operation symbols. The following are examples of algebraic expressions.

$$x^2 + 3x - 4 \qquad \frac{a^2 + 4}{6} \qquad 5(x + 6y) - 4 \qquad r - \sqrt{2t}$$

When we replace the variables in an algebraic expression with specific values, the expression becomes a numerical expression. Performing all indicated operations to determine the value of the expression is called **evaluating the expression.**

EXAMPLE 1 Evaluating Algebraic Expressions

Remind students to use the correct Order of Operations.

Evaluate each expression for the given value(s) of the variable(s).

(a) $x^2 + 3x - 4, \quad x = -2$

(b) $5x + 2(1 - x), \quad x = -3$

(c) $3x - |y - 2x|, \quad x = 3, y = 1$

(d) $\dfrac{a + 2b}{2a^2b}, \quad a = -1, b = 3$

Solution

(a) Replace each occurrence of x with -2. Then evaluate the resulting numerical expression.

$$x^2 + 3x - 4 = (-2)^2 + 3(-2) - 4 = 4 - 6 - 4 = -6$$

(b) Replace each occurrence of x with -3.

$$\begin{aligned} 5x + 2(1 - x) &= 5(-3) + 2[1 - (-3)] \\ &= -15 + 2[4] \\ &= -7 \end{aligned}$$

(c) Replace x with 3 and y with 1.

$$3x - |y - 2x| = 3(3) - |1 - 2(3)| = 9 - |-5| = 9 - 5 = 4$$

(d) $\dfrac{a + 2b}{2a^2b} = \dfrac{-1 + 2 \cdot 3}{2(-1)^2 \cdot 3} = \dfrac{5}{6}$ Replace a with -1 and b with 3.

Equations and Formulas

An **equation** is a statement that two expressions have the same value. If both expressions are numerical, then the truth of the equation can be determined by simply evaluating the expressions.

If one or both sides of an equation are algebraic expressions, then the equation may be true or false, depending on the replacement for the variable. A value of the variable that makes the equation *true* is called a **solution** of the equation.

EXAMPLE 2

Testing Solutions of an Equation

For each equation, determine whether the given number is a solution.

(a) $5 - x^2 = 3x - (2x + 1)$; 2 (b) $\sqrt{4 - x} - \sqrt{x + 6} = 2$; 3

Solution

(a) $5 - x^2 = 3x - (2x + 1)$

$5 - 2^2 = 3 \cdot 2 - (2 \cdot 2 + 1)$ Replace x with 2.

$1 = 1$ True

The last equation is true, so 2 is a solution.

(b) $\sqrt{4 - x} - \sqrt{x + 6} = 2$

$\sqrt{4 - 3} - \sqrt{3 + 6} = 2$ Replace x with 3.

$1 - 3 = 2$

$-2 = 2$ False

The last equation is false, so 3 is not a solution.

A **formula** is an equation that uses an expression to represent some specific quantity. For example, if the base of a triangle is b and the height is h, then the formula $A = \frac{1}{2}bh$ tells us how to determine the area A of the triangle.

An important formula is contained in the **Pythagorean Theorem.**

The Pythagorean Theorem is introduced now to prepare for the Distance Formula later.

The Pythagorean Theorem

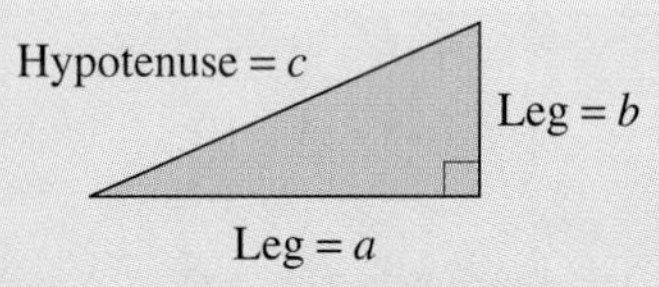

A triangle is a **right triangle** if and only if the sum of the squares of the **legs** is equal to the square of the **hypotenuse.** Symbolically, $a^2 + b^2 = c^2$.

We can use the formula in the Pythagorean Theorem to determine whether a triangle is a right triangle.

EXAMPLE 3

LEARNING TIP

When you work with the Pythagorean Theorem, begin by identifying the hypotenuse. It is the longest side, and it is located opposite the right angle.

Determining Whether a Triangle Is a Right Triangle

The three numbers given in each part are the lengths of the sides of a triangle. Determine whether the triangle is a right triangle.

(a) 2, 4, 5 (b) 5, 13, 12

Solution

Ask students why the assignment of c is critical, whereas the assignments of a and b are unimportant.

(a) We must determine whether the Pythagorean Theorem formula is true for the given lengths. The longest side is 5, so $c = 5$. It does not matter how we assign the values for a and b.

$$a^2 + b^2 = c^2$$

$$2^2 + 4^2 = 5^2 \quad \text{Let } a = 2, b = 4, \text{and } c = 5.$$

$$4 + 16 = 25$$

$$20 = 25 \quad \text{False}$$

The triangle is not a right triangle.

(b) The longest side is 13, so $c = 13$.

$$a^2 + b^2 = c^2$$

$$5^2 + 12^2 = 13^2 \quad \text{Let } a = 5, b = 12, \text{and } c = 13.$$

$$25 + 144 = 169$$

$$169 = 169 \quad \text{True}$$

The triangle is a right triangle.

Think About It

Is it necessary to use the Pythagorean Theorem to test 2, 3, and 7 to determine whether these numbers could be the lengths of the sides of a right triangle? Why?

Modeling with Real Data

We frequently model data with an expression in which the variable represents a year. For convenience, we usually select a base year and let the variable represent the number of years after the base year.

EXAMPLE 4

McDonald's Restaurants Around the World

Figure 2.1 shows the number of countries in which McDonald's restaurants were established during the period 1988–2000. (Source: McDonald's Corporation.)

Figure 2.1

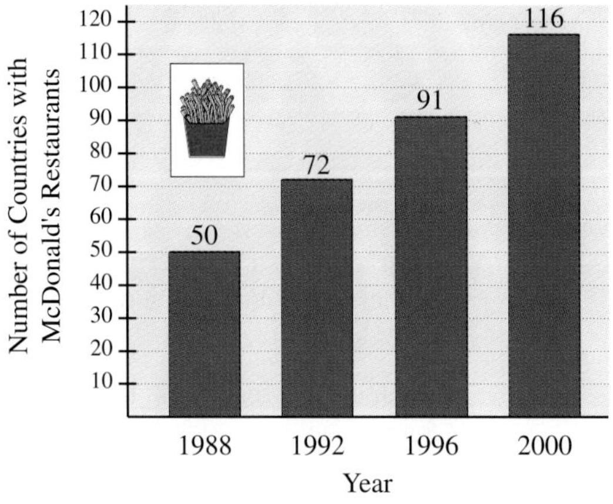

We select 1988 as our base year, and we let x represent the number of years *after* 1988. For 1988, $x = 0$; for 1992, $x = 4$; for 1996, $x = 8$; and so on.

With x defined in this way, suppose that we use the expression $5.425x + 49.7$ to *model* the number of restaurants x years after 1988.

(a) Evaluate the model expression for each year in the bar graph—that is, for $x = 0$, $x = 4$, $x = 8$, and so on. Then complete the following table. (The "Difference" column shows the difference between the actual data and the modeled data.)

Year	x	*Actual*	*Model*	*Difference*
1988				
1992				
1996				
2000				

(b) In what year is the model most accurate? least accurate?

Solution

(a)

Year	x	*Actual*	*Model*	*Difference*
1988	0	50	49.7	+0.3
1992	4	72	71.4	+0.6
1996	8	91	93.1	−2.1
2000	12	116	114.8	+1.2

(b) From the "Difference" column, we see that the model is most accurate for 1988 and least accurate for 1996.

Quick Reference 2.1

Evaluating Algebraic Expressions

- An **algebraic expression** is any combination of numbers, variables, grouping symbols, and operation symbols.
- To **evaluate** an algebraic expression, we replace each variable with a specific value and then perform all indicated operations.

Equations and Formulas

- An **equation** is a statement that two expressions have the same value.
- If an equation contains a variable, a value of the variable that makes the equation true is called a **solution.**
- A **formula** is an equation that uses an expression to represent some specific quantity.
- In a right triangle, the sides that form the right angle are called **legs**; the side opposite the right angle is called the **hypotenuse.**
- For a right triangle with legs a and b and hypotenuse c, the **Pythagorean Theorem** states that $a^2 + b^2 = c^2$.
- If the lengths of the three sides of a triangle satisfy the formula $a^2 + b^2 = c^2$, where c is the length of the longest side, then the triangle is a right triangle.

Speaking the Language 2.1

1. The longest side of a right triangle is called the ________.
2. We call $3x + 2$ a(n) ________ expression, whereas $3(5) + 2$ is an example of a(n) ________ expression.
3. A(n) ________ of an equation is any replacement for the variable that makes the equation true.
4. The sides of a right triangle are related by a rule known as the ________.

Exercises 2.1

Concepts and Skills

1. Explain why the algebraic expression $\dfrac{x+1}{x-2}$ cannot be evaluated for $x = 2$.

2. The expression $\dfrac{x-7}{7-x}$ has a value of -1 for any replacement of x except 7. Why?

In Exercises 3–18, evaluate each expression for the given value of the variable.

These are all one-variable expressions.

3. $5 - x$ for $x = -3$
4. $4 - 5x$ for $x = -2$
5. $-4x$ for $x = -1$
6. $3x - 2$ for $x - 4$
7. $5 - 3(x + 4)$ for $x = 2$
8. $x - 2(x - 1)$ for $x = -3$
9. $2x^2 - 4x - 9$ for $x = 3$
10. $x^2 - 5x$ for $x = -2$
11. $5 - |t + 3|$ for $t = -5$
12. $|3 - t| - 4$ for $t = 7$
13. $\dfrac{x+3}{5-x}$ for $x = 5$
14. $\dfrac{6}{x} + \dfrac{7}{x-2}$ for $x = 2$
15. $\dfrac{2x-3}{3-2x}$ for $x = 1$
16. $\dfrac{x^2-3}{x-4}$ for $x = -1$
17. $\sqrt{11 - 2x}$ for $x = 1$
18. $\sqrt{7x + 4}$ for $x = 3$

In Exercises 19–30, evaluate each expression for the given values of the variables.

Exercises 19–30 are all two-variable expressions.

19. $2x + 4y$ for $x = 3$ and $y = -2$
20. $x - y$ for $x = 4$ and $y = -3$
21. $2x - (x - y)$ for $x = 0$ and $y = -5$
22. $5 + x(x + y)$ for $x = -4$ and $y = 3$
23. $|2s - t|$ for $s = -4$ and $t = 2$
24. $5t - 2|t - s|$ for $s = 5$ and $t = 2$
25. $\dfrac{2x+y}{2x-y}$ for $x = 3$ and $y = 6$
26. $\dfrac{xy}{xy-12}$ for $x = 4$ and $y = 3$
27. $\dfrac{2x+3y}{x^2+y^2}$ for $x = -6$ and $y = 4$
28. $\dfrac{xy^2}{(x+y)^2}$ for $x = 0$ and $y = -5$
29. $\dfrac{(x-4)^2}{16} + \dfrac{(y+2)^2}{25}$ for $x = 2$ and $y = 3$
30. $\dfrac{4y^2}{9} + \dfrac{x^2}{4}$ for $x = -2$ and $y = 6$

In Exercises 31–34, evaluate the following expression for the given values of the variables.

$$\frac{-b + \sqrt{b^2 - 4ac}}{2a}$$

Although students are not expected to recognize the formulas in Exercises 31–38, they can still gain experience with evaluating them.

31. $a = 2$, $b = -3$, and $c = -2$
32. $a = 3$, $b = -5$, and $c = 2$
33. $a = 9$, $b = 12$, and $c = 4$
34. $a = 2$, $b = 3$, and $c = -14$

In Exercises 35–38, evaluate the following expression for the given values of the variables.

$$\frac{y_2 - y_1}{x_2 - x_1}$$

35. $x_1 = -2$, $x_2 = -4$, $y_1 = 3$, $y_2 = -1$
36. $x_1 = -12$, $x_2 = 4$, $y_1 = 5$, $y_2 = 5$

37. $x_1 = 0, x_2 = 5, y_1 = 2, y_2 = 0$

38. $x_1 = -3, x_2 = -3, y_1 = 4, y_2 = -2$

In Exercises 39–48, evaluate each expression for the given value(s) of the variable(s). Round decimal answers to the nearest hundredth.

39. $3x - 2$ for $x = -2.47$

40. $t^2 - 3t$ for $t = -4.56$

41. $-y + 3z$ for $y = 2.1, z = -3.7$

42. $m^2 - 2n^2$ for $m = -2.14$ and $n = 7.12$

43. $\sqrt{13y} + \sqrt{15x}$ for $x = 8$ and $y = 5$

44. $\sqrt{5a} - |-25b|$ for $a = 2$ and $b = 10$

45. $\sqrt{x^2 - 3x + 2}$ for $x = 2$

46. $\sqrt{x^2 - 3x}$ for $x = 2$

47. $|3 - 5x| - |4 - 7y|$ for $x = 4$ and $y = 2$

48. $\sqrt{1 - x}$ for $x = -5$

49. Suppose that you replace t with 5 in the expression $\sqrt{4 - t}$. Explain the result.

50. Suppose that someone evaluates the expression x^2 and tells you that the result is 9. Can you know for certain the value of x that the person used? Why?

CE In Exercises 51–56, suppose that you replace x with 5. With what value would you need to replace y in order for each expression to have a value of 9?

51. $x + y$

52. $x - y$

53. xy

54. $\frac{y}{x}$

55. $|y - x|$

56. y

57. Explain the difference between evaluating an expression and verifying that a number is a solution of an equation.

58. Explain how to determine which member of the set $A = \{-8, -1, 0, 2.3, 8, 37.92\}$ is a solution of the equation $3 \cdot (2x) = (3 \cdot 2)x$ without substituting any of the numbers in set A.

In Exercises 59–64, determine which member of the following set is a solution of the given equation.

$$\left\{-5, -3, -\frac{1}{2}, 2, \frac{5}{2}, 4\right\}$$

59. $7 - 3x = 1$

60. $2x + 3 = 2$

61. $3x - 7 = 5$

62. $5 - x = 8$

63. $13 - 4x = 3$

64. $3x + 17 = 2$

In Exercises 65–72, for each equation, determine whether the given number is a solution.

65. $3x - 12 = -3(4 - x)$; 0

66. $x - 7 = 7 - x$; 2

67. $x(x - 3) = 2$; 2

68. $(x - 5)(x + 2) = 0$; -2

69. $x^2 - x - 6 = 0$; -2

70. $2x^2 - 7x + 3 = 0$; 1

71. $\frac{x}{2} - 3 = 5$; 4

72. $\frac{x}{3} + \frac{x}{6} = \frac{3}{2}$; 3

Geometric Models

In Exercises 73–80, the three given numbers are the lengths of the sides of a triangle. Determine whether the triangle is a right triangle.

73. 5, 7, 4

74. 12, 13, 17

75. 17, 8, 15

76. 24, 25, 7

77. 9, 12, 15

78. 25, 24, 10

79. 1, 3, 3

80. 4, 3, 5

81. Some community volunteers have laid out a softball field at the local park. The distance between the bases is 60 feet. The distance from home plate to second base is 90 feet. (See figure.) Is the infield perfectly square? If not, what should be the distance from home plate to second base (to the nearest tenth)?

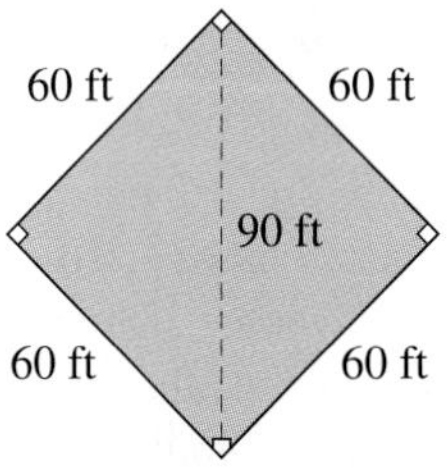

82. Standard letter paper is 8.5 inches wide and 11 inches long. To the nearest tenth, what is the diagonal distance d across the paper? (See figure.)

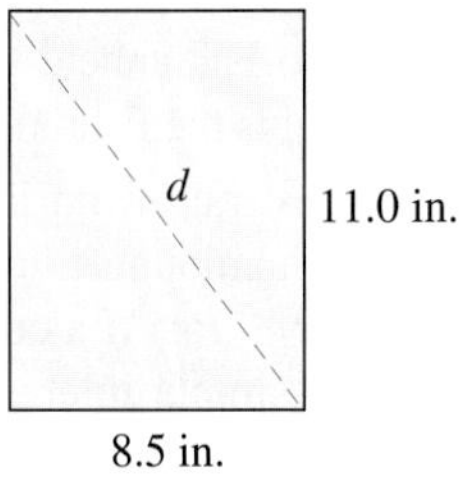

83. A building inspector is checking on an apartment building that is under construction. She finds that a window ledge, exactly 22 feet from the level ground, can be reached with a ladder that is 24 feet long. (See figure.) The bottom of the ladder is 6 feet away from the base of the wall, and the top of the ladder just reaches the window ledge. Is the wall perfectly vertical? Explain.

Figure for 83

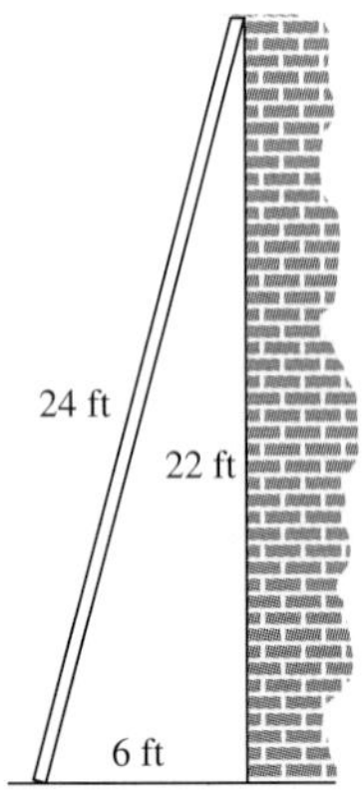

Figure for 84

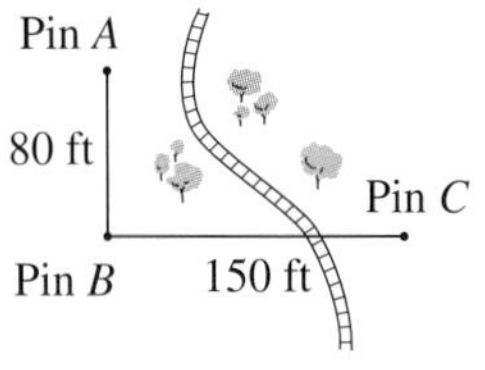

84. **Perpendicular lines** are lines that intersect to form right angles. A surveyor places pins at three corners of a property. (See figure.) The distance along one side of the property is 80 feet, and the distance along the other side is 150 feet. Explain what measurement the surveyor can use to determine whether the two sides of the property are perpendicular. What should the measurement be?

Real-Life Applications

85. A tourist is planning a January trip from his home in London to San Francisco. To plan his wardrobe, he learns that the average January temperature is 50°F. To convert to Celsius, the temperature scale in England, he uses the formula $C = \frac{5}{9}(F - 32)$. What is the corresponding Celsius temperature?

86. A radio station in Toronto, Canada, reports that the current temperature is $-10°$. Listeners who live in Buffalo, New York, know that this is a Celsius temperature. To convert to Fahrenheit, they use the formula $F = \frac{9}{5}C + 32$. What is the Fahrenheit temperature?

87. At a simple interest rate r, an investment of P dollars will grow to an amount A in t years according to the formula $A = P + Prt$. If a college fund of $1000 is invested at 6% simple interest, what will be the value of the investment after 5 years?

88. To signal the beginning of a yacht race, a flare is shot into the air. The height h (in feet) of the flare after t seconds is given by $h = 112t - 16t^2$. Determine the height of the flare after (a) 3.5 seconds and (b) 6 seconds.

Modeling with Real Data

89. The accompanying table shows the record high and low temperatures for selected states. (Source: National Oceanic and Atmospheric Administration.)

State	High (in °F)	Low (in °C)
Hawaii	100	−10
Oklahoma	121	−40
Florida	109	−19
Wisconsin	114	−48

Create a table with the headings "State," "High (in °C)," and "Low (in °F)." Then use the formulas $C = \frac{5}{9}(F - 32)$ and $F = \frac{9}{5}C + 32$ to fill in the table with the corresponding Fahrenheit F temperatures or Celsius C temperatures.

90. The number of letter carriers who were bitten by dogs decreased from 2787 in 1994 to 2541 in 1998. (Source: U.S. Postal Service.) With x representing the number of years since 1998, the expression $-61.5x + 2541$ is a model for the number of letter carriers who were bitten in year x. (Note that $x = -4$ for 1994.) Use the model to estimate the number of letter carriers who were bitten in 1996.

Data Analysis: *Bicycle Helmet Usage*

As more and more states enacted helmet laws for bicycle riders, the use of helmets has increased dramatically. The bar graph shows the percent of riders who wore helmets for selected years in the period 1992–1998. (Source: U.S. Bicycle Federation.)

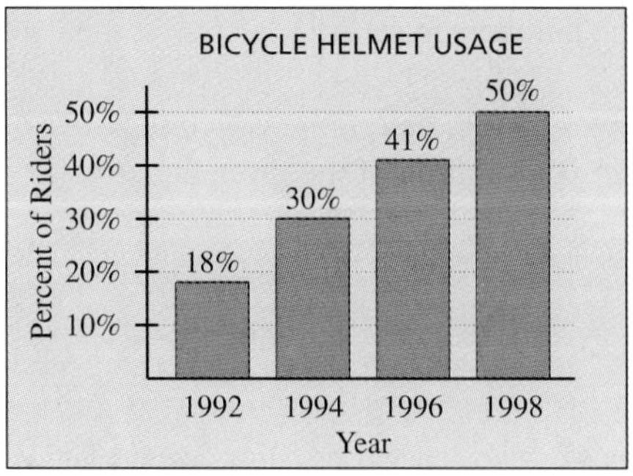

If we let t represent the number of years after 1992, then the percent of riders who wore helmets can be modeled by the expression $5.35t + 18.7$.

91. What values should you substitute for t in order to obtain approximations of the percents in the bar graph?

92. If you evaluate the model expression for $t = 10$, what is your interpretation of the result?

93. The bar graph does not give data for 1997. Use the model to estimate the percent for that year.

94. When $t = 16$, what is the value of the model expression? What does the result suggest about the validity of the model in the long-term future?

Challenge

95. The following equations illustrate sets of three integers that satisfy the Pythagorean Theorem formula $a^2 + b^2 = c^2$.

$3^2 + 4^2 = 5^2$

$5^2 + 12^2 = 13^2$

$7^2 + 24^2 = 25^2$

$9^2 + 40^2 = \square$

What pattern is revealed by these equations? Write the next two equations in the list and verify that they satisfy the Pythagorean Theorem formula.

96. In the accompanying figure, ΔABC is a right triangle. Squares I, II, and III are drawn so that one side of each square is also a side of the triangle. Explain how the area of Square I is related to the areas of the other two squares. (*Hint:* Consider the Pythagorean Theorem.)

Figure for 96

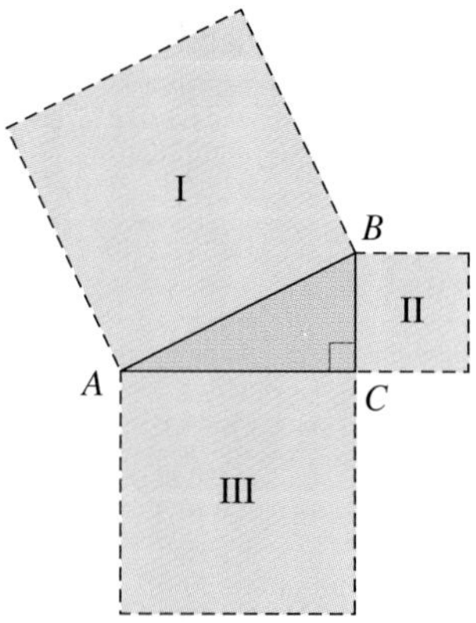

2.2 Simplifying Expressions

Terms and Coefficients • Simplifying Algebraic Expressions • Translations

Terms and Coefficients

We refer to the addends of an algebraic expression as **terms.** A term with no variable is called a **constant term.**

Because the expression $6x^2 - 8x + 7$ can be written as $6x^2 + (-8x) + 7$, the terms are $6x^2$, $-8x$, and 7. The constant term is 7. Because $x + y$ can be written as the expression $x + y + 0$, the constant term is 0.

Grouping symbols can affect the number of terms of an expression.

Expression	*Terms*
$x^2 + 5x + 3$	Three terms: x^2, $5x$, 3
$x^2 + 5(x + 3)$	Two terms: x^2, $5(x + 3)$

Note: An easy way to identify the terms of an expression is to determine the parts of the expression that are separated by plus or minus signs that are not inside grouping symbols.

Recall that in an expression such as $5xy$, the factors are 5, x, and y. The numerical factor in a term is called the **numerical coefficient** or simply the **coefficient.** Thus the coefficient of $5xy$ is 5. If the coefficient is not specifically indicated, it is understood to be 1 or -1.

Term	*Coefficient*	
y	1	Multiplicative Identity Property: $y = 1y$
$-ab$	-1	Multiplication Property of -1: $-ab = -1ab$

EXAMPLE 1

Identifying Terms and Coefficients

Identify the terms and coefficients in each expression.

Expression	*Terms*	*Coefficients*
(a) $4a^2 - 3a + 5$	$4a^2, -3a, 5$	$4, -3, 5$
(b) $y - (x - 3) - 2$	$y, -(x - 3), -2$	$1, -1, -2$
(c) $3y - x + \frac{y - 5}{2}$	$3y, -x, \frac{y - 5}{2}$	$3, -1, \frac{1}{2}$

Note that $\frac{y - 5}{2} = \frac{1}{2}(y - 5)$. Therefore, the coefficient is $\frac{1}{2}$.

Discuss why $-(x - 3)$ and $\frac{y - 5}{2}$ are considered terms. This concept is important for factoring by grouping.

Think About It

In the expression $5[(x + y + 2) - 7]$, how many terms are inside the parentheses? inside the brackets? in the entire expression?

Two terms are called **like terms** if they are both constant terms or if they have the same variable factors with the same exponents.

Like Terms	*Unlike Terms*
$2x, 7x$	$2x, 7$
$5y, 3y$	$5y, 3y^2$
$3a^2b, 6a^2b$	$3a^2b, 6ab^2$

Simplifying Algebraic Expressions

Equivalent expressions are expressions that have the same defined value regardless of the replacements for the variables. Two expressions may be equivalent even though some replacements for the variable are not permitted. For example, the expressions $\frac{x}{x}$ and 1 are equivalent because they have the same value for any replacement of x except 0.

To **simplify** an algebraic expression, we remove grouping symbols and combine like terms. The result is an expression that is equivalent to the original expression. Combining like terms is accomplished by applying the Distributive Property.

$$3a + 5a = a(3 + 5) = a \cdot 8 = 8a$$

Because the Commutative Property of Multiplication allows us to write factors in any order, we can save a step when we combine like terms. For example,

$$6n^2 - 7n^2 = (6 - 7)n^2 = -1n^2 = -n^2$$

EXAMPLE 2

Eventually students should be encouraged to perform the middle step mentally.

Combining Like Terms

(a) $2y - 3 - 5y + 1 = (2 - 5)y + (-3 + 1) = -3y - 2$

(b) $2a^2 + b - 2b - a^2 = (2 - 1)a^2 + (1 - 2)b = a^2 - b$

(c) $a^2b - ab^2 + 3a^2b - 5ab^2 = 4a^2b - 6ab^2$

We can also use the Distributive Property to remove certain grouping symbols. An expression of the form $a(b + c)$ is equivalent to the expression $ab + ac$.

An expression of the form $-(b + c)$ is equivalent to the expression $-1(b + c)$, which is equivalent to $-b - c$. The opposite sign in front of the group has the effect of changing the sign of each term inside the group.

EXAMPLE 3

These examples require careful attention to signs.

Simplifying Algebraic Expressions

(a) $(3 + 4x) - (5x - 1) = 3 + 4x - 5x + 1$ Remove parentheses.

$= -x + 4$ Combine like terms.

(b) $2(3x^2 - 2) - 3(x + 5) = 6x^2 - 4 - 3x - 15$ Distributive Property

$= 6x^2 - 3x - 19$ Combine like terms.

(c) $3x + x(5 - x) - 7x = 3x + 5x - x^2 - 7x$ Distributive Property

$= -x^2 + x$ Combine like terms.

(d) $7 - 3[4 - (x - 2)] = 7 - 3[4 - x + 2]$ Remove parentheses.

$= 7 - 3[6 - x]$ Combine like terms.

$= 7 - 18 + 3x$ Remove brackets.

$= -11 + 3x$ Combine like terms.

Translations

Students will be given much practice with translating throughout this book.

We can think of a *model* as a representation of an object, an idea, or information. When we use algebra to solve problems, it is often necessary to write the given information as an algebraic expression or equation that symbolically models the conditions of the problem and the operations to be performed.

The following are some typical translations involving the four basic operations.

Addition

Phrase	*Expression*
a number added to 5	$5 + x$
a number increased by 8	$x + 8$
7 more than a number	$x + 7$
the sum of two numbers	$x + y$

Subtraction

The order in subtraction is often a problem for students.

Phrase	*Expression*
-3 subtracted from a number	$x - (-3)$
a number decreased by 7	$x - 7$
6 less than a number	$x - 6$
the difference between 8 and a number	$8 - x$

Multiplication

Phrase	*Expression*
a number multiplied by -4	$-4x$
two-thirds of a number	$\frac{2}{3}x$
the product of two numbers	xy
twice a number	$2x$
6% of a number	$0.06x$

Division

Phrase	*Expression*
the quotient of -3 and a number	$\frac{-3}{x}$
the ratio of two numbers	$\frac{x}{y}$
a number divided by 5	$\frac{x}{5}$

When we translate information into an equation, we watch for words such as *is, equals, obtain,* and *results in,* which correspond to the equality symbol.

EXAMPLE 4

Translating into Equations

Translate the given information into an equation.

(a) The ratio of a number and 5 is 3.

(b) If a number is subtracted from -7, the result is 9.

(c) If you increase the product of two numbers by 1, you obtain 11.

LEARNING TIP

Avoid trying to translate all of the given information at once. Break the information down into words and phrases, and build your expression or equation a little bit at a time.

Solution

(a) Letting x represent the number, $\frac{x}{5} = 3$.

(b) Letting n represent the number, $-7 - n = 9$.

(c) Letting a and b represent the numbers, $ab + 1 = 11$.

In addition to knowing how to translate such typical phrases, we may need some common knowledge to write expressions.

EXAMPLE 5

Translating Word Phrases into Algebraic Expressions

Translate the given verbal expressions into algebraic expressions.

(a) The value in cents of n nickels and d dimes

(b) The difference between two numbers divided by the sum of their squares

(c) The reciprocal of a number x increased by two-thirds of the number

(d) The year-end value of y dollars invested at a 7% simple interest rate

Solution

(a) $5n + 10d$

(b) $\dfrac{x - y}{x^2 + y^2}$

(c) $\dfrac{1}{x} + \dfrac{2}{3}x$

(d) $y + 0.07y = 1y + 0.07y = 1.07y$

Quick Reference 2.2

Terms and Coefficients

- The addends of an algebraic expression are called **terms.** A term with no variable is a **constant term.**
- The numerical factor in a term is called the **coefficient.**
- Two terms are called **like terms** if they are both constant terms or if they have the same variable factors with the same exponents.

Simplifying Algebraic Expressions

- **Equivalent expressions** are expressions that have the same defined value regardless of the replacements for the variables.
- To **simplify** an algebraic expression, remove grouping symbols and combine like terms.
- We use the Distributive Property to combine like terms. The procedure is to combine the coefficients of the terms and retain the common variable factors.
- We also use the Distributive Property to remove grouping symbols.
  1. $a(b + c) = ab + ac$
  2. $-(b + c) = -b - c$

Speaking the Language 2.2

1. In the expression $2x + 3y$, 2 and 3 are called ________.
2. The property used to remove grouping symbols and to combine like terms is the ________ Property.
3. Because the expressions $3x + 2x$ and $5x$ have the same value for all replacements for x, we call the expressions ________.
4. The expressions $7x^2y$ and $4x^2y$ are examples of ________ terms.

Exercises 2.2

Concepts and Skills

1. Use the Additive Identity Property to explain why every algebraic expression has a constant term.
2. Use the Multiplicative Identity Property to explain why every term has a numerical coefficient.

In Exercises 3–10, identify the terms and coefficients in each expression.

3. $2x - y + 5$
4. $6a - 5b + c - 2d$
5. $3a^3 - 4b^2 + 5c - 6d$

Remind students to include the sign with the term.

6. $3x^2 + 5x - 6$

7. $7y^2 - 2(3x - 4) + \frac{2}{3}$

8. $\frac{5x}{3} - 2(y - 3)$

9. $\frac{x + 4}{5} - 5x + 2y^3$

10. $\frac{2x^2}{5} - y^2 - 8x + 2$

11. Explain why $2x + 3$ has two terms but $2(x + 3)$ is one term.

12. Explain why $5x^3y^2$ and $5x^2y^3$ are not like terms.

In Exercises 13–28, simplify the expressions by combining like terms.

13. $2x + 1 - 3x + 2$
14. $3 - x + 3x - 5$
15. $2x - 3y + x - 5$
16. $x - y - 3y + 4$
17. $2x + 4x - 7y$
18. $5y - 8y + 3x$
19. $-x + 3 - 4x + 1 + 5x$
20. $4 + 2x + x - 3 - x - 7$
21. $\frac{3}{2}x + \frac{2}{3}y - \left(-\frac{1}{3}\right)y - \frac{5}{2}x$
22. $\frac{3}{7}m - \frac{3}{8}n + \frac{4}{7}m + \frac{5}{8}n$
23. $8x^2 + 9x^2 - 7x + 10x$
24. $17x^3 - 4x^2 + 5x^3 - 7x^2$
25. $3ac^2 - 7ac + 7ac^2 - 6ac$
26. $10ax^2 + 10a^2x - 8ax^2 - 2a^2x$
27. $5.23x^2 - 5.23x^3 + 7.98x^3 - 9.63x^2$
28. $7.36a^2 - 8.97b^2 + 4.86a^2$

29. Explain why we can remove the parentheses in $-(2x - 5)$ and in $3 - (4a + 1)$ by changing the sign of each term inside the grouping symbols.

30. Do the expressions $2x^2z$ and $2xyz$ have the same number of factors? Why?

In Exercises 31–38, simplify each expression by removing the grouping symbols.

31. $-(a - b)$
32. $-(3 + n)$
33. $5(2x - 3)$
34. $-3(4 - x)$
35. $-(2a - 3b + 7d)$
36. $-(3h - 7k - 8g)$
37. $3(x - 2y) - (z - 2)$
38. $-3(2x + 4y) - 7(z - 2)$

In Exercises 39–44, match each expression in column A with its equivalent simplified expression in column B.

Column A	*Column B*
39. $2(x - 3) - 5x$	(a) $-(x - 1)$
40. $-(x - y) + x$	(b) $x^2y + xy^2$
41. $-x^2 + (-x)^2$	(c) $4x + 3$
42. $\frac{1}{2}(8x + 6)$	(d) $-3x - 6$
43. $-[-(1 - x)]$	(e) y
44. $xy^2 + x^2y$	(f) 0

In Exercises 45–68, simplify the expressions by removing grouping symbols and combining like terms.

45. $2x - (3x + 1)$
46. $5 + (3x - 4)$
47. $3(x - 4) - 2x$
48. $-2(3 - x) - 4$
49. $-(2a + 4b) - (5a - 2b)$
50. $-(3a - 7b) - (8b - a)$
51. $3x + 2(x - 4) - 5x$
52. $4 - 2(2x - 5) - x + 3$
53. $3(x + 5) + 2(1 - 3x)$
54. $5 - (x + 3) - 4$
55. $-(x - 1) + 2x + 1$
56. $5x - 1 + (x - 5)$
57. $(x + 1) + (2x - 1) + 5$
58. $-x - y - (x + y) + x$
59. $5(2t + 1) - 2(t - 4)$
60. $-(t + 1) - (2t - 1)$
61. $3x - 5y + 2x - (3x - 5y + 4)$
62. $7y - 6x + 9y - (7y - 3x - 8y)$
63. $6c - 5[6c - 5(c - 5)]$
64. $4a - 3[4a - 3(2a - 7)]$
65. $2(3x - 5) - 3(4x - 7) - 8(x + 6)$
66. $3(4x - 7) - 10(2x - 3) - 9(4 - x)$
67. $7 + 3[-(x - 2) - (3 - x)]$
68. $6 + 4[-2(5 - 3x) - (5x - 3)]$

In Exercises 69–72, simplify the expressions by removing grouping symbols and combining like terms.

69. $2x^2y^3 - 5x^3y^2 + 8x^3y^2 - x^2y^3$
70. $3x^4y - (x^2y^2 + x^4y)$
71. $-(x^5y^2 - x^2y^5) - (x^2y^5 - x^5y^2)$
72. $xyz + xy^2 - y(xz - xy)$

Modeling Conditions

In Exercises 73–76, translate the given verbal expressions into algebraic expressions. Let n represent the number. Simplify the expression, if possible.

73. The quotient of twice the number and 12

74. The product of the number and the number increased by 4

75. Three times the number plus the number itself

76. Eight more than the number less twice the number

In Exercises 77–80, column A contains word phrases describing a number n. Match each word phrase with the corresponding algebraic expression in column B.

Column A	*Column B*
77. Five less than a number	(a) $5 - n$
78. The difference between 5 and a number	(b) $n - 5$
79. The difference between a number and 5	
80. Five more than the opposite of a number	

In Exercises 81–90, translate the given information into an equation. Let x represent the unknown number, and let y represent any second unknown number.

81. Five less than a number is 3.

82. One more than twice a number is 7.

83. The product of 8 and a number is 11.

84. The quotient of a number and 4 is 3 times the number.

85. Twice a number is 3 less than the number.

86. If 25 is added to the square of a number, the result is 10 times the number.

87. Decreasing a number by 40% of the number results in 50.

88. If one number is increased by half of another number, the result is 16.

89. If you take the difference between a number and twice another number, you obtain -5.

90. If the numerator of the ratio $\frac{x}{3}$ is increased by 2, the value of the ratio becomes 7.

The following translations are important in applications in later chapters.

In Exercises 91–104, translate the verbal description into an algebraic expression.

91. The perimeter of a rectangle whose width is half its length L

92. The area of a triangle whose base is twice its height h

93. The value (in cents) of q quarters and d dimes

94. The distance (in miles) traveled in h hours at an average speed of 55 mph

95. The cost of online service for x hours if the monthly charge is \$15 plus \$2.50 per hour.

96. The portion of a job done in 1 hour if doing the entire job takes h hours

97. The sum of three consecutive integers, the smallest being n

98. The quotient of two consecutive odd integers, the smaller being n

99. The annual interest earned on d dollars if the simple interest rate is 8%

100. The amount needed to pay off a 1-year loan of d dollars if the simple interest rate is 9%

101. The hypotenuse of a right triangle if one leg is 2 units less than the other leg, whose length is a

102. The selling price of a television set purchased at a 35% discount off the original price c

103. A person's salary after a 5% raise if the old salary was s

104. The amount of acid in L liters of a 20% solution

105. Which is the correct translation of "the product of 3 and a number increased by 7"? Explain.

$$3x + 7 \quad \text{or} \quad 3(x + 7)$$

106. Evaluate the expression $\sqrt{a + b}$ and the expression $\sqrt{a} + \sqrt{b}$ for $a = 1$ and $b = 0$. Are these equivalent expressions? Use the definition of equivalent expressions to explain your answer.

Modeling with Real Data

107. For the period 1995–1999, the expression $1000(x + 6)$ represents the yearly number of passenger-service complaints about the airline industry, where x is the number of years since 1995. (Source: Department of Transportation.)

(a) Write the model expression without parentheses.

(b) Evaluate both the given expression and your expression in part (a) for $x = 3$.

(c) Interpret your result in part (b).

(d) If the model expression remains valid, how many complaints are predicted for the year 2002?

108. In 1999 the number of motor vehicle thefts per 100,000 people was 526. (Source: Federal Bureau of Investigation.) Write an algebraic expression that models the number of motor vehicle thefts in a city with population P.

Data Analysis: *Sport Fishing*

The Fish and Wildlife Service has estimated that a total of \$25 billion dollars is spent on sport (noncommercial) fishing. (Source: Fish and Wildlife Service, Bureau of the Census.)

109. If a total of d dollars is spent on bait, what algebraic expression represents the percent of the total spent on bait?

110. If x percent of the total dollars spent is for fishing equipment, what algebraic expression represents the total dollars spent for this fishing equipment?

111. In Exercises 109 and 110, suppose that $d = \$10$ million and $x = 26$. What total amount is spent on fishing gear and bait?

112. Suppose that, of the total of \$25 billion, the percents that are spent for travel, lodging, and food are $T\%$, $L\%$, and $F\%$, respectively. Write a *product* that represents the total amount spent in these three categories.

Challenge

In Exercises 113–118, determine whether the given expression is equivalent to (a) x^2 or (b) $-x^2$.

113. $(-x)^2$ **114.** $-(-x)^2$ **115.** $|x^2|$

116. $|-x^2|$ **117.** $-|x^2|$ **118.** $|x|^2$

In Exercises 119 and 120, simplify the given expression.

119. $x \cdot |x| + x^2$ where $x > 0$

120. $x \cdot |x| + x^2$ where $x < 0$

2.3 The Coordinate Plane

Rectangular Coordinate System • Distance Between Points • Midpoints of Line Segments • Real-Life Applications

Figure 2.2

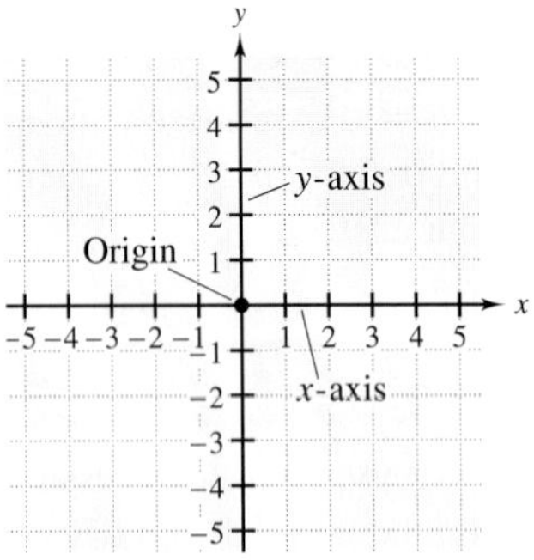

Rectangular Coordinate System

Sometimes it helps to compare a coordinate system to a road map where locations are given with a letter and a number.

To visualize certain sets of real numbers, we use a number line. To visualize pairs of numbers, we need to use two number lines. Conventionally, we represent the first number of the pair along a horizontal number line called the **x-axis,** and we represent the second number of the pair along a vertical number line called the **y-axis.** The two number lines (axes) intersect at a point called the **origin.** (See Fig. 2.2.)

This arrangement is called a **rectangular coordinate system,** a **Cartesian coordinate system,** or simply a **coordinate plane.**

Figure 2.3

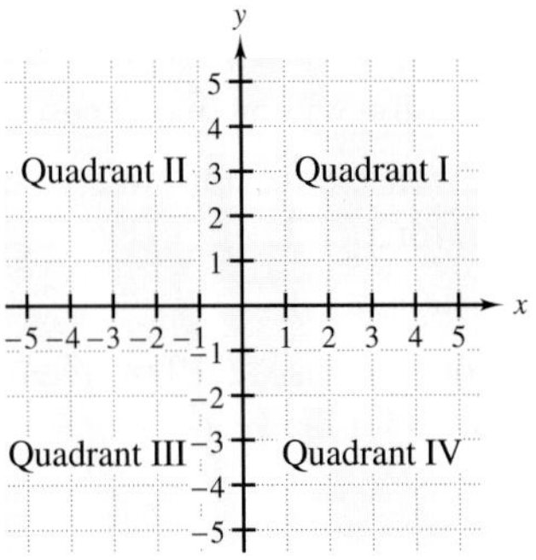

Note: Roughly speaking, you can think of a *plane* as the flat, unbounded surface (such as your paper or a chalk board) on which the axes are drawn.

In a coordinate plane, positive numbers are indicated with *tick marks* to the right of the origin on the x-axis and above the origin on the y-axis. Negative numbers are indicated with tick marks to the left of the origin on the x-axis and below the origin on the y-axis.

The axes divide the plane into four **quadrants** that are numbered counterclockwise. (See Fig. 2.3.)

Figure 2.4

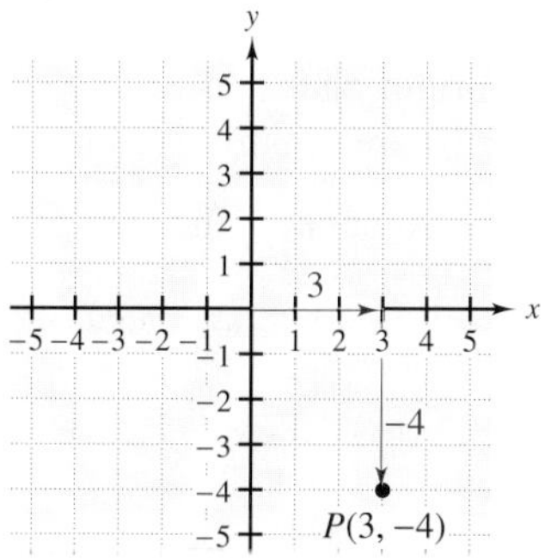

The location of each point in a coordinate plane can be described by a pair of numbers called an **ordered pair.** For example, the ordered pair (3, −4) describes a point P that is 3 units to the right of the origin and 4 units below the origin. (See Fig. 2.4.) In general, the ordered pair (x, y) is a pair in which the first coordinate is some number x, and the second coordinate is some number y.

Note: It is very important to remember that the first number of an ordered pair indicates horizontal direction and distance and that the second number of the pair indicates vertical direction and distance.

When we associate an ordered pair of numbers with a point in the coordinate plane, we call the numbers **coordinates** of the point. For the point $P(3, -4)$, we call 3 the ***x*-coordinate** of point P, and we call −4 the ***y*-coordinate** of point P. Each point in the coordinate system is associated with exactly one ordered pair of numbers, and each ordered pair of numbers is associated with exactly one point in the coordinate system. For this reason, we often use the terms *ordered pair* and *point* interchangeably.

LEARNING TIP

Think of the coordinate plane as a city laid out in blocks. The directions from the center of the city (origin) are given as a number of blocks east/west followed by a number of blocks north/south.

A point may lie in one of the four quadrants, or it may be a point of one of the axes, in which case the point does not belong to any quadrant.

EXAMPLE 1

Ordered Pairs and Points

Figure 2.5

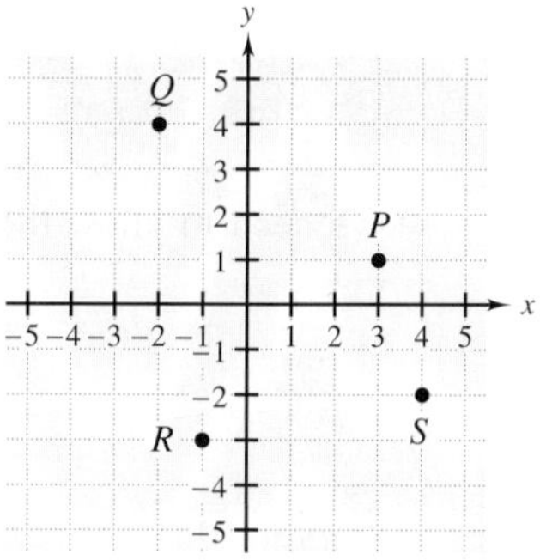

Use Figure 2.5 to answer each of the following questions.

(a) What are the coordinates of point P?

(b) Which point has the coordinates (−2, 4)?

(c) Which point is in Quadrant IV?

(d) What are the coordinates of the point in Quadrant III?

Solution

(a) (3, 1) (b) Point $Q(-2, 4)$ (c) Point $S(4, -2)$ (d) (−1, −3)

Note: In Example 1, the two different points Q and S have the same coordinates but in reverse order. Pairs of numbers are called *ordered* pairs because the order of the coordinates is significant.

Highlighting the point corresponding to an ordered pair is called **plotting** the point. Plotting the points corresponding to the ordered pairs of a set is called **graphing** the set.

EXAMPLE 2

Plotting Points

Plot the points associated with each of the following ordered pairs.

(a) (−6, 5)

(b) (−3, 0)

(c) (0, 4)

(d) (5, −3)

Solution

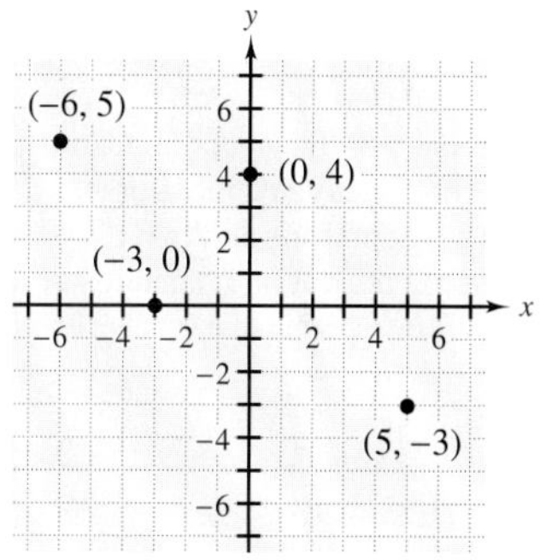

Think About It

Suppose that you wanted to represent ordered triples rather than ordered pairs. What kind of coordinate system could you use?

To plot a point in Quadrant I, we move right (+) from the origin and then up (+) to the point. Thus the coordinates of every point in Quadrant I are positive. Similarly, to plot a point in Quadrant III, we move left (−) from the origin and then down (−) to the point. Thus the coordinates of every point in Quadrant III are negative.

By using the same reasoning for Quadrants II and IV, we can summarize the sign patterns for all of the quadrants.

Quadrant I	*Quadrant II*	*Quadrant III*	*Quadrant IV*
(+, +)	(−, +)	(−, −)	(+, −)

Points of the axes do not belong to any quadrant.

A point of the x-axis is 0 units above or below that axis. Thus every point of the x-axis has a y-coordinate of 0.

Similarly, a point of the y-axis is 0 units to the right or left of that axis. Thus every point of the y-axis has an x-coordinate of 0.

Distance Between Points

We have seen that if the coordinates of two points of a number line are a and b, then the distance between the points can be found by calculating $|a - b|$ or $|b - a|$.

To find the distance between two points in a rectangular coordinate system, we use the same principle along with the Pythagorean Theorem.

Exploring the Concept

The Distance Formula

Figure 2.6 shows two points $A(-3, 2)$ and $B(5, 8)$. Suppose we need to know the distance AB between the points.

Figure 2.6

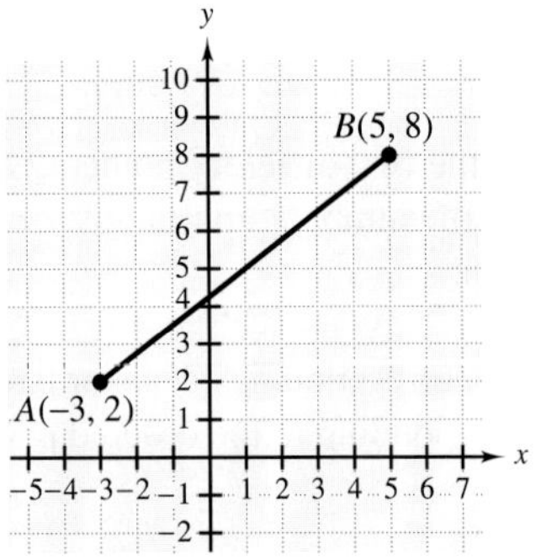

Figure 2.7

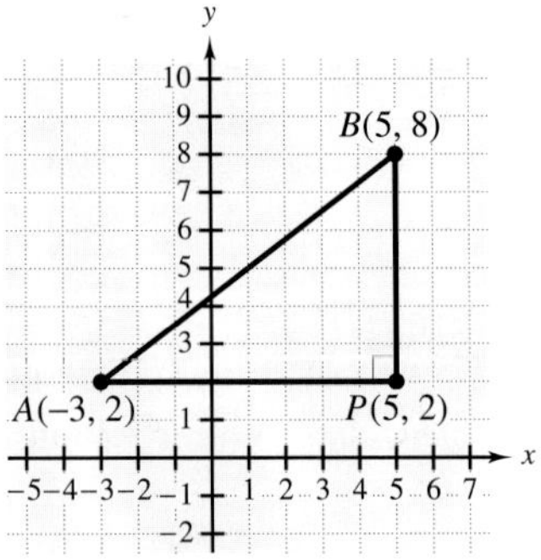

Observe that the y-coordinate does not change as you choose points of a horizontal line and that the x-coordinate does not change as you choose points of a vertical line.

We begin by adding a third point P, which is on a horizontal line containing point A and on a vertical line containing point B. (See Fig. 2.7.) Because point P has the same first coordinate as point B and the same second coordinate as point A, the coordinates of point P are (5, 2).

Because $\angle P$ is a right angle, ΔAPB is a right triangle, and we can use the Pythagorean Theorem to calculate the length AB of the hypotenuse.

$AP = |-3 - 5| = |-8| = 8$ Length of one leg

$BP = |8 - 2| = |6| = 6$ Length of the other leg

$$(AB)^2 = (AP)^2 + (BP)^2 \quad \text{Pythagorean Theorem}$$
$$= 8^2 + 6^2 = 64 + 36 = 100$$

Therefore, $AB = 10$.

Coordinates of points are sometimes written with *subscripts*. For example, point $P(x_2, y_1)$ has a first coordinate x_2 (read "x sub 2") and a second coordinate y_1 (read "y sub 1"). We often use this notation when we refer to general points rather than to points with specific coordinates.

Figure 2.8

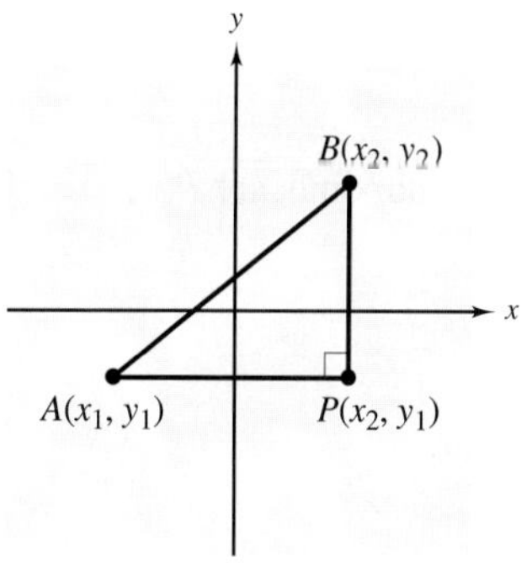

We can use the preceding technique to obtain a general method for finding the distance d between any two points $A(x_1, y_1)$ and $B(x_2, y_2)$. Figure 2.8 shows points A and B along with a third point $P(x_2, y_1)$ located so that ΔAPB is a right triangle.

The horizontal distance between A and B is $|x_2 - x_1|$, and the vertical distance is $|y_2 - y_1|$. According to the Pythagorean Theorem,

$$d^2 = |x_2 - x_1|^2 + |y_2 - y_1|^2$$

Therefore,

$$d = \sqrt{|x_2 - x_1|^2 + |y_2 - y_1|^2}$$

This formula is called the **distance formula** for the distance between two points in the Cartesian coordinate system. Because the differences are being squared, the order in which we subtract the coordinates is not important, and it is not necessary to take the absolute value of each difference.

The Distance Formula

If $A(x_1, y_1)$ and $B(x_2, y_2)$ are points in a Cartesian coordinate system, then the distance d between the points is given by

$$d = \sqrt{(x_2 - x_1)^2 + (y_2 - y_1)^2}$$

EXAMPLE 3

Using the Distance Formula

Find the distance between the points $E(-2, 1)$ and $F(3, 13)$.

You may want to illustrate that the same result is obtained from $\sqrt{[3 - (-2)]^2 + [13 - 1]^2}$.

Solution

$$d = \sqrt{(x_2 - x_1)^2 + (y_2 - y_1)^2}$$
$$d = \sqrt{(-2 - 3)^2 + (1 - 13)^2}$$
$$= \sqrt{(-5)^2 + (-12)^2}$$
$$= \sqrt{25 + 144}$$
$$= \sqrt{169}$$
$$= 13$$

Midpoints of Line Segments

The **midpoint** of a line segment is the point of the segment that is equidistant (the same distance) from the endpoints of the segment. It can be shown that the coordinates of the midpoint are the averages of the endpoint coordinates.

The Midpoint of a Line Segment

For the line segment with endpoints $A(x_1, y_1)$ and $B(x_2, y_2)$, the midpoint M has coordinates

$$x_m = \frac{x_1 + x_2}{2} \quad \text{and} \quad y_m = \frac{y_1 + y_2}{2}$$

EXAMPLE 4

Finding the Midpoint of a Line Segment

(a) Determine the midpoint M of the line segment whose endpoints are $P(4, 5)$ and $Q(-8, 9)$

(b) Use the distance formula to verify that $MP = MQ$.

Figure 2.9

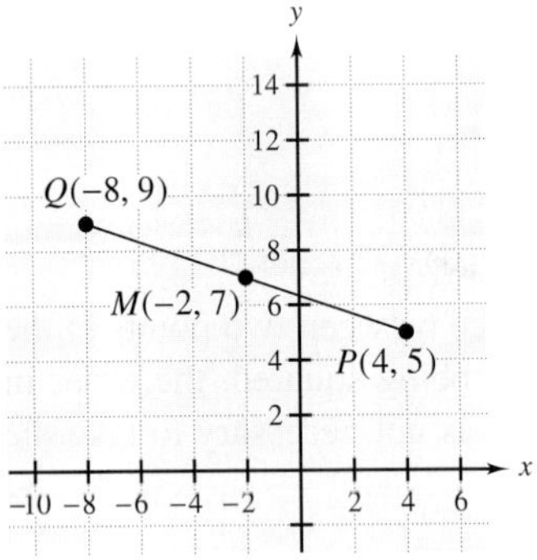

Solution

(a) $x_m = \frac{4 + (-8)}{2} = -2 \qquad y_m = \frac{5 + 9}{2} = 7$

The coordinates of midpoint M are $(-2, 7)$. (See Fig. 2.9.)

(b) $MP = \sqrt{(-2 - 4)^2 + (7 - 5)^2} = \sqrt{36 + 4} = \sqrt{40}$

$MQ = \sqrt{[-2 - (-8)]^2 + (7 - 9)^2} = \sqrt{36 + 4} = \sqrt{40}$

Therefore, $MP = MQ$.

Real-Life Applications

We can apply the distance and midpoint formulas to problems involving geometric figures drawn in a coordinate plane. In Example 5, we use the following definition: If two line segments have the same midpoint, then the line segments **bisect** each other.

EXAMPLE 5

The Diagonals of a Rectangle Bisect Each Other

Figure 2.10 shows a grid map of forest ranger facilities. Observation towers are located at points A, B, C, and D, which are the vertices of a rectangle. The main office is located at point P, where the diagonals of the rectangle intersect. Each tick mark represents 1 mile.

Figure 2.10

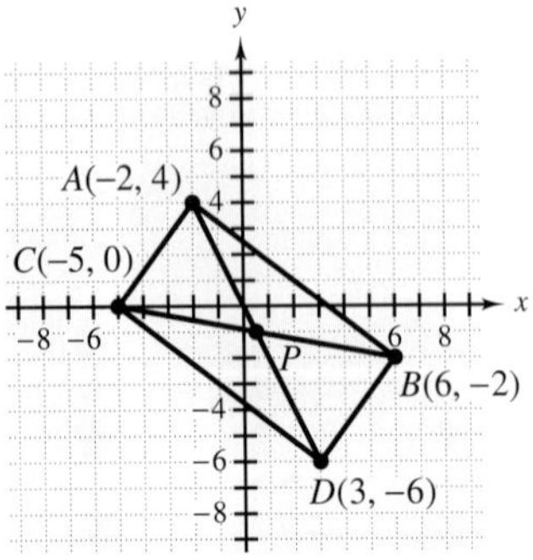

(a) Show that the main office is equidistant from the four observation towers by proving that the diagonals of the rectangle bisect each other.

(b) Find the distance from the main office to any one of the four towers.

Solution

(a) For the diagonal with endpoints A and D, the coordinates of the midpoint are as follows.

$$x_m = \frac{-2 + 3}{2} = \frac{1}{2} \qquad y_m = \frac{4 + (-6)}{2} = -1$$

For the diagonal with endpoints B and C, the coordinates of the midpoint are as follows.

$$x_m = \frac{-5+6}{2} = \frac{1}{2} \qquad y_m = \frac{0+(-2)}{2} = -1$$

Because $\left(\frac{1}{2}, -1\right)$ is the midpoint of both diagonals, the diagonals bisect each other. Therefore, the main office is equidistant from all four towers.

(b) The distance from the main office at $P\left(\frac{1}{2}, -1\right)$ to the tower at $A(-2, 4)$ is calculated with the distance formula.

$$d = \sqrt{[-2-0.5]^2 + [4-(-1)]^2} = \sqrt{6.25+25} = \sqrt{31.25} \approx 5.6 \text{ miles}$$

The distance from the office to any of the other towers is the same.

Quick Reference 2.3

Rectangular Coordinate System

- The **rectangular coordinate system** (or **Cartesian coordinate system** or simply **coordinate plane**) consists of a horizontal number line (***x*-axis**) and a vertical number line (***y*-axis**) that intersect at the **origin.**
- The axes divide the coordinate plane into four **quadrants.**
- The location of each point in a coordinate plane can be described by an **ordered pair** of numbers. The first and second numbers of the pair are called the ***x*-coordinate** and the ***y*-coordinate,** respectively.
- Highlighting the point that corresponds to an ordered pair is called **plotting** the point. Plotting the points that correspond to the ordered pairs of a set is called **graphing** the set.
- The following are the sign patterns for the coordinates in the four quadrants.

  Quadrant I: $(+, +)$ Quadrant II: $(-, +)$

  Quadrant III: $(-, -)$ Quadrant IV: $(+, -)$
- Every point of the x-axis has a y-coordinate of 0.
- Every point of the y-axis has an x-coordinate of 0.

Distance Between Points

- If $A(x_1, y_1)$ and $B(x_2, y_2)$ are points in a Cartesian coordinate system, then the distance d between the points is given by

  $$d = \sqrt{(x_2 - x_1)^2 + (y_2 - y_1)^2}$$

Midpoints of Line Segments

- For the line segment with endpoints $A(x_1, y_1)$ and $B(x_2, y_2)$, the midpoint M has coordinates

  $$x_m = \frac{x_1 + x_2}{2} \quad \text{and} \quad y_m = \frac{y_1 + y_2}{2}$$
- In words, the midpoint coordinates are the averages of the endpoint coordinates.

Speaking the Language 2.3

1. The axes of a coordinate plane divide the plane into regions called ______.
2. If a point represents an ordered pair (x, y), the number x is called the ______ of the point.
3. Ordered pairs of the form $(0, y)$ are represented by points of the ______.
4. The ______ is the point of intersection of the axes.

Exercises 2.3

Concepts and Skills

1. Explain in your own words how to plot the point associated with the ordered pair $A(-2, 4)$.
2. Give an example of a point that is not in any quadrant and describe its location.

In Exercises 3–6, plot the points associated with the given ordered pairs.

3. $A(4, -6)$, $B(-6, 4)$, $C(3, 5)$
4. $A(-2, -3)$, $B(-3, -2)$, $C(-5, 6)$
5. $A(0, -3)$, $B(-5, 0)$, $C(-2, -1)$
6. $A(3, 1)$, $B(3, 7)$, $C(4, -5)$

Recognizing patterns is important when we graph functions.

In Exercises 7–12, plot the points associated with the ordered pairs in the given set. From the resulting pattern, what do you think is the next ordered pair?

7. $\{(-2, -2), (0, -1), (2, 0), (4, 1), (6, 2), \ldots\}$
8. $\{(-5, 6), (-2, 4), (1, 2), (4, 0), (7, -2), \ldots\}$
9. $\{(-3, 5), (-2, 5), (-1, 5), (0, 5), (1, 5), \ldots\}$
10. $\{(4, 2), (4, 1), (4, 0), (4, -1), (4, -2), \ldots\}$
11. $\{(-4, 0), (-2, 3), (0, 0), (2, 3), (4, 0), \ldots\}$
12. $\{(-1, 1), (0, 0), (1, -1), (2, 0), (3, 1), \ldots\}$

In Exercises 13–18, without plotting the given point, name the quadrant in which the point lies.

13. $P\left(-2, -\frac{4}{5}\right)$
14. $B\left(\frac{3}{2}, -9\right)$
15. $P(2.4, 6.5)$
16. $B\left(-\sqrt{2}, -\sqrt{3}\right)$
17. $A\left(\sqrt{5}, -7\right)$
18. $P\left(-\sqrt{6}, 10\right)$

In Exercises 19–24, refer to the accompanying figure and determine the coordinates of the points.

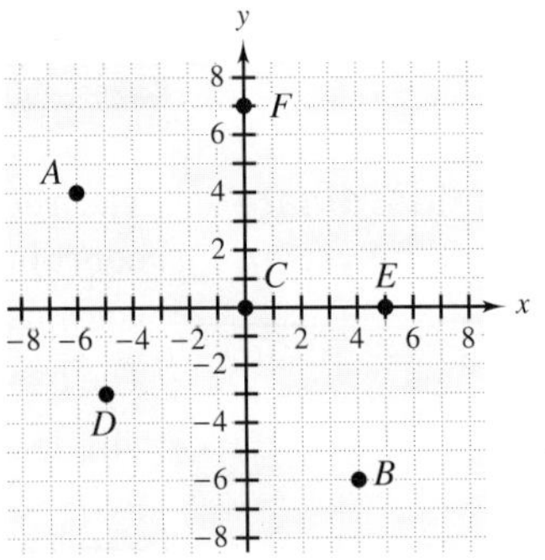

19. A
20. B
21. C
22. D
23. E
24. F
25. Describe the points whose x-coordinates are 0.
26. Describe the points whose y-coordinates are 0.

CE In Exercises 27–36, in which *possible* quadrants or on which axes could the point lie?

27. (x, y), $x > 0$
28. (x, y), $x < 0$
29. (x, y), $y < 0$
30. (x, y), $y > 0$
31. $(x, 0)$
32. $(0, y)$
33. $(-3, y)$
34. $(x, 5)$
35. (x, y), $xy < 0$
36. (x, y), $xy > 0$

CE In Exercises 37–40, what is true about x and y for the given location of point P?

37. $P(x, y)$ lies on the x-axis to the left of the origin.
38. $P(x, y)$ lies on the y-axis below the origin.
39. $P(x, y)$ lies on the y-axis above the origin.
40. $P(x, y)$ lies on the x-axis to the right of the origin.

CE **41.** Give the signs of a and b if $P(a, b)$ lies in

(a) Quadrant IV.

(b) Quadrant I.

(c) Quadrant II.

(d) Quadrant III.

CE **42.** If $P(x, y)$ is in Quadrant II, describe the location of the following points.

(a) $Q(-x, y)$

(b) $R(\ \ x,\ \ y)$

(c) $S(x, -y)$

CE **43.** In each part, the coordinates of a point are described. In which *possible* quadrants or on which axes could the point lie?

(a) $(a, 0), a > 0$

(b) $(0, b), b < 0$

(c) $(a, -a), a > 0$

(d) $(a, a), a \neq 0$

CE **44.** In each part, the location of $Q(b, a)$ is given. Determine the location of $P(a, b)$.

(a) $Q(b, a)$ is in Quadrant II.

(b) $Q(b, a)$ is in Quadrant IV.

(c) $Q(b, a)$ is in Quadrant I.

(d) $Q(b, a)$ is in Quadrant III.

CE In Exercises 45–50, the location of point P is described. Write the coordinates of point P.

45. Point P is on the y-axis 13 units above the x-axis.

46. Point P is on the x-axis directly below $(-14, 7)$.

47. Point P is directly above $(9, 0)$, and the coordinates are equal.

48. Point P is directly to the left of $(0, -11)$, and the coordinates are equal.

49. Point P is directly below $(-12, 13)$, and the y-coordinate is 6.

50. Point P is directly to the right of $(-17, -20)$, and the x-coordinate is -5.

CE In Exercises 51–54, start at point $A(12, -8)$ and move to point B according to the given directions. Then write the coordinates of point B.

51. Move 16 units left and then up 14 units.

52. Move 7 units right and then down 5 units.

53. Move 5 units right and then up 8 units.

54. Move 6 units left and then down 6 units.

55. In your own words, explain how to determine the distance between two points in a coordinate plane.

56. In your own words, explain how to determine the coordinates of the midpoint of a line segment drawn between two points.

In Exercises 57–66, determine the distance (to the nearest hundredth) between the given points P and Q.

For Exercises 57–66, you may want to suggest that students use a calculator to obtain a decimal approximation for radicands that are not perfect squares.

57. $P(5, 2), Q(-12, 2)$

58. $P(7, -4), Q(-17, -4)$

59. $P(3, 8), Q(3, -8)$

60. $P(-5, -14), Q(-5, 13)$

61. $P(-3, 2), Q(5, 8)$

62. $P(-5, -6), Q(4, 3)$

63. $P(-7, -2), Q(-3, -5)$

64. $P(-8, 5), Q(7, -9)$

65. $P(-4, 3), Q(-6, -7)$

66. $P(-5, -7), Q(5, -4)$

In Exercises 67–70, use the distance formula to determine whether the three given points are collinear. (Points are **collinear** if they are points of the same line.)

In Exercises 67–70, you may want to give students the additional fact that the points are collinear if the sum of the two shorter distances equals the longest distance.

67. $A(2, 4), B(0, -2), C(-1, -5)$

68. $A(-4, 3), B(2, 0), C(4, -1)$

69. $A(-4, 1), B(-1, 2), C(4, -1)$

70. $A(5, -4), B(3, -2), C(-1, 2)$

Geometric Models

CE In Exercises 71–74, calculate the perimeter and the area of the figure with the given vertices.

71. Rectangle: $(-4, 1), (-4, 6), (-1, 1), (-1, 6)$

72. Square: the endpoints of one side are $(3, 2)$ and $(6, 2)$.

73. Square: the endpoints of a diagonal are $(-3, 3)$ and $(0, 0)$.

74. Rectangle: the endpoints of a diagonal are $(-4, 0)$ and $(0, 5)$.

In Exercises 75–80, determine the midpoint of the line segment whose endpoints are given.

75. $(4, -3), (4, -1)$

76. $(-5, 6), (5, -6)$

77. $(3, 7), (-1, 3)$

78. $(3, -2), (-5, 5)$

79. $\left(2, \frac{1}{2}\right), (-2, 3)$

80. $\left(\frac{1}{3}, \frac{-11}{3}\right), \left(\frac{5}{3}, \frac{2}{3}\right)$

In Exercises 81 and 82, the given points are the four vertices (corners) of a rectangle. Show that the diagonals bisect each other.

81. $A(2, 5), B(8, 1), C(2, -8), D(-4, -4)$

82. $A(2, 4), B(6, -4), C(0, -7), D(-4, 1)$

In Exercises 83 and 84, show that the points A, B, C, and D are vertices of a parallelogram; that is, show that both pairs of opposite sides are equal in length.

83. $A(-2, 5), B(4, 3), C(-2, -3), D(-8, -1)$

84. $A(0, 7), B(8, 3), C(4, -1), D(-4, 3)$

In Exercises 85–88, determine whether the three given points are vertices of a right triangle.

85. $A(-2, 4), B(0, 9), C(3, 2)$

86. $A(-2, -1), B(4, -9), C(2, 2)$

87. $A(2, 4), B(-2, 1), C(3, -5)$

88. $A(-5, 0), B(2, 6), C(1, -7)$

Real-Life Applications

89. A train carrying hazardous waste derails and leaks poisonous gas. Authorities use a grid map to help them develop evacuation plans. They issue evacuation orders for all people living within a radius of 6 miles of the derailment. On the grid map, the scene of the accident is located at $(3, -2)$, and your house is located at $(-1, 4)$. Do you need to evacuate? (All coordinate units are in miles.)

90. The weather service is tracking Hurricane Jeffrey on a grid map with coordinate units representing 10 miles. At 10:00 A.M., the center of the hurricane is located at $(3, -4)$. Hurricane-force winds extend 70 miles out from the center. It is predicted that the eye of Jeffrey will be located at $(-1, -2)$ in two hours. Your cabin by Lake Helen is located at $(-7, -5)$ on the grid map. Will you feel the force of hurricane winds at noon? What is the predicted rate (in miles per hour) of the movement of the eye of Jeffrey?

91. When a rectangular coordinate system is superimposed on an aerial photograph, a church is at the origin and a school is at $(3, 2)$. Coordinate units are in miles. A cartographer uses the grid photograph to produce a map whose scale is 6 inches = 1 mile. To the nearest tenth of an inch, how far apart are the church and the school on the map?

92. On a grid map, a Navy cruiser is at $P(-20, 5)$ and is steaming toward Subic Bay located at $S(30, -15)$. To remain in the required sea lane, the cruiser must first travel to $Q(1, 1)$ and then, from there, head directly to port. How much farther will the cruiser travel than a helicopter that flies from point P directly to Subic Bay? (All coordinate units are in miles.)

Modeling with Real Data

93. About half of all employed women have children under age 18. The accompanying bar graph compares the job status of women with the number of children that they have. For example, we see that 15% of all mothers who are employed part-time have three children. (Source: Interep Research.)

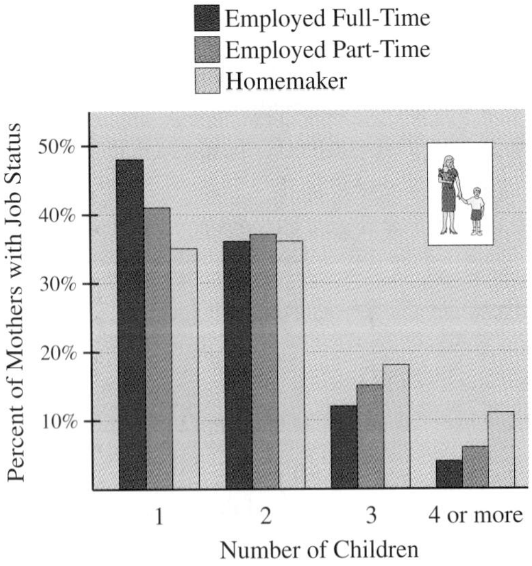

(a) Describe the woman's job and child status that is represented by the highest bar of any of the graphs.

(b) Describe the woman's job and child status that is represented by the lowest bar of any of the graphs.

94. Refer to the data in Exercise 93. For how many children are the percentages in each category about the same?

95. The following table gives the average high and low monthly temperatures in Omaha, Nebraska. The months are numbered beginning with 1 for January. (Source: National Climatic Data Center.)

Month	High	Low
1	31	11
2	37	17
3	49	28
4	64	40
5	74	51
6	84	60
7	88	66
8	85	63
9	77	54
10	66	41
11	49	29
12	35	16

(a) Using the ordered pairs of the form (month, high) and (month, low), plot these two sets of data in the same coordinate plane.

(b) What distance indicates the difference in the average high and low temperatures for any month?

96. The graphs in Exercise 95 indicate some gradual and some abrupt average temperature changes. During what 2-month period does the greatest change in the average high or low temperature occur?

Data Analysis: *Population Trends*

The accompanying figure shows a bar graph of population trends in the Midwest and South. (Source: U.S. Bureau of the Census.)

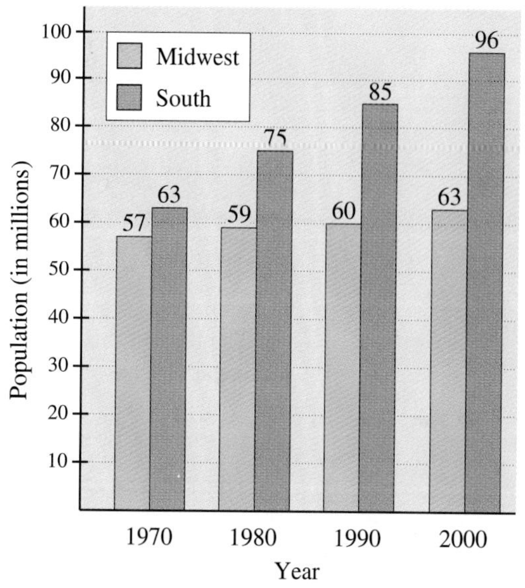

97. Write the four ordered pairs represented in the graph for the South.

98. Determine the coordinates of the midpoint of the line segment whose endpoints are (1970, 63) and (2000, 96).

99. What is a reasonable interpretation of your result in Exercise 98?

100. Using the data for 1970 and 2000, determine the average 10-year increase in the population in the South. Use your result to predict the population in the South in the year 2010.

Challenge

In Exercises 101 and 102, sketch the set of points that are the given distance from point C and describe the graph.

101. $C(1, 3)$, 2

102. $C(-2, 4)$, 4

In Exercises 103 and 104, point C is the midpoint of $\overline{AB}$. Determine the coordinates of point B.

103. $A(2, 4)$, $C(5, 6)$

104. $A(0, -3)$, $C(-2, -6)$

2.4 Relations and Functions

Relations • Functions • Graphs of Functions • Domain and Range • Real-Life Applications

Relations

"Burnable" CDs can be used to read and write large amounts of computer data. The price per disk (unit price) decreases as the number of disks purchased (volume) increases.

Volume	10	20	50	100
Unit Price	\$3.00	\$2.50	\$2.00	\$1.80

The data in this table can be written as ordered pairs. The first coordinate of each pair is the purchase volume, and the second coordinate is the unit price.

$$\{(10, 3.00), (20, 2.50), (50, 2.00), (100, 1.80)\}$$

This set is an example of a **relation.**

Definition of a Relation

A **relation** is a set of ordered pairs. The set of all first coordinates is the **domain** of the relation, and the set of all second coordinates is the **range** of the relation.

EXAMPLE 1 Determining the Domain and Range of a Relation

Consider writing sets in (a), (b), and (c) with one color highlighting the domain elements and another color highlighting the range elements.

State the domain and the range of each relation.

(a) $\{(3, 6), (-2, 7), (5, 3), (7, 3), (-5, -2)\}$

(b) $\{(3, 6), (3, 7), (5, 3), (7, 3)\}$

(c) $\{(1, 4), (2, 5), (3, 6), (4, 7), \ldots\}$

(d) The set of all ordered pairs in which the second coordinate is 3 more than the first coordinate.

Solution

(a) Domain: $\{-5, -2, 3, 5, 7\}$ — The set of all first coordinates.
Range: $\{-2, 3, 6, 7\}$ — The set of all second coordinates.

(b) Domain: $\{3, 5, 7\}$
Range: $\{3, 6, 7\}$

(c) Domain: $\{1, 2, 3, 4, \ldots\}$
Range: $\{4, 5, 6, 7, \ldots\}$

(d) The first coordinate x can be replaced by any real number, so the domain is **R**. The second coordinate y is simply 3 more than x, so the value of y is also any real number. Therefore, the range is **R**.

Rather than listing the ordered pairs of a relation or describing the relation in words, as in Example 1, we often find describing a relation with an algebraic expression to be more convenient. One way to do this is to use a special notation called **set-builder notation.** For example, the relation in which the y-coordinate is 3 more than the x-coordinate can be written in set-builder notation as follows.

$$\{(x, y) \mid y = x + 3\}$$

We read this relation as "the set of all ordered pairs (x, y) such that $y = x + 3$." Note that the vertical bar is read "such that."

In set-builder notation, the expression to the left of the vertical bar designates what type of members are in the set. For example, the members might be single elements x or ordered pairs (x, y). To the right of the vertical bar, we find the condition that must be met for an element to belong to the set. For example, $\{(x, y) \mid y = x + 3\}$ is the set of all ordered pairs of the form (x, y) such that the condition $y = x + 3$ is met.

Plotting the ordered pairs of a relation on a coordinate system is called **graphing the relation.**

EXAMPLE 2

Graphs of Relations

Graph the following relations.

(a) $A = \{(3, 2), (6, 2), (1, 5), (-1, 3), (0, -2)\}$

(b) $B = \{(1, -1), (2, -2), (3, -3), (4, -4), \ldots\}$

Solution

(a)

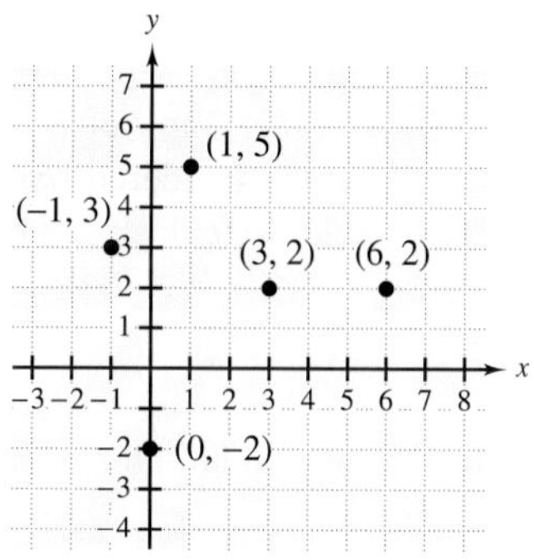

(b)

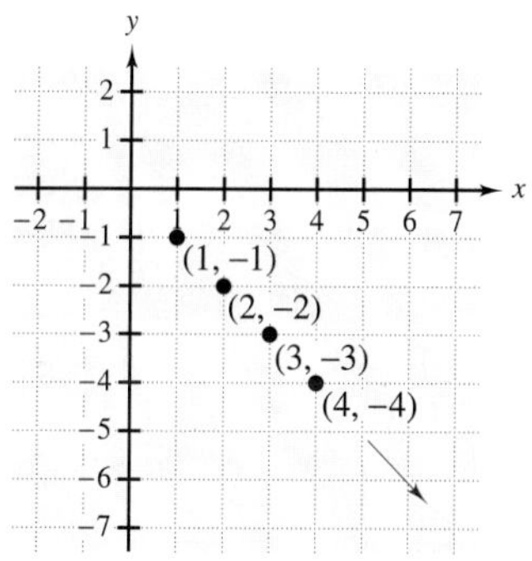

In the following exploration, we learn how to graph a relation that is described by an algebraic expression.

Exploring the Concept

The Graph of a Relation

Consider the relation $\{(x, y) \mid y = x + 3\}$. In the following table, we have selected some values for x and calculated the corresponding values of the expression $x + 3$.

x	-5	-3	0	1	4
$x + 3$	-2	0	3	4	7
$(x, x + 3)$	$(-5, -2)$	$(-3, 0)$	$(0, 3)$	$(1, 4)$	$(4, 7)$

As shown in the third row of the table, the ordered pairs of the relation have the form $(x, x + 3)$. In Figure 2.11, we have plotted those ordered pairs.

Observe that the plotted points appear to be points of a straight line. If we were to create more ordered pairs and plot additional points, they would also be points of the same line.

Figure 2.11

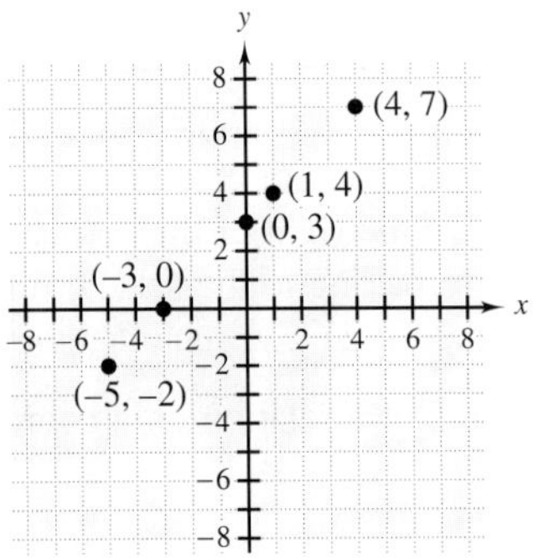

Figure 2.12

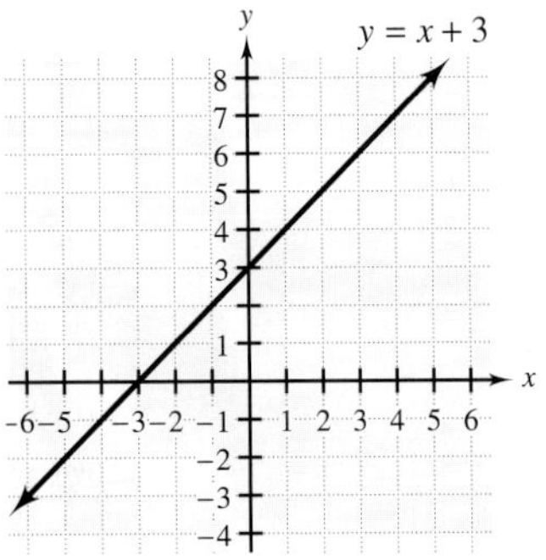

Because $x + 3$ can be evaluated for any real number x, there is no end to the number of ordered pairs and points. Thus we say that the relation contains *infinitely many* pairs $(x, x + 3)$. The graph of all such pairs is the graph of the relation $\{(x, y) \mid y = x + 3\}$, or simply the graph of $y = x + 3$. (See Fig. 2.12.)

Note that each point of the graph has a first coordinate that is the value of the variable x and a second coordinate that is the value of the expression $x + 3$.

In summary, we can sketch the graph of a relation that is described by an algebraic expression by creating a table of ordered pairs and plotting the corresponding points. Using those points as a guide, we can sketch the graph of the relation.

Functions

A **function** is a special kind of relation.

Definition of a Function

A **function** is a relation in which no two ordered pairs have the same first coordinate.

EXAMPLE 3

Recognizing Functions

Which of the following relations is a function?

(a) $A = \{(-2, 3), (1, 4), (4, 3), (5, -2)\}$

(b) $B = \{(1, 3), (2, 5), (-2, 5), (2, -1)\}$

(c) $C = \{(-1, 1), (-2, 2), (-3, 3), \ldots\}$

Solution

(a) Relation A is a function because no two ordered pairs have the same first coordinate. [There are two pairs, $(-2, 3)$ and $(4, 3)$, with the same *second* coordinate, but that does not violate the definition of a function.]

(b) Relation B is not a function because two pairs, $(2, 5)$ and $(2, -1)$, have the same first coordinate.

(c) If we assume that the indicated pattern continues, relation C is a function.

Think About It

Consider a relation whose ordered pairs are of the form (Person, Name of person). Is this relation a function? If the pairs were reversed, would the relation be a function?

Rather than listing the ordered pairs of a function, as in Example 3, we more often describe a function with an algebraic expression. For example, $\{(x, y) \mid y = 2x\}$ describes a function in which each y-coordinate is twice the x-coordinate.

When we describe a function with an expression, we can think of the function as a rule that states how to determine the y-coordinates given permissible x-coordinates. In effect, the rule indicates how to generate the ordered pairs of the function.

Alternative Definition of a Function

A **function** is a rule that assigns a value of the variable x to a unique value y. The **domain** of the function is the set of all permissible values of x. The **range** of the function is the set of all possible values of y.

Students usually have little difficulty identifying functions described as sets of ordered pairs, but expressions tend to be harder for them. The point made in this paragraph will probably need to be stressed many times.

In this alternative definition, the word *unique* is critical. Each permissible replacement for x must result in *exactly one* value of y. Informally, we can say that each *input* value of x results in exactly one *output* value of y.

EXAMPLE 4

Determining Whether an Equation Represents a Function

Determine whether the given equation represents a function.

(a) $y = x^2$ (b) $x = y^2$

Solution

LEARNING TIP

To judge whether an equation represents a function, ask yourself, "Can I choose any x-value that will result in more than one y-value?" If the answer is yes, then the equation does not represent a function.

(a) For each real number x, the expression x^2 has a unique value (exactly one value). Thus $y = x^2$ represents a function.

(b) For $x = 9$, for example, the equation $9 = y^2$ is true for two values of y: 3 and -3. That is, replacing x with 9 results in two different values for y. Therefore, the equation $x = y^2$ does not represent a function.

Graphs of Functions

Because a function is a relation, we graph a function just as we graph a relation: Plot the ordered pairs of the function. For example, we sketch the graph of $y = x^2 + 1$ by plotting the pairs (x, y) where $y = x^2 + 1$.

The graph of a relation provides an easy way to determine whether the relation is a function.

Remind students that all points of a vertical line have the same first coordinate.

Figure 2.13 shows a graph with a vertical line intersecting two points of the graph. The two points of intersection have the same first coordinate. This graph does not represent a function, because a function cannot have two ordered pairs with the same first coordinate.

Figure 2.13

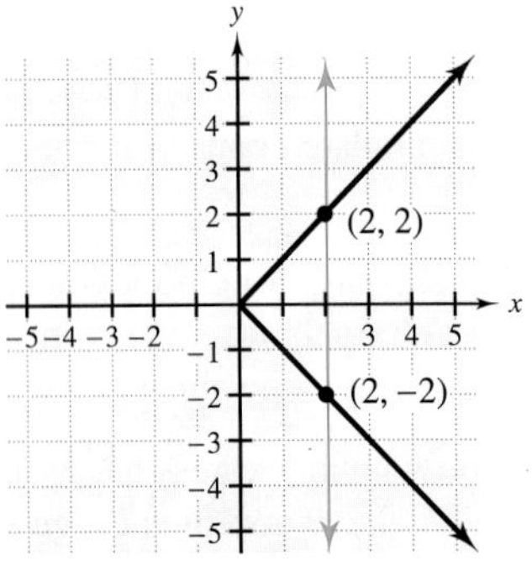

Figure 2.14

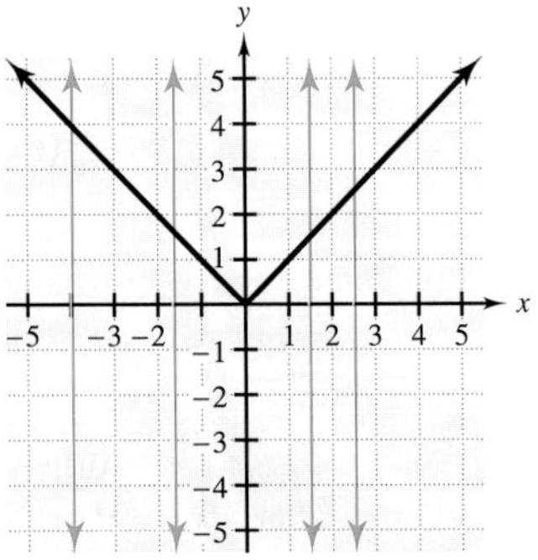

Figure 2.14 shows a graph for which it is not possible to draw a vertical line that would intersect the graph more than once. This graph represents a function.

This simple test for determining whether a graph represents a function is called the **Vertical Line Test.**

The Vertical Line Test

A graph represents a function if and only if no vertical line intersects the graph at more than one point.

EXAMPLE 5 Testing Relations for Functions

Determine whether the given graph represents a function.

Figure 2.15

(a)

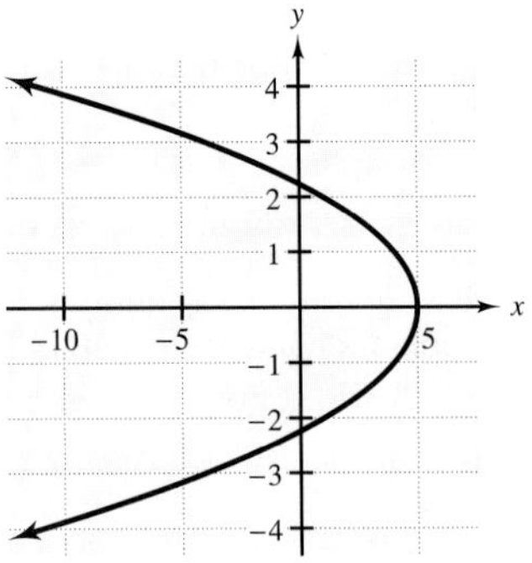

Figure 2.16

(b)

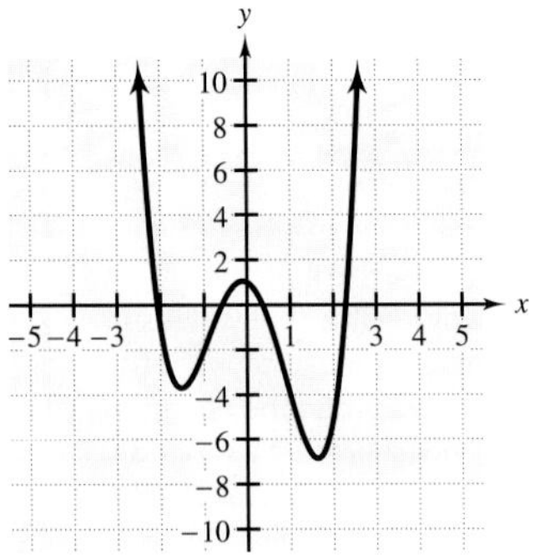

Solution

(a) Many vertical lines can be drawn that would intersect the graph at more than one point. Therefore, the graph does not pass the Vertical Line Test, and it does not represent a function.

(b) Assuming that the graph continues to widen on each side, no vertical line can be drawn that would intersect the graph at more than one point. The graph represents a function.

Domain and Range

We can think of the domain of a function as the set of all permissible values of the first coordinate. We have seen that when a function consists of a limited number of ordered pairs, we can write the domain simply by listing the first coordinates. In such cases, the range is also just a list of numbers—that is, the second coordinates.

When a function is described with an algebraic expression, then the domain and range might consist of *all* real numbers in one or more "intervals." In such cases, we obviously cannot write every member of the domain and range. Instead, we can use set-builder notation or another convenient method called **interval notation.**

Domain	*Number Line Graph*	*Set-Builder Notation*	*Interval Notation*
All numbers greater than 2		$\{x \mid x > 2\}$	$(2, \infty)$
All numbers greater than or equal to 2		$\{x \mid x \geq 2\}$	$[2, \infty)$
All numbers less than 5		$\{x \mid x < 5\}$	$(-\infty, 5)$
All numbers less than or equal to 5		$\{x \mid x \leq 5\}$	$(-\infty, 5]$
All numbers less than or equal to 6 and greater than 1		$\{x \mid 1 < x \leq 6\}$	$(1, 6]$
All numbers between 2 and 5		$\{x \mid 2 < x < 5\}$	$(2, 5)$
All numbers between 2 and 5, inclusive		$\{x \mid 2 \leq x \leq 5\}$	$[2, 5]$
All real numbers		**R**	$(-\infty, \infty)$

In interval notation, parentheses are associated with $<$ and $>$, and they are used to indicate that the endpoint of the interval is *not* included. Brackets are associated with $\leq$ and $\geq$, and they are used to indicate that the endpoint of the interval *is* included.

The *infinity* symbol indicates that the interval extends to the right (∞) or to the left ($-\infty$) without end. Note that parentheses are always used with the infinity symbol because such intervals do not have an endpoint.

Note: Be careful not to confuse, for example, the *ordered pair* (2, 5) with the *interval notation* (2, 5). You should be able to tell from the context whether the notation refers to ordered pairs or intervals.

Because domains and ranges cannot be inferred from graphs with certainty, we use words such as *suggest* and *estimate.*

As we see in Example 6, a graph of a relation (possibly a function) can be used to estimate the domain and the range.

In order to estimate the domain of a relation from its graph, we observe the x-coordinates of the points that are farthest to the left and right of the graph. The domain includes those x-values as well as the x-values of all points between those points. Similarly, to estimate the range, we observe the y-coordinates of the points that are the lowest and highest points of the graph. The range includes those y-values as well as the y-values of all points of the graph between the highest and lowest points.

EXAMPLE 6

Although holes and gaps can occur in graphs, our purpose at this point is to use graphs to make intelligent judgments about domains and ranges and to learn the geometric interpretation of domain and range.

Domain and Range

Determine whether the given graph represents a function. Then use the graph to estimate the domain and range.

Figure 2.17

(a)

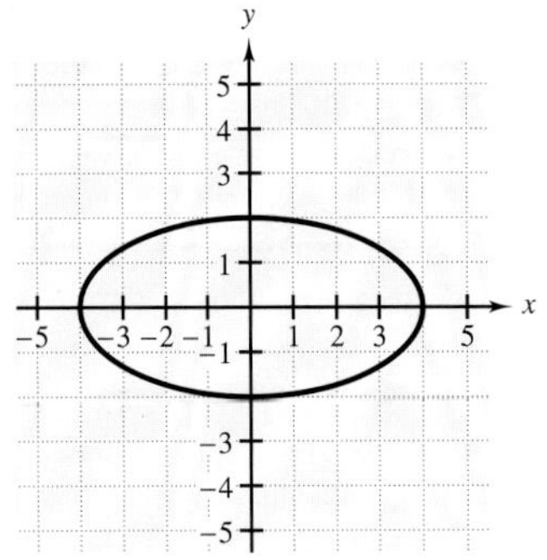

Figure 2.18

(b)

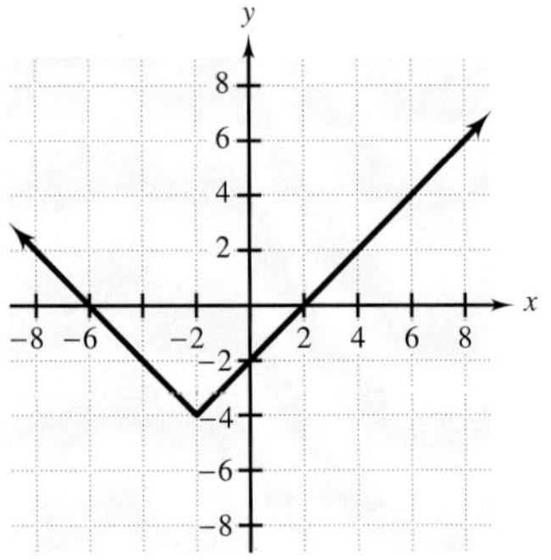

Solution

(a) We can draw many vertical lines that would intersect the graph at more than one point. Therefore, the graph does not represent a function.

As shown in Figure 2.19, the point farthest to the left of the graph has an x-coordinate of -4, and the point farthest to the right has an x-coordinate of 4. Therefore, the domain is $\{x \mid -4 \le x \le 4\}$ or $[-4, 4]$.

Similarly, the y-coordinate of the lowest point is -2 and the y-coordinate of the highest point is 2. Therefore, the range is $\{y \mid -2 \le y \le 2\}$ or $[-2, 2]$.

Figure 2.19

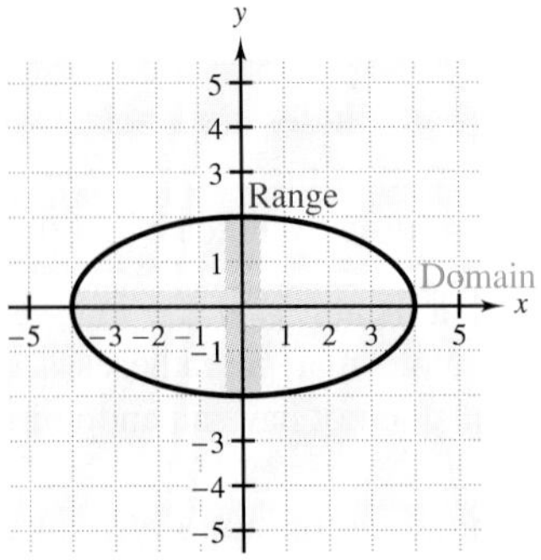

Figure 2.20

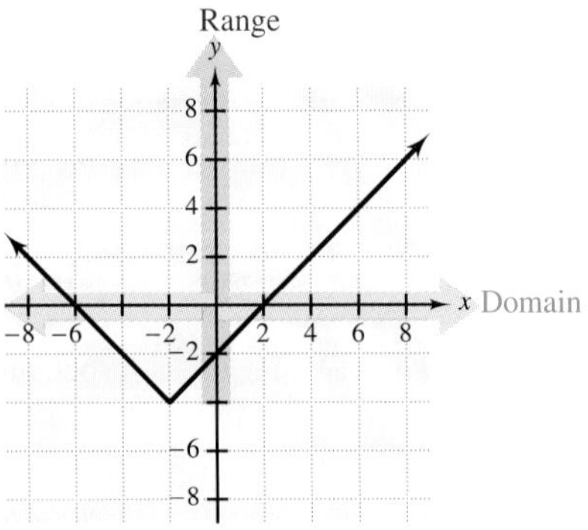

(b) The graph represents a function because no vertical line can be drawn that would intersect the graph at more than one point.

In Figure 2.20, the graph extends to the left and right without end. Therefore, the domain is **R** or $(-\infty, \infty)$.

The y-coordinate of the lowest point of the graph is -4, and the graph extends upward from there without end. Therefore, the range is $\{y \mid y \geq -4\}$ or $[-4, \infty)$.

Note: In Example 6, we give domains and ranges in both set-builder notation and interval notation. When it is possible to do so, we will most often use interval notation in the remainder of this book. However, for simplicity, we will use **R** for the set of all real numbers rather than the interval notation $(-\infty, \infty)$.

Real-Life Applications

Whenever lists of data can be paired, the ordered pairs form a relation. As Example 7 shows, the relation may or may not be a function.

EXAMPLE 7

Pairing Data to Form a Relation

Consider the following tables of information.

Elected Presidents, 1980–1996

Year	President
1980	Reagan
1984	Reagan
1988	Bush
1992	Clinton
1996	Clinton

World Series Winners, 1995–1999

Team	Year
Braves	1995
Yankees	1996
Marlins	1997
Yankees	1998
Yankees	1999

(a) If you were to write the data in the two tables as set $A = \{(\text{Year, President})\}$ and set $B = \{(\text{Team, Year})\}$, would the relations be functions? Why?

(b) Describe in words the domain and range of each relation.

Solution

(a) The relation A is a function because no two ordered pairs have the same first coordinate. However, the relation B is not a function because, for example, three ordered pairs have the first coordinate Yankees.

(b) For the relation A, the domain is the set of all presidential election years from 1980 to 1996, and the range is the set of all elected presidents during that period.

For the relation B, the domain is the set of all teams that won the World Series from 1995 to 1999, and the range is the set of all years in that period.

Quick Reference 2.4

Relations

- A **relation** is a set of ordered pairs.
- The set of all first coordinates is the **domain** of a relation, and the set of all second coordinates is the **range** of the relation.
- One way to describe a relation with an algebraic expression is to use a special notation called **set-builder notation.**
- Plotting the ordered pairs of a relation is called **graphing the relation.**

Functions

- A **function** is a relation in which no two ordered pairs have the same first coordinate.
- A function is often described by a rule that assigns any permissible value of the variable x to a unique value of y. The domain is the set of all permissible values of x.

Graphs of Functions

- When a relation or a function is defined by an expression, we graph the relation or function by plotting the pairs (x, y), where the first coordinate, x, represents the value of the variable and the second coordinate, y, is the value of the expression.
- The **Vertical Line Test** states that a graph represents a function if and only if no vertical line intersects the graph at more than one point.

Domain and Range

- The graph of a function can be used to estimate the domain and range of the function.
- Domains and ranges are often written with set-builder notation or with **interval notation.**

Speaking the Language 2.4

1. Any set of ordered pairs is called a(n) ________.
2. If (2, 3) and (2, 7) are pairs of a set, then the set is not a(n) ________.
3. The ________ describes a geometric way of determining whether a graph represents a function.
4. The ________ of a function in x is the set of all permissible values of x.

Exercises 2.4

Concepts and Skills

1. In your own words, explain the difference between a relation and a function.
2. Explain the meaning of *domain* and *range*.

In Exercises 3–8, use the given description to write a set of ordered pairs. Then determine the domain and range of the relation.

3. The first coordinate is an integer between -3 and 3, inclusive; the second coordinate is the square of the first coordinate.

4. The first coordinate is a positive integer less than 8; the second coordinate is the sum of the first coordinate and -4.
5. The first coordinate is the length of the side of a square, where the length is a positive integer not exceeding 5; the second coordinate is the perimeter of the square.
6. The first coordinate is the radius of a circle, where the radius is a positive integer not exceeding 4; the second coordinate is the area of the circle.
7. The first coordinate is the number of hours worked (10, 20, 30, 40); the second coordinate is the wages earned at $7 per hour.
8. The first coordinate is the number of toppings on a pizza (up to 5); the second coordinate is the price of a pizza, which is $7 plus $1 per topping.

In Exercises 9–18, sketch the graph of each relation and give the domain and range. Identify any relations that are functions.

9. $\{(1, 2), (3, 4), (5, 6), (7, 8), (9, 10)\}$
10. $\{(-1, 2), (-3, 4), (-5, 6), (-7, 8), (-9, 10)\}$
11. $\{(3, 1), (4, 1), (5, 1)\}$
12. $\{(2, 3), (5, 3), (6, 3)\}$
13. $\{(-2, 4), (-2, 5), (-2, 7)\}$
14. $\{(-1, -3), (-1, -5), (-1, -7)\}$
15. $\{(-2, 2), (-2, -3), (3, 4), (2, -3)\}$
16. $\{(4, -4), (-2, 3), (4, 3), (5, 0)\}$
17. $\{(-3, -4), (-4, -5), (-5, -6), \ldots\}$
18. $\{(-3, 4), (-4, 5), (-5, 6), \ldots\}$
19. Describe how to use the graph of a relation to estimate the domain.
20. Describe how to use the graph of a relation to estimate the range.
21. In your own words, state the Vertical Line Test and explain the reasoning behind it.
22. Are all functions relations? Are all relations functions? Explain.

In Exercises 23–26, write the set in interval notation.

23. $\{x \mid x < -4\}$
24. $\{x \mid x \geq -5\}$
25. $\{x \mid -3 < x \leq 7\}$
26. $\{x \mid 3 \leq x < 8\}$

In Exercises 27–30, write the given set in set-builder notation.

27. $[-3, 7)$
28. $[0, 5]$
29. $(-\infty, 0]$
30. $(-1, \infty)$

Remind students that the graphs given in Exercises 31–42 are to be considered *complete* and that we are estimating the domain and range.

In Exercises 31–42, estimate the domain and range of the relation whose graph is given. Identify any graphs that represent functions.

31.

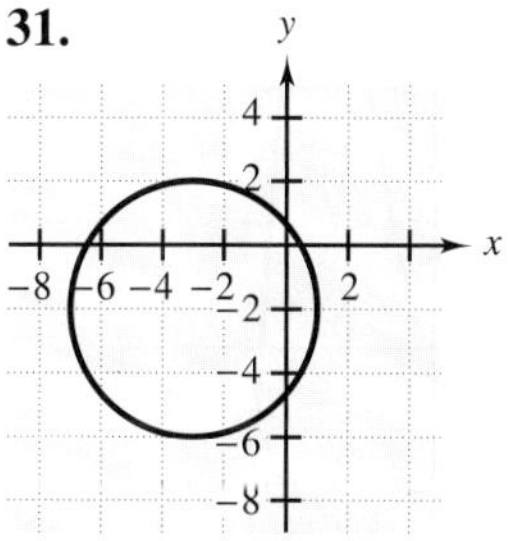

32.

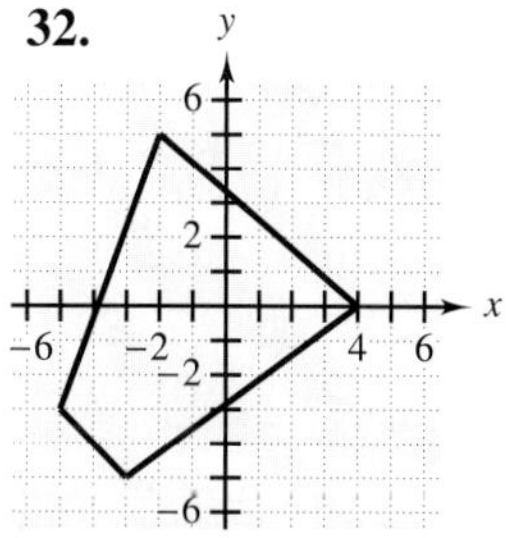

33.

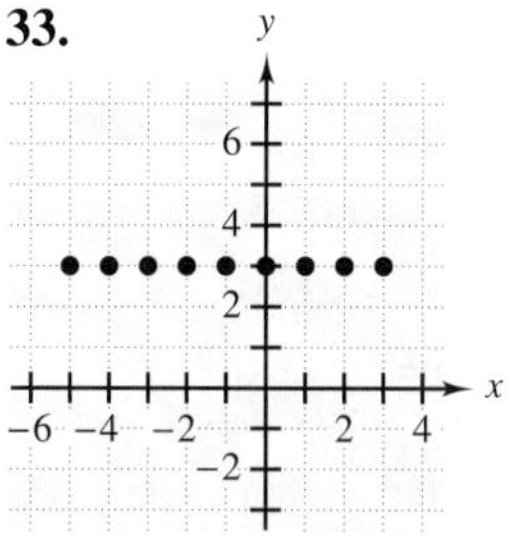

34.

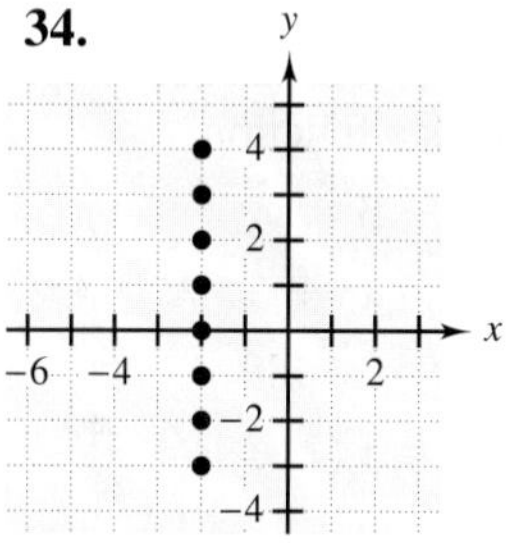

35.

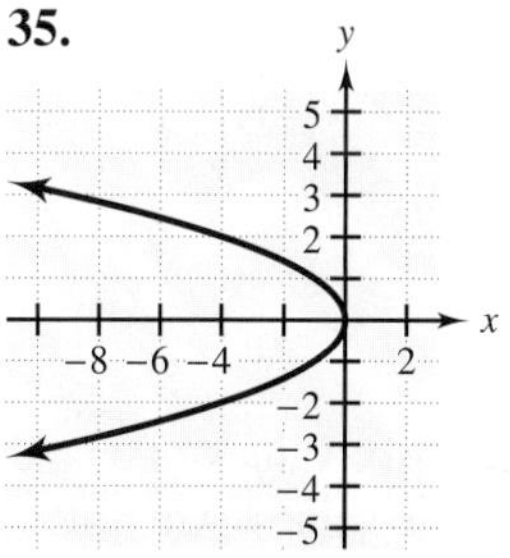

36.

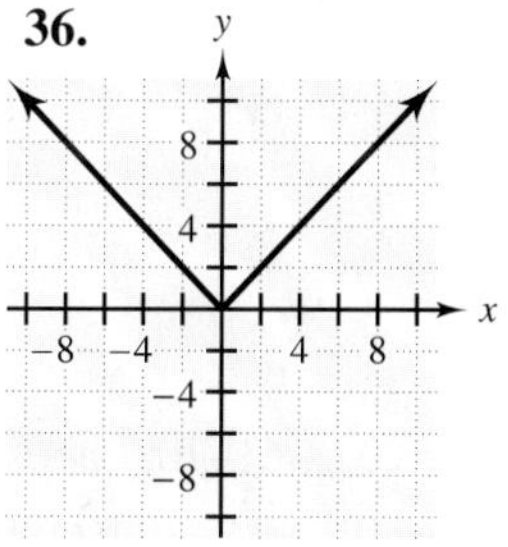

37.

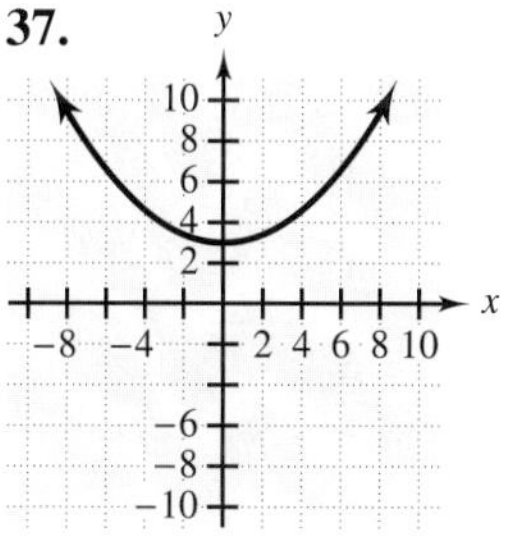

38.

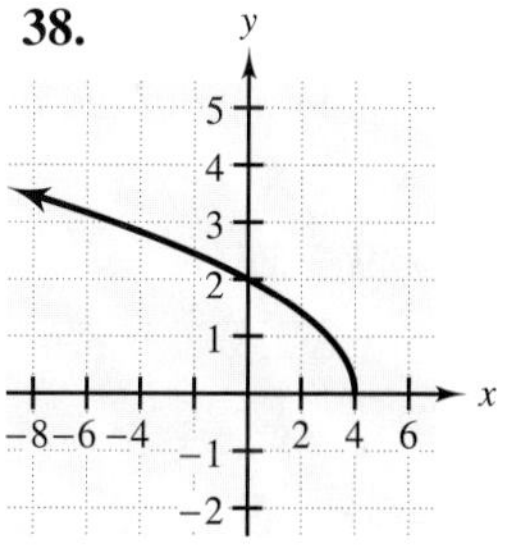

39.

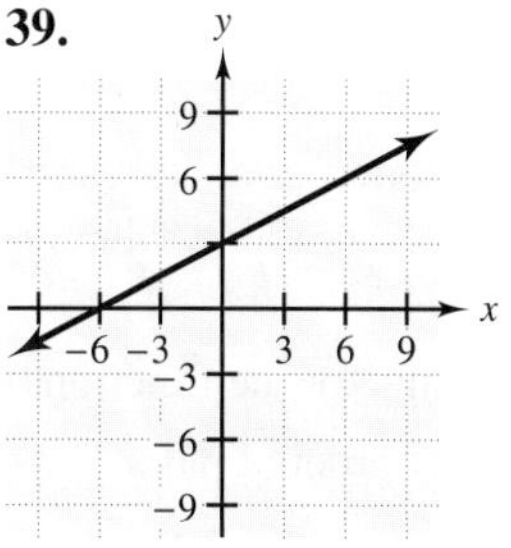

40.

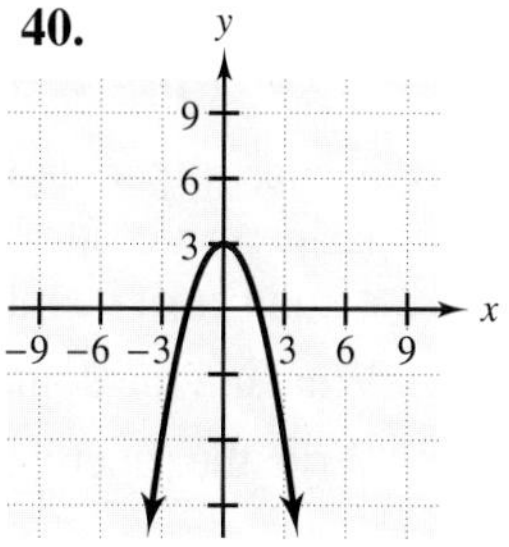

41.

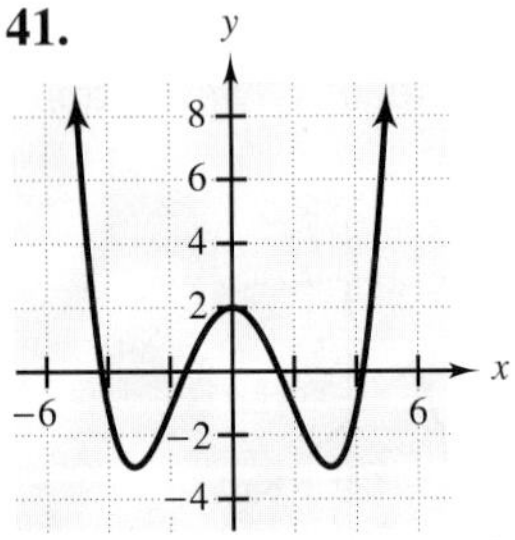

42.

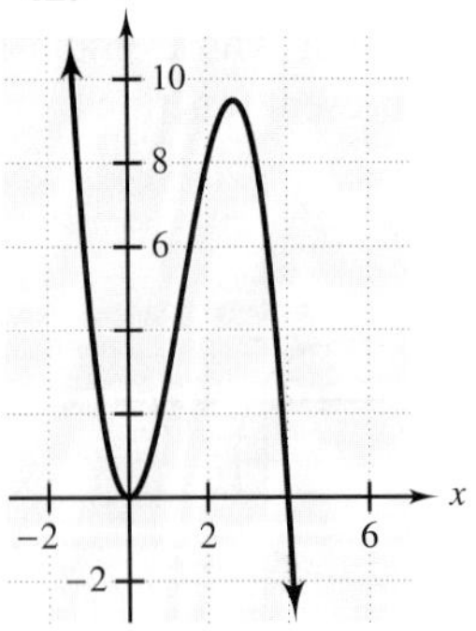

In Exercises 43–60, suggest that students choose a value of *x* and evaluate the expression that contains *x*. Then decide whether more than one value of *y* is associated with that value. Remind students that a number and its opposite have the same square and the same absolute value.

In Exercises 43–60, determine whether the equation represents a function.

43. $x - 3 = y$

44. $y = x - 5$

45. $y = x^2 + 5$

46. $y^2 = x + 1$

47. $y^2 = x + 9$

48. $y = (x - 4)^2$

49. $x = y^3$

50. $y = x^3$

51. $x = |y|$

52. $y = |x|$

53. $y = \sqrt{x}$

54. $x = \sqrt{y}$

55. $y^2 = 4 - x^2$

56. $y^2 = x^2 - 16$

57. $x^2 = y^2$

58. $|x| = |y|$

59. $x = \sqrt[3]{y}$

60. $y = \dfrac{1}{x}$

In Exercises 61–70, choose at least six values of *x* and evaluate the given function for those *x*-values. After plotting the points that represent the (*x*, *y*) pairs, sketch the graph of the function. Then use the graph to estimate the domain and range of the function.

61. $y = x - 8$

62. $y = 10 - x$

63. $y = 9 - x^2$

64. $y = x^2 - 5$

65. $y = \sqrt{9 - x}$

66. $y = \sqrt{x + 7}$

67. $y = x^2 + 4x + 3$

68. $y = 3 + 2x - x^2$

69. $y = |x - 2|$

70. $y = 3 - |x|$

Real-Life Applications

71. For at most 6 people to rent a condo at a ski resort, the per-person cost *c* is given by $c = 420/n$, where *n* is the number of people.

(a) What are the domain and range of the function?

(b) Write the function as a set of ordered pairs.

72. For a storage unit with up to 4 shelves, the price *p* is $p = 8 + 1.5t$, where *t* is the number of shelves.

(a) What are the domain and range of the function?

(b) Write the function as a set of ordered pairs.

73. A candy bar manufacturer must design a rectangular box to be used as a shipping container. The length of the base of the box is $20 - x$, where *x* is the width in inches.

(a) Write a function to describe the area of the base.

(b) Use the function to complete the table.

x	0	4	8	10	12	16	20
Area							

(c) Referring to the table, estimate the largest possible area of the base.

(d) On the basis of your estimate in part (c), what base dimensions produce the largest possible area?

(e) On the basis of the table, what are the dimensions of the base if the area is 64 square inches?

(f) What is the domain of the function?

74. At a local store, the price (in dollars) of a Rocky Colavito baseball card is $30 - x$, where *x* is the number of such cards that are available.

(a) Write a function for the income from the sale of *x* cards.

(b) Use the function to complete the table.

x	0	5	10	15	20	25	30
Income							

(c) Referring to the table, estimate the range of the function.

(d) Referring to the table, estimate the number of cards that should be sold for the most income.

(e) According to the table, how many cards are sold if the income is $200?

(f) What is the domain of the function?

Modeling with Real Data

75. The following table shows the most popular household pets in the United States. (Source: American Veterinary Medical Association.)

Household Pets	Number Owned (in millions)
Cats	59.0
Dogs	52.9
Birds	12.6
Rabbits	5.7

(a) Use the table to complete the following relation.

{(Birds, ____), (____ , 52.9), (____ , 5.7), (Cats, ____)}

(b) Does this relation represent a function? Why?

76. On the basis of the data in Example 7, would the relation {(President, Year)} be a function? Why?

77. A *mapping diagram* can be used to show how the entries in one list are paired with the entries of another list. (Source: *Landscape Plants of the Southeast.*)

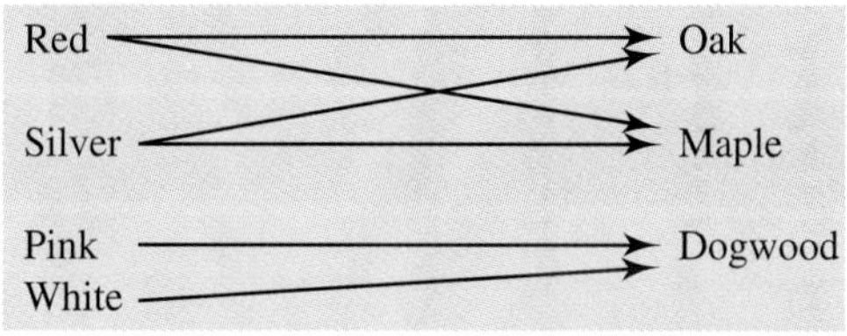

Does this mapping diagram represent a function? Why?

78. On the basis of the data in Example 7, would the relation {(Year, Team)} be a function? Why?

Data Analysis: *Day Care Costs*

Based on a monthly fee for a 3-year-old in a for-profit center 5 days per week, 8 hours per day, the median cost of day care nationwide is \$388. The bar graph shows the metropolitan areas that have the cheapest day care. (Source: Runzheimer International.)

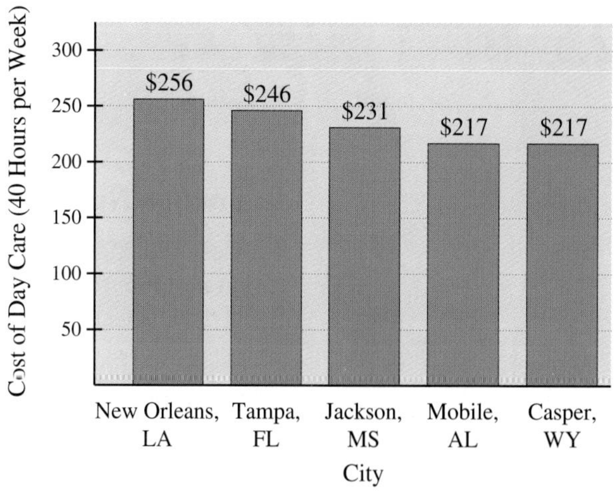

79. Write the information shown in the bar graph as a set A of ordered pairs of the form (City, Cost). Is set A a function?

80. Consider the set of ordered pairs (City, Cost), where the city is any city in the United States, and the cost is the average cost of day care for that city. Is this set a function?

81. Suppose that you selected a city and made a table of the costs at all day care centers in that city. Would the relation {(City, Cost)} be a function?

82. What minimum hourly wage would a full-time employee in New Orleans need to make just to pay for day care?

Challenge

In Exercises 83–88, the variables x and y are used to represent human beings. Determine whether the given relation is a function.

83. $A = \{(x, y) \mid y \text{ is an ancestor of } x\}$

84. $B = \{(x, y) \mid x \text{ is the mother of } y\}$

85. $C = \{(x, y) \mid y \text{ is the mother of } x\}$

86. $D = \{(x, y) \mid x \text{ is a brother of } y\}$

87. $E = \{(x, y) \mid y \text{ is an aunt of } x\}$

88. $F = \{(x, y) \mid x \text{ is a grandfather of } y\}$

In Exercises 89–92, determine the domain and range of the function.

89. $y = \dfrac{|x|}{x}$

90. $y = |x| + x$

91. $y = \sqrt{x^2}$

92. $y = \sqrt[3]{x^3}$

2.5 Functions: Notation and Evaluation

Function Notation • Evaluating Functions

Function Notation

When a function is described by an algebraic expression or rule, we usually use **function notation** to refer to the expression and to represent the value of the expression for a given replacement for the variable.

For the function $\{(x, y) \mid y = 2x\}$, we might name the function f and write $f(x) = 2x$. The notation $f(x)$ is read "f of x" and indicates that the name of the function is f and that the variable is x. We can use $f(x)$ and y interchangeably; that is, we can write $f(x) = 2x$ or $y = 2x$. In any case, the value of y is a function of (depends on) the value of x.

Just as we can use any variable when we write an expression, we can use letters other than f to name a function. The following are other examples of functions written with function notation.

$$g(x) = 2 - 3(x - 1) \qquad h(t) = |2t - 1| \qquad P(w) = \frac{w + 1}{w - 3} \qquad r(x) = \sqrt{x + 2}$$

Note: Be sure to read $f(x)$ as "f of x." Here the parentheses do *not* mean "f times x."

Evaluating Functions

Function notation is especially convenient when we want to indicate the value of a function for a given replacement for the variable. For example, if $f(x) = 2x^2 - 4x - 5$, then $f(3)$ represents the value of the function when x is replaced by 3.

Using function notation for evaluating functions usually helps to sell students on the value of the notation.

$$f(x) = 2x^2 - 4x - 5$$
$$f(3) = 2(3)^2 - 4(3) - 5 \qquad \text{Replace } x \text{ with 3.}$$
$$= 18 - 12 - 5$$
$$= 1$$

Note that because $f(3) = 1$, $(3, 1)$ is a point of the graph of the function. The value $f(3)$ is the y-coordinate of the point whose x-coordinate is 3.

EXAMPLE 1

Evaluating a Function

Let $g(t) = -t^2 + 2t + 6$. Evaluate $g(-4)$.

LEARNING TIP

Keep relating a function and its value to points of a graph. If $f(3) = 5$, then $(3, 5)$ is a point of the graph of f. As you study mathematics, go slowly and take the time to visualize.

Solution

$$g(t) = -t^2 + 2t + 6$$
$$g(-4) = -(-4)^2 + 2(-4) + 6 \qquad \text{Replace } t \text{ with } -4.$$
$$= -16 - 8 + 6 \qquad -(-4)^2 = -1(-4)^2 = -1 \cdot 16 = -16$$
$$= -18$$

EXAMPLE 2

Evaluating Functions

Let $h(x) = |3 - 2x|$ and $r(x) = \dfrac{12}{x}$. Evaluate each of the following.

(a) $h(5)$ (b) $r(4)$ (c) $h(0) - r(-4)$

Solution

(a)
$$\begin{aligned} h(x) &= |3 - 2x| \\ h(5) &= |3 - 2 \cdot 5| && \text{Replace } x \text{ with 5.} \\ &- |3 - 10| \\ &= |-7| \\ &= 7 \end{aligned}$$

(b)
$$\begin{aligned} r(x) &= \frac{12}{x} \\ r(4) &= \frac{12}{4} && \text{Replace } x \text{ with 4.} \\ &= 3 \end{aligned}$$

(c)
$$h(0) = |3 - 2 \cdot 0| = |3 - 0| = |3| = 3$$
$$r(-4) = \frac{12}{-4} = -3$$

Therefore, $h(0) - r(-4) = 3 - (-3) = 6$.

Students sometimes have trouble with the transition from evaluating a function for a number to evaluating for a variable expression.

As the next two examples illustrate, a function can be evaluated for a variable expression instead of a specific value.

EXAMPLE 3

Evaluating a Function for a Variable Expression

Let $g(x) = \sqrt{x^2 - 4}$. Find $g(t)$ and $g(3t)$.

Solution

$$g(t) = \sqrt{(t)^2 - 4} = \sqrt{t^2 - 4} \qquad \text{Replace } x \text{ with } t.$$
$$g(3t) = \sqrt{(3t)^2 - 4} = \sqrt{9t^2 - 4} \qquad \text{Replace } x \text{ with } 3t.$$

EXAMPLE 4

Operations with Functions

These operations may prepare students for Newton's quotient in later courses.

Let $f(x) = 3x - 4$. Determine each of the following.

(a) $f(5)$

(b) $f(5 + h)$

(c) $f(5 + h) - f(5)$

(d) $\dfrac{f(5 + h) - f(5)}{h}$

Think About It

Suppose that $f(x) = x + 1$. Show that $f(a + b) \neq f(a) + f(b)$.

Solution

(a) $f(5) = 3(5) - 4 = 15 - 4 = 11$

(b) $f(5 + h) = 3(5 + h) - 4$ — Substitute $5 + h$ for x.

$= 15 + 3h - 4$

$= 11 + 3h$

(c) $f(5 + h) - f(5) = (11 + 3h) - 11$ — Use the results of parts (a) and (b).

$= 3h$

(d) $\dfrac{f(5 + h) - f(5)}{h} = \dfrac{3h}{h}$ — Use the result of part (c).

$= 3$

Quick Reference 2.5

Function Notation

- A special notation called **function notation** can be used to represent a function and to represent its value for a given value of the variable.
- The notation $f(x)$ is read "f of x" and indicates that the name of the function is f and that the variable in the function is x.

Evaluating Functions

- Function notation is especially convenient when we want to indicate the value of a function for a particular value of the variable.
- If $f(x)$ represents a certain function and $f(3) = 5$, then $(3, 5)$ is a point of the graph of the function.

Speaking the Language 2.5

1. The symbol $f(z)$ indicates that the ________ in the function is z.
2. If $y = f(x)$, then $f(3)$ represents the ________ when $x = 3$.
3. If the graph of $y = f(x)$ has only one point whose y-coordinate is $f(3)$, then the ________ is 3.
4. If f is a function, then the process of determining $f(5)$ is called ________ the function when x is 5.

Exercises 2.5

Concepts and Skills

1. If $f(x) = 2x + 1$, explain how to evaluate $f(3)$.
2. If $g(x) = 1 - 3x$, explain how to evaluate $g(t + 4)$.

In Exercises 3–18, fill in the blank.

3. $f(x) = 5x + 4$; $f(3) = 5(\ \ \ ) + 4$
4. $h(x) = 7x - 6$; $h(1) = 7(\ \ \ ) - 6$
5. $R(x) = 4 - x^2$; $R(2) = 4 - (\ \ \ )^2$

6. $P(x) = 25 - x^2;\quad P(0) = 25 - (\quad)^2$

7. $h(x) = -x^2 + 2x + 5;$

$h(-2) = -(\quad)^2 + 2\,(\quad) + 5$

8. $g(x) = -x^4 - x^2;\quad g(-1) = -(\quad)^4 - (\quad)^2$

9. $f(x) = 2x^3 - 4x + 5;\quad f(\quad) = 2(2)^3 - 4(2) + 5$

10. $g(x) = 4x^3 + 3x^2 - 1;$

$g(\quad) = 4(1)^3 + 3(1)^2 - 1$

11. $h(x) = \sqrt{3x+4};\quad h(-1) = \sqrt{3(\quad) + 4}$

12. $g(x) - \sqrt{5 - 4x};\quad g(1) = \sqrt{5 - 4(\quad)}$

13. $a(x) = |x - 5|;\quad a(\quad) = |1 - 5|$

14. $b(x) = |2x + 9|;\quad b(\quad) = |2(4.5) + 9|$

15. $g(x) = 2x - 7;\quad g(3 + t) = 2(\quad) - 7$

16. $f(x) = 9x + 2;\quad f(b - 4) = 9(\quad) + 2$

17. $h(x) = -4x^2 + 3x - 2;\quad h(\quad) = -4a^2 + 3a - 2$

18. $P(x) = 7x^2 - 9x + 1;\quad P(\quad) = 7a^2 - 9a + 1$

In Exercises 19–24, evaluate the function as indicated.

19. $f(x) = 6 - x$

$f(3), f(-2)$

20. $f(x) = -3x + 5$

$f(-1), f(2)$

21. $g(x) = 2x + 9$

$g(-3), g(4)$

22. $h(x) = 5 - 6x$

$h(5), h(-4)$

23. $R(x) = -x^2 + x + 3$

$R(-2), R(3)$

24. $G(x) = 3x^2 - x + 7$

$G(0), G(1)$

25. Suppose that a certain expression is represented by $f(x)$ and that $f(4) = -7$. What is one point of the graph of the function? Explain how you know this.

26. Suppose that the graph of a function contains the point $(-3, 5)$. If the function is represented by $g(x)$, what is $g(-3)$? Explain how you know this.

In Exercises 27–32, evaluate the functions as indicated.

27. $f(x) = 7x - 2$ and $g(x) = 2x - 5$

(a) $g(-4)$ (b) $f(-3)$

(c) $g(3) - f(2)$ (d) $f(0) + g(1)$

28. $P(x) = x^2 - 3x + 5$ and $Q(x) = x^2 + 4x - 8$

(a) $P(-1)$ (b) $Q(1)$

(c) $Q(2) + P(0)$ (d) $P(2) - Q(0)$

29. $g(x) = -x^2 - 3x$ and $h(x) = x^4 - x$

(a) $g(-1)$ (b) $h(-1)$

(c) $h(-2)$ (d) $g(-3)$

30. $s(x) = -x^3 - 5x + 3$ and $T(x) = x^3 + 2x^2 - 5$

(a) $s(1)$ (b) $T(1)$

(c) $T(-2)$ (d) $s(-2)$

31. $h(x) = \sqrt{3x + 4}$ and $g(x) = |3 - 2x|$

(a) $h(-1)$ (b) $h(4)$

(c) $g(-2)$ (d) $g(2)$

32. $f(x) = |2x - 1|$ and $g(x) = \sqrt{5 - 4x}$

(a) $f(0)$ (b) $f(-1)$

(c) $g(1)$ (d) $g(-5)$

In Exercises 33–36, evaluate the functions as indicated.

33. $g(x) = 3x - 5$ and $f(x) = 8x - 3$

(a) $g(-9.3)$ (b) $g\left(\frac{3}{4}\right)$

(c) $f\left(\frac{5}{8}\right)$ (d) $f(50)$

34. $f(x) = 7 - 4x$ and $g(x) = \frac{3}{4}x + 1$

(a) $f(-2.7)$ (b) $f\left(\frac{2}{3}\right)$

(c) $g(25)$ (d) $g(0.8)$

35. $p(x) = -6x^2 - 8x + 3$ and $h(x) = -4x^2 + 3x - 5$

(a) $h\left(\frac{2}{5}\right)$ (b) $h(-11.4)$

(c) $p(25.7)$ (d) $p\left(-\frac{1}{7}\right)$

36. $f(x) = 2x^3 - 4x - 15$ and $g(x) = 4x^3 + 3x - 1$

(a) $f(-2.3)$ (b) $f(-1.9)$

(c) $g\left(\frac{3}{7}\right)$ (d) $g\left(-\frac{1}{7}\right)$

37. Suppose that $h(x) = \sqrt{3x - 6} \div 1 - x$ and $k(x) = \sqrt{3x - 6} \div (1 - x)$.

(a) Evaluate $h(5)$ and $k(5)$.

(b) Explain the role of the parentheses.

38. Suppose that $B(x) = 4x + 8 \div 2 - x$ and $C(x) = (4x + 8) \div (2 - x)$.

(a) Evaluate $B(1)$ and $C(1)$.

(b) Explain the role of the parentheses.

In Exercises 39–46, evaluate the given function as indicated.

39. $f(x) = x^2 - (4 - x);\quad f(-2)$

40. $g(z) = -z^2 - (3 - z);\quad g(-2)$

41. $h(c) = 5 - 5(5 - c);\quad h(5)$

42. $f(t) = 3 - 3(8 - t);\quad f(1)$

43. $k(m) = 7(3m - 1) - (8 - m);\quad k(-2)$

44. $f(p) = 6(2p - 3) - (1 - 3p);\quad f(2)$

45. $m(s) = [4 - (s - 1)] + 8s;\quad m(3)$

46. $n(t) = -[4 - (t - 2)] + 8t;\quad n(3)$

Exercises 47 and 48 illustrate the advantage of simplifying before evaluating.

47. Suppose $G(x) = 3\{x - [x - (x - 1)]\} - 3x$.

(a) Evaluate $G(1)$, $G(-4)$, and $G(2.3)$.

(b) What is your conjecture about $G(x)$ for any value of x?

(c) Show that your conjecture is correct by simplifying the expression.

48. Suppose $F(x) = 2[(x - 1) - 2] - 2(x - 2)$.

(a) Evaluate $F(3)$, $F(1)$, and $F(-3.7)$.

(b) What is your conjecture about $F(x)$ for any value of x?

(c) Show that your conjecture is correct by simplifying the expression.

In Exercises 49–52, use your calculator to evaluate each function as indicated. Round your results to the nearest hundredth.

49. $h(x) = \sqrt{3x + 4}$

(a) $h(20.8)$ (b) $h(0.9)$

50. $g(x) = \sqrt{5 - 4x}$

(a) $g(-2.4)$ (b) $g(-4.9)$

51. $k(x) = \sqrt{x^2 - 2x - 15}$

(a) $k(11.3)$ (b) $k(-4.5)$

52. $A(x) = \sqrt{2x^2 - x - 1}$

(a) $A(3.4)$ (b) $A(-5.7)$

In Exercises 53–56, use your calculator to evaluate each function as indicated. Round results to the nearest hundredth.

53. $c(x) = |x^2 + 2x - 7|$

(a) $c(106)$ (b) $c(-89)$

54. $d(x) = |2 - x - x^2|$

(a) $d(57)$ (b) $d(-104)$

55. $f(x) = \dfrac{x + 1}{2x - x^2}$

(a) $f(-1)$ (b) $f(0.7)$

56. $g(x) = \dfrac{x^2 + x}{3x + 2}$

(a) $g(1)$ (b) $g(-0.3)$

57. Given $f(x) = -x^2$, $g(x) = x^2$, and $h(x) = (-x)^2$, determine the values in parts (a) and (b). Then answer the questions in parts (c) and (d).

(a) $f(7)$, $g(7)$, and $h(7)$

(b) $f(-7)$, $g(-7)$, and $h(-7)$

(c) Let A represent the step "multiply by -1"; let B represent the step "square." Using A and B, write the sequence to be followed in evaluating each of the three functions.

(d) Will any of the three functions have the same graph? Explain.

58. Given $f(x) = -x^3$, $g(x) = x^3$, and $h(x) = (-x)^3$, determine the values in parts (a) and (b). Then answer the questions in part (c).

(a) $f(2)$, $g(2)$, and $h(2)$

(b) $f(-2)$, $g(-2)$, and $h(-2)$

(c) Which of the following is/are suggested by the results? Explain your answer.

(i) For all x, $f(x) = g(x)$.

(ii) For all x, $g(x) = h(x)$.

(iii) For all x, $f(x) = h(x)$.

59. Suppose $B(x) = \sqrt{x^2 + 25}$ and $C(x) = x + 5$.

Exercise 59 deals with an error commonly made by students.

(a) Evaluate $B(0)$ and $C(0)$.

(b) Evaluate $B(3)$ and $C(3)$.

(c) Evaluate $B(-6)$ and $C(-6)$.

(d) Would you expect the two functions to have the same graph? Explain.

60. Suppose $f(x) = \sqrt{x^2}$, $g(x) = x$, and $h(x) = |x|$.

(a) Evaluate $f(3)$, $g(3)$, and $h(3)$.

(b) Evaluate $f(-5)$, $g(-5)$, and $h(-5)$.

(c) Which two of these three functions do you think have the same graph? Why?

61. Suppose $f(x) = |x + 7|$ and $g(x) = |x| + 7$.

(a) If we replace x with only positive values, will the graphs of the two functions be the same?

(b) If we replace x with any real number, will the graphs of the two functions be the same?

62. Suppose $A(x) = \left(\sqrt{x - 5}\right)^2$ and $B(x) = |x - 5|$.

(a) Evaluate $A(14)$ and $B(14)$.

(b) Evaluate $A(1)$ and $B(1)$.

(c) Would you expect the two functions to have the same graph?

In Exercises 63–70, evaluate the function as indicated.

63. $f(x) = 5x - 7$

(a) $f(2a)$ (b) $f(a + 2)$

64. $g(x) = 6x + 2$

(a) $g(3 + z)$ (b) $g(3z)$

65. $h(x) = x^2 - 5x + 6$

(a) $h(t)$ (b) $h(2t)$

66. $g(x) = x^2 + 4x - 8$

(a) $g(-3t)$ (b) $g(t)$

67. $f(x) = \sqrt{x^2 + 5}$

(a) $f(a)$ (b) $f(-2a)$

68. $g(x) = \sqrt{6 - x^2}$

(a) $g(-3b)$ (b) $g(m)$

69. $f(x) = |x^3 - 2|$

(a) $f(-s)$ (b) $f(3s)$

70. $h(x) = |x^3 + 4|$

(a) $h(2r)$ (b) $h(-r)$

71. Let $f(x) = 2x - 5$. Evaluate f as indicated.

(a) $f(3)$ (b) $f(3 + h)$

(c) $f(3 + h) - f(3)$

(d) $\dfrac{f(3 + h) - f(3)}{h}$

72. Let $g(x) = 7 + 4x$. Evaluate g as indicated.

(a) $g(-8)$

(b) $g(-8 + h)$

(c) $g(-8 + h) - g(-8)$

(d) $\dfrac{g(-8 + h) - g(-8)}{h}$

73. Let $f(x) = 5 - 3x$. Evaluate f as indicated.

(a) $f(-2)$

(b) $f(-2 + h)$

(c) $f(-2 + h) - f(-2)$

(d) $\dfrac{f(-2 + h) - f(-2)}{h}$

74. Let $g(x) = 5x + 7$. Evaluate g as indicated.

(a) $g(4)$

(b) $g(4 + h)$

(c) $g(4 + h) - g(4)$

(d) $\dfrac{g(4 + h) - g(4)}{h}$

Real-Life Applications

75. A person walking a treadmill finds that her number of paces p per minute can be modeled by the function $p(m) = 60 - \sqrt{2m}$, where m is the number of minutes. To the nearest whole number, what is her pace rate halfway through a 30-minute workout?

76. A "square" of shingles covers 100 square feet. For a roof of x square feet, the function $N(x) = \dfrac{x}{100}$ models the number N of squares that will be needed. Evaluate N to determine the number of squares required for a roof of 5100 square feet.

77. The number B of bacteria present 24 hours after treatment with x milligrams of a drug is given by $B(x) = -2x^2 - 10x + 1500$. If the bacteria are treated with 8 milligrams of the drug, how many bacteria remain after 24 hours?

78. A person's average weekly expenditure G for groceries is related to the age x of the person by $G(x) = 3.25x - 20 - 0.0375x^2$. What is the average weekly grocery bill for a person who is

(a) 20 years old?

(b) 40 years old?

(c) 60 years old?

Modeling with Real Data

79. From 1995 to 1999, the number of semiprivate golf courses (clubs that also open to the public for a daily fee) rose while the number of private clubs declined. (Source: National Golf Foundation.)

Year	Semiprivate	Private
1995	8142	4759
1999	9012	4708

With x representing the number of years since 1995, the following functions model the data, where S is the number of semiprivate clubs and P is the number of private clubs.

$$S(x) = 217.5x + 8142$$

$$P(x) = -12.75x + 4759$$

Use functions S and P to complete the following table.

Year		1999		2002
x	2		6	
Semiprivate				
Private				

80. Veterinarians use the drug lufenuron for control of fleas in cats. The dosage depends on the weight of the cat. For example, a 15-pound cat receives 270 milligrams of lufenuron. (Source: Ciba Animal Health.)

You might use the function $D(w) = 13.5w + 67.5$ to estimate the dosage D (in milligrams) for a cat that weighs w pounds.

(a) If you owned a 12-pound cat, what dosage of lufenuron would you administer?

(b) Create a table to show the dosages for cats that weigh 6, 8, 10, 12, and 14 pounds. Then estimate the weight of a cat that receives 202 milligrams of lufenuron.

Data Analysis: *Gray Wolf Population*

Since the Endangered Species Act of 1973 became law, the gray wolf population in the continental United States has increased. The bar graph shows the estimated populations of gray wolves between 1960 and 2000. (Source: U.S. Fish and Wildlife Service.)

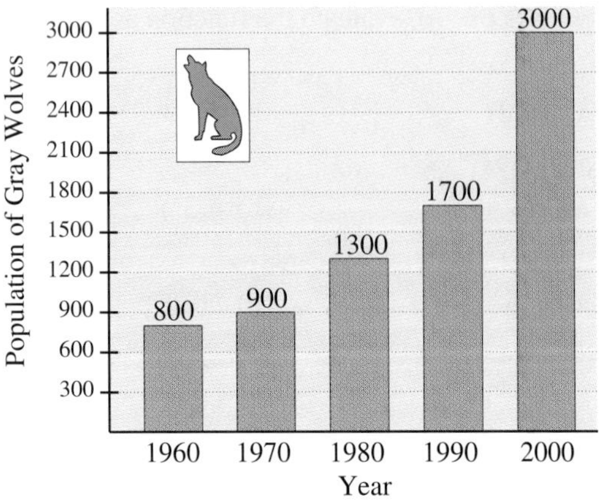

The following are two functions that could be used to model the data in the bar graph. For both functions, x is the number of years since 1950.

$$f(x) = 52x - 20$$

$$g(x) = 1.7x^2 - 50.9x + 1180$$

81. Use the model $f(x)$ to complete the following table. The column "f Difference" is the difference between the actual data and the corresponding value of f.

Year	$f(x)$	f Difference
1960		
1970		
1980		
1990		
2000		

82. Use the model $g(x)$ to complete the following table. The column "g Difference" is the difference between the actual data and the corresponding value of g.

Year	$g(x)$	g Difference
1960		
1970		
1980		
1990		
2000		

83. On the basis of the difference columns in the tables in Exercises 81 and 82, which function appears to model the data better?

84. Use each model to predict the populations in 2010 and 2020. On the basis of these predictions, compare the trend predicted by the models.

Challenge

85. Suppose $f(x) = \dfrac{3}{x}$.

(a) Evaluate each of the following.

(i) $f(1)$

(ii) $f(0.1)$

(iii) $f(0.01)$

(iv) $f(0.001)$

(v) $f(0.0001)$

(b) As the x-values become closer to zero, what is happening to the values of $f(x)$?

86. Suppose $g(x) = x^2$.

(a) Evaluate each of the following.

(i) $g(11)$

(ii) $g(111)$

(iii) $g(1111)$

(iv) $g(11{,}111)$

(b) On the basis of the pattern in part (a), predict the value of $g(11{,}111{,}111)$.

2 • Chapter Project **Bicycle Helmet Usage**

In Section 2.1, Exercises 91–94, we modeled the percent of bicycle riders who wore helmets with the expression $5.35t + 18.7$, where t is the number of years after 1992.

1. Complete the following table.

t	Actual Percent A	Modeled Percent M	$M - A$	$(M - A)^2$
0	18%			
2	30			
4	41			
6	50			

2. The columns headed "$M - A$" and "$(M - A)^2$" give some indication of how accurate the model is for a given value of t. Add the entries in the $(M - A)^2$ column, and then take the square root of the result. Call the resulting number D.

   The following are four other expressions that could be used to model the trend in helmet usage.

   (i) $6x + 18$ (ii) $5.5x + 19$

   (iii) $5x + 20$ (iv) $5.75x + 18$

   Form four groups of students, with each group selecting one of these alternative models. Your group will use the model expression that you selected to answer Questions 3–5.

3. Create and complete a table similar to that in Question 1.
4. Calculate the value of D as described in Question 2.
5. Along with the other groups, report your value of D. Assess whether your model is better or worse than the original model and whether your model is better or worse than the models used by the other groups. Explain your reasoning.

Chapter 2 Review Exercises

Section 2.1

1. Explain how to evaluate $x^2 + x + 1$ for $x = 3$.

2. What do we mean by a *solution* of an equation?

In Exercises 3–5, evaluate the expression for the given value(s) of the variable(s).

3. $\dfrac{2(x+1)}{x}$ for $x = -2$

4. $5a - 4b$ for $a = 3$ and $b = -2$

5. $2x^2 - 3y + 4xy$ for $x = -3$ and $y = 2$

6. Evaluate the expression $\sqrt{x^2 + 3y - 2xy}$ for $x = 5$ and $y = -2$. Round your result to the nearest hundredth.

In Exercises 7 and 8, determine which member of the following set is a solution of the given equation.

$$\left\{-7, -4, -\frac{2}{3}, 3, \frac{5}{4}, 6\right\}$$

7. $6x + 5 = 1$
8. $15 - 2x = 3$

9. Suppose that the value of x is -3. With what value would you need to replace y in order for the expression $\frac{x}{y}$ to have a value of 3?

10. The formula for converting Fahrenheit temperatures to Celsius is $C = \frac{5}{9}(F - 32)$. Convert 95°F to the corresponding Celsius temperature.

11. The three numbers given in each part are the lengths of the sides of a triangle. Determine whether the triangle is a right triangle.
    (a) 5, 4, 3
    (b) 12, 15, 18

12. A 15-foot post is supported by a 17-foot guy wire that is attached to the ground 8 feet from the base of the post. Show whether the post is perpendicular to the ground.

Section 2.2

13. Explain why $3x^2y$ and $3xy^2$ are not like terms.

14. Explain how to identify the terms of an expression.

In Exercises 15 and 16, identify the terms and coefficients of the expression.

15. $3x - 2y - z$
16. $3x^2 - 7(h - 5)$

In Exercises 17 and 18, state whether the given terms are like terms.

17. $7cb$, $7cd$
18. $4x^2y^3$, $-2x^2y^3$

In Exercises 19 and 20, simplify the expression by removing the grouping symbols.

19. $-(2a + 3b - 4c)$

20. $-2(3x - 4y) + 3(2h - 5k)$

In Exercises 21 and 22, simplify the expressions by combining like terms.

21. $14x^2 - 3x^3 + 2x^2 - 7x^3 + x^3$

22. $12ab^2 - 10a^2b + 5ab^2 - 10a^2b$

23. Translate the given information into an algebraic expression. Let n represent the number. Simplify the expression, if possible.
    (a) The product of -5 and the sum of 4 and 3 times the number
    (b) Ten more than the number less 4 times the number

24. (a) Translate the following into an equation: 3 less than twice a number is 7.
    (b) Show whether -2 is a solution of your equation.

In Exercises 25 and 26, simplify the given expression.

25. $3 - 3(2x - 1) + 4x$
26. $4y - 2[y - 5(6 - y)]$

Section 2.3

In Exercises 27–32, determine the coordinates of the points shown in the accompanying figure.

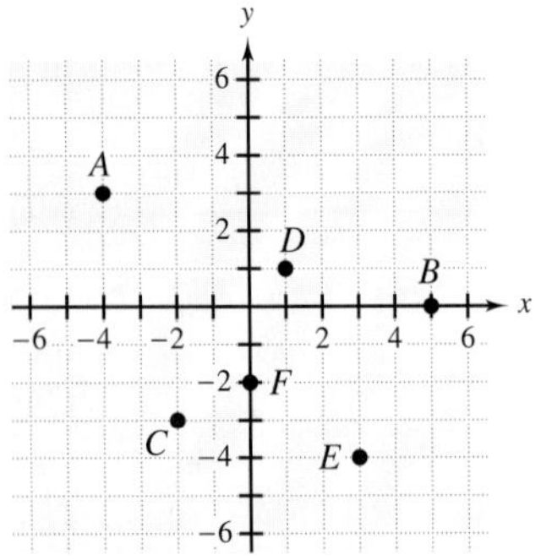

27. A
28. B
29. C
30. D
31. E
32. F

In Exercises 33–36, state in what quadrant the given point lies.

33. $P(-\pi, \pi)$ **34.** $Q(5, -6)$

35. $R(a, b)$ if $a < 0$ and $b > 0$

36. $S(3.25, 4.56)$

37. On which axis is $P(c, d)$ if $c = 0$?

38. In which quadrant is $P(a, b)$ if $Q(b, a)$ is known to be in Quadrant IV?

39. Graph the following set.

$$\{(-3, 5), (0, -6), (2, -4), (-5, -3), (7, 0)\}$$

40. If point P is directly to the right of $(-3, 4)$ and directly above $(2, -7)$, what are the coordinates of point P?

In Exercises 41 and 42, determine each of the following.

(a) The distance (to the nearest hundredth) between the points A and B.

(b) The midpoint of the line segment whose endpoints are A and B.

41. $A(-5, 4)$ and $B(-8, -5)$

42. $A(7, -1)$ and $B(13, -9)$

43. Find the perimeter of the parallelogram $ABCD$ if the vertices are $A(-6, 1)$, $B(-3, 5)$, $C(9, 0)$, and $D(6, -4)$.

44. Determine whether $A(-3, -2)$, $B(-1, 8)$, and $C(3, 4)$ are vertices of a right triangle.

Section 2.4

In Exercises 45 and 46, state the domain and range for each relation. Identify any relation that is a function.

45. $\{(0, 1), (3, 5), (-4, 7), (5, -6)\}$

46.

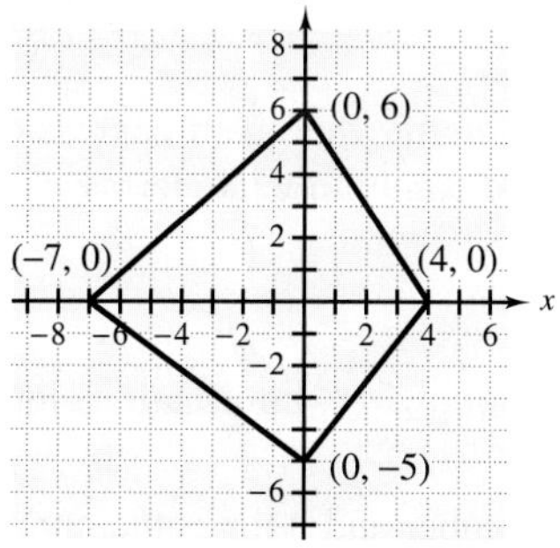

In Exercises 47–49, determine whether the given equation represents a function.

47. $y = |x + 2|$ **48.** $y = \sqrt{x + 1}$

49. $y^2 = x + 1$

50. Which of the following statements is true? Why?

(i) Every relation is a function.

(ii) Every function is a relation.

In Exercises 51 and 52, estimate the domain and range of the function whose graph is given.

51.

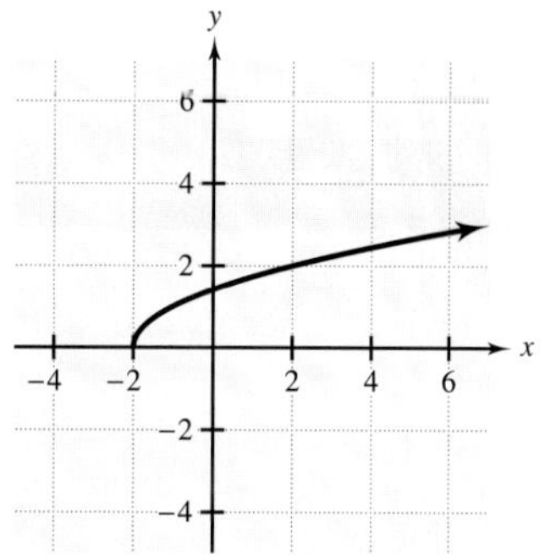

52.

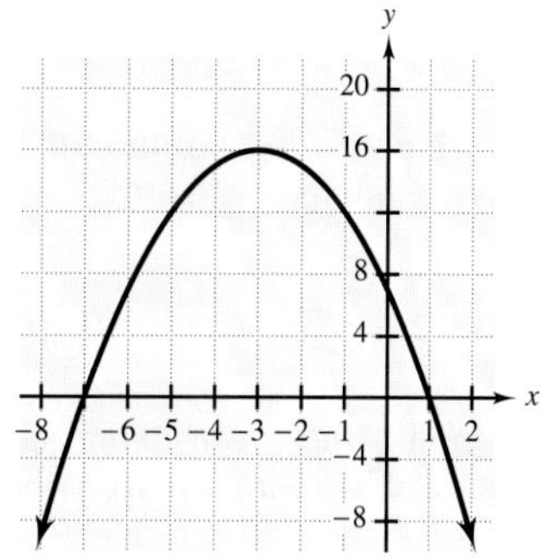

53. Which of the following graphs, if either, represents a function?

(i) A line containing $(-3, 2)$ and $(4, 0)$

(ii) A circle with its center at the origin

54. The sum of the base b and the height h of a parallelogram is 24 inches.

(a) Write a function that describes the area of the parallelogram.

(b) Use the function to complete the table.

b	2	4	8	12	16	20	22
Area							

(c) Use the table to estimate the greatest possible area of the parallelogram.

(d) What is the domain of the function?

Section 2.5

In Exercises 55–60, evaluate the function as indicated.

55. $f(x) = -x^2 - 5x + 2;\ f(-2)$

56. $g(a) = a^3 - 6a;\ g(-1)$

57. $h(c) = \sqrt{c^2 + 4c};\ h(3)$

58. $A(x) = |2x - 3|;\ A(-5)$

59. $Q(a) = 5a - 3(5a - 3) - a;\ Q(-1)$

60. $R(d) = \sqrt{5d + 4} - d^2;\ R(9)$

61. Let $f(x) = |2x - 5|$ and $g(x) = \dfrac{3x}{x^2 + 1}$. Evaluate the given expression.

(a) $f(1) + g(0)$

(b) $g(1) - f(-2)$

62. Determine $f(2 + h)$ if $f(x) = 7 - 3x$.

Looking Back

The following exercises review concepts and skills that you will need in Chapter 3.

In Exercises 1 and 2, evaluate the given expression for the given value of the variable.

1. $3 - \dfrac{3}{4}t;\ 8$

2. $4 - 2(3x + 2);\ -5$

In Exercises 3 and 4, for which member of the following set does the expression have the given value?

$\{-2, 0, 2, 5, 7\}$

3. $2 - 3x;\ -4$

4. $3(x - 2) - 5;\ 10$

In Exercises 5 and 6, simplify the given expression by removing grouping symbols and combining like terms.

5. $11x + 2(4 - 3x)$

6. $10\left(\dfrac{2x}{5} - \dfrac{2x - 1}{2}\right)$

In Exercises 7 and 8, determine whether the given number is a solution of the equation.

7. $2x + 1 = 4 - x;\ 1$

8. $x = 2 - 3x;\ -1$

9. Translate the given verbal description into a simplified algebraic expression.

(a) The perimeter of a rectangle whose width is 9 feet less than its length L

(b) The value in cents of n nickels and $9 - n$ dimes

(c) The year-end value of an investment of P dollars at 8% simple interest

10. Use set-builder notation and interval notation to describe all numbers that are less than 3.

11. Compare the value of the expression $|2x - 1|$ when x is replaced by 3 and -2.

12. Determine the member of $\{-3, 0, 3, 6\}$ that is a solution of the equation $|x - 3| = 0$.

Chapter 2 Test

In Questions 1 and 2, evaluate the expression for the given value(s) of the variable(s).

1. $x^2 + 1$ for $x = 2$

2. $\frac{1}{2}bh$ for $b = 5.32$ and $h = 7.6$

3. Identify the elements of $\{-2, 0, 2, 3\}$ that are solutions of the equation $3x - 4 = -10$.

4. Determine whether the numbers 5, 12, and 13 are the lengths of the sides of a right triangle.

5. Explain the difference in meaning between the words *term* and *factor*.

In Questions 6 and 7, simplify the given expression by removing parentheses and combining like terms.

6. $(2s - 5t) + 2(s + 2t)$

7. $4 - 2[3x - (x - 1)]$

8. (a) Translate the following information into an equation.

The sum of a number and 3 times the number is 2 less than twice the number.

(b) Determine whether -1 is a solution of your equation.

9. Determine the coordinates of the points in the accompanying figure.

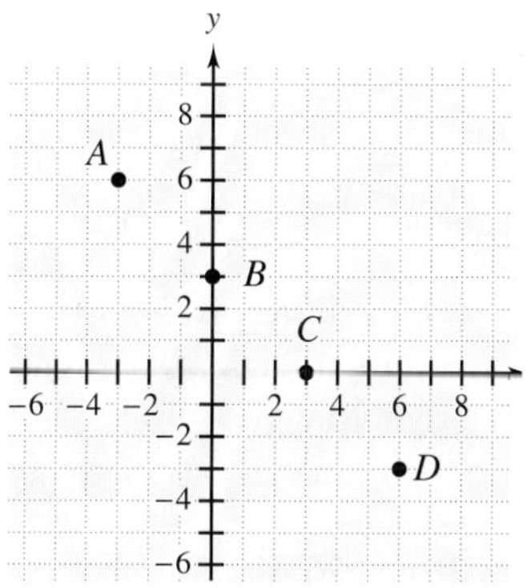

10. Use the given conditions to determine in which quadrant(s) or on what axis the point $P(a, b)$ could be located.

(a) $a > 0$

(b) $Q(b, a)$ is in Quadrant IV

11. Consider the points $A(-2, 1)$, $B(3, -2)$, and $C(2, 4)$.

(a) Determine the midpoint of the line segment $\overline{BC}$.

(b) If $\overline{AB}$ is a side of a triangle, how long is that side?

12. Suppose a weather radar system can detect a tornado within 75 miles of the radar location. The radar is located 30 miles east and 20 miles north of your home, and a tornado is located 40 miles west and 10 miles south of your home. (See figure.)

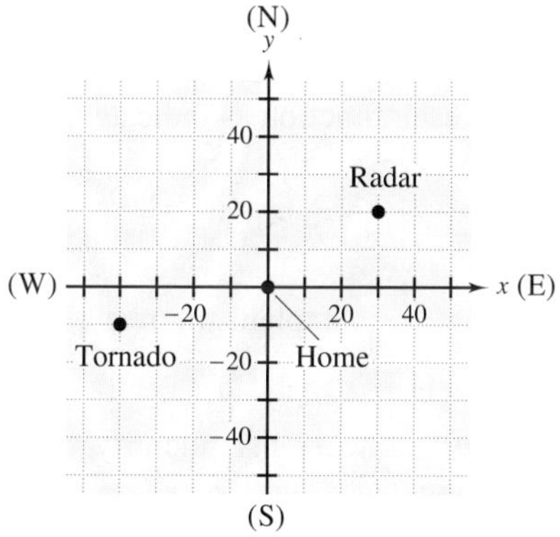

(a) How far is the tornado from the radar site? Is it within range of the radar?

(b) Show whether your home is on the line connecting the tornado and the radar location.

13. If the graph of $3x - 2$ contains a point whose x-coordinate is -4, what is the y-coordinate? Why?

14. For which member of $\{-1, 1, 5\}$ does the expression $6 - 2x$ have a value of -4?

In Questions 15–18, evaluate as indicated.

15. $f(x) = \sqrt{2x - 1};\ f(5)$

16. $f(x) = 2x + 1,\ g(x) = x^2 - 5;\ f(3) - g(1)$

17. $h(x) = \frac{1}{2}x - 8;\ h(6t)$

18. $h(x) = 3x + 2;\ h(1 + t) - h(1)$

19. Determine the domain and range for the given relation.

(a)

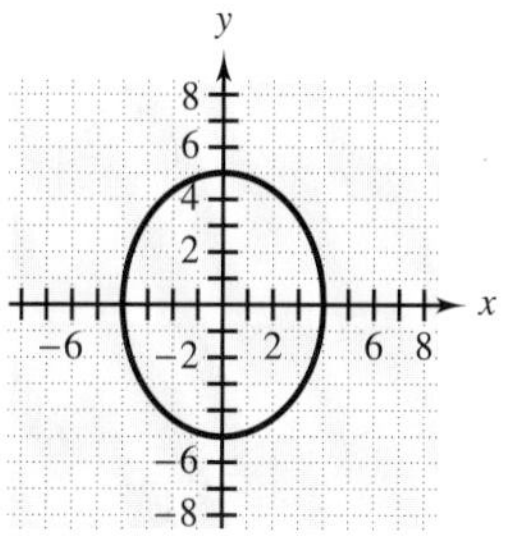

(b) $\{(1, 3), (2, 7), (3, -1), (5, 3)\}$

20. Determine whether the given graph represents a function and explain your answer.

(a)

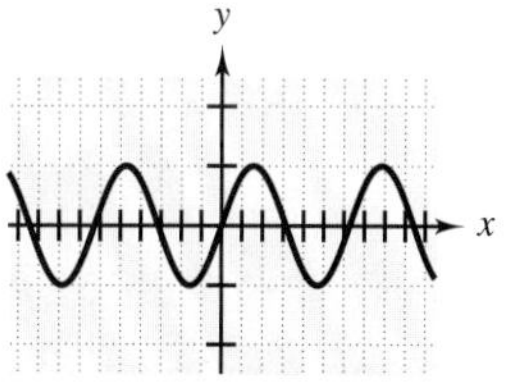

(b)

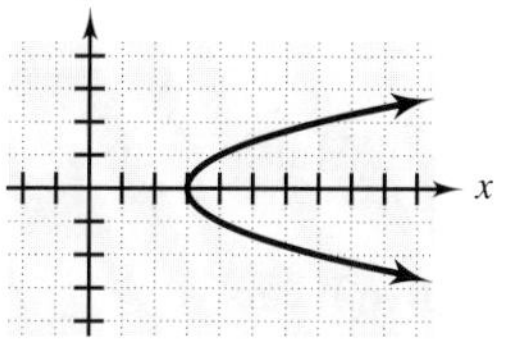

Cumulative Test, Chapters 1–2

1. Describe the set $\{x \mid -2 < x \le 3\}$ with interval notation.

2. For each of the following, determine whether the statement is true or false.

(a) All decimal numbers are irrational numbers.

(b) The inequalities $x < 4$ and $4 > x$ have the same meaning.

(c) The distance between -3 and 5 on the number line can be represented either by $|-3 - 5|$ or by $|5 - (-3)|$.

(d) If $n < 0$, then $-n$ represents a negative number.

In Questions 3–10, perform the indicated operations, if possible.

3. $-6 + [5 - (-4)]$

4. $\dfrac{3}{7} + \left(-\dfrac{2}{3}\right)$

5. $2 - |-5| + (-7)$

6. $-\dfrac{2}{5} - \dfrac{2}{3}$

7. $\dfrac{1}{2}(-1)(-4)(3)(-2)$

8. $\left(-\dfrac{1}{4}\right)\left(-\dfrac{6}{5}\right)\left(-\dfrac{15}{2}\right)$

9. $-1 \div 0$

10. $-24 \div 6 \div (-2)$

In Questions 11–16, evaluate the given expression, if possible.

11. -5^2

12. $15 \div \sqrt{16 - 7}$

13. $8 - 2 \cdot 3 + 10 \div 5$

14. $-1 - \sqrt{-1}$

15. $\dfrac{2^2 + 4^2}{2 - 2(5 - 6)}$

16. $\dfrac{-2 + \sqrt{(-2)^2 - 4(1)(-8)}}{2(1)}$

In Questions 17–20, use the given property of real numbers to write the expression in an alternative form.

17. Distributive Property: $9x + 12 =$ ______

18. Multiplication Property of -1: $-1x =$ ______

19. Commutative Property of Multiplication: $x(2 + 5) =$ ______

20. Property of Multiplicative Inverses: $\left(\dfrac{1}{5} \cdot 5\right)y =$ ______

In Questions 21 and 22, evaluate the expression for the given value(s) of the variable(s).

21. $\dfrac{\sqrt{20 - 2x}}{x^2 + 4}$ for $x = 2$

22. $\dfrac{x^2 y}{(x - y)^2}$ for $x = 5$, $y = -2$

23. Which member of the set $\{-2, 0, 3, 6\}$ is a solution of the equation $\dfrac{x}{2} = -(3 - x)$?

24. In each part, the three given numbers are the lengths of the sides of a triangle. Determine whether the triangle is a right triangle.

(a) 10, 24, 27 (b) 10, 8, 6

In Questions 25 and 26, simplify the given expression.

25. $-2(x - 5) - (4 - x)$

26. $x^2y - xy^2 - x(xy - y^2)$

27. Translate the given information into an algebraic expression or equation. Use x for your variable.

(a) Three more than the quotient of a number and 7

(b) If the difference of 6 and a number is multiplied by 2, the result is twice the number.

28. If $a < 0$ and $b > 0$, name the quadrant or axis that contains the given point.

(a) $P(a, b)$ (b) $Q(b, 0)$

(c) $R(0, -a)$ (d) $S(b, a)$

29. For $P(-6, 3)$ and $Q(8, -7)$, determine the following.

(a) PQ (b) The midpoint of $\overline{PQ}$

30. Suppose that the point (4, 5) belongs to the graph of the function $f(x) = 3x - 7$. What is your interpretation of the coordinates of that point?

31. Evaluate each function as indicated.

(a) $g(x) = 4(3x - 1) - (2 - x)$; $g(-1)$

(b) $f(x) = \sqrt{2x + 7} - x^2$; $f(9)$

32. If $f(x) = \dfrac{1}{x - 1}$, explain the result of attempting to evaluate $f(1)$.

33. From the following list, identify the relations that are functions.

(i) $\{(0, 1), (1, 2), (2, 3), \ldots\}$

(ii) The ordered pairs $(3, y)$, where y is any real number

(iii) The ordered pairs $(x, 3)$, where x is any real number

34. Explain how to use the Vertical Line Test to determine whether the graph of a relation represents a function.

Chapter 3

Linear Equations and Inequalities

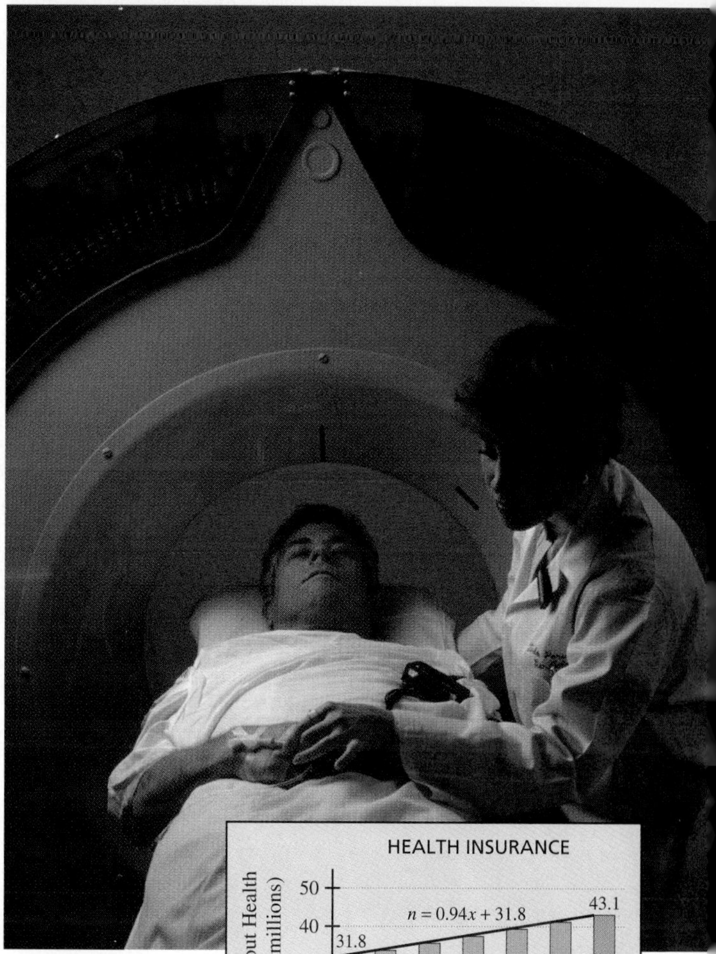

In 1999, about 43.1 million Americans were without health care insurance. This represents a 36% increase from 1987. The trend can be modeled by the ***equation*** $n = 0.94x + 31.8$*, where* n *is the number of uninsured persons and* x *is the number of years since 1987. We can also think of this equation as a* ***formula*** *for estimating the number of uninsured persons for any number of years after 1987. (For more on this topic, see Exercises 99–102 at the end of Section 3.8, and see the Chapter Project.)*

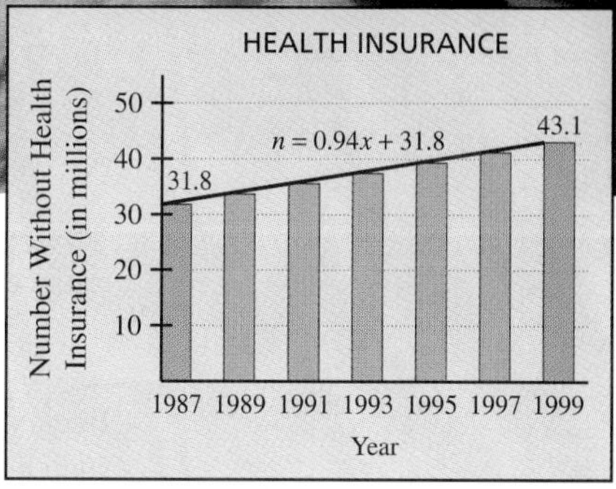

(Source: USA Today.)

In this chapter, we introduce linear equations and formulas as models, and we learn methods for solving and using them in various applications. We then consider solution techniques and applications for linear and compound inequalities, and also absolute value equations and inequalities.

3.1 Introduction to Linear Equations

Equations • Properties of Equations

Equations

In Chapter 2 we briefly considered equations and methods for testing whether numbers were solutions of equations. Now we present a more formal treatment of these topics along with some related vocabulary.

Definition of Equation

An **equation** is a statement that two algebraic expressions have the same value.

The expressions to the left and right of the equality symbol are called the *left side* and *right side* of the equation.

$$\underbrace{2x - 5}_{\text{left side}} = \underbrace{15 + 4x}_{\text{right side}}$$

Students often learn to solve equations without understanding the significance of their results.

If an equation contains a variable, we try to find replacements for that variable that make the equation true. We call such replacements **solutions** of the equation.

Definitions of Solution and Solution Set

A **solution** of an equation is a replacement for the variable that makes the equation true. A **solution set** is the set of all solutions of an equation.

To **solve** an equation means to determine each value of the variable that is a solution.

An equation may be true for any permissible replacement for the variable. Such equations are called **identities.** The following equations are all examples of identities.

$6(3x) = (6 \cdot 3)x$ Associative Property of Multiplication

$y + 3 = 3 + y$ Commutative Property of Addition

$3(x - 5) = 3x - 15$ Distributive Property

For each of these identities, every real number is a solution, so the solution set is **R**.

Some equations, called **inconsistent equations** (or **contradictions**), are false for all replacements of the variable. The following are examples of contradictions.

$$0 \cdot y = 4 \qquad x + 1 = x + 5$$

Because these contradictions have no solution, the solution set is "empty." We call such a set the **empty set,** which we denote with the symbol Ø.

A **conditional equation** is an equation that is true for at least one value of the variable but is not an identity. For example, $5x + 1 = 11$ is true if x is replaced with 2. However, the equation is not an identity because any other replacement for x makes the equation false.

There are many different types of equations. In this chapter we will focus on just one type, called a **linear equation in one variable.**

Definition of a Linear Equation in One Variable

A **linear equation in one variable** is an equation that can be written in the form $Ax + B = 0$, where A and B are real numbers and $A \neq 0$.

Similarly, a **linear function** is a function of the form $f(x) = ax + b$, where $a \neq 0$.

The following are linear equations.

$$3x + 5 = 0 \qquad 5(x + 3) = 4 - x$$

The first equation is in the form $Ax + B = 0$. It can be shown that the other equation can also be written in that form.

The following are examples of equations that are *not* linear equations.

$\frac{3}{x} = 9$ The variable cannot appear in a denominator.

$x + 2 = \sqrt{2x} - 1$ The variable cannot appear under a radical symbol.

$x^2 - 25 = 0$ The variable cannot have an exponent other than 1.

Verifying a solution involves testing whether a replacement for the variable makes the equation true. *Determining* a solution requires other methods. One approach is to estimate the solution by using a table of values.

Exploring the Concept

Estimating Solutions of Linear Equations

Consider the equation $x + 1 = 7 - x$. The following is a partial table of values for the expressions $x + 1$ and $7 - x$.

x	0	1	2	3	4	5	6	7
$x + 1$	1	2	3	4	5	6	7	8
$7 - x$	7	6	5	4	3	2	1	0

Observe that as x increases, the values of $x + 1$ increase and the values of $7 - x$ decrease. When $x = 3$, both expressions have a value of 4. Because the left and right sides of the equation are equal when $x = 3$, we conclude that 3 is a solution of the equation. On the basis of the table, the equation appears to be a conditional equation with one solution. For a contradiction, such as $x + 3 = 1 + x$, a table of values would suggest that the left and right sides *never* have the same value, and so the equation has no solution. For an identity, such as $x + 3 = 3 + x$, the table would indicate that the left and right sides have the same value for *all* values of x, and so every real number is a solution of the equation.

Think About It

Suppose that you create the following table to estimate the solution of $3x + 1 = x + 2$.

x	−1	0	1	2
$3x + 1$	−2	1	4	7
$x + 2$	1	2	3	4

In what interval does the solution lie? Why do you think so?

Although a table of values provides an interesting numerical method for estimating a solution of an equation, a table is limited in efficiency and accuracy. For example, the solution of $5x = 2x + 13$ is $\frac{13}{3}$, an x-value that you would be unlikely to include in your table. Algebraic methods enable us to solve such equations quickly and accurately.

Properties of Equations

Consider the following three equations.

$$2x - 1 = 7 \qquad 2x = 8 \qquad x = 4$$

It is easy to verify that 4 is the solution of all three equations. In fact, we say that the equation $x = 4$ can be solved *by inspection* because the solution is obvious. These equations are examples of **equivalent equations.**

Definition of Equivalent Equations

Equivalent equations are equations that have exactly the same solution sets.

The following properties of equations may be applied to produce equivalent equations.

Properties of Equations

Suppose A, B, and C represent algebraic expressions.

1. If $A = B$, then $A + C = B + C$. Addition Property of Equations

   Adding (or subtracting) the same quantity to (from) both sides of an equation produces an equivalent equation.

2. If $A = B$ and $C \neq 0$, then $AC = BC$. Multiplication Property of Equations

   Multiplying (or dividing) both sides of an equation by the same nonzero quantity produces an equivalent equation.

3. If $A = B$, then $B = A$. Symmetric Property of Equations

   The two sides of an equation can be swapped to produce an equivalent equation.

4. If $A = B$, then A may be replaced with B in any equation. Substitution Property of Equations

   Any expression in an equation may be replaced with an equivalent expression.

LEARNING TIP

You can think of the properties of equations in terms of a pan balance. For example, you can add the same weight to each pan, or you can swap the contents of the two pans, or you can replace any weight with an equivalent weight—all of these actions maintain the balance.

Solving an equation *algebraically* means using algebraic operations to produce equivalent equations until we obtain one that can be solved by inspection. Usually the goal is to find an equivalent equation in which the variable stands alone (is isolated) on one side of the equation so that the solution is obvious. To accomplish this, we use the Addition Property of Equations to isolate the variable term and the Multiplication Property of Equations to isolate the variable itself.

Our first example requires a single application of either the Addition or the Multiplication Property of Equations.

EXAMPLE 1

Solving with the Addition or Multiplication Property of Equations

Solve the following equations.

(a) $x + 7 = -4$

(b) $-5x = 35$

Solution

(a)
$$x + 7 = -4$$
$$x + 7 + (-7) = -4 + (-7) \quad \text{Addition Property of Equations}$$
$$x = -11$$

The solution is -11. Note that we would have obtained the same result by subtracting 7 from both sides of the equation.

(b) $-5x = 35$
$$\frac{-5x}{-5} = \frac{35}{-5} \quad \text{Multiplication Property of Equations}$$
$$x = -7$$

The solution is -7. Note that we would have obtained the same result by multiplying both sides of the equation by $-\frac{1}{5}$.

Except for very simple equations, such as those in Example 1, both properties are usually needed in the solving process. We apply the Addition Property first to isolate the variable term, which may include eliminating a variable term from one side of the equation. Then we apply the Multiplication Property to isolate the variable itself.

EXAMPLE 2

Applying Both Properties

Solve the following equations and verify the solutions by substitution.

(a) $2x + 3 = 11$ (b) $7x + 8 = -22 - 3x$

Solution

(a)
$$2x + 3 = 11$$
$$2x + 3 - 3 = 11 - 3 \quad \text{Addition Property of Equations}$$
$$2x = 8$$
$$\frac{2x}{2} = \frac{8}{2} \quad \text{Multiplication Property of Equations}$$
$$x = 4$$

We verify that 4 is the solution.

$$2x + 3 = 11$$
$$2(4) + 3 = 11 \quad \text{Replace } x \text{ with 4.}$$
$$11 = 11 \quad \text{True}$$

The Substitution Property enables us to replace one side with a simplified expression.

(b)
$$7x + 8 = -22 - 3x$$
$$7x + 8 + 3x = -22 - 3x + 3x \quad \text{Add } 3x \text{ to both sides of the equation.}$$
$$10x + 8 = -22 \quad \text{Combine like terms.}$$
$$10x + 8 - 8 = -22 - 8 \quad \text{Subtract 8 from both sides of the equation.}$$
$$10x = -30$$
$$\frac{10x}{10} = \frac{-30}{10} \quad \text{Divide both sides of the equation by 10.}$$
$$x = -3$$

To verify by substitution, replace x with -3 in the original equation.

$$7x + 8 = -22 - 3x$$

$$7(-3) + 8 = -22 - 3(-3) \quad \text{Replace } x \text{ with } -3.$$

$$-21 + 8 = -22 + 9$$

$$-13 = -13 \quad \text{True}$$

Quick Reference 3.1

Equations

- An **equation** is a statement that two algebraic expressions have the same value.
- A **solution** of an equation is a replacement for the variable that makes the equation true.
- The **solution set** of an equation is the set of all solutions of the equation.
- We verify a solution by replacing every occurrence of the variable by the proposed solution. If the replacement makes the equation true, it is a solution.
- A **linear equation in one variable** is an equation that can be written in the standard form $Ax + B = 0$, where A and B are real numbers and $A \neq 0$.
- A **linear function** is a function of the form $f(x) = ax + b$, where $a \neq 0$.
- An **identity** is an equation that is true for every permissible replacement for the variable.
- An **inconsistent equation** (or **contradiction**) is false for all replacements of the variable.
- A **conditional equation** is an equation that is true for at least one value of the variable but is not an identity.

Properties of Equations

- **Equivalent equations** are equations that have exactly the same solution sets.
- An equivalent equation results when
  1. the same quantity is added to or subtracted from both sides of an equation (Addition Property of Equations).
  2. both sides of an equation are multiplied or divided by the same nonzero quantity (Multiplication Property of Equations).
  3. the two sides of an equation are swapped (Symmetric Property of Equations).
  4. an expression in an equation is replaced with an equivalent expression (Substitution Property of Equations).
- Solving an equation algebraically means using algebraic operations to produce equivalent equations until we obtain one that can be solved by inspection. The goal is to isolate the variable.
- The Addition Property of Equations can be used to eliminate a variable term from one side of an equation and to isolate a variable term.
- The Multiplication Property of Equations can be used to isolate a variable by eliminating the coefficient of the isolated variable term.

Speaking the Language 3.1

1. The equation $x = x + 1$ is an example of a(n) ______.
2. To isolate the variable term in $2x + 3 = 8$, we use the ______ Property of Equations.
3. A(n) ______ is an equation that is true for all permissible replacements for the variable.
4. Equations of the form $Ax + B = 0$, where $A \neq 0$, are called ______.

Exercises 3.1

Concepts and Skills

1. In your own words, state the definition of an equation.
2. State the definition of a linear equation in one variable. Give two examples of equations that are not linear equations in one variable.

In Exercises 3–12, determine whether each of the following is an equation.

3. $x - 3 + 2x = 5$
4. $10 = x$
5. $4(x - 5) - 7(2x + 3)$
6. $\frac{9}{5} \cdot 5 - 6 + 7x$
7. $0 = -7$
8. $3 + 4 - 5 = 6$
9. $3x - 5 + 7x + 12$
10. $5t - 6 = 7$
11. $x^2 = 5x + 6$
12. $t^2 - 5t + 6$

In Exercises 13–20, state whether each equation is a linear equation in one variable.

13. $\frac{4}{7}x + 3 = 0$
14. $3x - 2 = 7$
15. $x^3 - 2x = 4$
16. $2x^2 + 5x = 6$
17. $2\pi^2 + 5x = 6$
18. $3\sqrt{2} - 7x = 0$
19. $\sqrt{x} = 5$
20. $6x = \sqrt{5}$
21. Explain how to verify that 7 is a solution of the equation $2x - 3 = x + 4$.
22. Suppose that you create a table of values for the expression on the left side of $2x - 3 = 5$. Explain how to recognize the solution of the equation.

In Exercises 23–32, verify that the given number is a solution of the given equation.

23. $3x - 5 = 7x + 7;\ -3$
24. $15 - 2x = 3x;\ 3$
25. $4(x - 2) + 5 = 5(3 - x) + 3(x + 4);\ 5$
26. $29 + 4(3x - 5) = 2(6x + 7) - 5;\ \frac{2}{3}$
27. $-2[2x - 3(1 - x)] = -5(x + 3) + 11;\ 2$
28. $\frac{2}{3}(x - 4) + \frac{3}{4}(x + 5) = 2(x - 1);\ \frac{37}{7}$
29. $\frac{3}{5}y = -2;\ -\frac{10}{3}$
30. $\frac{x}{2} - \frac{2x}{5} + \frac{1}{2} = \frac{7}{10}x - \frac{2}{5};\ \frac{3}{2}$
31. $6x - (5x - 8) = 1;\ -7$
32. $-0.2t = 5.4;\ -27$

In Exercises 33–40, determine which of the given numbers is a solution of the given equation.

33. $3x - 4 = 11;\ 3, 5, 7$
34. $4x + 7 = 17 - x;\ -2, 1, 2$
35. $\frac{x}{2} + \frac{1}{3} = \frac{x}{3} - \frac{1}{2};\ -1, -4, -5$
36. $\frac{t}{4} - 2 = 3 - \frac{3}{4}t;\ 2, 5, 8$
37. $3(x - 2) + 4 = 3x - 2;\ -3, 0, 2$
38. $4 - 2(3x + 2) = -4x - 2x;\ -5, -1, 4$
39. $5(t + 2) = 3(t - 4);\ -11, -1, 0$
40. $2y - 5 = 5(y - 1) - y;\ -2, 0, 2$

In Exercises 41–48, use the x-values $-3, -2, -1, 0, 1, 2$, and 3 to make a table of values for the functions associated with each side of the given equation. Use the result to estimate the solution of the equation.

41. $2x - 3 = x - 4$
42. $2 - 2x = x - 4$
43. $2x + 1 = 2x$
44. $5 - x = -x - 3$
45. $3x = x - 6$
46. $4x - 3 = x - 3$

47. $4 - x = -x + 4$

48. $1 + 2x = 2x + 1$

49. Consider the following two equations.

(i) $x + 2 = 3$ (ii) $-2x = 6$

When we solve these equations, explain why we *add* -2 to both sides of the first equation and *divide* both sides by -2 in the second equation.

50. What are equivalent equations?

In Exercises 51–62, simplify when necessary and use the Addition Property of Equations to solve the equations.

51. $x + 3 = 7$

52. $r + 4 = -10$

53. $z - 3 = -11$

54. $a - 5 = 8$

55. $9x = 8x$

56. $-6y = 8 - 7y$

57. $7 - 2s = 10 - 3s$

58. $14 - 3s = 19 - 4s$

59. $8 + 3z = 2z - 9$

60. $4k + 4 = 4 + 3k$

61. $4x - 3 + 3x = 7 + 6x - 4$

62. $7x - 7 - 4x - 2x + 12 = 8 - 3$

In Exercises 63–72, simplify when necessary and use the Multiplication Property of Equations to solve the given equation.

63. $7x = 0$

64. $3x = 12$

65. $-x = 9$

66. $-5m = 30$

67. $-3x = \frac{2}{5}$

68. $\frac{2}{5}x = \frac{9}{10}$

69. $\frac{y}{3} = 12$

70. $4 = \frac{-x}{5}$

71. $x + 3x + x = 30$

72. $9x - 4x + 3x = -72$

In Exercises 73–84, solve the equation algebraically and verify the solution.

73. $5x + 8 = 23$

74. $4x - 7 = -35$

75. $x - 6 = -2x$

76. $5x = 3x + 4$

77. $-5 = 4x + 13$

78. $4x - 3 = 5$

79. $7x - 5 = 8x + 7$

80. $7 + 8x = 4x - 13$

81. $4 - x = 3x + 12$

82. $2x - 3 = 7 - 3x$

83. $6x - 5 = 4x - 2$

84. $2x + 5 = 1 - x$

Real-Life Applications

85. An Internet provider charges a flat fee of \$20 per month plus \$2 per hour of online time. An alternative plan has a \$10 flat fee per month plus \$2.25 per hour online.

(a) Letting h represent the number of online hours, write functions for the total monthly cost of each plan.

(b) Complete the following table to compare the costs of the two plans.

h	10	20	30	40	50	60
Plan 1						
Plan 2						

(c) Write an equation to determine the number of hours for which the total charge is the same for both plans.

(d) Use the table to estimate the solution of the equation in part (c).

86. A price-sensitive item is one for which a small change in price results in a large change in demand. Suppose that the expression $-5x + 500$ represents the number of calculators that can be sold per month when the price is x dollars.

(a) Write an equation indicating that the number of calculators sold per month was 100.

(b) Write an equation indicating that the number of calculators sold per month was 50.

(c) Complete the following table and use the result to estimate the solutions to the equations in parts (a) and (b).

x	60	70	80	90	100
$-5x + 500$					

(d) What price change caused sales to be cut in half?

(e) According to this model, at what price would no calculators be sold?

Modeling with Real Data

87. The bar graph shows the number (in millions) of elementary and secondary teachers for selected years. (Source: National Center for Educational Statistics.)

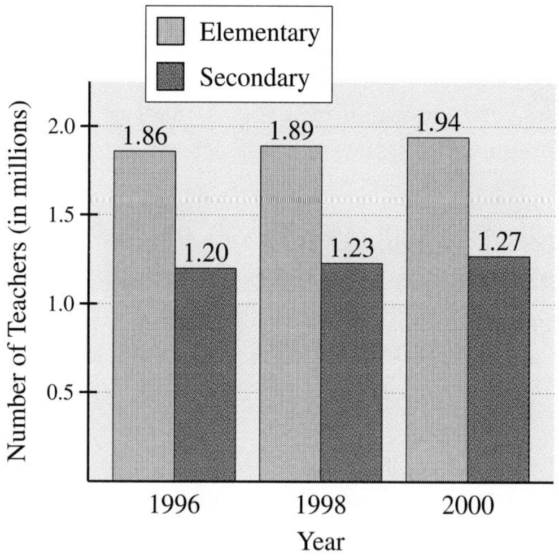

With x representing the number of years since 1990, the numbers of teachers can be modeled with the following functions.

Elementary: $E(x) = 0.020x + 1.74$

Secondary: $S(x) = 0.018x + 1.09$

Write an equation indicating that $E(x) - S(x) = 0.7$. Then solve the equation and interpret the solution.

88. Referring to the data in Exercise 87, write an equation indicating that the total number of teachers is 3.5 million. Then solve the equation and interpret the solution.

89. The percentage of people who pay off all their credit card bills every month increases with the age of the cardholder. For instance, 49% of those in the 30–45 age group pay their bills monthly, whereas 87% of those age 60 and over pay monthly. (Source: *USA Today.*) The function $P(a) = 1.16a + 9.27$ models the percentage P of people age a who pay off their bills monthly. Write and solve an equation to estimate the age at which 40% of cardholders pay their bills monthly.

90. At what age does the model in Exercise 89 predict that *all* people of that age pay off their credit cards monthly?

Challenge

In Exercises 91 and 92, solve the equation. Round results to the nearest hundredth.

91. $\sqrt{3}x - 12 = \pi x$

92. $\dfrac{x}{\sqrt{2}} = 8 - x$

In Exercises 93 and 94, use the x-values $-3, -2, -1, 0, 1, 2,$ and 3 to make a table of values for the function associated with the left side of the given equation. Use the result to estimate the solution(s) of the equation.

93. $|x - 1| = 2$

94. $x^2 + 3x = 0$

In Exercises 95 and 96, solve for x.

95. $b - ax = c$

96. $ax + b = cx + d$

97. If B and C are constants, is $0 \cdot x + B = C$ a linear equation in one variable? Why? Under what condition would the equation be an identity?

98. Show that a linear equation in one variable cannot be an inconsistent equation.

3.2 Solving Linear Equations

The Algebraic Solving Routine • Simplifying Both Sides • Clearing Fractions • Special Cases

The Algebraic Solving Routine

In Section 3.1 we began using the properties of equations to solve linear equations. In this section we illustrate a general strategy for solving linear equations in one variable.

The following is a general summary of the steps for solving a linear equation in one variable algebraically.

Algebraic Routine for Solving a Linear Equation In One Variable

1. If necessary, clear fractions by multiplying every term of both sides of the equation by the LCD of all the fractions.
2. Simplify both sides of the equation by
   (a) removing grouping symbols and/or
   (b) combining like terms.
3. If the variable is on both sides, use the Addition Property of Equations to eliminate the variable term from one side or the other.
4. Use the Addition Property of Equations to isolate the variable term.
5. Use the Multiplication Property of Equations to isolate the variable itself.
6. Verify the solution.

Simplifying Both Sides

The following illustrates how the expressions on the two sides of an equation may need to be simplified. Recall that the Distributive Property enables us to remove parentheses and combine like terms.

EXAMPLE 1

Simplifying Both Sides of an Equation

Solve the following equations.

(a) $4 - 2(x - 5) = 3x - 4(x + 2)$

(b) $1.09 - 0.3t = 0.4t - 1.7 + 0.2t$

Solution

(a) $4 - 2(x - 5) = 3x - 4(x + 2)$

$4 - 2x + 10 = 3x - 4x - 8$ Remove parentheses.

$14 - 2x = -x - 8$ Combine like terms.

$14 - 2x + x = -x - 8 + x$ Add x to both sides of the equation.

$14 - x = -8$ Combine like terms.

$14 - x - 14 = -8 - 14$ Subtract 14 from both sides of the equation.

$-1x = -22$

$\frac{-1x}{-1} = \frac{-22}{-1}$ Divide both sides of the equation by -1.

$x = 22$

(b) We can eliminate (or *clear*) the decimals from the original equation by multiplying both sides by 100.

$1.09 - 0.3t = 0.4t - 1.7 + 0.2t$

$100(1.09 - 0.3t) = 100(0.4t - 1.7 + 0.2t)$ Multiply both sides by 100.

$109 - 30t = 40t - 170 + 20t$ Remove parentheses.

$109 - 30t = 60t - 170$ Combine like terms.

Think About It

By what number would you multiply both sides of each of the following equations in order to clear the decimals?

(a) $2 - 0.1x = 3$

(b) $4x + 0.237 = 1.58$

Write a general rule for determining the multiplier to clear decimals from an equation.

$$109 - 30t - 60t = 60t - 170 - 60t \quad \text{Subtract } 60t \text{ from both sides.}$$
$$109 - 90t = -170 \quad \text{Combine like terms.}$$
$$109 - 90t - 109 = -170 - 109 \quad \text{Subtract 109 from both sides.}$$
$$-90t = -279$$
$$\frac{-90t}{-90} = \frac{-279}{-90} \quad \text{Divide both sides by } -90.$$
$$t = 3.1$$

Clearing Fractions

When equations involve fractions, it is often helpful to clear the fractions by using the Distributive Property to multiply both sides of the equation by the least common denominator (LCD).

Note: Remember to multiply each term by the LCD—even terms that are not fractions.

EXAMPLE 2

Clearing Fractions from an Equation

Solve the following equations.

(a) $\frac{3}{5}x - \frac{2}{3} = \frac{9}{10} + \frac{1}{15}x$

(b) $\frac{1}{2} - \frac{1}{3}(x - 2) = 2(x - 1) + \frac{2}{3}$

Solution

(a) $\frac{3}{5}x - \frac{2}{3} = \frac{9}{10} + \frac{1}{15}x$

Because the LCD is 30, multiply each term of both sides by 30.

LEARNING TIP

Clearing fractions and decimals from an equation is a matter of convenience. If you try to solve the equations in Example 2 without clearing fractions, you will see how useful this step is.

$$30 \cdot \frac{3}{5}x - 30 \cdot \frac{2}{3} = 30 \cdot \frac{9}{10} + 30 \cdot \frac{1}{15}x$$
$$\left(30 \cdot \frac{3}{5}\right)x - 30 \cdot \frac{2}{3} = 30 \cdot \frac{9}{10} + \left(30 \cdot \frac{1}{15}\right)x \quad \text{Associative Property of Multiplication}$$
$$18x - 20 = 27 + 2x$$
$$18x - 20 - 2x = 27 + 2x - 2x \quad \text{Subtract } 2x \text{ from both sides.}$$
$$16x - 20 = 27 \quad \text{Combine like terms.}$$
$$16x - 20 + 20 = 27 + 20 \quad \text{Add 20 to both sides.}$$
$$16x = 47$$
$$\frac{16x}{16} = \frac{47}{16} \quad \text{Divide both sides by 16.}$$
$$x = \frac{47}{16}$$

Stress the use of the Associative Property of Multiplication. Students often multiply $6 \cdot 2(x - 1)$ to obtain $12(6x - 6)$.

(b)

$\frac{1}{2} - \frac{1}{3}(x - 2) = 2(x - 1) + \frac{2}{3}$	The LCD is 6.
$6 \cdot \frac{1}{2} - 6 \cdot \frac{1}{3}(x - 2) = 6 \cdot 2(x - 1) + 6 \cdot \frac{2}{3}$	Multiply both sides by 6.
$6 \cdot \frac{1}{2} - \left(6 \cdot \frac{1}{3}\right)(x - 2) = (6 \cdot 2)(x - 1) + 6 \cdot \frac{2}{3}$	Associative Property of Multiplication
$3 - 2(x - 2) = 12(x - 1) + 4$	
$3 - 2x + 4 = 12x - 12 + 4$	Distributive Property
$7 - 2x = 12x - 8$	
$7 - 2x - 12x = 12x - 8 - 12x$	Subtract $12x$ from both sides.
$7 - 14x = -8$	Combine like terms.
$7 - 14x - 7 = -8 - 7$	Subtract 7 from both sides.
$-14x = -15$	
$\frac{-14x}{-14} = \frac{-15}{-14}$	Divide both sides by -14.
$x = \frac{15}{14}$	

Special Cases

In Section 3.1 we observed that equations such as $x + 3 = 1 + x$ are false for all replacements of the variable and thus have no solution. Other equations, such as $x + 3 = 3 + x$, are true for all replacements of the variable, so every real number is a solution. Solving these equations algebraically confirms that they are special cases.

$$x + 3 = 1 + x$$
$$x + 3 - x = x + 1 - x$$
$$3 = 1 \quad \text{False}$$

Solution set: ∅

$$x + 3 = 3 + x$$
$$x + 3 - x = 3 + x - x$$
$$3 = 3 \quad \text{True}$$

Solution set: **R**

EXAMPLE 3

Special Cases

Solve each equation.

(a) $3(2x - 1) = 2(4 + 3x)$

(b) $2(3x + 6) = 7x - (x - 12)$

Solution

(a)

$3(2x - 1) = 2(4 + 3x)$	
$6x - 3 = 8 + 6x$	Remove grouping symbols.
$6x - 3 - 6x = 8 + 6x - 6x$	Eliminate the variable term from one side.
$-3 = 8$	False

Because the resulting equation is false, the original equation is a contradiction, and the solution set is ∅.

(b)

$$2(3x + 6) = 7x - (x - 12)$$

$$6x + 12 = 7x - x + 12 \quad \text{Remove grouping symbols.}$$

$$6x + 12 = 6x + 12 \quad \text{Combine like terms.}$$

$$6x + 12 - 6x = 6x + 12 - 6x \quad \text{Eliminate the variable term from one side.}$$

$$12 = 12 \quad \text{True}$$

Because the resulting equation is true, the original equation is an identity, and the solution set is **R**.

Summary of Special Cases

If solving a linear equation leads to an equivalent equation with no variable, then

1. if the resulting equation is false, the original equation is a contradiction, and the solution set is Ø;
2. if the resulting equation is true, the original equation is an identity, and the solution set is **R**.

Quick Reference 3.2

Algebraic Solving Routine

- The following are the suggested steps for solving a linear equation in one variable algebraically.
  1. If necessary, clear fractions by multiplying every term of both sides of the equation by the LCD of all the fractions.
  2. Simplify both sides of the equation by
     (a) removing grouping symbols and/or
     (b) combining like terms.
  3. If the variable is on both sides, use the Addition Property of Equations to eliminate the variable term from one side or the other.
  4. Use the Addition Property of Equations to isolate the variable term.
  5. Use the Multiplication Property of Equations to isolate the variable itself.
  6. Verify the solution.

Special Cases

- The algebraic routine for solving an equation will reveal special cases.
  1. If a resulting equivalent equation is a false statement, then the original equation is inconsistent, and the solution set is Ø.
  2. If a resulting equivalent equation is a true statement, then the original equation is an identity, and the solution set is **R**.

Speaking the Language 3.2

1. A good first step in solving $2(x - 3) = -(x + 1)$ is to __________ both sides.

2. If the result of solving an equation is the equivalent equation $3 = -3$, then the original equation is a(n) __________.

3. To clear fractions from an equation, we multiply both sides by __________.

4. To solve $-3x = 12$, we use the __________ Property of Equations.

Exercises 3.2

Concepts and Skills

1. The following are possible first steps in solving the equation $\frac{2}{3}x = 10$. Explain why each approach is correct.

(i) Multiply both sides by $\frac{3}{2}$.

(ii) Divide both sides by $\frac{2}{3}$.

(iii) Multiply both sides by 3 and then divide both sides by 2.

2. What is an important advantage of the algebraic method over the table method?

In Exercises 3–14, solve the equation algebraically and verify the solution. For special cases, write the solution set.

3. $3(3 - 2x) = 33 - 2x$

4. $2(3y + 8) = -2(2 - y)$

5. $7w - 5 = 11w - 5 - 4w$

6. $9w + 7 = 16w + 7(1 - w)$

7. $9x + 5 = 5x + 3(x - 1)$

8. $11x + 2(4 - 3x) = 3(3 - x) + 15$

9. $16 + 7(6 - x) = 15 - 4(x + 2)$

10. $7 - 4(t - 3) = 6t - 3(t + 3)$

11. $2(3a + 2) = 2(a + 1) + 4a$

12. $8t - 7 - 3t = 5t + 2$

13. $-2(x + 5) = 5(1 - x) + 3(7 - x)$

14. $2y - 3(2y - 3) = 2y - 5(3y - 4)$

15. Consider solving the equation $\frac{1}{3} - 2x = \frac{2}{5}$. To clear fractions, we multiply both sides by 15. Why is it necessary to multiply $2x$ by 15?

16. What is the solution set of $3x = 2x$? Now divide both sides of $3x = 2x$ by x. What is the solution set of the resulting equation? Why are the equations not equivalent?

In Exercises 17–26, solve the given equation.

17. $2x - \frac{3}{4} = -\frac{5}{6}$

18. $\frac{2}{15}y + \frac{3}{5} = 2 - \frac{2}{3}y$

19. $\frac{y}{2} + \frac{y}{4} = \frac{7}{8} - \frac{y}{8}$

20. $\frac{5 + 3x}{2} + 7 = 2x$

21. $\frac{5}{4}(x + 2) = \frac{x}{2}$

22. $\frac{1}{4}(x - 4) = \frac{1}{3}(x + 6)$

23. $x + \frac{3x - 1}{9} = 4 + \frac{3x + 1}{3}$

24. $\frac{2x}{5} - \frac{2x - 1}{2} = \frac{x}{5}$

25. $\frac{1}{4} + \frac{1}{6}(4a + 5) = 2(a - 3) + \frac{5}{12}$

26. $\frac{5}{6}(1 - t) - \frac{t}{2} = -\frac{1}{3}(1 - 3t)$

In Exercises 27–30, solve the given equation.

27. $2.6 = 0.4z + 1$

28. $1 - 0.3c = -0.5$

29. $0.08x + 0.15(x + 200) = 7$

30. $0.6(2x + 1) - 0.4(x - 2) = 1$

CE In Exercises 31–36, determine whether the given equations are equivalent equations.

31. $x + 4 = 7$ and $3x - 5 = 4$

32. $\frac{1}{3}x = 2$ and $x + 3 = 9$

33. $x + 3 = 4 + x$ and $2x - 3 = 4 + 2x$

34. $x + 3 = 3 + x$ and $2x - 5 = -(5 - 2x)$

35. $\frac{1}{4}x = -2$ and $8 - x = 0$

36. $3x = 0$ and $x - 3 = 0$

37. Describe the solution set of an inconsistent equation.

38. Describe the solution set of an identity.

In Exercises 39–46, solve the equation. State whether the equation is an identity, an inconsistent equation, or a conditional equation.

39. $x - 6 = 6 - x$

40. $2x - 7 = x + 4$

41. $x + 6 = 6 + x$

42. $x + 3 = 5 + x$

43. $2x - 5 = 7 + 3x - (x - 3)$

44. $x + 3 = 5x - [2x - (3 - 2x)]$

45. $\frac{1}{5}x - 2 = \frac{1}{3}x - 4$

46. $-\frac{1}{3}x + 8 = 0$

47. Use the Distributive Property to verify that

(a) $-(x + 1) = -x + 1$ is an inconsistent equation.

(b) $-(3 - x) + 5 = x + 2$ is an identity.

48. Although the equation $|x| = x$ has infinitely many solutions, it is a conditional equation. Why?

Exercises 49–60 help the student to understand the nature of a solution.

CE In Exercises 49–52, determine a value of c so that the equations are equivalent.

49. $5 - 3x = c + 2x$ and $2x - 3 = 5$

50. $7 - 2x = 4x - c$ and $3x - 8 = 7x$

51. $2x + c = 3x + 2$ and $4x + 7 = 6x - 5$

52. $3x - c = 5x - 3$ and $6 - 5x = 3x + 14$

CE In Exercises 53–56, determine a value of k so that the equations are inconsistent.

53. $3x - 2 = 4 + kx$

54. $5 - kx = 7x + 2$

55. $2x - 4 = x - 1 + kx$

56. $kx - 3x = 2x + 1$

CE In Exercises 57–60, determine a value of c so that the equations are identities.

57. $2x + 3 = c + 2x$

58. $x - 4 = -(c - x)$

59. $cx + 2 = 2 + 5x$

60. $4 - 2x = 3x - (cx - 4)$

CE In Exercises 61–64, determine a value of m so that the solution of the equation is 7.

61. $3x - 5 = m$

62. $8 - 2x = m + 4$

63. $3x + m - 2 = 4 - x$

64. $mx - 7 = 13$

In Exercises 65–82, solve the equations.

Exercises 65–82 provide additional practice in solving equations.

65. $2x - 5 = 5 - 2x$

66. $19 - 5z = 6z + 41$

67. $17 - 4z = 3z + 52$

68. $3y + 2 = 4 + 3y$

69. $3(x + 1) - 2 = 2(x + 1) + x - 1$

70. $x + 3(2 - x) = 6 - 2x$

71. $5 + 3[t - (1 - 4t)] = 7$

72. $13 - 3(x + 2) = 5x - 9 + 2(x - 7)$

73. $3(2x + 1) = 8x - 2(x - 2)$

74. $3 + 2[3 - (7t - 4)] = 2t - [3 - (3t + 5)]$

75. $\frac{4}{9}t - \frac{1}{6} = \frac{1}{3}t + 1$

76. $\frac{t}{5} + \frac{7}{15} = 2t + \frac{2}{3}$

77. $\frac{3}{4}(x - 3) - \frac{1}{2}(3x - 5) = 2(3 - x)$

78. $\frac{3t - 8}{2} = -8 - \frac{t}{3}$

79. $0.12 - 0.05y = 0.04y - 0.15$

80. $1 - 0.3x = 0.6x - 1.7$

81. $x(x + 6) = 5(x + 1) + x^2$

82. $x^2 - 4 = x(x + 1)$

Real-Life Applications

83. The volume (in decibels) of a digital alarm clock depends on the loudness setting, where a setting of 1 produces 30 decibels and each increment from 1 to 9 produces an increase of 8 decibels.

(a) Write an expression that models the volume of the alarm for any setting s.

(b) Write an equation that indicates a volume of 62 decibels, and solve the equation to determine the setting.

84. At a medical facility, the parking lot is designed so that one-third of the spaces are for small cars, one-half of the spaces are standard size, and the remaining 35 spaces are reserved for professional staff.

(a) Letting x represent the total number of parking spaces, write an expression that models the sum of the three categories.

(b) Write an equation showing that the sum of the three categories is equal to the total number of spaces.

(c) Solve the equation in part (b) and determine the number of parking spaces for small cars.

Modeling with Real Data

85. The accompanying bar graph shows the percentage of the United States population who were enrolled in a health maintenance organization (HMO) in 1990 and 1998. (Source: National Center for Health Statistics.)

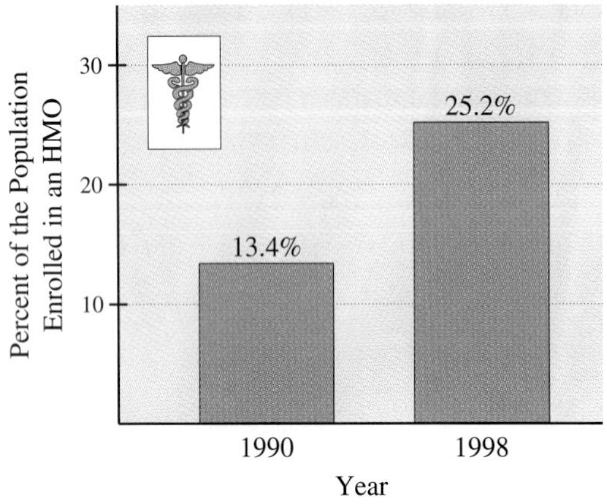

With x representing the number of years since 1990, the percentage can be modeled with the function $P(x) = 1.475x + 13.4$. Write and solve an equation to estimate the year in which half of the United States will be enrolled in an HMO.

86. In 2002, 1.2 million students received a Bachelor's degree. (Source: National Center for Educational Statistics.) With x representing the number of years since 2000, the function $D(x) = 0.01x + 1.20$ models the number of degrees D (in millions).

(a) How many Bachelor's degrees does this model predict will be awarded in 2005?

(b) Write and solve an equation that describes the year in which 1.3 million Bachelor's degrees will be granted.

87. The Social Security Act was passed in 1935. The accompanying table shows that the combined tax rate for employers and employees for Social Security and Medicare has risen steadily. (Source: Department of Health and Human Services.)

Year	Combined Tax Rate
1945	2.00%
1955	4.00%
1965	7.25%
1975	11.70%
1985	14.10%
1995	15.30%

The function $T(x) = 0.28x - 1.07$ models the tax rate T, where x is the number of years since 1935.

(a) Write an equation for estimating the year in which the tax rate will reach 10 times the rate in 1945.

(b) Determine and interpret the solution of the equation.

88. Refer to the data in Exercise 87. During the first 60 years of its existence, the Social Security tax rate climbed to 15.3%. What tax rate does the model function project after another 60 years?

Data Analysis: *Minority Populations*

The line graph shows trends in two minority populations (in millions) in the United States. (Source: U.S. Census Bureau.)

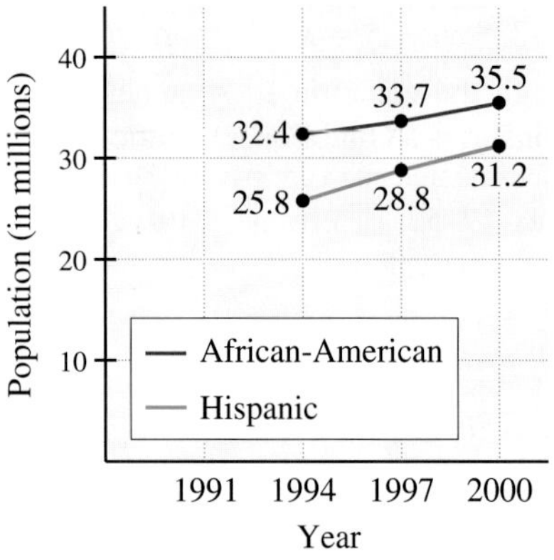

With t representing the number of years since 1990, the following functions model the populations (in millions).

African-American: $f(t) = 0.52t + 30.25$

Hispanic: $g(t) = 0.9t + 22.3$

89. Write a function that models the total African-American and Hispanic population in any given year since 1990. Evaluate the function to estimate the total in 2005.

90. Solve the equation $f(t) = g(t)$. Interpret the solution.

91. What does the equation $g(t) - f(t) = 2$ represent? Solve the equation and interpret the solution.

92. Write an equation to estimate the year in which the Hispanic population is estimated to be 10% more than the African-American population. Solve the equation and interpret the result.

Challenge

In Exercises 93 and 94, solve for x.

93. $a(x - b) = c$

94. $\dfrac{x + a}{3} = b$

In Exercises 95 and 96, solve the equation.

95. $\dfrac{x}{2} + \dfrac{x + 1}{3} + \dfrac{x + 2}{4} + \dfrac{x + 3}{5} = 1$

96. $x - 2\{(x - 1) - 3[(x - 2) - 4(x - 3)]\} = 19$

In Exercises 97 and 98, determine whether the given equations are equivalent.

97. $h^2 = 25$ and $h = 5$

98. $\dfrac{3}{x} = 0$ and $x = 0$

3.3 Formulas

Solving Formulas • Real-Life Applications

Solving Formulas

In Section 2.1 we described a **formula** as an equation that uses an expression to represent a specific quantity. For example, the formula for the perimeter P of a rectangle is $P = 2L + 2W$, where L represents the length of the rectangle and W represents the width. For given values of L and W, it is easy to calculate P. On the other hand, if the values of P and W are given, the task of calculating L involves solving an equation. If we must perform this task repeatedly, it is easier if we have a formula for L in terms of P and W.

Try to persuade students that this is not a new topic. Solving a formula is the same as solving an equation.

The process of isolating a variable in a formula is called *solving the formula* for the variable. Because a formula is simply an equation, the procedure for solving a formula is exactly the same as the procedure for solving equations.

EXAMPLE 1 Solving the Perimeter Formula

(a) Solve the formula $P = 2W + 2L$ for L.

(b) Use the formula from part (a) to find the length of a rectangle whose perimeter is 110 inches and whose width is 20 inches.

Solution

(a) Because the task is to solve for L, we treat L as the variable and W and P as constants. (For convenience, we swap the sides of the formula so that L appears on the left side.)

$$2W + 2L = P$$

$$2W - 2W + 2L = P - 2W \qquad \text{Subtract } 2W \text{ from both sides.}$$

$$2L = P - 2W$$

$$\frac{2L}{2} = \frac{P - 2W}{2} \qquad \text{Divide both sides by 2.}$$

$$L = \frac{P - 2W}{2}$$

LEARNING TIP

When you solve a formula for a particular variable, consider writing the variable in color or circling it in order to distinguish it from the other letters.

Students are often surprised that both forms are correct. Make sure they understand why.

In the last step, we used the Multiplication Property of Equations to divide both *sides* by 2. We might choose to divide *each term* by 2. Therefore, an alternative way to write the formula is

$$L = \frac{P}{2} - \frac{2W}{2} = \frac{P}{2} - W$$

(b) We verify that the formula and the alternative formula both give the same value for L.

$$L = \frac{P - 2W}{2} \qquad\qquad L = \frac{P}{2} - W$$

$$L = \frac{110 - 2(20)}{2} \qquad \text{Substitute 110 for } P \text{ and 20 for } W. \qquad L = \frac{110}{2} - 20$$

$$L = 35 \qquad\qquad L = 35$$

Figure 3.1

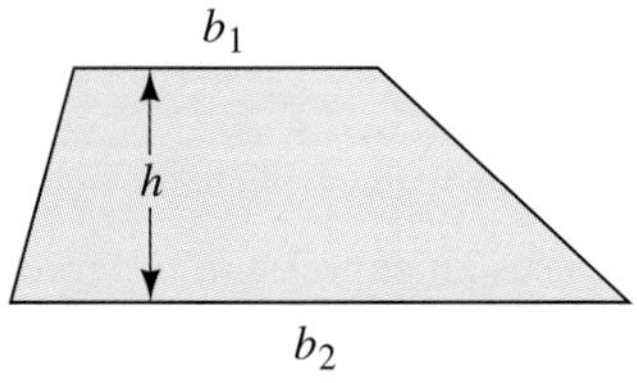

In Example 1 solving the formula for L, in general, involves exactly the same steps as solving the equation for L using specific values of P and W. By solving the formula for L, we have to go through the steps only once, and the new formula provides an arithmetic means of calculating L.

A **trapezoid** is a four-sided figure with one pair of sides parallel. (See Fig. 3.1.) The formula for calculating the area of a trapezoid is

$$A = \frac{1}{2}(b_1 + b_2)h$$

From previous experience, some students may think you want them to memorize all formulas. Relieving them of this concern will help put them at ease.

EXAMPLE 2

A Formula for the Height of a Trapezoid

Solve the formula for the area of a trapezoid for h.

Solution

In the formula $\frac{1}{2}(b_1 + b_2)h = A$, treat h as the variable and regard all other variables as constants.

$$2 \cdot \frac{1}{2}(b_1 + b_2)h = 2 \cdot A \qquad \text{Multiply both sides by 2 to clear the fraction.}$$

$$(b_1 + b_2)h = 2A \qquad \text{The coefficient of } h \text{ is } (b_1 + b_2).$$

$$\frac{(b_1 + b_2)h}{(b_1 + b_2)} = \frac{2A}{(b_1 + b_2)} \qquad \text{Divide both sides by } (b_1 + b_2).$$

$$h = \frac{2A}{b_1 + b_2}$$

Think About It

Suppose the circumference of any circle is increased by 1 foot. By how much is the radius increased?

Note: You will find a summary of common geometry formulas on the inside cover of the book. In particular, you will need the circle formulas for some of the exercises.

The formula for converting Fahrenheit temperatures F to Celsius temperatures C is $C = \frac{5}{9}(F - 32)$.

EXAMPLE 3

Converting Celsius Temperatures to Fahrenheit Temperatures

(a) Solve the formula $C = \frac{5}{9}(F - 32)$ for F.

(b) Use the formula for F to convert 0°C and 100°C to Fahrenheit temperatures.

Solution

(a) $\frac{5}{9}(F - 32) = C$ Swap sides to place F on the left.

$\frac{9}{5} \cdot \frac{5}{9}(F - 32) = \frac{9}{5} \cdot C$ Multiply both sides by $\frac{9}{5}$ to clear the fraction.

$F - 32 = \frac{9}{5}C$

$F = \frac{9}{5}C + 32$ Add 32 to both sides.

(b) $F = \frac{9}{5}C + 32$

$F = \frac{9}{5}(0) + 32$ Substitute 0 for C.

$F = 32$

$F = \frac{9}{5}C + 32$

$F = \frac{9}{5}(100) + 32$ Substitute 100 for C.

$F = 212$

Note that 32°F, the freezing temperature of water, corresponds to 0°C and that 212°F, the boiling temperature of water, corresponds to 100°C.

Graphing a function is often easier when the function is in the form $y =$ expression. In other words, the equation may need to be solved for y.

EXAMPLE 4

Solving an Equation for *y*

Solve the equation $3x - y = 4$ for y.

Solution

This skill will be important in the next chapter.

We treat x as though it were a constant and isolate the variable y.

$3x - y = 4$

$3x - y - 3x = -3x + 4$ Subtract $3x$ from both sides.

$-1y = -3x + 4$

$\frac{-1y}{-1} = \frac{-3x}{-1} + \frac{4}{-1}$ Divide both sides by -1.

$y = 3x - 4$

Real-Life Applications

Example 5 involves the volume V of a rectangular solid. The formula is $V = LWH$, where L, W, and H are the length, width, and height, respectively.

EXAMPLE 5

A Concrete Slab in the Form of a Rectangular Solid

In order to install an underground electric cable beneath an existing parking lot, a strip of concrete 200 feet long and 4 feet wide had to be removed. (See Fig. 3.2.) After the installation, the contractor ordered 8 cubic yards of concrete to repair the parking lot.

Figure 3.2

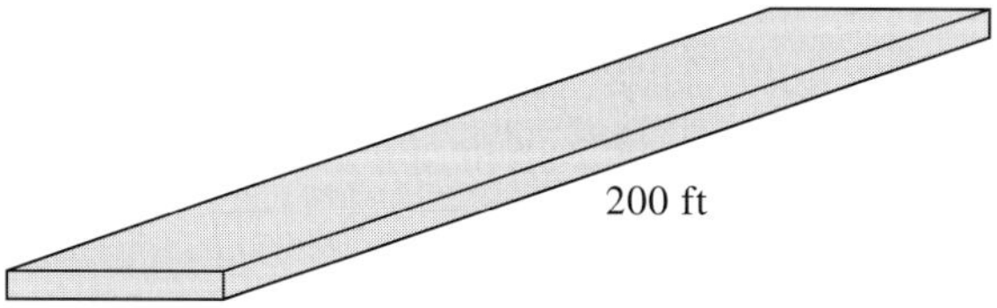

(a) Given the formula $V = LWH$, for which variable does the formula need to be solved in order to determine the thickness of the new slab of concrete? Solve the formula for that variable.

(b) Use the new formula to determine the thickness of the slab of concrete.

Solution

(a) The thickness of the concrete is represented by H, so we solve the formula $V = LWH$ for H.

$$LWH = V$$ The coefficient of H is LW.

$$\frac{LWH}{LW} = \frac{V}{LW}$$ Divide both sides by LW.

$$H = \frac{V}{LW}$$

(b) Because the length and width are expressed in feet, the volume of 8 cubic yards of concrete ordered must be expressed in cubic feet.

Many scientists keep track of units by "cancellation," which seems to offend some mathematicians. Whatever your method is, it should probably be stressed because students often have difficulty with units.

$$8 \text{ cubic yards} \cdot 27 \frac{\text{cubic feet}}{\text{cubic yard}} = 216 \text{ cubic feet}$$

$$H = \frac{216}{(200)(4)} = 0.27 \text{ feet}$$ $V = 216, L = 200, W = 4$

The concrete slab will be 0.27 foot, or about 3 inches, thick.

It is impossible to be a good equation solver and a poor formula solver. The procedures are identical in both cases.

No matter what your current or future line of work may be, you may very well find that being able to solve a formula is a valuable skill.

Quick Reference 3.3

Solving Formulas

- We previously described a **formula** as an equation that uses an expression to represent a specific quantity.
- Formulas are written in the form in which they are most commonly used. However, we sometimes need to know the value of another variable in the formula. If we need

to calculate that variable repeatedly, it is usually best to **solve the formula** for that variable.

- When solving a formula for a particular variable, treat all other variables as constants. The methods for solving a formula are identical to the algebraic methods for solving any other equation.

Speaking the Language 3.3

1. A formula is a specific kind of ______.
2. To solve $P = 2L + 2W$ for W, we treat P and L as ______.
3. A four-sided figure with one pair of parallel sides is called a(n) ______.
4. Writing the formula $A = bh$ as $h = \frac{A}{b}$ is called ______ for h.

Exercises 3.3

Concepts and Skills

1. In your own words, explain what is meant by solving a formula for a given variable.
2. If we solve the formula $I = Prt$ for r, which letter do we regard as the variable? How do we treat the other letters?

In Exercises 3–32, solve the formula for the given symbol.

These formulas are from various fields, including science, mathematics, and business.

3. $F = ma$ for m
4. $C = \pi d$ for d
5. $I = Prt$ for r
6. $V = LWH$ for W
7. $v = \frac{s}{t}$ for t
8. $r = \frac{d}{t}$ for t
9. $A = \frac{1}{2}bh$ for b
10. $V = \frac{1}{3}Bh$ for h
11. $A = \frac{1}{2}h(a + b)$ for a
12. $A = \frac{1}{2}h(a + b)$ for h
13. $A = \pi r^2$ for π
14. $x^2 = 4py$ for y
15. $E = mc^2$ for m
16. $V = \frac{4}{3}\pi r^3$ for π
17. $a = \frac{v - w}{t}$ for v
18. $z = \frac{x - u}{s}$ for u
19. $A = P + Prt$ for r
20. $IR + Ir = E$ for r
21. $P = \frac{nRT}{V}$ for R
22. $P = \frac{nRT}{V}$ for V
23. $R_T = \frac{R_1 + R_2}{2}$ for R_1
24. $A = \frac{r^2\theta}{2}$ for θ
25. $ax + by + c = 0$ for y
26. $y = mx + b$ for m
27. $s = \frac{1}{2}gt^2 + vt$ for v
28. $s = \frac{1}{2}gt^2 + vt$ for g
29. $A = a + (n - 1)d$ for n
30. $A = a + (n - 1)d$ for d
31. $S = \frac{n}{2}[2a + (n - 1)d]$ for a
32. $S = \frac{n}{2}[2a + (n - 1)d]$ for d
33. The following shows two different ways of solving the equation $2x + 3y = 6$ for y. Explain why both methods are correct.

$2x + 3y = 6$	$2x + 3y = 6$
$3y = -2x + 6$	$3y = -2x + 6$
$y = -\frac{2}{3}x + 2$	$y = \frac{-2x + 6}{3}$

34. In terms of graphing, what might be our reason for solving the equation $2x + 3y = 6$ for y?

In Exercises 35–52, solve the given equation for y.

35. $2x + 5y = 10$

36. $3x - y = 9$

37. $x - y + 5 = 0$

38. $x - 5y + 10 = 0$

39. $2x + 3y - 9 = 0$

40. $5x - 6y + 12 = 0$

41. $3x = 4y + 11$

42. $5x = 7y - 15$

43. $3x - y = 5x + 3y$

44. $3(2x - y) = 4x - y - 2$

45. $0.4x - 0.3y + 12 = 0$

46. $10 = 0.4y - 0.5x$

47. $y - 4 = \frac{1}{2}(x + 6)$

48. $y + 7 = -\frac{3}{4}(x - 12)$

49. $\frac{4}{5}x - \frac{2}{3}y + 8 = 0$

50. $\frac{y}{8} - \frac{x}{4} = 0$

51. $\frac{y - 2}{3} = \frac{x + 3}{4}$

52. $\frac{3 - x}{2} + 5 = \frac{y + 1}{3}$

In Exercises 53 and 54, round all decimal results to the nearest hundredth.

53. The surface area A of a right circular cylinder is given by the formula $A = 2\pi r^2 + 2\pi rh$, where r is the radius of the circular base and h is the height of the cylinder.

(a) What is the surface area of a right circular cylinder if the radius is 2.7 inches and the height is 5.83 inches?

(b) Solve the formula for h.

(c) What is the height of a right circular cylinder if its surface area is 2858.85 square centimeters and its radius is 13 centimeters?

54. The volume V of a right circular cone is given by the formula $V = \frac{1}{3}\pi r^2 h$, where r is the radius of the circular base and h is the height of the cone.

(a) What is the volume of a right circular cone if the radius of the base is 2.7 meters and the height is 11.3 meters?

(b) Solve the formula for h.

(c) What is the height of a right circular cone if the volume is 1256.64 cubic feet and the radius is 10 feet?

55. Which of the following formulas are equivalent to the formula $A = \frac{1}{2}bh$?

(i) $b = \frac{2A}{h}$ (ii) $b = 2A \cdot \frac{1}{h}$ (iii) $b = \frac{2}{h} \cdot A$

(iv) $b = \frac{A}{\frac{1}{2}h}$ (v) $b = \frac{A}{2h}$

56. A student drove 20 miles from home to college to take her math test. Her average speed was 40 mph. When she arrived at the college, she realized she had forgotten her calculator. Traveling at an average speed of 50 mph, she returned home to get her calculator. Finally, driving at an average speed of 60 mph, she drove back to the college to take the test. Because her speeds were 40, 50, and 60, was her average speed 50 mph? Explain. Most students are surprised that the answer is not 50 mph.

Geometric Models

57. A runner jogged 4 miles in ten laps around a circular track. What is the greatest distance straight across the track?

58. A person needs to buy a tablecloth for a circular table that has a diameter of 5 feet. The tablecloth is to hang over the edge of the table by 6 inches. A store advertises circular tablecloths with an area of 28.3 square feet. Is the tablecloth the proper size?

59. The perimeter of a rectangular swimming pool is 180 feet. If the length is twice the width, how far is it across the diagonal of the pool? (*Hint:* Use the Pythagorean Theorem.)

60. Suppose you want to put fringe around a circular tablecloth whose diameter is 6 feet. If you already have 6 yards of fringe, do you need to buy more? Explain.

61. Suppose you have two fish tanks. One tank is 5 feet by 2 feet, and the depth of the water is 18 inches. Will the water be deeper if you pour it into the other tank, which is 6 feet by 1.5 feet? Explain.

62. A regulation baseball has a circumference between 9 and $9\frac{1}{4}$ inches. Between what two values is the radius of the baseball?

63. The front of a concrete dam is in the shape of a trapezoid and has dimensions as shown in the accompanying figure. If the surface area of the front of the dam is 482,850 square feet, what is the height of the dam?

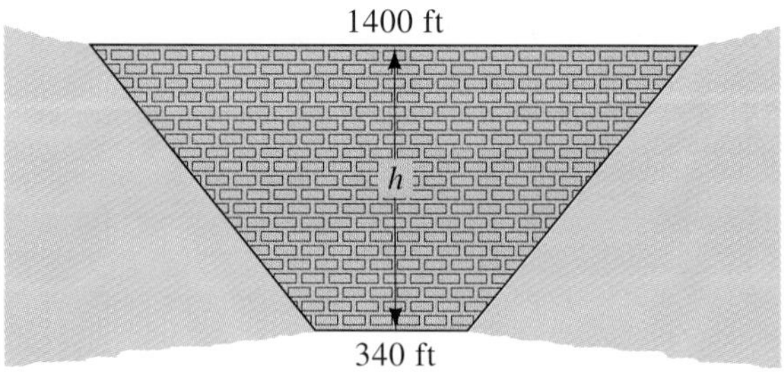

64. A garden in the shape of a trapezoid with the dimensions shown in the accompanying figure has an area of 1386 square feet. What is the length of the side perpendicular to the bases?

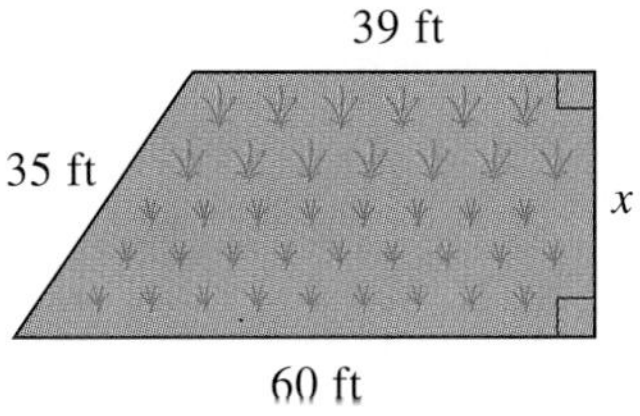

65. Determine a formula for the surface area A of the rectangular box shown in the accompanying figure. (The surface area is the sum of areas of the six sides.)

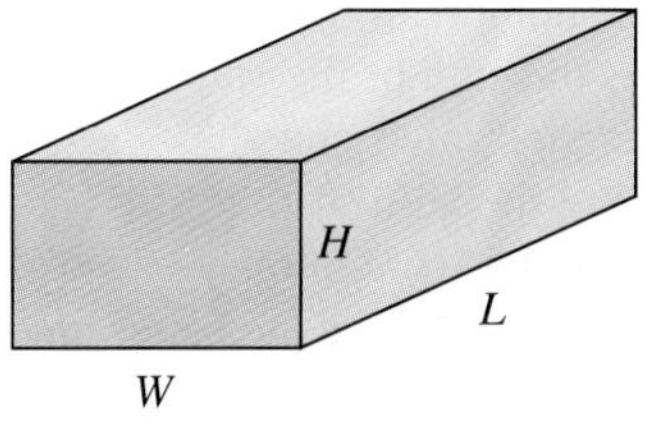

66. If a couple decides to paddle their canoe directly across (through the center of) a circular pond whose circumference is 200 meters, how far will they travel?

Real-Life Applications

In Exercises 67 and 68, the formula for an item's retail price R is $R = W + rW$, where W is the wholesale price. The formula for an item's discounted price D is $D = R - rR$. In these formulas, r is the percent of markup or discount.

67. As the Tool Department manager of a builders' supply company, you are responsible for markups and retail prices.

(a) Use the formula for R to determine the retail price for an item if the wholesale price was \$78 and the percent markup was 32%.

(b) Solve this formula for r and use the result to determine the percent markup if the retail price is \$448.50 and the wholesale price was \$390.

68. Suppose that a store owner decides to hold a holiday sale on all goods.

(a) Use the formula for D to determine the discounted price for an item if the retail price was \$96 and the percent of discount was 30%.

(b) Solve this formula for r and use the result to determine the percent of discount if the original retail price was \$56 and the discounted price is \$30.80.

In Exercises 69 and 70, solve the formula for the appropriate variable. Use your calculator for the repeated evaluations.

69. A satellite travels in a circular orbit of the earth. The circumference of the earth is approximately 25,000 miles. To the nearest mile, how high above the earth is the satellite if the length of one orbit is

(a) 25,800 miles?

(b) 26,000 miles?

(c) 26,500 miles?

70. Simple interest I on an investment P is calculated by the formula $I = Prt$, where r is the interest rate and t is the time in years. As a financial advisor, you have gathered the information shown in the following table to assist a client in deciding on an investment. Which investment has the best interest rate?

	Amount Invested	*Time*	*Interest*
(i)	\$10,000	6 months	\$375.00
(ii)	\$ 5,000	5 months	\$170.83
(iii)	\$ 5,000	3 months	\$100.00
(iv)	\$ 7,500	8 months	\$350.00
(v)	\$ 8,000	1 year	\$640.00

Modeling with Real Data

71. The liquid hydrogen tank of the space shuttle has a volume of approximately 1450 cubic meters and has a diameter of approximately 8.4 meters. (Source: NASA.)

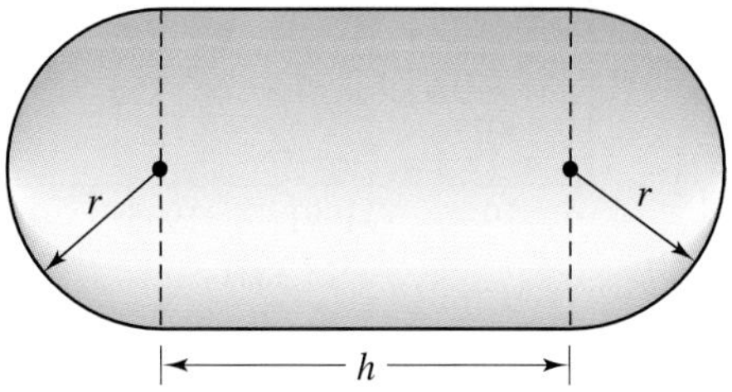

Use the formula $V = \frac{4}{3}\pi r^3 + \pi r^2 h$ to estimate the height h of the cylindrical part of the tank.

72. The wind speed of a storm is often given in *nautical* miles per hour. The accompanying table shows some notable storms with wind speeds given in either nautical or statute miles per hour. (Source: National Oceanic and Atmospheric Administration.)

Storm	Nautical Miles per Hour	Statute Miles per Hour
Beulah		109
Carla	126	
Donna		130
Juan	55	

The formula $N \approx 0.869M$ relates statute miles M to nautical miles N. Use this formula to complete the table (to the nearest tenth of a mile).

Data Analysis: *Wal-Mart Stores*

From a modest beginning, Wal-Mart Stores has risen to become one of the largest retailing chains in America. The accompanying bar graph shows the number of stores at the end of selected years in the period 1987–1999. (Source: Wal-Mart Stores, Inc.)

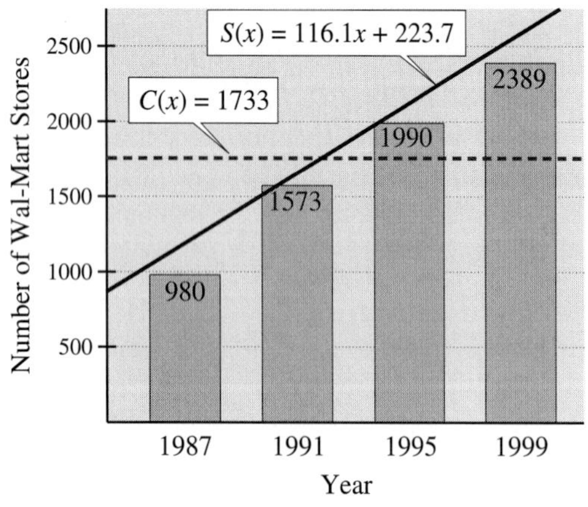

With x representing the number of years since 1980, the following functions are possible models for the data.

(i) $C(x) = 1733$

(ii) $S(x) = 116.1x + 223.7$

When more than one function may be considered as a model for data, various criteria may be considered to assess the validity of the model.

73. The dashed line in the figure shows the graph of $C(x)$, and the solid line shows the graph of $S(x)$. On the basis of a visual inspection of the graphs, which function appears to be the better model?

74. A numerical measure of the validity of a model is the sum of the differences $M - A$, where M is the model value and A is the actual data value. Determine the sum of the differences for model (i) and for model (ii). On the basis of the result, which function appears to be the better model?

75. Another numerical measure of the validity of a model is the sum of the values of $|M - A|$, where M is the model value and A is the actual data value. Determine the sum of the differences for model (i) and for model (ii). On the basis of the result, which function appears to be the better model?

76. On the basis of your results in Exercises 73–75, which numerical measure of validity of a model appears to be better? Why?

Challenge

In Exercises 77 and 78, solve the given equation for x.

77. $a(x + c) = b(x - c)$

78. $\dfrac{2x - 1}{x + 3} = y$

In Exercises 79–84, solve the given formula for the indicated variable.

79. $S = \dfrac{a}{1 - r}$ for r

80. $F = \dfrac{mv^2}{r}$ for r

81. $\dfrac{P_1V_1}{T_1} = \dfrac{P_2V_2}{T_2}$ for V_2

82. $E = IR + Ir$ for I

83. $A = 2LW + 2LH + 2WH$ for L

84. $A = P + Prt$ for P

3.4 Modeling and Problem Solving

A General Approach to Problem Solving • Numbers • Consecutive Integers • Piece Lengths • Angles of a Triangle • Rectangles • Things of Value

A General Approach to Problem Solving

From a practical standpoint, it makes little sense to learn mathematics and not apply it to anything. This section and the next give you an opportunity to bring together several skills and to use them in solving various kinds of application problems.

Many of the problems involve common, everyday situations. Others, though not as realistic, are important because they help you obtain a conceptual understanding of how mathematics can be used to model a problem and lead to the development of problem-solving strategies.

This is an important point. Students must be able to generalize to various types of problems.

Although we will illustrate various types of application problems, real-life problems rarely fall into a specific category. It is extremely important to concentrate on the general approach and to develop skills and strategies that can be applied to any problem.

The following is a summary of a general approach to problem solving. We will elaborate on these strategies as we discuss examples.

A General Approach to Problem Solving

1. On your first reading of the problem, make sure you understand what the problem is about and what question(s) you need to answer.
2. If appropriate, draw a figure or diagram or make a chart or table to organize information.
3. Assign a variable to the unknown quantity. Be specific about what the variable represents. If there are other unknown quantities, represent them in terms of the same variable.
4. Write an equation that describes, in symbols, the information given in the problem.
5. Solve the equation.
6. Use the solution to answer the question(s) asked in the problem.
7. Check the answer(s) by confirming that the conditions stated in the problem are met.

In the examples in this section, we include the step numbers to show how the general approach is used in solving the problem. (Step 2 will be shown only when a figure or table is used.)

The following examples illustrate how this general approach can be used in a variety of applications. Naturally, there is no way to include every possible type of problem that could ever be encountered. But these examples will provide a good start in developing methods that will generally work well for any problem.

Numbers

In a sense, all application problems are number problems, because the described conditions and the answer are all numerical. However, we begin with examples of problems that involve only numbers, with no reference to units.

EXAMPLE 1

Determining an Unknown Number

If 9 less than twice a number is -1, what is the number?

Solution

① The goal is to determine the described number.

③ Let $x =$ the number.

④ $2x - 9 = -1$ Nine less than twice the number is -1.

⑤ $2x - 9 + 9 = -1 + 9$ Add 9 to both sides.

$$2x = 8$$

$$\frac{2x}{2} = \frac{8}{2}$$ Divide both sides by 2.

$$x = 4$$

⑥ The number is 4. Write a sentence that answers the question.

⑦ Check: Twice the number is 8, and 9 less than 8 is -1.

Note: Because the equation you have written in step 4 might be in error, it is better to check an answer by confirming that it meets the conditions described in the problem than by substituting the answer into the equation.

Consecutive Integers

Two or more integers that appear immediately after one another on the number line are called **consecutive integers.** Problems can also involve **consecutive even integers** or **consecutive odd integers.** Here are examples.

Consecutive integers: 2, 3, 4, 5, 6, 7
Consecutive even integers: $-4, -2, 0, 2, 4, 6$
Consecutive odd integers: $-5, -3, -1, 1, 3, 5$

If we let n represent the first integer of a sequence, the rest of the representation looks like this.

Consecutive integers: $n, n + 1, n + 2, n + 3, \ldots$
Consecutive even integers: $n, n + 2, n + 4, n + 6, \ldots$
Consecutive odd integers: $n, n + 2, n + 4, n + 6, \ldots$

Students often need to be convinced that the two representations are the same.

Note that consecutive even integers and consecutive odd integers are represented the same way, but the variable n represents a different kind of number in the two cases.

EXAMPLE 2

Consecutive Odd Integers

The lockers on the right side of the hall are numbered consecutively with even integers; the lockers on the left side of the hall are numbered consecutively with odd integers. Two people have adjoining lockers on the left side of the hall. They note that their locker numbers have a sum of 60. What are their locker numbers?

Solution

① The goal is to determine the two locker numbers. Because the lockers are on the left side of the hall, the numbers are consecutive odd integers.

LEARNING TIP

When you assign a variable, always be precise. If the unknown quantity is the price of a rose, writing x = rose is not meaningful and can be misleading. Remember that a variable represents a number, not a thing.

③ Let x = the first odd integer and $x + 2$ = the second odd integer.

④ $x + (x + 2) = 60$ The sum of the locker numbers is 60.

⑤ $x + x + 2 = 60$ Remove parentheses.

$2x + 2 = 60$ Combine like terms.

$2x + 2 - 2 = 60 - 2$ Subtract 2 from both sides.

$2x = 58$

$\frac{2x}{2} = \frac{58}{2}$ Divide both sides by 2.

$x = 29$

⑥ Because x (the first locker number) is 29, $x + 2$ (the second locker number) is 31. The locker numbers are 29 and 31.

⑦ Check: The locker numbers are consecutive odd integers, and their sum is $29 + 31 = 60$.

Piece Lengths

In this type of problem, material of some known length is to be cut into two or more pieces. We are given the relative sizes of the pieces, and the task is to determine the actual length of each piece.

EXAMPLE 3

Encourage students to make a sketch for any problem involving a geometric figure.

Lengths of Pieces of Rope

A rope is 72 feet long and is to be cut into three pieces. The first piece must be 8 feet shorter than the second piece, and the third piece must be twice as long as the second piece. How long should the first piece be?

Solution

① Although the goal is to determine how long the first piece should be, we will determine all three lengths so that the results can be verified.

② **Figure 3.3**

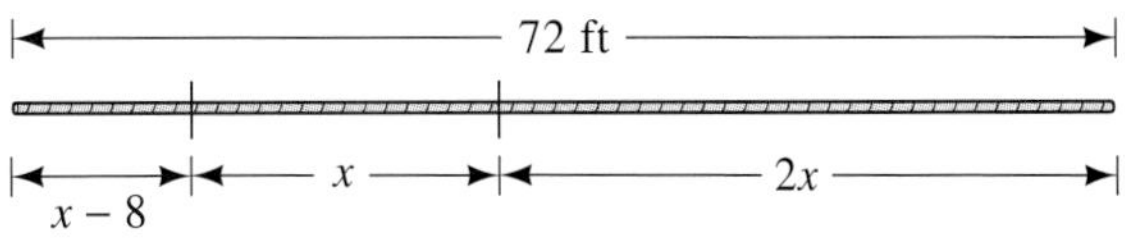

③ Because the first and third pieces are described in comparison to the second piece, we let the variable x represent the length of the second piece. Then we describe the lengths of the other two pieces in terms of that variable. (See Fig. 3.3.)

④ $(x - 8) + x + 2x = 72$ — The sum of the piece lengths is 72 feet.

⑤
$$x - 8 + x + 2x = 72 \quad \text{Remove parentheses.}$$
$$4x - 8 = 72 \quad \text{Combine like terms.}$$
$$4x - 8 + 8 = 72 + 8 \quad \text{Add 8 to both sides.}$$
$$4x = 80$$
$$\frac{4x}{4} = \frac{80}{4} \quad \text{Divide both sides by 4.}$$
$$x = 20$$

⑥ The length of the first piece is $x - 8 = 20 - 8 = 12$ feet. Note that the lengths of the second and third pieces are $x = 20$ feet and $2x = 40$ feet, respectively.

⑦ Check:

First piece	12	The first piece is 8 feet shorter than the second piece.
Second piece	20	
Third piece	40	The third piece is twice as large as the second piece.
Total length	72	

Think About It

Suppose that a 28-foot wire is to be cut into four pieces and that the lengths of the first three pieces are consecutive integers. Can you determine the lengths of the four pieces? Could you do it if the length of the wire were 7 feet?

LEARNING TIP

Try to establish connections among applications. For example, problems involving three piece lengths and angles of a triangle are essentially the same in that the sum of the lengths or measures is known.

Angles of a Triangle

Problems of this kind usually describe the relative measures of the three angles of a triangle. The task is to determine the measure of each angle. The basis for the equation in such problems is the theorem from geometry that states that the sum of the measures of the angles of a triangle is 180°.

EXAMPLE 4

Help students to see that the structure of this type of problem is identical to that of the piece lengths example.

The Measures of the Angles of a Triangle

A surveyor is mapping a plot of land in the form of a triangle. The second angle of the triangle is 21° greater than the first angle. The third angle is 11° more than twice the first angle. What are the measures of the three angles?

Solution

① The goal is to determine the measures of all three angles of the triangle.

② **Figure 3.4**

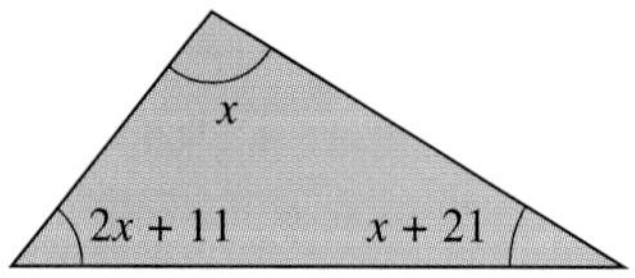

③ Because the measures of the second and third angles are described in comparison to the measure of the first angle, we assign the variable to the measure of the first angle. Figure 3.4 shows how the measures of the other two angles are represented.

④ $x + (x + 21) + (2x + 11) = 180$ — The sum of the measures of the angles of a triangle is 180°.

⑤
$$\begin{aligned} x + x + 21 + 2x + 11 &= 180 && \text{Remove parentheses.}\\ 4x + 32 &= 180 && \text{Combine like terms.}\\ 4x + 32 - 32 &= 180 - 32 && \text{Subtract 32 from both sides.}\\ 4x &= 148\\ \frac{4x}{4} &= \frac{148}{4} && \text{Divide both sides by 4.}\\ x &= 37 \end{aligned}$$

⑥ The measure of the first angle is 37°. The second angle is $x + 21 = 37 + 21$ or 58°. The third angle is $2x + 11 = 2 \cdot 37 + 11$ or 85°.

⑦ Check:

First angle	37°	
Second angle	58°	The second angle is 21° greater than the first angle.
Third angle	85°	The third angle is 11° more than twice the first angle.
Total	180°	

Rectangles

The basic quantities involving a rectangle are the length, width, perimeter, and area. Rectangle problems typically describe the relationship between the length and width and provide either the perimeter or the area.

EXAMPLE 5

Carpet for a Rectangular Room

A carpet layer has been to a home to provide an estimate of the cost of carpeting a rectangular dining room. When he returns to his office, he finds that he has misplaced the dimensions. He remembers that the perimeter is 52 feet and that the width is 4 feet less than the length. If the carpet costs \$12.95 per square yard, how can he still provide the estimate, and what is it?

Solution

① To provide a cost estimate, the carpet layer must determine the dimensions of the room.

② **Figure 3.5**

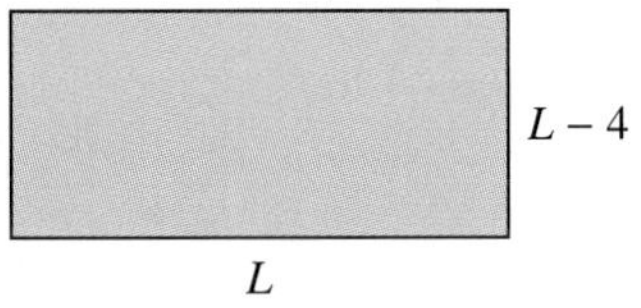

③ As shown in Figure 3.5, we let L represent the length and $L - 4$ the width. Note that we could have let W represent the width and $W + 4$ the length.

$$2L + 2W = P \quad \text{Formula for the perimeter of a rectangle}$$

④ $$2L + 2(L - 4) = 52 \quad \text{Replace } W \text{ with } L - 4. \text{ The perimeter } P \text{ is 52 feet.}$$

⑤
$$\begin{aligned} 2L + 2L - 8 &= 52 && \text{Distributive Property}\\ 4L - 8 &= 52 && \text{Combine like terms.}\\ 4L - 8 + 8 &= 52 + 8 && \text{Add 8 to both sides.}\\ 4L &= 60 \end{aligned}$$

$$\frac{4L}{4} = \frac{60}{4}$$ Divide both sides by 4.

$$L = 15$$

⑥ The length of the dining room is 15 feet, and the width is 15 − 4 or 11 feet. The area A is $A = LW = 15(11) = 165$ square feet or about 18.3 square yards. At \$12.95 per square yard, the cost estimate is 12.95(18.3) or about \$237.

⑦ Check: Length 15

Width 11 — The width is 4 feet less than the length.

Perimeter $= 2(15) + 2(11) = 52$

Things of Value

There are many applications involving things of value. The objects may be coins, tickets to a play, or pounds of bananas.

The problems usually involve more than one of an item. Typically, the per-unit value is given, as is the total value of all the items. The total value for a collection of items is determined by multiplying the number of items times their per-unit value. To organize the information in the problem, a table is particularly useful.

EXAMPLE 6 Two Kinds of Tickets with Different Values

A teacher organized a field trip to a zoo for students and their parents. The admission was \$2.00 for students and \$4.75 for adults. If there were six fewer adults than children on the trip and the total cost was \$93.00, how many adults went on the trip?

Solution

① The question asks for the number of adults who went on the trip, but to check our results, we will also determine the number of children who went.

The table helps organize the information and allows us to see the relationships among the quantities.

②

	Number of Tickets	Per-Ticket Cost	Cost of Tickets
Children	x	2.00	$2.00x$
Adults	$x - 6$	4.75	$4.75(x - 6)$
Total			93.00

③ The table shows how we have assigned the variable and how we have represented other unknown quantities.

④ $2.00x + 4.75(x - 6) = 93.00$ — The total cost of the tickets is \$93.00.

⑤ $2.00x + 4.75x - 28.50 = 93.00$ — Distributive Property

$6.75x - 28.50 = 93.00$ — Combine like terms.

$6.75x - 28.50 + 28.50 = 93.00 + 28.50$ — Add 28.50 to both sides.

$6.75x = 121.50$

$\frac{6.75x}{6.75} = \frac{121.50}{6.75}$ — Divide both sides by 6.75.

$x = 18$

⑥ The number of children who went to the zoo is 18. The number of adults is $18 - 6 = 12$.

⑦ Check: There were six fewer adults (12) than children (18).

Cost of children's tickets	$18(2.00) = \$36.00$
Cost of adults' tickets	$12(4.75) = \$57.00$
Total cost	$93.00

Students tend to ignore units. Emphasize the need to be consistent.

Note that the monetary units used in Example 6 were dollars. The units could have been cents to avoid the decimals, but the arithmetic involved in solving the equation is virtually the same either way. When solving applications involving monetary units, just be sure your units are the same in all expressions.

An application problem can have explicit or implied conditions that cannot be met. The following are some examples of applications with solutions that must be disqualified.

Application Problem	*Disqualified Solution*
Consecutive integers	Not an integer
Consecutive even integers	Odd integer
Consecutive odd integers	Even integer
Dimensions of a figure	Not a positive number
Number of things of value	Not a whole number
Time, rate, or distance	Not a positive number

Remember that a solution of an equation is not necessarily the answer to the question. Make sure your answer meets *all* the conditions of the problem.

Exercises 1–36 focus on more traditional concepts and skills. All other exercises deal with modeling and real-life applications.

Exercises 3.4

Concepts and Skills

Numbers

1. Suppose the total of two numbers is T. If one of the numbers is n, how can you represent the other number in terms of n? Show that the sum of these representations is T.
2. Explain why we should check the answer to an application problem by verifying that the answer meets the stated conditions of the problem rather than by substituting into the original equation.
3. The sum of a number and 6 more than the number is 32. What is the number?
4. The sum of a number and twice the number is 57. What is the number?
5. One number is 12 more than another number. If one-half of the smaller number equals one-third of the larger number, what is the larger number?
6. Increasing a number by 3 and multiplying the result by 4 is the same as decreasing the number by 2 and multiplying the result by 9. What is the number?

Consecutive Integers

7. Explain why x and $x + 2$ can be used to represent either two consecutive *even* numbers or two consecutive *odd* integers.
8. For three consecutive integers, the sum of the smallest and largest integers equals twice the middle integer. What are the integers? Explain the result.
9. The sum of two consecutive integers is 95. What are the two integers?
10. The sum of three consecutive integers is 3. What are the three integers?
11. The sum of two consecutive odd integers is 0. What are the two integers?

12. The sum of two consecutive even integers is 174. What are the two integers?

Piece Lengths

Remind students about the value of making a sketch whenever possible.

13. A 45-foot rope is to be cut into three pieces. The second piece must be twice as long as the first piece, and the third piece must be 9 feet longer than three times the length of the second piece. How long should each of the three pieces be?

14. A 54-foot log is to be cut into three smaller logs whose lengths are consecutive even integers. How long should each piece be?

15. A 105-yard bolt of cloth is to be cut into three lengths. The second piece is to be 5 yards shorter than the first piece, and the third piece is to be twice the length of the second piece. What are the lengths of the three pieces?

16. A 32-foot pipe is cut into three pieces. The first two lengths are consecutive integers, and the third piece is 6 feet longer than the second. How long are the three pieces?

Angles of a Triangle

17. One angle of a triangle is 20° greater than the first angle. The third angle is twice as large as the first angle. What are the measures of the three angles?

18. The second angle of a triangle is 7° more than three times the first angle. The third angle is 12° greater than the second angle. What are the measures of the three angles?

19. Three walkways intersect to enclose a triangular grassy area. The second angle of the triangular area is twice the first angle, and the third angle is one and one-half times the second angle. Determine whether the triangle is a right triangle.

20. An **isosceles triangle** is a triangle with at least two angles having the same measure. (See figure.) If the first angle of a triangular highway warning sign is 30° smaller than the third angle, and the third angle is 20° less than twice the second angle, determine whether the triangle is an isosceles triangle.

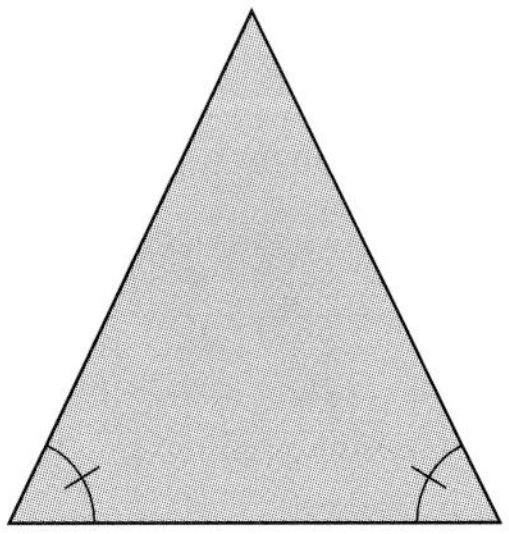

Rectangles

21. The width of a rectangular garage is 8 feet less than the length. If the perimeter of the garage is 64 feet, what are the dimensions?

22. The width of a rectangular real estate sign is 16 inches less than the length. If the perimeter of the sign is 88 inches, what are the dimensions?

23. The length of a rectangular piece of carpet is 4 yards greater than the width. If 60 yards of fringe border is required for the edge of the carpet, what are the dimensions of the carpet?

24. Suppose that 134 meters of fencing are required to enclose a rectangular animal feed lot. If the length of the lot is 7 meters greater than the width, what are the dimensions of the lot?

Things of Value

Remind students about the value of a table and about the need to be consistent with units.

25. A Salvation Army collection kettle contains \$2.25 in nickels and dimes. If the kettle contains nine more nickels than dimes, how many nickels are in the kettle?

26. The coin box of a gum vending machine, which accepts only dimes and quarters, contains \$5.20. If the number of quarters is four more than the number of dimes, how many quarters are in the box?

27. If a change dispenser contains 32 coins consisting of only dimes and quarters worth \$5.15, how many dimes are in the dispenser?

28. If a Lions Club mint box contains 36 coins consisting of only nickels and dimes worth \$2.50, how many nickels are in the box?

Miscellaneous Applications

29. For any three consecutive integers, show that the average of the smallest and largest integers is the middle integer.

30. Show that the absolute value of the difference between any two consecutive integers is always 1.

31. The sum of three consecutive even integers is 19. What are the integers? Could you have anticipated your answer without solving an equation?

32. The sum of three consecutive odd integers is 19. What are the integers? Can the conditions of the problem be met? Explain.

33. Increasing a number by 5 and multiplying the result by 7 is the same as multiplying the number by -5 and subtracting 1 from the result. What is the original number?

34. A child has saved \$3.55 in nickels and dimes. If there are ten fewer nickels than dimes, how many nickels does the child have?

35. The first angle of a triangle is 10° smaller than the second angle. The third angle is 10° less than twice the second angle. Show that this is a right triangle.

36. The width of a rectangle is 2 feet more than half the length. If the perimeter of the rectangle is 88 feet, what is the width?

Geometric Models

37. A supporting guy wire is attached to the top of a utility pole and is fastened at the ground 10 feet from the base of the pole. (See figure.) If the angle between the guy wire and the ground is eight times the angle between the guy wire and the pole, what is the measure of the angle between the guy wire and the ground?

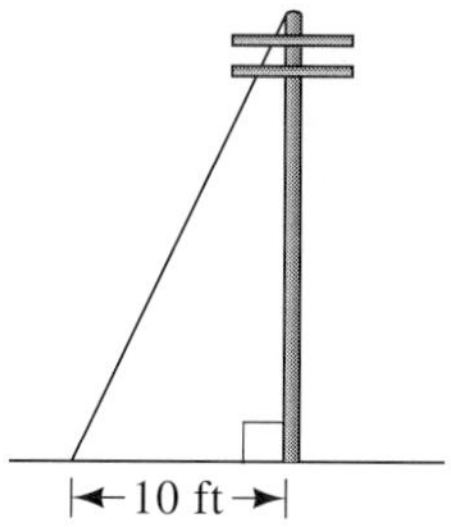

38. An Atlanta Braves pennant is a triangle with two of the angles having the same measure. The third angle is one-fifth the sum of the two equal angles. What is the measure of the smallest angle?

39. The height of a book is 4 inches greater than its width. If the perimeter of the book is 48 inches, can the book be placed upright on a bookcase shelf 12 inches high?

40. The width of one poster is 7 inches more than half the length. The width of a second poster is 9 inches less than the length of the first poster, and the length of the second poster is 10 inches greater than its width. If the two posters have the same perimeter, which poster is wider?

41. An art student drew a preliminary sketch for a painting on a small rectangular pad that was 5 inches longer than it was wide. To improve the scale of her drawing, she switched to another sketch pad that was twice as long as the smaller pad and 8 inches wider. If the perimeter of the larger pad was 60 inches, how long was the smaller pad?

42. The width of a sheet of paper is 2.5 inches less than its length. A photocopy is made that is 90% of the original dimensions. The reduced copy has a perimeter of 35.1 inches. How wide was the original paper?

43. Curtains 40 inches long are to be hung on a rectangular window. The height of the window is one and a half times its width, and the perimeter of the window is 150 inches. Will the curtains cover the full height of the window?

44. A computer is 2 inches taller than it is wide. The perimeter of the front of the computer is 60 inches. Will the computer fit under a shelf that is 18 inches above the top of the desk?

45. An **obtuse triangle** is a triangle with one angle having a measure that is greater than 90°. (See figure.) If one angle of a triangle is four times the second angle, and the third angle is 2° less than twice the second angle, determine whether the triangle is an obtuse triangle.

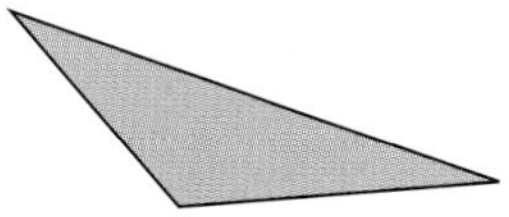

46. An **acute triangle** is a triangle with all three angles having measures that are less than 90°. (See figure.) If the second angle of a triangle is six-sevenths as large as the third angle, and the first angle is five-sixths as large as the second angle, determine whether the triangle is an acute triangle.

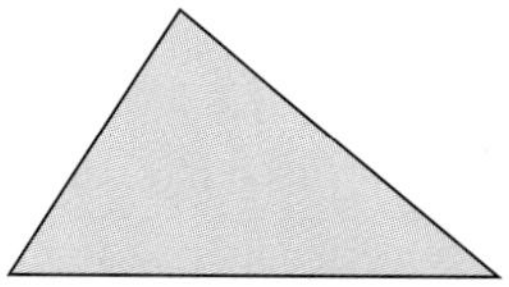

47. Two angles are **supplementary angles** if the sum of their measures is 180°. What is the measure of the smaller of two supplementary angles if the measure of the larger is 10.8° more than twice the measure of the smaller?

48. Two angles are **complementary angles** if the sum of their measures is 90°. What is the measure of the larger of two complementary angles if the measure of the smaller angle is 15° less than half the measure of the larger angle?

49. A ladder rests against a wall. (See figure.) The angle made by the ladder and the ground is 6° less than seven times the angle made by the ladder and the wall. What is the measure of the angle made by the ladder and the wall?

50. A 9-by-12-inch picture is to be mounted in a frame. The width of the frame is 5 inches less than the length, and the perimeter of the frame is 42 inches. Will the picture fit in the frame?

Real-Life Applications

Numbers

51. There are 39 students in a political science class. The instructor notices that there are half as many men as women. How many women are in the class?

52. An uncle wills \$3000 to his two nieces and one nephew. The older niece is to receive twice what the younger niece receives, and the nephew is to receive an amount equal to the sum of the nieces' amounts. How much will each person receive?

53. In a poker game, two people began with equal sums of money. When the game ended, the loser had lost \$50 more than a quarter of his money, and the winner had twice as much money as the loser. How much money did each person have at the end of the game?

54. An appliance company had four more washers than dryers in its inventory. On Friday half the dryers and one-third of the washers were shipped out. If a total of eight appliances were shipped, how many were dryers?

55. A retail store ordered some sofas and chairs from a manufacturer. The number of chairs ordered was four more than the number of sofas. The store received a partial shipment consisting of half of the sofas and one-third of the chairs that were ordered. If the total number of pieces was the same as the number of sofas originally ordered, how many sofas did the store receive?

56. Some guests had arrived at a party by 8:00. At 8:30, two more arrived. By 9:00 the 21 guests in attendance were three times the number who were there at 8:30. How many had arrived by 8:00?

57. By 10:00 A.M., the Dow Jones average had dropped from the opening average. By noon the Dow Jones average had slipped another 8 points. At the close of trading, the 50-point drop was five times the amount by which the average had dropped at noon. By how much had the average dropped at 10:00 A.M.?

58. The government's farm price support program pays a farmer not to plant any crops this year, so he must decide how not to plant his 100 acres. He decides not to plant corn on three times as many acres as he will not plant strawberries. The number of acres on which he will not harvest hay is equal to the total number of acres on which he will plant neither corn nor strawberries. On how many acres will the farmer not plant corn?

Consecutive Integers

59. At the sports club, two people have adjacent lockers, and the sum of their locker numbers is 434. Show how you know that the lockers are not numbered consecutively.

60. A precinct worker issues voting passes to three voters consecutively. The serial numbers on the passes total 237. What was the serial number of the second voter's pass?

61. After three college basketball games, a forward, center, and guard have a total of 143 points. The forward has scored the most points, and the guard has scored the fewest. The forward's and center's points are consecutive integers. The guard's and center's points are consecutive even integers. How many points has the forward scored?

62. A bookmark is placed in a book between two pages whose sum is 193. On what page did the reader stop reading? (There are two possible answers.)

Piece Lengths

63. A scout leader is on a camping trip with his troop. He wants to cut a 40-meter rope into three lengths. The second length must be 2 meters longer than the first length. The third length must be 6 meters longer than twice the first length. What are the three lengths of rope?

64. A fabric store worker has a 106-yard bolt of cloth that is to be cut into three lengths. The second piece is to be 4 yards longer than the first. The third piece is to be three times as long as the second piece. What is the length of each piece?

Things of Value

65. The sponsors of a Boys and Girls Club took some members to a play and bought seven more tickets for members than for adults. The members' tickets cost \$1.25 and the adults' tickets cost \$2.75. If a total of \$28.75 was paid for all the tickets, how many members went to the play?

66. The total receipts for a college basketball game were \$674, with a student ticket selling for \$1.25 and a nonstudent ticket selling for \$2.75. If there were 136 more students than nonstudents in attendance, how many people were at the game?

67. Hamburgers sell for \$2 each and hot dogs sell for \$1.25 each. On a certain day, the number of hamburgers sold was twice the number of hot dogs sold. Determine the number of hot dogs sold if the total income was \$304.50.

68. Mama's Pizzeria sells individual pizzas for \$8 each and orders of bread sticks for \$1 each. During a 2-week period, the number of pizzas sold was four times the number of bread stick orders sold. How many pizzas were sold if the total revenue was \$2970?

Miscellaneous Applications

69. A person invested one-fourth of an inheritance in bank stock, one-fifth in bonds, and one-half in a mutual fund. If the sum of her investments was \$38,000, how much did she inherit?

70. A commercial developer reads the zoning laws on how he must allot his 50 acres of land. The number of acres of lawn and landscaping must equal the number of acres used for buildings. The number of acres of parking must be two-fifths the number of acres used for buildings. The number of acres of roads and utilities must be one-tenth the number of acres of lawn and landscaping. How many acres of parking lot must the developer provide?

71. There are 266 men, women, and children at the company picnic. There are four times as many men as children and twice as many women as children. How many of each came to the picnic?

72. Three adjacent offices have office numbers that are consecutive odd integers. The manager is in a fourth office with a number that is 13 more than the number of the middle office. If the sum of the four office numbers is 97, what is the manager's office number?

73. A member of a three-person bowling team tells his wife that he bowled over 200 last night, but he will not give the actual score. His wife asks another member of the team about it, but she learns only that the team's three scores were consecutive odd integers. His wife then checks with the third member of the team, who says that the team score was 597. Could her husband have bowled over 200? Can she know for sure? Explain.

74. Soft drinks sell for 75 cents each and milk shakes sell for \$1.25 each. On a certain day, the number of soft drinks sold was three times the number of milk shakes sold. How many soft drinks were sold if the total income was \$280?

Modeling with Real Data

75. The number of college students aged 18–21 rose to 6.3 million in 2000. (Source: National Center for Educational Statistics.)

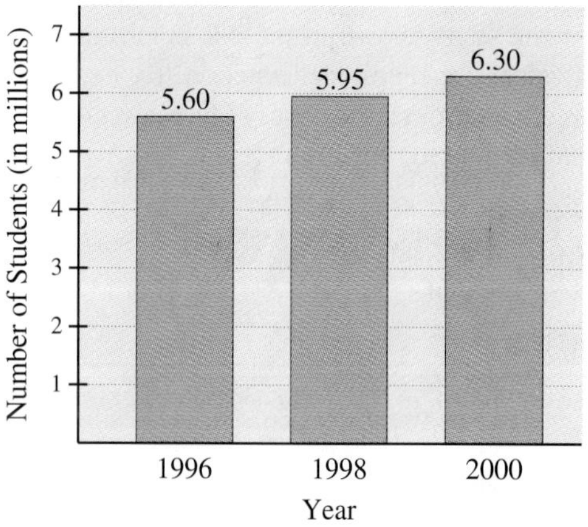

(a) Assuming that the average number of students continues to increase by 0.175 million per year, write an expression to model the number of students t years after 1996.

(b) Write and solve an equation to estimate the year in which the number of students in this age group will reach 8.4 million.

76. By the year 2050, the number of people over age 65 living in the United States is projected to be 78.9 million. This represents about a 125% increase from 2000. Letting p represent the population of the United States who were over 65 in 2000, write and solve an equation that models the given information.

Data Analysis: *Carbon Monoxide Emissions*

As a result of various clean-air measures, the level of carbon monoxide emissions in the atmosphere has declined. The bar graph shows the number of metric tons of carbon monoxide emitted for selected years. (Source: Environmental Protection Agency.) With x representing the number of years since 1990, the data can be modeled with the function $C(x) = -1.77x + 99.37$.

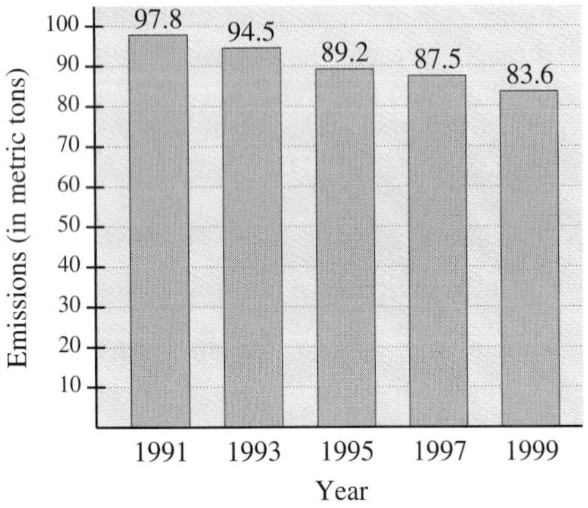

77. For the years shown in the bar graph, make a table of values for the model function. By examining the table, determine the yearly decrease in emissions predicted by the function.

78. By writing and solving an equation, estimate the year in which the emissions are predicted to decline to approximately 70 metric tons.

79. Write an equation that you can use to predict the year in which emissions are expected to be 20% less than the actual 1991 level.

80. Solve the equation in Exercise 79.

Challenge

81. George P. Burdell enjoys entertaining friends with number games. In one game he asks a friend to choose any number. Next he tells his friend to subtract one from the number, multiply the result by three, and then subtract twice the number increased by one. Finally, he tells his friend to add five to the result. George asks what the resulting number is and tells his friend that it is the chosen number. Write an equation to describe George's trick. Show that George's trick works for any number.

82. In one game George asks a friend to choose any number and add eight to it; multiply the result by three and then subtract three from the result; multiply by two and then divide by six; and finally, subtract the original number. George knows that the result is 7. Write an equation to describe George's trick. Show that George's trick works for any number.

3.5 Applications

Percentages • Fixed and Variable Costs • Simple Interest • Distance, Rate, and Time • Mixtures • Liquid Solutions

For most of the application problems in the previous section, equations were based on information provided by the problem itself.

This section deals with problems for which certain common knowledge is needed. In a sense, the use of this common knowledge makes the applications in this section more practical.

Although checking answers continues to be an important step in the general approach to problem solving, we will leave this step to you in the examples that follow.

Percentages

Ad valorem taxes are taxes based on the value of property. The valuation of the property, for tax purposes, is usually some percentage of the market or list value of the property. The ad valorem tax is then some percentage of the valuation.

EXAMPLE 1

Ad Valorem Tax on an Automobile

Suppose that the valuation of an automobile for tax purposes is 40% of the list value of the car and that the ad valorem tax is 2%. If the tax due on an automobile is \$126, what is the list value of the car?

Solution

Let x = the list value of the car.

$$\text{Valuation} = 0.40x \qquad \text{Valuation = 40\% of list value}$$

$$\text{Tax due} = (0.02)(0.40x) = 0.008x \qquad \text{Tax due = 2\% of the valuation}$$

$$0.008x = 126.00 \qquad \text{The tax due is \$126.}$$

$$\frac{0.008x}{0.008} = \frac{126.00}{0.008}$$

$$x = 15{,}750$$

The list value of the car is \$15,750.

Help the students translate information into word equations.

Original Price − Discount = Sale Price
Wholesale Price + Markup = Retail Price

When an item is placed on sale, the amount by which the price is reduced is called the **discount,** which is expressed as a percentage of the **original price.**

Retail store owners buy their goods at **wholesale prices.** Then, to cover overhead and the profit they want to make, they add an amount called **markup,** which is expressed as a percentage of the wholesale price. The result is the **retail price.**

EXAMPLE 2

Wholesale Price Plus Markup Equals Retail Price

All the goods at a hobby store are marked up 40%. If the retail price of one item is \$32.13, what was the wholesale cost to the store owner?

LEARNING TIP

Consider writing a word equation before you attempt to write an algebraic equation. For example, if you know or can represent the three quantities, then writing "original price − discount = sale price" gives you a guide for writing the actual equation.

Solution

Let w = the wholesale cost.

$$\text{Markup} = 0.40w \qquad \text{Markup = 40\% of wholesale cost}$$

$$w + 0.40w = 32.13 \qquad \text{Wholesale Cost + Markup = Retail Price}$$

$$1.00w + 0.40w = 32.13$$

$$1.40w = 32.13$$

$$\frac{1.40w}{1.40} = \frac{32.13}{1.40}$$

$$w = 22.95$$

The wholesale cost was \$22.95.

Think About It

Suppose that the wholesale price of an item is w and that the retail price reflects a 30% markup. If the item is later sold at a 20% discount, will the seller make money or lose money?

Fixed and Variable Costs

The cost of doing business can be separated into two components. One is the **fixed cost,** which is a cost that is incurred even if no products are sold or no service is rendered. The other component is the **variable cost,** which is a cost that depends on the number of products produced or sold or on the amount of service rendered.

EXAMPLE 3

The Cost of an Auto Rental

An employee submits an expense account for a 1-day round trip to a town 40 miles away. The expense account includes a car rental for \$82.00. The fixed daily rate is \$28.00 and the mileage rate is \$0.20 per mile. Is there reason to question the expense of the car rental?

Solution

Let M = the number of miles driven.

$$\text{Fixed cost} = 28.00 \qquad \text{Fixed daily rate is \$28.00.}$$
$$\text{Variable cost} = 0.20M \qquad \text{Mileage rate is \$0.20 per mile.}$$
$$28.00 + 0.20M = 82.00 \qquad \text{Fixed Cost + Variable Cost = Total Cost}$$
$$28.00 + 0.20M - 28.00 = 82.00 - 28.00$$
$$0.20M = 54$$
$$\frac{0.20M}{0.20} = \frac{54}{0.20}$$
$$M = 270$$

The employee has been charged for 270 miles when the trip should have been only 80 miles. Either the employee or the car rental agency has made a mistake.

Simple Interest

When money is deposited in a bank account, the bank pays for the use of the money. The amount deposited is called the **principal** (P). The amount paid by the bank is called the **interest** (I). Interest is expressed as a percentage (r) of the principal.

Simple interest is interest paid just once at the end of a given period of time. To calculate the simple interest that accumulates over a period of t years, we use the formula $I = Prt$.

EXAMPLE 4

Loan Plus Interest Equals Loan Payoff

A used car buyer borrows money for 1 year at a simple interest rate of 12%. If the payoff of the loan is \$5040, what was the amount of the original loan?

Solution

Let P = the amount of the loan.

$$\text{Interest} = 0.12P \qquad \text{Interest is 12\% of the loan amount.}$$
$$P + 0.12P = 5040 \qquad \text{Loan Amount + Interest = Loan Payoff}$$
$$1.00P + 0.12P = 5040$$
$$1.12P = 5040$$
$$\frac{1.12P}{1.12} = \frac{5040}{1.12}$$
$$P = 4500$$

The original loan was \$4500.

In a *dual-investment* problem, a given amount of money is divided into two different investments, each yielding a given interest percentage. With the total amount of interest earned given, the problem is to determine the individual investment amounts.

A table can be very useful in arranging the information for such problems.

EXAMPLE 5

Investments in Two Bond Funds

A total of \$10,000 was invested in two bond mutual funds, a junk bond fund and a government bond fund. The junk bond fund is risky and yields 11% interest. The safer government bond fund yields only 5%. The year's total income from the two investments was \$740. How much was invested in each fund?

Solution

Let x = the amount of money invested in the junk bond fund. The remaining money, $10{,}000 - x$, is the amount invested in the government bond fund.

	Amount Invested	Interest Rate	Interest Earned
Junk Bonds	x	0.11	$0.11x$
Government Bonds	$10{,}000 - x$	0.05	$0.05(10{,}000 - x)$
Totals	10,000		740

$$0.11x + 0.05(10{,}000 - x) = 740$$

The total interest earned is the sum of the interest earned from each fund.

$$0.11x + 500 - 0.05x = 740$$
$$0.06x + 500 = 740$$
$$0.06x + 500 - 500 = 740 - 500$$
$$0.06x = 240$$
$$\frac{0.06x}{0.06} = \frac{240}{0.06}$$
$$x = 4000$$

The amount invested in the junk bond fund was \$4000, and 10,000 − 4000, or \$6000, was invested in the government bond fund.

Distance, Rate, and Time

Distance d is related to the speed or rate r and time t by the formula $d = rt$. Again, a table is often useful for organizing the information.

EXAMPLE 6

Remind students to be consistent with units in distance, rate, and time problems.

Distance, Rate, and Time

Two friends drove from Deadwood, South Dakota, to Cody, Wyoming, a distance of 370 miles. The average driving speed of one driver was 16 mph slower than that of the other driver. The faster driver drove for 2.5 hours, and the other driver drove half an hour longer. What was the average speed of each driver?

LEARNING TIP

As you represent unknown quantities and write equations, make sure that your units are consistent. For example, if a rate is in miles per hour, then the distance must be in miles and the time must be in hours.

Solution

Let r = the average speed for the faster driver.

	Rate	Time	Distance
Fast Driver	r	2.5	$2.5r$
Slow Driver	$r - 16$	3	$3(r - 16)$
Total			370

$$2.5r + 3(r - 16) = 370 \quad \text{The sum of the distances is 370 miles.}$$

$$2.5r + 3r - 48 = 370$$

$$5.5r - 48 = 370$$

$$5.5r = 418$$

$$\frac{5.5r}{5.5} = \frac{418}{5.5}$$

$$r = 76$$

The faster driver averaged 76 mph, and the slower driver averaged 60 mph.

Mixtures

Mixture problems usually involve the mixing or blending of two items, each with its own unit price or cost. The resulting mixture has a unit value that is *between* the unit prices of the individual items.

Note: The unit price of a mixture is not the sum of the unit prices of the items that are mixed.

EXAMPLE 7

A Mixture of Nursery Stock

The owner of a small nursery specializes in iris and day lily starter plants. He has been selling irises at a price of \$3.00 per dozen and day lilies at a price of \$8.00 per dozen.

A landscaper wants to buy 15 dozen of a mixture of irises and day lilies, and she is willing to pay \$5.00 per dozen for the mixture. How many dozen of each type should the nursery owner include in the mixture?

Solution

Let x = the number of dozen of irises to include. Then

$15 - x$ = the number of dozen of day lilies.

The table summarizes the information in the problem.

	Number of Dozen	Price per Dozen	Total Cost
Iris	x	3.00	$3.00x$
Day Lily	$15 - x$	8.00	$8.00(15 - x)$
Totals	15	5.00	75.00

$$3x + 8(15 - x) = 75 \quad \text{The total cost of the mixture is \$75.}$$
$$3x + 120 - 8x = 75$$
$$-5x + 120 = 75$$
$$-5x = -45$$
$$x = 9$$

The mixture should consist of 9 dozen irises and $15 - 9 = 6$ dozen day lilies.

Liquid Solutions

When two liquids are mixed together, the result is a **liquid solution.**

Figure 3.6

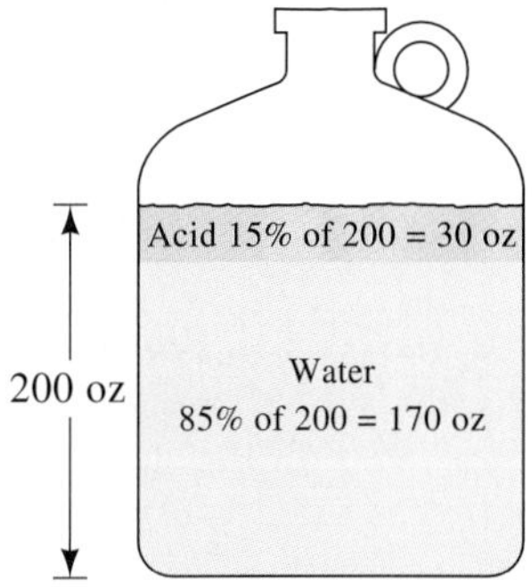

This preparatory work on *concentration* is essential. Students often have difficulty with liquid solution problems because they do not fully grasp what concentration means.

Figure 3.6 shows a bottle containing 200 ounces of a 15% acid solution, which means that 15% of the volume is pure acid. We say that the **concentration** of acid in the solution is 15%.

Because 15% of the total volume is acid, the number of ounces of acid in the bottle is $(0.15)(200) = 30$ ounces. The remaining 85% of the solution is water, which means that $(0.85)(200) = 170$ ounces of water are in the bottle. The total is $30 + 170 = 200$ ounces.

Liquid solution problems often involve the mixing of two solutions of different concentrations. The result is a solution whose concentration is between the concentrations of the original two solutions.

Note: When two solutions are mixed, the concentration of the result is *not* the sum of the concentrations of the two solutions.

In Figure 3.7, 50 liters of a 30% acid solution are mixed with 100 liters of a 75% solution. The result is a 150-liter solution.

Figure 3.7

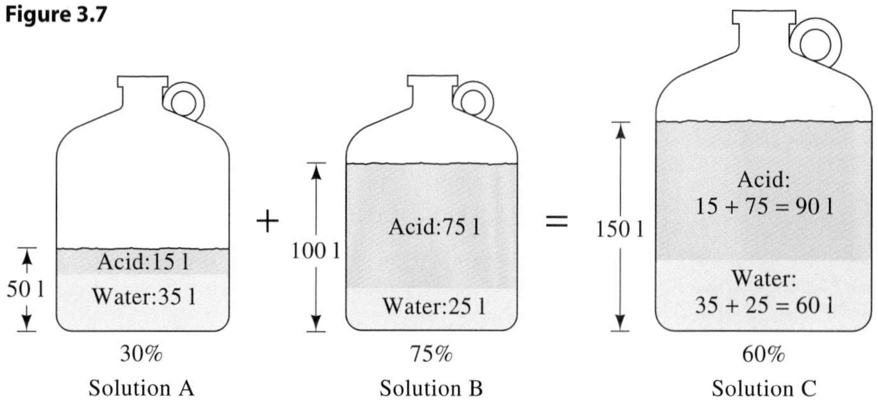

Note that the concentration of the result is between the concentrations of the two parts.

Solution A has $(0.30)(50) = 15$ liters of acid, and solution B has $(0.75)(100) = 75$ liters of acid. The resulting solution C will have $15 + 75 = 90$ liters of acid. The resulting concentration is $90 \div 150 = 0.60$ or 60%.

EXAMPLE 8

Mixing Two Acid Solutions

An embossing shop etches metal with acid solutions. The owner is discussing a certain job with the production manager, and they decide to use 30 ounces of a 40% acid solution.

Upon checking the acid solution supplies, the production manager has found some 25% acid solution and some 50% acid solution, but there is no 40% acid solution. The owner explains how the two existing solutions can be mixed to obtain the 40% solution. How much of each solution is needed?

Solution

Let $x =$ the number of ounces of the 25% solution. The remaining $30 - x$ ounces will be 50% solution.

The table summarizes the information in the problem.

	Ounces of Solution	Concentration of Acid	Ounces of Acid
25% Solution	x	0.25	$0.25x$
50% Solution	$30 - x$	0.50	$0.50(30 - x)$
Mixed Solution	30	0.40	12

$$
\begin{aligned}
0.25x + 0.5(30 - x) &= 12 \\
0.25x + 15 - 0.5x &= 12 \\
-0.25x + 15 &= 12 \\
-0.25x + 15 - 15 &- 12 - 15 \\
-0.25x &= -3 \\
\frac{-0.25x}{-0.25} &= \frac{-3}{-0.25} \\
x &= 12
\end{aligned}
$$

The total amount of acid in the mixture is the sum of the amounts of acid in each individual solution.

By mixing 12 ounces of the 25% solution and $30 - 12$, or 18, ounces of the 50% solution, the production manager obtains the required 30 ounces of 40% solution.

Although the sample problems in these sections have been organized into typical problem categories, they are offered only as illustrations of the general problem-solving strategy presented in the previous section.

The examples are not meant to suggest that every problem can be wedged into some known category and solved according to a fixed routine. It is more important to *think* about the problem rather than to seek out quick, ready-made prescriptions.

These exercises are initially organized by category. In the miscellaneous group, the problems are mixed.

Exercises 3.5

Real-Life Applications

Percentages

1. With respect to percents, how do we translate the word *of*? Translate the phrase *15% of 30* into a mathematical expression.

2. Suppose there is a 5% sales tax in your area. In which equation below does x represent the cost of an item *before* the sales tax? In which equation does x represent the cost *including* the sales tax?

   (i) $(1.05)(32.00) = x$

   (ii) $1.05x = 32.00$

3. A boy delivers newspapers for The Daily News, Inc., and he keeps 12% of all the money he collects. One week he kept \$28. How much money did he turn in to The Daily News, Inc.?

4. Washington County collects a 6% sales tax. If the tax on a new car is \$929.88, what is the total amount paid for the car? (Ignore all other fees associated with the purchase.)

5. On Monday an investor bought some stock. By Wednesday the stock had increased in value by 4%. However, on Friday the stock dropped 4% from Wednesday's value. The investor immediately sold the stock for \$349.44. What did the investor pay for the stock? Did she make money, lose money, or break even?

6. A student correctly answered 15 test questions of equal value. If his percentage grade was 83% (rounded off), how many questions were on the test?

7. Each month a worker's salary is reduced by 7% for social security, 20% for federal taxes, 6% for her pension, and 7% for state taxes. If the worker's monthly take-home pay is \$1200, what is her monthly salary?

8. A person's weekly wages are reduced by 36% for taxes and other deductions. He decides to spend his entire week's take-home pay on some stereo equipment costing \$220 plus a 4% sales tax. What is this person's weekly pay?

9. A real estate agent receives a 7% commission on the first \$120,000 of a sale and a 9% commission on the amount over \$120,000. If the agent received a commission of \$13,980 on the sale of a house, what was the selling price of the house?

10. A salesperson receives a base salary of \$360 a week and a commission of 4% on all sales. What was the amount of his sales during the first week of December if his weekly check was for \$500?

11. A teapot is on sale for \$17.50. If the teapot is marked down 30%, what was the original price?

12. A coffee maker is on sale for \$30.60. If the price of the coffee maker is marked down 15%, what was the original price?

13. A sign in a store window reads, "Everything marked down 20%." A customer found a suit on sale for \$239.20. What was the price of the suit before the markdown?

14. Cardigans Galore and The Sweater Shop both claim that their goods are priced exactly the same. When The Sweater Shop held a "20% Off" sale, a shopper found a sweater on sale for \$68.00. Cardigans Galore sells the same sweater for \$80.00. Was the original price at The Sweater Shop the same as the current price at Cardigans Galore?

Fixed and Variable Costs

15. Classify each of the following costs of running a business as a fixed cost or a variable cost.

(a) expenditures for materials

(b) annual insurance

(c) electric bill

(d) wages of hourly employees

(e) annual salary of the manager

16. If the cost C of producing x items is given by the function $C(x) = ax + b$, which term represents the fixed cost? Which term represents the variable cost?

17. Ace Construction Company rents a bulldozer for \$100 per day plus \$50 for each hour that the bulldozer is in use. If the bulldozer rental bill on April 18 was \$480, how many hours was the bulldozer used?

18. Sport Car Rentals rents a car for \$35 per day and 12 cents per mile. After 2 days, a bill (before tax) was \$127.84. How many miles were driven?

19. EMH Corporation has leased a photocopier. Because the company must pay \$300 per month plus 3 cents per copy made, it makes a policy that no more than 3600 copies can be made per month. At the end of the first month, the accounting department receives a photocopying bill for \$414. By how many copies has the company exceeded its limit?

20. Last year, Evans Company produced furnace vents with a variable cost of \$5.00. This year, the fixed cost has risen by 11%. If the cost of producing 100 furnace vents is now \$508.88, what was the fixed cost last year?

Simple Interest

21. If P represents the amount of an investment at 7% simple interest, which of the following represents the investment value after 1 year?

(i) $P + 0.07$ (ii) $0.07P$

(iii) $P + 0.07P$ (iv) $0.07 + 0.07P$

22. A big spender has two credit cards, one that carries an 18% finance charge and another that carries a 20% finance charge. This person uses only one of the cards to avoid paying 38% in finance charges. What is your opinion of this reasoning?

23. Suppose that you plan to save for one year to buy a sofa that costs \$600. What amount must you invest at 5% simple interest in order to have enough to buy the sofa?

24. Your father-in-law will sell his rider mower to you, but you do not have to pay him now. Instead, he will charge 5.5% simple interest, and you will owe him \$1000 at the end of one year. For how much is he selling the mower?

25. One year ago a college sophomore received a grant that she did not need until this year. She invested the money at 7% simple interest, and she now has \$1177. How much was the grant?

26. Last year a friend borrowed money from you at 9% simple interest, and he promised to repay the loan plus interest a year later. Now he has paid you \$1000, but he could not pay the remaining \$362.50 that he still owes. How much did he borrow from you?

27. The owner of a sporting goods store had \$15,000 cash that he could apply toward the cost of expanding his store. He needed to borrow the rest of the cost at 8% simple interest for 1 year. If his total cash outlay at the end of the year was \$36,600, what was the amount of his loan?

28. A homeowner took out an 8% simple interest equity loan on his home. At the end of the first year of the loan, he paid the interest due plus \$3800 toward the principal. At the end of the second year, before making any payment, he still owed \$33,000. What was the amount of the equity loan?

29. In 2000 you invested money in low-income mortgage loans paying 9% simple interest, while your more conservative sister placed her money in a safe 6% simple interest investment. If the total of your investments was \$6500 and the total of your interest incomes one year later was \$465, how much did each of you invest?

30. Two sons borrowed a total of \$26,000 from their father for one year. One son was charged 6% simple interest, but the less reliable son was charged 10% simple interest. If both pay off their loans on time and the father receives a total of \$2000 in interest, how much did each son borrow?

31. To purchase a car costing \$10,000, the buyer borrowed part of the money from the bank at 9% simple interest and the rest from her mother-in-law at 12% simple interest. If her total interest for the year was \$1080, how much did she borrow from the bank?

32. The owner of a small business, hoping to maintain good relations with two banks, invested a total of \$5000. Part of the money went to UltraBank, which pays 8% simple interest. The remainder went to Capital Savings, which pays 8% compounded continuously, yielding the equivalent of 8.3% simple interest. At the end of one year, the owner received a combined income of \$408.10. How much was invested at each bank?

Distance, Rate, and Time

33. An editor left home at 7:15 A.M. and took a bus to the office. The bus averaged 27 mph and arrived at the office building at 7:55 A.M. At 5:00 P.M., a co-worker drove the editor home, where she arrived at 5:24 P.M. What was the average speed for the drive home?

34. Two travelers left a restaurant in Salina, Kansas, and traveled in opposite directions on Interstate 70. If one driver averaged 65 mph and the other averaged 60 mph, how long was it before they were 400 miles apart?

35. A student left her college campus by bus to travel home, a distance of 472 miles. At the same time, her father left home in his car to meet the bus. Her father met the bus in 4 hours, and his average speed was 6 mph faster than that of the bus. Determine the speeds of the bus and the car.

36. Two cars left the Ranch Motel at the same time and traveled in opposite directions around a circular scenic loop. The first car averaged 40 mph, and the other car averaged 35 mph. Determine the distance around the loop if the first car returned to the Ranch Motel 30 minutes ahead of the second car.

37. One car left Valdosta averaging 55 mph, and another car left Atlanta averaging 70 mph. They traveled toward each other on Interstate 75. Atlanta is north of Valdosta by 300 miles. When did they meet if they both left at noon? How far south of Atlanta did they meet?

38. A mother jogs at a rate of 4.5 mph, and her daughter jogs at a rate of 5 mph. The mother begins her jog, and 15 minutes later her daughter begins her jog following the same route. How long will it take the daughter to overtake her mother, and how far will they have jogged?

Mixtures

39. The owner of The Nuttery wishes to mix 20 pounds of pecans selling for $1.80 per pound with some cashews selling for $2.40 per pound to make a mixture that will sell for $2 per pound. How many pounds of cashews should be used?

40. A seed company has fescue grass seed worth 55 cents per pound and bluegrass seed worth 95 cents per pound. How many pounds of each should be mixed to produce 100 pounds of seed mix that will sell for 70 cents per pound?

41. The Nature Center sells bird seed for 75 cents per pound and sunflower seed for $1.10 per pound. To the nearest pound, how many pounds of sunflower seed should be used to create a 50-pound mixture that sells for $0.90 per pound?

42. The Indian Tea Company makes a tea blend of cinnamon tea worth $2.50 per kilogram and black tea worth $2.00 per kilogram. How many kilograms of each should be used to produce a 20-kilogram blend whose total worth is $46.00?

43. A special grade of lime worth 90 cents per pound is to be mixed with 5 pounds of standard lime worth 50 cents per pound. If the mixture is to be worth 65 cents per pound, how many pounds of the special grade of lime should be used?

44. Two parts of pea gravel worth $10.00 per ton are mixed with one part of sand worth $8.00 per ton to form a base for road paving. If the bill for a project is $728.00, what is the approximate price per ton for the mixture?

Liquid Solutions

45. Cagle's Dairy mixed two grades of milk containing 3% and 4.5% butterfat, respectively, to obtain 150 gallons of milk that contained 4% butterfat. How many gallons of each were used in the mixture?

46. A precious-metals firm received an order for 160 pounds of an alloy that is 35% silver. The firm melted together two blocks of alloy, one that was 50% silver and one that was 25% silver. What was the weight of each original block?

47. A pharmacist has two decongestants, one containing 0.8% pseudoephedrine HCl, the other containing 1.4% of the same substance. A prescription calls for a decongestant containing 1.1% pseudoephedrine HCl. How much of each should be used to produce 4 ounces of a 1.1% solution?

48. A biologist wishes to have 15 gallons of an 80% formaldehyde solution. In her inventory, she has pure formaldehyde and some 50% formaldehyde solution. How many gallons of each should she mix to obtain the desired solution?

49. A worker has 64 ounces of a trisodium phosphate (TSP) solution to clean a wooden deck. However, the 20% concentration is too strong, so he must add water to dilute the solution to a 10% concentration. How many ounces of water should he add?

50. A leather finisher determines that his 8-ounce solution of 12% boric acid is too weak. How many ounces of pure boric acid must be added to increase the concentration to 20%?

Miscellaneous Applications

51. A person who wants to buy a computer decided to save for one year by investing money at 6% simple interest. However, at the end of the year, the price of the computer had risen to \$1800, which was \$236.50 more than the value of the investment. What was the original amount of the person's investment?

52. A furniture store advertises that no payment is due on any purchase for one year. However, customers must pay 11% simple interest on the amount of their purchases. If a certain customer buys a bedroom suite and pays \$1470.75 one year later, what was the cost of the furniture?

53. A college student bought his textbooks at the bookstore for \$209. He spent \$4 more on his biology book than on his English book. His math book cost \$16 more than his English book. How much did he spend on each book?

54. The cash register of Soft Hardware contains \$6.65 in just nickels and dimes. If there is a total of 85 coins, how many nickels are in the drawer?

55. Party Ideas has two grades of confetti. How many pounds of plain confetti worth \$4.50 per pound should be mixed with 6 pounds of sparkling confetti worth \$6 per pound in order to obtain a mixture worth \$5.40 per pound?

56. Candy Corner has two kinds of candy creams and wishes to make a mixture of 100 pounds to sell at \$3 per pound. If the two types are priced at \$2.50 per pound and \$3.75 per pound, how many pounds of each must be mixed to produce the desired amount?

57. In the figure, Elm and Maple are parallel streets that are intersected by 3rd Avenue. By a theorem in geometry, $\angle A$ and $\angle B$ are supplementary. The measure of the smaller angle is 20° less than one-fourth the measure of the larger angle. What is the measure of the angle between Elm Street and 3rd Avenue?

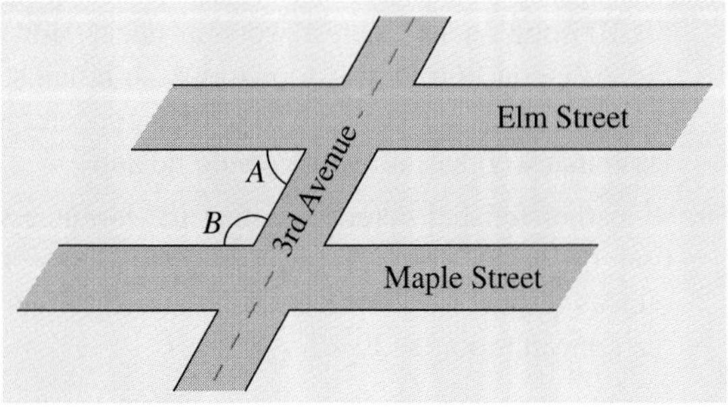

58. The fire-control computer on an M1-A1 tank plots a solution based on a right triangle in which the acute angles are complementary. The measure of one angle is 30° less than twice the measure of the other angle. What is the measure of each of the acute angles?

59. A chef has a 12% vinegar solution and a 25% vinegar solution. How much of each should he use to make 2 cups of a 15% vinegar solution?

60. A chemist added a solution that was 90% acid to an 18-gallon solution that was 60% acid. She obtained a solution that was 66% acid. How many gallons of 90% acid did she add?

61. The small digital computers used on road bikes measure time and distance, but the clock is active only when the bike is actually moving. The computer also calculates and displays the average speed for the entire trip. A cyclist rides 15 miles from 2:00 to 3:00 and stops for a sundae. At 3:30, she leaves for home and arrives at 4:20. To the nearest tenth, what average speed will the computer display?

62. A college student leaves her dorm at 10:00 A.M. and walks to the campus bookstore 2 miles away. She arrives at 10:30 A.M., buys her books, and leaves at 11:15 A.M. She then returns to the dorm, but she passes it and keeps walking an additional mile to the sorority house. What time does she arrive?

63. To obtain an \$80,000 mortgage on their home, a couple had to pay 2.5 discount points. A **discount point** is 1% of the mortgage amount. If the points were added to the mortgage amount, how much was the total loan?

64. To be able to buy a home, the purchaser had to borrow 90% of the purchase price. If her mortgage loan was \$108,000, what was the purchase price of the home?

Modeling with Real Data

65. The following table lists the speeds (in miles per hour) at which various animals can run. (Source: *World Almanac*.)

Animal	Speed (in mph)
Rabbit	35
Cat	30
Elephant	25
Coyote	43
Tortoise	0.17
Snail	0.03

Use the formula $d = rt$ to determine the time for each animal to complete a 100-yard dash.

66. In 1998 the average starting salary for graduates with a Bachelor's degree in nursing was \$32,927, which was a 1.7% decrease from 1995. (Source: Employment Research Institute.) What was the amount of decrease (to the nearest dollar) in average salaries from 1995 to 1998?

Data Analysis: *Court Case Dispositions*

The two pie graphs compare the dispositions of civil cases that were brought before the California Superior Courts in 1988–1989 and 1998–1999. (Source: Judicial Council of California.)

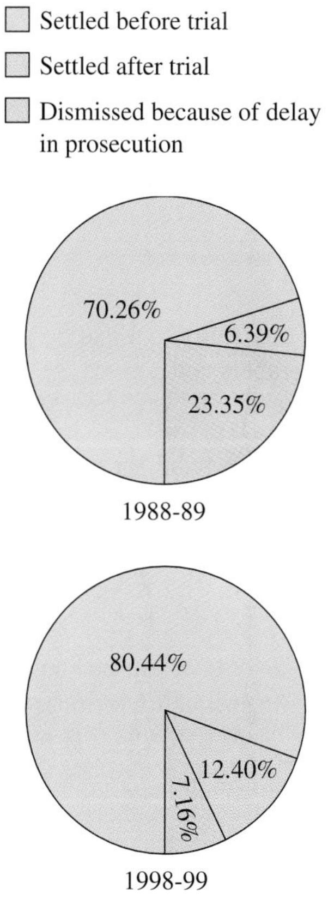

In the following exercises, percents and number totals may reflect some small rounding errors.

67. In 1998–1999, a total of 163,511 civil cases were filed with the California Superior Courts. Determine how many cases were filed in each of the categories shown in the pie graphs.

68. The total number of civil cases filed in 1998–1999 represents a 33.8% reduction of the total cases filed in 1988–1989. How many more cases were settled after trial in 1998–1999 than in 1988–1989?

69. The pie graphs indicate a reduction in the number of cases that were dismissed because of a delay in prosecution. What was the percent reduction?

70. The pie graphs indicate an increase in the percentage of cases that were settled before trial. Did the actual number of such cases increase? Explain.

Challenge

71. You have decided to sell your business. Your buyer has agreed to purchase your inventory at what it cost you to buy (wholesale price). The buyer asks, "What was your markup on all of your goods?" You reply that you marked up everything 25%. The buyer says, "Just give me a 25% discount on everything, and we'll be even." Will you agree to this plan? Why or why not?

72. Big Ed's Car Rental rents cars for \$25 a day plus \$0.40 per mile driven. Pop's Car Rental rents cars for \$33 a day plus \$0.30 per mile driven.

(a) Which agency provides the lowest cost for a one-day 100-mile trip?

(b) Let f represent the total cost of renting a car from Big Ed's, and let g represent the total cost of renting a car from Pop's. Let m represent the number of miles driven. Write f and g as functions of m.

(c) Suppose that you sketched the graphs of f and g. What is the significance of the point of intersection?

(d) Set the two functions equal to each other, and solve the resulting equation for m.

(e) What are the values of the left and right sides of the equation in part (d) for the solution? What do these values represent?

(f) From parts (d) and (e), what are the coordinates of the point of intersection of the graphs in part (c)?

(g) How would you use the information you obtained in parts (d) and (e) to make decisions about which car rental agency to use for *any* trip?

3.6 Linear Inequalities

Linear Inequalities in One Variable • Properties of Inequalities • Solving Inequalities Algebraically • Special Cases • Real-Life Applications

Linear Inequalities in One Variable

When we write an equation, we use an equality symbol to indicate that two algebraic expressions have the same value. In this section we use an inequality symbol to indicate that one algebraic expression is larger than or smaller than another expression. We call such a relation an **inequality.**

As is true for equations, an inequality may be true or false.

$-2 > -5$ True

$x < 9$ False if, for example, $x = 15$, but true if, for example, $x = -2$

$z \geq 0$ True if, for example, $z = 0$, but false if, for example, $z = -5$

The last inequality is true if z is replaced with 0 or any number greater than 0. We say that these replacements **satisfy** the inequality.

There are many different types of inequalities. In this section we discuss one type, a **linear inequality in one variable.**

Definition of Linear Inequality in One Variable

A **linear inequality in one variable** is an inequality that can be written in the form $Ax + B < 0$, where A and B are real numbers and $A \neq 0$.

This definition and all others in this section are also valid for the inequality symbols $>$, $\leq$, and $\geq$.

The following are linear inequalities.

$3x + 6 \geq 0$ $\quad t > -3 \quad$ $3(x - 5) \leq 6 - 3x$

The first example is written in the form $Ax + B \geq 0$. It can be shown that the other two can also be written in the defined form. An inequality written with the symbol $<$ or $>$ is called a **strict inequality.**

The following are inequalities that are *not* linear inequalities.

$x^2 - 5 < 0$ No exponent can exceed 1.

$\frac{3}{x} + 2x > 5$ No variable can be in a denominator.

$\sqrt{x} - 9 > 0$ There can be no variable in a radicand.

The definitions of **solution** and **solution set** for inequalities are similar to those for equations.

Definition of Solution and Solution Set of an Inequality

A **solution** of an inequality is any replacement for the variable that makes the inequality true. The **solution set** of an inequality is the set of all solutions of the inequality.

A solution of an inequality can be verified by substitution. However, inequalities typically have infinitely many solutions, so you will not be able to verify all of them.

Figure 3.8

The solutions of the inequality $x \geq 3$ are 3 and all numbers greater than 3. We can use a number line graph to help us visualize these infinitely many solutions. (See Fig. 3.8.)

LEARNING TIP

Be aware of the visual advantages of representing sets of numbers with a number line graph. Even when you are not asked to do so, sketch these number line graphs so that you can see what the numbers are.

We can also describe the solution set with the set-builder notation $\{x \mid x \geq 3\}$ or with the interval notation $[3, \infty)$. Recall that the bracket indicates that 3 is included in the set, and the symbol ∞ indicates that the interval extends without end. For $x > 3$, we write $(3, \infty)$, where the left parenthesis indicates that 3 is not included in the solution set.

We can use a table to estimate solutions of inequalities. The techniques are similar to those used in estimating solutions of equations.

Exploring the Concept

Make sure students understand the difference between the solution of the inequality and the solution of the associated equation.

Estimating Solutions of Linear Inequalities

The inequality $5 - 2x \leq 7$ states that the value of $5 - 2x$ must be 7 or less. The following table shows some values of $5 - 2x$ for selected values of x.

x	-3	-2	-1	0	1	2	3
$5 - 2x$	11	9	7	5	3	1	-1

From the table, we see that $5 - 2x$ is *equal* to 7 when $x = -1$, so -1 is a solution of the inequality. Moreover, the value of $5 - 2x$ is *less* than 7 for the x-values 0, 1, 2, 3, and all other real numbers that are greater than -1. Thus the solution set of the inequality is $[-1, \infty)$.

Figure 3.9

Note: Sketching a number line graph of a solution set provides a useful aid in visualizing the solutions. The graph in Figure 3.9 depicts the solution set of $5 - 2x \leq 7$.

A table can give us a good start in estimating the solutions of an inequality. However, because inequalities typically have infinitely many solutions, creating tables is usually not practical or accurate. Algebraic methods are needed to solve inequalities efficiently and precisely.

Students are usually relieved to hear that the algebraic method for solving inequalities is nearly identical to the algebraic method for solving equations.

Properties of Inequalities

As was true for equations, inequalities are **equivalent** if they have the same solution set. The Addition and Multiplication Properties of Inequalities may be applied to produce equivalent inequalities.

Addition Property of Inequalities

If $A < B$, then $A + C < B + C$.

Adding (or subtracting) the same quantity to (from) both sides of an inequality produces an equivalent inequality.

Before stating the Multiplication Property of Inequalities, we conduct an experiment.

Exploring the Concept

The Multiplication Property of Inequalities

Consider the following true inequalities.

$3 < 7 \qquad 4 > -3 \qquad -5 < -2$

First, we multiply both sides of each inequality by 3 and insert the correct inequality symbol in each result.

$3 < 7$	$4 > -3$	$-5 < -2$
$3 \cdot 3 \quad 7 \cdot 3$	$4 \cdot 3 \quad -3 \cdot 3$	$-5 \cdot 3 \quad -2 \cdot 3$
$9 < 21$	$12 > -9$	$-15 < -6$

Note that multiplying both sides by a positive number did not affect the direction of the inequality symbol.

Now we multiply both sides of each inequality by -3 and insert the correct inequality symbol in each result.

$3 < 7$	$4 > -3$	$-5 < -2$
$3(-3) \quad 7(-3)$	$4(-3) \quad -3(-3)$	$-5(-3) \quad -2(-3)$
$-9 > -21$	$-12 < 9$	$15 > 6$

Multiplying both sides by a negative number has reversed the direction of the inequality symbol.

An experiment involving division would yield the same results, which are summarized in the following property.

Multiplication Property of Inequalities

If $A < B$ and $C > 0$, then $AC < BC$.

Multiplying or dividing both sides of an inequality by the same *positive* number produces an equivalent inequality.

If $A < B$ and $C < 0$, then $AC > BC$.

Multiplying or dividing both sides of an inequality by the same *negative* number produces an equivalent inequality if the inequality symbol is reversed.

If $3x < -6$, then $x < -2$. Sometimes students want to reverse the inequality because they see a negative number on the right side.

Think About It

If $a = b$, then $b = a$. However, suppose that the weight in one pan of a balance is greater than the weight in the other pan. What happens if you exchange the weights in the two pans? What can you conclude about $a < b$ and $b < a$?

Note that multiplying both sides of an inequality by zero does *not* result in an equivalent inequality, and, of course, division by zero is undefined.

Solving Inequalities Algebraically

We can use the Addition and Multiplication Properties of Inequalities to solve inequalities algebraically. The solving routine is nearly identical to the solving routine for equations. The one exception is the need to reverse the inequality symbol when we multiply or divide both sides of an inequality by a negative number.

EXAMPLE 1

Solving an Inequality Algebraically

Use algebraic methods to solve the inequality $15 - 2x < -11$.

Solution

$$15 - 2x < -11$$

$$15 - 2x - 15 < -11 - 15 \quad \text{Addition Property of Inequalities}$$

$$-2x < -26$$

$$\frac{-2x}{-2} > \frac{-26}{-2} \quad \text{Multiplication Property of Inequalities—dividing both sides by } -2 \text{ reverses the inequality symbol.}$$

$$x > 13$$

The solution set is $(13, \infty)$. (See Fig. 3.10.)

Figure 3.10

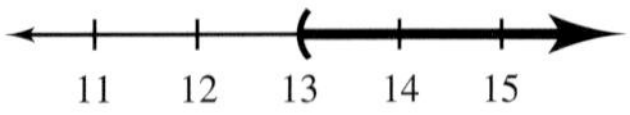

As with equations, it may be necessary to simplify both sides of an inequality by removing grouping symbols and combining like terms.

EXAMPLE 2

Simplifying Both Sides of an Inequality

Solve $2 + 3(x - 4) > x - 10$.

Solution

Compare the steps with those used in solving an equation.

$$2 + 3(x - 4) > x - 10 \quad \text{Simplify the left side.}$$

$$2 + 3x - 12 > x - 10 \quad \text{Distributive Property}$$

$$3x - 10 > x - 10$$

$$3x - 10 - x > x - 10 - x \quad \text{Addition Property of Inequalities}$$

$$2x - 10 > -10 \quad \text{Combine like terms.}$$

$$2x - 10 + 10 > -10 + 10 \quad \text{Addition Property of Inequalities}$$

$$2x > 0$$

$$\frac{2x}{2} > \frac{0}{2} \quad \text{Multiplication Property of Inequalities}$$

$$x > 0$$

Note that dividing both sides by 2 did not affect the inequality symbol. The solution set is $(0, \infty)$. (See Fig. 3.11.)

Figure 3.11

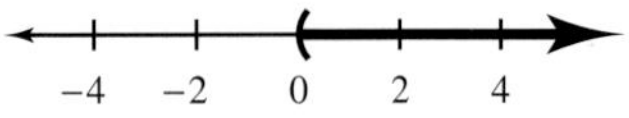

When solving equations, we often find it useful to clear fractions. The same is true for solving inequalities.

EXAMPLE 3

Clearing Fractions from an Inequality

Solve $\frac{1}{2}x + \frac{2}{3} < \frac{5}{6}x + 1$.

Solution

The LCM of all of the denominators is 6. To clear the fractions, multiply every term of both sides by 6.

Because we are multiplying by a positive number, the inequality symbol is not changed.

$$\frac{6}{1} \cdot \frac{1}{2}x + \frac{6}{1} \cdot \frac{2}{3} < \frac{6}{1} \cdot \frac{5}{6}x + 6 \cdot 1$$ Multiplication Property of Inequalities

$$3x + 4 < 5x + 6$$

$$3x + 4 - 5x < 5x + 6 - 5x$$ Addition Property of Inequalities

$$-2x + 4 < 6$$ Combine like terms.

$$-2x + 4 - 4 < 6 - 4$$ Addition Property of Inequalities

$$-2x < 2$$

$$\frac{-2x}{-2} > \frac{2}{-2}$$ Multiplication Property of Inequalities—dividing by -2 reverses the inequality symbol.

$$x > -1$$

The solution set is $(-1, \infty)$. (See Fig. 3.12.)

Figure 3.12

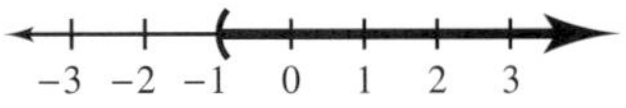

Special Cases

In certain special cases, the solution set of an equation is **R** or Ø. Similar special cases can occur with inequalities.

EXAMPLE 4

Special Cases

(a) $x + 2 < x - 1$

$$x - x + 2 < x - x - 1$$ Addition Property of Inequalities

$$2 < -1$$ False Inequality

The resulting false inequality is equivalent to the original inequality, so we conclude that there is no replacement that makes the original inequality true. The solution set is Ø.

(b) $x + 2 > x - 1$

$$x - x + 2 > x - x - 1$$ Addition Property of Inequalities

$$2 > -1$$ True inequality

The resulting true inequality is equivalent to the original inequality, so we conclude that every replacement makes the original inequality true. The solution set is **R**.

Real-Life Applications

Example 5 illustrates an application that can be solved with a linear inequality.

EXAMPLE 5

Minimum Bowling Average

To qualify for the semifinals in a bowling tournament, a bowler must have at least a 160 average for five games. Suppose the scores for your first four games were 172, 150, 148, and 162. What is the lowest score you can bowl in your fifth game and qualify for the semifinals?

Solution

Let $x =$ the fifth score. Your final average will be determined by adding your five scores and dividing by 5.

$$\frac{172 + 150 + 148 + 162 + x}{5} \geq 160$$ Your average must be at least 160.

$$\frac{632 + x}{5} \geq 160$$

$$\frac{5}{1} \cdot \frac{632 + x}{5} \geq 5 \cdot 160$$ Multiply both sides by 5.

$$632 + x \geq 800$$

$$632 + x - 632 \geq 800 - 632$$ Subtract 632 from both sides.

$$x \geq 168$$

You must bowl *at least* 168 for your fifth game in order to qualify.

Quick Reference 3.6

Linear Inequalities in One Variable

- A **linear inequality in one variable** is an inequality that can be written in the form $Ax + B < 0$, where A and B are real numbers and $A \neq 0$. The definitions involving $>$, $\geq$, and $\leq$ are similar.
- A **solution** of an inequality is any replacement for the variable that makes the inequality true. The **solution set** of an inequality is the set of all solutions.
- Set-builder notation, number line graphs, and interval notation can all be used to represent numbers that satisfy an inequality.

Properties of Inequalities

- Inequalities are **equivalent** if they have the same solution set.
- Addition Property of Inequalities: If $A < B$, then $A + C < B + C$. Adding (or subtracting) the same quantity to (from) both sides of an inequality produces an equivalent inequality.
- Multiplication Property of Inequalities:
  1. If $A < B$ and $C > 0$, then $AC < BC$. Multiplying or dividing both sides of an inequality by the same *positive* number produces an equivalent inequality.

2. If $A < B$ and $C < 0$, then $AC > BC$. Multiplying or dividing both sides of an inequality by the same *negative* number produces an equivalent inequality if the inequality symbol is reversed.

Solving Inequalities Algebraically

- The routine for solving inequalities is nearly identical to that for solving equations. The exception is the need to reverse the inequality symbol when we multiply or divide both sides by a negative number.
- As with equations, simplifying both sides and clearing fractions are steps in the solving process.

Special Cases

- Special cases occur when solving an inequality results in an equivalent inequality that is false (in which case the solution set is Ø) or true (in which case the solution set is **R**).

Speaking the Language 3.6

1. To indicate that an interval extends without end, we use the ______ symbol.
2. If $A < B$ and $C \neq 0$, then either $AC < BC$ or $AC > BC$, depending on whether C is ______ or ______.
3. In addition to set-builder notation, the solution set of an inequality can be represented with ______ or with ______.
4. Inequalities are ______ if they have the same solution set.

Exercises 3.6

Concepts and Skills

1. Explain the difference between $[a, b]$ and $(a, b]$.

2. For $\{x \mid x \geq 2\}$, is it correct to write $[2, \infty]$ or $[2, \infty)$?

In Exercises 3–8, represent the solution set with interval notation, set notation, and a number line graph.

3. $x < 2$
4. $x > -5$
5. $x \geq 3$
6. $x \leq -4$
7. $-3 < x \leq 2$
8. $-2 \leq x < 4$

In Exercises 9–16, solve the inequality algebraically.

9. $x + 3 < 8$
10. $x - 7 > -2$
11. $-7 + x \leq -7$
12. $x + 6 \geq -9$
13. $5x + 12 < 5 + 4x$
14. $4x \geq 3x + 2$
15. $3 - 6x \leq 2 - 7x$
16. $7 + 4x \geq 3x + 7$

17. In each part, the result of solving an inequality is given. What is the solution set? Why?

 (a) $-2 < 3$ (b) $3 < -5$ (c) $4 > 4$
 (d) $-7 \leq -7$ (e) $8 \geq -8$

18. If $-x < -z$, what is true about x? Explain.

CE In Exercises 19–26, assume that $a < b$ and insert the correct inequality symbol.

19. $2a$ ___ $2b$
20. $a - 6$ ___ $b - 6$
21. b ___ a
22. $-\frac{1}{2}a$ ___ $-\frac{1}{2}b$
23. $-a$ ___ $-b$
24. $\frac{a}{4}$ ___ $\frac{b}{4}$
25. $-3 + a$ ___ $-3 + b$
26. $a \div (-13)$ ___ $b \div (-13)$

CE In Exercises 27–32, assume that $a < b$ and determine the values of c for which the inequality is true.

27. $ac > bc$ **28.** $ac < bc$

29. $a + c < b + c$ **30.** $a - c < b - c$

31. $\frac{a}{c} < \frac{b}{c}$ **32.** $\frac{a}{c} > \frac{b}{c}$

33. Compare the algebraic method for solving an inequality with the algebraic method for solving an equation.

34. Explain the difference between the solution sets of $x < a$ and $x \leq a$.

In Exercises 35–42, solve the inequality algebraically.

35. $3x \leq 12$ **36.** $5x > 35$

37. $-x < 5$ **38.** $-4t \geq -32$

39. $-0.2y \geq 1.6$ **40.** $0.3n < -2.7$

41. $-\frac{2}{5}t > -6$ **42.** $6x < \frac{2}{3}$

In Exercises 43–76, solve the inequality algebraically. Except for special cases, write the solution set with interval notation.

43. $3x - 4 < 14$ **44.** $7x + 1 \geq 15$

45. $7 - 3t \leq -8$ **46.** $5 - y > 0$

47. $5x + 3 < 7 + 5x$ **48.** $3 - 2x \geq -2x + 1$

49. $4x - 4 \leq 7x - 13$ **50.** $2x + 7 < 7 + 5x$

51. $8 - x > x - 8$ **52.** $5 - t < t + 5$

53. $3 - x \geq 4x + 1$ **54.** $6 + 4x < 11 - 2x$

55. $3x - x < 2x$ **56.** $1 + 2x \leq 2x + 1$

57. $-(x - 2) \leq 2(x + 1)$ **58.** $3(x + 4) > 2x - 3$

59. $5 - 4x \geq 4(2 - x)$ **60.** $2 - 5x < -5(1 + x)$

61. $x - 4(x - 4) - 4x > x - 4$

62. $3(x - 1) - 4(2x + 3) > 0$

63. $5 + 4(x - 3) \geq 2x + 3(x - 1) + 4$

64. $5 - 5(5 - x) \leq 5(x - 1) + 2x$

65. $\frac{2}{3}x - 5 < \frac{5}{6}$ **66.** $\frac{5}{4}x - \frac{1}{2} \geq x - \frac{5}{4}$

67. $\frac{2x + 1}{-3} < 6$ **68.** $\frac{5x - 1}{-5} > \frac{3}{10}$

69. $x - \frac{1}{3} + \frac{5}{6}x \leq \frac{1}{2} - 3x$

70. $\frac{5}{12}x + \frac{1}{3} - x \leq \frac{7}{12} - \frac{x}{2}$

71. $\frac{3}{4}(x + 3) + 2x < 1$ **72.** $\frac{1}{4}(x - 2) < \frac{1}{2}x - \frac{2}{3}$

73. $0.2x + 1.7 \geq 2.94 - 5.3x$

74. $0.75x + 0.25 \leq x - 1.5$

75. $2.8(1.3x - 0.9) \leq 1.2 - 9.97x$

76. $0.5(0.6x + 3.1) \geq x - 1.73$

Real-Life Applications

77. The lengths of the sides of a triangle are represented by a, b, and c. What must be true about any two sides with respect to the third side? Write three true inequalities involving a, b, and c.

78. The length of a rectangle is 6 inches more than the width. The perimeter of the rectangle can be no more than 48 inches. What is the maximum width?

79. A pipe is at least 21 feet long, and you want to cut it into three pieces. The second piece is to be twice as long as the first piece, and the third piece is to be 1 foot longer than the second piece. What is the minimum length of the first piece?

80. You know that a rope is no more than 100 feet long. You need to cut the rope into three pieces. The second piece is to be three times as long as the first piece, and the third piece must be 18 feet long. What is the maximum length of the second piece?

81. Cross-country runners are awarded these quality points according to how the runners finish in a race.

Finish Position	Points
1	10
2	7
3	5
4	3
5	1

To win a trophy at the end of ten races, a runner must have an average of at least 7 points.

Here are your finish positions for your first nine races.

1 4 1 2 2 1 2 2 3

What is the minimum position in which you can finish the tenth race in order to win a trophy?

82. Your grades on tests 1, 2, and 4 were 82, 76, and 90. Unfortunately, you cut the third test and received a 0. If you have one test left to take, and if the passing grade for the course is 70, can you still pass the course?

83. A car rental agency rents cars for \$26.20 per day plus \$0.22 per mile driven. If your travel budget is \$200, what is the maximum number of miles you can drive during a 1-day rental?

84. Suppose that you are running a concession stand. A customer gives you \$18 and asks for six soft drinks and as many hot dogs as the remaining money will buy. If soft drinks are \$1.00 and hot dogs are \$1.75, what is the maximum number of hot dogs the person can buy?

85. A rich, eccentric aunt leaves some money in her will for Calvin, John, and Carrie. Calvin is to receive the least; Carrie is to receive the most. The aunt requires the three relatives to solve a puzzle before they can receive their inheritance. The will reads as follows.

> Calvin and John are given consecutive even integers. Carrie's number is 5 more than Calvin's number. If one-seventh of Calvin's number, one-third of John's number, and one-sixth of Carrie's number are added together, the sum exceeds 5. Determine the smallest integers that satisfy these conditions and multiply them by \$1000. The result is your inheritance.

How much will Calvin inherit?

86. Suppose you have a gift certificate worth \$20 for one long-distance phone call. If the charge is \$1.10 for the first minute and \$0.42 for each additional minute, what is the longest that you can talk?

Modeling with Real Data

87. Candidates for the United States Senate have been spending increasing amounts for their campaigns. The bar graph shows the average campaign expenses (in millions of dollars) for both winning and losing candidates in 1996 and 1998. (Source: Center for Responsive Politics.)

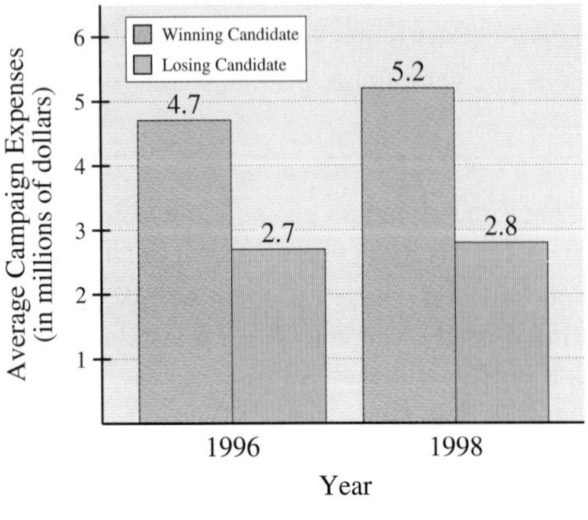

With x representing the number of years since 1996, the following functions can be used to model the data in the bar graph.

Winners: $W(x) = 0.25x + 4.7$

Losers: $L(x) = 0.05x + 2.7$

(a) Write an inequality to describe the years when an average of at least \$7.2 million will be spent by the winning candidates.

(b) Determine and interpret the solution set of your inequality in part (a).

(c) What visual evidence suggests that $W(x) \geq L(x)$ will always be true?

88. During the Christmas–New Year's Day holiday period in 1999, 44.6 million Americans traveled at least 100 miles from home. This figure was up 63% from 1990. (Source: AAA.) With t representing the number of years since 1990, the number of such travelers (in millions) can be modeled by $N(t) = 1.91t + 27.4$.

(a) Solve the inequality $N(t) \geq 40$.

(b) Interpret the solution set in part (a).

89. The following table shows the average numbers of runs scored per game in the American League and the National League for selected years in the period 1993–1999. (Source: ESPN.)

Year	*National*	*American*
1993	8.98	9.41
1996	9.66	11.03
1999	10.00	10.34

The average number of runs scored per game can be modeled by the following functions, where x is the number of years since 1993.

National: $N(x) = 0.17x + 9.04$

American: $A(x) = 0.16x + 9.80$

(a) Solve the inequality $N(x) \geq A(x)$.

(b) Interpret your solution set in part (a).

90. Referring to Exercise 89, write and solve an inequality to determine the years in which the difference between the average number of runs scored in the American League and that scored in the National League is at most 0.6.

Challenge

In Exercises 91–94, solve the inequality for x.

91. $5x + a \le b$

92. $cx < 10$ where $c < 0$

93. $a^2x \ge 8, a \ne 0$

94. $b - x > a$

95. Your instructor offers a one-million-dollar prize to anyone who can solve this puzzle: "There are three consecutive even integers, and at least two of them are positive. The sum of the first and third integers is no more than the second integer." Show why your instructor will never have to pay.

96. Suppose that solving an inequality results in an equivalent inequality in which both sides are 7. Explain your answers to the following questions.

(a) If you correctly conclude that the solution set is Ø, what is the inequality symbol? (There are two possibilities.)

(b) If you correctly conclude that the solution set is **R**, what is the inequality symbol? (There are two possibilities.)

3.7 Compound Inequalities

Double Inequalities • Conjunctions • Disjunctions • Real-Life Applications

Double Inequalities

A **compound inequality** consists of two or more inequalities connected with the word *and* or *or*.

$$2x < 9 \quad \textit{and} \quad x - 1 > 1$$

$$2 + x \ge 1 \quad \textit{or} \quad 3 \le 2 - x$$

$$-7 \le 2x + 1 \quad \textit{and} \quad 2x + 1 \le 3$$

Emphasize that the inequality symbols must be in the same direction and that the inequality must be correct if the middle component is omitted. Thus $3 < x > 5$ and $5 < x < -2$ are both incorrect.

The third example means that the value of the expression $2x + 1$ is between -7 and 3, inclusive. We can write this special case of a compound inequality in the more compact form $-7 \le 2x + 1 \le 3$. We call this form a **double inequality.**

Exploring the Concept

Solutions of Double Inequalities

The solutions of $-7 \le 2x + 1 \le 3$ are those replacements for x that make the double inequality true. Once again, a table can provide some guidance in estimating the solutions.

x	−5	−4	−3	−2	−1	0	1	2	3
$2x + 1$	−9	−7	−5	−3	−1	1	3	5	7

From the table, we observe that the value of $2x + 1$ is -7 when $x = -4$ and that the value of $2x + 1$ is 3 when $x = 1$. Thus -4 and 1 are solutions. We also see that the value of $2x + 1$ is between -7 and 3 for all x-values between -4 and 1. Thus the solution set can be described by the inequality $-4 \le x \le 1$, or, in interval notation, $[-4, 1]$.

To solve this same double inequality algebraically, our goal is to isolate the variable in the middle.

EXAMPLE 1

Solving a Double Inequality Algebraically

$$-7 \le 2x + 1 \le 3$$

$$-7 - 1 \le 2x + 1 - 1 \le 3 - 1 \qquad \text{Addition Property of Inequalities}$$

$$-8 \le 2x \le 2$$

$$\frac{-8}{2} \le \frac{2x}{2} \le \frac{2}{2} \qquad \text{Multiplication Property of Inequalities}$$

$$-4 \le x \le 1$$

Figure 3.13

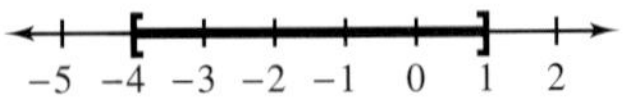

The last double inequality states that all numbers between -4 and 1, inclusive, are solutions. The solution set is $[-4, 1]$. (See Fig. 3.13.)

Conjunctions

A double inequality is a special case of a broader category of compound inequalities called **conjunctions** whose connective is the word *and.*

Suppose that A and B are the solution sets for the two inequalities of a conjunction. Then the solution set for the conjunction is the set S of numbers that belong to both A and B. We call the set S the **intersection** of sets A and B. In symbols, we write $A \cap B$.

EXAMPLE 2

LEARNING TIP

When you use number line graphs to assist you in forming unions and intersections, align the tick marks. This allows you to run your pencil vertically through the graphs to see where numbers are included or overlap.

Intersection of Solution Sets

Describe the solution set of the compound inequality

$$x < 4 \quad \textit{and} \quad x > -3$$

Solution

The solution set is the intersection of the individual sets. The graph of the solution set is as shown in Figure 3.14. In interval notation, the solution set is $(-3, 4)$.

Figure 3.14

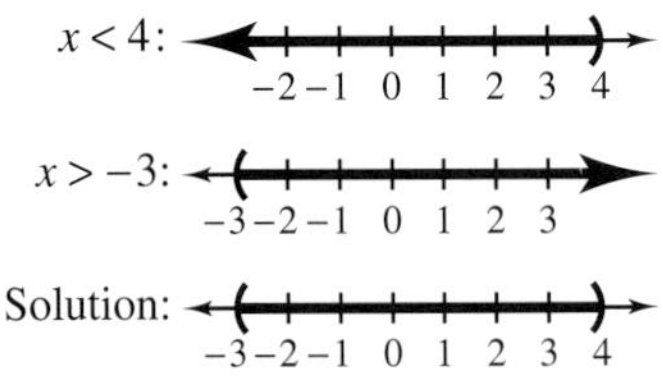

To solve a conjunction, we solve each individual inequality and then form the intersection of the individual solution sets.

EXAMPLE 3

For intersection, we look for the interval where the graphs overlap.

Solving a Conjunction

Solve the conjunction $3 - x < 5$ and $2x - 3 \le 7$.

Solution

Solve each inequality individually.

$$\begin{aligned} 3 - x &< 5 &\quad \text{and} \quad && 2x - 3 &\le 7 \\ 3 - x - 3 &< 5 - 3 &\quad \text{and} \quad && 2x - 3 + 3 &\le 7 + 3 \\ -x &< 2 &\quad \text{and} \quad && 2x &\le 10 \\ x &> -2 &\quad \text{and} \quad && x &\le 5 \end{aligned}$$

The connective is *and,* so the solution set for the compound inequality is the intersection of the two sets. The solution set is $(-2, 5]$. (See Fig. 3.15.)

Figure 3.15

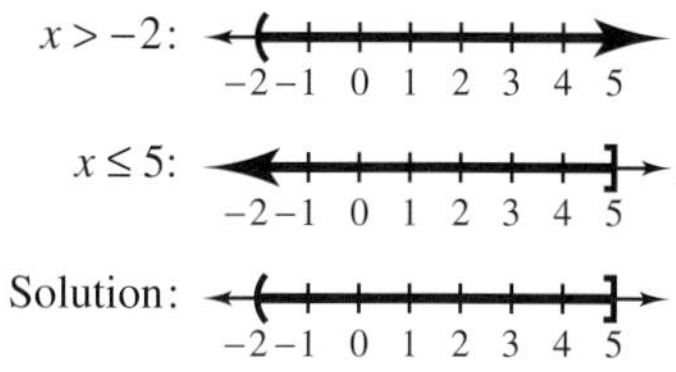

EXAMPLE 4

Solving a Conjunction

Solve the conjunction $x - 4 \ge -3$ and $2x - 6 > 0$.

Again, we look for the interval of overlap. This solution set is different from those in previous examples.

Solution

Solve each inequality individually.

$$\begin{aligned} x - 4 &\ge -3 &\quad \text{and} \quad && 2x - 6 &> 0 \\ x - 4 + 4 &\ge -3 + 4 &\quad \text{and} \quad && 2x - 6 + 6 &> 0 + 6 \\ x &\ge 1 &\quad \text{and} \quad && 2x &> 6 \\ x &\ge 1 &\quad \text{and} \quad && x &> 3 \end{aligned}$$

From the graphs, we can visually determine the intersection by noting the points common to both graphs. The solution set is $(3, \infty)$. (See Fig. 3.16.)

Figure 3.16

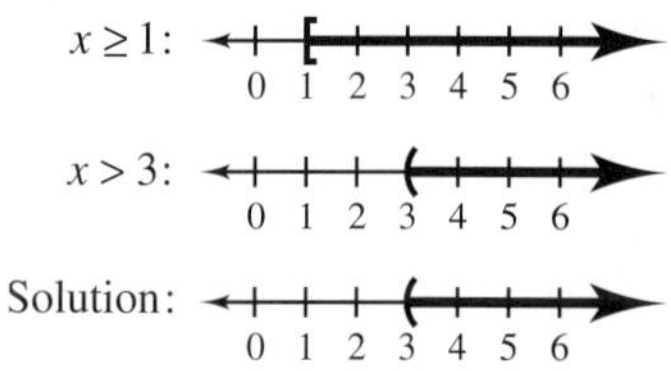

EXAMPLE 5

Solving a Conjunction

Solve the conjunction $x + 2 \le 0$ and $5 - x \le 3$.

Solution

Solve each inequality individually.

$$\begin{aligned} x + 2 &\le 0 &\quad \text{and} \quad && 5 - x &\le 3 \\ x + 2 - 2 &\le 0 - 2 &\quad \text{and} \quad && 5 - x - 5 &\le 3 - 5 \\ x &\le -2 &\quad \text{and} \quad && -x &\le -2 \\ x &\le -2 &\quad \text{and} \quad && x &\ge 2 \end{aligned}$$

Because there are no points common to both graphs in Figure 3.17, there are no solutions of the compound inequality. The solution set is Ø.

Figure 3.17

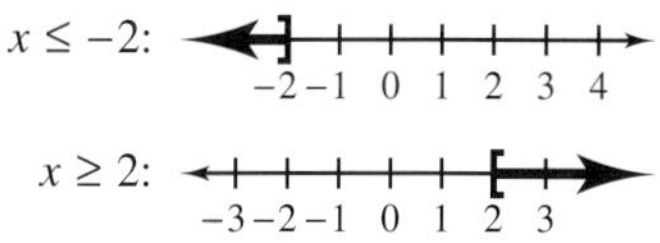

Disjunctions

A compound inequality whose connective is *or* is called a **disjunction.**

Suppose that A and B are the solution sets for the two inequalities of a disjunction. Then the solution set for the disjunction is the set S of numbers that belong to A or B (or both). We call the set S the **union** of sets A and B. In symbols, we write $A \cup B$.

EXAMPLE 6 Union of Solution Sets

Describe the solution set of the compound inequality

$$x > 1 \quad or \quad x < -2$$

For union, we look for all points that are shaded in either graph.

Solution

The solution set is the union of the individual sets. The graph of the solution set is as shown in Figure 3.18. In interval notation, the solution set is $(-\infty, -2) \cup (1, \infty)$.

Figure 3.18

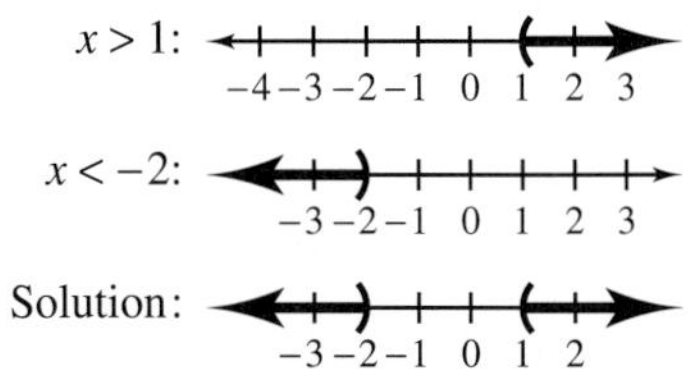

To solve a disjunction, we solve each individual inequality and then form the union of the individual solution sets.

EXAMPLE 7 Solving a Disjunction

Solve the disjunction $3x - 2 < 1$ or $x - 3 \geq 1$. That is, find all values of x for which at least one of the inequalities is true.

Students sometimes overlook the fact that the word *or* can mean "both."

Solution

Solve each inequality individually.

$$\begin{aligned} 3x - 2 &< 1 && \text{or} & x - 3 &\geq 1 \\ 3x - 2 + 2 &< 1 + 2 && \text{or} & x - 3 + 3 &\geq 1 + 3 \\ 3x &< 3 && \text{or} & x &\geq 4 \\ x &< 1 && \text{or} & x &\geq 4 \end{aligned}$$

Figure 3.19

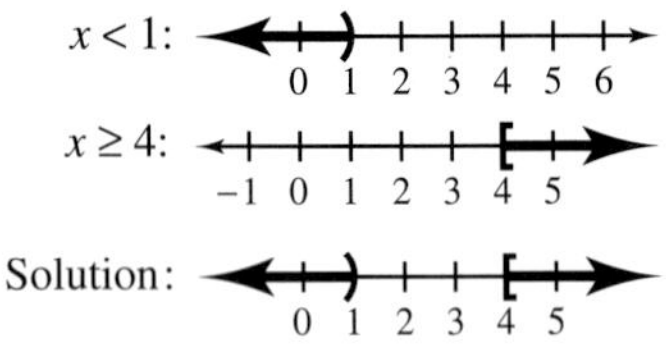

The connective is *or,* so the solution set for the compound inequality is the union of the two sets. From the graphs in Figure 3.19, we see the points that belong to at least one of the sets. The solution set is $(-\infty, 1) \cup [4, \infty)$.

EXAMPLE 8 Solving a Disjunction

This is an example of one of the special cases.

Solve the disjunction $x + 4 \geq 4$ or $x - 5 \leq 0$.

Solution

Solve each inequality individually.

$$\begin{aligned} x + 4 &\geq 4 && \text{or} & x - 5 &\leq 0 \\ x + 4 - 4 &\geq 4 - 4 && \text{or} & x - 5 + 5 &\leq 0 + 5 \\ x &\geq 0 && \text{or} & x &\leq 5 \end{aligned}$$

Think About It

What symbol would you use to describe the following sets?
(a) $\mathbf{R} \cap \varnothing$ (b) $\mathbf{R} \cup \varnothing$

We can see from the graphs in Figure 3.20 that every real number belongs to at least one of the two sets. The solution set for the compound inequality is **R**.

Figure 3.20

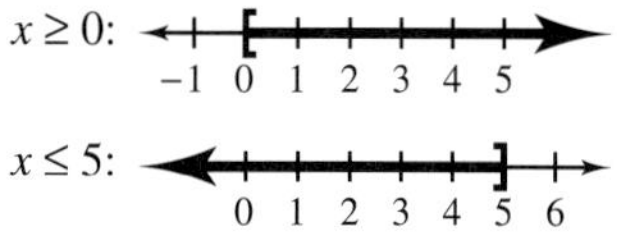

Real-Life Applications

In Example 9 we consider an application involving a compound inequality.

EXAMPLE 9

Fencing a Pasture

Farm Resources specializes in installing fencing. A customer who cannot spend more than \$6000 needs a 4-foot-high fence around a rectangular field. The width of the field must be 60% of the length, and the perimeter must be at least 800 feet. If Farm Resources charges \$5.00 per linear foot for this fencing, what are the minimum and maximum lengths that the field can be?

Solution

The minimum amount of fencing is 800 feet, and the maximum is the customer's budget divided by the cost per linear foot: $\frac{6000}{5} = 1200$. Thus the perimeter must be between 800 feet and 1200 feet, inclusive.

Figure 3.21

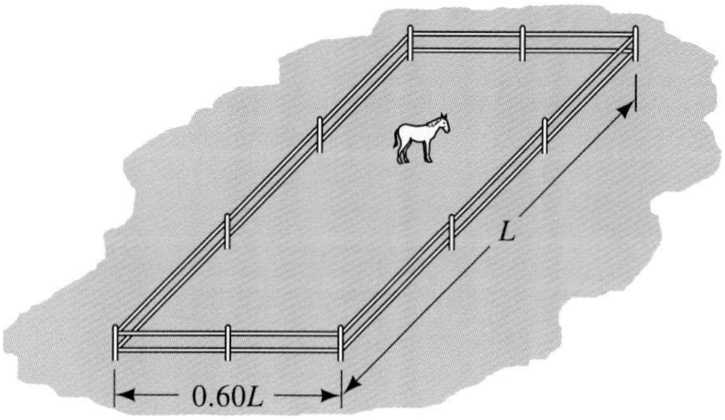

As shown in Figure 3.21, we let $L =$ the length of the field. Then the width is 60% of L, or $0.60L$.

$800 \le 2L + 2W \le 1200$ The perimeter of a rectangle is $2L + 2W$.

$800 \le 2L + 2(0.60L) \le 1200$ Replace W with $0.60L$.

$800 \le 2L + 1.2L \le 1200$

$800 \le 3.2L \le 1200$

$\frac{800}{3.2} \le \frac{3.2L}{3.2} \le \frac{1200}{3.2}$

$250 \le L \le 375$

Thus the length of the field will be at least 250 feet and at most 375 feet.

Quick Reference 3.7

Double Inequalities

- A **compound inequality** consists of two inequalities connected with the word *and* or *or*.
- A **double inequality** is a statement that the value of an expression is between two values. It is a special kind of compound inequality with the implied connective *and*.
- To solve a double inequality algebraically, apply the properties of inequalities to all three components at once.

Conjunctions

- A **conjunction** is a compound inequality whose connective is *and*. To solve a conjunction, solve each individual inequality and then form the *intersection* of the individual solution sets.

Disjunctions

- A **disjunction** is a compound inequality whose connective is *or*. To solve a disjunction, solve each individual inequality and then form the *union* of the individual solution sets.

Speaking the Language 3.7

1. The compound inequality $2 < x + 1 < 5$ is called a(n) ______.
2. The compound inequality $x < 3$ or $x > 7$ is an example of a(n) ______.
3. A conjunction consists of two or more inequalities connected with the word ______.
4. To solve a disjunction, we form the ______ of the individual solution sets; to solve a conjunction, we form the ______ of the individual solution sets.

Exercises 3.7

Concepts and Skills

1. Explain the use of the word *and* as it is used in compound inequalities.
2. Explain the use of the word *or* as it is used in compound inequalities.

In Exercises 3–10, determine whether the given numbers are solutions of the compound inequality.

For Exercises 3–10, remind students that for *and*, both parts must be true but that for *or*, only one part must be true.

3. $5x < 12$ and $3x > 3$; 2, 1
4. $-3x < 15$ or $2x > -5$; $-8, 4$
5. $2x + 7 < -3$ or $3x + 5 > 11$; $5, -7$
6. $4 - 3x < 2$ and $3 + 2x > -1$; $3, -6$
7. $0.5x > 4.5$ or $0.3x > 1.8$; 7, 11
8. $3 - 0.2x > 7$ and $1.2x > 14$; 8, 12
9. $0.5 \le 2x - 3 < 5$; 3.75, 5
10. $-3 < x + 2 \le 5$; $-5, 3, 0$
11. What is the difference between a conjunction and a disjunction?
12. For each of the given pairs of compound inequalities, explain the difference between their solution sets.

    (a) $x < -7$ or $x \ge 1$ $\quad$ $x < -7$ and $x \ge 1$

    (b) $x < 1$ or $x \ge -7$ $\quad$ $x < 1$ and $x \ge -7$

In Exercises 13–18, graph the solution set of each conjunction.

13. $x \ge -5$ and $x < 3$
14. $x \ge -6$ and $x \le 2$
15. $x < 5$ and $x \le 1$
16. $x \ge 0$ and $x > -3$

17. $x \le 4$ and $x > 6$

18. $x > -1$ and $x \le -7$

In Exercises 19–24, graph the solution set of each disjunction.

19. $x < -5$ or $x > 4$

20. $x \le -7$ or $x \ge 8$

21. $x \ge -3$ or $x > 4$

22. $x \le 2$ or $x < 7$

23. $x \le 4$ or $x > 0$

24. $x \ge -3$ or $x < 5$

25. For any real number c, describe the solution set of each compound inequality.

(a) $x \le c$ and $x \ge c$ (b) $x > c$ and $x < c$

(c) $x \le c$ or $x \ge c$ (d) $x > c$ or $x < c$

26. Tell whether each of the following pairs of compound inequalities are equivalent or not equivalent and explain why.

(a) $x > -3$ and $x < 7$ $\quad -3 < x < 7$

(b) $3 < x$ or $x < -2$ $\quad 3 < x < -2$

(c) $1 \le x$ and $x \ge 5$ $\quad 1 \le x \ge 5$

In Exercises 27–40, solve the inequalities algebraically.

27. $-4 < x + 3 < 7$

28. $-3 < x - 2 < 5$

29. $2 \le -3t \le 18$

30. $-14 \le -2y < -6$

31. $9 \ge 2x + 1 \ge -2$

32. $7 > 3x - 2 \ge -5$

33. $-5 \le -2x - 7 < 2$

34. $-3 < 5 - x < 4$

35. $-\frac{7}{4} \le \frac{3}{4}x - 1 < 2$

36. $3 > \frac{t}{2} + 1 \ge 0$

37. $-\frac{17}{6} < \frac{2}{3}x - \frac{1}{6} \le \frac{11}{3}$

38. $-2 < \frac{1}{2}x - \frac{1}{4} < \frac{7}{4}$

39. $6.5 \ge 0.75x + 0.5 > -2.5$

40. $-3.25 < 0.75 - 0.8x < 4.75$

41. Describe how the solution set of a disjunction can be **R** but the solution set of the corresponding conjunction can be Ø.

42. Explain why the solution set of a disjunction of linear inequalities can never be Ø.

In Exercises 43–48, solve each conjunction algebraically and write the solution set with interval notation.

43. $-x \le 5$ and $-x \ge -7$

44. $-x \ge -12$ and $-x \le 10$

45. $x + 1 \ge 5$ and $x - 2 < 12$

46. $x - 2 > 3$ and $x - 5 < 4$

47. $2x - 5 \le 7$ and $x + 4 > -7$

48. $2x + 3 \ge -1$ and $3x + 6 < 21$

In Exercises 49–54, solve each disjunction algebraically and write the solution set with interval notation.

49. $-x \ge 4$ or $-x \le -3$

50. $-x \ge 8$ or $-x \le -2$

51. $x + 4 < -3$ or $x + 4 > 3$

52. $x - 5 < -4$ or $x - 5 > 4$

53. $-4x - 5 > 0$ or $3x - 2 \ge 1$

54. $-3x - 6 > 3$ or $4x + 3 \ge 5$

In Exercises 55–70, solve each compound inequality and write the solution set with interval notation.

55. $2x - 5 \le -17$ and $3x > -27$

56. $-8x \le 24$ and $-2x \ge -8$

57. $x - 2 > 3$ or $x + 5 < 4$

58. $4 - x \ge 5$ or $2x - 3 > 5$

59. $7 - x \ge 10$ or $3x - 5 > 7$

60. $x + 3 > 5$ or $x + 4 < 0$

61. $-4x - 5 \ge 0$ and $3x - 2 \ge 1$

62. $2x - 3 \ge 5$ and $3x + 4 < -2$

63. $\frac{1}{2}x - 2 \ge 1$ or $\frac{2}{3}x - 1 \ge \frac{5}{3}$

64. $-\frac{1}{3}x + 1 \ge \frac{5}{3}$ or $\frac{4}{9}x - 1 < \frac{1}{3}$

65. $\frac{2}{5}x + \frac{1}{2} \ge \frac{1}{2}$ and $\frac{1}{4}x - \frac{1}{3} \le \frac{1}{6}$

66. $\frac{1}{3}x - 1 > -\frac{5}{3}$ and $\frac{1}{4}x + \frac{1}{2} < \frac{1}{4}$

67. $0.8 - 0.2x < -0.4$ or $0.75x - 0.25 < -0.25$

68. $-0.2x + 0.6 < 0$ or $4.6x - 11.5 \ge 6.9$

69. $1.8x + 3.6 < 5.4$ and $2.35x + 7.05 > -4.7$

70. $-19.2 < 4.8x - 8$ and $8.1x - 13.5 \le -10.8$

In Exercises 71–86, solve the compound inequality.

71. $-2x > x - 6$ and $2x - 4 \ge 1 + x$

72. $2 - x > 2 + 2x$ and $-2x > -3x$

73. $5x - 7 \ge 8$ and $2 - 3x < -4$

74. $-4x > x - 10$ and $x + 7 \le 4$

75. $5 - 2x < -9$ or $6x - 2 > 5x$

76. $-5x \le 10$ or $-3x \ge -4x - 2$

77. $x - 3 \le 1$ and $x - 4 \ge 0$

78. $4x \le 0$ and $x - 5 \ge -5$

79. $7 - x < 0$ or $x + 2 > 0$

80. $2x \ge 0$ or $2x \le 12 + 5x$

81. $-2x + 6 \geq -4$ and $2x - 1 > 9$

82. $3x - 5 < -5$ and $2(x - 4) > -8$

83. $-x \leq 20 - 3x$ and $4 - x \leq -6$

84. $3(2x + 1) \geq 5x$ and $2x - 5 \leq -11$

85. $5 - x \leq 5$ or $3(2x + 1) < 3 + 5x$

86. $1 - 2x < 7$ or $2x + 8 < 7 + x$

87. Which of the following is a correct interpretation of $3 < x > 8$? Explain your answer.

(i) x is between 3 and 8.

(ii) x is greater than 3 and x is greater than 8.

88. If x is a real number such that $0.99 < x < 1$, would you describe the number of solutions as none, one, or infinitely many? Why?

Real-Life Applications

89. A doll manufacturer makes happy dolls and grumpy dolls. Each day, the number of grumpy dolls made is two-thirds the number of happy dolls made. If the total daily production is between 75 and 100, what is the least number of happy dolls made? What is the maximum number of grumpy dolls made?

90. A senior, a junior, and a freshman work on the student newspaper. Each month, the junior works two-thirds the number of hours that the senior works, and the freshman works one-fifth the number of hours that the senior works. If their combined hours for the month are at least 56, what is the minimum number of hours that the junior works?

91. An automobile repair shop charges a flat rate of \$50 plus \$32 per hour for the mechanic's time. If a customer receives an estimate of between \$130 and \$170 for fixing his car, what is the estimate of the mechanic's time?

92. The distance from a lawyer's home to the shopping center is twice the distance from her home to her office. The office and the shopping center are 4 miles apart. The lawyer leaves her home, goes to her office and to the shopping center, and returns home. If the total distance traveled was between 11.8 and 12.4 miles, how far is it from the lawyer's home to her office?

93. A drugstore sells toothbrushes for \$1.30 and toothpaste for \$3.00. For an average month the store sells twice as much toothpaste as toothbrushes, and combined sales for these items are between \$211.70 and \$255.50. How many toothbrushes are sold in an average month?

94. A builder constructs rectangular swimming pools. For every design, the width is one-half of the length. If the perimeter of the pool can range between 120 feet and 192 feet, what pool widths does the builder offer?

95. In each of a player's ten basketball games, the number of free throws he made was two-thirds the number of two-point field goals he made. The player never scored less than 8 points and never scored more than 16 points. What are the minimum and maximum number of two-point field goals the player made?

96. In a small community the number of registered Republicans is 1.2 times the number of registered Democrats. It is known that 60% of the Democrats and 40% of the Republicans will vote in the special election. If the expected voter turnout is between 216 and 243, how many registered Democrats are there?

97. A packet of marigold seeds has twice as many seeds as a packet of delphinium seeds. The germination rate for marigolds is 90%, and the germination rate for delphiniums is 50%. A gardener plants all the seeds and expects to have a total of between 92 and 115 plants. How many marigold seeds were planted?

98. A light-bulb manufacturer finds that the number of defective 60-watt bulbs averages 1.4 times the number of defective 100-watt bulbs. On a given day the total number of defective bulbs is between 36 and 48. How many 100-watt bulbs are defective?

Modeling with Real Data

99. During the period 1996–1999, violent incidents of "road rage" increased by an average of 7% every year. (Source: AAA.) For this period, the number N of road rage incidents can be modeled by $N(x) = 155x + 1835$, where x is the number of years since 1995.

(a) Write an inequality to determine the year in which the number of road rage incidents was *at least* 2300.

(b) Solve the inequality in part (a) and write the solution in interval notation.

(c) Interpret your solution in part (b).

100. Refer to Exercise 99.

(a) Solve the inequality $2145 \le N(x) \le 2455$. Write the solution set in interval notation.

(b) Interpret your solution in part (a).

101. "Venture capital" is money that very wealthy individuals and institutions invest in businesses that are just starting up. Although such investments tend to be risky, the bar graph shows dramatic increases in venture capital during the period 1996–1999. (Source: PricewaterhouseCoopers.)

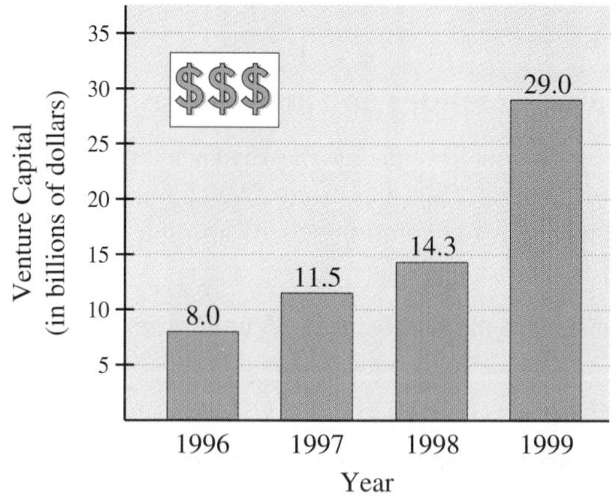

(a) For the period 1996–1998, you can verify that $V_1(x) = 3.15x + 4.97$ is quite accurate for modeling the amount V of venture capital x years after 1995. By what percent is the actual figure for 1999 higher than that predicted by the model?

(b) A similar model for the 1996 and 1999 data is $V_2(x) = 7x + 1$. Make a table of values for V_1 and V_2 for the years 1996–1999.

(c) Use the table in part (b) to determine the year for which V_1 and V_2 appear to be equally good models.

102. Refer to the table in part (b) of Exercise 101.

(a) According to the table, which model yields the lower estimate of venture capital investment in 1995?

(b) Which model appears to be the better model for 1997 and 1998?

Data Analysis: *ATM Transactions*

Increases in surcharges for the use of ATMs caused a decrease in the number of ATM transactions in the late 1990s. However, financial institutions are expected to continue to add terminals at a constant rate. The bar graph shows the number of ATM terminals (in thousands) in 1999 and the projections for the years 2000–2003. (Source: Speer and Associates, Inc.)

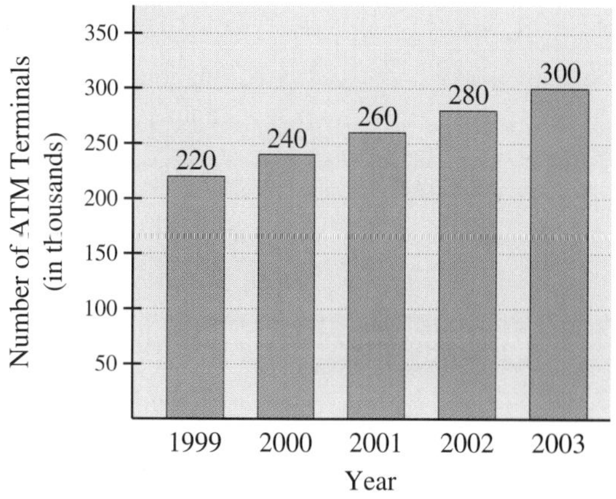

The function $T(x) = 20x + 200$ models the number T of terminals, where x is the number of years since 1998.

103. Solve the inequality $240 \le T(x) \le 280$ and write the solution in interval notation.

104. Interpret the solution of the inequality in Exercise 103.

105. In 1999 the average number of monthly ATM transactions per terminal was 3491. Suppose that this average remains constant for the period 1999–2003. Write a function $f(x)$ that represents the total number of transactions at all terminals for any year in the period.

106. To what level can the average number of monthly ATM transactions per terminal decrease in 2003 and there still be about the same total number of transactions at all terminals as there were in 1999?

Challenge

107. Given $x + k < 5$ and $k - x < 1$, for what values of k is the solution set $\varnothing$?

108. Given $x + k > -2$ or $k - x > -1$, for what values of k is the solution set $\mathbf{R}$?

In Exercises 109–112, solve the compound inequality algebraically and write the solution set with interval notation.

109. $3x + 2 \le x + 8 \le 2x + 15$

110. $2x - 1 < x + 1 < 7 - x$

111. $x < x + 1 < x + 2$

112. $x + 1 < x < x + 2$

3.8 Absolute Value: Equations and Inequalities

Graphical and Algebraic Interpretations • Algebraic Methods • Special Cases • Real-Life Applications

Graphical and Algebraic Interpretations

Equations and inequalities sometimes involve absolute values. The following are examples of **absolute value equations** and **absolute value inequalities.**

$$|2x - 3| = 5 \qquad |1 - x| \geq 9 \qquad |x^2 + 4x - 12| < 10$$

We can use a number line to develop generalizations about the solutions of absolute value equations and inequalities.

Exploring the Concept

Solutions of Absolute Value Equations and Inequalities

Recall that the symbol $|x|$ can be interpreted as the distance on a number line between the point whose coordinate is x and the origin. We can use this interpretation to develop generalizations about the solution sets of absolute value equations and inequalities.

1. The equation $|x| = 15$ means that the number x is 15 units from the origin. (See Fig. 3.22.)

   The solutions are -15 and 15.

Figure 3.22

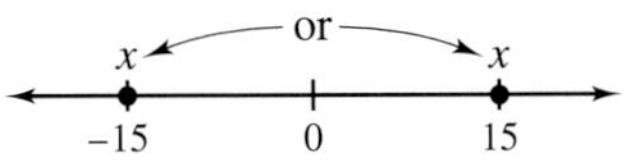

2. The inequality $|x| > 15$ means that the number x is more than 15 units from the origin. (See Fig. 3.23.)

   The solutions of $|x| > 15$ lie in the interval $(-\infty, -15) \cup (15, \infty)$.

Figure 3.23

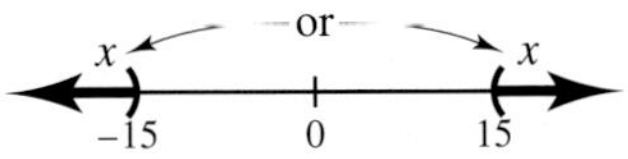

3. The inequality $|x| < 15$ means that the number x is less than 15 units from the origin. (See Fig. 3.24.)

   The solutions of $|x| < 15$ lie in the interval $(-15, 15)$.

Figure 3.24

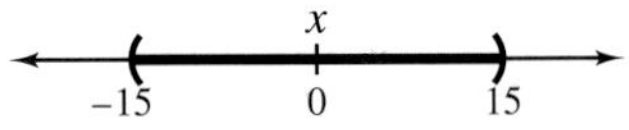

These methods can be generalized for any expression A and any positive number C.

When solving absolute value sentences, most students find these number line interpretations easier to use than the definition of absolute value.

	Graphical Interpretation	*Algebraic Interpretation*
1. $\|A\| = C$	A represents a number that is C units from the origin. (See Fig. 3.25.)	$A = C$ or $A = -C$
2. $\|A\| > C$	A represents a number that is more than C units from the origin. (See Fig. 3.26.)	$A < -C$ or $A > C$
3. $\|A\| < C$	A represents a number that is less than C units from the origin. (See Fig. 3.27.)	$A > -C$ and $A < C$

Figure 3.25

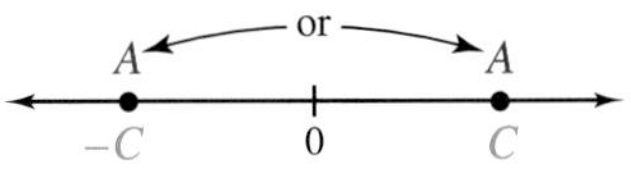

Figure 3.26

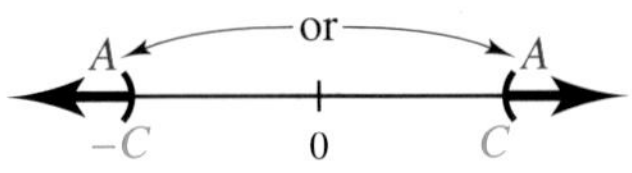

Figure 3.27

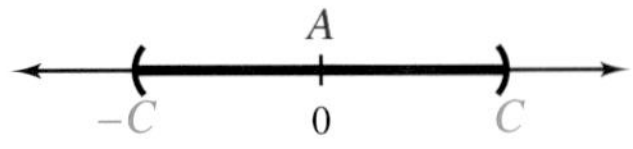

LEARNING TIP

Rather than memorizing the various possibilities for absolute value inequalities, use the definition of absolute value and graphs to see the associated inequalities.

In Example 1 we illustrate these algebraic interpretations with the assistance of number line graphs.

EXAMPLE 1

The absolute value equation leads to two linear equations; the absolute value inequalities lead to compound inequalities.

Combining Graphs and Algebraic Methods

Solve each absolute value equation or inequality.

(a) $|3 - 2x| = 17$ (b) $|2x + 1| > 9$ (c) $|3x - 5| < 4$

Solution

(a) Because $|3 - 2x| = 17$, the expression $3 - 2x$ represents a number that is 17 units from the origin. (See Fig. 3.28.)

Figure 3.28

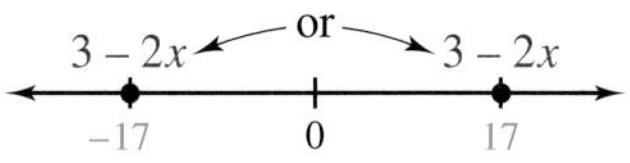

The figure suggests the following equations. Solve each one.

$$\begin{aligned} 3 - 2x &= -17 \quad \text{or} \quad & 3 - 2x &= 17 \\ -2x &= -20 \quad \text{or} \quad & -2x &= 14 \\ x &= 10 \quad \text{or} \quad & x &= -7 \end{aligned}$$

The solutions are 10 and -7.

(b) Because $|2x + 1| > 9$, the expression $2x + 1$ represents a number that is more than 9 units from the origin. (See Fig. 3.29.)

Figure 3.29

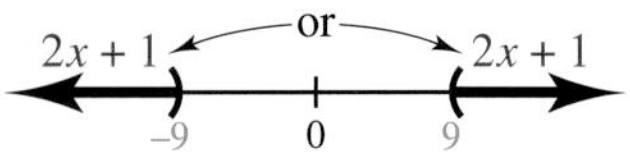

The figure suggests the following inequalities. Solve each one.

$$\begin{aligned} 2x + 1 &< -9 \quad \text{or} \quad & 2x + 1 &> 9 \\ 2x &< -10 \quad \text{or} \quad & 2x &> 8 \\ x &< -5 \quad \text{or} \quad & x &> 4 \end{aligned}$$

The solution set is $(-\infty, -5) \cup (4, \infty)$.

(c) Because $|3x - 5| < 4$, the expression $3x - 5$ represents a number that is less than 4 units from the origin. (See Fig. 3.30.)

Figure 3.30

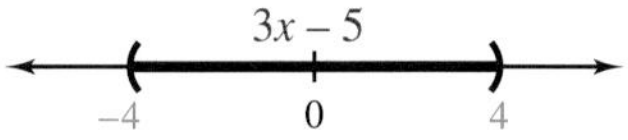

The figure suggests the following double inequality.

$$\begin{aligned} -4 < 3x - 5 &< 4 \\ 1 < 3x &< 9 \\ \frac{1}{3} < x &< 3 \end{aligned}$$

The solution set is $(\frac{1}{3}, 3)$.

Algebraic Methods

We can use algebraic methods to solve compound equations or inequalities arising from absolute value equations or inequalities. We normally begin by isolating the absolute value expression.

EXAMPLE 2

Always simplify each side (clear fractions if necessary) and isolate the absolute value expression.

Solving Absolute Value Equations and Inequalities Algebraically

Solve each of the following algebraically.

(a) $|2x + 5| - 4 = -2$

(b) $|1 - 3x| - 6 \geq 1$

(c) $9 - |x - 2| > 1$

Solution

(a)
$$|2x + 5| - 4 = -2$$
$$|2x + 5| - 4 + 4 = -2 + 4 \quad \text{Add 4 to both sides.}$$
$$|2x + 5| = 2 \quad \text{The equation is in the form } |A| = C.$$

Now write $|2x + 5| = 2$ as two linear equations and solve each one.

$$2x + 5 = 2 \quad \text{or} \quad 2x + 5 = -2 \qquad \text{If } |A| = C, \text{ then } A = C \text{ or } A = -C.$$
$$2x = -3 \quad \text{or} \quad 2x = -7$$
$$x = -\frac{3}{2} \quad \text{or} \quad x = -\frac{7}{2}$$

The solutions are $-\frac{3}{2}$ and $-\frac{7}{2}$.

(b)
$$|1 - 3x| - 6 \geq 1$$
$$|1 - 3x| - 6 + 6 \geq 1 + 6 \quad \text{Add 6 to both sides.}$$
$$|1 - 3x| \geq 7 \quad \text{The inequality is in the form } |A| \geq C.$$

The resulting inequality can be written as a disjunction.

$$1 - 3x \leq -7 \quad \text{or} \quad 1 - 3x \geq 7 \qquad \text{If } |A| \geq C, \text{ then } A \leq -C \text{ or } A \geq C.$$
$$-3x \leq -8 \quad \text{or} \quad -3x \geq 6$$
$$x \geq \frac{8}{3} \quad \text{or} \quad x \leq -2$$

Figure 3.31

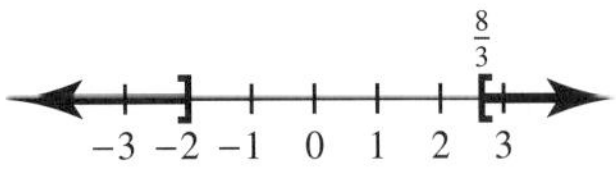

The solution set is $(-\infty, -2] \cup \left[\frac{8}{3}, \infty\right)$. (See Fig. 3.31.)

(c)
$$9 - |x - 2| > 1$$
$$9 - |x - 2| - 9 > 1 - 9 \quad \text{Subtract 9 from both sides.}$$
$$-|x - 2| > -8 \quad \text{Multiply both sides by } -1 \text{ and reverse the inequality symbol.}$$
$$|x - 2| < 8 \quad \text{The inequality is in the form } |A| < C.$$

The resulting inequality can be written as a conjunction.

$$x - 2 > -8 \quad \text{and} \quad x - 2 < 8 \qquad \text{If } |A| < C, \text{ then } A > -C \text{ and } A < C.$$
$$x > -6 \quad \text{and} \quad x < 10$$

Figure 3.32

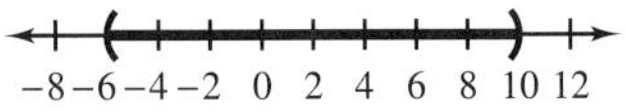

The solution set is $(-6, 10)$. (See Fig. 3.32.)

Think About It

Can $(-\infty, -2] \cup (2, \infty)$ represent the solutions of an absolute value inequality of the form $|Ax + B| > C$ or $|Ax + B| \geq C$?

If two expressions have the same absolute value, then either the expressions are equal or they are opposites. Thus an equation of the form $|A| = |B|$ is equivalent to the disjunction $A = B$ or $A = -B$.

EXAMPLE 3 **Solving Equations of the Form $|A| = |B|$**

Solve the following equations.

(a) $|4 - x| = |3 - 2(5 - x)|$ (b) $|4 - x| = |x - 7|$

Solution

(a)
$$|4 - x| = |3 - 2(5 - x)|$$
$$|4 - x| = |3 - 10 + 2x| \quad \text{Distributive Property}$$
$$|4 - x| = |-7 + 2x| \quad \text{The equation is in the form } |A| = |B|.$$

$$\begin{array}{rcl} 4 - x = -7 + 2x & \text{or} & 4 - x = -(-7 + 2x) \\ -3x = -11 & \text{or} & 4 - x = 7 - 2x \\ x = \dfrac{11}{3} & \text{or} & x = 3 \end{array}$$

If $|A| = |B|$, then $A = B$ or $A = -B$.

The solutions are $\frac{11}{3}$ and 3.

(b) $|4 - x| = |x - 7|$

$$\begin{array}{rcl} 4 - x = x - 7 & \text{or} & 4 - x = -(x - 7) \\ 4 - x = x - 7 & \text{or} & 4 - x = 7 - x \\ 11 = 2x & \text{or} & 4 = 7 \quad \text{False} \\ x = \dfrac{11}{2} & & \end{array}$$

The solution is $\frac{11}{2}$.

Special Cases

The absolute value of a number is never negative, and the absolute value of 0 is 0. These limitations can present some interesting special situations when we solve absolute value equations and inequalities.

EXAMPLE 4

Special Cases

Solve the following equations and inequalities. In each case, note the value of the constant on the right side. What general conclusions can you draw?

(a) $|x - 6| = -12$
(b) $|x - 6| > -12$
(c) $|2x + 14| = 0$
(d) $|2x + 14| < 0$

This example does not include all possible special cases. Ask students to think of others. An example is $|A| < C$, where $C < 0$.

Solution

(a) Note that $|x - 6|$ cannot be negative. Therefore, $|x - 6| = -12$ is false.

Generalization: An equation $|A| = C$, where $C < 0$, has no solution.

(b) Because $|x - 6| \geq 0$, the inequality $|x - 6| > -12$ is always true.

Generalization: An inequality $|A| > C$, where $C < 0$, is satisfied by all real numbers.

(c) Observe that $|2x + 14| = 0$ only if $2x + 14 = 0$, which implies that $x = -7$.

Generalization: An equation $|A| = 0$ implies that $A = 0$.

(d) Again, $|2x + 14|$ cannot be negative. Therefore, $|2x + 14| < 0$ is false.

Generalization: An inequality $|A| < 0$ has no solution.

Real-Life Applications

Absolute value inequalities typically arise in applied situations in which the difference between two quantities is less than or greater than a fixed amount. The following example illustrates how absolute value plays a role in matters of *tolerance* in measurements.

EXAMPLE 5

Temperature Tolerances

A nuclear reactor shuts down automatically if the temperature in the reactor differs from 1200°F by more than 30°. Write and solve an absolute value inequality that describes these conditions.

Solution

Let T = reactor temperature.

$|T - 1200| \leq 30$ The difference between T and 1200 is at most 30.

$T - 1200 \geq -30$ and $T - 1200 \leq 30$

$T \geq 1170$ and $T \leq 1230$

In interval notation, the solution set is [1170, 1230]. The reactor will operate as long as the temperature is between 1170° and 1230°, inclusive.

Quick Reference 3.8

Graphical and Algebraic Interpretations

- We can interpret the solutions of an absolute value equation or inequality in terms of their distance from 0 on the number line.
- If $C > 0$, then
  1. $|A| = C$ is equivalent to $A = C$ or $A = -C$.
  2. $|A| > C$ is equivalent to $A > C$ or $A < -C$.
  3. $|A| < C$ is equivalent to $A < C$ and $A > -C$.

Algebraic Methods

- To solve a compound absolute value equation or inequality algebraically:
  1. Isolate the absolute value expression.
  2. Translate the absolute value equation or inequality into an equivalent conjunction or disjunction.
  3. Solve each component of the compound equation or inequality and form the union or intersection of the solution sets.

Special Cases

- If $C = 0$, then
  1. $|A| = C$ is equivalent to $A = 0$.
  2. $|A| > C$ is true for every real number except for the solution of $A = 0$.
  3. $|A| < C$ has no solution.
- If $C < 0$, then
  1. $|A| = C$ has no solution.
  2. $|A| > C$ is true for every real number.
  3. $|A| < C$ has no solution.

Speaking the Language 3.8

1. The inequality $|x| < 6$ can be written as a(n) ________.
2. The inequality $|x| < a$ has no solution if a is ________ or if a is ________.
3. The equation $|x| = c$ has two solutions if c is ________.
4. The equation $|A| = |B|$ can be written as a(n) ________.

Exercises 3.8

Concepts and Skills

1. Interpret $|x - 2| = 5$ in terms of distance on a number line.

2. How does the solution set for $|x + 3| = 5$ differ from the solution set for $x + 3 = 5$?

In Exercises 3–14, solve the equation.

3. $|x| = 12$
4. $|x - 8| = 9$
5. $|2x - 5| = 10$
6. $|3 - 4x| = 12$
7. $|5x - 2| = 0$
8. $|5 - x| = 0$
9. $|2x - 10| + 7 = 7$
10. $|6 - x| - 9 = 5$
11. $|x + 4| - 3 = 7$
12. $6 + |3 - x| = 7$
13. $-|x + 10| = -15$
14. $10 - |x + 8| = 17$

15. What is the difference between the solution sets of $|x| \geq 5$ and $|x| \leq 5$?

16. Explain why the inequality $|x| < -5$ has no solution.

In Exercises 17–24, solve the inequality algebraically. Write the solution set with interval notation.

17. $|1 - 2t| < 9$
18. $|3y + 2| < 14$
19. $|7 + x| \leq 0$
20. $|x - 1| \leq 0$
21. $|4 - x| - 5 \leq 2$
22. $|2x - 1| - 3 \leq 10$
23. $|x + 5| < -3$
24. $|4 + 3x| < -5$

In Exercises 25–32, solve the inequality algebraically. Write the solution set with interval notation.

25. $|3x + 2| \geq 7$
26. $|5 - 2x| \geq 11$
27. $|x - 5| > 0$
28. $|3 + x| > 0$
29. $|3x - 2| \geq -5$
30. $|5 - 7x| \geq -1$
31. $|4t + 3| + 2 > 10$
32. $|2 - 3t| - 9 > 8$

CE In Exercises 33–36, x represents any real number. Write an absolute value inequality to describe each situation.

33. The distance from 5 to x is less than 7.
34. The distance from -5 to x is less than or equal to 4.
35. The distance from -7 to x is at least 10.
36. The distance from 7 to x is greater than 12.

In Exercises 37–42, solve the inequality.

37. $|x| < 12$
38. $|x - 8| \geq 9$
39. $15 - |x - 6| > 6$
40. $-|x + 10| \leq -15$
41. $|2x - 10| + 7 \geq 7$
42. $10 - |x + 8| \leq 17$

CE In Exercises 43–46, write an absolute value inequality in x to describe each graph.

Think of these in terms of distance from the middle of the interval.

43. 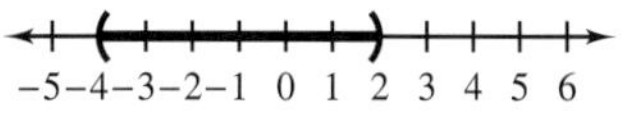

44. 

45. 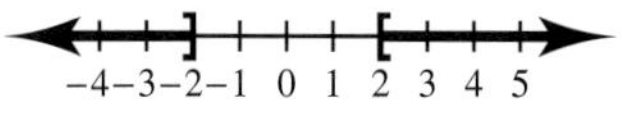

46. 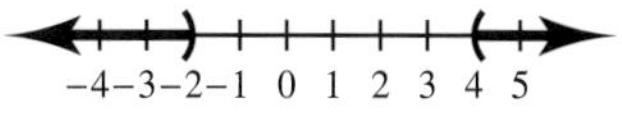

In Exercises 47–62, solve the equation.

47. $|2k - 3| = 8$
48. $|4 - 3m| = 15$
49. $|y + 1| - 3 = 7$
50. $|2x + 3| - 5 = 12$
51. $|t| + 8 = 5$
52. $|-t| - 7 = -12$

53. $5 - 3|x - 5| = -13$

54. $2 - |x + 1| = -7$

55. $|t + 4| - 3 = -23$

56. $|m - 23| + 9 = 4$

57. $|3x - 4| - 4 = -4$

58. $|5 - 2x| + 5 = 5$

59. $5 + 3|2x + 1| = 20$

60. $7 - 2|1 - 3x| = -13$

61. $\left|\frac{5 - 3x}{4}\right| = 3$

62. $\left|x - \frac{2}{3}\right| = \frac{3}{4}$

In Exercises 63–68, solve the given equation.

63. $|x - 1| = |x - 19|$ **64.** $|x + 3| = |x - 3|$

65. $|2x - 5| = |x|$ **66.** $|t| = |3 - 4t|$

67. $|x - 2| = |2 - x|$ **68.** $|1 - 2x| = |2x - 1|$

69. What is the first step in solving $-3|x - 2| \geq 12$?

70. Explain why each given pair of inequalities are equivalent or not equivalent.

(a) $|x| > 5$ $\quad x > 5$

(b) $|x - 1| \leq 3$ $\quad -3 \leq x - 1 \leq 3$

In Exercises 71–88, solve each inequality algebraically. Write the solution set with interval notation.

71. $-5|2x + 3| < -35$

72. $-2|7 - x| > -16$

73. $5 - |4 - 3x| > 2$

74. $12 - |6 - 5p| < 9$

75. $|4x + 3| + 9 \geq 4$

76. $|3x + 2| \leq -3$

77. $|3y - 5| + 4 \geq 10$

78. $|2x - 7| - 3 \leq 8$

79. $|3y - 5| + 6 \leq 6$

80. $|2k + 7| - 7 \geq 8 - 15$

81. $|5 - 3x| + 7 > 9$

82. $|4 - 3x| - 5 < 2$

83. $5 - |x - 2| \geq 7$

84. $24 - |m + 3| \leq 27$

85. $5 - 2|2x - 1| > -1$

86. $4 - 3|x + 5| < -8$

87. $\left|\frac{2x + 1}{4}\right| > 1$ **88.** $\left|\frac{4x - 3}{2}\right| + 5 \leq 7$

89. If x_1 and x_2 are two numbers on a number line, the distance between them is $|x_1 - x_2|$. Why is it necessary to take the absolute value of the difference?

90. Solving the inequality $|-x| < 3$, a student states that $|-x| = x$ and therefore $x < 3$. Explain the error in the student's reasoning.

Modeling Conditions

In Exercises 91–94, describe the given conditions with an absolute value inequality and write the corresponding double inequality.

91. A carpenter cuts a board to a length of 12 feet with no more than a 2% error. (Let L = the length of the cut board.)

92. A scale is never off its measurement by more than 3 pounds. The scale indicates that a person weighs 135 pounds. (Let w = the person's weight.)

93. A pipe 3 inches in diameter must have a tolerance of one-sixteenth of an inch so that the pipe can fit into a specific hole in the wall. (Let d = the diameter of the pipe.)

94. A freezer is most efficient when the temperature is 5°F plus or minus 6°. (Let t = temperature of the freezer.)

Real-Life Applications

In Exercises 95–98, write and solve an absolute value inequality that describes the given conditions.

95. On a psychological test, the actual score s differs from the expected score of 50 by no more than 6 points. What is the range of actual scores?

96. For submarine duty, a sailor's height h can differ from 6 feet by no more than 2 inches. What range of heights is required?

97. A medical clinic considers a patient's body temperature t to be normal if it deviates from 98.6°F by no more than 0.5°F. What is the normal temperature range?

98. A housing inspector approves residential wiring if the incoming voltage v deviates from 220 volts by less than 5 volts. What range of voltages would pass inspection?

Data Analysis: *Health Insurance*

In 1999 about 43.1 million Americans were without health care insurance. This represents a 36% increase from 1987. The trend can be modeled by $n = 0.94x + 31.8$, where n is the number of uninsured persons and x is the number of years since 1987.

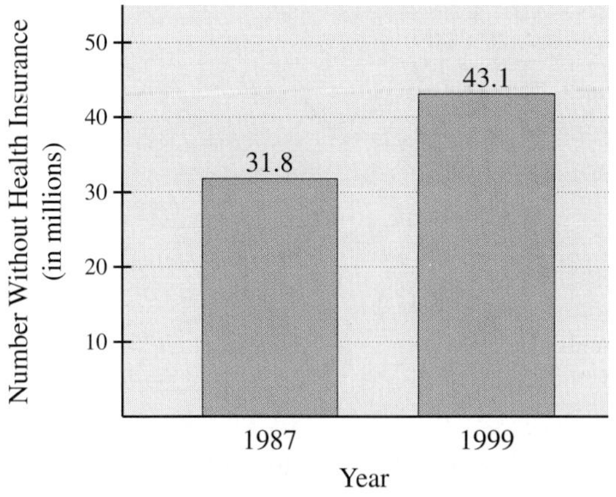

99. Use the model equation to estimate the number of uninsured persons for the years 1989, 1991, 1993, 1995, and 1997. (Round results to the nearest hundred thousand.)

100. If the given trend continues, what percent of the population will be uninsured in 2001? (Assume that the U.S. population in 2001 is 280 million.)

101. Using your result in Exercise 100, determine whether the percent of uninsured persons has increased or decreased since 1987. (Assume that the U.S. population in 1987 was 241 million.)

102. In 2001, how many more people would need to be insured in order for the percentage of Americans who are uninsured to be reduced to 10%?

Challenge

103. For what values are the following true? Explain your reasoning.

(a) $|x - 2| = x - 2$

(b) $|x - 2| = -(x - 2)$

(c) $|x - 2| = |2 - x|$

In Exercises 104–107, solve the given equation.

104. $|x - 4| = x + 2$

105. $|t - 3| = t + 3$

106. $|x - 1| = 7 - (x - 1)$

107. $|t + 5| = -t$

In Exercises 108 and 109, determine the values of x for which the given equation is true.

108. $x + |x| = 2x$

109. $x + |x| = 0$

110. Show that there are four possible equations that can be derived from the absolute value equation $|x - 2| = |2x + 1|$. Explain why two of the equations are equivalent to the other two.

3 • Chapter Project — Health Insurance

In a 1999 survey of voters, 61% stated that Congress should move quickly to help the uninsured obtain health insurance. (Source for all data in this project: American Medical Association.)

In this project, we will look more closely at the uninsured population and try to draw some conclusions about who they are. To do this, consider dividing the class into groups, with each group working on a specific set of data.

Group 1: Uninsured Families

As the bar graph shows, some of the 43.1 million uninsured persons are members of families headed by an employed person, and some are members of families in which the head of the family is not employed.

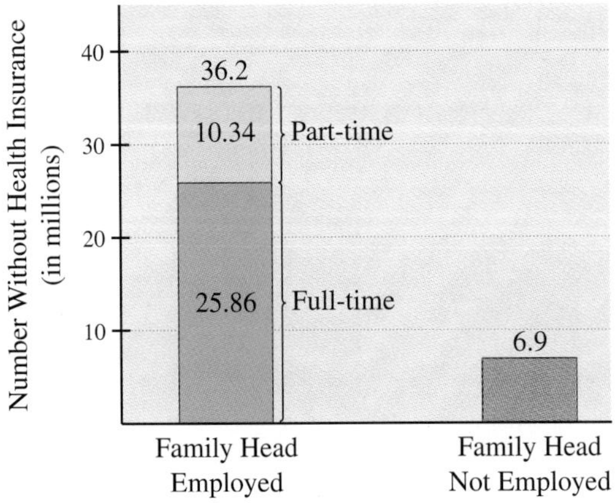

1. Assuming an average family size of four, how many uninsured families are headed by full-time workers?
2. What percentage of all uninsured persons live in families in which the family head is not employed?

Group 2: Age and Income

Of the 43.1 million uninsured persons, 32.4 million are "nonelderly adults" in the age range 18–64 with a median family income of \$20,000. The bar graph shows the percentages of these uninsured people who fall into various age groups.

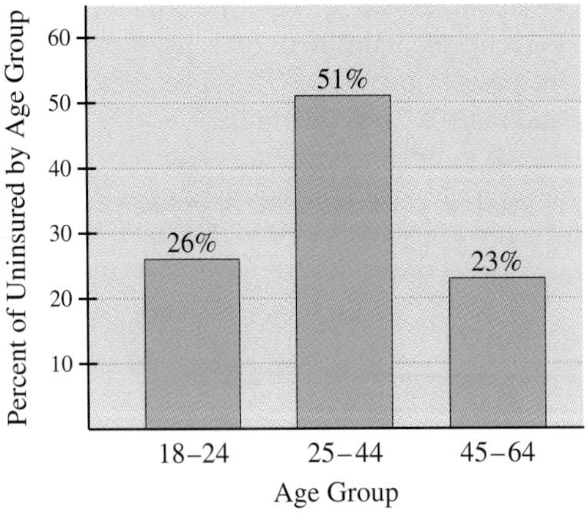

3. Calculate the number of uninsured persons in each age range.
4. Compare the number of uninsured persons in the 25–44 age group to the other age groups combined.

Group 3: Employer Size

Of the 32.4 million nonelderly adults (ages 18–64) who are uninsured, 24.6 million have either full-time or part-time jobs. The bar graph shows the sizes of the businesses for which these uninsured persons work.

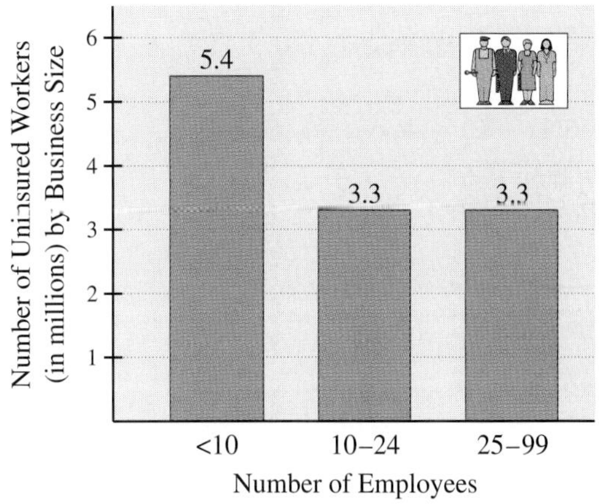

5. What percentage of uninsured workers are employed by businesses that have fewer than 100 employees?
6. If affordable health insurance were offered to all workers in businesses that have at least 100 employees, what total number of nonelderly adults would remain uninsured?

Conclusions

Bringing together the analysis of data from each group, develop a "profile" of the typical uninsured nonelderly adult. Consider employment status, age and income, and size of business for those who work.

Chapter 3 Review Exercises

Section 3.1

In Exercises 1 and 2, state whether each equation is a linear equation.

1. $3x^2 - 4x = 7$

2. $\frac{2}{3}x - 5 = 9$

In Exercises 3 and 4, verify that the given number is a solution of the equation.

3. $3x - 7 = 2(x - 5);\ -3$

4. $\frac{2}{7}(x - 5) + \frac{1}{3}(x + 7) = x + 2;\ -\frac{23}{8}$

5. What is the goal as you solve a linear equation algebraically?

6. Explain the use of the Addition and Multiplication Properties of Equations in solving a linear equation.

In Exercises 7–12, solve the given equation.

7. $y - 3 = 8$

8. $6t + 3 = 3 + 5t$

9. $-b = 8$

10. $4x + x = 20$

11. $4 - x = -2$

12. $2x - 5 = x + 3$

Section 3.2

In Exercises 13–20, solve each equation.

13. $\frac{-y}{3} = 10$

14. $4(5x - 6) - 6(3x - 2) = 4$

15. $0.2n + 0.3 = 0.3n - 5.2$

16. $\frac{3}{2}p + \frac{1}{6} = \frac{5}{3}$

17. $7x - 5 = 12x - 5 - 4x$

18. $\frac{x - 2}{6} = \frac{x}{3}$

19. $4(2s + 1) = s + 3(2s - 1)$

20. $\frac{2}{3}(k + 2) + \frac{1}{4}(k - 4) = k - \frac{1}{6}$

In Exercises 21–24, solve the equation. Then state whether the equation is an identity, an inconsistent equation, or a conditional equation.

21. $3x - 2(1 - 4x) = 6x + 5$

22. $3c - 4(1 + c) = 5 - c$

23. $2 - (2y - 3) = 5 - 2y$

24. $0.4q - 1.4 + 0.6q = 1 - 0.8q + 1.4$

In Exercises 25 and 26, determine k so that the equation meets the given condition.

25. $3 - 2x = kx - 2 + 3x$
The equation is inconsistent.

26. $-2(k - x) = 2x + 6$
The equation is an identity.

Section 3.3

In Exercises 27–32, solve each formula for the given variable.

27. $V = LWH$ for H

28. $R_T = \frac{R_1 + R_2}{2}$ for R_2

29. $\frac{ac}{x} + \frac{cd}{y} = 6$ for d

30. $A = \frac{1}{2}h(a + b)$ for b

31. $E = IR$ for I

32. $h = vt - 16t^2$ for v

In Exercises 33 and 34, solve each equation for y.

33. $3x - 4y - 25 = 0$

34. $\frac{4}{5}x = \frac{2}{3}y - \frac{2}{5}$

35. Solve the formula $A = P + Prt$ for r. Then use the formula to determine the simple interest rate (r) required for an investment of \$2400 (P) to be worth \$2556 (A) at the end of one year (t).

36. To obtain a building permit for a small office building, the developer must have at least 4500 square feet of paved parking area. (See the accompanying figure.) If the trapezoidal area in front of the building is paved for parking, will the building permit requirement be met? What is the actual paved area?

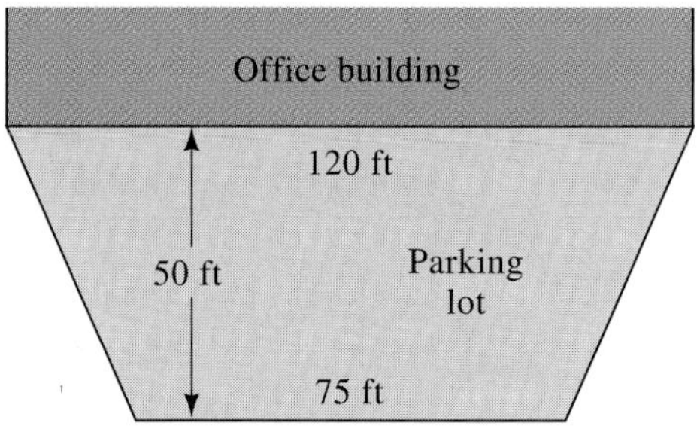

37. A circular goldfish pool is surrounded by a sidewalk. (See figure.) The circumference of the outside of the sidewalk is 43.98 feet, and the circumference of the inside of the sidewalk is 31.42 feet. What is the width of the sidewalk?

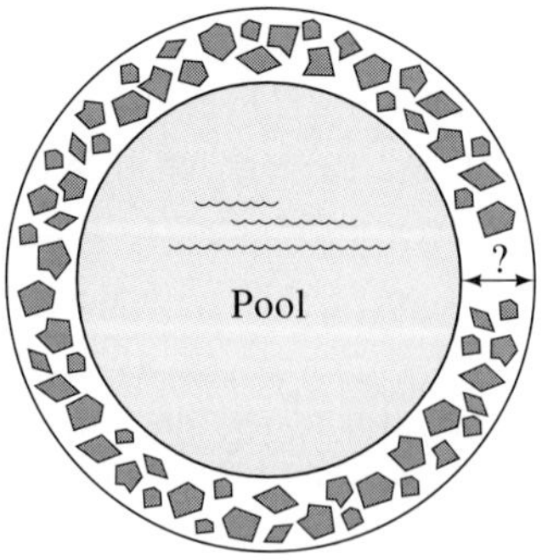

Section 3.4

38. One number is 3 less than another number. If one-fifth of the larger number is added to one-half of the smaller number, the result is 9. What is the smaller number?

39. A 100-foot rope is to be cut into three pieces whose lengths are consecutive integers. One foot of rope will be left over. How long should each piece of rope be?

40. A sailboat jib is not quite a right triangle. Angle A is $10°$ less than one-half of angle B. Angle C is $15°$ more than angle B. What are the measures of the three angles? (See figure.)

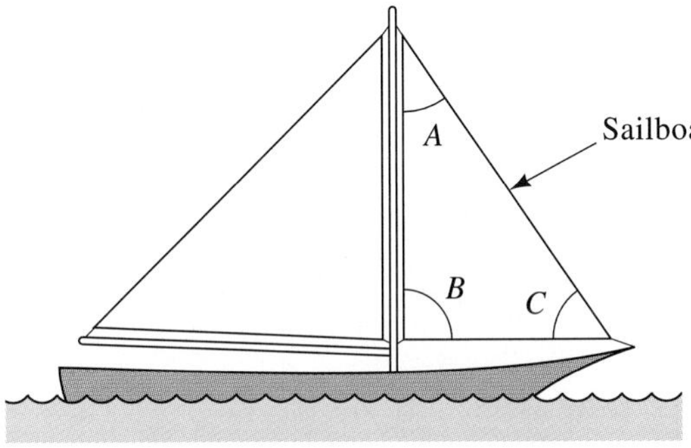

41. A coin collection jar is full of just nickels and dimes. If there are 355 coins valued at \$24.25, how many nickels are in the jar?

42. Determine two consecutive odd integers so that the difference between twice the first and 20 more than the second is 31.

Section 3.5

43. Menswear Unlimited advertised, "All suits are 40% off." If a suit is on sale for \$93, what was the original price?

44. How much money was invested at 6% simple interest if the value of the investment was \$1653.60 at the end of 1 year?

45. Stewart Construction Company rents a large crane for \$275 per day and provides an operator for \$20 per hour. The crane rental bill for July 29 was \$455. How long did the operator work?

46. A 10% salt solution is to be mixed with a 20% salt solution to obtain 15 gallons of a 12% salt solution. How many gallons of each should be used?

47. Two trucks leave Kansas City at midnight and travel in opposite directions. The truck traveling east is averaging 6 mph less than the truck going west. At 3:30 A.M. the trucks are 469 miles apart. What is the average rate of each truck?

48. Two investments yield 7.5% and 9.3%, respectively. If a total of \$8000 is invested and the income for the year is \$695.40, how much is invested at each rate?

49. Two angles are complementary. The difference between one-third of the larger angle and the smaller angle is $10°$. What is the measure of each angle?

50. At a chemical plant, a mixing vat receives water from a large pipe and an alkaline solution from a smaller pipe. The larger pipe delivers liquid 1.5 times as fast as the small pipe. The small pipe is opened at noon, and the larger pipe is opened at 1:45 P.M. At 5:30 P.M., the tank contains 133.5 gallons of liquid. At what rate does each pipe deliver liquid?

Section 3.6

In Exercises 51–53, for each graph write the set with interval notation.

51. 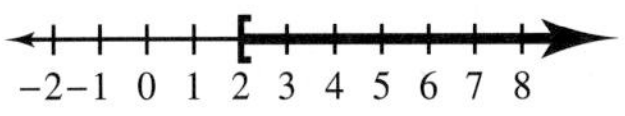

52. 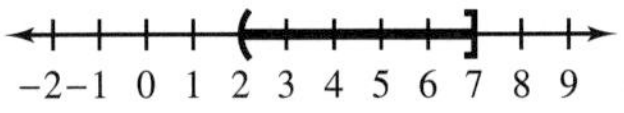

53. 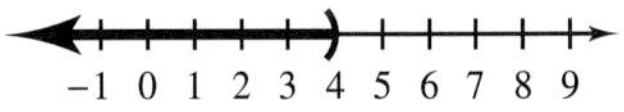

In Exercises 54–56, illustrate each set on the number line.

54. $(-\infty, 7]$

55. $[0, 3)$

56. $(-2, \infty)$

In Exercises 57–68, solve the inequality.

57. $3x - 4 < 14$

58. $7 - 3t \leq -8$

59. $4x - 4 \leq 7x - 13$

60. $5x - 3 \geq 6x + 5$

61. $-17 < 4x - 7 \leq -8$

62. $6 + 3(2x - 5) \geq 2x + 3(x - 2) - 7$

63. $-\frac{2}{3}(x - 4) \leq \frac{2x - 3}{-4}$

64. $-3 < \frac{2x - 1}{-1} < 5$

65. $7 > x + 3 > -4$

66. $\frac{1}{6}t + \frac{2}{3} < \frac{1}{3}t - \frac{1}{2}$

67. $-8 \leq -4x < 20$

68. $\frac{t}{-3} \geq -9$

69. The width of a rectangular lot is 7 meters less than the length. The perimeter of the lot can be no more than 254 meters. What is the maximum width?

70. A shipping company requires that the total of the length, width, and height of a box not exceed 153 inches. If the height is 75% of the length and the length is twice the width, what are the possible values for the dimensions?

71. Frugal Car Rental charges \$20 per day plus 10 cents per mile. Big C Car Rental charges \$26 per day plus 9 cents per mile. For what range of miles is Frugal Car Rental the best choice for a 1-day rental?

72. If 3 is subtracted from twice a positive integer, the result is negative. Show that this problem has only one solution.

Section 3.7

In Exercises 73–84, solve each compound inequality and write the solution set with interval notation.

73. $-t \geq -2$ or $\frac{t}{4} > -1$

74. $-5 \leq t - 7 < 5$

75. $\frac{x + 2}{4} \leq 1$ and $\frac{x + 2}{4} \geq -1$

76. $2x - 3 \geq 5$ and $x > 0$

77. $x + 1 > 0$ or $3x - 4 < 0$

78. $t + 3 \leq 5$ and $4t < -4$

79. $-5x + 2 \geq 12$ or $3x + 4 \geq 25$

80. $x - 6 > 4$ or $x + 5 \leq 20$

81. $3x + 4 < x$ or $x + 2 > 8 - x$

82. $4x - 3 \leq 3x$ and $5 - 2x \leq 1$

83. $3x - 4 > 5x - 2$ and $3x - 2 < 2x + 3$

84. $x + \frac{2}{3} < -\frac{1}{3}$ and $\frac{3}{2}x + 1 > \frac{1}{2}$

85. How much money must be invested at 8% simple interest to realize a yearly income between \$150 and \$200?

86. Equal amounts of money are invested in two funds that pay 7% and 8.5% simple interest. The first investment must earn at least \$119 per year, and interest from the other investment must not exceed \$238 per year. How much should be invested in each fund?

Section 3.8

In Exercises 87–92, solve each equation.

87. $|-y| = 8$

88. $|3 + 2x| = 18$

89. $|2x - 3| - 7 = 5$

90. $|3x + 5| = -2$

91. $2 + |x - 2| = 3 - |x - 2|$

92. $|-x| + 5 = 1$

In Exercises 93–98, solve each absolute value inequality. Write the solution set with interval notation.

93. $|3x + 5| < 14$

94. $|4 - x| > 5$

95. $7 - |c - 5| \leq 4$

96. $8 - |5 - d| \leq 12$

97. $|-t| \geq 7$

98. $|t| \leq -7$

99. Write a double inequality that is equivalent to the inequality $|x - 3| \leq 5$.

100. Write an absolute value inequality whose solution set is $(-\infty, -4) \cup (8, \infty)$.

101. Two brothers were born exactly 3 years apart. If one boy is 9 years old, how old is his brother? Write an absolute value equation and solve it.

102. A wrestler can remain in his weight class as long as his present weight does not vary by more than 5 pounds. If his current weight is 136 pounds, what is the range of weights in his weight class?

103. If a number is reduced by 1, the absolute value of the result is less than the original number. Show that the number cannot be negative.

Looking Back

The following exercises review concepts and skills that you will need in Chapter 4.

1. Evaluate each expression for the given value(s) of the variable(s).

(a) $5x - 2y$; $x = 2, y = \frac{3}{2}$

(b) $-x + 11$; -7

2. Which pairs of numbers are represented by points of the graph of the function $f(x) = 1 - 2x$?

(i) $(3, -5)$

(ii) $(-1, 1)$

(iii) $(0, 1)$

(iv) $(0.5, 0)$

3. Suppose that $f(x) = \frac{1}{2}x - 10$.

(a) Evaluate $f(0)$.

(b) Write and solve an equation to determine x so that $f(x) = 0$.

4. (a) For $c \neq 0$, which axis contains the point $(c, 0)$? Why?

(b) Describe the points of the y-axis.

In Exercises 5 and 6, solve for y.

5. $x - y = 5$

6. $2x + 3y = 12$

7. Evaluate $\frac{y_2 - y_1}{x_2 - x_1}$ for $x_1 = -2$, $y_1 = 4$, $x_2 = 5$, and $y_2 = 8$.

8. Identify the coefficient of x and the constant term in the expression $5 - \frac{3x}{4}$.

9. Write an expression or equation to model the given information.

(a) The cost in dollars of shipping a package at \$2.00 plus \$1.50 for each pound p.

(b) The length L is 4 meters less than the width W.

10. Solve for x.

(a) $3 - 2x \geq 11$

(b) $\frac{2}{3}x + 8 = 0$

11. Determine whether the given number is a solution of the equation.

(a) $2x - 15 = 7$; 11

(b) $1 - 3x = x - 9$; -2

12. Show that the triangle with vertices $A(-4, 4)$, $B(6, 9)$, and $C(2, -8)$ is a right triangle.

Chapter 3 Test

1. State the property that justifies each step.

$$2x - 3 = 7$$

$$2x - 3 + 3 = 7 + 3 \quad \text{(a) \_\_\_\_\_\_}$$

$$2x = 10 \quad \text{(b) \_\_\_\_\_\_}$$

$$\frac{2x}{2} = \frac{10}{2} \quad \text{(c) \_\_\_\_\_\_}$$

$$x = 5$$

In Questions 2–4, solve the equation. Then state whether the equation is an identity, an inconsistent equation, or a conditional equation.

2. $10 - 3x = 6x - 17$

3. $3x - 4 = 5x - 4 - 2x$

4. $3 - 2(x - 7) = 3(x + 1) - 5x$

5. Solve the formula $A = P + Prt$ for t.

6. Solve the equation $2x + 3y = 12$ for y.

7. What was the original investment at a simple interest rate of 8% if the value of the investment at the end of a year is \$2484?

8. A person has 200 meters of fencing to use to enclose a rectangular garden. If the width is three-fifths of the length, what are the dimensions of the garden?

9. A radiator with a 6-gallon capacity is filled with a 50% antifreeze solution. How much must be drained so that adding pure antifreeze gives a 60% antifreeze solution?

10. Two cars leave Wall, South Dakota, at the same time. The first car travels west while the other travels east at 12 mph less than that of the first car. After 1.5 hours they are 144 miles apart. What is the speed of each car?

11. Use interval notation to describe the set whose graph is given.

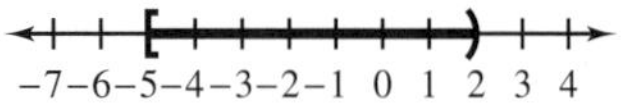

In Questions 12–15, solve the given inequality.

12. $3x + 1 > 13$

13. $4 - 2(x - 3) < 1 - 5x$

14. $\dfrac{2x + 13}{-2} \geq 1$

15. $x + \dfrac{2}{3}(x - 3) < \dfrac{1}{2}$

In Questions 16–19, solve the compound inequality.

16. $-3 \leq 2x + 1 \leq 5$

17. $3 - x < 5$ and $2x - 1 < 5$

18. $3 - 2x \leq 9$ or $x + 1 \geq 0$

19. $2x \geq 0$ or $x - 3 \leq 1$

20. Solve $|x + 2| - 5 = -3$.

In Questions 21–23, solve the given absolute value inequality.

21. $3 - |x - 1| < 1$

22. $|2x - 1| \leq 5$

23. $|x - 3| < -2$

24. The charge for cellular phone service is \$25 per month plus 50 cents per minute that the phone is used. If the phone bill is between \$31 and \$46 per month, what are the most and least numbers of minutes that the phone may have been used?

25. The average snowfall for the first week of January is 8.7 inches. If the snowfall totals for the first 6 days are 1.2, 0, 4.25, 1.5, 0, and 0.4 inches, how much snow could fall on January 7 without the total exceeding the average total?

Chapter 4

Properties of Lines

When Europeans first set foot on the North American continent, there were an estimated half-million bald eagles. By 1963, that number had dwindled to a mere 417 pairs. The accompanying graph shows the modest increases that have resulted from conservationists' efforts.

To visualize the trend, we can connect the data points for 1963 and 1998 with a line. This line has a ***slope*** *that represents the average* ***rate of change*** *in the bald eagle population. Moreover, the line can be described by a linear function, which can be used to model the number of pairs of bald eagles for any given year. (For more on this topic, see Exercises 101–104 at the end of Section 4.3, and see the Chapter Project.)*

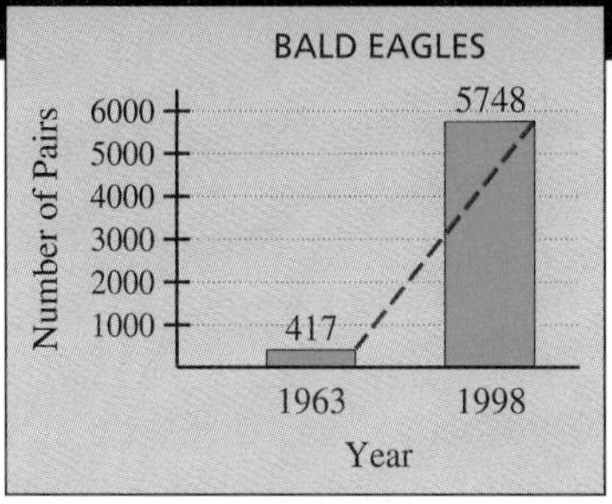

(Source: U.S. Fish and Wildlife Service.)

In Chapter 4 we consider linear equations in two variables and the properties of their graphs. In particular, we introduce the concept of slope and its many applications. We conclude with a discussion of linear inequalities and their graphs.

4.1 Linear Equations in Two Variables

The Standard Form • Solutions and Graphs • Intercepts • Special Cases

The Standard Form

Linear equations in one variable typically arise when a problem involves a single unknown quantity or when all unknown quantities can be easily represented with just one variable.

When problems involve two unknown quantities, it is often more convenient to use two different variables and to write equations in those variables. One such type of equation is a **linear equation in two variables.**

> **Definition of a Linear Equation in Two Variables**
>
> A **linear equation in two variables** is an equation that can be written in the *standard form* $Ax + By = C$, where A, B, and C are real numbers and A and B are not both zero.

Each of the following equations is a linear equation in two variables.

It is especially important to show that $y = 3$ and $x = -2$ can be written in standard form to convince students that these can be considered linear equations in *two* variables.

Equation	*Standard Form*	
$3x + 2y = 12$	$3x + 2y = 12$	$A = 3, B = 2, C = 12$
$y = 2x + 3$	$-2x + y = 3$	$A = -2, B = 1, C = 3$
$y = 3$	$0x + y = 3$	$A = 0, B = 1, C = 3$
$x = -2$	$x + 0y = -2$	$A = 1, B = 0, C = -2$

An equation may have two variables but not be a linear equation. The following are some examples of equations that *cannot* be written in the standard form $Ax + By = C$.

$x^2 + 3y = 5$ The exponents on both variables must be 1.

$\frac{2}{x} + y = -2$ No variable can appear in the denominator of a fraction.

$xy = 1$ The variables cannot be part of the same term.

Solutions and Graphs

To solve a linear equation in two variables, we must find a replacement for each variable so that the resulting equation is true. We say that such replacements *satisfy* the equation.

> **Definition of a Solution of a Linear Equation in Two Variables**
>
> A **solution** of a linear equation in two variables is a pair of numbers (x, y) that satisfies the equation.

We can use substitution to determine whether an ordered pair is a solution of an equation.

EXAMPLE 1

Testing Solutions of Linear Equations

Determine whether the following ordered pairs are solutions of $2y - 5x = 12$.

(a) $(2, 11)$ (b) $(6, 0)$ (c) $(-6, -9)$

Solution

For each ordered pair, replace x with the first coordinate and y with the second coordinate.

$$\begin{aligned} &(2, 11) \\ 2y - 5x &= 12 \\ 2(11) - 5(2) &= 12 \\ 12 &= 12 \quad \text{True} \end{aligned}$$

$$\begin{aligned} &(6, 0) \\ 2y - 5x &= 12 \\ 2(0) - 5(6) &= 12 \\ -30 &= 12 \quad \text{False} \end{aligned}$$

$$\begin{aligned} &(-6, -9) \\ 2y - 5x &= 12 \\ 2(-9) - 5(-6) &= 12 \\ 12 &= 12 \quad \text{True} \end{aligned}$$

The pairs $(2, 11)$ and $(-6, -9)$ are solutions of the equation.

In Example 1, we substituted values for both variables to test whether a given ordered pair is a solution of a linear equation. Determining solutions of a linear equation involves substituting values for just one variable and solving for the other variable.

Exploring the Concept

Graphs of Linear Equations

Consider the linear equation $2x + y = 8$. In the following table, the value of one variable is given. To complete the table, we can replace that variable with the given value and solve for the other variable. In the third row, we join the x- and y-values to form ordered pairs, which are solutions of the equation.

x	-4		0		4	
y		10		2		-10
(x, y)						

Calculations of the first two entries are shown.

$$\begin{aligned} 2x + y &= 8 \\ 2(-4) + y &= 8 \quad \text{Replace } x \text{ with } -4. \\ -8 + y &= 8 \\ y &= 16 \quad \text{Solve for } y. \end{aligned}$$

$$\begin{aligned} 2x + y &= 8 \\ 2x + 10 &= 8 \quad \text{Replace } y \text{ with } 10. \\ 2x &= -2 \\ x &= -1 \quad \text{Solve for } x. \end{aligned}$$

The other entries in the table are determined in a similar manner.

x	-4	-1	0	3	4	9
y	16	10	8	2	0	-10
(x, y)	$(-4, 16)$	$(-1, 10)$	$(0, 8)$	$(3, 2)$	$(4, 0)$	$(9, -10)$

You might also want to choose other points of the line that are not in the table and verify that they are solutions.

Each solution in the table can be represented by a point plotted in a coordinate system. (See Fig. 4.1.)

Figure 4.1

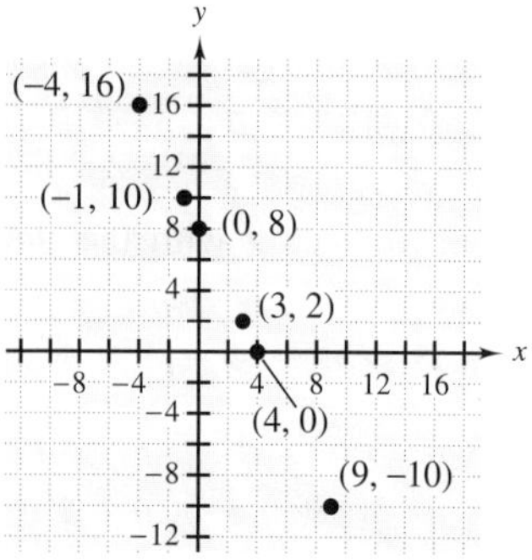

Figure 4.2

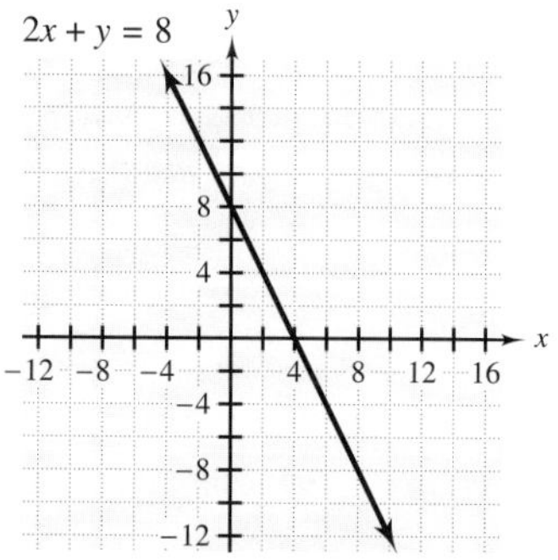

Because either variable can be replaced with any real number, there are infinitely many pairs (solutions) that could be formed. All of these solutions can also be represented by points in a coordinate system, and the points form a straight line. (See Fig. 4.2.) This line is called the **graph of the equation.**

We can make the following generalizations about a linear equation in two variables.

1. The graph is a straight line.
2. There is a one-to-one correspondence between the solutions of the equation and the points of the graph. In other words,
   (a) Every solution of the equation is represented by a point of the graph.
   (b) Every point of the graph represents a solution of the equation.

Because the graph of a linear equation is a straight line, two points of the line are all we need to sketch the graph. However, as a check, we usually determine three solutions of the equation.

Graphing a Linear Equation

1. Assign a value to one of the variables and solve for the other variable. The resulting pair of numbers is a solution of the equation.
2. Repeat step 1 to obtain three solutions of the equation.
3. Plot the points associated with the three solutions.
4. The graph is the straight line containing these three points.

EXAMPLE 2

Graphing a Linear Equation

Graph the equation $3x - 2y = 12$.

Figure 4.3

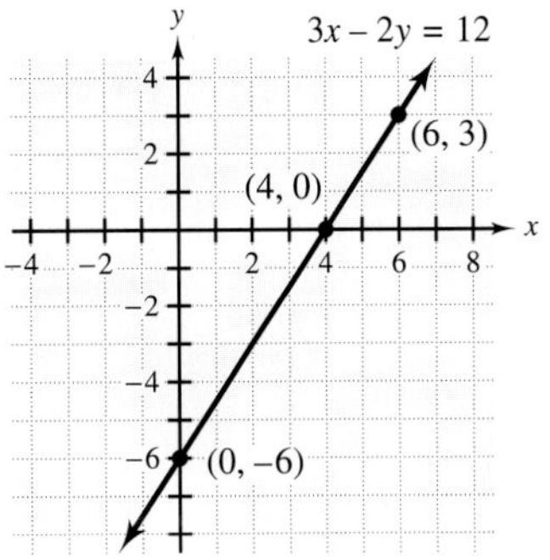

Solution

Choosing 0 for one variable is often a convenient way to begin. In this example we let $x = 0$ and obtain $y = -6$. Then we let $y = 0$ and obtain $x = 4$. For a third point, we let $x = 6$ and obtain $y = 3$. Our three ordered pairs are shown in the table.

x	0	4	6
y	-6	0	3
(x, y)	$(0, -6)$	$(4, 0)$	$(6, 3)$

Then we plot the associated points and draw the line containing the points. The line is the graph of $3x - 2y = 12$. (See Fig. 4.3.)

Intercepts

Students sometimes confuse these. To determine the y-intercept, they let $y = 0$, and to determine the x-intercept, they let $x = 0$. It helps to keep illustrating these points graphically.

Note that in Figure 4.3 the graph intersects the y-axis at $(0, -6)$ and intersects the x-axis at $(4, 0)$. We call the point $(0, -6)$ the ***y*-intercept** and the point $(4, 0)$ the ***x*-intercept.**

Definitions of Intercepts

A point where a graph intersects the y-axis is called a ***y*-intercept.** A point where a graph intersects the x-axis is called an ***x*-intercept.**

Although we define y-intercept and x-intercept as *points,* the words are often used to refer to the coordinates of the points.

LEARNING TIP

Visualizing the y-intercept as a point of the y-axis, where the x-coordinate is always 0, provides a way to remember that x is replaced with 0. Similarly, visualize the x-intercept as a point of the x-axis, where the y-coordinate is always 0.

We can determine the coordinates of the x- and y-intercepts algebraically.

1. The y-intercept is a point whose x-coordinate is 0. Replace x with 0 and calculate the corresponding value of y.
2. The x-intercept is a point whose y-coordinate is 0. Replace y with 0 and calculate the corresponding value of x.

EXAMPLE 3

Determining Intercepts Algebraically

Determine the intercepts for $y = 2x - 3$. Then graph the equation.

Solution

x-Intercept

$y = 2x - 3$

$0 = 2x - 3$ Replace y with 0.

$3 = 2x$ Solve for x.

$x = \dfrac{3}{2}$

The x-intercept is $\left(\frac{3}{2}, 0\right)$.

y-Intercept

$y = 2x - 3$

$y = 2(0) - 3$ Replace x with 0.

$y = 0 - 3$ Solve for y.

$y = -3$

The y-intercept is $(0, -3)$.

Formally, intercepts are described by *ordered pairs,* but it is common to say, for example, that the y-intercept is -3.

Use the two intercepts as two of the points to plot. For a third point, we let $x = 4$ and obtain $y = 5$. We use the three solutions shown in the table of values to sketch the graph. (See Fig. 4.4.)

x	0	$\frac{3}{2}$	4
y	-3	0	5

Figure 4.4

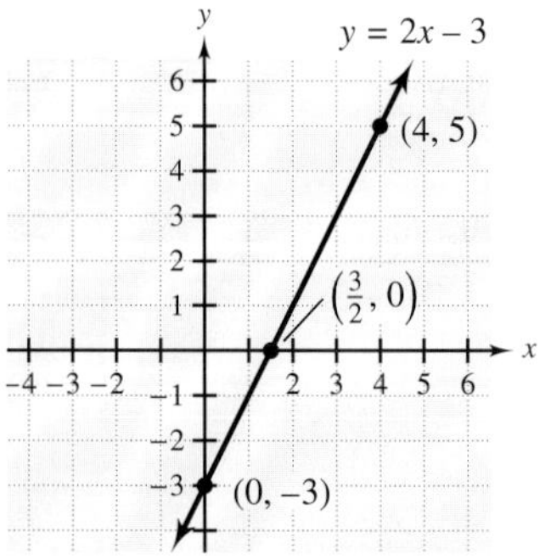

Exploring the Concept

Determining y-Intercepts from Equations

Each time we determine the y-intercept, we replace x with 0 and solve for y. Consider the results of determining the y-intercept of each of the following equations.

Equation	*Replace x with 0*	*y-Intercept*
$y = x + 12$	$y = 0 + 12 - 12$	$(0, 12)$
$y = x + 5$	$y = 0 + 5 = 5$	$(0, 5)$
$y = x + 0$	$y = 0 + 0 = 0$	$(0, 0)$
$y = x - 8$	$y = 0 - 8 = -8$	$(0, -8)$

Observe that in each case, the y-coordinate of the y-intercept is the same as the constant term of the equation.

This conclusion can be supported as follows. When we solve a linear equation in two variables for y, the resulting equation has the form $y = ax + b$, where a and b are real numbers. We determine the y-intercept by replacing x with 0.

$$y = ax + b$$

$$y = a(0) + b = 0 + b = b$$

Thus the y-intercept is $(0, b)$.

Think About It

The y-intercept of the graph of $y = ax + b$ is $(0, b)$. Can you write a general ordered pair for the x-intercept?

In summary, for the linear equation $y = ax + b$ or the linear function $f(x) = ax + b$, the y-intercept $(0, b)$ can be determined from the constant term of the equation.

EXAMPLE 4

Determining the y-Intercept from the Equation

Write each of the following equations in the form $y = ax + b$. Then determine the y-intercept in each case.

(a) $y = x - 4$ (b) $5x + 7y = 9$

Solution

(a) The equation is already solved for y. The constant term is -4. Therefore, the y-intercept is $(0, -4)$.

(b)
$$\begin{aligned} 5x + 7y &= 9 && \text{Solve the equation for } y. \\ 5x + 7y - 5x &= -5x + 9 && \text{Subtract } 5x \text{ from both sides.} \\ 7y &= -5x + 9 && \text{Combine like terms.} \\ \frac{7y}{7} &= \frac{-5x}{7} + \frac{9}{7} && \text{Divide each term on both sides by 7.} \\ y &= -\frac{5}{7}x + \frac{9}{7} \end{aligned}$$

The constant term is $\frac{9}{7}$. Therefore, the y-intercept is $\left(0, \frac{9}{7}\right)$.

If the task is only to determine intercepts, there is little value in writing the equation in the form $y = mx + b$.

Determining the y-intercept is easy when the equation is written in the form $y = ax + b$. However, the y-intercept can be determined from any equation form by replacing x with 0 and computing the corresponding y-value.

Consider again the equation in part (b) of Example 4.

$$\begin{aligned} 5x + 7y &= 9 \\ 5(0) + 7y &= 9 && \text{Replace } x \text{ with 0.} \\ 7y &= 9 && \text{Solve for } y. \\ y &= \frac{9}{7} \end{aligned}$$

The y-intercept is $\left(0, \frac{9}{7}\right)$.

Special Cases

At the beginning of this section, we stated that $y = 3$ and $x = -2$ are both examples of linear equations in two variables. In both equations, one of the variables is missing. For this reason, we consider the equations to be special cases.

EXAMPLE 5

Special Case: y = Constant

Consider the following equations.

(i) $y = 4$ (ii) $y = -5$

For each equation, complete the following table of values. Then sketch the graph of each equation.

x	-2	0	3
y			

Solution

To complete the table, note that equation (i) states that y is always 4. For equation (ii), y is always -5.

Plot the points associated with the tables of values and sketch the graphs. (See Fig. 4.5.)

(i) $y = 4$

x	-2	0	3
y	4	4	4

(ii) $y = -5$

x	-2	0	3
y	-5	-5	-5

Figure 4.5

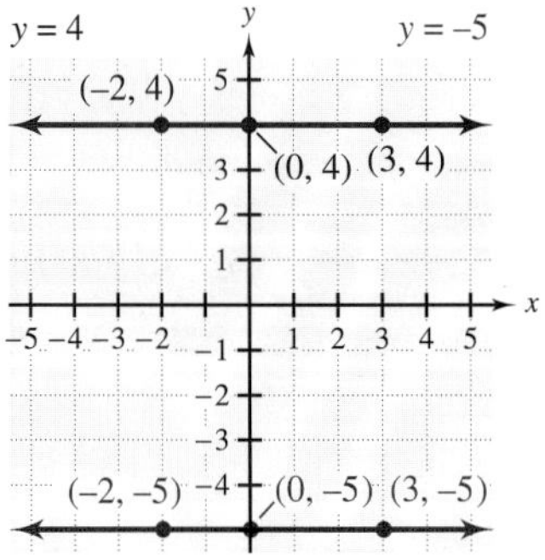

Note that in Example 5, the lines appear to be horizontal. The constant value of y indicates how far above or below the x-axis the line is drawn. By examining other points of the line, note that all the points of each line have the same y-coordinate. There is no x-intercept. The y-intercepts are, in order, $(0, 4)$ and $(0, -5)$.

EXAMPLE 6

Figure 4.6

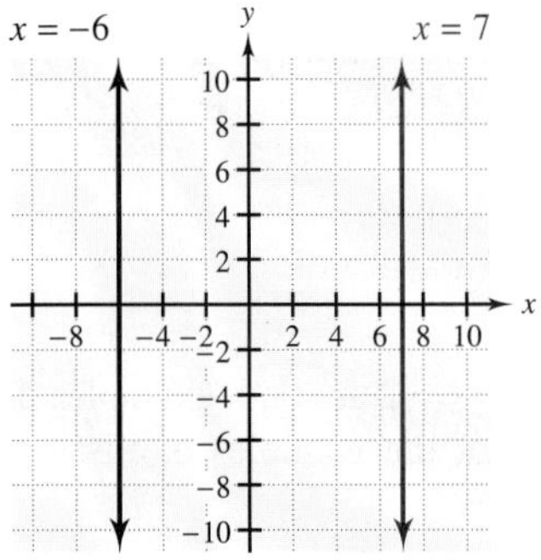

Special Case: x = Constant

Sketch the graphs of the following equations.

(a) $x = 7$ (b) $x = -6$

Solution

(a) For $x = 7$, the value of x is always 7. The following pairs are solutions of $x = 7$. Figure 4.6 shows the graph.

$$(7, -3), (7, 0), (7, 4), (7, 9)$$

(b) For $x = -6$, the value of x is always -6. The following pairs are solutions of $x = -6$. Figure 4.6 shows the graph.

$$(-6, -1), (-6, 3), (-6, 7), (-6, 12)$$

In Example 6, each line is a vertical line. The constant value x indicates how far to the left or right of the y-axis the line is drawn. The x-intercepts are $(7, 0)$ and $(-6, 0)$. Neither line has a y-intercept.

The following characteristics apply to all equations of the form $x = c$ and $y = b$, where c and b are constants.

1. For $x = c$ the first coordinate of all ordered pairs is c; for $y = b$ the second coordinate of all ordered pairs is b.
2. For $x = c$ the graph is a vertical line, where c indicates how far to the right or left of the y-axis to draw the line. For $y = b$ the graph is a horizontal line, where b indicates how far above or below the x-axis to draw the line.
3. The x-intercept of the graph of $x = c$ is $(c, 0)$, and if $c \neq 0$, then there is no y-intercept. The y-intercept of the graph of $y = b$ is $(0, b)$, and if $b \neq 0$, then there is no x-intercept.

Emphasize that the graph of $y = 0$ is the x-axis and illustrate the special nature of the intercepts.

Note: Because a vertical line does not pass the Vertical Line Test, the graph of $x = c$ does not represent a function.

It will be very useful to keep these two special cases in mind. As is true for all linear equations in two variables, the graph is a straight line. Knowing that the line is horizontal when y = constant and is vertical when x = constant simplifies sketching the graph.

Compare the intercepts of the graph of $x = 0$ with those of the graph of $y = 0$.

Here is a summary of what is known so far about the two special cases.

Summary of Special Cases of Linear Equations

Equation	*Graph*	*Intercepts*	*Function?*
$x = c$	vertical	x-intercept: $(c, 0)$ y-intercept: if $c \neq 0$, none.	no
$y = b$	horizontal	x-intercept: if $b \neq 0$, none. y-intercept: $(0, b)$	yes

Quick Reference 4.1

The Standard Form

- A **linear equation in two variables** is an equation that can be written in the *standard form* $Ax + By = C$, where A, B, and C are real numbers and A and B are not both zero.

Solutions and Graphs

- A **solution** of a linear equation in two variables is a pair of numbers (x, y) that satisfies the equation.
- Solutions can be determined by replacing one variable with some value and solving the equation for the other variable.
- There exists a one-to-one correspondence between the solutions of an equation and the points of its graph.
- The graph of a linear equation in two variables is a straight line.
- To sketch the graph of a linear equation, follow these steps.
  1. Assign a value to one of the variables. Solve for the other variable. The pair of numbers is a solution of the equation.
  2. Repeat step 1 to obtain three solutions of the equation.
  3. Plot the points associated with the three solutions.
  4. The graph is the straight line containing these three points.

Intercepts

- A point where a graph intersects the y-axis is called a ***y*-intercept.** A point where a graph intersects the x-axis is called an ***x*-intercept.**

- We determine the exact x-intercept algebraically by replacing y with 0 and solving for x. Similarly, we determine the exact y-intercept algebraically by replacing x with 0 and solving for y.
- If a linear equation is solved for y, the form of the equation is $y = ax + b$. From this, we can see that the y-intercept is $(0, b)$.

Special Cases

- The graph of the equation $y = b$ is a horizontal line. The constant b indicates how far the line is above or below the x-axis. Excluding the graph of $y = 0$, the graph has the y-intercept $(0, b)$ and no x-intercepts.
- The graph of the equation $x = c$ is a vertical line. The constant c indicates how far the line is to the right or left of the y-axis. Excluding the graph of $x = 0$, the graph has the x-intercept $(c, 0)$ and no y-intercepts.

Speaking the Language 4.1

1. The standard form of a linear equation in two variables is ________, whereas the form of a linear function is ________.
2. Points where a graph intersects the axes are called ________.
3. The graph of the equation $y = b$ is a(n) ________ line, whereas the graph of the equation $x = c$ is a(n) ________ line.
4. Any ordered pair (x, y) that satisfies a linear equation in two variables is called a(n) ________ of the equation.

Exercises 4.1

Concepts and Skills

1. Explain why $x = -5$ can be a linear equation in two variables even though there is no y-term.
2. How are the points of a graph of an equation related to the solutions of the equation?

In Exercises 3–8, state whether the equation is a linear equation in two variables.

3. $\frac{x}{2} + \frac{y}{3} - \frac{7}{8} = 0$
4. $y = \frac{2x}{3} + 5$
5. $y = x^2 - 3$
6. $4y = \frac{3}{x} + \frac{4}{5}$
7. $-6 = 2xy$
8. $4y + 5 = -7$

In Exercises 9–14, determine whether the given ordered pair is a solution of the given equation.

9. $(4, 5);\ y = 2x - 3$
10. $(-2, 2);\ y = 3x + 4$
11. $\left(2, \frac{3}{2}\right);\ 3x - 4y = 12$
12. $(4, 5);\ \frac{1}{4}x + \frac{2}{5}y = 3$
13. $(-2, -1);\ x - 5 = 3$
14. $(-2, -3);\ 7 + y = 4$

CE In Exercises 15–20, determine k so that the given ordered pair is a solution of the equation.

15. $2x - ky = 7;\ (1, -1)$
16. $y + 3x = k;\ (2, -5)$
17. $y = -\frac{2}{3}x + k;\ (12, -3)$
18. $kx + y = 0;\ (-2, 6)$
19. $y = -2x + k;\ (k, k + 1)$
20. $kx - y = 9;\ (3, k - 5)$

CE In Exercises 21–26, find the values of a, b, and c so that the ordered pairs are solutions of the given equation.

21. $y = -2x + 5$; $(-3, a)$, $(4, b)$, $(c, -7)$

22. $y = 4x - 3$; $(a, 5)$, $(b, -3)$, $(3, c)$

23. $x - 5 = -2$; $(a, 5)$, $(b, 9)$, $(c, -6)$

24. $5 - 2y = -1$; $(-1, a)$, $(4, b)$, $(-12, c)$

25. $x + 3y - 4 = 0$; $(a, 3)$, $(7, b)$, $(c, -4)$

26. $5y - 3x = -10$; $(a, -2)$, $(-5, b)$, $(10, c)$

27. For each equation or inequality in column A, a solution or solution set is given in column B. Explain how the differences in notation affect the meaning of each entry in column B.

Column A	*Column B*
$\|x\| = 2$	$\{-2, 2\}$
$\|x\| < 2$	$(-2, 2)$
$y = x + 4$	$(-2, 2)$

28. Explain the difference between the solution set of $x = 1$, a linear equation in one variable, and $x = 1$, a linear equation in two variables.

In Exercises 29–32, complete the table.

29. $y = 4 - x$

x	10	-6	
y			-3

30. $x = 2y + 3$

x	5	-13	
y			-6

31. $f(x) = 3x + 1$

x	0	3	-5
$f(x)$			
(x, y)			

32. $g(x) = -2x - 5$

x	-2	-7	3
$g(x)$			
(x, y)			

CE **33.** (a) Which of the following equations is graphed in the accompanying figure?

$x - 2y - 5 = 0$

$2x - 3y - 6 = 0$

$5x - 6y - 18 = 0$

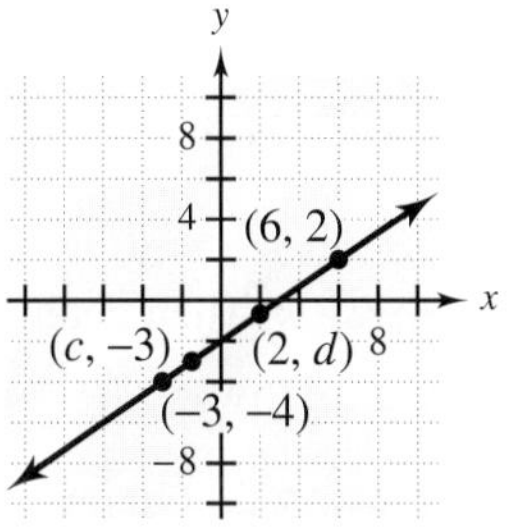

(b) What are the values of c and d in the figure?

CE **34.** (a) Which of the following equations is graphed in the accompanying figure?

$3x + 4y + 9 = 0$

$5x + 8y + 11 = 0$

$7x + 8y + 13 = 0$

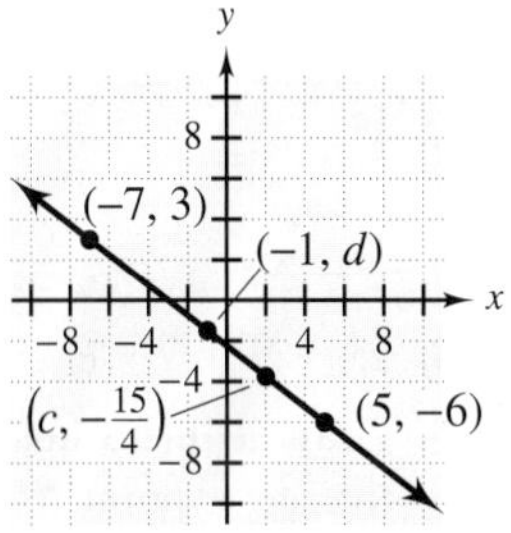

(b) What are the values of c and d in the figure?

In Exercises 35–48, determine three solutions of the given equation and sketch the graph.

35. $y = 3x - 4$

36. $y = 4x + 3$

37. $y = \frac{2}{3}x - 2$

38. $y = -\frac{3}{4}x + 5$

39. $3x - 5y = 15$

40. $2x - 3y = 6$

41. $y - 2 = 3$

42. $y + 5 = 7$

43. $x - 9 = -11$

44. $x + 7 = 9$

45. $y = x$

46. $y = -x$

47. $3x = 21 - 7y$

48. $5x + 2y - 10 = 0$

In Exercises 49–54, determine the intercepts. Then use the intercepts to sketch the graph of the equation.

49. $y = 12 - 3x$

50. $y = -12 + x$

51. $x = 3y + 30$

52. $x = 15 - y$

53. $3x - 4y = 36$

54. $2x + 3y = 30$

In Exercises 55 and 56, select from the given equations the one whose graph has the given y-intercept. (There could be more than one answer.)

55. The y-intercept is $(0, 3)$.

(i) $5x + 3y - 9 = 0$ (ii) $6x + 2y - 6 = 0$

(iii) $3x + 4y + 12 = 0$

56. The y-intercept is $(0, -2)$.

(i) $5x + 3y - 6 = 0$

(ii) $5x + 3y + 6 = 0$

(iii) $2x + 4y + 8 = 0$

In Exercises 57–60, write the given equation in the form $y = ax + b$. Then determine the y-intercept.

57. $2x + 5y = 14$

58. $4x - 3y = 15$

59. $y - 2 = -4(x + 3)$

60. $\frac{1}{3}x + \frac{3}{4}y - 9 = 0$

In Exercises 61–68, determine the intercepts algebraically.

61. $y = -2x$

62. $5x - 4y = 0$

63. $3x - 4y = 12$

64. $2x - 3y = 12$

65. $\frac{x}{3} + \frac{y}{4} = 1$

66. $\frac{x}{5} - \frac{y}{3} = 1$

67. $y + 2 = 17$

68. $2x + 9 = 7$

69. Draw a line whose x-intercept and y-intercept are the same point. If the equation of the line has the form $y = ax + b$, what is the value of b?

70. Describe the form of an equation whose graph is a

(a) vertical line.

(b) horizontal line.

Students often fail to see any connection among graphs such as those of $y = x, y = x + 4$, and $y = x - 3$.

CE In Exercises 71 and 72, sketch the graph of the given equation. Then determine the equations of the lines described in parts (a) and (b). To check your equations, sketch their graphs on the same axes.

71. $y = x$

(a) A line shifted upward 4 units from the given line

(b) A line shifted downward 3 units from the given line

72. $y = -1$

(a) A line shifted upward 5 units from the given line

(b) A line shifted downward 4 units from the given line

CE In Exercises 73–76, determine a, b, and c so that the given equation has the specified intercepts.

73. $ax + 4y = c$; $(0, 2), (-4, 0)$

74. $x - 2by = c$; $(0, 5), (-1, 0)$

75. $ax + by = 15$; $(0, -5), (3, 0)$

76. $\frac{x}{a} + \frac{y}{b} = 1$; $(0, 1), (1, 0)$

Remind students that if both variables appear in the equation with nonzero coefficients, then the graph is neither horizontal nor vertical.

In Exercises 77–82, determine whether the graph of the given equation is horizontal, vertical, or neither.

77. $3x = 12$

78. $x - 5 = 6$

79. $x - y = 0$

80. $y = x$

81. $-2y = -8$

82. $y + 3 = -2$

CE In Exercises 83–88, translate the given conditions into an equation. Then select several points of the graph of the equation and verify that their coordinates satisfy the conditions.

83. One number y is the opposite of another number x.

84. The numbers x and y are equal.

85. One number y is half of another number x.

86. One number y is 3 less than twice another number x.

87. The sum of two numbers x and y is 5.

88. The difference of two numbers x and y is 2.

Real-Life Applications

89. A window washer has 22 windows to wash.

(a) Write a linear function that models the number y of windows remaining to be washed as a function of the number x of windows that have been washed.

(b) Determine and interpret the intercepts of the graph of the function in part (a).

90. A package delivery service employs 40 drivers, of whom x are men.

(a) Write a linear function that models the number y of women drivers as a function of x.

(b) Determine and interpret the intercepts of the graph of the function in part (a).

91. A taxi fare is \$4.00 plus 15 cents per tenth of a mile. What are all of the possible fares for a trip of between 2 and 2.5 miles, inclusive?

92. When the daily high temperature is 75°F, the number of people using the local jogging path is 120 per day. For each 1°F temperature increase over 75°F (up to 100°F), the number of joggers decreases by 4. For what temperature range will the number of joggers be between 88 and 100, inclusive?

Modeling with Real Data

93. In the period 1992–1997, about \$911 million was spent nationwide on improvements to zoos and aquariums. The bar graph shows that these expenditures increased during the period. (Source: American Zoo and Aquarium Association.)

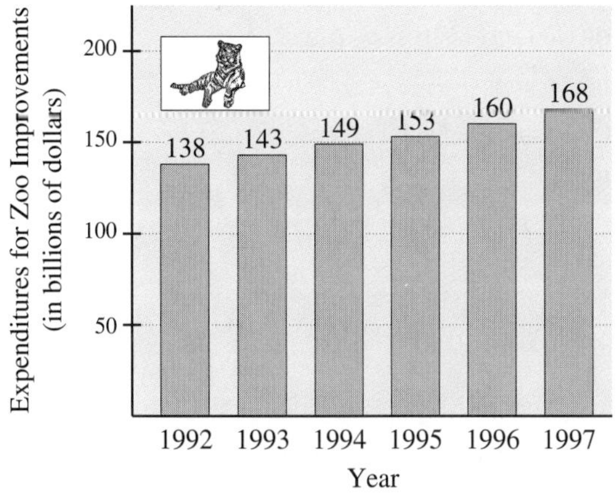

(a) What visual evidence does the bar graph provide to support modeling the data with a linear function?

(b) Let x represent the number of years since 1990. Evaluate each of the following linear functions for the years 1993, 1995, and 1997. Use your results to decide which of the functions appears to be the best model of the data.

(i) $E(x) = 152$

(ii) $E(x) = 5.9x + 125.5$

(iii) $E(x) = 5.9x + 168$

(c) If the trend in expenditures continues, what is the first year in which expenditures will be at least 50% more than they were in 1992?

94. (a) Determine the y-intercept for the graph of each of the functions in part (b) of Exercise 93.

(b) Explain how you could have used these y-intercepts to make your choice in part (b) of Exercise 93 without evaluating the functions.

95. In 2000, the life expectancy at birth was 76.7 years. This is an increase from the life expectancy of 73.7 years for people born in 1980. (Source: Centers for Disease Control.) With x representing the number of years since 1980, the linear function $L(x) = 0.15x + 73.7$ models the life expectancy L (in years).

(a) Determine and interpret the y-intercept of the graph of L.

(b) According to the model function, what is the average annual increase in our life expectancy?

(c) What is the x-intercept of the graph of function L? Explain why this point is meaningless in the context of this application.

96. In 2001, domestic consumer spending on books was \$33 billion. (Source: Book Industry Study Group, Inc.) The function $B(x) = 1.29x + 20.06$ can be used to model book sales B (in billions of dollars) x years after 1991.

(a) For the years 1998 to 2002, make a table of values for B and estimate the number by which B is increasing each year.

(b) Compare your result in part (a) to the coefficient of x in the model function.

Data Analysis: *Alternative Fuels*

The Clean Air Act of 1992 prompted increased attention to alternative fuels. The bar graph shows the production (in millions of gallons) of 85% ethanol (E85) fuel during the period 1996–1998. (Source: Energy Information Administration.)

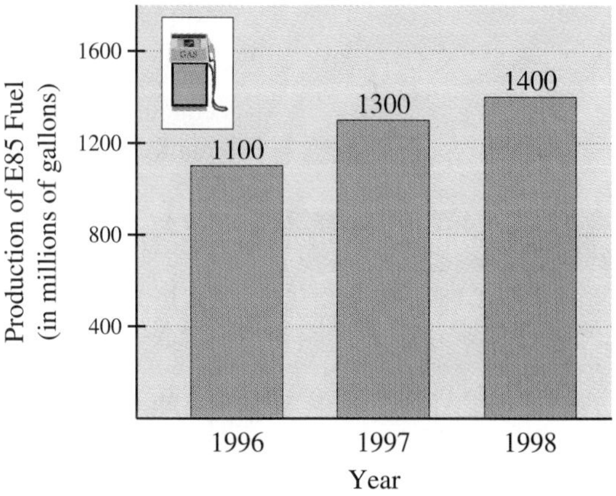

97. Suppose that we model the data with a linear function $y = ax + b$. If we let x represent the number of years since 1996, then what approximate value should we expect for b?

98. If we let x represent the number of years since 1995, then the linear function $P(x) = 150x + 967$ models the production P. Determine and interpret the y-intercept of the graph of P.

99. Let k represent the average annual production for the given 3-year period. Determine k and describe the graph of $y = k$. Explain why $y = k$ is not a realistic model of the data.

100. Suppose that the annual production of ethanol declined from 1999 to 2001. If the data for the period 1996–2001 were modeled by a linear function f, would you expect the y-intercept of the graph of f to be greater than or less than the y-intercept of the graph of P? Why?

Challenge

101. Interpret the roles of g and h in the equation

$$\frac{x}{g} + \frac{y}{h} = 1$$

(*Hint:* Consider where the graph crosses the axes.)

In Exercises 102–104, determine and plot the intercepts of the graph of the given equation. Be sure to take into account the conditions on a and b. Then sketch the graph.

102. $ax + by = ab;\ a > 0, b < 0$

103. $2x + by = 4b;\ b < 0$

104. $ax + 2y = -2a;\ a > 0$

In Exercises 105–108, graph using the specified domain.

105. $y = 2x + 1;\ x \geq 0$

106. $y = 1 - x;\ x \leq 0$

107. $y = x - 5;\ 10 \leq x \leq 15$

108. $y = 3x;\ -2 \leq x \leq 3$

4.2 Slope of a Line

The Slope Formula • The Slope–Intercept Form • Graphing with Slope

The Slope Formula

In the preceding section, we learned the significance of the constant term b when a linear equation is written in the form $y = ax + b$. The y-intercept of the graph of the equation is $(0, b)$. We now consider the coefficient a and learn how to interpret this number graphically.

Exploring the Concept

Effect of a on the Graph of $y = ax + b$

In Figures 4.7 and 4.8, all of the equations have the form $y = ax + 3$, so all of the graphs have the same y-intercept, $(0, 3)$.

Figure 4.7

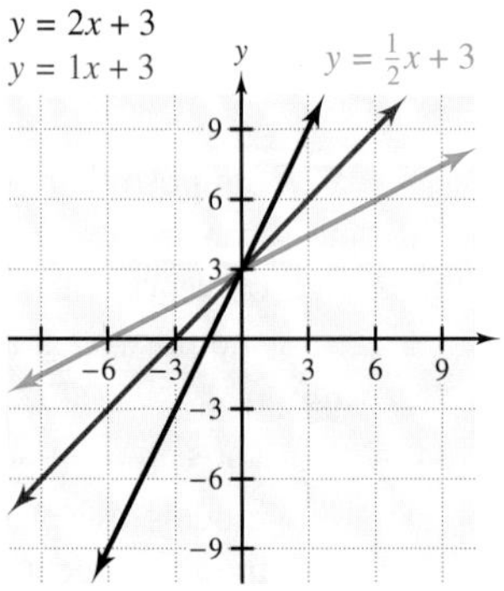

Figure 4.8

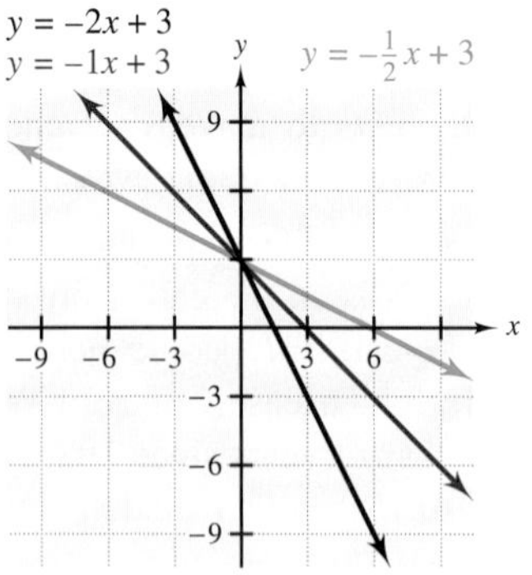

For the equations in Figure 4.7, the coefficient a is positive, and the lines rise from left to right. For the equations in Figure 4.8, the coefficient a is negative, and the lines fall from left to right.

We can also observe that the coefficient a influences the steepness of the line. In both figures, larger values of $|a|$ are associated with steeper lines.

Steepness is a relative word, and it is too vague to describe the orientation of a line specifically. What we need is a numerical measure of steepness. That measurement is called the **slope** of the line.

Exploring the Concept

Figure 4.9

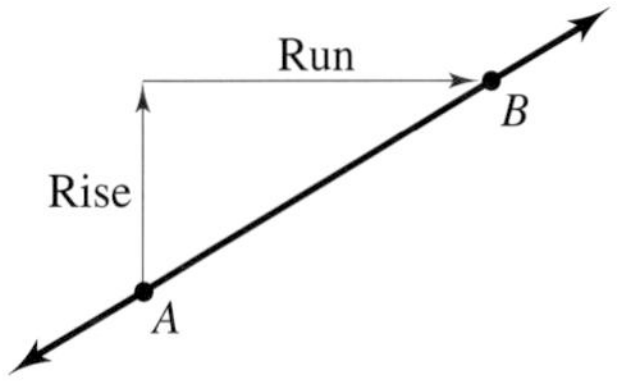

Figure 4.10

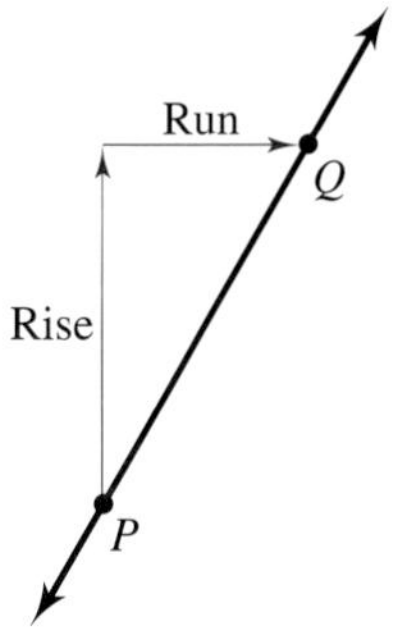

Students are not always aware that a ratio is a way of comparing two numbers.

The Slope of a Line

Consider the two lines in Figures 4.9 and 4.10.

One way to travel from point A to point B in Figure 4.9 is to start at A, travel vertically (rise) upward to the point directly across from B, and then travel horizontally (run) to the right to reach B. We can travel from point P to point Q in Figure 4.10 in a similar way.

We can see the difference in the steepness of the two lines by comparing the rise to the run in each case. For the line that is rising more gradually (Fig. 4.9), the rise is relatively small compared to the run. For the steeper line (Fig. 4.10), the rise is relatively large compared to the run.

This suggests a way of describing the steepness, or slope, of a line. We compare the rise to the run with a ratio.

In Figure 4.11, $P(x_1, y_1)$ and $Q(x_2, y_2)$ are two general points of a line. Note that point $R(x_1, y_2)$ is on a vertical line with P and on a horizontal line with Q. The triangle formed by points P, Q, and R is called a **slope triangle.**

Figure 4.11

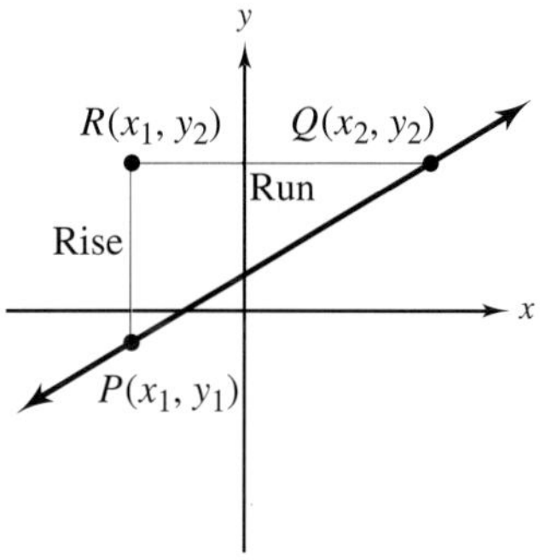

The **rise** PR is the difference between the y-coordinates.

$$\text{Rise} = y_2 - y_1$$

The **run** QR is the difference between the x-coordinates.

$$\text{Run} = x_2 - x_1$$

With these formulas for rise and run, we can now define the slope of a line.

Definition of Slope

The slope m of a line containing $P(x_1, y_1)$ and $Q(x_2, y_2)$ is defined as follows.

$$m = \frac{\text{rise}}{\text{run}} = \frac{y_2 - y_1}{x_2 - x_1}, \qquad x_2 \neq x_1$$

Note that the conventional symbol for slope is the letter m.

See whether students recall that the x-coordinates of all points of a vertical line are the same. Then show how the denominator of the slope formula would be zero in this case.

There are several important observations to be made from this definition. First, we can use the formula to calculate the slope of any line except a vertical line. The slope of a vertical line is undefined.

Next, the order in which we subtract the coordinates does not matter as long as we are consistent. We can choose to compute $y_2 - y_1$ for the rise, but then we must compute $x_2 - x_1$ for the run. Or we can choose to compute $y_1 - y_2$ for the rise, but then we must compute $x_1 - x_2$ for the run.

Finally, the definition refers to any two points of the line. No matter which two points we use, the calculated slope will be the same.

EXAMPLE 1

Determining Slope Given Two Points of a Line

Sketch the line determined by the given points and determine the slope of the line.

(a) $A(2, -1)$ and $B(4, 5)$ (b) $C(-2, 5)$ and $D(2, -3)$

Solution

(a) In Figure 4.12, the rise is $+6$ and the run is $+2$.

$$m = \frac{\text{rise}}{\text{run}} = \frac{6}{2} = 3$$

We can also use the slope formula to determine the slope.

$$m = \frac{y_2 - y_1}{x_2 - x_1} = \frac{5 - (-1)}{4 - 2} = \frac{6}{2} = 3$$

Figure 4.12

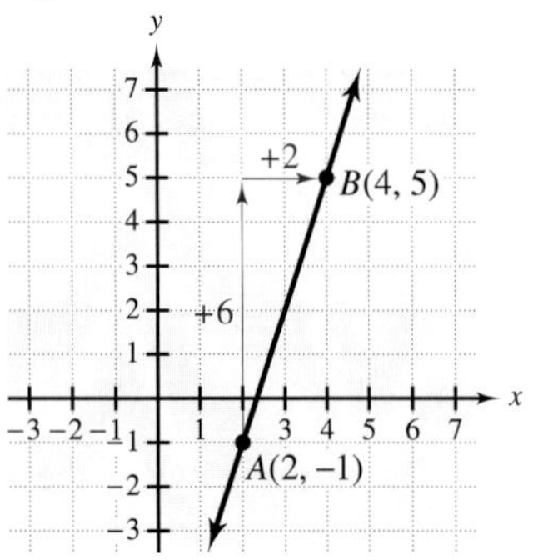

Figure 4.13

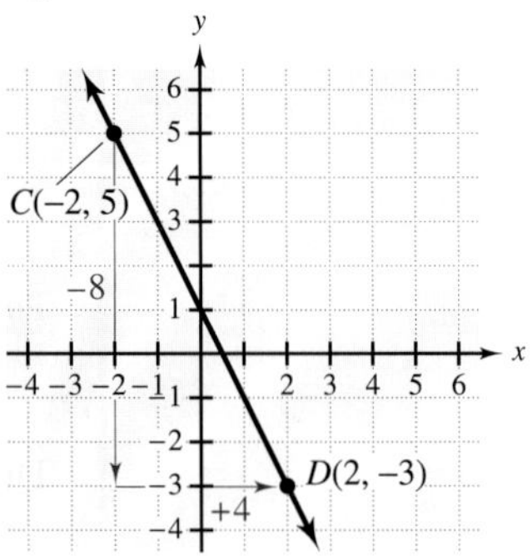

(b) In Figure 4.13, the rise is -8 and the run is $+4$.

$$m = \frac{\text{rise}}{\text{run}} = \frac{-8}{4} = -2$$

Using the slope formula, we obtain the same result.

$$m = \frac{-3 - 5}{2 - (-2)} = \frac{-8}{4} = -2$$

EXAMPLE 2

Special Cases

Determine the slope of the following lines.

(a) The equation of the line is $y = 5$.

(b) The line contains $E(1, 3)$ and $F(1, 7)$.

Solution

(a) Two points of the line $y = 5$ are shown in Figure 4.14. Note that points A and B have the same y-coordinate.

$$m = \frac{y_2 - y_1}{x_2 - x_1} = \frac{5 - 5}{-3 - 2} = \frac{0}{-3 - 2} = \frac{0}{-5} = 0$$

Figure 4.14

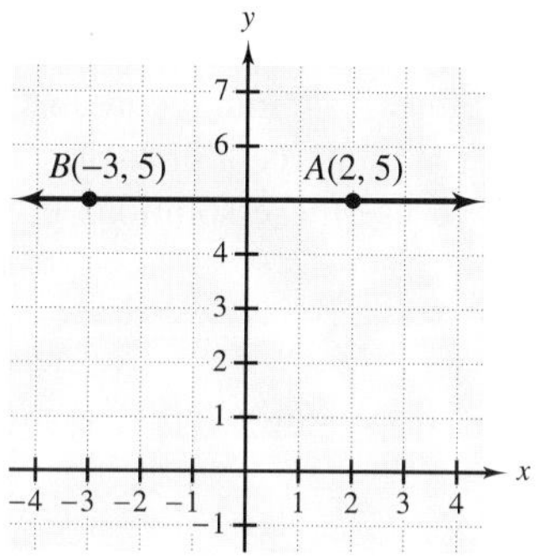

Figure 4.15

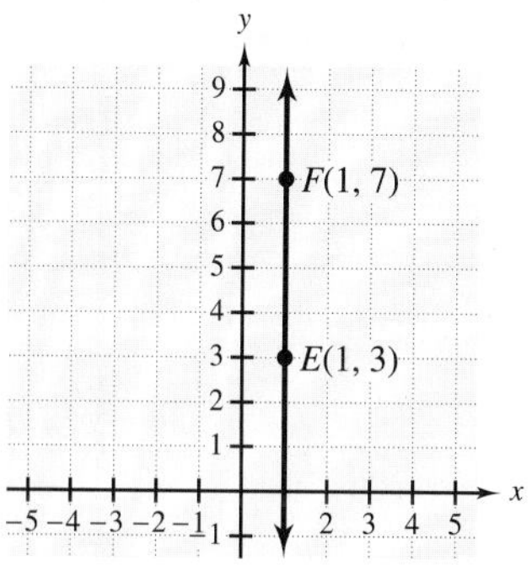

(b) In Figure 4.15, the line is vertical because points E and F have the same x-coordinate. The slope formula does not apply because the ratio is

$$\frac{y_2 - y_1}{x_2 - x_1} = \frac{7 - 3}{1 - 1} = \frac{4}{0}$$

which is undefined. In general, the slope of any vertical line is undefined.

Students tend to use words such as *zero*, *undefined*, and *no solution* interchangeably.

Note: For these special cases, it is important to know the difference between *zero* and *undefined*. A horizontal line has a defined slope, and it is 0. A vertical line does not have a defined slope. *Zero* does not mean the same thing as *undefined*.

One of the special cases, that of a vertical line, is handled in the definition of slope. The definition states that the slope of a vertical line is undefined.

In Example 2 we found that the slope of a horizontal line is 0. Because each point of a horizontal line has the same y-coordinate, $y_2 - y_1 = 0$ for any two points. Therefore,

$$m = \frac{y_2 - y_1}{x_2 - x_1} = \frac{0}{x_2 - x_1} = 0$$

LEARNING TIP

Compare the slope of a line to the *pitch* of a roof. A roof can have a small or a large pitch (slope). A flat roof has no pitch (zero slope), and there is no such thing as a vertical roof (undefined slope).

We can now add this new knowledge about slopes to our summary for special cases.

Summary of Special Cases

Equation	*Graph*	*Intercepts*	*Slope*
$x = c$	vertical	x-intercept: $(c, 0)$ y-intercept: If $c \neq 0$, none.	undefined
$y = b$	horizontal	y-intercept: $(0, b)$ x-intercept: If $b \neq 0$, none.	0

The Slope–Intercept Form

When a linear equation is written in the form $y = ax + b$, we have seen that the coefficient a affects the slope of the line. We now investigate this relationship in more detail.

Exploring the Concept

Determining Slopes from Equations

The following table lists four equations. For each equation, two solutions are given. Note that the y-intercept of each graph is $(0, 1)$, and so $(0, 1)$ is a solution of each equation. We use these solutions in the slope formula to calculate the slope of each line.

Equation	*First Solution*	*Second Solution*	*Slope*
$y = \frac{1}{2}x + 1$	$(0, 1)$	$(2, 2)$	$m = \frac{2-1}{2-0} = \frac{1}{2}$
$y = 2x + 1$	$(0, 1)$	$(1, 3)$	$m = \frac{3-1}{1-0} = \frac{2}{1} = 2$
$y = -1x + 1$	$(0, 1)$	$(1, 0)$	$m = \frac{0-1}{1-0} = \frac{-1}{1} = -1$
$y = -3x + 1$	$(0, 1)$	$(-1, 4)$	$m = \frac{4-1}{-1-0} = \frac{3}{-1} = -3$

In each case, we observe that the coefficient of x is the same as the slope of the line. These results suggest the following generalization: If an equation is written in the form $y = ax + b$, then the coefficient a is the same as the slope m. (The proof of this generalization is left as an exercise.)

Because $a = m$, we can write $y = ax + b$ in the more meaningful form $y = mx + b$, which is called the **slope–intercept form** of a linear equation.

Caution students that the slope is the number m, not the term mx.

Definition of Slope–Intercept Form

The **slope–intercept form** of a linear equation is the form $y = mx + b$. The slope of the graph is m and the y-intercept is $(0, b)$.

Graphing with Slope

If we know the slope and one point of a line, we can sketch the line. To do this, we use a starting point and the slope to construct a slope triangle.

EXAMPLE 3

Using Slope to Draw a Line

Draw the line that contains the point $P(-3, 4)$ and whose slope is -2.

Solution

Plot the point $P(-3, 4)$. We can write the slope as $\frac{-2}{1}$, which means that the rise is -2 and the run is 1.

You may want to locate additional points to show that all are points of the same line.

Starting at $P(-3, 4)$, move *down* 2 units and to the *right* 1 unit to locate a second point Q of the line. (See Fig. 4.16.) Points P and Q determine the line.

Figure 4.16

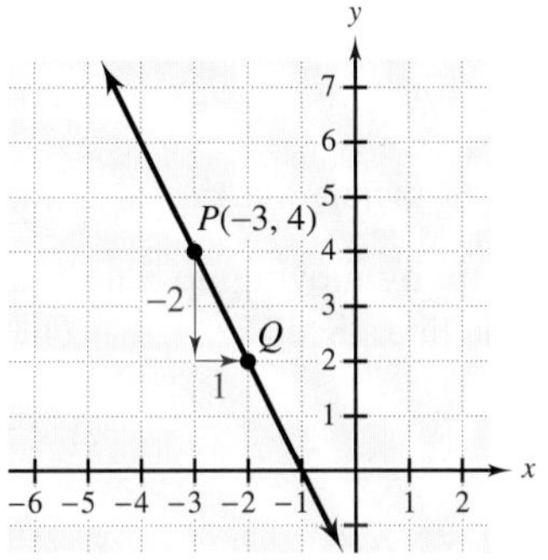

Figure 4.17

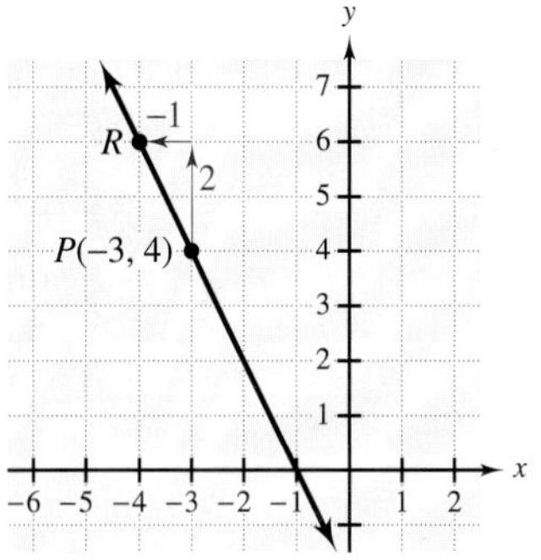

Note that the slope can also be written $\frac{2}{-1}$, which means that the rise is 2 and the run is -1. Starting at $P(-3, 4)$, move *up* 2 units and to the *left* 1 unit to locate a second point R of the line. (See Fig. 4.17.) The line determined by points P and R is the same as the line determined by points P and Q.

Note: When using a point and the slope to graph a line, always start the slope triangle at the plotted point, not at the origin.

Think About It

Why do the lines $y = -\frac{-2}{-3}x$, $y = -\frac{2}{3}x$, $y = \frac{-2}{3}x$, and $y = \frac{2}{-3}x$ all have the same slope?

If we are given just the equation of a line, we can write the equation in slope–intercept form, from which it is easy to determine the slope and y-intercept. Using this information, we can then draw the graph as we did in Example 3.

The following summarizes the approach to graphing a line with information about its slope and y-intercept.

Using a Line's Slope and y-Intercept to Graph the Line

1. Write the equation in the form $y = mx + b$.
2. Plot the y-intercept $(0, b)$.
3. Write the slope m as $\frac{\text{rise}}{\text{run}}$. Starting at the plotted point, move up or down as indicated by the rise. Then move left or right as indicated by the run. Plot the destination point, which is a second point of the line.
4. Draw the line through the two plotted points.

EXAMPLE 4

Using the Slope and y-Intercept to Draw a Line

Sketch the graph of $3x - 2y = 8$.

Solution

First, write the equation in slope–intercept form.

$$3x - 2y = 8$$
$$-2y = -3x + 8$$
$$y = \frac{3}{2}x - 4$$

Plot the y-intercept, $P(0, -4)$.

The slope is $\frac{3}{2}$, which means the rise is 3 and the run is 2. From the y-intercept, draw the slope triangle by moving up 3 units and right 2 units. The destination is a second point Q of the line. (See Fig. 4.18.) The points P and Q determine the line.

Figure 4.18

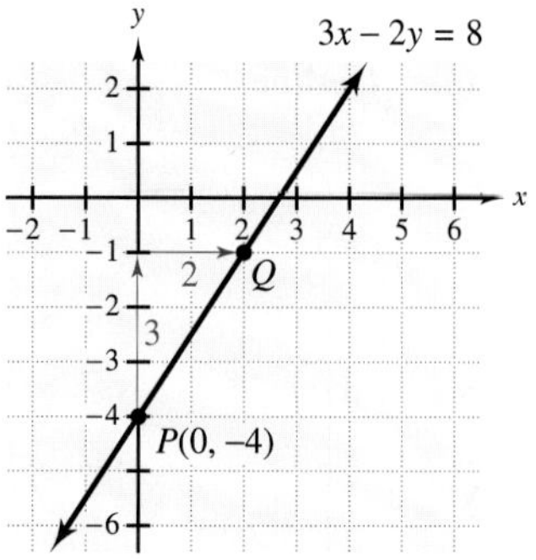

Quick Reference 4.2

The Slope Formula

- Given points A and B of a line, the steepness, or **slope,** of the line is measured by the ratio of the vertical change from A and B (**rise**) and the horizontal change from A and B (**run**).

$$\text{Slope} = \frac{\text{rise}}{\text{run}}$$

- The slope of a line does not depend on the points of the line used to calculate it or on the points designated as the starting and ending points.
- For two points (x_1, y_1) and (x_2, y_2) of a line, $x_1 \neq x_2$, the slope m is defined as follows.

$$m = \frac{y_2 - y_1}{x_2 - x_1} = \frac{y_1 - y_2}{x_1 - x_2}$$

- The slope of any horizontal line is 0; the slope of any vertical line is undefined.
- A line with a positive slope rises from left to right; a line with a negative slope falls from left to right.

The Slope–Intercept Form

- The **slope–intercept form** of a linear equation is the form $y = mx + b$.
- The constant term b is the y-coordinate of the y-intercept of the line. The coefficient m of the x-term is the slope of the line.

Graphing with Slope

- If we know the slope of a line and one of its points, we can use the following procedure to draw the line.
  1. Plot the given point.
  2. Interpreting slope as $\frac{\text{rise}}{\text{run}}$, and starting at the plotted point, draw a slope triangle. The destination point is another point of the line.
  3. Draw the line through the starting and ending points.
- If the equation is given, write it in the form $y = mx + b$ to determine the starting point (y-intercept) and slope.

Speaking the Language 4.2

1. The horizontal change and the vertical change from one point of a line to another are called the _____ and the _____, respectively.
2. A line with a(n) _____ slope falls from left to right.
3. If the equation of a line is $y = mx + b$, then b is the _____ and m is the _____.
4. The slope of a vertical line is _____, whereas the slope of a horizontal line is _____.

Exercises 4.2

Concepts and Skills

1. What is the geometric interpretation of the slope of a line?
2. Describe a line that has a
   (a) positive slope. (b) negative slope.

In Exercises 3–10, predict whether the slope is positive, negative, 0, or undefined. Then calculate the slope.

3.
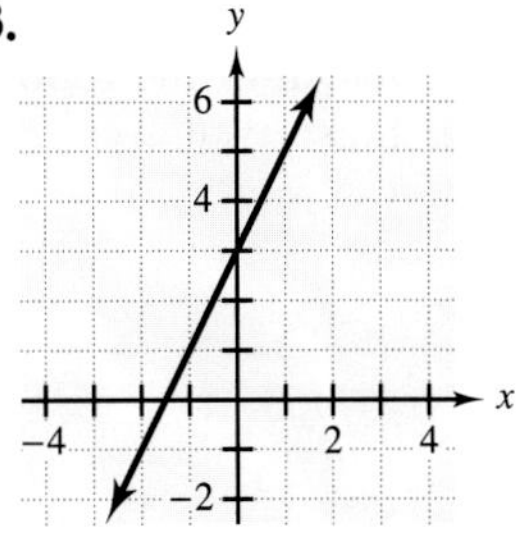

4.
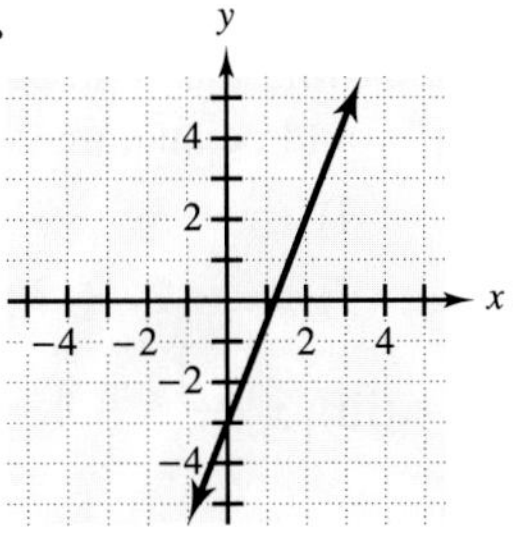

5.
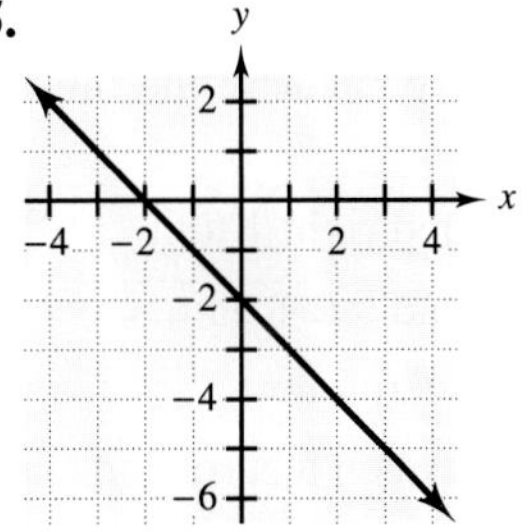

6.
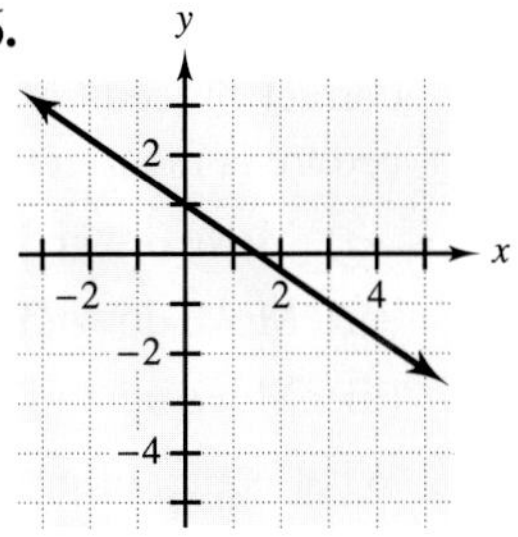

7.
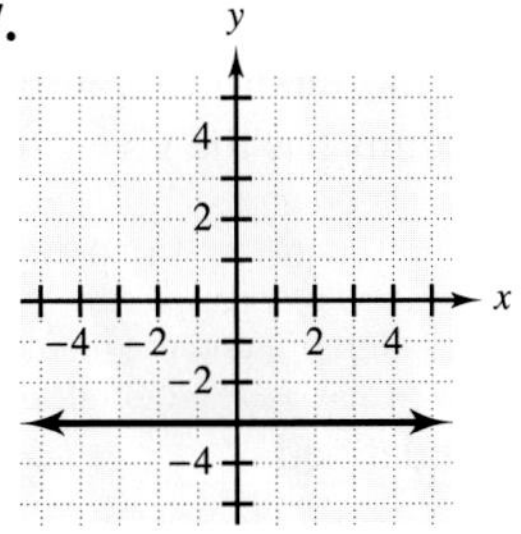

8.
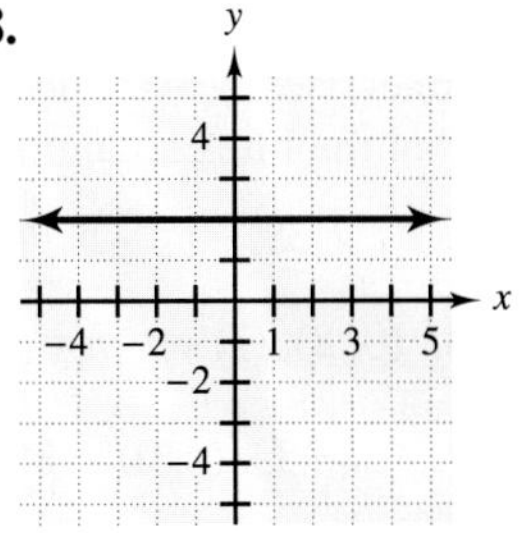

In Exercises 3–10, you might suggest to students that they try to locate a convenient slope triangle.

9.

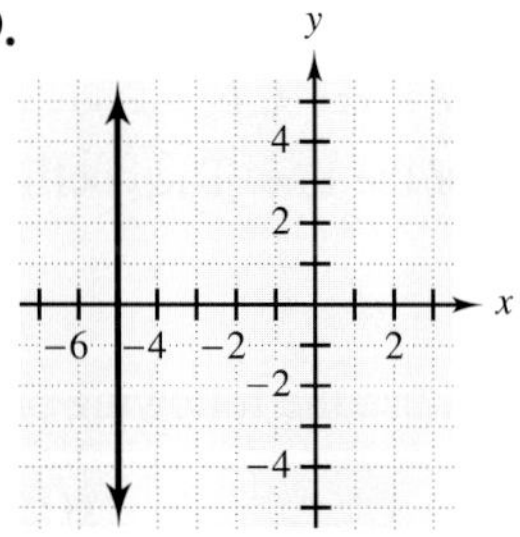

10.

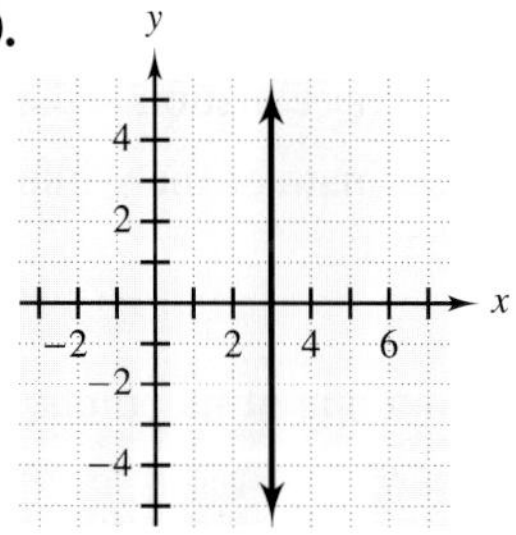

In Exercises 11–24, determine the slope of the line that contains the given points.

11. $(8, -2)$ and $(6, -12)$

12. $(-3, -4)$ and $(-6, -10)$

13. $(3, 4)$ and $(5, 11)$

14. $(-2, 4)$ and $(5, 8)$

15. $(5, -6)$ and $(5, 7)$

16. $(2, -4)$ and $(4, -4)$

17. $(2.3, -5.9)$ and $(-1.7, 5.4)$

18. $(0, 6.94)$ and $(2.45, -3.5)$

19. $(-2, 5)$ and $(8, -3)$

20. $(-1, -6)$ and $(-7, -4)$

21. $(-3, 7)$ and $(6, 7)$

22. $(-3, 7)$ and $(-3, 4)$

23. $\left(\frac{1}{2}, \frac{1}{3}\right)$ and $\left(\frac{2}{3}, \frac{3}{4}\right)$

24. $\left(\frac{1}{3}, \frac{2}{3}\right)$ and $\left(\frac{3}{4}, \frac{1}{2}\right)$

25. If the slope of a line is -3, two of the following procedures are correct ways to draw a slope triangle from a given point of the line. Identify the two correct procedures and explain why the third procedure is incorrect.

(i) Move down 3 units and right 1 unit.

(ii) Move down 3 units and left 1 unit.

(iii) Move up 3 units and left 1 unit.

26. What is the difference between a zero slope and an undefined slope?

CE In Exercises 27–32, determine the unknown coordinate so that the line containing the points has the given slope.

In Exercises 27–32, urge students to sketch the line and predict the result before determining the unknown coordinate algebraically.

27. $(-3, y), (2, -2);\ m = -1$

28. $(0, 8), (2, y);\ m = -\frac{5}{4}$

29. $(2, -1), (x, 0);$ slope is undefined

30. $(x, -3), (-5, 7);$ slope is undefined

31. $(-3, 4), (2, y);\ m = 0$

32. $(0, y), (6, 5);\ m = 0$

33. A line contains the points $P(3, -2)$ and $Q(-11, 6)$. Which of the following is a correct way to calculate the slope of the line? Why?

(i) $m = \dfrac{6 - (-2)}{3 - (-11)} = \dfrac{8}{14} = \dfrac{4}{7}$

(ii) $m = \dfrac{-2 - 6}{3 - (-11)} = \dfrac{-8}{14} = -\dfrac{4}{7}$

34. The slope of a line containing the points (x_1, y_1) and (x_2, y_2) is $\frac{4}{5}$. Does this mean that $y_2 - y_1 = 4$ and $x_2 - x_1 = 5$? Explain.

In Exercises 35–40, determine two solutions of the given equation and use the points to determine the slope of the line.

Remind students that *any* two solutions can be used to determine the slope.

35. $3x + 5y = 15$

36. $x - 2y - 6 = 0$

37. $3 + y = 12$

38. $7 + 3x = 22$

39. $y = \frac{3}{4}x - 3$

40. $y = \frac{2}{3}x + 2$

In Exercises 41–54, write the equation in slope–intercept form, if possible. Then determine the slope of the line.

41. $y + 2x = 17$

42. $y - 3x = 10$

43. $2x - y = 0$

44. $y + 3x = 0$

45. $y + 2 = 17$

46. $y - 3 = 5$

47. $x + 3y - 6 = 0$

48. $3x + 4y = 10$

49. $4x - 3 = 13$

50. $2x + 9 = 7$

51. $3y = 5x - 7$

52. $7x = 5y + 8$

53. $\frac{3}{4}x - \frac{2}{3}y = 12$

54. $\frac{1}{4}x + \frac{2}{5}y = 10$

CE **55.** If a linear equation is in the form $Ax + By = C$, $B \neq 0$, show that the slope of the graph is $-\frac{A}{B}$.

CE **56.** If a linear equation is in the form $Ax + By = C$, $B \neq 0$, show that the y-coordinate of the y-intercept of the graph is $\frac{C}{B}$.

57. A line has a slope of -2. Which of the following could be an equation of the line?

(i) $6x + 3y - 7 = 0$

(ii) $4x - 2y + 3 = 0$

(iii) $x + \frac{1}{2}y - 1 = 0$

58. A line has a slope of $\frac{1}{3}$. Which of the following could be an equation of the line?

(i) $\frac{1}{9}x = \frac{1}{3}y + 3$

(ii) $9x - 3y = 4$

(iii) $12x + 2 - 4y = 0$

59. What is the rise of a line if the line's slope is zero?

60. What is the run of a line if the line's slope is undefined?

In Exercises 61–72, draw a line that contains the given point and has the given slope.

61. $(0, 3);\ m = \frac{3}{4}$

62. $(0, -2);\ m = -\frac{3}{4}$

63. $(0, -1);\ m = -4$

64. $(0, 3);\ m = 2$

65. $(-5, 0);$ m is undefined

66. $(0, 2);\ m = 0$

67. $(-2, -3);\ m = \frac{2}{3}$

68. $(-3, -5);\ m = -\frac{3}{5}$

69. $(3, 7);\ m = 0$

70. $(-4, -3);$ m is undefined

71. $(2, 6);\ m = -5$

72. $(-5, 2);\ m = 4$

In Exercises 73–84, sketch the graph of the given equation.

73. $y = \frac{3}{4}x + 5$

74. $y = -\frac{2}{3}x - 2$

75. $y = -3x + 4$

76. $y = 2x + 3$

77. $y = 2$

78. $y = -2$

79. $y = x$

80. $y = -x$

81. $y = -\frac{3}{4}x$

82. $y = \frac{2}{5}x$

83. $x = 3$

84. $x = -4$

85. If an equation is in the form $y = 3x + b$, what is the effect on the graph if the value of b increases?

86. If an equation is in the form $y = mx + b$, where $m > 0$, what is the effect on the graph if the value of m increases?

In Exercises 87–96, if possible, write the given equation in the form $y = mx + b$. Then sketch the graph.

87. $2x + 5y = 10$

88. $4x - 5y = 20$

89. $x - 3 = 4$

90. $x + 6 = 2$

91. $2x - y - 6 = 2$

92. $x + 3y - 6 = 0$

93. $y + 3 = 10$

94. $y - 2 = 4$

95. $y - 2x = 0$

96. $y + 3x = 0$

Real-Life Applications

97. A pediatrician uses the rule of thumb that a boy grows about 2 inches per year between the ages of 6 and 14.

(a) If the average height of a boy at age 6 is 3 feet 9 inches, what is his expected height at age 14?

(b) Using the two data points in part (a), and assuming that the growth is linear, determine the slope of the graph.

(c) How does your answer in part (b) compare to the pediatrician's figure for average growth rate?

(d) Starting at the data point for age 6, suppose that you rise 8 units and run 4 units. What data would be represented by the destination point?

98. In recent years, math SAT scores have remained relatively constant. What does this imply about the slope of a line used to model the data?

99. The population of a small rural town was 1200 in 1990, but the population has decreased by an average of 75 people per year since then. Suppose that the slope–intercept form $y = mx + b$ is used to model the population y, where x represents the number of years since 1990.

(a) What is the value of m?

(b) What is the value of b?

(c) If you move from the data point for 1990 to the data point for 1997, what are the rise and run?

(d) In what year does the model equation predict that the town will be a ghost town?

100. A standard stairway has 9-inch treads. How long should the risers be in order for the stairway to have a slope of 0.8?

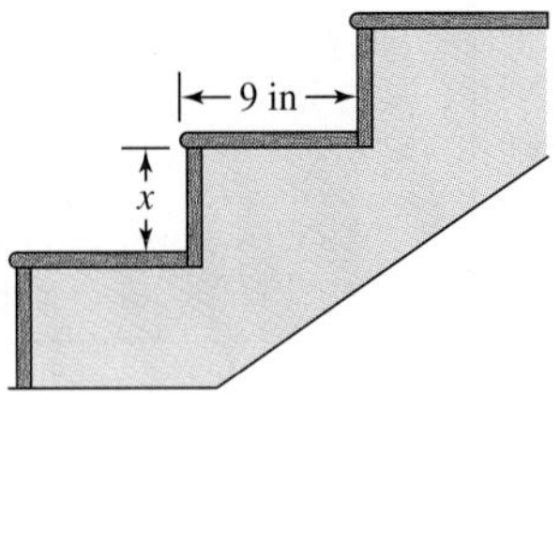

Modeling with Real Data

101. By the year 2000, 52% of all automobiles were projected to be equipped with airbags. The accompanying bar graph shows the data for 1991–2000. (Source: National Highway Traffic Safety Administration.)

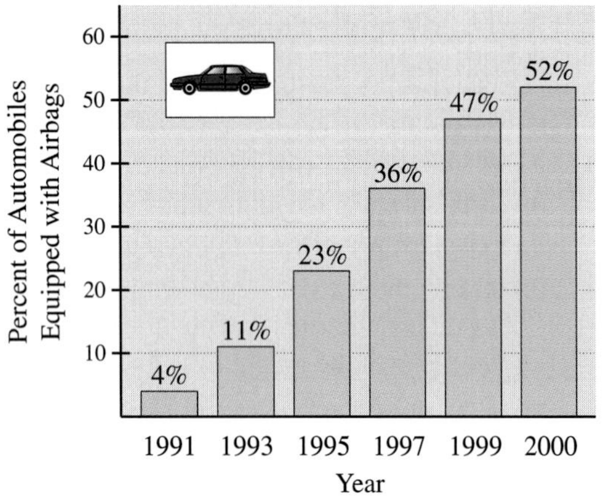

The percentage B of automobiles with airbags can be modeled by the linear function $B(t) = 5.6t - 3.6$, where t is the number of years since 1990.

(a) The model function is selected to fit all the data points as closely as possible. What is the slope of the graph of B?

(b) If a line were drawn through just the data points for 1991 and 2000, what would be the slope of that line?

102. In both parts (a) and (b) of Exercise 101, what does the slope indicate?

103. In Section 2.5 we observed the trends in the number of private and semiprivate golf courses from 1995 to 1999. The accompanying bar graph shows the data for those two years. (Source: National Golf Federation.)

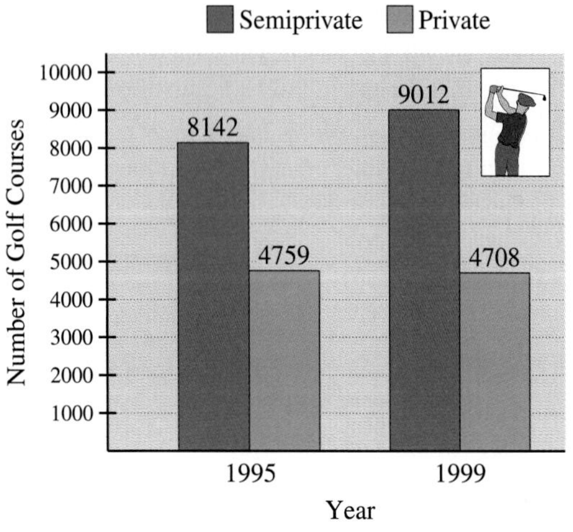

(a) Assuming that the number of courses of each type can be modeled with a linear function, what is the slope of each line?

(b) If you were given line graphs without the actual data, how would the slopes of the lines reveal the trends?

104. (a) Referring to Exercise 103, calculate the difference in the number of semiprivate and private golf courses for 1995 and 1999.

(b) Using the differences for 1995 and 1999 as data points, calculate the slope of the line that contains the points.

(c) What does the result in part (b) indicate?

Data Analysis: *Female State Legislators*

The bar graph shows the percentage of female state legislators during the period 1995–1999. (Source: Center for the American Woman and Politics.)

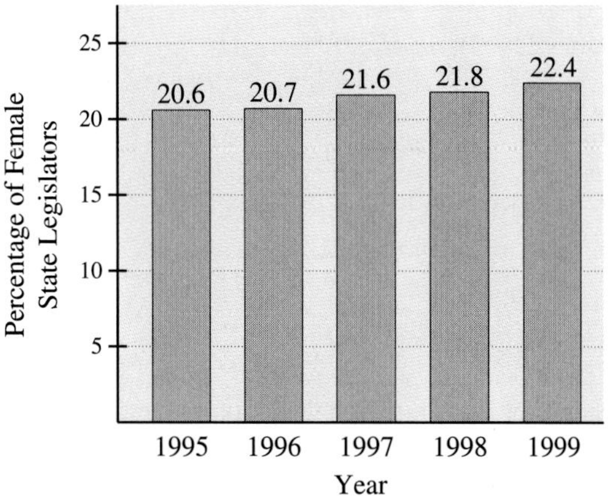

The data in the graph can be modeled by the linear function $P(x) = 0.47x - 917.17$, $x \geq 1995$, where $P(x)$ is the percentage of women in state legislatures in the year x.

105. Observe that the coefficient of x is positive. What does this tell you about the graph of the model function? Is this consistent with the shape of the bar graph?

106. Assuming that the model remains valid, by what year will the percentage of women in state legislatures reach 50%?

107. If a corresponding linear function Q were written to model the percentage of male state legislators during the same period, what could you predict about the coefficient of x? Why?

108. If functions P and Q were graphed together, what would be the coordinates of the point of intersection? How would you interpret those coordinates?

Challenge

109. $A(x, 3)$ and $B(5, 9)$ are points of a line. The midpoint of line segment $\overline{AB}$ is $M(1, y)$. Calculate the slope of the line.

110. Suppose that $P(x_1, y_1)$ and $Q(x_2, y_2)$ are two points of the graph of $y = ax + b$.

(a) Write an equation expressing the fact that (x_1, y_1) is a solution of $y = ax + b$.

(b) Write an equation expressing the fact that (x_2, y_2) is a solution of $y = ax + b$.

(c) Subtract the equation in part (b) from the equation in part (a) and solve for a.

(d) Explain why a is the slope of the graph of $y = ax + b$.

111. Complete the following table and use the results to sketch the graph of $y = 9 - x^2$.

x	-3	-2	-1	0	1	2	3
$9 - x^2$							

For each of the following points of the graph of $y = 9 - x^2$, draw a line through the point and $P(0, 9)$. Then determine the slope of the line.

(a) $A(-3, 0)$ (b) $B(-2, 5)$ (c) $C(-1, 8)$

112. In Exercise 111, points A, B, and C are closer and closer to point P. Calculate the slope of the line that contains points $P(0, 9)$ and $D(-0.1, 8.99)$. As we select points that are closer and closer to point P, what appears to be happening to the value of the slope?

4.3 Applications of Slope

Geometric Applications • Rate of Change • Direct Variation

Geometric Applications

In geometry, if two lines in the same plane do not intersect, then the lines are **parallel.** We begin this section by investigating the relationship between parallel lines and their slopes.

Exploring the Concept

Slopes of Parallel Lines

Figure 4.19 shows the graphs of several equations of the form $y = 2x + b$: $y_1 = 2x - 9$, $y_2 = 2x - 1$, $y_3 = 2x + 3$, $y_4 = 2x + 8$.

Figure 4.19

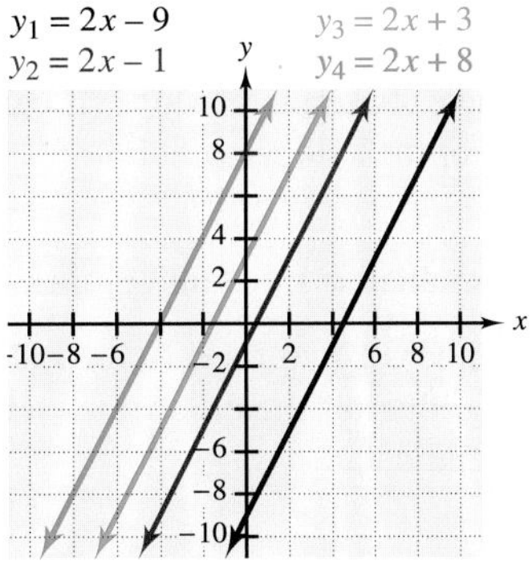

Because $m = 2$ for each equation, the slopes of all the lines are equal, and the lines appear to be parallel. These graphs suggest that distinct lines with the same slope are parallel.

Note: The word *distinct* means that the lines are different lines. If two lines have the same slope and the same y-intercept, then the lines coincide. Such lines are not parallel because they intersect everywhere.

The following relationship can be proved to be true.

> **Parallel Lines and Slopes**
>
> Two distinct, nonvertical lines are **parallel** if and only if they have the same slope. Also, any two distinct vertical lines are parallel.

Roughly speaking, a *quadrilateral* is a geometric figure with four sides. In Example 1, we consider a *parallelogram,* which is a quadrilateral whose opposite sides are parallel.

EXAMPLE 1 Determining Whether a Quadrilateral Is a Parallelogram

Figure 4.20

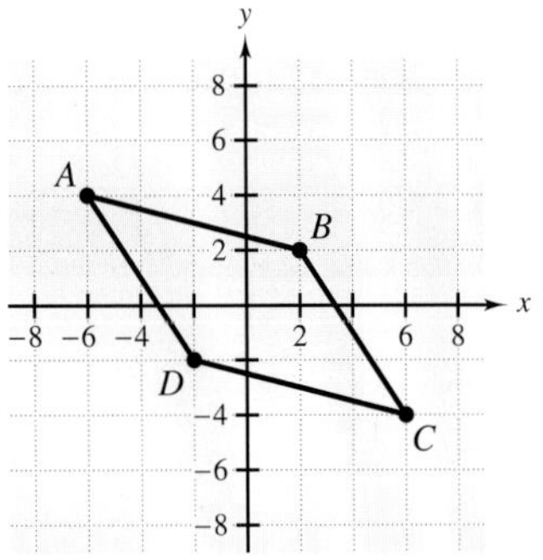

Figure 4.20 shows a quadrilateral drawn in a coordinate system. Show that the quadrilateral is a parallelogram.

Solution

Use the slope formula to determine the slopes of the sides of the quadrilateral.

Side	*Points*	*Slope*
$\overline{AB}$	$A(-6, 4), B(2, 2)$	$m = \dfrac{4-2}{-6-2} = \dfrac{2}{-8} = -\dfrac{1}{4}$
$\overline{DC}$	$D(-2, -2), C(6, -4)$	$m = \dfrac{-2-(-4)}{-2-6} = \dfrac{2}{-8} = -\dfrac{1}{4}$
$\overline{AD}$	$A(-6, 4), D(-2, -2)$	$m = \dfrac{4-(-2)}{-6-(-2)} = \dfrac{6}{-4} = -\dfrac{3}{2}$
$\overline{BC}$	$B(2, 2), C(6, -4)$	$m = \dfrac{2-(-4)}{2-6} = \dfrac{6}{-4} = -\dfrac{3}{2}$

The opposite sides are parallel because they have the same slope. Therefore, the quadrilateral is a parallelogram.

Think About It

A rhombus is a parallelogram with sides of equal length. To show that a quadrilateral is a rhombus, you would need to show that the opposite sides are parallel. What other formula would you need to complete the proof?

From geometry, we know that two lines are **perpendicular** if they intersect to form a right angle. In terms of slope, it can be shown that two lines are perpendicular if the product of their slopes is -1.

LEARNING TIP

If we know the slope of one line, then an easy way to determine the slope of a perpendicular line is to take the opposite of the reciprocal of the known slope. For instance, if the slope of a line is $-\frac{7}{5}$, then $\frac{5}{7}$ is the slope of a perpendicular line.

> **Perpendicular Lines and Slopes**
>
> If neither of two lines is vertical, then the lines are perpendicular if and only if the product of their slopes is -1. Also, a vertical line and a horizontal line are perpendicular.

Note: In symbols, if the slopes of two lines are m_1 and m_2, then the lines are perpendicular if and only if $m_1 m_2 = -1$. This relationship can also be expressed as $m_1 = -\dfrac{1}{m_2}$ or $m_2 = -\dfrac{1}{m_1}$.

EXAMPLE 2

Determining Whether Lines Are Perpendicular

Determine whether the graphs of the given equations are perpendicular.

(a) $y = \frac{1}{2}x + 3$
$y = -2x - 4$

(b) $7x - 12y = -36$
$11x + 6y = -24$

Solution

(a) From the equations, we see that the slopes of the lines are $m_1 = \frac{1}{2}$ and $m_2 = -2$. Because $m_1 m_2 = \frac{1}{2}(-2) = -1$, the lines are perpendicular.

(b) Write the equations in slope–intercept form.

$$y = \frac{7}{12}x + 3$$

$$y = -\frac{11}{6}x - 4$$

Multiply the slopes.

$$\frac{7}{12} \cdot \left(-\frac{11}{6}\right) = -\frac{77}{72} \approx -1.07$$

Because the product of the slopes is not -1, the lines are not perpendicular.

EXAMPLE 3

Parallel and Perpendicular Lines

Determine whether the lines in each pair are parallel, perpendicular, or neither.

(a) $y = -x + 3$ and $x - 1 + y = 0$

(b) $3x + 2y = 6$ and $-y = 2 - \frac{2}{3}x$

(c) $3x - y = 5$ and $x - 3y = 0$

Solution

	Equation	*Slope–Intercept Form*	*Slope*
(a)	$y = -x + 3$	$y = -x + 3$	-1
	$x - 1 + y = 0$	$y = -x + 1$	-1
(b)	$3x + 2y = 6$	$y = -\frac{3}{2}x + 3$	$-\frac{3}{2}$
	$-y = 2 - \frac{2}{3}x$	$y = \frac{2}{3}x - 2$	$\frac{2}{3}$
(c)	$3x - y = 5$	$y = 3x - 5$	3
	$x - 3y = 0$	$y = \frac{1}{3}x$	$\frac{1}{3}$

Because the lines in part (a) have the same slope and different y-intercepts, they are parallel. The product of the slopes in part (b) is -1, so the lines are perpendicular. Because the slopes in part (c) are not equal and their product is not -1, the lines are neither parallel nor perpendicular.

EXAMPLE 4

In Section 4.4 we use this technique in writing an equation of a line.

Slopes of Perpendicular Lines

Suppose the equation of line L is $y = \frac{2}{3}x - 5$. If line N is perpendicular to line L, what is the slope of line N?

Solution

From the equation of line L, the slope of line L is $\frac{2}{3}$.

Because line N is perpendicular to line L, the slope of line N is the opposite of the reciprocal of the slope of line L.

The opposite of the reciprocal of $\frac{2}{3}$ is $-\frac{3}{2}$. Therefore, the slope of line N is $-\frac{3}{2}$.

Previously, we determined whether a triangle was a right triangle by applying the Pythagorean Theorem. Now we can determine whether any two sides are perpendicular by examining their slopes.

EXAMPLE 5

Determining Whether a Triangle is a Right Triangle

The points $A(2, 5)$, $B(3, 7)$, and $C(6, 3)$ are the vertices of a triangle. (See Fig. 4.21.) Is the triangle a right triangle?

Figure 4.21

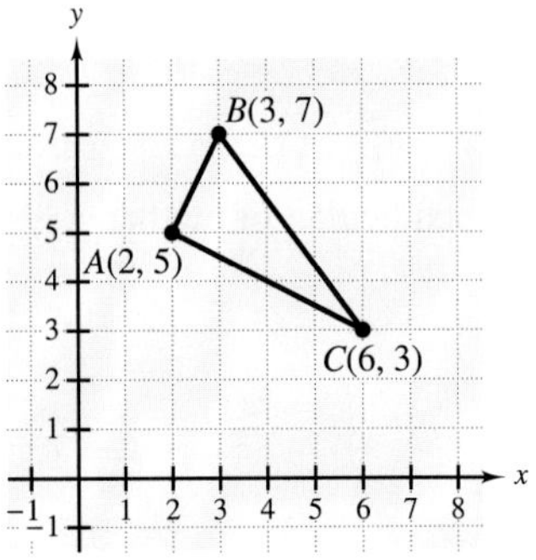

Solution

From Figure 4.21 it appears that line segments $\overline{AB}$ and $\overline{AC}$ are perpendicular. We use the slope formula to find the slopes of these line segments.

Points	*Slope*
$A(2, 5)$ and $B(3, 7)$	$m = \frac{7-5}{3-2} = \frac{2}{1} = 2$
$A(2, 5)$ and $C(6, 3)$	$m = \frac{3-5}{6-2} = \frac{-2}{4} = -\frac{1}{2}$

Because the product of the slopes is $2 \cdot \left(-\frac{1}{2}\right) = -1$, the line segments are perpendicular. Therefore, the triangle is a right triangle.

Rate of Change

This topic is optional and is not needed for any future material in this text.

An important concept in mathematics involves the rate at which a function changes. Specifically, for the function $y = mx + b$, the **rate of change** of y with respect to x is the amount by which y changes for a unit change in x.

Note: A *unit change in x* means a 1-unit increase or derease in the value of x. The corresponding change in y can be positive or negative.

Exploring the Concept

The technique in this illustration works well in class in relating rate of change and slope. You may want to repeat it with $y = -\frac{3}{2}x + 10$ or some other line with a negative slope.

Rate of Change and Slope

To see the rate of change for the function $y = 2x - 4$, consider the following table of values.

x	-1	0	1	2	3	4	5
y	-6	-4	-2	0	2	4	6

Because y increases by 2 units for each unit increase in x, the rate of change of y with respect to x is 2. Note that the rate of change is the same as the slope.

This result is true in general.

To show that the rate of change is not constant for every function, you might want to illustrate with a simple function such as $y = x^2$.

Slope as a Rate of Change

For a linear equation in two variables, the rate of change of y with respect to x is the slope of the graph of the equation. This rate of change, like the slope, is constant.

Many real-life situations can be modeled by an equation for which the rate of change is constant.

EXAMPLE 6

Hourly Pay as Rate of Change

A salesperson earns a base salary of \$300 per week plus a 5% commission on all sales.

(a) Write an equation for the salesperson's weekly income I as a function of weekly sales s.

(b) What is the rate of change in the weekly income with respect to the weekly sales?

Solution

(a) $I = 0.05s + 300$. The variables are I and s instead of y and x, but the equation is in slope–intercept form.

(b) The graph of the equation is a straight line with a slope of 0.05. The slope indicates that the rate of change of I with respect to s is 0.05. Note that the rate of change is simply the commission rate.

Direct Variation

If you do not intend to derive the slope–intercept and point–slope models for the equation of a line formally, this topic is optional.

Suppose that a neighborhood association collects dues of \$25 per residence. Then the amount y that is collected from x residences is $y = 25x$. In this case we say that y **varies directly** with x.

Definition of Direct Variation

The value of y **varies directly** with the value of x if there is a constant k such that $y = kx$.

Note that $y = kx$ is a special case of the slope–intercept form $y = mx + b$. The constant k is called the **constant of variation.** A direct variation can also be written $\frac{y}{x} = k$, $x \neq 0$. In this case we say that y is **directly proportional** to x, and k is called the **constant of proportionality.**

There are many examples of one variable varying directly with another variable. In each case, the y-intercept of the graph is the origin, and the slope m is the same as the constant k of variation.

EXAMPLE 7

Proportional Distances

The actual distance between two towns varies directly with the distance between them on a map. On a certain map, the distance between Tortilla Flat, Arizona, and Tombstone, Arizona, is $6\frac{3}{4}$ inches. Suppose that on this map, $\frac{1}{3}$ of an inch represents 7 miles. What is the actual distance between these towns?

Solution

Let d represent the actual distance in miles, and let x represent the map distance in inches. Because d varies directly with x, the relation is $d = kx$.

A map distance of $\frac{1}{3}$ of an inch corresponds to an actual distance of 7 miles. Replace x with $\frac{1}{3}$ and d with 7 to determine k.

$$d = kx$$

$$7 = k \cdot \frac{1}{3}$$

$$21 = k$$

The constant of variation is 21, so $d = 21x$. Because the map distance between the two towns is 6.75,

$$d = 21(6.75) = 141.75$$

Students should understand that determining the constant of variation is the same as determining the slope of the line.

The distance from Tortilla Flat to Tombstone is 141.75 miles.

Note that the slope of the graph of $d = 21x$ is the same as the constant of variation, $k = 21$. There are 21 miles of actual distance for each inch of map distance.

Quick Reference 4.3

Geometric Applications

- Two distinct, nonvertical lines are **parallel** if and only if they have the same slope. Any two distinct vertical lines are parallel.
- If neither of two lines is vertical, then the lines are **perpendicular** if and only if the product of their slopes is -1. A vertical line and a horizontal line are perpendicular.

Rate of Change

- For a linear function $y = mx + b$, the **rate of change** of y with respect to x is constant, and it is the slope of the graph of the function.

Direct Variation

- We say that y **varies directly** with x if there is a constant k such that $y = kx$. The linear function $y = kx$ is called a **direct variation,** where k is the **constant of variation.**
- If $x \neq 0$, a direct variation can be written $\frac{y}{x} = k$, where k is called the **constant of proportionality.**

Speaking the Language 4.3

1. The linear function $y = kx$ is called a(n) ________.
2. Two distinct vertical lines or two distinct nonvertical lines with the same slope are ________.
3. For a linear function, the rate of change of y with respect to x is the same as the ________ of the graph.
4. If the product of the slopes of two lines is -1, then the lines are ________.

Exercises 4.3

Concepts and Skills

Parallel and Perpendicular Lines

In all exercises in this section, L_1 refers to line 1, and L_2 refers to line 2. The symbol m_1 is used to represent the slope of line 1, and m_2 is used to represent the slope of line 2.

1. What is the geometric definition of perpendicular lines?
2. What is the geometric definition of parallel lines?

In Exercises 3–8, the slopes of L_1 and L_2 are given. Determine whether L_1 and L_2 are parallel, perpendicular, or neither.

3. $m_1 = 2;\ m_2 = -2$
4. $m_1 = \frac{1}{4};\ m_2 = 4$
5. $m_1 = \frac{1}{2};\ m_2 = 0.5$
6. $m_1 = -4;\ m_2 = 0.25$
7. $m_1 = \frac{3}{5};\ m_2 = -\frac{5}{3}$
8. $m_1 = \frac{7}{9};\ m_2 = \frac{9}{7}$

In Exercises 9–14, for the given slope, determine m_2 so that

(a) L_2 is parallel to L_1.
(b) L_2 is perpendicular to L_1.

9. $m_1 = -3$
10. $m_1 = \frac{5}{8}$
11. $m_1 = -\frac{3}{4}$
12. $m_1 = 2.5$
13. $m_1 = 0$
14. m_1 undefined

In Exercises 15–20, two points of L_1 and two points of L_2 are given. Determine whether the lines are parallel, perpendicular, or neither.

15. L_1: $(3, 5)$ and $(-2, 1)$
    L_2: $(-6, 9)$ and $(-1, 13)$
16. L_1: $(-6, 4)$ and $(-2, -8)$
    L_2: $(108, 120)$ and $(100, 144)$
17. L_1: $(0, 0)$ and $(8, 8)$
    L_2: $(-1, 4)$ and $(3, -7)$
18. L_1: $(4, 8)$ and $(1, -2)$
    L_2: $(-3, 5)$ and $(2, 2)$
19. L_1: $(-4, 2)$ and $(-4, 7)$
    L_2: $(0, 5)$ and $(3, 5)$
20. L_1: $(-2, 3)$ and $(1, 5)$
    L_2: $(0, 4)$ and $(2, 1)$
21. A vertical line and a horizontal line are perpendicular, but the relation $m_1 m_2 = -1$ is not valid. Why?
22. What can you conclude about lines L_1 and L_2 if $m_1 = -\frac{1}{m_2}$, where $m_2 \neq 0$? Why?

In Exercises 23–28, the equations of L_1 and L_2 are given. Determine whether the lines are parallel, perpendicular, or neither.

23. L_1: $3y = 5 - 2x$ $\quad L_2$: $y = -\frac{2}{3}x + 9$
24. L_1: $y = \frac{2}{3}x + 1$ $\quad L_2$: $3x + 2y = 0$
25. L_1: $x = -4$ $\quad L_2$: $y = -4$
26. L_1: $y - 3 = 12$ $\quad L_2$: $5 - y = 14$
27. L_1: $y = 5x - 3$ $\quad L_2$: $y = -5x + 3$

28. L_1: $\frac{1}{2}x - \frac{1}{3}y = 0$ L_2: $\frac{5}{6}x - \frac{5}{9}y = 1$

In Exercises 29–34, the equation of one line is given and two points of another line are given. Are the lines parallel, perpendicular, or neither?

29. L_1: $x - y = 2$ L_2: (5, 9) and (6, 8)

30. L_1: (−3, 1) and (4, 2) L_2: $y = \frac{1}{7}x - 2$

31. L_1: (−2, −5) and (0, 4) L_2: $y = \frac{9}{2}x - 4$

32. L_1: $-2x + 5y = 10$ L_2: (−6, 6) and (−4, 1)

33. L_1: $y = 5$ L_2: (2, 3) and (−6, 0)

34. L_1: (2, −1) and (−7, −1) L_2: $x = -1$

In Exercises 35–38, what can you conclude about the slope of line L?

35. L is parallel to the y-axis.

36. L is perpendicular to the x-axis.

37. L is perpendicular to the y-axis.

38. L is parallel to the x-axis.

In Exercises 39–44, determine the slope of a line that is

(a) parallel to the given line.

(b) perpendicular to the given line.

39. $2x - y = 6$

40. $3x + y - 7 = 0$

41. $4x + 3y - 6 = 0$

42. $2x = 5y - 5$

43. $3y - 9 = 0$

44. $2(x - 3) = 1$

In Exercises 45–48, determine the slope of the line and sketch the graph.

45. The line contains (−1, −4) and is parallel to the line containing (−2, 3) and (4, −1).

46. The line contains (−3, 1) and is perpendicular to the line $y = 2x - 3$.

47. The x-intercept of the line is (3, 0), and the line is parallel to the y-axis.

48. The y-intercept of the line is (0, −6), and the line is parallel to the x-axis.

Encourage students to sketch the lines and predict the value of k before working the problem algebraically.

CE In Exercises 49 and 50, information is given about two lines L_1 and L_2. Determine the value of k so that

(a) L_1 and L_2 are parallel.

(b) L_1 and L_2 are perpendicular.

49. L_1 contains the points (3, −4) and (−1, 2). L_2 contains the points (5, 0) and (−2, k).

50. The equation of L_1 is $3x + 12y = 15$. The equation of L_2 is $y = kx + 7$.

Geometric Models

In Exercises 51 and 52, show that the triangles in the figures are right triangles.

51.

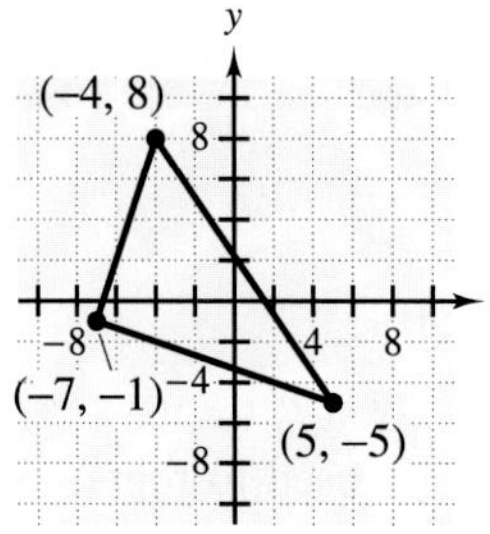

52.

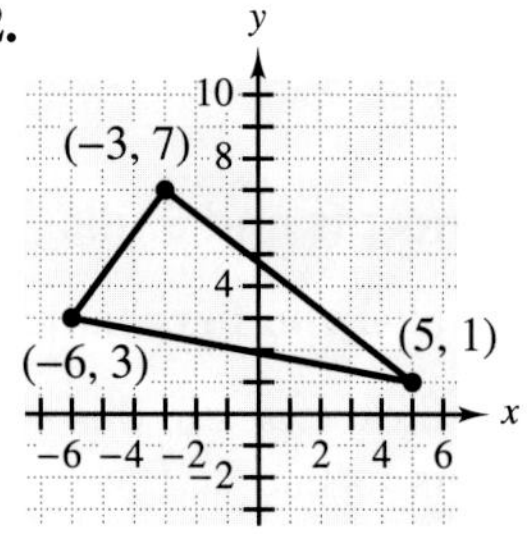

The intent of Exercises 53 and 54 is to determine whether the slopes are the same. However, you might want to discuss how the Distance Formula could be used.

CE In Exercises 53 and 54, determine whether the three given points are collinear. (Points are **collinear** if they are points of the same straight line.)

53. $P(-5, 8)$ $Q(-1, 3)$ $R(7, -7)$

54. $A(-7, -2)$ $B(0, 0)$ $C(21, 5)$

55. The consecutive vertices of a parallelogram are the points $A(3, -1)$, $B(-7, 9)$, $C(-2, 14)$, and $D(8, 4)$. Show that the parallelogram is a rectangle.

56. The consecutive vertices of a square are $A(-3, -2)$, $B(-3, 6)$, $C(5, 6)$, and $D(5, -2)$. Show that the diagonals of the square are perpendicular.

57. The consecutive vertices of a quadrilateral are $A(0, 0)$, $B(0, 5)$, $C(4, 7)$, and $D(4, 2)$. Show that the quadrilateral is a parallelogram.

58. A **trapezoid** is a quadrilateral with exactly one pair of parallel sides. The consecutive vertices of a quadrilateral are $A(11, 2)$, $B(-5, -4)$, $C(-2, 4)$, and $D(6, 7)$. Show that the quadrilateral $ABCD$ is a trapezoid.

Rate of Change

59. Describe the slope of a line in terms of rate of change.

60. For a linear function, explain why the rate of change is constant?

In Exercises 61–64, determine the rate of change of y with respect to x.

61. $y = \frac{1}{2}x + 3$

62. $y = -\frac{2}{3}x + 5$

63. $y = -4x + 9$

64. $y = 3x - 2$

In Exercises 65–68, determine the rate of change of y with respect to x for the line containing the given points.

65. $(-1, -2)$ and $(3, 4)$

66. $(-2, 3)$ and $(3, 1)$

67. $(2, -2)$ and $(-1, 5)$

68. $(-3, 1)$ and $(2, 5)$

CE In Exercises 69–76, the rate of change of a line whose equation is $y = mx + b$ is given. For the given increase/decrease of one variable, determine the corresponding increase/decrease of the other variable.

69. $m = 3$; x increased by 2

70. $m = -2$; y increased by 6

Sketching a slope triangle is especially helpful in Exercises 69–76.

71. $m = -\frac{5}{4}$; y increased by 5

72. $m = \frac{1}{3}$; x increased by 9

73. $m = 5$; x decreased by 2

74. $m = -4$; y decreased by 12

75. $m = -\frac{3}{2}$; y decreased by 6

76. $m = \frac{3}{5}$; x decreased by 20

Direct Variation

77. Explain the meaning of the statement "A varies directly with B."

78. Suppose y varies directly with x. If $x = a$ when $y = b$ and $x = c$ when $y = d$, explain why $\frac{a}{b} = \frac{c}{d}$ where a, b, c, and d are nonzero numbers.

In Exercises 79–84, determine whether the given equation represents a direct variation.

79. $y = 4x$

80. $4 = xy$

81. $\frac{y}{4} = x$

82. $y = 2x + 3$

83. $y = 5x - 3$

84. $y = 3$

CE **85.** Suppose that $y = kx$, $k > 0$. If x increases, then y ____, and if y decreases, then x ____.

CE **86.** Suppose that y varies directly with x and that k is the constant of proportionality. Then $y =$ ____ and $k =$ ____.

Real-Life Applications

Rate of Change

87. Ace Auto Rental rents a car for \$29 a day plus 10 cents per mile. For a 1-day rental, let m represent the number of miles driven, and let c represent the total cost.

(a) Write an equation for total rental cost.

(b) What is the rate of change in the total cost with respect to the number of miles driven?

88. A telemarketer earns \$100 per week plus a 30% commission on all sales. Let I represent the weekly income, and let s represent the weekly sales.

(a) Write an equation for the total weekly income.

(b) What is the rate of change in income with respect to sales?

In Exercises 89 and 90, assume that the described relationship is linear.

89. In 1950 the cost of a soft drink was 5 cents. In 2000 the cost of the same soft drink was 55 cents. During this period, what was the annual rate of change in the cost of the drink?

90. On a certain January day in Minnesota, the temperature at 6:00 A.M. was −15°F. By noon, the temperature was up to 12°F. What was the hourly rate of temperature change?

Direct Variation

91. The weight of a load of bricks varies directly with the number of bricks in the load. If a load of 500 bricks weighs 1175 pounds, what will a load of 1200 bricks weigh?

92. The weight that can be lifted by an automobile jack varies directly with the force exerted downward on the jack handle. If a force of 9 pounds will lift 954 pounds, what weight will be lifted by a force of 15 pounds?

93. If the height of a triangle remains constant, then the area of the triangle varies directly with the length of the base. If a triangular sail with a fixed height has an area of 165 square feet when the base is 12 feet, what would the area be if the base were increased to 18 feet?

94. In physics, Hooke's Law states the following: The distance d that a hanging spring is stretched varies directly with the weight of an attached object. If a weight of 4.5 pounds stretches a spring a distance of 3 inches, how far will a weight of 9.5 pounds stretch this spring?

95. The amount paid for typing varies directly with the number of pages typed. If \$100 is paid for typing 16 pages, how much will be paid for typing 25 pages?

96. The discount on a blouse varies directly with the marked (selling) price. If the discount on a blouse marked \$37 is \$11.84, what is the discount on a blouse marked \$50?

Modeling with Real Data

97. The average daily cost of a car vacation for a family of four is \$233. (Source: American Automobile Association.)

(a) Using data points of the form (days, cost), write a linear function to model the cost of a trip.

(b) Explain why your function in part (a) is a direct variation.

(c) What is the constant of variation?

98. Gateway manufactures and sells computers. This company's sales rose from \$5.0 billion in 1996 to \$10.4 billion in 2000. (Source: Value Line.) Assuming a constant rate of increase, find the average rate of change in sales during this period.

99. The bar graph shows the nuclear-based electricity generation (in billions of kilowatt-hours) by the United States for selected years. (Source: Energy Information Administration.) With x representing the number of years since 1990, a model function is $E(x) = 9.8x + 602.5$.

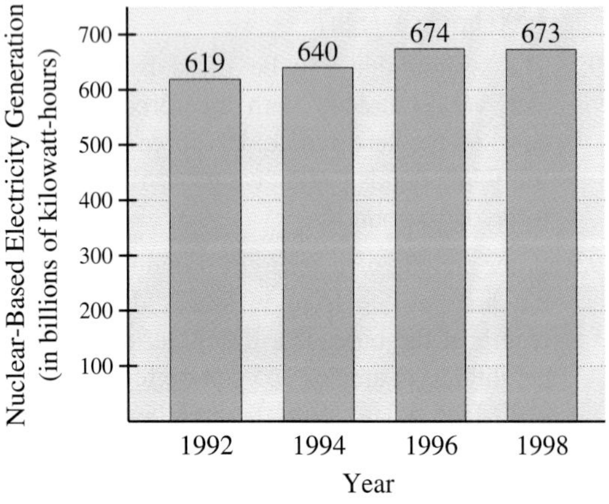

(a) What is the slope of the graph of E and what does it indicate?

(b) According to the model, what is the average rate of change in the nuclear-based production of electricity from 1992 to 1998? Compare your result to the slope in part (a).

100. The per-share dividends for Wachovia Corporation increased from \$1.38 in 1995 to \$2.06 in 1999. (Source: Wachovia Corporation.) Assume that the rate of change was constant.

(a) Using data points of the form (x, dividend), where x is the number of years since 1995, would a linear function describing this information be a direct variation? Why?

(b) If the function describing this information is in the slope–intercept form, what is m?

(c) What is your interpretation of the value of m?

Data Analysis: *Bald Eagles*

Although they were once about to become extinct, the population of bald eagles has increased steadily since 1986. (Source: U.S. Fish and Wildlife Service.)

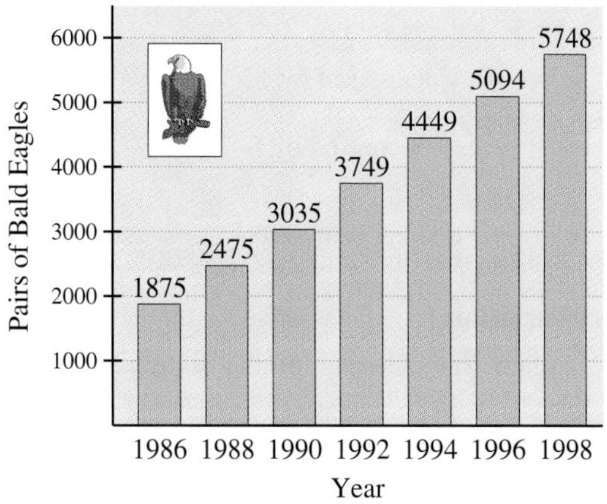

101. Letting x represent the number of years since 1984, plot the points whose coordinates are of the form (x, pairs). (Draw the x-axis from 0 to 16 in increments of 2; draw the y-axis from 0 to 6600 in increments of 600.)

102. Imagine a line containing the first and last points of the graph in Exercise 101. Calculate the slope of that line and interpret the slope in terms of rate of change.

103. Write the function $p(x) = mx + b$, where m is your result in Exercise 102 and b is calculated by using the data for 1986. Evaluate the function to verify the data for 1986 and 1998.

104. Use your function in Exercise 103 to predict in what year the population of bald eagles will reach 8976 pairs.

Challenge

In Exercises 105–110, the equation of a line is given. Determine k so that the given condition is met.

105. $kx + 3y = 10$; the slope of the line is -2.

106. $2x - ky = 2$; the line is parallel to the line $y = \frac{1}{2}x - 5$.

107. $3kx - 3y = 4$; the line is perpendicular to the line $x + y = 0$.

108. $5x + 2ky = 7$; the slope of the line is undefined.

109. $kx - 5y = 20$; the line is parallel to the x-axis.

110. $kx + ky = 4$; the slope of the line is -1.

111. A **rhombus** is a parallelogram whose sides are all the same length. The vertices of a rhombus are $A(0, 0)$, $B(3, 4)$, $C(8, 4)$, and $D(5, 0)$. Show that the diagonals of the rhombus are perpendicular.

112. If L_1, L_2, and L_3 are three lines in the same plane such that L_1 and L_2 are both perpendicular to L_3, show that L_1 and L_2 are parallel.

4.4 Equations of Lines

Slope–Intercept Form • Point–Slope Form • Special Cases • Analyzing Conditions • Real-Life Applications

Slope–Intercept Form

In previous sections of this chapter, we began with a given linear equation, and by writing the equation in the form $y = mx + b$, we easily determined the slope and y-intercept of the line and produced the graph.

In this section we begin with given information about the graph, such as the slope, the intercepts, or other points of the line. If we are given enough information to sketch the graph, then we can use this information along with the model $y = mx + b$ to write an equation of the line.

EXAMPLE 1 **Writing an Equation of a Line Given the Slope and the y-Intercept**

Write an equation of a line whose slope is 2 and whose y-intercept is $(0, -3)$.

Solution

We are given that $m = 2$ and $b = -3$.

$y = mx + b$ Slope–intercept form of the required equation

$y = 2x - 3.$ Replace m with 2 and b with -3.

The equation of the line is $y = 2x - 3$.

Suppose we are given the slope of a line along with some point other than the y-intercept. To determine an equation of the line, we use the key concept that relates equations and graphs:

Every point of a line has coordinates that are solutions of the equation of the line.

EXAMPLE 2

Writing an Equation of a Line Given the Slope and One Point

Write an equation of the line that contains the point $P(1, 3)$ and whose slope is 2.

Solution

Because $m = 2$, we can start the equation as follows.

$y = mx + b$ Slope–intercept form of a linear equation

$y = 2x + b$ Replace m with 2.

Because $P(1, 3)$ is a point of the line, the coordinates satisfy the equation.

$3 = 2(1) + b$ Replace x with 1 and y with 3.

$3 = 2 + b$

$1 = b$

Now we know that $b = 1$, so the equation can be completed.

$y = 2x + 1$ Replace b with 1.

Point–Slope Form

As we have seen, the form $y = mx + b$ can be used to write an equation of a line even if the given point is not the y-intercept.

Figure 4.22

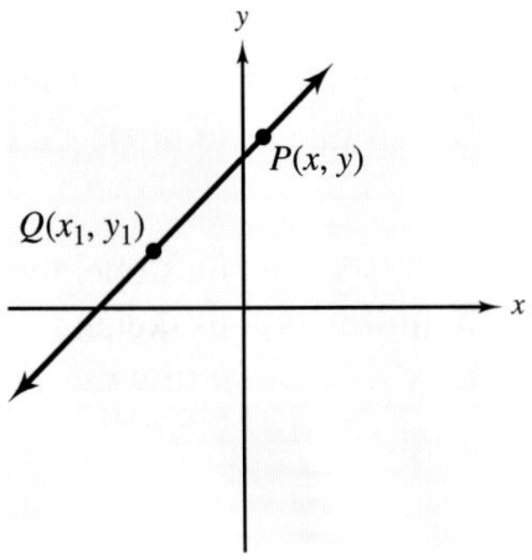

An alternative model can be used in such cases. Figure 4.22 shows a line with a *fixed* point $Q(x_1, y_1)$ and the point $P(x, y)$, which represents *any* other point of the line.

The slope of the line is given by

$$m = \frac{y - y_1}{x - x_1}$$

Because this equation is in the form of a direct variation, it is equivalent to

$$y - y_1 = m(x - x_1)$$

We use the definition of direct variation because we have not yet discussed solving rational equations.

This result is known as the **point–slope form** of a linear equation.

The Point–Slope Form of a Linear Equation

The **point–slope form** of a linear equation is

$$y - y_1 = m(x - x_1)$$

where m is the slope of the line and (x_1, y_1) is a fixed point of the line.

The point–slope form is particularly useful for writing an equation of a line when the slope and one point of the line are given.

EXAMPLE 3

Writing an Equation of a Line Given the Slope and One Point

Use the point–slope form to write an equation of a line that contains $P(-5, -3)$ and whose slope is $-\frac{2}{5}$.

Some instructors prefer to use the slope–intercept model exclusively and to ignore the point–slope model. The slope–intercept model has the advantage of being better for sketching a graph and for stressing the connection between points and solutions.

Solution

$$y - y_1 = m(x - x_1)$$ Point–slope form of a linear equation

$$y - (-3) = -\frac{2}{5}[x - (-5)]$$ Replace m with $-\frac{2}{5}$, x_1 with -5, and y_1 with -3.

$$y + 3 = -\frac{2}{5}(x + 5)$$

$$y + 3 = -\frac{2}{5}x - 2$$ Distributive Property

$$y = -\frac{2}{5}x - 5$$ Subtract 3 from both sides.

The slope–intercept form and the point–slope form both require that we know the slope of a line in order to write its equation. If we are given just two points of a line, we can calculate the slope and write an equation of the line with either of the forms.

EXAMPLE 4

Writing an Equation of a Line Given Two Points

Write an equation of the line containing the points $P(4, 2)$ and $Q(-3, -5)$.

Solution

We use the slope formula to compute the slope of the line.

$$m = \frac{y_2 - y_1}{x_2 - x_1} = \frac{-5 - 2}{-3 - 4} = \frac{-7}{-7} = 1$$

We use the point–slope form to begin writing the equation.

$$y - y_1 = m(x - x_1)$$

$$y - y_1 = 1(x - x_1)$$ Replace m with 1.

Because points P and Q both represent solutions, we can use either one as the fixed point. We use point $P(4, 2)$.

$$y - 2 = 1(x - 4)$$ Replace x_1 with 4 and y_1 with 2.

$$y - 2 = x - 4$$

$$y = x - 2$$ Add 2 to both sides.

Special Cases

As we have seen, the special cases of linear equations have graphs with the following characteristics.

Equation	*Graph Characteristics*
$y = b$	horizontal line, slope is 0
$x = c$	vertical line, slope is undefined

Students are usually persuaded about the value of knowing the special cases when they see that equations can be written immediately. Remind them that neither model can be used for vertical lines anyway.

If the slope is 0, we can use either the slope–intercept form or the point–slope form to write an equation of the line. However, it is easier to use the special-case form $y = b$.

If the slope is undefined, then neither model applies, and we must use the special-case form $x = c$.

EXAMPLE 5

Special Cases

(a) Write an equation of a line whose slope is 0 and that contains the point $R(2, 4)$.

(b) Write an equation of the line containing the points $C(4, 2)$ and $D(4, 7)$.

Solution

(a) Because the slope is 0, the line is horizontal. (See Fig. 4.23.) Therefore, the equation is of the form $y = b$, where b is the y-coordinate of every point of the line.

Because $R(2, 4)$ is a point of the line and its y-coordinate is 4, every point of the line has a y-coordinate of 4. The equation is $y = 4$.

Figure 4.23

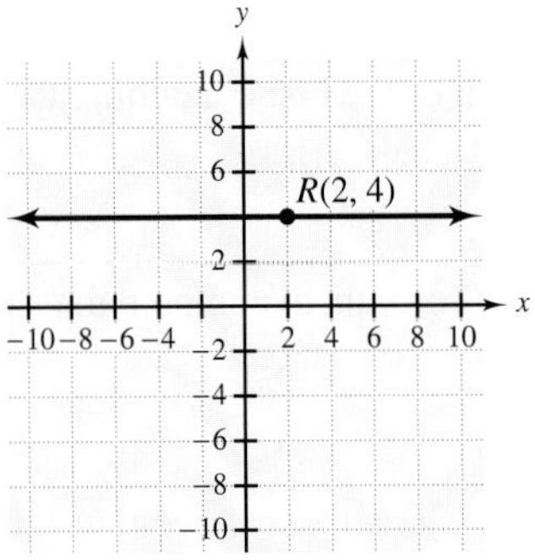

Figure 4.24

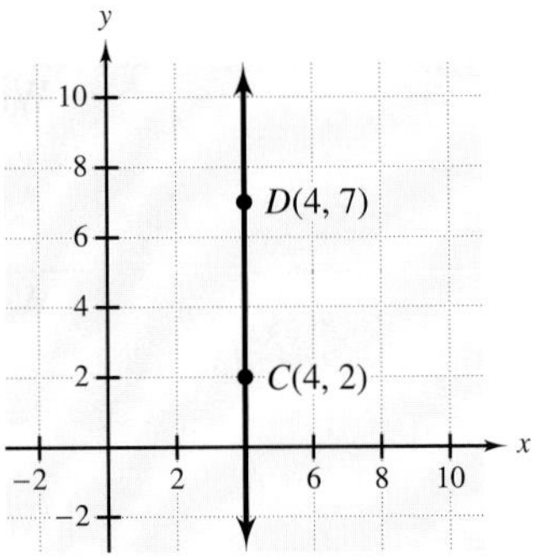

(b) Because the x-coordinates of points C and D are the same, the line is vertical. (See Fig. 4.24.) Therefore, the equation is of the form $x = c$. Points C and D and all other points of the line have an x-coordinate of 4. The equation is $x = 4$.

Analyzing Conditions

Sometimes a line is described in reference to one or more other lines. If enough information is furnished about the given line(s), we can deduce the features of the required line and write its equation.

EXAMPLE 6

Remind students about Example 4 in the previous section, which shows how to determine the slope of a line that is perpendicular to a given line. You can repeat this example with line *A* perpendicular to line *B* and having the same *y*-intercept as line *C*.

Describing a Line Relative to Other Lines

The equation of line B is $y = 2x - 1$ and the equation of line C is $y = x - 2$. Line A is parallel to line B and has the same y-intercept as line C. Write an equation of line A.

Solution

To visualize the information, we sketch the lines. (See Fig. 4.25.)

Figure 4.25

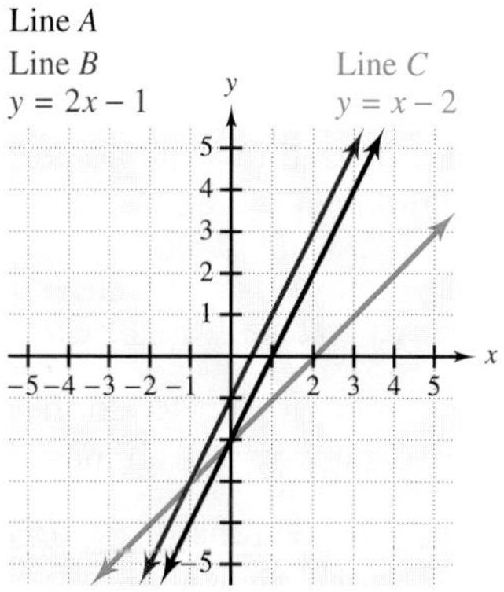

LEARNING TIP

When you work problems that involve two or more lines, sketching the lines and visualizing their relationship will be a great help in understanding what to do.

The equation of line B is $y = 2x - 1$, which means that the slope of line B is 2. Because line A is parallel to line B, the slope of line A is also 2. Thus we can begin writing the equation of line A.

$$y = mx + b$$

$$y = 2x + b \quad \text{Replace } m \text{ with 2.}$$

The equation of line C is $y = x - 2$, which means that the y-intercept of line C is $(0, -2)$. Because line A has the same y-intercept as line C, the y-intercept of line A is also $(0, -2)$. Now the equation of line A can be completed.

$$y = 2x - 2 \quad \text{Replace } b \text{ with } -2.$$

EXAMPLE 7

Horizontal and Vertical Lines

The equation of line L is $y = -2$. Line M is perpendicular to line L and contains the point $(3, 4)$. Write an equation of line M.

Solution

We can use the given information to sketch the lines. (See Fig. 4.26.)

Figure 4.26

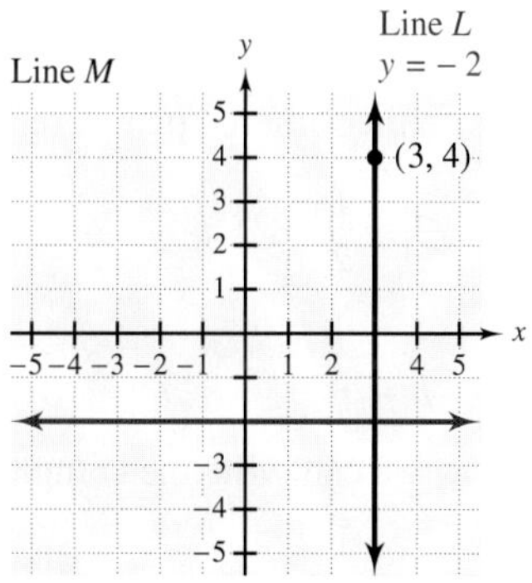

Think About It

Suppose that two lines are perpendicular and you know their point of intersection. Why is this information insufficient for writing the equations of the lines?

The equation of line L is of the form $y = b$. Therefore, line L is a horizontal line. Because line M is perpendicular to line L, line M must be a vertical line. Therefore, its equation is of the form $x = c$, where the constant is the x-coordinate of every point of the line.

Because $(3, 4)$ is a point of the line and its x-coordinate is 3, every point of the line has an x-coordinate of 3. The equation of line M is $x = 3$.

Real-Life Applications

Sometimes we need to write a linear equation to model all or part of the conditions in an applied problem.

EXAMPLE 8 **Modeling with a Linear Equation**

At age 8, a boy competed in the high jump and managed a personal best of 3 feet. By age 12, the boy was jumping 5 feet.

His coach decided that his progress in the high jump could be modeled by a linear equation. He let x represent the boy's age and y the height of his jump.

(a) What linear equation did the coach write?

(b) If he uses this equation to predict how high the boy can jump at age 20, what will his result be?

Solution

(a) If we write ordered pairs in the form (age, height) or (x, y), two solutions of the equation are $(8, 3)$ and $(12, 5)$.

The slope is $m = \dfrac{5 - 3}{12 - 8} = \dfrac{2}{4} = \dfrac{1}{2}$. Thus we can begin to write the equation.

$$y = \frac{1}{2}x + b$$

Use either ordered pair as replacements for x and y.

$$3 = \frac{1}{2} \cdot 8 + b \qquad \text{Replace } x \text{ with 8 and } y \text{ with 3.}$$

$$3 = 4 + b$$

$$b = -1$$

Now we can write the complete equation.

$$y = \frac{1}{2}x - 1$$

(b) The coach wants to predict the height y that the boy will jump when his age x is 20.

$$y = \frac{1}{2} \cdot 20 - 1 = 10 - 1 = 9 \text{ feet}$$

Using the linear equation as a model, the coach predicts that the boy will be able to jump 9 feet at age 20.

This is an opportunity to discuss a reasonable domain and to ask students to sketch a graph that would more closely model a person's ability to jump during a lifetime. Point out that a mathematical model can still be valuable even if the domain must be restricted.

Note: Try using the model in Example 8 to predict the height the boy will be able to jump at age 60. Because of the nature of a given problem, a mathematical model may be reasonably accurate only for a limited domain.

Quick Reference 4.4

Slope–Intercept Form

- The **slope–intercept** form $y = mx + b$ can be used as a model for writing an equation of a nonvertical line.
- If the slope m and the y-intercept b of a line are given, we can write an equation of the line by substituting directly into the slope–intercept form $y = mx + b$.
- We can also use the slope–intercept form if the slope and a point (x_1, y_1) of the line are given. Replace m with the given slope, and replace x and y with x_1 and y_1, respectively. Then solve for b.

Point–Slope Form

- An equivalent model for writing a linear equation of a nonvertical line is the **point–slope form** $y - y_1 = m(x - x_1)$. The point–slope form is particularly useful for writing an equation of a line for which the given point is not the y-intercept.
- If two points of a line are given, we can write an equation of the line by using the slope formula to calculate the slope. Then, using the point–slope form, we can replace m with the slope and replace x_1 and y_1 with the coordinates of either of the given points.

Special Cases

- The equation of a horizontal line can be found with either the slope–intercept form or the point–slope form, but it is much easier to use the model $y = b$. The constant is the y-coordinate of any point of the line.
- The equation of a vertical line cannot be found with the slope–intercept or point–slope model because the slope is undefined. Instead, use the model $x = c$, where the constant is the x-coordinate of any point of the line.

Analyzing Conditions

- If a nonhorizontal or nonvertical line is described in reference to one or more other lines, we use the given information to determine the slope of the required line and one of its points. Then we can write the equation of the line with the slope–intercept or point–slope model.
- If the given conditions indicate that a line is horizontal or vertical, then we can write the equation of the line with the simpler model $y = b$ or $x = c$.

Speaking the Language 4.4

1. The model $y - y_1 = m(x - x_1)$ is called the ________ form.
2. If a line is ________, then a form for writing its equation is $x = c$.
3. If the y-intercept and the slope of a line are given, a convenient model for writing an equation of the line is the ________ form.
4. Except in special cases, to write an equation of a line that is described in reference to a second line, we must know the ________ of the second line.

Exercises 4.4

Concepts and Skills

1. What information do we need to write an equation of a line with the slope–intercept form?
2. Why can the slope–intercept form be used to write an equation of a horizontal line but not an equation of a vertical line?

In Exercises 3–6, write the equation of the line.

3. $m = -4, b = -3$
4. $m = 3, b = 4$
5. $m = \frac{1}{2}, b = 3$
6. $m = -\frac{2}{3}, b = -2$

In Exercises 7–12, write an equation of the line that has the given slope and contains the given y-intercept.

7. $m = -5, (0, 1)$
8. $m = 6, (0, 6)$
9. $m = -\frac{3}{4}, (0, -5)$
10. $m = \frac{5}{8}, (0, 8)$
11. $m = 7, \left(0, -\frac{2}{3}\right)$
12. $m = -4, \left(0, \frac{1}{2}\right)$

In Exercises 13–18, write an equation for the line whose graph is shown in the figure.

Remind students to look for a convenient slope triangle.

13.

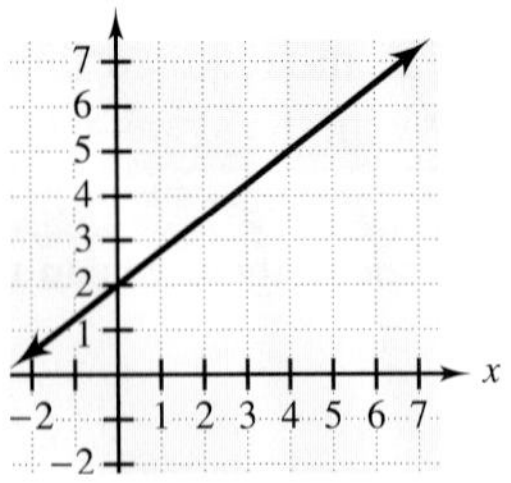

14.

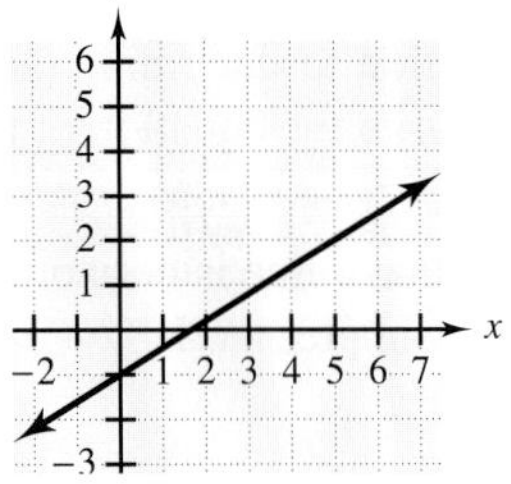

15.

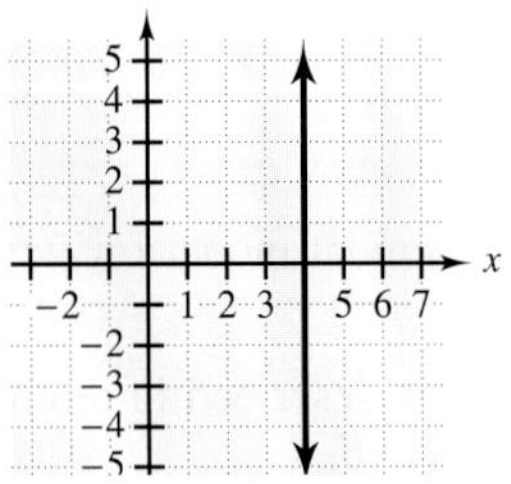

16.

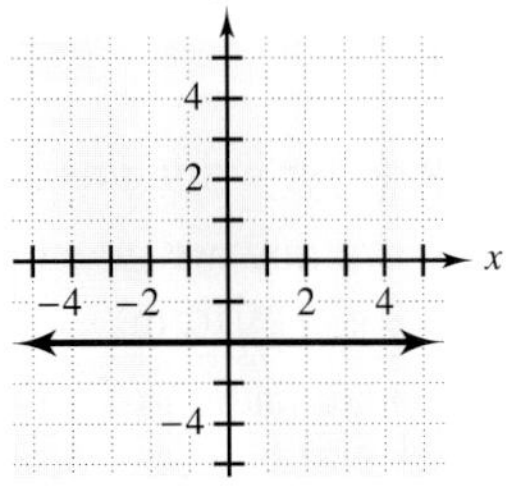

17.

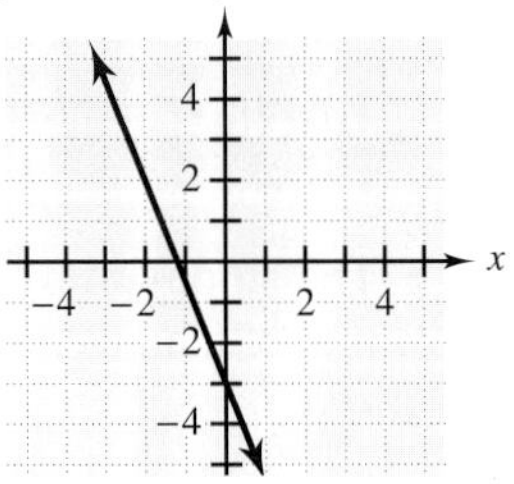

18.

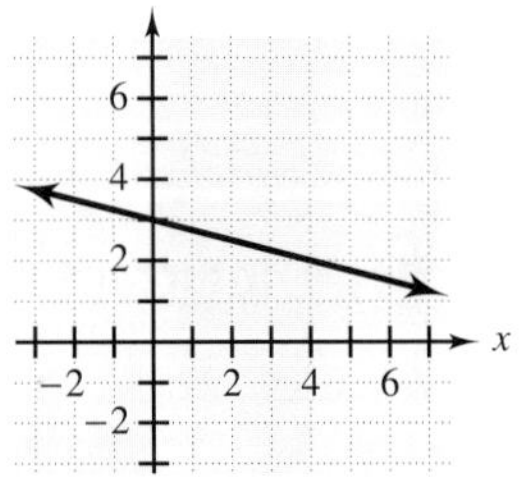

19. Why can the point–slope form be used to write an equation of a horizontal line but not an equation of a vertical line?
20. What information do we need to write an equation of a line with the point–slope form?

In Exercises 21–30, write an equation of the line that has the given slope and contains the given point. Express the equation in the form $y = mx + b$ if possible.

21. $m = 2, (1, 3)$
22. $m = -3, (2, 5)$
23. $m = \frac{2}{5}, (10, -4)$
24. $m = -\frac{1}{3}, (-6, 3)$
25. $m = -\frac{2}{3}, (-2, -5)$
26. $m = \frac{4}{5}, (-1, -2)$
27. $m = 0, (2, -5)$
28. $m = 0, (-3, -4)$
29. m is undefined, $(-3, 4)$
30. m is undefined, $(-4, 5)$

In Exercises 31–36, write an equation of the line that has the given slope and contains the given point. Express the equation in the form $Ax + By = C$.

31. $m = 1, (7, 4)$
32. $m = 2, \left(2, \frac{1}{2}\right)$
33. $m = \frac{5}{3}, (2, 0)$
34. $m = -2, \left(-1, \frac{2}{3}\right)$
35. $m = -\frac{1}{4}, \left(\frac{1}{2}, -1\right)$
36. $m = -\frac{5}{6}, (-3, 3)$
37. If we are given two points of a line but not the slope of the line, how can we use the point–slope or slope–intercept model to write an equation of the line?
38. If a line contains $A(2, -3)$ and $B(2, 5)$, why is the point–slope form not an appropriate model for writing an equation of the line?

In Exercises 39–52, write an equation of the line that contains the given pair of points.

39. (3, 4) and (1, 8)
40. (4, 7) and (6, 11)
41. (−2, 7) and (6, −9)
42. (5, −9) and (−1, 9)
43. (2, 3) and (−3, −2)
44. (12, 1) and (4, −1)
45. (2, −3) and (−4, 1)
46. (3, −2), and (−5, 4)
47. (−4, 2) and (−4, 5)
48. (2, −3) and (2, 4)
49. (−3, −4) and (1, −2)
50. (5, −4) and (2, −2)
51. (−3, 5) and (4, 5)
52. (2, −3) and (5, −3)

53. Describe the equation of a line that is
(a) perpendicular to a vertical line.
(b) parallel to the y-axis.

54. If the slope of L_1 is m_1, what additional information do we need in order to write an equation of L_1?

Encourage students to sketch the graph of the described line along with any given lines to see whether the given conditions appear to be met.

In Exercises 55–66, write an equation of the described line.

55. The line is horizontal and contains the point (3, −4).

56. The line is horizontal and contains the point (−5, 6).

57. The line contains the point (−2, 5) and has the same slope as the line whose equation is $y = 3$.

58. The line contains the point (5, −3) and has the same slope as the line whose equation is $y = -5$.

59. The line contains the point (−2, −3) and has the same slope as the line $2x + 3y = 1$.

60. The line contains the point (3, 5) and has the same slope as the line whose equation is $3x - 4y = 12$.

61. The line is vertical and contains the point (−2, 5).

62. The line is vertical and contains the point (7, −4).

63. The line contains the point (1, 5) and has the same y-intercept as the line whose equation is $2x + 3y = 6$.

64. The line contains the point (2, 3) and has the same y-intercept as the line $3x - 2y = 8$.

65. The line contains the point (−2, −3) and has an undefined slope.

66. The line contains the point (3, 5) and has an undefined slope.

In Exercises 67–70, use the given information to write an equation of L_1.

67. The equation of L_2 is $y = 4x - 7$. L_1 is parallel to L_2, and its y-intercept is (0, 2).

68. The equation of L_2 is $2x - 3y = 1$. L_1 is perpendicular to L_2, and its y-intercept is (0, −4).

69. The equation of L_2 is $y = 4$. L_1 is perpendicular to L_2 and contains the point (−2, 7).

70. The equation of L_2 is $x = -1$. L_1 is parallel to L_2 and contains the point (3, 5).

In Exercises 71 and 72, L_1 is perpendicular to L_2. Refer to the figures and write equations for L_1 and L_2.

71.

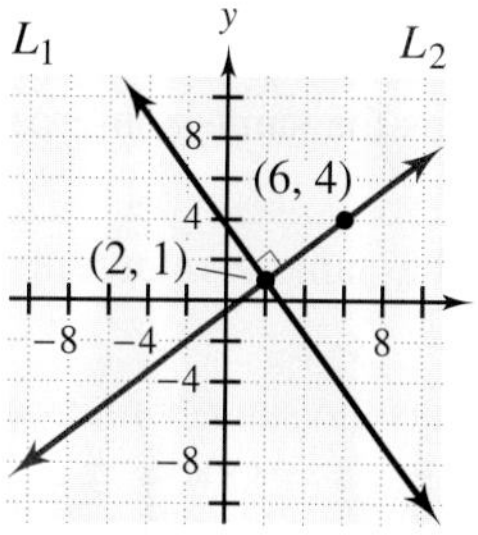

72.

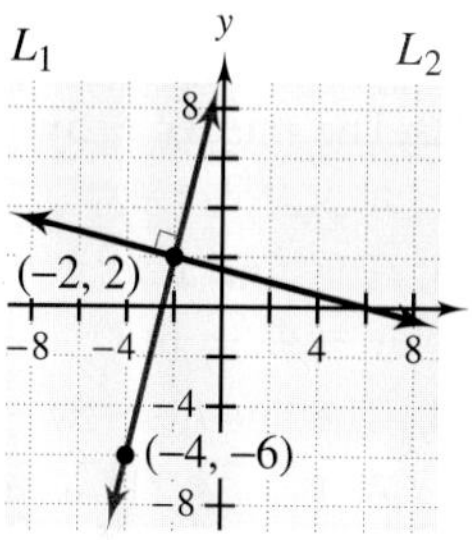

73. Consider the equations of L_1 and L_2.

L_1: $y = 3x - 7$ L_2: $x + y = 3$

Write an equation of L_3 where L_3 is parallel to L_2 and L_3 has the same y-intercept as L_1.

74. Consider the equations of L_1 and L_2.

L_1: $y = 3x - 7$ L_2: $x + y = 3$

Write an equation of L_3 where L_3 is perpendicular to L_1 and L_3 has the same y-intercept as L_2.

In Exercises 75 and 76, the equations of L_1 and L_2 are given. Write an equation of L_3 and an equation of L_4 according to the given conditions.

75. L_1: $3x + 4y = 12$ L_2: $x - y = 3$
(a) L_3 is perpendicular to L_2 and has the same y-intercept as L_1.
(b) L_4 is parallel to L_1 and has the same y-intercept as L_2.

76. L_1: $x - 2y = 6$ L_2: $4 - y = x$
(a) L_3 is parallel to L_1 and has the same y-intercept as L_2.
(b) L_4 is perpendicular to L_1 and has the same x-intercept as L_2.

CE In Exercises 77–84, determine the value of k for which the given condition is met.

77. The line $kx + 3y = 10$ is parallel to the line whose equation is $y = 3 - 2x$.

78. The line $3y + 2kx - 8 = 0$ is parallel to the line $y = -2 + x$.

79. The line $2x - ky = 2$ is perpendicular to the line $y = \dfrac{3 - 4x}{2}$.

80. The line $x = 5 - 3ky$ is perpendicular to the line $y = \dfrac{15 - 2x}{-3}$.

81. The line $x + 2y = 6k$ has the same y-intercept as $y = \dfrac{1 - x}{-2}$.

82. The line $3x - 5y - k = 0$ has the same y-intercept as $y = 2 - 4x$.

83. The line $x - y = 2k$ has the same x-intercept as $y = 4x - 8$.

84. The line $kx - 5y - 20 = 0$ has the same x-intercept as $y = 3x + 12$.

Real-Life Applications

85. In 2000 a person bought a computer for \$1500. She estimated the useful life of this computer to be 7 years with a remaining value of \$100 at the end of this period. She used the straight-line (linear) method of depreciation to estimate the loss in value of this equipment over time.

(a) Write the equation for the value v of the computer t years from 2000. (Let $t = 0$ for year 2000.)

(b) Use the equation in part (a) to determine the value of the computer in the year 2004.

86. The population of a small town grew from 35 in 1992 to 62 in 2002. Assuming the growth is linear, write an equation to describe the population P for any year x. What is the rate of change in the population?

87. The net loans at a certain bank increased from \$92.1 million in 1998 to \$105.7 million in 2000. Using a linear equation, predict the net loans in 2007. What is the rate of change of the net loans?

88. A library needs one children's librarian for each 1000 children attending story hour during a calendar quarter. The number of children attending story hour increased from 1153 during the first quarter of 2000 to 1732 during the same quarter of 2001. Using a linear equation, determine the year in which the library can fully support three children's librarians.

Modeling with Real Data

89. The bar graph shows the numbers of students (in millions) projected to receive Bachelor's degrees in the period 2000–2008. (Source: National Center for Educational Statistics.)

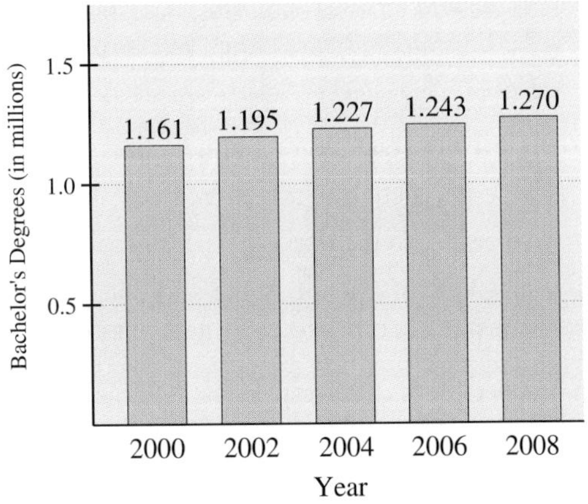

(a) Using the data for 2000 and 2008, write a model function in the slope–intercept form, where $N(x)$ is the number of degrees awarded (in millions) and x is the number of years since 1998.

(b) How would your function in part (a) change if you let x represent the number of years since 2000?

(c) From the actual data, (4, 1.195) is a data point, but $N(4) = 1.188$. Explain the discrepancy.

90. Leasing rather than purchasing automobiles has become increasingly popular. The percentage of car deals that were leases increased from 10% in 1984 to 35% in 1999. (Source: CNW Marketing.)

Write a linear function to model the percentage P for any year x after 1984.

91. From 1995 to 1998, the amount of energy used in the United States increased from 87.2 quadrillion BTUs to 90.5 quadrillion BTUs. (BTU stands for British Thermal Unit, which is a measure of heat energy. In the United States, one quadrillion is the number 1 followed by 15 zeros!)

(a) Suppose that we model the data with a function of the form $E(x) = mx + b$. How must we define x in order for the y-intercept of the graph to be (0, 87.2)?

(b) Write a function in the form described in part (a) to model the data.

(c) Use the model to predict the year in which energy use will reach 96 quadrillion BTUs.

92. In 1990 the circulation of *TV Guide* was 16 million. By 1999 the circulation had dropped to 13.1 million. (Source: PubList.com.)

(a) If a linear function were used to model the given data, what could you predict about the value of m? Why?

(b) Write a linear equation to model the circulation (in millions), where x is the number of years after 1990. Interpret the coefficient of x.

Data Analysis: *Single-Parent Families*

In recent years, the number of families with a woman as the sole parent has increased. The bar graph shows the number of such families (in millions) for selected years. (Source: U.S. Department of Commerce.)

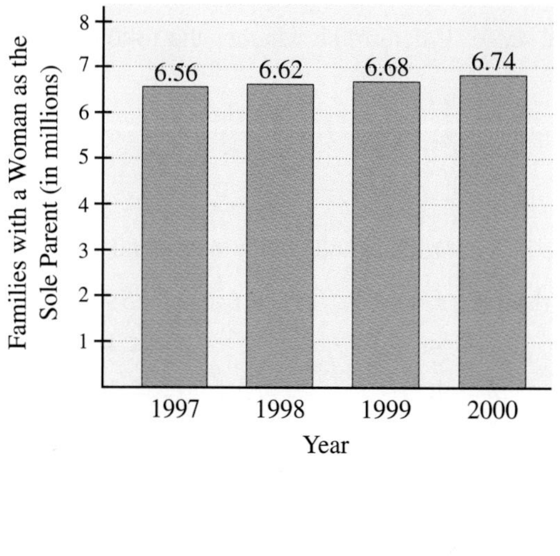

93. Let x represent the number of years since 1997. Choose any two data points and write a linear equation to model the number of families F for any year x. Do you obtain the same model if you choose two different points? Why?

94. Suppose that x is the number of years since 1995. Does the slope of the model function change? Does the y-intercept change?

95. The number of such families in 2010 is projected to be 7.19 million. If you were to create a new model function (of the same form) by using just the data for 1997 and 2010, would the coefficient of x be greater than or less than that of your model function in Exercise 93? Why?

96. Write a new model function (of the same form) by using just the data for 2000 and 2010. (Let x represent the number of years since 1997.)

Challenge

97. Show that if (x_1, y_1) and (x_2, y_2) are points of a line, an equation of the line can be written as

$$y - y_1 = \frac{y_2 - y_1}{x_2 - x_1}(x - x_1)$$

98. Show that the equation of the line with intercepts $(a, 0)$ and $(0, b)$ can be written as

$$\frac{x}{a} + \frac{y}{b} = 1, \quad a \neq 0 \text{ and } b \neq 0$$

In Exercises 99–102, use the results of Exercise 98 to determine the intercepts directly from the given equation.

99. $\frac{x}{3} + \frac{y}{2} = 1$

100. $\frac{x}{-4} - \frac{y}{2} = 1$

101. $\frac{2x}{5} - \frac{3y}{4} = 1$

102. $4x + 3y = 12$

103. Write an equation of the line that contains the point (a, b) and is perpendicular to the line whose equation is $ax + by = c$.

104. Write an equation of the line that contains the point (b, a) and is parallel to the line whose equation is $ax + by = c$.

4.5 Graphs of Linear Inequalities

Linear Inequalities • Graphs of Solution Sets • Graphs of Compound Inequalities

Linear Inequalities

The definition of a linear inequality in two variables is very similar to the definition of a linear equation in two variables.

Definition of a Linear Inequality in Two Variables

A **linear inequality in two variables** is an inequality that can be written in the form $Ax + By < C$, where A, B, and C represent any real numbers, and A and B are not both zero. Similar definitions can be stated with $>$, $\geq$, and $\leq$.

The following are some examples of linear inequalities along with their standard forms.

Inequality	*Standard Form*
$2x - 3y < 1$	$2x - 3y < 1$
$y \leq 4x - 1$	$-4x + y \leq -1$
$x > 3$	$1x + 0y > 3$
$y + 3 \geq 5$	$0x + 1y \geq 2$

Definitions of a Solution and a Solution Set

A **solution** of a linear inequality in two variables is a pair of numbers (x, y) that makes the inequality true. The **solution set** is the set of all solutions of the inequality.

As is true with equations, it is a simple matter to determine whether a particular pair of numbers is a solution of a linear inequality.

EXAMPLE 1 Verifying Solutions of an Inequality

Given the linear inequality $y \leq 4x - 3$, determine whether the following ordered pairs are solutions.

(a) $(0, -3)$ (b) $(1, 2)$ (c) $(1, -5)$

Solution

To be a solution, an ordered pair must satisfy either $y = 4x - 3$ or $y < 4x - 3$.

For each pair, replace x with the first coordinate and y with the second coordinate.

(a) $(0, -3)$

$$y \leq 4x - 3$$
$$-3 \leq 4(0) - 3$$
$$-3 \leq -3 \quad \text{True}$$

(b) $(1, 2)$

$$y \leq 4x - 3$$
$$2 \leq 4(1) - 3$$
$$2 \leq 1 \quad \text{False}$$

(c) $(1, -5)$

$$y \leq 4x - 3$$
$$-5 \leq 4(1) - 3$$
$$-5 \leq 1 \quad \text{True}$$

Both $(0, -3)$ and $(1, -5)$ are solutions.

Graphs of Solution Sets

The **graph** of the solution set of an inequality is a picture of all of the solutions of the inequality.

Exploring the Concept

Figure 4.27

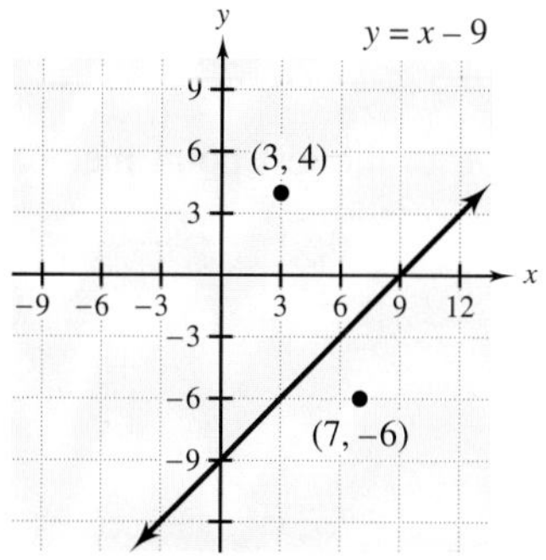

To involve students, have small groups investigate several points above and below the line and report their findings. Then use the results to form a conjecture about the points that represent solutions.

Graphs of Linear Inequalities

Consider the inequality $y \leq x - 9$. Every solution of the companion equation $y = x - 9$ is also a solution of the inequality. Thus the points of the line shown in Figure 4.27 represent solutions of the inequality $y \leq x - 9$.

Figure 4.27 also shows the point (3, 4), which is above the line, and the point (7, −6), which is below the line. We can easily determine whether these points represent solutions by substitution.

(3, 4)		*(7, −6)*	
$y \leq x - 9$		$y \leq x - 9$	
$4 \leq 3 - 9$		$-6 \leq 7 - 9$	
$4 \leq -6$	False	$-6 \leq -2$	True

Thus (3, 4) is not a solution of the inequality. In fact, no other point above the line represents a solution. However, (7, −6) is a solution, and all the points below the line represent solutions. In short, the solution set of $y \leq x - 9$ is represented graphically by the line and all points below the line. To depict these solutions, we shade the region below the line. (See Fig. 4.28.)

Figure 4.28

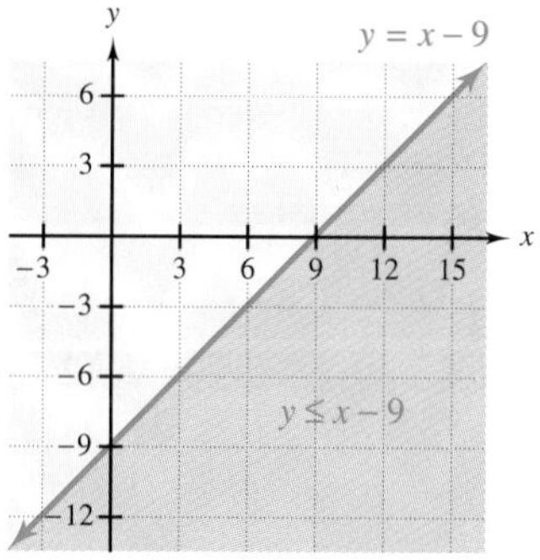

Figure 4.29

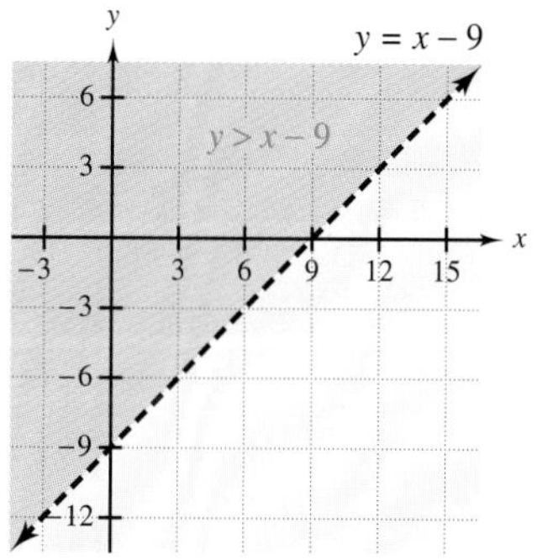

Figure 4.29 depicts the solution set of another inequality, $y > x - 9$. This time, points of the line $y = x - 9$ do not represent solutions, and we indicate this with a dashed line. The region above the line is shaded to show that all points in that region represent solutions.

The graph of the companion equation is called the **boundary line,** and the regions are called **half-planes.** We indicate that points of the boundary line represent solutions of $\leq$ and $\geq$ inequalities by drawing a solid line. (See Fig. 4.28.) For $<$ and $>$ inequalities, we draw a dashed boundary line to indicate that points of the line do not represent solutions. (See Fig. 4.29.)

To determine which half-plane contains points that represent solutions of an inequality, we can select any **test point** on either side of the boundary line and test the coordinates to see whether they satisfy the inequality.

The following is a summary of this method for sketching the graph of a linear inequality.

Sketching the Graph of a Linear Inequality

1. Sketch the graph of the companion equation. The boundary line is solid for $\leq$ and $\geq$ inequalities and dashed for $<$ and $>$ inequalities.
2. Select any test point in either half-plane (not on the boundary line). By substitution, determine whether the point represents a solution. If the test point represents a solution, shade the half-plane in which the test point lies. If the test point does not represent a solution, shade the other half-plane.

Emphasize that *any* point not on the boundary line will do. Some students fear that they will choose the wrong point.

The preceding discussion and figures suggest an alternative to the test point method for sketching the graph of a linear inequality.

Sketching the Graph of a Linear Inequality

1. Solve the inequality for y.
2. Draw the solid or dashed boundary line.
3. (a) For inequalities of the form $y < mx + b$ or $y \leq mx + b$, shade the region below the line.
   (b) For inequalities of the form $y > mx + b$ or $y \geq mx + b$, shade the region above the line.
4. For the special case of a vertical boundary line, shade the half-plane on the left for $x < c$ or $x \leq c$; shade the half-plane on the right for $x > c$ or $x \geq c$.

Some students are unsure about what "above" and "below" a line mean, especially if the line is very steep. Try selecting a point on the line and ask, "For smaller (greater) y-values, which way do we move?"

EXAMPLE 2

Graphing Linear Inequalities

Sketch the graph of each inequality.

(a) $y > 2x$ (b) $2x - 3y \geq 12$

(c) $y - 5 \leq 2$ (d) $x < -2$

Solution

To reinforce the concept, it is also a good idea to test some points to determine shading.

(a) The boundary line $y = 2x$ is dashed because the inequality symbol is $>$. Because the inequality is in the form $y > mx + b$, we shade above the boundary line. (See Fig. 4.30.)

Figure 4.30 also shows the point $(2, -3)$ that we can test to determine whether it represents a solution.

$$y > 2x$$
$$-3 > 2(2)$$
$$-3 > 4 \quad \text{False}$$

Because the test point $(2, -3)$ does not represent a solution, we shade the half-plane above the line.

Figure 4.30

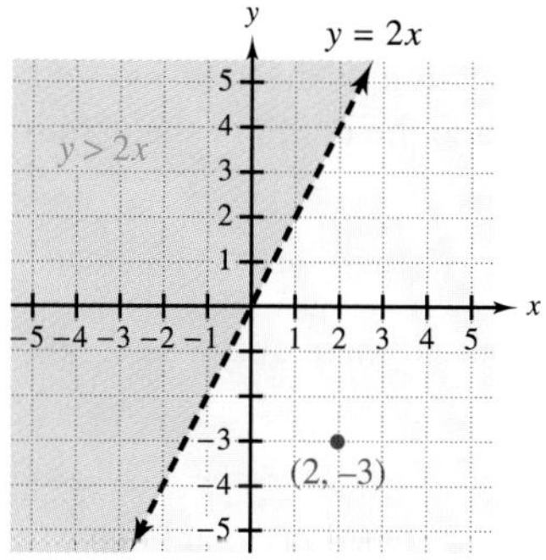

Figure 4.31

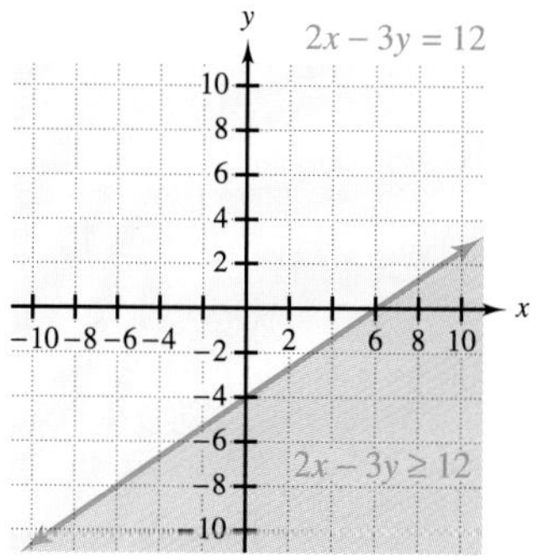

(b) $2x - 3y \geq 12$ Solve the inequality for *y*.

$-3y \geq -2x + 12$ Subtract 2*x* from both sides.

$\frac{-3y}{-3} \leq \frac{-2x}{-3} + \frac{12}{-3}$ Divide both sides by −3 and reverse the inequality symbol.

$y \leq \frac{2}{3}x - 4$

Because the inequality symbol is ≤, we draw the solid boundary line $y = \frac{2}{3}x - 4$. The inequality is in the form $y \leq mx + b$, so we shade below the line. (See Fig. 4.31.)

(c) $y - 5 \leq 2$ Solve the inequality for *y*.

$y \leq 7$

The boundary line is the solid horizontal line $y = 7$. Because the inequality symbol is ≤, we shade below the line. (See Fig. 4.32.)

Figure 4.32

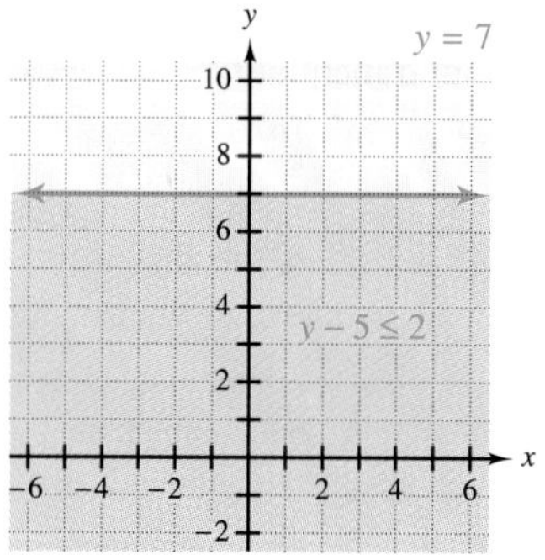

Figure 4.33

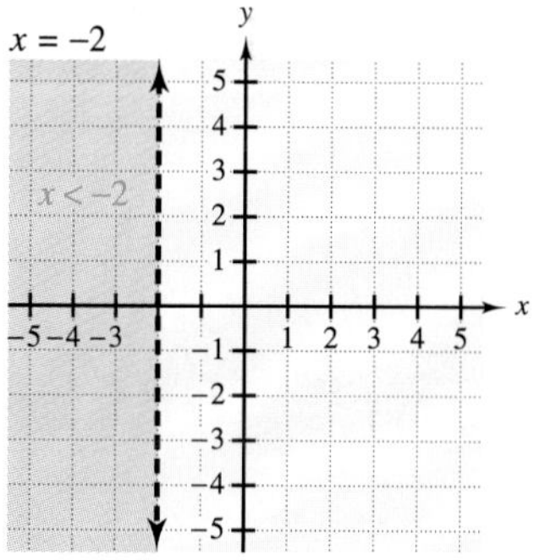

(d) Draw the dashed vertical line $x = -2$. Observe that the inequality symbol is <, so we shade to the left of the line. (See Fig. 4.33.)

Graphs of Compound Inequalities

In Section 3.7 we solved compound inequalities in one variable. Linear inequalities in two variables can also be connected with *and* or *or* to form compound inequalities.

To produce the graph of a compound inequality, we graph all the inequalities involved on the same coordinate system. For inequalities connected with *and* (a conjunction), we shade the intersection of the solution sets, and for inequalities connected with *or* (a disjunction), we shade the union of the solution sets.

EXAMPLE 3

Graphing a Conjunction

Graph the compound inequality $y \geq 2$ and $y \leq 2x + 5$.

Solution

LEARNING TIP

Think of the graph of each inequality in a compound inequality as being drawn on a transparent sheet with one sheet placed on top of the other. For a conjunction, the graph is the region where the individual graphs overlap. For a disjunction, the graph includes any region that is shaded on either graph.

The graph of $y \geq 2$ is the horizontal line $y = 2$ and the region above the line. The graph of $y \leq 2x + 5$ is the line $y = 2x + 5$ and the region below the line. (See Fig. 4.34.)

Figure 4.34

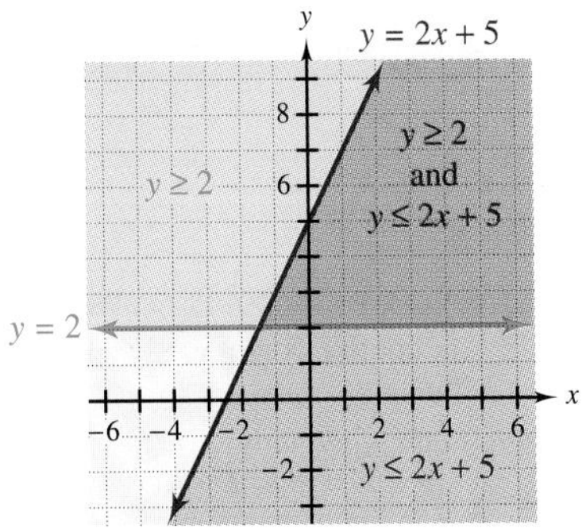

Figure 4.35

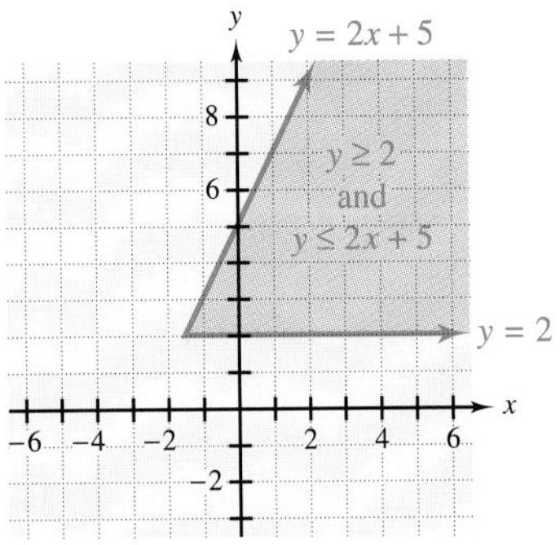

Because the inequalities are connected with *and,* the graph of the compound inequality is the intersection of the two graphs. (See Fig. 4.35.)

EXAMPLE 4

Graphing a Disjunction

Graph the compound inequality $y \geq x - 3$ or $2x + y < 1$.

Solution

The graph of $y \geq x - 3$ is the line $y = x - 3$ and the region above the line. The graph of $2x + y < 1$ is the region below the line $y = -2x + 1$. (See Fig. 4.36.)

Figure 4.36

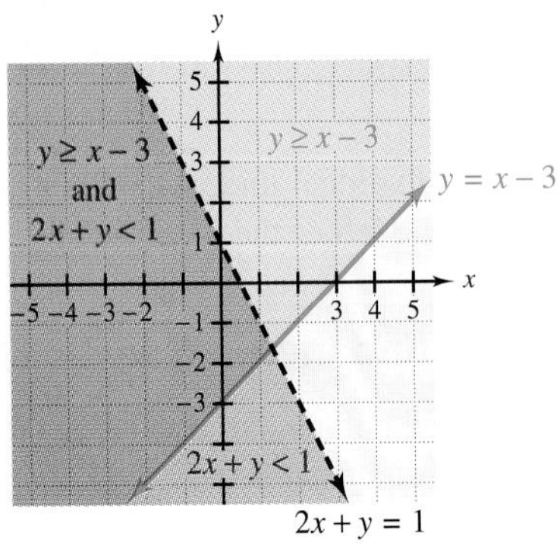

Figure 4.37

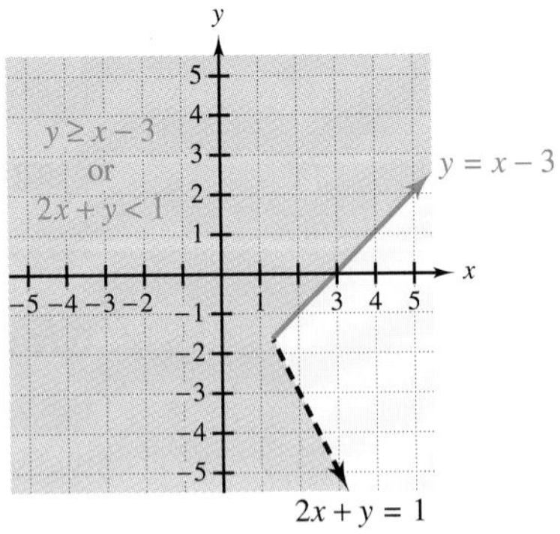

Because the inequalities are connected with *or,* the graph of the compound inequality is the union of the two graphs. (See Fig. 4.37.)

To graph inequalities involving absolute value, we must recall the following for any expression A and any positive number C:

Remind students that $A > -C$ *and* $A < C$ is equivalent to $-C < A < C$.

1. If $|A| < C$, then $A < C$ *and* $A > -C$.
2. If $|A| > C$, then $A > C$ *or* $A < -C$.

Similar rules hold for the symbols $\leq$ and $\geq$.

EXAMPLE 5

Graphing an Absolute Value Inequality

Graph the absolute value inequality $|x + y| > 2$.

Solution

The inequality means $x + y > 2$ or $x + y < -2$. We solve both inequalities for y to obtain $y > -x + 2$ or $y < -x - 2$.

The graph of $y > -x + 2$ is the region above the line $y = -x + 2$. The graph of $y < -x - 2$ is the region below the line $y = -x - 2$.

Because the inequalities are connected with *or,* the graph of the absolute value inequality is the union of the two graphs. (See Fig. 4.38.)

Figure 4.38

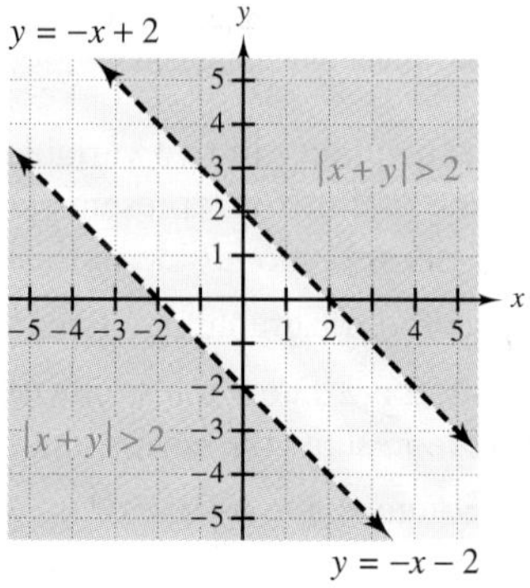

EXAMPLE 6

Graphing an Absolute Value Inequality

Graph the absolute value inequality $|y + x| \leq 4$.

Solution

The absolute value inequality means that $y + x \leq 4$ and $y + x \geq -4$. Solving each inequality for y, we obtain $y \leq -x + 4$ and $y \geq -x - 4$.

For $y \le -x + 4$ we shade below the line, and for $y \ge -x - 4$ we shade above the line. (See Fig. 4.39.)

Figure 4.39

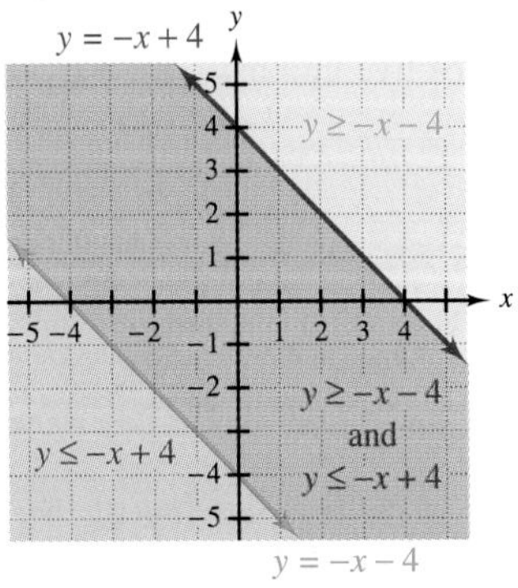

Figure 4.40

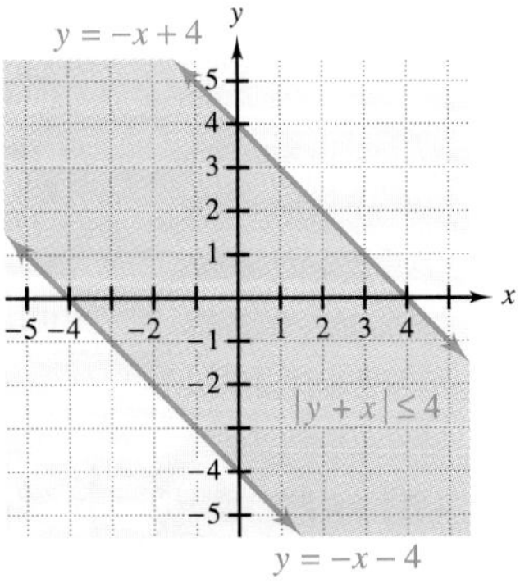

Think About It

Compare the graphs of $|x - y| < 0$ and $|x - y| \ge 0$.

Because the inequalities are connected with *and,* the graph of the absolute value inequality is the intersection of the two graphs. (See Fig. 4.40.)

Quick Reference 4.5

Linear Inequalities

- A **linear inequality in two variables** is an inequality that can be written in the form $Ax + By < C$, where A, B, and C represent any real numbers, and A and B are not both zero. A similar definition can be stated with $>$, $\ge$, and $\le$.
- A **solution** of a linear inequality in two variables is a pair of numbers (x, y) that makes the inequality true. The **solution set** is the set of all solutions of the linear inequality.

Graphs of Solution Sets

- The **graph** of a linear inequality in two variables is a picture of its solutions. It consists of a boundary line (solid if it represents solutions, dashed if it does not) and a shaded region (a **half-plane**) on one side of the boundary line.
- The recommended method for graphing a linear inequality is as follows.
  1. Solve the inequality for y. Draw the solid (for $\ge$ and $\le$) or dashed (for $>$ and $<$) boundary line that represents the corresponding equation.
  2. Determine the region to shade by one of these methods.
     (a) If the coordinates of a test point satisfy the inequality, shade the half-plane in which the test point lies; otherwise, shade the other half-plane.
     (b) For inequalities of the form $y < mx + b$ or $y \le mx + b$, shade below the boundary line; for inequalities of the form $y > mx + b$ or $y \ge mx + b$, shade above the boundary line.
  3. For the special case of a vertical boundary line, shade the half-plane on the left for $x < c$ or $x \le c$; shade the half-plane on the right for $x > c$ or $x \ge c$.

Graphs of Compound Inequalities

- Two or more linear inequalities in two variables can be connected by *and* (conjunction) or *or* (disjunction).

- To sketch a graph of a compound inequality, draw the graph of each component of the inequality on the same coordinate system. For a conjunction, shade the intersection of the two graphs; for a disjunction, shade the union of the two graphs.
- To sketch a graph of an absolute value inequality, translate the inequality into a conjunction or disjunction, and graph accordingly.

Speaking the Language 4.5

1. The graph of a linear inequality in two variables includes a region known as a(n) ________.
2. For a $<$ or $>$ linear inequality in two variables, the ________ is dashed.
3. Graphing a conjunction of two linear inequalities in two variables includes shading the ________ of the graphs of the inequalities.
4. When graphing a linear inequality in two variables, we can decide which region to shade by determining whether the coordinates of some point $P(x, y)$ satisfy the inequality. In this case, we call P a(n) ________.

Exercises 4.5

Concepts and Skills

1. How do the solution sets of the linear inequalities $x + 2y < 6$ and of $x + 2y \leq 6$ differ?
2. Does the inequality $y > x + 2$ define a function? Explain.

In Exercises 3–8, determine whether each ordered pair is a solution of the given inequality.

3. $y < 3x - 5$; $(1, 0), (0, -6), (2, 1)$
4. $y \geq 1 - 2x$; $(5, -9), (-3, 8), (2, -7)$
5. $2x - 7y \geq 12$; $(12, 1), (6, 0), (-4, -2)$
6. $4 - 3x - 5y < 0$; $(4, 3), (-5, -4), (7, 3.4)$
7. $y \leq 4$; $(1, 10), (0, 4), (4, 5)$
8. $x > -6$; $(-5, -7), (-5.99, 3), (-6.01, 4)$
9. For the inequality $x + 3 > 7$, we might write the solution set as $(4, \infty)$, or we might graph the solution set as a shaded half-plane. What assumptions would we be making in each case?
10. To sketch the graph of $x - 2y < 8$, you must draw the boundary line and then decide which half-plane to shade. A classmate tells you to shade below the line because the inequality symbol is $<$. Is your classmate correct? Explain.

In Exercises 11–16, the figure shows the boundary line for the graph of the given inequality. Complete the graph by shading the correct half-plane.

11. $y \leq 4x - 3$

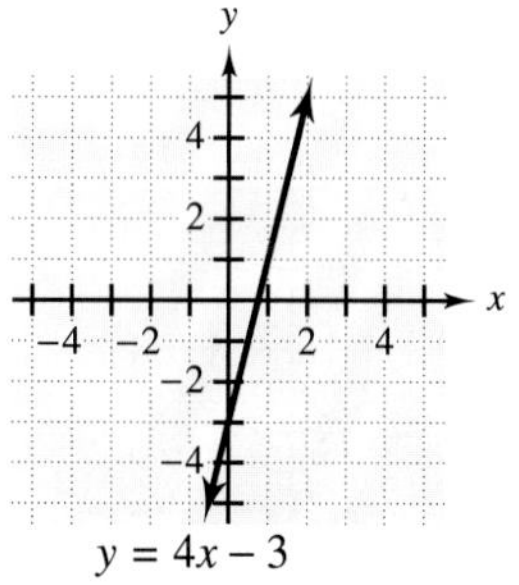

12. $y \geq -2x$

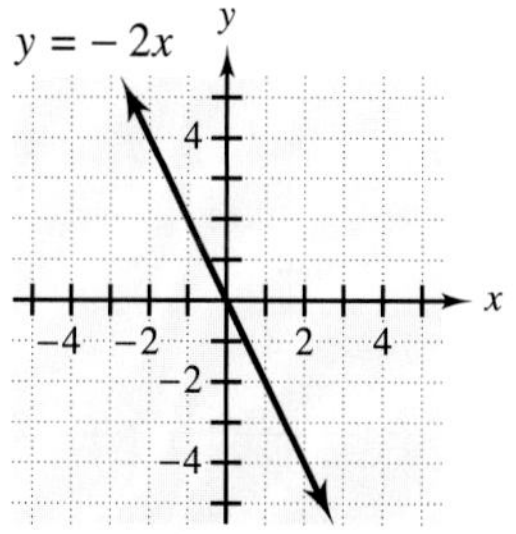

13. $2x + y > 6$

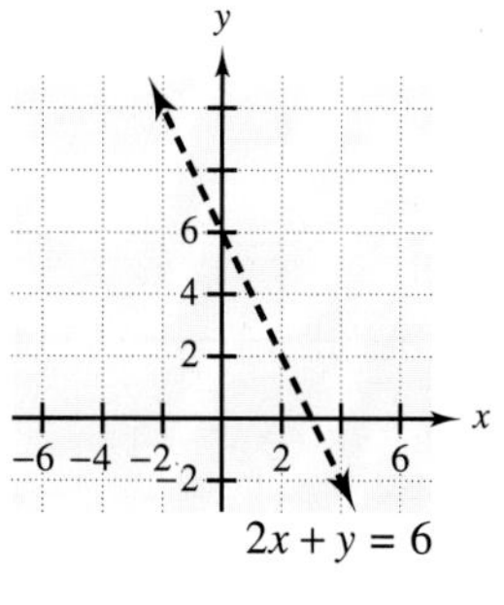

14. $3y - 2x < 9$

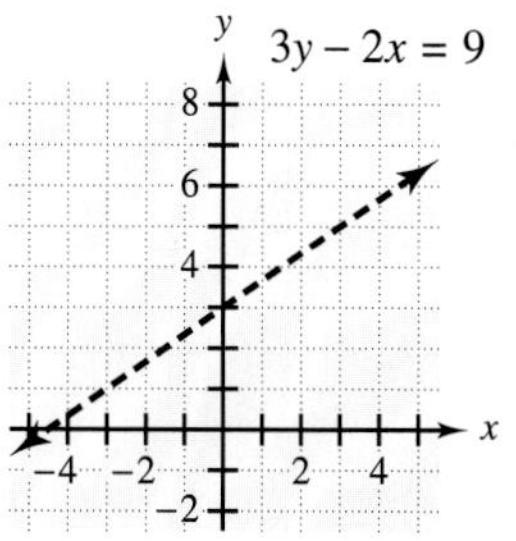

15. $y < -3$

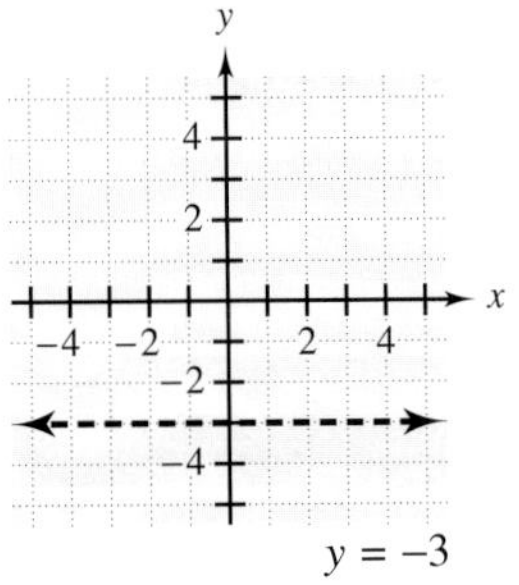

16. $2y - x \geq 0$

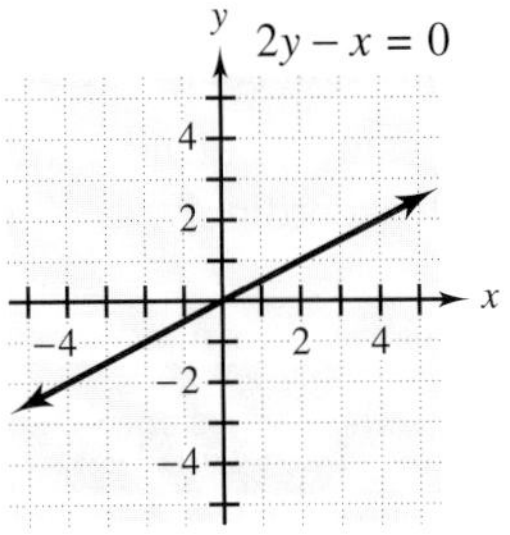

In Exercises 17–36, graph the inequality.

17. $y < 7 - 3x$

18. $y \geq \frac{1}{2}x - 5$

19. $-y \leq -2x$

20. $y \geq x$

21. $3x + 2y \leq 6$

22. $4x + 5y \geq 20$

23. $-2y \geq 4x - 7$

24. $2x \geq 3 - y$

25. $-6 \leq x - 2y$

26. $x - 3y \leq 0$

27. $2x - 5y \geq 10$

28. $x - 6 < -y$

29. $y - 3 \leq -2(x - 3)$

30. $y + 2 - 4(x - 1) > 0$

31. $\frac{x}{5} + \frac{y}{2} < 1$

32. $\frac{x}{3} - \frac{y}{2} \geq 1$

33. $y \geq 20$

34. $y - 3 \leq 2$

35. $x < 5$

36. $2x + 6 \geq 0$

CE In Exercises 37–42, write an inequality for the given graph.

37.

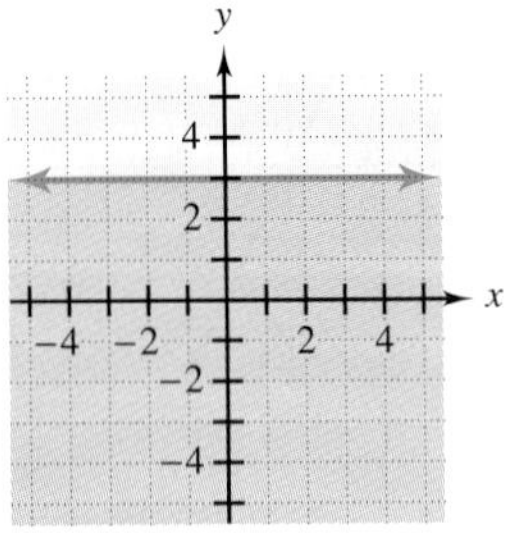

38.

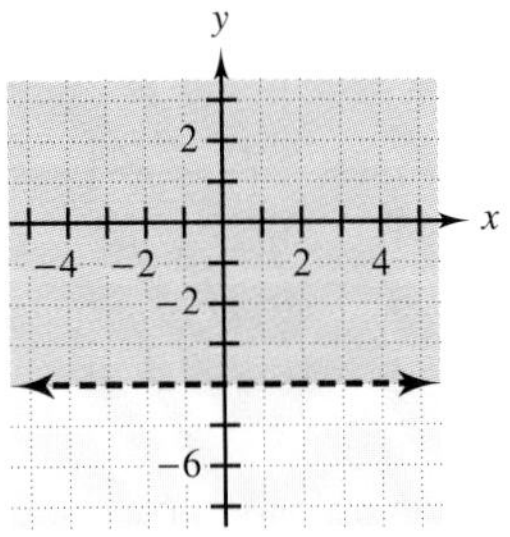

39.

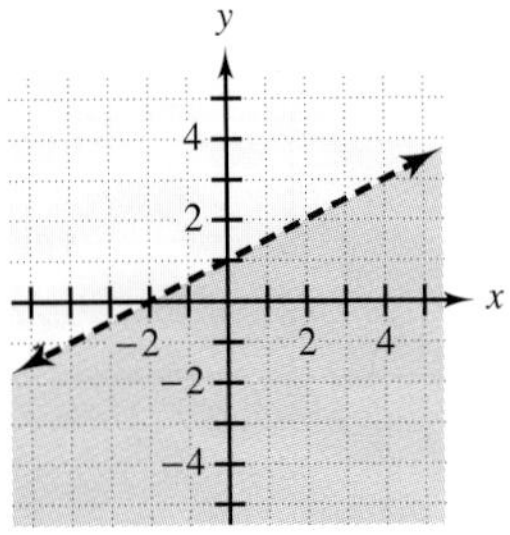

40.

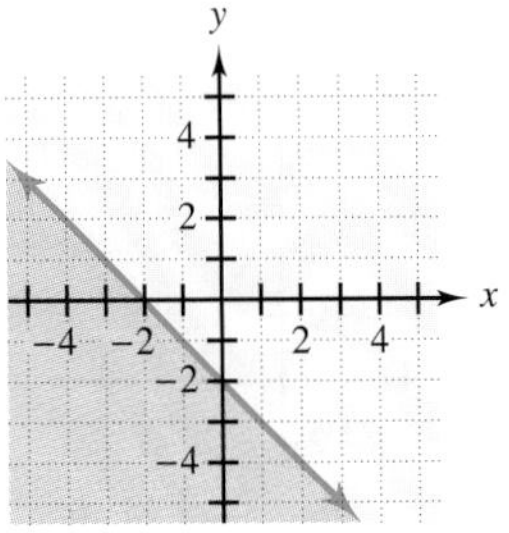

41.

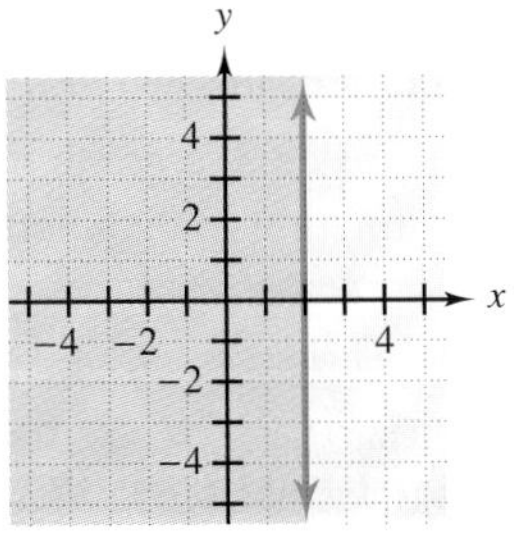

42.

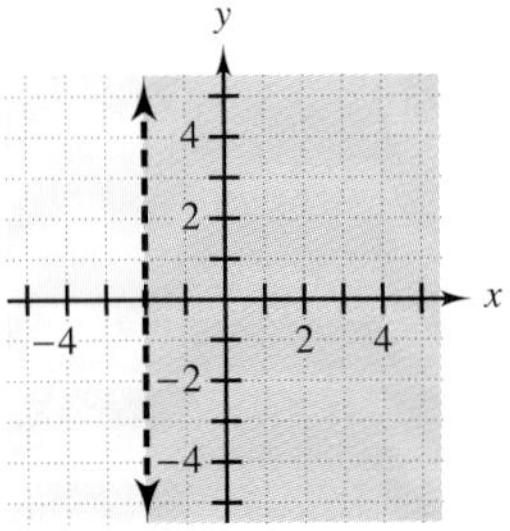

In Exercises 43–48, determine an ordered pair of numbers that satisfies the given condition. Then translate the given information into a linear inequality, sketch its graph, and test whether your pair is a solution.

43. One number x is at least 1 more than another number y.

44. One number y is at most twice another number x.

45. One number y is not less than 4 less than 3 times another number x.

46. One number x is no greater than half of another number y.

47. The sum of a number x and twice another number y exceeds 10.

48. The difference of a number x and two-thirds of another number y exceeds 7.

In Exercises 49–54, graph the conjunction.

49. $x \leq 2$ and $x \geq 0$

50. $y > -3$ and $x > 2$

51. $x + y \geq 3$ and $x \geq 0$

52. $y < -3x - 10$ and $y \leq 0$

53. $y < \frac{1}{2}x$ and $y \geq 3x$

54. $x + y \geq -3$ and $y > x - 5$

In Exercises 55–60, graph the disjunction.

55. $y > -1$ or $y \leq -4$

56. $x \geq 2$ or $y > -3$

57. $y > -\frac{2}{3}x$ or $y + 3x \leq 0$

58. $x + 2y > 2$ or $x \geq 0$

59. $y < x$ or $y \geq 0$

60. $3x < y + 2$ or $y > 4 - x$

In Exercises 61–72, sketch the graph of the given compound inequality.

61. $y \leq -x + 3$ and $y \leq 2x + 1$

62. $y \geq 4$ or $y < -2x + 3$

63. $3x + y < 5$ or $x - y < 2$

64. $3x + y < 5$ and $x - y < 2$

65. $2x - y - 5 < 0$ and $4x - 2y > 0$

66. $\frac{y}{2} - \frac{x}{3} \leq 1$ or $\frac{x}{3} + \frac{y}{2} \geq 1$

67. $5y + 3x < 30$ and $5y + 3x < -20$

68. $2x - 3y < 9$ or $2x - 3y < -12$

69. $x \geq y$ and $x \leq y$

70. $x - 3 > y$ or $x - 3 < y$

71. $x \geq 0$, $y \geq 0$, and $y \leq 10 - x$

72. $x \geq 0$, $y \geq 0$, and $y \geq 12 - 3x$

CE In Exercises 73–78, write a compound inequality that describes the graph.

73.

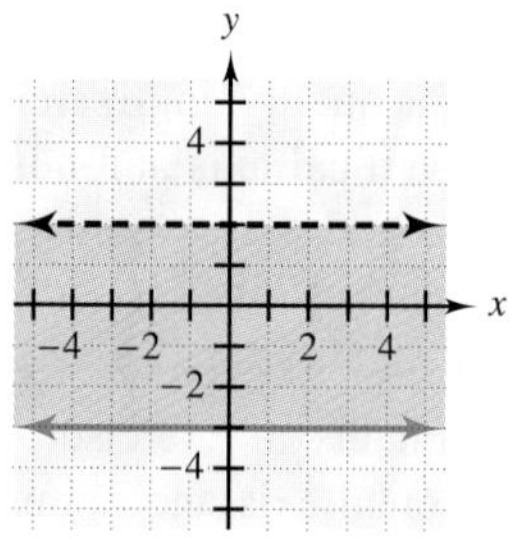

74.

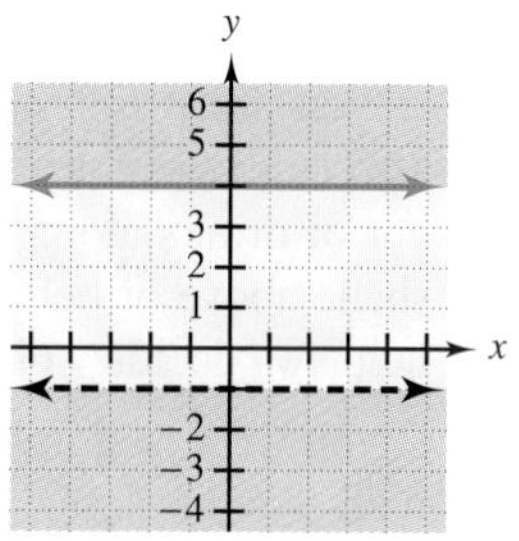

75.

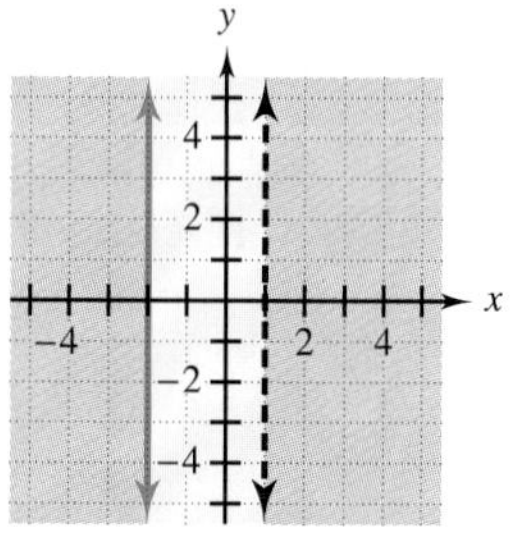

76.

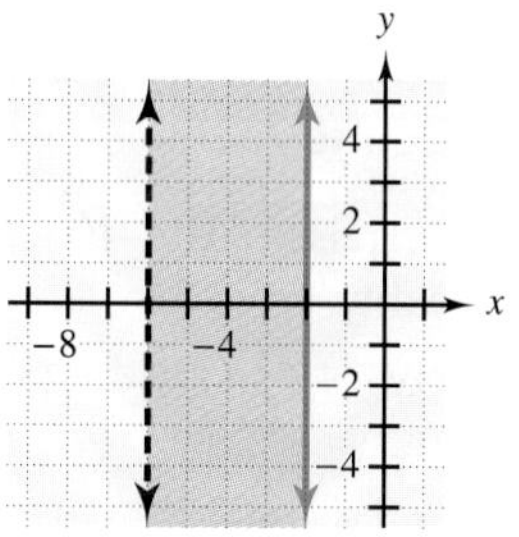

77.

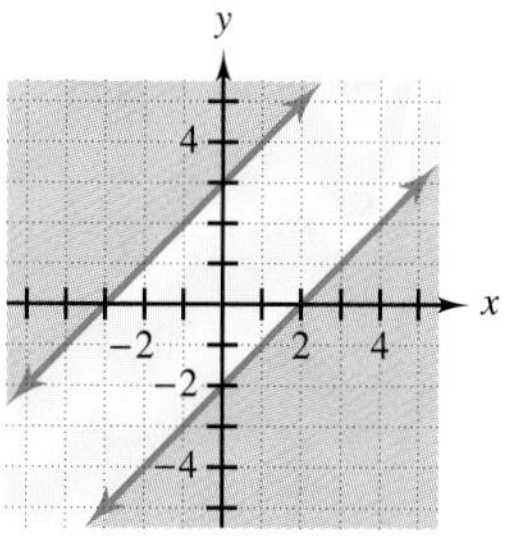

78.

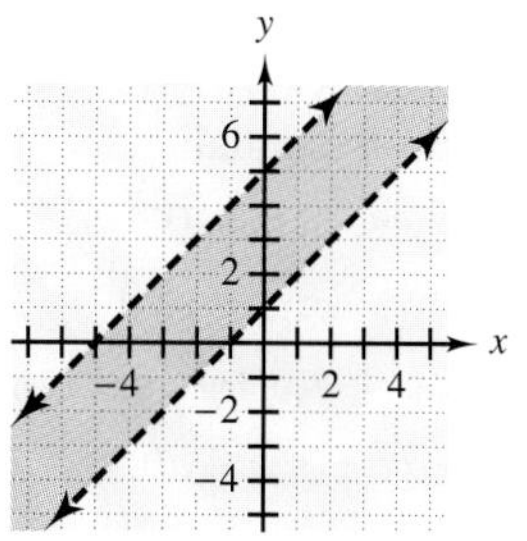

In Exercises 79–90, sketch the graph of the given absolute value inequality.

79. $|y| \leq 3$

80. $|x| \geq 4$

81. $|x - 2y| > 4$

82. $|2x + y| < 5$

83. $|y - x| \geq 3$

84. $|y + x| \leq 6$

85. $|x| \leq -2$

86. $|y| \geq -1$

87. $|x - y| \leq 0$

88. $|2x + 3y| \leq 0$

89. $|x + y| > 0$

90. $|y - 2x| > 0$

Real-Life Applications

91. The main hold of a supply ship has an area of 6000 square feet. Pallets of two sizes are used to store materials in the hold. One pallet is 4 feet by 4 feet, and the other is 3 feet by 5 feet. Write a linear inequality that describes the number of pallets that can be used, and produce its graph.

92. A hospital food service can serve at most 1000 meals per day. Patients on a normal diet receive three meals each day, and patients on special diets receive four meals each day. Write a linear inequality that describes the number of patients that can be served, and produce its graph.

93. Two basketball teams are considered evenly matched if the difference in their scores does not exceed 5 points. Write an absolute value inequality to describe this condition, and produce its graph.

94. A community organization has determined that social harmony among its members is greatest when the difference between the number of male and female members is less than 10. Write an absolute value inequality to describe this condition, and produce its graph.

95. A county animal shelter houses only cats and dogs. The maximum number of animals that can be kept is 100, but the county will close the shelter if the population falls below 20. Write a compound inequality that describes these conditions, and produce its graph.

96. To be eligible for certain development grants, a local arts council must have at least 50 members, and the difference between the number of male and female members must be at most 10. Write a compound inequality that describes these conditions, and produce the graph. Use the graph to show that not all nonnegative pairs of numbers satisfy the given conditions.

Modeling with Real Data

97. In 1996 the estimated number of registered voters in the United States was 190 million, and the estimated maximum voter turnout for the presidential election was 53.6%. (Source: Federal Elections Commission.)

(a) Let R represent the number of votes for the Republican candidate, and let D represent the number of votes for the Democratic candidate. Assuming that there are only two candidates, write an inequality that relates R and D to the total number of votes cast.

(b) Interpret the meaning of the points of the graph of the inequality in part (a).

98. (a) What change would you make to the inequality in Exercise 97 if there were other candidates in the race?

(b) How would the change in part (a) affect the graph in Exercise 97?

99. In 1999 taxpayers who were married and filing jointly, and who had a taxable income between \$43,050 and \$104,050, paid a tax of \$6457.50 plus 28% of the taxable income over \$43,050. (Source: Internal Revenue Service.)

(a) Suppose that a couple in this category had an adjusted gross income (before deductions) of x dollars. Write an inequality that describes the maximum tax y that the couple would pay.

(b) If you were to draw the graph of the inequality in part (a), would you shade above or below the boundary line?

100. In Exercise 99, if the base tax for the described tax bracket were to remain at \$6457.50 but the tax rate were to increase, what would be the effect on the graph of the inequality?

Data Analysis: *Blood Alcohol Level*

Research shows that when a person's blood alcohol level reaches 0.05, the chance of that person's having an automobile accident doubles. The blood alcohol level depends on the person's weight as well as on the amount of alcohol consumed. For example, a 100-pound person would reach a blood alcohol level of 0.05 with two drinks in a 2-hour period, whereas a 240-pound person could have four drinks in the same time period without reaching a blood alcohol level of 0.05. (Source: Alcohol Safety Action Project.)

The equation $0.0875D - 0.00125W = 0.05$ models combinations of weight W and number of drinks D that will produce a blood alcohol level of 0.05 in 2 hours.

101. To the nearest whole number, how many drinks in 2 hours will produce a blood alcohol level of 0.05 for a 180-pound person?

102. What does $0.0875D - 0.00125W > 0.05$ describe?

103. If you were to graph the inequality in Exercise 102, with weight along the vertical axis and number of drinks along the horizontal axis, would you shade above or below the boundary line? (*Hint:* To help you decide, solve the inequality for W.)

104. If you were to graph the inequality in Exercise 102 with weight along the horizontal axis and number of drinks along the vertical axis, would you shade above or below the boundary line?

Challenge

In Exercises 105–108, graph the given compound inequality.

105. $|x| > 3$ and $|y| \geq 2$

106. $|x| < 4$ or $|y| < 3$

107. $|x + y| > -2$ or $|x| < -5$

108. $|y + 2x| < -1$ and $|y| > -1$

In Exercises 109–112, draw an example of a graph of the given inequality.

109. $y < b, b > 0$

110. $y \geq mx, m < 0$

111. $y < mx + b, m > 0, b < 0$

112. $ax + by \leq ab, a > 0, b < 0$

4 • Chapter Project

Bald Eagle Population

The Bald Eagle Protection Act was passed in 1940. Alaska was exempted from this act because of pressure from the salmon-fishing industry and because the primary problem was in the lower 48 states. The Alaska exemption was removed in 1952.

Bald eagles were also protected by the Endangered Species Act, but in 1995, when the population had reached 4700 pairs, bald eagles were listed as "threatened" rather than "endangered." (Source: U.S. Fish and Wildlife Service)

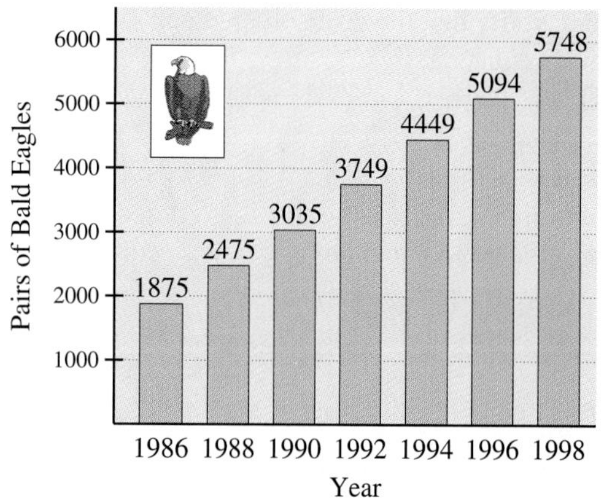

The linear function that best models the data is $E(x) = 326.27x + 1164.86$, where $E(x)$ is the number of pairs of bald eagles, and x is the number of years since 1984.

1. According to this model, what was the average yearly increase in the number of pairs of bald eagles during the given period?
2. In 1963 there were only 417 pairs of bald eagles. How well does function E model the data for that year? Explain.
3. As stated at the beginning of this chapter, when Europeans first arrived in North America, the bald eagle population was about half a million (250,000 pairs). According to the model function, how many years of protection will be needed to bring the bald eagle population back to this level?
4. Suppose that a goal of 10,000 pairs of bald eagles is set for the year 2006. Explain why this goal would require a greater average annual increase in the population than the slope of the graph of function E.
5. In 1996, the expenditures by private and public agencies for the recovery and protection of bald eagles were about \$1 million. What was the cost per pair of eagles that year?
6. Using the same rate of funding found in Exercise 5, and using function E as your model, determine what level of expenditure would be needed in the year 2004? Explain.

Chapter 4 Review Exercises

Section 4.1

In Exercises 1–4, state whether the equation is a linear equation in two variables.

1. $3x = \frac{2}{y} + \frac{3}{2}$ **2.** $3x - 8 = 15$

3. $x = \frac{y}{3} - \frac{4}{5}$ **4.** $5 = \frac{-3xy}{4}$

In Exercises 5 and 6, verify that the given ordered pair is a solution of the given equation.

5. $(5, -11)$; $y = -3x + 4$

6. $(-14, -6)$; $2x - 7y = 14$

In Exercises 7 and 8, find a, b, and c so that the ordered pairs are solutions of the given equation.

7. $y = 7 - x$; $(-2, a), (b, 14), (c, -5)$

8. $4y - 3x = 7$; $(1, a), (b, 4), (-9, c)$

In Exercises 9 and 10, complete the table of solutions of the given equation.

9. $2y - 3x = 9$

x	3	7	−7
y			

10. $2x - 4 = 3y$

x	2	11	−4
y			

In Exercises 11–14, determine the x- and y-intercepts.

11. $7x - 3y = 21$ **12.** $y = -2x + 3$

13. $y - 2 = 18$

14. $2x - 5 = 7$

In Exercises 15–18, write the given equation in the form $y = mx + b$. Then determine the y-intercept.

15. $15 - 3x + 6y = 0$ **16.** $4x - 5y = 19$

17. $2x - y = 8$ **18.** $\frac{1}{2}x - \frac{2}{5}y = 6$

In Exercises 19–22, determine three solutions of the equation and sketch the graph.

19. $5x + 3y = 15$ **20.** $\frac{1}{4}x - \frac{2}{5}y = 10$

21. $y + 3 = 9$ **22.** $2x + 1 = 7$

23. Which of the following equations does the graph represent?

(i) $3x + 4y = 12$

(ii) $3x + 4y = -12$

(iii) $3x - 4y = -12$

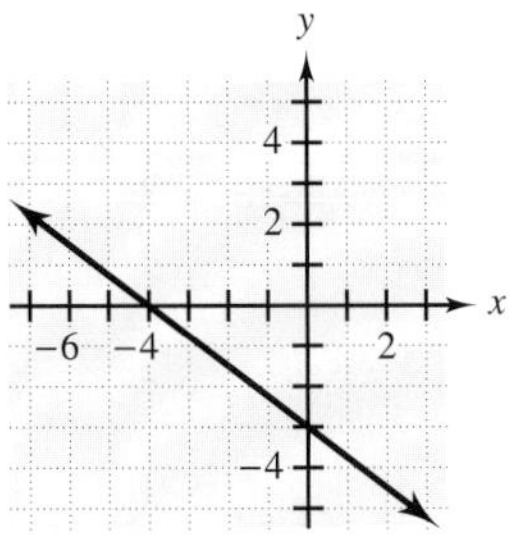

Section 4.2

24. In the accompanying figure, name the line whose slope is

(a) negative. (b) positive.

(c) undefined. (d) zero.

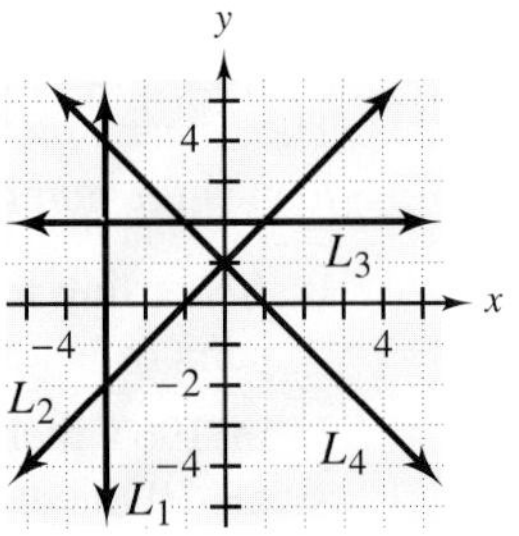

In Exercises 25–28, determine the slope of the line that contains the given points.

25. $(-5, -8)$ and $(2, -3)$ **26.** $(3, -4)$ and $(3, 5)$

27. $(2, -4)$ and $(-7, 6)$ **28.** $(-4, 5)$ and $(4, 5)$

In Exercises 29 and 30, write each equation in the form $y = mx + b$. Then determine the slope.

29. $3x - 4y = 15$

30. $x + 2y = 4$

In Exercises 31 and 32, draw a line through the given point with the given slope.

31. $(0, -4)$; $m = -\frac{3}{5}$ **32.** $(2, -3)$; $m = 0.5$

In Exercises 33–36, sketch the graph of the given equation.

33. $y = -x + 3$

34. $3y + 2x = 3$

35. $y = 3$

36. $x - 2 = 0$

Section 4.3

In Exercises 37–42, L_1 refers to line 1 with slope m_1, and L_2 refers to line 2 with slope m_2.

In Exercises 37–40, the slopes of two distinct lines, L_1 and L_2, are given. Determine whether L_1 and L_2 are parallel, perpendicular, or neither.

37. $m_1 = \frac{1}{4}$; $m_2 = 0.25$

38. $m_1 = 5$; $m_2 = -5$

39. $m_1 = \frac{1}{3}$; $m_2 = -3$

40. $m_1 = 0$; m_2 is undefined

In Exercises 41 and 42, the equations of L_1 and L_2 are given. Determine whether the lines are parallel, perpendicular, or neither.

41. L_1: $y = 0.5x - 3$ $\quad$ L_2: $y = -2x + 4$

42. L_1: $\frac{2}{3}x - \frac{3}{5}y = 1$ $\quad$ L_2: $15 - 10x = 9y$

43. Determine whether a triangle whose vertices are $A(-3, 6)$, $B(10, 0)$, and $C(2, -4)$ is a right triangle.

44. Determine whether the points $P(-4, 3)$, $Q(2, 1)$, and $R(5, -1)$ are collinear.

In Exercises 45 and 46, determine the rate of change of y with respect to x.

45. $y = -3x + 5$

46. $y = \frac{2}{3}x + 4$

In Exercises 47 and 48, the constant rate of change of a line is given, along with an increase or decrease in one variable. Determine the increase or decrease in the other variable.

47. $m = -5$; x increased by 3

48. $m = \frac{5}{3}$; y increased by 10

In Exercises 49 and 50, assume a constant rate of change.

49. The price of a ticket to a college football game increased from \$5.00 in 1988 to \$20.00 in 2000. What was the annual rate of change of the price of the ticket?

50. If the temperature fell from 20°F at 5:00 P.M. to −6°F at 6:00 A.M., what was the hourly rate of change in temperature?

In Exercises 51–54, determine whether the given equation represents a direct variation.

51. $y = 4x - 7$

52. $y = \frac{x}{-3}$

53. $y = 7x$

54. $xy = 4$

55. The number n of toothpicks produced by a machine varies directly with the number of hours h the machine is operating. If it produces 15,000 toothpicks in 6 hours, how many toothpicks are made in 45 hours by this one machine?

56. The per-share dividend that a company declares varies directly with the earnings per share. If a company pays a \$1.50 per-share dividend when earnings are \$4.00 per share, what would the earnings per share be for a dividend of \$2.00 per share?

Section 4.4

In Exercises 57 and 58, write an equation of the line having the given slope and containing the given point.

57. $m = -3$; $(-2, 5)$

58. $m = \frac{4}{3}$; $(1, 2)$

In Exercises 59–62, write an equation of the line containing the given pair of points.

59. $(2, -4)$ and $(-5, 3)$

60. $(2, -5)$ and $(2, 4)$

61. $(-3, -6)$ and $(1, -1)$

62. $(-2, -5)$ and $(4, -5)$

In Exercises 63–68, write an equation of the line with the following conditions.

63. The line is horizontal and contains the point $(-3, 6)$.

64. The line is vertical and contains the point $(6, -3)$.

65. The line contains the point $(1, 5)$ and is parallel to the line whose equation is $2x + 3y = 6$.

66. The line contains the point $(2, -3)$ and is perpendicular to the line $2x - 3y = 7$.

67. The line is graphed in the accompanying figure.

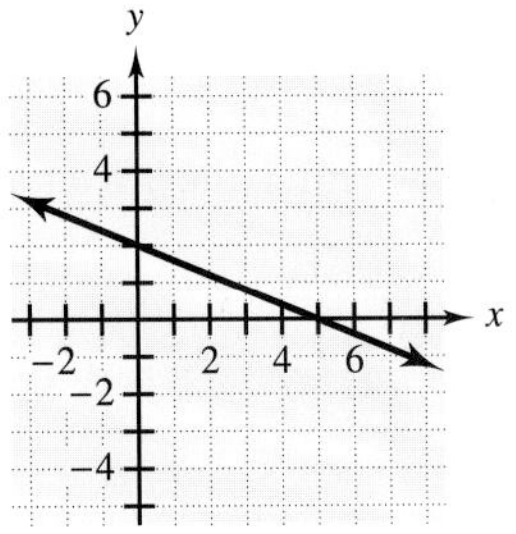

68. The line is perpendicular to the line whose equation is $y = 3x + 4$ and has the same y-intercept as the line whose equation is $y = 3x - 5$.

69. A store owner decides to expand the facilities when annual sales reach \$800,000. Sales were \$150,000 in 1998 and \$400,000 in 2000. Assuming the increase in sales is linear, predict the year in which the store owner will expand.

70. Suppose a public school system will lose its accreditation if its student–teacher ratio rises above 30 to 1. In 2000 the system had 1000 students and 40 teachers. During the following year, there were 1100 students and 45 teachers. Assuming a linear change in the student–teacher ratio, determine whether the system is headed for loss of accreditation.

Section 4.5

In Exercises 71–74, determine whether each of the given ordered pairs is a solution of the given inequality.

71. $4x - y \geq -2$; $(3, 14), (-2, 0), (-4, -20)$

72. $2x - 3y < 0$; $(0, 0), (5, 4), (-4, -4)$

73. $y \leq -3$; $(1, 10), (0, -4), (4, 5)$

74. $x > 5$; $(-5, -7), (4.99, 5), (5.01, 8)$

In Exercises 75–78, sketch the graphs of the following inequalities.

75. $3x - 4y \leq 12$

76. $x \leq 2y$

77. $y + 4 \leq 7$

78. $8 - 2x \geq 0$

In Exercises 79–82, sketch the graph of the compound inequality.

79. $5y - 2x > 7$ or $y + 3x < -2$

80. $y \leq 2x - 3$ and $y \geq 4 - x$

81. $y \leq 4$ and $x + 2 \leq 0$

82. $x \leq -3$ or $y < 2x - 1$

In Exercises 83–86, sketch the graph of the absolute value inequality.

83. $|y| \geq 2$

84. $|4x - 3y| \geq 5$

85. $|x - 2| - 2 < 5$

86. $|3x + 2y| \leq -1$

87. At a minor league stadium, grandstand seats cost \$6 and bleacher seats cost \$4. The total receipts from the ticket sales for a certain game were less than \$11,400. Write a linear inequality that describes the number of each kind of ticket sold.

88. An **isosceles triangle** is a triangle with two sides of equal length. The perimeter of a certain isosceles triangle is at most 42 inches. Write a linear inequality that describes the lengths of the sides of the triangle.

89. A community athletics council sponsors coed soccer teams. The number of boys and girls enrolled in the league cannot exceed 400, and there must be at least 100 girls. Write a compound inequality that describes the number of youngsters who can participate in the program.

90. In a survey a certain percentage of those polled were in favor of a local bond issue. The survey is known to be accurate to within 3 percentage points. Write an absolute value inequality that describes the relationship between the percentage of the sample who were in favor and the actual percentage of all voters who were in favor.

Looking Back

The following exercises review concepts and skills that you will need in Chapter 5.

1. (a) Graph the equation $2x - 3y = 6$.
   (b) Graph the inequality $2x - 3y \geq 6$.

2. Determine whether the given ordered pair is a solution of the equation $2x + y = 1$.

   (i) $(-2, 5)$ (ii) $(-1, 1)$ (iii) $\left(\frac{1}{2}, 0\right)$

3. Describe the graphs of $x = 3$ and $y = 3$.

In Exercises 4–6, solve the equation.

4. $2x - 2 = 13 - 3x$ 5. $3x + 2(4x - 13) = 29$

6. $2y - (y - 5) + 17 = 0$

7. For what value of x do the functions $f(x) = -5$ and $g(x) = 3x - 2$ have the same value?

In Exercises 8 and 9, the equations of line L_1 and line L_2 are given. Determine whether the lines are parallel.

8. L_1: $y + 3x = 6$ L_2: $3x - y + 8 = 0$

9. L_1: $y = -x - 3$ L_2: $x + y = 4$

10. (a) Suppose that you solve an equation and obtain $-2 = -2$. What is the equation called and what is the solution set?

(b) Suppose that you solve an equation and obtain $5 = 2$. What is the equation called and what is the solution set?

11. Evaluate each expression for $x = 4$, $y = \frac{1}{2}$, and $z = -3$.

(a) $-3x + 4y + z$ (b) $2x - 2y + 3z$

12. Write an expression that models the given information.

(a) The total cost in dollars of mailing x first-class letters at 34¢ each and y postcards at 20¢ each

(b) The number of gallons of boric acid in x gallons of a 20% solution and y gallons of a 9% solution

(c) The total distance traveled at x mph for 2 hours and y mph for 3 hours

Chapter 4 Test

1. Complete the ordered pairs so that they correspond to points of the graph of $2x - y = -2$.

(a) (___ , 0) (b) (0, ___)

(c) (−2, ___) (d) (___ , 3)

In Questions 2–5, sketch the graph of each equation. Give the x- and y-intercepts and the slope of each.

2. $y = x$

3. $y = -x - 2$

4. $2x - 3y = 9$

5. $x - 2 = 3$

6. Refer to the graph and fill in the blanks with the symbols < or >:

m ___ 0 and b ___ 0.

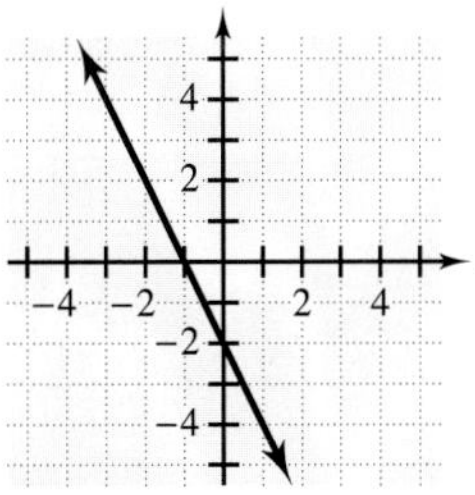

In Questions 7–12, determine an equation of the line that satisfies the given information.

7. The line contains the points $(2, -3)$ and $(-1, -5)$.

8. The line contains the point $(-1, 3)$ and has a slope of 2.

9. The line is vertical and contains the point $(2, 5)$.

10. The line contains the point $(-1, 3)$ and is parallel to the line whose equation is $y = -4$.

11. The line contains the point $(1, 4)$ and is perpendicular to the line whose equation is $y = -\frac{1}{2}x + 3$.

12. The graph of the line is shown in the figure.

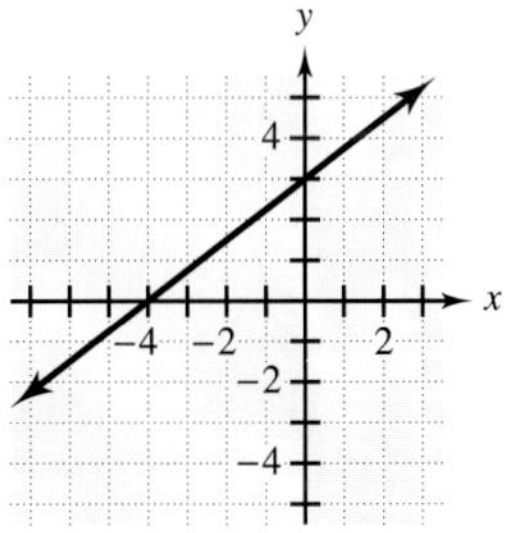

In Questions 13 and 14, graph the given inequalities.

13. $2x - y \le 5$

14. $y - 3 > 0$

In Questions 15 and 16, graph the given compound inequalities.

15. $y \ge 2x + 1$ and $x + y \ge 3$

16. $x - 2y < 2$ or $2y - x > 6$

In Questions 17 and 18, graph the given absolute value inequalities.

17. $|x + y| \leq 3$

18. $|x + 2y| > 5$

19. Show that the points $(1, -2)$, $(-2, 1)$, and $(5, 2)$ are the vertices of a right triangle.

20. Draw a line through $(-1, 3)$ with a slope of -2.

In Questions 21 and 22, determine whether the lines are perpendicular, parallel, or neither.

21. $2x + 3y = 3$ and $2y - 3x = -4$

22. $y = 3x + 1$ and $x - 3y = -1$

23. The velocity of an object dropped from rest varies directly with the time of the fall. After 0.5 second, the speed is 16 feet per second. Write the relation between the velocity and the time. What is the velocity after 2 seconds?

24. The earnings for a company dropped from \$1.35 per share in 1998 to \$0.45 in 2000. Assuming that the relationship is linear, what was the rate of change in per-share earnings per year?

25. At 6:00 A.M., the temperature in Death Valley was 88°F. At 10:00 A.M., the temperature was 98°F. Assuming that the rise in temperature is linear, write an equation to describe the rise in temperature. Use this equation to predict the temperature at 3:00 P.M.

Chapter 5

Systems of Linear Equations

The percentage of office communications conducted by e-mail is increasing rapidly, while the percentage conducted by hard copy (paper) is decreasing. (Source: Cognitive Communication Survey.) The trends in the use of e-mail and in the use of paper can each be modeled by a linear equation, and the two equations together form a ***system of equations.*** *The* ***solution*** *of this system is an ordered pair of numbers that indicates, for example, the year when the percentages for e-mail and paper are predicted to be the same. (For more on this topic, see Exercises 89–92 at the end of Section 5.2, and see the Chapter Project.)*

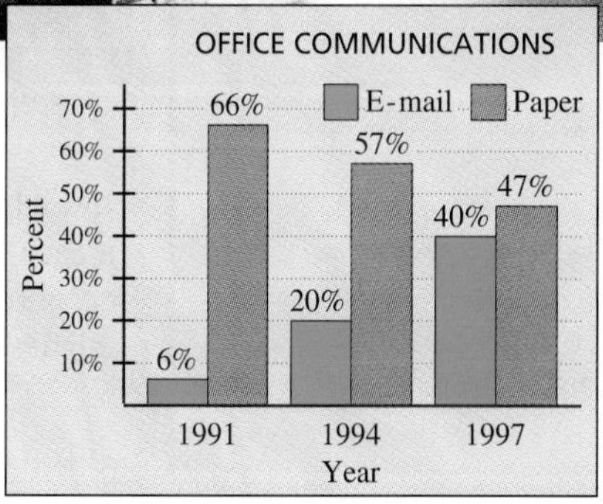

(Source: Cognitive Communication Survey.)

In this chapter we introduce systems of linear equations in two and three variables. We discuss graphical and algebraic methods for solving systems. These methods include Cramer's Rule, for which we need to learn about matrices and determinants. After using these new skills to solve application problems, we conclude by examining systems of linear inequalities.

5.1 Graphing and Substitution Methods

Systems of Equations • Estimating Solutions by Graphing • The Substitution Method • Special Cases • Applications

Systems of Equations

We write equations to describe the conditions and relationships given in applied problems. Up to this point, applications have led to a single equation in one or two variables. However, applications often lead to more than one equation.

Exploring the Concept

Systems of Equations

Suppose that we want to describe two numbers whose sum is 2. If x and y represent the two numbers, then an equation is $x + y = 2$. To sketch the graph, we solve for y and write $y = -x + 2$ and make a table of solutions. Note that there are infinitely many pairs of numbers whose sum is 2. The graph of this equation is shown in Figure 5.1.

Figure 5.1

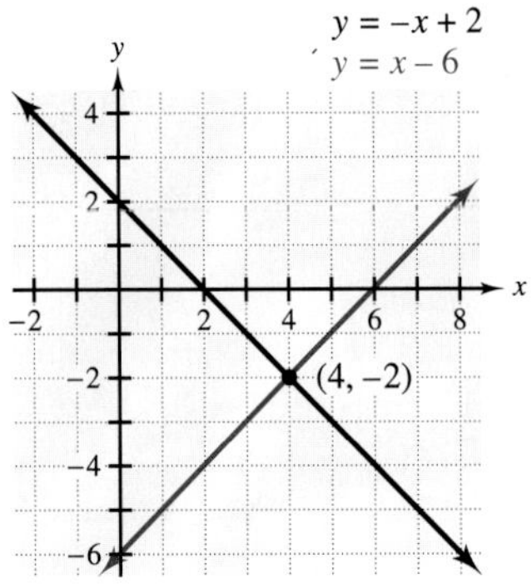

x	-1	0	1	2	3	4	5	6
$-x + 2$	3	2	1	0	-1	-2	-3	-4

Next, suppose that we want to describe two numbers whose difference is 6. If x is the larger number, then an equation is $x - y = 6$. We write $y = x - 6$ and complete a table of solutions. Again, there are infinitely many pairs of numbers whose difference is 6. The graph of this equation is also shown in Figure 5.1.

x	-1	0	1	2	3	4	5	6
$x - 6$	-7	-6	-5	-4	-3	-2	-1	0

Observe that $(4, -2)$ is the *only* pair that satisfies both conditions.

Now suppose that we want to find a pair of numbers whose sum is 2 *and* whose difference is 6. Because the point of intersection in Figure 5.1 belongs to both graphs, its coordinates $(4, -2)$ satisfy both equations. The two tables confirm that $(4, -2)$ satisfies both of the given conditions. Thus the required numbers are 4 and -2.

Considered simultaneously, the pair of equations $x + y = 2$ and $x - y = 6$ is an example of a **system of equations.** The ordered pair $(4, -2)$ is a **solution** of this system because it satisfies both equations.

Definition of a System of Equations and of Its Solution

Two or more equations considered simultaneously form a **system of equations.** A **solution** of a system of equations in two variables is an ordered pair (a, b) that satisfies all equations of the system.

A system of equations is a *conjunction*, but we usually write the system without explicitly including the word *and*.

Initially, you might want to write the word *and* to emphasize that *both* equations must be satisfied.

$$x + y = 2$$
$$x - y = 6$$

In certain applications there can be many equations and many variables. In this section we consider only systems of two linear equations in two variables.

EXAMPLE 1

Verifying Solutions of a System of Equations

For the following system of equations, determine whether the given pair of numbers is a solution.

$$x - y = 2$$
$$x - 2y = -2$$

(a) (5, 3) (b) (6, 4)

Solution

(a) In each equation replace x with 5 and y with 3.

$x - y = 2$		$x - 2y = -2$	
$5 - 3 = 2$		$5 - 2(3) = -2$	
$2 = 2$	True	$5 - 6 = -2$	
		$-1 = -2$	False

Although (5, 3) satisfies the first equation, it does not satisfy the second equation. Therefore, the pair is not a solution of the system of equations.

(b) In each equation replace x with 6 and y with 4.

$x - y = 2$		$x - 2y = -2$	
$6 - 4 = 2$		$6 - 2(4) = -2$	
$2 = 2$	True	$6 - 8 = -2$	
		$-2 = -2$	True

Because (6, 4) satisfies both equations, the pair is a solution of the system of equations.

Ask students to interpret this solution graphically.

Estimating Solutions by Graphing

The preceding discussion suggests a graphing method for estimating the solution of a system of equations.

Estimating Solutions of Systems by Graphing

1. Sketch the graphs of both equations in the same coordinate system.
2. Locate the point of intersection of the two graphs.
3. This point represents the estimated solution of the system of equations.

EXAMPLE 2

Estimating Solutions by Graphing

Use the graphing method to estimate the solution of the system of equations.

$$2x - y = 0$$
$$3y - 4x = 3$$

Solution

Solve each equation for y and sketch the graphs of both equations. (See Fig. 5.2.)

$$2x - y = 0 \rightarrow y = 2x$$
$$3y - 4x = 3 \rightarrow y = \frac{4}{3}x + 1$$

Figure 5.2

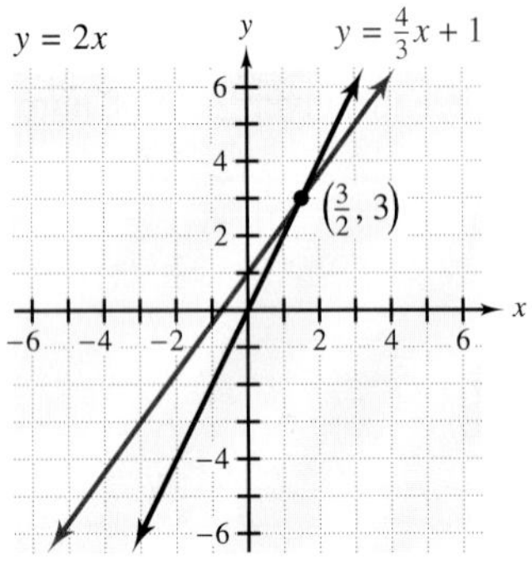

The point of intersection appears to be $\left(\frac{3}{2}, 3\right)$. The solution can be verified by substituting $\frac{3}{2}$ for x and 3 for y.

Substitute $x = \frac{3}{2}$ and $y = 3$

$2x - y = 0$	$3y - 4x = 3$
$2\left(\frac{3}{2}\right) - 3 = 0$	$3(3) - 4\left(\frac{3}{2}\right) = 3$
$3 - 3 = 0$	$9 - 6 = 3$
$0 = 0$ True	$3 = 3$ True

Note: As Example 2 illustrates, success with the graphing method depends on sketching a very accurate graph and then estimating the coordinates of the point of intersection.

The Substitution Method

When we use the graphing method to estimate the solution of a system of equations, the point of intersection of the two graphs corresponds to the solution of the system. At that point, one value of x gives the same value for y in each equation. The substitution method is an algebraic method that uses this same idea.

The substitution method is a particularly useful method when it is easy to solve one of the equations for a certain variable. In the following summary of the method, we begin by solving one equation for y. However, either x or y may be chosen.

The Substitution Method

1. Solve one equation for y to obtain $y =$ expression.
2. Replace y in the other equation with the expression. This results in an equation in one variable x.
3. Solve that equation.
4. Substitute the solution for x in either original equation and solve for y.
5. Check the solution.

EXAMPLE 3

Solving a System with the Substitution Method

Solve the following system with the substitution method.

$$\begin{aligned} 2x - y &= 9 \\ 3x - 4y &= -24 \end{aligned}$$

Solution

We choose to solve the first equation for y.

$2x - y = 9$	First equation
$-y = -2x + 9$	Subtract $2x$ from both sides.
$y = 2x - 9$	Multiply both sides by -1.

Substitute $2x - 9$ for y in the second equation.

$3x - 4y = -24$	Second equation
$3x - 4(2x - 9) = -24$	Replace y with $2x - 9$.
$3x - 8x + 36 = -24$	Distributive Property
$-5x + 36 = -24$	Combine like terms.
$-5x = -60$	Subtract 36 from both sides.
$x = 12$	Divide both sides by -5.

Students sometimes think they are finished when they have found the value of the first variable. Emphasize the need to substitute to find the value of the other variable.

Now determine the y-value of the solution by substituting 12 for x in the equation that is already solved for y.

$y = 2x - 9$

$y = 2(12) - 9$ Replace x with 12.

$y = 24 - 9$

$y = 15$

The solution is (12, 15).

LEARNING TIP

The solution of a system of equations in two variables is a pair of numbers. After solving for one variable, remember to substitute to determine the value of the other variable.

Think About It

The solution to the following system is easy to determine without the graphing method and without the substitution method.

$y = 2x - 7$
$y = -x - 7$

What is the solution? Why?

Special Cases

When we use the graphing method to estimate the solution of a system of two linear equations in two variables, one possible result is two lines that intersect at exactly one point. The pair of coordinates of that point of intersection is the solution of the system. However, two lines may not intersect at exactly one point.

See whether students can predict the special cases.

Exploring the Concept

Inconsistent Systems

Consider the following system of equations.

$$x - 2y = 5 \rightarrow y = \frac{1}{2}x - \frac{5}{2}$$

$$2x - 4y = 4 \rightarrow y = \frac{1}{2}x - 1$$

Figure 5.3

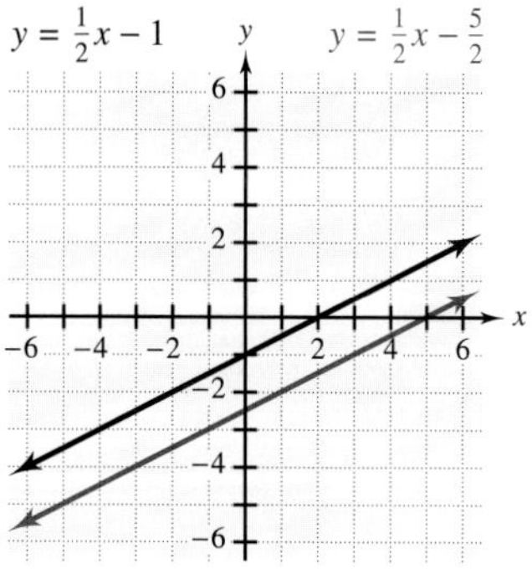

Figure 5.3 shows the graphs of the two equations.

Because the slope of each line is $\frac{1}{2}$, the lines are parallel. Thus there is no point of intersection, and the system has no solution.

To solve the system with the substitution method, we choose to solve the first equation for x.

$$x - 2y = 5 \quad \text{First equation}$$
$$x = 2y + 5 \quad \text{Add } 2y \text{ to both sides.}$$

Substitute $2y + 5$ for x in the second equation and solve for y.

$$2x - 4y = 4 \quad \text{Second equation}$$
$$2(2y + 5) - 4y = 4 \quad \text{Replace } x \text{ with } 2y + 5.$$
$$4y + 10 - 4y = 4 \quad \text{Distributive Property}$$
$$10 = 4 \quad \text{False}$$

Because the resulting equation is false, it has no solution. Therefore, the system of equations has no solution.

Systems of equations whose graphs do not intersect have no solution. Such systems are called **inconsistent.** If a system of equations has at least one solution, the system is called **consistent.**

Exploring the Concept

Dependent Equations

The following system of equations illustrates another special case.

$$2x + y = 3 \rightarrow y = -2x + 3$$
$$4x + 2y = 6 \rightarrow y = -2x + 3$$

When we solve for y, we see that the two equations are exactly the same. Because all points of the lines are points of intersection, the system has infinitely many solutions, and they are described by the equation $y = -2x + 3$.

The substitution method will reveal this special case.

$$2x + y = 3 \quad \text{First equation}$$
$$2x + (-2x + 3) = 3 \quad \text{Replace } y \text{ with } -2x + 3 \text{ from the second equation.}$$
$$3 = 3 \quad \text{True}$$

The resulting equation is an identity that has infinitely many solutions. Thus the solution set of this system contains the infinitely many solutions of the equation $y = -2x + 3$.

Systems are consistent or inconsistent; *equations* are dependent or independent.

Systems of equations whose graphs coincide have infinitely many solutions. Such equations are called **dependent.** If the lines do not coincide, the equations are called **independent.**

If an algebraic method of solving a system results in an identity, then the equations are dependent and the system has infinitely many solutions.

Encourage students to relate the number of solutions to the graph visually, not just to memorize the algebraic method.

Figure 5.4 summarizes the three possible outcomes when we solve a system of two linear equations in two variables.

Figure 5.4

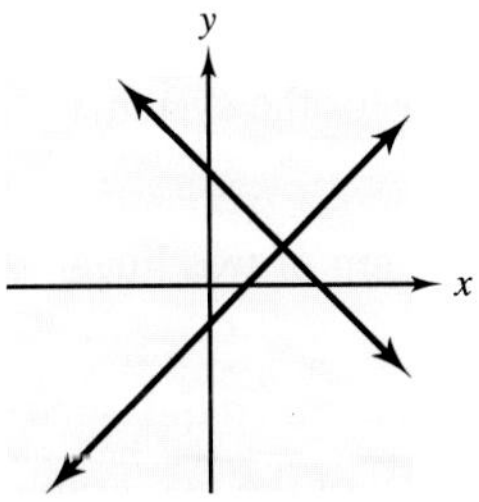

Solution is unique.
Equations are independent.
System is consistent.
The lines intersect.

(a)

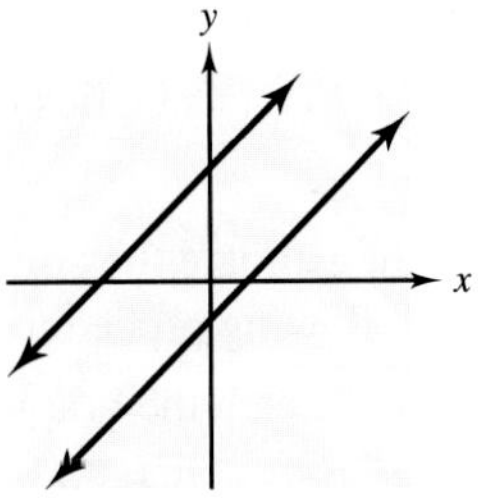

No solution.
Equations are independent.
System is inconsistent.
The lines are parallel.

(b)

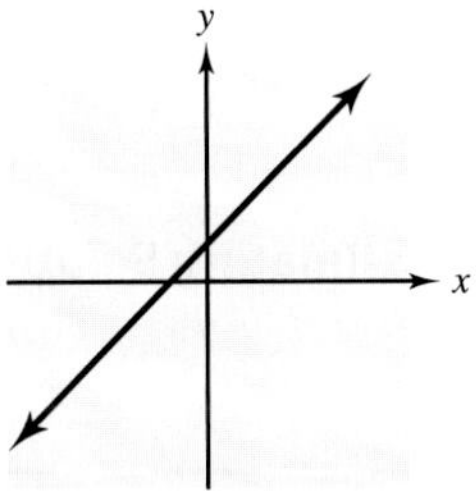

Infinitely many solutions.
Equations are dependent.
System is consistent.
The lines coincide.

(c)

Applications

Many applied problems can be modeled and solved with an equation that has a single variable. However, when the problem involves two unknown quantities, it is often more convenient to use two variables.

EXAMPLE 4

Using a System of Equations to Solve an Application Problem

Two angles are supplementary. One angle is 20° less than the other. What are the measures of the two angles?

This example and Exercises 67–78 are designed to ease the student into applying systems of equations. See Exercises 79–88 for real-life applications and modeling.

Solution

Let x = measure of one angle and
y = measure of the other angle.

The following system of equations describes the given conditions.

$x + y = 180$ The angles are supplementary.

$y = x - 20$ One angle is 20° less than the other angle.

Replace y in the first equation with $x - 20$.

$x + (x - 20) = 180$

$2x - 20 = 180$ Remove parentheses and combine like terms.

$2x = 200$ Add 20 to both sides.

$x = 100$ Divide both sides by 2.

Now we use the second equation of the system to determine y.

$y = x - 20$

$y = 100 - 20 = 80$ Replace x with 100.

The measures of the angles are 100° and 80°.

Quick Reference 5.1

Systems of Equations

- Two or more equations considered simultaneously form a **system of equations.**
- A **solution** of a system of two linear equations in two variables is an ordered pair (a, b) that satisfies both equations of the system.

Estimating Solutions by Graphing

- To estimate the solution of a system of two linear equations in two variables, use the following procedure.
  1. Sketch the graphs of both equations in the same coordinate system.
  2. Locate the point of intersection of the two graphs.
  3. This point represents the estimated solution of the system of equations.

Substitution Method

- The substitution method is an algebraic method for solving a system of equations.
- In the following summary of the method, we begin by solving one equation for y. However, either x or y may be chosen.
  1. Solve one equation for y to obtain $y =$ expression.
  2. Replace y in the other equation with the expression. This results in an equation in one variable x.
  3. Solve that equation.
  4. Substitute the solution for x in either original equation and solve for y.
  5. Check the solution.

Special Cases

- Two special cases may arise when we are solving a system of two linear equations in two variables.
  1. The system of equations is **inconsistent.** The graphs of the equations are parallel, and the system has no solution.
  2. The equations of the system are **dependent.** The graphs of the equations coincide, and the solution set of the system contains the infinitely many solutions of either equation.

Speaking the Language 5.1

1. If the graphs of a system of equations intersect at a single point, that point represents a(n) ________ of the system.

2. If a system of equations has no solution, then the system is ________.

3. If the graphs of a system of equations coincide, then the equations of the system are called ________.

4. The ________ method is an algebraic method for solving a system of equations.

Exercises 5.1

Concepts and Skills

1. Explain how to determine whether an ordered pair is a solution of a system of equations in two variables.

2. The pair (1, 2) does not satisfy the first equation of the following system.

$$2x + 3y = 7$$
$$x - 2y = 1$$

Why can you conclude, even without testing the pair in the second equation, that (1, 2) is not a solution of the system?

In Exercises 3–6, determine whether each ordered pair is a solution of the given system of linear equations.

3. $(0, 1), (2, -3), (4, 0)$
$2x + y = 1$
$3x - 2y = 12$

4. $\left(4, -\frac{5}{3}\right), (-0.5, 0), \left(\frac{1}{2}, \frac{2}{3}\right)$
$2x + 3y = 3$
$8x = 12y - 4$

5. $(1, 1), \left(-2, \frac{8}{7}\right), (2, 0)$
$2x + 7y = 4$
$14y + 4x = 8$

6. $(4, -2), (-2, 1), (-6, 13)$
$3x = 2(4 - y)$
$2y + 1 = -3(x - 3)$

CE In Exercises 7–10, determine the value of a, b, or c so that the given ordered pair is a solution of the system of equations.

7. $ax - 2y = 5$ $\quad (3, -1)$
$x + by = 1$

8. $ax + y = 5$ $\quad (1, 2)$
$x + by = 7$

9. $ax + 3y = 9$ $\quad (3, -1)$
$2x - y = c$

10. $3x - y = c$ $\quad (8, 2)$
$x + by = 2$

11. Explain why two linear equations in two variables cannot have exactly two solutions.

12. Describe the graph of a system that has
(a) no solution.
(b) a unique solution.
(c) infinitely many solutions.

In Exercises 13–18, determine the number of solutions of the system without graphing or solving.

13. $y = 3x - 4$
$y = 3x + 5$

14. $y = -2x - 3$
$y + 2x = 4$

15. $y + 2x = 6$
$y - 2x = 8$

16. $y = \frac{2}{3}x + 5$
$y = -\frac{3}{2}x - 4$

17. $y + x = 5$
$y = -x + 5$

18. $y + 3x = 4$
$y = -3x + 4$

19. (a) How do the graphs of the following systems of equations differ?
(i) The equations are independent.
(ii) The equations are dependent.
(b) Is it possible for a system of two independent equations to have no solution? Explain.

20. (a) How do the graphs of the following systems of equations differ?
(i) The system is consistent.
(ii) The system is inconsistent.
(b) Is it possible for a consistent system to have infinitely many solutions? Explain.

In Exercises 21–24, estimate the solution of the system of equations.

21.

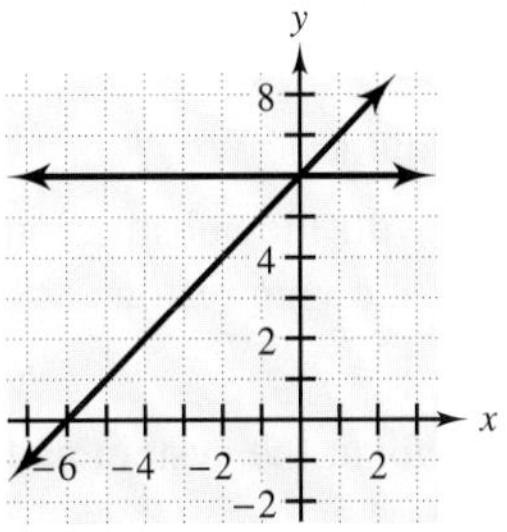

22.

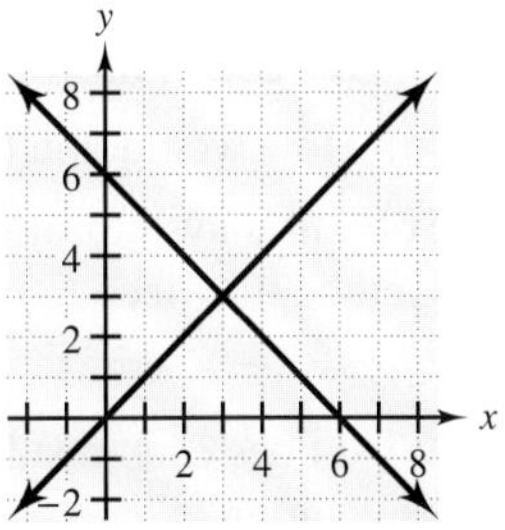

23.

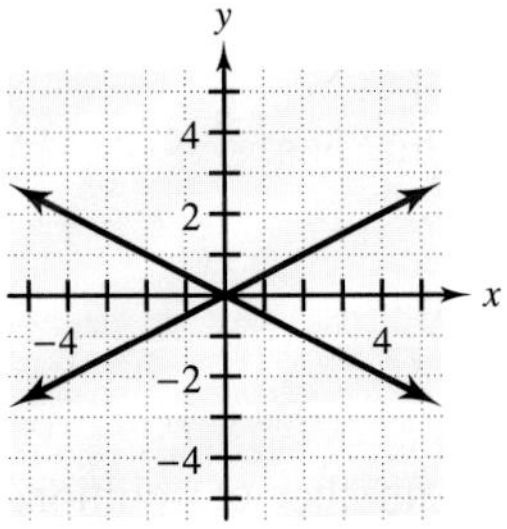

24.

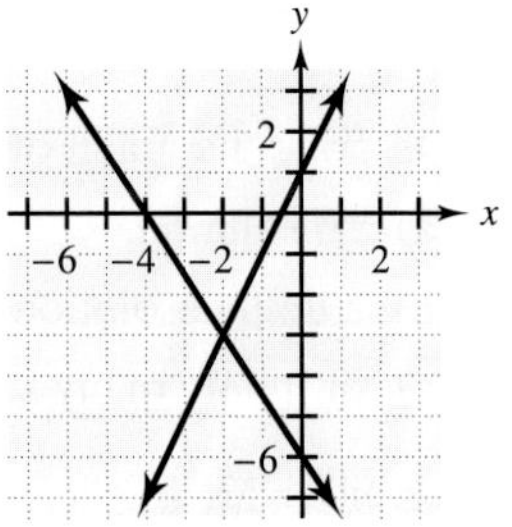

In Exercises 25–40, use the graphing method to estimate the solution of the given system of equations. Then use the substitution method to determine the solution.

25. $y = 2x - 2$
$y = -x + 7$

26. $y - x = 2$
$y = 2x - 3$

27. $y = 0.6x - 10$
$5y - 3x = 4$

28. $3x + y = 12$
$2y + 10 = -6x$

29. $2x + y = 8$
$y + 2 = 0$

30. $y = 3x - 2$
$y = 4$

31. $x + 3 = 0$
$2x + 3y = 6$

32. $x - 2 = 0$
$3x + 2y = 8$

33. $y + 4x - 7 = 0$
$2x + y - 5 = 0$

34. $3x - 4y + 24 = 0$
$2y + x - 2 = 0$

35. $x = -4$
$y = 7$

36. $x = 3$
$y = -5$

37. $y = -2(4 - x)$
$y + 8 = 2x$

38. $2y - 3x = 10$
$-1.5x = 5 - y$

39. $\frac{2}{3}x + \frac{1}{3}y = 1$
$\frac{1}{2}x + \frac{2}{3}y = -\frac{4}{3}$

40. $-\frac{5}{9}x + \frac{2}{3}y = 2$
$\frac{1}{6}x + \frac{1}{2}y = -2$

41. Suppose you plan to solve the following system of equations with the substitution method.

$3x - 4y = 16$
$2x + y = 7$

(a) Would it be correct to begin by solving one of the equations for x?

(b) How would you begin the solution? Why?

42. Solving a system of equations with the substitution method may lead to one of the following results.

(a) $-3 = -3$ (b) $13 = 7$

What does each result indicate about the system and its solution?

In Exercises 43–58, use the substitution method to solve the system of equations. Identify inconsistent systems and dependent equations.

43. $x - y - 5 = 0$
$2x - y + 17 = 0$

44. $2x + 3y + 15 = 0$
$x + 2y + 18 = 0$

45. $y + 3x - 4 = 0$
$12 = 9x + 3y$

46. $2x = y - 5$
$3y = 6x + 15$

47. $x = 3y + 9$
$3y - x = 6$

48. $x - 2y = 5$
$3x = 6y + 8$

49. $y = 2x - 5$
$4x - 2y - 10 = 0$

50. $x - 2y = 5$
$2x = 4y + 10$

51. $y = 2x - 7$
$3x + 2y = 0$

52. $x = 2y - 5$
$2x - 3y = -6$

53. $6x = 5y$
$12x + 7 = 10y$

54. $2x + 3y + 2 = 0$
$0 = 6x + 9y$

55. $y = -2x + 5$
$21x + 10y = 50$

56. $1.5x + y = 9$
$y = -1.52x + 9$

57. $x = y$
$11x - 10y = 80$

58. $y = \frac{7}{20}x$
$3y - x = -36$

CE In Exercises 59–62, determine the value of k so that the equations are dependent.

59. $y = \frac{2}{3}x + 1$
$kx + 3y = 3$

60. $3x + ky = 15$
$y = 5 - x$

61. $y = 3x + k$
$y + (2 - k)x = 5$

62. $kx - 4y = -2$
$4y - 3x = k - 1$

CE In Exercises 63–66, determine the value of k so that the system is inconsistent.

63. $kx + y = 2k$
$y = 2x + 5$

64. $2y - x = 6$
$x + ky = 0$

65. $kx + 3y = 12$
$y = \frac{4}{3}x + (k + 1)$

66. $x = ky - 6$
$3x + y = 9k$

In Exercises 67–70, translate the given information into a system of equations and solve the system to determine the two numbers.

67. The sum of two numbers is 16 and their difference is 4.

68. The difference of two numbers is one-third of the smaller number. The sum of the numbers is 6 less than twice the larger number.

69. One number exceeds a second number by 8. Three-fourths of the larger number is 3 more than the smaller number.

70. A number is 6 less than a larger number. Two-thirds of the larger number plus one-fourth of the smaller number is 15.

Geometric Models

71. Two angles are complementary. The difference between twice the larger angle and 5 times the smaller angle is 19°. What is the measure of each angle?

72. Two angles are complementary. The smaller angle is one-third of 10° less than the larger angle. What is the measure of each angle?

73. Two angles are supplementary. The larger angle is 24° less than twice the smaller angle. What is the measure of each angle?

74. Two angles are supplementary. Two-thirds of the larger angle is 30° more than the smaller angle. What is the measure of each angle?

75. The perimeter of a rectangle is 74 feet. The length is 7 feet more than twice the width. What are the dimensions of the rectangle?

76. The perimeter of a rectangle is 44 meters. The width is 4 meters less than the length. What are the dimensions of the rectangle?

77. The perimeter of a rectangle is 80 inches. The width is one-third of the length. What are the dimensions of the rectangle?

78. The perimeter of a rectangle is 32 meters. The length is 4 meters less than 3 times the width. What are the dimensions of the rectangle?

Real-Life Applications

79. Mills and Lopez were the only candidates in a county election. Of the 24,000 registered voters, only 30% turned out for the election, and they all voted for one of the candidates. If Lopez won the election by 200 votes, how many voted for Mills?

80. On an AAU swim team, the number of women was 1 more than the number of men. When two men and one woman left the team, there was a total of 26 swimmers. How many men were originally on the team?

81. A coed Little League team had 18 players. Even though it was raining, half the girls and one-third of the boys showed up for a game, but these seven players went home when the game was canceled. How many girls were on the team?

82. Perfecto and CTV were cable TV companies competing for a local area franchise. Perfecto offered the same channels as CTV plus two other channels. One-tenth of Perfecto's channels were Pay-Per-View, and one-eighth of CTV's channels were Pay-Per-View. CTV offered one more Pay-Per-View channel than Perfecto. How many channels did CTV offer?

Modeling with Real Data

83. Over the last century, the percentage of men over 65 in the work force has declined. The bar graph shows the percentages of men and women over 65 in the work force from 1950 to 2000. (Source: Bureau of the Census.)

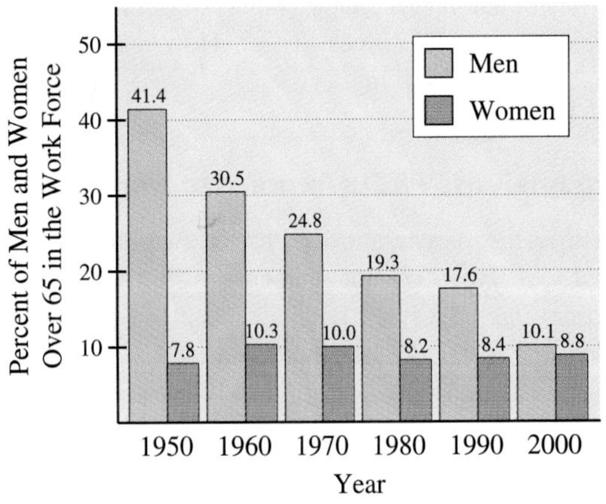

With x representing the number of years since 1950, the percentages can be modeled with the following functions.

Men: $M(x) = -0.57x + 38.29$

Women: $W(x) = -0.01x + 9.10$

(a) Treat these two equations as a system of equations. How do you know that the system of equations has one solution?

(b) Solve the system algebraically.

(c) Interpret the solution of the system of equations.

84. The table shows that tuition at both private and public four-year colleges continued to increase in the period 1997–1999. (Source: The College Board.)

	Tuition	
Year	**Private**	**Public**
1997	12,994	2,975
1998	13,664	3,111
1999	14,508	3,243

The following functions model the tuition y (in dollars), where x is the number of years since 1995.

Private: $y = 757x + 11{,}451$

Public: $y = 134x + 2708$

(a) Does the system of these model equations have a solution? If so, what does it represent?

(b) Suppose that you are interested only in years beyond 1997. Does the system have a solution for those values of x? How can you know without solving?

Data Analysis: *Ph.D.s Awarded to Women*

In 1967, the percentage of Ph.D.s awarded to women was 13%. By 1997, that percentage had risen to 41%. (Source: *Business Week.*)

85. Letting x represent the number of years since 1965, write a linear function $W(x)$ that models the percentage of Ph.D.s awarded to women.

86. (a) Interpret the function $y = 100 - W(x)$.

(b) Letting x represent the number of years since 1965, write a linear function $M(x)$ that models the percentage of Ph.D.s awarded to men.

(c) Compare your results in parts (a) and (b).

87. If you were to graph functions W and M, how could you know in advance what the y-coordinate of the point of intersection would be? Explain.

88. Solve the system algebraically to approximate the year when the same percentage of Ph.D.s will be awarded to men as to women.

Challenge

In Exercises 89–92, use the graphing method to estimate the solution of the given system of three equations in two variables.

89. $y = x - 12$

$2y + x = 0$

$2y = 3x - 32$

90. $x + y = 3$

$y = \frac{2}{5}x + 10$

$-3x + 2y = 31$

91. $y = 9 - x$

$3y + 3x = -21$

$2x = 18 - 2y$

92. $x - y = -7$

$x + y = 7$

$2y = x - 10$

93. Determine the vertices of a quadrilateral bounded by the graphs of $y = x - 5$, $y + x = 3$, $x + y + 2 = 0$, and $y - x = 7$. What specific kind of quadrilateral is this figure?

94. Determine the vertices of a triangle bounded by the graphs of $x + 3 = 0$, $y = 6$, and $6y - 7x = 15$. An **isosceles triangle** is a triangle in which two of the sides are of equal length. Is this triangle an isosceles triangle?

95. Show whether there is a value of k such that the given system has exactly one solution.

$x - y = k$

$3x - 3y = 2$

96. Is it possible to determine a and c so that the given system of equations has

(a) no solution?

(b) infinitely many solutions?

$2ax + ay = 1$

$y - x = c$

5.2 The Addition Method

Addition Method • Special Cases • Applications

Addition Method

The substitution method is not always the most convenient algebraic method for solving a system of equations. An alternative method is the **addition method,** which is sometimes called the **elimination method.**

The Addition Property of Equations allows us to add the same number to both sides of an equation: If $A = B$, then $A + C = B + C$. If $C = D$, then we can write the result as follows.

$$\begin{array}{r} A = B \\ \underline{C = D} \\ A + C = B + D \end{array}$$

We call this process *adding the equations.*

The goal of the addition method is to eliminate one of the variables in a system of equations by adding the two equations, with the result being a linear equation in one variable.

EXAMPLE 1

Solving a System of Equations with the Addition Method

Use the addition method to solve the following system of equations.

$$x + y = 6$$
$$x - y = 2$$

Solution

$x + y = 6$	Add the two equations.
$x - y = 2$	The coefficients of y are opposites.
$2x \quad = 8$	The variable y is eliminated.
$x = 4$	Solve for x.

Replace x with 4 in either of the original equations and solve for y.

$x + y = 6$	First equation
$4 + y = 6$	Replace x with 4.
$y = 2$	

The solution of the system is (4, 2).

Remind students not to stop after they have solved for the first variable. Students sometimes forget to solve for the second variable.

Adding two equations results in the elimination of a variable only when the variable appears in both equations and has coefficients that are opposites. If this is not the case, we can make them opposites by applying the Multiplication Property of Equations.

EXAMPLE 2

Using the Multiplication Property of Equations with the Addition Method

Solve the following system of equations.

$$\begin{aligned} 2x - y &= 9 \\ 3x - 4y &= -24 \end{aligned}$$

Solution

LEARNING TIP

Try to recognize the variable whose elimination requires the least amount of work. For instance, a variable with a coefficient of −1 or 1 is a good choice for a variable to eliminate.

If we choose to eliminate y, its coefficients must be opposites. Therefore, we multiply both sides of the first equation by −4 and then add the equations.

$$\begin{aligned} -4(2x - y) &= -4(9) &\rightarrow\quad -8x + 4y &= -36 &&\text{Multiply first equation by } -4. \\ 3x - 4y &= -24 &\rightarrow\quad 3x - 4y &= -24 &&\text{Second equation} \\ & & -5x &= -60 &&\text{Add the equations.} \\ & & x &= 12 &&\text{Solve for } x. \end{aligned}$$

Replace x with 12 in either equation to determine y.

$$\begin{aligned} 2x - y &= 9 &&\text{First equation} \\ 2(12) - y &= 9 &&\text{Replace } x \text{ with 12.} \\ 24 - y &= 9 \\ -y &= -15 \\ y &= 15 \end{aligned}$$

The solution is (12, 15).

Discuss ways to recognize the variable whose elimination requires the least work.

The decision of which variable to eliminate is arbitrary. Usually, we choose the variable whose elimination requires the least amount of work.

Writing both equations in the same form is a good idea when using the addition method, because then the like terms are lined up in columns. It is common to write both equations in the standard form $Ax + By = C$.

The Multiplication Property of Equations requires us to multiply both sides of an equation by the same number. However, we can multiply both sides of one equation by one number and both sides of the other equation by a *different* number. It is not necessary to multiply both equations by the same number.

EXAMPLE 3

Using the Multiplication Property of Equations with the Addition Method

Solve the following system.

$$\begin{aligned} 3x &= 33 - 5y \\ 3y &= 4x - 15 \end{aligned}$$

Solution

$$\begin{aligned} 3x + 5y &= 33 &&\text{Write each equation in standard form.} \\ -4x + 3y &= -15 \end{aligned}$$

Ask students how to eliminate y instead. Discuss general strategies.

To eliminate x, multiply the first equation by 4 and the second equation by 3. Then add the equations.

$$\begin{aligned} 4(3x + 5y) &= 4(33) & \rightarrow && 12x + 20y &= 132 \\ 3(-4x + 3y) &= 3(-15) & \rightarrow && -12x + 9y &= -45 \\ &&&& 29y &= 87 \\ &&&& y &= 3 \end{aligned}$$

Replace y with 3 in either equation.

$$\begin{aligned} 3x &= 33 - 5y && \text{First equation} \\ 3x &= 33 - 5(3) && \text{Replace } y \text{ with 3.} \\ 3x &= 18 \\ x &= 6 \end{aligned}$$

The solution is (6, 3).

When the equations involve fractions, we can multiply by the LCD to clear them.

EXAMPLE 4

Clearing Fractions

Solve the following system.

$$\begin{aligned} \frac{3}{2}x - \frac{3}{4}y &= \frac{15}{4} \\ \frac{4}{3}x - \frac{5}{3}y &= 3 \end{aligned}$$

Solution

Clear the fractions by multiplying each term by the LCD. For the first equation, the LCD is 4; for the second equation, the LCD is 3.

$$\begin{aligned} 4 \cdot \frac{3}{2}x - 4 \cdot \frac{3}{4}y &= 4 \cdot \frac{15}{4} & \rightarrow && 6x - 3y &= 15 \\ 3 \cdot \frac{4}{3}x - 3 \cdot \frac{5}{3}y &= 3 \cdot 3 & \rightarrow && 4x - 5y &= 9 \end{aligned}$$

Now use the Multiplication Property of Equations again, this time to adjust the coefficients. To eliminate x, multiply the first equation by 2 and the second equation by -3. Then add the equations.

Encourage students to write the system with the indicated multiplications, as shown here.

$$\begin{aligned} 2(6x - 3y) &= 2(15) & \rightarrow && 12x - 6y &= 30 \\ -3(4x - 5y) &= -3(9) & \rightarrow && -12x + 15y &= -27 \\ &&&& 9y &= 3 \\ &&&& y &= \frac{3}{9} = \frac{1}{3} \end{aligned}$$

There is an alternative to replacing y with $\frac{1}{3}$ in order to solve for x. We can return to the system in which the fractions have been cleared, eliminate y, and solve for x.

Some students prefer using this method in order to avoid fractions. However, you might want to show that using a calculator makes the arithmetic easy to perform.

$$\begin{aligned} 30x - 15y &= 75 && \text{Multiply the first equation by 5.} \\ -12x + 15y &= -27 && \text{Multiply the second equation by } -3. \\ \hline 18x &= 48 \\ x &= \frac{48}{18} = \frac{8}{3} \end{aligned}$$

The solution is $\left(\frac{8}{3}, \frac{1}{3}\right)$.

The following is a summary of the addition method for solving a system of equations.

The Addition Method

1. Write both equations in the form $Ax + By = C$.
2. If necessary, multiply one or both of the equations by appropriate numbers so that the coefficients of one of the variables are opposites.
3. Add the equations to eliminate a variable.
4. Solve the resulting equation.
5. Substitute that value in either of the original equations and solve for the other variable.
6. Check the solution.

Think About It

A system of linear equations has infinitely many solutions when an algebraic method results in an equation such as $0 = 0$. However, not *every* pair (x, y) is a solution. Why?

Special Cases

In Section 5.1 we saw how the substitution method detected the two special cases: lines that are parallel and lines that coincide. When both variables were eliminated, the result was either a false equation or an identity.

Example 5 shows that special cases can be identified algebraically with the addition method in exactly the same way that they are identified with the substitution method.

EXAMPLE 5

In both systems, both variables are eliminated. Stress the difference between the resulting *false* equation in one case and the resulting *true* equation in the other case.

Special Cases

Use the addition method to solve the following systems of equations.

(a) $6x - 9y = -34$
$2x - 3y = -12$

(b) $x - 0.2y = -0.4$
$5x - y = -2$

Solution

(a)
$$\begin{aligned} 6x - 9y &= -34 && \text{First equation} \\ -6x + 9y &= 36 && \text{Multiply the second equation by } -3. \\ \hline 0 &= 2 && \text{False} \end{aligned}$$

The system is inconsistent and has no solution.

(b)
$$\begin{aligned} -5x + y &= 2 && \text{Multiply the first equation by } -5. \\ 5x - y &= -2 && \text{Second equation} \\ \hline 0 &= 0 && \text{True} \end{aligned}$$

The equations are dependent, and the system has infinitely many solutions.

Write the equations in the slope–intercept form. Point out that in part (a) the graphs are parallel lines and the system has no solution and that in part (b) the graphs coincide and the system has infinitely many solutions.

Again, this example and Exercises 77–82 are more traditional to help the student get started. See Exercises 83–92 for real-life applications and modeling.

Applications

When we are solving applied problems that involve two variables, the initial work is always the same. We assign variables and write a system of equations. The only decision that remains is the choice of a method for solving the system.

EXAMPLE 6

A Collection of Coins

A collection jar for the Kidney Foundation contained 18 one-dollar bills and a total of 77 coins consisting only of quarters and dimes. If the value of all the money was \$30.95, how many quarters and dimes were in the jar?

Solution

Let q = the number of quarters and
d = the number of dimes.

The total value of the coins was \$30.95 − \$18.00 = \$12.95.

$$q + d = 77 \quad \text{There were 77 coins in all.}$$
$$25q + 10d = 1295 \quad \text{We choose cents as the monetary unit.}$$

$$-10q - 10d = -770 \quad \text{Multiply the first equation by } -10.$$
$$25q + 10d = 1295 \quad \text{Second equation}$$
$$15q = 525 \quad \text{Add the equations.}$$
$$q = 35 \quad \text{Solve for } q.$$

$$q + d = 77 \quad \text{First equation}$$
$$35 + d = 77 \quad \text{Replace } q \text{ with 35.}$$
$$d = 42$$

The jar contained 35 quarters and 42 dimes.

We now have three methods for solving a system of equations. The graphing method has the advantage of being visually helpful.

The substitution method and the addition method are both algebraic methods. Choose the one that is most convenient for the system. Either method will work for all linear systems.

The ideal solving process is a combination of graphing and algebraic methods. This will help you to remain aware of the conceptual nature of the problem and to avoid falling into a mere symbol manipulation routine.

Quick Reference 5.2

Addition Method

- The addition method is another algebraic method for solving a system of two equations. The following is a summary of the method.
  1. Write both equations in the form $Ax + By = C$.
  2. If necessary, multiply one or both of the equations by appropriate numbers so that the coefficients of one of the variables are opposites.

3. Add the equations to eliminate a variable.
4. Solve the resulting equation.
5. Substitute that value in either of the original equations and solve for the other variable.
6. Check the solution.

- It is usually best to clear fractions before using the addition method.

Special Cases

- After you add the equations to eliminate one variable, the resulting equation may be false, or it may be an identity.
  1. If it is false, then the system is inconsistent and has no solution.
  2. If it is an identity, then the two equations of the system are dependent, and the system has infinitely many solutions.

Speaking the Language 5.2

1. The addition method for solving a system of equations works only if the ______ of one of the variables are opposites.
2. If using the addition method results in both variables being eliminated, the graphs of the linear equations are ______, or they ______.
3. When using the addition method, the goal of adding the equations is to obtain an equation in ______.
4. The addition method is sometimes called the ______ method.

Exercises 5.2

Concepts and Skills

1. Suppose you wish to solve the following system.

$$3x - y = 8$$
$$x + 4y = 7$$

(a) Describe the operation you would perform to eliminate x.

(b) Describe the operation you would perform to eliminate y.

2. Suppose you wish to use the addition method to solve the following system. Describe the easiest way to eliminate one of the variables.

$$2x + y = 3$$
$$3x + 4y = -2$$

In Exercises 3–10, use the addition method to solve each system.

This group requires only addition.

3. $x + y = 10$, $x - y = 2$
4. $2x - y = 7$, $2x + y = 13$
5. $x + 2y = 4$, $-x - y = 3$
6. $3x - 3y = 14$, $3x + 3y = -2$
7. $3x + 2y = 4$, $4y = 3x + 26$
8. $x - 3y = 9$, $5y = x - 17$
9. $4x = 3y + 6$, $12 = 5x + 3y$
10. $2x = y + 9$, $3y = 2x - 19$

11. Suppose you solve a system of equations and the resulting equation is $0 = 0$. Does this mean that the solution of the system is (0, 0)? Explain.

12. Suppose you solve a system of equations in x and y, and you eliminate y to obtain $x = 4$. Is 4 the solution of the system? Explain.

In Exercises 13–20, use the addition method to solve each system. This group requires multiplication of just one of the equations.

13. $2x + 3y + 1 = 0$
$5x + 3y = 29$

14. $5x - 2y = 0$
$4y = 3x + 14$

15. $3x - 2y = 21$
$5x + 4y = 13$

16. $2x = 3y - 14$
$4x + 5y = 16$

17. $2x + 3y = 3$
$4x = 6y - 2$

18. $3x + 4y + 1 = 0$
$6x = 8y - 14$

19. $6x = 5y + 10$
$3x + 2y = 23$

20. $4x + 3y = 2$
$5x + 6y = 7$

21. Suppose the solution of a system of two linear equations is (3, 2). Try to visualize the general appearance of the graph of the system. Now explain why there are many other systems whose solutions are also (3, 2).

22. Explain why the addition method cannot be used to solve the following system of equations. What is the solution of the system?

$x = -3$
$y = 7$

In Exercises 23–30, use the addition method to solve each system. This group requires both equations to be multiplied.

23. $6x + 11y = 17$
$4x - 5y = -1$

24. $5x - 3y = 2$
$3x - 2y = -1$

25. $-6x + 5y = 10$
$5x + 4y = 8$

26. $-3x + 7y = 10$
$5x - 2y = -6$

27. $2x - 7y + 11 = 0$
$7x + 4y + 10 = 0$

28. $3x + 2y + 7 = 0$
$2x - 3y = 4$

29. $5y - 2x = 5$
$5x + 2y = 2$

30. $4x = 3y + 8$
$3x = 4y + 6$

31. When you use the addition method, how can you tell when the equations of a system are dependent?

32. When you use the addition method, how can you tell when a system is inconsistent?

In Exercises 33–44, use the addition method to solve the system. This group includes the special cases.

33. $2x + 3y + 18 = 0$
$5y = 6x - 2$

34. $y = 3x - 5$
$2x + 3y = 6$

35. $3x - 2y = 8$
$6x = 4y + 17$

36. $4x + 12y = 32$
$7 - x = 3y$

37. $3x - y = 7$
$2y = 6x - 14$

38. $4x + 12y = 36$
$9 - x = 3y$

39. $2 + 7y = 5x$
$15x = 21y - 6$

40. $5 = x - 4y$
$12y - 3x = -8$

41. $6y + 7x = 16$
$3x = 2y + 16$

42. $5y = 2(3x - 2)$
$8 + 3y = -2x$

43. $2y = 3x - 5$
$-12x = -20 - 8y$

44. $8x + 9y = -3$
$6 + 18y = -16x$

In Exercises 45–56, use the addition method to solve the system.

45. $4(x - 2) - 5(y - 3) = 14$
$3(x + 4) - 2(y + 5) = 13$

46. $3(x - 5) + (y + 7) = 7$
$2(x + 3) - 5(y + 1) = 11$

47. $\frac{1}{2}x + \frac{3}{10}y = \frac{1}{2}$
$-\frac{5}{3}x - y = \frac{4}{3}$

48. $-0.4x + 1.2y = 1$
$-0.6x + 1.8y = -1$

49. $x - 1.5y = 0.25$
$0.9y - 0.6x = -0.15$

50. $\frac{3}{2}x - y = \frac{3}{4}$
$2x - \frac{4}{3}y = 1$

51. $0.1x - 0.25y = 1.05$
$0.625x + 0.4y = 0.675$

52. $0.4x - 0.3y = -0.1$
$0.5x + 0.2y = 1.6$

53. $\frac{x+2}{9} + \frac{y+2}{6} = 1$
$\frac{x-1}{4} - \frac{y+1}{3} = -1$

54. $\frac{x-2}{12} - \frac{y-2}{4} = 1$
$\frac{x+3}{2} + \frac{y-4}{4} = 1$

55. $\frac{x-y}{6} + \frac{x+y}{3} = 1$
$y = 6 - 3x$

56. $\frac{x+y}{2} = \frac{1}{3} + \frac{x-y}{2}$
$\frac{x+2}{2} - \frac{y+4}{2} = 4$

CE In Exercises 57–60, determine values of a, b, or c so that the system has the given solution.

57. $x + by = -2a$
$ax - 4y = b$
$(-2, -2)$

Exercises 57–68 stress the meaning of a solution and the properties that lead to special cases. This helps students understand concepts, not just symbol manipulation.

58. $\begin{aligned} 5x + 2ay &= c \\ ax - 3y &= -6c \end{aligned}$
$(2, -2)$

59. $\begin{aligned} ax - by &= -17 \\ bx + ay &= -1 \end{aligned}$
$(-2, 5)$

60. $\begin{aligned} ax - y &= c + 6 \\ x - (3a + 4)y &= c \end{aligned}$
$(7, 1)$

CE In Exercises 61–64, determine k so that the system of equations is inconsistent.

61. $\begin{aligned} x + 2y &= 5 \\ 2x - ky &= 7 \end{aligned}$

62. $\begin{aligned} 3x - 2y &= 10 \\ kx + 6y &= 5 \end{aligned}$

63. $\begin{aligned} y &= -2x + 3 \\ kx - 3y &= 4 \end{aligned}$

64. $\begin{aligned} x &= 2y - 5 \\ 4x + ky &= 2 \end{aligned}$

CE In Exercises 65–68, determine a, b, or c so that the equations are dependent.

65. $\begin{aligned} x + 2y &= c \\ 3x + 6y &= 12 \end{aligned}$

66. $\begin{aligned} 2x + by &= 5 \\ -6x - 12y &= -15 \end{aligned}$

67. $\begin{aligned} ax + by &= 3 \\ x - y &= 1 \end{aligned}$

68. $\begin{aligned} 2x + 3y &= -4 \\ ax - 6y &= c \end{aligned}$

CE In Exercises 69–76, solve the given system and obtain a solution in terms of a, b, or c.

69. $\begin{aligned} x + y &= b \\ x - y &= c \end{aligned}$

70. $\begin{aligned} 2x + y &= 3 \\ x + y &= c \end{aligned}$

71. $\begin{aligned} ax + y &= 3 \\ x - y &= c \end{aligned}$

72. $\begin{aligned} x + by &= c \\ x - by &= c \end{aligned}$

73. $\begin{aligned} 2ax + y &= 3 \\ -ax + 3y &= 2 \end{aligned}$

74. $\begin{aligned} ax + by &= c \\ bx + ay &= c \end{aligned}$

75. $\begin{aligned} ax + by &= a \\ x - y &= 1 \end{aligned}$

76. $\begin{aligned} 2x + y &= 4 + c \\ x - 2 &= y \end{aligned}$

77. If 8 hamburgers and 4 milk shakes cost \$20 and 3 hamburgers and 2 milk shakes cost \$8.10, what are the prices of a hamburger and a milk shake?

78. If 5 root beers and 7 orders of French fries cost \$7.25 and 4 root beers and 5 orders of French fries cost \$5.50, what are the prices of one root beer and one order of French fries?

79. A parking meter contains \$3.85 in nickels and dimes. If there is a total of 50 coins, how many dimes are in the meter?

80. A coin jar contains 34 coins. If the coins are all quarters and dimes and their total value is \$6.55, how many quarters are in the jar?

81. There are 64 coins in a cash register and their total value is \$4.25. If the coins are all nickels and dimes, how many of each coin are there?

82. A vending machine attendant fills a coin dispenser with \$6.85 in nickels and quarters. If there is a total of 61 coins in the dispenser, how many of each coin are there?

Real-Life Applications

83. A subcontractor employs carpenters at \$22 per hour and unskilled labor at \$8 per hour. If the hourly cost for the company's total of 16 employees is \$212, how many carpenters are employed?

84. Tees Unlimited sold 35 more T-shirts than sweatshirts last Monday. T-shirts sell for \$11 each, and sweatshirts sell for \$20 each. In all, \$1718 worth of shirts were sold that day. How many of each type of shirt were sold?

85. A person working at home obtains contracts for word processing projects from small businesses. She charges \$2 per text page, but she adds \$3 to that charge for any pages that contain graphics. If her total charge for a 60-page report is \$147, what percent of the pages have graphics?

86. A farmer received 90 cents per bushel more for his wheat than he did for his corn. If he sold 6000 bushels of wheat and 4000 bushels of corn for a total of \$30,400, what was the price per bushel of each?

Modeling with Real Data

87. Adults—those age 18 and over—are spending more hours each year on the Internet. The table shows the number of hours per year that the average adult spends reading books and spends online. (Source: Communications Industry Report.)

	Number of Hours per Person per Year	
Year	**Reading**	**Internet**
1998	95	35
1999	95	39
2000	96	43
2001	97	46
2002	97	49

With x representing the number of years since 1995, the following equations model the data.

Reading: $y = 0.6x + 93$

Internet: $y = 3.5x + 24.9$

(a) Solve the system of these model equations.

(b) Interpret the solution.

88. The table shows that increasing numbers of both men and women were practicing M.D.s in the period 1994–1997. (Source: American Medical Association.)

	Number of Practicing MDs (in thousands)	
Year	**Male**	**Female**
1994	494.9	124.9
1995	505.9	140.1
1996	515.6	148.3
1997	527.0	157.7

With x representing the number of years since 1990, the data can be modeled with the following functions.

Male: $M(x) = 10.6x + 453.6$

Female: $F(x) = 10.6x + 84.1$

(a) What do the coefficients of x indicate?

(b) Solve the system of equations. What does the solution of the system indicate?

Data Analysis: *E-mail*

The accompanying bar graph shows that the percentage of office communications conducted by e-mail is increasing rapidly, while the percentage conducted by hard copy (paper) is decreasing. (Source: Cognitive Communication Survey.)

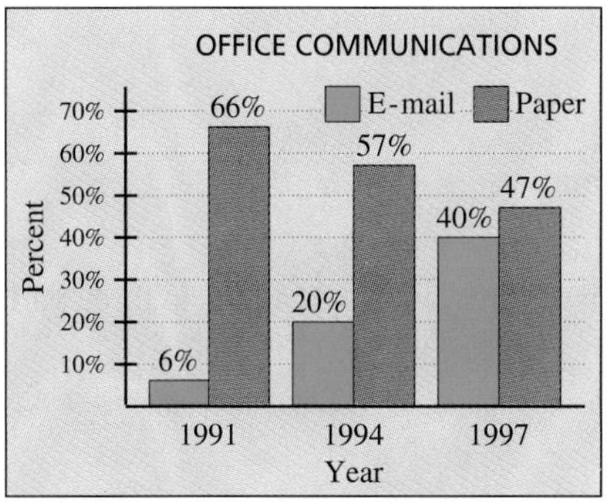

If we let y represent the percentage and x the number of years since 1990, the following equations model the use of e-mail and paper.

E-mail: $17x - 3y = 2$

Paper: $19x + 6y - 416 = 0$

89. Solve each equation for y. Round off all numbers to the nearest integer. Then use the graphing method to estimate the solution of the system of those two equations.

90. Solve the system algebraically and compare the result to your estimate in Exercise 89.

91. In the given data, the percentages for e-mail and paper do not total 100%. Why is this so?

92. Interpret the solution in Exercise 89.

Challenge

In Exercises 93–96, replace $\frac{1}{x}$ with u and $\frac{1}{y}$ with v. Solve the resulting system for u and v. Then use these values to determine x and y.

93. $\frac{3}{x} + \frac{2}{y} = 2$
$\frac{1}{x} - \frac{6}{y} = -1$

94. $\frac{3}{x} + \frac{2}{y} = 0$
$\frac{9}{x} - \frac{4}{y} = 5$

95. $\frac{5}{x} + \frac{10}{y} = 3$
$\frac{1}{x} + \frac{3}{y} = 1$

96. $\frac{1}{x} + \frac{1}{y} = -1$
$\frac{5}{x} - \frac{2}{y} = -1$

97. Using the technique described for Exercises 93–96, solve the following system. Explain the result.

$$\frac{3}{x} + \frac{5}{y} = 10$$
$$\frac{2}{x} + \frac{1}{y} = 2$$

In Exercises 98–101, determine a, b, or c so that the system is consistent.

98. $x - 2y = 3$
$ax - 2y = 4$

99. $-3x + 2y = 5$
$6x + by = 7$

100. $3x - 2y = c$
$-6x + 4y = -10$

101. $4x - 8y = 20$
$x - 2y = c$

In Exercises 102–105, determine a, b, or c so that the equations are independent.

102. $4x - 7y = -1$
$by - 8x = 2$

103. $ax + 3y = 2$
$2x + 6y = 4$

104. $x - 2y = c$
$6y - 3x = 3$

105. $x + y = 5$
$2x + 2y = c$

5.3 Systems of Equations in Three Variables

Linear Equations in Three Variables • Graphical Interpretation • Algebraic Methods • Special Cases • Real-Life Applications

Linear Equations in Three Variables

Although we have concentrated on linear equations in two variables, linear equations can be defined for more than two variables.

Definition of a Linear Equation in Three Variables

A **linear equation in three variables** is any equation that can be written in the *standard form* $Ax + By + Cz = D$, where A, B, C, and D are real numbers and A, B, and C are not all zero.

Similar definitions may be stated for linear equations in more than three variables. The following are some examples of linear equations in three variables.

Equation	*Standard Form*
$3x - y + 2z = 5$	$3x + (-1)y + 2z = 5$
$y = 2x + 1$	$-2x + 1y + 0z = 1$
$x = 1$	$1x + 0y + 0z = 1$

A solution of a linear equation in *two* variables is an ordered *pair* (x, y) that satisfies the equation. We can extend this definition for linear equations in three variables.

Definition of Solution

A **solution** of a linear equation in three variables is an **ordered triple** (x, y, z) of numbers that satisfies the equation.

EXAMPLE 1 Verifying a Solution

Verify that (5, 3, 6) is a solution of the equation $x + y - z = 2$.

Solution

$$x + y - z = 2$$

$$5 + 3 - 6 = 2 \quad \text{Replace } x \text{ with 5, } y \text{ with 3, and } z \text{ with 6.}$$

$$2 = 2 \quad \text{True}$$

The ordered triple (5, 3, 6) is a solution.

The following is an example of a system of three linear equations in three variables.

$$\begin{aligned} 3x - y + 2z &= 0 \\ 2x + 3y + 8z &= 8 \\ -x + y + 6z &= 0 \end{aligned}$$

A solution of such a system is an ordered triple (x, y, z) that satisfies each equation of the system.

In previous sections we developed graphical and algebraic methods for solving systems of two linear equations in two variables. We can extend those methods to solve systems of three linear equations in three variables.

Graphical Interpretation

We know that the graph of a linear equation in *two* variables is a *line* drawn in a *two*-dimensional (rectangular) coordinate system.

The graph of a linear equation in *three* variables is a *plane* drawn in a *three*-dimensional coordinate system, also called a **coordinate space.** (See Fig. 5.5.)

Note that a coordinate space includes three axes that are all perpendicular to each other. We must visualize the x-axis as coming out from the page.

Figure 5.6 shows the point $P(4, 2, 3)$ plotted in a coordinate space.

Figure 5.5

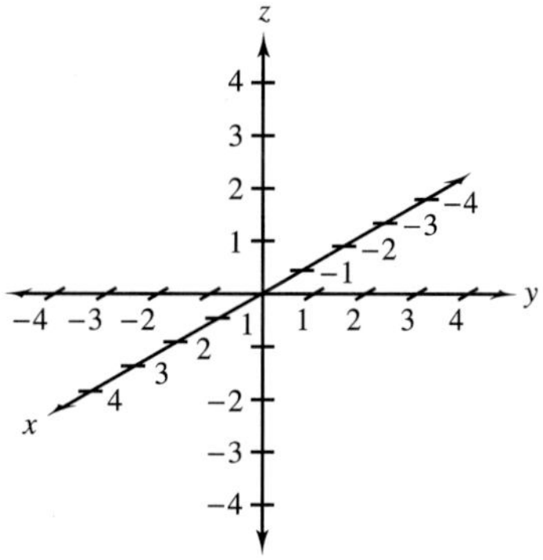

Figure 5.6

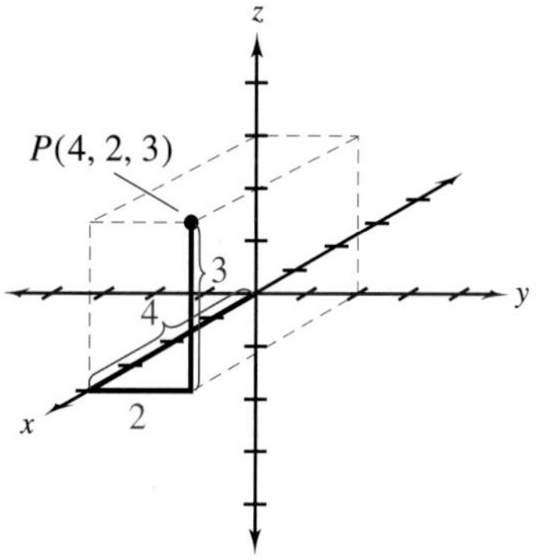

Figure 5.7 shows one of many planes that contains point P. Planes, like lines, extend without end. However, a plane in a coordinate space is conventionally drawn as a bounded figure so that its location can be visualized.

Although students are not expected to construct graphs in a coordinate space, it is helpful to visualize the geometric interpretation and to compare it with the two-variable case.

Figure 5.7

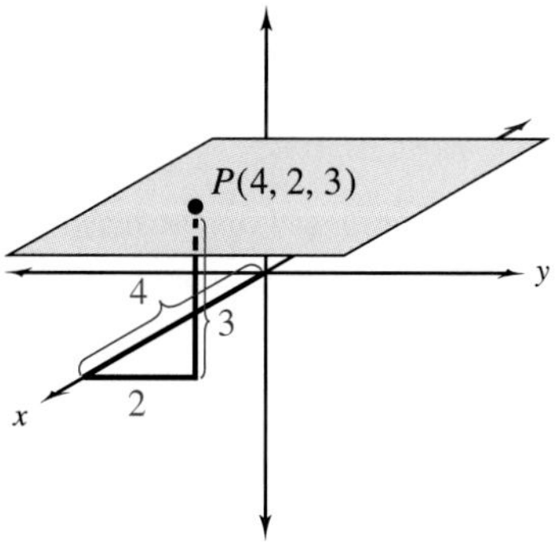

For a system of three linear equations in three variables, the graph of each equation of the system is a plane. The solution set of such a system is represented by the intersection

Figure 5.8

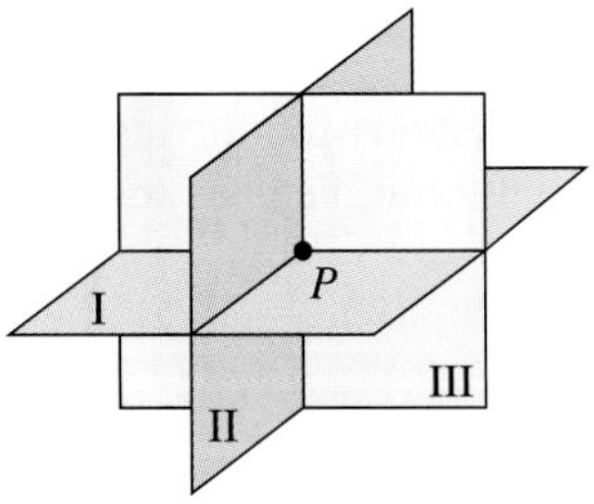

of the graphs (planes) of the three equations. Any point in common to all three planes represents a solution of the system. There are three possible outcomes.

First, the solution could be unique. This means that the three planes intersect at one point P. (See Fig. 5.8.)

Second, there could be no solution. This can occur when at least two of the planes are parallel (see Fig. 5.9) or when one plane is parallel to the line L of intersection of the other two planes (see Fig. 5.10).

Figure 5.9

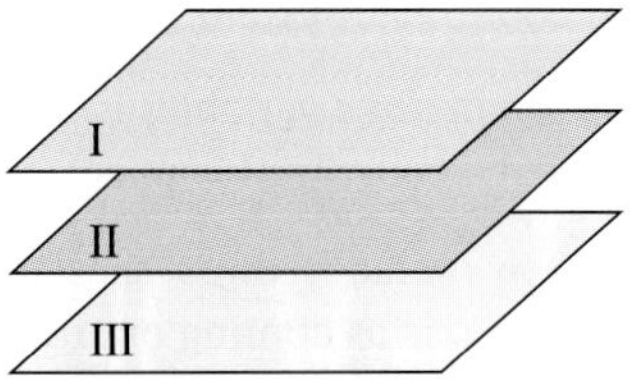

Figure 5.10

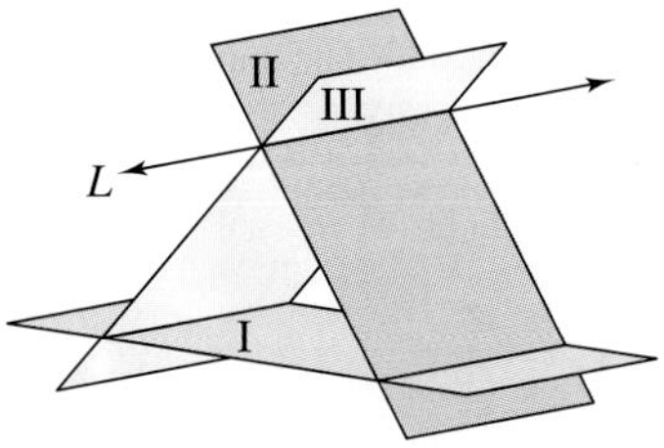

Third, there could be infinitely many solutions. This can occur when the planes coincide (see Fig. 5.11) or when they intersect in a common line L (see Fig. 5.12).

Figure 5.11

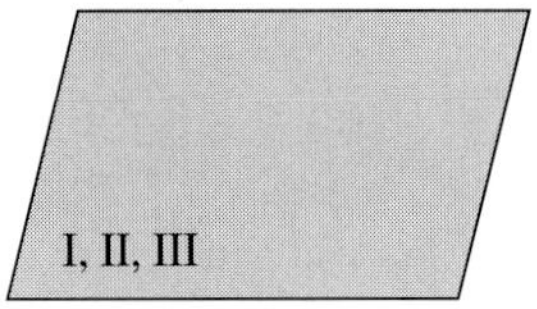

Figure 5.12

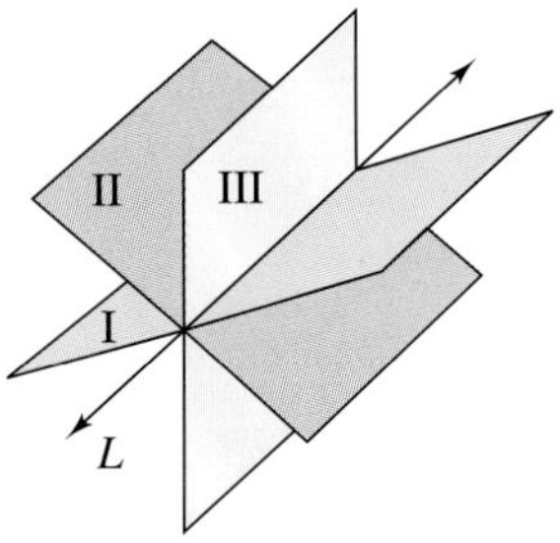

Think About It

How would you define a linear equation in four variables? Can you imagine the graph of such an equation?

Because of the difficulties inherent in drawing planes in coordinate space, we will not use graphing methods for estimating solutions of systems of three or more linear equations in three variables. Our focus will be on algebraic methods. Nevertheless, your ability to visualize the various ways in which planes can be oriented in space will provide you with a geometric interpretation of the system and its solution.

Algebraic Methods

The substitution method we used for solving systems of two linear equations in two variables can also be used to solve systems of three linear equations in three variables.

Although the substitution method can be used for any system, the addition method is more generally applicable. We will limit our discussion to that one algebraic method.

EXAMPLE 2 **Solving a System of Three Equations with the Addition Method**

Use the addition method to solve the following system.

$$\begin{aligned} 3x - y + 2z &= 0 \quad (1)\\ 2x + 3y + 8z &= 8 \quad (2)\\ x + y + 6z &= 0 \quad (3)\end{aligned}$$

LEARNING TIP

In some systems, it might be tempting to eliminate a variable from one pair of equations and to eliminate another variable from a second pair of equations. The addition method will not work unless you eliminate the same variable from each pair of equations.

Solution

The plan is to eliminate a variable from one pair of equations. Then we eliminate the same variable from another pair of equations. In the following, we eliminate y from equations (1) and (3) and from equations (2) and (3).

$$\begin{aligned} 3x - y + 2z &= 0 \quad (1)\\ x + y + 6z &= 0 \quad (3)\end{aligned} \qquad \begin{aligned} 2x + 3y + 8z &= 8 \quad (2)\\ x + y + 6z &= 0 \quad (3)\end{aligned}$$

In the first pair, eliminate y by adding the equations. In the second pair, multiply both sides of equation (3) by -3 and then add the equations.

$$\begin{aligned} 3x - y + 2z &= 0\\ x + y + 6z &= 0\\ \hline 4x \quad + 8z &= 0\end{aligned} \qquad \begin{aligned} 2x + 3y + 8z &= 8\\ -3x - 3y - 18z &= 0\\ \hline -x \quad - 10z &= 8\end{aligned}$$

The result is a system of two equations in two variables.

Point out that if we do not eliminate the same variable in each pair of equations, we will end up with two equations but there will still be three variables present.

$$\begin{aligned} 4x + 8z &= 0 \quad (4)\\ -x - 10z &= 8 \quad (5)\end{aligned}$$

Now we apply the addition method again to solve this system.

$$\begin{aligned} 4x + 8z &= 0 &&\\ -4x - 40z &= 32 && \text{Multiply equation (5) by 4.}\\ \hline -32z &= 32 && \text{Add the equations.}\\ z &= -1 &&\end{aligned}$$

Now substitute -1 for z in equation (4) to determine x.

$$\begin{aligned} 4x + 8z &= 0 && (4)\\ 4x + 8(-1) &= 0 && \text{Replace } z \text{ with } -1.\\ 4x - 8 &= 0 &&\\ 4x &= 8 &&\\ x &= 2 &&\end{aligned}$$

Finally, to determine the value of y, substitute 2 for x and -1 for z in any of the three original equations.

$$\begin{aligned} x + y + 6z &= 0 && (3)\\ 2 + y + 6(-1) &= 0 && \text{Replace } x \text{ with 2 and } z \text{ with } -1.\\ 2 + y - 6 &= 0 &&\\ y - 4 &= 0 &&\\ y &= 4 &&\end{aligned}$$

The solution is the ordered triple $(2, 4, -1)$.

Our work is simplified when one or more variables are missing from the equations of a system.

EXAMPLE 3

Systems with Missing Variables

Solve the following system of equations.

$$\begin{aligned} 3x + 4y \quad\quad &= 4 \quad (1)\\ -3x \quad\quad + 2z &= -3 \quad (2)\\ y + z &= 1 \quad (3) \end{aligned}$$

Solution

Note that x is missing from equation (3). We can easily eliminate x in equations (1) and (2).

$$\begin{aligned} 3x + 4y \quad\quad &= 4 \quad (1)\\ -3x \quad\quad + 2z &= -3 \quad (2)\\ \hline 4y + 2z &= 1 \quad (4) \end{aligned}$$

Now we use equations (4) and (3) to form a system of equations in two variables.

$$\begin{aligned} 4y + 2z &= 1 \quad (4)\\ y + z &= 1 \quad (3) \end{aligned}$$

$$\begin{aligned} 4y + 2z &= 1 \quad (4)\\ -2y - 2z &= -2 \quad \text{Multiply equation (3) by } -2.\\ \hline 2y \quad\quad &= -1\\ y &= -\frac{1}{2} \end{aligned}$$

$$-\frac{1}{2} + z = 1 \quad \text{Substitute } -\frac{1}{2} \text{ for } y \text{ in equation (3).}$$

$$z = \frac{3}{2} \quad \text{Solve for } z.$$

$$3x + 4\left(-\frac{1}{2}\right) = 4 \quad \text{Substitute } -\frac{1}{2} \text{ for } y \text{ in equation (1).}$$

$$x = 2 \quad \text{Solve for } x.$$

The solution is the ordered triple $\left(2, -\frac{1}{2}, \frac{3}{2}\right)$.

Special Cases

The addition method will detect the special cases in which the system has no solution or has infinitely many solutions.

EXAMPLE 4

An Inconsistent System of Three Linear Equations

Solve the following system.

$$\begin{aligned} 2x - y - 5z &= 3 \quad (1)\\ 5x - y + 14z &= -11 \quad (2)\\ 7x - 2y + 9z &= -5 \quad (3) \end{aligned}$$

Solution

We eliminate y in two pairs of equations.

$$\begin{aligned} 2x - y - 5z &= 3 \quad (1) \\ 7x - 2y + 9z &= -5 \quad (3) \end{aligned} \qquad \begin{aligned} 5x - y + 14z &= -11 \quad (2) \\ 7x - 2y + 9z &= -5 \quad (3) \end{aligned}$$

Multiply both sides of equations (1) and (2) by -2. Then add those equations to equation (3).

$$\begin{aligned} -4x + 2y + 10z &= -6 \\ 7x - 2y + 9z &= 5 \quad (3) \\ \hline 3x \qquad + 19z &= -11 \end{aligned} \qquad \begin{aligned} -10x + 2y - 28z &= 22 \\ 7x - 2y + 9z &= -5 \quad (3) \\ \hline -3x \qquad - 19z &= 17 \end{aligned}$$

When we add the two resulting equations, both variables are eliminated.

$$\begin{aligned} 3x + 19z &= -11 \\ -3x - 19z &= 17 \\ \hline 0 &= 6 \quad \text{False} \end{aligned}$$

The resulting false statement indicates that there is no solution of the system of equations. The solution set is Ø, and the system is inconsistent.

Students are not expected to determine the orientation of planes for these special cases.

Determining the orientation of the three planes in Example 4 is a matter that we must leave for more advanced courses. As a point of interest, the planes that represent equations (2) and (3) intersect in a line. The plane that represents equation (1) is parallel to that line. This situation was illustrated in Fig. 5.10.

EXAMPLE 5

A System of Three Dependent Equations

Solve the following system of equations.

$$\begin{aligned} 6x - 4y + 2z &= 12 \quad (1) \\ 18x - 12y + 6z &= 36 \quad (2) \\ -3x + 2y - z &= -6 \quad (3) \end{aligned}$$

Solution

Note that multiplying equation (1) by 3 results in equation (2). Multiplying equation (3) by -6 also results in equation (2).

In other words, the equations are dependent, and any one of them can be used to describe the solutions. Using equation (3), we say that the solution set is the set of all solutions to $-3x + 2y - z = -6$.

Graphically, it can be shown that the three planes coincide, as was illustrated in Figure 5.11.

Real-Life Applications

An applied problem involving three unknown quantities may be solved most conveniently with a system of three linear equations in three variables.

EXAMPLE 6

Three Different Exhibit Fees

Each fall the Riverfest Craft Fair rents three types of space to exhibitors. The first year, there were 40 spaces for food vendors, 50 spaces to sell crafts, and 20 spaces to demonstrate crafts. The total receipts were $3700.

The next year, there were 30 food vendors, 60 spaces for selling crafts, and 15 demonstration booths, and the total receipts were $3900. The third year, with 25 food vendors, 80 spaces for selling crafts, and 30 demonstration booths, the total receipts were $4925.

Assuming that the fees did not change over these 3 years, what was the fee for each type of space?

Solution

Let x = the fee for a food booth,
y = the fee for a craft booth, and
z = the fee for a demonstration booth.

The equations are as follows.

$$40x + 50y + 20z = 3700 \quad (1)$$
$$30x + 60y + 15z = 3900 \quad (2)$$
$$25x + 80y + 30z = 4925 \quad (3)$$

Because z has the smallest coefficients, we choose to eliminate z. We select equations (1) and (3) for one pair and equations (2) and (3) for the other pair.

$$40x + 50y + 20z = 3700 \quad (1) \qquad 30x + 60y + 15z = 3900 \quad (2)$$
$$25x + 80y + 30z = 4925 \quad (3) \qquad 25x + 80y + 30z = 4925 \quad (3)$$

In the first pair, multiply equation (1) by -3 and equation (3) by 2. In the second pair multiply equation (2) by -2.

$$\begin{array}{r} -120x - 150y - 60z = -11{,}100 \\ 50x + 160y + 60z = 9{,}850 \\ \hline -70x + 10y \qquad = -1{,}250 \end{array} \qquad \begin{array}{r} -60x - 120y - 30z = -7800 \\ 25x + 80y + 30z = 4925 \\ \hline -35x - 40y \qquad = -2875 \end{array}$$

Now we have a system of two equations in two variables.

$$-70x + 10y = -1250 \quad (4)$$
$$-35x - 40y = -2875 \quad (5)$$

To eliminate y, multiply equation (4) by 4.

$$\begin{array}{r} -280x + 40y = -5000 \\ -35x - 40y = -2875 \quad (5) \\ \hline -315x \qquad = -7875 \\ x = 25 \end{array}$$

Verify that the corresponding values for y and z are $y = 50$ and $z = 10$. The food vendors paid $25, the craft vendors paid $50, and the demonstrators paid $10.

Quick Reference 5.3

Linear Equations in Three Variables

- A **linear equation in three variables** is any equation that can be written in the standard form $Ax + By + Cz = D$, where A, B, C, and D are real numbers and A, B, and C are not all zero.
- A **solution** of a linear equation in three variables is an **ordered triple** (x, y, z) of numbers that satisfies the equation.
- A **solution** of a system of linear equations in three variables is an ordered triple (x, y, z) that satisfies each equation of the system.

Graphical Interpretation

- The graph of a linear equation in three variables is a plane drawn in a three-dimensional **coordinate space.**
- For a system of three linear equations in three variables, any point that is common to all three planes represents a solution.
- When we are solving a system of three linear equations in three variables, there are three possible outcomes.
  1. *Unique solution.* The three planes intersect at exactly one point.
  2. *No solution.* At least two of the planes are parallel or one plane is parallel to the line of intersection of the other two planes.
  3. *Infinitely many solutions.* The three planes coincide or the three planes intersect in a common line.

Algebraic Methods

- The addition method is the algebraic method discussed in this section. To use this method, we do the following.
  1. Write all equations in the form $Ax + By + Cz = D$.
  2. Add two pairs of equations to eliminate the same variable in each pair. It may be necessary to multiply one or both of the equations in each pair so that the variable will be eliminated when you add the equations.
  3. Solve the resulting system of two equations in two variables.
  4. Substitute to determine the values of the other two variables.

Special Cases

- When we use the addition method, we can detect special cases after the first elimination of a variable.
  1. If the resulting system of two linear equations in two variables is inconsistent, then the original system has no solution.
  2. If the two linear equations of the resulting system are dependent, then the original system has infinitely many solutions.

Speaking the Language 5.3

1. The graph of a linear equation in three variables is a(n) ________.

2. A solution of a linear equation in three variables has the form (x, y, z), which is called a(n) ________.

3. Suppose that in solving a system of three linear equations in three variables, you eliminate y from equations (1) and (2). Then the next step is to eliminate ______ from equations ______ or equations ______.

4. A system of three linear equations in three variables has a unique solution only if the graphs of the three equations intersect ______.

Exercises 5.3

Concepts and Skills

1. The equation $0 = 0$ can be written as follows.

$$0x + 0y + 0z = 0$$

Does this mean that $0 = 0$ is a linear equation in three variables? Explain.

2. Written beside each equation of the following system is a solution of that equation.

$$\begin{aligned} x + y + z &= 1 \qquad (0, 0, 1) \\ x - y + z &= 1 \qquad (1, 0, 0) \\ x + y - z &= 1 \qquad (1, 1, 1) \end{aligned}$$

Is $\{(0, 0, 1), (1, 0, 0), (1, 1, 1)\}$ the solution set of the system? Explain.

In Exercises 3–6, determine whether each of the given ordered triples is a solution of the system of equations.

3. $$\begin{aligned} 5x + 7y - 2z &= -1 \\ x - 2y + z &= 8 \\ 3x - y + 3z &= 14 \end{aligned}$$
$(3, -2, 1), (2, -1, 2)$

4. $$\begin{aligned} 3x - 2y - z &= 5 \\ x - y + 2z &= -5 \\ 2x + y - z &= 18 \end{aligned}$$
$(0, -1, -3), (5, 6, -2)$

5. $$\begin{aligned} 3x + 4y - z &= -12 \\ -x + 2y + 3z &= 8 \\ 2x + 6y + z &= -8 \end{aligned}$$
$(-3, 1, 7), (0, -2, 4)$

6. $$\begin{aligned} -3x + 4y + z &= -13 \\ 2x - 2y + 3z &= -2 \\ 5x + 6y - 2z &= 29 \end{aligned}$$
$\left(4, \frac{1}{2}, -3\right), (2, -1, -3)$

CE In Exercises 7–10, determine the value of a, b, c, or d so that the ordered triple is a solution to the system.

7. $$\begin{aligned} ax - 2y + z &= 7 \\ x - y + cz &= -2 \\ 2x + by - z &= 0 \end{aligned}$$
$(1, -1, 2)$

8. $$\begin{aligned} x + y + cz &= 0 \\ 3x + by + z &= 2 \\ ax - y - z &= 6 \end{aligned}$$
$(2, -2, -4)$

9. $$\begin{aligned} 2x + y + z &= d \\ x + by - z &= 7 \\ -3x + 2y + cz &= 4 \end{aligned}$$
$(0, 2, -1)$

10. $$\begin{aligned} ax - y + z &= 0 \\ x - 3y + z &= d \\ x + y + cz &= 8 \end{aligned}$$
$(3, 2, -1)$

11. Compare the graph of a linear equation in two variables with the graph of a linear equation in three variables.

12. If a system of three linear equations in three variables has a unique solution, describe the graph of the system.

13. Describe one way in which the graph of a system of three linear equations in three variables would indicate that the system has
   (a) no solution.
   (b) infinitely many solutions.

14. Suppose you are solving the following system.

$$\begin{aligned} 2x + y - 3z &= 12 \\ x - y + z &= 0 \\ 3x - 2y - z &= 7 \end{aligned}$$

It is easy to eliminate y in the first two equations and to eliminate z in the last two equations. If you do this, are you making progress toward solving the system? Explain.

In Exercises 15–18, use the addition method to solve the system of equations.

15. $$\begin{aligned} x + y + z &= 0 \\ 3x + y \quad &= 0 \\ y - 2z &= 7 \end{aligned}$$

Exercises 15–30 are straightforward and often require only one multiplication. Special cases are not included.

16. $\begin{aligned} 2x + 4y + 5z &= -3 \\ y + 4z &= 1 \\ 3x - y &= 9 \end{aligned}$

17. $\begin{aligned} x + 2y - 3z &= 5 \\ x + 2z &= 15 \\ 2y - z &= 6 \end{aligned}$

18. $\begin{aligned} x - y + z &= 0 \\ 2y + 3z &= 1 \\ x + 2z &= 0 \end{aligned}$

In Exercises 19–30, use the addition method to solve the system of equations.

19. $\begin{aligned} 2x - y + z &= 9 \\ 3x + 2y - z &= 4 \\ 4x + 3y + 2z &= 8 \end{aligned}$

20. $\begin{aligned} 2x + 3y - 3z &= 1 \\ x - y + 2z &= 7 \\ 3x + 2y + z &= 6 \end{aligned}$

21. $\begin{aligned} x + 2y - 3z &= 1 \\ x + y + 2z &= -1 \\ 3x + 3y - z &= 4 \end{aligned}$

22. $\begin{aligned} x + 3y - 5z &= -8 \\ -2x + y + 3z &= 9 \\ 3x - 2y + z &= 2 \end{aligned}$

23. $\begin{aligned} 4x + 5y - 2z &= 23 \\ -6x + 2y + 7z &= -14 \\ 8x + 3y + 3z &= 11 \end{aligned}$

24. $\begin{aligned} -x + y + z &= 0 \\ x - 2y + 3z &= 7 \\ -2x + y - 2z &= 9 \end{aligned}$

25. $\begin{aligned} 7x - 2y + 3z &= 19 \\ x + 8y - 6z &= 9 \\ 2x + 4y + 9z &= 5 \end{aligned}$

26. $\begin{aligned} x + y - 4z &= -3 \\ 2x - 3y - 2z &= 5 \\ x + y + z &= 2 \end{aligned}$

27. $\begin{aligned} 2x + 3y + 2z &= 2 \\ -3x - 6y - 4z &= -3 \\ -\frac{1}{6}x + y + z &= 0 \end{aligned}$

28. $\begin{aligned} 3x + 4y - 2z &= -4 \\ -5x + 3y + 4z &= 6 \\ -2x - 2y + 7z &= -3 \end{aligned}$

29. $\begin{aligned} 4x + 3y - 3z &= 5 \\ 2x - 6y + 9z &= -7 \\ 6x + 6y - 3z &= 7 \end{aligned}$

30. $\begin{aligned} 2x + 4y + 3z &= 1 \\ 2x - 8y - 9z &= 6 \\ 4x + 12y + 9z &= 0 \end{aligned}$

31. How do we determine algebraically if a system of equations has

(a) no solution?

(b) infinitely many solutions?

32. Describe a method other than the addition method to solve the following system of equations.

$$\begin{aligned} x + y - z &= 8 \\ 2y + z &= 1 \\ z &= 3 \end{aligned}$$

In Exercises 33–40, determine the solution of the given system. Describe the solution set for special cases.

Special cases are included in this group. Usually, only one multiplication is required.

33. $\begin{aligned} x + 2y - z &= 5 \\ x - 2y + z &= 2 \\ 2x + 4y - 2z &= 7 \end{aligned}$

34. $\begin{aligned} 3x - 3y - 4z &= -7 \\ 7x - 6y + 6z &= 4 \\ 4x - 3y + 10z &= -5 \end{aligned}$

35. $\begin{aligned} x + 2y - z &= 3 \\ -4x - 8y + 4z &= -12 \\ 3x + 6y - 3z &= 9 \end{aligned}$

36. $\begin{aligned} 2x - y + 4z &= 6 \\ 6x - 3y + 12z &= 18 \\ -4x + 2y - 8z &= -12 \end{aligned}$

37. $\begin{aligned} x + y &= 1 \\ 2x + 3y - z &= -1 \\ 3x - y - 4z &= 7 \end{aligned}$

38. $\begin{aligned} x - 2z &= 8 \\ 2x + y - z &= 5 \\ 3x - 2y - 4z &= 6 \end{aligned}$

39. $\begin{aligned} x - y + z &= 2 \\ 2x + y - 2z &= -2 \\ 3x - 2y + z &= 2 \end{aligned}$

40. $\begin{aligned} x + y - z &= 2 \\ 2x - 3y + z &= 5 \\ 3x + 2y - 4z &= 3 \end{aligned}$

In Exercises 41–52, solve the given system of equations. For special cases, describe the solution set.

This group includes special cases.

41. $\begin{aligned} x + y + z &= 0 \\ 2x + 3y + 2 &= 0 \\ y - 4z &= 0 \end{aligned}$

42. $$\begin{aligned} 3x + y + z &= 2 \\ x - 2z &= 7 \\ y + z + 1 &= 0 \end{aligned}$$

43. $$\begin{aligned} x + y \quad &= 1 \\ y - 2z &= 2 \\ x \quad - 3z &= 14 \end{aligned}$$

44. $$\begin{aligned} x + y \quad &= 1 \\ y - 2z &= 5 \\ x \quad - 3z &= 1 \end{aligned}$$

45. $$\begin{aligned} x + z &= y \\ \frac{1}{2}y + \frac{3}{4}z &= \frac{1}{4} \\ \frac{1}{2}x + z + \frac{1}{2} &= 0 \end{aligned}$$

46. $$\begin{aligned} x + y &= 3 + z \\ \frac{1}{4}y - \frac{3}{4}z &= 2 \\ \frac{1}{2}x - z &= \frac{3}{2} \end{aligned}$$

47. $$\begin{aligned} 3x + 2z &= y + 3 \\ 2(x + y) &= 2 - z \\ 4 + 3y &= x + z \end{aligned}$$

48. $$\begin{aligned} 6x + 3z &= 1 + 3y \\ 4 + y &= 2x + z \\ 4x - 2y &= 7 - 2z \end{aligned}$$

49. $$\begin{aligned} 4x + 6z &= 12 + 3y \\ 6 + 1.5y &= 2x + 3z \\ 0.5x - 0.375y + 0.75z &= 1.5 \end{aligned}$$

50. $$\begin{aligned} 5x - 2y + 15z &= 20 \\ x - 0.4y + 3z &= 4 \\ 2.5x - y + 7.5z &= 10 \end{aligned}$$

51. $$\begin{aligned} \frac{2}{3}x + \frac{5}{2}y - z &= -21 \\ x - \frac{1}{2}y + \frac{1}{8}z &= \frac{1}{2} \\ y + 2z &= 8 + 2x \end{aligned}$$

52. $$\begin{aligned} \frac{1}{4}x + \frac{1}{3}y - \frac{1}{6}z &= \frac{1}{3} \\ x + y + \frac{1}{2}z &= 1 \\ \frac{1}{3}x - \frac{1}{2}y + \frac{1}{12}z &= \frac{1}{4} \end{aligned}$$

In Exercises 53–56, extend the substitution method to solve the system of three equations in three variables.

53. $$\begin{aligned} 2x + y \quad &= -4 \\ x \quad + z &= 2 \\ 2y - z &= -1 \end{aligned}$$

54. $$\begin{aligned} x - y \quad &= 0 \\ 3x \quad + z &= 11 \\ y - z &= 1 \end{aligned}$$

55. $$\begin{aligned} x + y + z &= -2 \\ x - 3y &= 8 \\ z - 2y &= 8 \end{aligned}$$

56. $$\begin{aligned} 2x - 3y &= 0 \\ 2z + y &= -2 \\ 2x - 2y + 3z &= -4 \end{aligned}$$

In Exercises 57–62, translate the given information into a system of equations, solve the system, and answer the question.

57. In a triangle, angle B is 5° less than 3 times angle A. Angle C is 6° more than the sum of the other two angles. Determine the sizes of the three angles.

58. In a triangle, angle B is 8° more than twice angle A. Angle C is 11° less than angle B. Determine the sizes of the three angles.

59. A 100-yard rope is cut into three pieces. The second piece is 12 feet longer than the sum of the lengths of the other two pieces. The third piece is 6 yards less than the length of the first piece. How long is each piece?

60. A 14-foot board is cut into three pieces. The second piece is 2 feet less than twice the length of the first piece. The third piece is 36 inches less than the sum of the lengths of the other two pieces. How long is each piece?

61. A collection of 155 coins consists of nickels, dimes, and quarters. The total value of the coins is \$24.95. If the number of dimes is two more than twice the number of nickels, how many of each coin are in the collection?

62. A cash register contains a total of 90 coins consisting of pennies, nickels, dimes, and quarters. There are only three pennies, and the total value of the coins is \$9.28. If there are three times as many dimes as quarters, how many of each coin are in the cash register?

Real-Life Applications

63. At a small college of 1200 students, only 8% belonged to either the Republican Club or the Democratic Club, and all the remaining students call themselves Independents. There were 4 more Republican Club members than Democratic Club members.

(a) Using the variables R, D, and I, write a system of equations in three variables that models the given information.

(b) Solve the system to determine the number of students in each category.

64. A store has a sale on T-shirts, sweatshirts, and tank tops. A purchase of 3 T-shirts, 2 sweatshirts, and

5 tank tops costs \$77. It costs \$60 for 5 T-shirts and 1 sweatshirt or for 4 T-shirts and 6 tank tops. What is the sale price for each shirt?

65. Suppose that a marching band has 100 instruments consisting only of brass, woodwind, and percussion. There are twice as many brass as woodwind instruments, and there are three times as many woodwind as percussion instruments.

(a) Using the variables b, w, and p, write a system of equations in three variables that models the given information.

(b) Solve the system to determine the number of instruments in each category.

66. A total of \$10,000 is to be invested in three funds. The following table shows three possible combinations of investments, along with the total return for each combination.

Growth Fund	Income Fund	Money Market	Total Return
\$2000	\$5000	\$3000	\$460
5000	3000	2000	510
7000	2000	1000	550

What is the percentage rate of return of each fund?

Data Analysis: *Foreign-Born Population*

The accompanying bar graph shows that the percentage of the United States population that was foreign-born decreased from 11.6% in 1930 to 4.8% in 1970. Then the percentage began to increase and reached 9.5% by 2000. (Source: Bureau of Census.)

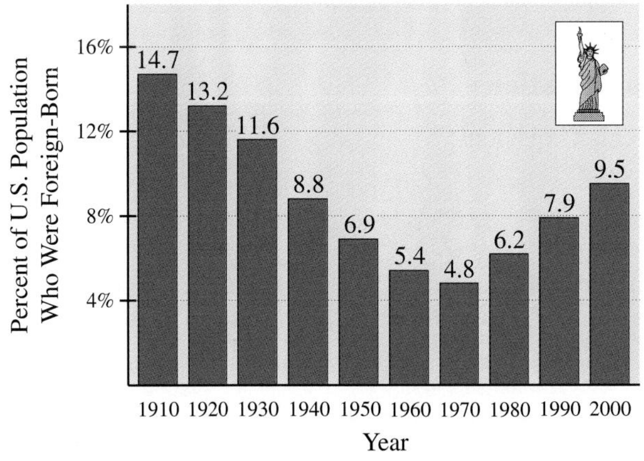

67. Write the data for 1930, 1970, and 2000 as ordered pairs of the form (t, p), where t is the number of years since 1900 and p is that year's percentage of the United States population that was foreign-born.

68. Assume that the data can be modeled with a function of the form $p(t) = at^2 + bt + c$. By substituting your three ordered pairs from Exercise 67, create a system of three linear equations in the variables a, b, and c.

69. Solve the system in Exercise 68. Then write the model function.

70. Evaluate the model function to estimate the percentage of the 2010 population that will be foreign-born.

Challenge

In Exercises 71 and 72 extend the addition method to solve a system of four linear equations in four variables.

71.
$$\begin{aligned} x + y + z - w &= -1 \\ 2x - y + 2z + w &= 1 \\ x - y - z + 2w &= 6 \\ x + 2y + 3z - 2w &= -4 \end{aligned}$$

72.
$$\begin{aligned} x + y + z &= 2 \\ y + z + w &= 1 \\ x + z + w &= -1 \\ x + y + w &= -2 \end{aligned}$$

In Exercises 73 and 74, make an appropriate substitution to solve the systems of equations. (See Section 5.2, Exercises 93–96.)

73.
$$\begin{aligned} \frac{3}{x} + \frac{2}{y} - \frac{1}{z} &= -1 \\ \frac{1}{x} + \frac{1}{y} + \frac{1}{z} &= 2 \\ \frac{1}{x} - \frac{1}{y} + \frac{2}{z} &= 6 \end{aligned}$$

74.
$$\begin{aligned} \frac{2}{x} + \frac{3}{y} &= 2 \\ \frac{3}{y} - \frac{1}{z} &= 0 \\ \frac{4}{x} - \frac{2}{z} &= 0 \end{aligned}$$

75. Solve the given system for x, y, and z. Express your solutions in terms of a, b, and c.

$$\begin{aligned} x + y &= a \\ y + z &= b \\ x + z &= c \end{aligned}$$

76. For what value(s) of a is the solution of the system unique?

$$\begin{aligned} ax + 2y + z &= 0 \\ 2x - y + 2z &= 0 \\ x + y + 3z &= 0 \end{aligned}$$

5.4 Matrices and Systems of Equations

Matrices • Row Operations • Special Cases

Matrices

A **matrix** is an array of numbers. As is true with ordered pairs, the position of each of the numbers is meaningful. The following are examples of matrices.

$$A = \begin{bmatrix} 3 & -5 & 0 \\ 2 & 4 & 1 \\ 2 & 1 & 6 \end{bmatrix} \qquad B = \begin{bmatrix} 2 & 4 & -5 & -1 \\ 0 & 2 & 7 & -2 \end{bmatrix}$$

The matrices are given letter names for easy reference. The numbers in the array are called the **elements** of the matrix. The **rows** of the matrix are read horizontally, and the **columns** are read vertically. In matrix B the elements in the first row are 2, 4, -5, and -1. The elements in the first column are 2 and 0.

We refer to the rows and columns of a matrix with the symbols R and C. For example, R_2 refers to row 2 and C_1 refers to column 1.

The **dimensions** of a matrix are the number of rows and the number of columns. Matrix A has 3 rows and 3 columns, and we refer to it as a 3×3 (read "3 by 3") matrix. Matrix B has 2 rows and 4 columns, and we refer to it as a 2×4 (read "2 by 4") matrix. In general, an $m \times n$ matrix has m rows and n columns.

If the number of rows and the number of columns is the same, the matrix is a **square matrix.** The 3×3 matrix A is an example of a square matrix.

Row Operations

We have discussed two algebraic methods for solving systems of equations. Another solving method involves the use of matrices.

Recall that we solve systems by using the properties of equations to produce equivalent equations. However, when we apply those properties, only the coefficients and constants change. Therefore, we can use matrices of these numbers along with techniques analogous to algebraic operations to solve the same systems.

Exploring the Concept

Matrices and Systems of Equations

Consider the following system of equations.

$$\begin{aligned} 2x + 3y &= -5 \\ 5x - 2y &= 16 \end{aligned}$$

We define the **coefficient matrix** to be a matrix whose elements are the coefficients of the variables in the equations. For this system, the coefficient matrix is

$$A = \begin{bmatrix} 2 & 3 \\ 5 & -2 \end{bmatrix}$$

The **constant matrix** is a matrix whose elements are the constants on the right side of the system of equations. For the given system, the constant matrix is

$$B = \begin{bmatrix} -5 \\ 16 \end{bmatrix}$$

The **augmented matrix** is the combination of the coefficient matrix and the constant matrix. For the given system, the augmented matrix is

Think About It

If the dimensions of a coefficient matrix are $n \times n$, what are the dimensions of the corresponding augmented matrix?

$$M = \left[\begin{array}{cc|c} 2 & 3 & -5 \\ 5 & -2 & 16 \end{array}\right]$$

To solve the original system of equations, our strategy is to try to transform the system's augmented matrix into a matrix of the form

$$\left[\begin{array}{cc|c} 1 & 0 & n_1 \\ 0 & 1 & n_2 \end{array}\right]$$

where n_1 and n_2 are numbers. If we can do this, then when we translate this matrix back into a system of equations, we obtain

$$1x + 0y = n_1$$
$$0x + 1y = n_2$$

Stress the relationship between the augmented matrix and the system of equations.

With the system written in this way, the solutions are obvious: $x = n_1$ and $y = n_2$.

Note: The augmented matrix $\left[\begin{array}{cc|c} 1 & 0 & n_1 \\ 0 & 1 & n_2 \end{array}\right]$ has 1's along the diagonal of the coefficient matrix and 0's elsewhere. We call this the **reduced row echelon form.**

In order to transform a matrix in this way, we use **row operations.** Row operations correspond to the algebraic operations we perform on the equations of a system. Just as the properties of equations allow us to produce equivalent systems, row operations can be used to produce **row-equivalent matrices.**

The following is a summary of permissible row operations. Included in this summary are examples of the notation we use to indicate each operation.

The Elementary Row Operations on Matrices

1. Two rows may be interchanged: $R_2 \leftrightarrow R_1$.
2. A row may be multiplied or divided by a nonzero number n and the original row may be replaced by the result: $n \cdot R_1 \rightarrow R_1$.
3. A row may be replaced by the sum of itself and another row: $R_1 + R_2 \rightarrow R_1$.
4. A row may be replaced by the sum of itself and a multiple of another row: $R_1 + n \cdot R_2 \rightarrow R_1$.

Although row operation 4 is a combination of row operations 2 and 3, it is convenient to treat it as a separate property.

EXAMPLE 1

Using Matrices to Solve a System of Two Equations in Two Variables

Solve the following system of equations.

$$2x + y = 1$$
$$x - y = 5$$

Solution

Row Operation	Row-Equivalent Matrix		System of Equations
	$\left[\begin{array}{cc|c} 2 & 1 & 1 \\ 1 & -1 & 5 \end{array}\right]$	Augmented matrix	$2x + 1y = 1$ $1x - 1y = 5$
$R_2 \leftrightarrow R_1$	$\left[\begin{array}{cc|c} 1 & -1 & 5 \\ 2 & 1 & 1 \end{array}\right]$	Interchange rows to obtain a 1 in row 1, column 1.	$1x - 1y = 5$ $2x + 1y = 1$
$-2 \cdot R_1 + R_2 \rightarrow$	$\left[\begin{array}{cc|c} 1 & -1 & 5 \\ 0 & 3 & -9 \end{array}\right]$	Multiply row 1 by -2 and add the result to row 2. Now we have a 0 in row 2, column 1.	$1x - 1y = 5$ $0x + 3y = -9$
$R_2 \div 3 \rightarrow$	$\left[\begin{array}{cc|c} 1 & -1 & 5 \\ 0 & 1 & -3 \end{array}\right]$	Divide row 2 by 3 to obtain a 1 in row 2, column 2.	$1x - 1y = 5$ $0x + 1y = -3$
$R_1 + R_2 \rightarrow$	$\left[\begin{array}{cc|c} 1 & 0 & 2 \\ 0 & 1 & -3 \end{array}\right]$	Replace row 1 with the sum of rows 1 and 2. Now we have a 0 in row 1, column 2.	$1x - 0y = 2$ $0x + 1y = -3$

The system of equations represented by this matrix is equivalent to the original system. The solution of the system is the ordered pair $(2, -3)$.

The matrix method for solving a system of three equations in three variables involves the same strategy as was illustrated in Example 1, but because the augmented matrix is bigger (3×4), more row operations are usually needed to achieve the result.

The goal remains the same. The coefficient matrix must have 1's down its diagonal and 0's elsewhere; that is, the matrix must be in reduced row echelon form. Then the solution (ordered triple) is determined from the numbers in the right-hand column.

EXAMPLE 2

Using Matrices to Solve a System of Three Equations in Three Variables

Before discussing Example 2, make sure students are clear on the goal. Students tend to flounder because they forget where they are headed.

Solve the following system of equations.

$$x + y + 2z = -1$$
$$3x + y - 6z = 7$$
$$-x + 2y + 2z = 0$$

Solution

$$\left[\begin{array}{ccc|c} 1 & 1 & 2 & -1 \\ 3 & 1 & -6 & 7 \\ -1 & 2 & 2 & 0 \end{array}\right]$$ Augmented matrix

LEARNING TIP

Unless you have a plan, it is easy to become lost when solving a system with the matrix method. Begin by obtaining a 1 in the first row, first column. Then perform row operations to obtain a 0 in each of the other positions in the first column. Next, obtain a 1 in the second row, second column. Then obtain a 0 in each of the other positions in the second column. Continue this process until the matrix is in the proper form.

$$-3 \cdot R_1 + R_2 \rightarrow \left[\begin{array}{ccc|c} 1 & 1 & 2 & -1 \\ 0 & -2 & -12 & 10 \\ -1 & 2 & 2 & 0 \end{array}\right]$$

Multiply row 1 by -3 and add the result to row 2.

$$R_1 + R_3 \rightarrow \left[\begin{array}{ccc|c} 1 & 1 & 2 & -1 \\ 0 & -2 & -12 & 10 \\ 0 & 3 & 4 & -1 \end{array}\right]$$

Replace row 3 with the sum of rows 1 and 3.

$$R_2 \div (-2) \rightarrow \left[\begin{array}{ccc|c} 1 & 1 & 2 & -1 \\ 0 & 1 & 6 & -5 \\ 0 & 3 & 4 & -1 \end{array}\right]$$

Divide row 2 by -2.

$$-1 \cdot R_2 + R_1 \rightarrow \left[\begin{array}{ccc|c} 1 & 0 & -4 & 4 \\ 0 & 1 & 6 & -5 \\ 0 & 3 & 4 & -1 \end{array}\right]$$

Multiply row 2 by -1 and add the result to row 1.

$$-3 \cdot R_2 + R_3 \rightarrow \left[\begin{array}{ccc|c} 1 & 0 & -4 & 4 \\ 0 & 1 & 6 & -5 \\ 0 & 0 & -14 & 14 \end{array}\right]$$

Multiply row 2 by -3 and add the result to row 3.

$$R_3 \div (-14) \rightarrow \left[\begin{array}{ccc|c} 1 & 0 & -4 & 4 \\ 0 & 1 & 6 & -5 \\ 0 & 0 & 1 & -1 \end{array}\right]$$

Divide row 3 by -14.

$$4 \cdot R_3 + R_1 \rightarrow \left[\begin{array}{ccc|c} 1 & 0 & 0 & 0 \\ 0 & 1 & 6 & -5 \\ 0 & 0 & 1 & -1 \end{array}\right]$$

Multiply row 3 by 4 and add the result to row 1.

$$-6 \cdot R_3 + R_2 \rightarrow \left[\begin{array}{ccc|c} 1 & 0 & 0 & 0 \\ 0 & 1 & 0 & 1 \\ 0 & 0 & 1 & -1 \end{array}\right]$$

Multiply row 3 by -6 and add the result to row 2.

The solution is $(0, 1, -1)$.

Special Cases

Matrix solution methods reveal inconsistent systems and dependent equations in the same way as algebraic methods.

EXAMPLE 3

An Inconsistent System

Solve the following system of equations.

$$x - 2y = 5$$
$$2x - 4y = 4$$

Solution

$$\left[\begin{array}{cc|c} 1 & -2 & 5 \\ 2 & -4 & 4 \end{array}\right]$$

Augmented matrix

$$-2R_1 + R_2 \rightarrow \left[\begin{array}{cc|c} 1 & -2 & 5 \\ 0 & 0 & -6 \end{array}\right]$$

Multiply row 1 by -2 and add the result to row 2.

The second row corresponds to the equation $0x + 0y = -6$ or $0 = -6$. This false equation indicates that the system has no solution.

EXAMPLE 4 **Dependent Equations**

Solve the following system of equations.

$$2x + y = 3$$
$$4x + 2y = 6$$

Solution

$$\left[\begin{array}{cc|c} 2 & 1 & 3 \\ 4 & 2 & 6 \end{array}\right] \quad \text{Augmented matrix}$$

$$-2R_1 + R_2 \rightarrow \left[\begin{array}{cc|c} 2 & 1 & 3 \\ 0 & 0 & 0 \end{array}\right] \quad \text{Multiply row 1 by } -2 \text{ and add the result to row 2.}$$

The second row corresponds to the equation $0x + 0y = 0$ or $0 = 0$, which is true. The system has infinitely many solutions.

Quick Reference 5.4

Matrices

- A **matrix** is an array of numbers called **elements.**
- The **dimensions** of a matrix are the number of rows and the number of columns. (An $m \times n$ matrix has m rows and n columns.) If the dimensions are the same, the matrix is a **square matrix.**

Row Operations

- For a system of equations written in standard form,
  1. the **coefficient matrix** is the matrix of coefficients of the variables,
  2. the **constant matrix** is the matrix of constants on the right sides of the equations, and
  3. the **augmented matrix** is the combination of the coefficient matrix and the constant matrix.
- For any matrix, the permissible row operations are as follows:
  1. Rows may be interchanged.
  2. A row may be multiplied or divided by a nonzero number.
  3. A row may be replaced by the sum of itself and another row.
  4. A row may be replaced by the sum of itself and a multiple of another row.
- When we are solving systems with matrix methods, the goal is to write the matrix in the **reduced row echelon form**—that is, with 1's along the diagonal of the coefficient matrix and 0's elsewhere. The solution is then determined from the numbers in the last column.

Special Cases

- When we are solving a system of equations with matrix methods, a special case exists if at any time a row of the coefficient matrix is all 0's.
  1. If the associated element of the constant matrix is not 0, then the system is inconsistent and has no solution.
  2. If the associated element of the constant matrix is 0, then at least two equations of the system are dependent, and the system has infinitely many solutions.

Speaking the Language 5.4

1. A 2×2 matrix is an example of a(n) ________ matrix.
2. The combination of a coefficient matrix and a constant matrix is called a(n) ________ matrix.
3. When we are solving a system of equations with matrix methods, the goal is to obtain 1's along the ________ of the coefficient matrix and 0's elsewhere.
4. If one row of an augmented matrix is $0\ 0\ 0 \mid 3$, then the system has ________ solution(s).

Exercises 5.4

Concepts and Skills

1. What is a 2×4 matrix?

2. For the matrix $\begin{bmatrix} 0 & 5 \\ 1 & 9 \end{bmatrix}$, are the elements of the second column 1 and 9 or 5 and 9? Why?

In Exercises 3–8, state the dimensions of the matrix.

3. $\begin{bmatrix} 2 & -3 & -5 & 4 \\ 1 & 0 & -6 & 7 \end{bmatrix}$

4. $\begin{bmatrix} 2 & -1 \\ -1 & 4 \\ 5 & -7 \end{bmatrix}$

5. $\begin{bmatrix} 2 & 4 \\ -5 & 0 \end{bmatrix}$

6. $\begin{bmatrix} 1 & 2 & -3 \\ 0 & -4 & 5 \\ 6 & 0 & -7 \end{bmatrix}$

7. $\begin{bmatrix} 2 & -3 & 4 & -5 \\ 0 & 1 & -7 & 6 \\ 5 & 8 & 0 & -9 \end{bmatrix}$

8. $\begin{bmatrix} 1 \\ 2 \\ -3 \\ 0 \end{bmatrix}$

9. Describe each of the following.
   (a) A coefficient matrix
   (b) A constant matrix
   (c) An augmented matrix

10. For a system of n linear equations in n variables, explain why the coefficient matrix must be a square matrix.

In Exercises 11–14, write the augmented matrix for the given system of linear equations.

11. $\begin{aligned} 3x + 2y &= 6 \\ x - 4y &= 9 \end{aligned}$

12. $\begin{aligned} 3x - 2y &= 8 \\ 6x &= 4y + 17 \end{aligned}$

13. $\begin{aligned} x + 2y - 3z &= 5 \\ x + 2z &= 15 \\ 2y - z &= 6 \end{aligned}$

14. $\begin{aligned} x + y - z &= 0 \\ 2y + 3z &= 1 \\ x + 2z &= 0 \end{aligned}$

In Exercises 15–18, write a system of linear equations associated with the augmented matrix.

15. $\left[\begin{array}{cc|c} 2 & 1 & 1 \\ 3 & -2 & 12 \end{array}\right]$

16. $\left[\begin{array}{cc|c} 2 & 1 & 8 \\ 0 & 1 & 2 \end{array}\right]$

17. $\left[\begin{array}{ccc|c} 1 & 1 & -1 & 2 \\ 2 & -3 & 1 & 5 \\ 3 & 2 & -4 & 3 \end{array}\right]$

18. $\left[\begin{array}{ccc|c} 1 & 1 & -1 & 3 \\ 0 & 1 & -3 & 8 \\ 1 & 0 & -2 & 3 \end{array}\right]$

19. When using matrices to solve a system of linear equations, what is your goal with respect to the coefficient matrix?

20. For a system of n linear equations in n variables, what are the dimensions of the constant matrix? Why?

In Exercises 21–24, state the row operation that has been performed in each step of the matrix solution of the given system of linear equations. Then state the solution.

21. $\begin{aligned} 3x + 6y &= 12 \\ 2x - 3y &= 1 \end{aligned}$

Matrix	*Row Operation*
$\left[\begin{array}{cc\|c} 3 & 6 & 12 \\ 2 & -3 & 1 \end{array}\right]$	Augmented matrix
$\left[\begin{array}{cc\|c} 1 & 2 & 4 \\ 2 & -3 & 1 \end{array}\right]$	(a) ________
$\left[\begin{array}{cc\|c} 1 & 2 & 4 \\ 0 & -7 & -7 \end{array}\right]$	(b) ________
$\left[\begin{array}{cc\|c} 1 & 2 & 4 \\ 0 & 1 & 1 \end{array}\right]$	(c) ________
$\left[\begin{array}{cc\|c} 1 & 0 & 2 \\ 0 & 1 & 1 \end{array}\right]$	(d) ________

(e) The solution is ________.

22. $2x - 6y = -18$
$3x + 4y = -\ 1$

Matrix	*Row Operation*
$\left[\begin{array}{cc|c} 2 & -6 & -18 \\ 3 & 4 & -1 \end{array}\right]$	Augmented matrix
$\left[\begin{array}{cc|c} 1 & -3 & -9 \\ 3 & 4 & -1 \end{array}\right]$	(a) ______
$\left[\begin{array}{cc|c} 1 & -3 & -9 \\ 0 & 13 & 26 \end{array}\right]$	(b) ______
$\left[\begin{array}{cc|c} 1 & -3 & -9 \\ 0 & 1 & 2 \end{array}\right]$	(c) ______
$\left[\begin{array}{cc|c} 1 & 0 & -3 \\ 0 & 1 & 2 \end{array}\right]$	(d) ______

(e) The solution is ______.

23. $x + y - z = 4$
$2x - y + z = -1$
$x + y - 2z = 5$

Matrix	*Row Operation*
$\left[\begin{array}{ccc|c} 1 & 1 & -1 & 4 \\ 2 & -1 & 1 & -1 \\ 1 & 1 & -2 & 5 \end{array}\right]$	Augmented matrix
$\left[\begin{array}{ccc|c} 1 & 1 & -1 & 4 \\ 0 & -3 & 3 & -9 \\ 0 & 0 & -1 & 1 \end{array}\right]$	(a) ______ (b) ______
$\left[\begin{array}{ccc|c} 1 & 1 & -1 & 4 \\ 0 & 1 & -1 & 3 \\ 0 & 0 & 1 & -1 \end{array}\right]$	(c) ______ (d) ______
$\left[\begin{array}{ccc|c} 1 & 1 & 0 & 3 \\ 0 & 1 & 0 & 2 \\ 0 & 0 & 1 & -1 \end{array}\right]$	(e) ______ (f) ______
$\left[\begin{array}{ccc|c} 1 & 0 & 0 & 1 \\ 0 & 1 & 0 & 2 \\ 0 & 0 & 1 & -1 \end{array}\right]$	(g) ______

(h) The solution is ______.

24. $x + y - z = 4$
$2y - 5z = 2$
$7z = 0$

Matrix	*Row Operation*
$\left[\begin{array}{ccc|c} 1 & 1 & -1 & 4 \\ 0 & 2 & -5 & 2 \\ 0 & 0 & 7 & 0 \end{array}\right]$	Augmented matrix
$\left[\begin{array}{ccc|c} 1 & 1 & -1 & 4 \\ 0 & 2 & -5 & 2 \\ 0 & 0 & 1 & 0 \end{array}\right]$	(a) ______
$\left[\begin{array}{ccc|c} 1 & 1 & 0 & 4 \\ 0 & 2 & 0 & 2 \\ 0 & 0 & 1 & 0 \end{array}\right]$	(b) ______ (c) ______
$\left[\begin{array}{ccc|c} 1 & 1 & 0 & 4 \\ 0 & 1 & 0 & 1 \\ 0 & 0 & 1 & 0 \end{array}\right]$	(d) ______
$\left[\begin{array}{ccc|c} 1 & 0 & 0 & 3 \\ 0 & 1 & 0 & 1 \\ 0 & 0 & 1 & 0 \end{array}\right]$	(e) ______

(f) The solution is ______.

In Exercises 25–28, fill in the blanks to indicate the results of performing row operations. State the solution.

25. $\left[\begin{array}{cc|c} 4 & 8 & 52 \\ 3 & -1 & -17 \end{array}\right]$

$\left[\begin{array}{cc|c} 1 & \text{(a)} & 13 \\ 3 & -1 & -17 \end{array}\right]$

$\left[\begin{array}{cc|c} 1 & \text{(b)} & 13 \\ 0 & -7 & \text{(c)} \end{array}\right]$

$\left[\begin{array}{cc|c} 1 & \text{(d)} & 13 \\ 0 & \text{(e)} & 8 \end{array}\right]$

$\left[\begin{array}{cc|c} 1 & 0 & \text{(f)} \\ 0 & \text{(g)} & 8 \end{array}\right]$

(h) Solution: ______

26. $\left[\begin{array}{cc|c} 6 & 2 & -2 \\ 1 & -1 & 1 \end{array}\right]$

$\left[\begin{array}{cc|c} \text{(a)} & -1 & \text{(b)} \\ 6 & 2 & -2 \end{array}\right]$

$\left[\begin{array}{cc|c} \text{(c)} & -1 & \text{(d)} \\ 0 & \text{(e)} & -8 \end{array}\right]$

$\left[\begin{array}{cc|c} \text{(f)} & -1 & \text{(g)} \\ 0 & 1 & \text{(h)} \end{array}\right]$

$\left[\begin{array}{cc|c} \text{(i)} & 0 & \text{(j)} \\ 0 & 1 & \text{(k)} \end{array}\right]$

(l) Solution: ______

27. $\left[\begin{array}{ccc|c} 2 & -3 & 1 & 0 \\ 1 & 1 & -1 & 3 \\ 3 & -2 & -2 & 7 \end{array}\right]$

$$\left[\begin{array}{ccc|c} 1 & 1 & -1 & \text{(a)} \\ 2 & \text{(b)} & 1 & 0 \\ 3 & -2 & -2 & 7 \end{array}\right]$$

$$\left[\begin{array}{ccc|c} 1 & 1 & -1 & 3 \\ 0 & -5 & \text{(c)} & -6 \\ 0 & -5 & 1 & \text{(d)} \end{array}\right]$$

$$\left[\begin{array}{ccc|c} 1 & \text{(e)} & -1 & 3 \\ 0 & -5 & 3 & -6 \\ 0 & 0 & \text{(f)} & 4 \end{array}\right]$$

$$\left[\begin{array}{ccc|c} 1 & 1 & -1 & 3 \\ 0 & -5 & 3 & \text{(g)} \\ 0 & 0 & 1 & \text{(h)} \end{array}\right]$$

$$\left[\begin{array}{ccc|c} 1 & 1 & 0 & \text{(i)} \\ 0 & -5 & \text{(j)} & 0 \\ 0 & 0 & 1 & -2 \end{array}\right]$$

$$\left[\begin{array}{ccc|c} 1 & \text{(k)} & 0 & 1 \\ 0 & 1 & 0 & \text{(l)} \\ 0 & 0 & 1 & -2 \end{array}\right]$$

$$\left[\begin{array}{ccc|c} 1 & 0 & \text{(m)} & 1 \\ \text{(n)} & 1 & 0 & 0 \\ 0 & 0 & 1 & -2 \end{array}\right]$$

(o) Solution:

28. $$\left[\begin{array}{ccc|c} 3 & 6 & -3 & 9 \\ 1 & -2 & 1 & 3 \\ 2 & -1 & -1 & 0 \end{array}\right]$$

$$\left[\begin{array}{ccc|c} \text{(a)} & 2 & -1 & 3 \\ \text{(b)} & -2 & 1 & 3 \\ 2 & -1 & -1 & 0 \end{array}\right]$$

$$\left[\begin{array}{ccc|c} 1 & 2 & -1 & 3 \\ 0 & \text{(c)} & 2 & 0 \\ 0 & \text{(d)} & 1 & -6 \end{array}\right]$$

$$\left[\begin{array}{ccc|c} 1 & -3 & \text{(e)} & -3 \\ 0 & 6 & 0 & \text{(f)} \\ 0 & -5 & 1 & -6 \end{array}\right]$$

$$\left[\begin{array}{ccc|c} 1 & -3 & 0 & -3 \\ 0 & \text{(g)} & 0 & 2 \\ 0 & -5 & 1 & \text{(h)} \end{array}\right]$$

$$\left[\begin{array}{ccc|c} 1 & 0 & 0 & \text{(i)} \\ 0 & 1 & 0 & 2 \\ 0 & 0 & 1 & \text{(j)} \end{array}\right]$$

(k) Solution:

29. Consider the system

$$2x = 3y + 2$$
$$y + 1 = x$$

What must be done before the coefficient matrix and the constant matrix can be written?

30. Each of the following shows the last row of an augmented matrix after row operations are performed. In each part, describe the solution set of the associated system of equations.

(a) $0 \quad 0 \mid 6$

(b) $0 \quad 0 \mid 0$

(c) $0 \quad 1 \mid 0$

In Exercises 31–42, use matrix methods to solve the given system of equations. Identify inconsistent systems and dependent equations.

31. $2x - 3y = 18$, $x + 2y = -5$

32. $3x + 4y = 0$, $x + 2y = 2$

33. $4x + 8y = 0$, $x - 2y = 1$

34. $y - x = 1$, $2x + y = 0$

35. $2x - 4y = 7$, $2y - x = 4$

36. $2x - 3y = 7$, $4x = 6y + 7$

37. $3x - 2y = 6$, $6y - 9x = -18$

38. $4x - 5y = 7$, $10y = 8x - 14$

39. $x + z = 1$, $y - z = 5$, $x - y = 2$

40. $x - y = 8$, $x + z = 7$, $y + 2z = 1$

41. $x + y - z = -4$, $-x - 2y + z = 7$, $2x + 2y + z = -2$

Remind students that a special case is indicated when any row of the coefficient matrix contains all zeros.

42. $-x + 2y - z = -7$, $3x - 7y - 2z = 23$, $-2x + 2y + 3z = -10$

In Exercises 43–46, write the augmented matrix that is associated with the given system of equations.

43. $2x - 5y = -5$, $2y + 5x = 2$

44. $x - 3y = 9$, $5y = x - 17$

45. $z + 3x = 4y + 2$, $1 - y = x + 3z$, $z = 2x + 3y$

46. $z = x + y$, $y - 2 = 2x + 3z$, $z - 1 = 3x + 2y$

In Exercises 47–54, solve the given system of equations.

Each system in this group has a unique solution.

47. $3x + 4y = -5$
$x + 3 = 0$

48. $y - 2 = 4$
$2x - y = -2$

49. $3x + y + z = 5$
$2x - y + z = 6$
$y = 2z + 2$

50. $x + 3y + z = 2$
$2y + z = 1$
$x - 2y - 3z = 4$

51. $x - 2y - 2z = -2$
$2x + 3y + z = -1$
$3x + y + z = 1$

52. $x + y + z = 0$
$2x - 2y + 3z = 6$
$5x - y + 2z = 0$

53. $x + y - z = 0$
$2x - 3y + 4z = 4$
$3x - 2y + z = 6$

54. $3x - 2y + 5z = -16$
$2x + 5y + 4z = 0$
$4x - 6y - 7z = 13$

Modeling with Real Data

Exercises 55 and 56 illustrate the fact that a matrix does not have to contain coefficients of equations. Any array of numbers is a matrix. The position of each element indicates the information.

55. The accompanying table shows the average number of minutes in the work day of employed men and women in the United States, in other industrialized countries, and in developing countries. (Source: *Human Development Report.*)

	Time (in minutes)	
	Men	**Women**
United States	428	453
Other industrialized countries	408	430
Developing countries	483	544

(a) Write the data as a 3×2 matrix. What does the entry in row 3, column 2 represent?

(b) Write the data as a 2×3 matrix. Having done so, in what row or column would you look to determine the difference between men's and women's times in the United States?

56. In part (a) of Exercise 55, suppose that you perform the following operations.

$$\frac{1}{60} \cdot R_i \to R_i \qquad \text{where } i = 1, 2, 3$$

What would the entries in the resulting matrix represent?

Challenge

In Exercises 57–60, solve the system.

57. $2x + y - z + w = -7$
$x - 2y + z - 2w = 16$
$-x + 2y + 3z + 4w = -15$
$3x - y - 2z - w = 3$

58. $x + y + z + w = -2$
$5x - 2y + 3z + w = 1$
$-x + 2y - z - 2w = -9$
$3x + 4y + 2z + 3w = -10$

59. $x - z = 6$
$y + w + 2 = 0$
$x - y + w + 6 = 0$
$y - x = 1$

60. $x + y = 0$
$y + 2z + 3 = 0$
$2z + 3w - 4 = 0$
$3w + 4x = 10$

61. Use matrix methods to solve the following system.

$x + y = 1$
$x + cy = 5$

For what value(s) of c is the solution unique? For what value(s) of c is there no solution?

62. Use matrix methods to solve the following system.

$x + 3y = 2$
$ax + y = 1$

For what value(s) of a is the solution unique? For what value(s) of a is there no solution?

63. Use matrix methods to determine p, q, and r so that the given system of equations has

(a) a unique solution.

(b) no solution.

(c) infinitely many solutions.

$x - y = p$
$y - z = q$
$z - x = r$

5.5 Determinants and Cramer's Rule

Determinant of a 2 × 2 Matrix • Determinant of a 3 × 3 Matrix • Cramer's Rule • Real-Life Applications

This section is optional.

Determinant of a 2 × 2 Matrix

We can write the general elements of a matrix by using subscripts. For example, we can write a 2 × 2 matrix as follows.

$$\begin{bmatrix} a_1 & b_1 \\ a_2 & b_2 \end{bmatrix}$$

LEARNING TIP

Be sure that you understand the difference between a matrix and a determinant. A matrix is an array or table of numbers, whereas a determinant is a *number*.

Note that the subscript on each element indicates the row of the matrix in which the element is located.

Associated with each square matrix is a real number called the **determinant** of the matrix.

Determinant of a 2 × 2 Matrix

The **determinant** of a 2 × 2 matrix is defined by

$$\begin{vmatrix} a_1 & b_1 \\ a_2 & b_2 \end{vmatrix} = a_1b_2 - a_2b_1$$

Students tend to write matrices and determinants interchangeably. Make sure they know the difference in notation.

Note: A matrix is enclosed by square brackets, whereas a determinant is enclosed by vertical lines.

As indicated in the diagram, an easy way to evaluate the determinant of a 2 × 2 matrix is to multiply the numbers on the diagonals and subtract.

$$\begin{vmatrix} a_1 & b_1 \\ a_2 & b_2 \end{vmatrix} = a_1b_2 - a_2b_1$$

EXAMPLE 1 Evaluating the Determinant of a 2 × 2 Matrix

Think About It

Consider the following matrices.

$$\begin{bmatrix} a_1 & b_1 \\ a_2 & b_2 \end{bmatrix} \qquad \begin{bmatrix} a_2 & b_2 \\ a_1 & b_1 \end{bmatrix}$$

Are the matrices equivalent? Describe the relationship between the associated determinants of these matrices.

Evaluate the determinant of the matrix $\begin{bmatrix} 2 & 1 \\ 1 & -1 \end{bmatrix}$.

Solution

$$\begin{vmatrix} 2 & 1 \\ 1 & -1 \end{vmatrix} = 2(-1) - 1(1) = -2 - 1 = -3$$

Determinant of a 3 × 3 Matrix

Defining the determinant for matrices larger than 2 × 2 requires the definition of a **minor** of a matrix.

Definition of a Minor of a Matrix

For each element of a matrix, the **minor** of the matrix corresponding to that element is the determinant of the matrix obtained by eliminating the row and column containing that element.

EXAMPLE 2 Determining Minors of a Matrix

For the following matrix, evaluate the minor corresponding to each of the given elements.

$$\begin{bmatrix} 5 & 2 & 9 \\ 3 & 7 & 4 \\ 6 & 1 & 8 \end{bmatrix}$$

(a) 2 (b) 8 (c) 3

Solution

	Element	*Matrix*	*Minor*
(a)	2	$\begin{bmatrix} \not{5} & \textcircled{2} & \not{9} \\ 3 & \not{7} & 4 \\ 6 & \not{1} & 8 \end{bmatrix}$	$\begin{vmatrix} 3 & 4 \\ 6 & 8 \end{vmatrix} = 3 \cdot 8 - 4 \cdot 6 = 24 - 24 = 0$
(b)	8	$\begin{bmatrix} 5 & 2 & \not{9} \\ 3 & 7 & \not{4} \\ \not{6} & \not{1} & \textcircled{8} \end{bmatrix}$	$\begin{vmatrix} 5 & 2 \\ 3 & 7 \end{vmatrix} = 5 \cdot 7 - 2 \cdot 3 = 35 - 6 = 29$
(c)	3	$\begin{bmatrix} \not{5} & 2 & 9 \\ \textcircled{3} & \not{7} & \not{4} \\ \not{6} & 1 & 8 \end{bmatrix}$	$\begin{vmatrix} 2 & 9 \\ 1 & 8 \end{vmatrix} = 2 \cdot 8 - 9 \cdot 1 = 16 - 9 = 7$

We define the determinant of a 3×3 matrix in terms of the minors of the elements of a row or a column. The following method for evaluating a determinant is called **expanding by minors along the first row.**

Note that the definition includes the 2×2 minors of a_1, b_1, and c_1.

$$\begin{vmatrix} a_1 & b_1 & c_1 \\ a_2 & b_2 & c_2 \\ a_3 & b_3 & c_3 \end{vmatrix} = a_1 \begin{vmatrix} b_2 & c_2 \\ b_3 & c_3 \end{vmatrix} - b_1 \begin{vmatrix} a_2 & c_2 \\ a_3 & c_3 \end{vmatrix} + c_1 \begin{vmatrix} a_2 & b_2 \\ a_3 & b_3 \end{vmatrix}$$

We can evaluate a determinant by expanding along any row or any column. To determine whether to add or subtract the product of an element and its minor, we use the following sign pattern.

$$\begin{vmatrix} + & - & + \\ - & + & - \\ + & - & + \end{vmatrix}$$

EXAMPLE 3

Evaluating the Determinant of a 3 × 3 Matrix

Evaluate the determinant of the matrix $\begin{bmatrix} 3 & -5 & 0 \\ 2 & 4 & 1 \\ 2 & 1 & 6 \end{bmatrix}$.

Solution

We expand along the first row.

$$\begin{vmatrix} 3 & -5 & 0 \\ 2 & 4 & 1 \\ 2 & 1 & 6 \end{vmatrix} = 3 \cdot \begin{vmatrix} 4 & 1 \\ 1 & 6 \end{vmatrix} - (-5) \cdot \begin{vmatrix} 2 & 1 \\ 2 & 6 \end{vmatrix} + 0 \cdot \begin{vmatrix} 2 & 4 \\ 2 & 1 \end{vmatrix}$$

$$= 3(23) + 5(10) + 0(-6) = 119$$

Expanding along another row or column gives the same result. For instance, we can expand along the second column. Note the sign pattern.

$$\begin{vmatrix} 3 & -5 & 0 \\ 2 & 4 & 1 \\ 2 & 1 & 6 \end{vmatrix} = -(-5) \cdot \begin{vmatrix} 2 & 1 \\ 2 & 6 \end{vmatrix} + 4 \cdot \begin{vmatrix} 3 & 0 \\ 2 & 6 \end{vmatrix} - 1 \cdot \begin{vmatrix} 3 & 0 \\ 2 & 1 \end{vmatrix}$$

$$= 5(10) + 4(18) - 1(3) = 119$$

LEARNING TIP

Another way to evaluate a 3 × 3 determinant is to use the following diagram.

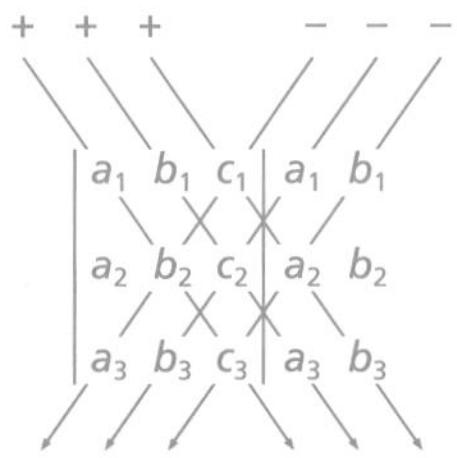

Cramer's Rule

When we solve a system of equations with the addition method, we perform the same operations repeatedly.

Exploring the Concept

Cramer's Rule

Consider the following general system of equations.

$$a_1x + b_1y = k_1$$
$$a_2x + b_2y = k_2$$

If we apply the addition method to this system of equations, we obtain the following formulas for the solution of the system.

$$x = \frac{k_1b_2 - k_2b_1}{a_1b_2 - a_2b_1} \quad \text{and} \quad y = \frac{a_1k_2 - a_2k_1}{a_1b_2 - a_2b_1}$$

Note that the formula for x is valid only if $a_1b_2 - a_2b_1 \neq 0$. If $a_1b_2 - a_2b_1 = 0$, then the solution is not unique.

(The derivation of these formulas is left as an exercise.) Note that the denominator of each formula is the determinant of the coefficient matrix.

$$D = \begin{vmatrix} a_1 & b_1 \\ a_2 & b_2 \end{vmatrix} = a_1b_2 - a_2b_1$$

The numerator of the formula for x is the determinant of the matrix obtained by replacing a_1 and a_2 (the coefficients of x in the coefficient matrix) with the constants k_1 and k_2. Similarly, the numerator of the formula for y is the determinant of the matrix obtained by replacing b_1 and b_2 (the coefficients of y in the coefficient matrix) with the constants k_1 and k_2.

$$D_X = \begin{vmatrix} k_1 & b_1 \\ k_2 & b_2 \end{vmatrix} = k_1b_2 - k_2b_1 \qquad D_Y = \begin{vmatrix} a_1 & k_1 \\ a_2 & k_2 \end{vmatrix} = a_1k_2 - a_2k_1$$

We can write the formulas for x and y with D, D_X, and D_Y.

$$x = \frac{D_X}{D} \quad \text{and} \quad y = \frac{D_Y}{D}$$

These conclusions are stated in Cramer's Rule.

Cramer's Rule for a 2 × 2 System

The solution of the system

$$a_1x + b_1y = k_1$$
$$a_2x + b_2y = k_2$$

is given by $x = \dfrac{D_X}{D}$ and $y = \dfrac{D_Y}{D}$,

where $D = \begin{vmatrix} a_1 & b_1 \\ a_2 & b_2 \end{vmatrix} \neq 0$, $\quad D_X = \begin{vmatrix} k_1 & b_1 \\ k_2 & b_2 \end{vmatrix}$, $\quad D_Y = \begin{vmatrix} a_1 & k_1 \\ a_2 & k_2 \end{vmatrix}$

Emphasize that if $D = 0$, Cramer's Rule does not apply. Other techniques must be used.

EXAMPLE 4

Using Cramer's Rule to Solve a 2 × 2 System

Use Cramer's Rule to solve the following system of equations.

$$2x + y = 1$$
$$x - y = 5$$

Solution

Evaluate D, D_X, and D_Y.

$$D = \begin{vmatrix} 2 & 1 \\ 1 & -1 \end{vmatrix} = 2(-1) - 1(1) = -3$$

$$D_X = \begin{vmatrix} 1 & 1 \\ 5 & -1 \end{vmatrix} = 1(-1) - 1(5) = -6$$

$$D_Y = \begin{vmatrix} 2 & 1 \\ 1 & 5 \end{vmatrix} = 2(5) - 1(1) = 9$$

Now use Cramer's Rule to calculate x and y.

$$x = \frac{D_X}{D} = \frac{-6}{-3} = 2 \quad \text{and} \quad y = \frac{D_Y}{D} = \frac{9}{-3} = -3$$

The solution is $(2, -3)$.

Note: If the determinant of the coefficient matrix is zero, then either the system has no solution or it has infinitely many solutions. Cramer's Rule cannot be applied.

Cramer's Rule can be extended to larger systems of equations.

Cramer's Rule for a 3 × 3 System

The solution of the system of equations

$$a_1x + b_1y + c_1z = k_1$$
$$a_2x + b_2y + c_2z = k_2$$
$$a_3x + b_3y + c_3z = k_3$$

is given by $x = \dfrac{D_X}{D}$, $y = \dfrac{D_Y}{D}$, and $z = \dfrac{D_Z}{D}$, where

$$D = \begin{vmatrix} a_1 & b_1 & c_1 \\ a_2 & b_2 & c_2 \\ a_3 & b_3 & c_3 \end{vmatrix} \neq 0,\ D_X = \begin{vmatrix} k_1 & b_1 & c_1 \\ k_2 & b_2 & c_2 \\ k_3 & b_3 & c_3 \end{vmatrix},\ D_Y = \begin{vmatrix} a_1 & k_1 & c_1 \\ a_2 & k_2 & c_2 \\ a_3 & k_3 & c_3 \end{vmatrix},\ D_Z = \begin{vmatrix} a_1 & b_1 & k_1 \\ a_2 & b_2 & k_2 \\ a_3 & b_3 & k_3 \end{vmatrix}$$

EXAMPLE 5

Using Cramer's Rule to Solve a 3 × 3 System

Use Cramer's Rule to solve the following system of equations.

$$x - y - z = 1$$
$$x + 2y - z = 1$$
$$x + y + z = 3$$

Solution

Evaluating D, D_X, D_Y, D_Z, we obtain the following results.

$$D = \begin{vmatrix} 1 & -1 & -1 \\ 1 & 2 & -1 \\ 1 & 1 & 1 \end{vmatrix} = 6 \qquad D_X = \begin{vmatrix} 1 & -1 & -1 \\ 1 & 2 & -1 \\ 3 & 1 & 1 \end{vmatrix} = 12$$

$$D_Y = \begin{vmatrix} 1 & 1 & -1 \\ 1 & 1 & -1 \\ 1 & 3 & 1 \end{vmatrix} = 0 \qquad D_Z = \begin{vmatrix} 1 & -1 & 1 \\ 1 & 2 & 1 \\ 1 & 1 & 3 \end{vmatrix} = 6$$

Now we use the formulas from Cramer's Rule to calculate x, y, and z.

$$x = \frac{D_X}{D} = \frac{12}{6} = 2 \qquad y = \frac{D_Y}{D} = \frac{0}{6} = 0 \qquad z = \frac{D_Z}{D} = \frac{6}{6} = 1$$

The solution is (2, 0, 1).

Real-Life Applications

In the following application, we use Cramer's Rule to solve the system of equations.

EXAMPLE 6

Frozen Yogurt Sales

In one day a frozen yogurt shop sold 120 waffle cones. There were 10 fewer strawberry cones sold than chocolate cones. The number of vanilla cones sold was 10 more than twice the number of chocolate cones sold. How many cones of each flavor were sold?

Solution

Let x = number of vanilla cones sold,
y = number of chocolate cones sold, and
z = number of strawberry cones sold.

The following system describes the conditions of the problem.

$x + y + z = 120$ A total of 120 cones were sold.

$z = y - 10$ There were 10 fewer strawberry cones sold than chocolate cones.

$x = 2y + 10$ The number of vanilla cones sold was 10 more than twice the number of chocolate cones sold.

This system can be written in the following form.

$$\begin{aligned} x + y + z &= 120 \\ y - z &= 10 \\ x - 2y &= 10 \end{aligned}$$

The associated determinants are as follows.

$$D = \begin{vmatrix} 1 & 1 & 1 \\ 0 & 1 & -1 \\ 1 & -2 & 0 \end{vmatrix} \qquad D_X = \begin{vmatrix} 120 & 1 & 1 \\ 10 & 1 & -1 \\ 10 & -2 & 0 \end{vmatrix}$$

$$D_Y = \begin{vmatrix} 1 & 120 & 1 \\ 0 & 10 & -1 \\ 1 & 10 & 0 \end{vmatrix} \qquad D_Z = \begin{vmatrix} 1 & 1 & 120 \\ 0 & 1 & 10 \\ 1 & -2 & 10 \end{vmatrix}$$

Evaluate the four determinants.

$$D = -4 \qquad D_X = -280 \qquad D_Y = -120 \qquad D_Z = -80$$

Finally, we use Cramer's Rule to determine the values of x, y, and z.

$$x = \frac{D_X}{D} = \frac{-280}{-4} = 70$$

$$y = \frac{D_Y}{D} = \frac{-120}{-4} = 30$$

$$z = \frac{D_Z}{D} = \frac{-80}{-4} = 20$$

There were 70 vanilla cones, 30 chocolate cones, and 20 strawberry cones sold.

Quick Reference 5.5

Determinant of a 2 × 2 Matrix

- Associated with each square matrix is a real number called a **determinant.**
- For the matrix $\begin{bmatrix} a_1 & b_1 \\ a_2 & b_2 \end{bmatrix}$, we define the determinant by $\begin{vmatrix} a_1 & b_1 \\ a_2 & b_2 \end{vmatrix} = a_1b_2 - a_2b_1$.

Determinant of a 3 × 3 Matrix

- The **minor** corresponding to an element of a matrix is the determinant of the matrix obtained by deleting the row and column containing that element.
- We can evaluate a determinant of a matrix by expanding by minors along any row or along any column.

Cramer's Rule

- For a system of equations written in standard form, the following notation is used in connection with Cramer's Rule.
  1. The determinant D is the determinant of the coefficient matrix.
  2. The determinants D_X, D_Y, and D_Z are the determinants of the coefficient matrix after the x-coefficients, y-coefficients, and z-coefficients, respectively, have been replaced by the constants.
- Cramer's Rule states that if $D \neq 0$, then
  1. the solution of a 2×2 system is given by $x = \frac{D_X}{D}$, $y = \frac{D_Y}{D}$, and
  2. the solution of a 3×3 system is given by $x = \frac{D_X}{D}$, $y = \frac{D_Y}{D}$, and $z = \frac{D_Z}{D}$.

Speaking the Language 5.5

1. A(n) ________ is a real number that is associated with a square matrix.

2. In a matrix, if we delete the row and column that contain the element a_1, then the determinant of the remaining matrix is called the ________ corresponding to a_1.

3. The method for solving a system of equations with determinants is given by ________ Rule.

4. In the formulas $x = \frac{D_X}{D}$ and $y = \frac{D_Y}{D}$, the symbol D represents the determinant of the ________.

Exercises 5.5

Concepts and Skills

1. What is the difference between a matrix and a determinant?

2. Note that matrix B is obtained by exchanging the rows of A. Are the determinants of the two matrices the same? Explain.

$$A = \begin{bmatrix} p & q \\ r & s \end{bmatrix} \qquad B = \begin{bmatrix} r & s \\ p & q \end{bmatrix}$$

In Exercises 3–6, evaluate the determinant of the given matrix.

3. $\begin{bmatrix} 0 & 1 \\ -7 & 5 \end{bmatrix}$

4. $\begin{bmatrix} 3 & 1 \\ -2 & 4 \end{bmatrix}$

5. $\begin{bmatrix} 3 & 2 \\ 9 & 6 \end{bmatrix}$

6. $\begin{bmatrix} 5 & -2 \\ -1 & -3 \end{bmatrix}$

In Exercises 7–10, evaluate the given determinant.

7. $\begin{vmatrix} 4 & 2 \\ 2 & 1 \end{vmatrix}$

8. $\begin{vmatrix} 0 & -1 \\ 1 & 1 \end{vmatrix}$

9. $\begin{vmatrix} \frac{3}{4} & \frac{2}{3} \\ 3 & 4 \end{vmatrix}$

10. $\begin{vmatrix} \sqrt{18} & 2 \\ 3 & \sqrt{2} \end{vmatrix}$

In Exercises 11–16, evaluate the minor that corresponds to the given element in matrix A.

$$A = \begin{bmatrix} 3 & -2 & 5 \\ 1 & 2 & -1 \\ 6 & -4 & -7 \end{bmatrix}$$

11. -2
12. 1
13. -1
14. -4
15. 2
16. -7

In Exercises 17–24, evaluate the determinant.

17. $\begin{vmatrix} 2 & 0 & 0 \\ 0 & 1 & 0 \\ 0 & 0 & -4 \end{vmatrix}$

18. $\begin{vmatrix} 1 & 3 & -2 \\ 0 & 1 & 5 \\ 0 & 0 & 2 \end{vmatrix}$

19. $\begin{vmatrix} 2 & 0 & 3 \\ -5 & 0 & 1 \\ 2 & 0 & -5 \end{vmatrix}$

20. $\begin{vmatrix} 3 & -1 & 2 \\ 5 & 2 & -4 \\ 0 & 0 & 0 \end{vmatrix}$

21. $\begin{vmatrix} 1 & 3 & 1 \\ 1 & 2 & 4 \\ 1 & -1 & -3 \end{vmatrix}$

22. $\begin{vmatrix} 2 & 4 & -1 \\ 1 & 1 & 1 \\ 3 & -2 & 7 \end{vmatrix}$

23. $\begin{vmatrix} -1 & -3 & 2 \\ 2 & 6 & -4 \\ 1 & 0 & -2 \end{vmatrix}$

24. $\begin{vmatrix} -3 & 1 & 0 \\ 6 & -2 & 2 \\ 3 & -1 & 1 \end{vmatrix}$

In Exercises 25–28, use the method for evaluating a determinant to write an equivalent algebraic expression.

25. $\begin{vmatrix} x & 9 \\ 1 & x \end{vmatrix}$

26. $\begin{vmatrix} 2 & y \\ y & 8 \end{vmatrix}$

27. $\begin{vmatrix} x-y & 2 \\ x+y & 3 \end{vmatrix}$

28. $\begin{vmatrix} x+2 & 1 \\ 3-x & -1 \end{vmatrix}$

In Exercises 29–32, solve for a.

29. $\begin{vmatrix} 4 & a \\ 5 & 7 \end{vmatrix} = -7$

30. $\begin{vmatrix} 2 & 5 \\ 3 & a \end{vmatrix} = 3$

31. $\begin{vmatrix} a & 4 \\ -1 & 3 \end{vmatrix} = 7$

32. $\begin{vmatrix} 2 & -2 \\ a & 7 \end{vmatrix} = -2$

In Exercises 33 and 34, use the definition of the determinant of a 3×3 matrix to solve for a.

33. $\begin{vmatrix} 1 & -2 & 0 \\ 0 & 2 & -1 \\ 3 & 0 & a \end{vmatrix} = 14$

34. $\begin{vmatrix} 1 & 0 & 4 \\ -1 & a & 0 \\ 0 & 3 & -2 \end{vmatrix} = 10$

35. Explain why we cannot use Cramer's Rule if the determinant of the coefficient matrix is zero.

36. When solving a 3×3 system with Cramer's Rule, why is it wise to calculate D before calculating D_x, D_y, and D_z?

Remind students that equations must be in standard form before Cramer's Rule can be applied.

In Exercises 37–42, determine whether Cramer's Rule can be used to solve the system of equations.

37. $x + 2y = 7$
$6x = -12y + 28$

38. $x - 2y = 8$
$3x - 5y = 10$

39. $x = 3y - 4$
$4x - 12y = 28$

40. $y = 4x + 5$
$12x - 3y = 11$

41. $y = 2x - 9$
$3x - 2y = 8$

42. $x = 4y + 3$
$x - 8y = 4$

In Exercises 43–50, use Cramer's Rule to solve the system of equations.

43. $3x + 2y = 16$
$5x + 4y = 30$

44. $x - 3y = 7$
$3x + y = -9$

45. $y = 3x + 5$
$3x + 2y = 1$

46. $x = y + 12$
$2x + y = 3$

47. $2x - \frac{4}{3}y = 6$
$0.3x - y = -1.5$

48. $3x + 1.5y = 0.5$
$3x + 2y = \frac{11}{6}$

49. $1.2x + 2.7y = 9$
$0.4x + 3y = 7.2$

50. $4.5x - 1.2y = -12$
$3.7x + 5.4y = 54$

In Exercises 51–58, solve the given system only if Cramer's Rule applies. If Cramer's Rule does not apply, so indicate.

51. $2x + 3y = 16$
$4x + 5y = 30$

52. $x - 2y = 7$
$3x + y = 7$

53. $2y - \frac{x}{3} = 0$
$2x + 6y = 9$

54. $x + \frac{2y}{3} = 1$
$4x = y - 7$

55. $y = 3x + 4$
$1.5x - 0.5y = 2$

56. $x = 0.2y + 1.4$
$0.5y = 2.5x - 3.5$

57. $\frac{y}{2} - \frac{2x}{3} = \frac{1}{2}$
$\frac{4x}{3} - 2y = -\frac{5}{3}$

58. $\frac{x}{5} - \frac{y}{6} = 1$
$x + \frac{2y}{3} = -1$

In Exercises 59–70, solve the given system only if Cramer's Rule applies. If Cramer's Rule does not apply, so indicate.

59. $\begin{aligned} x + y + z &= 6 \\ 3x - 2y + 4z &= 5 \\ 2x - y + z &= 1 \end{aligned}$

60. $\begin{aligned} x - 2y + z &= 15 \\ 2x + 3y + z &= -3 \\ 3x - 2y - 3z &= -1 \end{aligned}$

61. $\begin{aligned} x &= y + 2z - 7 \\ y &= z - 7 \\ 3x - 6y - z &= 8 \end{aligned}$

62. $\begin{aligned} x + y &= z + 6 \\ y + 2z &= x - 3 \\ 2x + 3z &= 3 \end{aligned}$

63. $\begin{aligned} x - 2y + z &= 7 \\ x + 2y - 3z &= 5 \\ 4x - 8y + 4z &= 14 \end{aligned}$

64. $\begin{aligned} x &= 2y + z + 5 \\ y &= 3z + 7 \\ 3x - 6y - 3z &= 15 \end{aligned}$

65. $\begin{aligned} x - y &= 3 \\ x - 2z &= 6 \\ z - 3y &= 1 \end{aligned}$

66. $\begin{aligned} 2x + y &= 1 \\ -3z - y &= 0 \\ z + x &= -3 \end{aligned}$

67. $\begin{aligned} 0.6x - 2y - 1.2z &= 3.8 \\ 1.5x - 0.5y + z &= 5 \\ x - y + 0.3z &= 4 \end{aligned}$

68. $\begin{aligned} 1.5x - y + 2z &= 10 \\ x - y + 2.4z &= 8.8 \\ 2.5x + y + 1.6z &= 13.2 \end{aligned}$

69. $\begin{aligned} \frac{2}{3}x - \frac{1}{6}y - \frac{1}{2}z &= -2 \\ \frac{5}{4}x + \frac{1}{4}y + \frac{3}{2}z &= 1 \\ \frac{3}{2}x - \frac{1}{4}y - \frac{3}{4}z &= -\frac{7}{2} \end{aligned}$

70. $\begin{aligned} \frac{1}{5}x - \frac{1}{2}y + \frac{3}{4}z &= -7 \\ \frac{1}{3}x + 2y - \frac{1}{2}z &= 25 \\ x - \frac{1}{4}y + \frac{1}{2}z &= 9 \end{aligned}$

71. In the first determinant, the number n is a common factor in the first row. In the second determinant, n is a common factor in the first column.

$$\begin{vmatrix} na & nb \\ c & d \end{vmatrix} \qquad \begin{vmatrix} na & b \\ nc & d \end{vmatrix}$$

Show that the number n can be factored out of either determinant; that is, show that

$$n\begin{vmatrix} a & b \\ c & d \end{vmatrix} = \begin{vmatrix} na & nb \\ c & d \end{vmatrix} = \begin{vmatrix} na & b \\ nc & d \end{vmatrix}$$

72. Show that the slope–intercept form of the equation of a line can be written

$$y = \begin{vmatrix} m & b \\ -1 & x \end{vmatrix}$$

In Exercises 73–76, determine the value of c so that Cramer's Rule will not apply.

73. $\begin{aligned} 2x + 3y &= 6 \\ 6x - cy &= 18 \end{aligned}$

74. $\begin{aligned} 3x - 4y &= 12 \\ cx + 12y &= 36 \end{aligned}$

75. $\begin{aligned} 5x - 4y &= 7 \\ 20x - cy &= 8 \end{aligned}$

76. $\begin{aligned} 4x - 7y &= 3 \\ cx - 21y &= 4 \end{aligned}$

In Exercises 77–80, use the definition of the determinant of a 3×3 matrix to determine the value of c so that Cramer's Rule does not apply.

77. $\begin{aligned} 2x - y + z &= 2 \\ y - 2z &= -5 \\ x + y + cz &= 3 \end{aligned}$

78. $\begin{aligned} cx + y &= 0 \\ -2x + 4y + z &= 4 \\ 3y + z &= -7 \end{aligned}$

79. $\begin{aligned} 2x + y &= 1 \\ x + cy - 3z &= 2 \\ cx + y + 2z &= 6 \end{aligned}$

80. $\begin{aligned} 2x + cy + z &= -2 \\ x - 2y + cz &= 3 \\ y - z &= -1 \end{aligned}$

Real-Life Applications

Exercises 81–84 are variations on "age" problems.

81. A local grocery store has been in business 4 years longer than the pharmacy. Six years ago, the grocery store had been in business twice as long as the pharmacy had been. How long has each store been in business?

82. A home was built 9 years before the home next door was built. Five years ago, the home was 4 times as old as the home next door was. How old is each home?

83. A checking account currently has twice as much money in it as a savings account. If $800 is deposited into each account next month, the checking account will have 1.5 times as much money in it as the savings account. How much money is currently in each account?

84. One spring is 3 centimeters shorter than a second spring. If each spring is stretched by 4 centimeters, the shorter spring will be $\frac{4}{5}$ as long as the second spring. How long is each spring before it is stretched?

85. A collection of 41 coins consisting of nickels, dimes, and quarters has a value of $3.50. If there are three times as many nickels as dimes in the collection, how many of each coin are in the collection?

86. In a triangle the measure of angle B is twice the measure of angle A. The measure of angle C is 20° more than the measure of angle B. Determine the measures of the three angles.

Challenge

In Exercises 87 and 88, solve the system of equations.

87. $\sqrt{3}x - 4\sqrt{2}y = 14$
$3\sqrt{3}x + \sqrt{2}y = 16$

88. $2\sqrt{5}x + \sqrt{6}y = -18$
$3\sqrt{5}x + 4\sqrt{6}y = 3$

In Exercises 89–92, write the given expression as a determinant.

89. $2x - 3y$

90. $x + 4y$

91. $ax + by$

92. $mx + b$

93. Consider the following system of equations.

$$a_1x + b_1y = k_1$$
$$a_2x + b_2y = k_2$$

Use the addition method to derive the following formulas for the solution of the system.

$$x = \frac{k_1b_2 - k_2b_1}{a_1b_2 - a_2b_1} \quad \text{and} \quad y = \frac{a_1k_2 - a_2k_1}{a_1b_2 - a_2b_1}$$

94. Explain why there is not a unique solution to the following problem.

The total value of 82 coins consisting of nickels, dimes, and quarters is \$8.20. There are three times as many nickels as quarters. Determine the number of each type of coin in the collection.

95. In each of the following matrices, either two rows or two columns are the same. Show that the determinant of each is 0.

$$\begin{bmatrix} a & b & c \\ a & b & c \\ x & y & z \end{bmatrix} \qquad \begin{bmatrix} a & a & x \\ b & b & y \\ c & c & z \end{bmatrix}$$

96. In each of the following matrices, either one row or one column consists of zeros. Show that the determinant of each is 0.

$$\begin{bmatrix} a & b & c \\ x & y & z \\ 0 & 0 & 0 \end{bmatrix} \qquad \begin{bmatrix} 0 & a & x \\ 0 & b & y \\ 0 & c & z \end{bmatrix}$$

5.6 Applications

2 × 2 Systems • 3 × 3 Systems

2 × 2 Systems

In this section we use our ability to solve systems of equations for solving a variety of application problems. Although our discussion focuses on specific examples, our goal remains to develop a general understanding of problem-solving strategies. We begin with examples of applications that lead to systems of two equations in two variables.

Mixture problems typically involve two or more items, each with a different cost. The items are combined to create a mixture whose cost is between the costs of the individual items.

EXAMPLE 1 **A Mixture of Cashews and Peanuts**

A snack food distributor provides cashews for \$6 per pound and peanuts for \$2 per pound. The distributor determines that the total sales would increase if a mixture of the two nuts for \$3 per pound were offered. How many pounds of each nut should be used for a 10-pound mixture?

The mixture's unit cost is closer to the unit cost for peanuts than to the unit cost for cashews. This implies that there will be more peanuts than cashews in the mixture. Help students develop this ability to predict sensible results.

Solution

Let c = the number of pounds of cashews and
p = the number of pounds of peanuts.

We can use a table to organize the information.

	Weight in Pounds	×	Cost per Pound	=	Total Cost
Cashews	c		6		$6c$
Peanuts	p		2		$2p$
Mixture	10		3		30

From this table we write the following system of equations.

$c + p = 10$ The total weight of the mixture is 10 pounds.

$6c + 2p = 30$ The total cost of the mixture is \$30.00.

LEARNING TIP

Beside each equation in the examples, we indicate the information that the equation models. As you write equations, consider making such notes to yourself as reminders of what the equations represent.

Multiply the first equation by -2 and add the equations.

$$\begin{aligned} -2c - 2p &= -20 \\ 6c + 2p &= 30 \\ \hline 4c &= 10 \\ c &= \frac{10}{4} = 2.5 \end{aligned}$$

$c + p = 10$ First equation

$2.5 + p = 10$ Replace c with 2.5.

$p = 7.5$

The mixture consists of 2.5 pounds of cashews and 7.5 pounds of peanuts.

Dual investment problems involve different amounts of money in two or more investments, each paying a different simple interest rate.

EXAMPLE 2

Dual Investments with Simple Interest

A bank customer invested in two certificates of deposit, one paying 8% and the other paying 3.8%. The total of her two investments was \$10,000. If the certificates paid yearly simple interest and the total interest earned at the end of the year was \$674, how much did this customer invest in each certificate?

Solution

Let x = the amount invested in the 8% certificate and
y = the amount invested in the 3.8% certificate.

We use a table to organize the information.

	Amount Invested	×	Interest Rate	=	Interest Earned
CD Paying 8%	x		0.08		$0.08x$
CD Paying 3.8%	y		0.038		$0.038y$
Totals	10,000				674

From the table, we can write a system of equations.

$$x + \quad y = 10{,}000 \qquad \text{The total amount invested was \$10,000.}$$
$$0.08x + 0.038y = 674 \qquad \text{The total interest earned was \$674.}$$

Multiply the first equation by -0.08 and add the equations.

$$\begin{aligned} -0.08x - 0.080y &= -800 \\ 0.08x + 0.038y &= 674 \\ \hline -0.042y &= -126 \\ y &= 3000 \end{aligned}$$

$$x + y = 10{,}000 \qquad \text{First equation}$$
$$x + 3000 = 10{,}000 \qquad \text{Replace } y \text{ with 3000.}$$
$$x = 7000$$

The bank customer invested \$7000 in the 8% certificate and \$3000 in the 3.8% certificate.

Students often have difficulty with liquid solution problems. You may want to use the figures in Section 3.5 as background review.

Recall that a liquid solution is some substance, such as acid, mixed with water. The *concentration* of the substance is the amount of the substance compared to the total amount of solution. Refer to Section 3.5 to review the details concerning this type of problem.

EXAMPLE 3

Creating a Liquid Solution with a Specified Concentration

A nurse has a 20% alcohol solution and a 45% alcohol solution. How many liters of each should be used to make 5 liters of a 30% alcohol solution?

Solution

Let x = the number of liters of the 20% (weak) solution and
y = the number of liters of the 45% (strong) solution.

	Liters of Solution	×	Concentration of Alcohol	=	Liters of Alcohol
Weak	x		0.20		$0.20x$
Strong	y		0.45		$0.45y$
Mix	5		0.30		1.5

From the table we write the following system of equations.

$$x + y = 5$$ The total amount of solution is 5 liters.

$$0.20x + 0.45y = 1.5$$ The total amount of alcohol in the solution is 1.5 liters.

This time, we use the substitution method to solve the system.

$$x = 5 - y$$ Solve the first equation for x.

$$0.20(5 - y) + 0.45y = 1.5$$ Replace x in the second equation with $5 - y$.

$$1 - 0.20y + 0.45y = 1.5$$ Distributive Property

$$1 + 0.25y = 1.5$$ Combine like terms.

$$0.25y = 0.5$$

$$y = \frac{0.5}{0.25} = 2$$

$$x = 5 - y$$ Replace y with 2 in the equation solved for x.

$$x = 5 - 2$$

$$x = 3$$

The nurse should use 3 liters of the 20% solution and 2 liters of the 45% solution.

Distance d, rate r, and time t are related by the formula $d = rt$. Water current and wind speed have the effect of decreasing or increasing a traveler's rate.

EXAMPLE 4

Determining Boat Speed and Rate of Current

A river boat takes tourists from St. Louis up the Mississippi River 72 miles to Clarksville. The trip up the river takes 6 hours, and the return downstream takes 2 hours less time. What is the speed of the current and what is the speed of the river boat?

Solution

Let b = speed of the boat in still water and
c = rate of the current.

As it travels upstream, the boat is slowed down by the current. The net rate is the boat's rate minus the rate of the current and is given by the expression $b - c$.

The boat is helped by the current as it returns downstream. The net rate is the boat's rate plus the rate of the current and is given by the expression $b + c$.

	Rate (mph)	×	Time (hours)	=	Distance (miles)
Upstream	$b - c$		6		$6(b - c)$
Downstream	$b + c$		4		$4(b + c)$

Because the boat traveled 72 miles upstream and 72 miles downstream, we write the following system of equations.

$$6(b - c) = 72$$

$$4(b + c) = 72$$

$$b - c = 12$$ Divide both sides of the first equation by 6.

$$b + c = 18$$ Divide both sides of the second equation by 4.

We use the addition method to solve the system.

$$\begin{aligned} b - c &= 12 && \text{First equation} \\ b + c &= 18 && \text{Second equation} \\ \hline 2b \quad &= 30 && \text{Add the equations.} \\ b &= 15 \end{aligned}$$

$$\begin{aligned} b + c &= 18 && \text{Second equation} \\ 15 + c &= 18 && \text{Replace } b \text{ with 15.} \\ c &= 3 \end{aligned}$$

The speed of the boat in still water is 15 mph, and the rate of the current is 3 mph.

When we model the conditions of a problem with a system of equations, we may learn that the conditions cannot be met.

EXAMPLE 5

Inventory of New and Used Vehicles

An auto dealer sells new and used cars and trucks. He is planning his inventory for next month. His plan is to double the total number of cars and have a total of 190 cars and trucks. He also wants 25% of his cars and 25% of his trucks to be used, and the total of used cars and trucks is to be 42. How many cars and trucks will he have in his inventory?

Solution

Let $x =$ the current total number of new and used cars and
$y =$ the current total number of new and used trucks.

If the dealer doubles the number of cars, his total number of vehicles can be described by

$$2x + y = 190$$

The number of used cars and trucks can be described by

$$0.25(2x) + 0.25y = 42$$

We solve each equation for y.

$$\begin{aligned} 2x + y &= 190 \\ y &= -2x + 190 \end{aligned} \qquad \begin{aligned} 0.25(2x) + 0.25y &= 42 \\ 0.50x + 0.25y &= 42 \\ 0.25y &= -0.50x + 42 \\ y &= -2x + 168 \end{aligned}$$

From the equations in the form $y = mx + b$, we can see that the slopes of their graphs are the same and the y-intercepts are different. Therefore, the graphs will be parallel. This system of equations has no solution, and the auto dealer cannot carry out his plan.

Help students to understand the difference between solving a system and solving a problem. In this example, the system has no solution, but the problem has an answer. The answer is that the conditions cannot be met.

Think About It

Suppose that in Example 5, the dealer has a total of 168 cars and trucks. How many cars and how many trucks will he have in his inventory?

3 × 3 Systems

Example 6 shows how an application involving three unknowns can be modeled with a system of three linear equations in three variables.

EXAMPLE 6

Furniture Production Schedule

The Wood Shop makes wooden porch furniture. Rockers require 5 hours of sawing time, 3 hours of assembly time, and 3 hours of finishing time. Chairs require 4 hours of sawing time, 2.5 hours of assembly time, and 2 hours of finishing time. Tables require 3 hours of sawing time, 2 hours of assembly time, and 2 hours of finishing time. The shop employs four people full-time to saw, two full-time and one half-time for assembly, and two full-time and one part-time (10 hours per week) for finishing. Assume that full time is 40 hours per week. How many rockers, chairs, and tables can The Wood Shop produce per week?

Solution

Let r = the number of rockers produced per week,
c = the number of chairs produced per week, and
t = the number of tables produced per week.

Four full-time workers provide 4(40) = 160 hours available for sawing. Two full-time workers and one half-time worker provide 2.5(40) = 100 hours available for assembly. Two full-time workers and one part-time (one-fourth of full time) worker provide 2.25(40) = 90 hours available for finishing.

	Hours Required			
	Rocker	**Chair**	**Table**	**Available Hours**
Sawing	5	4	3	160
Assembly	3	2.5	2	100
Finishing	3	2	2	90

The system of equations is as follows.

$$5r + 4c + 3t = 160 \quad (1)$$
$$3r + 2.5c + 2t = 100 \quad (2)$$
$$3r + 2c + 2t = 90 \quad (3)$$

$$\begin{array}{rl} 3r + 2.5c + 2t = 100 & (2) \\ -3r - 2c - 2t = -90 & \text{Multiply (3) by } -1. \\ \hline 0.5c = 10 & \\ c = 20 & \end{array}$$

Now we can replace c with 20 in equations (1) and (3) to obtain equations (4) and (5), respectively.

$$\begin{array}{lll} 5r + 3t = 80 \quad (4) & \rightarrow \quad 10r + 6t = 160 & \text{Multiply (4) by 2.} \\ 3r + 2t = 50 \quad (5) & \rightarrow \quad -9r - 6t = -150 & \text{Multiply (5) by } -3. \\ & \hline \quad r = 10 & \end{array}$$

$$30 + 2t = 50 \quad \text{Replace } r \text{ with 10 in (5).}$$
$$t = 10$$

The shop can produce 10 rockers, 20 chairs, and 10 tables each week.

The first four problems in the following groups lead to 2 × 2 systems, and the last two in each group lead to 3 × 3 systems.

Exercises 5.6

Real-Life Applications

Mixtures

1. A candy store offers a mixture of candies for $3.25 per pound. A 20-pound mixture is produced by mixing assorted creams worth $2.95 per pound with chocolate-covered almonds worth $3.70 per pound. How many pounds of assorted creams are in the 20-pound mixture?

2. A caterer is able to purchase nut mixtures from a wholesale company. Preparing for a convention, the caterer bought a 45-pound nut mixture for $5.40 per pound. The mixture was produced by mixing cashews worth $6.00 per pound with pecans worth $5.00 per pound. How many pounds of each kind of nut were used to make the mixture?

3. A Kroger's shopper bought a total of 8 pounds of fruit for $6.12. The apples cost $0.45 per pound, and the grapes cost $1.29 per pound. How many pounds of apples did the shopper buy?

4. A landscaper specializes in azaleas and rhododendrons. A commercial customer has a budget of $2250.00 and wants the landscaper to plant a mixture of 100 azaleas and rhododendrons around a new office building. Including planting costs, the landscaper charges $40.00 per rhododendron and $15.00 per azalea. How many of each shrub can the landscaper plant?

5. During a bake sale, an organization sold all of its cakes and cookies and three-fourths of its pies. The cakes were priced at $3.50 each, the pies at $4.00 each, and the cookies at $1.40 per dozen. A sum of $361.00 was earned at the sale. How many baked items were donated if there were six more pies than cakes and if 128 items were sold? (A dozen cookies represent one baked item.)

6. A caterer made 30 pounds of a party mix to sell for $4.05 a pound. Peanuts selling for $2.95 a pound, cashews selling for $6.25 a pound, and coated chocolates selling for $3.50 a pound were all used in the party mix. If two more pounds of peanuts than of cashews were used, how many pounds of each item were in the party mix?

Tickets

7. Arthur Miller's *Death of a Salesman* was the spring production at a certain college. The opening of the play was attended by 456 people. Patron's tickets cost $2, and all other tickets cost $3. If the total box office receipts were $1131, how many of each kind of ticket were sold?

8. A college basketball game was attended by 1243 people. Student tickets cost $2 and all other tickets cost $8. If the total receipts were $5018, how many students and how many nonstudents attended the game?

9. The attendance at an Atlanta Braves and Los Angeles Dodgers game was 42,245. All seats were either box seats costing $30 or reserved seats costing $20. The total receipts for the game were $987,510. How many box seats were sold?

10. Tickets for a charity ball were $10 apiece or $15 per couple. If 802 people attended the dance and a total of $6530 was collected from ticket sales, how many couples attended?

11. The attendance at an opera was 5500. Orchestra seats were $65 each, loge seats were $50 each, and balcony seats were $30 each. The total receipts for the performance were $246,750. If there were 100 more loge seats than orchestra seats sold, how many of each type of seat were sold?

12 At a state fair the total receipts for one evening were $17,000 for the 9000 people who attended. The entrance tickets were $3.00 for each adult, $1.50 for each student, and $1.00 for each child under 10 years old. If there were twice as many students as children under 10, how many adults attended the state fair that evening?

Investments

13. A financial adviser recommended that her client's $10,000 investment be diversified, part at 8.5% simple interest and the remainder at 6.9% simple interest. How much did her client invest at each rate if the total interest income at the end of 1 year was $745.84?

14. A lottery winner invested $11,000, part at 6.5% simple interest and the remainder at 9% simple interest. How much was invested at each rate if the total interest income at the end of 1 year was $830.85?

15. A used-car buyer borrowed $7000 from two sources. One loan was to be repaid at 6% simple interest and the other loan at 7% simple interest. If the borrower paid a total of $452 in interest for the year, how much was borrowed at each rate?

16. Newlyweds purchased kitchen appliances for \$8500. They borrowed the money, part at 9.5% and the remainder at 8% simple interest. If the couple owed a total of \$741.50 in interest at the end of the year, how much did they borrow at each rate?

17. On behalf of a client, a stockbroker invested a total of \$16,500 in a stock fund, a bond fund, and a mutual fund with yields of 6%, 7.5%, and 8% simple interest, respectively. At the end of 1 year, the total income from all three funds was \$1191.50. If the broker invested \$1200 less in the mutual fund than twice the amount that was invested in the bond fund, how much was invested in each fund?

18. A retiree placed his \$60,000 lump-sum pension distribution into three investments, each paying simple interest. He placed some of the money in a high-risk growth fund that paid 12%, some in a GNMA income fund that paid 6.4%, and the rest in a conservative money market account that paid 3.7%. The high-risk investment was \$1000 less than half the money market investment. If the total interest earned at the end of one year was \$3806, how much money was invested in each account?

Liquid Solutions

19. Suppose that a 20-ounce acid solution has a concentration of 10% acid. Explain how to determine the number of ounces of acid in the solution.

20. Suppose that each of three beakers contains 10 ounces of an alcohol solution. The concentrations are 40%, 1%, and 2%. If the three solutions are mixed together, would you expect the resulting alcohol concentration to be less than or greater than 20%? Why?

21. A brick mason has a 34% solution of muriatic acid and a 55% solution of muriatic acid. How much of each solution should be mixed together to produce 70 liters of a 40% acid solution?

22. An exterminator needed 26 liters of a 49% insecticide solution. To obtain this, he mixed solution A, which had a 34% concentration, with solution B, which had a 60% concentration. How much of each solution was required?

23. A water purification research facility needed 600 gallons of a 2.5% chlorine solution. How much 3.5% chlorine solution had to be added to a 2% chlorine solution to obtain the required result?

24. A nurse needs 15 liters of a 40% alcohol solution. A solution of 20% alcohol and another solution of 70% alcohol are available. How much of each solution should be mixed?

25. An infirmary has solutions of 10%, 20%, and 50% alcohol content. A nurse needs 18 liters of a 30% alcohol solution. How much of each solution is needed if twice as much of the 50% alcohol solution as the 20% alcohol solution is to be used?

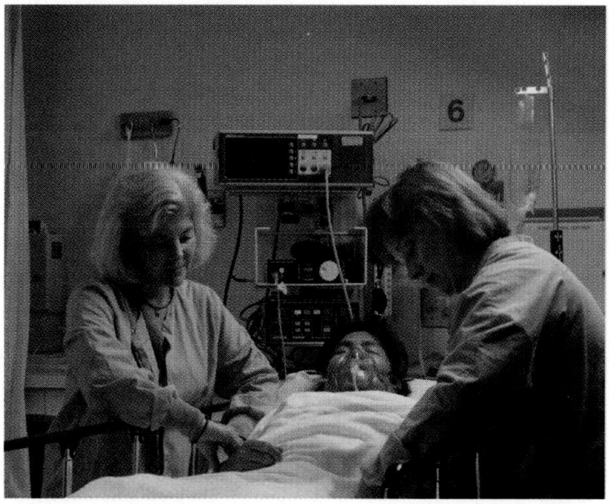

26. A chemist is analyzing three acid solutions labeled A, B, and C. Each solution was formed by mixing together two of three other acid solutions labeled 1, 2, and 3. The following table summarizes the amounts of solutions 1, 2, and 3 that were used to make solutions A, B, and C. All amounts are measured in cubic centimeters. The numbers in parentheses indicate the acid concentrations in solutions A, B, and C.

	Amount of Solution 1	Amount of Solution 2	Amount of Solution 3
A (28%)	2	8	0
B (31%)	0	12	3
C (46%)	9	0	6

What are the acid concentrations in solutions 1, 2, and 3?

Distance, Rate, and Time

27. If the still-water rate of a boat is r and the rate of the current is c, then the downstream rate of the boat is $r + c$. Explain why this is so.

28. Suppose that a boater travels upstream 5 miles and then returns downstream to the starting point. If the river has a current and the time was the same for both trips, compare the rates for the two trips.

29. A man paddles his canoe 21 miles upstream in 7 hours. Then he paddles downstream to his starting point in 3 hours. What is the speed of the canoe in still water and what is the speed of the current?

30. A motor boat travels 137.5 miles upstream in 12.5 hours. The next day, it travels downstream to its starting point in 5.5 hours. What is the speed of the boat in still water and what is the speed of the current?

31. The distance from Cincinnati to Dublin is 3360 miles. Suppose that an airplane made a flight with the wind in 7 hours and returned against the wind in 8 hours. What was the speed of the airplane in calm air and what was the speed of the wind?

32. An airplane flew from Minneapolis to San Juan, a distance of 2860 miles. Because there was a tail wind, the flight took 5.5 hours. There was a head wind on the return flight, so the trip took 6.5 hours. What was the speed of the airplane in calm air and what was the speed of the wind?

33. A vacationer paddled his canoe downstream in 2 hours and paddled upstream back to his starting point in 6 hours. At noon the next day, he decided to drift downstream. At 5:00 P.M. he fell asleep. He woke up at 7:00 P.M. to discover that he had drifted 2 miles past his turn-around point of the previous day. Determine the rate of the current and the rate of the canoe in still water.

34. An athlete who is training for the triathlon has discovered that her rates seem to be staying the same. On Monday she jogged for 30 minutes and rode the bike for 30 minutes and covered 21 miles. On Tuesday she rode the bike for one-half hour and swam for one hour and covered 18.4 miles. On Wednesday she jogged for 15 minutes and biked for 45 minutes and covered 25.5 miles. What was her rate for each sport?

Miscellaneous Applications

35. When a father and his son play golf, the father agrees to pay twice as much as his son pays. The total cost of one round of golf is $51. How much does each pay for one round of golf?

36. In *The Joy of Cooking,* a torte recipe calls for three times as much flour as sugar. If 6 cups of the mixture are to be used, how many cups of flour are needed?

37. At the close of business, a store owner had $505 in five- and ten-dollar bills in the cash register. If there was a total of 70 of these bills, how many of each denomination were there?

38. A biology class had 5 more students than a French class. After four more students enrolled in each class, the French class had three-fourths as many students as the biology class. Determine the original enrollment in each class.

39. Your building-supply store stocks two brands of electric drills. The retail price of the heavy-duty drill is $46.50, and the price of the homeowner's drill is $18.95. You have hired a part-time employee to help with your annual inventory. You neglect to tell the employee that you need to know how many of each kind of drill you have in your inventory. Therefore, one line of your inventory report is

Item	*Quantity*	*Retail Value*
Drills	31	$918.05

From this report, determine how many of each kind of drill you have.

40. A city school system wants to ensure a balance of minority students and teachers in the schools. At Hamilton Middle School, there is a combined total of 435 students and faculty of whom 167 are minorities. If 40% of the students and 20% of the faculty are minorities, how many students attend the school?

41. A paint store has a total of 75 one-gallon cans of paint with high-gloss enamel selling at $19.95 per gallon and flat latex selling at $10.95 per gallon. If the total value of the one-gallon cans is $1109.25, how many of each type are in stock?

42. The owner of a precious metals shop wishes to produce 69.8 ounces of an alloy that is 25% gold. Some 20% gold alloy and some 28% gold alloy are available. How many ounces of each alloy should be melted and mixed to produce the desired alloy?

43. A florist sells orchid corsages for $23 each and carnation corsages for $14 each. If the total receipts for 95 corsages were $1663, how many of each type of corsage were sold?

44. At a concession stand, students sell candy bars for 75 cents each and popcorn for $1.25 a box. If the total receipts for 365 items were $352.75, how many of each item did they sell?

45. At a local election, 536 of the 700 voters voted in favor of a bond referendum. If 88% of the Democrats and 72% of the Republicans supported the referendum, how many Republicans voted against it?

46. A computer programmer has produced two programs, one in the BASIC language and one in the Pascal language. Together, the programs total 950 lines of code, of which a total of 15 lines contain errors. If 1% of the BASIC program lines contain errors, and 2% of the Pascal program lines contain errors, how many lines of code were written in Pascal?

Exercises 47–54 involve three variables.

47. A hiking club of 40 members climbed a mountain on three different trails. Part of the group climbed an 8-mile trail, another part climbed a rougher 6-mile trail, and the third group climbed the steepest trail of 5.5 miles. The second group had two more people than the third group. If the members of the club walked a total of 280 person-miles, how many hikers climbed each trail? (*Note:* If a group of 25 people walks 10 miles, the group has walked a total of $10 \cdot 25$ or 250 person-miles.)

48. The church secretary, the organist, and a volunteer addressed envelopes for a bulk mailing. The organist can address 120 envelopes per hour, the volunteer 116 per hour, and the secretary 140 per hour. They worked a total of 6 person-hours and addressed a total of 740 envelopes. The volunteer worked 1 hour longer than the secretary. How long did each work?

49. At a certain college, A's are worth 4 points, B's are worth 3 points, and C's are worth 2 points. A student's *quality points* are determined by multiplying the number of credit hours for each course times the point value of the grade earned in the course. One semester, a student earned 52 quality points for 17 credit hours of course work. The number of credit hours in which the student received B's was twice the number of credit hours in which the student received C's. If none of the grades was below C, what was the number of credit hours the student earned for each grade?

50. The sum of four digits on a license plate is 18. The first and last digits are the same, and the sum of the first and the second digits is 10. The second digit is twice the third. What is the number on the license plate?

51. A bookshelf 4 feet in length holds exactly three sets of books totaling 35 volumes. The books in the first set are 1 inch thick, those in the second set are 1.5 inches thick, and those in the third set are 2 inches thick. If the number of books in the first set is one more than the number in the second set, how many volumes are in each set?

52. The perimeter of a triangle is 40 inches. The length of the shortest side is 4 inches more than the positive difference between the lengths of the other two sides. The length of the longest side is 4 inches less than the sum of the lengths of the other two sides. Which of the three sides can you determine from this information? What can you say about the other two sides?

53. The following shows three orders placed at a fast-food restaurant.

Order	*Cost*
2 hamburgers, 3 fries, 2 shakes	\$ 9.75
4 hamburgers, 2 shakes	\$10.80
5 fries, 3 shakes	\$ 9.25

What is the price of each item?

54. Suppose that your hardware store carries 4-penny, 6-penny, and 8-penny nails and you have a total of 85 pounds in your inventory. The nails sell for \$3.50 per pound, \$4.00 per pound, and \$4.50 per pound, respectively, and the total retail value of your inventory is \$350. How many more pounds of 8-penny nails than of 4-penny nails do you have in stock?

Modeling with Real Data

55. The table shows, for 1998 and 1999, what percentages of the stories covered on local TV news were crime and human interest stories. (Source: Project for Excellence in Journalism.)

	1998	1999
Crime	28%	22%
Human interest	5%	11%

(a) Let x represent the number of years since 1998, and write linear functions $C(x)$ and $H(x)$ to model the data.

(b) Treat these two equations as a system of equations and solve the system.

(c) Interpret the solution.

56. The accompanying table shows the cost of college as a percent of the median family income for the given year. (Source: U.S. Department of Education.) For example, the median income in 2000 was \$53,350, and the cost of attending a public college was 20.2% of that income.

	Percent of Median Income Required to Pay for College	
Year	Public	Private
1979	9.1%	20.7%
2000	20.2%	40.0%

(a) Let x represent the number of years since 1974, and write linear equations to model the percentages for public and private schools.

(b) Treating your two equations in part (a) as a system of equations, how can you tell, just by looking at the equations, that the system has no solution for years beyond 1974?

(c) If the model equations were to remain valid indefinitely, by what year would attending a private college cost half of the median family income?

Data Analysis: *Small Business Loans*

One function of the Small Business Administration (SBA) is to make or guarantee loans to minority business owners. The accompanying bar graph shows the trend in such loans during the period 1988–1996. (Source: Small Business Administration.)

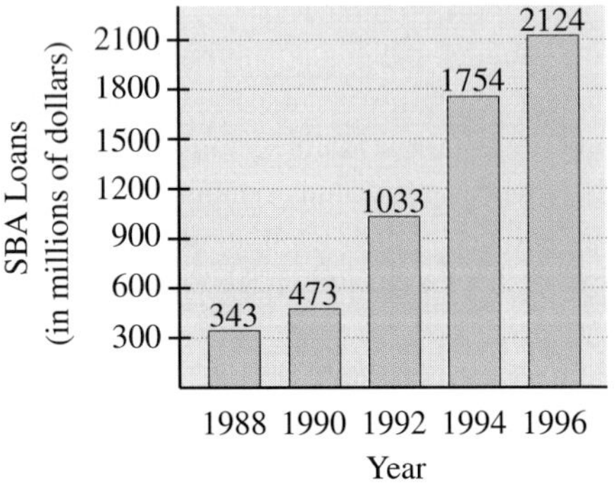

We can model the data in the bar graph with an equation of the form $y = ax^2 + bx + c$.

57. For the data given for 1988 ($x = 2$), 1992 ($x = 6$), and 1996 ($x = 10$), replace y with the amount of the loan and x with the year number. Write the result as a system of equations in three variables a, b, and c.

58. Use the matrix method to solve the system of equations in Exercise 57. Then write the equation that models the three data points by replacing a, b, and c in the equation $y = ax^2 + bx + c$ with their respective values.

59. For the years for which data are given, make a table of values for the model function that you wrote in Exercise 58.

60. Judging on the basis of your table of values in Exercise 59 and the given data, how well does the equation appear to model the data?

Challenge

61. A two-cycle engine requires an oil and gas mixture that is 2.5% oil. A man mixed 6 gallons of gas and oil but then discovered he had mistakenly made a 2% mixture. How many *ounces* of oil does he need to add to the mixture to correct the problem? (There are 128 ounces in a gallon.)

62. On a pan balance, two baseballs and three softballs balance with one softball and five baseballs. In another experiment, two softballs and one baseball balance with four baseballs. How many softballs are needed to balance with three baseballs?

63. A track inspector and a crew chief were standing in a tunnel whose length is 0.6 mile. They departed in opposite directions to leave the tunnel before the next train came through. Both exited their ends of the tunnel in 3 minutes, and the inspector reached the end of the tunnel just as the train arrived. The train passed the crew chief, who had walked 0.12 mile beyond the other end of the tunnel. If the train was traveling at 30 mph, how fast did each person walk?

64. A woman is applying to be a salesperson for Ace Auto Company. She must decide whether she wants to have a base salary of $20,000 per year with a 12% commission on annual sales or to have a base salary of $30,000 per year with an 8% commission on annual sales. If x represents the annual sales and y_1 and y_2 represent the first and second annual earnings options, respectively, write and solve a system of equations.

At what level of annual sales will the annual earnings be the same? For what range of sales should the salesperson choose the first option? For what range of annual sales should she choose the second option?

5.7 Systems of Linear Inequalities

Systems of Inequalities • Real-Life Applications

Systems of Inequalities

When a problem involving inequalities requires that two or more conditions be satisfied, a system of inequalities is needed.

Definition of a System of Inequalities, Solution, and Solution Set

Two or more inequalities considered simultaneously form a **system of inequalities.** A **solution** of a system of inequalities in two variables is an ordered pair that satisfies each inequality. The **solution set** is the set of all solutions.

In this section we consider only systems of linear inequalities in two variables. For brevity, we refer to such systems simply as *systems of inequalities.*

The inequalities of a system are implicitly joined by the word *and.* Therefore, a system of inequalities is an instance of a compound inequality, a concept that was introduced in Section 4.5.

EXAMPLE 1 Verifying Solutions of a System of Inequalities

Determine whether each ordered pair is a solution of the given system of inequalities.

$$x - 2y < 5$$
$$y \geq 3 - x$$

(a) $(6, 11)$ (b) $(-5, 3)$

Solution

(a) Replace x with 6 and y with 11.

$x - 2y < 5$		$y \geq 3 - x$	
$6 - 2(11) < 5$		$11 \geq 3 - 6$	
$-16 < 5$	True	$11 \geq -3$	True

Because $(6, 11)$ satisfies both inequalities, it is a solution of the system.

(b) Replace x with -5 and y with 3.

$x - 2y < 5$		$y \geq 3 - x$	
$-5 - 2(3) < 5$		$3 \geq 3 - (-5)$	
$-11 < 5$	True	$3 \geq 8$	False

Although $(-5, 3)$ satisfies the first inequality, it does not satisfy the second inequality. Therefore, $(-5, 3)$ is not a solution of the system.

A graph of a system of inequalities is a picture of all the solutions of the system. Because a solution of a system of inequalities satisfies each inequality, the graph of the solution set of the system is the intersection of the graphs of the inequalities.

Graphing the Solution Set of a System of Inequalities

1. Graph each inequality.
2. Determine the *intersection* of the solution sets.

EXAMPLE 2 The Graph of a System of Inequalities

Graph the following system of inequalities.

$$3x - y \leq 9$$
$$y + 6 > -x$$

Solution

To sketch the graphs, we solve each inequality for y.

$$y \geq 3x - 9$$
$$y > -x - 6$$

The graphs of these inequalities are the half-planes shown in Figures 5.13 and 5.14.

You may need to remind students how to determine which half-plane to shade.

LEARNING TIP

After graphing the solution set, test a couple of points inside and outside the shaded region. It is impossible to check every solution, but checking a few may expose errors.

Think About It

Can the entire coordinate plane represent the solution set of a system of linear inequalities?

Figure 5.13

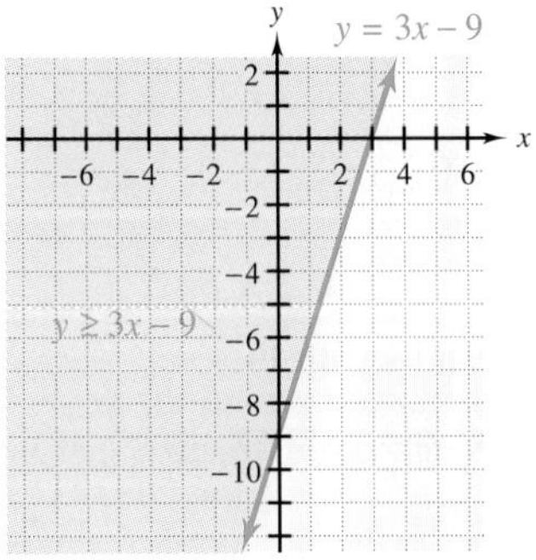

Figure 5.14

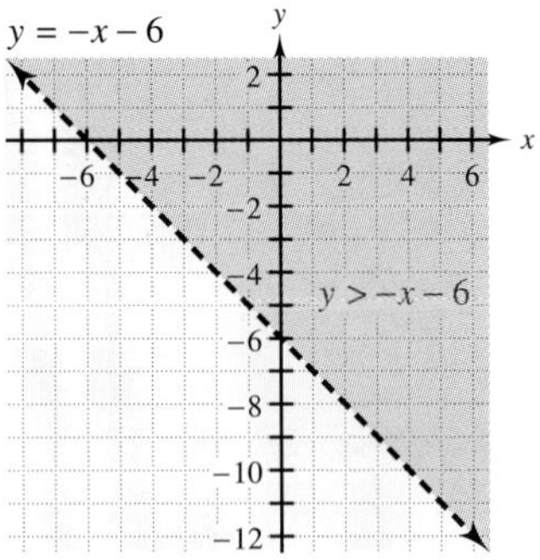

Figure 5.15 shows the two half-planes in the same coordinate system. Their intersection is shown in Figure 5.16.

Figure 5.15

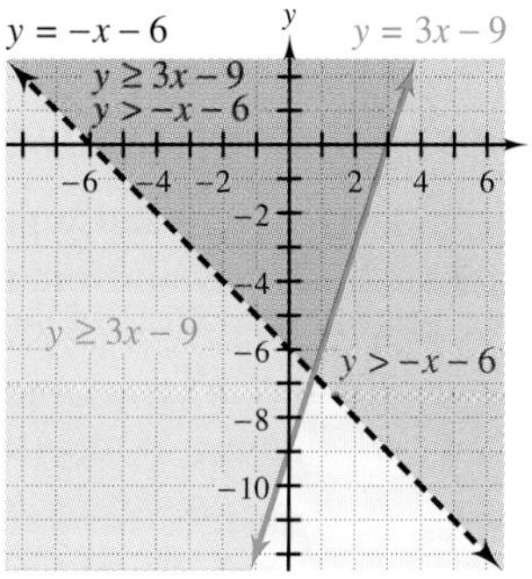

Figure 5.16

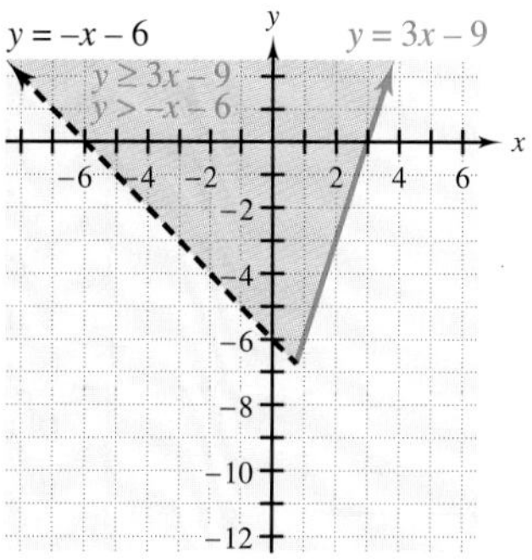

EXAMPLE 3

The Graph of a System of Inequalities

Graph the following system of inequalities.

$$y \le x - 7$$
$$x \le y + 3$$

Figure 5.17

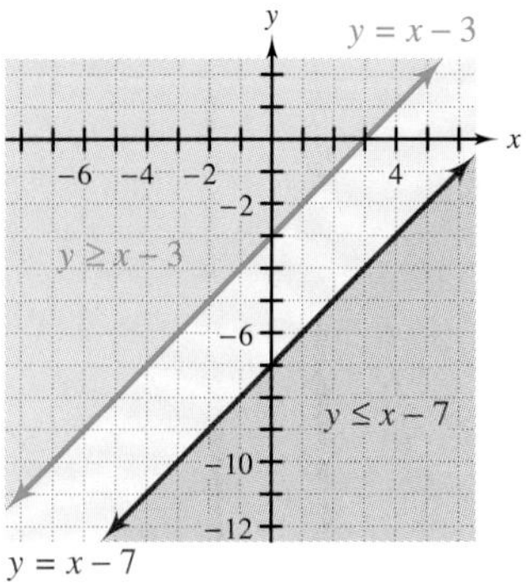

Solution

To sketch the graphs, we solve each inequality for y.

$$y \le x - 7$$
$$y \ge x - 3$$

The graph of each inequality is shown in Figure 5.17. Because the graphs do not intersect, the system of inequalities has no solution.

EXAMPLE 4

A Graph of a System of Three Inequalities

Graph the following system of inequalities.

$$x - y < 3$$
$$2x + y > 1$$
$$x - 4 \le 0$$

Figure 5.18

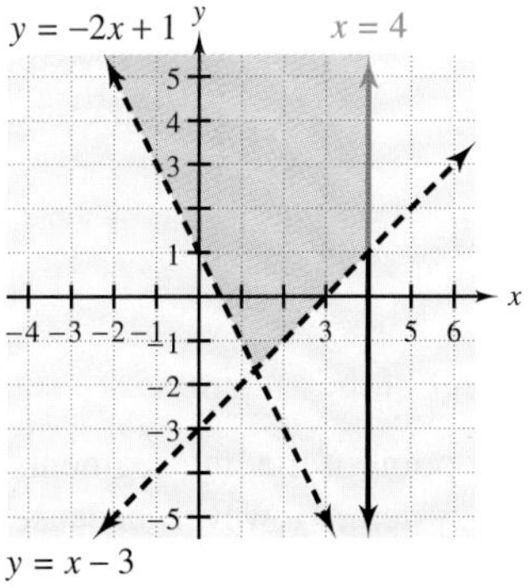

Solution

To sketch the graphs, we solve the first two inequalities for y and the third inequality for x.

$x - y < 3$	$2x + y > 1$	$x - 4 \le 0$
$y > x - 3$	$y > -2x + 1$	$x \le 4$

The graph of each inequality is a half-plane. The solution set of the system of inequalities is the intersection of the three half-planes. (See Fig. 5.18.)

Real-Life Applications

Application problems may require two or more inequalities to describe the conditions of the problem.

EXAMPLE 5

Feasible Production of Recycled Products

A recycling company collects type 1 and type 2 plastics for the manufacture of recycling bins and litter boxes. A recycling bin requires 5 units of type 1 plastic and 1 unit of type 2 plastic. A litter box requires 3 units of type 1 plastic and 2 units of type 2 plastic. If there are 75 units of type 1 plastic available and 36 units of type 2 plastic available, describe the number of recycling bins and litter boxes that can be manufactured.

Solution

The given information is summarized in the following table.

	Units Required	
	Type 1	Type 2
Recycling Bins	5	1
Litter Boxes	3	2
Available	75	36

Let x = the number of recycling bins manufactured and
y = the number of litter boxes manufactured.

The amount of plastic available gives two inequalities.

$$5x + 3y \le 75, \quad \text{or} \quad y \le -\frac{5}{3}x + 25$$

$$x + 2y \le 36, \quad \text{or} \quad y \le -\frac{1}{2}x + 18$$

Because it is impossible to make a negative number of items, each variable must be nonnegative. Thus the complete system of inequalities is

$$y \le -\frac{5}{3}x + 25$$

$$y \le -\frac{1}{2}x + 18$$

$$x \ge 0$$

$$y \ge 0$$

The shaded region in Figure 5.19 represents the solution set of the system.

Figure 5.19

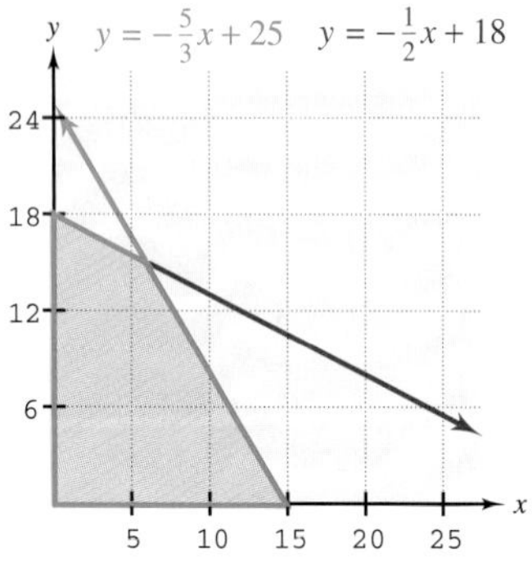

Advise students to be aware of such implicit inequalities in application problems.

Note that every point in the shaded area, as well as every point of the line segments that bound the shaded area, represents a solution of the system. However, because of the nature of the problem, we can consider only those points whose coordinates are integers.

Quick Reference 5.7

Systems of Inequalities

- A **system of inequalities** consists of two or more inequalities considered simultaneously. In this section we consider systems of linear inequalities in two variables.
- A **solution** of a system of inequalities is a pair of numbers that satisfies each inequality of the system.
- To graph the solution set of a system of inequalities,
  1. graph each inequality and
  2. determine the intersection of the solution sets.

Speaking the Language 5.7

1. Two or more inequalities considered simultaneously form a(n) ________.
2. The solution set of a linear inequality in two variables is a half-plane. The solution set of a system of linear inequalities in two variables is the ________ of the half-planes.
3. Because the inequalities of a system are implicitly joined by the word *and,* the system is a(n) ________.
4. If the variables in a system of inequalities represent positive numbers, then the solution set of the system is represented by points that lie entirely in the ________ quadrant.

Exercises 5.7

Concepts and Skills

1. Use the word *half-plane* to describe the graph of the solution set of a system of linear inequalities.
2. Why is a system of inequalities a conjunction?

In Exercises 3–14, graph the solution set of the given system of linear inequalities.

3. $x \geq 3$
$y \geq 2$

4. $y > 1$
$x < -2$

5. $16 \geq 3x - 4y$
$y < 6$

6. $5y + 3x > -20$
$x \geq -2$

7. $3y - 4x < 15$
$y > 0$

8. $x \geq 0$
$x + 3y \leq 12$

9. $y \geq \frac{1}{2}x$
$y \leq 3x$

10. $y < 2x - 5$
$y > -\frac{2}{3}x - 5$

11. $6 \leq 2x + 3y$
$2y - 5x \geq 14$

12. $x - y > 3$
$2x + y \leq 3$

13. $y - 2x > 7$
$y + 3 < 2x$

14. $2x + 3y \leq -15$
$y \geq -\frac{2}{3}x - 1$

CE In Exercises 15–18, write the system of linear inequalities whose solution set is represented by the shaded region of the graph.

15.

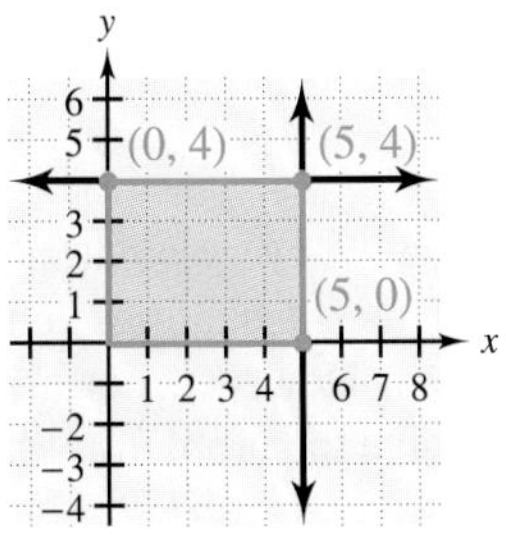

16.

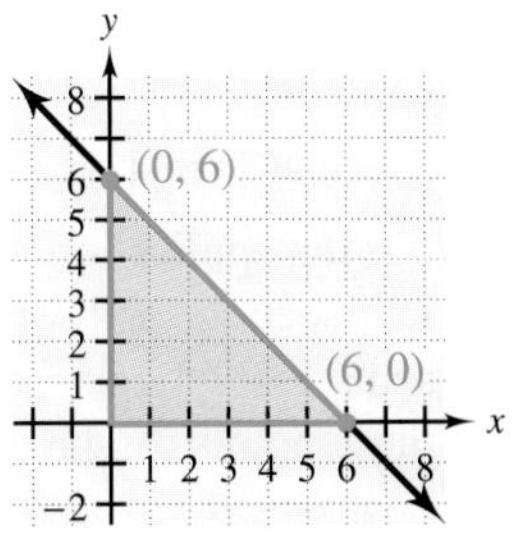

17.

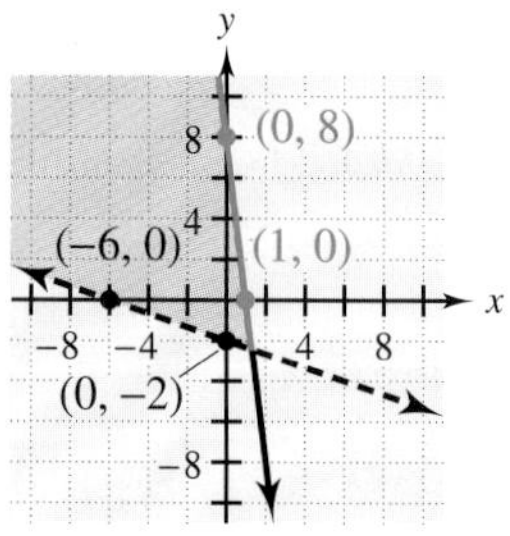

18.

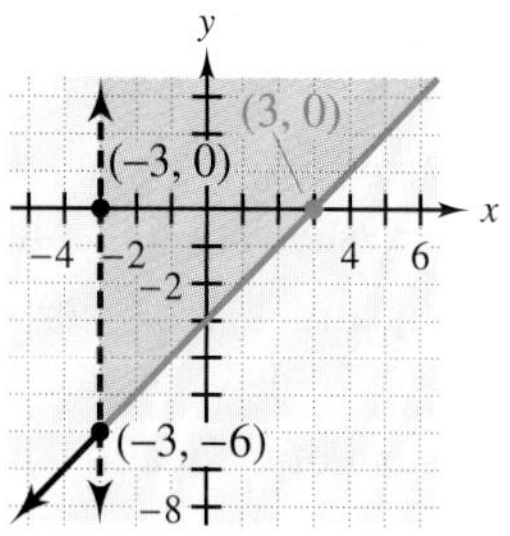

19. Describe the graph of the solution set of the following system of linear inequalities.

$$y \le 5$$
$$y \ge -5$$
$$x \le 5$$
$$x \ge -5$$

20. Suppose you want to determine whether (a, b) is a solution of a system of three inequalities, and you find that the pair does not satisfy the first inequality. Why is it not necessary to test the pair in the other two inequalities?

In Exercises 21–26, graph the given system of linear inequalities.

21. $y + 3x \le 3$, $y + x + 1 \ge 0$, $y < x + 7$

22. $y \le x + 5$, $y - 3x - 2 > 0$, $y + x + 2 \ge 0$

23. $x + 5 \ge 0$, $y - x > 1$, $2x + y < 6$

24. $x + y \le 7$, $2y > x$, $y < 5x$

25. $y - 2x < 1$, $y + x + 2 \ge 0$, $y + 7 \ge 0$

26. $y \ge 0$, $x + 2y \le 8$, $3x - 2y < 4$

27. For an application involving a number x of men and a number y of women, suppose the given conditions can be translated into the following two linear inequalities. State two additional implied inequalities.

$$2x + y \le 30$$
$$x + y \ge 10$$

28. For the application in Exercise 27, every point in the intersection of the graphs of the inequalities represents a solution of the system. However, not all of these points can be used to solve the problem. Why?

In Exercises 29–40, graph the given system of linear inequalities.

29. $2x + 3y \le 18$, $x \ge 0$, $y \ge 0$

30. $3x + y \le 10$, $x \ge 0$, $y \ge 0$

31. $y \ge x$, $y \le 8$, $x \ge 0$, $y \ge 0$

32. $x \le 6$, $y \le 4$, $x \ge 0$, $y \ge 0$

33. $-2x + y \le 7$, $x \le 5$, $x \ge 0$, $y \ge 0$

34. $2x - y \le 5$, $y \le 7$, $x \ge 0$, $y \ge 0$

35. $-x + y \le 1$, $3x + 4y \ge 32$, $x \ge 0$, $y \ge 0$

36. $2x - y \le 7$, $x + y \ge 2$, $x \ge 0$, $y \ge 0$

37. $3x + y \ge 5$, $x + y \ge 3$, $y \ge \frac{1}{2}x$, $x \ge 0$, $y \ge 0$

38. $4x + 3y \ge 18$, $x - y \le 8$, $2x + 3y \ge 12$, $x \ge 0$, $y \ge 0$

39. $-x + y \ge 3$, $x + y \le -2$, $x \ge 0$, $y \ge 0$

40. $x - y \ge 5$, $x + y \le 0$, $x \ge 0$, $y \ge 0$

Exercises 41–46 are precursors to linear programming problems that students might encounter in future courses.

Real-Life Applications

41. A bank invests at most \$10 million either in short-term notes or in mortgages. If the amount in mortgages cannot exceed \$7 million, describe the distribution of funds with a system of inequalities and graph the system. If \$2 million is invested in short-term notes, what is the most that can be invested in mortgages? If \$6 million is invested in short-term notes, what is the most that can be invested in mortgages?

42. A farmer allocates at most 2000 acres of land for either corn or soybeans. At most 40% of the total available acres can be used for soybeans. The number of acres of corn cannot exceed the number of acres of soybeans by more than 700 acres. Write and graph the system of inequalities that describes these conditions. If 1100 acres are in corn, what is the greatest and what is the least number of acres that can be planted in soybeans?

43. A restaurant prepares tuna fruit salads and chicken Caesar salads for luncheon parties. The number of people who request tuna fruit salad never exceeds two-thirds of the number who order chicken Caesar salads. The smallest group that the restaurant books is 50. Write a system of inequalities that describes the number of salads to be made and graph the system. Describe the number of tuna fruit salads that should be prepared if 60 chicken Caesar salads are prepared?

44. A coffee company produces at least 100 pounds of Irish cream coffee and Hawaiian macadamia coffee. The number of pounds of Irish cream cannot exceed twice the number of pounds of Hawaiian macadamia. Describe and graph the number of pounds of each kind of coffee that can be produced. What is the least number of pounds of Hawaiian macadamia coffee that can be produced?

45. A builder specializes in the construction of small barns and tool sheds. The following time-table shows the number of hours needed to frame and roof the buildings and the maximum number of hours available for each kind of construction.

	Hours to Frame	Hours to Roof
Barn	15	7
Tool Shed	5	6
Maximum Hours Available	90	84

(a) Write a system of linear inequalities and draw the graph of the system. Shade the region that represents the possible numbers of barns and tool sheds that can be built under these conditions.

(b) Can the builder construct 4 barns and 5 tool sheds?

(c) Can the builder construct 5 barns and 4 tool sheds?

46. A businesswoman runs a full-service catering business. She prepares food ahead of time, and on the day of the event, she sets up the tables and provides decorations. Her time requirements and constraints are summarized in the following table.

	Food Preparation Hours	Set-Up Hours
Banquet	13	8
Reception	3	5
Maximum Hours Available	39	40

(a) Write a system of linear inequalities and draw the graph of the system. Shade the region showing the possible number of banquets and receptions that can be catered.

(b) Can she cater 3 banquets and 2 receptions?

(c) Can she cater 2 banquets and 3 receptions?

Challenge

47. Refer to Exercise 41. If mortgages yield 7% and short-term notes yield 5%, how should the funds be invested to maximize the return to the bank?

48. Refer to Exercise 42. Suppose the profit on soybeans is 70 cents per bushel and the profit on corn is 40 cents per bushel. If the farmer can produce 35 bushels of soybeans per acre and 150 bushels of corn per acre, what land allocation would produce the greatest profit?

5 • Chapter Project

Office Communications

The following table shows the percentages of office communications conducted by e-mail and by paper for three selected years. (Source: Cognitive Communication Survey.)

	Percent of Office Communications	
Year	**E-mail**	**Paper**
1991	6%	66%
1994	20%	57%
1997	40%	47%

1. For 1994 and 1997, what was the percentage of office communications conducted *either* by e-mail *or* by paper?
2. Assume that all other office communications are conducted orally (telephone, personal contact, meetings). What trend in oral communication can be deduced from your data in Question 1?
3. In Section 5.2 we modeled the percentage of office communications conducted by e-mail with the equation $17x - 3y = 2$, where y represents the percentage and x represents the number of years since 1990. Use this model to predict the year when half of all office communications will be conducted by e-mail.
4. Suppose that an office has 1 million communications events in the year that you estimated in Question 3. Suppose further that 10% of all e-mail is printed and that each e-mail message occupies one sheet of paper. How many reams of paper (500 sheets per ream) would be used for printed e-mail?
5. Given your results in Question 4, do you think that the percentages in the table for paper are an accurate reflection of paper consumption in office communications? Why?
6. What does the model equation in Question 3 predict for the year 2008? How realistic do you think this prediction is?

Chapter 5 Review Exercises

Section 5.1

1. Suppose you draw the graphs of $y_1 = ax + b$ and $y_2 = cx + d$ on the same coordinate axes. If the graphs intersect only at point Q, what is the significance of that point?

2. Suppose you use the substitution method to solve the following system.

$$2x + 3y = 3$$
$$y = 7x + 1$$

What is the first equation in one variable that you would write? Why?

In Exercises 3 and 4, determine whether the given ordered pair is a solution of the system of equations.

3. (2, 1); $3x - 5y = 1$
$y = 2x - 3$

4. (−4, 5); $3x + y = -7$
$2x - 3y = 12$

In Exercises 5–8, use the graphing method to estimate the solution of the given systems of equations.

5. $y = 2x - 5$
$y = -x + 4$

6. $x - 3y = 7$
$3x = 9y + 2$

7. $2x - 5y = 10$
$10y - 4x + 20 = 0$

8. $2x + y + 2 = 0$
$3x + 2y - 1 = 0$

In Exercises 9–12, determine the number of solutions of the system without graphing or solving algebraically.

9. $y = -x + 4$
$y = x - 3$

10. $7x - 6y = 5$
$7x - y = 5(y + 1)$

11. $y = 5$
$x - 2 = 4$

12. $3x = y + 2$
$3y = 9x + 21$

In Exercises 13–18, use the substitution method to solve the given system of equations.

13. $y = x + 1$
$2x + 3y = 6$

14. $x = -2y + 3$
$2y + x = 4$

15. $y = x + 3$
$2x - 2y + 6 = 0$

16. $x - 2y = 6$
$2x + y = 4$

17. $3x + y = 2$
$4x + 8y = 1$

18. $\frac{5}{3}x + \frac{3}{2}y = 1$
$15x + 18y = 6$

Section 5.2

19. Suppose that as you solve a system of equations with the addition method, you obtain $x = \frac{17}{31}$. Explain how you can determine the y-value without using substitution.

20. Suppose that as you solve a system of equations with the addition method, you obtain an equation with no variable. Explain how to interpret this result.

In Exercises 21–28, use the addition method to solve the given system of equations.

21. $3x - 4y = 22$
$2x + 5y = -16$

22. $x - 2y + 4 = 0$
$3x + y + 5 = 0$

23. $y - 3x = 4$
$6x - 2y + 8 = 0$

24. $y = 2x + 3$
$4x - 2y - 6 = 0$

25. $\frac{3}{4}x + \frac{1}{3}y - 4 = 0$
$\frac{1}{2}x = \frac{2}{3}y + 8$

26. $\frac{3}{4}x + \frac{1}{3}y - 3 = 0$
$9x = 36 - 4y$

27. $\frac{1}{5}x - \frac{2}{3}y = 2$
$3x = 10y + 19$

28. $\frac{2}{5}x + \frac{1}{2}y = 2$
$\frac{3}{10}x + \frac{1}{4}y = 0$

In Exercises 29 and 30, determine c so that the given system of equations is inconsistent.

29. $x + 3y = 6$
$4x - cy = 8$

30. $y - 4x = c$
$y - (c + 1)x = -7$

In Exercises 31 and 32, determine c so that the equations of the given system are dependent.

31. $x - 2y = 5$
$cx + 6y + 15 = 0$

32. $\frac{1}{2}x + y - 3 = 0$
$cx + 4y - 12 = 0$

Section 5.3

33. What geometric figure represents the solutions of a linear equation in three variables? What do we call the coordinate system in which such a figure is drawn?

34. Suppose you want to solve the following system of equations.

$$2x + y + 3z = 7$$
$$3x - y + z = 1$$
$$x + 4y - z = 2$$

You can easily eliminate y from the first two equations and z from the last two equations. What is your opinion of this initial approach?

In Exercises 35 and 36, determine whether the given ordered triple is a solution of the given system of equations.

35. $x + y + z = 2$
$2x - 3y + 3z = 2$
$x - 2y + z = 1$
(−2, 1, 3)

36. $3x - z = 3$
$y + 2x - 2 = 0$
$x + y + z = 5$
(2, 0, 3)

In Exercises 37–46, use the addition method to solve the given system of equations.

37. $x + 2y + z = 5$
$2x + y - 2z = -3$
$4x + 3y - z = 0$

38. $y = x + 2$
$x + 2y + 4z = 9$
$x + y - z = 9$

39. $5x - 2y + 7z = -19$
$x + 3y - 2z = 3$
$4x - y - z = 16$

40. $x + 2y + 3z = -10$
$3x + 2y + z = -2$
$x + 3y + 2z = -8$

41. $2y = x + z - 3$
$7x + 5z = 3y + 1$
$2x + 2z + 5 = 4y$

42. $2x - 5y + 7z = -3$
$x + 2y - 4z = 0$
$2x - 5y + 3z = -7$

43. $2x - 2y + 10z = 12$
$x - 2y + 3z = -1$
$4x - 3y + z = -32$

44. $3x + 4z = 2y - 6$
$2(y - x) = 4z + 1$
$2(z - 2) = y - x$

45. $x + y - z = 8$
$y + 2z + 8 = 0$
$x - 2y + 3 = 0$

46. $x - y + z = 9$
$2x + y - z = 6$
$5x + 2y - 2z = -4$

Section 5.4

47. What is the difference between a coefficient matrix and an augmented matrix?

48. Describe the row operation used to transform

$$\begin{bmatrix} 2 & 5 \\ -1 & 3 \end{bmatrix} \text{ into } \begin{bmatrix} 0 & 11 \\ -1 & 3 \end{bmatrix}.$$

In Exercises 49 and 50, write the augmented matrix for the given system of equations.

49. $4x - 3y = 12$
$x - 2y + 3 = 0$

50. $y - 2z - 7 = -11$
$2z - 2x = 5$
$2y - 6x = 1$

In Exercises 51 and 52, write a system of linear equations associated with the given augmented matrix.

51. $\left[\begin{array}{ccc|c} 1 & 2 & -3 & 4 \\ 3 & 1 & -2 & 5 \\ 1 & 1 & 0 & 0 \end{array}\right]$

52. $\left[\begin{array}{cc|c} 5 & -2 & -2 \\ 6 & 0 & 7 \end{array}\right]$

In Exercises 53–62, use the matrix method to solve the given system of equations.

53. $2x - y = 5$
$x - 2y = 4$

54. $y + 9 = 1.2x$
$15 - x = 0$

55. $x - y - 5 = 0$
$2x - y + 17 = 0$

56. $4x - 3y - 6z = 6$
$x - 3y + 6z = -1$
$x - 6y + 12z = -1$

57. $x - y - 2z = -2$
$x + y + 6z = 6$
$y - z = -1$

58. $3x - 4y + 2z = 1$
$-3x + 3y - z = 0$
$-6x + 9y - 3z = 3$

59. $x + 10 = 0$
$y + 0.3x = 6$

60. $4x - 3y = -12$
$5x + 2y = 1$

61. $1.6x + 1.2y - z = 3.2$
$x - 3y - 0.75z = -13$
$0.5x + y + z = 3.5$

62. $x - y = 1$
$x + 3z = 6$
$y - 2z = 0$

Section 5.5

63. Suppose that you apply Cramer's Rule to solve a system of equations and obtain the following expression for x. What can you conclude?

$$x = \frac{1}{\begin{vmatrix} 2 & 4 \\ 3 & 6 \end{vmatrix}}$$

64. The number 13 is associated with the following matrix. What do we call this number and how is it calculated?

$$\begin{bmatrix} 2 & -1 \\ 3 & 5 \end{bmatrix}$$

In Exercises 65 and 66, evaluate the determinant.

65. $\begin{vmatrix} 7 & 20 \\ 3 & 10 \end{vmatrix}$

66. $\begin{vmatrix} 0 & 0 \\ 0 & 1 \end{vmatrix}$

67. For the following matrix, evaluate the minor for the element 6.

$$\begin{bmatrix} 1 & 2 & 3 \\ 4 & 5 & 6 \\ 7 & 8 & 9 \end{bmatrix}$$

68. Evaluate the determinant associated with the matrix in Exercise 67.

In Exercises 69 and 70, use Cramer's Rule to solve the system of equations.

69. $7x + 3y = -4$
$-11x - 13y = -2$

70. $2x - 5y + z = -1$
$x + y - 4z = 13$
$3x - 2y + 5z = -12$

Section 5.6

In Exercises 71–76, assign variables to the unknown quantities and write a system of equations to describe the situation. Use any method to solve the system.

71. A cash register drawer contains \$705 in five- and twenty-dollar bills. If there is a total of 60 bills, how many bills of each denomination are in the drawer?

72. The difference between one number and twice a smaller number is 20. The smaller number plus 10 is one-half the larger number. What are the numbers?

73. A hardware store stocks two types of hammers. The store has a total of 42 hammers, with sledge hammers selling at \$22.95 each and claw hammers selling at \$10.95 each. If the total value of the hammers is \$639.90, how many of each type are in stock?

74. A forestry ranger had to hike from the park office up into the mountains to reach the newly planted seedlings that she had to inspect. She hiked at an average rate of 3.5 mph to reach this area. The ranger then averaged 2.5 mph through the inspection area. On her return to the park office, she again averaged 3.5 mph. The total length of her hike was 17 miles, which she covered in 6 hours. How long was she in the planted area? How far from the park office is the inspection site?

75. The sum of the three digits on a license plate is 12. The sum of the first two digits is 7, and the sum of the last two digits is 8. What is the three-digit number on the license plate?

76. A tree removal service charges \$80 for oak trees, \$40 for sourwood trees, and \$30 for pines. A customer had four times as many pines as oaks, and the number of oaks was one more than the number of sourwoods. If the total cost of removing the customer's trees was \$680, how many of each tree were there?

Section 5.7

In Exercises 77–86, graph the given system of inequalities.

77. $y < 4$
$2x - 5y < 10$

78. $x \geq -3$
$2x + 3y \leq 5$

79. $y < 3x + 1$
$2y + 3x \leq 10$

80. $y + x < -5$
$y - x \geq -5$

81. $y \leq 3x$
$3y \geq x$
$x + y \leq 9$

82. $4y - 3x < 16$
$y > -5$
$6y < 5x + 36$

83. $y > 2$
$y \geq -3$
$x \geq -5$

84. $y - x > 3$
$y < x - 5$
$3x + 7y < 14$

85. $x + y \leq 10$
$x + y \geq 2$
$x \geq 0$
$y \geq 0$

86. $2x - 3y \leq 6$
$y \leq 7$
$x \leq 9$
$x \geq 0$
$y \geq 0$

87. A certain pickup truck can carry no more than 1000 pounds and no more than 20 bags of mortar mix and/or lime. A bag of mortar mix weighs 80 pounds, and a bag of lime weighs 40 pounds. Write a system of inequalities that describes the given conditions and graph the solution set.

Looking Back

The following exercises review concepts and skills that you will need in Chapter 6.

1. Evaluate each exponential expression.
(a) 9^2 (b) $(-2)^5$ (c) 10^4

2. Evaluate the expression $3x^2 - 2x - 7$ for $x = 2$.

3. Let $x = -2$ and $y = 3$. Evaluate $4y^2 + \frac{1}{2}x^3$.

4. Use the commutative and associative properties to rewrite the expression $(3x)(-5y)$ in simpler form.

In Exercises 5 and 6, use the Distributive Property to remove the parentheses.

5. $5(2x - 1)$

6. $-1(4 + y)$

In Exercises 7 and 8, simplify the given expression.

7. $3x + 1 - x + 4$

8. $12x^3 - x^2 + x^3 - 6x^2$

In Exercises 9 and 10, simplify the given expression.

9. $3(t - 2) - t$

10. $(a + 3) - (2a - 1) + 6$

11. Write the product in exponential form.
(a) $x \cdot x \cdot x \cdot x \cdot x$ (b) $(a - 1)(a - 1)$

12. Write the exponential form in factored form.
(a) $3y^2$ (b) $(t + 2)^3$

Chapter 5 Test

1. Use the graphing method to estimate the solution of the following system of equations. Show your verification of the solution by substitution.

$$y = -2x + 3$$
$$x - y = 3$$

In Questions 2 and 3, use the substitution method to solve the system of equations.

2. $x + 2y = 10$
$3x + 4y = 8$

3. $2x + y = 7$
$x - y = 8$

In Questions 4 and 5, use the addition method to solve the system of equations.

4. $x - 2y = 3$
$3x = 6y + 9$

5. $x - \frac{3}{2}y = 13$
$\frac{3}{2}x - y = 17$

In Questions 6 and 7, use the addition method to solve the system of equations.

6.
$$\begin{aligned} x - 2y + 3z &= 7 \\ 2x + y + z &= 4 \\ -3x + 2y - 2z &= -10 \end{aligned}$$

7.
$$\begin{aligned} x + 2y - z &= -3 \\ 2x - 4y + z &= -7 \\ -2x + 2y - 3z &= 4 \end{aligned}$$

In Questions 8 and 9, use matrix methods to solve the system of equations.

8. $3x - 6y = 24$
$5x + 4y = 12$

9.
$$\begin{aligned} x + y - z &= 6 \\ 3x - 2y + z &= -5 \\ x + 3y - 2z &= 14 \end{aligned}$$

In Questions 10–12, classify the given system of equations as a consistent system with a unique solution, an inconsistent system, or a system of dependent equations.

10. $2x + 3y = 5$
$4x + 6y = 10$

11. $2x + y = 2$
$8x + 4y = 7$

12. $x + y = 6$
$x - y = 4$

13. Describe the graph of a system of two linear equations in two variables for each of the following.

(a) The equations are dependent.

(b) The system has a unique solution.

(c) The system is inconsistent.

14. Determine k so that the following equation is true.

$$\begin{vmatrix} 2 & -1 \\ 4 & 6 \end{vmatrix} = \begin{vmatrix} -3 & k \\ -1 & -5 \end{vmatrix}$$

15. Use Cramer's Rule to solve the system of equations.

$$x - 3y = -8$$
$$2x + 7y = -3$$

In Questions 16 and 17, graph the solution set of the system of inequalities.

16. $y < -2x$
$y > -\frac{2}{3}x$

17. $2x + y \le 12$
$y \le 8$
$x \ge 0$
$y \ge 0$

In Questions 18–21, define variables, translate the problem into a system of equations, solve the system of equations, and state the answer to the question.

18. With a tail wind an airplane flies 600 miles in 3 hours, and against a head wind the plane flies the same distance in 4 hours. What are the wind speed and the airspeed of the plane?

19. Two angles are complementary. The measure of one angle is 5° less than twice the measure of the other. What is the measure of each angle?

20. A chemistry lab has two containers of hydrochloric acid. One is a 15% solution and the other a 40% solution. How many liters of each are required to make 100 liters of a 30% solution?

21. For a seafood party, the caterer knows that the number of pounds of shrimp must be half the combined weight of the crabs and lobster. The number of pounds of crabs should be 5 less than the number of pounds of lobster. The cost of shrimp is \$5 per pound, the cost of crab is \$7 per pound, and lobster costs \$9 per pound. How many pounds of each can the caterer purchase for \$960.50?

22. On a vacant lot, the total number of pines and dogwoods does not exceed 20. There are at least twice as many pines as dogwoods. Write a system of inequalities that describes the given conditions, and graph the solution set.

Cumulative Test, Chapters 3–5

1. To solve $5x - 3 = 2x + 9$, we graph $y_1 = 5x - 3$ and $y_2 = 2x + 9$ and note that the point of intersection of the two graphs is (4, 17). Explain the significance of the 4 and 17.

2. Solve each equation algebraically.
   (a) $2(3x - 1) - 2x = -(x - 4)$
   (b) $\frac{1}{3}(x + 1) - \frac{1}{2} = \frac{3}{5}(2x - 3)$

3. Solve the formula $A = \frac{1}{2}bh$ for b.

4. After three tests of equal weight, a student has an 81 average. If the first and last test scores were 80 and 91, what was the second test score?

5. A car and a bicycle are 39 miles apart. The car driver and the cyclist leave at 1:00 P.M. and head directly toward each other. If the driver averages 40 mph and the cyclist averages 12 mph, at what time will they meet?

6. A person invests \$8000 in two funds, one at 5% simple interest and the other at 6.5% simple interest. If the total income for the year is \$475, how much was invested at each rate?

7. Solve each inequality algebraically and graph the solution set on a number line.
   (a) $x - 2(x - 2) < 5x + 8$ (b) $\frac{3 - x}{-4} \geq x$

8. Solve each compound inequality. Write the solution set with interval notation and graph the solution set on a number line.
   (a) $-5x + 1 > 6$ or $2x - 1 \geq 5$
   (b) $4 - x < 1$ and $\frac{1}{3}x < 2$

9. Solve each of the following.
   (a) $|x - 3| < 8$ (b) $2|x + 1| - 3 = 5$
   (c) $-2|3 + x| \leq -4$

10. The length of a rectangle is 3 more than twice the width. If the perimeter of the rectangle does not exceed 42 inches, what is the maximum length of the rectangle?

11. Find the values of a and b so that the ordered pairs are solutions of the given equation.
    $2x + 3y = 7$ $(3, a)$ $(b, 1)$

12. Determine the intercepts of the graph of the equation $5x + y = -10$.

13. For each equation, state whether the slope of the graph is positive, negative, zero, or undefined.
    (a) $x + 2y = 3$ (b) $x + 2 = 3$
    (c) $0 \cdot x + 2y = 3$ (d) $x = 2y + 3$

14. Write $x - 2y + 4 = 0$ in slope–intercept form. Use the slope and y-intercept to graph the equation.

15. Consider the equation $y = \frac{2}{3}x - 1$. Describe the rate of change of y with respect to x.

16. Line L_1 contains the point $P(-4, 3)$. Line L_2 is perpendicular to L_1. If L_1 and L_2 intersect at the point $Q(3, -2)$, what is the slope of L_2?

17. Write an equation of a line that contains the points $A(-2, -5)$ and $B(6, 3)$. Express the equation in the form $Ax + By = C$.

18. The equation of line L_1 is $2x + 3y = 6$. Line L_2 is parallel to L_1 and contains the point $P(1, -4)$. Write the equation of L_2 in slope–intercept form.

19. The graph of $x - y < 7$ is a shaded half-plane.
    (a) Describe the boundary line.
    (b) Describe two methods for determining which half-plane to shade.

20. Draw the graph of $|x - y| \leq 5$.

21. Each of the following is a graph of a system of two linear equations in two variables. From the graph, describe the solution set of the system.

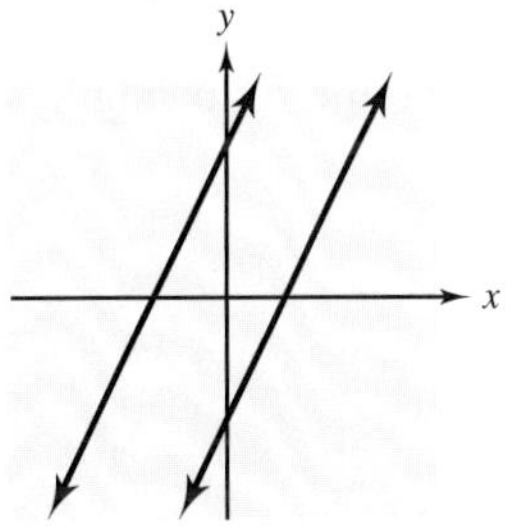

(a)

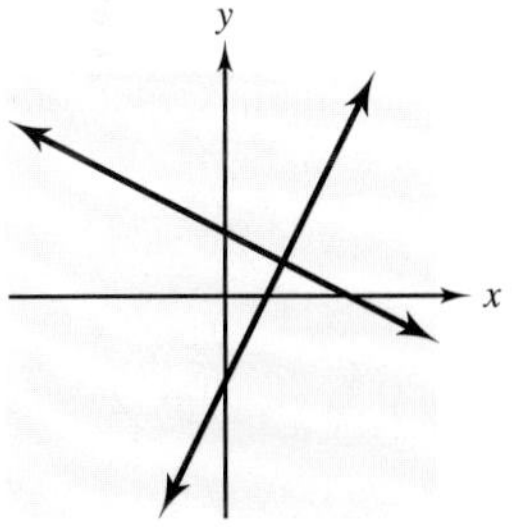

(b)

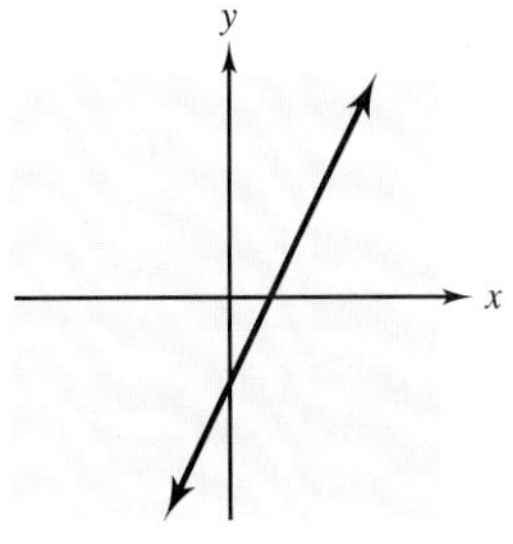

(c)

22. Use the substitution method to solve the following system of equations.

$$\begin{aligned} x + 7y &= 40 \\ 3x - 2y &= -18 \end{aligned}$$

23. Use the addition method to solve each of the following systems. Indicate if a system is inconsistent or if the equations are dependent.

(a) $2x - 3y = -21$

$y = \frac{2}{3}x + 7$

(b) $4x - y = -9$

$2y - x = 11$

(c) $y = 5x + 1$

$x - \frac{y}{5} = 0.2$

24. A college bookstore placed a textbook order for Mathematics 101 and English 122. The mathematics textbooks cost \$35 each, and the English textbooks cost \$28 each. If a total of 420 books were ordered and the total cost was \$13,160, how many of each book were ordered?

In Questions 25 and 26, use the addition method to solve the given system of equations.

25.
$$\begin{aligned} 2x + 3y + 5z &= 18 \\ 3x - 2y + 4z &= 13 \\ 4x - 3y + 3z &= 8 \end{aligned}$$

26.
$$\begin{aligned} x + y + z &= 2 \\ 2x - y + z &= 9 \\ -x + 2y + 2z &= -5 \end{aligned}$$

27. Use matrix methods to solve the following system of equations.

$$\begin{aligned} x - 2y + z &= 13 \\ 2x + y - z &= 5 \\ 3x - y + 2z &= 22 \end{aligned}$$

28. For what value of k is the following equation true?

$$\begin{vmatrix} 2 & -7 \\ k & -2 \end{vmatrix} = -25$$

29. Show why Cramer's Rule *cannot* be used to solve the following system.

$$\begin{aligned} -2x + y + 3z &= 1 \\ x - 2y + z &= 2 \\ 3x - 3y - 2z &= -2 \end{aligned}$$

30. A cyclist rides 32 miles against the wind in 2 hours. The return trip takes 40 minutes less time. What is the wind speed?

31. A cruise ship has a total of 500 living accommodations on decks A, B, and C. For a 4-day cruise, A-deck spaces cost \$1200, B-deck spaces cost \$900, and C-deck spaces cost \$500. The combined number of A- and B-deck spaces equals the number of C-deck spaces. If the total receipts from a sold-out cruise are \$380,000, how many B-deck spaces are there?

32. Graph the solution set of the following system of inequalities.

$$\begin{aligned} x + y &\le 9 \\ 2x - y &\le 0 \\ y &\ge 2 \end{aligned}$$

33. The perimeter of a rectangle cannot exceed 40 inches. The width of the rectangle can be no longer than half the length. Write a system of *all* inequalities that describe these conditions.

Chapter 6

Exponents and Polynomials

In 1996 the American space shuttle docked with the Russian space station Mir. Shannon Lucid, an American astronaut, stayed on Mir, where she set an endurance record in space.

The mathematics of space exploration typically involves very large numbers, which can be represented with ***exponents*** *and* ***scientific notation.*** *Equally large are the costs of NASA programs. The accompanying bar graph shows the NASA budgets proposed by the National Space Society for the years 2000–2003. (For more on this topic, see Exercises 83–86 at the end of Section 6.3, and see the Chapter Project.)*

We begin this chapter with the topic of exponents and their properties. We define the zero exponent and negative exponents, and we learn how to represent numbers with scientific notation. The chapter concludes with the introduction of polynomials and a discussion of polynomial addition, subtraction, and multiplication.

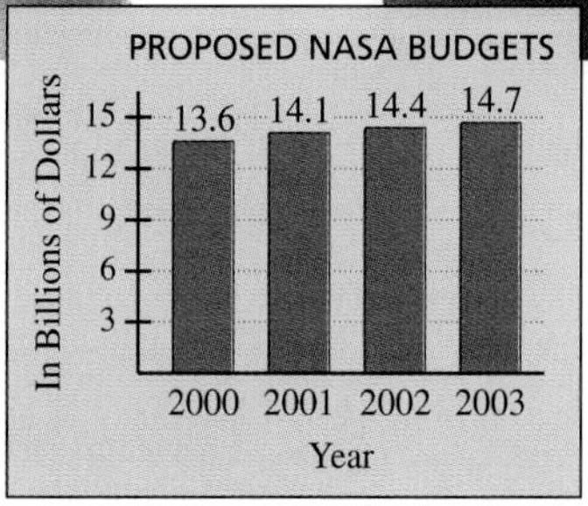

(Source: NASA.)

6.1 Properties of Exponents

Bases • Product Rule for Exponents • Power Rules for Exponents • Quotient Rule for Exponents

Bases

In Section 1.6 we used a positive integer **exponent** to represent repeated multiplication. The exponent indicates the number of times the **base** is used as a factor.

$$7 \cdot 7 \cdot 7 = 7^3 \qquad b^5 = b \cdot b \cdot b \cdot b \cdot b$$

When performing operations with exponents, students seem to benefit from writing in the understood exponent of 1.

In the **exponential expression** 7^3, the base is 7 and the exponent is 3. When no exponent is written, we assume it to be 1: $b = b^1$. When we work with exponential expressions, identifying the base is important.

The distinction between $(-3)^2$ and -3^2 is often difficult for students.

$(-3)^2 = (-3)(-3) = 9$	The base is -3.
$-3^2 = -1 \cdot 3^2 = -1 \cdot 3 \cdot 3 = -9$	The base is 3.
$2x^4 = 2 \cdot x \cdot x \cdot x \cdot x$	The base for the exponent 4 is x.
$(2x)^4 = (2x)(2x)(2x)(2x)$	The base for the exponent 4 is $2x$.

Think About It

What repeated exponential operation would result in exponents in the following sequence?

1, 2, 4, 8, 16, 32, 64, …

Product Rule for Exponents

In the remainder of this section, we will consider rules for performing operations with exponential expressions. In doing so, we will use the fact that *equivalent expressions* have the same value for all permissible values of the variable.

We begin with products of exponential expressions that have the same base. Consider, for example, the product $x^3 \cdot x^5$. Because the positive integer exponents indicate repeated multiplication, we can write this product as follows.

$$x^3 \cdot x^5 = \underbrace{x \cdot x \cdot x}_{x^3} \cdot \underbrace{x \cdot x \cdot x \cdot x \cdot x}_{x^5} = x^8$$

(A total of eight factors of x)

This example illustrates the fact that $x^3 \cdot x^5$ and x^8 are equivalent expressions and leads to the following rule.

> **Product Rule for Exponents**
>
> For any real number b, and for any positive integers m and n,
>
> $$b^m \cdot b^n = b^{m+n}$$
>
> In words, to multiply exponential expressions with like bases, retain the base and add the exponents.

Note: The bases can be numbers, variables, or any expressions, but they must be the same.

EXAMPLE 1

In part (a), students sometimes write 9^{16}.

To emphasize that the bases must be the same, compare part (d) to $(x + 3)^2 \cdot (x - 3)^5$, where the bases are not the same and the Product Rule for Exponents does not apply.

Using the Product Rule for Exponents

(a) $3^7 \cdot 3^9 = 3^{7+9} = 3^{16}$

(b) $x^{18} \cdot x^{13} = x^{18+13} = x^{31}$

(c) $(2x^2y)(5x^5y^4) = (2 \cdot 5)(x^2 \cdot x^5)(y^1 \cdot y^4) = 10x^{2+5}y^{1+4} = 10x^7y^5$

(d) $(x + 3)^2 \cdot (x + 3)^5 = (x + 3)^{2+5} = (x + 3)^7$

Note: For products such as $x \cdot x^4$, writing the understood exponent 1 helps you to apply the Product Rule for Exponents properly: $x \cdot x^4 = x^1 \cdot x^4 = x^{1+4} = x^5$.

Power Rules for Exponents

The expression $(x^5)^3$ is an example of an exponential expression raised to a power. We can use the Product Rule for Exponents to simplify this expression.

$$(x^5)^3 = x^5 \cdot x^5 \cdot x^5 \quad \text{The base is } x^5 \text{ and the exponent is 3.}$$
$$= x^{5+5+5} \quad \text{Product Rule for Exponents}$$
$$= x^{15}$$

Observe that the product of the exponents in the first expression is the exponent in the result.

Power to a Power Rule

For any real number b and positive integers m and n,

$$(b^m)^n = b^{mn}$$

In words, to raise an exponential expression to a power, retain the base and multiply the exponents.

At this point, students start to confuse when to multiply exponents and when to add exponents.

To simplify a product or quotient raised to a power, we can apply the definition of exponent.

$$(cd)^2 = (cd) \cdot (cd) \qquad \left(\frac{x}{y}\right)^3 = \left(\frac{x}{y}\right) \cdot \left(\frac{x}{y}\right) \cdot \left(\frac{x}{y}\right)$$

$$= c \cdot c \cdot d \cdot d \qquad = \frac{x \cdot x \cdot x}{y \cdot y \cdot y}$$

$$= c^2d^2 \qquad = \frac{x^3}{y^3}$$

Observe that each factor of the product is raised to the power 2. For the quotient, both the numerator and the denominator are raised to the power 3. These observations are the basis for the Product to a Power Rule and the Quotient to a Power Rule.

Raising Products and Quotients to a Power

For any real numbers a and b and a positive integer n,

$(ab)^n = a^n b^n$ Product to a Power Rule

$\left(\frac{a}{b}\right)^n = \frac{a^n}{b^n},\ b \neq 0$ Quotient to a Power Rule

In words, to raise a product to a power, raise each factor to that power. To raise a quotient to a power, raise both the numerator and the denominator to that power.

EXAMPLE 2

Simplifying with the Power Rules for Exponents

Simplify by removing the parentheses. Assume that no divisor has a value of 0.

(a) $(y^5)^4 = y^{5\cdot 4} = y^{20}$ Power to a Power Rule

(b) $(a^5)^3(a^4)^2 = a^{5\cdot 3} \cdot a^{4\cdot 2}$ Power to a Power Rule

$= a^{15} \cdot a^8$

$= a^{23}$ Product Rule for Exponents

(c) $(gt)^8 = g^8t^8$ Product to a Power Rule

Students often forget to raise the coefficient to the power, or else they multiply the coefficient by the exponent.

(d) $(-3y^2)^3 = (-3)^3(y^2)^3$ Product to a Power Rule

$= -27y^6$ Power to a Power Rule

(e) $\left(\frac{x}{y}\right)^{11} = \frac{x^{11}}{y^{11}}$ Quotient to a Power Rule

(f) $\left(\frac{2a^3}{b^4}\right)^4 = \frac{(2a^3)^4}{(b^4)^4}$ Quotient to a Power Rule

$= \frac{2^4(a^3)^4}{(b^4)^4}$ Product to a Power Rule

$= \frac{16a^{12}}{b^{16}}$ Power to a Power Rule

Quotient Rule for Exponents

We conclude our list of basic exponent rules by considering quotients of exponential expressions.

Consider the quotient $\frac{x^7}{x^4}$. We can simplify this quotient by writing x^7 as x^{4+3} and then dividing out the common factor x^4.

$$\frac{x^7}{x^4} = \frac{x^{4+3}}{x^4} = \frac{x^4 \cdot x^3}{x^4} = \frac{x^4}{x^4} \cdot \frac{x^3}{1} = 1 \cdot x^3 = x^3$$

The equivalent expression x^3 could have been obtained by subtracting the exponents. The following is the general rule.

In this section we restrict our attention to the case where $m > n$.

Quotient Rule for Exponents

For any nonzero real number b and natural numbers m and n,

$$\frac{b^m}{b^n} = b^{m-n}$$

In words, to divide exponential expressions with the same base, retain the base and subtract the exponents.

EXAMPLE 3 Using the Quotient Rule for Exponents

Use the Quotient Rule for Exponents to perform the indicated operations. Assume that no divisor has a value of 0.

(a) $\dfrac{20^{12}}{20^8} = 20^{12-8} = 20^4 = 160{,}000$

(b) $\dfrac{6x^9}{9x^6} = \dfrac{3}{3} \cdot \dfrac{2}{3} \cdot \dfrac{x^9}{x^6} = 1 \cdot \dfrac{2}{3} \cdot x^{9-6} = \dfrac{2}{3}x^3$

You might want to point out that the result can also be written $\dfrac{2x^3}{3}$.

(c) $\dfrac{(x-2)^9}{(x-2)^5} = (x-2)^{9-5} = (x-2)^4$

(d) $\dfrac{5a^7b^6}{10a^6b^2} = \dfrac{5}{5} \cdot \dfrac{1}{2} \cdot a^{7-6}b^{6-2} = \dfrac{1}{2}ab^4$

LEARNING TIP

In these quotients, we treat the coefficients and the exponents differently. We write the coefficients as a fraction and reduce it, but we subtract the exponents.

The only exponents we have used have been natural numbers. The following examples illustrate how applying the Quotient Rule for Exponents may lead to exponents other than natural numbers.

$$\frac{b^3}{b^3} = b^{3-3} = b^0 \qquad \frac{b^4}{b^7} = b^{4-7} = b^{-3}$$

We will consider these situations in the next section.

Quick Reference 6.1

Bases

- A positive integer **exponent** is used to represent repeated multiplication. The exponent indicates the number of times the **base** is used as a factor.

Product Rule for Exponents

- For any real number b and for positive integers m and n, $b^m \cdot b^n = b^{m+n}$.
- To multiply exponential expressions with like bases, retain the base and add the exponents.

Power Rules for Exponents

- Power to a Power Rule
  1. For any real number b and for positive integers m and n, $(b^m)^n = b^{mn}$.
  2. To raise an exponential expression to a power, retain the base and multiply the exponents.
- Product to a Power Rule
  1. For any real numbers a and b and for a positive integer n, $(ab)^n = a^n b^n$.
  2. To raise a product to a power, raise each factor of the product to the power.
- Quotient to a Power Rule
  1. For any real numbers a and b, $b \neq 0$, and for a positive integer n, $\left(\frac{a}{b}\right)^n = \frac{a^n}{b^n}$.
  2. To raise a quotient to a power, raise the numerator and the denominator to the power.

Quotient Rule for Exponents

- For any nonzero real number b and natural numbers m and n, $\frac{b^m}{b^n} = b^{m-n}$.
- To divide exponential expressions with like bases, retain the base and subtract the exponents.

Speaking the Language 6.1

1. In the expression c^4, c is called the ________ and 4 is called the ________.
2. To simplify $(y^2)^3$, ________ the exponents.
3. The ________ and ________ Rules for Exponents require that the bases be the same.
4. In this section, the only exponents that are used are ________ numbers.

Exercises 6.1

Concepts and Skills

1. Explain why we can multiply $2x^2$ by $3x$ but we cannot combine the terms in the sum $2x^2 + 3x$.
2. Explain why -5^3 and $(-5)^3$ have the same value and why -5^2 and $(-5)^2$ do not have the same value.

In Exercises 3 and 4, simplify, if possible. Then identify the parts that are equivalent.

3. (a) $-(-b)^3$ (b) $-b^3$ (c) $(-b)^3$ (d) b^3 (e) $-(b)^3$
4. (a) $-(-b)^2$ (b) $-b^2$ (c) b^2 (d) $(-b)^2$

In Exercises 5–18, use the Product Rule for Exponents and express the result in exponential form.

5. $2^4 \cdot 2^7$
6. $t \cdot t^4$
7. $-5x^4 \cdot x^2$
8. $y^3 \cdot 2y^5$
9. $6t \cdot 3t^7 \cdot t^2$
10. $-2y^5 \cdot y^2 \cdot y^3$
11. $(6 + t)^5(6 + t)^4$
12. $(x - y)^2(x - y)^3$
13. $y^5(3y^2)$
14. $2x^2(3x)$
15. $(-2x^7)(3x)$
16. $(-8t^9)(5t^7)$

17. $(2xy^2)(-3x^3y)$

18. $(-x^3y^7)(-2x^4y^2)$

CE In Exercises 19–24, perform the following operations.

(a) If possible, add the given expressions.

(b) Multiply the given expressions.

19. $8y^5, 8y^4$

20. $6t^5, -7t^5$

21. $n, 3n, -6n$

22. $m^2, 2m^2, -3m^2$

23. $y^3, 5y^3, -4y^3$

24. x, x^2, x^5

CE In Exercises 25 and 26, simplify the given expression by using the

(a) Order of Operations.

(b) Distributive Property.

25. $x^2(3x^3 + 5x^3)$

26. $3y(y^2 + 3y^2)$

27. When performing operations with exponential expressions, sometimes we multiply exponents, sometimes we add exponents, and sometimes we do nothing with the exponents. Use the following expressions to explain the differences.

$x^3 \cdot x^3 \qquad x^3 + x^3 \qquad (x^3)^3$

28. From the following list, identify the two expressions that are equivalent to $(2x^3)^5$ and explain why the third expression is not equivalent to $(2x^3)^5$.

(i) $2^5(x^3)^5$ (ii) $2x^{15}$ (iii) $32x^{15}$

In Exercises 29–44, simplify by removing the parentheses. Assume that no divisor has a value of 0.

29. $(x^3)^4$

30. $(t^5)^2$

31. $[(x - 3)^3]^4$

32. $[(2 - 5x)^6]^2$

33. $(xy)^5$

34. $(-2t)^3$

35. $(-x^2y)^3$

36. $(a^5b^9)^6$

37. $(-2x^2y^4)^3$

38. $(3x^3y^5)^4$

39. $\left(\dfrac{a}{b}\right)^6$

40. $\left(-\dfrac{3}{v}\right)^5$

41. $\left(\dfrac{4x^6}{y^5}\right)^2$

42. $\left(\dfrac{xy^2}{z}\right)^3$

43. $\left(\dfrac{7x^3}{5y^5}\right)^2$

44. $\left(\dfrac{5a^5}{-2b^6}\right)^4$

In Exercises 45–60, use the Quotient Rule for Exponents and express the result in exponential form. Assume that all expressions are defined.

45. $\dfrac{-7^{11}}{7^5}$

46. $\dfrac{z^5}{z^2}$

47. $\dfrac{x^{12}}{x}$

48. $\dfrac{y^8}{y^7}$

49. $\dfrac{(x+1)^5}{(x+1)^3}$

50. $\dfrac{(2x-1)^6}{(2x-1)^5}$

51. $\dfrac{-5x^4}{10x^3}$

52. $\dfrac{-9y^4}{-3y}$

53. $\dfrac{-3z^5}{-4z^2}$

54. $\dfrac{24x^3}{-40x}$

55. $\dfrac{a^4b^5}{a^2b^4}$

56. $\dfrac{x^9y^{12}}{x^6y^5}$

57. $\dfrac{a^4b^5}{b^2}$

58. $\dfrac{x^{15}y^{10}}{x^{12}y^2}$

59. $\dfrac{6x^{12}y^8}{12x^6y^4}$

60. $\dfrac{48a^{17}b^{10}}{6a^4b^2}$

CE In Exercises 61–64, insert grouping symbols so that the expression has the given value.

These are good class discussion exercises.

61. $7 - 2^2$; 25

62. $9 \div 3^2$; 9

63. $8 \cdot 5^2$; 1600

64. $5 + 3^2 \div 4$; 16

In Exercises 65–70, determine whether the statement is true or false. Assume that x and y are both positive.

65. $\dfrac{x^{12}}{x^3} = x^4$

66. $(x^2)^3 = x^8$

67. $(xy)^2 = x^2y^2$

68. $(x^3)^7 = (x^7)^3$

69. $\dfrac{x^{20}}{x^{19}} > 0$

70. $x^6 \cdot x^2 = x^{12}$

In Exercises 71–88, perform the indicated operations. Assume that all expressions are defined.

71. $(-5x^3)(4x^5)$

72. $(x^3y^5)(x^2y)(x^7y^9)$

73. $(3s^3t^8)^3$

74. $(-2ab^7c^4)^5$

75. $(-t)^3(t^2)^4$

76. $(x^3)^2 \cdot x^2$

77. $(-2a^4)(a^2)^3$

78. $(-2x^3)^4(3x^5)$

79. $(xy^2)^2(xy)^2$

80. $3x^5 \cdot (2x^3)^3$

81. $\dfrac{-3x^{12}}{12x^4}$

82. $\dfrac{6s^7t^8}{4s^4t^3}$

83. $\dfrac{(-2x^3)^4}{8x^6}$

84. $\dfrac{9y^{14}}{(3y^3)^3}$

85. $\dfrac{(2x^3)^2}{10x^2}$

86. $\dfrac{8y^{13}}{(2y^4)^3}$

87. $\dfrac{(a^5b^7)^2}{(a^2b^4)^3}$

88. $\dfrac{(5x^3y^4)^2}{x^5y^6}$

89. If $x^2 = 5$, what is x^6? Explain your reasoning.

90. We know that $(cd)^2 = c^2d^2$ because of the Product to a Power Rule. However, there is no corresponding exponent rule for a sum raised to a power. For example, how can you show that $(c + d)^2$ is not the same as $c^2 + d^2$?

CE In Exercises 91–94, determine the value of n.

91. $(x^n)^3 = x^{12}$

92. $x^n \cdot x^3 = x^{12}$

93. $\dfrac{x^n}{x^4} = x^2$

94. $(x^5)^n = x^{20}$

Data Analysis: *The Price of Government*

The price of government increased significantly during the twentieth century. The table shows the taxes (in dollars) that every person in the United States needed to pay to support federal, state, and local governments. (Source: Tax Foundation.)

Year	Federal Taxes (dollars)	State and Local Taxes (dollars)
1900	137	211
1910	116	243
1920	447	233
1930	213	501
1940	694	670
1950	2067	725
1960	2630	1101
1970	3554	1850
1980	4646	2179
1990	5478	2785
2000	7826	3435

The functions $F(t) = 0.008t^3$ and $S(t) = 0.35t^2$ model the data for federal and state/local taxes, respectively, as functions of the number of years since 1900.

95. Make a table of values for each function for even-numbered years from 2002 to 2010. What do the tables of values indicate about federal taxes relative to state and local taxes?

96. Write a simplified function $R(t)$ that models the ratio of federal taxes to state and local taxes.

97. Make a table of values for the function R for even-numbered years from 2002 to 2010. What trend does the table suggest?

98. Another possible model for federal taxes is the function $f(t) = 0.7t^2$. Using this model, write a simplified function $r(t)$ that models the ratio of federal taxes to state and local taxes. What trend does r suggest?

Challenge

The Quotient Rule for Exponents cannot be applied to $\dfrac{4^8}{8^4}$ because the bases are not the same. However, in cases like this, we can sometimes change all of the bases to the same base.

For example: $4^8 = (2^2)^8 = 2^{16}$ and $8^4 = (2^3)^4 = 2^{12}$.

$$\text{Then } \frac{4^8}{8^4} = \frac{2^{16}}{2^{12}} = 2^4 = 16.$$

In Exercises 99–102, use this technique to perform the indicated operations.

99. $9^4 \cdot 3^5$

100. $8^3 \cdot 4^5$

101. $\dfrac{8^5}{2^{10}}$

102. $\dfrac{9^5}{3^8}$

In Exercises 103–106, show that the statement is true. Assume that x represents a nonzero number.

103. $\dfrac{x^5}{x^4} + \dfrac{x^4}{x^3} + \dfrac{x^3}{x^2} + \dfrac{x^2}{x} + x = 5x$

104. $(x^3)^4 + x^7 \cdot x^5 + \dfrac{x^{20}}{x^8} = 3x^{12}$

105. $(2x)^3 - x^3 \cdot x^2 - 8x^3 + x^5 = 0$

106. $(-x)^2 + (-x^2) - x^2 + (x^1)^2 = 0$

In Exercises 107–110, use the exponent rules to simplify the expression.

107. $y^{3n} \cdot y^2$

108. $\dfrac{y^{5m+1}}{y^{2m}}$

109. $(a^{3n})^{4n}$

110. $\left(\dfrac{b^{2n}}{a^m}\right)^n$

6.2 Zero and Negative Exponents

Zero Exponent • Negative Exponents • Properties of Exponents

Zero Exponent

In the previous section, we applied the Quotient Rule for Exponents to write $\frac{b^3}{b^3} = b^{3-3} = b^0$. An alternative approach to the problem uses the fact that any nonzero number divided by itself is 1. Therefore, $\frac{b^3}{b^3} = 1$.

Because b^0 and 1 are both results of simplifying $\frac{b^3}{b^3}$, b^0 and 1 must be equivalent. This is the basis for the following definition.

Definition of the Zero Exponent

For any nonzero real number b, $b^0 = 1$.

Note: The base can be any quantity except 0; that is, 0^0 is undefined. In all expressions involving a variable base and zero exponent, we will assume that the base is not zero.

EXAMPLE 1 Exponential Expressions with Zero Exponents

Discuss the roles of the parentheses in these examples. In particular, in part (e) the exponent on 4 is 1: $4y^0 = 4^1y^0$. Only y is raised to the 0 power.

(a) $5^0 = 1$ — The base is 5.

(b) $(-5)^0 = 1$ — The base is -5.

(c) $-5^0 = -1 \cdot 5^0 = -1 \cdot 1 = -1$ — The base is 5.

(d) $(4y)^0 = 1$ — The base is $4y$.

(e) $4y^0 = 4 \cdot y^0 = 4 \cdot 1 = 4$ — The base is y.

Negative Exponents

In the previous section, we applied the Quotient Rule for Exponents to write $\frac{b^4}{b^7} = b^{4-7} = b^{-3}$. An alternative approach to the problem uses the Product Rule for Exponents and the fact that any nonzero number divided by itself is 1.

$$\frac{b^4}{b^7} = \frac{1 \cdot b^4}{b^3 \cdot b^4} = \frac{1}{b^3} \cdot \frac{b^4}{b^4} = \frac{1}{b^3} \cdot 1 = \frac{1}{b^3}$$

Because b^{-3} and $\frac{1}{b^3}$ are both the results of simplifying $\frac{b^4}{b^7}$, b^{-3} and $\frac{1}{b^3}$ must be equivalent. This conclusion suggests the following definition.

Definition of the Negative Exponent

For any nonzero real number b and any integer n, $b^{-n} = \frac{1}{b^n}$.

Note: The base cannot be zero. In all expressions involving a variable base and a negative exponent, we will assume that the base is not zero.

Think About It

Show why $\left(\frac{a}{b}\right)^{-2} = \left(\frac{b}{a}\right)^2$.

According to the definition of an integer exponent, all of the following are true.

$$y^{-5} = \frac{1}{y^5} \qquad a^{-1} = \frac{1}{a} \qquad 6^{-7} = \frac{1}{6^7} \qquad (-8)^{-5} = \frac{1}{(-8)^5} \qquad \left(\frac{3}{4}\right)^{-5} = \left(\frac{4}{3}\right)^5$$

EXAMPLE 2

Emphasize that the sign of the base does not change. In part (a), students sometimes write $(-4)^{-2} = \frac{1}{4^2}$.

Exponential Expressions with Negative Exponents

(a) $(-4)^{-2} = \frac{1}{(-4)^2} = \frac{1}{16} = 0.0625$ — The sign of the exponent changes, but the sign of the base does not.

(b) $-4^{-2} = -1 \cdot 4^{-2} = -1 \cdot \frac{1}{4^2} = -\frac{1}{16} = -0.0625$

(c) $-(-5)^{-1} = -1 \cdot (-5)^{-1} = \frac{-1}{(-5)^1} = \frac{1}{5}$

(d) $3^{-1} + 2^{-1} = \frac{1}{3} + \frac{1}{2} = \frac{2}{6} + \frac{3}{6} = \frac{5}{6}$

(e) $(3 + 2)^{-1} = 5^{-1} = \frac{1}{5}$

(f) $\left(\frac{2}{5}\right)^{-2} = \left(\frac{5}{2}\right)^2 = \frac{25}{4}$ $\qquad \left(\frac{a}{b}\right)^{-n} = \left(\frac{b}{a}\right)^n$

Because n can be any integer, the definition may be expressed in the following way.

$$b^{-n} = \frac{1}{b^n} \quad \text{or} \quad b^n = \frac{1}{b^{-n}}$$

EXAMPLE 3

In part (c), $3k^{-5} = 3^1k^{-5}$. Only k has a negative exponent. Students are inclined to write $3k^{-5} = \frac{1}{3k^5}$.

Writing Exponential Expressions with Positive Exponents

(a) $x^{-5} = \frac{1}{x^5}$

(b) $(3k)^{-5} = \frac{1}{(3k)^5}$ — The base is $3k$.

(c) $3k^{-5} = 3 \cdot \frac{1}{k^5} = \frac{3}{k^5}$ — The base is k.

(d) $\frac{1}{b^{-3}} = b^3$

(e) $\frac{3}{p^{-8}} = 3 \cdot \frac{1}{p^{-8}} = 3p^8$

(f) $\frac{7^{-2}}{2^{-5}} = \frac{7^{-2}}{1} \cdot \frac{1}{2^{-5}} = \frac{1}{7^2} \cdot \frac{2^5}{1} = \frac{2^5}{7^2}$

Example 3 suggests that we can move exponential factors in a fraction from the numerator to the denominator or from the denominator to the numerator if we change the sign of the exponent. This method applies only to exponential *factors,* not to exponential terms.

EXAMPLE 4

Writing Exponential Expressions with Positive Exponents

Write each expression with positive exponents.

(a) $\frac{s^{-6}}{t^{-9}} = \frac{t^9}{s^6}$

Students tend to move every factor. Emphasize that only factors with negative exponents are moved. Factors with positive exponents remain where they are.

(b) $\frac{c^2d^{-3}}{5k^{-2}} = \frac{c^2k^2}{5d^3}$ The factors c^2 and 5 have positive exponents and are not moved.

(c) $6x^{-1}y = \frac{6x^{-1}y}{1} = \frac{6y}{x}$ The factors 6 and y have positive exponents and are not moved.

(d) $x^{-2}y + xy^{-2} = \frac{y}{x^2} + \frac{x}{y^2}$ Apply the definition of negative exponent to each term.

Properties of Exponents

Assuming that the expression is defined, all the properties of exponents discussed in Section 6.1 are valid for an expression with any integer exponents.

EXAMPLE 5

Products and Quotients with Integer Exponents

Perform the indicated operations and express the results with positive exponents. Assume that all variables represent nonzero numbers.

(a) $(3x^5)(-2x^{-3}) = -6x^{5+(-3)} = -6x^2$

(b) $(-3x^{-5}y^5z^{-3})(5x^4yz^4) = -15x^{-5+4}y^{5+1}z^{-3+4} = -15x^{-1}y^6z^1 = \frac{-15y^6z}{x}$

In part (b), students sometimes move the -15 to the denominator because it is negative. Emphasize that the exponent of -15 is $+1$.

(c) $\frac{4x^{-2}}{6x^{-5}} = \frac{4x^5}{6x^2} = \frac{2}{3} \cdot x^{5-2} = \frac{2}{3} \cdot x^3 = \frac{2x^3}{3}$

(d) $\frac{x^4y^{-3}z^0}{x^{-1}y} = x^{4-(-1)}y^{-3-1}z^0 = x^5y^{-4} \cdot 1 = \frac{x^5}{y^4}$

EXAMPLE 6

Using the Power Rules for Integer Exponents

Perform the indicated operations and express the results with positive exponents. Assume that all variables represent nonzero numbers.

(a) $(2r^{-3}t^2)^{-3} = 2^{-3}(r^{-3})^{-3}(t^2)^{-3} = 2^{-3}r^9t^{-6} = \frac{r^9}{2^3t^6} = \frac{r^9}{8t^6}$

LEARNING TIP

When more than one exponent rule is needed, there is often more than one way to simplify an expression. To become aware of the choices that you have, practice varying your approaches and confirm that the results are the same.

(b) $\left(\frac{m^2}{n^{-3}}\right)^{-4} = \frac{(m^2)^{-4}}{(n^{-3})^{-4}} = \frac{m^{-8}}{n^{12}} = \frac{1}{m^8n^{12}}$

(c) $\frac{2cd^{-1}}{(2cd)^{-1}} = \frac{2c(2cd)^1}{d^1} = \frac{4c^2d}{d} = 4c^2$

(d) $\frac{(3x^{-1}y)^3(4y^{-3})^{-2}}{2x^2} = \frac{(3x^{-1}y)^3}{2x^2(4y^{-3})^2} = \frac{27x^{-3}y^3}{2x^2 \cdot 16y^{-6}} = \frac{27y^3y^6}{32x^2x^3} = \frac{27y^9}{32x^5}$

Quick Reference 6.2

Zero Exponent

- A zero exponent arises when we use the Quotient Rule for Exponents to divide $\frac{b^n}{b^n} = b^{n-n} = b^0, b \neq 0.$
- For any nonzero real number b, $b^0 = 1$.

Negative Exponents

- A negative exponent arises when we use the Quotient Rule for Exponents to divide $\frac{b^m}{b^n}$, $b \neq 0$, where $m < n$.
- For any nonzero real number b and any integer n, $b^{-n} = \frac{1}{b^n}$ and $b^n = \frac{1}{b^{-n}}$.

Properties of Exponents

- Assuming that the expression is defined, all the properties of exponents discussed in Section 6.1 are valid for an expression with any integer exponents.

Speaking the Language 6.2

1. When we write $(-x)^{-3}$ with a positive exponent, the sign of the ______ changes, but the sign of the ______ does not.
2. If b is a nonzero number, then b^{-n} can be written as the ______ of b^n.
3. The ______ exponent arises when we apply the Quotient Rule for Exponents to $\frac{b^m}{b^n}$, where $m = n$.
4. In this section, we have extended the properties of exponents to expressions with ______ exponents.

Exercises 6.2

Concepts and Skills

1. In the definition of b^0, why must b be a nonzero number?
2. Explain why $2x^0$ and $(2x)^0$ have different values.

In Exercises 3–20, evaluate each expression.

3. 4^0
4. $\frac{1}{9^0}$
5. $3t^0$
6. $-7x^0$
7. -6^0
8. $(2y)^0$
9. 5^{-2}
10. 7^{-1}
11. $(-2)^{-3}$
12. $(-4)^{-2}$
13. $\frac{1}{15^{-1}}$
14. $\frac{1}{-3^{-2}}$
15. $\left(\frac{1}{4}\right)^{-1}$
16. $\left(-\frac{2}{3}\right)^{-3}$
17. $\frac{2^{-2}}{2}$
18. $\frac{8 \cdot 4^{-2}}{3^{-3}}$
19. $3^{-1} + 2^{-2}$
20. $2^{-1} - 5^{-1}$
21. In the definition of b^{-n}, why must b be a nonzero number?
22. Explain how to evaluate a number that has a negative exponent.

In Exercises 23–36, write the expressions with positive exponents.

23. x^{-7}
24. $\frac{1}{y^{-3}}$

25. $4x^{-4}$

26. $-5y^{-1}$

27. $2^{-5}x^3$

28. $4u^{-3}v^2$

29. $-x^{-3}y^{-4}$

30. yz^{-3}

31. $\dfrac{1}{-5k^{-6}}$

32. $\dfrac{1}{3t^{-5}}$

33. $(2 + y)^{-3}$

34. $\dfrac{3}{(x + 1)^{-5}}$

35. $\dfrac{-3a^{-1}}{2b^{-1}}$

36. $\dfrac{-2a^{-4}}{b^{-3}}$

In Exercises 37–40, insert <, =, or > to make the statement true.

37. $-3^0 \; \underline{\qquad} \; (-3)^0$

38. $(-8)^0 \; \underline{\qquad} \; 8^0$

39. $-5^{-2} \; \underline{\qquad} \; (-5)^{-2}$

40. $\left(\dfrac{2}{3}\right)^{-1} \; \underline{\qquad} \; \dfrac{3}{2}$

CE In Exercises 41 and 42, list the given expressions in order from least to greatest. Assume x represents a nonzero number.

41. $(-10)^0$, 0, $-|8|^0$, $-1 - 4^0$, $4^0 + x^0$

42. $-x^0 + 1$, $(-x)^0 + 1$, $|-6|^0$, $(-2)(-x)^0$

In Exercises 43–48, use the Product Rule for Exponents and express the result in exponential form with positive exponents.

43. $x^5 \cdot x^{-5}$

44. $y^{-6} \cdot y^5 \cdot y^0$

45. $(-4x^{-4})(3x)$

46. $(x^{-4})(-3x^{-6})(-x^7)$

47. $7x^{-4}y^5(2x^{-3}y^{-4})$

48. $(x^{-4}y^7)(xy^2)(x^8y^{-10})$

In Exercises 49–62, use the Quotient Rule for Exponents and express the result in exponential form with positive exponents. Assume that all expressions are defined.

49. $\dfrac{z^{-2}}{z^5}$

50. $\dfrac{x^6}{x^{-7}}$

51. $\dfrac{y^{-8}}{y^{-3}}$

52. $\dfrac{-2t^5}{t^{-7}}$

53. $\dfrac{3z^{-8}}{2z^{-5}}$

54. $\dfrac{21t^4}{7t^{10}}$

55. $\dfrac{x^{-3}y^8}{x^2y^{-2}}$

56. $\dfrac{x^{-2}y^{-3}}{x^{-3}y^{-1}}$

57. $\dfrac{4^{-1}x^{-3}}{-4^2x^{-2}}$

58. $\dfrac{3^{-2}y^{-5}}{3^{-1}y^{-7}}$

59. $\dfrac{3^{-2}x^{-5}y^{-7}}{4x^5y^{-9}}$

60. $\dfrac{10x^{-5}y^3}{5x^{-7}y^7}$

61. $\dfrac{x^{21}y^{-22}}{x^{-18}y^{26}}$

62. $\dfrac{8x^{-3}y^0}{24x^2y^{-5}}$

CE In Exercises 63–66, determine the unknown expression.

63. The quotient of the expression and x^{-3} is x^{-5}.

64. The quotient of the expression and x^{-2} is 3.

65. The product of the expression and x^{-4} is 1.

66. The product of the expression and x^5 is x.

In Exercises 67–76, simplify and express the result with positive exponents. Assume that all expressions are defined.

67. $(x^{-6})^5$

68. $(y^2)^{-4}$

69. $(2a^{-5})^{-4}$

70. $(xy^{-3})^2$

71. $(-2x^{-2}y^4)^3$

72. $(2x^6y^{-8})^{-2}$

73. $\left(-\dfrac{2}{k}\right)^{-5}$

74. $\left(\dfrac{x^3}{y^{-4}}\right)^{-3}$

75. $\left(\dfrac{-4x^{-3}}{5^{-1}y^4}\right)^3$

76. $\left(\dfrac{5x^{-3}}{3y^2}\right)^{-2}$

CE In Exercises 77–82, insert grouping symbols in the expression on the left side of the equation to make the statement true.

77. $4 - 3^{-9} = 1$

78. $5 \cdot 2^{-1} = 0.1$

79. $2^5 + 1^0 = 1$

80. $5 \div 3^{-1} = \dfrac{3}{5}$

81. $2xy^{-2} = \dfrac{2}{x^2y^2}$

82. $x^5y^{-3}z^0 = x^5$

In Exercises 83–98, perform the indicated operations and express the result with positive exponents. Assume that all expressions are defined.

83. $(3x^{-2})^0(2x^{-3})^{-4}$

84. $(-5x^{-2}y)^{-1}$

85. $-4x^{-2}(x^3)^{-1}$

86. $(-t)^3(t^{-2})^4$

87. $-6x^3(2x^{-3}y^{-2})$

88. $(2x^{-2}y^3)(-3x^4y^{-7})$

89. $(a^{-5}b^{-6})^5(a^{-2}b^5)^{-4}$

90. $(-2s^{-3}t^5)^4(-s^{-1}t^{-15})$

This group includes a mix of all the exponent operations.

91. $\dfrac{-4x^2y^{-2}}{8y^5}$

92. $\dfrac{6x^{-3}y^4}{-2x^2y^{-3}}$

93. $\left(\dfrac{-2x^{-2}}{y^5}\right)^{-3}$

94. $\left(\dfrac{3x^{-3}}{x^5}\right)^{-3}$

95. $\dfrac{(x^{-2})^0x^{-3}}{x^5}$

96. $\dfrac{(t^{-6})^2}{t^4(t^3)^{-4}}$

97. $\dfrac{(a^2b^{-3})^{-4}}{(a^{-3}b^5)^2}$

98. $\dfrac{9a^2b^{-2}}{(3a^{-2}b)^{-1}}$

Challenge

In Exercises 99 and 100, determine the number or expression that satisfies the given conditions.

99. The number raised to the -2 power is $\frac{9}{16}$.

100. When the expression is squared and then multiplied by $3x^{-2}$, the result is $3x^6$.

101. Show that $\dfrac{x^{-m}}{y^{-n}} - \dfrac{y^n}{x^m} = 0$.

102. Show that $\dfrac{(ab)^n}{a^n b^n} = x^0$.

103. Show that $\dfrac{x^{-4}}{x^{-5}} - \dfrac{x^{-3}}{x^{-4}} + \dfrac{x^{-2}}{x^{-3}} - \dfrac{x^{-1}}{x^{-2}} + \dfrac{x^0}{x^{-1}} - x = 0$

104. Show that $\dfrac{b^{m+n}}{b^{m-n}} = b^{2n}$.

In Exercises 105 and 106, determine the values of m and n that make the statement true. Assume that all expressions are defined.

105. $\dfrac{a^3b^n}{a^5b^2} \cdot \dfrac{a^4b^3}{a^mb^{-5}} = 1$

106. $x^6(x^n)^3 = x^{12}$

6.3 Scientific Notation

Powers of 10 • Decimal Form to Scientific Notation • Real-Life Applications

Powers of 10

The design of the decimal number system makes multiplying a number by a positive or negative power of 10 easy to do.

Exploring the Concept

Multiplying by Powers of 10

Observe the position of the decimal point in the following products. We begin with positive powers of 10.

$$2.587 \cdot 10^1 = 2.587 \cdot 10 = 25.87$$
$$2.587 \cdot 10^2 = 2.587 \cdot 10 \cdot 10 = 258.70$$
$$2.587 \cdot 10^3 = 2.587 \cdot 10 \cdot 10 \cdot 10 = 2587.00$$

Now consider these negative powers of 10.

$$2.587 \cdot 10^{-1} = \frac{2.587}{10} = 0.2587$$
$$2.587 \cdot 10^{-2} = \frac{2.587}{10 \cdot 10} = 0.02587$$
$$2.587 \cdot 10^{-3} = \frac{2.587}{10 \cdot 10 \cdot 10} = 0.002587$$

Multiplying a number by a positive power of 10 moves the decimal point to the right. Multiplying a number by a negative power of 10 moves the decimal point to the left. The number of places that the decimal point moves is the same as the absolute value of the exponent.

Each product in the first column is in the form $2.587 \cdot 10^p$, which is an example of a number that is written in **scientific notation.**

Think About It

The definition of scientific notation is stated for positive numbers. Restate the definition for negative numbers.

General Scientific Notation

The scientific notation for a positive number is $n \cdot 10^p$ where $1 \le n < 10$ and p is an integer.

To convert a number from scientific notation $n \cdot 10^p$ to **decimal** form, we multiply n by the indicated power of 10. We can do this mentally simply by moving the decimal point to the right or left, depending on the exponent p.

Scientific Notation	*Decimal Form*	
$5.8 \cdot 10^6$	5,800,000	Move the decimal point 6 places to the right.
$2.36 \cdot 10^{-4}$	0.000236	Move the decimal point 4 places to the left.

LEARNING TIP

When numbers are written in the form $n \cdot 10^p$, at least be aware that the number is large when p is positive and small when p is negative. Also, try to develop a number sense by being familiar with certain powers of 10: 10^3 means one thousand, 10^6 means one million, and so on.

Decimal Form to Scientific Notation

To write a number in scientific notation $n \cdot 10^p$, we must determine the number n, where $1 \le n < 10$, and the exponent p.

Writing a Number in Scientific Notation

1. Place the decimal point after the first nonzero digit. The result is n.
2. The number of places from the decimal point to its original position is the absolute value of p.
3. If the original position of the decimal point is to the right, then p is positive; if the original position of the decimal point is to the left, then p is negative.

EXAMPLE 1

Writing Large Numbers in Scientific Notation

Write the following numbers in scientific notation.

(a) 250,000 (b) 63,420,000

Solution

Note that when the decimal point is moved to the right to its original location, p is positive.

(a) $n = 2.50000$ — Place the decimal point after the first nonzero digit, 2.

2.50000. — The original location of the decimal point is 5 places to the right: $p = 5$.

$250{,}000 = 2.5 \cdot 10^5$

(b) $n = 6.3420000$ — Place the decimal point after the first nonzero digit, 6.

6.3420000. — The original location of the decimal point is 7 places to the right: $p = 7$.

$63{,}420{,}000 = 6.342 \cdot 10^7$

EXAMPLE 2

Writing Small Numbers in Scientific Notation

Write each of the following numbers in scientific notation.

(a) 0.007 (b) 0.00000058

Solution

Note that when the decimal point is moved to the left to its original location, p is negative.

(a) $n = 7.0$ — Place the decimal point after the first nonzero digit, 7.

$.007.0$ — The original location of the decimal point is 3 places to the left: $p = -3$.

$0.007 = 7.0 \cdot 10^{-3}$

(b) $n = 5.8$ — Place the decimal point after the first nonzero digit, 5.

$.0000005.8$ — The original location of the decimal point is 7 places to the left: $p = -7$.

$0.00000058 = 5.8 \cdot 10^{-7}$

Note: For consistency, we use the multiplication dot when writing numbers in scientific notation. However, $\times$ is also commonly used as the multiplication symbol. The result in part (b) of Example 2 can be written 5.8×10^{-7}.

A scientific or graphing calculator will automatically use scientific notation to display very large or very small numbers. Typically, the letter E is used to indicate the value of the exponent on 10. For example, 3.1E4 means $3.1 \cdot 10^4$.

EXAMPLE 3

Scientific Notation on a Calculator

Use a calculator to perform the following operations. Write the result in scientific notation $n \cdot 10^p$ with n rounded to the nearest tenth.

(a) 500^{20} (b) $0.00005 \div 600{,}000$ (c) $857{,}000 \cdot 26{,}984$

Solution

Examples 3–6 are intended to be solved with a calculator.

(a) $500^{20} \approx 9.5 \cdot 10^{53}$ (b) $0.00005 \div 600{,}000 \approx 8.3 \cdot 10^{-11}$

(c) $857{,}000 \cdot 26{,}984 \approx 2.3 \cdot 10^{10}$

EXAMPLE 4

Calculations with Numbers in Scientific Notation

(a) $(4.2 \cdot 10^5)(7.23 \cdot 10^{-9}) \approx 3.04 \cdot 10^{-3}$

(b) $\dfrac{4.59 \cdot 10^{-5}}{1.3 \cdot 10^4} \approx 3.53 \cdot 10^{-9}$

(c) $(8.45 \cdot 10^3)^6 \approx 3.64 \cdot 10^{23}$

Real-Life Applications

As the name implies, scientific notation is used in the sciences as a compact way of working with very large and very small numbers.

EXAMPLE 5

Water Use in the United States

In the United States, an estimated 1620 gallons of water per day per person are withdrawn from both ground and surface sources.

(a) Based on a U.S. population of 280 million people, how many gallons are withdrawn each day?

(b) If the volume of fresh water in the world is approximately $9.9 \cdot 10^{18}$ gallons, what percentage of the world's fresh water supply does the United States withdraw in a year?

Solution

(a) Each day $(1.62 \cdot 10^3)(2.80 \cdot 10^8) \approx 4.54 \cdot 10^{11}$ gallons are withdrawn in the United States.

(b) Each year $365(4.54 \cdot 10^{11}) \approx 1.66 \cdot 10^{14}$ gallons are withdrawn.

$$\frac{1.66 \cdot 10^{14}}{9.9 \cdot 10^{18}} \approx 1.68 \cdot 10^{-5} = 0.0000168 \approx 0.002\%$$

EXAMPLE 6

The Doomsday Gap

If the sun were suddenly to burn out, how long would it take for earthlings to find out about it? (Use 93,000,000 miles as the distance from the sun to the earth, and use 186,000 miles per second as the speed of light.)

Solution

$$\begin{aligned}
\text{Time} &= \frac{\text{distance from sun to earth}}{\text{speed of light}} \\
&= \frac{9.3 \cdot 10^7 \text{ miles}}{1.86 \cdot 10^5 \dfrac{\text{miles}}{\text{second}}} \\
&= \frac{9.3}{1.86} \cdot 10^{7-5} \text{ seconds} \\
&= 5 \cdot 10^2 \text{ seconds} \\
&= 500 \text{ seconds}
\end{aligned}$$

It would take 500 seconds, or approximately 8.3 minutes, before disaster would strike the earth.

Quick Reference 6.3

Powers of 10

- A positive number in **scientific notation** has the form $n \cdot 10^p$, where n is a number such that $1 \le n < 10$ and the exponent p is an integer.
- To convert a number from scientific notation $n \cdot 10^p$ to decimal form, we multiply n by the indicated power of 10.

Decimal Form to Scientific Notation

- To write a number in scientific notation $n \cdot 10^p$, we must determine the number n, where $1 \le n < 10$, and the exponent p.
  1. Place the decimal point after the first nonzero digit. The result is n.
  2. The number of places from the decimal point to its original position is the absolute value of p.
  3. If the original position of the decimal point is to the right, then p is positive; if the original position of the decimal point is to the left, then p is negative.
- Because a calculator can display only a limited number of digits, scientific notation is used to display very large and very small numbers. You can also enter numbers in scientific notation, and you can set the calculator to display results in scientific notation.

Speaking the Language 6.3

1. Multiplying a number by 10 shifts the decimal point to the ________, whereas dividing a number by 10 shifts the decimal point to the ________.
2. The number form $n \cdot 10^p$, where p is an integer and $1 \le n < 10$, is called ________.
3. To convert $6.2 \cdot 10^5$ to ________ form, we move the decimal point 5 places to the right.
4. On a calculator, the symbol E means ________.

Exercises 6.3

Concepts and Skills

1. A number is in scientific notation when it is written in the form $n \cdot 10^p$. What must be true about the numbers n and p?
2. From the following list, identify the one number that is in scientific notation. Explain why the other two numbers are not in scientific notation.

   (i) $35 \cdot 10^{-4}$ (ii) $1 \cdot 10^{-9}$ (iii) $10 \cdot 10^3$

In Exercises 3–16, write the given number in scientific notation.

3. 125,000
4. 15,000
5. 2570.4
6. 110.25
7. −537,600,000
8. 1,000,000,000
9. 456,700,000,000
10. −35,795,000
11. −0.4
12. 0.025
13. 0.00000645
14. −0.00000001
15. 0.0000003749
16. 0.000000246

In Exercises 17–28, write the number in decimal form.

17. $1.34 \cdot 10^6$
18. $7.358 \cdot 10^{-6}$
19. $-4.214 \cdot 10^{-7}$
20. $6.87 \cdot 10^4$
21. 10^{-6}
22. 10^9
23. $3 \cdot 10^4$
24. $-5 \cdot 10^{-2}$
25. $-5.72 \cdot 10^9$
26. $6.3 \cdot 10^{-9}$
27. $8.9 \cdot 10^{-4}$
28. $-4.83 \cdot 10^7$

In Exercises 29–36, use a calculator to perform the indicated operations. Write the result in scientific notation $n \cdot 10^p$ where n is rounded to the nearest tenth.

29. 428^{20}
30. $(-249)^{15}$
31. -376^{-9}
32. 426^{-23}

33. $0.0000006 \div 50{,}000$

34. $0.0000047637 \div 423{,}300$

35. $8{,}765{,}432 \cdot 654{,}321$

36. $-2{,}763{,}742 \cdot 2{,}060{,}000$

In Exercises 37–40, insert $<$, $=$, or $>$ to make the statement true.

37. $2.84 \cdot 10^9$ ___ $2.84 \cdot 10^8$

38. $1.73 \cdot 10^5$ ___ $17.3 \cdot 10^4$

39. $6.58 \cdot 10^{-3}$ ___ $6.58 \cdot 10^{-4}$

40. 10^7 ___ $1.0 \cdot 10^7$

In Exercises 41–44, write the given expression as a single number in decimal form.

41. $7 \cdot 10^3 + 1 \cdot 10^2 + 5 \cdot 10^1 + 3$

42. $5 \cdot 10^3 + 4 \cdot 10^2 + 9 \cdot 10^1 + 1$

43. $2 + 3 \cdot 10^{-1} + 4 \cdot 10^{-2}$

44. $9 + 2 \cdot 10^{-1} + 8 \cdot 10^{-2}$

You may want to discuss a couple of these in class to point out the structure of a decimal number.

In Exercises 45–50, perform the indicated operations mentally.

Encourage students to use the properties of exponents on the powers of 10.

45. $(2 \cdot 10^2)(3 \cdot 10^1)$

46. $(2.5 \cdot 10^{-5})(2 \cdot 10^3)$

47. $\dfrac{4.2 \cdot 10^8}{2.1 \cdot 10^6}$

48. $\dfrac{2 \cdot 10^{-3}}{4 \cdot 10^{-5}}$

49. $(2 \cdot 10^{-1})^{-1}$

50. $(4 \cdot 10^{-2})^{-1}$

In Exercises 51–60, use a calculator to perform the indicated operations. Express the result in scientific notation.

51. $(3.72 \cdot 10^7)(9.8 \cdot 10^3)$

52. $(5.87 \cdot 10^{-4})(6.73 \cdot 10^7)$

53. $\dfrac{4.23 \cdot 10^8}{5.2 \cdot 10^{-5}}$

54. $\dfrac{8.6 \cdot 10^{-4}}{4.9 \cdot 10^6}$

55. $(5.7 \cdot 10^3)^4$

56. $(6.89 \cdot 10^{-5})^3$

57. $(8.31 \cdot 10^5)^{-1}$

58. $(4.5 \cdot 10^{-5})^{-1}$

59. $\dfrac{(4.9 \cdot 10^5)(3 \cdot 10^{-7})}{3.79 \cdot 10^8}$

60. $\dfrac{(9.5 \cdot 10^5)(6.52 \cdot 10^{-7})}{7.3 \cdot 10^{12}}$

CE In Exercises 61–66, write the given number in scientific notation.

Some students believe that any number followed by a power of 10 is in scientific notation.

61. $3456 \cdot 10^6$

62. $423 \cdot 10^9$

63. $0.0001234 \cdot 10^{12}$

64. $0.00258 \cdot 10^{-10}$

65. $0.0000975 \cdot 10^{-7}$

66. $75{,}394 \cdot 10^{-9}$

Real-Life Applications

67. The total area of Georgia is $3.77 \cdot 10^7$ acres. Write this number in decimal form.

68. The distance between an electron and a proton in a hydrogen atom is $5.3 \cdot 10^{-11}$ meter. Write this distance in decimal form.

69. The radius of the nucleus of a heavy atom is 0.000000000007 millimeter. Write the radius in scientific notation.

70. In 2010, the population of India is predicted to be 1,631,800,000. Write this number in scientific notation.

71. If a computer can do one addition problem in $6 \cdot 10^{-6}$ second, how many such problems can the computer do in one-half minute?

72. Property tax rates are expressed in mils, where one mil is one dollar for each thousand dollars of assessed property value. If the millage rate for a county is 43 mils and the total property value in the county is $5.3 \cdot 10^7$ dollars, what will the annual tax revenue be for the year? Express your answer in decimal notation and in scientific notation.

73. A report indicates that a total of 50,455,000 cable TV subscribers pay an average of $17.58 per month. What is the total amount paid by all subscribers in one year?

74. A report indicates that approximately 3.2 million gallons of ice cream are consumed per day in the United States. How many gallons are consumed in a year?

75. Placed side by side, about 4000 typical cells of the human body would span 1 inch. How wide is a typical cell?

76. What percentage of the U.S. population plays NBA basketball? (Assume that there are 29 teams with 11 players per team and that the population of the United States is 280,000,000.)

Modeling with Real Data

In Exercises 77–82, write your answers in scientific notation.

77. In one particular year, AIDS and measles each caused the deaths of 1 million people worldwide. There were 110% more deaths from malaria than from AIDS, and 210% more people died from tuberculosis than from measles. (Source: World Health Report.) What was the total number of deaths due to these four diseases?

78. Americans make 30 billion visits to fast-food restaurants annually. (Source: *USA Today*.) If the average sale per visit is \$5, what are the total annual sales?

79. About 60 million people in the 26–44 age group commute to work, and they average 22 minutes of travel. (Source: Bureau of Census.) How many person-hours per week are spent commuting?

80. The typical American family watches television for an average of 7 hours and 42 minutes each day. (Source: Nielsen Media Research.) What is the number of family-hours per year that the 97 million households in America spend watching television?

81. The hummingbird produces the smallest bird egg. A total of 78 such eggs would weigh only 1 ounce. (Source: *A Guide to Birds*.) What is the weight in pounds of a single hummingbird egg?

82. The most accurate clock is the Hewlett-Packard atomic clock, which is accurate to within 1 second in 1.6 million years. (Source: *Guinness Book of Records*.) How much time each year does the clock lose?

Data Analysis: *NASA and Space Shuttle Budgets*

In 2000, the total budget of the National Aeronautics and Space Administration (NASA) was \$13,653,000,000. The budget for the space shuttle program was \$2,986,000,000. (Source: NASA.)

83. Write the total NASA budget and the space shuttle budget in scientific notation.

84. Using the numbers in scientific notation, write the ratio of the space shuttle budget to the total NASA budget. Use this expression to determine the percentage of the total budget that is represented by the space shuttle budget.

85. Suppose that Congress increased the NASA budget by 2% but decreased the shuttle budget by 1.5%. How many dollars would be available for NASA programs other than the space shuttle?

86. The National Space Society proposed a 4% increase in the total NASA budget for 2001. How many additional dollars would have had to be added to the space shuttle budget in order to maintain the percentage in Exercise 84?

Challenge

In Exercises 87–90, determine the value of n that makes the statement true.

87. $\dfrac{(4 \cdot 10^{7})(8 \cdot 10^{-3})}{(16 \cdot 10^{-2})(2 \cdot 10^{n})} = 1$

88. $\dfrac{(3 \cdot 10^{4})(8 \cdot 10^{n})}{12 \cdot 10^{-2}} = 2$

89. $\dfrac{(15 \cdot 10^{-8})(4 \cdot 10^{n})}{(24 \cdot 10^{12})(5 \cdot 10^{-7})} = \dfrac{1}{2}$

90. $\dfrac{(18 \cdot 10^{5})(5 \cdot 10^{-3})}{(8 \cdot 10^{n})(15 \cdot 10^{7})} = \dfrac{3}{4}$

In **engineering notation,** numbers are written in the form $n \cdot 10^{p}$, where $1 \le n < 1000$ and p is an integer that is a multiple of 3.

In Exercises 91–96, perform the operation and write the result in engineering notation $n \cdot 10^{p}$ with n rounded to the nearest tenth.

91. $80{,}210 \div 65$

92. $625 \cdot 80$

93. 3^{-28}

94. $1 \div 1{,}000{,}000$

95. 5^{30}

96. $8745 \cdot 9579$

6.4 Polynomials

Definitions • Addition of Polynomials • Subtraction of Polynomials • Polynomial Functions

Definitions

Recall that **terms** are parts of an expression that are separated by plus signs. In this section we consider one kind of term called a **monomial,** and we consider expressions called **polynomials** whose terms are monomials.

Students tend to be skeptical about a constant being a monomial. Writing, for example, 2 as $2x^0$ is usually only mildly persuasive.

> **Definition of Monomial and Polynomial**
>
> A **monomial** is a number or a product of a number and one or more variables with nonnegative integer exponents. A **polynomial** is a finite sum of monomials.

An example of a polynomial is $4x^5 - 5x^3 + 7x + 2$. We can write this expression as a *sum* of monomials: $4x^5 + (-5x^3) + 7x + 2$. The terms are $4x^5$, $-5x^3$, $7x$, and 2 (or $2x^0$).

Remind students to include the sign with the coefficient.

The **numerical coefficient** (usually just called the **coefficient**) of a term is the numerical factor of the term. In our example, the coefficients are 4, −5, 7, and 2. We also say that 2 is the **constant term.**

The following are some examples of algebraic expressions that are not polynomials.

$5x^2 + 3x^{-5}$ Exponents must be nonnegative integers.

$\dfrac{x-4}{x^2+1}$ Variables cannot be in the denominator of a fraction.

$\sqrt{x} + 3$ Variables cannot appear in a radicand.

A polynomial consists of one or more terms (monomials). When a polynomial has only one term, we usually refer to the expression simply as a monomial. A polynomial with two nonzero terms is called a **binomial,** and a polynomial with three nonzero terms is called a **trinomial.**

Polynomial	*Number of Terms*	*Name*
$t^2 - 6 + 3t$	3	trinomial
$16 - x^2$	2	binomial
$5x^4$	1	monomial
-15	1	monomial

Except for 0, the **degree of a monomial** in one variable is the value of the exponent on the variable. The degree of a monomial in more than one variable is the sum of the exponents on the variables. For instance, the degree of $5x^4$ is 4, and the degree of a^2b^5 is 2 + 5, or 7. Because a constant term such as 2 can be written $2x^0$, the degree of a nonzero constant is 0. The degree of the monomial 0 is not defined because 0 can be written $0x^n$ for any nonnegative integer n.

The **degree of a polynomial** is the degree of the term with the highest degree.

EXAMPLE 1

Determining the Degree of a Polynomial

Polynomial	*Degree*	
$2y$	1	$2y = 2y^1$
$7 - 3x^4$	4	The degree of $-3x^4$ is 4.
$4t^5 - 5t^3 + 7t + 2$	5	The largest exponent is 5.
$3x^2y^4 - 4xy^3 + x^3y^5$	8	The degree of x^3y^5 is $3 + 5$, or 8.

The conventional way to write a polynomial in one variable is to write the terms in order of descending (decreasing) degrees. We say that the polynomial is written in **descending order.**

In some instances, it is convenient to write the polynomial in **ascending** (increasing) **order.** In such cases, we begin with the term with the lowest exponent. Then the terms that follow have increasing degrees.

When rearranging the order of terms, keep the signs with the terms.

Polynomial	*Descending Order*	*Ascending Order*
$4x^5 - 5x^3 + 7x + 2$	$4x^5 - 5x^3 + 7x + 2$	$2 + 7x - 5x^3 + 4x^5$
$-2x - x^5 + 7x^3$	$-x^5 + 7x^3 - 2x$	$-2x + 7x^3 - x^5$

In the first polynomial in this example, there is no term with x^4 or x^2. We say that these are *missing terms.* The polynomial could have been written with the missing terms:

$$4x^5 + 0x^4 - 5x^3 + 0x^2 + 7x + 2$$

We evaluate polynomials in the same way as we evaluate any expression.

EXAMPLE 2

Evaluating a Polynomial in Two Variables

Evaluate the polynomial $3x^2y^4 - 4xy^3 + x^3y^5$ for $x = 2$ and $y = -1$.

Solution

$$\begin{aligned} 3x^2y^4 - 4xy^3 + x^3y^5 &= 3(2)^2(-1)^4 - 4(2)(-1)^3 + (2)^3(-1)^5 \\ &= 3(4)(1) - 4(2)(-1) + 8(-1) \\ &= 12 + 8 - 8 \\ &= 12 \end{aligned}$$

Addition of Polynomials

Students should be reminded that the Distributive Property allows us to combine like terms.

Terms of a polynomial are **like terms** if they are both constant terms or if they have the same variables with the same exponents. To add polynomials, we simply combine the like terms.

EXAMPLE 3

Adding Polynomials

Add the polynomials $5x^3 - 4x + 2$ and $-x^3 + 5x^2 - 2x + 5$.

Solution

$$(5x^3 - 4x + 2) + (-x^3 + 5x^2 - 2x + 5)$$
$$= 5x^3 - 4x + 2 - x^3 + 5x^2 - 2x + 5 \quad \text{Remove parentheses.}$$
$$= 4x^3 + 5x^2 - 6x + 7 \quad \text{Combine like terms.}$$

Sometimes we write a polynomial addition problem as column addition. This approach has the advantage that like terms are aligned in columns. When you use the column approach, it is helpful to write any missing terms with a coefficient of 0.

$$\begin{array}{rl} 5x^3 + 0x^2 - 4x + 2 & \text{First polynomial} \\ -1x^3 + 5x^2 - 2x + 5 & \text{Second polynomial} \\ \hline 4x^3 + 5x^2 - 6x + 7 & \text{Add the polynomials.} \end{array}$$

Students are less likely to make sign errors if they can see that adding and subtracting polynomials is just a matter of simplifying expressions—that is, removing parentheses and combining like terms.

Subtraction of Polynomials

The definition of subtraction of real numbers applies to the subtraction of polynomials. If P and Q represent polynomials, then $P - Q = P + (-Q)$.

EXAMPLE 4

Subtracting Polynomials

Subtract the polynomial $2x^3 - x + 4$ from $-2x^3 + x^2 - 4x + 1$.

LEARNING TIP

When you subtract polynomials, write the intermediate step that shows the result of removing grouping symbols. This will help you to avoid the sign errors that commonly occur in subtraction.

When subtracting in columns, students generally do not want to write the problem twice. Instead, they change signs by crossing out the original signs and inserting the new ones. This often leads to careless sign errors. One suggestion is that the student circle the new signs to make them prominent.

Solution

$$(-2x^3 + x^2 - 4x + 1) - (2x^3 - x + 4)$$
$$= -2x^3 + x^2 - 4x + 1 - 2x^3 + x - 4 \quad \text{Remove parentheses.}$$
$$= -4x^3 + x^2 - 3x - 3 \quad \text{Combine like terms.}$$

Polynomial subtraction problems can also be done with columns. Again, it is helpful to write any missing terms with the coefficient of 0.

$$\begin{array}{rcr} -2x^3 + 1x^2 - 4x + 1 & & -2x^3 + 1x^2 - 4x + 1 \\ 2x^3 + 0x^2 - 1x + 4 & \text{Change signs.} & -2x^3 - 0x^2 + 1x - 4 \\ \hline & \text{Add the polynomials.} & -4x^3 + 1x^2 - 3x - 3 \end{array}$$

Examples 3 and 4 show that we add and subtract polynomials simply by removing parentheses and combining like terms. We can simplify any polynomial expression in the same way.

EXAMPLE 5

Simplifying Polynomial Expressions

(a) $2(x^3 - 2x^2 - 6x - 1) - 3(-2x^3 + x^2 - 4x - 2)$
$$= 2x^3 - 4x^2 - 12x - 2 + 6x^3 - 3x^2 + 12x + 6 \quad \text{Distributive Property}$$
$$= 8x^3 - 7x^2 + 4 \quad \text{Combine like terms.}$$

(b) $(2x^3y^4 - x^2y^3 + 3xy - 7) + (x^3y^4 + 3x^2y^3 + 5x - 2)$
$$= 2x^3y^4 - x^2y^3 + 3xy - 7 + x^3y^4 + 3x^2y^3 + 5x - 2 \quad \text{Remove parentheses.}$$
$$= 3x^3y^4 + 2x^2y^3 + 3xy - 9 + 5x \quad \text{Combine like terms.}$$

(c) $(x^2y^3 - 2xy^2 - 5xy - 7) - (x^2y^3 - xy^2 - 5xy + 3)$

$= x^2y^3 - 2xy^2 - 5xy - 7 - x^2y^3 + xy^2 + 5xy - 3$ Remove parentheses.

$= -xy^2 - 10$ Combine like terms.

Polynomial Functions

We can use function notation to give a polynomial a name and to indicate the variable used in the polynomial. For example, to write a polynomial in t, we give the polynomial a name such as p and use the function notation $p(t)$. (We can also use function notation for polynomials in more than one variable, but that topic is beyond the scope of this book.)

You may want to refer students to Sections 2.1 and 2.5 for a review of notation and evaluation.

Function notation is especially useful when we evaluate a polynomial. We can write "the value of the polynomial p when t has a value of 3" as $p(3)$.

EXAMPLE 6

Evaluating a Polynomial

For $p(t) = -t^3 + 3t - 6$, calculate $p(-2)$.

Solution

$$\begin{aligned} p(t) &= -t^3 + 3t - 6 \\ p(-2) &= -1(-2)^3 + 3(-2) - 6 \\ &= -1(-8) - 6 - 6 \\ &= 8 - 6 - 6 \\ &= -4 \end{aligned}$$

Think About It

We use $(p + q)(x)$ and $(p - q)(x)$ to represent the sum and difference of two polynomial functions. How could we represent the product of two polynomial functions? the quotient?

When two polynomial *expressions* are added or subtracted, the result is another polynomial expression. If $p(x)$ and $q(x)$ are two polynomial *functions,* then $p(x) + q(x)$, the sum of the two functions, is also a function. The function notation used for $p(x) + q(x)$ is $(p + q)(x)$. Similarly, we write $p(x) - q(x) = (p - q)(x)$.

EXAMPLE 7

Evaluating Sums of Polynomial Functions

Given $p(x) = x^2 - 5$ and $q(x) = 3x + 2$, write $(p + q)(x)$. Then evaluate the functions to verify that $(p + q)(4) = p(4) + q(4)$.

Solution

$$\begin{aligned} (p + q)(x) &= p(x) + q(x) \\ &= (x^2 - 5) + (3x + 2) \\ &= x^2 + 3x - 3 \end{aligned}$$

Evaluate the functions at 4.

$$(p + q)(4) = 4^2 + 3(4) - 3 = 16 + 12 - 3 = 25$$

$$p(4) = 4^2 - 5 = 16 - 5 = 11$$

$$q(4) = 3(4) + 2 = 12 + 2 = 14$$

Observe that $(p + q)(4)$ and $p(4) + q(4)$ are both 25.

Quick Reference 6.4

Definitions

- A **monomial** is a number or the product of a number and one or more variables raised to nonnegative integer powers. A **polynomial** is a finite sum of monomials.
- The **numerical coefficient** of a term is the numerical factor of the term. A **constant term** of a polynomial is a term that is just a number.
- Polynomials with one, two, and three terms are called **monomials, binomials,** and **trinomials,** respectively.
- Except for 0, the **degree of a monomial** in one variable is the value of the exponent on the variable. The degree of a monomial in more than one variable is the sum of the exponents on the variables. The **degree of a polynomial** is the degree of the term with the highest degree.

Addition of Polynomials

- Terms of a polynomial are **like terms** if they are both constant terms or if they have the same variables with the same exponents.
- To add polynomials, we remove parentheses and combine the like terms. Column addition has the advantage that like terms are aligned in columns.

Subtraction of Polynomials

- To subtract polynomials, remove parentheses by changing the sign of each term of the polynomial that is being subtracted. Then combine like terms.
- Column subtraction has the advantage that like terms are aligned in columns. However, remember to change the sign of each term of the polynomial being subtracted. Then add the like terms.
- All polynomial expressions are simplified by removing parentheses and combining like terms.

Polynomial Functions

- We can use function notation to give a polynomial a name and to indicate the variable used in the polynomial. Function notation is especially useful when we evaluate a polynomial.
- When two polynomial functions $p(x)$ and $q(x)$ are added or subtracted, the result is another polynomial function, which can be written as $(p + q)(x)$ for addition or $(p - q)(x)$ for subtraction.

Speaking the Language 6.4

1. The expression $5x^2$ is called a(n) ________; the expression $5x^2 - 6x + 3$ is called a(n) ________.
2. In the expression $7x^2y^3$, 7 is the ________ and 5 is the ________.
3. The expression $x^3 - 4x + 11$ is written in ________ order, and x^2 is a(n) ________ term.
4. We call $p(x) = x^2 - x + 1$ a(n) ________ function.

Exercises 1–30 deal primarily with vocabulary.

Exercises 6.4

Concepts and Skills

1. Explain why neither of the following is a polynomial.

(a) $\dfrac{5}{x+2}$ (b) $\dfrac{1}{3}x^{-4}$

2. Two of the following statements are true. Explain why the other two statements are not true.
   (i) The term $5x^2$ is both a monomial and a polynomial.
   (ii) The coefficients of $2x^3 + 5x - 8$ are 2 and 5.
   (iii) An example of a third-degree term is 2^3.
   (iv) The expression $5 + 4x + 3x^2 + 2x^3$ is in ascending order.

In Exercises 3–6, state whether the expression is a polynomial.

3. $5x^2 + 4x - 6$
4. $\dfrac{5}{x^2} + 4x - 6$
5. $\sqrt{7} - 4x$
6. $4x^3 - 8x + 5x^2$

In Exercises 7–12, give the coefficient and degree of each monomial.

7. 5
8. $-y$
9. $-4x^3$
10. $2\pi z^2$
11. xy
12. $4ab^2$

13. What is the difference between the degree of a term and the degree of a polynomial with more than one term?

14. The expression $x^2 + 5$ can be written in the form $x^2 + 0x + 5$. Does this mean that $x^2 + 5$ is a trinomial? Why?

In Exercises 15–18, determine the degree of the given polynomial.

15. $2x + 3y$
16. $4xy - 3y$
17. $2x^4 - xy^2 - x^2y^3$
18. $7a^4 - 3a^2b$

In Exercises 19–24, simplify each polynomial and determine its degree. Classify the simplified polynomial as a monomial, binomial, trinomial, or none of these.

19. $7x + 3x^5 - 9$
20. $5x^6$
21. $5x^3 + 7x - 2x^2 - 3$
22. $2x^2 + 3x - 1 - 2x^2$
23. $x^2y^3 + xy^2$
24. $3xy^2 + 2^3 - 3xy^2$

In Exercises 25–30, determine the degree of the polynomial. Then write the polynomial in descending order and in ascending order.

25. $6 - 3y^2 + 6y$
26. $7 - 2y^3$
27. $x^3 - 5x^6 + 4x^9 - 7x^8$
28. $3x^4 + 5x^7 - 6x^2 + 4x - 7x^{11}$
29. $4n^8 - 3n^4 + n^{12} - 2n^6$
30. $8p^5 - p^9 + 22p^4 - 5^7$

In Exercises 31–36, evaluate the polynomial for the given values of the variables.

31. $2x^2 + 3xy$
    $x = -1, y = 2$
32. $x^2y - 3y^2$
    $x = 2, y = -1$
33. $ab^2 - 2ab$
    $a = 1, b = -3$
34. $ab^2 + a^2b$
    $a = -2, b = 2$
35. $x^2y + 3xy - 2y^2$
    $x = -2, y = 3$
36. $x^3y^2 - 4x^2y + 5y^3$
    $x = 1, y = -1$

In Exercises 37–40, add the polynomials.

37. $(-a^2 + 2a - 1) + (a^2 + 2a + 1)$
38. $(x^2 - 5x) + (3x^2 - 4x - 2)$
39. $(2xy^2 - y^3) + (y^3 + xy^2)$
40. $(m^2n + mn) + (mn + 1)$

In Exercises 41–46, subtract the polynomials.

41. $(-5x - 3) - (2x - 9)$
42. $(5x - 6) - (-4x + 8)$
43. $(2x^2 - 4x + 7) - (-2x^2 - 3x - 7)$
44. $(x^2 + 6x + 3) - (4x - x^2)$

45. $(-3t + 7) - (2t^2 + 6)$ **46.** $(m^2 + 3) - (4 - 3m)$

In Exercises 47–52, perform the indicated operations.

47. $(4mn + n^2) - (m^2 - 3mn)$

48. $(x^2y + 3) - (2 - x^2y)$

49. $(x^2y + 2xy + 3y^2) + (-x^2y - 4xy + 2y^2)$

50. $(ab^2 + a^2b - 3ab) + (3ab^2 - 2ab - a^2b)$

51. $(9a^2b + 2ab - 5a^2) - (2ab - 3a^2 + 4a^2b)$

52. $(4y^3 - 3xy^2 + 4) - (2xy^2 - 3y^3 - 2)$

In Exercises 53–60, add or subtract as indicated.

53. Subtract
$$\begin{array}{l} 4x^3 - 10x \\ \underline{6x^3 - 2x} \end{array}$$

54. Subtract
$$\begin{array}{r} 3y - 6 \\ \underline{-5y + 2} \end{array}$$

55. Add
$$\begin{array}{r} x^2 + 3x - 7 \\ \underline{-5x^2 + 2x - 1} \end{array}$$

56. Add
$$\begin{array}{rrrr} 4x^3 & -\,5x^2 & & +\,6 \\ \hline & 2x^2 & -\,5x & +\,7 \end{array}$$

57. Subtract
$$\begin{array}{l} 4x^2 - x - 6 \\ \underline{3x^2 - x + 6} \end{array}$$

58. Subtract
$$\begin{array}{llll} 2x^4 & +\,3x^2 & & -\,6 \\ \underline{3x^4} & & -\,2x & +\,5 \end{array}$$

59. Add
$$\begin{array}{rrr} x^2 & +\,3x & -\,2 \\ -2x^2 & -\,x & \\ 3x^2 & & +\,4 \\ \hline \end{array}$$

60. Add
$$\begin{array}{rrrr} y^3 & +\,2y^2 & & -\,5 \\ -4y^3 & & -\,3y & +\,2 \\ & 6y^2 & +\,2y & -\,5 \\ \hline \end{array}$$

61. For the problem "Subtract $x - 2$ from $3x + 4$," which of the following is correct? Why?

(i) $(x - 2) - (3x + 4)$

(ii) $(3x + 4) - (x - 2)$

(iii) $3x + 4 - x - 2$

62. To solve the equation $\frac{1}{2}x - \frac{1}{3} = \frac{3}{4}$, we clear fractions by multiplying by the LCD. Can we do the same for the polynomial $\frac{1}{2}x - \frac{1}{3} - \frac{3}{4}$? Why?

63. Find the sum of $3x^2 - 6x$ and $4x + 7$.

64. Find the sum of $y^3 - 2y + 4$ and $y^2 + 2y + 1$.

65. Subtract $7x^2 - 2x + 9$ from $-5x - 2x^2$.

66. Subtract $2x^3 + 3x - 1$ from $x^3 + 3x - 1$.

CE **67.** What must be added to $x - 7$ to obtain 0?

CE **68.** What must be added to $3x + 5$ to obtain $x + 9$?

CE **69.** What must be subtracted from $x - 3$ to obtain 0?

CE **70.** What must be subtracted from $5x - 9$ to obtain $3x + 2$?

In Exercises 71–80, simplify. For Exercises 71–80, remind students about the Order of Operations.

71. $(2x - 3) - (3x^2 + 4x - 5) - (3x - 4x^2)$

72. $(-x^2 - x - 1) + (2x^2 + 3x + 4) - (2x - 3x^2)$

73. $-[3x - (2x - 1)]$

74. $-(2y + 4) - (2y - 4)$

75. $(x - 3) + [2x - (3x - 1)]$

76. $(2 - 3x) - [(3x + 1) - (1 - 2x)]$

77. $[3 - (y - 1)] - (2y + 1)$

78. $[(a^2 - 2) + (a + 2)] - (a^2 - 3a)$

79. $(3x^2 - 2xy + y^2) - (4xy + 3x^2 - y^2) + (4xy^2 - 2x^2)$

80. $(4a^2 + 5ab - 7b^2) + (5ab - a^2 + 3b^2) - (2a^2 - 7ab + 3a^2b)$

In Exercises 81–86, evaluate the function as indicated.

81. $P(x) = 3 - x$
$P(-2), P(3)$

82. $P(x) = -2x + 3$
$P(0), P(-1)$

83. $Q(x) = -x^2 + 3x - 1$
$Q(0), Q(-2)$

84. $P(x) = 3x^3 - 2x^2 + 6x + 4$
$P(-3), P(1)$

85. $R(x) = x^3 - 4x^5 + 5x^2 + 2$
$R(2.1), R(-1.7)$

86. $R(x) = x^4 + 5x^3 - 2x^2 - 5x$
$R(-3.1), R(1.2)$

In Exercises 85 and 86, encourage students to evaluate with a scientific calculator.

In Exercises 87–90, let $P(x) = 4x + 3$, $Q(x) = x^2 - 5$, and $R(x) = 2x^2 - 7x - 8$. Perform the indicated operations.

87. $P(x) + Q(x)$ **88.** $Q(x) - R(x)$

89. $R(x) - [P(x) - Q(x)]$

90. $[R(x) - P(x)] - Q(x)$

Real-Life Applications

91. The attendance for a game at a minor league baseball stadium is modeled by $A(t) = -2t^2 + 280t - 8000$, where t is the noonday temperature (Fahrenheit). For each of the following noonday temperatures, what would be the expected attendance?

(a) 80°F (b) 50°F (c) 98°F

(d) 40°F (e) 100°F

(f) Using the results in parts (d) and (e), and taking into account what the function represents, determine the domain of $A(t)$.

92. The number of people swimming at a certain Florida beach is modeled by $N(t) = -t^2 + 140t - 4500$, where t is the Fahrenheit temperature of the water. For each of the following water temperatures, what would be the expected number of people swimming?

(a) 60°F (b) 75°F

(c) 86°F (d) 50°F

(e) 90°F

(f) Using the results in parts (d) and (e), and taking into account what the function represents, determine the domain of $N(t)$.

93. The following functions describe costs and revenue (in dollars) from the production and sale of x units of light bulbs.

Cost C of Manufacturing	$C(x) = 20x + 12$
Cost M of Marketing	$M(x) = x^2 - 3x + 10$
Revenue R from Sales	$R(x) = 2x^2 + 8x$

(a) What is the cost of manufacturing 100 units of bulbs?

(b) What is the cost of marketing 100 units of bulbs?

(c) Write a function $T(x) = C(x) + M(x)$ that represents the total cost of manufacturing and marketing x units of bulbs.

(d) What is the total cost of manufacturing and marketing 100 units of bulbs?

(e) Show that $C(100) + M(100) = T(100)$.

(f) What is the revenue from selling 100 units of bulbs?

(g) Assuming profit is revenue minus total cost, write a function $P(x) = R(x) - T(x)$ that represents the profit from selling x units of bulbs.

(h) What is the profit from selling 100 units of bulbs?

(i) Show that $P(100) = R(100) - T(100)$.

94. The following functions describe costs and revenue (in dollars) from the production and sale of x units of radios.

Cost C of Manufacturing	$C(x) = 25x + 700$
Cost M of Marketing	$M(x) = x^2 - 2x + 15$
Revenue R from Sales	$R(x) = 4x^2 + 3x$

(a) What is the cost of manufacturing 70 units of radios?

(b) What is the cost of marketing 70 units of radios?

(c) Write a function $T(x) = C(x) + M(x)$ that represents the total cost of manufacturing and marketing x units of radios.

(d) What is the total cost of manufacturing and marketing 70 units of radios?

(e) Show that $C(70) + M(70) = T(70)$.

(f) What is the revenue from selling 70 units of radios?

(g) Assuming profit is revenue minus total cost, write a function $P(x) = R(x) - T(x)$ that represents the profit from selling x units of radios.

(h) What is the profit from selling 70 units of radios?

(i) Show that $P(70) = R(70) - T(70)$.

Challenge

In Exercises 95–98, determine a, b, and c so that the sum or difference on the left is equal to the polynomial on the right.

95. $(ax^2 + bx - 3) + (x^2 + 2x + c) = 3x + 1$

96. $(by + c) - (ay^3 - 2y + 5) = y^3 - 1$

97. $(x^2 + bx - c) + ax^2 = x^2 + 6x + 9$

98. $(2x^2 + bxy + 3y^2) - (ax^2 - xy - cy^2)$
$= x^2 + xy - 4y^2$

99. If $(ax^2 + 5) + (bx^2 - 7) = 4x^2 - 2$, explain why it is not possible to determine unique values for a and b.

100. If the degree of $p(x)$ is $n > 1$, what is the degree of $p(3)$?

101. Suppose that $P(x) = x^3 - 5x^2 + 6x + 10$ and $P(0) = a + b + c$, where a, b, and c are constants. If c is the sum of a and b, and if b is one less than twice a, what are a, b, and c?

6.5 Multiplication and Special Products

Multiplying by a Monomial • Multiplying Polynomials • Multiplying Binomials • Special Products • Simplifying Expressions

Multiplying by a Monomial

In Section 6.1 we learned the Product Rule for Exponents.

$$b^m \cdot b^n = b^{m+n}$$

In words, if the bases are the same, then we add the exponents. To multiply polynomials, we use this rule, often repeatedly.

Remind students that they already know how to multiply monomials. Also, students are usually heartened when you tell them that all polynomial multiplication comes down to this one basic skill.

To multiply two monomials, we use the associative and commutative properties of multiplication with the Product Rule for Exponents.

$$(4x^4)(5x^2) = (4 \cdot 5)(x^4 \cdot x^2) = 20(x^{4+2}) = 20x^6$$

To multiply a monomial times a polynomial, we use the Distributive Property.

EXAMPLE 1

The Product of a Monomial and a Polynomial

Multiply $3x^2(5x^3 - x^2 - 4x + 2)$.

Solution

We use the Distributive Property to multiply the monomial times each term of the polynomial.

$$\begin{aligned} &3x^2(5x^3 - x^2 - 4x + 2) \\ &\quad = 3x^2(5x^3) - 3x^2(x^2) - 3x^2(4x) + 3x^2(2) \\ &\quad = 15x^5 - 3x^4 - 12x^3 + 6x^2 \end{aligned}$$

Multiplying Polynomials

Multiplying a polynomial times another polynomial requires repeated use of the Distributive Property.

EXAMPLE 2

The Product of Two Polynomials

Multiply $(2x - 3)(x^2 + 3x - 5)$.

Solution

Multiply the second polynomial by $2x$ and then by -3.

$$\begin{aligned} &(2x - 3)(x^2 + 3x - 5) \\ &\quad = 2x(x^2 + 3x - 5) - 3(x^2 + 3x - 5) \\ &\quad = 2x(x^2) + 2x(3x) - 2x(5) - 3(x^2) - 3(3x) + 3(5) \\ &\quad = 2x^3 + 6x^2 - 10x - 3x^2 - 9x + 15 \\ &\quad = 2x^3 + 3x^2 - 19x + 15 \end{aligned}$$

Again, you can frame the task in terms of something students already know how to do: simplifying an expression. The most prevalent errors are in the signs.

Another approach to the problem in Example 2 is column multiplication. This method makes it easier to align and combine like terms.

$$
\begin{array}{rrrrl}
 & x^2 + & 3x - & 5 & \\
 & & 2x - & 3 & \\
\hline
 & -3x^2 - & 9x + & 15 & \quad -3(x^2 + 3x - 5) \\
2x^3 + & 6x^2 - & 10x & & \quad 2x(x^2 + 3x - 5) \\
\hline
2x^3 + & 3x^2 - & 19x + & 15 & \quad \text{Add the results.}
\end{array}
$$

EXAMPLE 3

The Product of Two Trinomials

Multiply $(x^2 + 3x - 1)(2x^3 - x + 1)$.

Solution

Use the Distributive Property to remove the parentheses. Then combine the like terms.

$$
\begin{aligned}
(x^2 + 3x - 1)&(2x^3 - x + 1) \\
&= x^2(2x^3 - x + 1) + 3x(2x^3 - x + 1) - 1(2x^3 - x + 1) \\
&= 2x^5 - x^3 + x^2 + 6x^4 - 3x^2 + 3x - 2x^3 + x - 1 \\
&= 2x^5 + 6x^4 - 3x^3 - 2x^2 + 4x - 1
\end{aligned}
$$

With practice, you will soon find that you do not need to write the first step. Also, you may find that column multiplication is especially appealing when the polynomials are large.

EXAMPLE 4

The Product of Two Binomials

Multiply $(2x - 1)(3x + 5)$.

Solution

Again, we use the Distributive Property.

$$
\begin{aligned}
(2x - 1)(3x + 5) &= 2x(3x + 5) - 1(3x + 5) \\
&= 6x^2 + 10x - 3x - 5 \\
&= 6x^2 + 7x - 5
\end{aligned}
$$

Multiplying Binomials

To multiply binomials efficiently, shortcuts or mnemonic devices are often used.

Exploring the Concept

The FOIL Method

Look carefully at the result in Example 4.

$$(2x - 1)(3x + 5) = 6x^2 + 10x - 3x - 5$$

1. The product of the *first terms* ($2x$ and $3x$) of the binomials is the first term of the result ($2x \cdot 3x = 6x^2$).
2. The products of the *outer terms* ($2x$ and 5) and the *inner terms* (-1 and $3x$) are the next two terms of the result $[2x \cdot 5 + (-1)(3x) = 10x - 3x]$.
3. The product of the *last terms* (-1 and 5) of the binomials is the last term of the result $[(-1)(5) = -5]$.

Using the letters in the word FOIL might help you remember which terms are multiplied and allow you to compute the product quickly.

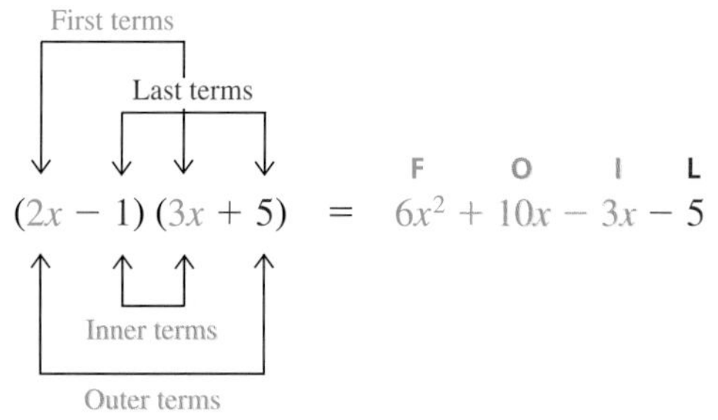

Color coding is a very effective way of emphasizing patterns, and it can be carried over later into factoring.

With practice, you should be able to combine like terms mentally.

EXAMPLE 5

Using the FOIL Pattern to Multiply Binomials

(a) $(x - 3)(4x - 1) = \overset{F}{4x^2} \overset{O}{- x} \overset{I}{- 12x} \overset{L}{+ 3}$

$= 4x^2 - 13x + 3$

(b) $(x + 4)^2 = (x + 4)(x + 4)$

$= \overset{F}{x^2} + \overset{O}{4x} + \overset{I}{4x} + \overset{L}{16}$

$= x^2 + 8x + 16$

Unless you do not want to mention it, this is a good opportunity to point out that $(a + b)^2 \neq a^2 + b^2$.

Note that the product $(x + 4)^2$ is the product of the same two binomials.

(c) $(3x + 5)(3x - 5) = \overset{F}{9x^2} \overset{O}{- 15x} \overset{I}{+ 15x} \overset{L}{- 25}$

$= 9x^2 - 25$

Note that the product $(3x + 5)(3x - 5)$ is the product of the sum of two terms and the difference of the same two terms.

Special Products

Parts (b) and (c) of Example 5 illustrate two special products of binomials.

If A and B represent terms (monomials), the special product of the sum and difference of those two terms can be written as $(A + B)(A - B)$. We can use the FOIL method to find the product.

$$(A + B)(A - B) = A^2 - AB + AB - B^2$$
$$= A^2 - B^2$$

In words, the product of the sum and difference of the same two terms is the difference of their squares.

Product of the Sum and Difference of the Same Two Terms

$$(A + B)(A - B) = A^2 - B^2$$

EXAMPLE 6

The Product of a Sum and Difference of the Same Two Terms

Multiply.

(a) $(3x + 5)(3x - 5)$ (b) $(5 + x^2)(5 - x^2)$

Solution

(a) $(A + B)(A - B) = A^2 - B^2$

$\downarrow \quad \downarrow \; \downarrow \quad \downarrow \qquad \downarrow \quad \downarrow$

$(3x + 5)(3x - 5) = (3x)^2 - 5^2 = 9x^2 - 25$

(b) $(5 + x^2)(5 - x^2) = 5^2 - (x^2)^2 = 25 - x^4$

The square of a binomial is another special product. Letting A and B represent terms and using the FOIL method, we find that the square of a binomial is a trinomial of a specific form.

$$(A + B)^2 = (A + B)(A + B) = A^2 + AB + AB + B^2 = A^2 + 2AB + B^2$$

$$(A - B)^2 = (A - B)(A - B) = A^2 - AB - AB + B^2 = A^2 - 2AB + B^2$$

Encourage students to verbalize the special product patterns.

The first term of the result is the square of the first term of the binomial. The last term of the result is the square of the last term of the binomial. The middle term of the result is double the product of the two terms of the binomial.

The Square of a Binomial

$$(A + B)^2 = A^2 + 2AB + B^2$$

$$(A - B)^2 = A^2 - 2AB + B^2$$

EXAMPLE 7

The Square of a Binomial

Determine the following products.

(a) $(x + 4)^2$ (b) $(2x^3 + 5)^2$ (c) $(3 - 2y)^2$

Solution

(a) $(A + B)^2 = A^2 + 2 \cdot A \cdot B + B^2$

$\downarrow \quad \downarrow \qquad \downarrow \quad \downarrow \quad \downarrow \quad \downarrow \quad \downarrow$

$(x + 4)^2 = x^2 + 2 \cdot (x) \cdot (4) + 4^2 = x^2 + 8x + 16$

(b) $(2x^3 + 5)^2 = (2x^3)^2 + 2(2x^3)(5) + 5^2 = 4x^6 + 20x^3 + 25$

(c) $(3 - 2y)^2 = 9 - 12y + 4y^2$

Think About It

Suppose that you create a table of values for $(x + 3)^2$, $x^2 + 9$, and $x^2 + 6x + 9$. Which columns should be the same? Why?

Simplifying Expressions

Sometimes we must multiply polynomials as part of a larger simplification problem. In the following examples, we use the Distributive Property, we use the FOIL pattern, and we recognize special products in simplifying the expressions.

EXAMPLE 8

Simplifying Expressions

Simplify the following expressions.

(a) $(x + 3)(2x + 1) - (x + 3)(x - 3)$

(b) $2x(x - 5)(x + 2)$

(c) $(x + 5)(x - 3)(x - 5)$

LEARNING TIP

For an expression such as $x(x + 1)(x + 2)$, avoid the error of multiplying x times both $x + 1$ and $x + 2$. Multiplying the binomials first and then multiplying the result by x is generally the easiest approach.

Solution

(a) The first product can be found with the FOIL method. The second product is the sum of two terms times the difference of the same two terms, and the result is the difference of their squares.

$$\begin{aligned}&(x + 3)(2x + 1) - (x + 3)(x - 3)\\&\quad= (2x^2 + 7x + 3) - (x^2 - 9)\\&\quad= 2x^2 + 7x + 3 - x^2 + 9\\&\quad= x^2 + 7x + 12\end{aligned}$$

In the first step, the product $(x + 3)(x - 3)$ is the *quantity* $(x^2 - 9)$. The quantity must remain in parentheses until the subtraction is performed.

This is a very common error, and the correct method needs emphasis.

(b) We can multiply $2x$ and $x - 5$ and then multiply the result by $x + 2$.

$$2x(x - 5)(x + 2) = (2x^2 - 10x)(x + 2) = 2x^3 - 6x^2 - 20x$$

Alternatively, we can multiply the binomials and then multiply the result by $2x$.

$$2x(x - 5)(x + 2) = 2x(x^2 - 3x - 10) = 2x^3 - 6x^2 - 20x$$

(c) Look for ways to rearrange the factors so that special product patterns can be used.

$(x + 5)(x - 3)(x - 5)$	
$= (x + 5)(x - 5)(x - 3)$	Commutative Property of Multiplication
$= (x^2 - 25)(x - 3)$	$(A + B)(A - B) = A^2 - B^2$
$= x^3 - 3x^2 - 25x + 75$	FOIL

Quick Reference 6.5

Multiplying by a Monomial

- To multiply a monomial by a monomial, multiply the coefficients and the bases.
- To multiply a monomial by a polynomial, use the Distributive Property to multiply each term of the polynomial by the monomial.

Multiplying Polynomials

- To multiply two polynomials, multiply each term of one polynomial by each term of the other polynomial and then combine like terms.
- For larger polynomials, multiplying in columns helps to align like terms for easier addition.

Multiplying Binomials

- The letters in the word FOIL (**F**irst–**O**uter–**I**nner–**L**ast) give us a pattern for remembering how to multiply two binomials quickly.

Special Products

- The product of the sum and difference of two terms has the form $(A + B)(A - B)$.

  $(A + B)(A - B) = A^2 - B^2$
- The square of a binomial has the form $(A + B)^2$ or $(A - B)^2$. The products have the following forms.

  $(A + B)^2 = A^2 + 2AB + B^2$

  $(A - B)^2 = A^2 - 2AB + B^2$

Simplifying Expressions

- The Distributive Property, the FOIL pattern, and the special product forms are useful in simplifying algebraic expressions that involve products of polynomials.

Speaking the Language 6.5

1. The ________ method is a shortcut for multiplying binomials.
2. All multiplication of polynomials involves the ________ Property.
3. If A and B are monomials, then $A^2 + 2AB + B^2$ is the result of ________ the binomial $A + B$.
4. The product $(A + B)(A - B)$ is called the product of the ________ and ________ of A and B.

Exercises 6.5

Concepts and Skills

1. Which of the following statements is false? Why?

   (i) The simplified product of a nonzero monomial and a binomial is a binomial.

   (ii) The simplified product of a binomial and a binomial is a trinomial.

2. Give an example of two binomials whose simplified product has the following number of terms.

   (a) 4 (b) 3 (c) 2

In Exercises 3–8, multiply the monomial times the polynomial.

3. $3x^2(2x - 3)$
4. $2x^3(3x + 4)$
5. $-4x^3(2x^2 + 5)$
6. $a^4(a^2 - 4a + 3)$
7. $-xy^2(x^2 - xy - 2)$
8. $-6x^2y(x^2 - 2xy + 7xy^2)$

In Exercises 9–28, multiply the binomials.

9. $(x - 6)(x + 4)$
10. $(x + 7)(x - 3)$
11. $(y - 7)(y - 2)$
12. $(x - 6)(x - 5)$

13. $(y + 5)(3 - y)$
14. $(8 - x)(x + 2)$
15. $(5x - 6)(4x + 3)$
16. $(2x + 3)(4x - 5)$
17. $(9x + 2)(x + 3)$
18. $(4y - 7)(7 - 4y)$
19. $(4 - 5y)(2y + 1)$
20. $(1 - 4x)(2 + 7x)$
21. $(x - 9y)(x - 4y)$
22. $(x - 5y)(x - 3y)$
23. $(a + 5b)(a - 2b)$
24. $(m + 4n)(m - n)$
25. $(6x - 5y)(2x - 3y)$
26. $(3x + 2y)(4x + y)$
27. $(2x^2 - 3)(3x^2 - 2)$
28. $(4x^2 - 5)(3x^2 - 7)$

In Exercises 29–38, multiply the polynomials.

Encourage students to experiment with the horizontal and vertical methods to see which they like better.

29. $(x + 2)(x^2 - 4x + 5)$
30. $(x - 3)(x^2 + 5x - 3)$
31. $(2x - 3)(3x^2 + x - 1)$
32. $(3x + 2)(2x^2 - 2x + 1)$
33. $(x + y)(x^2 - xy + y^2)$
34. $(x - 2y)(x^2 + 2xy + 4y^2)$
35. $(2x - 1)(2x^3 - x^2 + 3x - 4)$
36. $(3x + 2)(3x^3 - 2x^2 + 5x - 2)$
37. $(x^2 - 2x + 1)(x^2 + 2x + 1)$
38. $(x^2 + 4x + 4)(x^2 - 4x + 4)$

In Exercises 39–42, multiply the polynomials.

39. $\begin{array}{r} x^2 + 2x + 4 \\ x - 2 \\ \hline \end{array}$
40. $\begin{array}{r} x^2 - 3x + 9 \\ x + 3 \\ \hline \end{array}$
41. $\begin{array}{r} 2x^2 - 4x + 5 \\ x^2 - 2x + 3 \\ \hline \end{array}$
42. $\begin{array}{r} 2x^2 + 3x - 2 \\ x^2 - 2x + 1 \\ \hline \end{array}$

43. Can the FOIL method be used to determine the product $(x + 3)(x - 3)$? What is the advantage of recognizing this as a special product instead?

44. Can the FOIL method be applied to determine the product $x - 3(x - 3)$? Why?

In Exercises 45–54, square the binomial.

45. $(x + 3)^2$
46. $(x - 5)^2$
47. $(2 - y)^2$
48. $(4 + x)^2$
49. $(7x - 3)^2$
50. $(9 + 2a)^2$
51. $(3a - b)^2$
52. $(m - 6n)^2$
53. $(2x + 5y)^2$
54. $(3x - 4y)^2$

In Exercises 55–64, multiply the binomials.

55. $(x - 4)(x + 4)$
56. $(x + 3)(x - 3)$
57. $(7 - y)(7 + y)$
58. $(3 + n)(3 - n)$
59. $(2x - 3)(2x + 3)$
60. $(6 + 5a)(6 - 5a)$
61. $(4x + 5y)(4x - 5y)$
62. $(2x - 7y)(2x + 7y)$
63. $(9n - 4m)(9n + 4m)$
64. $(a - 2b)(a + 2b)$

65. For $x^2(x + 1)(x - 3)$, two approaches are
(i) multiply the two binomials and then multiply the result by x^2.
(ii) multiply $x + 1$ by x^2 and then multiply the result times $x - 3$.
Explain why both approaches are correct. Which approach is easier?

66. Explain why the following two products are the same. Which is easier to multiply?
(i) $(x - a)[(x + a)(x + a)]$
(ii) $[(x - a)(x + a)](x + a)$

In Exercises 67–86, multiply and simplify.

This group mixes the special products and uses the Order of Operations.

67. $(a^3 - b^3)(a^3 + b^3)$
68. $(a^2 + b^2)(a^2 - b^2)$
69. $(4 - pq^2)(4 + p^2q)$
70. $(a^2b^2 - 1)(a^2b^2 - 3)$
71. $(x^2y^2 + 2)^2$
72. $(x^3 + 4yz^2)^2$
73. $2x^3(2x - 5)(3x + 1)$
74. $-3y(y^2 + 1)(y^2 - 1)$
75. $5x - 6(4x + 3)$
76. $2x + 3(4x - 5)$
77. $(b - 3)(2b + 5)(b - 2)$
78. $(a - 5)(3a + 7)(a + 2)$
79. $(2x + 1)(4x^2 + 1)(2x - 1)$
80. $(1 + 3a)(9a^2 - 1)(3a - 1)$
81. $(2x + 5)(2x - 5) - (2x + 5)^2$
82. $(x + 7)(x - 7) - (x + 3)(x - 3)$
83. $(x + 6)(x - 4) - (2x - 3)(x + 2)$
84. $(3x + 5)(2x + 4) - (x - 6)(4x - 5)$
85. $(x + 3)^2 - (x - 3)^2$
86. $(x - 2)^2 + (5 - 2x)(1 + x)$
87. Multiply the square of the quantity $4x - 3$ by $2x^3$.
88. Multiply the square of the quantity $2 + 3y$ by $y + 1$.
89. Subtract the square of the quantity $5 - 2y$ from the product of the sum and difference of $3y$ and 2.
90. Subtract the product of $3 - 5x$ and $x + 4$ from $7x^2 + 5x + 6$.

CE In Exercises 91–94, determine a and b so that the product is equal to the polynomial.

91. $(ax + 3)(2x + b) = 8x^2 - 2x - 6$
92. $(ax + b)(3x + 1) = 6x^2 - 7x + b$
93. $(ax - 2)(3x + b) = 9x^2 - 4$

94. $(ax - b)^2 = 49 + 14x + x^2$

95. At noon, a car left a gas station and traveled north at an average of r miles per hour. The car reached its destination t hours later. At 1:00 P.M., a van left the same gas station and traveled south. Its average speed was 5 mph greater than the car's average speed. Write a polynomial expression (in two variables) that represents the total distance between the car and the van at the time that the car reached its destination.

96. Write a polynomial expression that represents the product of three consecutive integers.

97. A gardener uses 300 feet of fencing to enclose three sides of a rectangular garden. With w representing the width of the garden, write a polynomial expression that represents the area of the garden.

98. The lengths of the two legs of a right triangle are $(3x + 2)$ and $(x - 5)$. Use the Pythagorean Theorem to derive a polynomial expression for the square of the hypotenuse.

Data Analysis: *College Textbook Sales*

The following table shows the number (in millions) of college textbooks sold during the period 1991–1997. Also shown are the average wholesale prices (in dollars) for college textbooks during the same period. (Source: Book Industry Study Group, Inc.)

Year	Number Sold (millions)	Average Price (dollars)
1991	137	17.33
1993	136	18.37
1995	142	19.07
1997	154	20.19

99. Use the concept of rate of change to explain whether you think a linear model would be appropriate for the number of books sold.

100. Letting x represent the number of years since 1990 and using the data for 1991 and 1997, write a linear function P to model the average wholesale price of textbooks. (Round coefficients to the nearest hundredth.)

101. The function $B(x) = 0.81x^2 - 3.65x + 139.79$ models the number of books sold as a function of the number of years since 1990. Referring to Exercise 100, write a simplified expression for $E(x) = B(x) \cdot P(x)$. (Round coefficients to the nearest hundredth.)

102. Referring to Exercise 101, explain what the function $E(x)$ represents.

Challenge

In Exercises 103–106, multiply.

103. $(x^n + 2)(x^n + 4)$

104. $(x^n - 2)^2$

105. $(x^n + 4)(x^n - 4)$

106. $(x + 1)(x^{2n} - x^n + 1)$

107. In the accompanying figure, the total area of the largest rectangle is the sum of the areas of the four inner rectangles. Use this fact to show the following.

$$(a + b)^2 = a^2 + 2ab + b^2$$

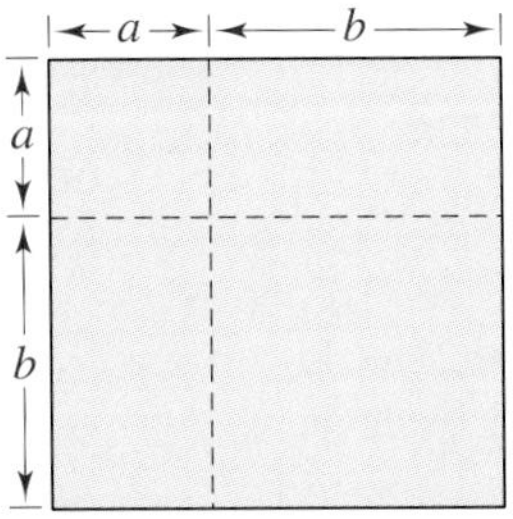

108. In the accompanying figure, the total area of the largest rectangle is the sum of the areas of the three inner rectangles. Use this fact to show the following.

$$(a - b)(a + b) = a^2 - b^2$$

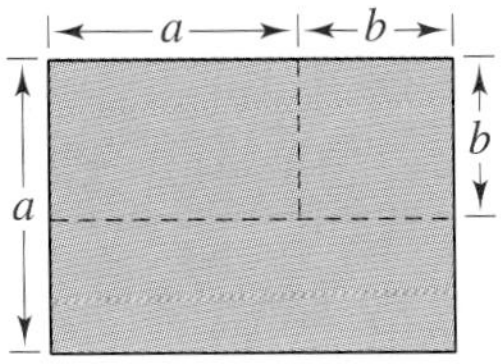

109. Explain how to use a special product to find the product of the trinomials.

$$(x + 3y - 4)(x + 3y + 4).$$

In Exercises 110 and 111, decide how to rearrange terms and to group within the factors to obtain a special product. Then perform the multiplication.

110. $(x + y + 2)(2 + x + y)$

111. $(x^2 + 5x + 3)(x^2 + 5x - 3)$

In Exercises 112 and 113, multiply.

112. $[2y - (x - 3)]^2$

113. $[(x - 3) + 5y][(x - 3) - 5y]$

6 • Chapter Project — Microgravity and Space Flight

We can think of "microgravity" as a small fraction of the gravitational pull that we experience on the surface of the earth.

1. To reach a point where the earth's gravitational pull is one-millionth of that at the surface, you would have to travel into space a distance of 6,370,000 kilometers. Write this distance in scientific notation.
2. The closest that the moon comes to the earth is approximately 221,000 miles. Given that 1 mile $\approx$ 1.61 kilometers, how many times farther is the distance in Exercise 1 than the distance from the earth to the moon?
3. A specially equipped airplane, the KC-135, is used to create microgravity conditions. When this airplane rises at a 45° angle, pushes over the top of an arc, and then descends at 45°, the gravitational pull on the passengers is about $1.0 \cdot 10^{-2}$ the pull at the earth's surface. How much would a 170-pound person weigh during this condition?
4. To remain in orbit, the space shuttle must travel at a velocity of about 4.9 miles per second. Convert this velocity to miles per hour, and write the result in both scientific and decimal notation.
5. The *period* of an orbit is the time required to travel one orbit. Use the formula $d = rt$ and the result in Exercise 4 to calculate the period (in minutes) of an orbit of 24,900 miles.
6. The required orbiting speed depends on the altitude of the orbiter. If the shuttle is at a 350-kilometer altitude, it must travel at 27,740 kilometers per hour to achieve a stable orbit. Using the unit conversion given in Exercise 2, write this speed in miles per hour, and express your result in scientific notation.

Chapter 6 Review Exercises

Section 6.1

1. Explain why we simplify the product $x^2 \cdot x^3$ by adding the exponents rather than by multiplying them.

2. To simplify $(x^2y^3)^4$, we use the Product to a Power Rule. Explain why the Power to a Power Rule must also be used.

In Exercises 3 and 4, use the Product Rule for Exponents to simplify. Express the result in exponential form.

3. $-3 \cdot 3^4 \cdot 3^7$

4. $k^4 \cdot k^5 \cdot k \cdot k^2$

In Exercises 5 and 6, use the Quotient Rule for Exponents to simplify. Express the result in exponential form.

5. $\dfrac{4^{11}}{4^7}$

6. $\dfrac{x^5y^9}{xy^6}$

In Exercises 7–12, simplify.

7. $\left(-\dfrac{2}{5}\right)^3$

8. $(c^7)^6$

9. $\left(\dfrac{2h^3}{k^2}\right)^3$

10. $(-x^3)^3(2x^7)$

11. $(-7x^5y)(5x^7y^8)$

12. $\dfrac{30c^{18}}{6c^6}$

Section 6.2

13. If $b \neq 0$, we can use the Quotient Rule for Exponents to write $\frac{b^n}{b^n} = b^{n-n} = b^0$. Explain what prompted us to define $b^0 = 1$ for $b \neq 0$.

14. If $b^{-n} = a$, where $n > 0$, explain why neither a nor b can be 0.

In Exercises 15–20, evaluate the expression.

15. 1776^0 16. -2^{-2}

17. $(-3)^{-3}$ 18. $5^{-1} + 3^{-1}$

19. $(5 + 3)^{-1}$ 20. $7^0 \cdot 4^{-2}$

In Exercises 21 and 22, write the expression with positive exponents.

21. $\left(-\frac{h}{k}\right)^{-1}$ 22. $\frac{-4h^{-2}}{3k^{-3}}$

In Exercises 23 and 24, use the Product Rule for Exponents to simplify. Express the result in exponential form with positive exponents.

23. $r^0 \cdot r^{-3} \cdot r \cdot r^{-7}$ 24. $(-4x^{-2})(x^{-8})(x^7)$

In Exercises 25 and 26, use the Quotient Rule for Exponents to simplify. Express the result in exponential form with positive exponents.

25. $\frac{y^{-7}}{y^{-5}}$ 26. $\frac{2^{-3}x^{-4}}{2^2x^{-5}}$

In Exercises 27–30, simplify. Express the result with positive exponents.

27. $\left(\frac{c^2}{d}\right)^{-7}$ 28. $(2x^4y^{-5})^{-3}$

29. $\left(\frac{2h^{-5}}{k^3}\right)^{-4}$ 30. $\frac{y^4(y^3)^{-5}}{y^{-7}}$

Section 6.3

31. We can write 23,000 as $2.3 \cdot 10^p$. Explain how to determine p.

32. Use your calculator to compute 5^{25}. Explain how to interpret the displayed result.

In Exercises 33 and 34, write the given number in scientific notation.

33. 24,500,000,000 34. 0.00000563

In Exercises 35 and 36, write the given number in decimal notation.

35. $6.83 \cdot 10^{-7}$ 36. $8.49 \cdot 10^5$

In Exercises 37 and 38, use your calculator to perform the indicated operations. Write the result in scientific notation $n \cdot 10^p$, where n is rounded to the nearest tenth.

37. $(-439)^{33}$ 38. $\frac{7.4 \cdot 10^{-9}}{4.3 \cdot 10^7}$

39. The land area of Missouri is 68,945 square miles. The population of Missouri in 1995 was 5,137,804.
    (a) Write both of these numbers in scientific notation $n \cdot 10^p$ with n rounded to the nearest tenth.
    (b) Use the numbers in part (a) to calculate the average population per square mile.

40. The cost to the U.S. Fish and Wildlife Service of restoring the hooded bobwhite to the Southwest was $23,000 per bird. If 55 birds were known to survive, and the population of the United States was $2.6 \cdot 10^8$, what was your share of the total cost?

Section 6.4

In Exercises 41–44, answer each of the following.

(a) State whether the given polynomial is a monomial, a binomial, a trinomial, or none of these.
(b) State the degree of the polynomial.
(c) Write the polynomial in descending order.
(d) State the coefficient of the x^2 term.

41. $3x - 2x^2$ 42. x^4

43. $x^2 + 9x^5 - 3$ 44. $x^6 - 7x^2 + 5x^5 - 4x$

In Exercises 45 and 46, determine the degree of the polynomial.

45. $4x^2 + y^2 + x^2y^4$

46. $a^3b - b^3$

In Exercises 47 and 48, evaluate the given function as indicated.

47. $P(x) = x^5 - x^3 + 2x^2$; $P(1)$

48. $P(x) = -x^2 - 3x - 5$; $P(-2)$

In Exercises 49 and 50, evaluate the given polynomial as indicated.

49. $x^5y^3 - 4x^3y^4 + x^2$ for $x = 1$ and $y = 2$

50. $a^3 - a^2b + ab^2 - b^3$ for $a = 1$ and $b = -1$

In Exercises 51–54, perform the indicated operations.

51. $(5x^2 - 3x + 4) + (-5x - 3x^2 + 9)$

52. $(3x - 2) - (x^2 - 3x + 2) + (x^2 - 4x - 5)$

53. Find the sum of $3x^2 - 6x$ and $4x + 7$.

54. Subtract $-2x^2 + 3x - 7$ from $5x^2 - x$.

Section 6.5

In Exercises 55–62, multiply.

55. $4x^5(-2x^3)$

56. $2x^3(3x^2 - 4)$

57. $2a^2b(5 - b^2 + 3b)$

58. $(x - 4)(x + 3)$

59. $(2x + 3)(8 - 3x)$

60. $(x + 2)(2x^2 - 3x + 1)$

61. $(x + 3)(2x - 5) - (3x + 1)(x + 4)$

62. $3x(x - 1)(x + 2)$

63. Write a polynomial expression for the area of a rectangle with a length of $4x - 5$ and a width of $2x - 3$.

64. If three numbers are represented by $x + y$, $y - 3$, and $4 - x$, write a polynomial expression for the product of the numbers.

65. In words, describe how to recognize a difference of two squares.

66. For what special product is the following template used for multiplying?

$$(\quad)^2 + 2(\quad)(\quad) + (\quad)^2$$

In Exercises 67–72, perform the indicated operations.

67. $(3x - 2)(3x + 2)$

68. $(1 - 3y)(1 + 3y)$

69. $(3y - 2)^2$

70. $(ab + 4)^2$

71. $(2x - 1)(x - 3) - (x - 3)^2$

72. $[(x + 2y) + 5][(x + 2y) - 5]$

73. Write a polynomial expression for the product of three consecutive integers where x represents the second integer.

74. The width and height of a rectangular carton are the same. The length is 1 foot longer than the height. This carton is then placed inside a second carton in the shape of a cube with the same length as the first carton. (See figure.) Write a polynomial expression for the space that remains inside the second carton.

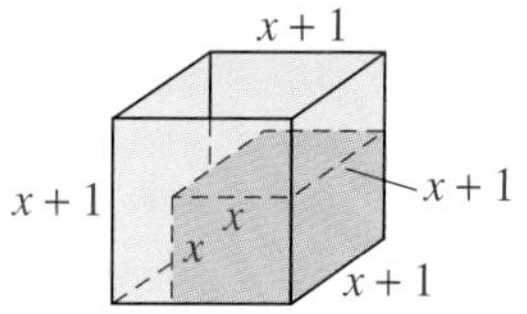

Looking Back

The following exercises review concepts and skills that you will need in Chapter 7.

1. Use the Distributive Property to rewrite each of the given expressions.

(a) $5(2 - 3y)$ (b) $3x + 3y$

2. Write each polynomial in descending order.

(a) $2x - 15 + x^2$ (b) $7 - 15x^2 - 8x$

In Exercises 3–8, multiply the polynomials.

3. $(y + 5)(y - 5)$

4. $(1 + 4x)(1 - 4x)$

5. $(a + 1)^2$

6. $(2x - 3)^2$

7. $(3x + 1)(x - 5)$

8. $(x + 3)(x^2 - 3x + 9)$

9. Find two numbers whose sum is the first number and whose product is the second number.

(a) 9, 20 (b) $-5, -6$

(c) $-7, 12$ (d) $6, -16$

In Exercises 10 and 11, solve the given equation.

10. (a) $2x + 1 = 0$ (b) $-5x = 0$

11. $|x - 7| = 8$

12. Explain how to determine the x-intercept of the graph of the linear equation $y = mx + b$.

Chapter 6 Test

In Questions 1–3, simplify the given product. Express the results with positive exponents.

1. $(x^3y^2)(xy^4)$ **2.** $t^3 \cdot t^{-2} \cdot t$ **3.** $(2a^{-5}b^4)^2$

In Questions 4–6, simplify the given quotient. Express the results with positive exponents.

4. $\left(\dfrac{a^2}{b^{-3}}\right)^{-2}$ **5.** $\dfrac{4a^{-3}}{5b^{-2}}$ **6.** $\dfrac{c^{-3}d^4}{c^{-5}d^{-2}}$

In Questions 7–9, evaluate the given expression.

7. $\dfrac{(-2)^{-3}}{(-2)^2}$ **8.** $3x^0$ **9.** $(-2)^{-3}$

In Questions 10 and 11, write the given number in decimal notation.

10. $2.7 \cdot 10^7$

11. $4.56 \cdot 10^{-5}$

12. The wavelength of sunlight is approximately 0.00000057 meter. Write this number in scientific notation.

13. Assume that the salary of the President of the United States is \$0.0008 per person and that the population of the United States is 260,000,000. Write both of these numbers in scientific notation and then determine their product. What does the result represent?

14. Evaluate $-x^2y^3 + 2xy^2$ for $x = -1$ and $y = 2$.

15. Suppose that $p(x) = 3x^4 - 4x^3 - 5x^2 + 6x - 7$ and $q(x) = 5x^3 - 8x + 2$. Show two ways to evaluate each of the following.

(a) $(p + q)(1)$

(b) $(p - q)(-1)$

In Questions 16–21, determine the product.

16. $-2x^3(x^2 - 5x + 3)$ **17.** $(2t + s)(3t + 4s)$

18. $(x - 3)(x^2 + 3x - 2)$ **19.** $(x^2 + 3y)^2$

20. $(5x - 7z)(5x + 7z)$ **21.** $(-3a^2b^5)(2a^3b)$

22. Suppose $p(x) = x^2 + 1$ and $q(x) = x^3 - 1$. Write the polynomial $P(x)$, where $P(x) = p(x) \cdot q(x)$.

Chapter 7

Factoring

The accompanying table shows the number (in thousands) of foreign travelers to the United States during the period 1992–1997. The table also shows the average amount of money spent per trip. (Source: U.S. Department of Commerce.)

These data can be modeled with polynomial expressions. Various techniques for ***factoring*** *polynomials can be used to solve* ***quadratic equations*** *and to learn about the features of the graphs of polynomials. (For more on this topic, see Exercises 113–116 at the end of Section 7.2, and see the Chapter Project.)*

Most of this chapter is devoted to techniques for factoring a polynomial and to the application of factoring to solving equations. The chapter concludes with division of polynomials, including synthetic division.

Year	Number of Visitors (thousands)	Average Expenditures
1992	47,300	$1158
1993	45,800	$1264
1994	44,800	$1305
1995	43,300	$1464
1996	46,500	$1500
1997	47,800	$1534

(Source: U.S. Department of Commerce.)

7.1 Common Factors and Grouping

Greatest Common Factor • Common Factors in Polynomials • Factoring by Grouping

Greatest Common Factor

To factor a number, we write the number as a product of numbers. Sometimes there is more than one way to factor a number.

$$24 = 1 \cdot 24 \qquad 24 = 2 \cdot 12 \qquad 24 = 3 \cdot 8 \qquad 24 = 4 \cdot 6$$

The numbers on the right are numbers that can be divided into 24 with a remainder of 0. These numbers are called **factors** of 24.

A **prime number** is a natural number greater than 1 that is divisible only by itself and 1. Therefore, the only natural number factors of a prime number are the number itself and 1. The first ten prime numbers are

2, 3, 5, 7, 11, 13, 17, 19, 23, 29

To factor a number *completely* means to write the number as the product of prime numbers. For example, to factor 24 completely, we write

$$24 = 3 \cdot 2 \cdot 2 \cdot 2$$

The prime factors of 24 are 3 and 2.

The **greatest common factor (GCF)** of two numbers a and b is the largest integer that is a factor of both a and b.

EXAMPLE 1

Finding the Greatest Common Factor

Find the GCF for 36, 48, and 120.

Solution

$$36 = 3 \cdot 3 \cdot 2 \cdot 2 = 12 \cdot 3 \qquad \text{Factor each number completely.}$$
$$48 = 3 \cdot 2 \cdot 2 \cdot 2 \cdot 2 = 12 \cdot 4$$
$$120 = 3 \cdot 2 \cdot 2 \cdot 2 \cdot 5 = 12 \cdot 10$$

The GCF is 12.

We can determine the GCF of two or more monomials in the same way.

$$10x^3 = 2 \cdot 5 \cdot x^3$$

$$15x^5 = 3 \cdot 5 \cdot x^3 \cdot x^2$$

Because 5 and x^3 are the greatest factors that are common to both monomials, the GCF is $5x^3$. Note that for the variable factors, we can simply use the smallest exponent.

Determining the GCF of Two or More Monomials

To determine the GCF of two or more monomials, perform the following steps.

1. Determine the GCF of the numerical coefficients.
2. For each variable that is common to the monomials, determine the smallest exponent and write the variable with that exponent.
3. The GCF of the monomials is the product of the results in steps 1 and 2.

Note: The numbers 1 and -1 are factors of every monomial. When we determine the GCF, we usually ignore these factors.

EXAMPLE 2

Determining the GCF in Monomials

Determine the GCF of the given monomials.

(a) $6y^4, 9y^6$

(b) $75x^3y^4, 45x^2y^5, 225x^4y^7$

Solution

(a) $6y^4 = 2 \cdot 3 \cdot y^4$ Factor the monomials completely.

$9y^6 = 3 \cdot 3 \cdot y^6$

The GCF of 6 and 9 is 3, and the smallest exponent on y is 4. Thus the GCF of the monomials is $3y^4$.

The monomials can be written with the GCF as a factor.

$$6y^4 = 3y^4 \cdot 2 \qquad 9y^6 = 3y^4 \cdot 3y^2$$

(b) $75x^3y^4 = 3 \cdot 5 \cdot 5 \cdot x^3 \cdot y^4$

$45x^2y^5 = 3 \cdot 3 \cdot 5 \cdot x^2 \cdot y^5$

$225x^4y^7 = 3 \cdot 3 \cdot 5 \cdot 5 \cdot x^4 \cdot y^7$

The GCF of the numerical coefficients is $3 \cdot 5 = 15$. The smallest exponents on x and y are 2 and 4, respectively. Thus the GCF of the three monomials is $15x^2y^4$.

The monomials can be written with the GCF as one factor.

$$75x^3y^4 = 15x^2y^4 \cdot 5x$$
$$45x^2y^5 = 15x^2y^4 \cdot 3y$$
$$225x^4y^7 = 15x^2y^4 \cdot 15x^2y^3$$

Common Factors in Polynomials

Factoring a polynomial means writing the polynomial as a product of two or more polynomials. The first step is to determine whether the terms have a common factor other than 1 or -1.

To multiply a polynomial by a monomial, we use the Distributive Property. To factor a polynomial, we reverse the process. For example, to factor $12x^3 - 20x^2$, we begin by determining that the GCF of $12x^3$ and $20x^2$ is $4x^2$.

To relate factoring out a GCF to previous experience, point out that the same procedure is used to combine like terms.

$$12x^3 - 20x^2 = 4x^2 \cdot 3x - 4x^2 \cdot 5$$ Factor each term with the GCF as one factor.

$$= 4x^2(3x - 5)$$ Apply the Distributive Property.

If the polynomial to be factored has integer coefficients, then we try to factor so that the resulting expression has integer coefficients. If this is not possible, we consider the expression to be a **prime polynomial.**

A polynomial is **factored completely** if each factor other than a monomial factor is prime. For example, although $12x^3 - 20x^2$ can be written as $2x^2(6x - 10)$, we usually want to factor completely and write $4x^2(3x - 5)$.

Factoring Out a GCF from a Polynomial

1. Determine the GCF of the terms.
2. Factor each term with the GCF as one factor.
3. Use the Distributive Property to factor the polynomial with the GCF as one of the factors.
4. Verify the result by multiplying the factors. The product should be the original polynomial.

Note: Although we do not show the verification step in each example, you should always verify your result.

EXAMPLE 3

Factoring Out the Greatest Common Monomial Factor

Factor completely.

(a) $10x^3 - 5x^2$

(b) $-4x^3 - 8y^2$

(c) $-12x^2y^3 + 10xy^3$

Solution

(a) $10x^3 - 5x^2 = (5x^2)(2x) - (5x^2)(1)$ Determine the GCF of each term.

$= 5x^2(2x - 1)$ Distributive Property

With practice, students will not need to write the intermediate step. But they should be encouraged to write the step until they are comfortable with omitting it.

(b) $-4x^3 - 8y^2 = (-4)x^3 + (-4)(2y^2)$ We regard the common numerical factor as negative.

$= -4(x^3 + 2y^2)$ Distributive Property

(c) $-12x^2y^3 + 10xy^3 = (-2xy^3)(6x) - (-2xy^3)(5)$ $\text{GCF} = -2xy^3$

$= -2xy^3(6x - 5)$ Distributive Property

EXAMPLE 4

Factoring Out the Greatest Common Monomial Factor

Factor the trinomial $54x^3y^5 + 36x^4y^2 - 90x^6y^3$.

Solution

$54x^3y^5 = 18 \cdot 3 \cdot x^3 \cdot y^5$ Factor the monomials.

$36x^4y^2 = 18 \cdot 2 \cdot x^4 \cdot y^2$

$90x^6y^3 = 18 \cdot 5 \cdot x^6 \cdot y^3$

$\text{GCF} = 18x^3y^2$ Use the smallest exponent on each common factor.

In factored form, $54x^3y^5 + 36x^4y^2 - 90x^6y^3 = 18x^3y^2(3y^3 + 2x - 5x^3y)$.

EXAMPLE 5

Factoring Out the Greatest Common Binomial Factor

Factor each of the following.

(a) $3x(x + 4) - y(x + 4)$

(b) $5(2x - 1) - 2y(1 - 2x) + z(2x - 1)$

(c) $x^2(x + 5) + 3x^4(x + 5) - 6x^3(x + 5)$

(d) $6x^3(y - 1)^2 + 9x(y - 1)^5$

Solution

Some students will have trouble regarding a quantity, such as a binomial, as a common factor. This is an important lead-in to the grouping method.

(a) $3x(x + 4) - y(x + 4)$

$= (x + 4)(3x - y)$

The binomial $(x + 4)$ is common to both terms.

(b) $5(2x - 1) - 2y(1 - 2x) + z(2x - 1)$

$= 5(2x - 1) + 2y(2x - 1) + z(2x - 1)$

$= (2x - 1)(5 + 2y + z)$

Use the rule $a - b = -(b - a)$ to rewrite $1 - 2x$.

Now $(2x - 1)$ is common to all three terms.

(c) $x^2(x + 5) + 3x^4(x + 5) - 6x^3(x + 5)$

$= x^2(x + 5)(1 + 3x^2 - 6x)$

Note that x^2 and $(x + 5)$ are common to all three terms.

(d) $6x^3(y - 1)^2 + 9x(y - 1)^5$

$= 3x(y - 1)^2[2x^2 + 3(y - 1)^3]$

Note that 3, x, and $(y - 1)^2$ are common to both terms.

Factoring by Grouping

In some cases we must be creative to find the GCF.

Exploring the Concept

Factoring by Grouping

At first glance, the polynomial $x^3 - 3x^2 + 2x - 6$ does not appear to have a common factor. A factor of x appears in three of the terms but not in the last term.

However, suppose the first two terms and the last two terms are grouped together.

$$x^3 - 3x^2 + 2x - 6 = (x^3 - 3x^2) + (2x - 6)$$

Stress that a factored expression consists of just one term and that that term is a product.

The first group has a GCF of x^2; the second group has a GCF of 2.

$$x^3 - 3x^2 + 2x - 6 = x^2(x - 3) + 2(x - 3)$$

The polynomial is still not factored because it is a sum of two terms, not a product. Observe that the two terms have a common binomial factor $(x - 3)$.

$$\begin{aligned} x^3 - 3x^2 + 2x - 6 &= x^2(x - 3) + 2(x - 3) \\ &= (x - 3)(x^2 + 2) \end{aligned}$$

In factored form, $x^3 - 3x^2 + 2x - 6 = (x - 3)(x^2 + 2)$.

This method of factoring is called the **grouping method.**

Using color to highlight the two groups to be factored is effective. Whichever method is used, the sign change in the second group will be troublesome for many students, and they should be encouraged to check the factoring by multiplication.

Factoring by the Grouping Method

1. Group the terms so that each group has a common factor.
2. Factor out the GCF in each group.
3. If there is a common factor in the resulting terms, factor it out. If there is no common factor, a different grouping may produce the desired result.
4. Multiply the factors to verify that the original expression is obtained.

EXAMPLE 6

Factoring by the Grouping Method

Factor completely.

(a) $ax + bx + ay + by$ (b) $6x - 3y - 2ax + ay$

(c) $x^3 + 2x^2 - 5x - 10$

Solution

(a) $ax + bx + ay + by$

$= (ax + bx) + (ay + by)$ Group the expression in pairs.

$= x(a + b) + y(a + b)$ Factor out the GCF in each pair.

$= (a + b)(x + y)$ In the resulting two terms, the GCF is $(a + b)$.

In part (b), an alternative and sometimes preferred approach is to factor out $-a$ from the second group: $6x - 3y - 2ax + ay = 3(2x - y) - a(2x - y)$.

(b) Care must be taken when we group negative terms. Just as we change signs when we *remove* parentheses from $-(m - n) = -m + n$, we must change signs when we *insert* parentheses in $-m + n = -(m - n)$.

$6x - 3y - 2ax + ay$

$= (6x - 3y) - (2ax - ay)$ Change signs in the second group.

$= 3(2x - y) - a(2x - y)$ Factor out the GCF in each pair.

$= (2x - y)(3 - a)$ In the resulting two terms, the GCF is $(2x - y)$.

LEARNING TIP

In part (c) of Example 6, we insert parentheses and then factor. In such cases, you may find it easier just to factor out -5 from the last two terms. Either way, be careful to make the necessary sign changes.

(c) $x^3 + 2x^2 - 5x - 10$

$= (x^3 + 2x^2) - (5x + 10)$ Change signs in the second group.

$= x^2(x + 2) - 5(x + 2)$ Factor out the GCF in each pair.

$= (x + 2)(x^2 - 5)$ In the resulting two terms, the GCF is $(x + 2)$.

Think About It

Use the grouping method to factor the following expression.

$xy(x + 1) + x(x + 1) + y^2(x + 1) + y(x + 1)$

Note: These examples illustrate the method of grouping by *pairs*. In some cases, other grouping arrangements may be needed. Also, though the grouping method is a good technique to try, there is no guarantee that it will always work. For example, although $ax + ay + bx - by$ can be grouped and written as $a(x + y) + b(x - y)$, there is no common binomial factor with which to complete the factorization.

Quick Reference 7.1

Greatest Common Factor

- A **prime number** is a natural number greater than 1 that has exactly two different natural number factors, 1 and itself.
- To factor a number *completely* means to write the number as the product of prime numbers.
- The **greatest common factor (GCF)** of two numbers a and b is the largest integer that is a factor of both a and b.
- To determine the GCF of two or more monomials, perform the following steps.
  1. Determine the GCF of the numerical coefficients.
  2. For each variable that is common to the monomials, determine the smallest exponent and write the variable with that exponent.
  3. The GCF of the monomials is the product of the results in steps 1 and 2.

Common Factors in Polynomials

- **Factoring a polynomial** means writing the polynomial as a product of two or more polynomials.
- If the polynomial to be factored has integer coefficients, then we try to factor so that the resulting expression has integer coefficients. If this is not possible, we consider the expression to be a **prime polynomial.**
- A polynomial is **factored completely** if each factor other than a monomial factor is prime.
- To factor out a GCF from a polynomial, perform the following steps.
  1. Determine the GCF of the terms.
  2. Factor each term with the GCF as one factor.
  3. Use the Distributive Property to factor the polynomial with the GCF as one of the factors.
  4. Verify the result by multiplying the factors. The product should be the original polynomial.

Factoring by Grouping

- To factor with the grouping method, perform the following steps.
  1. Group the terms so that each group has a common factor.
  2. Factor out the GCF in each group.
  3. If there is a common factor in the resulting terms, factor it out. If there is no common factor, a different grouping may produce the desired result.
  4. Multiply the factors to verify that the original expression is obtained.

Speaking the Language 7.1

1. The largest integer that is a factor of two numbers is called the ______ of those numbers.
2. If n is a natural number greater than 1 that is divisible only by n and 1, then n is a ______ number.
3. The process of writing a polynomial expression as a product is called ______ the expression.
4. If an expression consists of four unlike terms, it might be possible to factor the expression with the ______ method.

Exercises 7.1

Concepts and Skills

1. For the monomials x^5y^3, x^7y^2, and x^4y^6, explain how you can determine the GCF by inspecting exponents.
2. Is the GCF of an expression always a monomial? Give an example that supports your answer.

In Exercises 3–10, determine the GCF.

3. 24, $60x$
4. $21x$, 35
5. $12x$, $16y$, $24z$
6. $25a$, $30b$, $15c$
7. $3x^5$, $4x^7$, $5x^3$
8. $2y^{10}$, $5y^4$, $6y^5$
9. m^5n^3, mn^6
10. p^5q, pq^5

In Exercises 11–14, complete the factorization.

11. $-6xy + 3y = -3y(\quad)$
12. $-8x^2 + 6x = -2x(\quad)$
13. $5x^2 + 10x + 15 = \quad(x^2 + 2x + 3)$
14. $30 - 15a + 9a^2 = \quad(10 - 5a + 3a^2)$

In Exercises 15–32, determine the GCF and factor the given expression completely.

15. $3x + 12$
16. $2x + 6$
17. $5x + 10y - 30$
18. $6 - 24a - 18b$
19. $12x^2 - 4x$
20. $6x^2 + 2x$
21. $7x^2 - 10y$
22. $5a^3 - 8b^2$
23. $x^3 - x^2$
24. $c^4 + 3c^3$
25. $3x^3 - 9x^2 + 12x$
26. $15b^5 + 5b^7 - 10b^4$
27. $x^5y^4 - x^4y^3 + xy^2$
28. $x^7y^8 + xy^5 - x^5y$
29. $3x(a + by) - 2y(a + by)$
30. $2x(x - 3) - 5y(x - 3)$
31. $4x^2(2x + 3) - 7y^2(2x + 3)$
32. $5x^5(x - 2) - 9y^3(x - 2)$
33. Determine whether each of the following is a correct factorization of $4x - 12$. If it is incorrect, explain why.
    (a) $4(x - 3)$
    (b) $-4(-x + 3)$
    (c) $4(-3 + x)$
    (d) $-4(3 - x)$
34. Explain how the Multiplication Property of -1 can be used to factor $x - 2y$ so that one of the factors is $2y - x$.

For Exercises 35–44, you may want to remind students about the rule $-(a - b) = b - a$.

In Exercises 35–38, factor out the GCF and the opposite of the GCF.

35. $4x - 12$
36. $10 - 4x$
37. $6y + 3xy - x^2y$
38. $-a^2b + 5b$

In Exercises 39–44, factor the given expression.

39. $y(y - 3) + 2(3 - y)$
40. $x(x - 2) - 3(2 - x)$
41. $5(3 - z) - z(z - 3)$
42. $4(6 - a) + a(a - 6)$
43. $(x + 3)(x - 4) - (x + 2)(4 - x)$
44. $(3 - a)(2a - 1) + (a - 2)(a - 3)$
45. If $(2x - 5)(x + 3)$ is the result of factoring an expression by grouping, what were the four terms in the original expression?
46. Which of the following is the correct way to group $x^3 + x^2 - x - 1$ in pairs? Explain.
    (i) $(x^3 + x^2) - (x - 1)$
    (ii) $(x^3 + x^2) - (x + 1)$

In Exercises 47–64, use the grouping method to factor the given expression completely.

47. $ab + bx + ay + xy$

48. $ax + 3a + 2x + 6$

49. $cd + 3d - 4c - 12$

50. $ac + 5a - 2c - 10$

51. $ab - ay - bx - xy$

52. $ab - by + ax + xy$

53. $ax^2 + 3x^2 + ay + 3y$

54. $ax^2 + 2a - x^2y - 2y$

55. $x^3 + 2x^2 + 3x + 6$

56. $x^3 + 4x^2 - 3x - 12$

57. $3y^3 + 9y^2 + 12y + 36$

58. $4y^3 - 12y^2 + 20y - 60$

59. $a^2b + a^2y + b + y$

60. $ab^2 - abx - b + x$

61. $x^3 - 15 + 3x - 5x^2$

62. $2x^3 + 12 - 3x - 8x^2$

63. $6x^3 - 12 + 8x - 9x^2$

64. $6x^3 + 6 + 9x + 4x^2$

In Exercises 65–68, factor completely.

65. $18x^3 + 6x^2 - 45x - 15$

66. $2a^3 + 2a^2b + 6ab + 6b^2$

67. $3xy^3 - 6xy^2 + 7xy - 14x$

68. $r^2t^2 + r^2t + rt^2 + rt$

CE In Exercises 69–72, complete the factorization of each expression.

In Exercises 69–76, the intent is to write the polynomial as a fraction times a prime polynomial with integer coefficients.

69. $\frac{2}{3}y - \frac{5}{6} = \frac{1}{6}(\qquad)$

70. $1 - \frac{3}{4}x = \frac{1}{4}(\qquad)$

71. $\frac{1}{2}x^2 + \frac{1}{3}x + 1 = \frac{1}{6}(\qquad)$

72. $x^2 + \frac{1}{4}x - \frac{1}{2} = \frac{1}{4}(\qquad)$

CE In Exercises 73–76, factor the given expression as in Exercises 69–72.

73. $\frac{1}{3}x + 5$

74. $\frac{1}{4}a - \frac{3}{2}b$

75. $\frac{1}{6}x^2 + \frac{1}{4}x + 3$

76. $x^2 - \frac{1}{2}x + \frac{1}{5}$

In Exercises 77–84, factor out the GCF in the expression.

77. $d^{20} - d^{15}$

78. $y^{12} + y^{20}$

79. $88x^{48} + 11x^{51}$

80. $100a^{50} + 10a^{25}$

81. $24a^5b^3 - 6ab^4 + 12a^3b^2$

82. $16c^6d^4 + 8c^2d^3 - 12c^2d$

83. $4a^2b^3(x - 4)^2 + 12a^3b^2(x - 4)^3$

84. $15c^4d^5(x + 3)^5 + 5c^5d^4(x + 3)^3$

In Exercises 85–88, write the given formula in completely factored form.

85. If P dollars are invested at a simple interest rate r for one year, the account value A is given by $A = P + Pr$.

86. If an object is propelled upward at an initial velocity of 560 feet per second, the height h of the object after t seconds is given by $h = 560t - 16t^2$.

87. The total surface area A of a right circular cylinder is given by $A = 2\pi rh + 2\pi r^2$, where h is the height of the cylinder and r is its radius.

88. The area A of a trapezoid is $A = \frac{1}{2}b_1h + \frac{1}{2}b_2h$, where h is the altitude of the trapezoid and b_1 and b_2 are the lengths of the bases.

Challenge

89. Explain how to verify the following.

$$3x^{-4} + 2x^{-6} = x^{-6}(3x^2 + 2)$$

90. Although the expression $2x^{-3} + 5x^{-4}$ is not a polynomial, can we still determine the GCF by using the smaller exponent on x? Is the GCF x^{-4} in this case? If so, what is the other factor?

In Exercises 91 and 92, complete the factorization of each expression.

In Exercises 91 and 92, the quantity in parentheses is to be a prime polynomial with positive Integer exponents.

91. $2x^{-3} - 5x^{-4} = x^{-4}(\qquad)$

92. $x^{-5} + x^5 = x^{-5}(\qquad)$

In Exercises 93–96, factor completely as in Exercises 91 and 92.

93. $2x^3 - 10x^{-4}$

94. $y^{-2} + 3y^{-4}$

95. $5x^{-4}y^7 - 15x^3y^{-2}$

96. $3x^{-1}y^{-2} + 9xy^{-3}$

97. Use the grouping method to factor the following expression.

$$a^2x + 3a^2 + bx + 3b - 5x - 15$$

98. If n is an integer, show that when $4n^2 + 2n$ is factored completely, one factor is an even number and the other factor is an odd number.

7.2 Special Factoring

Difference of Two Squares • Perfect Square Trinomials • Sum and Difference of Two Cubes

Difference of Two Squares

We can reverse the special product $(A + B)(A - B) = A^2 - B^2$ and write

$$A^2 - B^2 = (A + B)(A - B)$$

It is important to relate the special factoring patterns to the special product patterns discussed earlier.

We recognize a **difference of two squares** as two perfect square terms separated by a minus sign. If the binomial we want to factor matches the pattern $A^2 - B^2$, then the factors are $A + B$ and $A - B$.

Note: There is no special factoring pattern for a *sum* of two squares $A^2 + B^2$.

EXAMPLE 1 Factoring a Difference of Two Squares

LEARNING TIP

Remember to check your results by multiplying the factors to verify that the original expression is obtained.

Factor completely.

(a) $81 - 4y^2$

(b) $25x^2 - 49y^2$

(c) $(x - 1)^2 - 25$

Solution

The binomial in each part is in the form $A^2 - B^2$.

(a) $81 - 4y^2 = 9^2 - (2y)^2 = (9 + 2y)(9 - 2y)$

(b) $25x^2 - 49y^2 = (5x)^2 - (7y)^2 = (5x + 7y)(5x - 7y)$

(c) $(x - 1)^2 - 25 = (x - 1)^2 - 5^2$

$= [(x - 1) + 5][(x - 1) - 5]$

$= (x + 4)(x - 6)$

EXAMPLE 2 Combining Methods

Factor completely.

(a) $5x^4 - 80$

(b) $8x^3 - 2x$

Solution

(a) $5x^4 - 80 = 5(x^4 - 16)$ Factor out the GCF 5.

$= 5[(x^2)^2 - 4^2]$ The other factor is a difference of two squares.

$= 5(x^2 + 4)(x^2 - 4)$ $A^2 - B^2 = (A + B)(A - B)$

$= 5(x^2 + 4)(x + 2)(x - 2)$ The factor $(x^2 + 4)$ cannot be factored further.

(b) $8x^3 - 2x = 2x(4x^2 - 1) = 2x(2x + 1)(2x - 1)$

Perfect Square Trinomials

Previously, we have seen the results of squaring binomials.

$$(A + B)^2 = A^2 + 2AB + B^2$$
$$(A - B)^2 = A^2 - 2AB + B^2$$

The trinomials on the right are called **perfect square trinomials** because they are trinomials that are the squares of binomials. By reversing these special products, we can establish two factoring patterns.

$$A^2 + 2AB + B^2 = (A + B)^2$$
$$A^2 - 2AB + B^2 = (A - B)^2$$

EXAMPLE 3

Factoring Perfect Square Trinomials

Factor completely.

(a) $x^2 - 12x + 36$ (b) $25x^2 + 70xy + 49y^2$ (c) $x^6 + 10x^3y^2 + 25y^4$

Solution

Rather than write the second step, students should eventually be able to write just the squared binomial and check the middle term mentally.

$$A^2 - 2\,A\,B + B^2 = (A - B)^2$$
$$\downarrow \quad \downarrow \downarrow \downarrow \quad \downarrow \quad \downarrow \; \downarrow$$

(a) $x^2 - 12x + 36 = (x)^2 - 2(x)(6) + (6)^2 = (x - 6)^2$

(b) $25x^2 + 70xy + 49y^2 = (5x)^2 + 2(5x)(7y) + (7y)^2 = (5x + 7y)^2$

(c) $x^6 + 10x^3y^2 + 25y^4 = (x^3)^2 + 2(x^3)(5y^2) + (5y^2)^2 = (x^3 + 5y^2)^2$

Sum and Difference of Two Cubes

Sometimes a factoring pattern can be discovered by experimenting with multiplication. Consider the following products.

$$\begin{aligned}(A + B)(A^2 - AB + B^2) &= A^3 - A^2B + AB^2 + A^2B - AB^2 + B^3\\ &= A^3 + B^3\\ (A - B)(A^2 + AB + B^2) &= A^3 + A^2B + AB^2 - A^2B - AB^2 - B^3\\ &= A^3 - B^3\end{aligned}$$

Help students observe that the sign in the binomial is the same as the sign in the original expression and that the sign in front of *AB* is the opposite of the sign in the binomial.

The first result is a **sum of two cubes,** and the second result is a **difference of two cubes.** These patterns give us a way of factoring a sum of two cubes or a difference of two cubes.

$$A^3 + B^3 = (A + B)(A^2 - AB + B^2)$$
$$A^3 - B^3 = (A - B)(A^2 + AB + B^2)$$

EXAMPLE 4

Factoring Sums and Differences of Two Cubes

Factor completely.

(a) $y^3 + 64$

(b) $54x^3 + 2y^3$

(c) $t^6 - 125$

LEARNING TIP

Be aware of the difference between $A^2 + 2AB + B^2$, which is a perfect square trinomial pattern, and $A^2 + AB + B^2$, which is the second factor for a difference of two cubes and is prime.

Think About It

By combining like terms, $x^3 + x^3 = 2x^3$. Use the factoring pattern for the sum of two cubes on $x^3 + x^3$ to see whether you obtain the same result.

Solution

$$A^3 + B^3 = (A + B)(A^2 - A \cdot B + B^2)$$

$$\downarrow \quad \downarrow \quad \downarrow \; \downarrow \; \downarrow \quad \downarrow \; \downarrow \; \downarrow$$

(a) $y^3 + 64 = (y)^3 + (4)^3 = (y + 4)(y^2 - y \cdot 4 + 4^2)$

$= (y + 4)(y^2 - 4y + 16)$

(b) $54x^3 + 2y^3$

$= 2(27x^3 + y^3)$ — Note the GCF 2.

$= 2[(3x)^3 + y^3]$ — The other factor is a sum of two cubes.

$= 2(3x + y)[(3x)^2 - (3x)(y) + y^2]$ — $A^3 + B^3 = (A + B)(A^2 - AB + B^2)$

$= 2(3x + y)(9x^2 - 3xy + y^2)$

(c) $t^6 - 125 = (t^2)^3 - 5^3$ — The expression is a difference of two cubes.

$= (t^2 - 5)[(t^2)^2 + t^2 \cdot 5 + 5^2]$ — $A^3 - B^3 = (A - B)(A^2 + AB + B^2)$

$= (t^2 - 5)(t^4 + 5t^2 + 25)$

Quick Reference 7.2

Difference of Two Squares

- An expression consisting of two perfect square terms separated by a minus sign is called a **difference of two squares.**
- The special factoring pattern for a difference of two squares is

  $A^2 - B^2 = (A + B)(A - B)$

Perfect Square Trinomials

- A **perfect square trinomial** is a trinomial of the form

  $A^2 + 2AB + B^2 \quad \text{or} \quad A^2 - 2AB + B^2$
- The special factoring patterns for perfect square trinomials are as follows:

  $A^2 + 2AB + B^2 = (A + B)^2$

  $A^2 - 2AB + B^2 = (A - B)^2$

Sum and Difference of Two Cubes

- A **sum of two cubes** is a binomial of the form $A^3 + B^3$; a **difference of two cubes** is a binomial of the form $A^3 - B^3$.
- The special factoring patterns for the sum of two cubes and the difference of two cubes are as follows:

  $A^3 + B^3 = (A + B)(A^2 - AB + B^2)$

  $A^3 - B^3 = (A - B)(A^2 + AB + B^2)$

Speaking the Language 7.2

1. The pattern $A^2 - B^2$ is called a(n) ______.
2. The expression $x^2 + 6x + 9$ is an example of a special pattern called a(n) ______.
3. After we factor out the GCF 5 from $5x^3 + 5$, the resulting binomial factor is called a(n) ______.
4. We can check the factorization of an expression by ______ the factors.

Exercises 7.2

Concepts and Skills

1. In words, how do we recognize a difference of two squares?

2. Why can we factor $9x^2 + 36$ but not $9x^2 + 16$?

In Exercises 3–16, factor completely.

3. $x^2 - 9$ **4.** $9x^2 - 16$

5. $1 - 4x^2$ **6.** $25 - a^2$

7. $16x^2 - 25y^2$ **8.** $49x^2 - 64y^2$

9. $\frac{1}{4}x^2 - 1$ **10.** $\frac{1}{9}y^2 - \frac{1}{16}$

11. $a^6 - b^{16}$ **12.** $x^4 - y^2$

13. $(a + b)^2 - 4$ **14.** $(x - 3)^2 - y^2$

15. $16 - (x - y)^2$ **16.** $25 - (x + 4)^2$

17. In words, how do we recognize a perfect square trinomial?

18. For $x^2 + 6xy + 9y^2$, how do we check the suspected factorization $(x + 3y)^2$?

In Exercises 19–28, factor completely.

19. $a^2 - 4a + 4$ **20.** $x^2 + 8x + 16$

21. $4x^2 + 12x + 9$ **22.** $9x^2 - 12x + 4$

23. $x^2 - 14xy + 49y^2$ **24.** $x^2 + 12xy + 36y^2$

25. $x^2 + x + \frac{1}{4}$ **26.** $y^2 - \frac{4}{3}y + \frac{4}{9}$

27. $x^4 - 10x^2 + 25$ **28.** $y^4 + 6y^2 + 9$

In Exercises 29–38, factor completely.

29. $x^3 + 1$ **30.** $8 - a^3$

31. $27y^3 - 8$ **32.** $64y^3 + 27$

33. $125 - 8x^3$ **34.** $64 - 27z^3$

35. $27x^3 + y^3$ **36.** $8a^3 - b^3c^3$

37. $x^6 + 1$ **38.** $b^6 - 125$

Exercises 39–64 are a mixture of all the special factoring patterns.

In Exercises 39–64, factor completely.

39. $x^2 + 2x + 1$ **40.** $x^2 - 10x + 25$

41. $0.25x^2 - 0.49$ **42.** $1.44a^2 - 0.0016$

43. $x^3 - 1$ **44.** $27 + b^3$

45. $c^2 + 100 - 20c$ **46.** $d^2 + 81 + 18d$

47. $2x^3 + 54$ **48.** $3x^3 - 24$

49. $81 + y^2$ **50.** $32a^2 + 50$

51. $36 - 84y + 49y^2$ **52.** $25 + 60p + 36p^2$

53. $4x^2 - 64$ **54.** $16 - 4y^2$

55. $25x^2 - 40xy + 16y^2$ **56.** $16x^2 + 40xy + 25y^2$

57. $x^4 - 16$ **58.** $81a^4 - b^4$

59. $8 + 125x^6$ **60.** $125 - 27x^6$

61. $y^8 - 6y^4 + 9$ **62.** $x^8 + 4x^4 + 4$

63. $216x^3 - y^3$ **64.** $a^3b^3 + 8$

This is a mixed and slightly more difficult group.

In Exercises 65–92, factor completely.

65. $x^6 - 16x^3 + 64$ **66.** $x^6 + 64 + 16x^3$

67. $x^{12} - 4y^{20}$ **68.** $49 - 81x^4$

69. $x^6 + 4 - 4x^3$ **70.** $y^6 + 9 + 6y^3$

71. $98 - 2y^6$ **72.** $125x^2 - 5x^8$

73. $125x^5y^3 - x^2$ **74.** $27x^9y^7 - 8y^4$

75. $49x^7 - 70x^5 + 25x^3$ **76.** $9x^6 - 42x^4 + 49x^2$

77. $x^4 + 36y^4$ **78.** $16a^4 + 25b^6$

79. $64 + 27x^3y^9$ **80.** $1 - 27x^6y^3$

81. $x^4 - 8x^2 + 16$ **82.** $x^4 - 18x^2 + 81$

83. $16x^4 - 25y^2$ **84.** $c^8 - 1$

85. $64x^4 + 80x^2y + 25y^2$

86. $49x^6 + 126x^3y^2 + 81y^4$

87. $16a^4 - 81d^4$ **88.** $x^4 - c^4$

89. $2x^4 + 16x$ **90.** $3y^5 - 81y^2$

91. $125x^6 - 27y^{12}$ **92.** $27x^9 + 8y^6$

In Exercises 93–96, factor completely.

93. $4a(x^2 - y^2) + 8b(x^2 - y^2)$

94. $10x(a^2 - b^2) + 5y(a^2 - b^2)$

95. $x^2(a + b) - y^2(a + b)$

96. $4a^2(x - y) - b^2(x - y)$

This group combines a GCF with special factoring patterns.

CE In Exercises 97–100, factor the expression completely. (*Hint:* Replace the binomial with u and factor the resulting expression. Then replace u with the original binomial and simplify.)

97. $(2x + 1)^2 - 10(2x + 1) + 25$

98. $36(x - 1)^2 + 36(x - 1) + 9$

99. $(x^2 - 2)^2 - 4(x^2 - 2) + 4$

100. $(x^2 - 9)^2 + 16(x^2 - 9) + 64$

101. Match the expression $x^6 - y^6$ with the appropriate special factoring expression to show that the expression is a difference of two squares and is also a difference of two cubes.

102. The expression $x^6 + y^6$ is both a sum of two squares and a sum of two cubes. Which way should you regard the expression if you want to factor it? If you intend to cover completing the square, these are good seeding exercises.

CE In Exercises 103–108, determine a, b, or c so that the trinomial is a perfect square.

103. $9x^2 + bx + 1$ **104.** $x^2 + bx + 100$

105. $ax^2 - 10x + 25$ **106.** $ax^2 + 12x + 36$

107. $9x^2 - 24x + c$ **108.** $4x^2 - 20x + c$

Real-Life Applications

109. A circular swimming pool of radius r is surrounded by a circular deck of radius R. (See figure.) Write an expression for the area of the deck. Then write the expression in factored form.

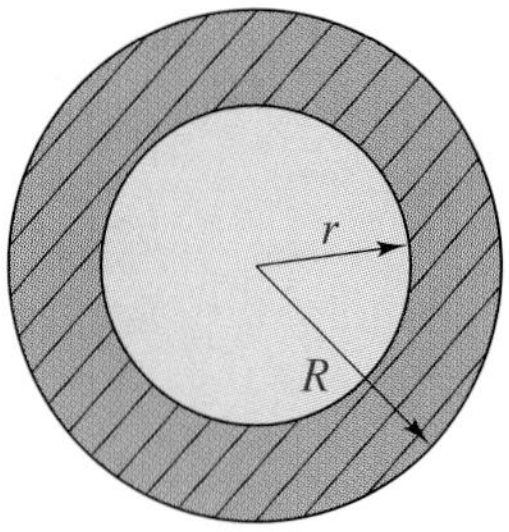

110. A Sea-Land shipping container for machine parts is a rectangular box with a volume that is represented by $x^3 + 10x^2 + 25x$. The length and width of the container are represented by binomials with integer coefficients. By factoring the expression for the volume, show that the height of the container is 5 units less than the length and width.

111. In physics, kinetic energy is the energy an object has by virtue of its motion. The formula for kinetic energy K is $K = \frac{1}{2}mv^2$, where m is the mass of the object and v is the velocity of the object. Suppose two objects of equal mass have velocities v_1 and v_2, respectively. Write an expression for the difference in their kinetic energies. Then write the expression in factored form.

112. A pharmaceutical company determines that a drug costing x dollars to develop and market will yield x^4 dollars in revenue. Write an expression for the net earnings (revenue minus cost) of the drug. Then write the expression in factored form.

Data Analysis: *Foreign Visitors to the United States*

The accompanying table shows the number (in thousands) of foreign visitors to the United States during the years 1992–1997. The table also shows the average amount of money (in dollars) each visitor spent. (Source: U.S. Department of Commerce.)

Year	Number of Visitors (thousands)	Average Expenditures
1992	47,300	$1158
1993	45,800	$1264
1994	44,800	$1305
1995	43,300	$1464
1996	46,500	$1500
1997	47,800	$1534

Let x represent the number of years before (negative) or after (positive) 1995. Then $V(x) = 560x^2 + 44{,}400$ models the number of visitors (in thousands), and $E(x) = 80x + 1420$ models the average expenditure (in dollars).

113. Write function V in factored form.

114. Write function E in factored form.

115. Use the factored form of function V to write the function $F(x)$ that models the number of visitors in millions.

116. Write the model function $C(x)$ (in factored form) that models the total expenditures in millions of dollars.

Challenge

In Exercises 117–120, factor completely. (*Hint:* In each case, begin by factoring the expression as a difference of two squares.)

117. $x^6 - 64$ **118.** $c^6 - 1$

119. $a^6 - b^6$ **120.** $x^6 - 64y^6$

In Exercises 121–124, factor completely. (*Hint:* Start with the grouping method.)

121. $4x^5 - x^3 - 32x^2 + 8$

122. $x^5 - 32 + 8x^2 - 4x^3$

123. $x^2y^3 - x^2z^3 - y^3 + z^3$

124. $y^2 - 9x^2 + 12x - 4$

7.3 Factoring Trinomials of the Form $ax^2 + bx + c$

Leading Coefficient $a = 1$ • Leading Coefficient $a \neq 1$ • General Strategy

Leading Coefficient $a = 1$

Recall that when we multiply binomials such as $(x - 3)(x + 5)$, the result is a trinomial.

$$(x - 3)(x + 5) = x^2 + 2x - 15$$

We call this trinomial a **quadratic polynomial.** In general, a quadratic polynomial in one variable can be written in the form $ax^2 + bx + c$, where $a \neq 0$. The term ax^2 is called the **quadratic term,** the term bx is called the **linear term,** and the term c is called the **constant term.** The coefficient a of the quadratic term is called the **leading coefficient.**

Note: The definition of a quadratic polynomial prohibits the leading coefficient from being 0; that is, there must be a quadratic term. However, b or c can be 0, so there need not be a linear term or a constant term. Thus all of the following are quadratic polynomials.

$$3x^2 - 4x + 2 \qquad y^2 - 16 \qquad x^2 + x \qquad 5x^2$$

We can factor a quadratic polynomial with no constant term simply by factoring out the GCF.

$$2x^2 - 7x = x(2x - 7)$$

We also know how to factor a quadratic polynomial with no linear term, provided that the polynomial is a difference of two squares.

$$9y^2 - 16 = (3y + 4)(3y - 4)$$

In this section, we will concentrate on factoring quadratic polynomials that have all three terms.

Exploring the Concept

To emphasize this, you may want to perform a FOIL multiplication or two.

Factoring a Quadratic Trinomial

When the leading coefficient a is 1, the trinomial has the form $x^2 + bx + c$. From our experience with multiplication, we may recall that $x^2 + bx + c$ was often the result of multiplying two binomials of the form $(x + m)(x + n)$, where m and n are constants.

Factors	F O I L	$x^2 + bx + c$	
$(x + 4)(x - 2)$	$= x^2 - 2x + 4x - 8$	$= x^2 + (-2 + 4)x - 8$	$b = -2 + 4, c = -8$
$(x - 5)(x - 3)$	$= x^2 - 3x - 5x + 15$	$= x^2 + (-3 - 5)x + 15$	$b = -3 - 5, c = 15$
$(x + 6)(x + 1)$	$= x^2 + x + 6x + 6$	$= x^2 + (1 + 6)x + 6$	$b = 1 + 6, c = 6$
$(x + m)(x + n)$	$= x^2 + nx + mx + mn$	$= x^2 + (m + n)x + mn$	$b = m + n, c = mn$

These patterns suggest that we can factor the trinomial $x^2 + bx + c$ if we can find numbers m and n such that $mn = c$ and $m + n = b$.

EXAMPLE 1

Factoring a Quadratic Trinomial $x^2 + bx + c$

Factor each trinomial.

(a) $x^2 + 8x + 12$ (b) $x^2 - 5x + 6$

Solution

(a) Because $c = 12$ and $b = 8$, we find pairs of numbers m and n whose product is 12 and whose sum is 8.

Factors of 12	*Sums*
1, 12 and −1, −12	13 and −13
2, 6 and −2, −6	8 and −8
3, 4 and −3, −4	7 and −7

The required values for m and n are 2 and 6. Thus,

$$x^2 + 8x + 12 = (x + 2)(x + 6)$$

Even though we used a systematic way of determining 2 and 6, it is important for students to get into the habit of multiplying as a check on the factoring.

We can (and should) verify that the factorization is correct by confirming that $(x + 2)(x + 6) = x^2 + 8x + 12$.

(b) We look for a pair of numbers m and n whose product is 6 and whose sum is -5. The required numbers are -2 and -3.

$$x^2 - 5x + 6 = (x - 2)(x - 3)$$

In both parts of Example 1, $c > 0$. In Example 2, we consider quadratic trinomials in which $c < 0$.

EXAMPLE 2

Factoring a Quadratic Trinomial $x^2 + bx + c$

Factor each trinomial.

(a) $x^2 - x - 20$ (b) $x^2 + 5x - 14$

Solution

(a) Because $c = -20$ and $b = -1$, we find pairs of numbers m and n whose product is -20 and whose sum is -1.

Students should soon be able to work through such lists mentally for easy problems.

Factors of −20	*Sums*
1, −20 and −1, 20	−19 and 19
2, −10 and −2, 10	−8 and 8
4, −5 and −4, 5	−1 and 1

The required values for m and n are 4 and -5. Thus,

$$x^2 - x - 20 = (x + 4)(x - 5)$$

(b) We look for a pair of numbers m and n whose product is -14 and whose sum is 5. The required numbers are 7 and -2.

$$x^2 + 5x - 14 = (x + 7)(x - 2)$$

With practice, you can determine the pair of numbers m and n mentally without making the list of factors and sums.

EXAMPLE 3

Extending the Methods

Factor each trinomial.

(a) $12 - x - x^2$ (b) $3x^2 + 18x - 21$

(c) $t^2 + 11st + 18s^2$ (d) $y^4 - 5y^2 - 14$

Solution

(a) The trinomial $12 - x - x^2$ is in ascending order, and $a = -1$.

Verify that $12 - x - x^2 = (4 + x)(3 - x)$.

An alternative approach is to write the trinomial in descending order and factor out -1.

$$12 - x - x^2 = -x^2 - x + 12 = -1(x^2 + x - 12)$$

To factor $x^2 + x - 12$, we determine numbers whose product is -12 and whose sum is 1. The required numbers are 4 and -3.

$$12 - x - x^2 = -1(x^2 + x - 12) = -1(x + 4)(x - 3)$$

Note that the two approaches lead to equivalent factorizations.

(b) $3x^2 + 18x - 21 = 3(x^2 + 6x - 7)$ Factor out the GCF 3.

$= 3(x + 7)(x - 1)$ Choose 7 and -1 because their product is -7 and their sum is 6.

(c) $t^2 + 11st + 18s^2 = (t + 2s)(t + 9s)$ Choose 2 and 9 because their product is 18 and their sum is 11.

(d) Although $y^4 - 5y^2 - 14$ is a fourth-degree polynomial, not a quadratic polynomial, the method still applies.

$y^4 - 5y^2 - 14 = (y^2 - 7)(y^2 + 2)$ Choose -7 and 2 because their product is -14 and their sum is -5.

Both factors have a degree greater than 1, so they should be inspected to see whether they can be factored further. In this case, neither of them can be.

Leading Coefficient $a \neq 1$

When the leading coefficient of a quadratic trinomial is a number other than 1, an approach to factoring the trinomial is the *trial-and-check method.*

Exploring the Concept

The Trial-and-Check Method

Suppose that we want to factor $6x^2 + 7x - 3$. To write the trinomial as the product of binomials, recall that the product of the first terms must be $6x^2$ and the product of the last terms must be -3.

Remind students about FOIL.

$6x^2$

$$6x^2 + 7x - 3 = (\quad\quad\quad)(\quad\quad\quad)$$

-3

Here are the possibilities.

$$6x^2 = 6x \cdot x \quad \text{or} \quad 2x \cdot 3x \qquad\qquad -3 = 1(-3) \quad \text{or} \quad -1 \cdot 3$$

Using all the combinations of these possibilities, we can list all the possible factors of $6x^2 + 7x - 3$. Then we check the middle term resulting from multiplying the factors.

Possible Factors	
$(6x + 1)(x - 3)$	The middle term is $-17x$.
$(6x - 3)(x + 1)$	The factor $6x - 3$ has a common factor 3.
$(6x - 1)(x + 3)$	The middle term is $17x$.
$(6x + 3)(x - 1)$	The factor $6x + 3$ has a common factor 3.
$(2x + 1)(3x - 3)$	The factor $3x - 3$ has a common factor 3.
$(2x - 3)(3x + 1)$	The middle term is $-7x$.
$(2x - 1)(3x + 3)$	The factor $3x + 3$ has a common factor 3.
$(2x + 3)(3x - 1)$	The middle term is $7x$.

Students should be able to eliminate many choices quickly and mentally instead of writing all of the possibilities.

Because $6x^2 + 7x - 3$ does not have a common factor, none of its factors has a common factor. This means that we can eliminate any factorization in which a binomial factor has a common factor.

The last possible factorization in the list produces the correct middle term $7x$.

$$6x^2 + 7x - 3 = (2x + 3)(3x - 1)$$

As always, factoring out the GCF, if any, is the first step in factoring. Once you have done so, it is not necessary to list or try any combination in which a binomial factor has a common factor.

EXAMPLE 4

Factoring Quadratic Trinomials by Trial and Check

Factor each trinomial.

(a) $4x^2 + 11x + 6$ (b) $12x^2 + 7x - 12$ (c) $2t^2 - 3t - 1$

Solution

LEARNING TIP

If a is positive, then the signs in a trinomial provide valuable clues to factoring. For example, if $c > 0$, then the binomial factors have the same sign as b. If $c < 0$, then the binomial factors have opposite signs.

(a) $4x^2 + 11x + 6$

Because $4x^2 = 4x \cdot x$ or $2x \cdot 2x$, the possible factorizations have the following forms.

$$(4x + \quad)(x + \quad) \quad \text{or} \quad (2x + \quad)(2x + \quad)$$

Because b and c are both positive, the factors of c must be positive: $6 \cdot 1$ or $3 \cdot 2$.

The only factorizations that do not have a common factor are as follows:

$(4x + 1)(x + 6)$ The middle term is $25x$.
$(4x + 3)(x + 2)$ The middle term is $11x$.

Because $b = 11$, the correct factorization is $(4x + 3)(x + 2)$.

(b) $12x^2 + 7x - 12$

The factors of 12 are $12 \cdot 1$, $6 \cdot 2$, and $4 \cdot 3$. Taking signs into account and eliminating combinations that have a common factor, we have the following possible factorizations.

$(12x + 1)(x - 12)$ The middle term is $-143x$.
$(12x - 1)(x + 12)$ The middle term is $143x$.
$(4x + 3)(3x - 4)$ The middle term is $-7x$.
$(4x - 3)(3x + 4)$ The middle term is $7x$.

Because $b = 7$, the correct factorization is $(4x - 3)(3x + 4)$.

(c) $2t^2 - 3t - 1$

The only possibilities are

$(2t + 1)(t - 1)$ The middle term is $-t$.
$(2t - 1)(t + 1)$ The middle term is t.

Because neither possibility gives us the correct middle term $-3t$, we conclude that the trinomial is prime.

EXAMPLE 5

Extending the Methods

Factor each trinomial.

(a) $30x^2 - 65x + 10$ (b) $3y^2 + yz - 10z^2$

Solution

(a) First, we factor out the GCF, 5.

$$30x^2 - 65x + 10 = 5(6x^2 - 13x + 2)$$

The quadratic trinomial factor may be factorable. Because c is positive and b is negative, the factors of the constant term must be negative. The only possibilities are

$(x - 2)(6x - 1)$ The middle term is $-13x$.
$(3x - 2)(2x - 1)$ The middle term is $-7x$.

Because $b = -13$, the correct factorization of the trinomial $6x^2 - 13x + 2$ is $(x - 2)(6x - 1)$. Thus the complete factorization of $30x^2 - 65x + 10$ is $5(x - 2)(6x - 1)$.

(b) $3y^2 + yz - 10z^2$

The only possible factorizations are

Possible Factors	*Middle Term*	*Possible Factors*	*Middle Term*
$(3y - 10z)(y + z)$	$-7yz$	$(3y + 10z)(y - z)$	$7yz$
$(3y - 5z)(y + 2z)$	yz	$(3y + 5z)(y - 2z)$	$-yz$
$(3y - z)(y + 10z)$	$29yz$	$(3y + z)(y - 10z)$	$-29yz$
$(3y - 2z)(y + 5z)$	$13yz$	$(3y + 2z)(y - 5z)$	$-13yz$

Because $b = 1$, the correct factorization is $(3y - 5z)(y + 2z)$.

You may want to illustrate that the placement of the values for a and c makes a difference. For example, we obtain different middle terms for $(3y - 10z)(y + 1z)$ and $(3y + 1z)(y - 10z)$.

A disadvantage of the trial-and-check method is that the number of possible combinations can sometimes be quite large. An alternative method is to rewrite the polynomial so that the grouping method can be applied. The following is a summary of the procedure.

Think About It

If you forget the GCF, you might factor $6x^2 + 6x - 36$ as $(3x + 9)(2x - 4)$. What can you do now to factor completely?

Factoring $ax^2 + bx + c$ with the Grouping Method

1. Find the product of the leading coefficient a and the constant term c.
2. List the pairs of numbers whose product is ac.
3. Choose the pair whose sum is b, the coefficient of the middle term.
4. Use these numbers to write the middle term as a sum of two terms.
5. Factor by grouping.

EXAMPLE 6

Factoring Trinomials with the Grouping Method

Factor $6x^2 - x - 2$.

Solution

Because $ac = -12$ and $b = -1$, we look for pairs of numbers whose product is -12 and whose sum is -1.

Factors of −12	*Sums*
$1, -12$ and $-1, 12$	-11 and 11
$2, -6$ and $-2, 6$	-4 and 4
$3, -4$ and $-3, 4$	-1 and 1

We use 3 and −4 to rewrite the middle term as $3x - 4x$.

$$
\begin{aligned}
6x^2 - x - 2 &= 6x^2 + 3x - 4x - 2 && \text{Write } -x \text{ as } 3x - 4x.\\
&= (6x^2 + 3x) - (4x + 2) && \text{Group in pairs.}\\
&= 3x(2x + 1) - 2(2x + 1) && \text{Factor each group.}\\
&= (2x + 1)(3x - 2) && \text{Factor out } 2x + 1.
\end{aligned}
$$

The factorization of $6x^2 - x - 2$ is $(2x + 1)(3x - 2)$.

EXAMPLE 7

Using the Grouping Method

Factor $10x^2 - 21x + 9$.

Solution

We see that $ac = 90$ and $b = -21$. Because the numbers we are seeking must have a positive product and a negative sum, the numbers must be negative. The required numbers are -15 and -6.

$$
\begin{aligned}
10x^2 - 21x + 9 &= 10x^2 - 15x - 6x + 9 && \text{Write } -21x \text{ as } -15x - 6x. \\
&= (10x^2 - 15x) - (6x - 9) && \text{Group in pairs.} \\
&= 5x(2x - 3) - 3(2x - 3) && \text{Factor each group.} \\
&= (2x - 3)(5x - 3) && \text{Factor out } 2x - 3.
\end{aligned}
$$

Remember that you should always verify your factorization.

$$(2x - 3)(5x - 3) = 10x^2 - 21x + 9$$

General Strategy

As we solve problems in algebra, we sometimes must factor a polynomial and need to decide which of the methods of factoring we have studied is appropriate. It is helpful to have a general strategy to guide us in making the decision.

A polynomial is factored when it is written as the product of polynomials. A polynomial is factored completely if each factor other than a monomial factor is prime. Sometimes more than one of these methods is needed to factor the polynomial completely:

A General Strategy for Factoring

1. When factoring a polynomial, always look for a greatest common factor first. Even if other methods of factoring are required, factoring out the GCF will make other factoring methods easier to use.
2. When factoring a polynomial with just two terms, determine whether it matches one of the special product patterns.
3. When factoring a polynomial that has three terms, check whether it is a perfect square trinomial, and if so, use the special product for the square of a binomial.
4. When factoring a trinomial that is not a perfect square trinomial, use the grouping method or the trial-and-check method.
5. When factoring a polynomial that has more than three terms, try the grouping method.
6. Multiply the factors to verify that the original expression is obtained.

Quick Reference 7.3

Leading Coefficient $a = 1$

- A trinomial of the form $ax^2 + bx + c$ is called a **quadratic trinomial.** The number a is called the **leading coefficient.**
- To factor a quadratic trinomial $x^2 + bx + c$ (the leading coefficient is 1), one approach is as follows:
  1. List the pairs of numbers whose product is the constant term c.
  2. Choose the pair whose sum is b, the coefficient of the middle term.
  3. These numbers are the constant terms of the binomial factors.

Leading Coefficient $a \neq 1$

- To factor a quadratic trinomial $ax^2 + bx + c$ by the trial-and-check method, follow these steps.
  1. List all the pairs of numbers whose product is the leading coefficient a.
  2. List all the pairs of numbers whose product is the constant term c.
  3. Use these two sets of numbers to form possible binomial factors of the trinomial. (Eliminate all possibilities that would produce a common factor in the terms of either of the binomial factors.)
  4. Multiply the remaining possibilities to determine the correct factorization.
- An alternative approach is to use the grouping method.
  1. Find the product of the leading coefficient a and the constant term c.
  2. List the pairs of numbers whose product is ac.
  3. Choose the pair whose sum is b, the coefficient of the middle term.
  4. Use these numbers to write the middle term as a sum of two terms.
  5. Factor by grouping.

General Strategy

- The first step in factoring is to factor out the greatest common factor.
- If the polynomial has just two terms, determine whether it matches the pattern of one of the special products.
  1. Difference of two squares
     $$A^2 - B^2 = (A - B)(A + B)$$
  2. Difference of two cubes
     $$A^3 - B^3 = (A - B)(A^2 + AB + B^2)$$
  3. Sum of two cubes
     $$A^3 + B^3 = (A + B)(A^2 - AB + B^2)$$
- If the polynomial has three terms, determine whether it is a perfect square trinomial.
  1. $A^2 + 2AB + B^2 = (A + B)^2$
  2. $A^2 - 2AB + B^2 = (A - B)^2$
- If the trinomial to be factored is not a perfect square trinomial, use the grouping method or the trial-and-check method.
- If the polynomial has more than three terms, try the grouping method. It might be necessary to rearrange terms or to group in ways other than in pairs.
- Multiply the factors to verify that the original expression is obtained.

Speaking the Language 7.3

1. In the expression $x^2 + 4x - 7$, x^2 is called the ______ term, and $4x$ is called the ______ term.
2. In the expression $3x^2 - 5x + 4$, the number 3 is called the ______ coefficient.
3. Listing all possible combinations of factors is known as the ______ method for factoring an expression.
4. The recommended first step in factoring is to determine whether the expression has a(n) ______.

Exercises 7.3

Concepts and Skills

1. When we factor $x^2 + 3x + 2$, does it matter whether we write $(x + 1)(x + 2)$ or $(x + 2)(x + 1)$? Why?
2. After you factor an expression, how can you check the accuracy of your work?

In Exercises 3–12, factor the given quadratic trinomial completely.

3. $x^2 + 3x + 2$
4. $x^2 - 4x + 3$
5. $x^2 + 18 - 9x$
6. $x^2 - 18 + 3x$
7. $14 + 9x + x^2$
8. $21 + 10x + x^2$
9. $x^2 + 2x + 2$
10. $x^2 + 3x + 5$
11. $x^2 - 6xy - 16y^2$
12. $x^2 - 4xy - 45y^2$

In Exercises 13–22, factor the trinomial completely.

13. $2x^2 + 5x + 2$
14. $2x^2 + 7x + 6$
15. $12 + 4x - x^2$
16. $4 + 3x - x^2$
17. $3x^2 - 8x - 3$
18. $3x^2 - 4x - 15$
19. $4x^2 - 17x - 15$
20. $12 + 7x - 10x^2$
21. $12x^2 + 19x + 21$
22. $12x^2 + 13x + 14$

This group mixes $a = 1$ and $a \neq 1$.

In Exercises 23–58, factor the trinomial completely.

23. $c^2 + c - 2$
24. $x^2 - 2x - 3$
25. $6x^2 + 13x - 15$
26. $15x^2 + 8x - 7$
27. $x^2 + 5x - 24$
28. $3x + x^2 - 40$
29. $24 + 2x - x^2$
30. $15 + 2x - x^2$
31. $b^2 - 2b - 15$
32. $c^2 - 4c - 12$
33. $3x^2 + 5x + 2$
34. $3x^2 - 7x - 6$
35. $15 + 6x^2 - 19x$
36. $6 + x - 12x^2$
37. $x^2 + 25 + 10x$
38. $16 + y^2 + 8y$
39. $2 + x - 15x^2$
40. $2 - 11x + 15x^2$
41. $18 - 11x - 10x^2$
42. $3x^2 + 10x - 8$
43. $24 + a^2 - 10a$
44. $b^2 - 2b - 24$
45. $9x^2 - 9x - 10$
46. $6x^2 + x - 15$
47. $x^2 - 16x + 63$
48. $x^2 - 2x - 63$
49. $9x^2 - 9x - 18$
50. $9x^2 - 9x - 54$
51. $12x^2 + 16x - 3$
52. $13x + 35x^2 - 12$
53. $8x^2 - 2xy - 21y^2$
54. $8x^2 - 2xy - 15y^2$
55. $2x^2 - 12 + 2x$
56. $3x^2 - 12 + 9x$
57. $24x + 36 + 4x^2$
58. $48x + 40 + 8x^2$
59. Why is factoring $17x^2 + 392x + 23$ easier than factoring $24x^2 + 11x - 18$?
60. Which of the following is a correct factorization of $6 - 5x + x^2$? Explain.

    (i) $(2 - x)(3 - x)$ (ii) $(x - 2)(x - 3)$

In Exercises 61–66, factor completely.

61. $8x^3y - 24x^2y^2 - 80xy^3$
62. $6x^6 - 60x^5 + 96x^4$
63. $45x + 12x^2 - 12x^3$
64. $10y^6 + 55y^5 + 75y^4$
65. $28x^2y - 30xy - 18y$
66. $15y^3 - 84y^2 - 36y$

In Exercises 67–78, factor the expression completely.

67. $a^2b^2 - 3ab - 10$
68. $c^2d^2 - 4cd - 21$
69. $12x^2y + 36xy + x^3y$
70. $28x^2 + 98x + 2x^3$
71. $x^2y^2 - 2xy - 48$
72. $36 - 9ab - a^2b^2$
73. $18x^2 - 12xy + 2y^2$
74. $8x^2y^2 + 18xy - 5$
75. $6m^2 + 5mn - n^2$
76. $16m^2n^4 + 12mn^5 + 2n^6$
77. $16x^3y^2 - 12x^2y^3 + 2xy^4$
78. $30ab^4 + 24a^2b^3 - 6a^3b^2$

In this group, the GFC is a binomial. Then other techniques are required.

CE In Exercises 79–84, factor the expression completely.

79. $x^3(x - 1) + 11x^2(x - 1) - 42x(x - 1)$

80. $x^3(2x - 1) + 4x^2(2x - 1) - 21x(2x - 1)$

81. $x^4(x^2 - 25) - 10x^2(x^2 - 25) + 9(x^2 - 25)$

82. $3x^4(y^2 - 4) + 7x^2(y^2 - 4) - 6(y^2 - 4)$

83. $2x^2y^2(xy + 1) + 3xy(xy + 1) - 9(xy + 1)$

84. $2x^2(2 - y) + xy(2 - y) - 10y^2(2 - y)$

85. If we write $x^4 - x^2 - 12 = (x^2 + 3)(x^2 - 4)$, have we factored completely? Explain.

86. How can you tell at a glance that $x^4 + 7x + 10$ cannot be factored with our current methods for factoring trinomials?

In Exercises 87–98, factor completely.

87. $x^4 + 28x^2 - 60$

88. $x^4 - 17x^2 - 60$

89. $x^6 + 17x^3 + 60$

90. $x^6 - 16x^3 + 60$

91. $x^4 - 13x^2 + 36$

92. $x^4 - 5x^2 + 4$

93. $5x^8 + 3x^4 - 14$

94. $9x^8 + 77x^4 - 36$

95. $4x^4 - 37x^2 + 9$

96. $4x^4 + 11x^2 - 3$

97. $9x^4 - 13x^2 + 4$

98. $9x^4 - 31x^2 - 20$

CE In Exercises 99–104, factor completely. (*Hint:* Replace the binomial with u and factor the resulting expression. Then replace u with the binomial and simplify.)

You may have chosen to introduce this method previously.

99. $(x - 1)^2 + 11(x - 1) + 28$

100. $(x + 2)^2 + 11(x + 2) + 24$

101. $(2x + 1)^2 - 15(2x + 1) + 56$

102. $12(2 - 3y)^2 - 29(2 - 3y) + 14$

103. $2x^2(x + 2)^2 + x(x + 2) - 6$

104. $2x^2(2 + y)^2 - x(2 + y) - 15$

Geometric Models

105. Real data applications can rarely be modeled with polynomials whose coefficients are integers. For example, an architect's design shows a building with a rectangular floor plan for which the area is modeled by $1.01x^2 + 4.99x - 24.05$, a polynomial that cannot be factored with our present methods. Round off the coefficients of this expression and approximate a factorization.

106. A sheet metal fabricator produces a triangular piece of aluminum with area $\frac{1}{2}x^2 + 7x + 20$, where the base and height are polynomials with integer coefficients. Factor the expression to determine the difference between the base and the height of the piece of aluminum.

Data Analysis: *Food Costs*

The accompanying bar graph shows the average annual food costs (in dollars) for couples 20–50 years old during the period 1990–1998. (Source: U. S. Department of Agriculture.)

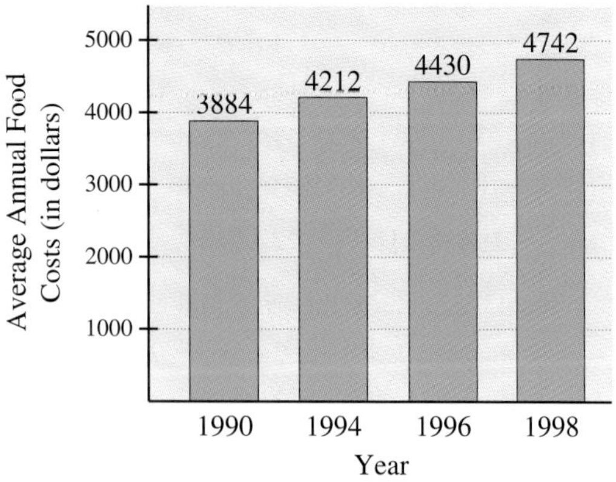

A model that can be used for the average annual food cost is $f(t) = 78t + 3900$, where t is the number of years since 1990.

107. Write $f(t)$ in factored form with 52 as a common factor.

108. What does the other factor represent?

109. What does the coefficient of t in the original function indicate about the rate of increase in annual food costs? How can you interpret the coefficient of t in the factored form of the function?

110. What does the constant term in the original function represent? How can you interpret the constant term in the factored form of the function?

Challenge

In Exercises 111–114, determine all the integers k such that the trinomial can be factored.

111. $x^2 + kx + 10$ **112.** $x^2 + kx - 12$

113. $3x^2 + kx + 5$ **114.** $7x^2 + kx - 2$

In Exercises 115 and 116, determine a positive number k such that the polynomial can be factored.

115. $x^2 + 3x + k$ **116.** $x^2 + 2x + k$

In Exercises 117–123, factor completely.

117. $x^{2n} + 11x^n + 24$ **118.** $y^{2n} + 11y^n + 30$

119. $y^{4n} - 20y^{2n} + 99$ **120.** $x^{4n} - 17x^{2n} + 72$

121. $5x^{2n} + 2x^n - 3$ **122.** $7x^{2n} + 16x^n - 15$

123. $3x^{4n} + 5x^{3n} + 2x^{2n}$

124. If x is an odd integer, show that $3x^2 + 12x + 20$ is the sum of the squares of three consecutive odd integers. (*Hint:* Write $3x^2$ as $x^2 + x^2 + x^2$, write $12x$ as $4x + 8x$, and write 20 as $4 + 16$.)

This supplementary group mixes the methods of factoring and gives students an opportunity to practice the general strategy.

Supplementary Exercises 7.3

In Exercises 1–56, factor the expression completely.

1. $5x^2 - 7x - 6$ **2.** $5x^2 - 31x + 6$

3. $25 - 81y^2$ **4.** $9x^2 - 81$

5. $2x^3 + x^2 - 6x - 3$ **6.** $3x^3 + 9x^2 + x + 3$

7. $3x(x - 2) + 5y(x - 2)$ **8.** $y^2(y - 3) - 5(y - 3)$

9. $125x^3 + 1$ **10.** $64 - 27a^3$

11. $6x + x^2 + 8$ **12.** $x^2 - x - 12$

13. $x^2 + 9$

14. $x^2 - 5x + 36$

15. $4x^2 + 20x + 25$

16. $45a^2 - 30a + 5$

17. $ax - 3a - bx + 3b$

18. $xy + 3y - 5x - 15$

19. $25y^3 - 25$ **20.** $a^3b^3 + 8$

21. $y^4 - 16x^4$ **22.** $5x^2 - 20$

23. $a^2b^7 + 3a^3b^5 - a^2b^4$ **24.** $3x^4 + 15x^3 - 3x^2$

25. $x + 21x^2 - 2$ **26.** $3 - 22y + 7y^2$

27. $16 - 8a + a^2$ **28.** $x^2 + 6xy + 9y^2$

29. $50 - 18x^4$ **30.** $3x^4 + 14x^2 + 8$

31. $2y + 10 - 3xy - 15x$

Exercises 29–56 often use more than one factoring technique in one problem.

32. $2x^3 - 3x^2 - 18x + 27$

33. $24x^4 - 24x^2 + 20x^3$

34. $8a^2b - 27ab + 9b$

35. $c^{20} - d^8$ **36.** $81 - t^4$

37. $x^2 + xy + y^2$ **38.** $x^2 + 4x + 16$

39. $a^2x^3 - a^2y^3 - 4x^3 + 4y^3$

40. $18a^3 - 8a + 45a^2 - 20$

41. $a^5b^4 - a^7b^8 + a^6b^3$

42. $(a + b)(c + d) - 2(a + b)(c - d)$

43. $m^2 - 10m + 16$ **44.** $8 + 7y - y^2$

45. $4x^4 - 64y^4$ **46.** $8x^2 - 98y^2$

47. $(a + b)^3 - 8$ **48.** $(c + d)^2 - 16$

49. $20x^2 - 60x + 45$

50. $x^2y^2 + 3x^2y - 88x^2$

51. $30a^6 - 104a^5 + 90a^4$

52. $12a^2 - 22a - 42$

53. $c^5 - 4c^3 - c^2 + 4$

54. $16x^5 - 54x^2$

55. $x^4 + 2x^2y^2 - 99y^4$

56. $a^2 - 13a + 40$

7.4 Solving Equations by Factoring

Quadratic Equations • Solving by Factoring

Quadratic Equations

In Chapter 3 we developed a routine for solving linear equations in one variable. Equations involving polynomials of degree 2 or higher require different solution techniques.

An equation involving a polynomial of degree 2 is called a **second-degree** or **quadratic equation.**

Definition of a Quadratic Equation

A **quadratic equation** is an equation that can be written in the *standard form*

$$ax^2 + bx + c = 0$$

where a, b, and c are real numbers and $a \neq 0$.

When a quadratic equation is written in standard form, the term ax^2 is called the **quadratic term,** the term bx is called the **linear term,** and the term c is called the **constant term.**

In Chapter 3 we used a table to estimate the solutions of first-degree equations. We can use a similar approach to estimate the solutions of a quadratic equation.

Exploring the Concept

Estimating Solutions of Quadratic Equations

Recall that solving an equation such as $x^2 - 2x - 5 = 3$ involves finding values of x for which the expression $x^2 - 2x - 5$ has a value of 3. Consider the following table of values for the expression $x^2 - 2x - 5$.

x	−3	−2	−1	0	1	2	3	4	5
$x^2 - 2x - 5$	10	3	−2	−5	−6	−5	−2	3	10

This is the same way we estimated solutions of first-degree equations.

From the table, the expression $x^2 - 2x - 5$ appears to have a value of 3 for two different x-values, −2 and 4. Thus the equation $x^2 - 2x - 5 = 3$ appears to have two solutions, −2 and 4.

The table suggests that $x^2 - 2x - 5$ has a value of −6 for only one x-value, 1. Thus the equation $x^2 - 2x - 5 = -6$ appears to have only one solution, 1.

As the x-values in the table increase from −3 to 5, the values of $x^2 - 2x - 5$ decrease to −6 and then increase. Because the expression apparently does not have a value less than −6, we conclude that the equation $x^2 - 2x - 5 = -10$ has no solution.

On the basis of these observations, we might suspect that a quadratic equation can have zero, one, or two real number solutions.

Solving by Factoring

Even though we can use a table to estimate solutions of a quadratic equation, we want to develop an algebraic method to determine the solutions exactly, just as we did for linear equations.

A simple but important property of real numbers is central to the method we will develop. The Zero Factor Property states that if the product of two numbers is zero, then one of the numbers must be zero.

> **Zero Factor Property**
>
> For real numbers a and b, if $ab = 0$, then $a = 0$ or $b = 0$.

Note: As stated, the Zero Factor Property refers to the product of two numbers. The property can easily be extended to products with any number of factors.

For the Zero Factor Property to apply, two conditions must be met.

1. One side of the equation must be 0.
2. The other side of the equation must be a product.

EXAMPLE 1

Using the Zero Factor Property to Solve an Equation

Solve the equation $(x - 5)(x + 3) = 0$.

Solution

$(x - 5)(x + 3) = 0$ — The equation meets the conditions of the Zero Factor Property.

$x - 5 = 0 \quad \text{or} \quad x + 3 = 0$ — One of the factors must be 0.

$x = 5 \quad \text{or} \quad x = -3$ — Solve each first-degree equation.

The solutions are 5 and -3.

Sometimes students do not realize that these are first-degree equations that are solved with previous methods.

If one side of an equation is not a product, then we can create a product by factoring the expression.

$x^2 - 2x - 15 = 0$ — The Zero Factor Property does not apply.

$(x - 5)(x + 3) = 0$ — Factor the expression.

Now the Zero Factor Property applies, and we can proceed as in Example 1.

The following summarizes the procedure for solving a quadratic equation by factoring.

Solving a Quadratic Equation by Factoring

1. Write the equation in standard form.
2. Factor the polynomial completely.
3. Set each factor equal to 0 (Zero Factor Property).
4. Solve the resulting first-degree equations.
5. Check the solutions in the original equation.

Note: In our examples, we usually leave step 5 to you. However, we strongly encourage you to check all solutions.

EXAMPLE 2

Solving Quadratic Equations by Factoring

Solve each equation.

(a) $2x^2 - 9x - 5 = 0$ (b) $9x^2 + 4 = 12x$ (c) $x^2 + 4x - 3 = 4$

Solution

(a)

$$2x^2 - 9x - 5 = 0 \quad \text{The equation is in standard form.}$$

$$(2x + 1)(x - 5) = 0 \quad \text{Factor the expression on the left side.}$$

$$2x + 1 = 0 \quad \text{or} \quad x - 5 = 0 \quad \text{Zero Factor Property}$$

$$x = -\frac{1}{2} \quad \text{or} \quad x = 5 \quad \text{Solve each case for } x.$$

(b)

$$9x^2 + 4 = 12x$$

$$9x^2 - 12x + 4 = 0 \quad \text{Write the equation in standard form.}$$

$$(3x - 2)(3x - 2) = 0 \quad \text{The factors are identical.}$$

$$3x - 2 = 0 \quad \text{We obtain only one first-degree equation.}$$

$$x = \frac{2}{3} \quad \text{Solve for } x.$$

Naturally, you want to avoid discussing the technical motivation for calling this solution a "double root." You can temporize by noting that two identical factors give rise to two roots, but the roots are equal.

Because the two factors are the same, we obtain just one distinct solution (or root). We say that $\frac{2}{3}$ is a **double root.**

(c)

$$x^2 + 4x - 3 = 4$$

$$x^2 + 4x - 7 = 0 \quad \text{Write the equation in standard form.}$$

Because we cannot factor $x^2 + 4x - 7$, we are unable to use our current algebraic method for solving the equation. However, this does not necessarily mean that the equation has no solution. Using solving methods that we will learn in Chapter 10, we can show that the equation has two irrational number solutions.

EXAMPLE 3

Solving Quadratic Equations Algebraically

Solve each equation.

(a) $12x^2 = 3x$ (b) $x(6x + 1) = 12$

Explain to students why they should not divide both sides by x. Students are sometimes unsure about what to do with the $3x$. Emphasize that the equation must be written in standard form.

Solution

(a)

$12x^2 = 3x$	
$12x^2 - 3x = 0$	Write the equation in standard form.
$3x(4x - 1) = 0$	Factor the left side.
$3x = 0$ or $4x - 1 = 0$	Zero Factor Property
$x = 0$ or $x = \frac{1}{4}$	Solve each case for x.

The two solutions are 0 and $\frac{1}{4}$.

A student who wants to write "$x = 12$ or $6x + 1 = 12$" does not understand the required conditions of the Zero Factor Property. These conditions must be stressed repeatedly.

(b)

$x(6x + 1) = 12$	
$6x^2 + x = 12$	Distributive Property
$6x^2 + x - 12 = 0$	Write the equation in standard form.
$(2x + 3)(3x - 4) = 0$	Factor the left side.
$2x + 3 = 0$ or $3x - 4 = 0$	Zero Factor Property
$x = -\frac{3}{2}$ or $x = \frac{4}{3}$	Solve each case for x.

The two solutions are $-\frac{3}{2}$ and $\frac{4}{3}$.

LEARNING TIP

Because the left side is factored in part (b) of Example 3, you might be tempted to apply the Zero Factor Property right away. Keep in mind, however, that the Zero Factor Property applies only when one side of the equation is 0.

While the factoring method is a classic technique for solving quadratic equations, it can also be used for other types of equations.

EXAMPLE 4

Solving Other Types of Equations by Factoring

Solve the following equations.

(a) $4x^3 + 12x^2 - x - 3 = 0$ (b) $|x^2 - 13| = 12$

Solution

(a) Note that this is a third-degree equation, not a quadratic equation. However, if we can factor the left side, the factoring method can still be used to solve the equation.

$4x^3 + 12x^2 - x - 3 = 0$	
$(4x^3 + 12x^2) - (x + 3) = 0$	Group the left side in pairs.
$4x^2(x + 3) - (x + 3) = 0$	Factor out the GCF in the first pair.
$(x + 3)(4x^2 - 1) = 0$	Factor out the GCF $x + 3$.
$(x + 3)(2x + 1)(2x - 1) = 0$	Factor the difference of two squares.
$x + 3 = 0$ or $2x + 1 = 0$ or $2x - 1 = 0$	Zero Factor Property
$x = -3$ or $x = -\frac{1}{2}$ or $x = \frac{1}{2}$	Solve each case for x.

The solutions are -3, $-\frac{1}{2}$, and $\frac{1}{2}$.

(b) The equation $|x^2 - 13| = 12$ states that $x^2 - 13$ represents a number whose absolute value is 12. Therefore, $x^2 - 13$ represents either 12 or -12.

$$x^2 - 13 = 12 \quad \text{or} \quad x^2 - 13 = -12$$

The absolute value equation results in two quadratic equations, which, in turn, result in two first-degree equations.

Think About It

Use the factoring method to solve $x^3 - x + 2 = 2x^2$.

Each possibility is a quadratic equation that we will try to solve by the factoring method.

$$
\begin{array}{lll}
x^2 - 13 = 12 & \qquad & x^2 - 13 = -12 \\
x^2 - 25 = 0 & & x^2 - 1 = 0 \\
(x + 5)(x - 5) = 0 & & (x + 1)(x - 1) = 0 \\
x + 5 = 0 \quad \text{or} \quad x - 5 = 0 & & x + 1 = 0 \quad \text{or} \quad x - 1 = 0 \\
x = -5 \quad \text{or} \quad x = 5 & & x = -1 \quad \text{or} \quad x = 1
\end{array}
$$

The four solutions are 5, -5, 1, and -1.

Quick Reference 7.4

Quadratic Equations

- A **quadratic equation** is an equation that can be written in the *standard form*

  $ax^2 + bx + c = 0$

  where a, b, and c are real numbers and $a \neq 0$.
- The term ax^2 is the **quadratic term,** the term bx is the **linear term,** and the term c is the **constant term.**

Solving by Factoring

- The **Zero Factor Property** is as follows:

  For real numbers a and b, if $ab = 0$, then $a = 0$ or $b = 0$.

  In words, if a product of two or more factors is equal to zero, then one of the factors must equal zero.
- For the Zero Factor Property to apply,
  1. one side of the equation must be a product, and
  2. the other side of the equation must be 0.
- To solve a quadratic equation by factoring:
  1. Write the equation in standard form.
  2. Factor the polynomial completely.
  3. Set each factor equal to zero.
  4. Solve each resulting equation.
  5. Check the solution(s).
- In the solving of a quadratic equation, if two factors are the same, each results in the same solution. That solution is called a **double root.**

Speaking the Language 7.4

1. The equation $x^2 + 6x - 10 = 0$ is an example of a(n) ________ equation.

2. In the equation $x^2 + 13x + 40 = 0$ we call x^2 the ________ term and $13x$ the ________ term.

3. If $7x = 0$, then the ________ Property guarantees that $x = 0$.

4. Because -3 is the only solution of $x^2 + 6x + 9 = 0$, we call -3 a(n) ________ root.

Exercises 7.4

Concepts and Skills

1. Applying the Zero Factor Property to solve the equation $3x(x + 5) = 0$, one person writes $3 = 0$, $x = 0$, or $x + 5 = 0$. Another person writes $3x = 0$ or $x + 5 = 0$. Explain why both approaches are correct.

2. Explain why the Zero Factor Property does not apply to the equation $(x + 2)(x - 3) = -6$.

In Exercises 3–10, use the Zero Factor Property to solve the equation.

3. $(x - 4)(x + 3) = 0$
4. $3(2x + 5)(x + 1) = 0$
5. $5x(6 - x) = 0$
6. $x(1 - 2x) = 0$
7. $(x - 5)^2 = 0$
8. $4(x + 4)^2 = 0$
9. $(x + 5)(2x - 6)(3x + 3) = 0$
10. $x(3 + 2x)(1 - x) = 0$

11. How do we distinguish between a first-degree equation and a second-degree equation?

12. Describe the factors of a second-degree equation that has only one distinct solution.

Remind students that fractions can be cleared as we did for first-degree equations.

In Exercises 13–20, solve.

13. $x^2 + 7x = 0$
14. $2c^2 - 5c = 0$
15. $2g^2 - 18 = 0$
16. $h^2 - 49 = 0$
17. $x^2 - 5x + 6 = 0$
18. $-3b^2 - 7b + 6 = 0$
19. $\frac{2}{5}x^2 - \frac{3}{5}x - 1 = 0$
20. $x^2 - \frac{7}{6}x - \frac{1}{2} = 0$

21. To solve $(x - 1)(2x - 1) = 1$, a person concludes that because the product is 1 and $1 \cdot 1 = 1$, each factor must be 1. Thus the person writes $x - 1 = 1$ and $2x - 1 = 1$. Explain why this reasoning is not valid.

22. Verify that solving $x^2 = 7x$ results in two solutions. Explain why dividing both sides of the equation by x to obtain $x = 7$ does not reveal both solutions.

In Exercises 23–38, solve the given equation.

23. $a^2 = 5a$
24. $x^2 = 25$
25. $(x - 2)^2 = 25$
26. $(a + 3)^2 - 64 = 0$
27. $x^2 + 5x = 6$
28. $y^2 - 2y = 8$
29. $x^2 + 35 = 12x$
30. $x^2 + 12 = 7x$
31. $12 = 11d - 2d^2$
32. $15 = 26g - 7g^2$
33. $-8 = -9y^2 - 14y$
34. $-x - 6 = -12x^2$
35. $x(x + 1) = 30$
36. $x(x - 1) = 42$
37. $a(a - 32) + 60 = 0$
38. $c(c + 23) + 60 = 0$

In Exercises 39–48, solve.

The equations in this group require significant algebraic manipulation to be written in standard form.

39. $3x(2 + x) = 24$
40. $x(2 - 3x) = \frac{1}{3}$
41. $\frac{2}{5}x(x - 2) = x - 2$
42. $3(x + 1)^2 = 9 - x$
43. $(x - 3)(x + 2) = 6$
44. $(x - 7)(x - 1) = -8$
45. $(3x - 2)(x + 2) = (x - 2)(x + 1) + 10$
46. $3(x - 3)(x + 1) = (2x - 1)(x - 1) + 8$
47. $(4x - 3)^2 + (4x - 3) = 6$
48. $5(x + 2)^2 - 17(x + 2) - 12 = 0$

In Exercises 49–64, solve the equation.

49. $(x - 5)(x^2 + 3x - 18) = 0$
50. $(2x - 5)(8x^2 + 14x - 15) = 0$
51. $3x^3 - 27x = 0$
52. $8x^2 - 50x^4 = 0$
53. $6x^3 + 2x^2 = 4x$
54. $6y^2 = 3y^3 + 9y^4$
55. $x^3 - 5x^2 - x + 5 = 0$
56. $x^3 + 3x^2 - x - 3 = 0$
57. $2x^3 - 3x^2 - 8x + 12 = 0$
58. $3x^3 + 4x^2 - 27x - 36 = 0$

Although the equations in this group are not quadratic, they can be solved by factoring.

59. $x^4 + 36 = 13x^2$
60. $x^4 - 26x^2 + 25 = 0$
61. $6(y - 2)^3 - (y - 2)^2 - (y - 2) = 0$
62. $y^2(4y^2 - 9) - 16y(4y^2 - 9) + 48(4y^2 - 9) = 0$
63. $3x^2(3 - x) + 3x(3 - x) + 18(x - 3) = 0$
64. $10(3 - 2x)x^2 - 27(3 - 2x)x - 5(2x - 3) = 0$

CE In Exercises 65–68, one solution of the equation is given. Determine a, b, or c and another solution of the equation.

65. $2x^2 + bx + 2 = 0; \quad x = -2$
66. $(2a + 1)x^2 + x + 10 = 0; \quad x = 2$
67. $2x^2 + x + (3 - 4c) = 0; \quad x = 3$
68. $-x^2 + x + c = 0; \quad x = -1$

CE In Exercises 69–74, write an equation with the given solutions.

You may want to suggest to students that they reverse the steps used to solve a quadratic equation—that is, that they start with the solutions, determine the factors that led to those solutions, write the equation in factored form, and then write it in standard form.

69. 3, −4
70. −2, 5

71. 4 (double root)

72. -2 (double root)

73. $0, \frac{1}{2}$

74. $\frac{2}{3}, -2$

CE In Exercises 75–80, solve for x.

75. $x^2 - cx - 2c^2 = 0$

76. $x^2 + xy - 2y^2 = 0$

77. $3y^2 - 4xy = 4x^2$

78. $cx + 2c^2 = 6x^2$

79. $x^2 + 2ax = 3a^2$

80. $4a^2 = 10x^2 + 3ax$

In Exercises 81–86, solve the given equation.

81. $|x^2 + 5x| = 6$

82. $|3x^2 - 5x| = 2$

83. $|x^2 + 10x + 15| = 6$

84. $|x^2 + 7x + 8| = 2$

85. $|x^2 - 3x - 1| = 3$

86. $|x^2 - 5x + 5| = 1$

In Exercises 87–94, write an equation describing the conditions of the problem. Then solve the equation to determine the number(s).

87. The sum of a number and its square is 72.

88. The sum of twice a number and its square is 63.

89. The sum of the squares of three consecutive integers is 50.

90. The sum of the squares of three consecutive integers is 77.

91. The sum of the cube of a number and 6 times the number is equal to 5 times the square of the number.

92. The sum of the cube of a number and 12 times the number is equal to 8 times the square of the number.

93. Two numbers have a sum of 3 and a product of -10.

94. The sum of two numbers is 12 and the sum of their squares is 74.

Geometric Models

95. The length of a rectangle is 5 meters more than the width. The area of the rectangle is 594 square meters. What are the dimensions of the rectangle?

96. The width of a rectangular living room is 7 feet less than the length. The area of the room is 294 square feet. What are the dimensions of the rectangle?

97. One leg of a right triangle is 7 inches more than the other leg. The area of the triangle is 60 square inches. What are the lengths of the three sides?

98. One leg of a right triangle is 5 feet more than the other leg. The area of the triangle is 150 square feet. What are the lengths of the three sides?

Real-Life Applications

99. A landscape designer decided to surround a rectangular garden with a path of uniform width. The garden was 14 feet wide and 20 feet long. If the total area of the path was 152 square feet, how wide was the path?

100. A farmer wants to use 500 feet of fencing to enclose a rectangular area of 15,000 square feet. To meet these conditions, what should be the dimensions of the fenced area?

101. A product-packaging firm manufactures cardboard trays for holding 24 12-ounce cans of soft drinks. The process begins with a rectangular piece of cardboard that is 5 inches longer than it is wide. Two-inch squares are cut from each corner, and the flaps are folded up to form the tray.

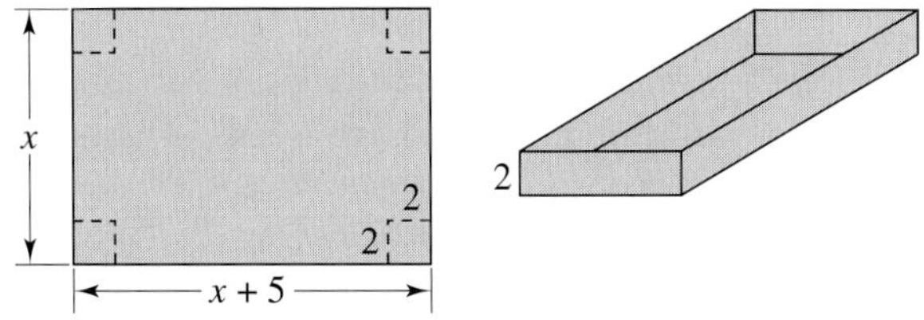

If the area of the bottom of the tray must be 176 square inches, what are the dimensions of the original cardboard rectangle?

102. A small software development firm finds that its profits (in hundreds of thousands of dollars) can be modeled by the expression $-0.4x^2 + 4x$, where x is the number of sales representatives. For how many sales representatives (other than none) will the company earn no profit?

Modeling with Real Data

103. A survey of coffee drinkers revealed that 62% of those in the 35–49 age group become upset if they do not have coffee at a regular time. The accompanying table gives the results for the three age groups that were surveyed. (Source: LMK Associates.)

Age Group	Percent
18–34	50%
35–49	62%
50 and over	56%

The function $C(x) = -0.05x^2 + 4.4x - 32$ models the percents as a function of age x.

(a) Write $C(x)$ with a common factor of -0.05.

(b) Write $C(x)$ as a completely factored polynomial.

(c) Solve the equation $C(x) = 0$.

(d) What do the solutions represent?

104. In a recent year the average annual cost of utilities was \$1984. However, the accompanying table shows that the average annual cost varies with age. (Source: Bureau of Labor Statistics.)

Age Group	Average Cost
Under 25	\$1024
25–34	1797
35–44	2232
45–54	2375
55–64	2255
65 and over	1816

The function $f(x) = -1.4x^2 + 142.8x - 1288$ models the average costs as a function of age x.

(a) Write the model function in factored form. (*Hint:* Begin by factoring out the common factor -1.4.)

(b) Use the factoring method to solve the equation $f(x) = 0$.

(c) Interpret the solutions and state whether you think that they are meaningful.

Challenge

105. Solve the following equation.

$$(x + y + 1)(1 - x) + (x + y - 1)(x - 1) = 0$$

In Exercises 106 and 107, solve the equation. (*Hint:* Recall the substitution technique: $u = 1/x$.)

106. $3\left(\frac{1}{x}\right)^2 + 14\left(\frac{1}{x}\right) - 24 = 0$

107. $\frac{2}{x^2} - \frac{1}{x} - 1 = 0$

108. A rectangular swimming pool is surrounded by a concrete deck. The diagram shows the relative dimensions of the pool and deck.

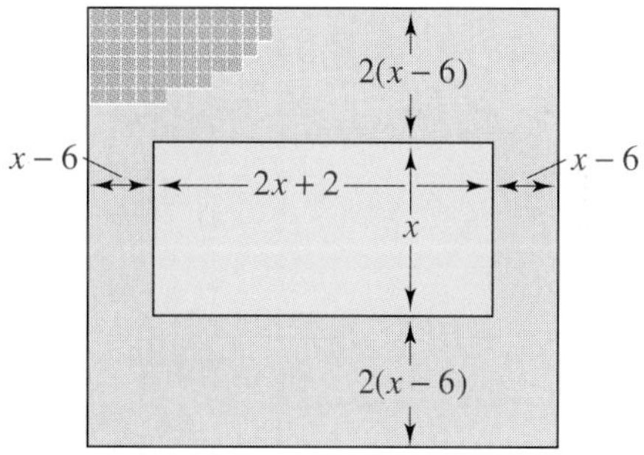

The combined area of the pool and deck is 2116 square feet. The owner plans to cover the concrete deck with tile that costs \$80 per hundred square feet. Determine the total cost of the job.

7.5 Division

Quotient of a Polynomial and a Monomial • Long Division • Synthetic Division

Quotient of a Polynomial and a Monomial

The most basic example of a quotient of polynomials is the quotient of two monomials. We use properties of exponents as well as the properties of real numbers to simplify the quotient.

$$\frac{15x^5}{9x^3} = \frac{15}{9} \cdot \frac{x^5}{x^3} = \frac{5}{3} \cdot \frac{x^{5-3}}{1} = \frac{5}{3} \cdot \frac{x^2}{1} = \frac{5x^2}{3}$$

$$\frac{4y^3}{20y^5} = \frac{4}{20} \cdot \frac{y^3}{y^5} = \frac{1}{5} \cdot \frac{y^{3-5}}{1} = \frac{1}{5} \cdot \frac{y^{-2}}{1} = \frac{1}{5y^2}$$

Recall the following rule for adding fractions.

For real numbers a, b, and c, $c \neq 0$, $\frac{a}{c} + \frac{b}{c} = \frac{a+b}{c}$.

To determine the quotient of a polynomial and a monomial, we use this same rule but in reverse order.

Dividing a Polynomial by a Monomial

For monomials a, b, and c, $c \neq 0$,

$$\frac{a+b}{c} = \frac{a}{c} + \frac{b}{c}$$

In words, divide each term of the numerator by the denominator.

EXAMPLE 1

Dividing a Polynomial by a Monomial

Find the following quotients.

(a) $\frac{12x^3 - 8x^2 + 6x}{2x}$ (b) $\frac{15x^6 + 21x^4 - 2x^2 + 3x}{3x^2}$

Solution

Encourage students to write the middle step until they are comfortable omitting it.

(a) $\frac{12x^3 - 8x^2 + 6x}{2x} = \frac{12x^3}{2x} - \frac{8x^2}{2x} + \frac{6x}{2x}$ Write the quotient as a sum of quotients.

$= 6x^2 - 4x + 3$ Simplify each quotient.

(b) $\frac{15x^6 + 21x^4 - 2x^2 + 3x}{3x^2} = \frac{15x^6}{3x^2} + \frac{21x^4}{3x^2} - \frac{2x^2}{3x^2} + \frac{3x}{3x^2}$

$= 5x^4 + 7x^2 - \frac{2}{3} + \frac{1}{x}$

Long Division

In this section, the vocabulary associated with division will be important.

To divide a polynomial by a polynomial with more than one term, we use a procedure similar to long division in arithmetic.

$$\begin{array}{rrl} & 5 & \leftarrow \text{Quotient} \\ \text{Divisor} \rightarrow & 41\overline{)211} & \leftarrow \text{Dividend} \\ & \underline{205} & \\ & 6 & \leftarrow \text{Remainder} \end{array}$$

The result can be written in the following form.

$$\frac{\text{Dividend}}{\text{Divisor}} = \text{Quotient} + \frac{\text{Remainder}}{\text{Divisor}}$$

$$\frac{211}{41} = 5 + \frac{6}{41} = 5\frac{6}{41}$$

To check the result, we use the following relation.

Quotient	·	Divisor	+	Remainder	=	Dividend
↓		↓		↓		↓
5	·	41	+	6	=	211

EXAMPLE 2

Dividing a Polynomial by a Binomial

Divide $(2x^3 - 3x^2 - x + 2)$ by $(x - 2)$.

Remind students about the care they must take with subtraction. Circling the changed signs can help prevent careless errors.

LEARNING TIP

In long division of polynomials, the most frequent error is in the subtraction step. Rather than scribbling out signs, you might want to circle sign changes to make them more prominent.

Solution

$$x - 2\overline{)2x^3 - 3x^2 - x + 2}$$

Write the problem in long-division format.

$$\begin{array}{r} 2x^2 \\ x - 2\overline{)2x^3 - 3x^2 - x + 2} \end{array}$$

Divide the first term of the dividend $2x^3$ by the first term of the divisor x.

$$\begin{array}{r} 2x^2 \qquad\qquad \\ x - 2\overline{)2x^3 - 3x^2 - x + 2} \\ 2x^3 - 4x^2 \qquad\qquad \end{array}$$

Multiply the result $2x^2$ times the divisor $x - 2$. Write the result under the dividend.

$$\begin{array}{r} 2x^2 \qquad\qquad \\ x - 2\overline{)2x^3 - 3x^2 - x + 2} \\ \underline{2x^3 - 4x^2} \qquad\qquad \\ x^2 - x \qquad \end{array}$$

Subtract (by changing signs and adding) and then bring down the next term of the dividend $-x$.

$$\begin{array}{r} 2x^2 + x \qquad \\ x - 2\overline{)2x^3 - 3x^2 - x + 2} \\ \underline{2x^3 - 4x^2} \qquad\qquad \\ x^2 - x \qquad \end{array}$$

Divide the first term of the result x^2 by the first term of the divisor, x.

$$\begin{array}{r} 2x^2 + x \qquad \\ x - 2\overline{)2x^3 - 3x^2 - x + 2} \\ \underline{2x^3 - 4x^2} \qquad\qquad \\ x^2 - x \qquad \\ x^2 - 2x \qquad \end{array}$$

Multiply the result, x, times the divisor, $x - 2$.

Write the result under the dividend.

$$\begin{array}{r} 2x^2 + x \qquad \\ x - 2\overline{)2x^3 - 3x^2 - x + 2} \\ \underline{2x^3 - 4x^2} \qquad\qquad \\ x^2 - x \qquad \\ \underline{x^2 - 2x} \qquad \\ x + 2 \end{array}$$

Subtract (by changing signs and adding) and then bring down the next term of the dividend, 2.

$$\begin{array}{r} 2x^2 + x + 1 \\ x - 2\overline{)2x^3 - 3x^2 - x + 2} \\ \underline{2x^3 - 4x^2} \qquad\qquad \\ x^2 - x \qquad \\ \underline{x^2 - 2x} \qquad \\ x + 2 \\ \underline{x - 2} \\ 4 \end{array}$$

Divide the first term of the result, x, by the first term of the divisor, x. Then multiply the result, 1, by the divisor, $x - 2$.

Then subtract.

The quotient is $2x^2 + x + 1$ and the remainder is 4. We can write the result in the following form.

$$\frac{2x^3 - 3x^2 - x + 2}{x - 2} = 2x^2 + x + 1 + \frac{4}{x - 2}$$

Students may need to be reminded about the relation (Quotient)(Divisor) + Remainder = Dividend

We check the result with the following relation.

Quotient	·	Divisor	+	Remainder	=	Dividend
$(2x^2 + x + 1)$	·	$(x - 2)$	+	4	=	$2x^3 - 3x^2 - x + 2$

Note: When we use long division to divide polynomials, the dividend and the divisor should be in descending order. We continue the division operation until the remainder is 0 or the degree of the remainder is less than the degree of the divisor.

EXAMPLE 3

Missing Terms in the Dividend

Divide $x^3 + 27$ by $x + 3$.

Solution

As we write the problem in the long-division format, it is helpful to include the missing terms in the dividend with zero coefficients.

$$\begin{array}{r|l} x^2 - 3x + 9 & \\ x + 3\overline{)x^3 + 0x^2 + 0x + 27} & \text{Include the missing } x^2 \text{ and } x \text{ terms.} \\ \underline{x^3 + 3x^2} \qquad\qquad\quad & x^2(x + 3) = x^3 + 3x^2 \\ -3x^2 + 0x \qquad & \text{Subtract and bring down } 0x. \\ \underline{-3x^2 - 9x} \qquad & -3x(x + 3) = -3x^2 - 9x \\ 9x + 27 & \text{Subtract and bring down 27.} \\ \underline{9x + 27} & 9(x + 3) = 9x + 27 \\ 0 & \text{Subtract. The remainder is 0.} \end{array}$$

The quotient is $x^2 - 3x + 9$ and the remainder is 0.

$$\frac{x^3 + 27}{x + 3} = x^2 - 3x + 9$$

To check, multiply $(x^2 - 3x + 9)$ by $(x + 3)$. The result is $x^3 + 27$.

When we divide 20 by 4, we obtain a quotient of 5. Because the remainder is 0, we say that 20 is *divisible* by 4, and we can write the factorization $20 = 5 \cdot 4$.

Similarly, in Example 3 the remainder is 0, and so we say that $x^3 + 27$ is divisible by $x + 3$, and we can write the dividend in factored form.

Dividend	=	Quotient	·	Divisor
$x^3 + 27$	=	$(x^2 - 3x + 9)$	·	$(x + 3)$

EXAMPLE 4

Missing Terms in the Dividend and the Divisor

Divide $x^3 + 2x - 1$ by $2x^2 + 4$.

Solution

Write the problem in the long-division format with zero for the coefficients of the missing terms in the dividend and the divisor.

$$2x^2 + 0x + 4\overline{)x^3 + 0x^2 + 2x - 1}$$

$$\begin{array}{r} \tfrac{1}{2}x \\ 2x^2 + 0x + 4\overline{)x^3 + 0x^2 + 2x - 1} \\ \underline{x^3 + 0x^2 + 2x\phantom{{}-1}} \\ -1 \end{array}$$

Divide x^3 by $2x^2$.

$\frac{1}{2}x(2x^2 + 0x + 4) = x^3 + 0x^2 + 2x$

Subtract and bring down -1.

We stop dividing because the degree of the remainder is less than the degree of the divisor. The quotient is $\frac{1}{2}x$; the remainder is -1.

Synthetic Division

Synthetic division is a numerical method for dividing a polynomial by a binomial of the form $x - c$. The technique involves writing only the essential parts of a long division problem.

Exploring the Concept

Synthetic Division

Consider the following long-division problem on the left and compare it to the numerical arrangement on the right.

$$\begin{array}{r} 2x^2 + 3x + 8 \\ x - 3\overline{)2x^3 - 3x^2 - 1x + 2} \\ \underline{2x^3 - 6x^2\phantom{{}-1x+02}} \\ 3x^2 - x\phantom{{}+02} \\ \underline{3x^2 - 9x\phantom{{}+02}} \\ 8x + 2 \\ \underline{8x - 24} \\ 26 \end{array}$$

$$\begin{array}{c|rrrr} ③ & 2 & -3 & -1 & 2 \\ & & 6 & 9 & 24 \\ \hline & 2 & 3 & 8 & \boxed{26} \end{array}$$

We can make the following observations about the numbers on the right.

1. The circled number 3 is the same as the 3 in the divisor, $x - 3$. We refer to this number as the *divider.*
2. The numbers 2, -3, -1, and 2 along the first row are the coefficients of the dividend, $2x^3 - 3x^2 - x + 2$.
3. The numbers 2, 3, and 8 along the bottom row are the coefficients of the quotient, $2x^2 + 3x + 8$.
4. The boxed number 26 is the remainder.
5. Each number in the bottom row is the sum of the two numbers above it.
6. Each number in the middle row is the product of the divider 3 and the number in the bottom row of the preceding column.

The following illustrates the use of synthetic division to divide $2x^3 - 3x^2 - x + 2$ by $x - 3$.

$$\begin{array}{r|rrrr} 3 & 2 & -3 & -1 & 2 \\ & & & & \\ \hline & & & & \end{array}$$

Write the divider 3 and the coefficients of the dividend.

$$\begin{array}{r|rrrr} 3 & 2 & -3 & -1 & 2 \\ & & & & \\ \hline & 2 & & & \end{array}$$

Bring the first number of the top row down to the bottom row.

$$\begin{array}{r|rrrr} 3 & 2 & -3 & -1 & 2 \\ & & 6 & & \\ \hline & 2 & & & \end{array}$$

Multiply the first number in the bottom row, 2, by the divider 3 to obtain 6, the first number in the middle row.

$$\begin{array}{r|rrrr} 3 & 2 & -3 & -1 & 2 \\ & & 6 & & \\ \hline & 2 & 3 & & \end{array}$$

Add the −3 and 6 to obtain the second number in the bottom row, 3.

$$\begin{array}{r|rrrr} 3 & 2 & -3 & -1 & 2 \\ & & 6 & 9 & \\ \hline & 2 & 3 & 8 & \end{array}$$

Similarly, multiply the 3 in the bottom row by the divider 3 to obtain 9, the second number in the middle row. Then add the column to obtain 8.

$$\begin{array}{r|rrrr} 3 & 2 & -3 & -1 & 2 \\ & & 6 & 9 & 24 \\ \hline & 2 & 3 & 8 & 26 \end{array}$$

Finally, multiply this 8 by the divider 3 to obtain 24, the last number in the middle row. Then add the column to obtain 26.

The entries in the bottom row are the coefficients of the quotient and the remainder.

Quotient: $2x^2 + 3x + 8$ Remainder: 26

Note that the degree of the quotient is 1 less than the degree of the dividend. Because the degree of the dividend is 3, the degree of the quotient is 2. Thus we begin the quotient with $2x^2$.

Synthetic division is a valid shortcut to long division, provided that certain requirements are met.

You probably need to stress this. Students often use the wrong divider.

1. The divisor must be of the form $x - c$. For a divisor such as $x - 7$, the divider is 7. For a divisor such as $x + 5$, we write $x - (-5)$, and the divider is -5.
2. The dividend must be written in descending order.
3. We write 0 as the coefficient of any missing term.

The advantages of synthetic division are the ease and speed with which it can be performed. The method eliminates much of the repetition involved in long division. For example, it is unnecessary to keep writing the variable factors of the terms.

EXAMPLE 5

Dividing with Synthetic Division

Use synthetic division to divide $x^4 - 3x^3 + 2x - 4$ by $x - 3$.

Solution

3	1	−3	0	2	−4	Include a 0 as the coefficient of the missing x^2 term.
		3	0	0	6	
	1	0	0	2	2	The remainder is 2.
	↓	↓	↓	↓		
	$1x^3 +$	$0x^2 +$	$0x +$	2		Quotient

The degree of the dividend is 4, so the degree of the quotient is 3. The quotient is $x^3 + 2$ and the remainder is 2.

We can also write the result in the following form.

$$\frac{x^4 - 3x^3 + 2x - 4}{x - 3} = x^3 + 2 + \frac{2}{x - 3}$$

To check, multiply the quotient by the divisor and add the remainder. The result is the dividend.

$$(x^3 + 2)(x - 3) + 2 = x^4 - 3x^3 + 2x - 4$$

Think About It

Consider the quotient $\frac{x^2 + 5x - 4}{2x - 1}$. Synthetic division may be used to perform the division if we first divide the numerator and the denominator by 2. Do this and then perform the synthetic division.

EXAMPLE 6

Dividing with Synthetic Division

Divide $x^3 - 6x^2 - x + 30$ by $x + 2$. Then use the result to factor $x^3 - 6x^2 - x + 30$ completely.

Solution

The divisor must be written as $x - (-2)$ with $c = -2$.

−2	1	−6	−1	30	
		−2	16	−30	
	1	−8	15	0	The remainder is 0.
	↓	↓	↓		
	$1x^2 -$	$8x +$	15		Quotient

Because the remainder is 0, we can write the dividend in factored form.

$$\begin{aligned} x^3 - 6x^2 - x + 30 &= (x + 2)(x^2 - 8x + 15) \\ &= (x + 2)(x - 5)(x - 3) \end{aligned}$$

Complete the factorization by factoring the trinomial.

Quick Reference 7.5

Quotient of a Polynomial and a Monomial

- To divide a monomial by a monomial, use the Quotient Rule for Exponents.
- To divide a polynomial by a monomial, divide each term of the polynomial by the monomial.

Long Division

- To divide a polynomial by a polynomial, we can use a method similar to long division in arithmetic.

- To keep like terms in columns when dividing polynomials, use coefficients of 0 for missing terms.
- When dividing polynomials, continue the division operation until the remainder is 0 or the degree of the remainder is less than the degree of the divisor.
- To check the answer, multiply the quotient times the divisor and add the remainder. The result should be the dividend.

Synthetic Division

- To divide a polynomial (written in descending order) by $x - c$, where c is a constant called the *divider,* follow these steps.
  1. In the first row, write the coefficients of the dividend with zero as the coefficient of any missing term.
  2. Write the divider c to the left of the first row in step 1.
  3. Bring the first coefficient in the top row down to the bottom row.
  4. Multiply the divider c times the new entry in the bottom row and place the result in the middle row of the next column.
  5. Add the column and write the result underneath in the bottom row.
  6. Repeat steps 4 and 5 until the bottom row is completely filled.
  7. The bottom row then gives the coefficients of the quotient. The last number in the bottom row is the remainder. The degree of the quotient will be 1 less than the degree of the dividend.

Speaking the Language 7.5

1. The quantities involved in a division problem are related by the form

Quotient · Divisor + ________ = ________

2. In a long-division problem, we write any ________ term in the dividend with a coefficient of 0.

3. Performing long division with only coefficients is a numerical method called ________.

4. If the remainder in a long-division problem is not 0, then we continue the division process until the degree of the ________ is less than the degree of the ________.

Exercises 7.5

Concepts and Skills

1. Identify the dividend, divisor, quotient, and remainder in the following long-division problem.

$$\begin{array}{r} 1 \\ x\overline{)x + 7} \\ \underline{x} \\ 7 \end{array}$$

2. Suppose that you divide $x^2 + 5x + 9$ by $x + 1$ and obtain the quotient $x + 4$ and a remainder of 5. Explain how you can check your work. Then perform the check to determine whether this result is correct.

In Exercises 3–12, determine the quotient.

3. $\dfrac{10x^4 + 20x^3}{5x^2}$

4. $\dfrac{4x^3 - 2x^2}{2x^2}$

5. $\dfrac{4x^5 + 6x^3 + x^2}{2x^2}$

6. $\dfrac{6x^6 - 3x^5 + 2x^3}{3x^3}$

7. $\dfrac{16a^4 + 12a^3 - 6a^2 + 8a}{2a}$

8. $\dfrac{18a^4 - 12a^3 - 15a^2 + 9a}{3a}$

9. $\dfrac{20x^5 - 12x^4 + 6x^2 - 8x}{4x^2}$

10. $\dfrac{20x^5 + 10x^4 - 3x^2 + 15x}{5x^2}$

11. $\dfrac{15c^5d^7 + 6c^4d^5 - 9cd^3}{3c^2d^2}$

12. $\dfrac{18c^4d^7 - 10c^2d^5 + 8cd^2}{2cd^3}$

In Exercises 13–16, determine the quotient. (*Hint:* In Exercises 15 and 16, begin by simplifying the numerator.)

13. $\dfrac{6x^3(3x + 1) + 4x^2(2x - 3)}{2x^2}$

14. $\dfrac{9x^2(2x - 1) - 3x(x - 2)}{3x}$

15. $\dfrac{(x + 4)^2 - (x - 4)^2}{2x}$

16. $\dfrac{(3 - x)^2 - (3 + x)^2}{3x}$

17. When we perform long division, when do we stop dividing?

18. If a polynomial of degree $n > 2$ is divided by a polynomial of degree 2, what are the possible degrees of the remainder? Explain.

In Exercises 19–28, use long division to determine the quotient and remainder.

19. $\dfrac{2x + 1}{x - 3}$

20. $\dfrac{4x + 3}{x + 2}$

21. $(x^2 + 4x + 3) \div (x + 2)$

22. $(x^2 - x - 6) \div (x + 3)$

23. $\dfrac{3x^2 + 2}{x - 1}$

24. $\dfrac{x^2 - 2}{x + 1}$

25. $(4a^3 - a^2 + 5a + 4) \div (4a + 3)$

26. $(3x^4 + 2x^3 - 4x^2 + 7x - 5) \div (3x - 1)$

27. $\dfrac{12x^3 + 13x^2 - 1}{4x - 1}$

28. $\dfrac{10x^3 + x^2 + 12x + 9}{5x + 3}$

It may be necessary to remind students about the importance of including missing terms.

In Exercises 29–34, use long division to determine the quotient and remainder.

29. $\dfrac{x^2 + 3}{x^2 - 1}$

30. $\dfrac{x^2 - 4}{x^2 + 1}$

31. $\dfrac{x^2 + 3x - 4}{x^2 + 2x}$

32. $\dfrac{3x^3 + 2x - 1}{x^2 + 2}$

33. $(a^4 + 7a - a^2 - 3) \div (a^2 - 2a + 3)$

34. $(2c^4 + 3 - 8c) \div (c^2 - 3c + 1)$

In Exercises 35–38, use long division to determine the quotient and remainder.

35. $\dfrac{x^2 + 4xy + 3y^2}{x - 2y}$

36. $\dfrac{4x^2 - 2xy - y^2}{2x + y}$

37. $\dfrac{4x^3 - 3xy^2 - 2y^3}{x + y}$

38. $\dfrac{x^3 - x^2y - y^3}{x - 2y}$

39. To use synthetic division to divide a polynomial by $x + 3$, show how to write the divisor in order to determine the divider, and state what the divider is.

40. Suppose that you want to use synthetic division to divide $\dfrac{x - x^5}{x - 2}$. How should you write the row of dividend coefficients?

In Exercises 41–50, use synthetic division to determine the quotient and remainder.

41. $\dfrac{3x^2 - 10x - 8}{x - 4}$

42. $\dfrac{x^2 - 2x - 15}{x + 3}$

43. $(6x^2 + x + 1) \div \left(x - \dfrac{1}{2}\right)$

44. $(2x^2 - 3x - 10) \div \left(x + \dfrac{5}{2}\right)$

45. $(2x^4 - 3x^3 + x^2 - 3x + 4) \div (x - 2)$

46. $(3x^3 - 4x^2 - 2x - 5) \div (x - 3)$

47. $(x^2 + x^4 - 14) \div (x + 2)$

48. $(x^2 - 18x + 2x^3 + 42) \div (x + 4)$

49. $\dfrac{2x^2 - 6 + x^3 - 3x}{x - 2}$

50. $\dfrac{4x + x^3 - 13}{x - 3}$

51. Can you use synthetic division to divide the polynomial $x^4 + x - 2$ by $x^2 - 1$? Why?

52. When we say, "Determine the quotient of $x + 2$ and $x^2 + 5x - 3$," do we divide $x + 2$ by $x^2 + 5x - 3$ or do we divide $x^2 + 5x - 3$ by $x + 2$?

CE In Exercises 53 and 54, by inspecting the given synthetic division problem, determine a, b, c, d, and e.

Exercises 53–56 require students to think about the synthetic-division process.

53. $$\begin{array}{r|rrrrr} 2 & a & b & c & d & e \\ & & 2 & 0 & 0 & 2 \\ \hline & 1 & 0 & 0 & 1 & 0 \end{array}$$

54. $$\begin{array}{r|rrrrr} -3 & a & b & c & d & e \\ & & -6 & 15 & 3 & -12 \\ \hline & 2 & -5 & -1 & 4 & -6 \end{array}$$

CE In Exercises 55 and 56, the given synthetic-division problem represents the division of a third-degree polynomial by a divisor of the form $x - c$. State the dividend and the divisor.

55. $$\begin{array}{r|rrrr} & 1 & 3 & -2 & -4 \\ & & -1 & -2 & 4 \\ \hline & 1 & 2 & -4 & 0 \end{array}$$

56. $$\begin{array}{r|rrrr} & 1 & -4 & 3 & -11 \\ & & 4 & 0 & 12 \\ \hline & 1 & 0 & 3 & 1 \end{array}$$

In Exercises 57–68, divide.

57. $\dfrac{x^2 - 2x - 9}{x - 6}$

58. $\dfrac{3x^2 + 14x + 5}{x + 3}$

59. $(3x^2 + 5x + 2) \div (x - 3)$

60. $(2x^2 - 4x - 1) \div (x + 2)$

61. $\dfrac{x^2 + 9x + 20}{x + 4}$

62. $\dfrac{x^2 - 8x + 16}{x - 4}$

63. $(6a^3 + 5a^2 + 3) \div (3a - 2)$

64. $(4x^4 + 5x^2 - 7x) \div (2x + 1)$

65. $(2x^3 + 5x^2 - 3x + 2) \div (x + 2)$

66. $(3x^4 + x^3 - 5x^2 + 2x - 3) \div (x + 1)$

67. $\dfrac{x + 5 - 2x^3 + x^5}{x + 2}$

68. $\dfrac{2x^4 - 25x + x^5 - 5}{x + 3}$

In Exercises 69–74, divide the first polynomial by the given binomial. Then use the result to factor the polynomial completely.

69. $x^3 - 13x + 12;\quad x - 1$

70. $x^3 - 49x + 120;\quad x + 8$

71. $x^3 - 3x^2 - 10x + 24;\quad x + 3$

72. $x^3 - 6x^2 + 11x - 6;\quad x - 2$

73. $2x^3 - 3x^2 + 1;\quad 2x + 1$

74. $3x^3 - x^2 - 8x - 4;\quad 3x + 2$

In Exercises 75–78, perform the indicated operations.

75. $(2x + 1)(x - 5) \div (x + 2)$

76. $(x^2 + 7x + 12) \div (x + 3) \div (x - 2)$

77. $(3x + 1)^2 \div (x - 1)$

78. $(x^2 + 2)^2 \div (x^2 + 1)$

As a hint for Exercises 79–82, remind students about the relation (Quotient)(Divisor) + Remainder = Dividend

CE In Exercises 79–82, determine the missing quantity.

	Quotient	*Divisor*	*Remainder*	*Dividend*
79.	$x + 3$	$2x - 1$	2	
80.	$x^2 + 1$	$x + 1$	-3	
81.	$x - 3$		-1	$x^2 - x - 7$
82.	$x - 2$		1	$x^3 - 2x^2 + x - 1$

83. Divide $x - 3$ into $x - 12 - 17x^2 + x^5 - 3x^3$.

84. Divide $x^4 + x^2 + 2$ by $x + 1$.

85. Find the quotient of $x^3 + 2x - 3$ and the product of x and $x - 3$.

86. Subtract $4x - 1$ from the quotient of $6x^2 + 11x - 7$ and $2x - 1$.

CE **87.** What polynomial divided by $3x^2$ results in the quotient $3x^3 + 4x^2 + x - 4$?

CE **88.** When $x^2 - 2x - 3$ is divided by $P(x)$, the quotient is $x + 2$ and the remainder is 5. What is $P(x)$?

Real-Life Applications

89. A physicist found that a particle traveled a distance of $x^2 + 5x - 24$ meters in $x - 3$ seconds. What expression represents the average speed of the particle in meters per second?

90. A moving company loads cubic boxes into a moving van. Suppose that $w + 8$ boxes can be placed from the front to the back of the van, and w boxes can be stacked on each other. If the capacity of the truck is $w^3 + 10w^2 + 16w$ boxes, what expression represents the number of boxes that can be placed along one end of the truck?

91. The length in feet of a roll of fencing is represented by $x^3 - 65x - 6$. The fencing is to be stretched across $x + 9$ posts, which are evenly spaced.

(a) Write an expression that represents the distance between the posts. (*Hint:* The number of spans between the posts is 1 less than the number of posts.)

(b) How much fencing will be left over?

92. A production engineer determined that the cost of producing n rolls of fiber optic cable is $n^2 + 52n$ dollars. If 50 additional rolls are produced, the total cost is increased by \$100.

(a) What is the unit cost of the original n rolls?
(b) Does producing 50 additional rolls increase or decrease the unit cost? By how much?

Challenge

In Exercises 93 and 94, determine k so that the quotient has the specified remainder.

93. $(x^2 + 4x + k) \div (x - 2), \quad R = 4$

94. $(3x^2 + 7x + k) \div (3x + 1), \quad R = 0$

In Exercises 95 and 96, determine the value of k so that the remainder is zero.

95. $(x^3 + kx^2 - 2k + 3) \div (x - 1)$

96. $(4x^3 + kx^2 + 7) \div (x + 1)$

In Exercises 97 and 98, determine the quotient.

97. $\dfrac{16x^{4n} - 64x^{2n}}{4x^{2n}}$ **98.** $\dfrac{6x^{6n} + 12x^{5n} - 9x^{4n}}{3x^{2n}}$

In Exercises 99 and 100, divide.

99. $(x^{4n} + 3x^{3n} - x^{2n} + 2x^n + 1) \div (x^n + 3)$

100. $(x^{2n} + 2x^n - 4) \div (x^n + 1)$

101. For $(x^2 + bx + b^2) \div (x - b)$, determine the positive value of b for which the remainder is 12.

102. Suppose that when $ax^2 + 3x + c$ is divided by $x + 1$, the remainder is 0. If a is twice c, what are the values of a and c?

7 • Chapter Project Foreign Currency Exchange

In Section 7.2, we saw that 47,800,000 foreigners visited the United States in 1997 and spent an average of \$1534 per trip. In most cases, these dollars were purchased with the currency of the traveler's home country. The following table lists five countries along with the number of monetary units, in each country's currency, needed to purchase one dollar.

Country	One Dollar Exchange
England	0.65 pound
Italy	1517.00 lire
Germany	1.48 marks
Norway	6.41 crowns
Spain	125.00 pesetas

1. For each country in the table, convert \$1534 to the monetary units of that country.
2. If c represents the number of crowns, d represents the number of dollars, and p represents the number of pounds, write equations of the form $c = k_1d$ and $p = k_2d$, where k_1 and k_2 are determined from the table.
3. Solve the two equations in Exercise 2 for d. Then, using substitution, write an equation of the form $c = k_3p$. What does this formula model?
4. Suppose that you leave for England with \$2000 and you exchange this money for pounds. While in England, you spend 400 pounds. You then travel to Norway, where you exchange your pounds to crowns. How many crowns will you have?
5. Continuing with Exercise 4, if you spend 6000 crowns while in Norway, how many dollars will you receive when you return to the United States and exchange crowns for dollars?
6. Arrange a visit with an economics professor and learn the meanings of a "strong dollar" and a "weak dollar." On the basis of what you learn, prepare an argument for or against traveling to another country when the dollar is "strong."

Chapter 7 Review Exercises

Section 7.1

1. Suppose you wanted to factor $-x^3y^2 - 2x^2y^3$ so that one of the factors is a binomial with a positive leading coefficient. What would you select for the GCF?
2. Explain why $2x + 4y + 3x + 9y$ cannot be factored with the grouping method.

In Exercises 3–10, factor completely.

3. $x^{25} + x^{35}$
4. $6x^3 - 36x^2$
5. $x^6y^3 - x^7y^4 - x^3y^5$
6. $3x(x - a) - 4y(x - a)$
7. $x^3 + 3x^2 - 2x - 6$
8. $ac - af + bf - bc$
9. $14x^7 - 35x^5 + 42x^3$
10. $(3x - 2)^4 + 2(3x - 2)^3$
11. An odd number can be represented by $2n - 1$, where n is an integer. Using this representation, show that the difference between the square of an odd number and the odd number itself can be written $2(2n - 1)(n - 1)$.
12. The total cost of z items is given by $xy^2z - 4z$. What binomial represents the unit cost of the items?

Section 7.2

13. How can you quickly tell that $9x^2 - 30x + 24$ is not a perfect square trinomial?
14. Suppose that you factor an expression and obtain the following: $(x + 1)(x^2 - x + 1)$. What expression did you factor?

In Exercises 15–22, factor completely.

15. $49x^2 - 16y^2$
16. $a^{20} - b^{10}$
17. $16 - a^4$
18. $8a^3 - 1$
19. $27b^6 + 8a^3$
20. $4x^2 + 12xy + 9y^2$
21. $c^2 + 64 + 16c$
22. $c^4 + 6c^2 + 9$
23. The base and height of a parallelogram are represented by completely factored binomials. If the area of the parallelogram is given by $a^2 - 16$ square feet, show that the base and height differ by 8 feet.
24. The total revenue from selling n items is given by $4x^2 + 9y^2$. If the total cost of the n items is $12xy$, write a completely factored expression for the profit.

Section 7.3

In Exercises 25–28, factor completely.

25. $x^2 - 5x - 6$
26. $x^2 + 5x + 4$
27. $2x^2 + 12x - 14$
28. $x^4 + 7x^2 - 60$
29. What is a disadvantage of using the trial-and-check method to factor $6x^2 + 25x + 24$?
30. One way to factor $x(2x - 1) + 3(2x - 1)$ is to simplify the expression to obtain $2x^2 + 5x - 3$. Then, using the trial-and-check method, we can write $2x^2 + 5x - 3 = (2x - 1)(x + 3)$. Is this approach correct? What is a more efficient approach?

In Exercises 31–38, factor completely.

31. $3x^2 + 14x + 15$
32. $8x^2 - 3x - 5$
33. $12x^2 + 8x - 15$
34. $12x^2 - 12x - 9$
35. $2x - 8 + x^2$
36. $125x^2 - 80y^2$
37. $125x^3 - 64$
38. $3x^2 - 3x - 60$
39. While factoring $4x^2 + 16xy + 16y^2$, suppose you overlook the GCF of 4 and write $(2x + 4y)^2$. Can you now factor out 2 and write $2(x + 2y)^2$? Why?
40. Last year, a billboard designer created a sign L feet long and W feet wide. The area of the sign was A square feet. This year, the designer changed the dimensions of the sign, and the new area is given by $A - 5W + 2L - 10$ square feet. If the dimensions of the new sign can be represented by binomials, what changes did the designer make to the sign?

In Exercises 41–48, factor completely.

41. $3x^2 + 4x - 15$
42. $2x^3 + 5x^2 - 18x - 45$
43. $x^9 + 27y^6$
44. $y^4 - x^4$
45. $x^3y^4 - x^4y^3$
46. $50x^3 - 25x^4$
47. $6x^3 + 25x^2 + 21x$
48. $12(2x - 3)^2 + 15(3 - 2x) - 18$

Section 7.4

49. In which of the following equations is the value of a or b definitely known?

$$a + b = 0 \qquad ab = 0$$

50. The left side of $5(x - 2)(x + 4) = 0$ has three factors, but the equation has only two solutions. Why?

In Exercises 51–60, solve the equation algebraically.

51. $x^2 = 8x$
52. $x^2 - 81 = 0$
53. $x^2 + x = 42$
54. $4x^2 - 23x + 15 = 0$
55. $x(x + 2) = 24$
56. $(x + 3)(x - 2) = 6$
57. $x^3 + 3x^2 - 4x = 12$

58. $(x - 1)^2 - 2(x - 1) = 15$

59. $|3x^2 + 7x - 2| = 4$ **60.** $|5x^2 + 2x + 2| = 5$

61. The sum of twice a number and its square is 63. Determine the number.

62. Standing on the roof of an eight-story building, a person throws a ball upward with an initial velocity of 64 feet per second. If the release point is 80 feet above the ground, then the height h of the ball after t seconds is determined by $h(t) = -16t^2 + 64t + 80$. Use this equation to find the time it takes for the ball to reach the ground.

Section 7.5

63. Explain how to check the result of a long division problem.

64. Interpret the bottom row of numbers in the following synthetic division problem.

$$\begin{array}{r|rrrr} -1 & 2 & 1 & 0 & 4 \\ & & -2 & 1 & -1 \\ \hline & 2 & -1 & 1 & 3 \end{array}$$

In Exercises 65 and 66, determine the quotient.

65. $\dfrac{9x^5 - 12x^3}{3x^2}$ **66.** $\dfrac{6a^5b^4 - 4a^3b^3 + 8a^2b}{2a^2b^3}$

In Exercises 67–70, use long division to find the quotient and remainder.

67. $(9x^3 - 18x^2 - 7x + 5) \div (3x - 2)$

68. $(12x^4 - 4x^3 + 2x - 17) \div (2x^2 + 3)$

69. $(x^3 - 3x^2y + 3xy^2 - y^3) \div (x - 3y)$

70. $(x^4 - 1) \div (x - 1)^2$

71. If a polynomial $P(x)$ is divided by $x + 1$, the quotient is $2x^2 - x + 1$ and the remainder is 3. Write a polynomial expression for $P(x)$.

72. For the following synthetic division problem, determine a, b, c, and d.

$$\begin{array}{r|rrrr} 3 & a & b & c & d \\ & & 3 & 3 & 6 \\ \hline & 1 & 1 & 2 & 13 \end{array}$$

In Exercises 73–76, use synthetic division to find the quotient and remainder.

73. $(2x^3 - 3x^2 + 5) \div (x - 1)$

74. $(3x - 4x^3 + x^5 + 2) \div (x + 2)$

75. $(x^5 - 32) \div (x - 2)$

76. $(2x^3 + 3x^2 - 5) \div (x - 0.5)$

Looking Back

The following exercises review concepts and skills that you will need in Chapter 8.

1. Perform the indicated operations.

(a) $\dfrac{-2}{3} + \dfrac{1}{6}$ (b) $\dfrac{5}{8} - \dfrac{3}{10}$

(c) $\dfrac{-3}{14} \cdot \dfrac{4}{5}$ (d) $\dfrac{9}{5} \div 6$

2. Evaluate the numerical expression.

$$\left(\frac{1}{2} + \frac{3}{4}\right) \div \left(\frac{1}{3} - 2\right)$$

3. Evaluate $\dfrac{x - 3}{x + 6}$ for the given value of x.

(a) 3 (b) -6 (c) -5

4. Simplify and express the result with positive exponents. Assume that all expressions are defined.

(a) $3x^{-1}$ (b) $\dfrac{8x^3y^4}{12x^5y}$

In Exercises 5–7, factor completely.

5. $2x^2 + 6x$

6. $x^2 - 81$

7. $6 - x - 2x^2$

8. Explain how to factor $2 - 2x$ so that one factor is

(a) $1 - x$ (b) $x - 1$

In Exercises 9–11, solve the given equation.

9. $x^2 + 2x - 8 = 0$

10. $2 - \dfrac{2}{15}t = \dfrac{3}{5} + \dfrac{2t}{3}$

11. $\dfrac{x}{2} + \dfrac{5}{6}(x - 1) = \dfrac{3x - 1}{3}$

12. Write an expression to model the given information.

(a) The time required to travel 250 miles at r mph

(b) The price per foot of molding that costs $84 dollars for f feet

Chapter 7 Test

In Questions 1–6, factor completely.

1. $18r^3s^5 - 12r^5s^4$

2. $xy + 3y - x - 3$

3. $27 - 12x^2$

4. $4y^2 - 20xy + 25x^2$

5. $1 - 8x^3$

6. $6x^2 + 11x + 4$

In Questions 7–10, solve.

7. $3x^2 - 4x = 0$

8. $5x^2 + 13x = 6$

9. $2x^3 - x^2 - 8x + 4 = 0$

10. $|2x^2 - 3x - 2| = 3$

11. If $P(x)$ is divided by $D(x) = x + 3$, the quotient $Q(x)$ is $2x^2 - x + 3$, and the remainder R is -10. Determine $P(x)$. Then write an equation (with no fractions) that relates $P(x)$, $Q(x)$, $D(x)$, and R.

In Questions 12–15, divide by any method to determine the quotient and remainder.

12. $\dfrac{3x^3 - 2x^2 + 5x - 7}{x + 2}$

13. $\dfrac{15x^6y^4 - 21x^5y^8}{-3xy^2}$

14. $\dfrac{x^3 + 3x^2 - 3x + 1}{x^2 - 1}$

15. $\dfrac{x^5 - 1}{x + 2}$

16. For the following synthetic division calculation, what is the significance of the number k?

$$\begin{array}{r|rrrr} -2 & a & b & c & d \\ & & e & f & g \\ \hline & a & h & j & k \end{array}$$

17. Show that when the sum of the squares of three consecutive even integers is divided by the second integer, the remainder is 8.

18. A rectangular garden is 3 feet longer than it is wide. If the garden were widened by 3 feet and lengthened by 5 feet, its area would be 10 square feet more than twice the area of the original garden. Find the dimensions of the original garden.

19. A city lot is in the shape of a trapezoid with two right angles. The figure shows the relative lengths of three sides of the lot. If the area of the lot is 5500 square feet, what are the lengths of the parallel sides? [The area A of a trapezoid is $A = \frac{1}{2}h(b_1 + b_2)$, where h is the height and b_1 and b_2 are the lengths of the parallel sides.]

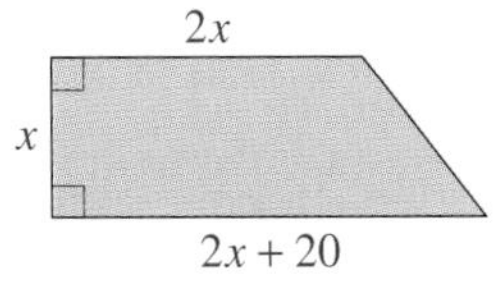

Chapter 8

Rational Expressions

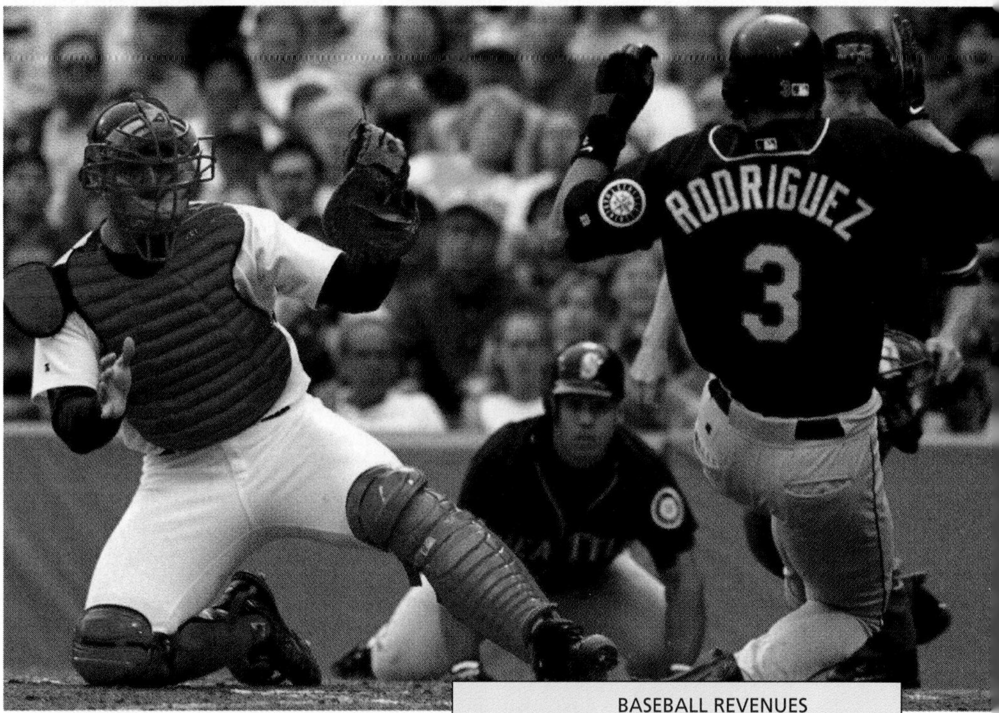

The accompanying bar graph shows the dramatic increase in the average salaries of major league baseball players from 1995 to 1999. Also shown are attendance figures for the same period. Both sets of data can be modeled with a ***rational expression,*** *which is a special kind of fraction.*

The analysis of such data is central to the resolution of salary disputes among owners, players, and their agents. For example, does the trend in attendance justify the trend in salaries? What effect do salaries have on ticket prices? (For more on this topic, see Exercises 71–74 at the end of Section 8.4, and see the Chapter Project.)

In this chapter we learn how to perform the basic operations with rational expressions, and we develop methods for solving equations and application problems that involve rational expressions.

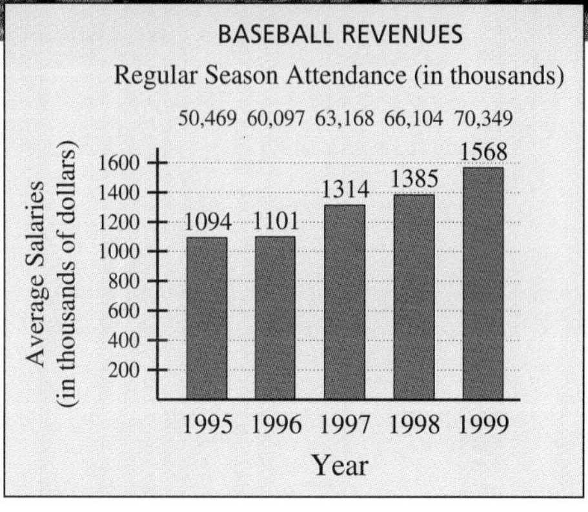

(Source: Major League Baseball Players Association.)

8.1 Introduction to Rational Expressions

Definitions • Restricted Values • Simplifying Rational Expressions • Equivalent Rational Expressions

Definitions

LEARNING TIP

Operations with rational expressions are similar to operations with fractions in arithmetic. You may benefit from a review of fractions before beginning this chapter.

Recall that a **rational number** is a number that can be written as $\frac{p}{q}$, where p and q are integers and $q \neq 0$. In algebra, **rational expressions** play the same role as rational numbers, or fractions, in arithmetic. In fact, we sometimes refer to rational expressions as *algebraic fractions.*

Definition of a Rational Expression

A **rational expression** can be written as $\frac{P(x)}{Q(x)}$, where P and Q are polynomials and $Q(x) \neq 0$.

Throughout this chapter, students will probably benefit from a review of the basics of fractions.

The following are examples of rational expressions.

$$\frac{3+x}{x-5} \qquad \frac{x^2-2x-3}{3x^3-6x^2} \qquad x^3$$

Note that x^3 is a rational expression because it can be written $\frac{x^3}{1}$.

Restricted Values

A **rational function** is a function of the form $y = R(x)$, where $R(x)$ is a rational expression.

We evaluate a rational function in the same manner as any other function.

EXAMPLE 1

Evaluating a Rational Function

Evaluate the rational function $R(x) = \frac{3+x}{x-5}$ as indicated.

(a) $R(7)$ (b) $R(-3)$ (c) $R(5)$

Solution

(a) $R(7) = \frac{3+7}{7-5} = \frac{10}{2} = 5$

(b) $R(-3) = \frac{3+(-3)}{-3-5} = \frac{0}{-8} = 0$

(c) $R(5) = \frac{3+5}{5-5} = \frac{8}{0}$

Because division by 0 is undefined, 5 is not a permissible replacement for x. Thus $R(5)$ is undefined.

You may want to remind students that the numerator can be 0.

Recall that the domain of a function is the set of all permissible replacements for the variable. In general, a rational function is defined for all values of the variable except those for which the denominator is 0. Thus the domain of the rational function in Example 1 is $\{x \mid x \neq 5\}$.

Restricted values will be important later when we solve equations with rational expressions.

A value that is not a permissible replacement for the variable is sometimes called a *restricted value*. When we determine restricted values for rational functions, only the denominator needs to be checked. Any replacement is permissible in the numerator.

Determining Restricted Values

Use the following steps to determine the restricted value(s), if any, of a rational function.

1. Set the denominator equal to 0.
2. Solve the resulting equation.
3. Any solution of the equation is a restricted value.

EXAMPLE 2

Determining the Domain of a Rational Function

Determine the domain of each rational function.

(a) $f(x) = \dfrac{2x}{x + 3}$ (b) $g(x) = \dfrac{x^2 - 5}{x^2 + 4x}$ (c) $h(x) = \dfrac{x}{x^2 + 1}$

Solution

(a) $x + 3 = 0$ Set the denominator equal to 0.

$x = -3$ Solve for x and exclude the result from the domain.

The domain of function f is $\{x \mid x \neq -3\}$.

Students are familiar with solving equations to determine what values to *include* in a solution set. However, here we are solving equations to determine what values to *exclude* from a domain.

(b) $x^2 + 4x = 0$ Set the denominator equal to 0.

$x(x + 4) = 0$ Solve with the factoring method.

$x = 0$ or $x + 4 = 0$ Zero Factor Property

$x = 0$ or $x = -4$ Solve each equation for x and exclude the results from the domain.

The domain of g is $\{x \mid x \neq 0 \text{ and } x \neq -4\}$.

(c) Because $x^2 + 1$ is never 0, the domain of h is **R**.

Simplifying Rational Expressions

The process of simplifying a rational expression is similar to that of reducing a fraction to lowest terms.

Property of Equivalent Fractions

If a, b, and c are real numbers, where $b \neq 0$ and $c \neq 0$, then

$$\frac{ac}{bc} = \frac{a}{b}$$

If we let a, b, and c represent algebraic expressions (with the same restrictions), then the Property of Equivalent Fractions is also valid for rational expressions.

We use the phrase *divide out the common factor* rather than the commonly used word *cancel* because the phrase is more descriptive of the operation being performed.

In order for us to apply the Property of Equivalent Fractions, the numerator and denominator must be products; that is, they must be in factored form. Then, if the numerator and the denominator have a common factor, we can divide the numerator and denominator by that factor. We will refer to this process as "dividing out the common factor."

$$\frac{10}{16} = \frac{2 \cdot 5}{2 \cdot 8} \quad \text{Factor.} \quad \frac{9x^2y}{15xy^3} = \frac{3xy \cdot 3x}{3xy \cdot 5y^2}$$

$$= \frac{5}{8} \quad \text{Divide out common factors.} \quad = \frac{3x}{5y^2}$$

You may also want to show how properties of exponents can be used.

When the numerator or the denominator contains more than one term, we factor in order to divide out any common factors.

$$\frac{2x - 6}{5x - 15} = \frac{2(x - 3)}{5(x - 3)} = \frac{2}{5}$$

Note: Because 3 is a restricted value, $\frac{2x - 6}{5x - 15} = \frac{2}{5}$ only if $x \neq 3$. In our discussion of operations with rational expressions, we will usually not specifically state restricted values. It is understood that the equalities hold only for those values for which the expressions are defined.

A rational expression is considered simplified if no factor other than 1 or -1 is common to the numerator and the denominator. The following summarizes the procedure for simplifying a rational expression.

Simplifying Rational Expressions

1. Factor the numerator and the denominator completely.
2. Divide out any factor that is common to both.

EXAMPLE 3

Simplifying a Rational Expression

(a) $$\frac{5x^4 + 10x^3}{15x^2} = \frac{5x^3(x + 2)}{15x^2} \quad \text{Factor.}$$

$$= \frac{x(x + 2)}{3} \quad \text{Divide out the common factors.}$$

(b) $$\frac{x^2 - 9}{x^2 - 8x + 15} = \frac{(x + 3)(x - 3)}{(x - 5)(x - 3)} \quad \text{Factor.}$$

$$= \frac{x + 3}{x - 5} \quad \text{Divide out the common factor.}$$

Think About It

Find two values of k such that $\frac{x^2 + 8x + 15}{x^2 + 2kx + k^2}$ can be simplified.

If we wish to list restricted values, the best time to do that is when the rational expression is in factored form. If we looked only at the simplified result in part (b) of Example 3, we might mistakenly identify 5 as the only restricted value. From the factored form of the expression, we can see that 5 and 3 are restricted values.

When simplifying rational expressions, we sometimes need to use the fact that $b - a = -1(a - b)$.

EXAMPLE 4

Simplifying a Rational Expression by Factoring Out −1

(a) $\dfrac{y-4}{4-y} = \dfrac{1(y-4)}{-1(y-4)}$ Factor.

$= \dfrac{1}{-1}$ Divide out the common factor.

$= -1$

(b) $\dfrac{t^2-25}{5-t} = \dfrac{(t+5)(t-5)}{-1(t-5)}$ Factor.

$= \dfrac{t+5}{-1}$ Divide out the common factor.

$= -(t+5)$ or $-t-5$

Because the main step in simplifying a rational expression is the factoring step, it is necessary to remember the many techniques we have developed for factoring.

EXAMPLE 5

Simplifying Rational Expressions

(a) $\dfrac{4x^4+6x^3-10x^2}{6x^4-6x^2} = \dfrac{2x^2(2x^2+3x-5)}{6x^2(x^2-1)}$ Factor out the GCF.

$= \dfrac{2x^2(2x+5)(x-1)}{6x^2(x+1)(x-1)}$ Factor the polynomials.

$= \dfrac{2x+5}{3(x+1)}$ Divide out the common factors.

(b) $\dfrac{x^3-1}{x^2-1} = \dfrac{(x-1)(x^2+x+1)}{(x+1)(x-1)}$ Difference of two cubes and difference of two squares

$= \dfrac{x^2+x+1}{x+1}$ Divide out the common factors.

Equivalent Rational Expressions

Remind students that equivalent fractions are fractions that have the same value. Later, you can point out that equivalent rational expressions have the same value for all values of the variable for which the expressions are defined.

We use the rule $\dfrac{ac}{bc} = \dfrac{a}{b}$ to simplify a fraction. In reverse, the rule is $\dfrac{a}{b} = \dfrac{ac}{bc}$. This means that we can multiply the numerator and the denominator of a fraction by any nonzero number c. For example, to write $\dfrac{2}{5}$ with a denominator of 15, we multiply the numerator and the denominator by 3.

$$\frac{2}{5} = \frac{2 \cdot 3}{5 \cdot 3} = \frac{6}{15}$$

Although renaming a fraction is basic arithmetic, it is useful to illustrate this method. Most students use the "goes into" method of renaming a fraction.

This process is sometimes called *renaming the fraction*. Note that the appearance of the fraction is changed but its value has not; that is, the fractions are **equivalent.**

We use a similar method to write an equivalent rational expression with a specified denominator. Determine the factors in the specified denominator that are not in the original denominator. Then multiply the numerator and the denominator of the original fraction by those factors.

EXAMPLE 6 **Writing Equivalent Rational Expressions**

Rewrite the given rational expression with the specified denominator.

(a) $\dfrac{2}{3x} = \dfrac{}{12x^2y^4}$

(b) $\dfrac{a+2}{a^2-2a-3} = \dfrac{}{(a+1)^2(a-3)}$

Solution

(a) Factor the specified denominator $12x^2y^4$ so that the original denominator $3x$ is one of the factors.

$$12x^2y^4 = 3x \cdot 4xy^4$$

Because $4xy^4$ is a factor of the specified denominator, but not of the original denominator, we multiply the numerator and the denominator of the original fraction by $4xy^4$.

$$\frac{2}{3x} = \frac{2(4xy^4)}{3x(4xy^4)} = \frac{8xy^4}{12x^2y^4}$$

(b) Original denominator: $a^2 - 2a - 3 = (a+1)(a-3)$
Specified denominator: $(a+1)(a+1)(a-3)$

There is one factor of $(a+1)$ in the specified denominator that is not in the original denominator. Therefore, we multiply the numerator and the denominator of the original fraction by $(a+1)$.

$$\frac{a+2}{a^2-2a-3} = \frac{a+2}{(a-3)(a+1)} = \frac{(a+1)(a+2)}{(a+1)(a+1)(a-3)}$$

$$= \frac{a^2+3a+2}{(a+1)^2(a-3)}$$

Quick Reference 8.1

Definitions

- A **rational expression** can be written as $\dfrac{P(x)}{Q(x)}$, where P and Q are polynomials and $Q(x) \neq 0$.

Restricted Values

- A **rational function** is a function of the form $y = R(x)$, where $R(x)$ is a rational expression.
- A restricted value is a number that is not a permissible replacement for a variable. Use the following steps to determine the restricted value(s), if any, of a rational function.
  1. Set the denominator equal to 0.
  2. Solve the resulting equation.
  3. Any solution of the equation is a restricted value.

Simplifying Rational Expressions

- The Property of Equivalent Fractions is used to simplify a rational expression. If a, b, and c are real numbers, where $b \neq 0$ and $c \neq 0$, then $\frac{ac}{bc} = \frac{a}{b}$.
- To simplify a rational expression,
  1. factor the numerator and the denominator completely, and
  2. divide out any factor that is common to both.

Equivalent Rational Expressions

- Multiplying the numerator and the denominator of a rational expression by a nonzero number produces an **equivalent** expression.
- To write a rational expression with a specified denominator,
  1. determine the factors in the specified denominator that are not in the original denominator, and
  2. multiply both the numerator and the denominator of the original fraction by those factors.

Speaking the Language 8.1

1. If $\frac{P(x)}{Q(x)}$ is a rational expression, then $P(x)$ and $Q(x)$ are ________.

2. In the expression $\frac{x+3}{x-3}$, we call 3 a(n) ________.

3. According to the Property of ________, the expressions $\frac{5x}{3x}$ and $\frac{5}{3}$ are equivalent provided that $x \neq 0$.

4. Simplifying a rational expression begins with ________ the numerator and the denominator.

Exercises 8.1

Concepts and Skills

1. In general, what numbers are excluded from the domain of a rational function?

2. What is the first step in determining the domain of

$$f(x) = \frac{x-6}{x^2 - 2x - 15}?$$

In Exercises 3–6, evaluate the rational expression, if possible, for each of the given values of the variable.

3. $\frac{x}{x-3}$; $-3, 0, 3$

4. $\frac{x-2}{2x+4}$; $-2, 0, 2$

5. $\frac{t-3}{4-t}$; $-4, 3, 4$

6. $\frac{t-6}{6-2t}$; $-3, 3, 6$

In Exercises 7–18, determine the domain of the given rational function.

7. $y = \frac{2x+3}{x-4}$

8. $y = \frac{x+3}{x+3}$

9. $y = \frac{10}{x-5}$

10. $y = \frac{x^2-4}{x-2}$

11. $y = \frac{x-4}{(x+2)^2}$

12. $y = \frac{3x-1}{x^2}$

13. $y = \frac{2x-6}{x^2-9}$

14. $y = \frac{x+1}{2x^2+6x}$

15. $y = \frac{3}{x^2+2x-8}$

These exercises will help students to be aware of restricted values when we solve equations later.

16. $y = \dfrac{4x - 3}{5x^2 - 9x - 2}$

17. $y = \dfrac{2x - 1}{x^2 + 4}$

18. $y = \dfrac{1 - 2x}{1 + 2x^2}$

CE In Exercises 19–22, fill in the blank so that the given number(s) is/are not in the domain of the function.

19. $y = \dfrac{5}{x \underline{\hspace{1cm}} 2}$; -2

20. $y = \dfrac{x}{x \underline{\hspace{1cm}} 3}$; 3

21. $y = \dfrac{x}{x^2 \underline{\hspace{1cm}}}$; $5, -5$

22. $y = \dfrac{2x}{x^2 \underline{\hspace{1cm}}}$; $3, -3$

Exercises 23 and 24 emphasize that we divide out factors, not terms. Using proper terminology helps reduce confusion.

23. Explain why the first of the following expressions can be reduced but the second cannot.

(i) $\dfrac{2(x + 3)}{(x + 3)}$ (ii) $\dfrac{2x + 3}{x + 3}$

24. If $x \neq 3$, then we know that $\dfrac{3(x - 3)}{x - 3} = 3$ is true. Is $\dfrac{3x - 3}{x - 3} = 3$ also true? Why?

In Exercises 25–32, simplify the rational expression.

25. $\dfrac{6x^3 - 3x^2}{9x^2}$

26. $\dfrac{4a^5 + 8a^7}{12a^4}$

27. $\dfrac{2(3x + 1)}{2 + 6x}$

28. $\dfrac{-3(2x - 1)}{3 - 6x}$

29. $\dfrac{x^2 - 81}{x + 9}$

30. $\dfrac{x + 8}{x^2 - 64}$

31. $\dfrac{a^2 + 5a}{4a - 20}$

32. $\dfrac{3b - 9}{b^2 + 3b}$

In Exercises 33–44, simplify the rational expression.

33. $\dfrac{2 - x}{2 + 9x - 5x^2}$

34. $\dfrac{x + 3}{12 + x - x^2}$

35. $\dfrac{x^2 - 16}{x^2 - x - 12}$

36. $\dfrac{10x^2 - x - 2}{6x^2 + x - 2}$

37. $\dfrac{a^3 - a}{6a - 6}$

38. $\dfrac{5a + 10}{a^3 - 4a}$

39. $\dfrac{x - 3}{2x^2 - 5x - 3}$

40. $\dfrac{2 + x}{6 - x - 2x^2}$

41. $\dfrac{x^2 - 14x + 45}{x^2 - 17x + 72}$

42. $\dfrac{4a^2 - 13a - 12}{a^2 - a - 12}$

43. $\dfrac{2x^2 + 6x - 1}{2x(x + 3) - 1}$

44. $\dfrac{a + 3}{a(a + 6) + 9}$

In Exercises 45–50, simplify the rational expression.

45. $\dfrac{ab - 3a}{b^2 - 9}$

46. $\dfrac{5ab + 25b}{a^2 - 25}$

47. $\dfrac{x + 3y}{x^2 + 2xy - 3y^2}$

48. $\dfrac{a + 4b}{a^2 + 3ab - 4b^2}$

49. $\dfrac{c^2 - 5cd + 6d^2}{2c^2 - 5cd - 3d^2}$

50. $\dfrac{2x^2 + 13xy + 20y^2}{2x^2 + 17xy + 30y^2}$

CE In Exercises 51–54, determine the unknown numerator or denominator.

51. $\dfrac{2}{a - b} = \dfrac{\underline{\hspace{1cm}}}{b - a}$

52. $\dfrac{2x - 3}{3x - 2} = \dfrac{\underline{\hspace{1cm}}}{2 - 3x}$

53. $\dfrac{x}{x - 3} = -\dfrac{x}{\underline{\hspace{1cm}}}$

54. $\dfrac{x + 2}{2 - x} = -\dfrac{x + 2}{\underline{\hspace{1cm}}}$

In Exercises 55–58, determine whether the given expression has a value of 1, -1, or neither.

55. (a) $\dfrac{x + 3}{x - 3}$ (b) $\dfrac{3 + x}{x + 3}$ (c) $\dfrac{x - 3}{3 - x}$

56. (a) $\dfrac{x - y}{-x + y}$ (b) $\dfrac{x + y}{-y + x}$ (c) $-\dfrac{y - x}{x - y}$

57. (a) $\dfrac{-2x + 5}{-(5 - 2x)}$ (b) $\dfrac{2x - 5}{-5 + 2x}$ (c) $\dfrac{2x - 5}{2x + 5}$

58. (a) $\dfrac{x + y}{y + x}$ (b) $\dfrac{x - y}{y - x}$ (c) $\dfrac{x - y}{x + y}$

In Exercises 59–68, simplify the rational expression.

59. $\dfrac{5x - 2}{2 - 5x}$

60. $\dfrac{x - y}{y - x}$

61. $\dfrac{5x - 3}{-15x + 9}$

62. $\dfrac{-3x + 6}{4x - 8}$

63. $\dfrac{a^2 - 49}{7 - a}$

64. $\dfrac{x^2 - 2x - 8}{4 - x}$

65. $\dfrac{z - 5w}{25w^2 - z^2}$

66. $\dfrac{x^2 - 4}{(2 - x)^2}$

67. $\dfrac{-2x^2 - 3x + 20}{2x^2 + x - 15}$

68. $\dfrac{16 - x^2}{2x^2 - 9x + 4}$

In Exercises 69–74, simplify the rational expression.

This group involves factoring by grouping and factoring sums and differences of cubes.

69. $\dfrac{x^3 + 3x^2 - 9x - 27}{x^2 + 6x + 9}$

70. $\dfrac{8 - 27y^3}{2 - 3y + 2x - 3xy}$

71. $\dfrac{x^3 + 8}{x^2 - 4}$ **72.** $\dfrac{1 + x + x^2}{1 - x^3}$

73. $\dfrac{18x^5 - 39x^4 + 18x^3}{8x^4 - 18x^2}$ **74.** $\dfrac{16 - x^4}{x^3 + 2x^2 + 4x + 8}$

In Exercises 75–86, supply the missing numerator.

75. $\dfrac{3}{2xy^2} = \dfrac{\square}{8x^3y^3}$ **76.** $\dfrac{-3}{5ab^2} = \dfrac{\square}{15a^2b^5}$

77. $\dfrac{5}{x - 5} = \dfrac{\square}{5x - 25}$ **78.** $\dfrac{-5}{x - 6} = \dfrac{\square}{7x - 42}$

79. $x + 5 = \dfrac{\square}{x + 5}$ **80.** $3x = \dfrac{\square}{2x^2}$

81. $\dfrac{5}{4x + 12} = \dfrac{\square}{4x^2 - 36}$ **82.** $\dfrac{9}{5x + 5} = \dfrac{\square}{5x^2 - 5}$

83. $\dfrac{-3}{x + 1} = \dfrac{\square}{x^2 + 5x + 4}$

84. $\dfrac{5}{x + 1} = \dfrac{\square}{x^2 - 3x - 4}$

85. $\dfrac{9}{x^2 - 5x - 6} = \dfrac{\square}{(x - 6)(x - 5)(x + 1)}$

86. $\dfrac{8}{x^2 - 7x + 6} = \dfrac{\square}{(x - 6)(x^2 + 3x - 4)}$

CE In Exercises 87–90, supply the denominator.

87. $\dfrac{5t}{6} = \dfrac{-10t^2}{\square}$

88. $\dfrac{4}{5x^2} = \dfrac{12x^3y}{\square}$

89. $\dfrac{2x - 1}{1 - x} = \dfrac{1 - x - 2x^2}{\square}$

90. $\dfrac{x + 2}{2 - 3x} = \dfrac{3x^2 + 4x - 4}{\square}$

Real Life Applications

91. Suppose that you are in charge of cleaning up an oil spill from a tanker and that one of your jobs is to estimate the total cost. The model for estimating the total cost C (in millions of dollars) is

$$C(x) = \frac{100}{100 - x}, \qquad 0 \le x < 100$$

where x is the percentage of the oil that is cleaned up.

(a) What is the cost of cleaning up 50% of the oil spill?

(b) What is the cost of cleaning up 90% of the oil spill?

(c) Describe the cost of cleaning up 95% of the oil spill compared to the cost of cleaning up 90% of the oil spill.

92. In diving board competitions, scoring is on a scale from 1 to 10, with 10 being perfect. A certain diver uses the function $s(x) = \dfrac{10x}{x + 5}$ to predict the score on a particular dive, where x is the number of training dives per day.

(a) Suppose that the diver performs 20 training dives per day. What is the predicted score on the competition dive?

(b) Use a table of values of s to estimate the number of training dives per day needed to achieve a predicted score of 8.5.

(c) What is the predicted increase in score if a diver who currently does 30 dives per day doubles the number of dives per day?

Data Analysis: *Net Worth of Households*

The accompanying table shows the percentages of heads of households who are in various age groups, along with their average net worth. (Source: Federal Reserve Bulletin.)

Age Group	Percent	Net Worth
34 and under	25.9	\$ 60,200
35–44	22.7	157,000
45–54	16.2	304,500
55–64	13.1	371,000
65–74	12.7	369,800
75 and over	9.4	257,600

The function $N(x) = -0.293x^2 + 37.077x - 808.444$ models the net worth $N(x)$, where x is the age of the head of household.

93. Use long (or synthetic) division to simplify the rational expression $\dfrac{N(x)}{x - 28}$.

94. You should have obtained a remainder of 0 in Exercise 93. What does this imply about the expression $x - 28$?

95. Write $N(x)$ in factored form.

96. Solve the equation $N(x) = 0$ and interpret the result.

Challenge

In Exercises 97–100, simplify the given rational expression.

97. $\dfrac{2x^{-2} - 3x^{-1}}{3x^2 - 2x}$

98. $\dfrac{3x^{-2} + 5x^{-4}}{3x + 5x^{-1}}$

99. $\dfrac{x^2 + 3x}{3x^{-4} + x^{-3}}$

100. $\dfrac{2x^5 - x^2}{2x - x^{-2}}$

In Exercises 101–104, determine the domain of the rational function where n is a positive integer.

101. $\dfrac{3}{x^n - 1}$

102. $\dfrac{4}{x^n + 1}$

103. $\dfrac{x}{x^{2n} + 1}$

104. $\dfrac{x + 3}{x^{2n} - 1}$

105. Evaluate the expression without doing any significant arithmetic. (*Hint:* Let $x = 5250$.)

$$\frac{5251^2 - 5248^2}{5250(5254) - 5251^2}$$

8.2 Multiplication and Division

Multiplication • Division

Multiplication

To multiply fractions, recall that we multiply the numerators, multiply the denominators, and then reduce the answer.

$$\frac{10}{21} \cdot \frac{15}{8} = \frac{150}{168} = \frac{25 \cdot 6}{28 \cdot 6} = \frac{25}{28}$$

An easier method is to reduce first and then multiply the numerators and the denominators.

$$\frac{10}{21} \cdot \frac{15}{8} = \frac{5 \cdot 2}{7 \cdot 3} \cdot \frac{5 \cdot 3}{4 \cdot 2} = \frac{25}{28}$$

In arithmetic, reducing before multiplying is a convenience. For rational expressions, it is nearly a necessity.

The definition of multiplication of rational expressions is similar to the definition of multiplication of arithmetic fractions.

Definition of Multiplication of Rational Expressions

If a, b, c, and d represent algebraic expressions, where b and d are not 0, then

$$\frac{a}{b} \cdot \frac{c}{d} = \frac{ac}{bd}$$

We are guided by our experience with arithmetic in stating the routine for multiplying rational expressions.

To Multiply Rational Expressions

1. Factor each numerator and each denominator completely.
2. Divide out factors that are common to both a numerator and a denominator.
3. Multiply the numerators and multiply the denominators.

Students usually know that they can do this, but few know why. You can demonstrate that this is so by reversing steps 2 and 3 in the previous box.

As in arithmetic, we can divide out common factors from the numerator of one fraction and the denominator of another fraction.

$$\frac{xy}{y} \cdot \frac{y-1}{1}\frac{x}{y} = \frac{x}{1} = x$$

EXAMPLE 1

Multiplying Rational Expressions

Determine the product of the rational expressions.

LEARNING TIP

Rewrite the problem with the numerators and denominators in factored form. Attempting to skip steps by writing the factors above the polynomial invites errors.

(a) $\dfrac{8x^2 - 4x}{2x^2 + 5x - 3} \cdot \dfrac{x^2 - 9}{4x^2}$

$= \dfrac{4x(2x-1)}{(2x-1)(x+3)} \cdot \dfrac{(x+3)(x-3)}{4 \cdot x \cdot x}$ Factor numerator and denominator.

$= \dfrac{x-3}{x}$ Divide out common factors.

(b) $(x^2 - 4xy + 4y^2) \cdot \dfrac{6}{2x^2 - 8y^2}$

Encourage students to write 1 for the denominator. This has the visual benefit of lining up the fractions.

$= \dfrac{(x-2y)(x-2y)}{1} \cdot \dfrac{2 \cdot 3}{2(x+2y)(x-2y)}$ Factor.

$= \dfrac{3(x-2y)}{x+2y}$ Divide out common factors.

(c) $\dfrac{x^2 - y^2}{y - x} \cdot \dfrac{1}{x^3 + y^3}$

$= \dfrac{(x+y)(x-y)}{-1(x-y)} \cdot \dfrac{1}{(x+y)(x^2 - xy + y^2)}$ Factor.

$= \dfrac{1}{-(x^2 - xy + y^2)}$ or $-\dfrac{1}{x^2 - xy + y^2}$ Divide out common factors.

Division

Although this is a proper definition of reciprocals, there is little harm in using the word *invert* in describing how to form a reciprocal.

Two rational expressions are **reciprocals** of each other if their product is 1. The following are some examples of rational expressions and their reciprocals.

Rational Expression	*Reciprocal*
$\dfrac{x}{2x+3}$	$\dfrac{2x+3}{x}$
x^2	$\dfrac{1}{x^2}$
$\dfrac{1}{x-1}$	$x - 1$

From arithmetic, we know that to divide by a fraction, we multiply by the reciprocal of the divisor.

$$10 \div \frac{2}{3} = \frac{10}{1} \cdot \frac{3}{2} = \frac{5 \cdot 2}{1} \cdot \frac{3}{2} = \frac{15}{1} = 15$$

Division of rational expressions is similar to division of fractions in arithmetic.

Definition of Division of Rational Expressions

If a, b, c, and d represent algebraic expressions, where b, c, and d are not 0, then

$$\frac{a}{b} \div \frac{c}{d} = \frac{a}{b} \cdot \frac{d}{c}$$

EXAMPLE 2

Division of Rational Expressions

(a) $\frac{3x^2 - 12}{x^2} \div \frac{3x + 6}{x} = \frac{3x^2 - 12}{x^2} \cdot \frac{x}{3x + 6}$ Definition of division

$= \frac{3(x + 2)(x - 2)}{x \cdot x} \cdot \frac{x}{3(x + 2)}$ Factor.

$= \frac{x - 2}{x}$ Divide out common factors.

(b) $\frac{6x^2 + 3x}{3x - 4} \div (6x^2 - 5x - 4)$

$= \frac{6x^2 + 3x}{3x - 4} \cdot \frac{1}{6x^2 - 5x - 4}$ Definition of division

$= \frac{3x(2x + 1)}{3x - 4} \cdot \frac{1}{(3x - 4)(2x + 1)}$ Factor.

$= \frac{3x}{(3x - 4)^2}$ Divide out common factors.

(c) $\frac{x^3 - x^2 + 2x - 2}{2x^2 - 5x + 3} \div \frac{x^4 - 4}{2x^3 + 3x^2 - 4x - 6}$

$= \frac{x^3 - x^2 + 2x - 2}{2x^2 - 5x + 3} \cdot \frac{2x^3 + 3x^2 - 4x - 6}{x^4 - 4}$ Definition of division

$= \frac{x^2(x - 1) + 2(x - 1)}{(2x - 3)(x - 1)} \cdot \frac{x^2(2x + 3) - 2(2x + 3)}{(x^2 + 2)(x^2 - 2)}$ Factor.

$= \frac{(x - 1)(x^2 + 2)}{(2x - 3)(x - 1)} \cdot \frac{(2x + 3)(x^2 - 2)}{(x^2 + 2)(x^2 - 2)}$

$= \frac{2x + 3}{2x - 3}$ Divide out common factors.

Think About It

How does the domain of $f(x) = \frac{x - 5}{x + 5} \div \frac{x - 5}{x + 5}$ compare to the domain of $g(x) = 1$? Explain the differences.

Multiplication and division operations can appear in the same expression.

EXAMPLE 3 **Combined Operations**

$$\frac{y-3}{y^2+y-2} \div (y+2) \cdot \frac{y^2+2y-3}{9-y^2}$$

$$= \frac{y-3}{y^2+y-2} \cdot \frac{1}{y+2} \cdot \frac{y^2+2y-3}{9-y^2}$$ Definition of division

$$= \frac{-1(3-y)}{(y+2)(y-1)} \cdot \frac{1}{y+2} \cdot \frac{(y+3)(y-1)}{(3+y)(3-y)}$$ Factor.

$$= \frac{-1}{(y+2)^2}$$ Divide out common factors.

Quick Reference 8.2

Multiplication

- If a, b, c, and d represent algebraic expressions, where b and d are not 0, then

$$\frac{a}{b} \cdot \frac{c}{d} = \frac{ac}{bd}$$

- Follow these steps to multiply rational expressions.
  1. Factor each numerator and denominator completely.
  2. Divide out any factors that are common to both a numerator and a denominator.
  3. Multiply the numerators and multiply the denominators.

Division

- Two rational expressions are **reciprocals** of each other if their product is 1.
- If a, b, c, and d represent algebraic expressions, where b, c, and d are not 0, then

$$\frac{a}{b} \div \frac{c}{d} = \frac{a}{b} \cdot \frac{d}{c}$$

- Follow these steps to divide rational expressions.
  1. Determine the reciprocal of the divisor.
  2. Change the operation from division to multiplication by the reciprocal.
  3. Follow the steps for multiplication.
- In expressions involving both multiplication and division of rational expressions, follow the Order of Operations.

Speaking the Language 8.2

1. If the product of two expressions is 1, then the expressions are called ________.
2. To divide by a rational expression, we multiply by the reciprocal of the ________.
3. Multiplying rational expressions is usually more efficient if we divide out ________ before we multiply.
4. In the expression $\frac{P}{Q} \div \frac{R}{S}$, ________ cannot be 0.

Exercises 8.2

Concepts and Skills

1. Consider the following multiplication problem.

$$\frac{x+1}{x^2-2x-15}\cdot\frac{x^2-5x}{x^2-1}$$
$$=\frac{(x+1)(x^2-5x)}{(x^2-2x-15)(x^2-1)}$$
$$=\frac{x^3-4x^2-5x}{x^4-2x^3-16x^2+2x+15}$$

The next step is to simplify the result, but the numerator and denominator will be difficult to factor. How could the problem have been done more easily?

2. Suppose $R(x) = x^{-1} \cdot x$. Write $R(x)$ as a product of rational expressions. Explain why the domain of $y = R(x)$ is not the same as the domain of $y = 1$.

In Exercises 3–14, multiply and simplify.

3. $\dfrac{5a^2b}{b^3c^4}\cdot\dfrac{b^2c^4}{25a^3}$

4. $\dfrac{4x^3y^2}{z^3}\cdot\dfrac{y^3z^4}{2x^5}$

5. $(a+4)\cdot\dfrac{a-4}{3a+12}$

6. $\dfrac{2x+1}{x-5}\cdot(15-3x)$

7. $\dfrac{5x-7}{12-3x}\cdot\dfrac{3x-12}{7-5x}$

8. $\dfrac{5p-25}{2p^2}\cdot\dfrac{6p}{p-5}$

9. $\dfrac{4c^2}{5c+30}\cdot\dfrac{c^2-36}{16c^3}$

10. $\dfrac{2v}{v+7}\cdot\dfrac{v^2+11v+28}{8v^2}$

11. $\dfrac{7w^2-14w}{w^2+3w-10}\cdot\dfrac{w+5}{21w}$

12. $\dfrac{z^2-11z+18}{6z^3}\cdot\dfrac{z^2+2z}{z^2-4}$

13. $\dfrac{y^2-1}{y^2+14y+48}\cdot\dfrac{y^2+5y-6}{(y-1)^2}$

14. $\dfrac{z^2+5z-14}{z^2-8z+12}\cdot\dfrac{z^2+z-42}{z^2+9z+14}$

CE In Exercises 15–18, fill in the blank to make the resulting equation true.

15. $\dfrac{x-3}{3x-21}\cdot\dfrac{7-x}{\underline{\qquad}}=\dfrac{1}{3}$

16. $\dfrac{\underline{\qquad}}{x+1}\cdot\dfrac{x+1}{x-2}=x+2$

17. $\dfrac{3x^2+8x-3}{2x+1}\cdot\dfrac{\underline{\qquad}}{3x^2-4x+1}=x+3$

18. $\dfrac{x^2-16}{\underline{\qquad}}\cdot\dfrac{x^2+3x-10}{x^2-6x+8}=\dfrac{x+4}{x+2}$

19. What is the difference between the *opposite* of $x + 3$ and the *reciprocal* of $x + 3$?

20. For $R(x) = \dfrac{5}{x-1} \div \dfrac{x-2}{3}$, why is 2 excluded from the domain of R?

In Exercises 21–32, divide and simplify.

21. $\dfrac{24x^3y^7}{16x^4y^3}\div\dfrac{9y^6x}{x^5}$

22. $\dfrac{25x^5y^4}{15y}\div\dfrac{x^3y^3}{3x}$

23. $\dfrac{2x+6}{5x^2}\div\dfrac{x+3}{15x}$

24. $\dfrac{(x+1)^2}{8x+8}\div\dfrac{7x^2+7x}{8x^2}$

25. $\dfrac{5x-x^2}{x+2}\div\dfrac{x^2-6x+5}{x^2-3x+2}$

26. $\dfrac{x^2-64}{x^2-9}\div\dfrac{6x^2+48x}{2x-6}$

27. $\dfrac{x^2+3x-4}{8x}\div\dfrac{x^2+10x+24}{x+6}$

28. $\dfrac{3x-12}{x^2-9x+20}\div\dfrac{9x+54}{x^2+x-30}$

29. $\dfrac{x^2-3x-10}{x^2-25}\div\dfrac{x^2-4}{x^2-x-30}$

30. $\dfrac{x^2-7x-8}{x^2-2x-48}\div\dfrac{6x^2+6x}{x^2+7x+6}$

31. $\dfrac{x^2-16}{2x^2-8x}\div(3x^2+10x-8)$

32. $\dfrac{x^2+12x+36}{x^2+3x-18}\div(x^2+9x+18)$

CE In Exercises 33–36, fill in the blank to make the resulting equation true.

33. $\dfrac{\underline{\qquad}}{5}\div\dfrac{3x-4}{10}=4x-2$

34. $\dfrac{x^2+7x}{x-x^2}\div\dfrac{\underline{\qquad}}{7x-7}=-1$

35. $\dfrac{2x^2+4x}{x^2+2x+4}\div\dfrac{x^2-4}{\underline{\qquad}}=2x$

36. $\dfrac{x^3-5x^2}{\underline{\qquad}}\div\dfrac{x^2-4x-5}{x^2+2x}=\dfrac{x+2}{x^2+x}$

In Exercises 37–54, perform the indicated operation and simplify.

37. $\dfrac{32x^5}{3y^4}\cdot\dfrac{5y^3}{-8x^2}$

38. $\dfrac{6x^3}{7y^4}\div\dfrac{36x^7}{49y^5}$

39. $-2x \cdot \dfrac{x+3}{4x-x^2}$

40. $\dfrac{x-3x^2}{3-x} \div -3x^2$

41. $\dfrac{x^2-4}{x^2-1} \div \dfrac{12x^2-24x}{4x+4}$

This is a mixed group that requires a variety of factoring techniques.

42. $\dfrac{b^2-16}{b^2-9} \cdot \dfrac{4b+12}{8b^2+32b}$

43. $\dfrac{4x-8}{x^2-5x+6} \div \dfrac{8x+48}{x^2+3x-18}$

44. $\dfrac{b^2-7b+10}{2b-10} \cdot \dfrac{4b^2+8b}{b^2-4}$

45. $\dfrac{a^2-81}{a^2-5a-36} \cdot \dfrac{a^2+10a+24}{a^2-3a-54}$

46. $\dfrac{x^2-5x-24}{x^2-3x-18} \div \dfrac{x^2-64}{3x-18}$

47. $\dfrac{x^2-3x-4}{x^2-10x+24} \div \dfrac{x^2+5x+4}{x^2-5x-6}$

48. $\dfrac{t^2+t-20}{t^2+2t-24} \cdot \dfrac{t^2+11t+30}{t^2+9t+20}$

49. $\dfrac{3x^3+6x^2}{2x^2+x-6} \div (6x^2-15x)$

50. $(w^2-w-12) \cdot \dfrac{9w^2}{3w^4-27w^2}$

51. $\dfrac{xy+3x-2y-6}{x^2-6x+8} \cdot \dfrac{xy-4y+2x-8}{y^2+5y+6}$

52. $\dfrac{ab-ad+bc-cd}{b^2+bd-2d^2} \div \dfrac{a^2-ac-2c^2}{ab+2ad-bc-2cd}$

53. $\dfrac{x^3-27}{x^2-9} \div \dfrac{x^2+3x+9}{x^2+8x+15}$

54. $\dfrac{x^3+y^3}{x^2-y^2} \cdot \dfrac{x^2-3xy+2y^2}{x-2y}$

In Exercises 55–62, perform the indicated operations.

55. $\dfrac{6x-18}{3x-6} \cdot \dfrac{5+5x}{25x+25} \div \dfrac{12-4x}{9x-18}$

56. $\dfrac{x^2-2x}{4y} \cdot \dfrac{28xy^2}{3x-6} \div \dfrac{2x^2+2x}{21x^2y^2}$

57. $\dfrac{2x+6}{x-5} \div \left[\dfrac{x}{4x+12} \cdot \dfrac{2(x+3)^2}{x^2-5x}\right]$

58. $(2x^2+3x-5) \div \left(\dfrac{6x+15}{x-2} \div \dfrac{3}{x-2}\right)$

59. $\dfrac{x^3-1}{x+1} \cdot \dfrac{x^2+2x+1}{x^2+x+1} \div \dfrac{x+1}{3x^2}$

60. $\dfrac{x^3-8}{x^2-4} \div \dfrac{x^2+2x+4}{x^3-3x^2} \cdot \dfrac{x^2+6x+8}{x^2+x-12}$

Remind students that multiplication and division have equal rank in the Order of Operations. Unless there are grouping symbols, we multiply and divide from left to right.

61. $\dfrac{x^2+x-2}{x^2-1} \div (x+2) \cdot \dfrac{x^2-9}{x^2+3x}$

62. $\dfrac{x^2+x-6}{x^2-16} \cdot \dfrac{x^2-4x}{x^2-4} \div (x^2+3x)$

Modeling with Real Data

63. British psychologists studied children's preferences for teddy bears with adult features and with baby features. Older children were more likely than younger children to prefer baby features.

The following table shows the number of boys and girls, by age, who chose baby-featured bears. The combined data are the averages for the two genders. (Source: *Discover.*)

Age	Boys	Girls	Combined
4	6	8	7
6	8	10	9
8	10	14	12

The following functions model the data, where x is the age.

Boys: $B(x) = x + 2$

Girls: $G(x) = \dfrac{3}{2}x + \dfrac{5}{3}$

(a) Write a rational expression that models the combined data. Then simplify the expression.

(b) Sketch the graphs of the three model expressions and describe the graph of the expression in part (a) relative to the other two graphs.

(c) Use the three graphs to estimate the number of children age 5 who preferred baby-featured bears.

64. The following table shows the number (in millions) of U.S. travelers to foreign countries. The table also shows the total amount (in millions of dollars) spent by those travelers. (Source: U.S. Department of Commerce.)

Year	Number of travelers (in millions)	Expenditures (in millions of dollars)
1993	44.4	40,713
1995	50.8	44,916
1997	52.7	51,220

The number T of travelers and the expenditures E can both be modeled as functions of the number of years since 1990.

$T(x) = 2x + 38$

$E(x) = 2626x + 32{,}482$

The average amount spent by each traveler can be modeled by the rational expression $\frac{E(x)}{T(x)}$.

(a) Write a simplified rational expression $A(x)$ for the average amount spent.

(b) Using a table of values for $A(x)$, compare the averages predicted by A to the averages determined from the actual data. For which year is the model average closest to the actual average?

Challenge

In Exercises 65–68, answer the question by determining the unknown quantity.

65. What rational expression multiplied by

$$\frac{x+3}{x-5} \text{ is } \frac{x^2+x-6}{x^2-2x-15}?$$

66. The product of $x^2 + x - 1$ and what rational expression is 1?

67. What rational expression divided by

$$\frac{2x^2+x-15}{2x^2-3x-5} \text{ is } \frac{2x-3}{x+3}?$$

68. What rational expression divided into

$$\frac{x+4}{x^2-1} \text{ is } \frac{1}{x^2-5x+4}?$$

For Exercises 69 and 70, create a table of values for the function $f(x) = \frac{8}{x}$, where $x = 0.0001, 0.001, 0.01, 0.1, 10, 100, 1000,$ and $10{,}000$.

69. For x-values closer and closer to 0, what is true about the corresponding y-values? Describe how you can select x-values to make the y-values as large as you want.

70. For larger and larger x-values, what is true about the corresponding y-values? Describe how you can select x-values to make the y-values as close to 0 as you want.

8.3 Addition and Subtraction

Like Denominators • Least Common Multiples • Unlike Denominators • Combined Operations

Like Denominators

Unlike multiplication and division, addition and subtraction of fractions require that the fractions have a common denominator. Recall that to add fractions with the same denominator, we simply add the numerators, retain the common denominator, and reduce the result to lowest terms.

Some students want to add the denominators.

$$\frac{2}{9}+\frac{4}{9}=\frac{6}{9}=\frac{2\cdot 3}{3\cdot 3}=\frac{2}{3}$$

The procedure for subtracting fractions with like denominators is similar. We use the same principle for addition and subtraction of rational expressions with the same denominator.

Definition of Addition and Subtraction of Rational Expressions

If a, b, and c are algebraic expressions, where $c \neq 0$, then

$$\frac{a}{c} + \frac{b}{c} = \frac{a+b}{c} \quad \text{and} \quad \frac{a}{c} - \frac{b}{c} = \frac{a-b}{c}$$

The following are the steps for adding or subtracting rational expressions with like denominators.

To Add or Subtract Rational Expressions with Like Denominators

1. Add (or subtract) the numerators.
2. Retain the common denominator.
3. Simplify the result.

When we add or subtract rational expressions, each numerator must be treated as though it were enclosed in parentheses. For a subtraction problem, in particular, we will insert parentheses around each numerator. Then, when we subtract the numerators, we will remove the parentheses to simplify.

Encourage the use of parentheses around numerators to avoid common sign errors.

EXAMPLE 1

Combining Rational Expressions with Like Denominators

Perform the indicated operations.

(a) $\dfrac{3r-2}{r^2s} + \dfrac{2}{r^2s} = \dfrac{3r-2+2}{r^2s} = \dfrac{3r}{r^2s} = \dfrac{3 \cdot r}{rs \cdot r} = \dfrac{3}{rs}$

(b) $\dfrac{3x+1}{x+3} - \dfrac{3x+2}{x+3}$

$= \dfrac{(3x+1)-(3x+2)}{x+3}$ Insert parentheses.

$= \dfrac{3x+1-3x-2}{x+3}$ Subtract by removing parentheses.

$= \dfrac{-1}{x+3}$ or $-\dfrac{1}{x+3}$ Combine like terms.

(c) $\dfrac{x}{x^2-1} - \dfrac{4x-3}{x^2-1}$

$= \dfrac{x-(4x-3)}{x^2-1}$ Insert parentheses.

$= \dfrac{x-4x+3}{x^2-1}$ Subtract by removing parentheses.

$= \dfrac{-3x+3}{x^2-1}$ Combine like terms.

$= \dfrac{-3(x-1)}{(x+1)(x-1)}$ Factor the numerator and the denominator.

$= \dfrac{-3}{x+1}$ Divide out common factors.

(d) $\frac{6x}{x^2 - x - 12} - \frac{4x + 7}{x^2 - x - 12} + \frac{x - 5}{x^2 - x - 12}$

$= \frac{6x - (4x + 7) + (x - 5)}{x^2 - x - 12}$ Insert parentheses.

$= \frac{6x - 4x - 7 + x - 5}{x^2 - x - 12}$ Simplify by removing parentheses.

$= \frac{3x - 12}{x^2 - x - 12}$ Combine like terms.

$= \frac{3(x - 4)}{(x - 4)(x + 3)}$ Factor.

$= \frac{3}{x + 3}$ Divide out the common factor.

Least Common Multiples

A **common multiple** of two numbers is a number that is divisible by both numbers. For example, 20, 40, 60, . . . are all common multiples of 4 and 5. Of these infinitely many common multiples, 20 is the smallest and is called the **least common multiple (LCM)** of 4 and 5. Factoring is useful in determining the LCM of two or more numbers or expressions. Suppose that we need to know the LCM of 6 and 8.

Students are sometimes unsure about how to determine the prime factors. Remind them that they can use *any* factors and then factor those factors.

$6 = 2 \cdot 3$

$8 = 2 \cdot 2 \cdot 2 = 2^3$

LCM: $2^3 \cdot 3 = 24$

Note that the LCM consists of each factor that appears in either factorization, and the exponent on each factor of the LCM is the largest one that appears on that factor in the factorizations of the numbers. We determine the LCM of polynomials in a similar way.

Determining the LCM of Polynomials

1. Factor each polynomial completely and write the result in exponential form.
2. Include in the LCM each factor that appears in at least one polynomial.
3. For each factor, use the largest exponent that appears on that factor in any polynomial.

EXAMPLE 2

Determining the LCM of Polynomials

Determine the LCM of the given polynomials.

(a) $3xy^4, 4x^3y, 6x^5y^2$

(b) $y, y - 2$

(c) $3x^2 - 3x, 6x - 6$

(d) $2x^2 - x - 3, 4x^2 - 4x - 3$

(e) $t^2 + 4t + 4, t^2 - t - 6$

Solution

(a) $3xy^4 = 3xy^4$ Factor completely.

$4x^3y = 2^2x^3y$ The bases that appear in at least one expression are 2, 3, *x*, and *y*.

$6x^5y^2 = 2 \cdot 3x^5y^2$

$\text{LCM} = 2^2 \cdot 3x^5y^4 = 12x^5y^4$

For each base, use the largest exponent that appears in any expression.

(b) $\text{LCM} = y(y - 2)$

If the polynomials cannot be factored, then the LCM is simply the product of the polynomials.

(c) $3x^2 - 3x = 3x(x - 1)$

Factor completely.

$6x - 6 = 6(x - 1) = 2 \cdot 3(x - 1)$

$\text{LCM} = 2 \cdot 3x(x - 1) = 6x(x - 1)$

The bases that appear in at least one expression are 2, 3, x, and $x - 1$.

(d) $2x^2 - x - 3 = (2x - 3)(x + 1)$

Factor completely.

$4x^2 - 4x - 3 = (2x + 1)(2x - 3)$

$\text{LCM} = (2x - 3)(x + 1)(2x + 1)$

(e) $t^2 + 4t + 4 = (t + 2)^2$

Factor completely.

$t^2 - t - 6 = (t - 3)(t + 2)$

The bases that appear in at least one expression are $t + 2$ and $t - 3$.

$\text{LCM} = (t + 2)^2(t - 3)$

For the base $t + 2$, the largest exponent that appears in either expression is 2.

In part (d), $2x - 3$ is a factor of both polynomials, but because the largest exponent is 1, there is only one factor of $2x - 3$ in the LCM. In part (e), $t + 2$ is a factor of both polynomials, but the largest exponent is 2. Help students to see the difference between these two problems.

Unlike Denominators

Think About It

Suppose that $a = 2^m \cdot 3^n$ and $b = 2^n \cdot 3^m$, where $m < n$. What is the GCF of a and b? What is the LCM of a and b?

If the rational expressions to be added or subtracted have unlike denominators, then the first task is to write equivalent rational expressions with the same denominator.

The routine begins with determining the **least common denominator (LCD)**—that is, the LCM of the denominators. Then we rename the fractions so that each one has the LCD as its denominator. Finally, we perform the addition or subtraction.

EXAMPLE 3

Combining Rational Expressions with Unlike Denominators

Perform the indicated operations.

(a) $\dfrac{2}{r^2s} - \dfrac{3}{rs^2}$

The LCD is r^2s^2.

$= \dfrac{2 \cdot s}{r^2s \cdot s} - \dfrac{3 \cdot r}{rs^2 \cdot r}$

Write equivalent fractions.

$= \dfrac{2s}{r^2s^2} - \dfrac{3r}{r^2s^2}$

$= \dfrac{2s - 3r}{r^2s^2}$

Combine the numerators.

(b) $\dfrac{x - 1}{6} + \dfrac{2x + 1}{9}$

The LCD is 18.

$= \dfrac{(x - 1) \cdot 3}{6 \cdot 3} + \dfrac{(2x + 1) \cdot 2}{9 \cdot 2}$

Write equivalent fractions.

$= \dfrac{3x - 3}{18} + \dfrac{4x + 2}{18}$

$= \dfrac{3x - 3 + 4x + 2}{18}$

Combine the numerators.

$= \dfrac{7x - 1}{18}$

Combine like terms.

(c) $\dfrac{1}{x-3} - \dfrac{2}{3-x}$

$= \dfrac{1}{x-3} - \dfrac{(-1)\cdot 2}{(-1)\cdot(3-x)}$ $\quad -1(a-b) = b-a$

Observe that the denominators are opposites and that multiplying by −1 will result in like denominators.

$= \dfrac{1}{x-3} - \dfrac{-2}{x-3}$ $\quad$ Now the denominators are the same.

$= \dfrac{1-(-2)}{x-3}$ $\quad$ Combine the numerators.

$= \dfrac{3}{x-3}$

(d) $x + \dfrac{3}{x+2} = \dfrac{x}{1} + \dfrac{3}{x+2}$ $\quad$ The LCD is $x+2$.

$= \dfrac{x(x+2)}{1(x+2)} + \dfrac{3}{x+2}$ $\quad$ Write equivalent fractions.

$= \dfrac{x^2+2x}{x+2} + \dfrac{3}{x+2}$

$= \dfrac{x^2+2x+3}{x+2}$ $\quad$ Combine the numerators.

In Example 3 it was not necessary to factor any of the denominators in order to determine the LCD. In the next example, both of the problems require factoring.

EXAMPLE 4

Factoring to Determine the LCD

(a) $\dfrac{3x-2}{3x^2+3x} - \dfrac{x}{x^2+2x+1}$ $\quad$ Factor the denominators.

$= \dfrac{3x-2}{3x(x+1)} - \dfrac{x}{(x+1)^2}$ $\quad$ The LCD is $3x(x+1)^2$.

$= \dfrac{(3x-2)(x+1)}{3x(x+1)(x+1)} - \dfrac{x\cdot 3x}{3x(x+1)^2}$ $\quad$ Write equivalent fractions.

$= \dfrac{3x^2+x-2}{3x(x+1)^2} - \dfrac{3x^2}{3x(x+1)^2}$ $\quad$ Simplify the numerators.

$= \dfrac{3x^2+x-2-3x^2}{3x(x+1)^2}$ $\quad$ Combine the numerators.

$= \dfrac{x-2}{3x(x+1)^2}$ $\quad$ Combine like terms.

LEARNING TIP

After you rename rational expressions so that they have a common denominator, avoid the temptation to divide out common factors. The simplifying step occurs after the addition or subtraction has been performed.

(b) $\dfrac{x+4}{x^2+x-6} - \dfrac{2x+1}{x^2-6x+8}$ $\quad$ Factor the denominators.

$= \dfrac{x+4}{(x+3)(x-2)} - \dfrac{2x+1}{(x-4)(x-2)}$ $\quad$ The LCD is $(x+3)(x-2)(x-4)$.

$= \dfrac{(x+4)(x-4)}{(x+3)(x-2)(x-4)} - \dfrac{(2x+1)(x+3)}{(x-4)(x-2)(x+3)}$ $\quad$ Write equivalent fractions.

$$= \frac{x^2 - 16}{(x-4)(x-2)(x+3)} - \frac{2x^2 + 7x + 3}{(x-4)(x-2)(x+3)}$$ Simplify the numerators.

$$= \frac{(x^2 - 16) - (2x^2 + 7x + 3)}{(x-4)(x-2)(x+3)}$$ Combine the numerators.

$$= \frac{x^2 - 16 - 2x^2 - 7x - 3}{(x-4)(x-2)(x+3)}$$ Remove parentheses.

$$= \frac{-x^2 - 7x - 19}{(x-4)(x-2)(x+3)}$$ Combine like terms.

Combined Operations

We may find it necessary to perform both addition and subtraction in the same problem.

EXAMPLE 5

Combined Operations with Rational Expressions

$$\frac{-2x}{x+3} + \frac{3}{3-x} - \frac{8x-12}{x^2-9}$$

$$= \frac{-2x}{x+3} + \frac{-1(3)}{-1(3-x)} - \frac{8x-12}{(x+3)(x-3)}$$ $-1(a - b) = b - a$

$$= \frac{-2x}{x+3} + \frac{-3}{x-3} - \frac{8x-12}{(x+3)(x-3)}$$ The LCD is $(x + 3)(x - 3)$.

$$= \frac{-2x(x-3)}{(x+3)(x-3)} + \frac{-3(x+3)}{(x+3)(x-3)} - \frac{8x-12}{(x+3)(x-3)}$$ Write equivalent fractions.

$$= \frac{-2x^2 + 6x}{(x+3)(x-3)} + \frac{-3x-9}{(x+3)(x-3)} - \frac{8x-12}{(x+3)(x-3)}$$ Simplify the numerators.

$$= \frac{(-2x^2 + 6x) + (-3x - 9) - (8x - 12)}{(x+3)(x-3)}$$ Combine the numerators.

$$= \frac{-2x^2 + 6x - 3x - 9 - 8x + 12}{(x+3)(x-3)}$$ Remove parentheses.

$$= \frac{-2x^2 - 5x + 3}{(x+3)(x-3)}$$ Combine like terms.

$$= \frac{-1(2x^2 + 5x - 3)}{(x+3)(x-3)}$$ Factor the numerator.

$$= \frac{-1(2x-1)(x+3)}{(x+3)(x-3)}$$

$$= \frac{-(2x-1)}{x-3} \quad \text{or} \quad -\frac{2x-1}{x-3}$$ Divide out the common factor.

You may want to review the definition of a negative exponent.

Expressions with negative exponents sometimes lead to addition or subtraction of rational expressions.

EXAMPLE 6

Expressions with Negative Exponents

Add $x^{-1} + y^{-1}$. Express the result with positive exponents.

Students tend to write $x^{-1} + y^{-1} = \frac{1}{x+y}$.

Solution

$$x^{-1} + y^{-1} = \frac{1}{x} + \frac{1}{y}$$ Definition of negative exponent

$$= \frac{y}{xy} + \frac{x}{xy}$$ The LCD is xy.

$$= \frac{x+y}{xy}$$ Combine the numerators.

Quick Reference 8.3

Like Denominators

- If a, b, and c are algebraic expressions, where $c \neq 0$, then

$$\frac{a}{c} + \frac{b}{c} = \frac{a+b}{c} \quad \text{and} \quad \frac{a}{c} - \frac{b}{c} = \frac{a-b}{c}$$

- To add or subtract rational expressions with like denominators, follow these steps.
  1. Add or subtract the numerators.
  2. Retain the common denominator.
  3. Simplify the result.
- When subtracting rational expressions, be sure to treat each numerator as a quantity in parentheses.

Least Common Multiples

- To determine the **least common multiple (LCM)** of two or more polynomials, follow these steps.
  1. Factor each polynomial completely and write the result in exponential form.
  2. Include in the LCM each factor that appears in at least one polynomial.
  3. For each factor, use the largest exponent that appears on that factor in any polynomial.

Unlike Denominators

- Follow these steps when adding or subtracting rational expressions with unlike denominators.
  1. Determine the **least common denominator (LCD),** which is the LCM of the denominators.
  2. Rewrite each rational expression as an equivalent expression with the LCD as its denominator.
  3. Combine and simplify the numerators.
  4. Simplify the result.

Combined Operations

- The procedure for adding and subtracting rational expressions can be extended to addition and subtraction involving three or more rational expressions.
- Using the definition of a negative exponent, we can write expressions involving negative exponents as rational expressions.

Speaking the Language 8.3

1. To add rational expressions with the same denominator, we add the ________ and retain the ________.
2. The smallest number that is a multiple of both a and b is called the ________ of a and b.
3. The LCM of three polynomials includes each ________ that appears in at least one polynomial.
4. By applying the definition of a negative exponent, we can write $(x + 1)^{-1}$ as a(n) ________ expression.

Exercises 8.3

Concepts and Skills

1. Suppose that two rational expressions have the same denominator. When we add the expressions, how do we determine the numerator and denominator of the sum?

2. So far, the following work is correct.

$$\frac{x^2 - 15}{x - 3} + \frac{2x}{x - 3} = \frac{x^2 - 15 + 2x}{x - 3}$$

Explain why the problem is not yet finished.

In Exercises 3–14, perform the indicated operations.

3. $\dfrac{5}{a} + \dfrac{3}{a}$

4. $\dfrac{5}{t} - \dfrac{1}{t}$

5. $\dfrac{5x - 4}{x} - \dfrac{8x + 5}{x}$

6. $\dfrac{8a - 2b}{b} + \dfrac{4a - 5b}{b}$

7. $\dfrac{4x - 45}{x - 9} + \dfrac{2x - 9}{x - 9}$

8. $\dfrac{3x - 63}{x - 6} + \dfrac{8x - 3}{x - 6}$

9. $\dfrac{3c + 51d}{c - 8d} - \dfrac{9c + 3d}{c - 8d}$

10. $\dfrac{3x - 24y}{x + 7y} - \dfrac{6x - 3y}{x + 7y}$

11. $\dfrac{2x + 5}{x^2 - 2x - 8} + \dfrac{x + 1}{x^2 - 2x - 8}$

12. $\dfrac{3t + 4}{t^2 + 10t + 25} - \dfrac{2t - 1}{t^2 + 10t + 25}$

13. $\dfrac{x^3 + 6x^2 - 20x}{x(x - 2)} - \dfrac{x^2 - 6x}{x(x - 2)}$

14. $\dfrac{x(x - 1)}{(2x + 3)(x + 1)} - \dfrac{2(2x + 3)}{(2x + 3)(x + 1)}$

In Exercises 15–20, perform the indicated operations.

15. $\dfrac{x}{x^2 - x - 6} - \dfrac{x + 1}{x^2 - x - 6} + \dfrac{x - 2}{x^2 - x - 6}$

16. $\dfrac{2x + 1}{x^2 + x - 2} + \dfrac{3x + 4}{x^2 + x - 2} - \dfrac{4x + 3}{x^2 + x - 2}$

17. $\dfrac{6}{2x + 3} - \left(\dfrac{x}{2x + 3} - \dfrac{3 - x}{2x + 3}\right)$

18. $\dfrac{x}{4 - x} - \left(\dfrac{x + 4}{4 - x} + \dfrac{x}{4 - x}\right)$

19. $\left(\dfrac{3x + 1}{x - 2} - \dfrac{2x - 2}{x - 2}\right) \div (x + 3)$

20. $\left(\dfrac{3x + 1}{5 - x} + \dfrac{1 - x}{5 - x}\right) \cdot \dfrac{x - 5}{x + 1}$

Remind students that addition and subtraction have equal rank in the Order of Operations. Unless there are grouping symbols, we add and subtract from left to right.

CE In Exercises 21–26, fill in the blank to make the resulting equation true.

21. $\dfrac{2x - 5}{x + 4} + \dfrac{\_\_\_\_}{x + 4} = \dfrac{3x + 1}{x + 4}$

22. $\dfrac{x - 4}{2x - 3} - \dfrac{\_\_\_\_}{2x - 3} = \dfrac{2}{2x - 3}$

23. $\dfrac{\_\_\_\_}{t^2 + 3} - \dfrac{3t - t^2}{t^2 + 3} = \dfrac{t^2 - 3}{t^2 + 3}$

24. $\dfrac{x^2 - 5x + 2}{x^2 + x - 1} + \dfrac{\_\_\_\_}{x^2 + x - 1} = \dfrac{3x^2 + x - 4}{x^2 + x - 1}$

25. $\dfrac{\_\_\_\_}{(x + 2)(x - 5)} - \dfrac{x - x^2}{(x + 2)(x - 5)} = \dfrac{3x^2 - 2x}{(x + 2)(x - 5)}$

26. $\dfrac{\square}{a-2b} + \dfrac{a^2+b}{a-2b} = \dfrac{a^2+ab}{a-2b}$

27. Explain how to rewrite the following expression so that the fractions have the same denominator.

$$\frac{15}{x} + \frac{7}{-x}$$

28. Explain how to rewrite the following expression so that the fractions have the same denominator.

$$\frac{x}{2x-1} - \frac{1}{1-2x}$$

CE In Exercises 29–32, fill in the blanks so that the resulting equation is true.

29. $\dfrac{a}{3} - \dfrac{5}{-3} = \dfrac{a}{3} + \dfrac{\square}{3} = \dfrac{\square}{3}$

30. $\dfrac{7}{c} + \dfrac{2}{-c} = \dfrac{7}{c} + \dfrac{\square}{c} = \dfrac{\square}{c}$

31. $\dfrac{3x+9}{x-2} - \dfrac{8x}{2-x} = \dfrac{3x+9}{x-2} + \dfrac{\square}{x-2}$

$= \dfrac{\square}{x-2}$

32. $\dfrac{2x-1}{x-2} + \dfrac{7x}{2-x} = \dfrac{2x-1}{x-2} + \dfrac{\square}{x-2}$

$= \dfrac{\square}{x-2}$

In Exercises 33–38, perform the indicated operations.

33. $\dfrac{2x+9}{7x-8} - \dfrac{7x-6}{8-7x}$

34. $\dfrac{5x+3}{9x-2} + \dfrac{6x-6}{2-9x}$

35. $\dfrac{x^2}{x-4} + \dfrac{16}{4-x}$

36. $\dfrac{x^2}{x-2y} + \dfrac{4y^2}{2y-x}$

37. $\dfrac{a^2}{a-b} - \dfrac{b^2}{b-a}$

38. $\dfrac{x-4}{x^2-25} - \dfrac{x-6}{25-x^2}$

In Exercises 39–48, determine the LCM of the given expressions.

39. $5x^2y,\ 10xy^2,\ 15x^3y$

40. $3x^5y,\ 6xy^4,\ 9x^3y^7$

41. $6x,\ 3x-12$

42. $2t,\ 4t+8$

43. $t+5,\ -3t-15,\ 10t$

44. $x-7,\ 4x-28,\ 6x$

45. $3x^2+x-2,\ 4x^2+5x+1$

46. $3x^2+4x-15,\ x^2+5x+6$

47. $x^2+3x+2,\ x^2-4,\ 3x+6$

48. $y^2+2y-3,\ y^2+y-2,\ 3y-3y^2$

49. Suppose the LCD of two fractions is $x(x-4)$. If one of the fractions is $\dfrac{x}{x-4}$, explain how to rewrite the fraction so that its denominator is the LCD.

50. Suppose the GCF of two denominators is 1. Explain how to determine the LCD for the two fractions.

In Exercises 51–64, perform the indicated operations.

Remind students to write 1 as the denominator for expressions that are not fractions.

51. $\dfrac{5}{a^3b} + \dfrac{7}{ab^2}$

52. $\dfrac{3}{5c^3d^2} - \dfrac{4}{15c^2d^4}$

53. $x - \dfrac{3}{x}$

54. $3 + \dfrac{5x}{x-2}$

55. $\dfrac{3}{x} + \dfrac{8}{x-4}$

56. $\dfrac{3}{t+3} - \dfrac{2}{t}$

57. $\dfrac{1}{x-5} - \dfrac{1}{x+7}$

58. $\dfrac{3}{x-8} + \dfrac{x}{x-3}$

59. $\dfrac{2}{t+1} + t + 1$

60. $1 - 2t - \dfrac{2t}{t+3}$

61. $\dfrac{3}{6(t-5)} - \dfrac{2-t}{9(t-5)}$

62. $\dfrac{2x+1}{3(2x-1)} + \dfrac{2-3x}{2(1-2x)}$

63. $\dfrac{5t+2}{t(t+1)^2} - \dfrac{3}{t(t+1)}$

64. $\dfrac{3}{y(y-2)} + \dfrac{2y-1}{y^2(y-2)}$

In Exercises 65–78, perform the indicated operations.

65. $\dfrac{2}{3t^2} + \dfrac{3}{2t^2+t-1}$

66. $\dfrac{3}{2x} - \dfrac{x-3}{2x^2-x-6}$

67. $\dfrac{2x^2-7}{10x^2-3x-4} - \dfrac{x-3}{5x-4}$

68. $\dfrac{9}{x-3} + \dfrac{x}{x^2-6x+9}$

69. $\dfrac{x+9}{4x-36} - \dfrac{x-9}{x^2-18x+81}$

70. $\dfrac{x-3}{2x+6} - \dfrac{x+3}{x^2+6x+9}$

71. $\dfrac{7}{x^2+15x+56} + \dfrac{1}{x^2-64}$

72. $\dfrac{5}{x^2-4x-12} + \dfrac{9}{x^2-15x+54}$

73. $\dfrac{x}{x+6} - \dfrac{72}{x^2-36}$

74. $\dfrac{x}{x+3} - \dfrac{18}{x^2-9}$

75. $\dfrac{x}{x^2+4x+3} - \dfrac{2}{x^2-2x-3}$

76. $\dfrac{x}{x^2+5x+4} + \dfrac{3}{x^2+11x+10}$

77. $\dfrac{x}{x^2 + 8x + 15} + \dfrac{15}{x^2 + 4x - 5}$

78. $\dfrac{x}{x^2 + 4x + 3} - \dfrac{3}{x^2 - 4x - 5}$

79. Write each of the following as a single rational expression with no negative exponents. Are the two expressions equivalent?

$(x + y)^{-1}$ $\qquad x^{-1} + y^{-1}$

80. Write each of the following as a single rational expression with no negative exponents. Are the two expressions equivalent?

$(xy)^{-1}$ $\qquad x^{-1}y^{-1}$

In Exercises 81–84, simplify and express the result as a single rational expression with no negative exponents.

81. $x^{-1} - y^{-1}$

82. $3x^{-1} + 4y^{-1}$

83. $x(x + 3)^{-1} + 3(x - 2)^{-1}$

84. $4(x - 3)^{-1} + x(x - 3)^{-2}$

Real-Life Applications

85. A quality control manager has measured the average number of errors that employees make while assembling electronic circuit boards. She has modeled her data with $E_0(x) = \dfrac{5}{x}$, where $E_0(x)$ is the average number of errors made after x hours of training ($x \geq 1$).

(a) Using integer values of x between 1 and 10, make a table of values for E_0. Does the table suggest that increasing the hours of training reduces errors? Can the average number of errors be reduced to zero if enough training is provided?

(b) By using certain incentives along with training, the manager believes she can reduce the average number of errors by $\dfrac{5}{x + 1}$. The revised model is

$$E_1(x) = \frac{5}{x} - \frac{5}{x + 1}.$$

Write this new function as a single rational expression. Using the same x-values as in part (a), create a table of values for E_1. Do the tables in parts (a) and (b) suggest that the average number of errors would be reduced even further with the incentive plan?

86. Suppose that one *cycle* of a sound wave can be modeled with the function $y_1 = \dfrac{8x}{x^2 + 1}$, where $-3 \leq x \leq 3$. Roughly speaking, the *amplitude* is the vertical distance between the highest and lowest points of the wave.

(a) The graph of y_1 is shown in the figure. Use the graph to estimate the amplitude of the wave.

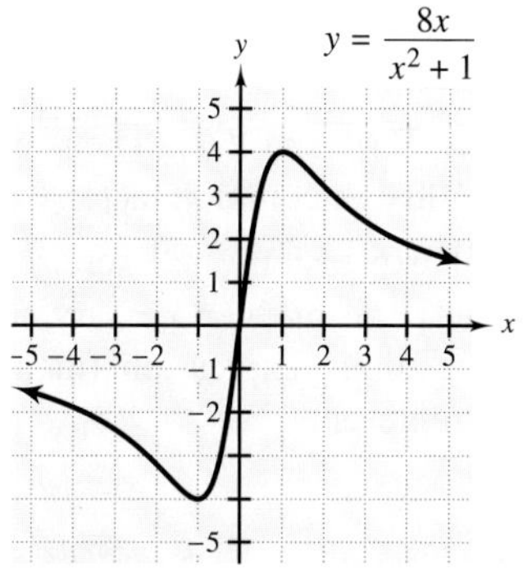

(b) Suppose that a second sound wave modeled by the function $y_2 = \dfrac{2x}{x^2 + 1}$ is mixed with the first so that the resulting wave is modeled by the sum of y_1 and y_2. Write a simplified expression for the resulting sound wave. Estimate the amount by which the amplitude is increased. (*Hint:* Make a table of values for y_2 with $-3 \leq x \leq 3$.)

Data Analysis: *Automobile Operating Costs*

The following table summarizes the costs associated with operating an automobile. Variable costs (in cents per mile) refer to such items as fuel, oil, and maintenance. Fixed costs (in dollars per year) refer to such items as insurance, licenses, finance charges, and depreciation. (Source: American Automobile Manufacturers Association.)

Year	Variable Costs (in cents/mile)	Fixed Costs (in dollars/year)
1993	9.3	3584
1994	9.2	3745
1995	10.0	3891
1996	10.1	4133
1997	10.8	4228

The costs can be modeled by the following functions, where t is the number of years since 1990.

$$\text{Variable costs: } V(t) = \frac{28t + 157}{t + 25}$$

$$\text{Fixed costs: } F(t) = \frac{1000(9t + 73)}{t + 25}$$

87. Write a new function $V_1(t)$ that models the annual variable costs (in dollars) associated with driving 10,000 miles per year.

88. Write the sum of $V_1(t)$ and $F(t)$. Then rewrite the expression with 100 as a common factor.

89. Add the rational expressions in Exercise 88. Use your calculator to perform the necessary arithmetic. What does the resulting expression (including the common factor) represent?

90. In 1997, the Internal Revenue Service allowed a deduction of 31.5 cents per mile for business-related travel by automobile. Referring to the table, did this deduction fully cover the total costs associated with operating the automobile for 10,000 miles?

Challenge

In Exercises 91 and 92, perform the indicated operations.

91. $\left(\dfrac{1}{x+h} - \dfrac{1}{x}\right) \div h$ **92.** $\left[\dfrac{1}{(x+h)^2} - \dfrac{1}{x^2}\right] \cdot \dfrac{1}{h}$

In Exercises 93 and 94, perform the indicated operations.

93. $\dfrac{\dfrac{1}{x} - \dfrac{2x}{x^2+1}}{\dfrac{1}{x}}$ **94.** $\dfrac{\dfrac{x}{x+1} + \dfrac{1}{x}}{\dfrac{1}{x} - 1}$

8.4 Complex Fractions

Order of Operations • Multiplying by the LCD • Expressions with Negative Exponents

Order of Operations

LEARNING TIP

When you write a complex fraction, make sure that you draw the main fraction bar longer than the other fraction bars in order to separate the numerator and the denominator clearly.

A **complex fraction** is a fraction that contains a fraction in the numerator, in the denominator, or in both.

$$\frac{\dfrac{x+5}{5}}{\dfrac{x^2-25}{10}} \qquad \frac{y + \dfrac{3}{y-2}}{y+1} \qquad \frac{\dfrac{3}{y} + 2}{\dfrac{5}{y} - 3} \qquad \frac{\dfrac{3}{x+3}}{\dfrac{1}{x+3} - \dfrac{1}{x}}$$

To simplify a complex fraction, we eliminate all fractions in the numerator and in the denominator.

The simplification methods discussed in this section are valid for all values of the variable except restricted values.

There are two common methods for simplifying a complex fraction. Because a fraction bar indicates division, one method is to write the complex fraction with a division symbol. Then we carry out the division in accordance with the normal Order of Operations. The following example is from arithmetic.

$$\frac{\dfrac{3}{4}}{\dfrac{12}{7}} = \frac{3}{4} \div \frac{12}{7} = \frac{3}{4} \cdot \frac{7}{12} = \frac{3}{4} \cdot \frac{7}{3 \cdot 4} = \frac{7}{16}$$

EXAMPLE 1 **Writing a Complex Fraction with a Division Symbol**

$$\frac{\frac{x+5}{5}}{\frac{x^2-25}{10}} = \frac{x+5}{5} \div \frac{x^2-25}{10}$$ The fraction indicates division.

$$= \frac{x+5}{5} \cdot \frac{10}{x^2-25}$$ Definition of division

$$= \frac{x+5}{5} \cdot \frac{5 \cdot 2}{(x+5)(x-5)}$$ Factor.

$$= \frac{2}{x-5}$$ Divide out common factors.

Recall that a fraction bar is also a grouping symbol. When the numerator or the denominator is a sum or a difference, we combine the terms and then carry out the division.

EXAMPLE 2 **Using the Order of Operations to Simplify a Complex Fraction**

$$\frac{\frac{3}{y}+2}{\frac{5}{y^2}-\frac{1}{y}} = \frac{\frac{3}{y}+\frac{2y}{y}}{\frac{5}{y^2}-\frac{y}{y^2}}$$ The LCD of the numerator is y; the LCD of the denominator is y^2.

$$= \frac{\frac{3+2y}{y}}{\frac{5-y}{y^2}}$$ Combine the fractions.

$$= \frac{3+2y}{y} \div \frac{5-y}{y^2}$$ The fraction indicates division.

$$= \frac{3+2y}{y} \cdot \frac{y^2}{5-y}$$ Multiply by the reciprocal of the denominator.

$$= \frac{y(3+2y)}{5-y}$$ Divide out the common factor y.

The following summarizes the Order of Operations method for simplifying a complex fraction.

Simplifying a Complex Fraction with the Order of Operations

1. Perform any addition or subtraction that is indicated in the numerator or the denominator.
2. Write the resulting complex fraction with a division symbol and perform the division.

Multiplying by the LCD

Emphasize that all terms in the numerator and denominator, not just the fractions, are multiplied by the LCD.

Another way to simplify a complex fraction is to multiply the numerator and denominator by the LCD of *all the fractions*. Because multiplying the numerator and denominator of the complex fraction by the same number is equivalent to multiplying by 1, the resulting expression will be equivalent to the original complex fraction.

To compare the two methods, we use this approach on the same complex fraction that we simplified in Example 2.

EXAMPLE 3 Using the LCD to Simplify a Complex Fraction

$$\frac{\frac{3}{y}+2}{\frac{5}{y^2}-\frac{1}{y}} = \frac{y^2\cdot\left(\frac{3}{y}+2\right)}{y^2\cdot\left(\frac{5}{y^2}-\frac{1}{y}\right)}$$

Multiply the numerator and the denominator by the LCD, y^2.

$$= \frac{y^2\cdot\frac{3}{y}+y^2\cdot 2}{y^2\cdot\frac{5}{y^2}-y^2\cdot\frac{1}{y}}$$

Applying the Distributive Property results in multiplying each term by the LCD.

$$= \frac{3y+2y^2}{5-y}$$

Simplify each product.

LEARNING TIP

Write the LCD beside each term of the complex fraction. Attempting to multiply mentally often leads to errors.

Note: Unlike the Order of Operations method, in which we can use one LCD for the numerator and a different LCD for the denominator, this second method requires just one LCD, which is the LCD of every fraction in the numerator and the denominator.

Generally, multiplying both the numerator and the denominator by the LCD is the easier of the two methods. It is the method that we will use in the remaining examples.

Simplifying a Complex Fraction by Multiplying by the LCD

1. Determine the LCD of every fraction in the complex fraction.
2. Multiply the numerator and the denominator of the complex fraction by the LCD.
3. Simplify the result, if possible.

EXAMPLE 4 Simplifying a Complex Fraction

$$\frac{\frac{1}{p-1}+p}{\frac{1}{p-1}-p} = \frac{(p-1)\cdot\left(\frac{1}{p-1}+p\right)}{(p-1)\cdot\left(\frac{1}{p-1}-p\right)}$$

Multiply the numerator and denominator by the LCD, $p-1$.

$$= \frac{(p-1)\cdot\frac{1}{p-1}+p(p-1)}{(p-1)\cdot\frac{1}{p-1}-p(p-1)}$$

Distributive Property

$$= \frac{1+p^2-p}{1-p^2+p} \quad \text{or} \quad \frac{1-p+p^2}{1+p-p^2}$$

Simplify the numerator and denominator.

EXAMPLE 5

Simplifying a Complex Fraction

$$\frac{\frac{3}{x+3}}{\frac{1}{x+3}-\frac{1}{x}} = \frac{x(x+3)\cdot\frac{3}{x+3}}{x(x+3)\cdot\left(\frac{1}{x+3}-\frac{1}{x}\right)}$$ Multiply the numerator and denominator by the LCD, $x(x+3)$.

$$= \frac{x(x+3)\cdot\frac{3}{x+3}}{x(x+3)\cdot\frac{1}{x+3} - x(x+3)\cdot\frac{1}{x}}$$ Distributive Property

$$= \frac{3x}{x-(x+3)}$$ Simplify.

$$= \frac{3x}{x-x-3}$$ Remove parentheses.

$$= \frac{3x}{-3}$$ Combine like terms.

$$= -x$$ Divide out the common factor 3.

Think About It

Simplify $\dfrac{\frac{bx+a}{b}}{\frac{a}{ax+b}}$.

Then show that the result simplifies to a perfect square trinomial if $a = b$.

Note: The complex fraction in Example 5 has two restricted values, 0 and -3. The very simple expression $-x$ is equivalent to the original complex fraction only if x is not replaced with 0 or -3.

Expressions with Negative Exponents

When an expression with negative exponents is rewritten with positive exponents, the result is sometimes a complex fraction.

EXAMPLE 6

Simplifying Expressions with Negative Exponents

Rewrite the following expression with positive exponents and simplify the result.

$$\frac{x^{-1}+y^{-1}}{x^{-1}y^{-1}}$$

Instead of writing the expression with positive exponents, you may want to discuss the alternative of multiplying the numerator and denominator of the original expression by xy.

Solution

$$\frac{x^{-1}+y^{-1}}{x^{-1}y^{-1}} = \frac{\frac{1}{x}+\frac{1}{y}}{\frac{1}{xy}}$$ Definition of negative exponent

$$= \frac{xy\cdot\left(\frac{1}{x}+\frac{1}{y}\right)}{xy\cdot\frac{1}{xy}}$$ Multiply the numerator and denominator by the LCD, xy.

$$= \frac{xy\cdot\frac{1}{x}+xy\cdot\frac{1}{y}}{xy\cdot\frac{1}{xy}}$$ Distributive Property

$$= \frac{y+x}{1} = x+y$$

Quick Reference 8.4

Order of Operations

- A **complex fraction** is a fraction that contains a fraction in the numerator, in the denominator, or in both.
- Use the following steps to simplify a complex fraction with the Order of Operations method.
  1. Perform any addition or subtraction that is indicated in the numerator or the denominator.
  2. Write the resulting complex fraction with a division symbol and perform the division.

Multiplying by the LCD

- Use the following steps to simplify a complex fraction by multiplying by the LCD.
  1. Determine the LCD of every fraction in the complex fraction.
  2. Multiply the numerator and the denominator of the complex fraction by the LCD.
  3. Simplify the result, if possible.

Expressions with Negative Exponents

- When expressions with negative exponents are rewritten with positive exponents, the result is sometimes a complex fraction.

Speaking the Language 8.4

1. If the expression Y is a fraction whose denominator is also a fraction, we call Y a(n) ______.
2. To use the Order of Operations in simplifying $\dfrac{\frac{a}{b}+\frac{c}{d}}{c}$, we begin by ______.
3. To use the LCD method in simplifying $\dfrac{\frac{a}{b}+\frac{c}{d}}{c}$, we begin by ______.
4. The result in Example 6 is true only if neither x nor y is ______.

Exercises 8.4

Concepts and Skills

1. Do the following expressions mean the same thing? Why? Evaluate each expression to support your answer.

$$\frac{\frac{2}{3}}{4} \quad \text{and} \quad \frac{2}{\frac{3}{4}}$$

2. How can the following expression be written as a complex fraction?

$$\frac{a+2}{a-1} \div \frac{a-3}{a+5}$$

In Exercises 3–16, simplify the complex fraction.

3. $\dfrac{\frac{3}{x}}{\frac{9}{x^3}}$

4. $\dfrac{-\frac{42}{y^5}}{\frac{35}{y^2}}$

5. $\dfrac{\frac{2x}{x+3}}{\frac{6x^2}{x+3}}$

6. $\dfrac{\frac{c+d}{4}}{\frac{c-d}{3}}$

7. $\dfrac{\frac{5}{x+5}}{\frac{10}{x^2-25}}$

8. $\dfrac{-\frac{2}{x^2-1}}{\frac{2}{1-x}}$

9. $\dfrac{\frac{4x^2}{6x+54}}{\frac{32x^3}{x^2-81}}$

10. $\dfrac{\frac{6x}{x-5}}{\frac{12x^2}{x^2+3x-40}}$

11. $\dfrac{\frac{8x}{x+5}}{\frac{16x^2}{x^2+x-20}}$

12. $\dfrac{\frac{7x-35}{4x^3}}{\frac{x^2-25}{4x^2+20x}}$

13. $\dfrac{\frac{5x}{x^2-4}}{\frac{35x^2}{x^2-x-6}}$

14. $\dfrac{\frac{2x}{x^2-4}}{\frac{8x^2}{x^2-7x-18}}$

15. $\dfrac{\frac{2x^2+3x-2}{x^2-x-6}}{\frac{1-x-2x^2}{x^2+3x+2}}$

16. $\dfrac{\frac{3x^2-13x-30}{6x^2+13x+5}}{\frac{x^2-4x-12}{2x^2+x}}$

17. One method for simplifying a complex fraction is to use the Order of Operations. Describe this method.

18. We can write the expression $\left(\frac{3}{4}+\frac{2}{3}\right)\div\frac{5}{8}$ as a complex fraction.

$$\dfrac{\frac{3}{4}+\frac{2}{3}}{\frac{5}{8}}$$

Explain why there is no need for parentheses in the complex fraction.

In Exercises 19–30, simplify the complex fraction.

19. $\dfrac{\frac{2}{3}+\frac{4}{t}}{\frac{1}{t}+\frac{1}{2}}$

20. $\dfrac{\frac{3}{x}-\frac{5}{9}}{\frac{4}{3}-\frac{4}{x}}$

21. $\dfrac{\frac{2}{y^2}+\frac{1}{y}}{\frac{2}{y}+1}$

22. $\dfrac{4-\frac{1}{x^2}}{\frac{2}{x}-\frac{1}{x^2}}$

23. $\dfrac{4-\frac{1}{x^2}}{2+\frac{1}{x}}$

24. $\dfrac{1+\frac{3}{x}}{x-\frac{9}{x}}$

25. $\dfrac{\frac{1}{c}-\frac{1}{cd}}{\frac{1}{cd}-\frac{1}{c}}$

26. $\dfrac{\frac{1}{x}+\frac{1}{y}}{\frac{1}{x}+\frac{1}{xy}}$

27. $\dfrac{\frac{5}{x^4}-\frac{2}{x^3}}{\frac{5}{x^3}-\frac{2}{x^2}}$

28. $\dfrac{4-\frac{5}{x}}{\frac{4}{x}-\frac{5}{x^2}}$

29. $\dfrac{x+y}{\frac{1}{x}+\frac{1}{y}}$

30. $\dfrac{\frac{3}{a^2}-\frac{2}{b}}{a+2b}$

In Exercises 31–42, simplify the complex fraction.

31. $\dfrac{\frac{2}{x+3}-\frac{3}{x-3}}{2+\frac{1}{x^2-9}}$

32. $\dfrac{\frac{4}{4-x^2}-1}{\frac{1}{x+2}+\frac{1}{x-2}}$

33. $\dfrac{\frac{3}{2x+1}-(x+1)}{(x-3)-\frac{2}{2x+1}}$

34. $\dfrac{(2x-1)+\frac{1}{x-2}}{x+\frac{1}{x-2}}$

35. $\dfrac{\frac{1}{1-y}-\frac{1}{1+y}}{\frac{1+y}{1-y}-\frac{1-y}{1+y}}$

36. $\dfrac{\frac{x+4}{x-4}+\frac{x-4}{x+4}}{\frac{1}{x-4}+\frac{1}{x+4}}$

37. $\dfrac{3+\frac{1}{x}-\frac{14}{x^2}}{6+\frac{11}{x}-\frac{7}{x^2}}$

38. $\dfrac{6-\frac{1}{x}-\frac{35}{x^2}}{2-\frac{11}{x}+\frac{15}{x^2}}$

39. $\dfrac{5+\frac{2x}{y}+\frac{2y}{x}}{\frac{2x}{y}-\frac{2y}{x}-3}$

40. $\dfrac{\frac{12x}{y^2}+\frac{13}{xy}-\frac{35}{x^3}}{\frac{10}{x^3}-\frac{3}{xy}-\frac{4x}{y^2}}$

41. $\dfrac{x+\frac{8}{x^2}}{1-\frac{2}{x}+\frac{4}{x^2}}$

42. $\dfrac{\frac{8}{y^3}-27}{\frac{4}{y^3}-\frac{12}{y^2}+\frac{9}{y}}$

43. Explain how to determine the domain of the following rational function, where P, Q, R, and S are polynomials.

$$f(x) = \frac{P/Q}{R/S}$$

44. Explain why $R(3)$ is undefined for the following function.

$$R(x) = \frac{\dfrac{x^2}{x+1}}{\dfrac{x-3}{2-x}}$$

In Exercises 45–54, simplify the complex fraction.

45. $\dfrac{\dfrac{4y^2}{2y+x} + (3x - y)}{\dfrac{5x^2}{2y+x} - (3y + 2x)}$

46. $\dfrac{x + 1 + \dfrac{2}{2x-3}}{x + 2 + \dfrac{3}{2x-3}}$

47. $\dfrac{\dfrac{1}{x+h} - \dfrac{1}{x}}{h}$

48. $\dfrac{\dfrac{1}{x+3} - \dfrac{1}{3}}{x}$

49. $\dfrac{a+b}{1 + \dfrac{2b}{a-b}}$

50. $\dfrac{1 + \dfrac{a+6}{a}}{\dfrac{a}{9} - \dfrac{1}{a}}$

51. $\dfrac{x + 3 + \dfrac{4}{x-2}}{x^2 + 3x + 2}$

52. $\dfrac{1 + \dfrac{c}{a+c}}{1 + \dfrac{3c}{a-c}}$

53. $\dfrac{\dfrac{1}{a^3} + \dfrac{1}{b^3}}{b^2 - ab + a^2}$

54. $\dfrac{\dfrac{1}{27} - \dfrac{1}{a^3}}{a^2 + 3a + 9}$

55. Why is it incorrect to multiply the following expression by the LCD as we do when we simplify a complex fraction?

$$\frac{x}{x+3} + \frac{5}{x-3}$$

You may want to discuss Exercise 56 in class and then suggest the method in part (a) as a way of simplifying the expressions in Exercises 57–64.

56. Simplify the following expression with the methods described in parts (a) and (b). Then answer part (c).

$$\frac{x^{-3} - x^{-2}}{2x^{-3} + x^{-2}}$$

(a) Multiply the numerator and denominator by x^3.

(b) Write the expression with positive exponents and then simplify the result.

(c) Which of the two methods did you find easier?

In Exercises 57–64, rewrite the expression with positive exponents and simplify.

57. $\dfrac{a^{-1}b^{-1}}{a^{-1} + b^{-1}}$

58. $\dfrac{a^{-1} - b^{-1}}{a^{-2} - b^{-2}}$

59. $\dfrac{x^{-1} + x}{x^{-2} + x^2}$

60. $\dfrac{a^{-1} + b}{a + b^{-1}}$

61. $\dfrac{1 - x^{-2}}{3 - x^{-1} - 4x^{-2}}$

62. $\dfrac{1 - 4x^{-2}}{2 + 5x^{-1} + 2x^{-2}}$

63. $\dfrac{xy^{-3} + x^{-2}y}{x - y}$

64. $\dfrac{-2a}{a^{-2} - a^{-1}}$

65. Write and simplify a complex fraction that is the reciprocal of $x + \dfrac{5}{x+1}$.

66. Write and simplify a complex fraction that is the average of $\dfrac{1}{a}$ and $\dfrac{1}{a-1}$.

67. Consider any three consecutive integers.

(a) Write a complex fraction whose numerator is the ratio of the first integer to the second integer and whose denominator is the ratio of the second integer to the third integer.

(b) Simplify the complex fraction and show that the denominator of the result is always 1 more than the numerator.

68. One number is 5 greater than another. If the reciprocal of the sum of the numbers is divided by the reciprocal of the difference of the numbers, the result is $\frac{5}{13}$. What is the larger number?

Modeling with Real Data

69. The accompanying table gives voter information for a recent presidential election. (Source: U.S. Bureau of the Census.)

Age Group	Voting Age Population (in millions)	Percent Registered	Percent Who Voted
18–20	9.7	48.3	38.5
21–24	14.6	55.3	45.7
25–34	41.6	60.6	53.2
35–44	39.7	69.2	63.6
45–64	49.1	75.6	70.0
65 and over	30.8	75.2	70.1

The following functions model the data, where t represents age.

Population (in millions):

$$p(t) = -0.004t^2 + 0.27t - 0.48$$

Percent registered: $r(t) = \dfrac{65t}{0.7t + 10}$

Percent who voted: $v(t) = \dfrac{58t}{0.7t + 10}$

Write a model function for the number of registered voters by age.

70. Referring to Exercise 69, write a model function for the number of people who voted by age.

Data Analysis: *Baseball Attendance and Salaries*

In professional sports, revenue comes from television rights, promotions, and other marketing ventures as well as from ticket sales. The accompanying bar chart shows the average salaries for major league baseball players, along with the total attendance figures, for the 1995–1999 seasons. (Source: Major League Baseball Players Association.)

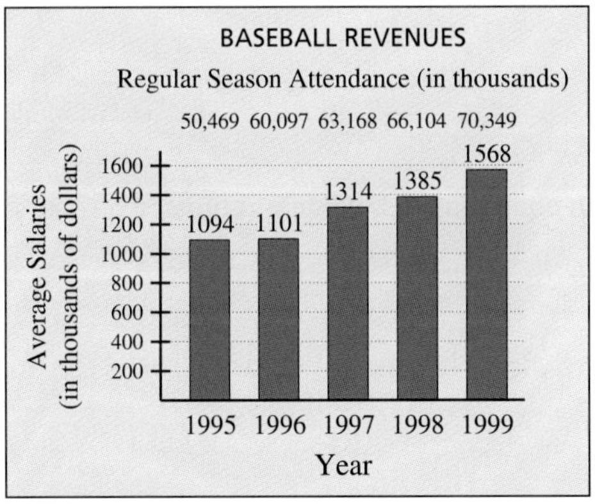

The attendance $A(t)$ and the average salary $S(t)$, both in thousands, are modeled by the following functions, where t is the number of years since 1990.

$$A(t) = \frac{1000(102t - 75)}{t + 3}$$

$$S(t) = \frac{34t^3 - 390t^2 + 2857t + 7}{t + 3}$$

Assume that there were 650 major league players in 1998.

71. Write and simplify the complex fraction $\dfrac{650 \cdot S(t)}{A(t)}$.

72. Assume that ticket sales were the only source of funds for players' salaries and that all revenue from ticket sales was used to pay the players. Use the model expression in Exercise 71 to estimate the cost of a ticket in 1998.

73. Calculate the percentage increases in attendance and in average salaries from 1998 to 1999. How would an owner use this information to argue against higher salaries?

74. Use the assumptions in Exercise 72 to estimate the percentage increase in the price of a ticket from 2000 to 2010.

Challenge

In Exercises 75–78, use the given function to write

$$\frac{f(x + h) - f(x)}{h}$$

Then simplify the expression.

75. $f(x) = \dfrac{1}{2x - 1}$

76. $f(x) = \dfrac{x}{x + 2}$

77. $f(x) = \dfrac{x + 1}{x - 2}$

78. $f(x) = \dfrac{2 - 3x}{x}$

In Exercises 79–82, simplify the complex fraction.

79. $2x - \dfrac{5 + \frac{3}{x}}{7 - \frac{2}{x}}$

80. $3x - \dfrac{4 + \frac{2}{x}}{\frac{3}{x} - 5}$

81. $1 + \dfrac{1}{1 + \dfrac{1}{1 + \frac{1}{2}}}$

82. $\dfrac{1 - \dfrac{1}{1 - \frac{1}{3}}}{1 + \dfrac{1}{1 + \frac{1}{3}}}$

8.5 Equations with Rational Expressions

Solving Equations • Solving Formulas

Solving Equations

In Chapter 3 we solved linear equations containing one or more fractions with constant denominators. The solving technique includes multiplying both sides of the equation by the LCD to eliminate fractions. Example 1 is a review of that method.

EXAMPLE 1

Using the Technique of Clearing Fractions to Solve an Equation

Solve the equation $\frac{t}{8} + \frac{t+1}{2} = 3$.

Solution

$$\frac{t}{8} + \frac{t+1}{2} = 3$$

$$8 \cdot \frac{t}{8} + 8 \cdot \frac{t+1}{2} = 8 \cdot 3 \quad \text{Multiply both sides by the LCD, 8.}$$

$$t + 4(t+1) = 24 \quad \text{Simplify each term.}$$

$$t + 4t + 4 = 24 \quad \text{Distributive Property}$$

$$5t + 4 = 24 \quad \text{Combine like terms.}$$

$$5t = 20 \quad \text{Solve for } t.$$

$$t = 4$$

We will proceed in a similar way with equations that contain rational expressions with variables in the denominators.

EXAMPLE 2

In preparation for equations with extraneous solutions, you may want to encourage students to list restricted values at the beginning of the problem.

Solving an Equation That Contains a Rational Expression

Solve the equation $\frac{15}{x} + 5 = 2$.

Solution

$$\frac{15}{x} + 5 = 2 \quad \text{Note that 0 is a restricted value.}$$

$$x \cdot \frac{15}{x} + 5 \cdot x = 2 \cdot x \quad \text{Multiply both sides by the LCD, } x.$$

$$15 + 5x = 2x \quad \text{Simplify each term.}$$

$$15 = -3x \quad \text{Solve for } x.$$

$$x = -5$$

Using substitution we can verify that the solution of the equation is -5.

EXAMPLE 3

Solving an Equation That Contains a Rational Expression

Solve the equation $\frac{2}{x} = x + 1$.

Solution

$$\frac{2}{x} = x + 1 \qquad \text{Note that 0 is a restricted value.}$$

$$x \cdot \frac{2}{x} = x \cdot (x + 1) \qquad \text{Multiply both sides by the LCD, } x.$$

$$2 = x^2 + x \qquad \text{The result is a quadratic equation.}$$

$$0 = x^2 + x - 2 \qquad \text{Write the equation in standard form.}$$

$$0 = (x + 2)(x - 1) \qquad \text{Factor the trinomial.}$$

$$x + 2 = 0 \quad \text{or} \quad x - 1 = 0 \qquad \text{Zero Factor Property}$$

$$x = -2 \quad \text{or} \quad x = 1 \qquad \text{Solve each case for } x.$$

The solutions are -2 and 1, which we can verify by substitution.

After clearing fractions, we may have a first-degree equation (Example 2), a second-degree equation (Example 3), or a higher-degree equation. Point out that we must be prepared to solve whatever the resulting equation is.

When a rational expression has a variable in its denominator, we must be careful not to replace that variable with numbers that would cause the denominator to be 0.

EXAMPLE 4

Apparent Solutions That Are Restricted Values

Solve $\frac{x^2 + 3x - 10}{x - 2} - x = x - 1$.

Solution

$$\frac{x^2 + 3x - 10}{x - 2} - x = x - 1 \qquad \text{Note that 2 is a restricted value.}$$

$$\frac{x^2 + 3x - 10}{x - 2} \cdot (x - 2) - x(x - 2) = (x - 1)(x - 2) \qquad \text{Multiply by the LCD, } x - 2.$$

$$x^2 + 3x - 10 - x^2 + 2x = x^2 - 3x + 2 \qquad \text{Distributive Property}$$

$$x^2 - 8x + 12 = 0 \qquad \text{Quadratic equation in standard form}$$

$$(x - 6)(x - 2) = 0 \qquad \text{Factor.}$$

$$x - 6 = 0 \quad \text{or} \quad x - 2 = 0 \qquad \text{Set each factor equal to 0.}$$

$$x = 6 \quad \text{or} \quad x = 2 \qquad \text{Solve for } x.$$

There appear to be two solutions. However, when we look at the denominator of the original equation, we see that 2 is a restricted value and cannot be a solution. The only solution is 6.

Extraneous solution is correct but unfortunate terminology. Emphasize that an extraneous solution is *not* a solution.

As we saw in Example 4, an apparent solution of an equation may be a restricted value, which must be disqualified. Such values are called **extraneous solutions.**

To clear fractions, it may be necessary to factor the denominators in order to determine the LCD.

EXAMPLE 5 Factoring Denominators to Determine the LCD

Solve the equations.

LEARNING TIP

Listing restricted values before solving the equation may help identify an extraneous solution.

(a) $\dfrac{4}{x+1} - \dfrac{5}{x} = \dfrac{20}{x^2+x}$ Begin by factoring the denominators.

$\dfrac{4}{x+1} - \dfrac{5}{x} = \dfrac{20}{x(x+1)}, \quad x \neq 0, -1$ Note that 0 and −1 are restricted values.

Now we see that the LCD is $x(x + 1)$. Multiply both sides by the LCD.

$$x(x+1) \cdot \frac{4}{x+1} - x(x+1) \cdot \frac{5}{x} = x(x+1) \cdot \frac{20}{x(x+1)}$$

$$4x - 5(x+1) = 20 \quad \text{Simplify each term.}$$

Students often skip this step and incorrectly write $4x - 5x + 5 = 20$.

$$4x - 5x - 5 = 20 \quad \text{Distributive Property}$$

$$-x - 5 = 20 \quad \text{Combine like terms.}$$

$$-x = 25 \quad \text{Solve for } x.$$

$$x = -25$$

Because it is not a restricted value, −25 is the solution.

(b) $\dfrac{2}{x+3} - \dfrac{1}{x} = \dfrac{-6}{x^2+3x}$ Begin by factoring the denominators.

$\dfrac{2}{x+3} - \dfrac{1}{x} = \dfrac{-6}{x(x+3)}, \quad x \neq 0, -3$ Note that 0 and −3 are restricted values.

The LCD is $x(x + 3)$. Multiply both sides of the equation by the LCD to clear the fractions.

$$x(x+3) \cdot \frac{2}{x+3} - x(x+3) \cdot \frac{1}{x} = x(x+3) \cdot \frac{-6}{x(x+3)}$$

$$2x - (x+3) = -6 \quad \text{Simplify each term.}$$

$$2x - x - 3 = -6 \quad \text{Distributive Property}$$

$$x - 3 = -6 \quad \text{Combine like terms.}$$

$$x = -3 \quad \text{Solve for } x.$$

Our solution process leads us to −3 as the apparent solution, but −3 is a restricted value. Therefore, −3 cannot be a solution, and the equation has no solution.

Reinforce this point by substituting −3 into the original equation to show that two of the denominators are 0.

EXAMPLE 6 Solving an Equation with Rational Expressions

Solve the equation $\dfrac{1}{x} - \dfrac{8}{2x - x^2} = \dfrac{x+2}{x-2}$.

Solution

$$\frac{1}{x} - \frac{(-1)8}{(-1)(2x - x^2)} = \frac{x+2}{x-2} \quad -1(a-b) = b-a$$

$$\frac{1}{x} - \frac{-8}{x^2 - 2x} = \frac{x+2}{x-2} \quad \text{Now the denominators are in descending order.}$$

Think About It

For the equation $\dfrac{kx}{x-1} = \dfrac{x+1}{x-1}$, 1 is an extraneous solution. What is the value of k?

$$\frac{1}{x} - \frac{-8}{x(x-2)} = \frac{x+2}{x-2}, \qquad x \neq 0, 2 \qquad \text{Factor the denominators.}$$

Note that the LCD is $x(x - 2)$ and the restricted values are 0 and 2. Multiply both sides of the equation by the LCD.

$$\frac{1x(x-2)}{x} - \frac{-8x(x-2)}{x(x-2)} = \frac{x(x-2)(x+2)}{x-2}$$

$$(x-2) - (-8) = x(x+2) \qquad \text{Simplify each term.}$$

$$x - 2 + 8 = x^2 + 2x \qquad \text{Distributive Property}$$

$$x + 6 = x^2 + 2x \qquad \text{The result is a quadratic equation.}$$

$$0 = x^2 + x - 6 \qquad \text{Write the equation in standard form.}$$

$$0 = (x+3)(x-2) \qquad \text{Factor the trinomial.}$$

$$x + 3 = 0 \quad \text{or} \quad x - 2 = 0 \qquad \text{Zero Factor Property}$$

$$x = -3 \quad \text{or} \quad x = 2 \qquad \text{Solve each case for } x.$$

Point out that none, some, or all of the apparent solutions could be extraneous.

The apparent solutions are -3 and 2. However, because 2 is a restricted value, it is an extraneous solution. Therefore, the only solution is -3.

Solving Formulas

When solving a formula for a specified variable, we regard the specified variable as the only variable and treat all other variables as though they were constants.

We treat the other variables as constants for the purpose of solving the formula. However, these variables may have restricted values.

Sometimes, formulas contain rational expressions. The method for solving such formulas for specified variables is the same as the method for solving equations containing rational expressions.

EXAMPLE 7

Solving a Formula with a Rational Expression

In electronics, a *resistor* is any device that offers resistance to the flow of electric current. When a light bulb is on, it acts as a resistor. Suppose two light bulbs are placed in a parallel circuit. (See Fig. 8.1.) Let r_1 and r_2 represent their resistances.

Figure 8.1

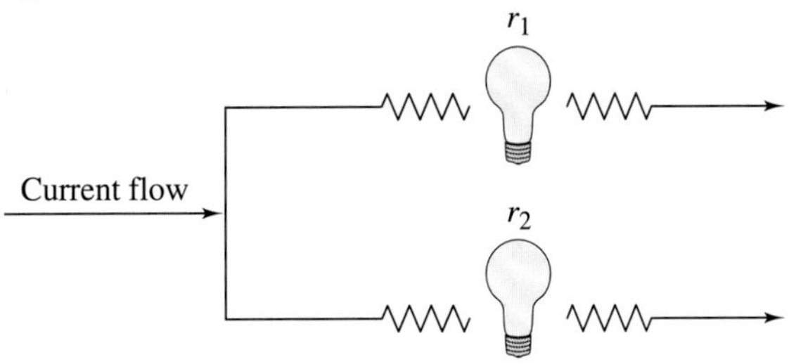

The formula for the total resistance R in this parallel circuit is

$$\frac{1}{R} = \frac{1}{r_1} + \frac{1}{r_2}$$

Solve this formula for r_1.

Solution

$\frac{1}{R} = \frac{1}{r_1} + \frac{1}{r_2}$	The LCD is Rr_1r_2.
$Rr_1r_2 \cdot \frac{1}{R} = Rr_1r_2 \cdot \frac{1}{r_1} + Rr_1r_2 \cdot \frac{1}{r_2}$	Multiply both sides by the LCD.
$r_1r_2 = Rr_2 + Rr_1$	Simplify each term.
$r_1r_2 - Rr_1 = Rr_2$	Place the terms containing r_1 on the same side.
$(r_2 - R)r_1 = Rr_2$	Factor out the common factor r_1.
$r_1 = \frac{Rr_2}{r_2 - R}$	Divide both sides by $r_2 - R$.

Quick Reference 8.5

Solving Equations

- If a replacement for a variable in a rational expression causes the denominator of the expression to be zero, the replacement is a restricted value.
- To solve an equation containing rational expressions, follow these steps.
  1. Factor each denominator completely and note the restricted values, if any.
  2. Determine the LCD for the fractions.
  3. Clear the fractions by multiplying each term of both sides of the equation by the LCD.
  4. Simplify each side of the equation.
  5. Use previously discussed techniques to solve the resulting equation.
  6. Verify the solution in the original equation.
- If the value obtained in step 5 is a restricted value, it cannot be a solution, and it is called an **extraneous solution.**

Solving Formulas

- To solve a *formula* containing rational expressions, treat the specified variable as the only variable and all other variables as constants. Solve the formula for the specified variable using the same steps as outlined for solving an *equation* containing rational expressions.

Speaking the Language 8.5

1. To solve an equation containing rational expressions, we factor the _______ and then multiply both sides by the LCD of the denominators in order to _______ the fractions.

2. If an equation has an apparent solution, but that solution is a restricted value, we call the solution a(n) _______ solution.

3. The technique for solving a formula that contains rational expressions is identical to the technique for solving a(n) _______ that contains rational expressions.

4. If $R(x) = 0$ is an equation that contains rational expressions, and if k is a restricted value, then $x - k$ is a factor in at least one _______.

Exercises 8.5

Concepts and Skills

1. Why are -3 and 0 not permissible replacements for x in the following equation? What do we call -3 and 0 in this case?

$$\frac{1}{x+3} - \frac{1}{x} = 5$$

2. If you solve $\frac{x}{3} + \frac{3}{x} = 0$ by clearing fractions, how can you tell immediately that the equation has no real number solution?

In Exercises 3–6, state the restricted value(s) for the given equation.

3. $\frac{x}{x+3} - \frac{2}{x-4} = \frac{1}{x}$

4. $\frac{3}{a^2} + \frac{a}{a-2} = 4$

5. $\frac{y}{y^2-9} + \frac{7}{y} = \frac{3y+2}{y+3}$

6. $\frac{2x+1}{x^2+2x-15} - \frac{x-3}{x^2-25} = \frac{x+5}{2}$

7. If we solve $\frac{x^2-4}{x-2} = 12$ by clearing fractions, we obtain $x^2 - 12x + 20 = 0$, which has two solutions, 2 and 10. If instead we begin by simplifying the left side, we obtain $x + 2 = 12$, which has just one solution, 10. Explain the discrepancy.

8. The equation $\frac{x}{x+1} = \frac{2}{x}$ has no x^2 terms in it. Does this mean that when we clear the fractions, the result will not be a quadratic equation?

In Exercises 9–20, solve the equation.

In this group, all lead to first-degree equations and none have extraneous solutions.

9. $\frac{1}{8} = \frac{1}{2t} + \frac{2}{t}$

10. $\frac{5}{4} + \frac{4}{3} = \frac{7}{x}$

11. $\frac{5}{x-2} = 3 - \frac{1}{x-2}$

12. $4 + \frac{6}{5-x} = \frac{2x}{5-x}$

13. $\frac{7}{12} + \frac{1}{2y-10} = \frac{5}{3y-15}$

14. $\frac{4}{3x+2} = 1 - \frac{2}{9x+6}$

15. $\frac{5}{x-2} = \frac{1}{x+2} + \frac{6}{x^2-4}$

16. $\frac{2}{t-5} + \frac{3}{t+5} = \frac{10}{t^2-25}$

17. $\frac{t}{t+4} - \frac{2}{t-3} = 1$

18. $\frac{1}{x} + \frac{1}{x-1} = \frac{2}{x+2}$

19. $\frac{3}{x+3} = \frac{2}{x-4} - \frac{10}{x^2-x-12}$

20. $\frac{3}{x+1} - \frac{4}{2x-1} = \frac{5}{2x^2+x-1}$

In Exercises 21–28, solve the equation. In this group, all lead to second-degree equations and none have extraneous solutions.

21. $2x - \frac{15}{x} = 7$

22. $12 - \frac{72}{x^2} = \frac{5}{x}$

23. $\frac{x}{4} - \frac{5}{x} = \frac{1}{4}$

24. $\frac{x}{2} - \frac{2}{x} = 0$

25. $\frac{t}{t-2} + 1 = \frac{8}{t-1}$

26. $\frac{8}{t} + \frac{3}{t-1} = 3$

27. $\frac{3}{x+2} + 1 = \frac{6}{4-x^2}$

28. $\frac{x+2}{x-12} = \frac{1}{x} + \frac{14}{x^2-12x}$

29. Neither of the following equations has a solution, but the reason there is no solution is different in each case. Explain why each equation has no solution.

(i) $\frac{x-4}{x-9} = \frac{5}{x-9}$ (ii) $\frac{4}{x+1} = \frac{4}{x-1}$

30. Verify that multiplying both sides of the following equation by the LCD, $x + 1$, results in the equation $1 = 1$.

$$\frac{x}{x+1} - \frac{x-1}{x+1} = \frac{1}{x+1}$$

Because $1 = 1$ is true, should we conclude that all real numbers are solutions of the original equation? Why?

In Exercises 31–54, solve the equation.

This is a mixed group leading to first- or second-degree equations. Some have extraneous solutions.

31. $\frac{1}{x+1} + \frac{2}{3x+3} = \frac{1}{3}$

32. $\frac{4}{x-2} + \frac{3}{x} = \frac{15}{x^2-2x}$

33. $\frac{x+2}{x+3} = \frac{x-1}{x+1}$

34. $\frac{-2}{3x+2} = \frac{3}{1-2x}$

35. $\frac{5}{x+5} = 4 - \frac{x}{x+5}$

36. $\frac{4}{5x} = \frac{1}{3x}$

37. $\frac{10}{(2x-1)^2} = 4 + \frac{3}{2x-1}$

38. $\frac{1}{x+3} + \frac{21}{(x+3)^2} = 2$

39. $\frac{2}{x^2+x-2} + \frac{3x}{x^2+5x+6} = \frac{5x}{x^2+2x-3}$

40. $\frac{4}{2x^2-7x-15} - \frac{2}{2x^2+13x+15} = \frac{-2}{x^2-25}$

41. $2x + 15 = \frac{8}{x}$

42. $x = \frac{4-11x}{3x}$

43. $\frac{x}{x-3} + \frac{2}{x} = \frac{3}{x-3}$

44. $\frac{x+8}{x+2} + \frac{12}{x^2+2x} = \frac{2}{x}$

45. $\frac{8r}{r-4} - \frac{7}{r^2-16} = 8$

46. $\frac{8}{q-3} + \frac{9}{q^2-11q+24} = \frac{2}{q-8}$

47. $\frac{5}{x-7} + \frac{2}{x+5} = \frac{1}{x^2-2x-35}$

48. $\frac{3}{x+3} = \frac{4}{x+1} - \frac{5}{x^2+4x+3}$

49. $\frac{x+4}{6x^2+5x-6} + \frac{x}{2x+3} = \frac{x}{3x-2}$

50. $\frac{2-6x}{x^2-x-6} = \frac{x-1}{x+2} - \frac{x+1}{3-x}$

51. $\frac{2}{x+1} - \frac{1}{x-3} = \frac{-8}{x^2-2x-3}$

52. $\frac{6}{z-5} + \frac{6}{z^2-11z+30} = \frac{2}{z-6}$

53. $\frac{x}{x+2} + \frac{2}{x+3} + \frac{2}{x^2+5x+6} = 0$

54. $\frac{1}{9x^2+3x-2} = \frac{x}{3x-1} - \frac{1}{9x+6}$

55. To solve the equation

$$\frac{x}{x-1} + \frac{x+3}{x+2} = 3$$

and to perform the addition

$$\frac{x}{x-1} + \frac{x+3}{x+2} + 3$$

we begin in the same way: Determine the LCD. However, what we do with the LCD is completely different for the two problems. Without actually doing the problems, explain what to do with the LCD in each case.

56. In the following problems, identify the one for which an LCD is not needed and explain why.

(i) $\frac{2}{x} - \frac{x^2}{7-x}$ (ii) $\frac{2}{x} = \frac{x^2}{7-x}$

(iii) $\frac{2}{x} \div \frac{x^2}{7-x}$ (iv) $\frac{2}{x} + \frac{x^2}{7-x}$

CE In Exercises 57–68, solve the equation or simplify the expression.

57. $\frac{3}{x-8} + \frac{x}{x-3}$

58. $\frac{3}{x+4} + \frac{6}{x} + \frac{12}{x^2+4x}$

Sometimes students overlook the difference between a rational expression and an equation with rational expressions, and they confuse the techniques for dealing with the two.

59. $4 + \frac{19}{2t-3} = \frac{t+2}{3-2t}$

60. $\frac{2x+1}{x-2} = \frac{2x-3}{x-3}$

61. $\frac{y^2-49}{8y+56} \cdot \frac{y^2-7y}{(y-7)^2}$

62. $\frac{t^2-4}{t^2-25} \cdot \frac{3t-15}{6t^2-12t}$

63. $1 - 2t - \frac{2t}{t+3}$

64. $\frac{3}{6(t-5)} - \frac{2-t}{9(t-5)}$

65. $\frac{x+5}{x-1} = \frac{8}{x-3} - \frac{7}{x^2-4x+3}$

66. $9 + \frac{3}{x+1} = \frac{-4}{x^2+x}$

67. $\frac{x^2+x-6}{x^2+9x+18} \div \frac{x^2-4}{4x+24}$

68. $\frac{3x^3+6x^2}{2x^2+x-6} \div (6x^2-15x)$

69. The current I flowing in an electrical circuit is given by $I = E/R$, where E is the voltage and R is the resistance. Solve this formula for R.

70. The electrostatic force F between two charged bodies bodies is given by $F = \frac{q_1q_2}{kr^2}$, where q_1 and q_2 are the electric charges on the two bodies, r is the distance between the two bodies and k is a constant. Solve this formula for q_1.

71. In chemistry, Charles's law is $\frac{P_1V_1}{T_1} = \frac{P_2V_2}{T_2}$, where P is pressure, T is absolute temperature, and V is

volume. The subscript 1 might be used for initial conditions, and the subscript 2 might be used for final conditions. Solve this formula for T_1.

72. When a metal bar is heated, its length increases according to the formula $L_t = L_0(kt + 1)$, where L_0 is the length of the bar at 0°C, L_t is the length of the bar at t°C, and k is the coefficient of linear expansion. Solve this formula for k.

73. An invested amount P can be calculated with the formula $P = \frac{I}{rt}$, where I is interest earned, r is the simple interest rate, and t is time. Solve this formula for r.

74. In the accompanying figure, triangle ABC is a right triangle with altitude $\overline{AP}$. In geometry, it can be shown that $\frac{a}{h} = \frac{h}{b}$. Solve this equation for b.

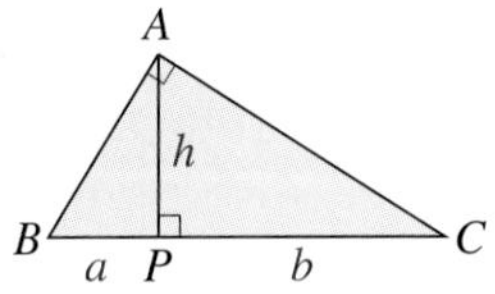

In Exercises 75–80, solve the formula for the specified variable.

75. $F = \frac{m_1 m_2}{\mu r^2}$; m_2

76. $p = \frac{2T}{r}$; r

77. $S = \frac{a(1 - r^n)}{1 - r}$; a

78. $\frac{1}{f} = (k - 1)\left(\frac{1}{p} + \frac{1}{q}\right)$; f

79. $\frac{x}{a} + \frac{y}{b} = 1$; a

80. $\frac{x}{a} + \frac{y}{b} + \frac{z}{c} = 1$; c

In Exercises 81–84, solve for y.

81. $\frac{y - 4}{x + 3} = -2$

82. $\frac{y + 2}{x - 5} = \frac{2}{3}$

83. $\frac{y}{x + 4} = -\frac{4}{5}$

84. $\frac{y - 1}{x} = -\frac{3}{4}$

Data Analysis: *DUI Arrests*

Statistics show that people in the 25–34 age group account for about half of the arrests made for driving under the influence of alcohol. (Source: U.S. Bureau of Justice Statistics.) The number $A(x)$ of arrests per 100,000 drivers can be modeled by the following function, where x represents a driver's age.

$$A(x) = \frac{27{,}720(x - 14)}{x^2 + 9} - 5x$$

85. The figure shows the graph of function A.

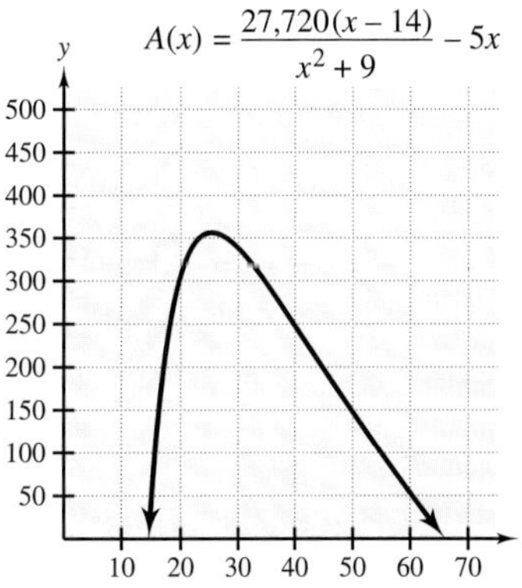

The graph is said to be *skewed* to the right. How can this property of the graph be interpreted?

86. Use the graph to estimate the solution of the equation $A(x) = 356$.

87. Interpret the solution in Exercise 86.

88. Use the graph to estimate the solution of the equation $A(x) = 400$. Interpret the solution.

Challenge

In Exercises 89–92, solve the equation.

89. $\dfrac{\frac{x + 3}{x - 2}}{\frac{2}{x - 2}} = 5$

90. $\dfrac{1 + \frac{6}{x}}{1 + \frac{1}{x}} = 2x$

91. $\frac{1}{18} + \frac{1}{9x} = \frac{1}{2x^2} + \frac{1}{x^3}$

92. $1 + \frac{2}{x} = \frac{1}{x^2} + \frac{2}{x^3}$

In Exercises 93 and 94, solve for y.

93. $\frac{y - y_1}{x - x_1} = m$

94. $\frac{y - b}{x} = m$

In Exercises 95 and 96, use substitution to solve the given equation.

95. $\left(\frac{1}{x + 1}\right)^2 + 3\left(\frac{1}{x + 1}\right) - 4 = 0$

96. $\left(\frac{x - 1}{x + 6}\right)^2 + 4\left(\frac{x - 1}{x + 6}\right) - 12 = 0$

8.6 Applications

Numbers • Ratio and Proportion • Percentages • Inverse Variation • Work Rate • Motion

In Chapter 3 we developed a general approach to application problem solving. This approach is appropriate regardless of the type of equation used to model the problem. In this section we present a variety of examples of applications that lead to equations with rational expressions.

Numbers

Application problems involving numbers and their reciprocals may give rise to equations with rational expressions.

EXAMPLE 1

Numbers and Their Reciprocals

The difference between a number and six times its reciprocal is -1. What is the number?

Solution

Let $x =$ the number. Then its reciprocal is $\frac{1}{x}$.

$x - \frac{6}{x} = -1$	Six times the reciprocal is $6 \cdot \frac{1}{x} = \frac{6}{x}$.
$x \cdot x - x \cdot \frac{6}{x} = -1 \cdot x$	Multiply by the LCD, x.
$x^2 - 6 = -x$	Simplify each term.
$x^2 + x - 6 = 0$	Write the quadratic equation in standard form.
$(x + 3)(x - 2) = 0$	Factor the trinomial.
$x + 3 = 0 \quad \text{or} \quad x - 2 = 0$	Zero Factor Property
$x = -3 \quad \text{or} \quad x = 2$	Solve each case.

There are two numbers that satisfy the conditions, -3 and 2.

Ratio and Proportion

The value of x can be compared to the value of y by means of a **ratio,** which is written as a quotient $\frac{x}{y}$ or in the form $x{:}y$.

You may want to show other ways in which the proportion can be written. For example, the ratios can be inverted or the inner and outer terms can be swapped.

The student is asked to justify this rule in Exercise 6.

A **proportion** is an equation stating that two ratios are equal. If the ratio of x to y is equal to the ratio of a to b, we can write $\frac{x}{y} = \frac{a}{b}$.

One method for solving a proportion, or any other equation of the form $\frac{A}{B} = \frac{C}{D}$, is to **cross-multiply.**

$$\text{If } \frac{A}{B} = \frac{C}{D}, \quad \text{then} \quad AD = BC.$$

Multiplying both sides of the equation by the LCD produces the same result, but clearing fractions by cross-multiplying is usually a more efficient method. We will illustrate this method in Example 2.

In plane geometry, **similar triangles** are triangles whose corresponding angles have the same measure. In the next example, we use the theorem from plane geometry that states that the corresponding sides of similar triangles are proportional.

EXAMPLE 2

Similar Triangles

To construct a scale model for wind tunnel testing, an engineer makes a plastic triangle DEF similar to the full-size component ABC. Determine the lengths (in feet) of sides $\overline{AB}$ and $\overline{DE}$.

LEARNING TIP

Sketching a figure allows you to translate the given information into a visual model and to assign variables and expressions to unknown quantities.

Figure 8.2

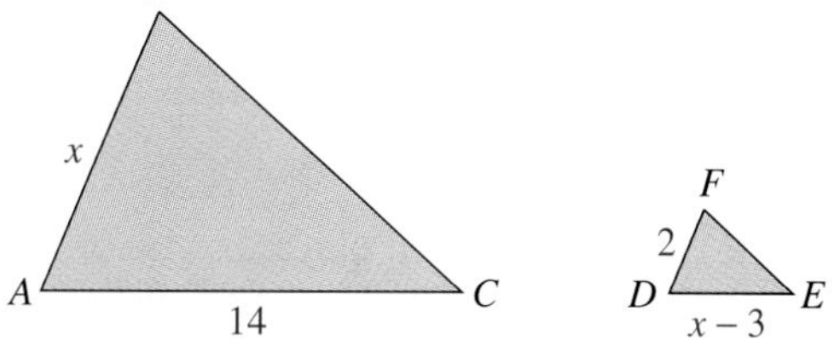

Solution

Because the triangles are similar, the lengths of the corresponding sides are proportional. In Figure 8.2, $\overline{AB}$ corresponds to $\overline{DF}$, and $\overline{AC}$ corresponds to $\overline{DE}$.

$\dfrac{x}{2} = \dfrac{14}{x - 3}$	$\dfrac{AB}{DF} = \dfrac{AC}{DE}$
$x(x - 3) = 28$	Cross-multiply.
$x^2 - 3x = 28$	Distributive Property
$x^2 - 3x - 28 = 0$	Write the quadratic equation in standard form.
$(x - 7)(x + 4) = 0$	Factor the trinomial.
$x - 7 = 0 \quad \text{or} \quad x + 4 = 0$	Zero Factor Property
$x = 7 \quad \text{or} \quad x = -4$	Solve each case for x.

Because length is not negative, the only solution is 7. Thus the side $\overline{AB}$ of the full-size component is 7 feet and the side $\overline{DE}$ of the model component is 4 feet.

Note: In Example 2, the solution -4 is not an extraneous solution; it is a permissible replacement for x. We disqualify -4 simply because it is not a meaningful solution for the application.

EXAMPLE 3

Win-to-Loss Ratio

At the All-Star break, the ratio of wins to losses for the Pittsburgh Pirates was 3 to 2. If the same ratio holds for the entire 162-game season, how many games will the Pirates win?

Solution

Let x = the number of games that will be won. Then $162 - x$ = the number of games that will be lost.

$$\frac{3}{2} = \frac{x}{162 - x}$$ The ratio of wins to losses is 3 to 2.

$$3(162 - x) = 2x$$ Cross-multiply.

$$486 - 3x = 2x$$ Distributive Property

$$486 = 5x$$ Add $3x$ to both sides.

$$97.2 = x$$ Solve for x.

If the 3 to 2 win–loss ratio holds, the Pirates' season record will be about 97 wins and 65 losses, probably good enough to win the division championship.

Percentages

For example, the ratio 1 to 4 is equal to 0.25. To write the ratio as a percent, we must multiply by 100 to obtain 25%.

Because percentages are actually ratios (multiplied by 100), applications involving percentages sometimes lead to equations with rational expressions.

EXAMPLE 4

Basketball Free Throw Percentage

A basketball player has made 32 free throws in 45 attempts, for a current success rate of 71%. How many consecutive free throws must the player make to raise the average to 75%?

Solution

Let x = the required number of consecutive, successful free throws. Then the total number of free throws made is $32 + x$, and the total number of free throws attempted is $45 + x$.

The success rate is the ratio of the number of free throws made to the number of free throws attempted.

$$\frac{\text{Free throws made}}{\text{Free throws attempted}} = \frac{32 + x}{45 + x} = 0.75$$ The goal is a 75% success rate.

$$32 + x = 0.75(45 + x)$$ Cross-multiply.

$$32 + 1.00x = 33.75 + 0.75x$$ Distributive Property

$$0.25x = 1.75$$ Solve for x.

$$x = 7$$

Seven consecutive successful free throws will raise the player's average to 75%.

Inverse Variation

In Chapter 4 we studied direct variation. Another type of variation is **inverse variation.**

Definition of Inverse Variation

A quantity y **varies inversely** with x if there is a constant k so that $y = \frac{k}{x}$.

We sometimes say that y is **inversely proportional** to x. The constant k is called the **constant of variation.**

Compare to the equation $y = kx$ for direct variation.

EXAMPLE 5

As one variable increases, the other decreases.

Inverse Variation

If the temperature is constant, the volume of a gas varies inversely with the pressure. If the pressure of 2 cubic feet of a gas is 12 pounds per square inch, what is the pressure when the volume is 1.6 cubic feet?

Solution

Let P = the pressure and V = the volume.

$$V = \frac{k}{P}$$ Volume varies inversely with pressure.

$$2 = \frac{k}{12}$$ Substitute the initial conditions $V = 2$ and $P = 12$.

$$k = 24$$ Solve for k.

Now we substitute the value of k into the original formula.

$$V = \frac{24}{P}$$ Replace k with 24.

$$1.6 = \frac{24}{P}$$ Substitute 1.6 for V.

$$1.6P = 24$$ Cross-multiply.

$$P = 15$$ Solve for P.

The pressure is 15 pounds per square inch.

Work Rate

You may want to give other examples of work rates. For example, in 3 hours, the machine completes $\frac{3}{5}$ of the task.

Work rate problems nearly always involve two or more people or machines performing tasks at different rates.

If a machine can perform a task in 5 hours, then it completes $\frac{1}{5}$ of the task each hour. This is the machine's *work rate*. The combined work rate of two or more machines is the sum of their individual work rates.

EXAMPLE 6

Work Rate

At a juice-processing plant, one pipe can fill a storage tank with cranberry juice in 2 hours. Another pipe can fill the same tank with apple juice in 3 hours. If both pipes are open, how long does filling the tank with cran-apple juice take?

Try to estimate a solution before solving the equation. Note that the number of hours should be less than the shortest time, 2 hours.

Solution

Let x = the number of hours to fill the tank with both pipes open.

	Hours to Fill Tank	Work Rate
Cranberry Juice	2	$\frac{1}{2}$
Apple Juice	3	$\frac{1}{3}$
Combined	x	$\frac{1}{x}$

The combined work rate is the sum of the individual work rates.

$$\frac{1}{2} + \frac{1}{3} = \frac{1}{x}$$

$$6x \cdot \frac{1}{2} + 6x \cdot \frac{1}{3} = 6x \cdot \frac{1}{x}$$ Multiply by the LCD $6x$.

$$3x + 2x = 6$$ Simplify each term.

$$5x = 6$$ Combine like terms.

$$x = \frac{6}{5} \text{ or } 1.2$$

The time needed to fill the tank is 1.2 hours.

Think About It

Suppose one person requires k times as long as another person to perform a task. How are the work rates related?

Motion

In our previous work with motion problems, we used the formula $d = rt$, where d, r, and t represent distance, rate, and time, respectively. If the focus of a motion problem is on time, then it is convenient to represent time in terms of distance and rate. We can do this by solving the formula $d = rt$ for t to obtain $t = \frac{d}{r}$.

EXAMPLE 7

Distance, Rate, and Time

A canoeist can paddle 2 miles up the Olentangy River in the same time she can paddle 6 miles downstream. The speed of the current is 2 mph. What is her speed in still water?

Solution

Let r = rate of canoe in still water.

	d	r	$t = \frac{d}{r}$
Upstream	2	$r - 2$	$\frac{2}{r-2}$
Downstream	6	$r + 2$	$\frac{6}{r+2}$

$$\frac{2}{r-2} = \frac{6}{r+2}$$ The upstream and downstream times are equal.

$$2(r + 2) = 6(r - 2)$$ Cross-multiply.

$$2r + 4 = 6r - 12$$ Distributive Property

$$-4r = -16$$ Solve for r.

$$r = 4$$

The canoeist paddles at a rate of 4 mph in still water.

EXAMPLE 8

Distance, Rate, and Time

From Front Royal, Virginia, a couple drove 210 miles on the scenic route to Roanoke, Virginia. From Roanoke they traveled Interstate 81 for 120 miles to Wytheville, Virginia. The average speed on the interstate was 25 mph faster than along the scenic route. The entire trip took 8 hours. How long did they drive on the scenic route and how long on the interstate? What was their average speed on each?

Solution

Let s = the average speed on the scenic route. Then $s + 25$ is the average speed on the interstate.

	d	r	$t = \frac{d}{r}$
Scenic	210	s	$\frac{210}{s}$
Interstate	120	$s + 25$	$\frac{120}{s+25}$

$$\frac{210}{s} + \frac{120}{s+25} = 8$$ The total time is 8 hours.

$$s(s+25) \cdot \frac{210}{s} + s(s+25) \cdot \frac{120}{s+25} = 8 \cdot s(s+25)$$ Multiply by the LCD, $s(s + 25)$.

$$210(s + 25) + 120s = 8s(s + 25)$$ Simplify each term.

$$210s + 5250 + 120s = 8s^2 + 200s$$ Distributive Property

$$330s + 5250 = 8s^2 + 200s$$ Combine like terms.

$$8s^2 - 130s - 5250 = 0$$ Write the quadratic equation in standard form.

$$2(4s^2 - 65s - 2625) = 0$$ GCF is 2.

$$2(4s + 75)(s - 35) = 0$$ Factor the trinomial.

$$4s + 75 = 0 \quad \text{or} \quad s - 35 = 0$$ Zero Factor Property

$$4s = -75 \quad \text{or} \quad s - 35 = 0$$ Solve each case for s.

$$s = -18.75 \quad \text{or} \quad s = 35$$

Because speed is not negative, the only solution is $s = 35$. The average speed on the scenic route was 35 mph, and the average speed on the interstate was 60 mph.

The driving time on the scenic route was $\frac{210}{35}$ or 6 hours, and the driving time on the interstate was $\frac{120}{60}$ or 2 hours.

Exercises 1–30 are organized by category. The remaining exercises are mixed.

Exercises 8.6

Concepts and Skills

Numbers

1. Twice a number plus three times its reciprocal is 7. Find the number.
2. The difference between three times a number and eight times its reciprocal is 10. Find the number.
3. One number is 4 less than another. The sum of the reciprocal of the larger number and three times the reciprocal of the smaller number is $\frac{3}{5}$. What are the numbers?
4. One number is three more than another number. The sum of the reciprocal of the smaller number and two times the reciprocal of the larger number is $\frac{9}{20}$. What are the numbers?

Real-Life Applications

Ratio and Proportion

5. Consider the following two problems.
   (i) Find the sum of $\frac{1}{3}$ and the reciprocal of a number.
   (ii) If the sum of $\frac{1}{3}$ and the reciprocal of a number is 2, what is the number?

   For both problems, the LCD is $3x$. Compare the two different ways in which the LCD is used to solve the problems.
6. An arithmetic rule states that if $\frac{a}{b} = \frac{c}{d}$, then the **cross-products** ad and bc are equal. Explain why this is so.
7. The center of a stained glass window in the shape of an isosceles triangle contains a blue glass piece that is also an isosceles triangle and similar to the window. If the ratio of a side of the window to the base is 3 to 2 and the side of the blue piece is 2 inches longer than the base of the blue piece, what is the length of the base of the blue triangle?
8. A large, modern sculpture includes a bronze plate component in the form of an obtuse triangle. At a certain time of the day, the sun casts a shadow of this component such that the plate and the shadow are similar triangles. (See figure.) Determine the lengths of the sides $\overline{AC}$ and $\overline{DE}$.

Figure for 8

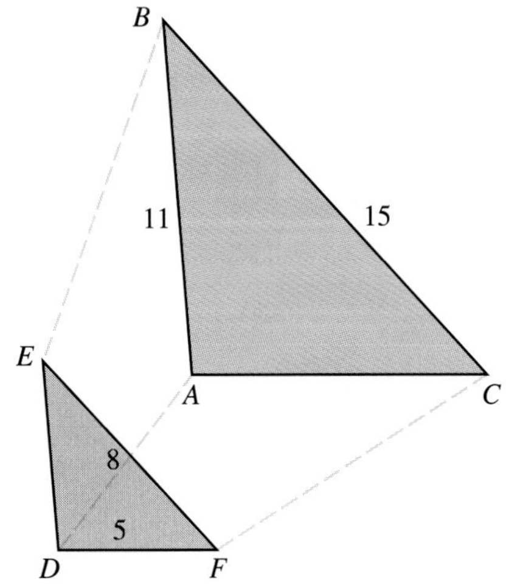

9. At a certain small college, the ratio of traditional age students to nontraditional age students is 4 to 3. If the college enrollment is 1260, how many nontraditional age students attend the college?
10. To investigate a crime, 7 more state than federal agents were assigned. If the ratio of federal agents to the total number of agents was 2 to 5, how many state agents were assigned to the case?

Percentages

11. A softball player has a batting average of 0.252 with 112 hits out of 444 times at bat. If the batter goes on a hitting streak, how many hits in a row would raise the batting average to 0.267?
12. A baseball player has a batting average of 0.295 with 105 hits out of 356 times at bat. The batter then goes into a hitless streak, and the batting average drops to 0.269. How many consecutive at bats without a hit did the batter have?
13. In 1980, 360 male students attended a certain college. By 1990, the number of male students had increased by 300, and the number of female students had increased by 700. If the percentage of male students was the same in 1990 as it was in 1980, what was the total enrollment in 1980?
14. A group of people suffering from a particular ailment agreed to participate in an experimental testing of a new medication. Some of the people received the medication, whereas the others received a placebo

(a preparation containing no medicine and given for its psychological effect). Of the group receiving the medication, 35 claimed it helped. Ten people in the group that received the placebo also said they felt much better. For both groups combined, 50% reported no beneficial effect. How many people were tested?

Inverse Variation

15. If y varies inversely as x, and if $y = 9$ when $x = 5$, what is y when $x = 3$?

16. If a varies inversely as c, and if $a = 12$ when $c = 2$, what is a when $c = 10$?

17. If w varies inversely as z^2, and if $z = 2$ when $w = 24$, what is w when $z = 4$?

18. If y varies inversely as x^2, and if $x = 2$ when $y = 150$, what is y when $x = 10$?

19. Suppose that four workers take 9 hours to roof a house. If the time required to do the job varies inversely as the number of people working on it, how long do five workers take to finish a roof of the same size?

20. Suppose that 5 workers take three 8-hour days to landscape a yard completely. If the time required to do the job varies inversely as the number of people working on it, how long would 16 workers take to landscape the same yard?

21. Suppose the price of oil varies inversely with the supply. If an OPEC nation can sell oil for $26.00 per barrel when daily production is 3 million barrels, what will the price of oil be if the daily production is increased to 4 million barrels?

22. The average number of words that can be printed on standard paper varies inversely with the font size. If 320 words can be printed with a 10-point font, how many words can be printed with a 12-point font?

Work Rate

23. Working together, two people can wash their car in 10 minutes. One person, working alone, can wash the car in a half hour. How long does the other person take to do the job working alone?

24. Using a 21-inch mower, a person can mow the front lawn in 45 minutes. Mowing takes only a half hour with a 36-inch mower. If two people work together, one with the smaller mower and the other with the larger mower, how long will they take to mow the lawn?

25. A professional painter can paint the outside trim on a certain house as fast as two apprentices working together. One apprentice can do the painting alone in 15 hours; the other apprentice can finish the job in 10 hours. How long would the painter take to paint the trim without help?

26. Working together, two front-end loaders can fill a railroad car with gravel in 20 minutes. Working alone, the smaller machine would take 9 minutes longer than the larger machine would take working alone. How long would each of the loaders take working alone?

Motion

27. A small plane travels 90 mph faster than a train. The plane travels 525 miles in the same time it takes the train to travel 210 miles. Determine the rate of each.

28. A private plane takes 4 hours less time than a car to make a trip of 495 miles. If the plane's rate is $\frac{9}{5}$ of the car's rate, what is the rate of each?

29. Two cyclists competed in a 5-mile bicycle race. The more experienced cyclist gave the novice cyclist a head start of 0.6 miles and still won the race by 2 minutes. If the experienced cyclist's average speed was 3 mph faster than the novice's average speed, what was the novice's average speed?

30. A pitcher throws a fastball to a catcher standing 66 feet away. At the same instant, the shortstop throws a ball to the first baseman standing 124 feet away. The first baseman catches the throw from the shortstop a half second after the catcher catches the pitch. If the pitcher can pitch 8 feet per second faster than the shortstop can throw, how long does a batter have to react to the pitcher's fastball?

Miscellaneous

31. Consider two data items A and B. If an increase in A causes a corresponding decrease in B, or if a decrease in A causes a corresponding increase in B, we say that A and B are inversely related. Which of the following would you suspect are inversely related?

(a) Humidity (A) and drying time of varnish (B)

(b) Number of amateur skiers (A) and the price of ski equipment (B)

(c) Illiteracy rate (A) and library circulation (B)

(d) Number of auto accidents (A) and insurance premiums (B)

(e) Unemployment (A) and home ownership (B)

32. A soft-drink bottling company uses a vat that can be filled with carbonated water in 1 hour. The vat can be filled with syrup in 9 hours. The vat of soft-drink mixture can be emptied to the bottling apparatus in 3 hours. If the operator opens all three valves and then takes a 90-minute lunch break, describe the operator's state of mind when he returns from lunch and show why.

33. In the figure, a yardstick is placed perpendicular to the ground, and it casts a shadow 15 inches long. If the nearby oak tree casts a 25-foot shadow, how tall is the oak tree? You may need to point out that the triangles are similar triangles.

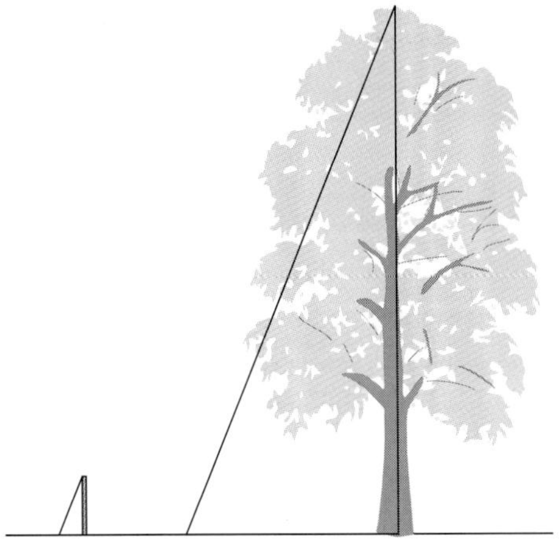

34. To estimate the bass population of a pond, a fish biologist used a net to capture 20 bass. She tagged these fish and released them back into the pond. A few days later, the biologist netted 21 bass and found that 3 of the fish had tags. What was the estimated bass population of the pond?

35. One person takes twice as long to shovel snow from the driveway as another person takes using a snow blower. If the two of them together can clear the driveway in 8 minutes, how long does the person shoveling alone take?

36. Two self-employed house cleaners can wash the windows of a particular client in a half hour. Working alone, one cleaner could wash all the windows in 11 minutes less time than the other would take, but the faster cleaner refuses to do windows. Working alone, how long will the other cleaner take to wash the windows?

37. The number of hairs in an animal's coat varies inversely with the average winter temperature. One year, the average winter temperature was 10° lower than the previous year, and an animal was found to have 1.5 times as many hairs as it had the previous year. What was the average winter temperature during the previous year?

38. A motorist took the interstate from Cincinnati to Columbus, a distance of 80 miles. On the return trip the motorist took some back roads, which reduced the average speed by 15 mph. If the total driving time for the round trip of 160 miles was $3\frac{1}{9}$ hours, what was the average speed on the interstate?

39. A sprinter ran the 100-yard dash for the college track team. It took 1 second longer to run the final 50 yards than it took to run the first 50 yards. If the sprinter's average speed for the total distance was $9\frac{1}{11}$ yards per second, what was the time for the final 50 yards?

40. Two people inherited a total of \$76,000. The will specified that the inheritance was to be divided in a 3 to 2 ratio. How much did each receive?

41. When the wind speed is 40 mph, a plane can fly 200 miles against the wind in the same time it can fly 300 miles with the wind. What is the speed of the plane?

42. A 1200-bushel grain bin can be filled in 3 hours and emptied in 2 hours. If grain is added to a full bin at the same time that the bin is being emptied, how long will it be before there is no grain in the bin?

43. An elevated water tank serves both commercial and residential customers. The tank can be filled in 14 hours. Commercial usage alone can empty the tank in 7 hours and residential usage alone can empty the tank in 9 hours. If the intake valve is open at the same time both outtake valves are open, how long will exhausting the water supply take if the tank is initially full?

44. A boat can travel 15 miles upstream and 15 miles back downstream in a total of 4 hours. If the speed of the current is 5 mph, what is the speed of the boat in still water?

45. Two cats can eat a 7-pound bag of cat food in 24 days. One cat can eat a 3.5-pound bag of food in 3 weeks. How long would the other cat take to eat a 3.5-pound bag?

46. A clothing chain has noticed that the number of down parkas sold by any one of their stores is inversely proportional to the average January temperature in the city where the store is located. If the store in Houston, Texas, where the average January temperature is 50°F, sells 500 parkas, how many parkas are sold in Great Falls, Montana, where the average January temperature is 21°F? What would the average January temperature be in a city that sold 4000 parkas?

Modeling with Real Data

47. The accompanying table shows that some progress was made in the recycling of aluminum and glass from 1980 to 2000. All data are in millions of tons. (Source: Franklin Associates Ltd.)

	Waste Generated (millions of tons)		Waste Recovered (millions of tons)	
Year	**Aluminum**	**Glass**	**Aluminum**	**Glass**
1980	1.8	15.0	0.3	0.8
1990	2.8	13.1	1.0	2.6
2000	3.1	11.8	1.1	2.8

The following functions model these data (in millions of tons), where x is the number of years since 1980.

Aluminum generated:
$A_1(x) = 0.07x + 1.9$

Aluminum recovered:
$A_2(x) = 0.04x + 0.4$

Glass generated:
$G_1(x) = -0.16x + 14.9$

Glass recovered:
$G_2(x) = 0.12x + 1.02$

(a) Write a rational expression to model the percentage of aluminum that was recovered.

(b) Write a rational expression to model the percentage of glass that was recovered.

48. Refer to the models given in Exercise 47.

(a) Write equations to determine in what years the percentages are predicted to be 40%.

(b) Solve the equations to determine the year for each.

Challenge

49. In Woodville Notch, all registered voters are either Democrats, Republicans, or Independents. The numbers of Democrats and Republicans are equal. There are 11 Independents, of whom 9 are women. In a recent election, five-sixths of the Democrats voted, seven-ninths of the Republicans voted, and all the Independents voted. Half of the Democrats are women, and two-thirds of the Republicans are women. All the women voted, which means that 75% of those who voted were women. How many registered voters are there in Woodville Notch?

50. Pecans worth \$30.00 are mixed with cashews worth \$45.00. There are 3 pounds more of cashews than of pecans. If the unit price of the cashews is 50 cents greater than the unit price of pecans, how many pounds of each kind of nut were used? Exercise 50 has two solutions.

51. A builder bought three sizes of lag bolts to build a gazebo. The builder needed 4 pounds more of medium bolts than of small bolts and 3 pounds fewer of large bolts than of small bolts. The sales invoice showed the following.

Lag bolts, small	\$3.20
Lag bolts, medium	8.40
Lag bolts, large	5.50

If the sum of the unit prices for the small and medium bolts is equal to the unit price for the large bolts, how many pounds of each size did the builder buy?

52. The two legs of a right triangle are represented by $\frac{3}{x}$ and $\frac{12}{x+1}$. If the hypotenuse is represented by $\frac{5}{x}$, determine the lengths of the sides of the triangle.

53. In the figure, the square on the left and the rectangle on the right have dimensions as shown. If the area of the rectangle is 3 square inches greater than the area of the square, find the dimensions of each.

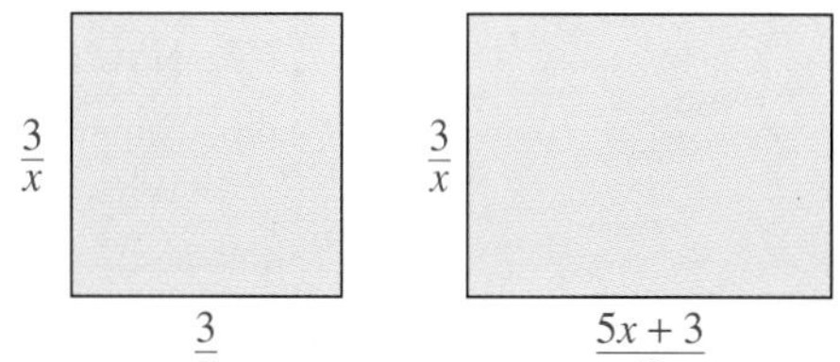

8 • Chapter Project

Major League Salaries

In Section 8.4, we observed that the average salaries of major league baseball players increased by about 50% from 1995 to 1999.

The following table shows the number of players who made at least \$1 million per year in selected years. (Source: *USA Today.*)

Year	Number of Players
1990	138
1994	265
1995	215
1996	241
1999	342

The function $N(x) = \dfrac{1.9x^4 - 24.0x^3 + 76.0x^2 + 240.3x + 138.4}{x + 1}$ models the number of million-dollar players, where x is the number of years since 1990.

1. The following is a table of values of $N(x)$ for various values of x.

x	$N(x)$
0	138
4	253
5	238
6	228
9	343

According to the model function N, how many million-dollar players were there in 1996?

2. Does the model function underestimate or overestimate the number of million-dollar players in 1995?
3. Use function N to estimate the number of million-dollar players in 2001. Assuming that the number of major league players remains 650, how valid do you think this estimate is?
4. Use long or synthetic division to simplify the model function, and call the result $P(x)$.
5. How does the domain of $P(x)$ differ from the domain of $N(x)$?
6. Suppose you wish to create a table of values of function P for $x = 0, 4, 5, 6$, and 9. Why would evaluating P for those values of x be unnecessary?

Chapter 8 Review Exercises

Section 8.1

1. For $f(x) = \dfrac{x-7}{x^2-2x-15}$, find two numbers a_1 and a_2 for which $f(a_1)$ and $f(a_2)$ are undefined. What do we call a_1 and a_2 in this case?

2. Consider the function $g(x) = \dfrac{1}{x-5}$.

(a) Evaluate each of the following.

(i) $g(6)$

(ii) $g(5.5)$

(iii) $g(5.01)$

(iv) $g(5.0001)$

(v) $g(5)$

(b) Describe the values of $g(x)$ as x becomes closer to 5.

(c) Explain the result in part (v).

In Exercises 3 and 4, determine the domain of the rational function.

3. $y = \dfrac{5x-2}{2x-5}$

4. $y = \dfrac{3x-4}{3x^2+11x-20}$

5. Explain why 3 can be divided out of the numerator and denominator of $\dfrac{3(x+1)}{3(x+5)}$ but not of $\dfrac{3x+1}{3x+5}$.

In Exercises 6–9, simplify the rational expression.

6. $\dfrac{2x-3}{4x^2-9}$

7. $\dfrac{x^2+x-12}{2x^2-9x+9}$

8. $\dfrac{9-x^2}{x^2+x-12}$

9. $\dfrac{ab+4a-3b-12}{ab+2a-3b-6}$

10. General Appliance Company has x refrigerators in its inventory. After the company receives a shipment of five refrigerators, the total value of the inventory (in thousands of dollars) is $x^2-2x-35$. Write a simplified expression for the average unit value of the refrigerators in the inventory.

In Exercises 11 and 12, supply the missing numerator.

11. $\dfrac{5}{3x+12} = \dfrac{\qquad}{3x^2-48}$

12. $\dfrac{3}{x^2+2x-8} = \dfrac{\qquad}{(x-2)(x^2-x-20)}$

Section 8.2

In Exercises 13 and 14, multiply and simplify.

13. $\dfrac{8x-40}{4x^3} \cdot \dfrac{4x^2+20x}{x^2-25}$

14. $\dfrac{x^2-x-56}{x^2-49} \cdot \dfrac{8x+64}{x^2-64}$

15. What is the product of $\dfrac{x^2+3x+2}{x^2+8x+12}$ and $\dfrac{x^2+7x+6}{x^2-x-2}$?

16. Multiply $\dfrac{2x^3}{4x^4-36x^2}$ by $x^2-2x-15$.

17. Consider two nonzero numbers x and y.

(a) What is the reciprocal of their product?

(b) What is the product of their reciprocals?

(c) Are the results found in parts (a) and (b) equivalent?

In Exercises 18 and 19, divide and simplify.

18. $\dfrac{3x}{x+9} \div \dfrac{15x^2}{x^2+7x-18}$

19. $\dfrac{x^2-5x+6}{x^2+3x-18} \div \dfrac{x^2-4}{6x+36}$

20. Divide $\dfrac{3x-3}{8x^3}$ by $\dfrac{x^2-1}{8x^2+8x}$.

21. What is the quotient of

$\dfrac{x^2+3x-4}{x^2-2x-24}$ and $\dfrac{x^2-5x+4}{x^2-7x+6}$?

22. List all the restricted values in Exercise 21.

23. The length and width of a rectangle are, respectively, $\dfrac{x^2+5x+6}{x^2-3x-18}$ and $\dfrac{2x-12}{x^2-4}$.

Write a simplified expression for the area of the rectangle.

24. If a and b are nonzero numbers, explain why dividing by $\dfrac{a^{-1}}{b^{-1}}$ is the same as multiplying by $\dfrac{a}{b}$.

Section 8.3

In Exercises 25 and 26, determine the LCD.

25. $\dfrac{1}{x^2-49}, \dfrac{1}{9x-63}$

26. $\dfrac{1}{x^2+3x+2}, \dfrac{1}{x^2-4}, \dfrac{1}{x^2-x-2}$

27. To add $\frac{2}{3}$ and $\frac{5}{3}$, we use the Distributive Property as follows:

$$\frac{2}{3}+\frac{5}{3}=\frac{1}{3}\cdot 2+\frac{1}{3}\cdot 5=\frac{1}{3}(2+5)=\frac{1}{3}\cdot 7=\frac{7}{3}$$

With this as background, explain why it is not possible to add fractions that have unlike denominators.

In Exercises 28–31, perform the indicated operations.

28. $\dfrac{5x+3}{x-1}-\dfrac{9x-1}{x-1}$

29. $\dfrac{2x-1}{5x-4}+\dfrac{6x-6}{4-5x}$

30. $\dfrac{3}{10x}+\dfrac{8}{5x^2}$

31. $\dfrac{7}{x^2+x-12}+\dfrac{3}{x^2-16}$

32. Determine the unknown numerator.

$$\frac{x}{5x+10}=\frac{\square}{5x^2-25x-70}$$

33. Write $2x^{-1}+3(x+2)^{-1}$ as a single rational expression with no negative exponents.

34. What is the sum of

$$\frac{x}{x^2+8x+15} \text{ and } \frac{15}{x^2+4x-5}?$$

35. Subtract $\dfrac{1}{x^2-x-30}$ from $\dfrac{7}{x^2+2x-48}$.

36. Consider any two consecutive odd integers. Find the sum of their reciprocals and show that the numerator of the sum is always an even number.

Section 8.4

37. Describe two different methods for simplifying a complex fraction.

In Exercises 38 and 39, simplify the complex fraction.

38. $\dfrac{9-\dfrac{1}{x^2}}{3+\dfrac{1}{x}}$

39. $\dfrac{\dfrac{a}{9}-\dfrac{1}{a}}{\dfrac{1}{3}+\dfrac{a+4}{a}}$

40. The area of a rectangle is $1+\dfrac{2}{x}-\dfrac{3}{x^2}$. If the length of the rectangle is $1-\dfrac{9}{x^2}$, write a simplified expression for the width.

41. For nonzero numbers m and n, show that

$$\frac{\dfrac{m^{-1}}{n^{-1}}}{\dfrac{n}{m}}=1$$

42. Rewrite $\dfrac{1-9x^{-2}}{2+3x^{-1}-9x^{-2}}$ with positive exponents and simplify.

In Exercises 43 and 44, simplify the complex fraction.

43. $\dfrac{\dfrac{x+4}{x-4}+\dfrac{x-4}{x+4}}{\dfrac{1}{x+4}+\dfrac{1}{x-4}}$

44. $\dfrac{\dfrac{12x}{y^2}+\dfrac{13}{xy}-\dfrac{35}{x^3}}{\dfrac{10}{x^3}-\dfrac{3}{xy}-\dfrac{4x}{y^2}}$

Section 8.5

45. Suppose that you make no errors in solving an equation containing rational expressions. How is it possible that the apparent solution does not check? What do we call the apparent solution in this case?

46. Suppose that you are solving an equation containing rational expressions. After you clear the fractions and simplify, you obtain an equation of the form $a=b$, where a and b are constants. How do you interpret this result?

In Exercises 47–54, solve.

47. $10x-\dfrac{2}{x}=1$

48. $\dfrac{x}{5}+\dfrac{5}{4x}=\dfrac{5}{4}$

49. $\dfrac{x+3}{x-5}=\dfrac{8}{x-5}$

50. $\dfrac{5}{x+3}+\dfrac{8}{x-3}=\dfrac{5}{x^2-9}$

51. $\dfrac{7}{x+1}-\dfrac{2}{x-1}=\dfrac{1}{x^2-1}$

52. $\dfrac{9}{3-x}+\dfrac{9}{x+3}=\dfrac{-54}{x^2-9}$

53. $\dfrac{2x}{x-2}-\dfrac{16}{x^2-4}=2$

54. $\dfrac{6}{x-1}+\dfrac{7}{x^2-5x+4}=\dfrac{2}{x-4}$

55. Solve $\dfrac{1}{x}+\dfrac{1}{y}=\dfrac{1}{3}$ for x.

Section 8.6

56. If each is working alone, a helper takes twice as long to do a welding job as an experienced welder would take. Working together, the two welders take 8 hours to complete the job. How long does the helper take working alone?

57. Seven members of the Future Farmers of America painted a small barn in 6 hours. If two additional members had helped, how long would painting the barn have taken?

58. A chemical tank truck can be filled through an intake pipe in 2.5 hours. With the outtake pipe open, the tank can be emptied in 3 hours. If the tank is initially empty, how long does filling the tank take with both pipes open?

59. A baseball player has a batting average of 0.244 with 42 hits out of 172 times at bat. If he goes on a hitting streak, how many hits in a row would raise his batting average to 0.278?

60. The total travel time for a family to reach a campsite is 2 hours. They drive 81 miles and hike 2 miles. If they drive 50 mph faster than they hike, what is their average driving speed?

61. During the first half of the season, a basketball player missed 7 free-throw attempts. During the second half, she made twice as many free throws as she made during the first half, but she had 16 more attempts than she had during the first half. If her free-throw percentage at the end of the season was 75%, how many free throws did she make for the season?

Looking Back

The following exercises review concepts and skills that you will need in Chapter 9.

In Exercises 1–4, simplify and express the result with positive exponents. Assume that all expressions are defined.

1. (a) $(x^2)^4$ (b) $(a^{-3})^5$
2. (a) $a^3 \cdot a^5$ (b) $x^{-5} \cdot x^3$
3. (a) $(a^2b^3)^4$ (b) $\left(\dfrac{3}{x^5}\right)^2$
4. (a) $\dfrac{3x^{-2}}{y^{-4}}$ (b) $\dfrac{40a^4b^7}{5a^9b^4}$

5. Evaluate each expression.
   (a) 5^{-2} (b) $-3 \cdot 6^0$ (c) $|-8|$

In Exercises 6 and 7, factor completely.

6. $y^2 + 6y + 9$
7. $a^4b^6 + a^2b^8$
8. Use the Distributive Property to show how to combine like terms in the expression $2n + 5n$.

In Exercises 9–11, solve the given equation.

9. $x^2 = 16$
10. $3x - 1 = 2^3$
11. $x^2 - 3x = 2^2$
12. Show that a triangle with sides 9, 12, and 15 is a right triangle.

Chapter 8 Test

1. For $\dfrac{x-3}{x^2+5x-24}$, explain why 3 is a restricted value.

2. Determine the domain of the function $y = \dfrac{2x-1}{x^2+3x}$.

For Questions 3–5, simplify.

3. $\dfrac{6x^2 + 12x}{3x^3 - 6x^2}$
4. $\dfrac{(x+5)^2}{x^2 - 4x - 45}$
5. $\dfrac{x^2 + xy - 2y^2}{y^2 - x^2}$

For Questions 6 and 7, perform the indicated multiplication or division.

6. $\dfrac{t^2 + 3t - 4}{t^2 + 2t - 3} \cdot \dfrac{t^2 - t - 6}{t^2 + t - 12}$

7. $\dfrac{y^2 - 25}{2y + 10} \div (y^2 - 10y + 25)$

8. Determine the missing numerator.

$$\frac{3x}{x^2 - x - 2} = \frac{\quad\quad}{(x - 2)^2(x + 1)}$$

9. Explain why the following subtraction is easier to perform than it first appears and show that the result does not involve r.

$$\frac{(m + n)^{-1}}{r^{-1}} - \frac{r - 1}{m + n}$$

For Questions 10–12, perform the indicated addition or subtraction.

10. $\dfrac{7x - 1}{3x + 4} - \dfrac{4x - 5}{3x + 4}$

11. $\dfrac{3}{t^2} - \dfrac{5}{t}$

12. $\dfrac{x - 7}{2x^2 + 9x - 5} + \dfrac{4 - x}{4x^2 + 23x + 15}$

13. Which of the following is a complex fraction? Show that two of the expressions are equivalent.

(i) $\dfrac{\frac{m}{n}}{p}$ (ii) $\dfrac{m}{\frac{n}{p}}$ (iii) $\dfrac{\frac{m}{n}}{\frac{1}{p}}$

For Questions 14 and 15, simplify the complex fraction.

14. $\dfrac{\dfrac{x + 3y}{y}}{\dfrac{x^2 - 9y^2}{6y}}$

15. $\dfrac{x - 3 + \dfrac{x - 3}{x^2}}{x + 2 + \dfrac{1}{x} + \dfrac{2}{x^2}}$

16. Write $\dfrac{a^{-2}}{a^{-2} - b^{-2}}$ as a single rational expression with positive exponents.

For Questions 17–20, solve the equation.

17. $\dfrac{3}{a + 3} = \dfrac{2}{a - 2}$

18. $1 - \dfrac{3}{x} = \dfrac{10}{x^2}$

19. $\dfrac{x}{x - 2} + \dfrac{2}{3} = \dfrac{2}{x - 2}$

20. $\dfrac{8t}{t^2 - 16} = \dfrac{3}{t + 4} + \dfrac{5}{4 - t}$

21. For different reasons, the solution sets of the following equations are empty. Explain why this is so for each equation.

(a) $\dfrac{x - 2}{x - 1} = \dfrac{1}{1 - x}$

(b) $\dfrac{x - 2}{x - 1} = \dfrac{x + 3}{x - 1}$

22. Solve the formula $\dfrac{1}{A} + \dfrac{1}{B} = 2$ for B.

23. In a target-shooting competition, scores are found to vary inversely with the wind speed. If the success rate is 60% when the wind speed is 20 mph, what is the success rate when the wind speed is 30 mph?

24. In the spring a cat sheds so fast that the carpet is covered with hair in 4 hours. If the cat is out of the house, cleaning the carpet takes 1.5 hours. How long does cleaning the carpet take with the cat in the house?

25. The time a person takes to paddle a kayak 2 miles downstream is the same as the time to paddle half a mile upstream. If the rate of the current is 3 mph, what is this person's paddling rate in still water?

Cumulative Test, Chapters 6–8

In Questions 1–6, use the properties of exponents to simplify the given expression. Write the result with positive exponents. Assume all expressions are defined.

1. $(x - 3)^0$

2. $\dfrac{x^{-3} \cdot x^5}{x^3}$

3. $(2x^2)^{-3}$

4. $\left(\dfrac{x^{-1}}{y}\right)^{-2}$

5. $\dfrac{18x^{-3}}{27z^{-2}}$

6. $\dfrac{(2x^{-2})^2(3x^4)}{(3x^2)^{-2}}$

7. (a) Write $2.7 \cdot 10^{-5}$ in decimal form.

(b) Write 0.0000000083 in scientific notation.

8. At the equator, the circumference of the earth is approximately 25,000 miles. Write this number in scientific notation.

9. Perform the indicated operations.

$(3x^2 - 6x + 8) - (x^3 - 2x^2 - 7) +$
$(2x^3 - 4x^2 + 7x - 16)$

10. Multiply and simplify.

$(x - 3)^2 - (x + 3)(x - 3)$

11. Factor completely.

(a) $x^4y^3 - x^3y^4$

(b) $ax + 3ay - 7bx - 21by$

12. Factor completely.

(a) $(2x + 1)^2 - 16$

(b) $2x^2 - 12x + 18$

(c) $y^3 + 8$

13. Factor completely.

(a) $z^2 + 2z - 63$

(b) $a^2 - 4ab - 45b^2$

(c) $12 + 9x - 3x^2$

14. Explain how the Zero Factor Property can be used to solve $x^2 - 3x = 0$.

15. Solve.

(a) $(x - 4)(x - 3) = 12$

(b) $a^4 + 25 = 26a^2$

16. If the square of a number is decreased by 24, the result is 5 times the number. What is the number?

17. Find the quotient and remainder for the following.

$(4c^4 + 5c^2 - 7c) \div (2c - 1)$

18. Divide synthetically.

$(x^5 - 3x^3 - 17x^2 + x - 13) \div (x - 3)$

19. Simplify $\dfrac{x^2 - x - 12}{4x^2 - 13x - 12}$.

20. Multiply $\dfrac{a^2 - 16}{a^2 - 3a - 4} \cdot \dfrac{a^2 + 3a + 2}{a^2 - 2a - 8}$.

21. Add $\dfrac{3}{y^2 + 11y + 10} + \dfrac{y}{y^2 + 5y + 4}$.

22. Simplify $\dfrac{\dfrac{1}{x+3}}{\dfrac{1}{x} - \dfrac{1}{x+3}}$.

23. Solve $\dfrac{2}{x} - \dfrac{x+8}{x+2} = \dfrac{12}{x^2 + 2x}$.

24. A landscape worker can lay sod in a yard in 6 hours. A helper would take 8 hours to do the same job. How long would sodding the yard take if both people worked together?

25. A driver of a car averaged 5 mph faster during the final 120 miles of a 340-mile trip. If the entire trip took 6 hours, at what average speed did the driver travel during the first 220 miles?

Chapter 9

Radical Expressions

Suppose that you are a staff member with the National Park Service. Faced with budgetary cuts, you assemble data on the number of visitors to all areas that are managed by the Service (see graph) and on the cost of administering those areas.

The data in the graph can be modeled by an expression with a ***rational exponent.*** *This model, along with the cost data, can be used to show the cost per visitor and to project budgetary requirements for future years. (For more on this topic, see Exercises 99–102 at the end of Section 9.3, and see the Chapter Project.)*

In Chapter 9 we discuss expressions with radicals and rational exponents. We learn about the properties of such expressions and about how to simplify them and perform operations with them. Next, we develop methods for solving equations containing radicals and exponents. We conclude by extending the number system to the set of complex numbers.

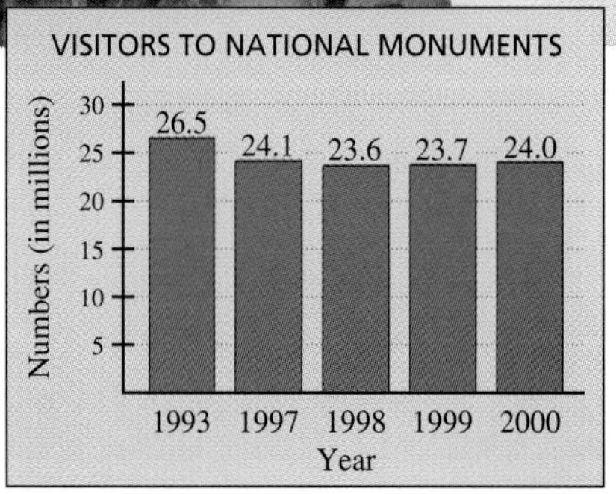

(Source: National Park Service.)

9.1 Radicals

Higher-Order Roots • Simplifying $(\sqrt[n]{b})^n$ and $\sqrt[n]{b^n}$

Higher-Order Roots

In Section 1.6 we defined a to be a **square root** of b if $a^2 = b$. For example, 5 and -5 are both square roots of 25 because $5^2 = 25$ and $(-5)^2 = 25$.

Explaining the definition of *n*th root will require some patience, because students almost always associate roots with radical symbols.

The concept of square root can be generalized to higher-order roots. For example, 3 is the **third root** (or **cube root**) of 27 because $3^3 = 27$; 2 and -2 are **fourth roots** of 16 because $2^4 = 16$ and $(-2)^4 = 16$.

Definition of *n*th Root

For any real numbers a and b and any integer $n > 1$, a is an ***n*th root** of b if and only if $a^n = b$.

EXAMPLE 1

Determining *n*th Roots

(a) The third root of 8 is 2. $2^3 = 8$

(b) The fifth root of -32 is -2. $(-2)^5 = -32$

(c) The fourth roots of 81 are 3 and -3. $3^4 = 81$ and $(-3)^4 = 81$

(d) The fourth root of -81 is undefined. There is no real number a such that $a^4 = -81$.

Example 1 leads to the following observations.

1. Any real number has one odd root.
2. A positive number has two even roots that are opposites, but a negative number has no real number even root.
3. Because $0^n = 0$ for all integers $n > 1$, 0 has one nth root, 0.

In Section 1.6 we defined the **principal square root** of b to be the *nonnegative* square root, and we represented this number with $\sqrt{b}$. We can extend this definition to higher-order even roots by defining the **principal *n*th root** of b to be the nonnegative nth root of b, and we represent this number with $\sqrt[n]{b}$. For odd roots, there is only one nth root, and we represent this number with $\sqrt[n]{b}$.

Definition of $\sqrt[n]{b}$

For any even integer $n > 1$ and any nonnegative real number b, $\sqrt[n]{b}$ represents the nonnegative nth root of b.

For any odd integer $n > 1$ and any real number b, $\sqrt[n]{b}$ represents the nth root of b.

LEARNING TIP

For an even index, the radicand must be positive or zero. For an odd index, the radicand can be negative, positive, or zero.

In the expression $\sqrt[n]{b}$, the symbol $\sqrt{\ }$ is called the **radical symbol,** b is called the **radicand,** n is called the **index,** and the entire symbol $\sqrt[n]{b}$ is called a **radical.**

Note: For an even index n, $\sqrt[n]{b}$ is a real number only if $b \geq 0$. For an odd index n, $\sqrt[n]{b}$ is defined for all real numbers b.

EXAMPLE 2

Evaluating *n*th Roots

(a) $\sqrt[3]{-64} = -4$ $\quad (-4)^3 = -64$

(b) $\sqrt{-64}$ is not a real number. $\quad$ Even root of a negative number

(c) $-\sqrt{64} = -8$ $\quad -1 \cdot \sqrt{64} = -1 \cdot 8 = -8$

(d) $\sqrt[4]{81} = 3$ $\quad 3^4 = 81$

(e) $\sqrt[5]{32} = 2$ $\quad 2^5 = 32$

(f) $\sqrt[6]{-1}$ is not a real number. $\quad$ Even root of a negative number

(g) $\sqrt[9]{-1} = -1$ $\quad (-1)^9 = -1$

Note: Most scientific calculators have square root keys. Keys for higher-order roots can also be found on some calculators. A typical selection is $\sqrt[x]{\ }$.

EXAMPLE 3

Calculating Higher-Order Roots

Use your calculator to evaluate the given radical. Round your results to the nearest hundredth.

(a) $\sqrt[3]{96} \approx 4.58$

(b) $\sqrt[5]{-6397} \approx -5.77$

(c) $\sqrt[4]{0.012} \approx 0.33$

Simplifying $\left(\sqrt[n]{b}\right)^n$ and $\sqrt[n]{b^n}$

Sometimes we encounter radical expressions of the form $\left(\sqrt[n]{b}\right)^n$ and $\sqrt[n]{b^n}$.

Exploring the Concept

Simplifying $\left(\sqrt[n]{b}\right)^n$ and $\sqrt[n]{b^n}$

To evaluate $\left(\sqrt{25}\right)^2$, we take the square root of 25 and then square the result: $\left(\sqrt{25}\right)^2 = (5)^2 = 25$. Similarly, $\left(\sqrt[3]{8}\right)^3 = (2)^3 = 8$. However, before we conclude that $\left(\sqrt{b}\right)^2$ is always b, we must recall that $\sqrt{b}$ is a real number only if $b \geq 0$. In fact, this provision applies to all even roots. With this limitation in mind, we state our conclusion as follows.

If n is odd, then $\left(\sqrt[n]{b}\right)^n = b$.

If n is even, then $\left(\sqrt[n]{b}\right)^n = b$ only if $b \geq 0$.

Now consider the quantities $\sqrt{7^2}$ and $\sqrt[3]{4^3}$. To evaluate these quantities, we begin with the exponent and then take the root.

$$\sqrt{7^2} = \sqrt{49} = 7 \qquad \sqrt[3]{4^3} = \sqrt[3]{64} = 4$$

These results might suggest that $\sqrt[n]{b^n}$ is always b, and this conclusion is indeed true if n is odd. However, a different result is obtained when n is even and b is negative.

$$\sqrt{(-10)^2} = \sqrt{100} = 10 \qquad \sqrt[4]{(-3)^4} = \sqrt[4]{81} = 3$$

In each of these instances, b is a negative number, but the result is positive because $\sqrt[n]{\ }$ indicates the *principal* (nonnegative) nth root when n is even. Regardless of the sign of b, we can conclude that $\sqrt[n]{b^n} = |b|$ when n is even.

The following summary gives the rules for simplifying expressions of the form $\left(\sqrt[n]{b}\right)^n$ and $\sqrt[n]{b^n}$.

Think About It

Raising $\sqrt[n]{b}$ to the nth power reverses or undoes the operation of taking the nth root. Give other examples of an operation that reverses an original operation.

Simplifying $\left(\sqrt[n]{b}\right)^n$ and $\sqrt[n]{b^n}$

Let b represent any real number.

1. If $\sqrt[n]{b}$ is defined, then $\left(\sqrt[n]{b}\right)^n = b$.
2. For even integers n, $\sqrt[n]{b^n} = |b|$. For odd integers n, $\sqrt[n]{b^n} = b$.

Note that for $b < 0$ and even integers n, $\left(\sqrt[n]{b}\right)^n$ is undefined.

EXAMPLE 4

Evaluating $\sqrt[n]{b^n}$ with Numerical Radicands

Sometimes students have trouble understanding why $\sqrt{(-5)^2} \neq -5$.

(a) $\sqrt{(-5)^2} = |-5| = 5$ — Because n is even, absolute value symbols are needed.

(b) $\sqrt[5]{(-4)^5} = -4$ — Because n is odd, absolute value symbols are not used.

(c) $\sqrt[4]{3^4} = |3| = 3$ — Because n is even, absolute value symbols are needed.

When the radicand is a variable expression and the index is even, we must use absolute value symbols around the result if there is a possibility that the result could be negative.

EXAMPLE 5

Simplifying $\sqrt[n]{b^n}$ with Variable Radicands

(a) $\sqrt[4]{a^4} = |a|$ — The index is even and a could be negative. Therefore, absolute value symbols are needed.

(b) $\sqrt[3]{t^3} = t$ — Because the index is odd, absolute value symbols are not used.

(c) $\sqrt{(x-3)^2} = |x-3|$ — Absolute value symbols are needed because the index is even and $x - 3$ could be negative.

The rules for simplifying $\sqrt[n]{b^n}$ can be useful for simplifying expressions in which the exponent on b is a multiple of the index.

EXAMPLE 6

Simplifying Radicals

(a) $\sqrt{c^6} = \sqrt{(c^3)^2}$ — Write the expression in the form $\sqrt[2]{b^2}$.

$= |c^3|$ — The index is even and c^3 could be negative.

(b) $\sqrt[3]{y^{15}} = \sqrt[3]{(y^5)^3}$ — Write the expression in the form $\sqrt[3]{b^3}$.

$= y^5$ — The index is odd, so absolute value symbols are not used.

(c) $\sqrt[4]{x^8} = \sqrt[4]{(x^2)^4}$ — Write the expression in the form $\sqrt[4]{b^4}$.

$= |x^2|$

$= x^2$ — The index is even, but x^2 is not negative.

Initial results should always be enclosed in absolute value symbols when the index is even. Then the symbols can be removed if they are not needed. Compare parts (a) and (c).

Note: The techniques used in Example 6 are applicable only if the exponent on the radicand is a multiple of the index. For expressions such as $\sqrt{x^5}$, other methods are needed.

Quick Reference 9.1

Higher-Order Roots

- For any real numbers a and b and any integer $n > 1$, a is an ***n*th root** of b if and only if $a^n = b$.
- The following generalizations apply to nth roots.
  1. Any real number has one odd root.
  2. A positive number has two even roots that are opposites, but a negative number has no real number even root.
  3. Because $0^n = 0$ for all integers $n > 1$, 0 has one nth root, 0.
- For any even integer $n > 1$ and any nonnegative real number b, $\sqrt[n]{b}$ represents the nonnegative nth root of b.

  For any odd integer $n > 1$ and any real number b, $\sqrt[n]{b}$ represents the nth root of b.
- In the expression $\sqrt[n]{b}$, the symbol $\sqrt{}$ is called the **radical symbol,** b is called the **radicand,** n is called the **index,** and the entire symbol $\sqrt[n]{b}$ is called a **radical.**

Simplifying $\left(\sqrt[n]{b}\right)^n$ and $\sqrt[n]{b^n}$

- Let b represent any real number.
  1. If $\sqrt[n]{b}$ is defined, then $\left(\sqrt[n]{b}\right)^n = b$.
  2. For even integers n, $\sqrt[n]{b^n} = |b|$. For odd integers n, $\sqrt[n]{b^n} = b$.

Speaking the Language 9.1

1. In the expression $\sqrt{x+1}$, the _____ is 2 and $x + 1$ is called the _____.

2. Because $2^4 = 16$ and $(-2)^4 = 16$, we say that 2 and -2 are _____ of 16.

3. Although 25 has two square roots, $\sqrt{25}$ indicates the _____ square root of 25.

4. If b is positive, then the process of writing $\sqrt[n]{b^n}$ as b is called _____ the radical.

Remind students that the instructions for Exercises 3–14 are not the same as those for Exercises 15–30.

Exercises 9.1

Exercise 1 distinguishes between the words *square root* and the meaning of the $\sqrt{\ }$ symbol.

Concepts and Skills

1. Why do the following two questions have different answers?
   (a) What are the square roots of 9?
   (b) What is $\sqrt{9}$?
2. In the expression $\sqrt[n]{b}$, what do we call n and b?

In Exercises 3–14, determine the specified root(s), if any, of the given number.

3. Square root of 36
4. Square root of $\frac{16}{25}$
5. Square root of -4
6. Square root of -25
7. Third root of 27
8. Third root of -64
9. Fourth root of 625
10. Fourth root of 16
11. Fifth root of -243
12. Fifth root of -32
13. Sixth root of -64
14. Sixth root of 64

In Exercises 15–30, evaluate.

15. $\sqrt{16}$
16. $\sqrt{64}$
17. $\sqrt[3]{-64}$
18. $\sqrt[5]{32}$
19. $\sqrt{\frac{4}{9}}$
20. $-\sqrt{\frac{25}{144}}$
21. $\sqrt[4]{-16}$
22. $\sqrt[6]{-64}$
23. $-\sqrt{16}$
24. $-\sqrt{36}$
25. $4\sqrt{25}$
26. $-2\sqrt{9}$
27. $\sqrt[3]{64}$
28. $\sqrt[4]{16}$
29. $-\sqrt[5]{-32}$
30. $-\sqrt[3]{-125}$

In Exercises 31–36, evaluate the given radical. Round results to the nearest hundredth.

31. $\sqrt[3]{343}$
32. $\sqrt[3]{-3.375}$
33. $\sqrt[4]{52}$
34. $\sqrt[6]{5600}$
35. $\sqrt[5]{-0.65}$
36. $\sqrt[3]{15}$
37. Why are absolute value symbols required for one of the following results but not for the other?
$$\sqrt{x^6} = |x^3| \qquad \sqrt{x^4} = x^2$$
38. Describe the conditions under which the following is true.
$$\left(\sqrt[n]{x}\right)^n = \sqrt[n]{x^n} = x$$

In Exercises 39–44, evaluate the radical expression.

39. $\sqrt{(-4)^2}$
40. $\sqrt[4]{(-3)^4}$
41. $\left(\sqrt[4]{7}\right)^4$
42. $\left(\sqrt[5]{-3}\right)^5$
43. $\sqrt[3]{10^3}$
44. $\sqrt[3]{(-5)^3}$

In Exercises 45–48, evaluate each expression.

45. $\sqrt{4^2 - 4(3)(1)}$
46. $\sqrt{9^2 - 4(2)(9)}$
47. $\sqrt{9^2 - 4(10)(-9)}$
48. $\sqrt{19^2 - 4(6)(-7)}$

In Exercises 49–54, translate the given phrase into a radical expression. Assume that all variables represent positive numbers.

49. Twice the square root of x
50. Five less than the cube root of y
51. The difference of the fifth root of a and 9
52. The square root of c plus b
53. The square root of the quantity $c + b$
54. The fourth root of the product of x and y

In Exercises 55–70, determine the indicated root. Assume that all variables represent positive numbers.

In this group, as in other later groups, we assume that variables represent positive numbers so as not to be distracted from the fundamental task. Compare to Exercises 73–86.

55. $\sqrt{x^{10}}$
56. $\sqrt{x^{16}}$
57. $4\sqrt{y^6}$
58. $-3\sqrt{z^8}$
59. $\sqrt[3]{x^{21}}$
60. $\sqrt[3]{y^{12}}$
61. $-\sqrt[3]{y^{15}}$
62. $4\sqrt[3]{a^6}$
63. $\sqrt[5]{x^{35}}$
64. $\sqrt[7]{x^{35}}$
65. $-\sqrt[4]{x^8}$
66. $5\sqrt[6]{y^{18}}$
67. $\sqrt{(3x)^6}$
68. $\sqrt{(5ac)^4}$
69. $\sqrt{(x+3)^{10}}$
70. $\sqrt{(x-2)^{16}}$
71. In each part, determine whether the functions are equivalent and explain why or why not.
   (a) $f(x) = \sqrt{x^2}$, $g(x) = x$
   (b) $f(x) = \sqrt[3]{x^3}$, $g(x) = x$
72. Which of the following radicals cannot be simplified with our present methods? Why?
   (i) $\sqrt{x^{12}}$
   (ii) $\sqrt[3]{x^{12}}$
   (iii) $\sqrt[4]{x^{12}}$
   (iv) $\sqrt[5]{x^{12}}$

In Exercises 73–86, determine the indicated root. Assume that all variables represent any real number.

In this group, no assumption is made about the variables, and absolute value symbols are needed in some of the answers.

73. $\sqrt{x^6}$ **74.** $\sqrt{t^{10}}$

75. $-\sqrt{x^8}$ **76.** $5\sqrt{y^{14}}$

77. $\sqrt[5]{x^{15}}$ **78.** $\sqrt[7]{y^{21}}$

79. $\sqrt[4]{a^4}$ **80.** $\sqrt[6]{y^6}$

81. $\sqrt{(x+2)^2}$ **82.** $\sqrt{(x^2-9)^2}$

83. $-2\sqrt{(x-1)^2}$ **84.** $\sqrt{(3x^3-5x)^2}$

85. $\sqrt[3]{(1-x)^3}$ **86.** $\sqrt[8]{(x-7)^8}$

In Exercises 87–90, insert =, <, or > to describe the correct relationship between the two given quantities. Use your calculator to evaluate the quantities.

87. $\sqrt{7-5} \quad \square \quad \sqrt{7}-\sqrt{5}$

88. $\sqrt{10+3} \quad \square \quad \sqrt{10}+\sqrt{3}$

89. $\sqrt{25-9} \quad \square \quad \sqrt{25}-\sqrt{9}$

90. $\sqrt{9}+\sqrt{16} \quad \square \quad \sqrt{25}$

CE In Exercises 91–96, supply the missing number to make each statement true.

91. $\sqrt{x^{\square}} = x^2$ **92.** $\sqrt[3]{y^{\square}} = y^4$

93. $\sqrt[\square]{y^{24}} = |y^3|$ **94.** $\sqrt[\square]{y^4} = |y|$

95. $\square\sqrt[4]{16} = -6$ **96.** $\square\sqrt[3]{-8} = 2$

Modeling with Real Data

97. Total political action committee (PAC) contributions to candidates rose from \$189.6 million in 1994 to \$219.9 million in 1998. (Source: Federal Election Commission.) The function $P(x) = 148\sqrt[5]{x}$ models PAC contributions (in millions of dollars), where x is the number of years since 1990.

(a) What are the predicted PAC contributions for 2004?

(b) Use the model to predict the percentage increase in PAC contributions from 2000 to 2010.

98. The senior citizen population in Arizona is expected to increase dramatically. From 1993 to 2020, the over-65 age group is predicted to rise from 529,000 to 1,121,000 and the over-85 age group from 46,000 to 146,000. The functions $S(x) = 368\sqrt[3]{x}$ (over 65) and $E(x) = 26.5\sqrt{x}$ (over 85) model the age group populations (in thousands) as functions of the number of years since 1990.

(a) What is the projected population of each age group in 2010? (Round results to the nearest thousand.)

(b) For each age group, what is the projected percent increase from 2000 to 2010?

Data Analysis: *Wind-chill Temperatures*

The following table lists the average January temperatures (in Fahrenheit degrees) and wind speeds (in miles per hour) for four cities. (Source: U.S. National Oceanic and Atmospheric Administration.)

City	Temperature	Wind Speed
Cheyenne, WY	26.5	15.3
Boston, MA	28.6	13.8
Seattle, WA	40.1	9.7
Norfolk, VA	39.1	8.1

The U.S. Meteorological Service uses the following formula to calculate wind-chill temperatures.

$$Y = 91.4 - (91.4 - T)\left[0.478 + 0.301\left(\sqrt{x} - 0.02x\right)\right]$$

In this formula, T is the actual air temperature in Fahrenheit degrees, x is the wind speed in miles per hour, and Y is the wind-chill temperature.

99. Use the formula to compute the average January wind-chill temperatures for each of the four cities in the table.

100. Suppose that the temperature is 32°F. Replacing T with 32 and simplifying results in the formula

$$Y = 63.007 - 17.879\sqrt{x} + 0.358x$$

which gives the wind-chill temperature as a function of wind speed. Use the graph of this function (see figure) to estimate the wind speed for which the wind-chill temperature would be 0°F.

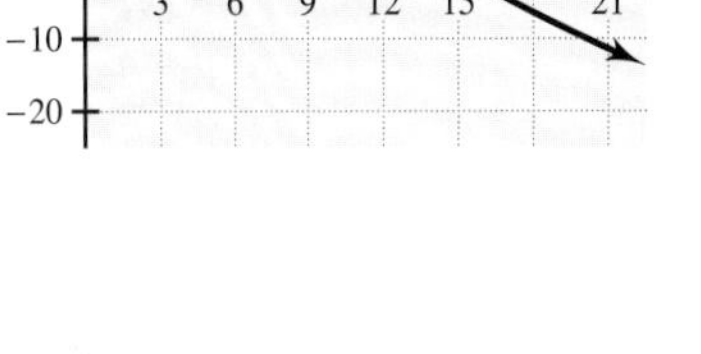

101. Suppose that a wind speed of 9 mph in Norfolk creates a wind-chill temperature of 20°F. Substitute these values into the wind-chill formula and solve the equation for T to determine the actual air temperature.

102. When the wind speed is 3 mph, Y and T are approximately equal. Does this mean that when the wind speed is less than 3 mph, it will feel warmer than it actually is? What restrictions would you place on x to make the wind-chill formula valid?

Challenge

In Exercises 103–108, evaluate the given expression, if possible.

103. $\sqrt[3]{-\sqrt[4]{1}}$

104. $\sqrt[4]{-\sqrt[3]{1}}$

105. $\sqrt[3]{\sqrt{64}}$

106. $\sqrt[3]{-\sqrt{64}}$

107. $\sqrt{\sqrt[3]{64}}$

108. $\sqrt{-\sqrt[3]{64}}$

109. We define $\sqrt[n]{b}$ for $n > 1$.

(a) Why do we exclude $n = 1$? What meaning would $\sqrt[1]{b}$ have?

(b) Why do we exclude $n = 0$? What meaning would $\sqrt[0]{b}$ have? (*Hint:* Consider the two cases $b = 1$ and $b \neq 1$.)

9.2 Rational Exponents

Definition of $b^{1/n}$ • Definition of $b^{m/n}$ • Negative Rational Exponents

Definition of $b^{1/n}$

Students usually benefit from an overview. Once we decided to define the zero and negative exponents, we had already departed from the original intent of an exponent as a shorthand way of writing repeated multiplication. The idea of a rational exponent is simply a next step in the evolution of exponents.

Initially, we used an exponent as a convenient way to write repeated multiplication: $5 \cdot 5 \cdot 5 \cdot 5 = 5^4$. However, to perform operations with exponents, we had to define a zero exponent and a negative exponent.

In this section we extend the definition of exponent to include all rational numbers. An application of the Power to a Power Rule extended to rational exponents provides the basis for the definition.

$$(9^{1/2})^2 = 9^{2 \cdot (1/2)} = 9^1 = 9$$

Because $(9^{1/2})^2 = 9$, the definition of square root implies that $9^{1/2}$ is a square root of 9.

In general, $(b^{1/n})^n = b^{n \cdot (1/n)} = b^1 = b$. Thus, by the definition of nth root, $b^{1/n}$ is an nth root of b.

Students often find rational exponents to be an abstract concept. Stress that they are just another way of writing roots of numbers.

Definition of $b^{1/n}$

For any integer $n > 1$ and for any real number b for which $\sqrt[n]{b}$ is defined, $b^{1/n}$ is defined as $\sqrt[n]{b}$.

Note: Observe that the denominator of the exponent is the index of the radical. Also, the definition implies that b cannot be negative when n is even.

EXAMPLE 1

Evaluating Expressions with Rational Exponents

(a) $25^{1/2} = \sqrt{25} = 5$ — $5^2 = 25$

(b) $(-25)^{1/2}$ is not a real number. — n is even and $b < 0$.

(c) $-25^{1/2} = -\sqrt{25} = -5$ — $-1 \cdot \sqrt{25} = -1 \cdot 5 = -5$

(d) $\left(\frac{4}{9}\right)^{1/2} = \sqrt{\frac{4}{9}} = \frac{2}{3}$ — $\left(\frac{2}{3}\right)^2 = \frac{4}{9}$

(e) $8^{1/3} = \sqrt[3]{8} = 2$ — $2^3 = 8$

(f) $(-8)^{1/3} = \sqrt[3]{-8} = -2$ — $(-2)^3 = -8$

(g) $256^{1/4} = \sqrt[4]{256} = 4$ — $4^4 = 256$

(h) $32^{1/5} = \sqrt[5]{32} = 2$ — $2^5 = 32$

In Section 9.1 we calculated higher-order roots with the $\sqrt[x]{\ }$ key. If your calculator does not have such a key, you can calculate a higher-order root by writing the radical as an expression with a rational exponent.

EXAMPLE 2

Finding Higher-Order Roots with a Calculator

Write each radical in exponential form and use your calculator to evaluate the resulting expression. Round the results to two decimal places.

(a) $\sqrt[5]{16{,}807}$ (b) $\sqrt[4]{10}$ (c) $\sqrt[7]{-17}$

Solution

(a) $\sqrt[5]{16{,}807} = 16{,}807^{1/5} = 7$

(b) $\sqrt[4]{10} = 10^{1/4} \approx 1.78$

(c) $\sqrt[7]{-17} = (-17)^{1/7} \approx -1.50$

Although $(-17)^{1/7} = -17^{1/7}$, it is good practice always to enclose negative bases in parentheses.

Definition of $b^{m/n}$

So far, the rational exponents we have used have had 1 as the numerator. However, extending the Power to a Power Rule to rational exponents suggests a logical definition of *any* rational exponent.

Although the Power to a Power Rule is not officially stated until the next section, anticipating it serves as a rationale for the definition.

Suppose m and n are positive integers with no common factor except 1. Then for all real numbers b for which $b^{1/n}$ is defined,

$$b^{m/n} = (b^{1/n})^m \qquad \text{Power to a Power Rule}$$

$$= \left(\sqrt[n]{b}\right)^m \qquad \text{Definition of } b^{1/n}$$

LEARNING TIP

Often the best approach to evaluating $b^{m/n}$ is to find the nth root of b first. Then raise the result to the mth power.

Definition of $b^{m/n}$

If m and n are positive integers with no common factor other than 1, then $b^{m/n} = \left(\sqrt[n]{b}\right)^m$ and $b^{m/n} = \sqrt[n]{b^m}$ for all real numbers b for which $b^{1/n}$ is defined.

Emphasize these roles of m and n.

Note that when we write $b^{m/n}$ as a radical, n indicates the index of the radical and m indicates the power to which the radical is raised.

EXAMPLE 3

Evaluating Expressions with Rational Exponents

(a) $8^{2/3} = \left(\sqrt[3]{8}\right)^2 = (2)^2 = 4$

(b) $(-32)^{4/5} = \left(\sqrt[5]{-32}\right)^4 = (-2)^4 = 16$ With parentheses, the base is -32.

(c) $-32^{4/5} = -\left(\sqrt[5]{32}\right)^4 = -1 \cdot (2)^4 = -16$ Without parentheses, the base is 32.

(d) $36^{3/2} = \left(\sqrt{36}\right)^3 = (6)^3 = 216$

Think About It

Consider $(-1)^{2/4}$ and explain why $b^{m/n} = \left(\sqrt[n]{b}\right)^m$ only if m and n have no common factors.

Exponential expressions with numerical bases can be evaluated without a calculator as long as the roots and powers are familiar. In Example 4 we do need a calculator.

EXAMPLE 4

Using a Calculator to Evaluate Exponential Expressions

Evaluate the following expressions to two decimal places.

(a) $6^{2/3}$ (b) $(-6)^{2/3}$ (c) $6^{3/2}$ (d) $(-6)^{3/2}$

Solution

(a) $6^{2/3} \approx 3.30$

(b) $(-6)^{2/3} \approx 3.30$

(c) $6^{3/2} \approx 14.70$

(d) The expression $(-6)^{3/2}$ is not a real number because $(-6)^{1/2}$ is not defined.

Even if we cannot evaluate an expression, we can use the definition of $b^{m/n}$ to write the expression in the form of a radical.

EXAMPLE 5

Writing Exponential Expressions as Radicals

Write the exponential expressions as radicals.

(a) $x^{3/5} = \left(\sqrt[5]{x}\right)^3$ or $\sqrt[5]{x^3}$

(b) $(5x)^{3/4} = \sqrt[4]{(5x)^3} = \sqrt[4]{125x^3}$ The base of the exponent is $5x$.

(c) $5x^{3/4} = 5\sqrt[4]{x^3}$ The base of the exponent is x.

Compare parts (b) and (c). Make sure students understand why 5 is not part of the radicand in part (c).

Negative Rational Exponents

We can extend the definition of a negative integer exponent to include negative rational exponents.

Definition of a Negative Rational Exponent

If m and n are positive integers with no common factor other than 1, and if $b \neq 0$, then $b^{-m/n} = \dfrac{1}{b^{m/n}}$ for all real numbers b for which $b^{1/n}$ is defined.

EXAMPLE 6

In part (b), emphasize that only the sign of the exponent changes, not the sign of the base.

Evaluating Exponential Expressions with Negative Exponents

Evaluate each of the following.

(a) $16^{-3/4}$ (b) $(-125)^{-2/3}$ (c) $\left(\dfrac{16}{25}\right)^{-1/2}$

Solution

The rule $\left(\dfrac{a}{b}\right)^{-\frac{m}{n}} = \left(\dfrac{b}{a}\right)^{\frac{m}{n}}$ was introduced in Section 6.2. Although we do not formally extend the properties of exponents to rational exponents until the next section, this provides a straightforward way to evaluate part (c).

(a) $16^{-3/4} = \dfrac{1}{16^{3/4}} = \dfrac{1}{\left(\sqrt[4]{16}\right)^3} = \dfrac{1}{2^3} = \dfrac{1}{8}$

(b) $(-125)^{-2/3} = \dfrac{1}{(-125)^{2/3}} = \dfrac{1}{\left(\sqrt[3]{-125}\right)^2} = \dfrac{1}{(-5)^2} = \dfrac{1}{25}$

(c) $\left(\dfrac{16}{25}\right)^{-1/2} = \left(\dfrac{25}{16}\right)^{1/2} = \sqrt{\dfrac{25}{16}} = \dfrac{5}{4}$ $\quad \left(\dfrac{a}{b}\right)^{-m/n} = \left(\dfrac{b}{a}\right)^{m/n}$

Quick Reference 9.2

Definition of $b^{1/n}$

- For any integer $n > 1$ and for any real number b for which $\sqrt[n]{b}$ is defined, $b^{1/n}$ is defined as $\sqrt[n]{b}$.
- If n is an odd integer, then b can be any real number. If n is an even integer, then b must be nonnegative.
- By writing a radical as an expression with a rational exponent, we can use a calculator to find higher-order roots.

Definition of $b^{m/n}$

- If m and n are positive integers with no common factor other than 1, then for all real numbers b for which $b^{1/n}$ is defined $b^{m/n} = \left(\sqrt[n]{b}\right)^m$ and $b^{m/n} = \sqrt[n]{b^m}$.
- The denominator n indicates the index of the radical, and the numerator m indicates the power to which the radical is raised.

Negative Rational Exponents

- If m and n are positive integers with no common factor other than 1, and if $b \neq 0$, then $b^{-m/n} = \dfrac{1}{b^{m/n}}$ for all real numbers b for which $b^{1/n}$ is defined.

Speaking the Language 9.2

1. In the expression $b^{1/n}$, we call $1/n$ a(n) ________ exponent.
2. When we write $b^{1/n}$ as a radical, n is the ________ of the radical.
3. To evaluate $4^{3/2}$, we take the square root of 4 and then ________.
4. When we apply the definition of a negative exponent to $(-3)^{-3/5}$, the sign of the ________ changes, but the sign of the ________ does not.

Exercises 9.2

Concepts and Skills

1. One of the following is false. Identify the false statement and explain why it is not true.
   (i) The square roots of 16 are 4 and −4.
   (ii) $\sqrt{16} = 4$
   (iii) $16^{1/2} = 4$ or -4
   (iv) $64^{1/3} = 4$

2. Only one of the following expressions represents a real number. Identify and evaluate this expression. Then explain why the other expressions do not represent real numbers.
   (i) $(-4)^{1/2}$ (ii) $(-4)^{-1/2}$
   (iii) $4^{-1/2}$ (iv) $\left(-\frac{1}{4}\right)^{-1/2}$

In Exercises 3–18, evaluate the given expression.

3. $49^{1/2}$ **4.** $(-32)^{1/5}$
5. $(-64)^{1/3}$ **6.** $81^{1/4}$
7. $(-16)^{1/4}$ **8.** $(-36)^{1/2}$
9. $\left(\frac{8}{27}\right)^{1/3}$ **10.** $\left(\frac{1}{32}\right)^{1/5}$
11. $(-27)^{1/3}$ **12.** $16^{1/4}$
13. $-32^{1/5}$ **14.** $-64^{1/3}$
15. $-(-625)^{1/4}$ **16.** $(-64)^{1/2}$
17. $-625^{1/4}$ **18.** $-36^{1/2}$

In Exercises 19–24, evaluate the given expression. Round results to two decimal places.

19. $15^{1/4}$ **20.** $23^{1/3}$
21. $(-12)^{1/2}$ **22.** $(-7)^{1/4}$
23. $-4^{1/4}$ **24.** $-20^{1/2}$

25. When you convert $9^{2/5}$ to a radical, how do you interpret the numerator and denominator of the exponent?

26. Which of the following is the correct way to write $\left(\sqrt[3]{2}\right)^5$ in exponential form? Use the words *index*, *base*, and *power* to explain how you know.
(i) $2^{5/3}$ (ii) $5^{2/3}$ (iii) $3^{5/2}$

In Exercises 27–34, write the given exponential expression as a radical. Assume all variables represent positive numbers.

27. $3x^{2/3}$ **28.** $-4y^{5/6}$
29. $(3x)^{2/3}$ **30.** $(2ab)^{3/4}$
31. $(2x + 3)^{2/3}$ **32.** $(t + 2)^{5/4}$
33. $3x^{1/2} + (2y)^{1/2}$ **34.** $(-2a)^{1/3} - 2b^{1/3}$

In Exercises 35–44, write the given radical expression in exponential form. Assume all variables represent positive numbers.

35. $\sqrt[3]{y^2}$ **36.** $\sqrt[5]{z^3}$
37. $\sqrt[4]{3x^3}$ **38.** $\sqrt[5]{7x^2y}$
39. $\sqrt{x^2 + 4}$ **40.** $\sqrt{x + 3}$
41. $\sqrt{t} - \sqrt{5}$ **42.** $\sqrt{2y} + 1$
43. $\sqrt[3]{(2x - 1)^2}$ **44.** $\sqrt[4]{(x + 1)^3}$

In Exercises 45–50, evaluate each quantity. Round your results to two decimal places.

45. $\sqrt[5]{-90}$ **46.** $\sqrt[4]{20}$
47. $-\sqrt[6]{740}$ **48.** $4\sqrt[7]{4020}$
49. $\sqrt[8]{230^5}$ **50.** $\sqrt[7]{49^3}$

51. Describe two ways to use a calculator to evaluate $\sqrt[5]{470}$.

52. Explain why $8^{2/(-3)}$ can be written as $\left(\sqrt[3]{8}\right)^{-2}$.

In Exercises 53–74, evaluate the given expression.

53. $9^{-1/2}$ **54.** $(-27)^{-1/3}$
55. $\left(\frac{1}{9}\right)^{-1/2}$ **56.** $\left(\frac{1}{27}\right)^{-1/3}$

Remind students that they cannot directly evaluate an expression with a negative exponent. They must change to a positive exponent first.

57. $27^{2/3}$ **58.** $36^{3/2}$
59. $(-27)^{-2/3}$ **60.** $-16^{3/2}$
61. $32^{3/5}$ **62.** $25^{3/2}$
63. $(-16)^{-1/2}$ **64.** $(-4)^{3/2}$
65. $8^{-2/3}$ **66.** $8^{-4/3}$
67. $\left(\frac{8}{27}\right)^{-2/3}$ **68.** $\left(\frac{32}{243}\right)^{-4/5}$
69. $-(-8)^{-4/3}$ **70.** $(-8)^{5/3}$
71. $(-36)^{5/2}$ **72.** $-64^{5/6}$
73. $-4^{-5/2}$ **74.** $(-8)^{-2/3}$

In Exercises 75–80, evaluate the given expression.

75. $16^{1/2} + 49^{1/2}$ **76.** $16^{3/2} - 16^{3/4}$
77. $27^{1/3} + 4^{-1/2}$ **78.** $81^{-3/4} + 81^{-3/2}$
79. $(-8)^{2/3}(16^{1/4})$ **80.** $(9^{-3/2})(-27^{1/3})$

CE In Exercises 81–84, evaluate the given expression.

81. $16^{0.25}$ **82.** $(-27)^{0.\overline{3}}$

83. $9^{-0.5}$ **84.** $49^{1.5}$

In Exercises 85–90, evaluate the given expression. Round results to two decimal places.

85. $16^{2/3}$ **86.** $16^{-2/3}$ **87.** $\left(\frac{5}{9}\right)^{5/9}$

88. $10^{4/5}$ **89.** $(-5)^{-2/5}$ **90.** $(-9)^{2/3}$

CE In Exercises 91–96, supply the missing number to make each statement true.

91. $(\square)^{-1/2} = \frac{1}{4}$ **92.** $(\square)^{-1/4} = \frac{1}{2}$

93. $(\square)^{5/3} = -32$ **94.** $(\square)^{2/3} = 9$

95. $(-27)^{(\square)} = 9$ **96.** $(-8)^{(\square)} = \frac{1}{16}$

CE **97.** The figure shows the graphs of the three functions $f(x) = x^{1/2}$, $g(x) = x^{1/4}$, and $h(x) = x^{1/8}$.

(a) Where do the three graphs intersect? Why would *all* graphs of $y = x^{1/n}$, $n > 1$, intersect at these points?

(b) For what values of x is $x^{1/2} < x^{1/4} < x^{1/8}$?

(c) For what values of x is $x^{1/2} > x^{1/4} > x^{1/8}$?

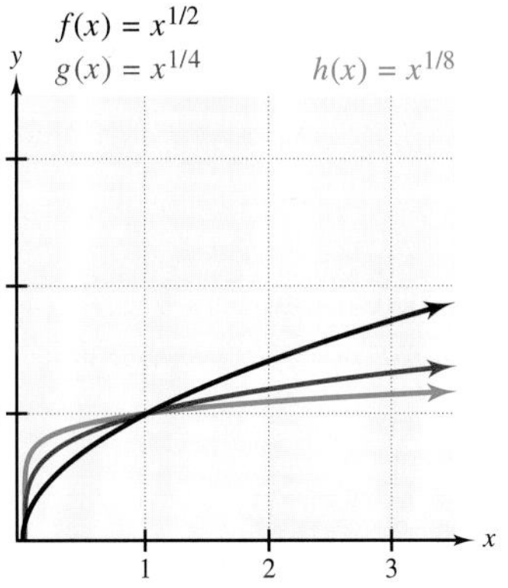

98. Show that $16^{(1/2)/(2/3)} = 8$.

Real-Life Applications

99. A researcher has developed a mathematical model for predicting a final exam grade: $G = 60T^{2/9}$. In this model, G is the predicted grade, and T is the number of hours spent studying for the exam. Create a table of T-values from 1 to 10 and calculate the corresponding G-values, rounded to the nearest integer.

(a) If the passing grade is 70, how many hours of study are needed to pass the exam?

(b) How many additional hours of study would be needed to earn a 92 on the exam?

(c) What grade is predicted if a student does not study at all? According to the model, can a student earn a 100? How realistic is the model at these extreme points of the graph?

100. Suppose that the Department of Labor uses a mathematical model to relate annual federal funding F (in billions of dollars) for jobs creation to the unemployment rate R. The function F is given by $F = 10R^{-1/3}$. (If the current unemployment rate is 6%, $R = 6$.)

(a) Approximately what amount of annual funding would be needed to maintain an unemployment rate of 5%?

(b) What funding would be required to reduce the unemployment rate to 3.18%?

(c) Why is the model not valid for zero funding?

Modeling with Real Data

101. Over 70% of taxpayers file their tax returns early—that is, at least a month before the tax deadline. The accompanying bar graph shows the percentage of various age groups who file early. (Source: Bruskin/Goldring Research.)

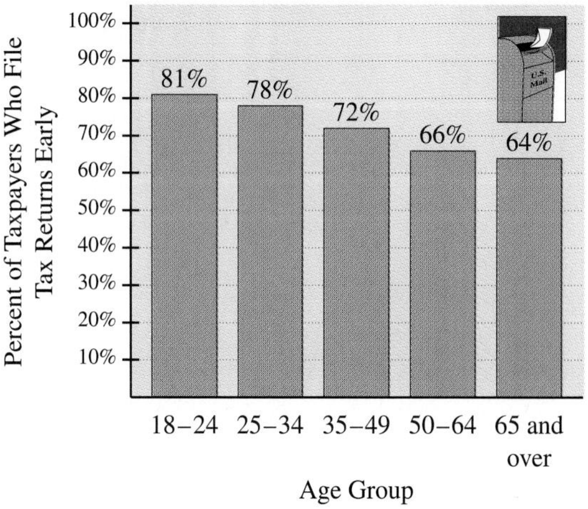

The function $f(x) = 154x^{-1/5}$ models the percentage of those age x who are expected to file early.

(a) What percentage of those age 25 are expected to file early?

(b) The data indicate that 78% of those in the 25–34 group file early. According to the model, what is the difference in the percentages for those age 25 and those age 34?

102. The number of music CDs shipped by manufacturers increased from 22.6 million in 1985 to 847 million in 1998. (Source: Recording Industry Association of America.) The function $C(x) = 5.2x^{9/5}$ models the number (in millions) of CDs shipped as a function of the number of years since 1985.

(a) According to the model, how many CDs are projected to be shipped in 2004?

(b) What is the predicted increase in the number of CDs shipped from 2002 to 2004?

Challenge

In Exercises 103–106, evaluate.

103. $\sqrt{16^{-1/2}}$

104. $\left(\sqrt{81}\right)^{-1/2}$

105. $\left(\sqrt[3]{64}\right)^{-1/2}$

106. $\sqrt[3]{729^{-1/2}}$

107. Consider the function $f(x) = (-2)^{1/x}$. Describe the integer values of x that are in the domain of the function.

108. Consider $f(x) = (x^2)^{1/2}$ and $g(x) = (x^3)^{1/3}$. For each function, create a table of values for integer values of x between -5 and 5. Then use your tables to sketch the graphs of functions f and g. On the basis of your graphs, for what values of x are the functions (a) equal and (b) not equal?

9.3 Properties of Rational Exponents

Simplifying Exponential Expressions • Simplifying Radical Expressions

Simplifying Exponential Expressions

The properties of exponents that we stated for integer exponents in Sections 6.1 and 6.2 are valid for rational exponents, provided the expressions are defined.

Properties of Rational Exponents

Let m and n be rational numbers. For nonzero real numbers a and b for which the expressions are defined, the following are properties of exponents.

$a^m a^n = a^{m+n}$	Product Rule for Exponents
$\dfrac{a^m}{a^n} = a^{m-n}$	Quotient Rule for Exponents
$a^{-n} = \dfrac{1}{a^n}$	Definition of Negative Exponent
$(a^m)^n = a^{mn}$	Power to a Power Rule
$(ab)^n = a^n b^n$	Power of a Product Rule
$\left(\dfrac{a}{b}\right)^n = \dfrac{a^n}{b^n}$	Power of a Quotient Rule
$\left(\dfrac{a}{b}\right)^{-n} = \left(\dfrac{b}{a}\right)^n$	Negative Power of a Quotient Rule

EXAMPLE 1

Evaluating Exponential Expressions

Use the properties of exponents to evaluate the expression.

(a) $(5^6)^{1/2}$ (b) $2^{1/2} \cdot 8^{1/2}$

(c) $\left(\dfrac{3^6}{5^3}\right)^{-2/3}$ (d) $16^{1/3} \cdot 16^{5/12}$

Solution

(a) $(5^6)^{1/2} = 5^{6 \cdot (1/2)}$ Power to a Power Rule

$= 5^3 = 125$

(b) $2^{1/2} \cdot 8^{1/2} = (2 \cdot 8)^{1/2}$ Power of a Product Rule

$= 16^{1/2} = \sqrt{16} = 4$

(c) $\left(\dfrac{3^6}{5^3}\right)^{-2/3} = \dfrac{3^{6\cdot(-2/3)}}{5^{3\cdot(-2/3)}}$ Power of a Quotient Rule

$= \dfrac{3^{-4}}{5^{-2}} = \dfrac{5^2}{3^4} = \dfrac{25}{81}$

LEARNING TIP

You can change negative exponents to positive exponents at the beginning or at the end of the problem. You will develop a preference as you gain experience.

The following is an alternative method.

$\left(\dfrac{3^6}{5^3}\right)^{-2/3} = \left(\dfrac{5^3}{3^6}\right)^{2/3}$ $\left(\dfrac{a}{b}\right)^{-n} = \left(\dfrac{b}{a}\right)^{n}$

$= \dfrac{5^{3\cdot(2/3)}}{3^{6\cdot(2/3)}}$ Power of a Quotient Rule

$= \dfrac{5^2}{3^4} = \dfrac{25}{81}$

Compare the products in parts (b) and (d). When the exponents are the same, we can multiply the bases; when the bases are the same, we add the exponents.

(d) $16^{1/3} \cdot 16^{5/12} = 16^{(1/3)+(5/12)}$ Product Rule for Exponents

$= 16^{(4/12)+(5/12)}$

$= 16^{9/12} = 16^{3/4} = \left(\sqrt[4]{16}\right)^3 = 2^3 = 8$

Recall that $\sqrt[n]{b^n} = |b|$ if n is even and $\sqrt[n]{b^n} = b$ if n is odd. But $\sqrt[n]{b^n} = (b^n)^{1/n}$. Thus we can state the following generalization.

> **Simplifying Expressions of the Form $(b^n)^{1/n}$**
>
> For an even positive integer n, $(b^n)^{1/n} = |b|$.
>
> For an odd integer n, $n > 1$, $(b^n)^{1/n} = b$.

EXAMPLE 2

Simplifying Expressions of the Form $(b^n)^{1/n}$

(a) $[(-3)^2]^{1/2} = |-3| = 3$ Because n is even, absolute value symbols are needed.

(b) $[(-3)^5]^{1/5} = -3$ Because n is odd, absolute value symbols are not used.

(c) $(y^3)^{1/3} = y$ Because n is odd, absolute value symbols are not used.

(d) $(z^8)^{1/8} = |z|$ Because n is even, absolute value symbols are needed.

Unless otherwise indicated, the variables in the remainder of this section will represent positive numbers. This will allow us to perform operations with exponential and radical expressions without regard to absolute value in the results.

EXAMPLE 3

Simplifying Exponential Expressions

Use the properties of exponents to simplify each of the following. Write all results with positive exponents.

(a) $\dfrac{x^{1/3}}{x^{4/3}} = x^{1/3 - 4/3}$ Quotient Rule for Exponents

$= x^{-3/3} = x^{-1} = \dfrac{1}{x}$ Definition of negative exponent

(b) $y^{1/3} \cdot y^{1/6} = y^{1/3 + 1/6}$ Product Rule for Exponents

$= y^{2/6 + 1/6} = y^{3/6} = y^{1/2}$

(c) $(a^3b^{-1})^{-1/2} = a^{3 \cdot (-1/2)}b^{-1 \cdot (-1/2)}$ Power of a Product Rule

$= a^{-3/2}b^{1/2} = \dfrac{b^{1/2}}{a^{3/2}}$ Definition of negative exponent

(d) $\left(\dfrac{x^{1/2}}{y^3}\right)^{-1/3} = \left(\dfrac{y^3}{x^{1/2}}\right)^{1/3}$ $\left(\dfrac{a}{b}\right)^{-n} = \left(\dfrac{b}{a}\right)^n$

$= \dfrac{y^{3 \cdot (1/3)}}{x^{(1/2) \cdot (1/3)}}$ Power of a Quotient Rule

$= \dfrac{y}{x^{1/6}}$

(e) $a^{1/3}(a^{5/3} - a^{-2/3}) = a^{1/3} \cdot a^{5/3} - a^{1/3} \cdot a^{-2/3}$ Distributive Property

$= a^{1/3 + 5/3} - a^{1/3 - 2/3}$ Product Rule for Exponents

$= a^{6/3} - a^{-1/3}$

$= a^2 - \dfrac{1}{a^{1/3}}$ Definition of negative exponent

Think About It

Factor $x - 9$ by writing x as $(x^{1/2})^2$. Use rational exponents to factor $x^{2/3} - 4$.

Simplifying Radical Expressions

If the index of a radical and the exponent of the radicand have a common factor, the expression can be written with a smaller index. We call this process **reducing the index.**

As we will see in Example 4, the technique begins with writing the radical in exponential form.

EXAMPLE 4

Reducing the Index of a Radical

Write the given expression with a smaller index.

(a) $\sqrt[12]{9^6} = 9^{6/12} = 9^{1/2} = \sqrt{9} = 3$

(b) $\sqrt[15]{t^{10}} = t^{10/15} = t^{2/3} = \sqrt[3]{t^2}$

(c) $\sqrt[6]{8y^3} = \sqrt[6]{(2y)^3} = (2y)^{3/6} = (2y)^{1/2} = \sqrt{2y}$

An alternative approach is to factor the index and the exponent. Then $\sqrt[kn]{b^{km}} = \sqrt[n]{b^m}$.

The properties of exponents can also be used to write a term with more than one radical as a term involving only one radical.

EXAMPLE 5 **Reducing the Number of Radicals in a Term**

Write each expression with only one radical symbol.

(a) $\sqrt[3]{\sqrt{2}} = \left(2^{1/2}\right)^{1/3} = 2^{1/6} = \sqrt[6]{2}$

(b) $\sqrt[3]{x} \cdot \sqrt[4]{x} = x^{1/3} \cdot x^{1/4} = x^{4/12} \cdot x^{3/12} = x^{7/12} = \sqrt[12]{x^7}$

Quick Reference 9.3

Simplifying Exponential Expressions

- The properties of exponents that we stated for integer exponents in Sections 6.1 and 6.2 are valid for rational exponents, provided the expressions are defined.
- Expressions of the form $(b^n)^{1/n}$, where n is an integer, $n > 1$, can be simplified according to the following rules.

  If n is even, $(b^n)^{1/n} = |b|$.

  If n is odd, $(b^n)^{1/n} = b$.

Simplifying Radical Expressions

- If the index of a radical and the exponent of the radicand have a common factor, we can **reduce the index.**
  1. Write the radical in exponential form and reduce the exponent to lowest terms.
  2. Write the exponential expression as a radical.
- The properties of exponents can be used to write a term with more than one radical as a term involving only one radical.

Speaking the Language 9.3

1. We can use the ________ Rule to simplify $(x^{-1/3})^6$.
2. For rational exponents, the Product and Quotient Rules for Exponents apply only if ________.
3. We can reduce the ________ of $\sqrt[n]{b^m}$ if m and n have a common factor.
4. To write $\sqrt[5]{\sqrt{3}}$ as an expression with only one radical symbol, we begin by writing the expression with ________.

Exercises 9.3

Concepts and Skills

1. If a and b are positive real numbers, only one of the following statements is true.

   (i) $(a^2b^2)^{1/2} = ab$ (ii) $(a^2 + b^2)^{1/2} = a + b$

   Identify the statement that is true and state a rational exponent rule that supports your claim. For the statement that is not true, select values for a and b for which the left and right sides are not equal.

2. Explain why it is easier to simplify $[(x^{1/3})^{1/5}]^{1/2}$ than it is to simplify $x^{1/3} \cdot x^{1/5} \cdot x^{1/2}$.

In Exercises 3–22, use the properties of exponents to evaluate the given expression.

3. $(6^3)^{2/3}$
4. $(3^{-6})^{1/3}$
5. $(2^{2/7})^{14}$
6. $(3^{2/5})^{10}$
7. $12^{1/2} \cdot 3^{1/2}$
8. $2^{2/3} \cdot 4^{2/3}$
9. $\left(\dfrac{7^3}{2^6}\right)^{-2/3}$
10. $\left(\dfrac{5^6}{3^9}\right)^{1/3}$
11. $65^{1/3} \cdot 65^{2/3}$
12. $7^{1/4} \cdot 7^{-1/4}$
13. $49^{7/10} \cdot 49^{-1/5}$
14. $16^{-1} \cdot 16^{1/2}$
15. $8^{2/9} \cdot 8^{1/9}$
16. $16^{5/8} \cdot 16^{-3/8}$
17. $6^{0.6} \cdot 6^{0.4}$
18. $5^{0.22} \cdot 5^{1.78}$
19. $\dfrac{27^{1/3}}{27^{2/3}}$
20. $\dfrac{16^{-3/8}}{16^{-1/8}}$
21. $8^{1/6} \div 8^{1/2}$
22. $25^{1/4} \div 25^{3/4}$

23. Which of the following is true? Explain your choice.

(i) $x^{2/5} \cdot x^{5/2} = x^1 = x$ (ii) $x^{2/5} \cdot x^{5/2} = x^{29/10}$

24. Describe the conditions under which each of the following is true.

(a) $(x^n)^{1/n} = x$ (b) $(x^n)^{1/n} = |x|$

In Exercises 25–40, use the properties of exponents to simplify the given expression. Write the result with only positive exponents.

25. $x^{2/3} \cdot x^{1/3}$
26. $a^{-2/5} \cdot a^{4/5}$
27. $t \cdot t^{5/7}$
28. $b^{-1/3} \cdot b^{1/4} \cdot b^{5/6}$
29. $(a^{-1/2}b)(a^{3/4}b^{1/2})$
30. $(x^{1/3}y^{-1/3})(x^{2/3}y^{-2/3})$
31. $\dfrac{z^{1/2}}{z^{1/3}}$
32. $\dfrac{m^{1/2}}{m^2}$
33. $\dfrac{a^{1/3}b}{b^{-1/3}}$
34. $\dfrac{x^{1/2}y^{-2}}{xy^{1/2}}$
35. $(a^{-2})^{1/2}$
36. $(y^{-3/4})^{-4}$
37. $(y^{2/3})^{3/2}$
38. $(z^{3/4})^{-4/3}$
39. $(a^6b^4)^{3/2}$
40. $\left(\dfrac{a^6}{b^8}\right)^{-1/2}$

CE In Exercises 41–44, supply the missing number.

41. $(9x^4)^{(\square)} = 3x^2$
42. $x^{1/2} \cdot x^{(\square)} = x^{3/2}$
43. $(x^{2/3})^{(\square)} = x$
44. $\dfrac{8^{2/3}}{8^{(\square)}} = 32$

45. Explain why $\sqrt[6]{x^4}$ can be simplified but $\sqrt[6]{x^5}$ cannot.

46. If $(x^{1/n})^n = x$, explain what must be true about x and n.

In Exercises 47–54, simplify the expression. All variables represent any real number.

This group requires absolute value symbols.

47. $\left[(-2)^6\right]^{1/2}$
48. $\left[(-3)^3\right]^{1/3}$
49. $(a^{15})^{1/5}$
50. $(x^4)^{1/4}$
51. $(a^4b^2)^{1/2}$
52. $(x^3y^6)^{1/3}$
53. $\left[(x+y)^2\right]^{1/2}$
54. $\left[(2x-1)^2\right]^{1/2}$

In Exercises 55–72, use the properties of exponents to simplify the given expression. Write the result with only positive exponents.

55. $(a^{-4}b^{1/3})^{1/2}$
56. $(x^2y^{-1})^{-1/2}$
57. $(a^6b^{-9})^{2/3}$
58. $(25a^4b^6)^{1/2}$
59. $(-3^{2/3}x^{1/2}y^{-1/3})^3$
60. $(x^{3/4}y^{-3/8}z^{-1/4})^8$
61. $\left(\dfrac{a^{4/5}}{b^{8/9}}\right)^{-15/4}$
62. $\left(\dfrac{x^6}{y^{-3/2}}\right)^{-2/3}$
63. $\left(\dfrac{x^{-1/3}}{y^{-1/2}}\right)^{-6}$
64. $\left(\dfrac{a^{-4}}{b^{-2}}\right)^{-5/2}$
65. $\left(\dfrac{16x^{-12}y^8}{z^4}\right)^{1/4}$
66. $\left(\dfrac{27x^{-12}y^9}{z^{-6}}\right)^{1/3}$
67. $(a^{-2/3}b^3)^6(a^{5/7}b^{-2})^7$
68. $(-27x^9)^{1/3}(2x^{-1/2})^2$
69. $\dfrac{(a^{5/6}b^{-2/5}c^{7/60})^{30}}{a^{21}b^{-5}}$
70. $\dfrac{(x^{3/4}y^{-5/6}z^{-5/24})^{12}}{x^7z^{1/2}}$
71. $\dfrac{(9x^4y^{-2})^{-1/2}}{(x^6y^{-3})^{1/3}}$
72. $\left(\dfrac{x^{-2}y^{2/3}}{z^{-1}}\right)\left(\dfrac{8x^3y}{z^6}\right)^{1/3}$

In Exercises 73–80, find the product.

73. $a^{3/4}(a^{5/4} + a^{-3/4})$
74. $y^{-2/3}(y^{5/3} + y^{2/3})$
75. $x^{-1/4}(x^{9/4} - x^{2/3})$
76. $b^{7/5}(b^{-2/5} - b^{4/15})$
77. $(x^{1/2} + y^{1/2})(x^{1/2} - y^{1/2})$
78. $(3^{1/2} + 2^{1/2})(3^{1/2} - 2^{1/2})$
79. $(x^{1/2} - 3^{1/2})^2$
80. $(x^{1/2} + y^{1/2})^2$

In Exercises 81–88, reduce the index.

81. $\sqrt[6]{x^3}$
82. $\sqrt[4]{x^2}$
83. $\sqrt[9]{8x^3}$
84. $\sqrt[8]{81x^4}$
85. $\sqrt[10]{x^4y^6}$
86. $\sqrt[8]{x^2y^4}$
87. $\left(\sqrt[4]{a}\right)^2$
88. $\left(\sqrt[6]{a}\right)^2$

In Exercises 89–96, write the given expression with just one radical symbol.

89. $\sqrt[3]{x^2}\sqrt[4]{x}$ **90.** $\sqrt[3]{x}\sqrt[5]{x}$

91. $\dfrac{\sqrt[4]{x}}{\sqrt[3]{x^2}}$ **92.** $\dfrac{x}{\sqrt[3]{x}}$

93. $\sqrt[4]{\sqrt[3]{7}}$ **94.** $\sqrt[4]{\sqrt[5]{x^3}}$

95. $\sqrt{2}\sqrt[3]{5}$ **96.** $\sqrt[3]{7}\sqrt{3}$

Real-Life Applications

97. A criminal justice expert developed a model to predict crime in a city. According to the model, for every 1000 people in a city, $16p^{1/4}$ people would be crime victims, where p is the population of the city (in thousands).

(a) Suppose that a city's population is 90,000. According to the model, for each 1000 people, how many would be crime victims?

(b) Write a simplified expression for the total number of crimes in a city with a population of p thousand.

(c) How many total crimes does the model predict for a city with a population of 1 million?

98. Outdoor Enthusiast is a large mail-order company that sells clothes and equipment for outdoor activities. The company predicts that, following the issue of a new catalog, the total number of daily orders (in thousands) will be $3w^{-1/3}$, where w is the number of weeks since the catalog was issued, and that the total revenue (in thousands of dollars) will be $1200w^{-4/5}$.

(a) Write a simplified expression for the average dollar amount of an order.

(b) For each order, about 1% of the price of the order covers the cost of shipping the items. What is the average shipping cost per order four weeks after a catalog is issued?

Data Analysis: *National Park Service*

The following table shows the number of visitors (in millions) to all National Park Service sites during the period 1996–2000. Also shown are the expenditures (in millions of dollars) for the National Park Service during the same period. (Source: U.S. National Park Service.)

Year	Number of Visitors (in millions)	Expenditures (in millions of dollars)
1996	265.8	1391
1997	275.3	1473
1998	286.7	1605
1999	286.6	1700
2000	291.2	1800

With t representing the number of years since 1994, the function $E(t) = 1150t^{1/4}$ models the expenditures, and $V(t) = 250t^{1/12}$ models the number of visitors.

99. Use the model functions to estimate the expenditures and the number of visitors expected in 2005.

100. Write the function $C(t) = \dfrac{E(t)}{V(t)}$ and simplify the expression.

101. What is your interpretation of $C(t)$?

102. Make a table of values for $C(t)$ for the years 2005, 2010, 2015, and 2020. Using your interpretation in Exercise 101, what can you conclude?

Challenge

Sometimes expressions with rational exponents have a common factor that can be factored out of the expression. Remember to look for the smallest exponent.

Example: $5x^{-2/3} - 8x^{1/3} = x^{-2/3}(5 - 8x)$

In Exercises 103–106, use the preceding technique to factor the expression.

103. $3x^{-1/2} + 5x^{3/2}$ **104.** $a^{-1/2} + a^{-5/2}$

105. $5a^{1/2} - 10a^{-1/2}$ **106.** $6x^{1/3} - 3x^{4/3}$

In Exercises 107–110, note that the given expression is in the form $ax^2 + bx + c$ and use the trial-and-check method to factor the expression.

107. $t^{2/5} + 4t^{1/5} - 5$

108. $3x^{4/3} + 2x^{2/3} - 5$

109. $2x^{1/3} - 7x^{1/6} - 15$

110. $6y^{2/3} - 5y^{1/3} + 1$

In Exercises 111 and 112, write the given expression with just one radical symbol.

111. $\sqrt{\sqrt{\sqrt{x}}}$ **112.** $\sqrt[4]{\sqrt[3]{\sqrt{x}}}$

9.4 The Product Rule for Radicals

Multiplying Radicals • Simplifying with the Product Rule

Multiplying Radicals

We begin our discussion of operations with products of radicals.

Exploring the Concept

Multiplying Radicals

Consider the following product of two radicals.

$$\sqrt{4}\sqrt{9} = 2 \cdot 3 = 6 = \sqrt{36}$$

Thus $\sqrt{4}\sqrt{9} = \sqrt{36}$.

This simple experiment might lead us to the generalization that $\sqrt{a}\sqrt{b} = \sqrt{ab}$. We can show that this conclusion is true by writing the radicals with rational exponents and then applying the Product to a Power Rule.

$$\sqrt{a}\sqrt{b} = a^{1/2} \cdot b^{1/2} = (ab)^{1/2} = \sqrt{ab}$$

We can generalize even further to higher-order roots.

$$\sqrt[n]{a}\sqrt[n]{b} = a^{1/n} \cdot b^{1/n} = (ab)^{1/n} = \sqrt[n]{ab}$$

Thus we see that the product of two nth root radicals is the nth root of the product of the radicands.

Students usually expect this operation, as well as the quotient, to be harder than they are. This discovery comes as a nice surprise.

The Product Rule for Radicals

For all real numbers a and b for which the operations are defined, $\sqrt[n]{a} \cdot \sqrt[n]{b} = \sqrt[n]{ab}$.

Note: The Product Rule for Radicals applies only if the indices are the same.

So that all radicals are defined and results can be expressed without absolute value, we will assume that all variables in this section represent positive numbers.

EXAMPLE 1

Using the Product Rule for Radicals

Use the Product Rule for Radicals to multiply the radical expressions.

(a) $\sqrt{2a} \cdot \sqrt{7b} = \sqrt{2a \cdot 7b} = \sqrt{14ab}$

(b) $\sqrt[4]{\dfrac{1}{x}} \cdot \sqrt[4]{\dfrac{2}{y}} = \sqrt[4]{\dfrac{1}{x} \cdot \dfrac{2}{y}} = \sqrt[4]{\dfrac{2}{xy}}$

(c) $\sqrt[3]{4y} \cdot \sqrt[3]{3y} = \sqrt[3]{12y^2}$

Think About It

For what values of n are the expressions in each part equal?

(i) $\sqrt{n}\sqrt{n}, n$ (ii) $\sqrt{-n}\sqrt{n}, -n$ (iii) $\sqrt{-n}\sqrt{-n}, n$

Simplifying with the Product Rule

In Section 9.1 we found that we could determine the indicated root of any radical that can be written in the form $\sqrt[n]{b^n}$.

$$\sqrt{36x^8} = \sqrt{(6x^4)^2} = 6x^4$$
$$\sqrt[3]{8y^9} = \sqrt[3]{(2y^3)^3} = 2y^3$$

The radicand $36x^8$ is a *perfect square* because each factor can be written with an exponent of 2: $36x^8 = 6^2 \cdot (x^4)^2$. Similarly, $8y^9$ is a *perfect cube* because each factor can be written with an exponent of 3: $8y^9 = 2^3 \cdot (y^3)^3$.

LEARNING TIP

Observe that a^m is a perfect nth power if m is divisible by n.

In general, an expression is a **perfect *n*th power** if each of its factors can be written with an exponent of n.

Another method for determining the indicated roots of $\sqrt{36x^8}$ and $\sqrt[3]{8y^9}$ is to use the Product Rule for Radicals.

$$\sqrt{36x^8} = \sqrt{36} \cdot \sqrt{x^8} = 6x^4 \qquad \sqrt{ab} = \sqrt{a}\sqrt{b}$$
$$\sqrt[3]{8y^9} = \sqrt[3]{8} \cdot \sqrt[3]{y^9} = 2y^3 \qquad \sqrt[3]{ab} = \sqrt[3]{a}\sqrt[3]{b}$$

Because the radicand in each case is a perfect nth power, either of the preceding methods can be used to determine the indicated root or write the expression without a radical symbol. This process is an instance of *simplifying* a radical.

We say that a radical is *simplified* if it satisfies three conditions. In this section we consider the first condition.

Simplifying a Radical: Condition 1

The radicand of a simplified nth root radical must not contain a perfect nth power factor.

If the radicand is not a perfect nth power, we may still be able to use the Product Rule for Radicals to satisfy condition 1. If the radicand is an integer, write it as a product of the largest perfect nth power factor of the integer and another factor.

EXAMPLE 2 Using the Product Rule to Simplify a Radical

Simplify each radical.

(a) $\sqrt{75} = \sqrt{25 \cdot 3}$ — The largest perfect square factor of 75 is 25.

$= \sqrt{25} \cdot \sqrt{3}$ — Product Rule for Radicals

$= 5\sqrt{3}$

(b) $\sqrt{32} = \sqrt{16 \cdot 2}$ — The largest perfect square factor of 32 is 16.

$= \sqrt{16} \cdot \sqrt{2}$ — Product Rule for Radicals

$= 4\sqrt{2}$

Students sometimes write $\sqrt{5}\sqrt{3}$ as the result in part (a). Emphasize that when they take the square root of 25, the result is 5.

In part (b), it is easier to use the largest perfect square factor. Show students that writing $\sqrt{32} = \sqrt{4 \cdot 8}$ eventually leads to the same result but requires more steps.

(c) $\sqrt[3]{24} = \sqrt[3]{8 \cdot 3}$ — The largest perfect cube factor of 24 is 8.

$= \sqrt[3]{8} \cdot \sqrt[3]{3}$ — Product Rule for Radicals

$= 2\sqrt[3]{3}$

(d) $\sqrt[4]{48} = \sqrt[4]{16 \cdot 3}$ — The largest perfect fourth power factor of 48 is 16.

$= \sqrt[4]{16} \cdot \sqrt[4]{3}$ — Product Rule for Radicals

$= 2\sqrt[4]{3}$

When the radicand is negative, we usually choose the negative perfect power.

(e) $\sqrt[3]{-54} = \sqrt[3]{-27 \cdot 2}$ — We choose the negative perfect cube factor -27.

$= \sqrt[3]{-27} \cdot \sqrt[3]{2}$ — Product Rule for Radicals

$= -3\sqrt[3]{2}$

A similar procedure is used when the radicand contains a variable raised to a power.

EXAMPLE 3

Using the Product Rule to Simplify a Radical

(a) $\sqrt{y^7} = \sqrt{y^6 \cdot y}$ — The largest perfect square factor of y^7 is y^6.

$= \sqrt{y^6} \cdot \sqrt{y}$ — Product Rule for Radicals

$= y^3\sqrt{y}$

(b) $\sqrt[3]{a^4b^8} = \sqrt[3]{a^3 \cdot a \cdot b^6 \cdot b^2}$ — The largest perfect cube factors are a^3 and b^6.

$= \sqrt[3]{a^3b^6} \cdot \sqrt[3]{ab^2}$ — Product Rule for Radicals

$= ab^2\sqrt[3]{ab^2}$

(c) $\sqrt{x^2 + 12x + 36} = \sqrt{(x + 6)^2}$ — Factor the radicand.

$= x + 6$

To save a step in the simplifying process, we can just write the given radical as a product of two radicals with one of the radicands containing all of the perfect *n*th power factors.

EXAMPLE 4

Using the Product Rule to Simplify a Radical

Remind students that the exponents of perfect powers should be divisible by the index.

(a) $\sqrt{49a^7} = \sqrt{49a^6} \cdot \sqrt{a} = 7a^3\sqrt{a}$

(b) $\sqrt{300r^{12}s^9} = \sqrt{100r^{12}s^8} \cdot \sqrt{3s} = 10r^6s^4\sqrt{3s}$

(c) $\sqrt[3]{24x^9y^{11}} = \sqrt[3]{8x^9y^9} \cdot \sqrt[3]{3y^2} = 2x^3y^3\sqrt[3]{3y^2}$

(d) $\sqrt[5]{64p^{12}q^{23}} = \sqrt[5]{32p^{10}q^{20}} \cdot \sqrt[5]{2p^2q^3} = 2p^2q^4\sqrt[5]{2p^2q^3}$

(e) $\sqrt{2x^2 + 12x + 18} = \sqrt{2(x^2 + 6x + 9)}$ — Factor out the GCF 2.

$= \sqrt{2(x + 3)^2}$ — Factor the trinomial.

$= \sqrt{(x + 3)^2} \cdot \sqrt{2}$ — Product Rule for Radicals

$= (x + 3)\sqrt{2}$ — Note that parentheses are needed.

Sometimes we need to use the Product Rule for Radicals twice in the same problem. First, we use it to multiply the radicals. Then we use it again to simplify the result.

Note: Although it is often possible to do some simplifying before multiplying, it is generally less confusing to find the product first and then simplify.

EXAMPLE 5

Multiplying and Simplifying Radicals

(a) $\sqrt{12a^4b^4} \cdot \sqrt{3a^2b^6} = \sqrt{36a^6b^{10}}$ Product Rule for Radicals

$= 6a^3b^5$ The radicand is a perfect square.

(b) $\sqrt{6x^3y^7} \cdot \sqrt{8x^4y^3} = \sqrt{48x^7y^{10}}$ Product Rule for Radicals

$= \sqrt{16x^6y^{10}} \cdot \sqrt{3x}$ Product Rule for Radicals

$= 4x^3y^5\sqrt{3x}$

Quick Reference 9.4

Multiplying Radicals

- The Product Rule for Radicals

  For all real numbers a and b for which the operations are defined, $\sqrt[n]{a} \cdot \sqrt[n]{b} = \sqrt[n]{ab}$.

Simplifying with the Product Rule

- An expression is a **perfect nth power** if each of its factors can be written with an exponent of n.
- Simplifying a Radical: Condition 1

  The radicand of a simplified nth root radical must not contain a perfect nth power factor.
- To use the Product Rule for Radicals to simplify a radical, write the radical as a product of two radicals with one of the radicands containing all the perfect nth power factors.

Speaking the Language 9.4

1. The Product Rule for Radicals applies only if the ________ are the same.
2. If each factor of an expression can be written with an exponent of 3, we call the expression a(n) ________.
3. Because the radicand of $\sqrt{32}$ has a perfect square factor, the radical is not considered to be ________.
4. The Product Rule for Radicals is used to multiply $\sqrt{3} \cdot \sqrt{6} = \sqrt{18}$. Then the ________ Rule for Radicals is used to simplify $\sqrt{18} = \sqrt{9} \cdot \sqrt{2} = 3\sqrt{2}$.

Exercises 9.4

Concepts and Skills

1. Explain why the following is an incorrect application of the Product Rule for Radicals.

$$\sqrt{-3}\sqrt{-12} = \sqrt{36} = 6$$

2. Explain whether the Product Rule for Radicals applies to the problem $\sqrt[3]{2}\sqrt{2}$.

In Exercises 3–10, use the Product Rule for Radicals to multiply the radicals.

3. $\sqrt{2}\sqrt{3}$
4. $\sqrt{7}\sqrt{x}$
5. $\sqrt{3x}\sqrt{5y}$
6. $\sqrt{2b}\sqrt{ac}$
7. $(-4\sqrt{2})(3\sqrt{7})$
8. $(2\sqrt[5]{6})(\sqrt[5]{3})$
9. $\sqrt[3]{x-2}\sqrt[3]{x+3}$
10. $\sqrt[4]{x}\sqrt[4]{x+2}$

11. What is the first step in simplifying the radical $\sqrt{x^2+10x+25}$? Why is this step necessary? Can this same step be used to simplify the radical $\sqrt{x^2+10x+21}$? Explain.

12. Explain why $8x^6$ is a perfect cube.

In Exercises 13–26, determine the indicated root.

13. $\sqrt{25y^2}$
14. $\sqrt{16x^6y^{10}}$
15. $\sqrt{49w^8t^4}$
16. $\sqrt{36x^8y^{12}}$
17. $\sqrt[3]{8t^6}$
18. $\sqrt[4]{16x^8y^{16}}$
19. $\sqrt[4]{81x^{12}y^{20}}$
20. $\sqrt[3]{27x^{27}y^{12}}$
21. $\sqrt{x^2(y-3)^4}$
22. $\sqrt{y^4(2x+7)^2}$
23. $\sqrt{x^2+12x+36}$
24. $\sqrt{y^2+6y+9}$
25. $\sqrt{4y^2+4y+1}$
26. $\sqrt{9t^2+12t+4}$

27. Explain why $\sqrt{32}$ can be simplified but $\sqrt{30}$ cannot.

28. To simplify $\sqrt{16x^{16}}$, we treat the coefficient and the exponent differently even though they are the same number. Explain what to do with each.

In Exercises 29–40, simplify the radical.

29. $\sqrt{12}$
30. $\sqrt{20}$
31. $\sqrt{50}$
32. $\sqrt{18}$
33. $\sqrt[3]{54}$
34. $\sqrt[3]{192}$
35. $\sqrt[3]{-48}$
36. $\sqrt[3]{-54}$
37. $\sqrt[4]{32}$
38. $\sqrt[4]{162}$
39. $\sqrt[5]{256}$
40. $\sqrt[5]{128}$

In Exercises 41–46, simplify.

41. $\sqrt{a^5b^3}$
42. $\sqrt{x^9y^4z^7}$
43. $\sqrt[3]{a^9b^8}$
44. $\sqrt[3]{x^{17}y^{15}}$
45. $\sqrt[4]{x^{16}y^9}$
46. $\sqrt[4]{a^5b^7}$

In Exercises 47–60, simplify.

47. $\sqrt{60t}$
48. $\sqrt{20x^{10}}$
49. $\sqrt{25x^{25}}$
50. $\sqrt{18a^9}$
51. $\sqrt{500a^7b^{14}}$
52. $\sqrt{100a^3b}$
53. $-\sqrt{49x^6y^7}$
54. $-\sqrt{8r^5t^9}$
55. $\sqrt{28w^6t^9}$
56. $\sqrt{75x^5y^7z^9}$
57. $\sqrt[3]{81x^9y^{11}}$
58. $\sqrt[3]{54x^8}$
59. $\sqrt[5]{-32x^7y^9}$
60. $\sqrt[4]{32x^6y^8}$

61. It would be incorrect to simplify $\sqrt{4x^2+16}$ by taking the square root of each term. What step must first be taken before the radical can be simplified?

62. Use the following methods to simplify $\sqrt{8x}\sqrt{2x^3}$.
    (a) Simplify each radical; then multiply.
    (b) Multiply; then simplify the result.
    Which method seems easier?

In Exercises 63–68, simplify.

In this group, the radicands must be factored.

63. $\sqrt{12x+8y}$
64. $\sqrt{ab^2+b^3}$
65. $\sqrt{4x^2+16}$
66. $\sqrt{x^4+x^2}$
67. $\sqrt{9x^7y^6+9x^6y^7}$
68. $\sqrt{50a^3+75a^4b}$

In Exercises 69–74, simplify. Assume all variables represent any real number. Use absolute value symbols when needed.

69. $\sqrt{9y^2+9y^4}$
70. $\sqrt{t^6+t^4}$
71. $\sqrt{r^4t^2+r^6t^4}$
72. $\sqrt{a^4b^6+a^2b^8}$
73. $\sqrt{y^2-10y+25}$
74. $\sqrt{2t^2-28t+98}$

In Exercises 75–90, multiply and simplify.

75. $(-2\sqrt{5})(2\sqrt{5})$
76. $(4\sqrt{6x})(-2\sqrt{2x})$

77. $\sqrt[3]{3}\sqrt[3]{54}$

78. $\sqrt[3]{2}\sqrt[3]{20}$

79. $\sqrt{7x}\sqrt{14x}$

80. $\sqrt{3y^2}\sqrt{15y}$

81. $\sqrt{3x^3}\sqrt{2x^6}$

82. $\left(\sqrt{3x^2y}\right)^2$

83. $\left(2\sqrt{5}\right)^2$

84. $\left(a\sqrt{3}\right)^2$

85. $\sqrt{6xy^3}\sqrt{12x^3y^2}$

86. $\sqrt{3c^{11}d^{13}}\sqrt{15c^{12}d^{14}}$

87. $\sqrt[3]{25x^2y^2}\sqrt[3]{5x^3y^4}$

88. $\left(\sqrt[3]{9x^2y^4}\right)^2$

89. $\sqrt[4]{54x^3y}\sqrt[4]{3x^2y^4}$

90. $\sqrt[4]{27x^2y^5}\sqrt[4]{3x^3y^3}$

CE In Exercises 91–96, supply the missing numbers to make the statement true.

91. $\sqrt{\square a^{\square} b^{\square}} = 4a^2b\sqrt{b}$

92. $\square\sqrt{4x^{\square}y^{\square}} = -6x^3y^2\sqrt{xy}$

93. $\sqrt{3x^{\square}y^3}\sqrt{\square x^2y^{\square}} = 3x^2y^3$

94. $\sqrt{6a^{\square}b^3}\sqrt{\square ab^{\square}} = 3a^2b^4\sqrt{2a}$

95. $\sqrt[3]{\square x^{\square}y^{\square}} = 2xy\sqrt[3]{2x}$

96. $\sqrt[3]{\square x^{\square}y^{\square}} = 5x^2y\sqrt[3]{2y^2}$

Real-Life Applications

97. An auctioneer specializes in the sale of vintage automobiles. For cars more than 10 years old, he has developed a model to estimate the price as the product of the car's ratings for design, performance, and reliability. For a certain type of car, the ratings are $\sqrt{5y^3}$, $\sqrt{10y^2}$, and $\sqrt{y}$, respectively, where y is the age of the car.
   (a) Which rating has the greatest effect on potential price?
   (b) Write a simplified expression for the potential price.
   (c) What is the estimated price for a car of this type that is 15 years old?

98. An educator studying 6-year-old children estimated that a child's number of reading levels above the average could be modeled by $\dfrac{\sqrt{20h^5}}{1000}$, where h is the average number of hours per month that parents spent reading to the child before age 6.
   (a) How many reading levels above average is the child if the parents read to the child 20 hours per month?
   (b) Write the model expression in simplified form.

Modeling with Real Data

99. As shown in the accompanying table, the percentage of African-American and Hispanic households with computers increased during the period 1984–1998. (Source: National Telecommunications and Information Administration.)

Year	Percentage of Households with Computers: African-American	Hispanic
1984	4.0	4.0
1989	7.0	6.7
1994	11.0	12.3
1997	19.9	19.4
1998	23.8	25.7

The following are functions of the number of years since 1980.

Percentage of households with computers:

African-American: $C_1(x) = 0.7x\sqrt[5]{x}$

Hispanic: $C_2(x) = 0.31x\sqrt{x}$

Number of households (in millions):

African-American: $H_1(x) = 6.9\sqrt[5]{x}$

Hispanic: $H_2(x) = 2.1\sqrt{x}$

(a) Write a simplified function that models the actual number of African-American households (in millions) with computers.

(b) Write a simplified function that models the actual number of Hispanic households (in millions) with computers.

100. In the United States, the number of deaths due to motor vehicle accidents per 100,000 population declined from 23.4 in 1980 to 15.8 in 1997. (Source: National Safety Council.) The following are model functions, where x is the number of years since 1970.

U.S. population (in millions):

$$P(x) = \sqrt{1150x + 4150}$$

Motor vehicle deaths (per 100,000 population):

$$D(x) = \frac{82.3\sqrt{x}}{x}$$

(a) Use the model function to estimate the number of motor vehicle deaths per 100,000 population in 2005.

(b) Write a function N that models the actual number of deaths due to motor vehicle accidents.

Challenge

101. To simplify $\sqrt[6]{z^{10}}$, we can proceed as follows:

$$\sqrt[6]{z^{10}} = z^{10/6} = z^{5/3} = \sqrt[3]{z^5}$$

Is the radical now simplified? Why?

In Exercises 102–106, simplify.

102. $\sqrt[15]{x^{20}}$

103. $\sqrt[8]{16x^{12}y^{20}}$

104. $\sqrt[12]{27x^6y^9}$

105. $\sqrt[21]{64x^{18}y^{15}}$

106. $\sqrt[6]{8x^{15}y^9}$

These exercises require index reduction, often followed by further simplification.

107. Suppose you want to write a computer program that will factor any given positive integer n. One approach is to divide n successively by 2, then 3, then 4, and so on, until a number is found that divides n. Explain why it would not be necessary to continue testing numbers beyond $\sqrt{n}$.

9.5 The Quotient Rule for Radicals

Dividing Radicals • Simplifying with the Quotient Rule • Rationalizing Monomial Denominators

Dividing Radicals

The following equalities show how a quotient of two square roots can be written as a single square root.

$$\frac{\sqrt{100}}{\sqrt{25}} = \frac{10}{5} = 2$$

$$\sqrt{\frac{100}{25}} = \sqrt{4} = 2$$

The fact that $\frac{\sqrt{100}}{\sqrt{25}} = \sqrt{\frac{100}{25}}$ suggests that the quotient of two square roots is the square root of the quotient of the radicands. We can use the Quotient to a Power Rule to generalize this rule to quotients of nth root radicals.

$$\frac{\sqrt[n]{a}}{\sqrt[n]{b}} = \frac{a^{1/n}}{b^{1/n}} = \left(\frac{a}{b}\right)^{1/n} = \sqrt[n]{\frac{a}{b}}$$

The Quotient Rule for Radicals

For all real numbers a and b for which the operations are defined, $\frac{\sqrt[n]{a}}{\sqrt[n]{b}} = \sqrt[n]{\frac{a}{b}}$.

Note: The Quotient Rule for Radicals applies only if the indices are the same.

In this section, we continue to assume that all variables represent positive numbers.

EXAMPLE 1

Using the Quotient Rule for Radicals

(a) $\dfrac{\sqrt[5]{64}}{\sqrt[5]{2}} = \sqrt[5]{\dfrac{64}{2}} = \sqrt[5]{32} = 2$

(b) $\dfrac{\sqrt{x^5}}{\sqrt{x^3}} = \sqrt{\dfrac{x^5}{x^3}} = \sqrt{x^2} = x$

(c) $\dfrac{\sqrt{6x^3}}{\sqrt{2x^2}} = \sqrt{\dfrac{6x^3}{2x^2}} = \sqrt{3x}$

Think About It

The Quotient Rule for Radicals does not apply to $\dfrac{\sqrt{8}}{\sqrt[3]{2}}$. Use another method to write this expression as a single radical.

After we divide radicals, the Product Rule for Radicals may be needed to simplify the result.

EXAMPLE 2

Remind students about the rules for simplifying exponential expressions.

Using the Quotient and Product Rules for Radicals

(a) $\dfrac{\sqrt{a^{11}b^8c^5}}{\sqrt{a^2bc}} = \sqrt{\dfrac{a^{11}b^8c^5}{a^2bc}}$ Quotient Rule for Radicals

$= \sqrt{a^9b^7c^4}$ Divide the numerator by the denominator.

$= \sqrt{a^8b^6c^4}\sqrt{ab}$ Product Rule for Radicals

$= a^4b^3c^2\sqrt{ab}$

(b) $\dfrac{\sqrt[3]{48x^9y^{12}}}{\sqrt[3]{3xy^2}} = \sqrt[3]{\dfrac{48x^9y^{12}}{3xy^2}}$ Quotient Rule for Radicals

$= \sqrt[3]{16x^8y^{10}}$ Divide the numerator by the denominator.

$= \sqrt[3]{8x^6y^9}\,\sqrt[3]{2x^2y}$ Product Rule for Radicals

$= 2x^2y^3\sqrt[3]{2x^2y}$

After applying the Quotient Rule for Radicals, we may find that the numerator is not divisible by the denominator. However, we may be able to simplify the radicand.

EXAMPLE 3

Using the Quotient Rule for Radicals

(a) $\dfrac{\sqrt{45xy}}{\sqrt{x^3}} = \sqrt{\dfrac{45xy}{x^3}}$ Quotient Rule for Radicals

$= \sqrt{\dfrac{45y}{x^2}}$ Simplify the radicand.

$= \dfrac{\sqrt{9}\,\sqrt{5y}}{\sqrt{x^2}}$ Product Rule for Radicals

$= \dfrac{3\sqrt{5y}}{x}$

(b) $\dfrac{\sqrt[3]{72x^2y^7}}{\sqrt[3]{x^{11}}} = \sqrt[3]{\dfrac{72x^2y^7}{x^{11}}}$ Quotient Rule for Radicals

$= \sqrt[3]{\dfrac{72y^7}{x^9}}$ Simplify the radicand.

$= \dfrac{\sqrt[3]{8y^6} \cdot \sqrt[3]{9y}}{\sqrt[3]{x^9}}$ Product Rule for Radicals

$= \dfrac{2y^2\sqrt[3]{9y}}{x^3}$

Simplifying with the Quotient Rule

The following is the second condition for a simplified radical.

Simplifying a Radical: Condition 2

The radicand of a simplified radical must not contain a fraction.

The easiest fractional radicand to simplify is one in which the denominator is a perfect power. Then the Quotient Rule for Radicals can be applied directly to simplify the radical.

EXAMPLE 4 **Simplifying a Radical with a Fractional Radicand**

(a) $\sqrt{\dfrac{7}{9}} = \dfrac{\sqrt{7}}{\sqrt{9}}$ $\sqrt[n]{\dfrac{a}{b}} = \dfrac{\sqrt[n]{a}}{\sqrt[n]{b}}$

$= \dfrac{\sqrt{7}}{3}$ Simplify the denominator.

(b) $\sqrt[3]{\dfrac{9z}{125}} = \dfrac{\sqrt[3]{9z}}{\sqrt[3]{125}}$ $\sqrt[n]{\dfrac{a}{b}} = \dfrac{\sqrt[n]{a}}{\sqrt[n]{b}}$

$= \dfrac{\sqrt[3]{9z}}{5}$ Simplify the denominator.

(c) $\sqrt[4]{\dfrac{a^5}{16x^8}} = \dfrac{\sqrt[4]{a^5}}{\sqrt[4]{16x^8}}$ $\sqrt[n]{\dfrac{a}{b}} = \dfrac{\sqrt[n]{a}}{\sqrt[n]{b}}$

$= \dfrac{\sqrt[4]{a^4} \cdot \sqrt[4]{a}}{2x^2}$ Product Rule for Radicals

$= \dfrac{a\sqrt[4]{a}}{2x^2}$ $\sqrt[4]{a^4} = a, a > 0$

If the radicand is a fraction, it may be possible to perform a simple division to eliminate the fraction. If so, then we can use previously discussed procedures to complete the simplification.

EXAMPLE 5

Simplifying a Radical with a Fractional Radicand

$$\sqrt{\frac{45x^9y}{x^3}} = \sqrt{45x^6y} \quad \text{Perform the indicated division in the radicand.}$$

$$= \sqrt{9x^6}\sqrt{5y} \quad \text{Product Rule for Radicals}$$

$$= 3x^3\sqrt{5y}$$

Rationalizing Monomial Denominators

The following is the third condition for a simplified radical.

Simplifying a Radical: Condition 3

A simplified radical must not contain a radical in the denominator.

The expression $\frac{3}{\sqrt{5}}$ is not considered simplified because the radical $\sqrt{5}$ appears in the denominator. If we multiply the numerator and the denominator by $\sqrt{5}$, the resulting denominator will be a rational number and the expression will be simplified.

$$\frac{3}{\sqrt{5}} = \frac{3 \cdot \sqrt{5}}{\sqrt{5} \cdot \sqrt{5}} = \frac{3\sqrt{5}}{\sqrt{25}} = \frac{3\sqrt{5}}{5}$$

Students sometimes divide out the 5's. Stress that 5 and $\sqrt{5}$ are different numbers.

This process is called **rationalizing the denominator.**

Note: We can save a step by recalling the fact that $\sqrt{n}\sqrt{n} = n$ for nonnegative numbers n. Also, remember that the $\sqrt{5}$ in the numerator and the 5 in the denominator are different numbers, and they cannot be divided out.

EXAMPLE 6

Rationalizing a Monomial Denominator

Rationalize the denominator.

(a) $\frac{\sqrt{7}}{\sqrt{3}}$ (b) $\frac{4\sqrt{3}}{\sqrt{2}}$ (c) $\frac{12}{\sqrt{27}}$

LEARNING TIP

When you simplify radicals, often the best approach is first to multiply or divide, then to simplify with the product or quotient rules, and finally to rationalize the denominator.

Solution

(a) Multiply the numerator and denominator by $\sqrt{3}$.

$$\frac{\sqrt{7}}{\sqrt{3}} = \frac{\sqrt{7} \cdot \sqrt{3}}{\sqrt{3} \cdot \sqrt{3}} = \frac{\sqrt{21}}{3}$$

(b) Multiply the numerator and denominator by $\sqrt{2}$.

$$\frac{4\sqrt{3}}{\sqrt{2}} = \frac{4\sqrt{3} \cdot \sqrt{2}}{\sqrt{2} \cdot \sqrt{2}} = \frac{4\sqrt{6}}{2} = 2\sqrt{6}$$

(c) We can simplify the expression before rationalizing the denominator.

$$\frac{12}{\sqrt{27}} = \frac{12}{\sqrt{9}\sqrt{3}}$$ Product Rule for Radicals

$$= \frac{12}{3\sqrt{3}}$$

$$= \frac{4}{\sqrt{3}}$$ Divide out the common factor 3.

$$= \frac{4 \cdot \sqrt{3}}{\sqrt{3} \cdot \sqrt{3}}$$ Multiply the numerator and denominator by $\sqrt{3}$.

$$= \frac{4\sqrt{3}}{3}$$

An alternative approach:

$$\frac{12}{\sqrt{27}} \cdot \frac{\sqrt{3}}{\sqrt{3}} = \frac{12\sqrt{3}}{\sqrt{81}} = \frac{12\sqrt{3}}{9} = \frac{4\sqrt{3}}{3}$$

In Example 7, instead of simplifying radicals first, we multiply the numerator and the denominator by a radical selected so that the resulting radicand in the denominator is a perfect square.

EXAMPLE 7

Rationalizing a Monomial Denominator

(a) $$\sqrt{\frac{45}{8}} = \frac{\sqrt{45} \cdot \sqrt{2}}{\sqrt{8} \cdot \sqrt{2}} = \frac{\sqrt{90}}{\sqrt{16}} = \frac{\sqrt{9}\sqrt{10}}{4} = \frac{3\sqrt{10}}{4}$$

(b) $$\frac{3\sqrt{2a}}{\sqrt{x^3}} = \frac{3\sqrt{2a} \cdot \sqrt{x}}{\sqrt{x^3} \cdot \sqrt{x}} = \frac{3\sqrt{2ax}}{\sqrt{x^4}} = \frac{3\sqrt{2ax}}{x^2}$$

(c) $$\frac{x}{\sqrt{x^2+4}} = \frac{x \cdot \sqrt{x^2+4}}{\sqrt{x^2+4} \cdot \sqrt{x^2+4}} = \frac{x\sqrt{x^2+4}}{x^2+4}$$

Again, watch for the common error of writing $\sqrt{x^2+4} = x + 2$.

If the index of the radical in a denominator is 2, we multiply by a number that makes the result a perfect square. If the index is 3, we multiply by a number that makes the result a perfect cube. A similar strategy is used for higher indices.

EXAMPLE 8

Rationalizing Denominators with Higher Indices

Rationalize the denominator.

(a) $\frac{10}{\sqrt[3]{4}}$ (b) $\frac{\sqrt[3]{2}}{\sqrt[3]{r^4}}$ (c) $\frac{-2}{\sqrt[4]{y^7}}$

Solution

Remind students about the value of knowing some of the perfect powers. The exponent of the radicand must be divisible by the index.

(a) Multiplying by $\sqrt[3]{4}$ would result in $\sqrt[3]{16}$ in the denominator, but 16 is not a perfect cube. Instead, we multiply by $\sqrt[3]{2}$ to obtain the perfect cube radicand 8.

$$\frac{10}{\sqrt[3]{4}} = \frac{10 \cdot \sqrt[3]{2}}{\sqrt[3]{4} \cdot \sqrt[3]{2}} = \frac{10\sqrt[3]{2}}{\sqrt[3]{8}} = \frac{10\sqrt[3]{2}}{2} = 5\sqrt[3]{2}$$

(b) Multiply by $\sqrt[3]{r^2}$ to obtain a perfect cube in the denominator.

$$\frac{\sqrt[3]{2}}{\sqrt[3]{r^4}} = \frac{\sqrt[3]{2} \cdot \sqrt[3]{r^2}}{\sqrt[3]{r^4} \cdot \sqrt[3]{r^2}} = \frac{\sqrt[3]{2r^2}}{\sqrt[3]{r^6}} = \frac{\sqrt[3]{2r^2}}{r^2}$$

(c) Multiply by $\sqrt[4]{y}$ to obtain a perfect fourth power in the denominator.

$$\frac{-2}{\sqrt[4]{y^7}} = \frac{-2 \cdot \sqrt[4]{y}}{\sqrt[4]{y^7} \cdot \sqrt[4]{y}} = \frac{-2\sqrt[4]{y}}{\sqrt[4]{y^8}} = \frac{-2\sqrt[4]{y}}{y^2}$$

Quick Reference 9.5

Dividing Radicals

- The Quotient Rule for Radicals
  For all real numbers a and b for which the operations are defined, $\dfrac{\sqrt[n]{a}}{\sqrt[n]{b}} = \sqrt[n]{\dfrac{a}{b}}$.

Simplifying with the Quotient Rule

- Simplifying a Radical: Condition 2
  The radicand of a simplified radical must not contain a fraction.
- The Quotient Rule for Radicals can be used to satisfy condition 2 if the denominator is a perfect nth power (possibly after simplifying the fraction) or if the numerator is divisible by the denominator.

Rationalizing Monomial Denominators

- Simplifying a Radical: Condition 3
  A simplified radical must not contain a radical in the denominator.
- We satisfy condition 3 by **rationalizing the denominator.** The method is to multiply the numerator and the denominator by a number or expression selected so that the resulting radicand in the denominator is a perfect nth power.

Speaking the Language 9.5

1. We use the ______ Rule for Radicals to write $\dfrac{\sqrt{x^8}}{\sqrt{x^3}} = \sqrt{x^5}$, and then we use the ______ Rule for Radicals to simplify the result.

2. A radical is not considered simplified if the radicand contains a(n) ______.

3. The expression $\dfrac{\sqrt{2}}{\sqrt{3}}$ is not considered simplified because it contains a radical in the ______.

4. The process of writing $\dfrac{1}{\sqrt{5}} = \dfrac{1}{\sqrt{5}} \cdot \dfrac{\sqrt{5}}{\sqrt{5}} = \dfrac{\sqrt{5}}{5}$ is called ______.

Exercises 9.5

Concepts and Skills

1. Explain whether the Quotient Rule for Radicals applies to the problem $\frac{\sqrt[3]{16}}{\sqrt{2}}$.

2. To simplify $\frac{\sqrt{15}}{\sqrt{9}}$, why is it unnecessary to apply the Quotient Rule for Radicals?

In Exercises 3–12, divide and simplify, if possible.

3. $\frac{\sqrt{30}}{\sqrt{5}}$

4. $\frac{\sqrt{42t}}{\sqrt{7}}$

5. $\frac{\sqrt{48}}{\sqrt{3}}$

6. $\frac{\sqrt[3]{54}}{\sqrt[3]{2}}$

7. $\frac{\sqrt{x^7}}{\sqrt{x}}$

8. $\frac{\sqrt{45a^{11}b^3}}{\sqrt{5ab}}$

9. $\frac{\sqrt{49a^7b^4}}{\sqrt{4a}}$

10. $\frac{\sqrt{81a^5b^7}}{\sqrt{16a^3b^5}}$

11. $\frac{\sqrt[3]{x^{11}}}{\sqrt[3]{x^8}}$

12. $\frac{\sqrt[4]{x^{15}}}{\sqrt[4]{x^{11}}}$

In Exercises 13–20, divide and simplify.

13. $\frac{\sqrt{x^5y^{12}}}{\sqrt{xy^3}}$

14. $\frac{\sqrt{60a^{10}b^9}}{\sqrt{3a^3b^9}}$

15. $\frac{\sqrt{45w^9t^8}}{\sqrt{w^2t^3}}$

16. $\frac{\sqrt{40a^{14}b^7}}{\sqrt{5a^9b^4}}$

17. $\frac{\sqrt{18x^9d^{13}}}{\sqrt{2x^4d^{10}}}$

18. $\frac{\sqrt{144r^{11}s^{12}}}{\sqrt{2r^4}}$

19. $\frac{\sqrt[3]{54x^{13}}}{\sqrt[3]{x^2}}$

20. $\frac{\sqrt[4]{32a^7b^{15}}}{\sqrt[4]{2a^2b^5}}$

In Exercises 21–30, simplify.

21. $\frac{\sqrt{10}}{\sqrt{360}}$

22. $\frac{\sqrt[4]{3}}{\sqrt[4]{48}}$

23. $\frac{\sqrt{t^3}}{\sqrt{t^9}}$

24. $\frac{\sqrt{12a^4b^7}}{\sqrt{3a^{10}b^5}}$

25. $\frac{\sqrt{3a^3}}{\sqrt{75a^5}}$

26. $\frac{\sqrt{18t^7}}{\sqrt{50t^{13}}}$

27. $\frac{\sqrt{16xy}}{\sqrt{9x^3y^5}}$

28. $\frac{\sqrt{ab^3}}{\sqrt{a^7b^5}}$

29. $\frac{\sqrt[4]{x^6}}{\sqrt[4]{16x^2}}$

30. $\frac{\sqrt[3]{24x^5}}{\sqrt[3]{3x^{20}}}$

31. Suppose that P and Q are monomials. Describe two circumstances in which simplifying $\sqrt{\frac{P}{Q}}$ would be possible without rationalizing the denominator.

32. We can satisfy condition 2 by writing $\sqrt{\frac{5x^4}{x}}$ as $\sqrt{5x^3}$. Explain why this result is still not simplified.

In Exercises 33–40, simplify.

33. $\sqrt{\frac{5}{49}}$

34. $\sqrt{\frac{17x}{100}}$

35. $\sqrt[3]{\frac{16x^2}{2x^{14}}}$

36. $\sqrt[4]{\frac{y^{11}}{81y^3}}$

37. $\sqrt{\frac{28}{9}}$

38. $\sqrt{\frac{125a}{144}}$

39. $\sqrt[3]{\frac{-16a}{1000}}$

40. $\sqrt[4]{\frac{243w}{16}}$

In Exercises 41–54, simplify.

41. $\sqrt{\frac{60x^{11}y^3}{3y}}$

42. $\sqrt{\frac{24x^{13}y^5}{2x^3}}$

43. $\sqrt{\frac{75x^7y^{13}}{x^3y^5}}$

44. $\sqrt{\frac{150x^{15}y^7}{2y^3}}$

45. $\sqrt{\frac{245x^7y^9z^{12}}{5x^3y^2z^5}}$

46. $\sqrt{\frac{54r^7s^5w^9}{3rw}}$

47. $\sqrt[3]{\frac{72a^9b^{10}}{3a^2b}}$

48. $\sqrt[3]{\frac{108a^5b^7c^9}{2abc}}$

49. $\sqrt{\frac{x^3y^5}{x^5y}}$

50. $\sqrt{\frac{ab^5}{a^5b^4}}$

51. $\sqrt{\frac{2a^{12}b^3}{18a^9b^9}}$

52. $\sqrt{\frac{8b^5c^4}{2b^7c^3}}$

53. $\sqrt[4]{\frac{x^7}{16x^2}}$

54. $\sqrt[3]{\frac{40x^{10}y^5}{5xy^{11}}}$

55. To simplify $\frac{5}{2\sqrt{3}}$, should you multiply the numerator and denominator by $2\sqrt{3}$ or by just $\sqrt{3}$? Will you obtain the correct result either way? If so, which method do you prefer?

56. To simplify $\frac{1}{\sqrt[n]{x^m}}$, we must multiply the numerator and denominator by a radical of the form $\sqrt[n]{x^k}$. For the method to succeed, what must be the relationship among m, k, and n?

In Exercises 57–66, rationalize the denominator.

57. $\frac{5}{\sqrt{5}}$ **58.** $\frac{6}{\sqrt{3}}$

59. $\frac{-\sqrt{5}}{\sqrt{3}}$ **60.** $\frac{1}{3\sqrt{2}}$

61. $\frac{15}{\sqrt{10}}$ **62.** $\frac{\sqrt{6}}{-3\sqrt{12}}$

63. $\frac{7}{\sqrt[3]{3}}$ **64.** $\frac{9}{\sqrt[3]{25}}$

65. $\frac{9}{\sqrt[4]{8}}$ **66.** $\frac{2}{\sqrt[5]{8}}$

In Exercises 67–82, simplify.

67. $\frac{6}{\sqrt{3x}}$ **68.** $\frac{3}{5\sqrt{t}}$

69. $\frac{2}{\sqrt{x+2}}$ **70.** $\frac{4}{\sqrt{4+y^2}}$

71. $\sqrt{\frac{a^3}{8}}$ **72.** $\sqrt{\frac{x^4}{20}}$

73. $\sqrt{\frac{5}{y^5}}$ **74.** $\frac{x^3}{\sqrt{x^3}}$

75. $\frac{\sqrt{5x^2}}{\sqrt{12x^7}}$ **76.** $\frac{\sqrt{a^4}}{\sqrt{18a^5}}$

77. $\frac{18x^2y^3}{\sqrt{3x^2y^5}}$ **78.** $\frac{\sqrt{8a^6b^4}}{\sqrt{32a^4b^5}}$

79. $\frac{9x}{\sqrt{12x^5}}$ **80.** $\frac{\sqrt{8xy^4}}{\sqrt{6x^2}}$

81. $\sqrt{\frac{x^5y^2}{20y^3}}$ **82.** $\sqrt{\frac{4xy}{16x^3y^2}}$

CE In Exercises 83–88, supply the missing number to make the statement true.

83. $\frac{\sqrt{3}}{\sqrt{5}} = \frac{\sqrt{15}}{\square}$ **84.** $\sqrt{\frac{3}{8}} = \frac{\sqrt{6}}{\square}$

85. $\sqrt{\frac{2}{7}} = \frac{\square}{7}$ **86.** $\frac{\sqrt{63}}{\sqrt{5}} = \frac{\square}{5}$

87. $\frac{3}{\square} = \sqrt{3}$ **88.** $\frac{5}{\square} = \frac{\sqrt{5x}}{x}$

In Exercises 89–96, simplify.

89. $\frac{2b}{\sqrt[3]{b^2}}$ **90.** $\frac{7}{\sqrt[4]{27}}$

91. $\sqrt[3]{\frac{5}{4}}$ **92.** $\frac{5}{\sqrt[4]{125}}$

93. $\frac{x}{\sqrt[4]{xy^3}}$ **94.** $\frac{5x}{\sqrt[5]{x^2}}$

95. $\sqrt[4]{\frac{4x^5}{64xy^3}}$ **96.** $\sqrt[5]{\frac{64a^4b^2}{a^6b^6}}$

CE In Exercises 97–102, determine the value of a.

97. $4\sqrt{x} = \sqrt{ax}$ **98.** $10\sqrt{x^2y} = \sqrt{ax^2y}$

99. $\frac{6}{\sqrt{x}} = \sqrt{\frac{a}{x}}$

100. $\frac{2}{\sqrt{y}} = \sqrt{\frac{a}{y}}$

101. $\sqrt{x^3} = \sqrt[a]{x^6}$

102. $\sqrt[3]{x^5} = \sqrt[a]{x^{10}}$

Real-Life Applications

103. A used car dealer planned an extensive advertising campaign for a Labor Day weekend sale. The dealer estimated that d dollars (in thousands) of advertising would result in $30\sqrt{d^3}$ sales (number of sales) with a total revenue of $90{,}000d^2$ dollars.

(a) Write a simplified expression for the estimated average price of each car sold.

(b) If the dealer spends \$4000 on advertising, what is the average price of a car?

104. A naturalist estimates that the number of birds that can be attracted to a preserve is $120\sqrt[5]{m^2}$, where m is the number of months after the preserve is established. The number of pounds of bird seed that must be provided per month per bird to supplement natural sources is predicted to be $\frac{1}{3\sqrt[5]{m}}$, where m is the number of months after the establishment of the preserve.

(a) What is the estimated bird population in the preserve after 6 months?

(b) Write a simplified expression for the number of pounds per month of bird seed required after m months.

(c) Does the seed requirement per bird increase or decrease? Does the total seed requirement increase or decrease?

Data Analysis: *Tax Payments and Refunds*

A survey showed that 59% of all taxpayers expected a tax refund, whereas 41% expected to pay more taxes than were withheld. However, the percentages varied according to age group. (Source: Bruskin/Goldring Research Survey for Turbo Tax.)

The function $R(x) = \frac{734}{x^{0.7}}$ can be used to model the percentage $R(x)$ of taxpayers expecting a refund, and the function $P(x) = \frac{0.2x(\sqrt{x} + 12)}{\sqrt{x}}$ can be used to model the percentage $P(x)$ of taxpayers expecting to pay more taxes. In both models, x represents the age of the taxpayer. The accompanying figure shows the graphs of R and P.

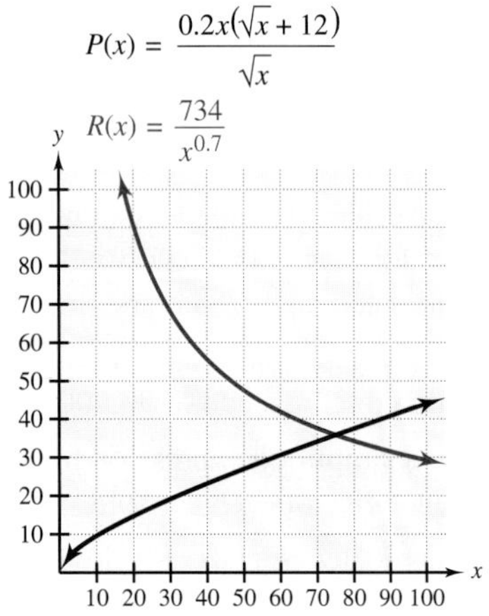

105. Describe the trend in refund expectations as taxpayers grow older.

106. Estimate the age at which 20% of the taxpayers at that age expect to pay more taxes.

107. Estimate the age at which the percentage of people expecting a refund equals the percentage of people expecting to pay more taxes.

108. On the basis of the age that you estimated in Exercise 107, what percentage of taxpayers of that age expect neither a refund nor a payment?

Challenge

109. Evaluate each of the following expressions.

$$\sqrt{2} \qquad \sqrt{2+\sqrt{2}} \qquad \sqrt{2+\sqrt{2+\sqrt{2}}}$$

$$\sqrt{2+\sqrt{2+\sqrt{2+\sqrt{2}}}}$$

$$\sqrt{2+\sqrt{2+\sqrt{2+\sqrt{2+\sqrt{2}}}}}$$

What number do the values of these expressions seem to be approaching? Do you think that this pattern holds true for any positive integer n? If so, repeat the experiment for $n = 3$.

In Exercises 110 and 111, rationalize the denominator.

110. $\frac{1}{\sqrt{\sqrt{x}}}$

111. $\frac{1}{\sqrt{\sqrt[3]{x}}}$

112. Consider three nonzero numbers x, y, and z. How many of these numbers must be positive for the following to be defined in the real number system?

(a) $\sqrt{x}\sqrt{y}\sqrt{z}$

(b) $\sqrt{xyz}$

113. Show that $\frac{7}{\sqrt{7}} = \sqrt{7}$.

From this, you may conclude that dividing a number by its square root always results in the square root of the number. Use rational exponents to prove that your conclusion is true for any positive number.

9.6 Operations with Radicals

Addition and Subtraction • Multiplication • Rationalizing Binomial Denominators

Students will be less inclined to add the radicands if you stress this use of the Distributive Property in combining terms.

LEARNING TIP

The familiar steps of removing parentheses and combining like terms apply to simplifying radical expressions. There is nothing new about the techniques used in this section.

Addition and Subtraction

Radicals with the same radicand and the same index are **like radicals.** In a manner similar to adding like terms, we use the Distributive Property to add or subtract like radicals.

$$3x + 5x = (3 + 5)x = 8x$$

$$3\sqrt{7} + 5\sqrt{7} = (3 + 5)\sqrt{7} = 8\sqrt{7}$$

Sometimes one or both of two unlike radicals can be simplified so that they become like radicals and can then be combined.

EXAMPLE 1

Adding and Subtracting Radicals

(a) $\sqrt{3} + 5\sqrt{3} = 1\sqrt{3} + 5\sqrt{3} = (1 + 5)\sqrt{3} = 6\sqrt{3}$ Distributive Property

(b) $\sqrt{45} - \sqrt{80} = \sqrt{9}\sqrt{5} - \sqrt{16}\sqrt{5}$ Product Rule for Radicals

$= 3\sqrt{5} - 4\sqrt{5}$

$= (3 - 4)\sqrt{5}$ Distributive Property

$= -\sqrt{5}$

(c) $3\sqrt{75} - 2\sqrt{48} = 3\sqrt{25}\sqrt{3} - 2\sqrt{16}\sqrt{3}$ Product Rule for Radicals

$= 3 \cdot 5\sqrt{3} - 2 \cdot 4\sqrt{3}$

$= 15\sqrt{3} - 8\sqrt{3}$

$= 7\sqrt{3}$ Combine the coefficients and retain the radical.

(d) $\sqrt[3]{32} + \sqrt[3]{4} = \sqrt[3]{8}\sqrt[3]{4} + \sqrt[3]{4}$ Product Rule for Radicals

$= 2\sqrt[3]{4} + 1\sqrt[3]{4}$

$= 3\sqrt[3]{4}$ Combine the coefficients and retain the radical.

Note: Radicals can be combined only if the radicands and the indices are the same. Thus the terms cannot be combined in either $4\sqrt{2} + 3\sqrt[3]{2}$ or $\sqrt{3} + \sqrt{2}$.

So that all radicals are defined and results can be expressed without regard to absolute value, we will assume that all variables in this section represent positive numbers.

EXAMPLE 2

Adding and Subtracting Radical Expressions

(a) $\sqrt{48x} - \sqrt{27x} = \sqrt{16} \cdot \sqrt{3x} - \sqrt{9} \cdot \sqrt{3x}$ Product Rule for Radicals

$= 4\sqrt{3x} - 3\sqrt{3x}$

$= \sqrt{3x}$ Combine the terms.

(b) $4\sqrt{a^3} + 2a\sqrt{9a} = 4\sqrt{a^2} \cdot \sqrt{a} + 2a\sqrt{9} \cdot \sqrt{a}$ Product Rule for Radicals

$= 4a\sqrt{a} + 2a \cdot 3\sqrt{a}$

$= 4a\sqrt{a} + 6a\sqrt{a}$

$= 10a\sqrt{a}$ Combine the terms.

(c) $3\sqrt[3]{16t^7} + 4t\sqrt[3]{2t^4} = 3\sqrt[3]{8t^6} \cdot \sqrt[3]{2t} + 4t\sqrt[3]{t^3} \cdot \sqrt[3]{2t}$ Product Rule for Radicals

$= 3 \cdot 2t^2\sqrt[3]{2t} + 4t \cdot t\sqrt[3]{2t}$

$= 6t^2\sqrt[3]{2t} + 4t^2\sqrt[3]{2t}$

$= 10t^2\sqrt[3]{2t}$ Combine the terms.

Note: In order for us to combine terms, the radical factors *and the variable factors* must be identical. [See parts (a) and (b) of Example 2.] In the expression $3\sqrt{4t} + 5t\sqrt{4t}$, the Distributive Property can be used to factor the expression as $(3 + 5t)\sqrt{4t}$, but the terms cannot be combined.

The following example illustrates the method for combining radicals whose radicands are fractions or rational expressions.

EXAMPLE 3

You may need to review the process of finding a least common denominator.

Adding and Subtracting Radical Expressions

(a) $\sqrt{\frac{8}{9}} - \sqrt{\frac{18}{49}} = \frac{\sqrt{8}}{\sqrt{9}} - \frac{\sqrt{18}}{\sqrt{49}}$ Quotient Rule for Radicals

$= \frac{\sqrt{4} \cdot \sqrt{2}}{3} - \frac{\sqrt{9} \cdot \sqrt{2}}{7}$ Product Rule for Radicals

$= \frac{2\sqrt{2}}{3} - \frac{3\sqrt{2}}{7}$

$= \frac{14\sqrt{2}}{21} - \frac{9\sqrt{2}}{21}$ Rewrite the fractions with 21 as the LCD.

$= \frac{5\sqrt{2}}{21}$ Subtract the numerators.

(b) $\sqrt{\frac{81}{a^6}} + \sqrt{\frac{49}{a^{10}}} = \frac{9}{a^3} + \frac{7}{a^5}$ Both radicands are perfect squares.

$= \frac{9a^2}{a^5} + \frac{7}{a^5}$ Rewrite the fractions with a^5 as the LCD.

$= \frac{9a^2 + 7}{a^5}$ Add the numerators.

Multiplication

We can use the Product Rule for Radicals along with the Distributive Property to multiply radical expressions that have more than one term.

EXAMPLE 4

Remind students about distributing a monomial, FOIL, and special products.

Multiplying Radical Expressions

(a) $3\sqrt{5}(2\sqrt{5} - \sqrt{2}) = (3\sqrt{5})(2\sqrt{5}) - (3\sqrt{5})(\sqrt{2})$ — Distributive Property

$= 6 \cdot 5 - 3\sqrt{10}$ — Product Rule for Radicals

$= 30 - 3\sqrt{10}$

(b) $\sqrt{6}(\sqrt{8} - \sqrt{2}) = \sqrt{6}\sqrt{8} - \sqrt{6}\sqrt{2}$ — Distributive Property

$= \sqrt{48} - \sqrt{12}$ — Product Rule for Radicals

$= \sqrt{16}\sqrt{3} - \sqrt{4}\sqrt{3}$ — Product Rule for Radicals

$= 4\sqrt{3} - 2\sqrt{3}$ — Combine the terms.

$= 2\sqrt{3}$

(c) $(\sqrt{2} - 3)(\sqrt{3} + \sqrt{2}) = \sqrt{2}\sqrt{3} + \sqrt{2}\sqrt{2} - 3\sqrt{3} - 3\sqrt{2}$ — FOIL

$= \sqrt{6} + 2 - 3\sqrt{3} - 3\sqrt{2}$ — Product Rule for Radicals

None of the four terms are like terms, so the result cannot be simplified further.

(d) $(\sqrt{3} + \sqrt{6})^2 = (\sqrt{3})^2 + 2 \cdot \sqrt{3}\sqrt{6} + (\sqrt{6})^2$ — $(A + B)^2 = A^2 + 2AB + B^2$

$= 3 + 2\sqrt{18} + 6$ — Product Rule for Radicals

$= 9 + 2\sqrt{9}\sqrt{2}$ — Product Rule for Radicals

$= 9 + 2 \cdot 3\sqrt{2}$

$= 9 + 6\sqrt{2}$

Rationalizing Binomial Denominators

Recall that the product of a sum of two terms and the difference of the same two terms is the difference of their squares.

$$(A + B)(A - B) = A^2 - B^2$$

The expressions $A + B$ and $A - B$ are called **conjugates** of each other.

Exploring the Concept

Square Root Conjugates

Consider the following products of square root conjugates.

$$(4 - \sqrt{5})(4 + \sqrt{5}) = 4^2 - (\sqrt{5})^2$$
$$= 16 - 5 = 11$$

$$(\sqrt{11} + 3\sqrt{2})(\sqrt{11} - 3\sqrt{2}) = (\sqrt{11})^2 - (3\sqrt{2})^2$$
$$= 11 - 9 \cdot 2 = 11 - 18 = -7$$

$$(-x + 2\sqrt{y})(-x - 2\sqrt{y}) = (-x)^2 - (2\sqrt{y})^2$$
$$= x^2 - 4y$$

In each case, multiplying square root conjugates results in an expression that has no square root.

Note: The conclusion holds only for square root conjugates. For example, in the following product of conjugates, the radical is not eliminated.

$$(\sqrt[3]{5} - 1)(\sqrt[3]{5} + 1) = \sqrt[3]{25} - 1$$

If the denominator of a fraction contains a binomial with at least one square root term, often we can rationalize the denominator by multiplying the numerator and the denominator by the conjugate of the denominator.

EXAMPLE 5

Rationalizing a Binomial Denominator

(a) $\dfrac{10}{\sqrt{6} - 2} = \dfrac{10 \cdot (\sqrt{6} + 2)}{(\sqrt{6} - 2) \cdot (\sqrt{6} + 2)}$ Multiply the numerator and denominator by the conjugate of $\sqrt{6} - 2$.

$= \dfrac{10(\sqrt{6} + 2)}{(\sqrt{6})^2 - 2^2}$ $(A + B)(A - B) = A^2 - B^2$

$= \dfrac{10(\sqrt{6} + 2)}{6 - 4}$

$= \dfrac{5 \cdot 2 \cdot (\sqrt{6} + 2)}{2}$ Factor the numerator.

$= 5(\sqrt{6} + 2)$ Divide out the common factor 2.

(b) $\dfrac{\sqrt{3} - \sqrt{2}}{\sqrt{5} + \sqrt{3}} = \dfrac{(\sqrt{3} - \sqrt{2})(\sqrt{5} - \sqrt{3})}{(\sqrt{5} + \sqrt{3})(\sqrt{5} - \sqrt{3})}$ Multiply the numerator and denominator by the conjugate of $\sqrt{5} + \sqrt{3}$.

$= \dfrac{\sqrt{15} - \sqrt{9} - \sqrt{10} + \sqrt{6}}{(\sqrt{5})^2 - (\sqrt{3})^2}$ FOIL

$= \dfrac{\sqrt{15} - 3 - \sqrt{10} + \sqrt{6}}{5 - 3}$

$= \dfrac{\sqrt{15} - 3 - \sqrt{10} + \sqrt{6}}{2}$

Simplified expressions often look simpler than the original expression, but not always. Remind students that we are conforming to the *conditions* for a simplified radical, not the appearance.

(c) $\dfrac{\sqrt{x}}{\sqrt{x} + \sqrt{2y}} = \dfrac{\sqrt{x}(\sqrt{x} - \sqrt{2y})}{(\sqrt{x} + \sqrt{2y})(\sqrt{x} - \sqrt{2y})}$ Multiply the numerator and denominator by the conjugate of $\sqrt{x} + \sqrt{2y}$.

$= \dfrac{\sqrt{x^2} - \sqrt{2xy}}{(\sqrt{x})^2 - (\sqrt{2y})^2}$ $(A + B)(A - B) = A^2 - B^2$

$= \dfrac{x - \sqrt{2xy}}{x - 2y}$ We cannot simplify further because the numerator and denominator cannot be factored.

Think About It

Rationalize the denominator of $\dfrac{1}{\sqrt{\sqrt{x} + 3}}$.

Quick Reference 9.6

Addition and Subtraction

- **Like radicals** have the same radicand and the same index. Only like radicals can be combined by addition or subtraction.
- Simplify all radicals before attempting to add or subtract them.
- Use the Distributive Property to add or subtract like radicals.

Multiplication

- To multiply radical expressions that have more than one term, use the Distributive Property, the FOIL method, and other familiar multiplication techniques that we used with polynomials.
- The special product patterns for polynomials also apply to multiplication of radicals.

$$(A + B)^2 = A^2 + 2AB + B^2$$
$$(A - B)^2 = A^2 - 2AB + B^2$$
$$(A + B)(A - B) = A^2 - B^2$$

Rationalizing Binomial Denominators

- Expressions of the form $A + B$ and $A - B$ are said to be **conjugates** of one another.
- The product of square root conjugates does not contain a square root.
- If the denominator of a fraction is a binomial with at least one square root term, often we can rationalize the denominator by multiplying the numerator and the denominator by the conjugate of the denominator.

Speaking the Language 9.6

1. Because $5\sqrt{x^3}$ and $-2\sqrt{x^3}$ have the same index and the same radicand, the radicals are called ______.
2. The expressions $x + \sqrt{y}$ and $x - \sqrt{y}$ are called square root ______.
3. To remove the parentheses in $\sqrt{x}(\sqrt{x} + \sqrt{a})$, we use the ______ Property.
4. Multiplying $\sqrt{x} - \sqrt{5}$ by its conjugate results in an expression that does not have a(n) ______ term. This fact can be used to ______ a denominator.

Exercises 9.6

Concepts and Skills

1. Explain the difference between $\sqrt{7} \cdot \sqrt{7}$ and $\sqrt{7} + \sqrt{7}$.
2. Consider the following two uses of the Distributive Property.

$$3\sqrt{5} + 7\sqrt{5} = (3 + 7)\sqrt{5} = 10\sqrt{5}$$
$$3\sqrt{5} + x\sqrt{5} = (3 + x)\sqrt{5}$$

Explain why parentheses are needed in the second result but not in the first.

In Exercises 3–10, perform the indicated operation.

In this group, the radicals do not need to be simplified in order for us to perform the operations.

3. $5\sqrt{13} + \sqrt{13}$
4. $\sqrt{7} - 2\sqrt{7}$
5. $6\sqrt{x} + 10\sqrt{x}$
6. $5\sqrt{ax} - 6\sqrt{ax}$
7. $9\sqrt[3]{y} - 12\sqrt[3]{y}$
8. $3\sqrt[3]{x^2} + 7\sqrt[3]{x^2}$
9. $\sqrt{2x} + 4\sqrt{3} - 5\sqrt{2x} - 3\sqrt{3}$
10. $ab\sqrt{a} + 2\sqrt{b} - 2ab\sqrt{a} + \sqrt{b}$
11. Explain why the terms of $\sqrt{18} + \sqrt{27}$ cannot be combined with the expression in its present form. Use the Distributive Property in your explanation. What step should you take before deciding whether the two terms can be combined?
12. Explain how the Commutative and Associative Properties of Multiplication are used to determine the product $(3\sqrt{2})(5\sqrt{7})$.

In Exercises 13–22, perform the indicated operation.

13. $\sqrt{27} - \sqrt{12}$
14. $\sqrt{45} + \sqrt{80}$

15. $\sqrt[3]{250} + \sqrt[3]{54}$ **16.** $\sqrt[3]{24} - \sqrt[3]{192}$

17. $\sqrt[4]{162} - \sqrt[4]{32}$ **18.** $\sqrt[4]{48} + \sqrt[4]{243}$

19. $5\sqrt{32} + 7\sqrt{72}$ **20.** $10\sqrt{20} - 3\sqrt{45}$

21. $4\sqrt{12} - 2\sqrt{27} + 5\sqrt{8}$

22. $3\sqrt{50} - 2\sqrt{20} + 6\sqrt{18}$

In Exercises 13–36, radicals must be simplified before we combine terms.

In Exercises 23–36, perform the indicated operation.

23. $\sqrt{72x} - \sqrt{50x}$ **24.** $\sqrt{80c} + \sqrt{20c}$

25. $x\sqrt{48} + 3\sqrt{27x^2}$ **26.** $3\sqrt{48x^2y} - 2x\sqrt{75y}$

27. $4a\sqrt{20a^2b} + a^2\sqrt{45b}$ **28.** $5\sqrt{a^5} + 7a\sqrt{4a^3}$

29. $x\sqrt{2x} - 3\sqrt{8x^2} + \sqrt{50x^3}$

30. $\sqrt{9x^4} + x\sqrt{12x} - \sqrt{27x^3}$

31. $\sqrt{9xy^3} + 4\sqrt{x^3y} - 5y\sqrt{4xy}$

32. $3\sqrt{x^2} + \sqrt{x^9} - x^3\sqrt{x^3}$

33. $2b\sqrt[3]{54b^5} - 3\sqrt[3]{16b^8}$ **34.** $\sqrt[4]{32x^3} + \sqrt[4]{2x^3}$

35. $\sqrt[3]{-54x^3} - \sqrt[3]{-16x^3}$ **36.** $x\sqrt[4]{xy^4} - y\sqrt[4]{x^5}$

In Exercises 37–44, multiply and simplify the result.

37. $4\sqrt{3}(2\sqrt{3} - \sqrt{7})$ **38.** $2\sqrt{2}(3\sqrt{8} + \sqrt{5})$

39. $5\sqrt{7}(4\sqrt{3} + 7)$ **40.** $6\sqrt{11}(5\sqrt{5} - 8)$

41. $\sqrt{5}(\sqrt{35} - \sqrt{15})$ **42.** $\sqrt{7x}(\sqrt{28} + \sqrt{63x})$

43. $\sqrt{3}(\sqrt{y} - \sqrt{3y})$ **44.** $\sqrt{5}(\sqrt{10y} - 2y)$

CE In Exercises 45–48, supply the missing number to make each statement true.

45. $\_\_\_\_(3 - \sqrt{2}) = 3\sqrt{7} - \sqrt{14}$

46. $\_\_\_\_(2 - \sqrt{3y}) = 2y\sqrt{3} - 4\sqrt{y}$

47. $4(\_\_\_\_) = 4x + 8\sqrt{3}$

48. $\sqrt{3}(\_\_\_\_) = 3\sqrt{5} - 5\sqrt{3}$

49. Explain how to combine $\sqrt{64} + \sqrt[3]{64} + \sqrt[6]{64}$ even though the indices are not the same.

50. Which of the following is not a rational number? Explain your choice.

(i) The square of $2\sqrt{5}$.

(ii) The square of $2 + \sqrt{5}$.

(iii) The product of $2 + \sqrt{5}$ and its conjugate.

In Exercises 51–60, multiply and simplify the result.

51. $(\sqrt{3} - 2)(\sqrt{2} + \sqrt{3})$

52. $(3\sqrt{5} - 2\sqrt{3})(\sqrt{3} - 5)$

53. $(\sqrt{7x} - 3)(\sqrt{3} + \sqrt{7x})$

54. $(\sqrt{3t} - 2)(\sqrt{3} + 4t)$

55. $(2\sqrt{10} + \sqrt{5})(2\sqrt{2} - 5)$

56. $(\sqrt{x} - 5)(2\sqrt{x} + 3)$

57. $(\sqrt{2} - \sqrt{6})^2$ **58.** $(\sqrt{3} + \sqrt{15})^2$

59. $(\sqrt{5} + \sqrt{10})^2$ **60.** $(\sqrt{10} - \sqrt{2})^2$

In Exercises 61–66, determine the product of the expression and its conjugate.

61. $2\sqrt{2} - 3$ **62.** $3\sqrt{2} + 2$

63. $3\sqrt{2} + 4\sqrt{6}$ **64.** $4\sqrt{3} - 3\sqrt{5}$

65. $10 - 5t\sqrt{2}$ **66.** $7w + 3\sqrt{3}$

In Exercises 67–70, determine the product.

67. $\dfrac{\sqrt{3} + 2}{\sqrt{3} - 2} \cdot \dfrac{\sqrt{3} + 2}{\sqrt{3} + 2}$

68. $\dfrac{\sqrt{5} - \sqrt{3}}{\sqrt{3} + \sqrt{2}} \cdot \dfrac{\sqrt{3} - \sqrt{2}}{\sqrt{3} - \sqrt{2}}$

69. $\dfrac{\sqrt{7}}{\sqrt{3} + \sqrt{7}} \cdot \dfrac{\sqrt{3} - \sqrt{7}}{\sqrt{3} - \sqrt{7}}$

70. $\dfrac{5}{\sqrt{3} - 1} \cdot \dfrac{\sqrt{3} + 1}{\sqrt{3} + 1}$

In Exercises 71–82, rationalize the denominator.

71. $\dfrac{3}{3 - \sqrt{6}}$ **72.** $\dfrac{12}{\sqrt{5} - \sqrt{2}}$

73. $\dfrac{\sqrt{5} - \sqrt{3}}{\sqrt{5} + \sqrt{3}}$ **74.** $\dfrac{\sqrt{5} - \sqrt{2}}{\sqrt{7} + \sqrt{3}}$

75. $\dfrac{\sqrt{y} + 5}{\sqrt{y} - 4}$ **76.** $\dfrac{5z}{\sqrt{z} + z}$

77. $\dfrac{\sqrt{x}}{\sqrt{x} + 2}$ **78.** $\dfrac{\sqrt{3}}{\sqrt{3} - \sqrt{x}}$

79. $\dfrac{1}{\sqrt{x + 1} - 2}$ **80.** $\dfrac{2}{3 + \sqrt{x + 2}}$

81. $\dfrac{\sqrt{x} - \sqrt{y}}{\sqrt{x} + \sqrt{y}}$ **82.** $\dfrac{\sqrt{xy}}{\sqrt{x} + \sqrt{y}}$

CE In Exercises 83–86, show that the given number and its conjugate are both solutions of the equation.

83. $x^2 - 4x + 1 = 0;\quad 2 + \sqrt{3}$

84. $x^2 - 2x - 4 = 0; \quad 1 - \sqrt{5}$

85. $4x^2 - 12x + 7 = 0; \quad \dfrac{3 - \sqrt{2}}{2}$

86. $9x^2 - 12x - 41 = 0; \quad \dfrac{2 + 3\sqrt{5}}{3}$

87. Explain why $\dfrac{1}{\sqrt{5}} + \dfrac{\sqrt{2}}{\sqrt{10}}$ cannot be performed with the fractions in their given form. Must the denominators of the two fractions be rationalized before the fractions can be added? Why?

88. The following shows the start of two different approaches to performing the given addition.

$$\frac{1}{\sqrt{2}} + \frac{1}{\sqrt{3}} = \frac{1}{\sqrt{2}} \cdot \frac{\sqrt{2}}{\sqrt{2}} + \frac{1}{\sqrt{3}} \cdot \frac{\sqrt{3}}{\sqrt{3}} = \cdots$$

$$\frac{1}{\sqrt{2}} + \frac{1}{\sqrt{3}} = \frac{1}{\sqrt{2}} \cdot \frac{\sqrt{3}}{\sqrt{3}} + \frac{1}{\sqrt{3}} \cdot \frac{\sqrt{2}}{\sqrt{2}} = \cdots$$

Complete the two problems and verify that the results are the same. Then comment on the approach that you believe to be more efficient.

CE In Exercises 89–98, simplify.

You may want to show students that they can rationalize first and then find the LCD, or they can find the LCD, combine, and rationalize last.

89. $\dfrac{\sqrt{x} - 3}{\sqrt{x}}$

90. $\dfrac{5 + \sqrt{2t}}{2\sqrt{t}}$

91. $\dfrac{1}{\sqrt{5}} + \sqrt{45}$

92. $\dfrac{\sqrt{2}}{\sqrt{3}} + \dfrac{1}{\sqrt{6}}$

93. $1 - \dfrac{3}{\sqrt{2}}$

94. $\dfrac{1}{\sqrt{2}} - 2\sqrt{8}$

95. $2\sqrt{x+1} - \dfrac{2x}{\sqrt{x+1}}$

96. $\dfrac{x}{\sqrt{x+3}} - \sqrt{x+3}$

97. $\dfrac{2}{\sqrt{3}+2} - \dfrac{1}{\sqrt{3}-2}$

98. $\dfrac{\sqrt{x}}{\sqrt{x}+5} + \dfrac{2\sqrt{x}}{\sqrt{x}-5}$

CE In Exercises 99–104, rationalize the *numerator*. (A rationalized *numerator* is not one of the conditions for a simplified radical. However, in more advanced studies, there are occasions when we must rationalize the numerator. The technique is the same as that for rationalizing denominators.)

99. $\dfrac{\sqrt{5} + \sqrt{7}}{4}$

100. $\dfrac{\sqrt{11} - \sqrt{3}}{8}$

101. $\dfrac{\sqrt{x} + 3}{\sqrt{x} - 3}$

102. $\dfrac{5 - \sqrt{a}}{5 + \sqrt{a}}$

103. $\dfrac{\sqrt{x+2} - \sqrt{x}}{2}$

104. $\dfrac{\sqrt{x+h} - \sqrt{x}}{h}$

In Exercise 105, it is not intended that the student write and solve an equation. From their familiarity with square roots, students should be able to realize that the only consecutive square roots are 0 and 1.

105. Determine two consecutive integers whose square roots are also consecutive integers.

106. Consider two positive consecutive integers. Show that the reciprocal of the sum of their square roots is equal to the difference of their square roots. Also, explain how you know that both expressions are positive.

Exercises 107–108 are precursors to the Even Root Property in Section 9.7. By definition of a square root, if $d^2 = x^2 + 50^2$, then d is a square root of $x^2 + 50^2$.

Geometric Models

107. A 50-foot pole is supported by two guy wires, one attached to the top of the pole and the other attached below the top of the pole. The two guy wires, one 10 feet shorter than the other, are attached at the same point on the ground. (See figure.)

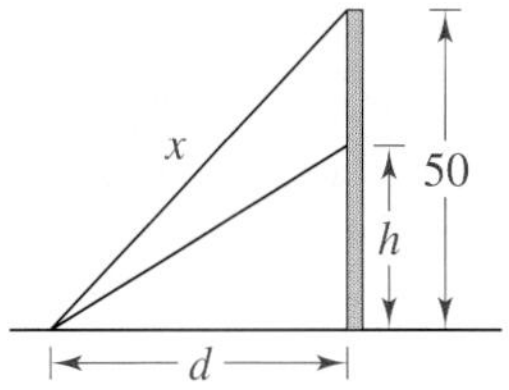

(a) Use the Pythagorean Theorem to write an expression in x for the distance d from the base of the pole to the point at which the wires are attached to the ground.

(b) Write an expression in x for the height h at which the shorter wire is attached to the pole.

108. A rectangular piece of paper is folded so that the top edge is along the side edge. (See figure.) Show that the length d of the diagonal is a multiple of $\sqrt{2}$ and that its value depends only on the width of the paper, not on the length.

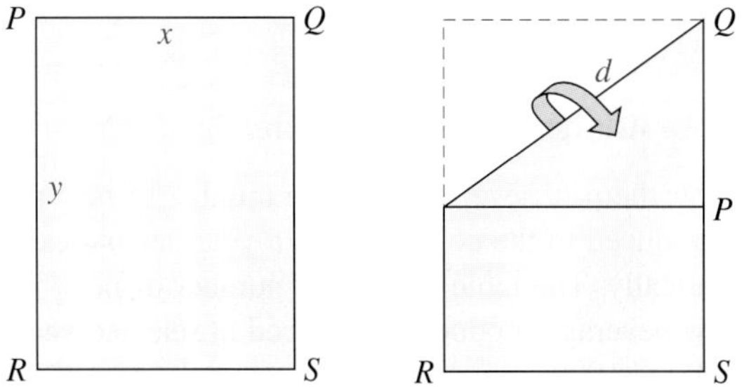

Real-Life Applications

109. Using T to represent the water temperature (in °F), a commercial fisherman determines that for each 100 pounds of fish in the nets, $\sqrt{T}$ pounds are not marketable fish and are returned to the water.

(a) Suppose that the water temperature is 60°F. What percentage of the catch is marketable?

(b) Write a simplified expression for the ratio of the number of pounds of fish that are not marketable to the number of pounds that are marketable.

110. A computer firm sells antivirus protection software. The probability that the consumer is not fully protected against a virus is $\dfrac{0.01\sqrt{D}}{0.01\sqrt{D}+1}$, where D is the number of days after the software product is developed.

(a) Write a simplified expression for the probability expression.

(b) What is the probability that a person is not protected after 30 days?

Data Analysis: *New Consumer Products*

Over the past several years, the number of new products introduced to the consumer each year has increased dramatically. The table shows the number of new food and new beverage products introduced in selected years from 1980 to 1999. (Source: Marketing Intelligence Service.)

	Number of New Products Introduced	
Year	**Food**	**Beverage**
1981	1192	256
1985	2327	585
1990	3453	630
1995	3891	809
1999	3793	1205

The following functions model the data as functions of the number of years since 1980.

Food: $F(t) = \dfrac{1252t}{\sqrt[5]{t^3}}$

Beverage: $B(t) = \dfrac{15.3t^2 + 135t}{\sqrt[5]{t^3}}$

111. Write the function $F(t)$ with a rationalized denominator.

112. Write the function $B(t)$ with a rationalized denominator.

113. Write a simplified function T to model the total number of new food and beverage products introduced.

114. On the basis of the model in Exercise 113, predict the total number of new food and beverage products that will be introduced in 2004.

Challenge

In Exercises 115 and 116, evaluate the expression.

115. $\sqrt[3]{3\sqrt[3]{3}}$

116. $\sqrt{3\sqrt{3\sqrt{3}}}$

In Exercises 117–120, perform the indicated operation.

117. $\sqrt[4]{x^2} + \sqrt{x}$

118. $\sqrt{12} + \sqrt[4]{9}$

119. $\sqrt[6]{8} - 2\sqrt[4]{4}$

120. $\sqrt[6]{x^2} - \sqrt[9]{x^3}$

121. Rationalize the denominator of

$$\frac{\sqrt{2}-\sqrt{3}}{\sqrt{2}+\sqrt{3}+\sqrt{5}}$$

122. Write the special factoring pattern for the difference of two cubes. Then explain how this pattern can be used to rationalize the denominator of

$$\frac{1}{\sqrt[3]{x}-\sqrt[3]{y}}$$

In Exercises 123 and 124, rationalize the denominator.

123. $\dfrac{3}{\sqrt[3]{x} - \sqrt[3]{2}}$

124. $\dfrac{5}{\sqrt[3]{x} - \sqrt[3]{3}}$

In Exercises 125 and 126, simplify the complex fraction and express the result with a rationalized denominator.

125. $\dfrac{\dfrac{1}{\sqrt{3}} + \dfrac{\sqrt{3}}{\sqrt{x}}}{\dfrac{\sqrt{x}}{\sqrt{3}} + \dfrac{3}{\sqrt{3x}}}$

126. $\dfrac{\dfrac{3}{\sqrt{y}} + \dfrac{3}{\sqrt{x}}}{\dfrac{1}{\sqrt{y}} + \dfrac{1}{\sqrt{x}}}$

9.7 Equations with Radicals and Exponents

Domain of a Radical Function • Radical Equations • Equations with Exponential Expressions

Domain of a Radical Function

A **radical function** is a function defined by a radical expression with a variable in the radicand. The domain of a radical function is the set of values of the variable for which the expression is defined.

Recall that $\sqrt[n]{b}$ is defined for any real number b if the index n is odd. However, if the index n is even, then b cannot be negative.

EXAMPLE 1 Determining the Domain of a Radical Function

Determine the domain of each function.

(a) $f(x) = \sqrt{1 - 2x}$ (b) $g(x) = \sqrt[3]{x + 3}$ (c) $h(x) = \sqrt[4]{x^2 + 1}$

Solution

(a) The expression $\sqrt{1 - 2x}$ is defined for all numbers for which $1 - 2x \geq 0$. To determine the domain, we solve the following inequality.

$$1 - 2x \geq 0$$
$$-2x \geq -1$$
$$x \leq \frac{1}{2}$$

The domain of the function is $\{x \mid x \leq \frac{1}{2}\}$.

For an odd root, the domain is all numbers for which the radicand is defined.

(b) Because the index is odd, $\sqrt[3]{x + 3}$ is defined for all real numbers. Therefore, the domain of the function is **R**.

For an even root, the domain is all numbers for which the radicand is defined and the radicand is nonnegative.

(c) Although the index is even, $x^2 + 1$ is positive for any replacement for x. Thus the domain of the function is **R**.

Radical Equations

An equation that contains a radical with a variable in the radicand is called a **radical equation.** The following are radical equations that we will solve in subsequent examples.

$$\sqrt{1 - 2x} = 3 \qquad \sqrt{6x + 13} = 2x + 1 \qquad \sqrt[3]{x + 1} - 3 = 1$$

The following property can be used to solve a radical equation that contains a square root radical.

Squaring Property of Equality

For real numbers a and b, if $a = b$, then $a^2 = b^2$.

Note: Although $a = b$ implies that $a^2 = b^2$, the reverse is *not* true. For example, $3^2 = (-3)^2$, but $3 \neq -3$. Therefore, $a^2 = b^2$ does not imply that $a = b$.

The Squaring Property of Equality can be used to eliminate a square root radical in a radical equation. For example, if $\sqrt{x} = 4$, then squaring both sides produces $x = 16$.

In general, if both sides of an equation are squared, then all solutions of the original equation will be solutions of the new equation. However, the new equation may have solutions that are not solutions of the original equation. For instance, the equation $x = 5$ has just one solution, 5. Squaring both sides produces $x^2 = 25$, which has two solutions, 5 and -5. In this case, -5 is an **extraneous solution.**

Note: Because extraneous solutions can occur when we solve a radical equation, we must check all solutions in the original equation.

EXAMPLE 2

Solving an Equation with One Radical

Solve the equation $\sqrt{1 - 2x} = 3$.

Solution

To solve the equation, we use the Squaring Property of Equality.

$$\sqrt{1 - 2x} = 3$$

$$\left(\sqrt{1 - 2x}\right)^2 = 3^2 \quad \text{Square both sides of the equation.}$$

$$1 - 2x = 9 \quad \text{The radical is eliminated.}$$

$$-2x = 8$$

$$x = -4$$

LEARNING TIP

A negative number may be a solution to a radical equation. However, we must reject any value that makes the radicand negative if the index of the radical is even.

We can check the solution by substitution.

When $x = -4$, $\sqrt{1 - 2x} = \sqrt{1 - 2(-4)} = \sqrt{9} = 3$.

EXAMPLE 3

Solving an Equation with One Radical

Solve the equation $\sqrt{6x + 13} = 2x + 1$.

Solution

$$\sqrt{6x + 13} = 2x + 1$$

$$\left(\sqrt{6x + 13}\right)^2 = (2x + 1)^2 \quad \text{Square both sides.}$$

$$6x + 13 = 4x^2 + 4x + 1 \quad \text{The radical is eliminated.}$$

Stress that we square both *sides*. We do not square each term.

$$0 = 4x^2 - 2x - 12$$ Write the resulting quadratic equation in standard form and factor.
$$0 = 2(2x^2 - x - 6)$$
$$0 = 2(2x + 3)(x - 2)$$
$$2x + 3 = 0 \quad \text{or} \quad x - 2 = 0$$ Zero Factor Property
$$2x = -3 \quad \text{or} \quad x = 2$$ Solve each case for x.
$$x = -\frac{3}{2} \quad \text{or} \quad x = 2$$

We can check the proposed solutions by substitution.

$$\sqrt{6x + 13} = 2x + 1 \qquad\qquad \sqrt{6x + 13} = 2x + 1$$
$$\sqrt{6\left(-\frac{3}{2}\right) + 13} = 2\left(-\frac{3}{2}\right) + 1 \quad x = -\frac{3}{2} \qquad\qquad \sqrt{6 \cdot 2 + 13} = 2 \cdot 2 + 1 \quad x = 2$$
$$\sqrt{4} = -3 + 1 \qquad\qquad \sqrt{25} = 4 + 1$$
$$2 = -2 \quad \text{False} \qquad\qquad 5 = 5 \quad \text{True}$$

Solving the equation produced two apparent solutions. However, $-\frac{3}{2}$ is an extraneous solution. Thus the equation has only one solution, 2.

The Squaring Property of Equality can be extended to any power: If $a = b$, then $a^n = b^n$. In words, both sides of an equation can be raised to the nth power. Remember, however, that doing this may produce extraneous solutions.

As illustrated in Example 4, an initial step in solving radical equations is to isolate the radical before raising both sides of the equation to a power.

EXAMPLE 4

Solving an Equation with a Third Root Radical

Solve the equation $\sqrt[3]{x + 1} - 3 = 1$.

You might want to explain why the radical must be isolated first. If this is not done, then the radical will not be eliminated.

Solution

$$\sqrt[3]{x + 1} - 3 = 1$$
$$\sqrt[3]{x + 1} = 4$$ Add 3 to both sides to isolate the radical.
$$\left(\sqrt[3]{x + 1}\right)^3 = 4^3$$ Raise both sides to the third power.
$$x + 1 = 64$$ The radical is eliminated.
$$x = 63$$

You should check that 63 is the solution.

To solve equations with two radicals, it may be necessary to square both sides more than once.

EXAMPLE 5

Solving an Equation with Two Radicals

Solve the equation $\sqrt{2a} = 2 + \sqrt{a - 2}$.

Solution

We need to isolate one of the radicals. It is usually better to isolate the one with the greater number of terms in the radicand.

Remind students about the special product for squaring a binomial. The middle term contains a radical.

$$\begin{aligned}
\sqrt{2a} &= 2 + \sqrt{a-2} && \\
\sqrt{2a} - 2 &= \sqrt{a-2} && \text{Isolate the radical on the right.} \\
\left(\sqrt{2a} - 2\right)^2 &= \left(\sqrt{a-2}\right)^2 && \text{Square both sides.} \\
2a - 4\sqrt{2a} + 4 &= a - 2 && \text{The radical on the right is eliminated.} \\
-4\sqrt{2a} &= -a - 6 && \text{Subtract } 2a \text{ and 4 to isolate the radical term.} \\
\left(-4\sqrt{2a}\right)^2 &= (-a-6)^2 && \text{Square both sides.} \\
16(2a) &= a^2 + 12a + 36 && \text{The radical is eliminated.} \\
32a &= a^2 + 12a + 36 && \\
0 &= a^2 - 20a + 36 && \text{Write the quadratic equation in standard form.} \\
0 &= (a-18)(a-2) && \text{Factor.}
\end{aligned}$$

$$a - 18 = 0 \quad \text{or} \quad a - 2 = 0 \qquad \text{Zero Factor Property}$$

$$a = 18 \quad \text{or} \quad a = 2$$

Check that 18 and 2 are both solutions.

EXAMPLE 6

Solving an Equation with Two Radicals

Solve the equation $1 = \sqrt{t-3} - \sqrt{2t-4}$.

Solution

The choice of radicals to isolate is arbitrary.

$$\begin{aligned}
1 &= \sqrt{t-3} - \sqrt{2t-4} && \\
1 + \sqrt{2t-4} &= \sqrt{t-3} && \text{Isolate one radical.} \\
\left(1 + \sqrt{2t-4}\right)^2 &= \left(\sqrt{t-3}\right)^2 && \text{Square both sides.} \\
1 + 2\sqrt{2t-4} + 2t - 4 &= t - 3 && \\
2\sqrt{2t-4} + 2t - 3 &= t - 3 && \\
2\sqrt{2t-4} &= -t && \text{Isolate the radical term.} \\
\left(2\sqrt{2t-4}\right)^2 &= (-t)^2 && \text{Square both sides.} \\
4(2t-4) &= t^2 && \\
8t - 16 &= t^2 && \\
0 &= t^2 - 8t + 16 && \text{Quadratic equation in standard form.} \\
0 &= (t-4)^2 && \text{Factor.} \\
0 &= t - 4 && \text{Zero Factor Property} \\
4 &= t &&
\end{aligned}$$

When $t = 4$, the left side of the equation has a value of 1, but the right side has a value of -1. Thus 4 is an extraneous solution. This equation has no solution.

Equations with Exponential Expressions

We now expand our equation-solving routines to include methods for solving equations that contain exponential expressions. There are two properties that we use to solve equations in the form $A^n = B$. The first is the Odd Root Property.

The Odd Root Property

Suppose that n is an odd integer, $n > 1$, and c is a real number. If $x^n = c$, then $x = \sqrt[n]{c}$.

EXAMPLE 7

Solving Equations with the Odd Root Property

Solve each equation.

(a) $y^5 = -32$

$y = \sqrt[5]{-32}$ Odd Root Property

$y = -2$

(b) $(x - 3)^7 = 1$

$x - 3 = \sqrt[7]{1}$ Odd Root Property

$x - 3 = 1$

$x = 4$

The other property that we use for solving equations that contain exponential expressions is the Even Root Property.

The Even Root Property

Suppose that n is a positive even integer and c is a real number.

1. If $c > 0$ and $x^n = c$, then $x = \sqrt[n]{c}$ or $x = -\sqrt[n]{c}$.
2. If $c = 0$ and $x^n = c$, then $x = 0$.
3. If $c < 0$, then $x^n = c$ has no real number solution.

Note: The symbol $\pm$ indicates two values, one positive and the other negative. Thus part 1 of the Even Root Property can be stated in a more compact form: If $c > 0$ and $x^n = c$, then $x = \pm\sqrt[n]{c}$.

EXAMPLE 8

Solving Equations with the Even Root Property

Solve each equation.

Because we stress that the symbol $\sqrt[n]{}$ refers to a nonnegative value when n is even, students tend to omit the $\pm$ sign.

(a) $x^4 = 81$

$x = \pm\sqrt[4]{81}$ Even Root Property with $c > 0$

$x = \pm 3$ Both 3 and -3 are solutions.

(b) $(x + 5)^6 = 0$

$x + 5 = 0$ Even Root Property with $c = 0$

$x = -5$

(c) $(x - 1)^2 - 4 = -1$ Isolate the variable term.

$(x - 1)^2 = 3$ Add 4 to both sides.

$x - 1 = \pm\sqrt{3}$ Even Root Property with $c > 0$

$x = 1 \pm \sqrt{3}$ Add 1 to both sides.

$x \approx 2.73$ or $x \approx -0.73$

To apply the Even Root Property or the Odd Root Property, we must isolate the exponential quantity.

(d) $2(t + 1)^2 + 3 = -3$

$2(t + 1)^2 = -6$ Subtract 3 from both sides.

$(t + 1)^2 = -3$ Divide both sides by 2.

Because $c < 0$, the Even Root Property states that the equation has no real number solution.

To solve equations of the form $A^{m/n} = B$ for A, raise both sides to the nth power to obtain an integer exponent on A.

EXAMPLE 9

Solving an Equation with a Rational Exponent

Solve the equation $(x - 2)^{2/3} = 4$.

Solution

$(x - 2)^{2/3} = 4$

$\left[(x - 2)^{2/3}\right]^3 = 4^3$ Raise both sides to the third power.

$(x - 2)^2 = 64$ Now the exponent is an integer.

$x - 2 = \pm 8$ Even Root Property

$x = 2 + 8 = 10$ or $x = 2 - 8 = -6$

You should check that both 10 and -6 are solutions.

Think About It

In Example 9, why not just raise both sides to the $\frac{3}{2}$ power and apply the properties of exponents?

Quick Reference 9.7

Domain of a Radical Function

- A **radical function** is a function defined by a radical expression with a variable in the radicand. The domain of a radical function is the set of values of the variable for which the expression is defined.
- The domain of an odd root radical function is the set of all values of the variable for which the radicand is defined. The domain of an even root radical function is the set of values of the variable for which the radicand is nonnegative.

Radical Equations

- Squaring Property of Equality

  For real numbers a and b, if $a = b$, then $a^2 = b^2$.

- The Squaring Property of Equality can be extended to any power:

  If $a = b$, then $a^n = b^n$.
- All solutions of radical equations must be checked because raising both sides of the equation to a power may produce extraneous solutions.

Equations with Exponential Expressions

- The Odd Root Property

  Suppose that n is an odd integer, $n > 1$, and c is a real number. If $x^n = c$, then $x = \sqrt[n]{c}$.
- The Even Root Property

  Suppose that n is an even positive integer and c is a real number.

  1. If $c > 0$ and $x^n = c$, then $x = \pm\sqrt[n]{c}$.
  2. If $c = 0$ and $x^n = c$, then $x = 0$.
  3. If $c < 0$, then $x^n = c$ has no real number solution.
- To solve equations of the form $A^{m/n} = B$ for A, raise both sides to the nth power to obtain an integer exponent on A. Then apply the Odd Root Property or the Even Root Property.

Speaking the Language 9.7

1. To eliminate the radicals in the equation $\sqrt{x} = \sqrt{3}$, we can use a property called the ________.
2. The solution of the equation $x = 7$ is 7. If we square both sides, the resulting equation has two solutions, 7 and -7. We call -7 a(n) ________.
3. The domain of an even root radical function is the set of all values of the variable for which the ________ is nonnegative.
4. Suppose that we are solving an equation of the form $(x + 2)^n = 5$. If $n = 3$, we apply the ________ Property, and if $n = 4$, we apply the ________ Property.

Exercises 9.7

Concepts and Skills

1. Consider the radical function $y = \sqrt{R}$, where R represents the radicand. Explain how to determine the domain of the function.
2. Explain how you can tell that $\sqrt{x - 3} = \sqrt{2 - x}$ has no solution without solving it. (*Hint:* Consider the domains of the radical expressions.)

In Exercises 3–14, determine the domain of the given radical function.

3. $y = \sqrt{x - 2}$
4. $y = \sqrt{x + 7}$
5. $y = \sqrt{2x - 3}$
6. $y = \sqrt{2x}$
7. $y = \sqrt{2 - x}$
8. $y = \sqrt{1 - 2x}$
9. $y = \sqrt{1 + 3x^2}$
10. $y = \sqrt{x^2 + 5}$
11. $y = \sqrt[3]{3 - 2x}$
12. $y = \sqrt[5]{5x + 7}$
13. $y = \sqrt[4]{x + 3}$
14. $y = \sqrt[6]{5 - x}$
15. If we square both sides of the radical equation $\sqrt{x} = -5$, we obtain $x = 25$. Are the two equations equivalent? Why? Explain why it is essential to check the values obtained when solving a radical equation.

16. Suppose you are solving the radical equation $\sqrt{x} + 1 = 5$ and you begin by squaring both sides. What is the resulting equation? Have you met your goal of eliminating the radical from the equation? If not, what should your first step have been?

In Exercises 17–42, solve the equation.

In Exercises 17–42, both sides are squared once to eliminate all radicals.

17. $\sqrt{5x + 1} = 4$

18. $\sqrt{2x + 1} = \frac{1}{2}$

19. $\sqrt[3]{3x - 1} = 2$

20. $\sqrt[3]{4x + 1} + 3 = 0$

21. $\sqrt{1 - 5x} + 6 = 3$

22. $\sqrt{x^2} = 3$

23. $\sqrt[4]{x + 1} = 2$

24. $\sqrt[4]{2x + 1} = \sqrt[4]{x + 6}$

25. $\sqrt{x^2 - 3x} = 2$

26. $\sqrt{3x^2 + 7x - 5} = 1$

27. $\sqrt{x}\sqrt{x - 5} = 6$

28. $x = 4 + \sqrt{32 - 2x}$

29. $x\sqrt{2} = \sqrt{5x - 2}$

30. $x\sqrt{3} = \sqrt{9x + 30}$

31. $\sqrt{4x + 13} = 2x - 1$

32. $\sqrt{2x - 3} = 5 - 2x$

33. $\sqrt[3]{x^2 + x + 2} = 2$

34. $\sqrt[3]{x^2 + 6x + 9} = 1$

35. $\sqrt[4]{x + 1} + \sqrt[4]{2x - 3} = 0$

36. $\sqrt[5]{x^2 - 7x + 9} = -1$

37. $\sqrt{x(x - 2)} - x = -10$

38. $\sqrt{x}\sqrt{x - 3} = x - 12$

39. $x - 9 = \sqrt{x^2 - x - 4}$

40. $x = \sqrt{45 - 11x} + 5$

41. $\sqrt{\frac{t}{3}} = \sqrt{\frac{t + 4}{2 + t}}$

42. $\sqrt{t + \frac{1}{t}} = \frac{\sqrt{17}}{2}$

43. In which of the following equations can we eliminate the radical by squaring both sides once?

(i) $\sqrt{2x} = x$

(ii) $\sqrt{2x} + 1 = x$

(iii) $\sqrt{2x} = \sqrt{x + 1}$

Describe the resulting equations in (i) and (ii).

44. For which one of the following equations will it be necessary to square both sides twice in order to solve it?

(i) $\sqrt{x + 1} = \sqrt{x} + 1$ (ii) $x + 1 = \sqrt{x + 1}$

Explain how the other equation can be solved by squaring both sides just once.

In Exercises 45–60, solve the radical equation.

In this group, both sides must be squared twice to eliminate radicals.

45. $\sqrt{x - 3} + 1 = \sqrt{x}$

46. $\sqrt{x + 5} = 1 + \sqrt{x}$

47. $\sqrt{4x - 3} - \sqrt{3x - 5} = 1$

48. $\sqrt{5x + 1} - \sqrt{3x - 5} = 2$

49. $\sqrt{5x - 9} + 3 = \sqrt{x}$

50. $\sqrt{x - 8} + \sqrt{x} = 2$

51. $\sqrt{7x + 4} - \sqrt{x + 1} = 3$

52. $\sqrt{7 - 2x} - 1 = \sqrt{2}\sqrt{x - 1}$

53. $\sqrt{x + 3} = 1 + \sqrt{x - 2}$

54. $\sqrt{x^2 + 3x - 1} - 1 = \sqrt{x^2 + x - 2}$

55. $\sqrt{5x - 6} = 1 + \sqrt{3x - 5}$

56. $\sqrt{2x - 1} - 1 = \sqrt{x - 1}$

57. $\sqrt{4x + 1} - \sqrt{3x - 2} = 1$

58. $\sqrt{2x + 3} - \sqrt{x - 2} = 2$

59. $\sqrt{7x - 3} - \sqrt{2x + 1} = 2$

60. $\sqrt{5x} - \sqrt{x + 4} = 2$

61. Suppose $x^{a/b} = k$. What is the first step in solving the equation for x? What is the purpose of this step?

62. Without solving, explain why the following equation has no solution: $\sqrt{x + 1} + 1 = 0$.

In Exercises 63–74, solve the given equation.

This group uses the Even Root Property or the Odd Root Property.

63. $x^2 = 6$

64. $\frac{1}{3}x^2 = 15$

65. $t^2 + 9 = 0$

66. $3x^2 - 27 = 0$

67. $(2x - 5)^2 = 16$

68. $25(x + 2)^2 = 16$

69. $(3x - 1)^2 = 3$

70. $(x + 2)^2 = 18$

71. $(x - 5)^5 = 0$

72. $(x - 3)^3 = -125$

73. $(2x - 3)^5 = -32$

74. $(2x + 3)^4 = -2$

In Exercises 75–82, solve the equations involving rational exponents.

75. $x^{2/3} = 9$

76. $y^{3/4} = -8$

77. $x^{-3/4} = 27$

78. $y^{-2/3} = 25$

79. $(x - 3)^{-2/3} = 4$

80. $(x + 5)^{-3/5} = 8$

81. $(3x - 2)^{3/4} = 8$

82. $(2x - 3)^{2/3} = 16$

You may want to show two methods for solving equations with negative exponents. Review solving equations with rational expressions, or raise both sides to the $-n$ power to leave a positive exponent on the variable term.

83. The square root of the product of a positive number and its square is twice the number. What is the number?

84. The fourth root of the cube root of the square root of a number is 2. What is the number?

In Exercises 85–88, determine the value of x.

85.

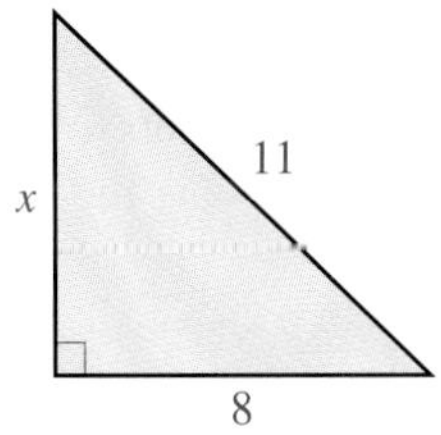

86.

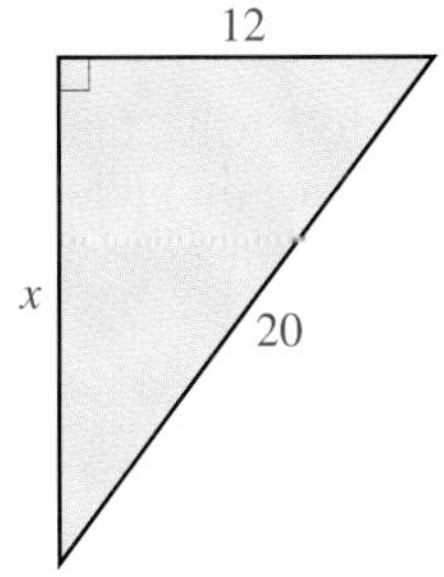

87.

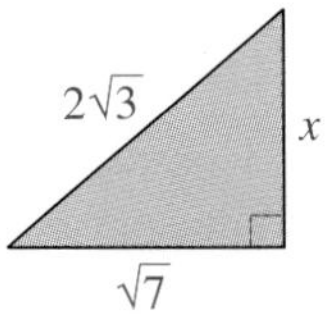

88.

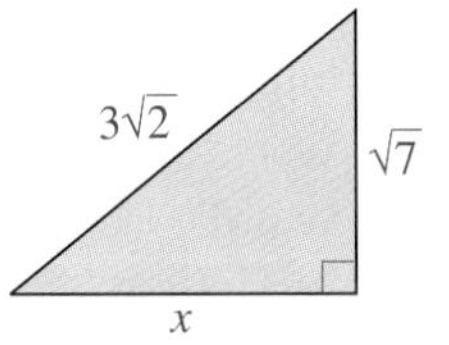

Geometric Models

89. A 26-foot ladder rests against a wall with the bottom of the ladder 7 feet from the base of the wall. How far up the wall does the top of the ladder rest? (See figure.)

Figure for 89

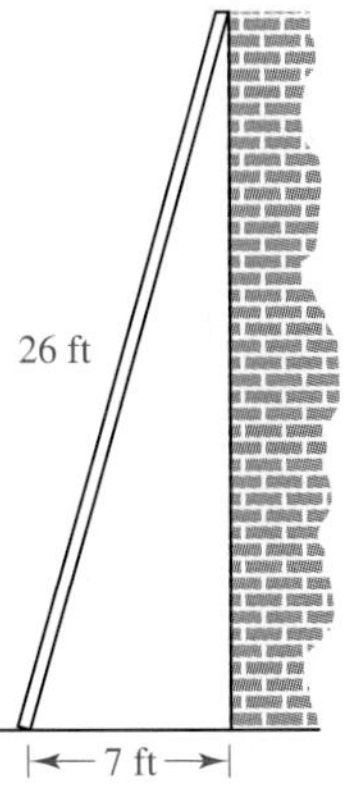

Figure for 90

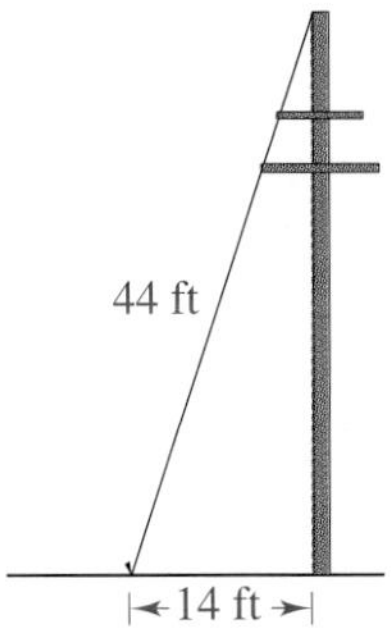

90. A guy wire 44 feet long reaches from the top of a telephone pole to a point on the ground 14 feet from the base of the pole. How tall is the telephone pole? (See figure.)

91. A billboard that is 30 feet high is supported by a brace that is anchored to the ground at a point 10 feet from the base of the sign. How long is the brace?

92. A 32-foot ramp connects a platform with a sidewalk that is at ground level. If the platform is 6 feet above ground level, what is the distance from the base of the platform to the sidewalk?

93. A surveyor needs to know the distance BC across the pond in the accompanying figure. She places a stake at point C so that the line segments $\overline{AC}$ and $\overline{BC}$ are at right angles. The trees are 150 feet apart, and the distance from the stake to the tree at point A is 125 feet. What is the distance from the stake to the tree at point B?

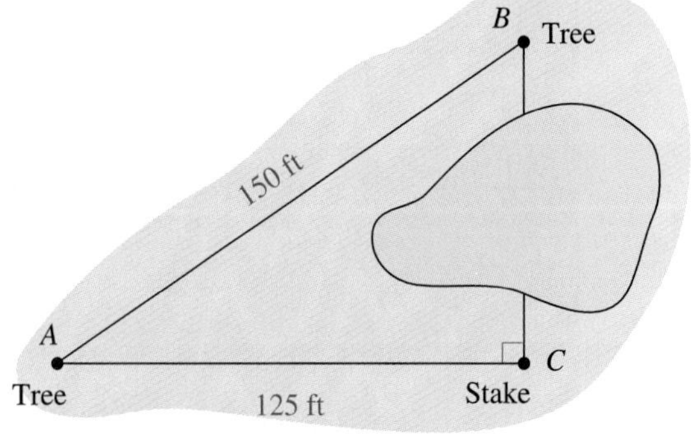

94. A car traveled south from Peoria at 46 mph. A second car left the same point at the same time and traveled west at 54 mph. How far apart were the cars after 1.5 hours?

Real-Life Applications

95. A circular rug was advertised as having an area of 95 square feet. Will the rug fit in a square room 11 feet on each side? Show how you know.

96. Clocks with pendulums can be adjusted to run faster or slower by changing the position of the weight at the end of the pendulum. The time t (in seconds) that it takes a pendulum to swing through one cycle is given by the formula $t = 2\pi\sqrt{\frac{L}{980}}$, where L is the length (in centimeters) of the pendulum. To the nearest tenth of a centimeter, what is the length of a pendulum that takes 1 second to swing through one cycle?

97. A dolphin trainer at Marine World estimated that the number of commands C that a dolphin learns to understand is a function of the number w of weeks of instruction: $C(w) = w - \sqrt{w}$. After how many weeks will the dolphin understand two commands?

98. The wildlife population of a certain area began to decrease as a result of development of the land. The population is modeled by the expression $270\sqrt{2y + 1} - 270\sqrt{3y - 5}$, where y is the number of years after development began. After how many years will the population reach 0?

Data Analysis: *College Financial Aid*

Loans have grown faster than grants as sources of federal government student financial assistance for a college education. The accompanying table compares the data (in billions of dollars) from the beginning to the end of the 1990s. (Source: U.S. Department of Education.)

Government Financial Aid for College (in billions of dollars)

	1991	1999
Grants	6.379	8.147
Loans	13.539	32.712

The following functions model the amounts of grants and loans (in billions of dollars), where x is the number of years since 1990.

Grants: $G(x) = \sqrt{3.2x + 37.4}$

Loans: $L(x) = 9.75\sqrt{x} + 3.45$

99. Write and solve an equation to predict in what year the amount from loans will be approximately \$40 billion.

100. Write and solve an equation to predict in what year the amount from grants will be approximately \$10 billion.

101. Write an equation to predict in what year the amount from loans will exceed the amount from grants by \$35 billion.

102. Verify that 18 is an approximate solution of the equation in Exercise 101. Interpret the solution.

Challenge

In Exercises 103–106, solve the equation.

103. $\sqrt[6]{3 + t - t^2} = \sqrt[3]{2t + 3}$

104. $\sqrt{\sqrt{t^2 + 3t + 12} + 3} = \sqrt{t + 5}$

105. $\sqrt{x + 2} - \sqrt{x - 3} = \sqrt{4x - 1}$

106. $\sqrt[4]{x + 3} = \sqrt{x + 1}$

107. If we square both sides of each of the following equations, the solution set of each appears to be the set of all real numbers. However, closer examination shows that the solution sets are all different. Solve the equations and compare solution sets.

(a) $\sqrt{x^2} = x$

(b) $\sqrt{x^2} = -x$

(c) $\sqrt{x^4} = x^2$

9.8 Complex Numbers

Imaginary Numbers • The Complex Number System • Addition and Subtraction • Multiplication and Division

Imaginary Numbers

Students find the concept of defining a set of numbers into existence very difficult to grasp. Their experience has been limited to real numbers, and they resist expanding the number system.

The equation $x^2 = -1$ has no solution in the set of real numbers. To be able to solve such equations, we begin expanding the number system by defining an **imaginary number** i.

Definition of the Imaginary Number i

The symbol i represents an **imaginary number** with the properties $i = \sqrt{-1}$ and $i^2 = -1$.

Using this new number i, we can define the square root of any real number.

Some attention to detail is probably needed here. We have always stressed that the symbol $-n$ does not necessarily represent a negative number. However, when we state that n represents a positive number, then $-n$ definitely represents a negative number.

Definition of $\sqrt{-n}$

For any positive real number n, $\sqrt{-n} = i\sqrt{n}$.

Note: The word *imaginary* is used to distinguish these new numbers from real numbers. It should not be inferred from their name that imaginary numbers do not really exist. Like all numbers in mathematics, imaginary numbers exist because we *define* them to exist.

EXAMPLE 1 **Using the Definition of $\sqrt{-n}$**

Write each square root in terms of i.

(a) $\sqrt{-49} = i\sqrt{49} = 7i$

(b) $\sqrt{-7} = i\sqrt{7}$

(c) $\sqrt{-\frac{4}{9}} = i\sqrt{\frac{4}{9}} = \frac{2}{3}i$

Note: In part (b) of Example 1, to avoid confusing $\sqrt{7}i$ with $\sqrt{7i}$, we usually write $i\sqrt{7}$.

Imaginary numbers are not real numbers, and the properties of real numbers do not apply to imaginary numbers. For example, we cannot apply the Product Rule for Radicals to $\sqrt{-12}\ \sqrt{-3}$ to obtain $\sqrt{36}$. However, some properties, such as the Distributive Property, do apply.

As we perform operations with imaginary numbers, we must begin with the definition of $\sqrt{-n}$.

EXAMPLE 2

Products of Imaginary Numbers

Determine the products.

(a) $\sqrt{-12}\sqrt{-3}$ (b) $\sqrt{-2}(\sqrt{-8}-\sqrt{-6})$

Solution

Stress that $\sqrt{-12}\sqrt{-3} \neq \sqrt{36}$.

(a) $\sqrt{-12}\sqrt{-3} = i\sqrt{12} \cdot i\sqrt{3}$ Definition of $\sqrt{-n}$

$= i^2 \cdot \sqrt{36}$ Product Rule for Radicals

$= -1 \cdot 6$ $i^2 = -1$

$= -6$

(b) $\sqrt{-2}(\sqrt{-8}-\sqrt{-6}) = i\sqrt{2}(i\sqrt{8} - i\sqrt{6})$ Definition of $\sqrt{-n}$

$= i^2\sqrt{16} - i^2\sqrt{12}$ Distributive Property

$= -1 \cdot 4 - (-1)\sqrt{4}\sqrt{3}$ Simplify the radicals.

$= -4 + 2\sqrt{3}$

We have defined $i = \sqrt{-1}$ and $i^2 = -1$. To facilitate computations, we need to investigate higher powers of i.

Exploring the Concept

Powers of i

Consider the following successive powers of i.

$$
\begin{aligned}
i &= && i\\
i^2 &= && -1\\
i^3 &= i^2 \cdot i = -1 \cdot i = && -i\\
i^4 &= (i^2)^2 = (-1)^2 = && 1\\
i^5 &= (i^2)^2 \cdot i = (-1)^2 \cdot i = 1 \cdot i = && i\\
i^6 &= (i^2)^3 = (-1)^3 = && -1\\
i^7 &= (i^2)^3 \cdot i = (-1)^3 \cdot i = -1 \cdot i = && -i\\
i^8 &= (i^2)^4 = (-1)^4 = && 1
\end{aligned}
$$

An alternative method is to observe that $i^n = i^r$, where r is the remainder when n is divided by 4, and to remember i^r for $r = 0, 1, 2,$ and 3.

By continuing these computations, we can see that all powers of i cycle through the numbers i, -1, $-i$, and 1. Furthermore, if we write powers of i as $(i^2)^n$ or $(i^2)^n \cdot i$, then the computations are reduced to raising -1 to an integer power.

EXAMPLE 3

Evaluating Powers of i

Evaluate.

(a) $i^{11} = (i^2)^5 \cdot i = (-1)^5 \cdot i = -1 \cdot i = -i$

(b) $i^{201} = (i^2)^{100} \cdot i = (-1)^{100} \cdot i = 1 \cdot i = i$

(c) $i^{48} = (i^2)^{24} = (-1)^{24} = 1$

The Complex Number System

We define an expanded number system, called the **complex number system,** to be numbers that can be written as the sum of a real number and a multiple of i.

Think About It

For what values of n is i^n a real number? an imaginary number? a complex number?

The word *complex* means consisting of more than one part. In the definition of a complex number, we refer to numbers consisting of a *real part* and an *imaginary part*.

Definition of a Complex Number

If a and b are real numbers, then $a + bi$ is the standard form of a **complex number.** If $b \neq 0$, then $a + bi$ is an imaginary number. If $b = 0$, then $a + bi$ is simply the real number a. For $a + bi$, a is called the **real part** and b is called the **imaginary part.**

Figure 9.1 shows the structure of the complex number system.

Sometimes the term *pure imaginary* is used to describe complex numbers with $a = 0$ and $b \neq 0$.

Figure 9.1

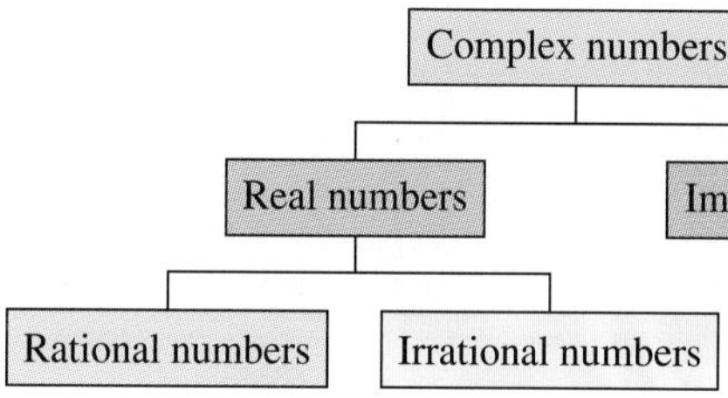

Just as real numbers are either rational or irrational, but not both, complex numbers are either real or imaginary, but not both.

The real number 5 is a complex number because it can be written $5 + 0i$. Similarly, the imaginary number $3i$ is a complex number because it can be written $0 + 3i$.

In Figure 9.2 we see that 3, $\sqrt{7}$, $4 + 0i$, $-\sqrt{5}$, and $\frac{7}{8}$ are real numbers. The numbers i, $0 - 3i$, $3 - 7i$, $\sqrt{-5}$, and $\frac{1}{2}i$ are imaginary numbers. Every number in the figure is a complex number.

Figure 9.2

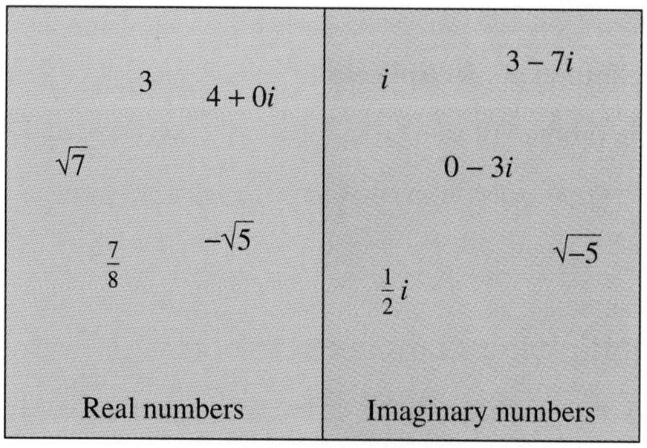

Complex Numbers

Two complex numbers are defined to be equal if their real parts are equal and their imaginary parts are equal.

Definition of Equality of Complex Numbers

$a + bi = c + di$ if and only if $a = c$ and $b = d$.

In the complex number system, addition and multiplication are commutative and associative. The Distributive Property also holds for complex numbers.

LEARNING TIP

Although i is not a variable, treating it as a variable in operations is often helpful.

When we add, subtract, and multiply, it is helpful to treat complex numbers as though they were binomials.

Addition and Subtraction

To add (or subtract) complex numbers, we add (or subtract) the real and imaginary parts.

Students should not feel that they must memorize such definitions. In words, add the real parts and add the imaginary parts.

Definitions of Addition and Subtraction

The sum and difference of two complex numbers $a + bi$ and $c + di$ are defined as follows:

$$(a + bi) + (c + di) = (a + c) + (b + d)i$$
$$(a + bi) - (c + di) = (a - c) + (b - d)i$$

EXAMPLE 4 Adding and Subtracting Complex Numbers

Compare this to combining like terms.

(a) $(3 - 2i) + (-4 + 7i) = (3 - 4) + (-2 + 7)i = -1 + 5i$

(b) $8 - (5 - 2i) = 8 - 5 + 2i = 3 + 2i$

(c) $(1 + i) - (1 - i) = 1 + i - 1 + i = 2i$

Multiplication and Division

To find the product of two complex numbers, use the Distributive Property or the FOIL method just as you did for binomials. Remember that $i \cdot i = i^2 = -1$.

Again, dissuade students from using this definition as a formula. Just use FOIL.

Definition of Multiplication

The product of two complex numbers $a + bi$ and $c + di$ is defined as follows:

$$(a + bi)(c + di) = (ac - bd) + (bc + ad)i$$

EXAMPLE 5 Multiplying Complex Numbers

(a) $5i(4 - 3i) = 20i - 15i^2$ Distributive Property

$= 20i - 15(-1)$

$= 15 + 20i$

(b) $(3 + 2i)(-1 + 4i) = -3 + 12i - 2i + 8i^2$ FOIL

$= -3 + 12i - 2i + 8(-1)$

$= -3 + 12i - 2i - 8$

$= -11 + 10i$

(c) $(2 + i)^2 = 2^2 + 4i + i^2$ Square of a binomial

$= 4 + 4i - 1$

$= 3 + 4i$

(d) $(5 + 4i)(5 - 4i) = 25 - 20i + 20i - 16i^2$ FOIL

$= 25 - 16(-1)$

$= 41$

In part (d) of Example 5, we call the pair of complex numbers $5 + 4i$ and $5 - 4i$ **complex conjugates.** Note that the product of these complex conjugates is the real number 41.

In general, the product of complex conjugates is the sum of the squares of the real parts and the imaginary parts, and the result is always a real number.

LEARNING TIP

Be careful not to confuse $(a + b)(a - b) = a^2 - b^2$ with $(a + bi)(a - bi) = a^2 + b^2$.

Product of Complex Conjugates

For any two complex conjugates $a + bi$ and $a - bi$,

$$(a + bi)(a - bi) = a^2 + b^2$$

To divide a complex number $a + bi$ by a nonzero real number c, we divide the real part and the imaginary part by c.

$$\frac{a + bi}{c} = \frac{a}{c} + \frac{b}{c}i$$

To divide a complex number by an imaginary number, we simply multiply the numerator and the denominator by the conjugate of the denominator.

EXAMPLE 6

Dividing Complex Numbers

(a) $\frac{4 + 6i}{2} = \frac{4}{2} + \frac{6}{2}i = 2 + 3i$

(b) $\frac{2 - i}{1 + 2i} = \frac{(2 - i)(1 - 2i)}{(1 + 2i)(1 - 2i)}$ Multiply the numerator and the denominator by the conjugate of the denominator.

$= \frac{2 - 4i - i - 2}{1 + 4}$ $(a + bi)(a - bi) = a^2 + b^2$

$= -\frac{5i}{5} = -i$

Compare this to the process of rationalizing a denominator.

(c) $\frac{5 + 2i}{2i} = \frac{(5 + 2i)(-2i)}{(2i)(-2i)}$ The conjugate of the denominator, $0 + 2i$, is $0 - 2i$, or $-2i$.

$= \frac{-10i + 4}{4}$ $(-2i)(2i) = -4i^2 = -4(-1) = 4$

$= 1 - \frac{5}{2}i$

Quick Reference 9.8

Imaginary Numbers

- The symbol i represents an **imaginary number** with the properties $i = \sqrt{-1}$ and $i^2 = -1$.
- For any positive real number n, $\sqrt{-n} = i\sqrt{n}$.
- To perform operations with radicals having negative radicands, use the definition of i to rewrite the radicals in the form of imaginary numbers.
- Powers of i can be evaluated by writing the expression in the form $(i^2)^n$ or $(i^2)^n \cdot i$, where n is an integer.

Complex Number System

- If a and b are real numbers, then $a + bi$ is the standard form of a **complex number.** If $b \neq 0$, then $a + bi$ is an imaginary number. If $b = 0$, then $a + bi$ is the real number a.
- For $a + bi$, a is called the **real part** and b is called the **imaginary part.**
- Imaginary numbers and real numbers are also complex numbers.
- The complex numbers $a + bi$ and $c + di$ are defined to be equal if and only if $a = c$ and $b = d$.

Addition and Subtraction

- The sum of two complex numbers is defined as

  $$(a + bi) + (c + di) = (a + c) + (b + d)i$$

- The difference of two complex numbers is defined as

  $$(a + bi) - (c + di) = (a - c) + (b - d)i$$

Multiplication and Division

- The product of two complex numbers can be found by treating the numbers as binomials and using the FOIL method. The definition is

  $$(a + bi)(c + di) = (ac - bd) + (bc + ad)i$$

- The complex numbers $a + bi$ and $a - bi$ are called **complex conjugates.**
- In general, $(a + bi)(a - bi) = a^2 + b^2$.
- If c is a nonzero real number, then $\frac{a + bi}{c} = \frac{a}{c} + \frac{b}{c}i$.
- To find the quotient of $a + bi$ and $c + di$, multiply the numerator and denominator by the conjugate of the denominator.

Speaking the Language 9.8

1. A number that can be written as the sum of a real number and a multiple of i is called a(n) ______ number.
2. The numbers 5 and $5i$ are both complex numbers. In particular, 5 is a real number, whereas $5i$ is a(n) ______ number.
3. The numbers $a + bi$ and $a - bi$ are called ______.
4. To add, subtract, and multiply complex numbers, we can informally treat the numbers as though they were ______.

Exercises 9.8

Concepts and Skills

1. To solve $x^2 = -1$, it is necessary to define a new type of number. What is the number and what two properties does it possess according to its definition?

2. To perform $\sqrt{-3}\,\sqrt{-5}$, we begin by writing $\sqrt{-3}\,\sqrt{-5} = i\sqrt{3} \cdot i\sqrt{5} = (i \cdot i) \cdot \left(\sqrt{3} \cdot \sqrt{5}\right)$. What two properties are used and for what purpose?

In Exercises 3–10, write the radical expression in the complex number form $a + bi$.

3. $\sqrt{-9}$ **4.** $\sqrt{-5}$

5. $\sqrt{-18}$ **6.** $\sqrt{-\dfrac{16}{25}}$

7. $\sqrt{25} + \sqrt{-36}$ **8.** $\sqrt{80} + \sqrt{-27}$

9. $\dfrac{-4 - \sqrt{-12}}{2}$ **10.** $\dfrac{-3 + \sqrt{-27}}{-3}$

11. Describe a method for simplifying i^{34}.

12. Explain how to add two complex numbers.

CE In Exercises 13–16, determine a and b.

This group uses the definition of equality of complex numbers.

13. $(a + 1) + 2bi = 3 - 2i$

14. $(2a + 1) + 3bi = 7 + 6i$

15. $2a + bi = \sqrt{-25} - 2$

16. $4a + 2bi = 3\sqrt{-4}$

17. Which of the following statements is true?

(a) The set of real numbers is a subset of the set of imaginary numbers.

(b) If a and b are both nonzero real numbers, then $a + bi$ is an imaginary number.

(c) The number 0 is both a real number and an imaginary number.

18. Classify the following as real numbers or imaginary numbers.

(a) $\sqrt{7}$ (b) $i\sqrt{7}$

(c) $\sqrt{-7}$ (d) $\sqrt{7} + \sqrt{-7}$

(e) $-\sqrt{7} + \sqrt{7}$ (f) $\sqrt{-7} - \sqrt{-7}$

In Exercises 19–30, perform the indicated operation.

19. $(5 + 6i) + (-4 - 5i)$

20. $(-8 + 4i) - (9 - 4i)$

21. $(7 + 3i) - (-8 - 7i)$

22. $(-5 + 2i) + (5 - 3i)$

23. $7i - 2(-3 - 7i)$ **24.** $8 - 3(5 + 2i)$

25. $(7 - 2i) - (3 - 4i) + 2(5 - i)$

26. $(3 - 4i) + 3(2 - 6i) - (3 + 2i)$

27. $\sqrt{-36} + \sqrt{-49}$ **28.** $\sqrt{-32} - \sqrt{-72}$

29. $3 \quad i \quad \sqrt{4} \quad \sqrt{\ \ 9}$

30. $\left(4 + \sqrt{-50}\right) - 4\left(3 - \sqrt{-72}\right)$

CE In Exercises 31–34, determine a and b.

31. $(a + 3i) + (2 + bi) = i$

32. $(a - 2i) - (7 + bi) = -4 - 8i$

33. $\left(5 - \sqrt{-4}\right) - (a + bi) = 6$

34. $(a + bi) + \left(\sqrt{4} - \sqrt{-16}\right) = -1 - 6i$

In Exercises 35–46, determine the product.

35. $(2i)(5i)$ **36.** $(-6i)(2i)$

37. $7i(3 - 4i)$ **38.** $-2i(-3 - 5i)$

39. $(2 - 3i)(3 - 4i)$ **40.** $(1 + i)(-3 + 5i)$

41. $(3 + i)^2$ **42.** $(2 - i)(-4 + 3i)$

43. $\sqrt{-8}\,\sqrt{-2}$ **44.** $\sqrt{-10}\,\sqrt{2}$

45. $\sqrt{-3}\left(\sqrt{-6} + \sqrt{-27}\right)$

46. $\left(\sqrt{-25} - \sqrt{4}\right)\left(\sqrt{-4} + \sqrt{25}\right)$

In Exercises 47–54, evaluate the given power of i.

47. i^{15} **48.** i^{45} **49.** i^{52}

50. i^{22} **51.** i^{30} **52.** i^{39}

53. i^{13} **54.** i^{24}

55. Suppose that a and b are nonzero real numbers. What is the difference in the results of the following operations?

(a) $(a + b)(a - b)$ (b) $(a + bi)(a - bi)$

56. To find the quotient $\dfrac{a + bi}{c + di}$, where a, b, c, and d are nonzero real numbers, what step must we take first?

In Exercises 57–62, write the complex conjugate of the given complex number. Then find the product of the complex number and its conjugate.

57. $7i$ **58.** $-4i$

59. $2 + 3i$ **60.** $3 - 4i$

61. $\sqrt{5} - 4i$

62. $7 - i\sqrt{2}$

In Exercises 63–74, determine the quotient.

63. $\dfrac{-3}{2i}$

64. $\dfrac{4}{5i}$

65. $\dfrac{9 + 6i}{3i}$

66. $\dfrac{-4 + 10i}{-2i}$

67. $\dfrac{3 - i}{1 + 2i}$

68. $\dfrac{4 + 5i}{3 - 2i}$

69. $\dfrac{3 - 4i}{-7 - 5i}$

70. $\dfrac{5 - 2i}{3 + 5i}$

71. $\dfrac{3}{2 + \sqrt{-3}}$

72. $\dfrac{-4}{1 - \sqrt{-2}}$

73. $\dfrac{\sqrt{-5} - \sqrt{3}}{\sqrt{-3} - \sqrt{9}}$

74. $\dfrac{\sqrt{6} + \sqrt{-3}}{3 - \sqrt{-2}}$

In Exercises 75–92, perform the indicated operations for each expression. This is a mixed group.

75. $(3 - 4i) + (5 + i)$

76. $(-2 + 3i) - (2 - 7i)$

77. $(4i)(-7i)$

78. $(-3i)(-i)$

79. $\dfrac{1 + 2i}{3i}$

80. $\dfrac{i - 4}{-2i}$

81. $3(2 + i) + 3i(1 - i)$

82. $5 - (4 - 5i)$

83. $(-6 + 5i)^2$

84. $(-2 - i)(3 + 2i)$

85. $\dfrac{3 + 4i}{3 - 4i}$

86. $\dfrac{2i}{5 - i}$

87. $7\sqrt{-25} - 4\sqrt{-49}$

88. $5\sqrt{-28} + 2\sqrt{-63}$

89. $\left(1 + \sqrt{-3}\right)\left(-1 - \sqrt{-12}\right)$

90. $\sqrt{-2}\left(\sqrt{18} - \sqrt{-8}\right)$

91. $\dfrac{\sqrt{-75}}{\sqrt{-3}}$

92. $\dfrac{\sqrt{6}\,\sqrt{-10}}{\sqrt{-2}}$

CE In Exercises 93–96, verify that the given number and its conjugate are both solutions of the given equation.

93. $x^2 - 4x + 13 = 0;\quad 2 + 3i$

94. $x^2 + 9 = 0;\quad -3i$

95. $x^2 - 4x + 5 = 0;\quad 2 - i$

96. $4x^2 + 3 = 0;\quad \dfrac{\sqrt{3}}{2}i$

The rule for the product of complex conjugates states that $(a + bi)(a - bi) = a^2 + b^2$. We can use this rule to factor expressions such as $x^2 + 4$ *over the complex numbers*.

Example: $x^2 + 4 = (x + 2i)(x - 2i)$

CE In Exercises 97–100, use the preceding method to factor the given expression over the complex numbers.

97. $x^2 + 9$

98. $x^2 + 25$

99. $4x^2 + 1$

100. $49x^2 + 16$

Challenge

101. For a nonzero complex number $a + bi$ and a positive integer n, $(a + bi)^{-n}$ is defined as

$$\frac{1}{(a + bi)^n}$$

Use this definition to evaluate the following.

(a) i^{-2}

(b) i^{-3}

(c) $(2 + 3i)^{-1}$

(d) $(1 - 2i)^{-2}$

102. Verify that $\dfrac{\sqrt{2}}{2}(1 + i)$ is a square root of i. What is your conjecture about another square root of i? Show that your conjecture is correct.

103. In the complex number system, a nonzero number has two square roots. For example, the square roots of 4 are 2 and -2; the square roots of -4 are $2i$ and $-2i$.

One cube root of -8 is -2. What is your conjecture about the total number of cube roots of a number in the complex number system? To support your conjecture, verify the following.

(a) $\left(1 + i\sqrt{3}\right)^3 = -8$

(b) $\left(1 - i\sqrt{3}\right)^3 = -8$

In Exercises 104 and 105, perform the indicated operations and express the result in the form $a + bi$.

104. $\dfrac{2i + 1}{i} - \dfrac{1 - 3i}{4 - i}$

105. $\dfrac{2 + 3i}{2 - 3i} + \dfrac{i}{1 - i}$

9 • Chapter Project

Visitors to National Parks and Monuments

James Bryce, a British ambassador to the United States, observed that the national parks are "the best idea America ever had." Theodore Roosevelt, who was instrumental in preserving the natural wonders in the United States, once said, "Leave it as it is. You cannot improve on it. The ages have been at work on it, and only man can mar it." Roosevelt's philosophy led to the establishment of the National Park Service in 1916 to preserve our natural and historic treasures. Today there are 357 sites and 80.7 million acres under the administration of the National Park Service.

The accompanying table shows the number of visitors (in millions) to national parks and to national monuments for selected years during the last decade. (Source: National Park Service.)

	Visitors (in millions)	
Year	**National Parks**	**National Monuments**
1993	59.8	26.5
1997	65.3	24.1
1998	64.5	23.6
1999	64.7	23.7
2000	65.5	24.0

The following functions model the number (in millions) of visitors to national parks and monuments as functions of the number of years since 1990.

National parks: $P(x) = 54.6x^{2/25}$

National monuments: $M(x) = 30.4x^{-3/25}$

1. Overcrowding in national parks has become such a serious problem that the number of visitors is sometimes restricted. If the National Park Service were to determine that 70 million is the maximum number of visitors that the national parks can absorb, in what year would that maximum be reached?
2. Our national monuments occupy 4.8 million acres. Write an expression in x that models the number of visitors per acre. Use this expression to predict the percent decrease in visitors per acre from 2000 to 2004.
3. Write a simplified expression (ratio) that compares the number of visitors to national parks to the number of visitors to national monuments. What is this ratio projected to be in the year 2005?
4. Wrangell–St. Elias National Park and Preserve in Alaska is the park with the largest land area, representing 9.95% of all the land administered by the National Park Service. Determine the number of acres in this park.
5. The smallest unit in the National Park Service system is Thaddeus Kosciuszko National Memorial in Pennsylvania. This unit occupies just 0.02 acre. If the land for this memorial were a square, what would be the length (in feet) of each side of the square? (1 acre = 43,560 square feet.)

6. State parks in the United States consist of 12,484,000 acres and had 783,400,000 visitors in 1997. After determining the total number N of acres of state parks in your state, calculate the ratio of your state's population to N for each of the last four census dates.

(a) What do these data suggest regarding the need for more state park land in your state?

(b) If the population of your state is increasing, how does that fact work against the establishment of new state park areas?

Chapter 9 Review Exercises

Section 9.1

1. If $x < 0$, then $\sqrt{x^2} = |x|$ and $\sqrt[3]{x^3} = x$. Explain why absolute value symbols are used in one result but not the other.

2. Because $3^2 = 9$ and $(-3)^2 = 9$, we say that 9 has two square roots, 3 and -3. Why, then, do we write $\sqrt{9} = 3$?

In Exercises 3–6, evaluate, if possible.

3. $-\sqrt{64}$
4. $-\sqrt[3]{-64}$
5. $-\sqrt[3]{64}$
6. $\sqrt[4]{-64}$

In Exercises 7–10, simplify the given expression.

7. $\sqrt[4]{x^8}$
8. $\sqrt{(3x)^4}$
9. $\sqrt{(x+4)^2}$
10. $7\sqrt[3]{x^9}$

11. Translate the given phrase into a radical expression.

(a) The square root of 10 minus a.

(b) The square root of the quantity 10 minus a.

12. The side of a cube is given by $s = \sqrt[3]{V}$, where V is the volume of the cube. What is the side of a cube whose volume is 4.096 cubic inches?

Section 9.2

13. Explain why $(-4)^{1/2}$ is not a real number.

14. When you convert $b^{m/n}$ into a radical, what are the interpretations of m and n?

In Exercises 15–19, evaluate the given expression, if possible.

15. $16^{3/2}$
16. $\left(-\dfrac{8}{27}\right)^{1/3}$
17. $(-27)^{-2/3}$
18. $-9^{-1/2}$
19. $(-9)^{-1/2}$

20. Write $4x^{6/7}$ and $(4x)^{6/7}$ as radical expressions.

21. Show how to evaluate $25^{1.5}$ without a calculator.

22. The length d of a diagonal of a rectangle is given by $d = (L^2 + W^2)^{1/2}$, where L and W are the length and width of the rectangle, respectively. To the nearest hundredth of a foot, how long is the diagonal of a rectangle whose width is 7 feet and whose length is 5 yards?

Section 9.3

23. Explain why the result of simplifying $(x^4)^{1/4}$ depends on whether x is nonnegative or is any real number.

24. Which of the following is a correct way to begin simplifying the given expression? Explain.

(i) $\left(\dfrac{4}{9}\right)^{-1/2} = \left(\dfrac{9}{4}\right)^{1/2} = \cdots$

(ii) $\left(\dfrac{4}{9}\right)^{-1/2} = \dfrac{4^{-1/2}}{9^{-1/2}} = \cdots$

25. Use the properties of exponents to evaluate each expression.

(a) $18^{1/2} \cdot 2^{1/2}$
(b) $8^{1/6} \cdot 8^{1/2}$

In Exercises 26–31, simplify the expression and write the result with positive exponents.

26. $\dfrac{f^{5/7}}{f^{3/7}}$ **27.** $(b^{-4/3}c^2)^9$

28. $(a^{-2})^{-1/4}$ **29.** $\left(\dfrac{27x^6z^{-9}}{y^{-3}}\right)^{1/3}$

30. $(2x^{-3/2})^{-1}$ **31.** $\dfrac{(3x^{-1})^{-2}}{x^{1/2}}$

32. Determine the values of a and b that make the statement true.

$$(x^a y^{12})^{1/4} = x^2 y^b$$

33. Simplify $x^{2/3}(x^{-5/3} - x^{7/3})$, $x \neq 0$. Write the result with positive exponents.

34. (a) Write $\sqrt[3]{\sqrt[5]{x}}$ with just one radical symbol.

(b) Write $\sqrt[10]{x^8y^6}$ with a smaller index.

Section 9.4

35. Explain why the Product Rule for Radicals cannot be used to multiply $\sqrt[3]{5}\sqrt{5}$. Show a method that can be used to perform the operation.

36. Explain why $\sqrt[3]{24}$ can be simplified but $\sqrt[3]{25}$ cannot.

In Exercises 37–44, simplify, if possible.

37. $(3\sqrt{5x})^2$ **38.** $\sqrt[3]{27a^6b^9}$

39. $\sqrt{6}\sqrt{7x}$ **40.** $\sqrt{a}\sqrt{a+1}$

41. $\sqrt{x^5}$ **42.** $\sqrt{45}$

43. $\sqrt{c^{17}d^{20}}$ **44.** $\sqrt{x^2+9}$

45. Assuming that x represents any real number, show that $\sqrt{x^2 - 6x + 9}$ represents the distance between x and 3 on a number line.

46. If the length and width of a rectangle are $2\sqrt{x}$ and $5\sqrt{x+3}$, find a simplified expression for the area.

Section 9.5

47. State the three conditions under which a radical is simplified.

48. Explain why we cannot rationalize the denominator of $\dfrac{1}{\sqrt[3]{x}}$ by multiplying the numerator and denominator by $\sqrt[3]{x}$.

In Exercises 49–52, simplify.

49. $\dfrac{\sqrt[3]{x^5}}{\sqrt[3]{x^2}}$ **50.** $\dfrac{\sqrt{9x^3y^4}}{\sqrt{4xy^2}}$

51. $\sqrt{\dfrac{5a^7}{9b^4}}$ **52.** $\sqrt{\dfrac{3x^{15}}{192x^5}}$

In Exercises 53–57, rationalize the denominator.

53. $\dfrac{6}{\sqrt{6}}$ **54.** $\sqrt{\dfrac{13}{3}}$

55. $\dfrac{3x}{\sqrt{x^5}}$ **56.** $\dfrac{1}{\sqrt[4]{8}}$

57. $\dfrac{\sqrt{18xy^5}}{\sqrt{2x^2y^3}}$

58. Show that $\dfrac{\sqrt[3]{x^2}}{\sqrt{x}}$ and $\dfrac{\sqrt{x}}{\sqrt[3]{x}}$ both have the same simplified form.

59. Use the Quotient Rule for Radicals to show that the square root of the reciprocal of a positive number is the reciprocal of the square root of the number.

60. One of the following expressions *cannot* be simplified with just one operation. Identify this expression and then simplify the other two expressions.

(a) $\dfrac{\sqrt{15}}{\sqrt{3}}$ (b) $\dfrac{\sqrt{16}}{\sqrt{3}}$ (c) $\dfrac{\sqrt{15}}{\sqrt{9}}$

Section 9.6

61. Explain why the terms in the given expression cannot be combined.

(a) $3\sqrt{x} + 3\sqrt{y}$

(b) $2\sqrt{x} + 2\sqrt[3]{x}$

62. Explain why we cannot rationalize the denominator of $\dfrac{1}{3+\sqrt{2}}$ by multiplying the numerator and denominator by $\sqrt{2}$. Describe the correct method.

In Exercises 63–65, perform the indicated operations.

63. $3\sqrt{45x} + 2\sqrt{20x}$

64. $\sqrt[3]{54} - \sqrt[3]{16}$

65. $(\sqrt{3} + \sqrt{5})^2$

66. Find the product of $\sqrt{x} + \sqrt{2}$ and its conjugate.

67. Supply the missing number to make the statement true.

$$(\quad)\left(\sqrt{3} + \sqrt{5x}\right) = \sqrt{6} + \sqrt{10x}$$

68. Determine which of the following is a rational number by evaluating each expression.

(a) $\left(3\sqrt{5}\right)^2$

(b) $\left(3 + \sqrt{5}\right)\left(3 - \sqrt{5}\right)$

(c) $\left(3 + \sqrt{5}\right)^2$

In Exercises 69 and 70, rationalize the denominator.

69. $\dfrac{5}{5 - \sqrt{3}}$

70. $\dfrac{\sqrt{x} + 7}{\sqrt{x} + 3}$

In Exercises 71 and 72, simplify.

71. $\dfrac{1}{\sqrt{3}} + \dfrac{5\sqrt{3}}{3}$

72. $\dfrac{1}{\sqrt{5} + a} - \dfrac{1}{\sqrt{5} - a}$

73. The lengths of the three sides of a triangle are $\sqrt{20}$, $\sqrt{125}$, and $\sqrt{180}$. Express the perimeter of the triangle as a single simplified radical.

74. The length of a rectangle is one more than the square root of a number x. If the area of the rectangle is $x + \sqrt{x}$, find a simplified expression for the width of the rectangle.

Section 9.7

75. What is an extraneous solution?

76. Describe the first step in solving an equation of the form $A^{m/n} = B$.

In Exercises 77–80, solve.

77. $(2x - 3)^3 = 27$

78. $\sqrt{6x + 1} = 2x - 3$

79. $\sqrt{5x - 4} - \sqrt{x + 3} = -1$

80. $(x + 2)^{2/3} = 9$

81. An old totem pole is supported by a 20-foot wire extending from the top of the pole to a point 13 feet from the base of the pole. How tall is the totem pole?

82. The square root of the sum of a number and 4 is 2 more than the square root of the difference of the number and 4. What is the number?

83. One number is 5 more than another number. If the product of their square roots is 6, what are the numbers?

84. A Texas ranch is in the shape of a square 60 miles on each side. If the ranch house is at the exact center of the square, how far is the house from any corner of the property?

Section 9.8

85. In the real number system, $\sqrt{-4}\sqrt{-9}$ cannot be performed. Why? What is the result in the complex number system?

86. What is the difference, if any, between the results of $(a + b)(a - b)$ and $(a + bi)(a - bi)$?

In Exercises 87–94, perform the indicated operations for each expression.

87. $(3 - 5i) - (5 - 7i)$

88. $\sqrt{-9} + (4 - 5i)$

89. $(6 - 7i)(2 + 3i)$

90. $(5 + i)^2$

91. $\dfrac{8 - 4i}{2}$

92. $\left(4 + \sqrt{-7}\right)\left(4 - \sqrt{-7}\right)$

93. i^{57}

94. $\dfrac{3 + 4i}{1 - 2i}$

95. If $i(a + bi) = 6$, determine a and b.

96. Which of the following statements is false?

(a) The number $a + bi$ is an imaginary number if b is not 0.

(b) The number $a + bi$ is a real number if b is 0.

(c) The number $a + bi$ is a complex number if b is 0.

(d) The number $a + bi$ is a real number, an imaginary number, and a complex number if neither a nor b is 0.

Looking Back

The following exercises review concepts and skills that you will need in Chapter 10.

1. Let $f(x) = 6 - x - x^2$.
   (a) Evaluate $f(0)$.
   (b) Determine x so that $f(x) = 0$.

In Exercises 2–5, solve the given equation by factoring.

2. $2x^2 - 3x - 5 = 0$
3. $9x^2 + 4 = 12x$
4. $5x - 15x^2 = 0$
5. $x^3 + 3x^2 - 4x - 12 = 0$
6. Use the Even Root Property to solve $(2x - 1)^2 = 16$.

In Exercises 7 and 8, use the given values of a, b, and c to evaluate the following expression.

$$\frac{-b + \sqrt{b^2 - 4ac}}{2a}$$

7. $a = 2, b = -7, c = 3$
8. $a = 4, b = -9, c = -3$

In Exercises 9 and 10, solve the given equation.

9. $\dfrac{12}{t^2 - 9} = \dfrac{2}{t - 3} - 1$
10. $x = 2 + \sqrt{2x - 1}$
11. Explain how to graph the inequality $y \le x - 6$.
12. Working together, an office worker and his assistant could complete the job of copying department personnel manuals in $2\frac{2}{5}$ hours. Working alone, the office worker could complete the job 2 hours sooner than the assistant. If the assistant began working alone at 8:00 A.M. and did not take a break, at what time would he complete the job?

Chapter 9 Test

1. Determine the domain of $f(x) = \sqrt{3 - x}$.
2. Simplify $(-8x^{-6}y^{12})^{1/3}$.
3. Explain the difference between $-25^{1/2}$ and $(-25)^{1/2}$.
4. Supply the missing number to make the statement true.
   $$\sqrt{\frac{\square}{49}} = \frac{2\sqrt{5}}{7}$$
5. Write $\sqrt[3]{\sqrt[4]{x^7}}$ as a single radical.
6. Write $\sqrt[3]{x} \cdot \sqrt{x}$ as a single radical expression.
7. Add $\dfrac{\sqrt{12}}{4} + \dfrac{\sqrt{27}}{6}$.
8. Multiply $(3\sqrt{5} - 4)(2\sqrt{5} + 1)$.
9. For the expression $\dfrac{\sqrt{b}}{\sqrt{a}}$, describe two conditions under which the denominator can be rationalized without multiplying the numerator and denominator by $\sqrt{a}$.
10. Rationalize the denominator of $\dfrac{3\sqrt{2} - 1}{\sqrt{2} + 3}$.
11. For the triangle in the accompanying figure, show that the length of the hypotenuse is $x + 2$.

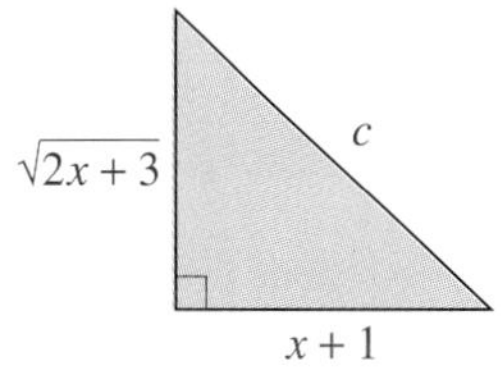

12. Simplify $\sqrt{98x^{11}y^{16}}$.
13. Evaluate $\left(\dfrac{16}{25}\right)^{-3/2}$.
14. Solve $(3t - 1)^{2/3} - 4 = 0$.
15. Explain why it is essential to check the numbers obtained when solving a radical equation.
16. Solve $3\sqrt{x} + 4 = 13$.

17. An **isosceles triangle** is a triangle with two sides of the same length. If the lengths of these two sides are $\sqrt{1 - 3x}$ and $\sqrt{3x + 1}$, how long are the two sides?

18. Subtract $(6 - 4i) - (7 - 10i)$.

19. What is the square of the complex number $3 + 2i$?

20. Divide $\frac{1 - 2i}{3 - i}$.

21. In what form must the numbers $\sqrt{-7}$ and $\sqrt{-3}$ be written before they can be multiplied?

22. Which *two* of the following statements are true?

(i) Every imaginary number is a complex number, but not every complex number is an imaginary number.

(ii) The product of complex conjugates is an imaginary number.

(iii) The radicals $\sqrt{4x^2}$, $\sqrt{x^2 + 4}$, and $\sqrt{x^2 + 4x + 4}$ can all be simplified.

(iv) The expressions $3 + \sqrt{5}$ and $3 - \sqrt{5}$ are both irrational numbers, but their product is a rational number.

23. The volume V of a cube is given by $V = s^3$, where s is the length of every side of the cube. Use a rational exponent to write a formula for s. Determine the side of a cube whose volume is 200 cubic inches. (Round to two decimal places.)

24. Simplify $\sqrt{3} + \sqrt{-3} + \sqrt{-3}\left(\sqrt{3} + \sqrt{-3}\right)$.

25. In each part, complete the sentence by stating the conditions under which the statement is true.

(a) If $A^n = B$ and n is an even positive integer, then the equation has no real number solution if ________.

(b) For a positive integer n, $\sqrt[n]{a^n} = a$ if ________ or if ________.

Chapter 10

Quadratic Equations

The bar graph shows that heart disease–related deaths increased significantly in the early 1990s. However, medical advances and good health practices brought about a decrease in such deaths after 1995.

By using a ***quadratic function*** *to model the data in the bar graph, we can graphically estimate the years when the number of deaths due to heart disease was above or below a specified level. And we can determine these years exactly by solving* ***quadratic inequalities.*** *(For more on this topic, see Exercises 95–98 at the end of Section 10.7, and see the Chapter Project.)*

In Chapter 10 we add the Quadratic Formula and other methods to our techniques for solving quadratic equations, equations that are quadratic in form, and applications. We learn about quadratic functions and properties of their graphs. The topic of quadratic inequalities concludes the chapter.

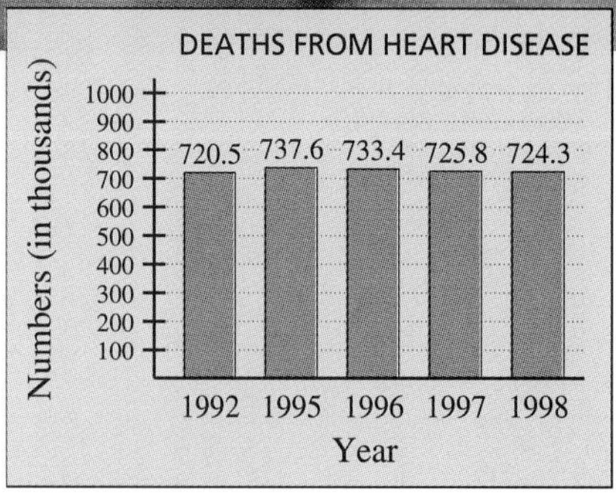

(Source: U.S. National Center for Health Statistics.)

10.1 Special Methods

Real Number Solutions • Complex Number Solutions

Real Number Solutions

In Section 7.4 we had our first experience with a class of equations known as **quadratic equations.** The following is a restatement of the definition of a quadratic equation.

Definition of a Quadratic Equation

A **quadratic equation** is an equation that can be written in the *standard form* $ax^2 + bx + c = 0$, where a, b, and c are real numbers and $a \neq 0$.

Associated with a quadratic equation is a **quadratic function.**

Definition of a Quadratic Function

A **quadratic function** is a function of the form

$$f(x) = ax^2 + bx + c$$

where a, b, and c are real numbers and $a \neq 0$.

Students tend to use the words *equation* and *function* interchangeably.

Note: Observe the difference between a quadratic *equation,* which can be written in the form $ax^2 + bx + c = 0$, and a quadratic *function,* which is written in the form $f(x) = ax^2 + bx + c$. The quadratic equation is the special case $f(x) = 0$.

The following example provides a review of previously discussed methods for solving a quadratic equation.

EXAMPLE 1 Quadratic Equations: Factoring Method and Even Root Property

Briefly review the available methods for solving quadratic equations and how to recognize the best method to use.

Solve the given equation.

(a) $x^2 - 2x - 8 = 0$ (b) $9x^2 - 12x + 4 = 0$

(c) $5 - x^2 = 0$ (d) $x^2 + 4 = 0$

Solution

(a) We use the factoring method.

$$x^2 - 2x - 8 = 0$$

$$(x - 4)(x + 2) = 0 \quad \text{Factor the trinomial.}$$

$$x - 4 = 0 \quad \text{or} \quad x + 2 = 0 \quad \text{Zero Factor Property}$$

$$x = 4 \quad \text{or} \quad x = -2 \quad \text{Solve each case for } x.$$

The solutions should be checked by substitution.

(b) We solve by factoring.

$$9x^2 - 12x + 4 = 0$$
$$(3x - 2)(3x - 2) = 0 \quad \text{Factor the trinomial.}$$
$$3x - 2 = 0 \quad \text{Zero Factor Property}$$
$$x = \frac{2}{3}$$

We can also write $(3x - 2)^2 = 0$ and use the Even Root Property.

The equation has only one solution, $\frac{2}{3}$, which is called a **double root.** The solution can be verified by substitution.

(c) To solve the equation, we use the Even Root Property:
If $x^2 = c$ and $c > 0$, then $x = \sqrt{c}$ or $x = -\sqrt{c}$.

Review the three cases for the number of real solutions when the Even Root Property is used. If $c > 0$, two solutions; if $c = 0$, one solution; if $c < 0$, no solutions.

$$5 - x^2 = 0$$
$$x^2 = 5 \quad \text{Write in the form } x^2 = c.$$
$$x = \sqrt{5} \approx 2.24 \quad \text{or} \quad x = -\sqrt{5} \approx -2.24 \quad \text{Even Root Property}$$

(d) The Even Root Property states that an equation of the form $x^2 = c$ has no real number solution if $c < 0$. Thus $x^2 = -4$ has no real number solution.

Complex Number Solutions

Think About It

Suppose that $(x + 3)^2 = k$. For what value(s) of k does the equation have the following solutions?
(i) $-3 \pm \sqrt{k}$ (ii) -3
(iii) no real number

In Example 1, the Even Root Property showed that the equation $x^2 = -4$ has no real number solution. An extension of the Even Root Property for equations of the form $x^2 = c$ enables us to determine complex number solutions of quadratic equations.

The Square Root Property

For any real number c, if $x^2 = c$, then $x = \sqrt{c}$ or $x = -\sqrt{c}$.

Note that for $c < 0$, x is an imaginary number. Thus the equation $x^2 = -4$ has two solutions: $\pm\sqrt{-4} = \pm 2i$.

Our next example illustrates how we can use the Square Root Property to solve other quadratic equations.

EXAMPLE 2

Using the Square Root Property to Solve Quadratic Equations

LEARNING TIP

The standard form $ax^2 + bx + c = 0$ is required for the factoring method, but the form $A^2 = c$, where A is an algebraic expression, is needed for us to apply the Square Root Property.

Use the Square Root Property to solve the given equation.

(a) $(2x + 1)^2 = 5$

$$2x + 1 = \pm\sqrt{5} \quad \text{Square Root Property with } c > 0$$
$$2x = -1 \pm\sqrt{5} \quad \text{Solve for } x.$$
$$x = \frac{-1 \pm \sqrt{5}}{2}$$

Recall that the symbol $\pm$ indicates two solutions—one evaluated with the plus symbol, the other evaluated with the minus symbol. Rounded to two decimal places, the solutions are 0.62 and -1.62.

(b) $y^2 + 9 = 0$

$y^2 = -9$ — Write in the form $A^2 = c$.

$y = \pm\sqrt{-9} = \pm i\sqrt{9} = \pm 3i$ — Square Root Property with $c < 0$

(c) $(t - 3)^2 = -5$

$t - 3 = \pm\sqrt{-5}$ — Square Root Property with $c < 0$

$t - 3 = \pm i\sqrt{5}$

$t = 3 \pm i\sqrt{5} \approx 3 \pm 2.24i$ — Add 3 to both sides to solve for t.

(d) $7 + 25(2x + 3)^2 = 0$ — Isolate the squared quantity.

$25(2x + 3)^2 = -7$

$(2x + 3)^2 = -\frac{7}{25}$

$2x + 3 = \pm\sqrt{-\frac{7}{25}}$ — Square Root Property with $c < 0$

$2x + 3 = \pm\frac{i\sqrt{7}}{5}$

$2x = -3 \pm \frac{i\sqrt{7}}{5}$ — Subtract 3 and divide by 2 to solve for x.

$x = -\frac{3}{2} \pm \frac{\sqrt{7}}{10}i \approx -1.5 \pm 0.26i$

Quick Reference 10.1

Real Number Solutions

- A **quadratic equation** is an equation that can be written in the *standard form* $ax^2 + bx + c = 0$, where a, b, and c are real numbers and $a \neq 0$.
- The previously discussed methods for solving a quadratic equation are as follows:
  1. Factoring Method
     (a) Write the quadratic equation in standard form and factor the polynomial.
     (b) Apply the Zero Factor Property and set each factor equal to zero.
     (c) Solve the resulting equations.
  2. Even Root Property
     If $x^2 = c$ and $c > 0$, then $x = \sqrt{c}$ or $x = -\sqrt{c}$.

Complex Number Solutions

- Square Root Property
  For any real number c, if $x^2 = c$, then $x = \sqrt{c}$ or $x = -\sqrt{c}$.
- In the Square Root Property, if $c < 0$, then x is an imaginary number.
- The Square Root Property enables us to determine complex number solutions of a quadratic equation.

Speaking the Language 10.1

1. The equation $2x^2 - 5x + 3 = 0$ is an example of a(n) ______ equation.
2. The equation $x^2 + 6x + 9 = 0$ has only one solution. We call that solution a(n) ______.
3. The easiest way to solve $(x + 4)^2 = 7$ is to use the ______ Property.
4. According to the ______ Property, the solutions of $(x + 4)^2 = -2$ are ______ numbers.

Exercises 10.1

Concepts and Skills

1. Explain why the Zero Factor Property cannot be used to solve all quadratic equations.
2. In order for us to use the factoring method to solve a quadratic equation, one side must be 0. Why?

In Exercises 3–12, use the factoring method to solve.

3. $x^2 + 7x + 6 = 0$
4. $4x^2 + 3 = 8x$
5. $x^2 = x$
6. $25 = 10x - x^2$
7. $\dfrac{3x}{4} = \dfrac{1}{2} + \dfrac{2}{x}$
8. $x = \dfrac{28}{x + 3}$
9. $x(x - 3) = x - 3$
10. $(3x - 1)^2 - 25x^2 = 0$
11. $8x^2 + x - 1 = 3x^2 - 2x + 1$
12. $2x(x + 3) = -5(3 + x)$
13. The equations $6 - x^2 = 0$ and $6 + x^2 = 0$ are very similar, but their solution sets are quite different. Use the Even Root Property to explain why one equation has two solutions and the other equation has no real number solution.
14. Suppose that $x^2 = c$, where $c < 0$. Describe the difference between the Even Root Property and the Square Root Property as each property applies to the solution(s) of the equation.

In Exercises 15–22, solve by using the Square Root Property.

This group has real number solutions.

15. $(5x - 4)^2 = 9$
16. $(3x - 2)^2 = 15$
17. $16(x + 5)^2 - 5 = 0$
18. $9(x + 3)^2 - 4 = 0$
19. $\left(x - \dfrac{3}{2}\right)^2 = \dfrac{9}{4}$
20. $(4 - x)^2 = \dfrac{36}{25}$
21. $x(x - 6) = 3(3 - 2x)$
22. $x(x + 12) = 12(3 + x)$

In Exercises 23–40, determine the complex number solutions of the given equation.

In Exercises 23–40, the equations have imaginary solutions as well as real solutions. All may be solved with the Square Root Property.

23. $t^2 + 16 = 0$
24. $25 + t^2 = 0$
25. $24 + y^2 = 0$
26. $y^2 + 18 = 0$
27. $(x - 3)^2 = -4$
28. $(x + 4)^2 = -9$
29. $(t - 1)^2 + 169 = 144$
30. $(2t + 1)^2 + 25 = 16$
31. $(2x + 5)^2 + 6 = 0$
32. $(3x + 7)^2 + 5 = 0$
33. $8(x + 1)^2 + 21 = 3$
34. $25(x + 2)^2 + 11 = 0$
35. $10 + \dfrac{4}{y^2} = 1$
36. $40 + \dfrac{25}{y^2} = 4$
37. $4 = \dfrac{9}{(1 - x)^2}$
38. $\dfrac{16}{(2 - x)^2} = 49$
39. $x(x + 4) = -4(1 - x)$
40. $x(8 - x) = 8(2 + x)$
41. In what form must a quadratic equation be written in order to solve the equation with (a) the factoring method and (b) the Square Root Property?
42. When solving a quadratic equation, how can we recognize when to use the factoring method and when to use the Square Root Property?

In Exercises 43–64, solve.

This group uses a variety of methods, and the solutions may be imaginary or real.

43. $3x = x^2$
44. $12x^2 = 21 + 4x$
45. $t^2 + 45 = 0$
46. $x^2 + 20 = 0$
47. $36x^2 + 12x + 1 = 0$
48. $9 - 6t + t^2 = 0$
49. $y^2 = -1.44$
50. $x^2 = 1.21$

51. $x(x-5) = 5(2-x)$

52. $x^2 + 6x = 6(3+x)$

53. $x - 2 = \frac{35}{x}$

54. $\frac{7}{x} + 4 = \frac{15}{x^2}$

55. $(4x+5)^2 + 49 = 0$

56. $(3x-4)^2 + 36 = 0$

57. $\left(\frac{3x+2}{4}\right)^2 = 12$

58. $\left(\frac{x-1}{2}\right)^2 = 6$

59. $t(t+3) = 3(t-2)$

60. $5(x^2+15) = 2x^2$

61. $x(x-4) = x - 4$

62. $6x^2 + 9x + 3 = 1 - 4x^2$

63. $(4x-1)(x-1) = 7x - 8$

64. $(4x+1)(5x+1) = x(-5x-1)$

Given two numbers, we can write a quadratic equation for which the numbers are solutions. For example, if 2 and -5 are solutions, then

$$x = 2 \quad \text{or} \quad x = -5$$
$$x - 2 = 0 \quad \text{or} \quad x + 5 = 0$$
$$(x-2)(x+5) = 0$$
$$x^2 + 3x - 10 = 0$$

Note that the equation is not unique. Multiplying both sides by any nonzero number produces an equivalent equation.

CE In Exercises 65–72, use the preceding technique to write a quadratic equation with integer coefficients that has the given pair of numbers as solutions.

65. 3, 5

66. $-1, -2$

67. 3, 3

68. 0, 0

69. $3, \frac{4}{3}$

70. $\frac{2}{3}, \frac{3}{4}$

71. $-4i, 4i$

72. $i\sqrt{3}, -i\sqrt{3}$

73. The product of two consecutive negative integers is 72. What are the two integers?

74. A book lies open so that the product of the facing page numbers is 1806. To what possible pages is the book open?

Real-Life Applications

75. A physical education major is doing his student teaching at an elementary school. One day, he lines up his fifth-grade students and asks them to count off by even numbers beginning with 2. Two students standing side by side have numbers whose product is 360. What possible numbers could they have?

76. As a project for an elementary science class, a teacher launches a model rocket, and the students use a stop watch to time the flight. The teacher knows that the height h after t seconds is given by $h = 96t - 16t^2$. What flight time does the teacher anticipate that the students will record?

77. A sentry walks back and forth from his guardhouse to the end of a building. His distance d from the guardhouse at any given time t (in minutes) is given by $d = 18t^2 - 90t$. How long does the sentry take to complete one round trip?

78. Two trucks left a truck stop together. One traveled north and the other traveled east. When the trucks were 50 miles apart, the eastbound truck had traveled 10 miles farther than the northbound truck. How far had each truck traveled?

Modeling with Real Data

79. The accompanying bar graph shows the number of cellular phone subscribers (in millions) for selected years in the period 1990 to 1998. (Source: Cellular Telecommunications Industry Association.)

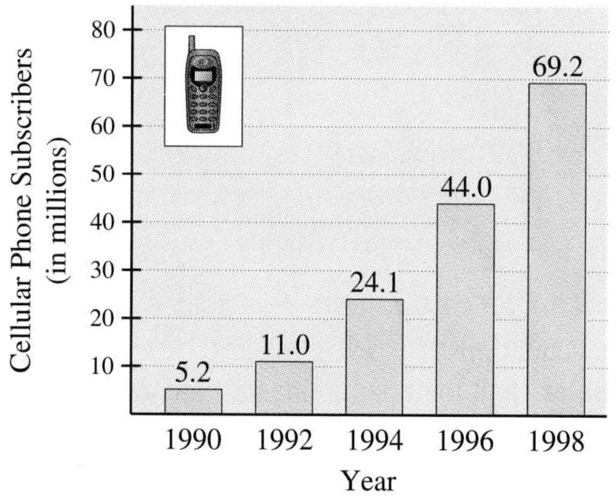

The function $C(t) = 0.82t^2 + 1.5t + 5$ models the number C of subscribers (in millions) as a function of the number of years since 1990.

(a) Write an equation that you can use to estimate the year in which 212 million subscribers are predicted.

(b) Verify that your equation in part (a) can be written as $(0.82t + 13.8)(t - 15) = 0$. Then determine the solution that corresponds to a year after 1990.

80. The accompanying table shows that the percentage of families who have a general-purpose credit card varies according to the age of the family head. (Source: Federal Reserve System.)

Age of Head of Family	Percentage of Age Group with Credit Cards
30	59.0
40	68.5
50	75.4
60	71.9
70	68.3
80	54.6

The function $P(x) = -0.028x^2 + 3x - 6.7$ models the data as a function of the age of the family head.

(a) Write an equation to estimate the age at which two-thirds of families have a credit card.

(b) Show that 38 and 69 are approximate solutions of your equation in part (a).

Challenge

In Exercises 81–84, solve for x. Assume all variables (other than x) represent positive numbers.

81. $(x + 2)^2 = -y$

82. $(x + a)^2 = 9$

83. $(x + c)^2 = -16$

84. $(ax + 1)^2 + 25 = 9$

85. Show that both $x - 1 + i$ and $x - 1 - i$ are factors of $x^2 - 2x + 2$. Then use the Zero Factor Property to solve $x^2 - 2x + 2 = 0$.

86. Show that both $x - 2 - 3i$ and $x - 2 + 3i$ are factors of $x^2 - 4x + 13$. Then use the Zero Factor Property to solve $x^2 - 4x + 13 = 0$.

In Exercises 87 and 88, use the Square Root Property to solve.

87. $(2x + 5)^2 = (x - 1)^2$

88. $(3 - x)^2 = (1 - 3x)^2$

10.2 Completing the Square

Square of a Binomial • Solving Quadratic Equations • Complex Number Solutions

Square of a Binomial

In the preceding section we used the Square Root Property to solve equations such as $(x - 3)^2 = 5$ and $(x + 2)^2 = -4$. Using a method called **completing the square,** we can write any quadratic equation in the form $(x \pm h)^2 = k$.

Eventually, this method will not be a desirable solving technique. However, the concept is important in deriving the Quadratic Formula.

The method is based on the special products for the square of a binomial.

$$(A + B)^2 = A^2 + 2AB + B^2$$
$$(A - B)^2 = A^2 - 2AB + B^2$$

Exploring the Concept

Completing a Perfect Square Trinomial

For the trinomial $x^2 + bx + c$, we begin with the first two terms and determine the constant c so that $x^2 + bx + c$ is a perfect square trinomial.

For instance, to determine c so that $x^2 + 8x + c$ is a perfect square, we compare the expression to the model for a perfect square trinomial.

$$x^2 + 8x + c$$
$$x^2 + 2Bx + B^2$$

For the coefficients of x to be equal, $2B = 8$ and so $B = 4$. Because $B^2 = 16$, the perfect square trinomial is $x^2 + 8x + 16$. In general, we can determine the constant term c so that $x^2 + bx + c$ is a perfect square by squaring half of the coefficient of x: $\left(\frac{b}{2}\right)^2$.

Stress that the leading coefficient must be 1.

Completing a Perfect Square Trinomial

If the first part of a perfect square trinomial is $x^2 + bx$, then the entire perfect square trinomial is $x^2 + bx + \left(\frac{b}{2}\right)^2$.

In words, the constant term is the square of half the coefficient of x.

EXAMPLE 1 Completing a Perfect Square Trinomial

Write a perfect square trinomial whose first two terms are given.

(a) $x^2 + 12x +$ ____

(b) $x^2 - 5x +$ ____

Note that the leading coefficient is 1 in both parts of the example.

Solution

(a) Half the coefficient of x is 6 and $6^2 = 36$. Thus the constant term is 36 and the trinomial is $x^2 + 12x + 36$.

(b) Half the coefficient of x is $-\frac{5}{2}$ and $\left(-\frac{5}{2}\right)^2 = \frac{25}{4}$. The constant term is $\frac{25}{4}$ and the trinomial is $x^2 - 5x + \frac{25}{4}$.

Think About It

To complete the square for $x^2 + ix$, would the constant term be a real number or an imaginary number?

Solving Quadratic Equations

By completing the square, we can write any quadratic equation in the form $(x \pm h)^2 = k$. Then we can use the Square Root Property to solve the equation.

Solving a Quadratic Equation by Completing the Square

1. If necessary, divide both sides by the leading coefficient to obtain a coefficient of 1 for the quadratic term.
2. Write the equation in the form $x^2 + bx = k$.
3. Add $\left(\frac{b}{2}\right)^2$ to both sides to complete the square.
4. Factor the resulting perfect square trinomial.
5. Solve with the Square Root Property.

Help students to see that it is the constant term that determines whether the trinomial is a perfect square. If it is not the "right" constant, we simply move it out of the way.

EXAMPLE 2 Solving Quadratic Equations by Completing the Square

(a)
$x^2 - 10x - 1 = 0$

$x^2 - 10x = 1$ Add 1 to both sides.

$x^2 - 10x + 25 = 1 + 25$ Add $\left(\frac{-10}{2}\right)^2 = (-5)^2 = 25$ to both sides to complete the square.

$(x - 5)^2 = 26$ Factor the perfect square trinomial.

$x - 5 = \pm\sqrt{26}$ Square Root Property

$x = 5 \pm \sqrt{26}$ Add 5 to both sides to solve for x.

$x \approx 10.10$ or $x \approx -0.10$

LEARNING TIP

Keep in mind that we complete the square in order to take advantage of the Square Root Property. However, if the expression can be factored, the factoring method is usually easier.

Emphasize this step. The purpose is to make the leading coefficient 1.

(b)

$$x(2x + 3) = 6$$

$$2x^2 + 3x = 6 \qquad \text{Remove parentheses.}$$

$$x^2 + \frac{3}{2}x = 3 \qquad \text{Divide both sides by 2.}$$

$$x^2 + \frac{3}{2}x + \frac{9}{16} = \frac{3}{1} + \frac{9}{16} \qquad \text{Add } \left(\frac{1}{2}\cdot\frac{3}{2}\right)^2 = \left(\frac{3}{4}\right)^2 = \frac{9}{16} \text{ to both sides to complete the square.}$$

$$\left(x + \frac{3}{4}\right)^2 = \frac{57}{16} \qquad \text{Add the fractions and factor.}$$

$$x + \frac{3}{4} = \pm\frac{\sqrt{57}}{4} \qquad \text{Square Root Property}$$

$$x = -\frac{3}{4} \pm \frac{\sqrt{57}}{4} \qquad \text{Subtract } \frac{3}{4} \text{ from both sides to solve for } x.$$

$$x \approx 1.14 \quad \text{or} \quad x \approx -2.64$$

(c)

$$x^2 + 6x + 8 = 0$$

$$x^2 + 6x = -8 \qquad \text{Subtract 8 from both sides.}$$

$$x^2 + 6x + 9 = -8 + 9 \qquad \text{Add } \left(\frac{6}{2}\right)^2 = 9 \text{ to both sides to complete the square.}$$

$$(x + 3)^2 = 1 \qquad \text{Factor the perfect square trinomial.}$$

$$x + 3 = \pm\sqrt{1} \qquad \text{Square Root Property}$$

$$x = -3 \pm 1 \qquad \text{Subtract 3 from both sides to solve for } x.$$

$$x = -2 \quad \text{or} \quad x = -4$$

Note: The equation in part (c) of Example 2 could have been solved by factoring. We used the method of completing the square to illustrate that the method is applicable to any quadratic equation.

Complex Number Solutions

Imaginary number solutions of a quadratic equation can be determined with the method of completing the square.

EXAMPLE 3

Quadratic Equations with Complex Number Solutions

Determine the solutions of each equation.

(a) $x^2 + 4x + 5 = 0$ (b) $3x^2 - 2x + 1 = 0$

Solution

(a) We solve by completing the square.

$$x^2 + 4x + 5 = 0$$

$$x^2 + 4x = -5 \qquad \text{Subtract 5 from both sides.}$$

$$x^2 + 4x + 4 = -5 + 4 \qquad \text{Add } \left(\frac{1}{2}\cdot 4\right)^2 = 2^2 = 4 \text{ to both sides.}$$

$$x^2 + 4x + 4 = -1$$

$$(x + 2)^2 = -1 \quad \text{Factor and write in exponential form.}$$
$$x + 2 = \pm\sqrt{-1} \quad \text{Square Root Property}$$
$$x + 2 = \pm i$$
$$x = -2 \pm i \quad \text{Subtract 2 from both sides to solve for } x.$$

(b) $3x^2 - 2x + 1 = 0$

$$3x^2 - 2x = -1 \quad \text{Subtract 1 from both sides.}$$
$$x^2 - \frac{2}{3}x = -\frac{1}{3} \quad \text{Divide both sides by 3.}$$
$$x^2 - \frac{2}{3}x + \frac{1}{9} = -\frac{1}{3} + \frac{1}{9} \quad \text{Add } \left[\frac{1}{2}\cdot\left(-\frac{2}{3}\right)\right]^2 = \left(-\frac{1}{3}\right)^2 = \frac{1}{9} \text{ to both sides.}$$
$$x^2 - \frac{2}{3}x + \frac{1}{9} = -\frac{3}{9} + \frac{1}{9} \quad \text{The LCD on the right side is 9.}$$
$$\left(x - \frac{1}{3}\right)^2 = -\frac{2}{9} \quad \text{Add the fractions and factor.}$$
$$x - \frac{1}{3} = \pm\sqrt{-\frac{2}{9}} \quad \text{Square Root Property}$$
$$x - \frac{1}{3} = \pm\frac{i\sqrt{2}}{3}$$
$$x = \frac{1}{3} \pm \frac{\sqrt{2}}{3}i \approx 0.33 \pm 0.47i$$

Quick Reference 10.2

Square of a Binomial

- If the first part of a quadratic trinomial is $x^2 + bx$, then the entire perfect square trinomial is $x^2 + bx + \left(\frac{b}{2}\right)^2$.
- In words, to determine the constant needed to form a perfect square trinomial (with 1 as the leading coefficient), take half the coefficient of x and square the result. This process is called **completing the square.**

Solving Quadratic Equations

- To solve a quadratic equation $ax^2 + bx + c = 0$ by completing the square, follow these steps.
  1. If necessary, divide both sides by the leading coefficient to obtain a coefficient of 1 for the quadratic term.
  2. Write the equation in the form $x^2 + bx = k$.
  3. Add $\left(\frac{b}{2}\right)^2$ to both sides to complete the square.
  4. Factor the resulting perfect square trinomial.
  5. Solve with the Square Root Property.

Complex Number Solutions

- The method of completing the square could result in an equation of the form $(x \pm h)^2 = k$, where $k < 0$. In this case, the solutions will be imaginary numbers.

Speaking the Language 10.2

1. To complete the square for $x^2 + kx$, we take ______ of k and ______ the result.
2. To solve $3x^2 + 5x - 7 = 0$ by completing the square, one step is to ______ by 3.
3. The purpose of completing the square is to create a(n) ______ on one side of the equation.
4. Writing $3 \pm i$ is a compact way of writing ______ or ______.

Exercises 10.2

Concepts and Skills

1. To solve a quadratic equation $ax^2 + bx + c = 0$ by the method of completing the square, we rewrite the equation in the form $A^2 = B$. Describe the form of the expression A.
2. For the expression $x^2 + bx + k$, describe in words how to determine k in order to complete the square.

In Exercises 3–8, determine k so that the trinomial is a perfect square.

3. $x^2 + 6x + k$
4. $x^2 - 8x + k$
5. $x^2 + 7x + k$
6. $x^2 - 9x + k$
7. $x^2 + \frac{4}{3}x + k$
8. $x^2 - \frac{5}{2}x + k$

9. Two students are solving $x^2 - 8x = -12$. One student rewrites the equation in standard form and solves with the factoring method. The other student adds 16 to both sides to complete the square and then solves with the Square Root Property. Which method is correct? Why?
10. Two students are solving $x^2 + x = -1$. One student rewrites the equation in standard form and tries to solve with the factoring method. The other student tries to solve with the method of completing the square. Which student will be successful? Why?

In Exercises 11–18, solve by factoring and then solve again by completing the square.

11. $x^2 + 4x + 3 = 0$
12. $x^2 + 6x + 8 = 0$
13. $t^2 + 2t = 15$
14. $t^2 = 6t + 55$
15. $x^2 - x - 12 = 0$
16. $x^2 + 5x - 6 = 0$
17. $6 = x^2 + x$
18. $x^2 = 5x + 14$

This is an opportunity to point out that completing the square always works, but factoring, if it can be done, is usually easier.

In Exercises 19–24, solve by completing the square.

19. $x^2 - 6x + 4 = 0$
20. $x^2 + 2x = 5$
21. $x^2 + 2x = 24$
22. $x^2 - 4x = 5$
23. $x^2 = 4x + 4$
24. $x^2 = 4x + 8$

25. If you want to solve the equation $3x^2 - 5x = 7$ with the method of completing the square, what is the first step you must take?
26. Although we regard the factoring method and the method of completing the square as two different methods for solving a quadratic equation, the method of completing the square does include factoring. At what step of the procedure do we need to factor?

In Exercises 27–30, solve by factoring and then solve again by completing the square.

27. $3y^2 + 4y + 1 = 0$
28. $5y^2 + 6y = 8$
29. $15 = 2x^2 + 7x$
30. $6x^2 + 3 = 11x$

In Exercises 27–38, the leading coefficient is not 1. All solutions are real numbers.

In Exercises 31–38, solve by completing the square.

31. $4x^2 = 8x + 21$
32. $4x^2 = 27 + 12x$
33. $4x^2 + 25 = 20x$
34. $16x^2 + 9 = 24x$
35. $x(2x - 3) = 1$

36. $4x^2 = 3(x + 1)$

37. $2x^2 + 7x - 1 = 0$

38. $4x^2 + 5x - 3 = 0$

CE **39.** Determine k so that $kx^2 + 6x + 1$ is a perfect square.

CE **40.** Determine k so that $4x^2 + kx + 25$ is a perfect square.

In Exercises 41–48, determine the imaginary number solutions by completing the square.

41. $x^2 - 4x + 8 = 0$

42. $x^2 + 6x + 10 = 0$

43. $x^2 + 6 = 4x$

44. $x^2 + 7 = 2x$

45. $x^2 + 3x + 5 = 0$

46. $x^2 + 5x + 8 = 0$

47. $2x^2 + x + 3 = 0$

48. $3x^2 + 2x + 7 = 0$

CE **49.** Show that the solutions of $x^2 + bx = k$ are imaginary numbers if $k < -\left(\frac{b}{2}\right)^2$.

CE **50.** For what value of k does $x^2 + bx = k$ have only one solution?

In Exercises 51–74, solve the equation.

51. $x^2 + 16 = 10x$

52. $2x^2 + x - 6 = 0$

53. $36x^2 - 49 = 0$

54. $\frac{1}{7}y^2 = 49$

55. $x(x - 3) = 10$

56. $b(b + 5) = 36$

57. $4(x + 3)^2 - 7 = 0$

58. $(3x - 2)^2 = 16$

59. $x(x + 1) + 9 = 0$

60. $2a(a + 2) = 3$

61. $(x - 3)(x - 1) = 5x - 6$

62. $(2x - 1)(8x - 1) = -1 - 2x$

63. $x^2 + \frac{3}{2}x = \frac{3}{4}$

64. $x^2 + \frac{4}{3}x = \frac{5}{9}$

65. $x^2 = x - 9$

66. $x^2 = x + 4$

67. $4x^2 + 7x = 3$

68. $3x^2 - x + 8 = 0$

69. $3x^2 + x = \frac{2}{3}$

70. $0.5x^2 - 2x = 1.75$

71. $(x - 2)^2 = -3(x - 2)$

72. $(x + 1)^2 = 3(x + 2)$

73. $4x^2 + 9 = 2x$

74. $2x^2 + 9 = 5x$

In this group, students can use a variety of solving techniques. Some solutions are imaginary numbers.

CE In Exercises 75–80, one solution is given. Determine a, b, or c and then determine another solution.

75. $ax^2 + x - 5 = 0;\ 1$

76. $ax^2 + 9x - 5 = 0;\ \frac{1}{3}$

77. $6x^2 + bx - 3 = 0;\ \frac{1}{3}$

78. $5x^2 + bx - 3 = 0;\ 1$

79. $x^2 + 2x + c = 0;\ 6$

80. $2x^2 + 7x + c = 0;\ -\frac{1}{2}$

81. The sum of the reciprocal of an integer and the square of the reciprocal is $\frac{6}{25}$. What is the integer?

82. The product of the reciprocal of a number and 20 is 8 less than the number. What two numbers satisfy this condition?

83. The square of a positive number is 3 more than 5 times the number. What is the number? (Round to two decimal places.)

84. Seven times a negative number is 1 less than the square of the number. What is the number? (Round to two decimal places.)

Real-Life Applications

85. A construction crew built a platform in the shape of a right triangle for a political campaign rally. If the lengths of the legs of the triangle differ by 8 feet, and the length of the longest side of the platform is 50 feet, how long is the shortest side of the platform? (Round to the nearest foot.)

86. Taxes on commercial property have increased by \$1.00 per frontage foot along Hayes Street. A store is on a lot in the shape of a right triangle with one leg along Hayes Street. The frontage along Minnow Road is 100 feet. (See figure.)

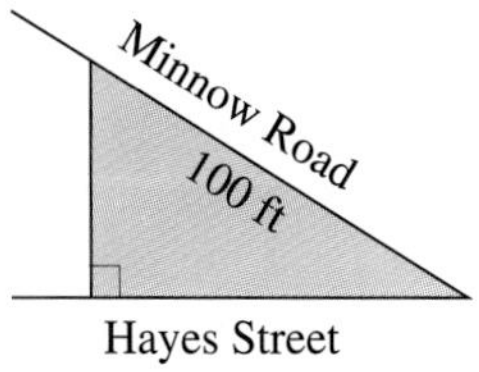

If the Hayes Street frontage is 30 feet more than twice the length of the other leg of the lot, what is the amount of the tax increase? (Round to the nearest dollar.)

87. A woman uses a rototiller to till her garden. Her neighbor has a larger tiller and could do the job in 30 minutes less time. How long does the woman take to till the garden if she and her neighbor, working together, can complete the job in 1 hour? (Round to the nearest minute.)

88. A bricklayer can complete a job in 2 hours less time than his apprentice. Working together, the two of them can complete the job in 5 hours. How long would the apprentice take working alone? (Round to the nearest hour.)

Data Analysis: *Cost of Dining Out*

As indicated in the bar graph, the average household's annual spending on eating away from home increased during the period 1994–1998. (Source: Bureau of Labor Statistics.)

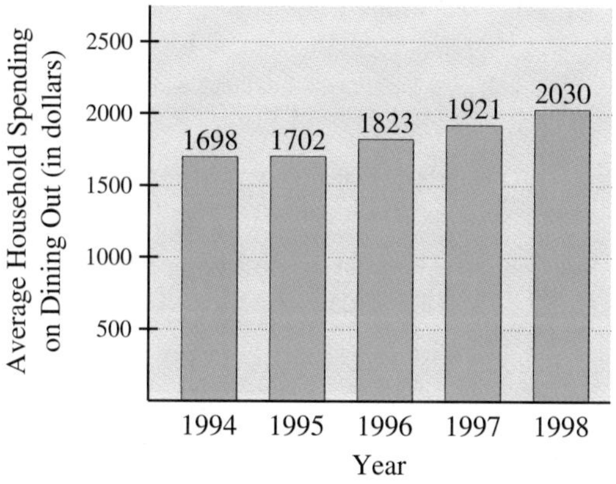

The function $A(t) = 13.4t^2 - 72t + 1759$ models the amount A spent annually on eating out as a function of the number of years since 1990.

89. Use a table of values for the years 2000–2005 to estimate a solution of the equation $A(t) = 3000$.

90. Explain why the factoring method is not convenient for solving the equation in Exercise 89.

91. Use the method of completing the square to solve the equation in Exercise 89.

92. Interpret the solutions to the equation in Exercise 89. Which solution is more likely to be a valid prediction? Why?

Challenge

In Exercises 93 and 94, use the method of completing the square to solve for x.

93. $x^2 - 2cx + 4c^2 = 0$

94. $ax^2 + 2x + a = 0, a \neq 0$

In Exercises 95 and 96, solve by completing the square.

95. $x^2 + 2\sqrt{5}x - 4 = 0$

96. $x^2 - 6\sqrt{3}x + 23 = 0$

In Exercises 97 and 98, solve for x. Determine the values of k for which (a) the solutions will be real numbers and (b) the solutions will be imaginary numbers.

97. $x^2 + kx + 1 = 0$

98. $x^2 + 2x + k = 0$

In Exercises 99 and 100, use the method of completing the square to solve the equation.

99. $x^2 + 4ix - 8 = 0$

100. $x^2 - 8ix - 25 = 0$

101. The method of completing the square can be illustrated geometrically. Consider the square (with a missing corner) in the accompanying figure.

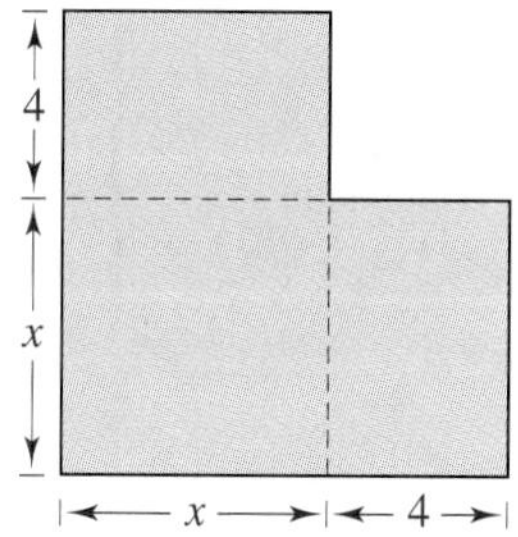

(a) Find the total area of the figure by adding the areas of the pieces. Simplify the result.

(b) What area needs to be added to "complete the square"? That is, what is the area of the missing corner in the figure?

(c) Find the area of the complete square by adding the areas found in parts (a) and (b).

(d) Use the fact that the lengths of the sides of the complete square are $x + 4$ to calculate the area of the complete square. Do not simplify.

(e) Compare the results in parts (c) and (d).

10.3 The Quadratic Formula

Derivation and Use • Complex Number Solutions • The Discriminant

Derivation and Use

Each time we solve a quadratic equation with the method of completing the square, we follow the same steps. By performing the steps just once for the general equation $ax^2 + bx + c = 0$, we can derive a formula for finding the solutions of *any* quadratic equation.

Exploring the Concept

The Quadratic Formula

We begin with the quadratic equation written in standard form, and we complete the square.

$ax^2 + bx + c = 0$	Standard form of a quadratic equation
$ax^2 + bx = -c$	Subtract c from both sides.
$x^2 + \frac{b}{a}x = -\frac{c}{a}$	The leading coefficient must be 1. Divide both sides by a.
$x^2 + \frac{b}{a}x + \frac{b^2}{4a^2} = -\frac{c}{a} + \frac{b^2}{4a^2}$	To complete the square, add $\left(\frac{1}{2} \cdot \frac{b}{a}\right)^2 = \frac{b^2}{4a^2}$ to both sides.
$x^2 + \frac{b}{a}x + \frac{b^2}{4a^2} = -\frac{4ac}{4a^2} + \frac{b^2}{4a^2}$	The LCD on the right side is $4a^2$.
$x^2 + \frac{b}{a}x + \frac{b^2}{4a^2} = \frac{b^2 - 4ac}{4a^2}$	Combine the fractions on the right side.
$\left(x + \frac{b}{2a}\right)^2 = \frac{b^2 - 4ac}{4a^2}$	Factor and write in exponential form.
$x + \frac{b}{2a} = \frac{\pm\sqrt{b^2 - 4ac}}{2a}$	Square Root Property
$x = \frac{-b}{2a} \pm \frac{\sqrt{b^2 - 4ac}}{2a}$	Subtract $\frac{b}{2a}$ to solve for x.
$x = \frac{-b \pm \sqrt{b^2 - 4ac}}{2a}$	Combine the fractions.

This formula is called the **Quadratic Formula.**

Remind students that $-b$ is read, "the opposite of b," not "negative b." In the Quadratic Formula, we replace $-b$ with the opposite of the number b.

The Quadratic Formula

For real numbers a, b, and c, with $a \neq 0$, the solutions of the quadratic equation $ax^2 + bx + c = 0$ are given by the **Quadratic Formula.**

$$x = \frac{-b \pm \sqrt{b^2 - 4ac}}{2a}$$

Think About It

In the Quadratic Formula, either b or c can be 0, but a cannot be 0. Why would this problem never arise?

Note: A quadratic equation must be written in standard form before the Quadratic Formula can be used to solve it.

EXAMPLE 1

Using the Quadratic Formula: Two Rational Number Solutions

Use the Quadratic Formula to solve $3x^2 - 13x = 30$.

Stress that *a, b,* and *c* cannot be listed until the equation is in standard form.

Solution

$$3x^2 - 13x = 30$$

$$3x^2 - 13x - 30 = 0 \qquad \text{Write the equation in standard form.}$$

Substitute $a = 3$, $b = -13$, and $c = -30$ into the Quadratic Formula.

$$x = \frac{-b \pm \sqrt{b^2 - 4ac}}{2a}$$

$$x = \frac{-(-13) \pm \sqrt{(-13)^2 - 4(3)(-30)}}{2(3)} \qquad \text{Note that the opposite of } b \text{ is 13.}$$

$$x = \frac{13 \pm \sqrt{169 + 360}}{6} \qquad \text{Simplify the radicand first.}$$

$$x = \frac{13 \pm \sqrt{529}}{6} \qquad \text{Note that 529 is a perfect square.}$$

$$x = \frac{13 \pm 23}{6} \qquad \text{Use the + sign for one solution and the − sign for the other solution.}$$

$$x = \frac{13 + 23}{6} = \frac{36}{6} \quad \text{or} \quad x = \frac{13 - 23}{6} = \frac{-10}{6}$$

$$x = 6 \quad \text{or} \quad x = -\frac{5}{3}$$

LEARNING TIP

Writing the Quadratic Formula each time you solve a problem will help you to learn the formula. List the numbers *a, b,* and *c* and substitute these values into the formula before you attempt to perform operations.

Sometimes students think that the Quadratic Formula cannot be used if the expression can be factored. Emphasize that the Quadratic Formula can be used to solve any quadratic equation, although other methods may be easier.

The equation in Example 1 has two real number solutions, and the equation could have been solved by factoring. Note that after substituting into the Quadratic Formula and simplifying, the radicand is a perfect square.

EXAMPLE 2

Using the Quadratic Formula: Double Root

Use the Quadratic Formula to solve $16t^2 + 9 = 24t$.

Solution

$$16t^2 + 9 = 24t$$

$$16t^2 - 24t + 9 = 0 \qquad \text{Write the equation in standard form.}$$

Substitute $a = 16$, $b = -24$, and $c = 9$ into the Quadratic Formula.

$$t = \frac{24 \pm \sqrt{(-24)^2 - 4(16)(9)}}{2(16)}$$

$$t = \frac{24 \pm \sqrt{576 - 576}}{32} \qquad \text{Simplify the radicand first.}$$

$$t = \frac{24 \pm 0}{32} = \frac{24}{32} = \frac{3}{4} \qquad \text{Note that the radicand is 0.}$$

The equation in Example 2 has only one distinct solution. The equation could have been solved by factoring, which would have revealed $\frac{3}{4}$ to be a double root. Because the radicand was 0, the symbol $\pm$ instructed us to add 0 for one root and subtract 0 for the other root. Of course, the result is the same in both cases.

EXAMPLE 3 Using the Quadratic Formula: Two Irrational Number Solutions

Use the Quadratic Formula to solve $3x^2 - 10 = 0$.

Either b or c can be zero. Remind students why a cannot be zero.

Solution

$$3x^2 - 10 = 0$$ Note that $b = 0$.

$$x = \frac{0 \pm \sqrt{0^2 - 4(3)(-10)}}{2(3)}$$ Substitute $a = 3$, $b = 0$, and $c = -10$ into the Quadratic Formula.

$$x = \frac{0 \pm \sqrt{0 + 120}}{6}$$

$$x = \pm\frac{\sqrt{120}}{6}$$ Note that 120 is not a perfect square.

$$x = \pm\frac{\sqrt{4}\sqrt{30}}{6} = \pm\frac{2\sqrt{30}}{6} = \pm\frac{\sqrt{30}}{3}$$

$x \approx 1.83$ or $x \approx -1.83$

The equation in Example 3 has two irrational number solutions and could have been solved with the Square Root Property. The fact that the radicand is not a perfect square indicates that the factoring method could not have been used.

EXAMPLE 4 Using the Quadratic Formula: Two Irrational Number Solutions

Use the Quadratic Formula to solve $4x = x^2 + 1$.

Solution

$$4x = x^2 + 1$$

$$-x^2 + 4x - 1 = 0$$ Write the equation in standard form.

$$x = \frac{-4 \pm \sqrt{4^2 - 4(-1)(-1)}}{2(-1)}$$ Substitute $a = -1$, $b = 4$, and $c = -1$ into the Quadratic Formula.

$$x = \frac{-4 \pm \sqrt{16 - 4}}{-2}$$

$$x = \frac{-4 \pm \sqrt{12}}{-2} = \frac{-4 \pm \sqrt{4}\sqrt{3}}{-2} = \frac{-4 \pm 2\sqrt{3}}{-2} = 2 \pm \sqrt{3}$$

$x \approx 0.27$ or $x \approx 3.73$

The equation in Example 4 has two irrational number solutions. The factoring method could not have been used. The equation could have been solved by completing the square, but the Quadratic Formula was derived from that method.

Complex Number Solutions

In the preceding examples, after we substituted a, b, and c into the Quadratic Formula, the radicand was never negative. Thus the solutions were always real numbers.

For other equations, it is possible for the radicand in the Quadratic Formula to be negative. In such cases the solutions are imaginary numbers. The Quadratic Formula can be used to determine all complex number solutions of any quadratic equation.

EXAMPLE 5

Using the Quadratic Formula: Complex Number Solutions

Use the Quadratic Formula to solve $3x^2 + 2x + 1 = 0$.

Solution

$$3x^2 + 2x + 1 = 0$$

$$x = \frac{-2 \pm \sqrt{2^2 - 4(3)(1)}}{2(3)}$$ Substitute $a = 3, b = 2,$ and $c = 1$ into the Quadratic Formula.

$$x = \frac{-2 \pm \sqrt{4 - 12}}{6}$$

$$x = \frac{-2 \pm \sqrt{-8}}{6}$$ The radicand is negative.

$$x = \frac{-2 \pm i\sqrt{8}}{6}$$

$$x = \frac{-2}{6} \pm \frac{\sqrt{8}}{6}i = -\frac{1}{3} \pm \frac{\sqrt{4}\sqrt{2}}{6}i = -\frac{1}{3} \pm \frac{2\sqrt{2}}{6}i = -\frac{1}{3} \pm \frac{\sqrt{2}}{3}i$$

$$x \approx -0.33 \pm 0.47i$$

In Example 5, the radicand is negative and the solutions are complex conjugates.

The Discriminant

In the Quadratic Formula, the radicand is $b^2 - 4ac$. This expression is called the **discriminant.** The discriminant provides information about the number and type of solutions of a quadratic equation with rational coefficients.

Information Obtained from the Discriminant $b^2 - 4ac$

If a, b, and c are rational numbers, the value of the discriminant determines the number and type of solutions of a quadratic equation as follows:

1. If the discriminant is positive and
   (a) a perfect square, there are two rational solutions.
   (b) not a perfect square, there are two irrational solutions.
2. If the discriminant is zero, there is one rational solution (double root).
3. If the discriminant is negative, the solutions are complex conjugates.

EXAMPLE 6

Using the Discriminant to Characterize Solutions of Quadratic Equations

Use the discriminant to determine the number and type of the solutions of each equation.

(a) $3x^2 + 1 = 2x$

$3x^2 - 2x + 1 = 0$ Write the equation in standard form.

$b^2 - 4ac = (-2)^2 - 4(3)(1) = 4 - 12 = -8$

Because the discriminant is negative, the solutions of the equation are complex conjugates.

(b) $x(5x + 1) = 12$

$5x^2 + x - 12 = 0$ Simplify and write the equation in standard form.

$b^2 - 4ac = 1^2 - 4(5)(-12) = 1 + 240 = 241$

The discriminant is positive, but it is not a perfect square. Thus the equation has two irrational solutions.

(c) $2 + 3x + x^2 = 0$

$x^2 + 3x + 2 = 0$

$b^2 - 4ac = 3^2 - 4(1)(2) = 9 - 8 = 1$

Because the discriminant is positive and a perfect square, the equation has two rational solutions.

(d) $25x^2 - 20x + 4 = 0$

$b^2 - 4ac = (-20)^2 - 4(25)(4) = 400 - 400 = 0$

The discriminant is zero. Thus the equation has one rational solution (double root).

We now have four methods available for solving a quadratic equation: the factoring method, the Square Root Property, the method of completing the square, and the Quadratic Formula.

When solving a quadratic equation, choose the method that is easiest for that particular equation. Calculating the discriminant can help you select a solution method. If the discriminant is a perfect square, the equation can be solved by factoring. Otherwise, our other techniques must be used. The following are some other criteria.

If	Use
$b = 0$	Square Root Property
$c = 0$	Factoring method
$b, c \neq 0$	Factoring method if it is easy; otherwise use the Quadratic Formula

The Quadratic Formula can be used to solve any quadratic equation, but the other methods are usually more efficient when they can be used.

Quick Reference 10.3

Derivation and Use

- The **Quadratic Formula** is derived by applying the method of completing the square to the general quadratic equation $ax^2 + bx + c = 0$.
- The solutions of the quadratic equation $ax^2 + bx + c = 0$ are given by the Quadratic Formula.

$$x = \frac{-b \pm \sqrt{b^2 - 4ac}}{2a}$$

- To use the Quadratic Formula, write the quadratic equation in standard form and substitute the values of a, b, and c into the formula.

Complex Number Solutions

- When the Quadratic Formula is used to solve a quadratic equation, the radicand may be negative. In such cases, the solutions are imaginary numbers.

The Discriminant

- In the Quadratic Formula, the expression $b^2 - 4ac$ is called the **discriminant.**
- If a, b, and c are rational numbers, then the value of the discriminant determines the number and type of solutions as follows:
  1. If the discriminant is positive and
     (a) a perfect square, there are two rational solutions.
     (b) not a perfect square, there are two irrational solutions.
  2. If the discriminant is zero, there is one rational solution (double root).
  3. If the discriminant is negative, the solutions are complex conjugates.

Speaking the Language 10.3

1. To use the Quadratic Formula, we begin by writing the equation in ________.
2. In the Quadratic Formula, the radicand is called the ________.
3. For a quadratic equation with rational number coefficients, if $b^2 - 4ac < 0$, then the solutions of the equation are ________.
4. Although the Quadratic Formula can be used to solve $x^2 - 2x = 0$, the ________ method is easier.

Exercises 10.3

Concepts and Skills

1. To use the Quadratic Formula to solve a quadratic equation, we must know the numbers a, b, and c. What must be true about the equation before these numbers can be determined?
2. What is your opinion about the best method for solving the equation $x^2 - 2x - 1088 = 0$? Can the factoring method be used? If so, what are the factors?

In Exercises 3–6, write the equation in standard form and then state the values of a, b, and c.

3. $x(3x - 2) = 7x + 6$
4. $(x - 2)^2 = -4x$
5. $(x + 1)^2 = 3x - 1$
6. $1 = 4x(2x - 5)$

In Exercises 7–14, solve with the factoring method and then solve again with the Quadratic Formula.

7. $x^2 - x - 2 = 0$
8. $x^2 - 3x + 2 = 0$

9. $x^2 + 16 = 8x$ **10.** $x^2 + 1 = 2x$

11. $2x^2 + 7x = 4$ **12.** $2x^2 = 7x - 3$

13. $x^2 = 5x$ **14.** $4x^2 - 3x = 3x$

In Exercises 15–22, use the Quadratic Formula to solve.

15. $4x^2 = 3 + 9x$ **16.** $4x^2 + 7x = 3$

17. $x^2 - 2x + 9 = 0$ **18.** $x^2 + 4x + 10 = 0$

19. $x(x + 2) = 5$ **20.** $x^2 = 2(2 - x)$

21. $x + 2x^2 + 4 = 0$ **22.** $9 + 3x^2 + 5x = 0$

In Exercises 23–48, solve by any method.

This is a mixed group. Students should solve with the most convenient method.

23. $6x - 9x^2 = 0$

24. $5 - 13x - 6x^2 = 0$

25. $9 - 49x^2 = 0$

26. $3 - 4x^2 = 0$

27. $49 = 14x - x^2$ **28.** $9x^2 - 6x = -1$

29. $(5x - 4)^2 = 10$ **30.** $(4x + 5)^2 = 49$

31. $0.25x^2 - 0.5x - 2 = 0$

32. $0.5x^2 + x = 0.875$

33. $x(x + 4) = 5$ **34.** $3 = x(x - 6)$

35. $5x^2 + 2x + 7 = 0$ **36.** $41 + 8x - 4x^2 = 0$

37. $x^2 - 3x + 8 = 0$ **38.** $x^2 - 5x - 5 = 0$

39. $6x - x^2 - 4 = 0$ **40.** $x^2 = 4 - 2x$

41. $x(5 + x) + 9 = 0$ **42.** $5 = 2x(1 - 3x)$

43. $\frac{1}{2}x^2 - \frac{2}{3}x + \frac{4}{3} = 0$ **44.** $\frac{1}{6}x^2 + \frac{3}{2} = \frac{1}{2}x$

45. $x^2 = \frac{5x}{2} + \frac{1}{2}$ **46.** $\frac{2}{3}x^2 + 2x = \frac{5}{9}$

47. $(x - 3)(2x + 1) = x(x - 4)$

48. $(2x + 1)^2 = 2(3x + 1)$

In Exercises 49–56, use the discriminant to determine the number and the nature of the solutions of the given equation.

49. $4x^2 + 12x = 7$ **50.** $x^2 + 12 = 8x$

51. $2x^2 + 5x + 7 = 0$ **52.** $4x^2 + x + 9 = 0$

53. $36x^2 + 1 = 12x$ **54.** $4x(x + 1) + 1 = 0$

55. $3x^2 + 4x - 8 = 0$ **56.** $1 + 6x - 9x^2 = 0$

CE In Exercises 57–60, determine the values of k so that the equation has two unequal real roots.

In Exercises 57–68, the value of k is determined by calculating the discriminant.

57. $x^2 + 8x + k = 0$

58. $3x^2 + 2x + k = 0$

59. $kx^2 - 4x + 1 = 0$

60. $kx^2 - 3x + 2 = 0$

CE In Exercises 61–64, determine the values of k so that the equation has two imaginary solutions.

61. $x^2 + 6x + k = 0$ **62.** $5x^2 + 4x + k = 0$

63. $kx^2 - 3x + 1 = 0$ **64.** $kx^2 - 5x + 4 = 0$

CE In Exercises 65–68, determine the values of k so that the equation has a double root.

65. $x^2 + kx + 9 = 0$ **66.** $x^2 + kx + 3 = 0$

67. $x^2 + kx + 2k = 0$

68. $kx^2 + (2k + 4)x + 9 = 0$

69. (a) Explain why $x^2 + ax - 5 = 0$ always has real solutions, regardless of the value of the real number a.

(b) For any positive real number k, does the equation $x^2 + ax - k = 0$ always have real solutions, regardless of the value of the real number a? Explain.

70. Explain how to use the discriminant to determine whether the factoring method can be used to solve the given equation.

(a) $2x^2 + 5x = 7$ (b) $2x^2 + 3x = 8$

CE In Exercises 71–74, use the Quadratic Formula to factor the polynomial.

Exercises 71–74 help to relate solutions, factors, and x-intercepts.

71. $x^2 + 4x - 480$

72. $x^2 + 32x + 240$

73. $x^2 + 7x - 450$ **74.** $x^2 + 4x - 672$

75. Use the Quadratic Formula to show that the sum of the roots of $ax^2 + bx + c = 0$ is $-\frac{b}{a}$. Test the conclusion by solving $2x^2 - 7x - 15 = 0$ and showing that the sum of the two roots is $-\frac{b}{a}$.

76. Use the conclusion in Exercise 75 to explain why 5 and -4 cannot possibly be solutions of the equation $2x^2 - 3x - 20 = 0$ even though the numbers are factors of -20.

77. Use the Quadratic Formula to show that the product of the roots of $ax^2 + bx + c = 0$ is $\frac{c}{a}$.

78. Use the conclusion in Exercise 77 to explain why 5 and -4 cannot possibly be solutions of the equation $2x^2 - 3x - 20 = 0$ even though the numbers are factors of -20.

In Exercises 79–82, determine the sum and the product of the roots of the given equation without solving the equation.

79. $3x^2 - 2x - 1 = 0$

80. $2x^2 - 2x + 1 = 0$

81. $x^2 + x + 1 = 0$

82. $x^2 - 4x - 1 = 0$

83. Show that $x^2 + kx + (k - 1) = 0$ has two real number solutions for any value of k except 2.

84. Show that $k^2x^2 - 4kx + 4 = 0$ has a double root for any nonzero value of k.

Real-Life Applications

85. A softball field should be laid out so that the distance between bases is 60 feet. Suppose the infield of a certain softball field is perfectly square and the distance between first base and third base is 84.9 feet. To the nearest foot, is this a regulation softball field?

86. An auto dealer wants to string a wire from the top of a pole to the ground and hang colorful pennants from it. The distance from the base of the pole to the point where the wire is attached to the ground is 3 times the height of the pole. If the wire is 63 feet long, how tall is the pole?

87. A firm that manufactures ready-to-assemble computer desks can sell 200 desks per day at a wholesale price of \$90. However, if the company can cut costs and reduce the price, sales will increase by 10 desks per day for each \$2 reduction in price. For what prices will the total revenue be \$20,000 per day?

88. A whitewater rafting outfitter provides equipment and pickup service for trips down the Ocoee River. When the price is \$40 per person, the outfitter can anticipate 140 customers per day on the weekend. For each \$5 increase in price, 12 customers will go to a competitor. For what price will the total revenue be \$4200 per weekend day?

Data Analysis: *Post-GED Education*

The general equivalency diploma is given to high school dropouts who pass the GED test. The accompanying bar graph shows a rise in the percentage of GED recipients who planned additional schooling. (Source: American Council on Education.)

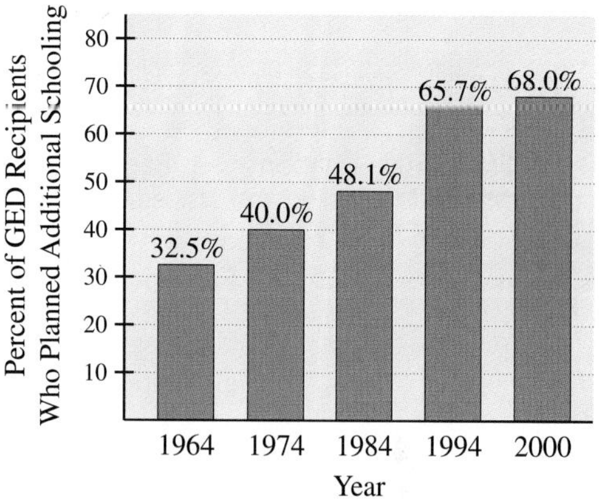

The function $p(x) = 0.01(0.96x^2 + 63x + 2950)$ models the data, where $p(x)$ is the percentage who planned more schooling and x is the number of years since 1960.

89. Use the model function to predict the percentage of GED recipients who will pursue more schooling in 2010.

90. Referring to the bar graph, indicate between what two years you would predict the percentage to be 57%.

91. In what year after 1964 does the model function estimate the percentage to be 57%? Compare your answer to your prediction in Exercise 90.

92. Determine and explain the significance of the sign of the discriminant in $p(x) = 0$.

Challenge

In Exercises 93 and 94, solve for x.

93. $x^2 + ax - 2a = 0$

94. $x^2 - yx - 2 = 0$

In Exercises 95 and 96, solve for y.

95. $x = y^2 + 2y - 8$

96. $x = 2y - 5 - y^2$

In Exercises 97 and 98, write a quadratic equation with the given solutions.

97. $-5 \pm 2\sqrt{5}$

98. $-2 \pm 3i$

In Exercises 99 and 100, use the Quadratic Formula to solve each equation.

99. $x^2 + \sqrt{3}x - 1 = 0$

100. $\sqrt{2}x^2 + \sqrt{5}x - \sqrt{8} = 0$

It can be shown that the Quadratic Formula holds for quadratic equations with imaginary coefficients. In Exercises 101 and 102, solve the equations and express the solutions in the complex number form $a + bi$.

101. $x^2 - 2ix - 3 = 0$ **102.** $ix^2 + 3x - 2i = 0$

10.4 Equations in Quadratic Form

Fractional Equations • Radical Equations • Quadratic Form • Complex Number Solutions

Fractional Equations

Clearing fractions in an equation with rational expressions sometimes leads to a quadratic equation.

EXAMPLE 1 Solving Equations with Rational Expressions

Solve the equation $\dfrac{2}{t} - \dfrac{t}{t-2} = 5$.

Remind students about restricted values and why we need to take note of them.

Solution

$$\frac{2}{t} - \frac{t}{t-2} = 5 \qquad \text{Note that 0 and 2 are restricted values.}$$

$$\frac{t(t-2)}{1} \cdot \frac{2}{t} - \frac{t(t-2)}{1} \cdot \frac{t}{t-2} = 5t(t-2) \qquad \text{Multiply by the LCD, } t(t-2).$$

$$2(t-2) - t^2 = 5t^2 - 10t \qquad \text{Simplify.}$$

$$2t - 4 - t^2 = 5t^2 - 10t$$

$$-6t^2 + 12t - 4 = 0 \qquad \text{Write the equation in standard form.}$$

Now use the Quadratic Formula to solve for t.

$$t = \frac{-12 \pm \sqrt{12^2 - 4(-6)(-4)}}{2(-6)} = \frac{-12 \pm \sqrt{144 - 96}}{-12} = \frac{-12 \pm \sqrt{48}}{-12}$$

$$= \frac{-12 \pm \sqrt{16}\sqrt{3}}{-12} = \frac{-12 \pm 4\sqrt{3}}{-12} = \frac{3 \pm \sqrt{3}}{3}$$

$$t \approx 1.58 \quad \text{or} \quad t \approx 0.42 \qquad \text{Neither solution is restricted.}$$

Radical Equations

To solve equations involving radicals, we raise both sides of the equation to a power equal to the index of the radical. The result may be a quadratic equation that we can solve with our present methods.

EXAMPLE 2

In this example, we need to square both sides only once. Remind students that some radical equations require us to square both sides twice.

Solving Equations with Radical Expressions

Solve the equation $\sqrt{x+1} = x - 2$.

Solution

$$\sqrt{x+1} = x - 2$$

$$\left(\sqrt{x+1}\right)^2 = (x-2)^2 \quad \text{Square both sides of the equation.}$$

$$x + 1 = x^2 - 4x + 4 \quad \text{Square the binomial.}$$

$$0 = x^2 - 5x + 3 \quad \text{Write the equation in standard form.}$$

Use the Quadratic Formula to solve for x.

$$x = \frac{5 \pm \sqrt{(-5)^2 - 4(1)(3)}}{2(1)} = \frac{5 \pm \sqrt{13}}{2}$$

$$x \approx 4.30 \quad \text{or} \quad x \approx 0.70$$

Checking the two possible solutions reveals that 0.70 is an extraneous solution. The only solution is $\frac{5 + \sqrt{13}}{2}$, or approximately 4.30.

Quadratic Form

An equation is in quadratic form if it can be written in the form $au^2 + bu + c = 0$ where u is an algebraic expression. To solve such equations, we make a substitution to write the equation in terms of u and then apply our methods for solving quadratic equations. The method is illustrated in Example 3.

EXAMPLE 3

Solving Equations That Are in Quadratic Form

Solve each equation.

(a) $x^4 - 6x^2 + 8 = 0$ (b) $16(z-1)^2 - 8(z-1) - 2 = 0$

(c) $28x^{-2} + x^{-1} - 2 = 0$

Solution

LEARNING TIP

To identify an appropriate substitution, look for a variable expression and its square, and let u represent the expression.

(a) The first term, x^4, is the square of the variable factor x^2 in the second term.

$$x^4 - 6x^2 + 8 = 0$$

$$(x^2)^2 - 6x^2 + 8 = 0$$

$$u^2 - 6u + 8 = 0 \quad \text{Let } u = x^2. \text{ Note that } u^2 = x^4.$$

$$(u-2)(u-4) = 0 \quad \text{Factor.}$$

$$u - 2 = 0 \quad \text{or} \quad u - 4 = 0 \quad \text{Zero Factor Property}$$

$$u = 2 \quad \text{or} \quad u = 4 \quad \text{Solve each case for } u.$$

Students often stop after determining u. Make sure they understand that they must continue until they determine the solutions for the original variable.

Now replace u with x^2.

$$x^2 = 2 \quad \text{or} \quad x^2 = 4$$

$$x = \pm\sqrt{2} \approx \pm 1.41 \quad \text{or} \quad x = \pm 2 \quad \text{Square Root Property}$$

The equation has four solutions: $\pm\sqrt{2}, \pm 2$.

(b) $16(z - 1)^2 - 8(z - 1) - 2 = 0$

$16u^2 - 8u - 2 = 0$ Let $u = z - 1$. Then $u^2 = (z - 1)^2$.

Use the Quadratic Formula to solve for u.

$$u = \frac{8 \pm \sqrt{(-8)^2 - 4(16)(-2)}}{2(16)} = \frac{8 \pm \sqrt{64 + 128}}{32} = \frac{8 \pm \sqrt{192}}{32}$$

$$= \frac{8 \pm \sqrt{64}\sqrt{3}}{32} = \frac{8 \pm 8\sqrt{3}}{32} = \frac{1 \pm \sqrt{3}}{4}$$

$u \approx 0.68$ or $u \approx -0.18$

To solve for z, replace u with the expression it represents.

$z - 1 \approx 0.68$ or $z - 1 \approx -0.18$

$z \approx 1.68$ or $z \approx 0.82$

(c) $28x^{-2} + x^{-1} - 2 = 0$

$28(x^{-1})^2 + x^{-1} - 2 = 0$

$28u^2 + u - 2 = 0$ Let $u = x^{-1}$. Then $u^2 = x^{-2}$.

$(4u - 1)(7u + 2) = 0$ Factor.

$4u - 1 = 0$ or $7u + 2 = 0$ Zero Factor Property

$4u = 1$ or $7u = -2$ Solve each case for u.

$u = \frac{1}{4}$ or $u = -\frac{2}{7}$

To solve for x, replace u with the expression it represents.

$x^{-1} = \frac{1}{4}$ or $x^{-1} = -\frac{2}{7}$ Solve for x.

$x = 4$ or $x = -\frac{7}{2}$

An equation involving rational exponents may be quadratic in form if one of the exponents is twice another exponent. Recognizing such patterns is essential to detecting quadratic forms.

EXAMPLE 4

Solving Equations with Rational Exponents

Solve the equation $x^{2/3} - 2x^{1/3} - 24 = 0$.

When rational exponents are involved, students are often unsure about how to assign the dummy variable. Another good example: $(x^{1/6})^2 = x^{1/3}$. Therefore, let $u = x^{1/6}$ and $u^2 = x^{1/3}$.

Solution

$x^{2/3} - 2x^{1/3} - 24 = 0$

$u^2 - 2u - 24 = 0$ Let $u = x^{1/3}$. Then $u^2 = (x^{1/3})^2 = x^{2/3}$.

$(u - 6)(u + 4) = 0$ Factor.

$u - 6 = 0$ or $u + 4 = 0$ Zero Factor Property

$u = 6$ or $u = -4$ Solve each case for u.

Think About It

Suppose that to solve the equation $(x + 1) + 5\sqrt{x + 1} + 6 = 0$, you let $u = \sqrt{x + 1}$ and write $u^2 + 5u + 6 = 0$. How can you tell immediately that the original equation has no solution?

Replace u with $x^{1/3}$.

$$x^{1/3} = 6 \quad \text{or} \quad x^{1/3} = -4$$
$$(x^{1/3})^3 = 6^3 \quad \text{or} \quad (x^{1/3})^3 = (-4)^3 \qquad \text{Raise both sides to the third power.}$$
$$x = 216 \quad \text{or} \quad x = -64$$

Complex Number Solutions

Whenever quadratic equations emerge in the process of solving an equation, the solutions may involve imaginary numbers.

EXAMPLE 5

Equations with Complex Number Solutions

Solve the equation $x^4 + x^2 = 12$.

Solution

$$x^4 + x^2 = 12$$
$$u^2 + u = 12 \qquad \text{Let } u = x^2. \text{ Then } u^2 = x^4.$$
$$u^2 + u - 12 = 0 \qquad \text{Write the equation in standard form.}$$
$$(u + 4)(u - 3) = 0 \qquad \text{Factor.}$$
$$u + 4 = 0 \quad \text{or} \quad u - 3 = 0 \qquad \text{Zero Factor Property}$$
$$u = -4 \quad \text{or} \quad u = 3 \qquad \text{Solve each case for } u.$$

The solutions for u are real numbers. To solve for x, replace u with the expression it represents.

Point out that the substitution variable may have only real solutions, but the original variable may have both real and imaginary solutions.

$$x^2 = -4 \quad \text{or} \quad x^2 = 3$$
$$x = \pm 2i \quad \text{or} \quad x = \pm\sqrt{3} \approx \pm\ 1.73$$

Two of the solutions for x are real (irrational) numbers, and two solutions are complex conjugates.

In this section we have seen that the ability to solve quadratic equations benefits us well beyond quadratic equations themselves. Solving a quadratic equation may be one part of solving a higher-degree equation; an equation involving rational expressions, radicals, or rational exponents; or any other equation that is quadratic in form.

Quick Reference 10.4

Fractional Equations

- To solve a fractional equation, clear the fractions by multiplying both sides of the equation by the LCD. This step may lead to a quadratic equation.
- Make sure that any solutions you obtain are not restricted values.

Radical Equations

- To solve a radical equation, raise both sides of the equation to a power equal to the index of the radical. The result may be a quadratic equation.
- Make sure that any solutions you obtain are not extraneous.

Quadratic Form

- An equation is in quadratic form if it can be written in the form $au^2 + bu + c = 0$, where u is an algebraic expression.
- To solve such equations, we make a substitution to write the equation in terms of u and then apply our methods for solving quadratic equations.

Complex Number Solutions

- Whenever quadratic equations emerge in the process of solving an equation, the solutions may involve imaginary numbers.

Speaking the Language 10.4

1. We say that the equation $5x^4 + 3x^2 - 1 = 0$ is in ________ form.
2. A fractional equation may lead to a quadratic equation, but a solution that is a(n) ________ value must be rejected.
3. Solving a radical equation may lead to a quadratic equation, but ________ solutions must be rejected.
4. If we use u as a substitute variable to write $x^{2/5} - 2x^{1/5} - 15 = 0$ in quadratic form, then $u =$ ________.

Exercises 10.4

Concepts and Skills

1. Consider the equation $x + \frac{1}{x-1} - \frac{5-4x}{x-1} = 0.$

   Show that clearing fractions results in the quadratic equation $x^2 + 3x - 4 = 0$, which has two solutions. Then show that if we perform the addition and subtraction first and simplify the results, we obtain a linear equation with one solution. Explain the apparent discrepancy.

2. Other than to find algebraic mistakes, why is it essential to check solutions for equations with rational expressions?

In Exercises 3–14, solve the given equation.

3. $\frac{3}{x} = 2 + \frac{4}{x^2}$

4. $x + \frac{5}{x} = 3$

5. $\frac{t-3}{t} = \frac{2}{t-3}$

6. $\frac{y}{y+1} = \frac{y+1}{3}$

7. $x + 2 = \frac{6}{1-x}$

8. $x + 1 = \frac{2}{2x+3}$

9. $\frac{x}{x+3} + \frac{3}{x} = 4$

10. $\frac{2}{x-2} + \frac{4}{x+3} = 1$

Some equations in this group have irrational or imaginary solutions.

11. $\frac{x}{x-2} - \frac{5}{x+2} = \frac{12}{x^2-4}$

12. $\frac{x}{x-5} - \frac{16}{x^2-25} = \frac{1}{x+5}$

13. $\frac{x+3}{x-2} - \frac{5}{2x^2-3x-2} = \frac{2}{2x+1}$

14. $\frac{x}{x-3} - \frac{1}{2x-1} = \frac{5x}{2x^2-7x+3}$

15. Consider the equation $\sqrt{x^2-5} + 2\sqrt{x} = 0$. Isolate the radicals, square both sides, and show that the result is $x^2 - 4x - 5 = 0$, which has two solutions. Explain how you know without substituting that both are extraneous solutions.

16. Other than to find algebraic mistakes, why is it essential to check your solutions for radical equations?

In Exercises 17–26, solve the radical equation.

There are no imaginary solutions in this group.

17. $5 = 3\sqrt{x} + \frac{2}{\sqrt{x}}$

18. $\sqrt{x+1} - \frac{2}{\sqrt{x+1}} = 1$

19. $2y + 1 = \sqrt{11y - 1}$

20. $2 + x = \sqrt{2x + 3}$

21. $x - 3 = \sqrt{x - 1}$

22. $2x + 5 = \sqrt{4x + 14}$

23. $\sqrt{3x - 2} = \sqrt{x + 2} - 1$

24. $\sqrt{3x + 4} = \sqrt{x - 2} - 3$

25. $\sqrt{x}\sqrt{x + 3} = 1$

26. $\sqrt{2x}\sqrt{x - 3} = 6$

27. Explain the role of substitution in solving equations that are quadratic in form. After you have solved the resulting equation, how do you determine the solution(s) of the original equation?

28. Suppose that an equation is quadratic in form. The two variable terms are $x^{1/6}$ and $x^{1/3}$, and you want to use a substitution to solve the equation. What is an appropriate substitution? Why?

In Exercises 29–48, solve by using an appropriate substitution to rewrite the given equation in quadratic form.

In this group, solutions include imaginary numbers.

29. $b^4 - 7b^2 + 12 = 0$

30. $x^4 - 14x^2 + 45 = 0$

31. $y^4 + 2y^2 = 63$

32. $x^4 + 17x^2 + 16 = 0$

33. $x - 6x^{1/2} + 8 = 0$

34. $c + 21 = 10c^{1/2}$

35. $y^{-2} - 2 = y^{-1}$

36. $1 - 6x^{-1} + 9x^{-2} = 0$

37. $y^{2/3} + 12 = 7y^{1/3}$

38. $3x^{1/10} + x^{1/5} = 4$

39. $(x^2 - 5x)^2 - 8(x^2 - 5x) = 84$

40. $(x^2 - x)^2 - 26(x^2 - x) + 120 = 0$

41. $(x + 1)^2 + 2(x + 1) = 2$

42. $(y - 2)^2 - 2(y - 2) = 4$

43. $(x^2 - 3)^2 - 5(x^2 - 3) - 6 = 0$

44. $(x^2 + 1)^2 - 2(x^2 + 1) - 15 = 0$

45. $3\left(\sqrt{x} - 2\right)^2 - 7\left(\sqrt{x} - 2\right) + 2 = 0$

46. $\left(\sqrt{x} - 8\right)^2 + 9\left(\sqrt{x} - 8\right) + 20 = 0$

47. $x - \sqrt{x} = 6$

48. $x + 3\sqrt{x} - 10 = 0$

In Exercises 49–78, determine all complex number solutions.

Exercises 49–78 are a mixed group.

49. $x^4 - 7x^2 + 12 = 0$

50. $x^4 - 10x^2 + 16 = 0$

51. $\dfrac{x}{x + 2} + \dfrac{4}{x} = 5$

52. $\dfrac{4}{x} + \dfrac{5x}{x + 3} = 2$

53. $2x + 5 = 2\sqrt{x + 3}$

54. $2x = 1 + \sqrt{x + 7}$

55. $a^4 + a^2 = 20$

56. $x^4 + 5x^2 + 6 = 0$

57. $x - 13x^{1/2} + 42 = 0$

58. $b = 4b^{1/2} + 21$

59. $\dfrac{x + 3}{x - 1} = \dfrac{5}{x}$

60. $x + 2 = \dfrac{6}{1 - x}$

61. $\sqrt{3x + 12} - \sqrt{x + 8} = 2$

62. $\sqrt{3x - 2} = \sqrt{x + 3} + 1$

63. $a^{2/3} = 2a^{1/3} + 15$

64. $2x^{1/6} + x^{1/3} = 8$

65. $\sqrt{3x} = \sqrt{2 - 7x}$

66. $2x = \sqrt{3}\sqrt{2x + 1}$

67. $\dfrac{x + 1}{3 - x} + \dfrac{12}{x^2 - 3x} + \dfrac{5}{x} = 0$

68. $1 = \dfrac{3}{x + 4} - \dfrac{5x}{x^2 + 3x - 4}$

69. $(3 - y)^2 - 3(3 - y) = 6$

70. $(x + 1)^2 + (x + 1) - 3 = 0$

71. $\sqrt{2x - 3} = 4 - \sqrt{x + 2}$

72. $\sqrt{3x - 5} = 2 - \sqrt{x + 1}$

73. $(3 - x)^{-2} - 3(3 - x)^{-1} = 4$

74. $4(x + 1)^{-1} + (x + 1)^{-2} - 21 = 0$

75. $\dfrac{x}{4(x - 4)} = \dfrac{1}{x + 2} + \dfrac{6}{x^2 - 2x - 8}$

76. $\dfrac{x + 4}{2x + 1} + \dfrac{1}{x + 1} = \dfrac{-1}{2x^2 + 3x + 1}$

77. $(x^2 - x)^2 + 3(x^2 - x) + 2 = 0$

78. $(2x^2 - 7x)^2 - 2(2x^2 - 7x) - 8 = 0$

Real-Life Applications

79. A small manufacturer makes motors for yard tractors. The cost to produce x hundred motors is $8000 + 8000x^2 - 80x^4$, and the total revenue from the sale of the motors is $1680x^2 - 4960$. How many motors must the company make and sell to break even?

80. Mail Express provides a mailing service for bulk mail advertising for small businesses. The company estimates that for x hundred brochures, preparation and mailing cost C (in dollars) will be modeled by $C = 20 + 20\sqrt{2x + 11}$, and the revenue R (in dollars) will be modeled by $R = 20\sqrt{5x + 1}$. For what size order will the company break even?

81. As part of a summer program, a college obtains a low bid of \$195,000 for a group of students to travel and study in South America. After initial registration, spaces for 40 students remain. If the remaining spaces are filled, the per-person cost will be reduced by \$100. How many students can the program accommodate, and what is the cost per person if all spaces are filled?

82. A business traveler left home at 6:00 A.M. and drove 50 miles to the airport for a flight to Pittsburgh. After the first 20 miles, traffic became congested and reduced the motorist's average speed by 24 mph. If the traveler arrived at the airport at 7:10 A.M., how many minutes of the trip were spent in heavy traffic?

Modeling with Real Data

83. The following bar graph shows the amount of coffee (in gallons) per person that is consumed annually in the United States. (Source: U.S. Department of Agriculture.)

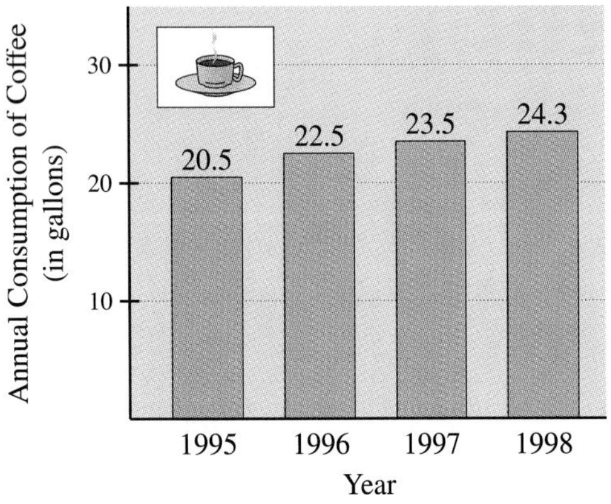

The following function models the amount of coffee (in gallons) consumed per person as a function of the number of years since 1995.

$$C(x) = -0.3(x + 5)^2 + 5.14(x + 5) + 2.34$$

(a) Write an equation to determine the year after 1998 in which the model predicts that the amount will decline to 15.5 gallons.

(b) Use a substitution to write the equation in part (a) in the standard form of a quadratic equation. Solve your equation.

84. A survey found that having a neat house is not a major concern for the majority of the 30–49 age group. The accompanying table shows the percentages of "yes" responses to the question "Is a neat house very important?" (Source: Soap and Detergent Association.)

Age Group	Percent
18–29	57%
30–49	48%
50 and over	62%

The following function models the percentage of people who think that a neat house is very important, where x is the number of years over (or under) age 40.

$$h(x) = 0.037(x + 40)^2 - 3.01(x + 40) + 109.14$$

(a) Write an equation to determine the age at which 50% of the respondents answered "yes."

(b) Use a substitution to write the equation in part (a) in the standard form of a quadratic equation. Then solve the equation and interpret the result.

Data Analysis: *National School Lunch Program*

The cost of the federally funded National School Lunch Program has risen steadily since its inception. The cost rose from about \$2.3 billion in 1980 to \$3.2 billion in 1990. More recent costs are shown in the accompanying table. (Source: U.S. Department of Agriculture.)

Year	Cost (in millions of dollars)
1994	4291
1995	4467
1996	4662
1997	4934
1998	5084

With t representing the number of years since 1990, the function $C(t) = 6.98(t + 10)^2 + 38.4(t + 10) + 2246$ models the cost C (in millions of dollars) of the school lunch program.

85. If the model function remains valid, what will be the national per capita cost for the school lunch program in 2006? (Assume a U.S. population of 285 million.)

86. Note that the expression for $C(t)$ is quadratic in form. Using a suitable substitution, solve the equation $C(t) = 10{,}000$. Interpret the result.

87. The form of the model function makes evaluating $C(-10)$ without a calculator easy. What previously given information does the result represent?

88. Suppose that in 2003, 28 million students are expected to participate in the lunch program. On the basis of the model, what is the predicted annual cost per student?

Challenge

In Exercises 89 and 90, rewrite the equation so that it is quadratic in form. Then make an appropriate substitution and solve.

89. $t = \sqrt{t+3}$ (*Hint:* Consider adding 3 to both sides.)

90. $x^2 + 6x - \sqrt{x^2 + 6x - 2} = 22$

In Exercises 91–96, determine all complex number solutions.

91. $x^6 - 7x^3 = 8$

92. $x^6 + 26x^3 = 27$

93. $x^2 + 2x = |2x + 1|$

94. $|x - 3| = x^2 - 3x$

95. $y^{-2} + 1 = y^{-1}$

96. $2\sqrt{x} - \dfrac{1}{\sqrt{x}} = 2$

10.5 Applications

Geometric Figures • Ratio and Proportion • Distance, Rate, Time • Work Rate • Investment

Students usually benefit from a brief review of the general approach to problem solving.

This section includes a sampling of applications that can be modeled with a variety of equations. As we have seen, even if the initial equation used to describe conditions is not quadratic, the solving process may lead to a quadratic equation.

Perimeter problems often lead to first-degree equations; area problems often lead to second-degree equations.

Geometric Figures

Problems involving the area of a rectangle can often be modeled with a quadratic equation.

EXAMPLE 1

Figure 10.1

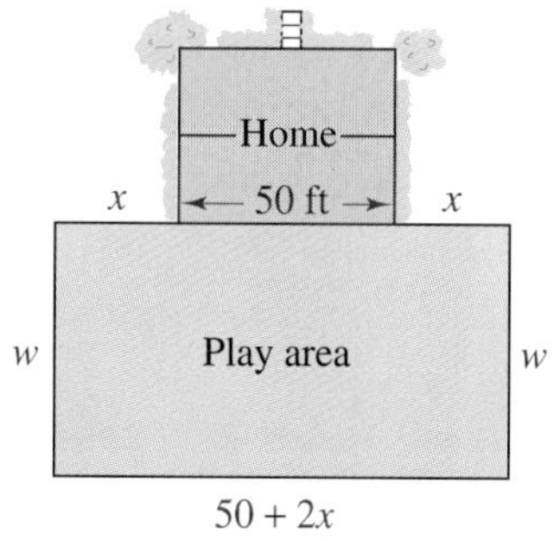

The temporary use of the variable w is to determine an expression for the width of the rectangle.

The Area of a Rectangular Play Area

A day care operator plans to use 170 feet of fencing to enclose a rectangular play area of 2800 square feet. She plans to construct the play area so that it is centered against the back of her home, which is 50 feet in length. Determine the dimensions necessary to enclose the required area.

Solution

Let x = the distance from the corner of the house to the corner of the fence. (See Fig. 10.1.) Then the length of the rectangle is $50 + 2x$.

Let w = the width of the play area. Note that no fencing is needed along the back of the home.

$$x + w + (50 + 2x) + w + x = 170 \quad \text{Total fencing used is 170 feet.}$$
$$2w + 50 + 4x = 170 \quad \text{Combine like terms.}$$
$$2w = 120 - 4x \quad \text{Solve for } w.$$
$$w = 60 - 2x$$
$$L \cdot W = A \quad \text{Formula for the area of a rectangle}$$
$$(50 + 2x)(60 - 2x) = 2800 \quad \text{Required area is 2800 square feet.}$$
$$3000 + 20x - 4x^2 = 2800 \quad \text{FOIL}$$
$$-4x^2 + 20x + 200 = 0 \quad \text{Write the equation in standard form.}$$
$$x^2 - 5x - 50 = 0 \quad \text{Divide both sides by } -4.$$
$$(x - 10)(x + 5) = 0 \quad \text{Factor.}$$

Think About It

It can be shown that the sum $1 + 2 + 3 + \cdots + n = \frac{n(n+1)}{2}$

If the sum of n numbers is 946, what is the last number in the sum?

$$x - 10 = 0 \quad \text{or} \quad x + 5 = 0 \qquad \text{Zero Factor Property}$$
$$x = 10 \quad \text{or} \quad x = -5 \qquad \text{Solve each case for } x.$$

Because distance cannot be negative, the only applicable solution is 10. The fence extends 10 feet on each side of the house, which means that the length of the play area is 70 feet. The width is $60 - 2(10) = 40$ feet.

Ratio and Proportion

The Golden Ratio was used extensively in Greek architecture.

According to early Greek mathematicians, the most perfect rectangles were those whose width W and length L satisfied the *Golden Ratio.*

$$\frac{W}{L} = \frac{L}{W + L}$$

We can use this formula in application problems in which a rectangle is to have dimensions that satisfy the Golden Ratio.

EXAMPLE 2

Rectangles That Satisfy the Golden Ratio

To create a miniature reproduction of the courtyard of an ancient Greek temple, an architect designed the courtyard so that it satisfied the Golden Ratio. If the perimeter of the miniature courtyard was 20 inches, what were the dimensions of the rectangle?

Solution

Let L = the length of the rectangle, W = the width of the rectangle, and P = the perimeter, which is given as 20 inches.

$$2L + 2W = P \qquad \text{Formula for the perimeter of a rectangle}$$
$$2L + 2W = 20 \qquad \text{The required perimeter is 20 inches.}$$
$$L + W = 10 \qquad \text{Divide both sides by 2.}$$
$$W = 10 - L \qquad \text{Solve for } W.$$

The Golden Ratio formula is $\frac{W}{L} = \frac{L}{W + L}$.

$$\frac{10 - L}{L} = \frac{L}{10} \qquad \text{Replace } W \text{ with } 10 - L.$$
$$\frac{10L}{1} \cdot \frac{10 - L}{L} = \frac{10L}{1} \cdot \frac{L}{10} \qquad \text{Multiply by the LCD } 10L \text{ to clear the fractions.}$$
$$100 - 10L = L^2 \qquad \text{Simplify.}$$
$$L^2 + 10L - 100 = 0 \qquad \text{Write the equation in standard form.}$$

We use the Quadratic Formula to solve for L.

$$L = \frac{-10 \pm \sqrt{10^2 - 4(1)(-100)}}{2(1)} = \frac{-10 \pm \sqrt{100 + 400}}{2} = \frac{-10 \pm \sqrt{500}}{2}$$
$$= \frac{-10 \pm \sqrt{100}\sqrt{5}}{2} = \frac{-10 \pm 10\sqrt{5}}{2} = -5 \pm 5\sqrt{5}$$

$L \approx 6.18 \quad \text{or} \quad L \approx -16.18$

Because length is positive, the negative solution is disqualified. The length is approximately 6.18 inches, and the width is approximately $10 - 6.18 = 3.82$ inches.

Distance, Rate, Time

Recall that distance D, rate R, and time T are related by the formula $D = RT$.

EXAMPLE 3

A Boating and Walking Trip

From a dock, it is 3 miles directly across a lake to a marina. A fisherman travels by boat at 5 mph from the dock to a pier on the opposite shore. Then he walks at 4 mph from the pier to a fishing spot that is 6 miles down the shoreline from the marina. (See Fig. 10.2.) If he made the trip in 1.5 hours, how far did he walk?

Figure 10.2

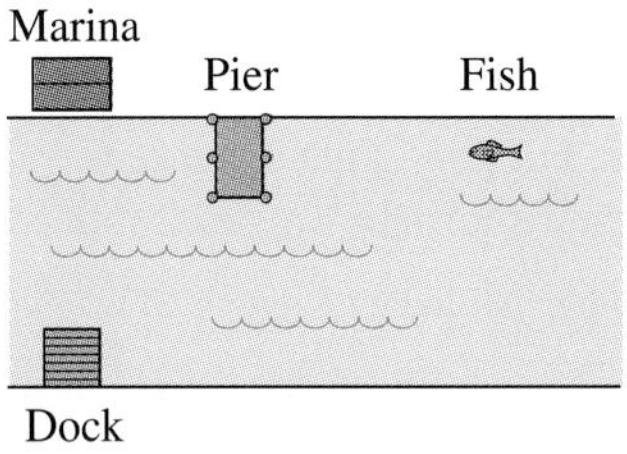

Solution

In Figure 10.3, we identify locations with letters.

D: Dock

M: Marina

P: Pier

F: Fishing spot

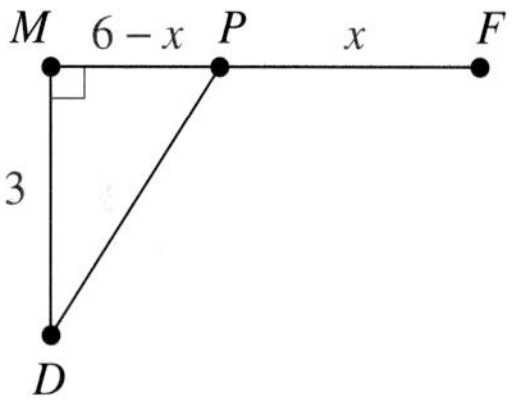

Figure 10.3

Let $x =$ the distance the fisherman walked from P to F. The distance from the marina M to the fishing spot F is 6 miles. Therefore, the distance from the marina M to the pier P is $6 - x$.

We use the Pythagorean Theorem on the right triangle DMP to determine the distance DP traveled by boat.

$$(DP)^2 = (DM)^2 + (MP)^2$$
$$(DP)^2 = 3^2 + (6 - x)^2$$
$$DP = \sqrt{9 + (6 - x)^2}$$

As with first-degree equations, encourage the use of a table for organizing information in distance, rate, and time problems.

	Distance (miles)	Rate (mph)	Time (hours)
Boat	$\sqrt{9 + (6 - x)^2}$	5	$\frac{\sqrt{9 + (6 - x)^2}}{5}$
Walking	x	4	$\frac{x}{4}$

$$\frac{\sqrt{9 + (6 - x)^2}}{5} + \frac{x}{4} = 1.5$$ The total travel time was 1.5 hours.

$$\frac{20}{1} \cdot \frac{\sqrt{9 + (6 - x)^2}}{5} + \frac{20}{1} \cdot \frac{x}{4} = 20 \cdot 1.5$$ Clear fractions by multiplying by the LCD, 20.

$4\sqrt{9 + (6 - x)^2} + 5x = 30$	Simplify.
$4\sqrt{9 + (6 - x)^2} = 30 - 5x$	Isolate the radical.
$\left[4\sqrt{9 + (6 - x)^2}\right]^2 = (30 - 5x)^2$	Square both sides.
$16[9 + (6 - x)^2] = (30 - 5x)^2$	$(ab)^2 = a^2b^2$
$16(9 + 36 - 12x + x^2) = (30 - 5x)^2$	Square the binomial $6 - x$.
$16(45 - 12x + x^2) = 900 - 300x + 25x^2$	Square $30 - 5x$.
$720 - 192x + 16x^2 = 900 - 300x + 25x^2$	Distributive Property
$9x^2 - 108x + 180 = 0$	Write the equation in standard form.
$x^2 - 12x + 20 = 0$	Divide both sides by 9.
$(x - 2)(x - 10) = 0$	Factor.
$x - 2 = 0$ or $x - 10 = 0$	Zero Factor Property
$x = 2$ or $x = 10$	Solve each case for x.

Even if 10 were not an extraneous solution, it would not be a valid solution to the problem. Remind students that when a problem calls for a positive solution, a positive solution for the variable is not necessarily valid.

By substitution, we find that 10 is an extraneous solution. For $x = 2$, the distance $6 - x$ would be 4, which is an acceptable value. Therefore, the fisherman walked 2 miles.

Work Rate

If T is the time needed to complete a task, then $\frac{1}{T}$ is the work rate. If two or more persons, groups, or machines work at the task, then the combined work rate is the sum of the individual work rates.

EXAMPLE 4

Combined Work Rates

Working together, two details of Air Force reservists can load a C-130 aircraft with relief supplies for hurricane victims in 6 hours. Working alone, the first detail can load the plane in 2 hours less time than the second detail can when working alone. How long would each detail, working alone, take to load the plane?

Solution

Let t = number of hours for the second detail, working alone, to load the plane. Then $t - 2$ = number of hours for the first detail, working alone, to load the plane.

	Time (hours)	Work Rate (loads/hour)
First Detail	$t - 2$	$\frac{1}{t-2}$
Second Detail	t	$\frac{1}{t}$
Together	6	$\frac{1}{6}$

$$\frac{1}{t-2}+\frac{1}{t}=\frac{1}{6}$$ Sum of the work rates equals the combined work rate.

$$\frac{6t(t-2)}{1}\cdot\frac{1}{t-2}+\frac{6t(t-2)}{1}\cdot\frac{1}{t}=\frac{6t(t-2)}{1}\cdot\frac{1}{6}$$ Multiply by the LCD, $6t(t-2)$.

$$6t+6(t-2)=t(t-2)$$ Simplify.

$$6t+6t-12=t^2-2t$$ Distributive Property

$$t^2-14t+12=0$$ Write the equation in standard form.

To solve for t, use the Quadratic Formula.

$$t=\frac{14\pm\sqrt{(-14)^2-4(1)(12)}}{2(1)}=\frac{14\pm\sqrt{196-48}}{2}=\frac{14\pm\sqrt{148}}{2}$$

$$=\frac{14\pm\sqrt{4}\sqrt{37}}{2}=\frac{14\pm2\sqrt{37}}{2}=7\pm\sqrt{37}$$

$t\approx13.08$ or $t\approx0.92$

Here is another positive but invalid solution.

Because the solution 0.92 makes the first detail's time negative, the only applicable solution is 13.08. The second detail would take approximately 13.08 hours to load the plane, and the first detail would take approximately 11.08 hours.

Investment

LEARNING TIP
Consider writing a "word equation" to relate the quantities involved in a problem. For example, (number of items) · (cost per item) = total cost.

In previous investment problems, we have used this relationship:

Number of shares · price per share = total investment

If the focus of an investment application is on price per share, then we can rewrite the relationship this way:

$$\text{Price per share}=\frac{\text{total investment}}{\text{number of shares}}$$

EXAMPLE 5

An Investment Application Problem

Six years ago a woman invested \$42,000 in stock. As a result of stock splits and dividend reinvestment, she now has 1100 more shares of the stock than she originally purchased, and the market value is \$73,600. However, the price per share is \$3 per share less than the price per share of the original purchase. How many shares did she purchase originally, and how many does she own now?

Solution

Let x = number of shares initially purchased and $x + 1100$ = number of shares currently owned. The price per share is the market value divided by the number of shares.

Help students understand that price per share is a fraction, so a problem involving price per share is likely to be described by an equation with rational expressions.

$$\frac{42{,}000}{x}=\text{price per share of initial purchase}$$

$$\frac{73{,}600}{x+1100}=\text{current price per share}$$

The difference between the initial price per share and the current price per share is \$3.

$$\frac{42{,}000}{x} - \frac{73{,}600}{x + 1100} = 3$$

Clear fractions by multiplying by the LCD, $x(x + 1100)$.

$$\frac{x(x + 1100)}{1} \cdot \frac{42{,}000}{x} - \frac{x(x + 1100)}{1} \cdot \frac{73{,}600}{x + 1100} = 3x(x + 1100)$$

$$42{,}000(x + 1100) - 73{,}600x = 3x(x + 1100)$$

$$42{,}000x + 46{,}200{,}000 - 73{,}600x = 3x^2 + 3300x$$

$$46{,}200{,}000 - 31{,}600x = 3x^2 + 3300x$$

$$3x^2 + 34{,}900x - 46{,}200{,}000 = 0$$

To solve for x, use the Quadratic Formula.

$$x = \frac{-34{,}900 \pm \sqrt{34{,}900^2 - 4(3)(-46{,}200{,}000)}}{2(3)}$$

$$x = \frac{-34{,}900 \pm \sqrt{1{,}772{,}410{,}000}}{6}$$

$$x = 1200 \quad \text{or} \quad x \approx -12{,}833.33$$

Because the number of shares must be positive, the only applicable solution is 1200. Thus the woman bought 1200 shares initially, and she now owns 2300 shares.

For applications that are geometric in nature, students should be urged to draw and label a diagram to organize information.

Exercises 10.5

Real-Life Applications

Geometric Figures

1. The dimensions of a rectangular front yard are 60 feet by 80 feet. If you begin mowing around the outside edge, how wide a border around the lawn must you mow to complete three-fourths of the job?

2. The dimensions of a rectangular hay field are 60 yards by 100 yards. A farmer is going to cut the hay around the outside edge. How wide a border around the field must the farmer cut to complete two-thirds of the job?

3. A triangular-shaped fountain area occupies one corner of a rectangular park. A low wall is to be constructed along the longest side, which is 160 feet long. A higher wall is to be constructed along the two perpendicular sides, whose lengths differ by 30 feet. What are the lengths of the two higher walls?

4. When a painter leans a 24-foot ladder against a house, the distance from the base of the ladder to the house is 8 feet less than the distance from the top of the ladder to the ground. How far up the house is the top of the ladder?

A key step in solving right triangle problems is identifying the hypotenuse. Remind students how this is done.

Ratio and Proportion

5. A woodworker is asked to build a rectangular shadow box with a perimeter of 56 inches. To make the shape aesthetically appealing, the rectangle should satisfy the Golden Ratio. What are the appropriate dimensions of the shadow box?

6. The design editor for a publisher created a front cover for a book. The cover satisfied the Golden Ratio and had a perimeter of 31 inches. What were the dimensions of the cover?

7. In the figure, triangles ABC and DEF are similar triangles. Determine the lengths AC and DE.

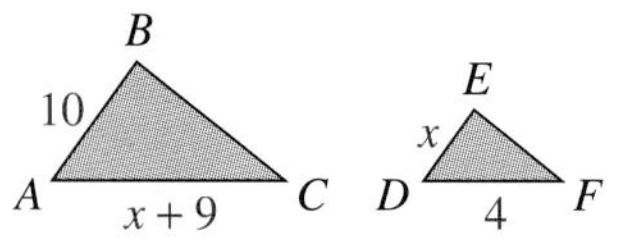

8. In the accompanying figure, triangles ABC and DEF are similar triangles. Determine the lengths AB and DF.

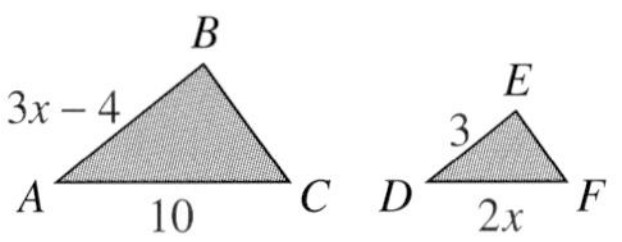

Remind students about the value of a table for distance, rate, and time problems.

Distance, Rate, Time

9. A boater drove a motorboat upstream a distance of 46 miles and then immediately drove back to the starting point. The total time of the trip was 3.5 hours. If the rate of the current was 4 mph, how fast would the boat have traveled in still water?

10. Two campers left camp at 8:00 A.M. to paddle their canoe upstream 10 miles to a fishing cove. They fished for 3.75 hours and then paddled back to the campsite and arrived at 6:00 P.M. If the rate of the current was 3 mph, how fast can the campers paddle in still water?

11. A family drove 45 miles and then hiked 1 mile to their campsite. The trip took 1 hour and 15 minutes. If they drove 55 mph faster than they hiked, how fast did they drive?

12. Two track team members ran a 200-meter relay in 29 seconds. (Each person runs 100 meters.) One sprinter ran 0.55 meter per second faster than the other. What was the speed of each runner?

Work Rate

13. Working together, two roommates can paint their apartment in 6 hours. Working alone, one person can complete the job in 2 hours less time than the other person working alone. How long would the job take each person, working alone?

14. Two landscape workers can complete a certain landscaping job in a total of 20 hours. Working alone, one can do this same job in 2 hours less time than the other. How long would the faster worker take to do the job if the slower worker were unable to work?

15. A backyard swimming pool has two fill pipes and a drain pipe. The first fill pipe takes 9 hours to fill the pool. The drain pipe takes 1 hour longer to drain the pool than the second fill pipe takes to fill the pool. If all three pipes are open at once, the pool can be filled in 8 hours. How long would the drain pipe take to drain the pool with the two fill pipes closed?

16. Both the inlet and outlet pipes to a cider barrel were mistakenly left open. Nevertheless, the barrel filled in 12 hours. With only the outlet pipe open, a full barrel of cider can be emptied in 1 hour more than the time to fill an empty barrel with only the inlet pipe open. How long would a full barrel of cider take to drain with the outlet pipe open and the inlet pipe closed?

Investment

In Exercises 17–20, two positive solutions will be obtained, but the conditions of the problem invalidate one of them.

17. Ten years ago, a man had less than 400 shares of stock worth \$16,095. Today, he has 405 more shares of stock than he had originally, and the investment is worth \$34,875. The price per share today is \$1.50 more than it was 10 years ago. How many shares of stock did the man have originally?

18. A woman invested \$13,500 to buy more than 200 shares of a certain stock. After 5 years she had 22.23 more shares of stock, and her investment was worth \$16,189.13. If each share increased in value by \$4, how many shares did the woman purchase initially?

19. A community service organization sells citrus fruit each year to raise money for the local library. Last year the total sales were \$8568. This year the organization sold 102 more boxes of fruit than last year, and the total sales were \$11,594. This year's selling price was up \$2 from last year's selling price. If fewer than 500 boxes were sold last year, how many boxes of fruit were sold last year and this year, and what was the price per box this year?

20. The community theater group performed a popular musical. The tickets for the Friday night performance were \$2 more than the Thursday night tickets. The total receipts on Thursday night were \$967.50, and the total receipts on Friday night were \$2034.00. On Friday 65 more people attended the play than had attended on Thursday night. How many people attended on Friday night?

Miscellaneous

21. The owner of an apartment complex collected \$24,000 per month in rent when all the apartments were occupied. Then the owner raised the rent \$80 per month, and ten apartments were vacated. However, the owner still collects \$24,000 a month in rent. How many apartments are in the complex?

22. An Olympics vendor purchased a box of souvenir key chains for \$70. When all but 12 were sold at a profit of 75¢ per key chain, he had recovered his original investment. How many key chains were in the box, and what was the selling price?

23. Find two consecutive odd integers such that the difference of their cubes is 866.

24. Find two consecutive integers such that the difference of their cubes is 169.

25. A plot of land next to a river is to be enclosed with 300 feet of fencing to form a rectangular area of 11,500 square feet. The river will serve as the fourth side. An 8-foot-wide gate is to be installed in the fence that is parallel to the river. (See figure.) How far should the fence extend from the river if the minimum distance is 70 feet?

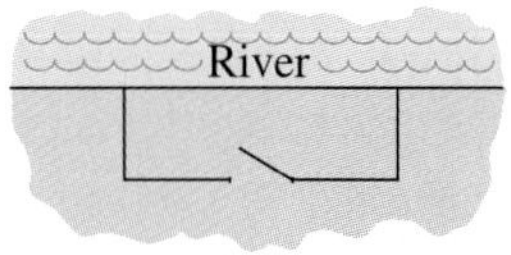

26. A woman bought a 21-square-yard rectangular piece of carpeting for a room that is 12 feet by 18 feet. She chose the dimensions of the carpet so that there would be a uniformly wide strip of uncovered floor around the outside of the carpet. How wide will this strip be?

27. If a rectangular sign is 5 feet wide, how long must it be in order to satisfy the Golden Ratio?

28. An architect's plans include a rectangular iron gate with a length that is 3 feet more than the width and that satisfies the Golden Ratio. What are the dimensions of the gate?

29. A polygon of n sides has $\frac{1}{2}n(n - 3)$ diagonals. If a polygon has 27 diagonals, how many sides does it have?

30. The height h of an object projected upward from an initial height h_0 with an initial velocity v_0 is given by $h = -16t^2 + v_0t + h_0$. The floor of the Golden Gate bridge is 220 feet above the water. If a rock is tossed upward from the floor of the bridge with an initial velocity of 30 feet per second, how many seconds later will the rock hit the water? (Round your answer to the nearest tenth.)

31. In geometry, the altitude h to the hypotenuse of a right triangle is related to the two segments a and b shown in the figure by the equation $\dfrac{a}{h} = \dfrac{h}{b}$.

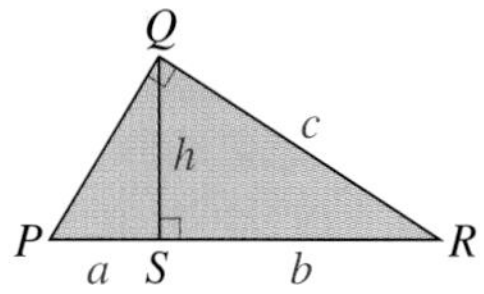

Suppose a home gardener uses the design shown in the figure to build a cold frame. The front segment b is 2.5 feet longer than the back segment a, and the height of the frame is 3 feet. How long is the front panel c?

32. In the figure, $\overline{AB}$ is a 4-inch mirror with two vertical shields represented by $\overline{AQ}$ and $\overline{BR}$. Light ray L_1 strikes the mirror at point P and is reflected as light ray L_2. By a law of optics, $\angle 1$ (the *angle of incidence*) has the same measure as $\angle 2$ (the *angle of reflection*). Given the relative dimensions in the figure, how tall is the shield $\overline{AQ}$?

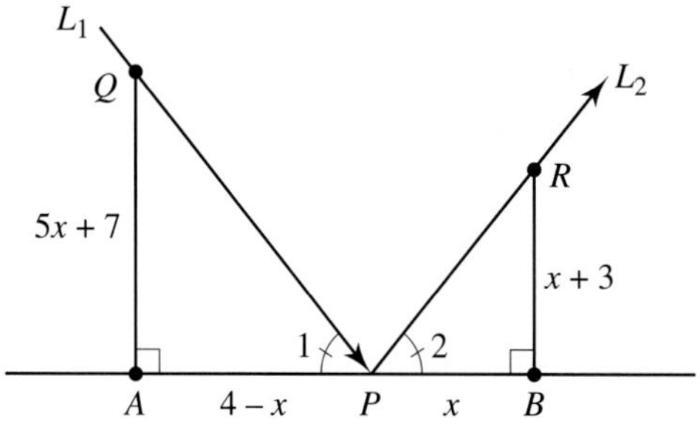

33. A farmer has two irrigation sprinklers that rotate about a fixed point to water circular areas. The circumference of the circle watered by the larger system is 4π yards more than the circumference of the circle covered by the smaller system. The total area irrigated by both systems is 164π square yards. What is the radius of the circular area irrigated by the smaller system?

34. An auto center has a large metal drum to store used motor oil. The height of the drum is 4 feet, and the total surface area is 24π square feet. What is the diameter of the drum? (*Hint:* The total surface area S of a right circular cylinder is $S = 2\pi rh + 2\pi r^2$, where r is the radius and h is the height.)

35. The area of a triangular mainsail is 440 square feet. If the height of the sail is 10 feet less than the base, what is the height?

36. A car and a truck left Luckenback, Texas, at noon. The car traveled due north, and the truck traveled due east. In 2 hours they were 155 miles apart, and the car had traveled 30 miles more than the truck. How far was each vehicle from Luckenback at 2:00 P.M.?

37. A shop for custom-made furniture has an order for a hardwood rectangular table that is 2 feet longer than it is wide. The customer wants the table to be designed so that an average adult can reach the center from any position at the table. The designer knows that the average person can reach 3 feet. What will be the width of the table?

38. To construct a paved walk of uniform width around a rectangular swimming pool, a contractor uses 5.5 cubic yards of concrete. The pool is 30 feet long and 20 feet wide. If the walk is to be 3 inches thick, how wide can it be?

39. When 1 is added to a number, the result is 2 more than the square root of the result. What is the number?

40. The square root of 7 more than a number is one less than twice the number. What is the number?

41. A person swims from a starting point S to a marker M that is 0.15 mile from the shore. From the marker, the person swims at a rate of 0.5 mph back to the shore at point B, which is a bicycle pickup point. From there, he rides the bike at 12 mph to the finish line F, which is 4.2 miles from the starting point. If the total time from the marker to the finish line is 50 minutes, how far did he ride the bicycle?

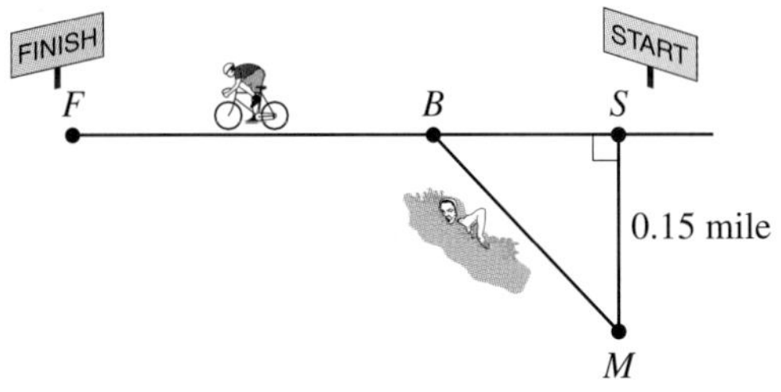

42. A tour bus traveled the first 200 miles of a trip at an average speed that was 6 mph faster than the average speed for the last 180 miles. If the entire trip took 6 hours and 40 minutes, what was the average speed for the first portion of the trip?

43. Working alone, one person can prepare a quarterly bank report in 4 hours less time than a less experienced person. If both begin working at 9:00 A.M. and stop at 5:00 P.M., but the less experienced worker takes a 2-hour lunch break and the experienced worker skips lunch, the job is 65% complete at the end of the day. If both begin working at 9:00 A.M. the next day, at what time will they complete the report?

44. Two air conditioning units can cool a building to 72°F in 3 hours. Working alone, the older unit takes 5 hours longer than the newer unit to cool the building. How many hours were required to cool the building before the new unit was installed?

45. As a result of a basketball injury, a senior forward played in three fewer games than a junior forward. The senior, who scored a total of 174 points, averaged 5.1 points per game more than the junior, who scored a total of 141 points. How many games did the senior play?

46. At 7:00 A.M. a trucker left Little Rock and traveled north. At 8:00 A.M. a second trucker left the same point and drove east, going 12 mph faster than the first. At noon, the two trucks were 310 miles apart. To the nearest tenth, what was the average speed of the second truck?

Data Analysis: *Credit Card Payoffs*

A majority of Americans have credit cards, but the percentage of card holders who pay the entire balance due each month varies with age. The accompanying table shows representative data for selected ages of card holders. (Source: Federal Reserve System.)

Age	Percent Having Credit Cards	Percent Who Always Pay the Balance Due
30	57.6	39.6
40	64.2	45.5
50	72.5	52.1
60	68.0	64.8
70	65.0	74.9
80	52.1	78.9

47. The function $N(x) = -0.025x^2 + 2.71x - 1.76$ can be used to model the percentage $N(x)$ of card holders at age x. Determine and interpret the x-intercepts of the graph of $N(x)$.

48. As is true with most models, the validity of $N(x)$ is restricted to the given x-values. To illustrate, use $N(x)$ to determine the percentage of 10-year-olds who have credit cards.

49. On the basis of the data in the table, which function, $y = ax + b$ or $y = ax^2 + bx + c$, would be the better model for the percentage who always pay the balance due? Use the data for ages 30 and 70 to write a model function for these data.

50. How would the graph of the function in Exercise 49 compare to the graph of a function that models the percentage who rarely pay the balance due?

Challenge

51. A campus co-op bought a shipment of calculators for \$10,000. The co-op first sold calculators to all students at the price needed to cover the total cost. The remaining 40 calculators were then sold to non-students at the same price that the students paid, and the profit was distributed to the students. If all of the calculators had been sold to students, the price would have been \$12.50 less than what they paid. What share of the profit did each student receive?

52. A fragile item is packed in a carton in the shape of a cube. This carton is then placed inside a larger carton, which is also in the shape of a cube and which has sides 2 feet longer than the sides of the smaller carton. The larger carton is then filled with 40 cubic feet of packing material around the smaller carton. How long is a side of the smaller carton?

53. A spring is attached at point A [see Fig. (a)] and is connected to the rim of a wheel at point B so that the spring is tangent to the wheel. In this position the spring is 26 inches long. The wheel is turned clockwise so that the spring stretches and the end of the spring is now at point D. [See Fig. (b).] If the distance between points C and D is 8 inches, how far did the spring stretch from its original position? (*Hint:* A theorem from geometry relates AB, AC, and AD as $AB/AD = AC/AB$.)

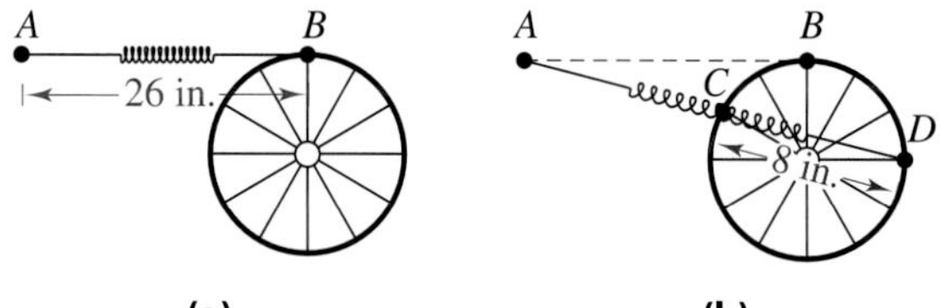

54. A ceremonial cannon, located on the bank of a river, is about to be fired. (See point C in the accompanying figure.) Observer 1 is directly across the river at point A, and Observer 2 is farther down the river at point B. Observer 1 is 2000 feet farther from Observer 2 than from the cannon. When the cannon is fired, Observer 2 hears the sound 3 seconds after she sees the flash. How wide is the river? (Assume that the speed of sound is 1100 feet per second.)

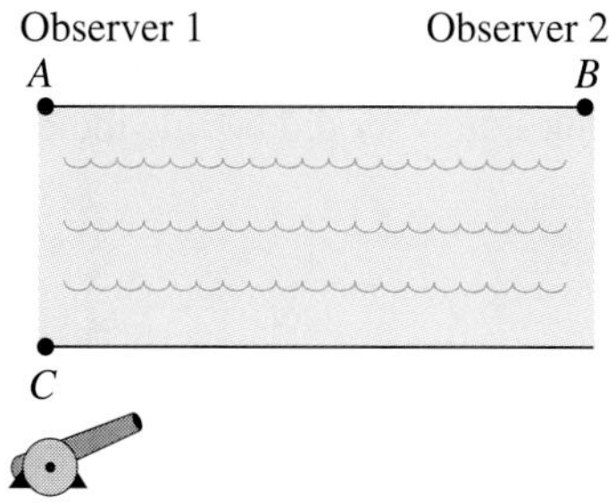

10.6 Quadratic Functions

Properties of Graphs • Variations of the Graph of $y = x^2$ • Vertex and Intercepts • Real-Life Applications

Properties of Graphs

Recall that a quadratic function is a function of the form $f(x) = ax^2 + bx + c$, where a, b, and c are real numbers and $a \neq 0$. In this section we consider quadratic functions and their graphs.

Exploring the Concept

Graphs of Quadratic Functions

We graph a quadratic function just as we graph other functions: We plot the pairs (x, y) where $y = f(x)$. For instance, to sketch the graph of the simplest quadratic function $y = x^2$, we can begin by creating a table of values for the function and plotting the points associated with the pairs. Then we use those points as a guide for sketching the graph. (See Fig. 10.4.)

x	-3	-2	-1	0	1	2	3
x^2	9	4	1	0	1	4	9

Figure 10.4

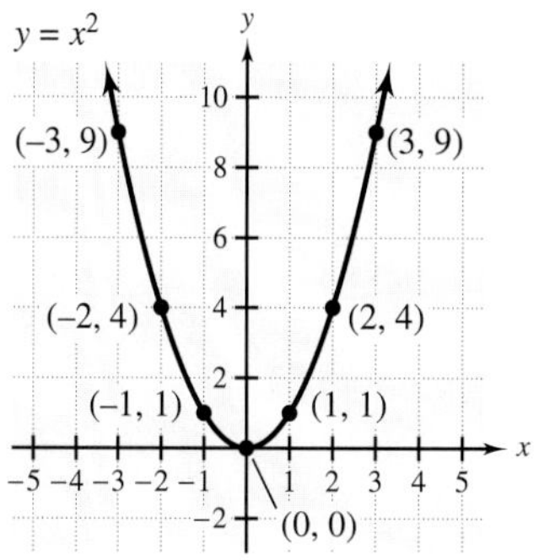

EXAMPLE 1

Sketching the Graph of a Quadratic Function

Sketch the graph of $f(x) = 9 - x^2$.

Solution

The following typical table of values can be used to sketch the graph of the function. (See Fig. 10.5.)

x	-4	-3	-1	0	1	3	4
$9 - x^2$	-7	0	8	9	8	0	-7

Figure 10.5

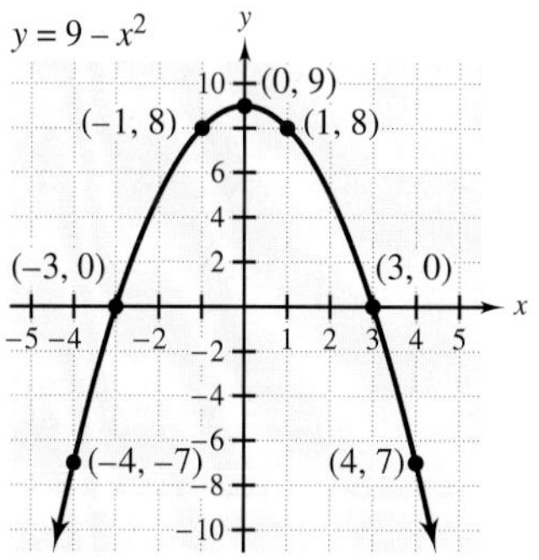

Figure 10.6

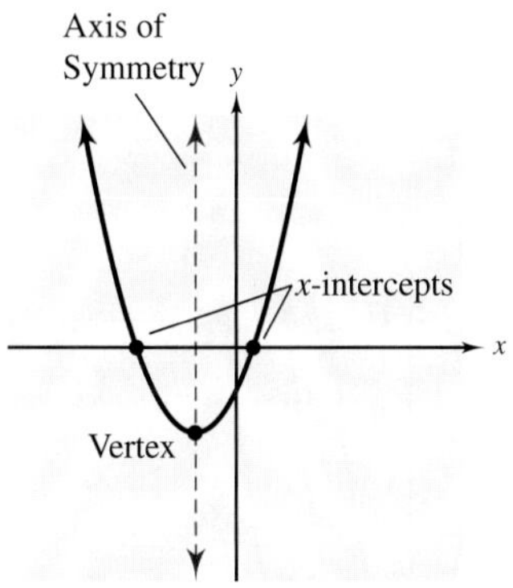

The graph of a quadratic function is called a **parabola.** When a parabola opens upward, as in Figure 10.6, it has a low point corresponding to the minimum value of the function. The parabola in Figure 10.7 opens downward and has a high point corresponding to the maximum value of the function. The highest or lowest point of a parabola is called the **vertex.**

Figure 10.7

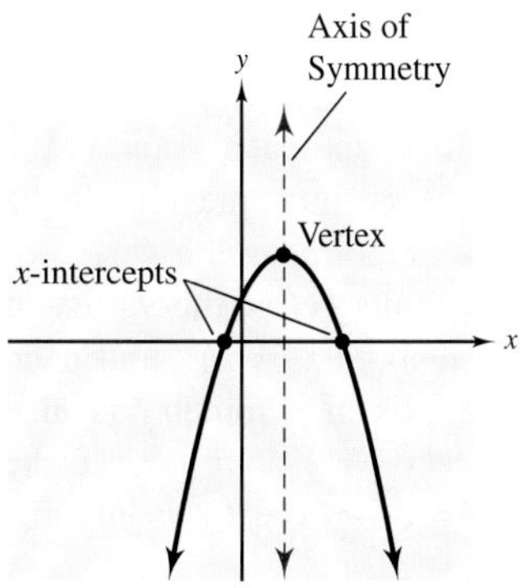

The parabolas in Figures 10.6 and 10.7 are *symmetric* with respect to a vertical line containing the vertex. This line is called the **axis of symmetry.** If the graph were folded together along the axis of symmetry, then the two sides of the parabola would coincide.

Note: Parabolas can be oriented in any direction in the coordinate plane. A parabola with a vertical axis of symmetry is sometimes called a *vertical parabola*. The graph of a quadratic function is always a vertical parabola.

Variations of the Graph of $y = x^2$

The graph of a quadratic function can often be sketched easily by comparing the function to the basic function $y = x^2$.

Exploring the Concept

Variations of the Graph of $y = x^2$

Consider the graphs of the functions of the form $y = ax^2$ shown in Figures 10.8 and 10.9.

Figure 10.8

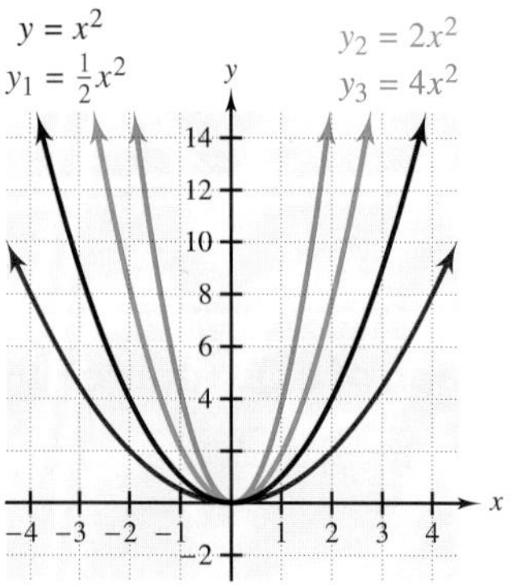

Figure 10.9

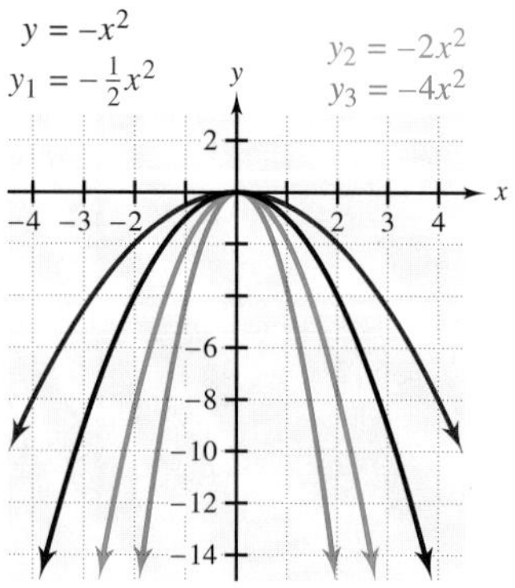

Observe that if a is positive, the graph opens upward and that if a is negative, the graph opens downward. As $|a|$ increases, the graph becomes narrower. Because the graph extends without bound to the left and right, the domain of each function is **R**. For $a > 0$, the range is $\{y \mid y \geq 0\}$; for $a < 0$, the range is $\{y \mid y \leq 0\}$. We see that the absolute value of a affects the width of a parabola and that the sign of a affects the direction in which a parabola opens. But the vertex for $y = ax^2$ is the origin for all values of a.

Now consider the graphs shown in Figures 10.10 and 10.11.

Figure 10.10

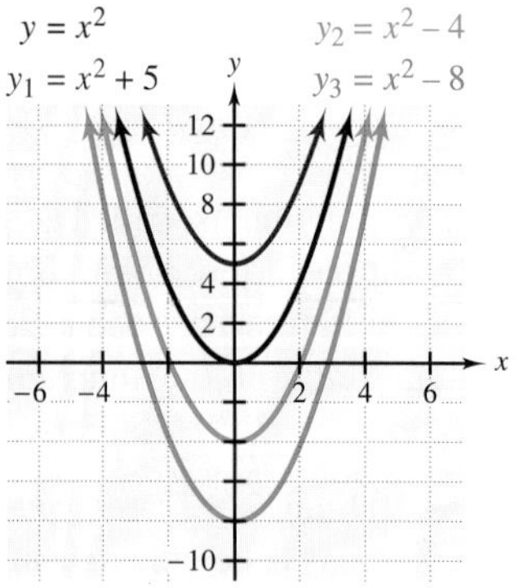

Figure 10.11

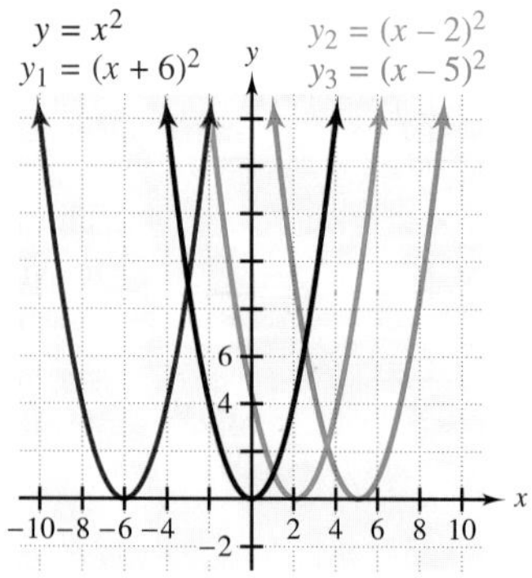

See whether students can understand the similarity between this experiment and the one for $y = mx + b$ in which we varied b.

In Figure 10.10, each function is in the form $f(x) = x^2 + k$, and each graph is the graph of $y = x^2$ shifted vertically. The axis of symmetry is the same for all the graphs. If $k > 0$, then the vertex of $f(x) = x^2 + k$ is k units *above* the vertex of $y = x^2$. If $k < 0$, then the vertex of $f(x) = x^2 + k$ is $|k|$ units *below* the vertex of $y = x^2$. In Figure 10.11, each function is in the form $g(x) = (x - h)^2$, and each graph is the graph of $y = x^2$ shifted horizontally. The axis of symmetry is also shifted in each case. If $h > 0$, then the vertex of $g(x) = (x - h)^2$ is h units to the *right* of the vertex of $y = x^2$. If $h < 0$, then the vertex of $g(x) = (x - h)^2$ is $|h|$ units to the *left* of the vertex of $y = x^2$.

A vertical or horizontal shift of $y = x^2$ does not affect the domain of the function, but it may affect the vertex, the axis of symmetry, and the range.

	Vertex	*Axis of Symmetry*	*Range*
$f(x) = x^2 + k$	$(0, k)$	$x = 0$	$\{y \mid y \geq k\}$
$g(x) = (x - h)^2$	$(h, 0)$	$x = h$	$\{y \mid y \geq 0\}$

Note: Our conclusions about $g(x) = (x - h)^2$ are valid only if the function is written in that form. To analyze $g(x) = (x + h)^2$, we first write $g(x) = [x - (-h)]^2$.

EXAMPLE 2

Graphing a Quadratic Function

Using the graph of $y = x^2$ as a guide, sketch the graph of each of the following functions. Describe the vertex, the axis of symmetry, and the range of the function.

(a) $f(x) = (x + 2)^2 - 3$ (b) $g(x) = (x - 4)^2 + 5$

Solution

(a) The graph of function f is the graph of $y = x^2$ with the vertex shifted to the left 2 units and down 3 units. The vertex is $(-2, -3)$ and the axis of symmetry is the line $x = -2$. The range of f is $\{y \mid y \geq -3\}$. (See Fig. 10.12.)

Figure 10.12

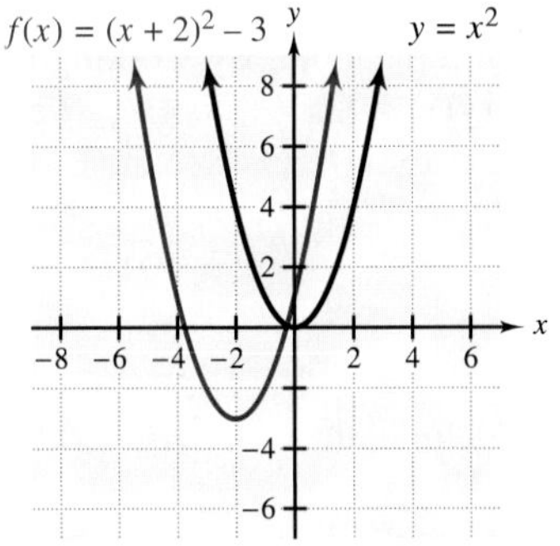

Figure 10.13

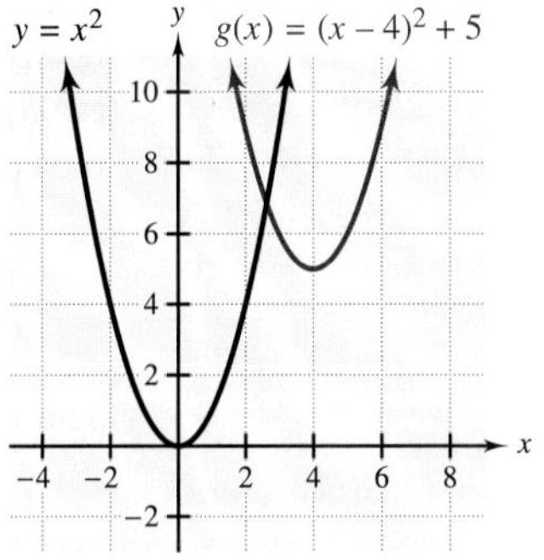

(b) The graph of function g is the graph of $y = x^2$ with the vertex shifted to the right 4 units and up 5 units. The vertex is $(4, 5)$, and the axis of symmetry is the line $x = 4$. The range of g is $\{y \mid y \geq 5\}$. (See Fig. 10.13.)

Example 2 prompts us to generalize as follows.

Properties of the Graph of $f(x) = a(x - h)^2 + k$

The graph of $f(x) = a(x - h)^2 + k$ is the graph of $y = ax^2$ with the vertex shifted to the point (h, k).

Vertex and Intercepts

Knowing the vertex of a parabola is important when we graph a quadratic function. To determine the coordinates of the vertex, we can use the method of completing the square to write the function in the form $f(x) = a(x - h)^2 + k$. The process is illustrated in Example 3.

EXAMPLE 3

Determining the Vertex of a Parabola by Completing the Square

Determine the vertex of the graph of $f(x) = x^2 - 8x + 10$.

Solution

As before, we wish to complete the square for $x^2 - 8x$. We subtract 10 from both sides to move the constant term out of the way.

By using the method of completing the square, we will write the function in the form $f(x) = (x - h)^2 + k$.

$$f(x) = x^2 - 8x + 10$$

$$f(x) - 10 = x^2 - 8x \qquad \text{Subtract 10 from both sides.}$$

$$f(x) - 10 + 16 = x^2 - 8x + 16 \qquad \text{To complete the square, add } \left(-\frac{8}{2}\right)^2 = 16 \text{ to both sides.}$$

$$f(x) + 6 = (x - 4)^2 \qquad \text{Factor and write in exponential form.}$$

$$f(x) = (x - 4)^2 + (-6) \qquad \text{Add } -6 \text{ to both sides to obtain the function.}$$

In this form we see that $h = 4$ and $k = -6$. Therefore, the vertex is $(4, -6)$.

The procedure illustrated in Example 3 is one we need to repeat each time we wish to determine the vertex of a parabola. By working through the same routine for the general quadratic function $f(x) = ax^2 + bx + c$, we can derive formulas for determining the coordinates of the vertex $V(h, k)$. (See the Exercise 109.)

LEARNING TIP

Observe that the x-coordinate of the vertex is midway between the x-coordinates of the x-intercepts. If you know the x-intercepts, then finding the average of the x-coordinates will give you the x-coordinate of the vertex.

$$h = -\frac{b}{2a} \quad \text{and} \quad k = \frac{4ac - b^2}{4a}$$

The vertex is $\left(-\frac{b}{2a}, \frac{4ac - b^2}{4a}\right)$.

Note: The coordinates of the vertex can be found by using these two formulas. However, in practice, it is easier to determine the first coordinate by using the formula $-\frac{b}{2a}$. Then the y-coordinate can be determined by evaluating $f\left(-\frac{b}{2a}\right)$.

EXAMPLE 4

Determining the Vertex Algebraically

Determine the vertex of the graph of $f(x) = 2x^2 + 12x + 17$.

Solution

The x-coordinate of the vertex is

$$x = -\frac{b}{2a} = -\frac{12}{2(2)} = -3$$

Because $f(-3) = -1$, the vertex is $(-3, -1)$.

In general, the y-intercept of a graph is a point whose x-coordinate is 0. For a quadratic function $f(x) = ax^2 + bx + c$, $f(0) = c$. Therefore, a vertical parabola always has one y-intercept $(0, c)$.

To determine the x-intercepts, we solve the quadratic equation $ax^2 + bx + c = 0$. We know this equation has zero, one, or two real number solutions. Thus a vertical parabola has zero, one, or two x-intercepts.

The following summarizes the characteristics of the graph of a quadratic function $f(x) = ax^2 + bx + c$.

Recall that the discriminant indicates the number and nature of the solutions of a quadratic equation.

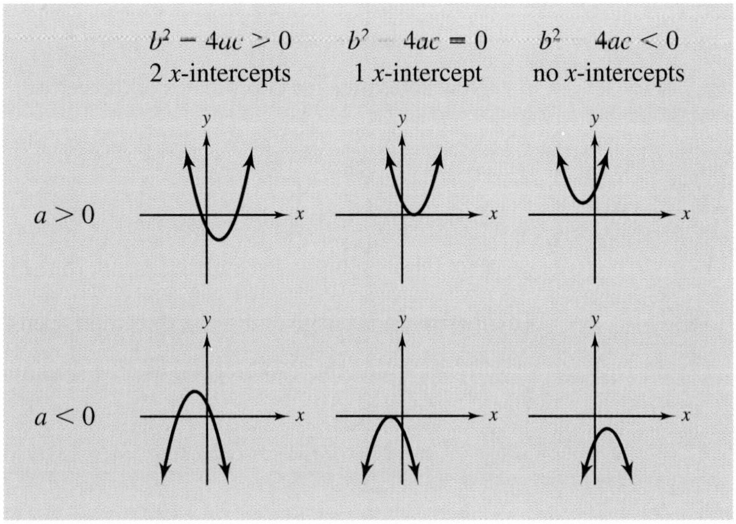

Graphing a quadratic function is facilitated by first determining the intercepts and vertex.

Graphing a Quadratic Function $f(x) = ax^2 + bx + c$

1. The y-intercept is $(0, c)$.
2. Determine the x-intercepts by solving $ax^2 + bx + c = 0$.
3. Use completing the square or the vertex formula to determine the coordinates of the vertex.
4. Plot the points determined in steps 1–3 and any additional points that are needed to sketch the graph.
5. Verify that the parabola opens upward for $a > 0$ and downward for $a < 0$.

EXAMPLE 5

Using the Vertex and Intercepts to Sketch the Graph of a Quadratic Function

Determine the vertex and the intercepts of the graphs of each of the following functions. Then sketch the graph of the function.

(a) $f(x) = 8 - 2x - x^2$ (b) $h(x) = x^2 - 4x + 1$

(c) $g(x) = 4x^2 - 12x + 9$ (d) $s(x) = -2x^2 + 12x - 19$

Figure 10.14

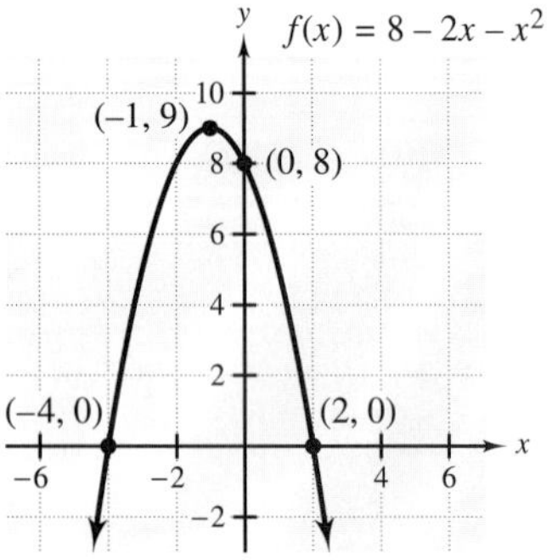

Figure 10.15

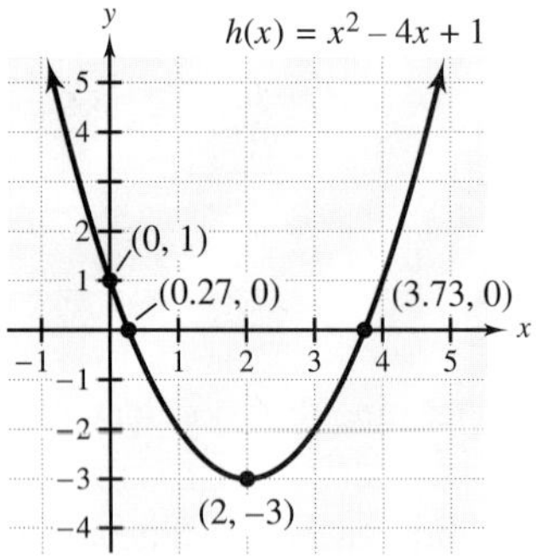

Solution

(a) Because a is negative ($a = -1$), the graph opens downward. The y-intercept is (0, 8).

To determine the x-intercepts, use factoring to solve the equation $8 - 2x - x^2 = 0$.

$$8 - 2x - x^2 = 0$$
$$(2 - x)(4 + x) = 0$$
$$2 - x = 0 \quad \text{or} \quad 4 + x = 0$$
$$x = 2 \quad \text{or} \quad x = -4$$

The x-intercepts are (2, 0) and (−4, 0).

Now determine the coordinates of the vertex.

$$x = -\frac{b}{2a} = -\frac{-2}{2(-1)} = -1$$

Because $f(-1) = 9$, the vertex is (−1, 9).

Plot these points and sketch the graph. (See Fig. 10.14.)

(b) Because a is positive ($a = 1$), the graph opens upward. The y-intercept is (0, 1).

To determine the x-intercepts, use the Quadratic Formula to solve the equation $x^2 - 4x + 1 = 0$.

$$x = \frac{4 \pm \sqrt{(-4)^2 - 4(1)(1)}}{2(1)} = \frac{4 \pm \sqrt{16 - 4}}{2} = \frac{4 \pm \sqrt{12}}{2}$$
$$= \frac{4 \pm \sqrt{4}\sqrt{3}}{2} = \frac{4 \pm 2\sqrt{3}}{2} = 2 \pm \sqrt{3}$$
$$x \approx 3.73 \quad \text{or} \quad x \approx 0.27$$

The (approximate) x-intercepts are (3.73, 0) and (0.27, 0). Now we determine the coordinates of the vertex.

$$x = -\frac{b}{2a} = -\frac{-4}{2(1)} = 2$$

Because $h(2) = -3$, the vertex is (2, −3).

Figure 10.15 shows the plotted points and the graph of the function.

(c) Because a is positive ($a = 4$), the graph opens upward. The y-intercept is (0, 9).

To determine the x-intercept, solve the associated equation.

$$4x^2 - 12x + 9 = 0$$
$$(2x - 3)^2 = 0$$
$$2x - 3 = 0$$
$$2x = 3$$
$$x = \frac{3}{2}$$

The x-intercept is $\left(\frac{3}{2}, 0\right)$.

Figure 10.16

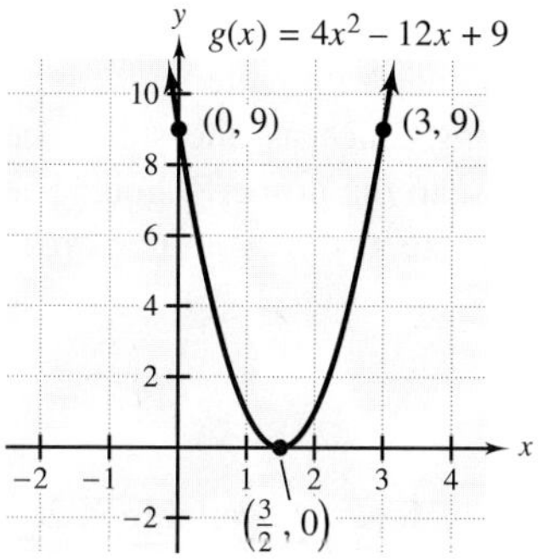

Now we determine the coordinates of the vertex.

$$x = -\frac{b}{2a} = -\frac{-12}{2(4)} = \frac{3}{2}$$

Because $g\left(\frac{3}{2}\right) = 0$, the vertex is $\left(\frac{3}{2}, 0\right)$. (Note that the vertex and the x-intercept are the same point.)

Because we have only two points to plot, we need an additional point to assist with sketching the graph. Because $g(3) = 9$, we can plot (3, 9) as a third point. Figure 10.16 shows the intercepts and the additional point.

(d) Because a is negative ($a = -2$), the parabola opens downward. The y-intercept is $(0, -19)$.

We determine the number of x-intercepts by evaluating the discriminant.

$$b^2 - 4ac = 12^2 - 4(-2)(-19) = 144 - 152 = -8$$

Because the discriminant is negative, there are no x-intercepts. Now we determine the coordinates of the vertex.

$$x = -\frac{b}{2a} = -\frac{12}{2(-2)} = 3$$

Because $s(3) = -1$, the vertex is $(3, -1)$.

Again we have only two points to plot. Figure 10.17 shows those points and one additional point that we determined to help us sketch the graph.

Think About It

Two points determine a line. What is the minimum number of points needed to determine a parabola? For a particular parabola, if you know that $c = 3$ and that $-\frac{b}{2a} = 6$, can you write the quadratic function?

Figure 10.17

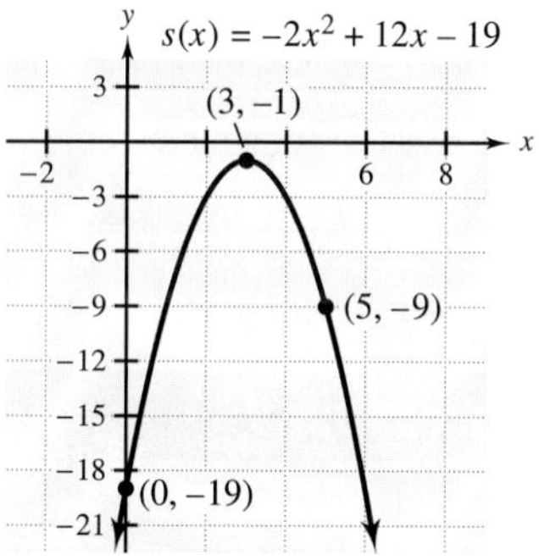

Real-Life Applications

Applications in which a quantity is to have an optimum (maximum or minimum) value can often be modeled with quadratic functions. The vertex of the graph is the point of interest, because it is at that point that the function has its maximum or minimum value.

In Section 10.5 we saw an example of a day care operator who wished to enclose a rectangular play yard of a specified area. In Example 6, we see a similar problem, but this time the goal is to *maximize* the area.

EXAMPLE 6

Maximizing a Rectangular Area

Figure 10.18

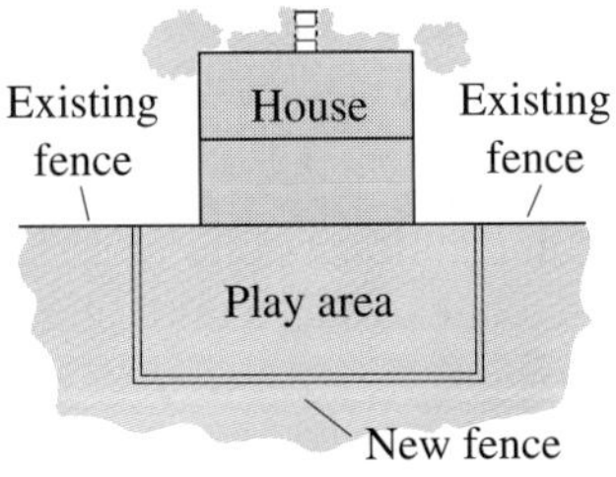

A day care operator wants to use 200 feet of fencing to enclose a rectangular play area for her day care center. She plans to use an existing fence as one side. The existing fence extends from each back corner of her home to the property line. (See Fig. 10.18.) What dimensions should she use to maximize the rectangular area? What is the maximum area?

Solution

Let x = the width of the rectangle. Because the amount of fencing available is 200 feet, the length of the rectangle is $200 - 2x$. (See Fig. 10.19.) The area is given by the function $A(x) = x(200 - 2x) = -2x^2 + 200x$.

Figure 10.19

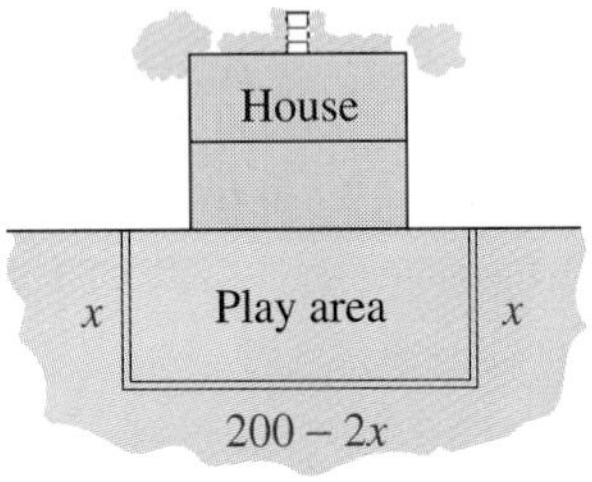

Because a is negative ($a = -2$), the parabola opens downward. The maximum value of the function occurs at the vertex.

The x-coordinate of the vertex is determined as follows.

$$x = -\frac{b}{2a} = -\frac{200}{2(-2)} = 50$$

Because x represents the width of the play area, the maximum area is achieved when the width is 50 feet. The length is

$$200 - 2x = 200 - 2(50) = 100$$

The area of the yard can be found either by evaluating the function for $x = 50$ or by using the formula $A = LW$.

$$A(50) = 5000$$
$$A = LW = 100 \cdot 50 = 5000$$

The maximum area of 5000 square feet is achieved when the width is 50 feet and the length is 100 feet.

Quick Reference 10.6

Properties of Graphs

- The graph of a quadratic function $f(x) = ax^2 + bx + c$ is called a **parabola.**
- The lowest point of a parabola (when $a > 0$) or the highest point (when $a < 0$) is called the **vertex.**
- The graph of a quadratic function is **symmetric** with respect to a vertical line containing the vertex. This line is called the **axis of symmetry.**

Variations of the Graph of $y = x^2$

- If $a > 0$, the parabola opens upward; if $a < 0$, the parabola opens downward. As $|a|$ increases, the parabola becomes narrower; as $|a|$ decreases, the parabola becomes wider.
- The domain of a quadratic function is **R**. If $V(h, k)$ is the vertex, then the range of the function is $\{y \mid y \geq k\}$ when $a > 0$ or $\{y \mid y \leq k\}$ when $a < 0$.
- For $f(x) = x^2 + k$, the vertex of the parabola is $V(0, k)$.

- For $f(x) = (x - h)^2$, the vertex of the parabola is $V(h, 0)$.
- The graph of $f(x) = a(x - h)^2 + k$ is the graph of $y = ax^2$ with the vertex shifted to $V(h, k)$.

Vertex and Intercepts

- To determine the vertex of the graph of a quadratic function, we can use the method of completing the square to rewrite the function in the form $f(x) = a(x - h)^2 + k$.
- In general, the x-coordinate of the vertex is $-\dfrac{b}{2a}$.

  The y-coordinate can be determined by evaluating $f\left(-\dfrac{b}{2a}\right)$.
- If $a > 0$, then the y-coordinate of the vertex represents the minimum value of the function; if $a < 0$, then the y-coordinate represents the maximum value of the function.
- The y-intercept of the graph of $f(x) = ax^2 + bx + c$ is $(0, c)$.
- To determine the x-intercepts, solve the quadratic equation $ax^2 + bx + c = 0$. The number of x-intercepts depends on the sign of the discriminant.
- To graph a quadratic function $f(x) = ax^2 + bx + c$, follow these steps.
  1. The y-intercept is $(0, c)$.
  2. Determine the x-intercepts by solving $ax^2 + bx + c = 0$.
  3. Use completing the square or the vertex formula to determine the coordinates of the vertex.
  4. Plot the points determined in steps 1–3 and any additional points that are needed to sketch the graph.
  5. Verify that the parabola opens upward for $a > 0$ and downward for $a < 0$.

Speaking the Language 10.6

1. The graph of a quadratic function is called a(n) ______.
2. The highest or lowest point of the graph of $f(x) = ax^2 + bx + c$ is called the ______.
3. If the lowest point of the graph of a quadratic function is $V(p, q)$, then $x = p$ is the equation of the ______.
4. The graph of $f(x) = x^2 + k$ is a ______ shift of the graph of $y = x^2$, whereas the graph of $g(x) = (x - h)^2$ is a ______ shift of the graph of $y = x^2$.

Exercises 10.6

Concepts and Skills

1. What do we call the graph of a quadratic function $f(x) = ax^2 + bx + c$? Describe how the value of a affects the shape and orientation of the graph.

2. (a) Describe the graph of $f(x) = x^2 + k$ in comparison to the graph of $y = x^2$.
   (b) Describe the graph of $g(x) = (x - h)^2$ in comparison to the graph of $y = x^2$.

In Exercises 3–8, match the function with its graph.

3. $f(x) = x^2 - 4$

4. $g(x) = x - 4$

5. $h(x) = -2x^2$

6. $f(x) = (x - 3)^2$

7. $g(x) = -\frac{1}{2}x - 2$

8. $h(x) = -\frac{1}{2}(x + 1)^2 - 2$

(a)

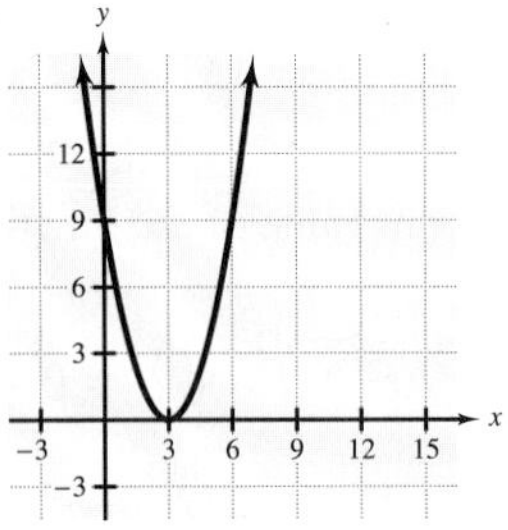

(b)

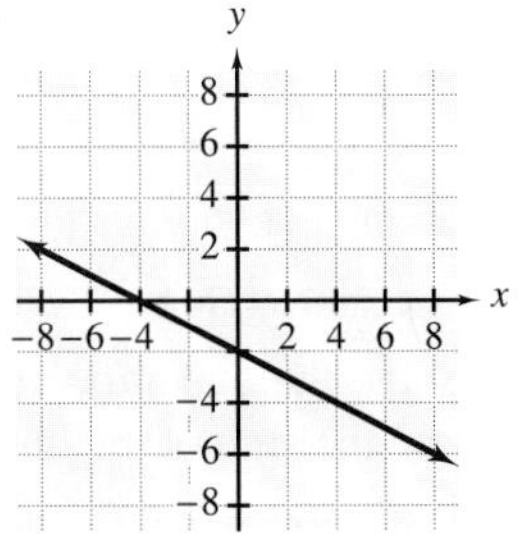

(c)

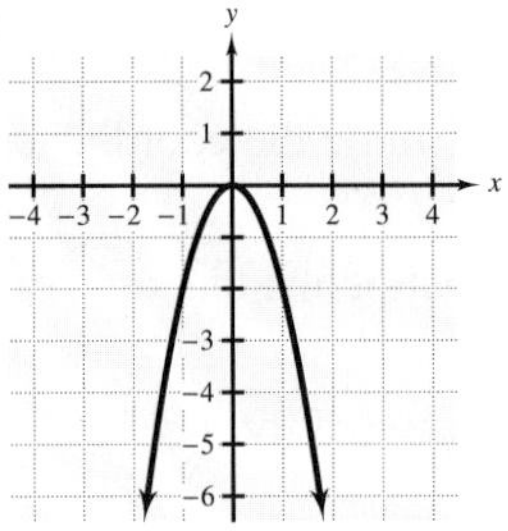

(d)

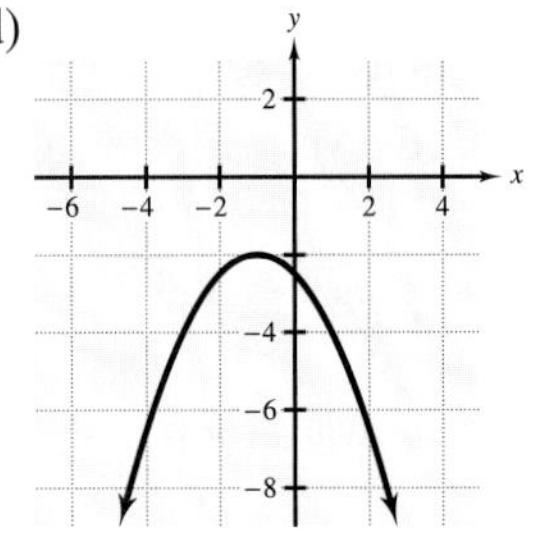

(e)

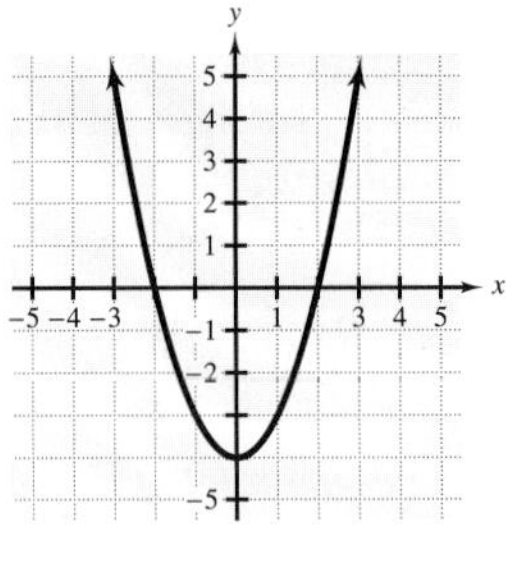

(f)

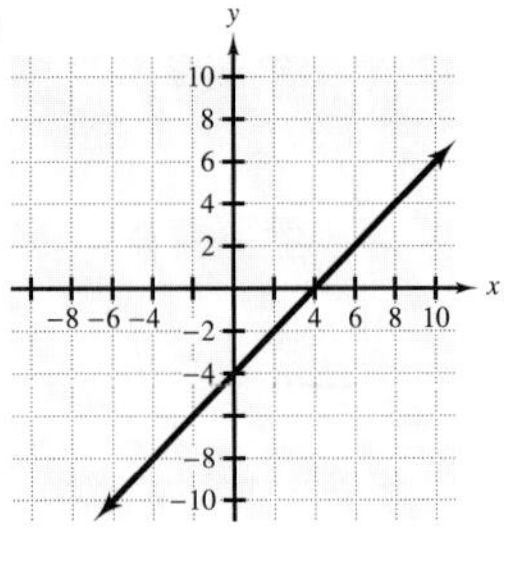

In Exercises 9–14, using the graph of $y = x^2$ as a guide, sketch the graphs of the given pair of functions. State whether the graphs are more or less narrow and describe any horizontal or vertical shifts.

9. (a) $y = -x^2$ (b) $y = -2x^2$

10. (a) $y = x^2 + 4$ (b) $y = x^2 - 4$

11. (a) $y = (x + 3)^2$ (b) $y = (x - 3)^2$

12. (a) $y = x^2 - 1$ (b) $y = (x - 1)^2$

13. (a) $y = (x - 3)^2 - 2$ (b) $y = (x + 1)^2 + 4$

14. (a) $y = -2(x + 1)^2 - 1$ (b) $y = 3(x - 1)^2 + 2$

15. Given the function $f(x) = ax^2 + bx + c$, we can determine the vertex of the graph by using the method of completing the square. Describe a faster method for determining the x-coordinate of the vertex. How do we then determine the y-coordinate?

16. How can we determine the minimum or maximum value of a quadratic function if we know the vertex of the graph of the function?

In Exercises 17–22, sketch the graph of the given function. Determine the vertex and state the minimum or maximum value of the function.

17. $f(x) = (x - 3)^2 + 4$

18. $g(x) = (x + 3)^2 - 4$

19. $h(x) = x^2 + 2x - 3$

20. $f(x) = x^2 - 6x + 5$

21. $g(x) = 12x - 9 - 4x^2$

22. $h(x) = -6 + 5x - 3x^2$

In Exercises 23–28, determine the vertex of the graph and the equation of the axis of symmetry.

23. $f(x) = x^2 - 6x + 9$

24. $g(x) = x^2 + 4x + 9$

25. $h(x) = 2x^2 + 5x - 7$

26. $g(x) = 3x^2 + 6x + 8$

27. $h(x) = 5 - 4x - x^2$

28. $f(x) = 5x - 2x^2$

29. Use what you know about the domain of a quadratic function to explain why its graph always has a y-intercept. If you know what the function is, what is an easy way to determine the y-intercept?

30. If we know that the solutions of $ax^2 + bx + c = 0$ are m and n, what information do we have about the graph of $f(x) = ax^2 + bx + c$?

In Exercises 31–38, determine the x- and y-intercepts of the graph of the given function.

31. $f(x) = x^2 - 3x - 28$

32. $h(x) = 2x^2 - 11x + 12$

33. $g(x) = 4x^2 - 5x + 9$

34. $f(x) = -3 + x - x^2$

35. $f(x) = 12x - 4 - 9x^2$

36. $h(x) = 4x^2 + 4x + 1$

37. $h(x) = 3x^2 + 5x$

38. $g(x) = 0.2x - 0.4x^2$

In Exercises 39–44, determine the following information for each function.

(a) the x- and y-intercepts

(b) the vertex and the equation of the axis of symmetry

(c) Sketch the graph of the function and state the domain and range.

39. $f(x) = x^2 - 6x + 8$

40. $g(x) = x^2 + 8x + 12$

41. $f(x) = 5 - 4x - x^2$

42. $g(x) = 12 - 4x - x^2$

43. $h(x) = 2x^2 + 4x + 9$

44. $g(x) = 2x - x^2 - 8$

45. What is the domain of any quadratic function? If you know the vertex $V(h, k)$ of the graph of a quadratic function, can you specify the range of the function? If so, state what the range is. If not, what additional information do you need?

46. (a) Explain how to write the equation of the axis of symmetry if the vertex $V(h, k)$ of a vertical parabola is known.

(b) If the equation of the axis of symmetry of a parabola is $x = a$, what can you conclude about the vertex of the parabola?

In Exercises 47–50, for $f(x) = ax^2 + bx + c$, the value of a is given along with the vertex V of the graph of the function. What is the range of the function?

47. $a = -2$; $V(1, 3)$

48. $a = 1$; $V(-2, 1)$

49. $a = \frac{3}{5}$; $V(0, 2)$

50. $a = -3$; $V(-4, 0)$

In Exercises 51–54, write the equation of the vertical axis of symmetry for a parabola with vertex V.

51. $V(0, -1)$

52. $V(-1, -5)$

53. $V(3, 0)$

54. $V(0, 4)$

CE In Exercises 55–58, the range of a quadratic function is given along with the equation of the axis of symmetry of the graph of the function. What is the vertex of the graph?

55. $\{y \mid y \geq 3\}$; $x = 0$

56. $\{y \mid y \leq -2\}$; $x = -3$

57. $\{y \mid y \leq -1\}$; $x = 5$

58. $\{y \mid y \geq 0\}$; $x = -2$

In Exercises 59–62, $f(x) = ax^2 + bx + c$, where a, b, and c are rational numbers. In each exercise the value of the discriminant is given for the associated equation $f(x) = 0$. How many x-intercepts does the graph of $f(x)$ have? Are the x-coordinates of these points rational or irrational numbers?

59. -4

60. 0

61. 16

62. 20

CE In Exercises 63–68, $f(x) = ax^2 + bx + c$. In each exercise the value of a is given along with the vertex of the graph of $f(x)$. Determine the number of x-intercepts of the graph.

63. -3, $V(-1, -2)$

64. -1, $V(3, 2)$

65. $\frac{1}{2}$, $V(2, -4)$

66. 2, $V(-2, 1)$

67. 3, $V(2, 0)$

68. -4, $V(0, 1)$

CE In Exercises 69–74, information is given about the graph of a quadratic function. Determine the number of x-intercepts.

69. The vertex is $(-2, -5)$, and the graph opens upward.

70. The vertex is $(0, 1)$, and the graph opens upward.

71. The vertex is $(-3, 0)$.

72. The vertex is $(1, 2)$, and the y-intercept is $(0, 7)$.

73. The maximum value of the function occurs at $(-4, 3)$.

74. The minimum value of the function occurs at $(3, 0)$.

CE In Exercises 75–82, you are given information about the location of the graph of $f(x) = (x - h)^2 + k$ relative to the location of the graph of $y = x^2$. Determine the values of h and k.

75. The graph of $f(x)$ is up 3 units.

76. The graph of $f(x)$ is down $\frac{1}{2}$ unit.

77. The graph of $f(x)$ is left $\frac{3}{5}$ unit.

78. The graph of $f(x)$ is right 1 unit.

79. The graph of $f(x)$ is up 2 units and left 4 units.

80. The graph of $f(x)$ is left 1 unit and down 4 units.

81. The graph of $f(x)$ is right 3 units and up 2 units.

82. The graph of $f(x)$ is right 2 units and down 3 units.

CE In Exercises 83–88, determine the values of b and c in the quadratic function $f(x) = x^2 + bx + c$ so that the graph of f has the given x-intercept(s).

83. $(5, 0)$ and $(-3, 0)$

84. $(-2, 0)$ and $(-5, 0)$

85. $(0, 0)$ and $(-2, 0)$

86. $(-4, 0)$ and $(4, 0)$

87. $(4, 0)$ only

88. $(-1, 0)$ only

CE In Exercises 89–92, use the information about the graph of the quadratic function to write the specific function.

89. The graph of $f(x) = x^2 + bx + c$ has a y-intercept $(0, 3)$ and an axis of symmetry whose equation is $x = 2$.

90. The graph of $f(x) = 2x^2 + bx + c$ has the vertex $V(2, -3)$.

91. The graph of $f(x) = ax^2 + x + c$ has an x-intercept $(-1, 0)$, and the first coordinate of the vertex is $-\frac{1}{6}$.

92. The graph of $f(x) = ax^2 - 8x + c$ has the vertex $V(1, -9)$.

93. The sum of two numbers is 30. Determine the numbers such that their product is a maximum.

94. The difference of two numbers is 36. Determine the numbers such that their product is a minimum.

Real-Life Applications

95. A springboard diver is propelled upward from the board, and after reaching the peak of the dive, she performs a double twist before entering the water. Her height h (in feet) after t seconds is given by $h = 10 + 30t - 16t^2$. After how many seconds will the diver be at the peak of the dive?

96. A convention setup crew had 150 linear feet of dividers to create five adjacent meeting rooms of identical size. For each room, a wall of the building served as one side, and the dividers were used to form the other three sides. What was the largest area that the rooms could have?

97. The manager of a resort condominium complex can rent each of the 80 units for \$140 per night. However, five units will remain vacant for each \$20-per-night increase in the rental price. For what price will the total income be the greatest?

98. An engineer wants to design a parabolic arch whose height is 60 feet. At ground level, the span of the arch is 100 feet. If one ground point of the arch is placed at the origin of a coordinate system and the other ground point is placed along the positive x-axis, find the quadratic function that models the arch.

Data Analysis: *Mutual Fund Accounts*

As the accompanying bar graph shows, the number of mutual fund accounts has increased as individuals have sought to balance their investments through mutual funds. (Source: Investment Company Institute.)

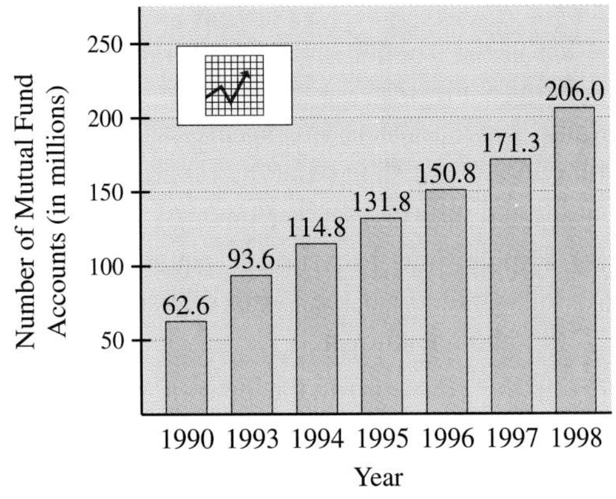

Let t represent the number of years since 1990.

99. Show that $I_1(t) = -2.1t^2 + 35.9t + 4.8$ is a reasonable model for the three years 1993, 1994, and 1995. Does the graph of the model function open upward or downward?

100. Show that $I_2(t) = 3.055t^2 - 11.125t + 99.48$ is a reasonable model for the three years 1993, 1997, and 1998. Does the graph of the model function open upward or downward?

101. Determine and interpret the coordinates of the vertices of the parabola in Exercise 99. What trend does the model predict?

102. Determine and interpret the coordinates of the vertices of the parabola in Exercise 100. What trend does the model predict?

Challenge

In Exercises 103–106, the graph of the quadratic function $f(x) = ax^2 + bx + c$ is a parabola containing the given points. Determine the vertex of the parabola.

103. $(-1, 0)$, $(1, -6)$, and $(2, -3)$

104. $(1, 6)$, $(-2, -12)$, and $(3, 8)$

105. $(1, 3)$, $(2, 7)$, and $(3, 13)$

106. $(0, 1)$, $\left(-\frac{1}{2}, 0\right)$, and $(1, 9)$

107. A commercial developer finds that the rental income I (in dollars) for a shopping mall is given by $I(x) = (700 + 25x)(30 - x)$, where x is the number of vacant stores.

(a) It seems reasonable to expect a maximum income when there are no store vacancies—that is, when $x = 0$. Evaluate $I(0)$ to determine the income when no stores are vacant.

(b) Now determine the vertex of the graph of the income function. For how many store vacancies is the income maximized?

(c) How many stores are in the shopping mall?

108. For a certain flight, a charter airline charges \$96 per person. Each person is charged an additional \$4 for each unsold seat. If the plane has 60 seats, find the number of unsold seats that produces the maximum revenue.

109. Using the following steps as a guide, derive the formulas for the coordinates of the vertex of a parabola.

(a) Begin with the general quadratic function $f(x) = ax^2 + bx + c$. Subtract c from both sides, and then divide both sides by a.

(b) By completing the square, write the right side as the square of a binomial.

(c) Write the function in the following form.

$$f(x) = a(x - h)^2 + k$$

10.7 Quadratic Inequalities

Algebraic Methods • Special Cases • Real-Life Applications

Algebraic Methods

The standard form of a quadratic equation is $ax^2 + bx + c = 0$. When the equality symbol is replaced by an inequality symbol, we obtain a **quadratic inequality.**

Definition of a Quadratic Inequality

A **quadratic inequality** is an inequality that can be written in the standard form $ax^2 + bx + c > 0$, where a, b, and c are real numbers and $a \neq 0$. The symbols $<$, $\leq$, and $\geq$ can also be used.

The **solution set** of a quadratic inequality is the set of all replacements for the variable that make the inequality true.

Exploring the Concept

Solutions of Quadratic Inequalities

Consider the quadratic inequality $x^2 + 2x - 8 < 0$. We can visualize the solution set of this inequality in reference to the graph of the associated quadratic function $y = x^2 + 2x - 8$. To sketch the graph, we use the methods in Section 10.6 to determine the vertex: $V(-1, -9)$. Then we determine the x-intercepts.

$$x^2 + 2x - 8 = 0$$

$$(x + 4)(x - 2) = 0$$

$$x + 4 = 0 \quad \text{or} \quad x - 2 = 0$$

$$x = -4 \quad \text{or} \quad x = 2$$

Figure 10.20

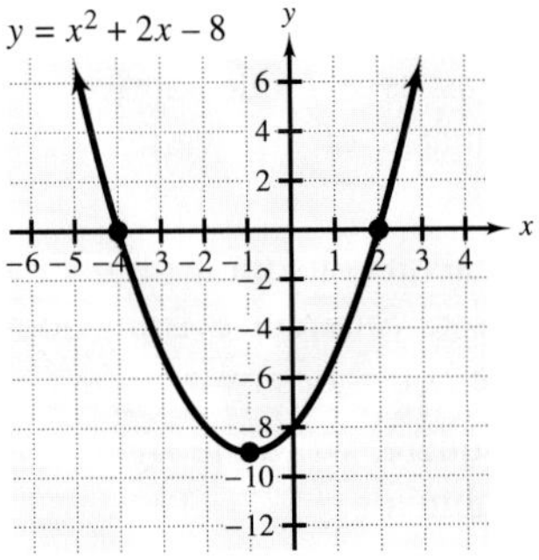

The x-intercepts are $(-4, 0)$ and $(2, 0)$. Figure 10.20 shows the graph of the function.

The solutions of the inequality are the x-values for which the graph lies below the x-axis. Note that this occurs between the two x-intercepts. Thus the solution set in interval notation is $(-4, 2)$. (See Fig. 10.21.)

Figure 10.21

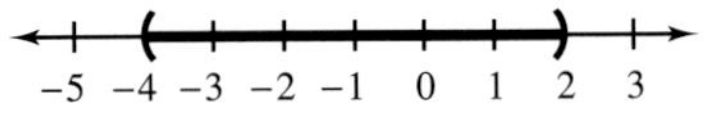

Similarly, for the inequality $x^2 + 2x - 8 \geq 0$, the solutions are the x-values for which the graph lies on or above the x-axis. Thus the solution set is $(-\infty, -4] \cup [2, \infty)$. (See Fig. 10.22.)

Figure 10.22

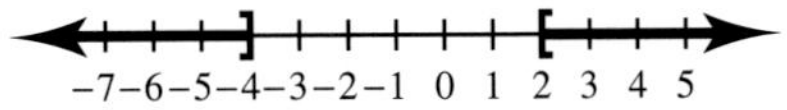

For a quadratic inequality, the graph of the associated function is a parabola. If the parabola has two x-intercepts, then these points divide the x-axis into three intervals. In each interval, the function is either always positive or always negative. Thus either the inequality is true in an entire interval or it is false in that interval.

Note: We will treat parabolas with only one x-intercept or with no x-intercept as special cases later in this section.

The key step is to determine the x-intercepts, which divide the x-axis into intervals. (We can plot these numbers on a number line without drawing the entire graph of the function.) Once these intervals are known, we can simply test values within those intervals to learn whether the inequality is true or false.

Solving a Quadratic Inequality

1. Write the inequality in standard form.
2. Solve the associated equation and plot each solution on a number line.
3. Choose a test point in each interval. Test the value in the inequality to determine whether the inequality is true in that interval.
4. The solution set for the inequality includes all x-values in each interval where the test value indicates that the inequality is true. For inequalities that include equality, the solutions of the associated equation are also included in the solution set.

EXAMPLE 1

Solving a Quadratic Inequality

Solve $x^2 - 3x \geq 10$.

Solution

$x^2 - 3x - 10 \geq 0$ Write the inequality in standard form.

Solve the associated quadratic equation.

$$x^2 - 3x - 10 = 0$$

$$(x - 5)(x + 2) = 0 \qquad \text{Factor the trinomial.}$$

$$x - 5 = 0 \quad \text{or} \quad x + 2 = 0 \qquad \text{Zero Factor Property}$$

$$x = 5 \quad \text{or} \quad x = -2 \qquad \text{Solve each equation for } x.$$

The numbers -2 and 5 divide the number line into three intervals, which we label as A, B, and C. (See Fig. 10.23.)

Figure 10.23

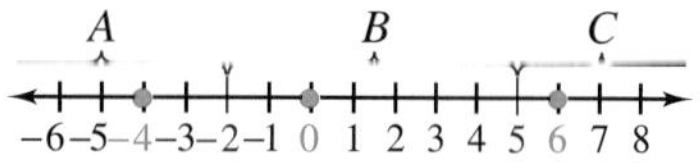

Students often fear that they will choose the wrong test point. Stress that any point of an interval is acceptable.

From these intervals we select the numbers -4, 0, and 6, and we test each number to determine whether it is a solution of the inequality.

Test Number: −4	*Test Number: 0*	*Test Number: 6*
$x^2 - 3x - 10 \geq 0$	$x^2 - 3x - 10 \geq 0$	$x^2 - 3x - 10 \geq 0$
$(-4)^2 - 3(-4) - 10 \geq 0$	$0^2 - 3 \cdot 0 - 10 \geq 0$	$6^2 - 3 \cdot 6 - 10 \geq 0$
$18 \geq 0$	$-10 \geq 0$	$8 \geq 0$
True	False	True

LEARNING TIP

If solving a quadratic inequality involves three intervals, "True" and "False" will alternate. Thus testing only one interval is sufficient.

Think About It

Suppose that two quadratic inequalities have the associated equations $x^2 + 2x - 15 = 0$ and $-x^2 - 2x + 15 = 0$. Note that -5 and 3 are the solutions of both equations. What are the quadratic inequalities if the solution set for both is $[-5, 3]$?

The results are summarized on the number line in Figure 10.24.

Figure 10.24

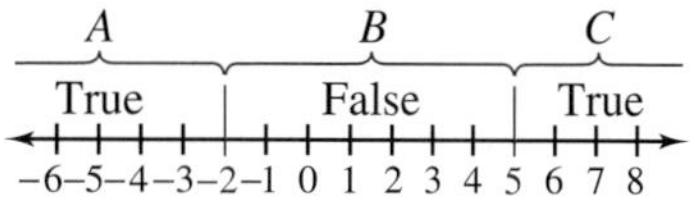

The solution set is the union of the two intervals A and C. Because the inequality symbol is $\geq$, the solutions of the equation are also solutions of the inequality. The solution set is $(-\infty, -2] \cup [5, \infty)$. (See Fig. 10.25.)

Figure 10.25

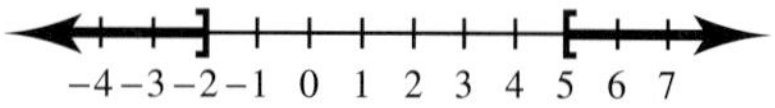

To solve a quadratic inequality, we use the factoring method, the Square Root Property, or the Quadratic Formula to determine the solutions of the associated equation.

EXAMPLE 2 Solving Quadratic Inequalities

Solve the following inequalities.

(a) $1 > x(4 + x)$ (b) $12x < 4x^2 + 7$

Solution

(a)

$$1 > x(4 + x)$$

$$1 > 4x + x^2 \qquad \text{Distributive Property}$$

$$-x^2 - 4x + 1 > 0 \qquad \text{Write the inequality in standard form.}$$

Use the Quadratic Formula to solve the associated equation $-x^2 - 4x + 1 = 0$.

$$x = \frac{4 \pm \sqrt{(-4)^2 - 4(-1)(1)}}{2(-1)} = \frac{4 \pm \sqrt{16 + 4}}{-2}$$

$$= \frac{4 \pm \sqrt{20}}{-2} = \frac{4 \pm \sqrt{4}\sqrt{5}}{-2} = \frac{4 \pm 2\sqrt{5}}{-2} = -2 \pm \sqrt{5}$$

$x \approx -4.24$ or $x \approx 0.24$

Locate the two solutions on the number line and test a number from each interval. (See Fig. 10.26.)

LEARNING TIP

When you divide the number line into intervals, an accurate scale is not necessary. Just make sure that the interval endpoints and the test points are in the proper order.

Figure 10.26

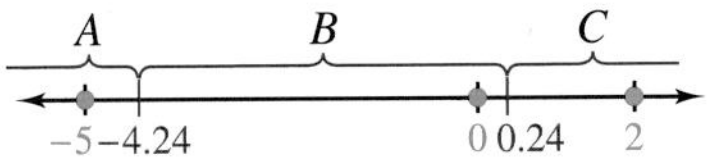

Test Number: −5	*Test Number: 0*	*Test Number: 2*
$-x^2 - 4x + 1 > 0$	$-x^2 - 4x + 1 > 0$	$-x^2 - 4x + 1 > 0$
$-(-5)^2 - 4(-5) + 1 > 0$	$-0^2 - 4 \cdot 0 + 1 > 0$	$-2^2 - 4 \cdot 2 + 1 > 0$
$-4 > 0$	$1 > 0$	$-11 > 0$
False	True	False

The (approximate) solution set is $(-4.24, 0.24)$. (See Fig. 10.27.)

Figure 10.27

(b) $12x < 4x^2 + 7$

$-4x^2 + 12x - 7 < 0$ Write the inequality in standard form.

Solve the associated quadratic equation $-4x^2 + 12x - 7 = 0$ with the Quadratic Formula.

$$x = \frac{-12 \pm \sqrt{12^2 - 4(-4)(-7)}}{2(-4)} = \frac{-12 \pm \sqrt{144 - 112}}{-8}$$

$$= \frac{-12 \pm \sqrt{32}}{-8} = \frac{-12 \pm \sqrt{16}\sqrt{2}}{-8} = \frac{-12 \pm 4\sqrt{2}}{-8} = \frac{3 \pm \sqrt{2}}{2}$$

$x \approx 0.79$ or $x \approx 2.21$

The solutions divide the number line into three intervals *A*, *B*, and *C*. Using the test points 0, 1, and 3, we obtain the results shown in Figure 10.28.

Figure 10.28

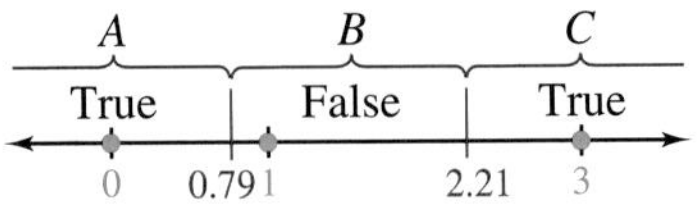

An approximation of the solution set is $(-\infty, 0.79) \cup (2.21, \infty)$. (See Fig. 10.29.)

Figure 10.29

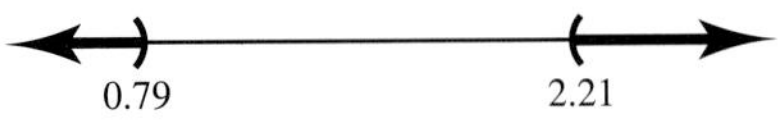

Special Cases

In each of the preceding examples, the solution set was an interval of x-values or the union of two intervals of x-values. The next example illustrates some special cases of quadratic inequalities that lead to different solution sets.

EXAMPLE 3

Special Cases of Quadratic Inequalities

Students should not try to memorize all of the special cases and their results. Encourage them to be confident about using test points to reveal the solutions.

Solve the inequalities.

(a) $x^2 + 3 \geq 0$ (b) $x^2 + 3 < 0$

(c) $x^2 + 6x + 9 \leq 0$ (d) $x^2 + 6x + 9 > 0$

Solution

(a) The Even Root Property tells us that the associated equation $x^2 + 3 = 0$ or $x^2 = -3$ has no real number solution. Thus there is only one interval, the entire real number line, and the inequality is either always true or always false. Choose a test point and check the inequality.

$$x^2 + 3 \geq 0 \quad \text{Original inequality}$$
$$0^2 + 3 \geq 0 \quad \text{Choose 0 as a test point.}$$
$$3 \geq 0 \quad \text{The inequality is true.}$$

The inequality is always true. The solution set is **R**.

(b) In part (a) we found that $x^2 + 3 \geq 0$ is always true. Thus $x^2 + 3 < 0$ is never true. The solution set is Ø.

(c) Solve the associated equation.

$$x^2 + 6x + 9 = 0$$
$$(x + 3)^2 = 0$$
$$x = -3$$

The solution divides the number line into two intervals. (See Fig. 10.30.) Choose a test point from each interval.

Figure 10.30

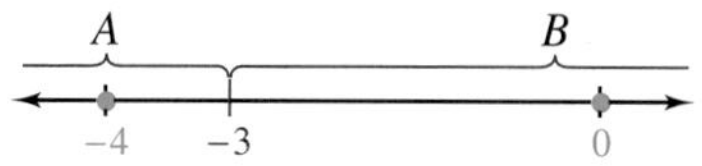

Test Number: −4

$$x^2 + 6x + 9 \leq 0$$
$$(-4)^2 + 6(-4) + 9 \leq 0$$
$$1 \leq 0$$
False

Test Number: 0

$$x^2 + 6x + 9 \leq 0$$
$$0^2 + 6 \cdot 0 + 9 \leq 0$$
$$9 \leq 0$$
False

The inequality is false in each interval. However, because the inequality symbol is ≤, we must include any solutions to the associated equation. The only solution of the inequality is −3.

(d) The associated equation is the same as in part (c). Using the same test points as in part (c), we see that the inequality is true in each interval. However, because the inequality is in the form $y > 0$, we do not include the solution to the associated equation. The solution set is $(-\infty, -3) \cup (-3, \infty)$.

Note: When we solve a quadratic inequality, we may find that the associated equation has no solution. However, as part (a) of Example 3 shows, this does not imply that the inequality has no solution.

Real-Life Applications

Applications involving quadratic inequalities are characterized by such words and phrases as *less than* and *at least*.

EXAMPLE 4

Minimum Area of a Rectangle

According to a local election code, candidates must be provided with at least 8000 square feet of space outside a polling place for campaigning activities. If the poll manager has 400 feet of rope with which to enclose a rectangular area, what is the length of the space?

Solution

Let $L =$ the length of the rectangle and $W =$ the width of the rectangle.

$2L + 2W = 400$ — The perimeter of the rectangle is 400 feet.

$2W = 400 - 2L$ — Solve for W.

$W = 200 - L$

Figure 10.31

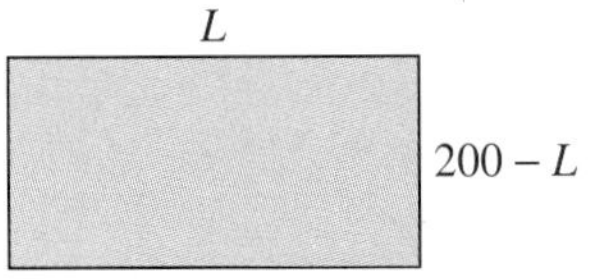

Figure 10.31 shows the rectangle with the dimensions labeled.

The area of the rectangle is given by the formula $A = LW$.

$A = LW = L(200 - L) = -L^2 + 200L$

$-L^2 + 200L \geq 8000$ — The area must be at least 8000 square feet.

$-L^2 + 200L - 8000 \geq 0$ — Write the inequality in standard form.

Use the Quadratic Formula to verify that the solutions of the associated equation $-L^2 + 200L - 8000 = 0$ are 55.28 and 144.72 (rounded to two decimal places). These numbers divide the number line into three intervals (see Fig. 10.32).

Figure 10.32

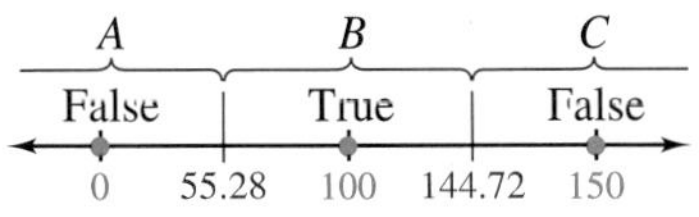

The test points 0, 100, and 150 indicate that the inequality is true in interval B. Therefore, the solution set of the inequality is approximately [55.28, 144.72].

To obtain an area of at least 8000 square feet, the length of the rectangle can be any value between 55.28 feet and 144.72 feet.

Quick Reference 10.7

Algebraic Methods

- A **quadratic inequality** is an inequality that can be written in the standard form $ax^2 + bx + c > 0$ (or $<, \leq, \geq$), where a, b, and c are real numbers and $a \neq 0$.
- The **solution set** of a quadratic inequality is the set of all replacements for the variable that make the inequality true.

- To solve a quadratic inequality, follow these steps.
  1. Write the inequality in standard form.
  2. Solve the associated equation and plot each solution on a number line.
  3. Choose a test point in each interval. Test the value in the inequality to determine whether the inequality is true in that interval.
  4. The solution set for the inequality includes all x-values in each interval where the test value indicates that the inequality is true. For inequalities that include equality, the solutions of the associated equation are also included in the solution set.

Special Cases

- When we solve a quadratic inequality, the associated equation may have only one solution a or it may have no solution.
- In such special cases, the solution set may be **R**, Ø, a set containing just the one number a, or the set of all real numbers except the number a.

Speaking the Language 10.7

1. The inequality $x^2 - 5x + 6 \geq 0$ is an example of a(n) ______.

2. If $f(x) = x^2 - 5x + 6$, then to solve $f(x) < 0$, we determine the ______ of $x^2 - 5x + 6 = 0$ and then plot those numbers on a number line to divide the line into ______.

3. In Question 2, the next step is to select a(n) ______ in each interval. If the coordinate of that point makes the inequality true, then all x-values in that interval belong to the ______.

4. When we solve a quadratic inequality, solutions of the associated equation are solutions of the inequality only if the inequality symbol is ______ or ______.

For Exercises 3–10 and 39–46, you may want to stress the meanings of the words *positive, nonpositive, negative,* and *nonnegative.*

Exercises 10.7

Concepts and Skills

1. Suppose that you are solving a quadratic inequality, and you have already determined the solutions of the associated equation. Describe the next step you will take in order to write the solution set.

2. To solve a quadratic inequality, we begin by determining the solutions of the associated equation. What methods can be used? Which of these methods will always work?

In Exercises 3–6, for what values of x is the given expression positive?

3. $(3 - x)(1 + x)$

4. $2x^2 + 5x - 3$

5. $5x^2 - 8x$

6. $6 - x^2 + 3x$

In Exercises 7–10, for what values of x is the given expression negative?

7. $x^2 + 10 - 7x$

8. $12 + 4x - x^2$

9. $3 + x - 2x^2$

10. $x^2 - 3x - 6$

11. Suppose that you are writing the solution set of a quadratic inequality. Explain how to determine whether the solutions of the associated equation are included in the solution set of the inequality.

12. For a quadratic inequality, suppose that the associated equation has two distinct solutions. Describe the possible intervals that make up the solution set.

In Exercises 13–18, sketch the graph of the associated function. After determining the x-intercepts algebraically, use the graph to write the solution set of the given inequality. There are no special cases in Exercises 13–26.

13. $9 - x^2 < 0$ **14.** $x(x + 3) \geq 0$

15. $5 - 4x^2 \geq 8x$ **16.** $x^2 > 2(x + 4)$

17. $x(x + 6) \leq 7$ **18.** $x(2 - x) \geq -3$

In Exercises 19–26, solve the inequality.

19. $(x - 2)(x + 5) > 8$ **20.** $5 + 4x - x^2 \geq 0$

21. $x^2 \leq 5x$ **22.** $4x^2 < 7$

23. $x(12x + 5) \geq 2$ **24.** $x(4x + 9) > 9$

25. $2x(x - 2) < 7$ **26.** $x^2 + 2x \geq 7$

In Exercises 27–30, match each inequality in column B with the correct sentence in column A.

Column A	*Column B*
27. A number does not exceed 5.	(a) $x > 5$
28. A number is at least 5.	(b) $x \leq 5$
29. A number is more than 5.	(c) $x - 5 < 0$
30. A number is less than 5.	(d) $5 \leq x$

In Exercises 31–38, translate the given information into an inequality and then solve it.

31. The product of x and $x + 1$ is at least 6.

32. The product of $x + 3$ and $x - 1$ is less than 12.

33. The product of x and $x - 1$ is not more than 6.

34. The sum of $3x^2$ and $5x$ is less than 2.

35. Six less than the product of x and $6x - 5$ is negative.

36. The difference of 6 and $5x^2 + 7x$ is positive.

37. Seven more than the product of $2x$ and $2 - x$ is not positive.

38. The difference of $4x + 7$ and $2x^2$ is not negative.

In Exercises 39–42, for what values of x is the function nonnegative?

39. $g(x) = (4 - x)(2 + x)$ **40.** $f(x) = x^2 + 3 - 4x$

41. $g(x) = 4x^2 - 7x$ **42.** $h(x) = 8 - 5x^2$

In Exercises 43–46, for what values of x is the function nonpositive?

43. $f(x) = 16 - x^2$

44. $h(x) = 3(1 - x^2) - 8x$

45. $g(x) = x(x - 8) - 5$

46. $f(x) = 9x^2 + 6x - 8$

CE **47.** Suppose that the associated equation of a quadratic inequality has no real number solution. List all possible solution sets of the inequality.

CE **48.** Suppose you find that the associated equation of a quadratic inequality has only one solution, 3. List all possible solution sets of the inequality.

49. Suppose that $x - 2$ is a factor of the expression $ax^2 + bx + c$. Explain whether 2 can be a solution of the inequality $ax^2 + bx + c > 0$.

50. The solution set of $x^2 - x - 6 < 0$ is the interval $(-2, 3)$. Explain how to determine the solution set of $x^2 - x - 6 \geq 0$.

In Exercises 51–76, solve the inequality.

51. $4x^2 + 23x + 15 \geq 0$ **52.** $x^2 + 7x + 6 < 0$

53. $5x - x^2 < 0$ **54.** $25 - x^2 \geq 0$

55. $x^2 + 3x + 6 < 0$ **56.** $3x - 2x^2 > 7$

57. $4x^2 + 20x + 25 \leq 0$ This group includes special cases.

58. $6x^2 - 41x + 70 \leq 0$

59. $x^2 - 4x \geq 4$ **60.** $x^2 + 2x + 5 \leq 0$

61. $x^2 + 9 \leq 6x$ **62.** $x^2 + 16 \leq 8x$

63. $2x^2 < 4x + 5$ **64.** $2x^2 \geq 5x + 4$

65. $x^2 + 5 \geq 6x$ **66.** $5x - x^2 < 8$

67. $0.5x - 0.25x^2 \leq 0.9$

68. $0.3x^2 - 0.7x + 0.9 \leq 0$

69. $x(6x + 17) \leq -7$ **70.** $3x(2 - 3x) \geq 1$

71. $x^2 + \frac{3}{4}x > \frac{5}{2}$ **72.** $\frac{x^2}{2} + x \geq 3$

73. $3x(x + 2) < 4$

74. $2x(3 - x) < 9$

75. $(2x + 1)(x + 2) \leq (x - 3)(x + 2)$

76. $(2x + 3)(x + 1) > 2(2x + 3)(1 - x)$

CE In Exercises 77–80, use the discriminant to determine all values of k so that the solution set of the quadratic inequality is **R**.

77. $2x^2 - 3x + k > 0$

78. $kx^2 + 1 - 3x > 0$

79. $9 + kx - x^2 > 0$

80. $7x^2 - kx > 0$

CE In Exercises 81–84, use the discriminant to determine all values of k so that the real number solution set of the quadratic inequality is $\varnothing$.

81. $x^2 + 5x + k < 0$

82. $kx^2 + 2x + 1 < 0$

83. $kx^2 + 8x - 5 > 0$

84. $5x^2 - 2x + k < 0$

CE In Exercises 85–88, use the discriminant to determine all values of k so that the real number solution set of the quadratic inequality is a set with only one element.

85. $x^2 + 12x + k \le 0$

86. $x^2 + 3x + k \le 0$

87. $x^2 + kx + 9 \le 0$

88. $4x^2 + kx + 1 \le 0$

89. Suppose that two numbers differ by 5. If the product of the numbers is at most 6, what is the largest possible value of the smaller number?

90. Consider a circle with radius r. For what values of r does the numerical value of the circumference of the circle exceed the numerical value of the area?

Real-Life Applications

91. A builder plans to construct a home on a lot that is 100 feet wide. The covenants for the subdivision include the following provisions.

(a) The front of a home must be at least 35 feet longer than its depth.

(b) No home can be closer than 15 feet to any property line.

(c) All homes must have more than 2556 square feet of ground-floor living space.

Show why the builder cannot construct the home.

92. The Woodstock Dairy Company wants to design a holding tank for chocolate milk. The tank is to be in the shape of a right circular cylinder 9.5 feet high. What should the radius of the tank be so that it holds at least 10,000 gallons? (1 gallon = 231 cubic inches.)

93. The Mathis Package Company wants to design a box that is 11 inches high with a base perimeter of 80 inches. The box's volume needs to be at least 2 cubic feet. What should be its length?

94. An army unit fired a mortar shell over a hill 112 feet high. The height h of the shell was given by $h = -16t^2 + 128t$, where t is the time of flight. To the nearest tenth of a second, during what time interval was the shell at least 30 feet higher than the hill?

Data Analysis: *Heart Disease*

The accompanying figure shows deaths (in thousands) due to heart disease in the United States for selected years in the period 1992–1998. (Source: U.S. National Center for Health Statistics.)

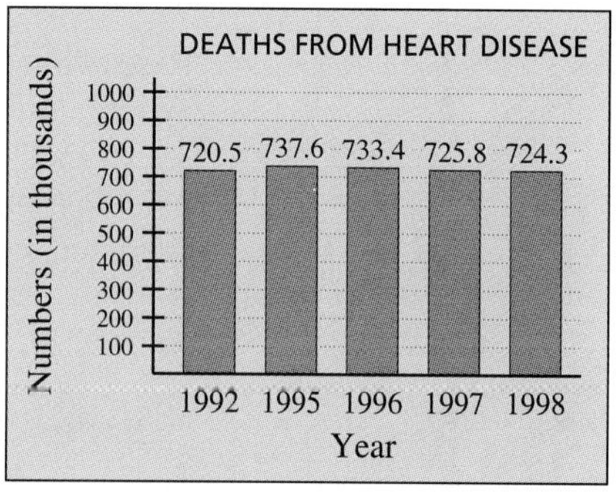

The number H of deaths (in thousands) can be modeled as a function of the number years t since 1990: $H(t) = -1.51t^2 + 15.3t + 696$.

95. Write an equation stating that the number of deaths due to heart disease was 680,000.

96. Solve the equation in Exercise 95 and interpret the solutions.

97. What does the inequality $H(t) \le 680$ model?

98. Solve the inequality $H(t) > 680$ and interpret the solutions.

Challenge

99. For $ax^2 > 0$ $(a \ne 0)$, use the discriminant to prove that the solution set cannot be **R**.

100. For nonzero real numbers a and b, use the discriminant to prove that the solution set of $ax^2 + bx > 0$ cannot be **R**.

In Exercises 101 and 102, determine all values of k so that the solution set of the quadratic inequality is **R**.

101. $x^2 + kx + 4 > 0$

102. $9x^2 + kx + 1 \ge 0$

In Exercises 103 and 104, determine all values of k so that the real number solution set of the quadratic inequality is $\emptyset$.

103. $3x^2 + kx - 1 > 0$

104. $-2x^2 + kx - 2 > 0$

In Exercises 105 and 106, determine all values of k so that the real number solution set of the quadratic inequality is a set with only one element.

105. $kx^2 + 3x + (k - 1) \ge 0$

106. $2kx^2 + x + (k - 1) \le 0$

10 • Chapter Project

Heart Attack Risks

Research indicates that lifestyle has a substantial effect on the risk of a heart attack. Weight, exercise, and blood cholesterol levels are three of the many factors on which medical research focuses in the prevention of heart disease. The *New England Journal of Medicine* reported the following factors and the amount by which heart attack risk is reduced by each.

Factor	Risk Reduction
Exercising regularly	45%
Maintaining ideal weight	35–55%
Decreasing cholesterol (per 1% decrease in cholesterol)	2–3%

1. One recommendation for exercise is to maintain a target heart rate for 20 minutes each day. The following formulas give the target heart rate T for a person age A. (Source: Kenneth H. Cooper, *The Aerobics Program for Total Well-Being*.)

   Women: $T = 0.8(220 - A)$

   Men: $T = 0.8(205 - 0.5A)$

   Using the formulas, make a table showing the target heart rates for men and women ages 20 through 70 in increments of 10 years.
2. Use algebraic methods to determine the age at which the target heart rates for men and women are the same.
3. The ideal weight W (in pounds) of a person with a medium frame can be modeled as a function of height h (in inches).

   Men: $W(h) = 0.077h^2 - 7.61h + 310.11$

   Women: $W(h) = 2.95h - 57.32$

   (a) Use the appropriate model to estimate the ideal weight for a man whose height is 5 feet 11 inches.

   (b) For women, what is the approximate increase in weight for each additional inch in height?
4. Exercise is also a factor in weight control. The number of calories C burned per hour by a 120-pound person walking at rate r (in mph) is modeled by $C(r) = 25.3r^{1.78}$.

   (a) How many calories are burned per hour by a 120-pound person walking 5 miles per hour? (Round to the nearest whole number.)

   (b) Determine the walking rate required to burn 700 calories per hour. (Round to the nearest tenth.)

5. A person's cholesterol level C increases with age a. Using the following model functions, what can you conclude about men and women age 60?

Women: $C(a) = 1.17a + 147.15$

Men: $C(a) = 0.76a + 171.4$

6. In groups, choose and investigate another factor in heart attack risk, such as alcohol consumption, smoking, blood pressure, or use of aspirin. Perhaps a representative of the nursing faculty, college wellness center, or college health education department can provide additional information and discuss efforts toward education to promote a healthful lifestyle on your campus or in your community. Report your findings and support your information with model functions and graphs.

Chapter 10 Review Exercises

Section 10.1

1. Describe two methods for solving the equation $4x^2 = 25$.

In Exercises 2 and 3, use the Zero Factor Property to solve.

2. $\left(4x - \frac{2}{7}\right)\left(6x - \frac{3}{5}\right) = 0$

3. $(d - 3)(d + 2) = 36$

In Exercises 4–6, use the factoring method to solve.

4. $16 + 9c^2 = 24c$

5. $x(x + 2) = 15$

6. $a^2 - 7a + 10 = 0$

In Exercises 7–9, solve by using the Square Root Property.

7. $25 - y^2 = 0$

8. $25(b + 4)^2 - 16 = 0$

9. $13 - 3z^2 = 0$

In Exercises 10 and 11, determine the complex number solutions.

10. $a^2 + 36 = 0$

11. $(b - 2)^2 + 37 = 0$

12. The sum of the squares of two consecutive odd integers is 130. What are the integers?

Section 10.2

13. Suppose you use the method of completing the square to solve $2x^2 + 5x + 2 = 0$. Explain why you will eventually add $\frac{25}{16}$ to both sides of the equation.

14. Determine the value of k so that $x^2 - 7x + k$ is a perfect square.

In Exercises 15 and 16, solve by completing the square.

15. $x^2 + 10x - 1 = 0$

16. $3y(y - 5) = 12$

In Exercises 17 and 18, determine the imaginary number solutions by completing the square.

17. $x^2 - 6x + 10 = 0$

18. $3x^2 - 2x + 4 = 0$

In Exercises 19–23, determine the complex number solutions.

19. $x^2 - 5x + 3 = 0$

20. $x^2 + 7 = 3x$

21. $3 - x - 2x^2 = 0$

22. $2x^2 + x + 1 = 0$

23. $1 = 2x^2 + 6x$

24. The sum of the reciprocals of two consecutive integers is $\frac{7}{12}$. What are the integers?

Section 10.3

25. Which of the following equations can be solved with the Quadratic Formula?

(i) $3x^2 = 12$

(ii) $(x + 1)(x - 5) = 0$

(iii) $x^2 + x + 1 = 0$

(iv) $(x + 3)^2 = 5$

From this list, identify those equations that can be solved with an alternative method and state the method.

In Exercises 26–29, solve the equation by using the Quadratic Formula.

26. $x(x + 3) = 7$

27. $5x^2 - 16 = 0$

28. $3 + \frac{4}{x^2} = \frac{7}{x}$

29. $4x^2 = 5x + 3$

In Exercises 30 and 31, determine the complex number solutions by using the Quadratic Formula.

30. $x^2 + 4x + 7 = 0$

31. $2x^2 + 9 = x$

In Exercises 32–35, use the discriminant to determine the number and the nature of the solutions of the given equation.

32. $8x = x^2 + 16$

33. $4x = x^2 - 12$

34. $8x = 2x^2 - 7$

35. $6x^2 - 2x + 5 = 0$

36. The sum of the square of a number and 5 times the number is 7. What is the number?

Section 10.4

37. If $x > 0$, which of the following equations are quadratic in form?

(a) $\frac{2}{x^2} + \frac{3}{x} - 5 = 0$

(b) $2x + \sqrt{2x} = 3$

(c) $(x + 1)^2 + 2(x + 1) + 1 = 0$

(d) $x^{1/4} - x^{1/2} + 3 = 0$

For the equations you selected, state the substitution you would use and rewrite the equation as a quadratic equation.

In Exercises 38–45, determine the real number solutions of the equation.

38. $\frac{3x}{x - 4} + \frac{3}{x + 3} = 4$

39. $\frac{x}{x + 1} = 2 + \frac{2}{x - 2}$

40. $\sqrt{3x - 4} = \sqrt{x + 1} - 2$

41. $\sqrt{1 - 2x} = x + 1$

42. $a^4 - 5a^2 + 6 = 0$

43. $x - 3x^{1/2} = 10$

44. $x^{1/6} + x^{1/3} = 2$

45. $(x^2 + x)^2 + 72 = 18(x^2 + x)$

In Exercises 46 and 47, determine all complex number solutions.

46. $t^4 + 8t^2 - 9 = 0$

47. $\frac{x}{x - 2} = \frac{2}{x + 5}$

48. Use an appropriate substitution to write the equation $3 + 10x^{-1} + 8x^{-2} = 0$ as a quadratic equation. Then determine the solutions of the original equation.

Section 10.5

49. Two tablecloths are to be cut from a rectangular piece of material. One tablecloth is a circle, and the other is a rectangle whose length is 3 feet longer than the width. (See figure.) The total area of the tablecloths is 90 square feet. How many square yards of material are needed?

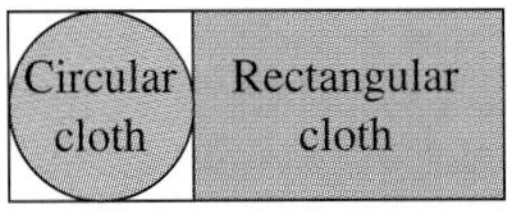

50. A fraternity sold tickets for its annual dinner-dance. Ten couples could not attend the dinner but did attend the dance. The fraternity collected \$1680 for dinner tickets and \$1300 for dance tickets. If the dinner tickets cost \$15 more than the dance tickets, how many couples attended the dance?

51. For Hanukkah an aunt sent \$6 to each child, and a grandmother sent \$60 to be divided equally among the children of the family. An uncle sent \$56 to be divided among the children and three other relatives. A grandfather sent each child \$5. If each child received a total of \$30, how many children were in the family?

52. A boat traveled up a river and back, a total distance of 40 miles, in 6 hours. If the rate of the current was 5 mph, what would the rate of the boat have been in still water?

53. Working together, two people can wash all the windows in their home in 6 hours. One person takes 5 hours longer than the other to wash the windows alone. How long would the slower worker take, working alone, to wash the windows?

54. A metal fabricator needs a piece of aluminum in the shape of a right triangle with an area of 150 square inches. If the length of one leg is 1 inch more than twice the length of the other leg, what is the length of the hypotenuse? (Round the result to the nearest tenth.)

Section 10.6

55. Explain how to determine the range of a quadratic function.

56. Explain how to determine the x-intercepts of the graph of a quadratic function.

In Exercises 57 and 58, determine the vertex and the intercepts of the graph of the function. Then sketch the graph of the function.

57. $f(x) = x^2 - 6x + 9$

58. $g(x) = 3x^2 - 4x - 7$

In Exercises 59–62, determine whether the graph of the given function opens upward or downward. Then sketch the graph to determine the range of the function.

59. $f(x) = 16 - x^2$

60. $g(x) = 2 + 5x - 3x^2$

61. $h(x) = 2x^2 - 5x$

62. $f(x) = 4x^2 - 6x + 7$

63. A dog owner has 250 feet of fencing to enclose a rectangular run area for his dogs. If he wants the maximum possible area, what should the length of the rectangle be?

64. If an object is thrown vertically upward, its height (in feet) after t seconds is given by $h(t) = 64t - 16t^2$.

(a) In how many seconds will the object be at its maximum height?

(b) What is the maximum height reached by the object?

Section 10.7

65. The solution set of $x^2 + 5x + 6 < 0$ is $(-3, -2)$. Explain how to determine the solution set of $x^2 + 5x + 6 \geq 0$.

In Exercises 66–71, solve the inequality.

66. $2x^2 + 6 \geq 7x$

67. $x(x - 6) + 9 \leq 0$

68. $4x(x + 5) + 9 \leq 0$

69. $x^2 + 5 \geq 3x$

70. $2x - x^2 + 2 > 0$

71. $x - x^2 - 1 > 0$

72. A dog owner has 250 feet of fencing to enclose a rectangular run area for his dogs. If he wants the area to be at least 3500 square feet, what can the length of the rectangle be?

In Exercises 73 and 74, determine the values of x for which the function satisfies the given condition?

73. $f(x) = 8 + 3x - 2x^2$; $f(x) > 0$

74. $g(x) = x^2 + 8x + 16$; $g(x) \geq 0$

In Exercises 75 and 76, use the discriminant to determine all values of k so that the quadratic inequality has the given solution set.

75. $3x^2 + 2x + k > 0$; $\mathbf{R}$

76. $kx^2 + 3x + 2 < 0$; $\emptyset$

Looking Back

The following exercises review concepts and skills that you will need in Chapter 11.

In Exercises 1 and 2, determine the domain of the given function.

1. $f(x) = \dfrac{x + 2}{x - 3}$ **2.** $g(x) = \sqrt{x + 5}$

3. Explain how the Vertical Line Test is used to determine whether a graph represents a function.

4. Determine whether the given equation represents a function.

(a) $y = x^2$ (b) $x = |y|$

5. Perform the indicated operations.

(a) Add the polynomials $x + 3$ and $x^2 + 4x - 3$.

(b) Subtract $4 - 6t$ from $3t + 1$.

6. Perform the indicated operations.

(a) $(x + 3)(x - 2)$

(b) $(x^2 - 5x + 6) \div (x - 2)$

In Exercises 7 and 8, evaluate the given function as indicated.

7. $f(x) = \sqrt{5 - x}$; $f(-4)$

8. $g(x) = \dfrac{x - 1}{x + 1}$; $g(2)$

In Exercises 9 and 10, evaluate the function $f(x) = 3x - 7$ as indicated.

9. $f(t + 4)$ **10.** $f(t^2)$

11. Let $g(x) = 3x - 7$. Determine the following values.

(a) $g(3)$

(b) x, if $g(x) = -22$

(c) x, if $g(x) = 17$

12. Evaluate the exponential expressions.

(a) 2^{-4} (b) $\left(\dfrac{1}{3}\right)^{-3}$ (c) 15^0

(d) $4^{3/2}$ (e) $125^{-2/3}$

Chapter 10 Test

1. Solve $3t^2 + 4t = 1$ by completing the square.

In Questions 2–8, solve the given equation.

2. $x(6x - 5) = 6$

3. $(2x - 1)^2 - 5 = 0$

4. $2 = \frac{2}{t} + \frac{1}{t^2}$

5. $(2y + 1)(y + 3) = 2$

6. $3x^2 - x + 2 = 0$

7. $9x^4 + 7x^2 = 2$

8. $\frac{2}{x+7} + \frac{16}{x^2 + 6x - 7} = \frac{x}{1 - x}$

9. The area A of a triangle is 20 square feet. The base b is 3 feet longer than the height h. What are the dimensions of the triangle? $\left(A = \frac{1}{2}bh.\right)$

10. At a dairy, a milk storage tank has an inlet pipe for receiving milk from the milking station. An outlet pipe takes the milk to the bottling room. The tank fills in 8 hours if both pipes are open. Emptying the tank with the outlet pipe takes 2 hours longer than filling the tank with the inlet pipe. How long does emptying a full tank take?

11. A gutter is formed from a 100-foot-long sheet of metal that is 12 inches wide. To make the gutter, the sides are folded up the same amount on each side. (See figure.) What should the height of the gutter be for maximum capacity?

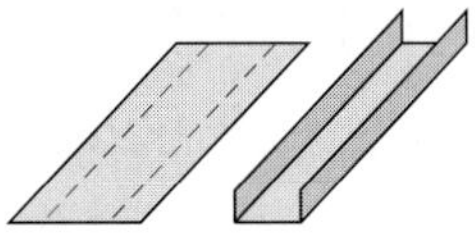

12. Suppose that $f(x)$ is a quadratic function. Explain how the equation $f(x) = 0$ can have no solution, but the solution set of the inequality $f(x) > 0$ is **R**.

13. Solve $6 + x - x^2 \leq 0$.

14. Suppose $f(x) = x^2 - 6x + 11$. Write the function in the form $f(x) = (x - h)^2 + k$. Then give the coordinates of the vertex of the graph of f and write the equation of the axis of symmetry.

15. The figure shows the graph of $f(x) = ax^2 + bx + c$.

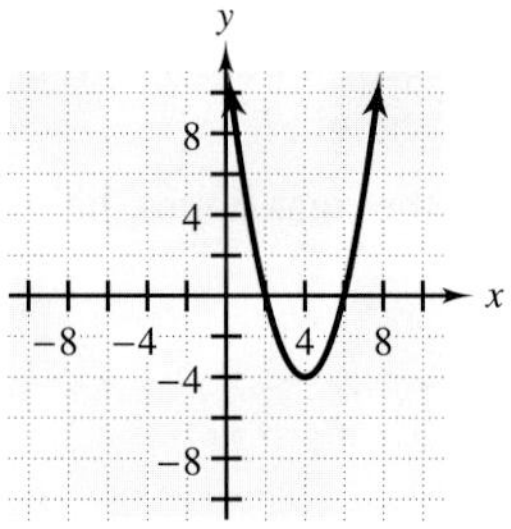

(a) What are the domain and range of the function?

(b) What do you know about the coefficient a?

(c) What are the estimated solutions of the equation $ax^2 + bx + c = 0$?

16. To the nearest hundredth, determine the x-intercepts of the graph of $f(x) = x^2 + 6x - 20$.

Chapter 11

Exponential and Logarithmic Functions

Work at home by way of computers, faxes, E-mail, and other forms of communications technology is becoming a reality for millions of workers.

In the early years when telecommuting first began, an ***exponential function*** *was a good model for the increasing numbers of at-home workers. However, as the bar graph shows, the rate of growth has slowed in recent years, and a better model for this data is a* ***logarithmic function.*** *(For more on this topic, see Exercises 109–112 at the end of Section 11.4, and see the Chapter Project.)*

In Chapter 11, we discuss the algebra of functions and their inverses. In particular, we study the properties and graphs of two inverse functions called exponential and logarithmic functions. Finally, we learn how to solve equations and applications that involve exponential and logarithmic expressions.

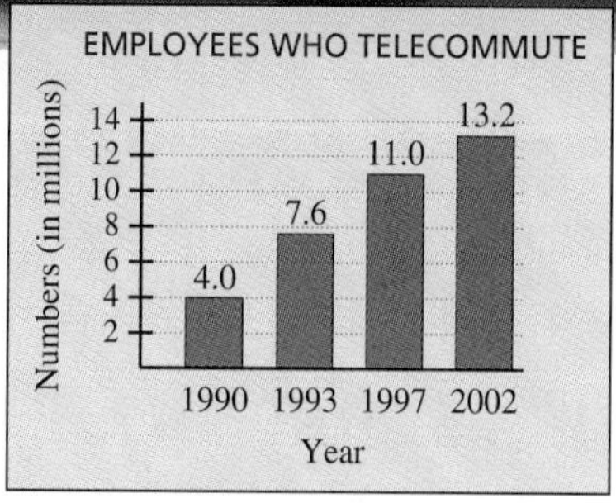

(Source: Gil Gordon Associates.)

11.1 Algebra of Functions

Review of Functions • Basic Operations on Functions • Composition of Functions • Real-Life Applications

Review of Functions

In Chapter 2 we defined a function, introduced function notation, and discussed methods for evaluating functions. We begin with a brief review of these topics.

Definition of a Function

A **function** is a set of ordered pairs in which no two different ordered pairs have the same first coordinate.

Figure 11.1

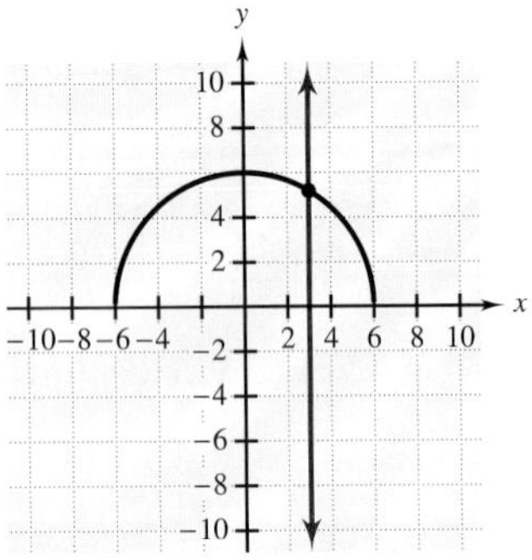

Figure 11.2

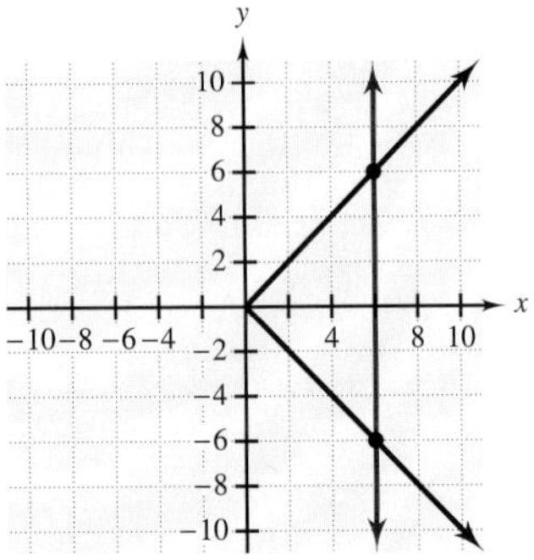

We can use the Vertical Line Test to determine whether a graph represents a function. Because no two ordered pairs of a function can have the same first coordinate, any vertical line will intersect the graph of a function at most once. For example, the graph in Figure 11.1 is the graph of a function; the graph in Figure 11.2 is not the graph of a function.

An equation in x and y describes a function if each value of x corresponds to a unique value of y. For example, because every real number has only one absolute value, the equation $y = |x|$ describes a function. However, the equation $x = |y|$ does not describe a function because its solution set includes, for instance, the pairs (4, 4) and (4, −4), which have the same first coordinate.

Although the definition of *function* refers specifically to a set of ordered pairs, we usually call equations such as $y = |x|$ functions because their solution sets are functions.

The **domain** of a function is the set of all first coordinates of the ordered pairs, and the **range** is the set of all second coordinates of the ordered pairs.

Note: Recall that a graph can be used to estimate the domain and range of a function.

Basic Operations on Functions

In previous chapters we have added, subtracted, multiplied, and divided algebraic expressions. Functions can also be combined with these operations.

Definitions of Basic Operations on Functions

If functions f and g have domains with at least one element in common, then the following operations are defined for each x in the domains of both f and g.

$(f + g)(x) = f(x) + g(x)$ Sum

$(f - g)(x) = f(x) - g(x)$ Difference

$(fg)(x) = f(x)g(x)$ Product

$\left(\frac{f}{g}\right)(x) = \frac{f(x)}{g(x)}$, for $g(x) \neq 0$ Quotient

EXAMPLE 1

Performing Basic Operations on Functions

Let $f(x) = x^2 + 2$ and $g(x) = |x - 1|$. Evaluate each of the following.

(a) $(f + g)(1)$ (b) $(f - g)(-1)$

(c) $(fg)(2)$ (d) $\left(\frac{f}{g}\right)(0)$

Solution

(a) Both f and g are defined at 1.

$$\begin{aligned}(f + g)(1) &= f(1) + g(1) && \text{Definition of sum}\\ &= (1^2 + 2) + |1 - 1| && \text{Evaluate each function.}\\ &= 3\end{aligned}$$

(b) Both f and g are defined at -1.

$$\begin{aligned}(f - g)(-1) &= f(-1) - g(-1) && \text{Definition of difference}\\ &= [(-1)^2 + 2] - |-1 - 1| && \text{Evaluate each function.}\\ &= (1 + 2) - |-2|\\ &= 1\end{aligned}$$

(c) Both f and g are defined at 2.

$$\begin{aligned}(fg)(2) &= f(2)g(2) && \text{Definition of product}\\ &= (2^2 + 2)\,|2 - 1| && \text{Evaluate each function.}\\ &= (4 + 2)\,|1|\\ &= 6\end{aligned}$$

(d) Both f and g are defined at 0 and $g(0) \neq 0$.

$$\begin{aligned}\left(\frac{f}{g}\right)(0) &= \frac{f(0)}{g(0)} && \text{Definition of quotient}\\ &= \frac{0^2 + 2}{|0 - 1|} && \text{Evaluate each function.}\\ &= \frac{2}{1}\\ &= 2\end{aligned}$$

EXAMPLE 2

Performing Basic Operations on Functions

Let $f(x) = 2x + 3$ and $g(x) = x - 5$. Determine each of the following.

(a) $(f + g)(x)$

(b) $(f - g)(x)$

(c) $(fg)(x)$

(d) $\left(\frac{f}{g}\right)(x)$

Solution

Note that the domain for both functions f and g is $\mathbf{R}$. Therefore, the operations can be performed for all values of x except as indicated in part (d).

(a) $(f + g)(x) = f(x) + g(x) = (2x + 3) + (x - 5) = 3x - 2$

(b) $(f - g)(x) = f(x) - g(x) = (2x + 3) - (x - 5) = x + 8$

(c) $(fg)(x) = f(x)g(x) = (2x + 3)(x - 5) = 2x^2 - 7x - 15$

(d) $\left(\frac{f}{g}\right)(x) = \frac{f(x)}{g(x)} = \frac{2x + 3}{x - 5} \quad (x \neq 5)$

Note that 5 is in the domain of both functions f and g, but because $g(5) = 0$, the domain of f/g does not include 5.

EXAMPLE 3 Determining the Domain of a Combined Function

Let $f(x) = \sqrt{x + 3}$ and $g(x) = \sqrt{2 - x}$. What is the domain of $(f + g)$?

Figure 11.3

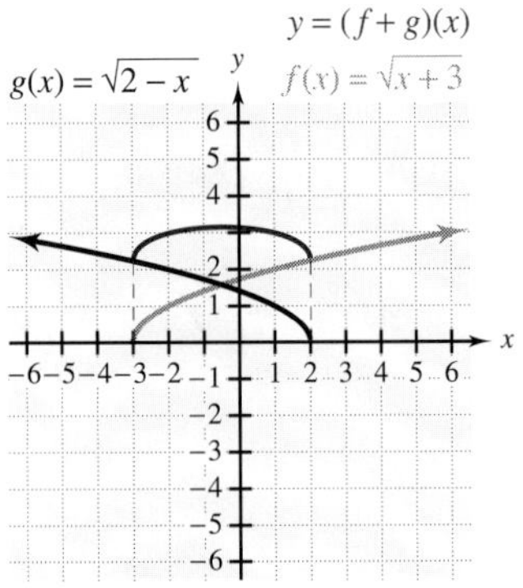

Solution

$$(f + g)(x) = f(x) + g(x) = \sqrt{x + 3} + \sqrt{2 - x}$$

The domain of $(f + g)$ is the intersection of the domains of f and g. From the graphs of f and g in Figure 11.3, we can estimate that the intersection of the domains of f and g is the interval of x-values between -3 and 2, inclusive.

Figure 11.3 also shows the graph of $(f + g)$, whose domain appears to be the same interval.

These estimates can be verified algebraically.

Function	*Requirement*	*Domain*
f	$x + 3 \geq 0$	$x \geq -3$
g	$2 - x \geq 0$	$x \leq 2$

In effect, we are solving the compound inequality $x + 3 \geq 0$ *and* $2 - x \geq 0$.

Because the intersection of the domains is $-3 \leq x \leq 2$, the domain of $f + g$ is $[-3, 2]$.

EXAMPLE 4 Determining the Domain of a Combined Function

Let $f(x) = x^2 - 16$ and $g(x) = 2x - 8$. What is the domain of f/g?

Solution

Although the domain of both functions f and g is the set of all real numbers, f/g is defined only if $g(x) \neq 0$. Because $g(x) = 0$ if $x = 4$, the domain of f/g is the set of all real numbers *except* 4.

The process of determining f/g will also reveal the domain of the function.

$$\left(\frac{f}{g}\right)(x) = \frac{f(x)}{g(x)}$$
$$= \frac{x^2 - 16}{2x - 8}$$

$$= \frac{(x+4)(x-4)}{2(x-4)} \qquad \text{Factor.}$$

$$= \frac{x+4}{2} \qquad \text{Divide out the common factor } x - 4.$$

Students often overlook the fact that 4 is not in the domain because $\frac{x+4}{2}$ does not appear to have a restricted value.

When we divide out the common factor $x - 4$, we are assuming that $x - 4$ is not 0, and so $x \neq 4$. That is why 4 must be excluded from the domain of f/g, even though the simplified expression for f/g appears to have no restricted values.

Composition of Functions

Another way to combine functions is to *compose* the functions.

Students may find the following diagram to be helpful in illustrating $(f \circ g)(x)$.

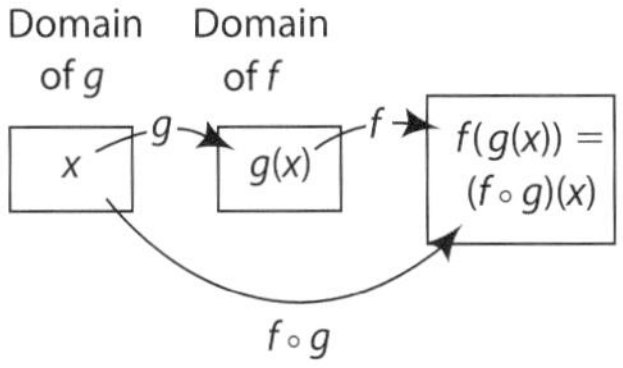

Definition of Composition

The **composition** of two functions f and g is another function $f \circ g$, where $(f \circ g)(x) = f(g(x))$. The domain of $f \circ g$ is the set of all x in the domain of g for which $g(x)$ is in the domain of f.

Note: The notation $f \circ g$, used for the *composition* of f and g, does not mean the same thing as fg, which is used for the *product* of f and g.

EXAMPLE 5

Composition of Two Functions

Let $f(x) = x^2 - 1$ and $g(x) = x + 3$. Determine each of the following.

(a) $(f \circ g)(-2)$ (b) $(g \circ f)(-2)$

(c) $(f \circ g)(x)$ (d) $(g \circ f)(x)$

LEARNING TIP

To evaluate the composite function $(f \circ g)(x)$, write $f(g(x))$ and begin by evaluating $g(x)$. Then substitute the value or expression for $g(x)$ for the variable in the definition of f.

Remind students about the correct order for evaluating $(f \circ g)(x)$ or $f(g(x))$. Evaluate g first and then evaluate f at the resulting value.

Solution

(a) Because $g(-2) = -2 + 3 = 1$,

$$(f \circ g)(-2) = f(g(-2)) = f(1) = 1^2 - 1 = 0$$

(b) Because $f(-2) = (-2)^2 - 1 = 4 - 1 = 3$,

$$(g \circ f)(-2) = g(f(-2)) = g(3) = 3 + 3 = 6$$

(c) We determine $(f \circ g)(x)$ by substituting $g(x)$ for all x in $f(x)$.

$$\begin{aligned}(f \circ g)(x) &= f(g(x))\\ &= f(x+3)\\ &= (x+3)^2 - 1\\ &= x^2 + 6x + 9 - 1\\ &= x^2 + 6x + 8\end{aligned}$$

(d) $(g \circ f)(x) = g(f(x)) = g(x^2 - 1) = (x^2 - 1) + 3 = x^2 + 2$

From Example 5, we observe that the operation of composition is not commutative. In general, $f \circ g \neq g \circ f$.

Sometimes we need to regard a function as the composition of simpler functions. For instance, $F(x) = (x + 1)^2$ consists of two operations (adding 1 and squaring), so F can be written as the composition of $g(x) = x + 1$ (adding 1) and $h(x) = x^2$ (squaring). We can write $F(x) = (h \circ g)(x)$ because

$$(h \circ g)(x) = h(g(x)) = h(x + 1) = (x + 1)^2 = F(x)$$

EXAMPLE 6

Writing a Function as a Composite Function

Use a composition of $f(x) = 2x - 1$ and $g(x) = \dfrac{1}{x}$ to write the given function h.

(a) $h(x) = \dfrac{2}{x} - 1$ (b) $h(x) = 4x - 3$

Solution

(a) $h(x) = \dfrac{2}{x} - 1 = 2 \cdot \dfrac{1}{x} - 1 = 2 \cdot g(x) - 1 = f(g(x)) = (f \circ g)(x)$

(b) $h(x) = 4x - 3 = 4x - 2 - 1$
$= 2(2x - 1) - 1 = 2 \cdot f(x) - 1 = f(f(x)) = (f \circ f)(x)$

Think About It

Suppose that the rule for function f is "take three steps forward" and that the rule for function g is "turn right 90°." If you follow the rule for $f \circ g$, will you arrive at the same point as if you follow the rule for $g \circ f$?

Real-Life Applications

In certain applications one variable may be a function of a second variable, which, in turn, is a function of a third variable. To describe how the first and third variables are related, a composite function can be used.

EXAMPLE 7

Distance, Time, and Fuel Consumption

The distance d (in miles) a race car has traveled after t minutes is given by $d(t) = 2.5t$. The number f of gallons of fuel remaining after d miles is given by $f(d) = 20 - 0.125d$.

(a) How far has the car traveled after 20 minutes?

(b) How much fuel remains after 20 minutes?

(c) Write a function to describe the amount of fuel remaining after t minutes.

(d) Use the function in part (c) to calculate the amount of fuel remaining after 20 minutes. Compare the result to that of part (b).

Solution

(a) $d(20) = 2.5(20) = 50$ Evaluate $d(t)$ for $t = 20$.

After 20 minutes, the car has traveled 50 miles.

(b) $f(50) = 20 - 0.125(50)$ From part (a), when $t = 20$, $d = 50$.

$= 13.75$ Evaluate $f(d)$ for $d = 50$.

After 20 minutes, 13.75 gallons of fuel remain.

(c) Fuel f is a function of distance d, which is a function of time t. To relate fuel and time directly, we compose functions f and d.

$$\begin{aligned}(f \circ d)(t) &= f(d(t))\\ &= f(2.5t)\\ &= 20 - 0.125[2.5t]\\ &= 20 - 0.3125t\end{aligned}$$

(d) $$\begin{aligned}(f \circ d)(20) &= 20 - 0.3125(20)\\ &= 20 - 6.25\\ &= 13.75\end{aligned}$$

As in part (b), 13.75 gallons of fuel remain after 20 minutes.

Quick Reference 11.1

Review of Functions

- A **function** is a set of ordered pairs in which no two different ordered pairs have the same first coordinate.
- The **domain** of a function is the set of all first coordinates of the ordered pairs; the **range** is the set of all second coordinates.
- A relation is a function if
  1. the graph of the relation passes the Vertical Line Test;
  2. each element of the domain corresponds to a unique element of the range.

Basic Operations on Functions

- If functions f and g have domains with at least one element in common, then the following operations are defined for each x in the domains of both f and g.
  1. $(f + g)(x) = f(x) + g(x)$ Sum
  2. $(f - g)(x) = f(x) - g(x)$ Difference
  3. $(fg)(x) = f(x)g(x)$ Product
  4. $\left(\dfrac{f}{g}\right)(x) = \dfrac{f(x)}{g(x)}$, for $g(x) \neq 0$ Quotient
- The domain of $f + g$, $f - g$, and fg is the intersection of the domains of f and g. The domain of f/g is the intersection of the domains of f and g, excluding any x-values for which $g(x) = 0$.

Composition of Functions

- The **composition** of functions f and g is a function $f \circ g$, where $(f \circ g)(x) = f(g(x))$. The domain of $f \circ g$ is the set of all x in the domain of g for which $g(x)$ is in the domain of f.
- In certain special cases, $(f \circ g)(x) = (g \circ f)(x)$, but in general, $f \circ g$ and $g \circ f$ are not equal.

Speaking the Language 11.1

1. A function is a set of ordered pairs in which no two different ordered pairs have the same ______ coordinate.
2. The set of all second coordinates of a function is called the ______ of the function.
3. The notation $f \circ g$ represents the ______ of functions f and g.
4. The domain of the function $f + g$ is the ______ of the domains of functions f and g.

Exercises 11.1

Concepts and Skills

1. Suppose that $f(x) = x - 4$ and $g(x) = 3x^2$. Describe two methods for determining $(f + g)(1)$.
2. The domain of both $f(x) = x$ and $g(x) = x + 1$ is **R**. Is **R** also the domain of $(f/g)(x)$? Explain.

In Exercises 3–10, determine the domain of the function.

3. $g(x) = |2x^2 - x - 1|$
4. $h(x) = 4 - 3|2 - x|$
5. $f(x) = \sqrt{1 - 2x}$
6. $g(x) = \dfrac{x}{\sqrt{x + 5}}$
7. $h(x) = \dfrac{x + 4}{x^3 + 3x^2 - 4x}$
8. $f(x) = \dfrac{x^2 - 9}{x^2 + 25}$
9. $g(x) = \sqrt{x^2 - 16}$
10. $f(x) = \sqrt{x^2 + 2x - 8}$

In Exercises 11–16, for the given functions f and g, determine each of the following.

(a) $(f + g)(-1)$ (b) $(f - g)(1)$

(c) $(fg)(0)$ (d) $\left(\dfrac{f}{g}\right)(0)$

11. $f(x) = 3x, \quad g(x) = x - 1$
12. $f(x) = \sqrt{x + 1}, \quad g(x) = 3$
13. $f(x) = |2x + 1|, \quad g(x) = x^2$
14. $f(x) = \dfrac{1}{x + 2}, \quad g(x) = \dfrac{1}{(x + 2)^2}$
15. $f(x) = 2^x, \quad g(x) = 2^{-x}$
16. $f(x) = 3^{x - 1}, \quad g(x) = 3^{1 - x}$

In Exercises 17–22, use the given functions f and g to evaluate the combined function in each part.

17. Let $f(x) = -4x$ and $g(x) = 3 - 2x$.

 (a) $(f + g)(-3)$ (b) $(f - g)(4)$

 (c) $(fg)(0)$ (d) $\left(\dfrac{f}{g}\right)(1)$

18. Let $f(x) = 3x + 1$ and $g(x) = x - 6$.

 (a) $(f - g)(-1.2)$ (b) $(f + g)\left(-\dfrac{1}{2}\right)$

 (c) $\left(\dfrac{f}{g}\right)(2.5)$ (d) $(fg)\left(\dfrac{2}{3}\right)$

19. Let $f(x) = |x - 3|$ and $g(x) = 2x - 3$.

 (a) $(f + g)(2)$ (b) $(f - g)(-2)$

 (c) $(fg)(0)$ (d) $\left(\dfrac{f}{g}\right)(-1)$

20. Let $f(x) = \sqrt{5 - x}$ and $g(x) = x - 5$.

 (a) $(f + g)(4)$ (b) $(f - g)(-4)$

 (c) $(fg)(5)$ (d) $\left(\dfrac{f}{g}\right)(1)$

21. Let $f(x) = x^2 - 2$ and $g(x) = \sqrt[3]{x}$.

(a) $(f + g)(0)$ (b) $(f - g)(1)$

(c) $\left(\frac{f}{g}\right)(27)$ (d) $(fg)(8)$

22. Let $f(x) = (x + 1)^{1/2}$ and $g(x) = x^3 + 2x$.

(a) $(f + g)(3)$ (b) $(f - g)(-1)$

(c) $(fg)(1)$ (d) $\left(\frac{g}{f}\right)(0)$

23. Suppose that $f(x) = x^2 + 3x$, $g(x) = x$, and $h(x) = x + 3$. Show that $(f/g)(x) = x + 3$. Is the domain of f/g the same as the domain of h? Explain.

24. Suppose that $A = \{x \mid x < 4\}$ is the domain of function f and that $B = \{x \mid x > -3\}$ is the domain of function g. Use interval notation to express the domain of $f + g$, $f - g$, and fg.

In Exercises 25–30, functions f and g are given. Determine the simplified combined function in each part and state its domain.

25. Let $f(x) = x^2 - 9$ and $g(x) = x + 3$.

(a) $(f - g)(x)$ (b) $(f + g)(x)$

(c) $\left(\frac{f}{g}\right)(x)$ (d) $(fg)(x)$

26. Let $f(x) = x + 3$ and $g(x) = x^2 + 4x + 3$.

(a) $(f + g)(x)$ (b) $(f - g)(x)$

(c) $(fg)(x)$ (d) $\left(\frac{f}{g}\right)(x)$

27. Let $f(x) = x^2 + 4$ and $g(x) = x^2 - 4$.

(a) $\left(\frac{f}{g}\right)(x)$ (b) $\left(\frac{g}{f}\right)(x)$

(c) $\left(\frac{f}{f}\right)(x)$ (d) $\left(\frac{g}{g}\right)(x)$

28. Let $f(x) = x^2 - 5x + 6$ and $g(x) = x - 2$.

(a) $\left(\frac{g}{f}\right)(x)$ (b) $\left(\frac{f}{g}\right)(x)$

(c) $\left(\frac{g}{g}\right)(x)$ (d) $\left(\frac{f}{f}\right)(x)$

29. Let $f(x) = \dfrac{1}{x + 2}$ and $g(x) = \dfrac{1}{x - 3}$.

(a) $(f - g)(x)$ (b) $(f + g)(x)$

(c) $\left(\frac{f}{g}\right)(x)$ (d) $(fg)(x)$

30. Let $f(x) = \dfrac{x + 1}{x + 3}$ and $g(x) = \dfrac{x + 3}{2 - x}$.

(a) $(f + g)(x)$ (b) $(f - g)(x)$

(c) $\left(\frac{g}{f}\right)(x)$ (d) $\left(\frac{f}{g}\right)(x)$

CE In Exercises 31–34, let $f(x) = x^2 - 4$ and $g(x) = x - 1$. Determine the given combined function.

31. $(f - g)(2t)$ **32.** $(f + g)(-t)$

33. $\left(\frac{f}{g}\right)\left(\frac{t}{2}\right)$ **34.** $(fg)(t + 2)$

CE In Exercises 35–38, let $f(x) = x + 5$ and $g(x) = x^2 - 3$. Determine the given combined function.

35. $(f + g)(3t)$ **36.** $(f - g)(t + 1)$

37. $(fg)(-2t)$

38. $\left(\frac{f}{g}\right)\left(\frac{t}{3}\right)$

39. Explain how you can use the graphs of functions f and g to estimate the domain of $f + g$.

40. Explain why we cannot add $f(x) = \sqrt{1 - x}$ and $g(x) = \sqrt{x - 5}$.

In Exercises 41–44, functions f and g are given. Determine the domain of the combined function given in each part.

41. Let $f(x) = \sqrt{2x - 3}$ and $g(x) = \sqrt{5 - x}$.

(a) $(f + g)(x)$ (b) $(f - g)(x)$

42. Let $f(x) = \sqrt{2 - x}$ and $g(x) = \sqrt{3 - x}$.

(a) $(f - g)(x)$ (b) $(f + g)(x)$

43. Let $f(x) = \sqrt{8 - x}$ and $g(x) = \sqrt{x - 1}$.

(a) $\left(\frac{f}{g}\right)(x)$ (b) $(fg)(x)$

44. Let $f(x) = \sqrt{2 - x}$ and $g(x) = \sqrt{x - 3}$.

(a) $(fg)(x)$ (b) $\left(\frac{g}{f}\right)(x)$

45. For two given functions f and g, determining the value of $(f \circ g)(5)$ requires two calculations. Describe these calculations in the order in which they are performed.

46. Suppose that $f = \{(1, 2), (2, 5), (4, -1)\}$ and $g = \{(1, 3), (2, 5)\}$. Explain why $(g \circ f)(1)$ is defined but $(f \circ g)(1)$ is not.

In Exercises 47–50, functions f and g are given. Evaluate the composite function given in each part.

47. Let $f(x) = x - 3$ and $g(x) = 2 - x^2$.

(a) $(f \circ g)(1)$ (b) $(f \circ g)(0)$
(c) $(g \circ f)(3)$ (d) $(g \circ f)(-1)$

48. Let $f(x) = 2x + 1$ and $g(x) = x^3$.

(a) $(f \circ g)(-1)$ (b) $(f \circ g)(2)$
(c) $(g \circ f)(0)$ (d) $(g \circ f)(-1)$

49. Let $f(x) = x^2$ and $g(x) = 2x - 3$.

(a) $(g \circ f)(-3)$ (b) $(f \circ g)\left(-\frac{1}{2}\right)$
(c) $(g \circ g)\left(\frac{3}{2}\right)$ (d) $(f \circ f)(-1)$

50. Let $f(x) = x^2 - 2x + 3$ and $g(x) = x + 4$.

(a) $(f \circ g)(-5)$ (b) $(g \circ f)(3)$
(c) $(f \circ f)(0)$ (d) $(g \circ g)(2)$

CE In Exercises 51 and 52, for the given functions f and g, determine each of the following.

(a) $(f \circ g)(2)$
(b) $(g \circ f)(-1)$
(c) $(f \circ g)(-3)$

51. $f = \{(-2, 3), (2, -4), (1, 5), (-3, -2), (-1, 3)\}$
$g = \{(-2, 5), (2, 1), (-3, 2), (3, -2)\}$

52. $f = \{(2, 5), (3, 6), (4, 7), (-2, 1), (-1, 2)\}$
$g = \{(-5, -1), (-3, -2), (0, 4), (-1, 3), (2, 3)\}$

In Exercises 53–60, for the given functions f and g, determine $(f \circ g)(x)$ and $(g \circ f)(x)$.

53. $f(x) = 3 - x^2, \quad g(x) = 2x + 1$

54. $f(x) = 3x - 2, \quad g(x) = 4x - x^2$

55. $f(x) = x^2 - 3x + 5, \quad g(x) = 3x - 2$

56. $f(x) = 2x - 5, \quad g(x) = x^2 - 4x - 5$

57. $f(x) = x^3, \quad g(x) = \sqrt[3]{x - 2}$

58. $f(x) = \sqrt[3]{2x - 3}, \quad g(x) = -x^3$

59. $f(x) = 5, \quad g(x) = 2x - 3$

60. $f(x) = x^2 - 3, \quad g(x) = 7$

In Exercises 61–68, for the given functions f and g, determine $(f \circ f)(x)$ and $(g \circ g)(x)$.

61. $f(x) = 3x - 4, \quad g(x) = x + 5$ Before assigning these, you may want to ask students for an interpretation of $f \circ f$.

62. $f(x) = x - 2, \quad g(x) = 4 - 3x$

63. $f(x) = |x - 3|, \quad g(x) = x^2 + 3$

64. $f(x) = x^2 - 2, \quad g(x) = |x + 3|$

65. $f(x) = 2x^2 - 3x, \quad g(x) = 2 - x$

66. $f(x) = x - 3, \quad g(x) = 3x^2 - 2x$

67. $f(x) = 3, \quad g(x) = 2$

68. $f(x) = -5, \quad g(x) = -2$

In Exercises 69–72, show that $(f \circ g)(x) = x$ for each given pair of functions.

This group prepares the student for the topic of inverse functions in the next section.

69. $f(x) = 2x + 3, \quad g(x) = \frac{x - 3}{2}$

70. $f(x) = 2x - 5, \quad g(x) = \frac{x + 5}{2}$

71. $f(x) = \frac{x}{3} + 2, \quad g(x) = 3x - 6$

72. $f(x) = \frac{x}{4} - 2, \quad g(x) = 4x + 8$

In Exercises 73–80, let $f(x) = x + 3$, $g(x) = 2x$, and $h(x) = x^2$. Use f, g, and h to write each function as a composition of two functions.

73. $F(x) = 2x + 6$

74. $H(x) = 4x^2$

75. $H(x) = 4x$

76. $G(x) = x + 6$

77. $G(x) = x^2 + 3$

78. $F(x) = (x + 3)^2$

79. $F(x) = 2x^2$

80. $G(x) = 2x + 3$

Real-Life Applications

81. On interstate trips, a driver averages 54 mph. The distance d (in miles) traveled in t hours is given by $d(t) = 54t$. Because the driver averages 25 miles per gallon, the number of gallons g used is given by $g(d) = d/25$. The cost per gallon is \$1.03, so the total fuel cost c is given by $c(g) = 1.03g$.

(a) Write a function describing the number of gallons used in t hours of travel.

(b) Write a function describing the total fuel cost in t hours of travel.

(c) Determine the total fuel cost of a 12-hour trip.

82. An oil tanker runs aground and springs a leak. The oil spreads out in a semicircular pattern from the shoreline. The distance r (in feet) from the tanker to the edge of the oil spill at time t (in minutes) is given by the function $r(t) = 20t$. If the area of the semicircle is given by $A(x) = \frac{1}{2}\pi x^2$, where x is the radius, write a function $A \circ r$ for the area covered by the oil at time t. What is the area of the oil spill after 5 minutes?

83. A party caterer charges $5.00 per person for 100 people. For a gathering of fewer than 100 people, the caterer increases the price per person by $0.05 for each person under 100 people. For example, 90 people is 10 less than 100, so the additional price per person is 10(0.05) = $0.50. Then the total cost is (90)(5.50) = $495.00.

(a) Write a function $f(x)$ to represent the number of people who attend the gathering.

(b) Write a function $c(x)$ to represent the per-person cost.

(c) Write a function $T(x) = (fc)(x)$ to represent the total cost of the gathering.

(d) Use the function in part (c) to determine the total cost for 84 people.

84. The distance between the bases of a baseball diamond (a square) is 90 feet. A ball is hit directly down the third-base line at a speed of 60 feet per second. If x is the distance from home plate to the ball, then x is a function of time t in seconds. (See figure.) Let $f(x)$ represent the distance from the ball to first base.

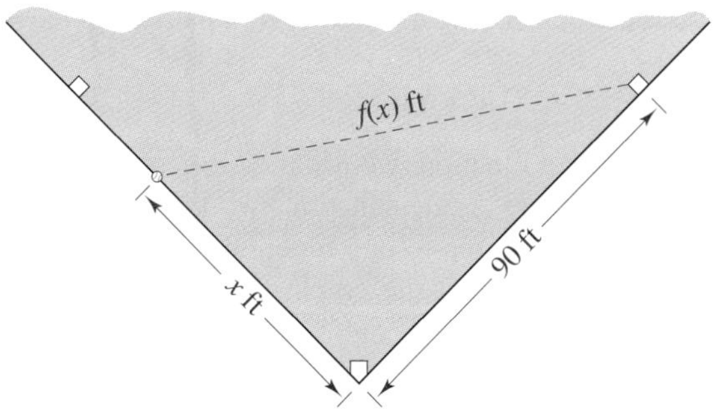

(a) Use the Pythagorean Theorem to write the function f.

(b) Write the function $x(t)$.

(c) Write the function $(f \circ x)(t)$. What does this function represent?

Data Analysis: *ATM Terminals*

The accompanying table shows that the number of automatic teller machine (ATM) terminals and the number of transactions at ATMs increased from 1993 to 1998. (Source: EFT Network Data Book.)

Year	ATM Terminals (in thousands)	ATM Transactions (in millions)
1993	94.8	7,705
1994	109.1	8,334
1995	122.7	9,689
1996	139.1	10,684
1997	165.0	10,920
1998	187.0	11,160

The number G of terminals (in thousands) can be modeled as a function of the number of years since 1990.

$$G(t) = 18.4t + 35$$

The number F of transactions (in millions) can be modeled as a function of x, where x is the number of terminals (in thousands).

$$F(x) = -0.55x^2 + 194x - 5879$$

85. Interpret $(F \circ G)(t)$.

86. Evaluate $G(8)$. Identify and describe the actual entry in the table for which $G(8)$ is an estimate.

87. Evaluate $F(187)$. Identify and describe the actual entry in the table for which $F(187)$ is an estimate.

88. Evaluate $(F \circ G)(8)$. Identify and describe the actual table entry for which $(F \circ G)(8)$ is an estimate.

Challenge

In Exercises 89–93, write each function $f(x)$ as a composition of two simpler functions g and h. State specifically what the functions g and h are.

89. $f(x) = \sqrt{x^2 - 5x + 6}$

90. $f(x) = \dfrac{1}{x^2 + 4x}$

91. $f(x) = (3x + 2)^4$

92. $f(x) = 3x^{1/2} - 5$

93. $f(x) = |2x + 5| + 8$

94. Let $f(x) = 3$ and $g(x) = \dfrac{1}{x - 3}$. Determine $(f \circ g)(x)$ and $(g \circ f)(x)$.

11.2 Inverse Functions

One-to-One Functions • Inverse Functions • Determining the Inverse

One-to-One Functions

Make sure students understand the difference between the definition of a *function* and the definition of a *one-to-one function.*

We have defined a function as a set of ordered pairs in which no two ordered pairs have the same *first* coordinate. If, in addition, we require that no two ordered pairs have the same *second* coordinate, then we have the special case of a **one-to-one function.**

Definition of a One-to-One Function

A function is a **one-to-one function** if no two different ordered pairs have the same second coordinate.

Think About It

Suppose a function associates each person with that person's mother. Is the function one-to-one?

This means that for any value of y in the range of the function f, there is a *unique* value of x such that $f(x) = y$.

EXAMPLE 1 One-to-One Functions

Determine whether each function is a one-to-one function.

(a) $f = \{(2, 3), (4, 5), (6, 3)\}$

(b) $g = \{(1, 2), (2, 3), (3, 4), (4, 5), (5, 6)\}$

Solution

(a) Function f is not a one-to-one function because (2, 3) and (6, 3) are different ordered pairs with the same second coordinate.

(b) Function g is a one-to-one function because no two ordered pairs have the same second coordinate.

To determine whether a graph represents a function, we use the Vertical Line Test. To determine whether a graph represents a one-to-one function, we use the Horizontal Line Test. Because no two ordered pairs of a one-to-one function have the same second coordinate, a horizontal line cannot intersect the graph of the function at more than one point.

The Horizontal Line Test for a One-to-One Function

A function is a one-to-one function if and only if no horizontal line intersects the graph of the function at more than one point.

A one-to-one function passes both the Vertical Line Test and the Horizontal Line Test. For example, the graph of $x = 3$ passes the Horizontal Line Test, but the graph does not represent a function.

The graphs in Figures 11.4 and 11.5 both pass the Vertical Line Test, so both represent functions.

Figure 11.4

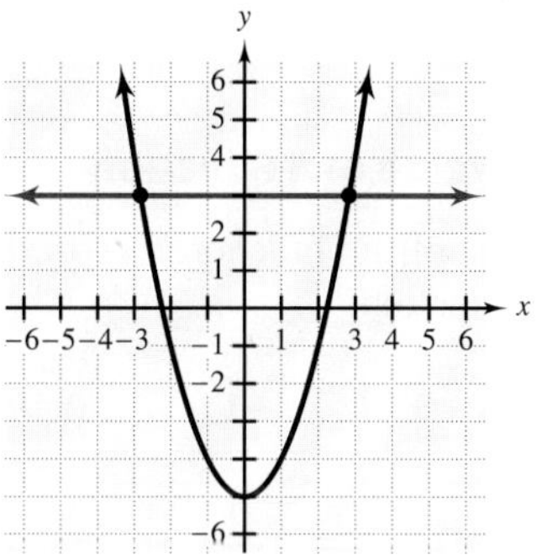

Figure 11.5

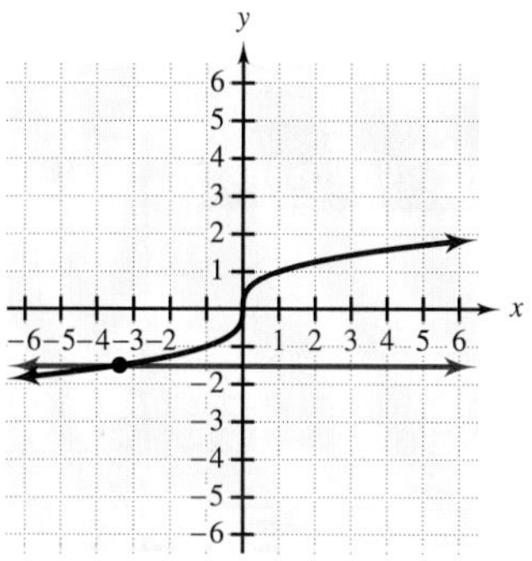

LEARNING TIP

Remember that a one-to-one function must pass both the Vertical Line Test and the Horizontal Line Test.

Using the Horizontal Line Test, we find that the graph in Figure 11.4 does not represent a one-to-one function, but the graph in Figure 11.5 does.

Inverse Functions

If the corresponding x- and y-coordinates of a one-to-one function f are interchanged, the resulting set of ordered pairs is also a function. This new function is called the **inverse function** and is written f^{-1} (read "f inverse").

Note: In the notation f^{-1}, the -1 is not an exponent. We read f^{-1} as a single symbol that represents an inverse function.

Exploring the Concept

Remind students that this set of ordered pairs means $f(-3) = 5$, $f(-1) = -4$, and so on.

Inverse Functions

Consider the following function.

$$f = \{(-3, 5), (-1, -4), (3, -2), (6, 1)\}$$

To write the inverse function, interchange the x- and y-coordinates of each ordered pair.

$$f^{-1} = \{(5, -3), (-4, -1), (-2, 3), (1, 6)\}$$

Note the domains and ranges of f and f^{-1}.

Domain of f:	$\{-3, -1, 3, 6\}$	Range of f:	$\{-4, -2, 1, 5\}$
Range of f^{-1}:	$\{-3, -1, 3, 6\}$	Domain of f^{-1}:	$\{-4, -2, 1, 5\}$

The domain of f is the same as the range of f^{-1}, and the range of f is the same as the domain of f^{-1}.

If we sketch the graphs of f and f^{-1} on the same coordinate system (see Fig. 11.6), we observe that the points lie the same distance on either side of the diagonal line whose equation is $y = x$. This line is called the **line of symmetry.** If we fold the graph along the line of symmetry, the points coincide.

Figure 11.6

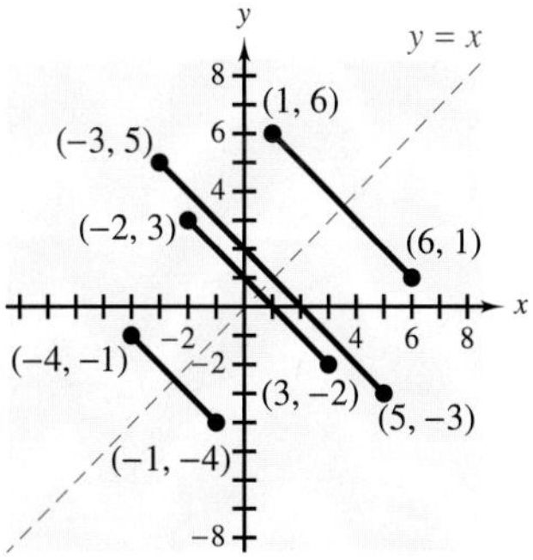

When a one-to-one function is described by a set of ordered pairs, forming the inverse function is simply a matter of exchanging coordinates.

When a function f is expressed as a rule, then the inverse function g often also can be expressed as a rule.

Exploring the Concept

Determining the Inverse of a Function

To determine the inverse of the function $f(x) = x + 2$, we begin by constructing a table of values for the function.

x	-3	-2	-1	0	1	2	3	4
$f(x)$	-1	0	1	2	3	4	5	6

Then, to write a table of values for the inverse function, we interchange the x- and y-coordinates of the pairs that belong to f.

x	-1	0	1	2	3	4	5	6
$f^{-1}(x)$	-3	-2	-1	0	1	2	3	4

In the table for the inverse function, each value of $f^{-1}(x)$ is 2 less than the value of x. Thus the inverse function appears to be described by $f^{-1}(x) = x - 2$.

Observe that the function f adds 2 to each value of x, whereas the function f^{-1} subtracts 2 from each value of x.

Encourage students to develop this intuitive idea about f^{-1}.

This observation suggests that an inverse function f^{-1} reverses the operations indicated in f. Thus, if $f(x) = 3x$, then $f^{-1}(x) = \frac{x}{3}$; if $f(x) = x^3$, then $f^{-1}(x) = \sqrt[3]{x}$.

The definition of an inverse function offers a way to verify that our conjecture is correct.

Definition of an Inverse Function

Functions f and g are **inverse functions** if $(f \circ g)(x) = x$ for each x in the domain of g and if $(g \circ f)(x) = x$ for each x in the domain of f. We write $g = f^{-1}$ and $f = g^{-1}$.

We can use this definition to verify that the functions $f(x) = x + 2$ and $g(x) = x - 2$ are inverse functions.

$$(f \circ g)(x) = f(g(x)) = f(x - 2) = (x - 2) + 2 = x$$

$$(g \circ f)(x) = g(f(x)) = g(x + 2) = (x + 2) - 2 = x$$

EXAMPLE 2

Verifying Inverse Functions

Verify that the given pair of functions are inverses.

(a) $f(x) = \dfrac{x - 2}{3}$ and $g(x) = 3x + 2$

(b) $f(x) = \sqrt[3]{x + 2}$ and $g(x) = x^3 - 2$

Solution

(a) We apply the definition of inverse functions.

$$(f \circ g)(x) = f(g(x))$$
$$= f(3x + 2) = \frac{(3x + 2) - 2}{3} = \frac{3x}{3} = x$$
$$(g \circ f)(x) = g(f(x))$$
$$= g\left(\frac{x - 2}{3}\right) = 3 \cdot \frac{x - 2}{3} + 2 = (x - 2) + 2 = x$$

Because $(f \circ g)(x) = x$ and $(g \circ f)(x) = x$, f and g are inverse functions. Figure 11.7 shows their graphs and the line of symmetry.

LEARNING TIP

Rotating a figure so that the line $y = x$ is vertical may help you to see the symmetry of the graphs of a function and its inverse.

Figure 11.7

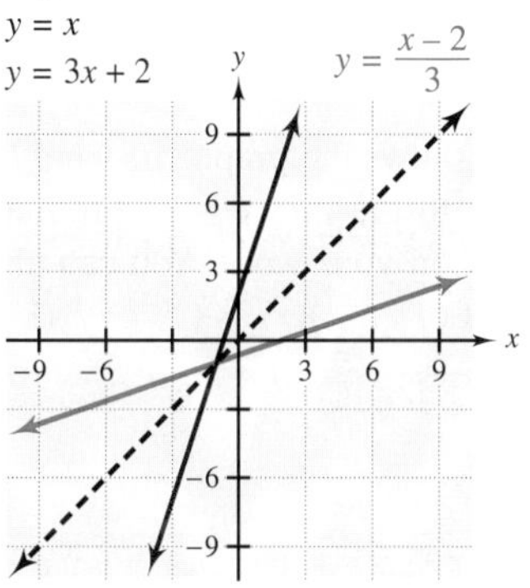

Figure 11.8

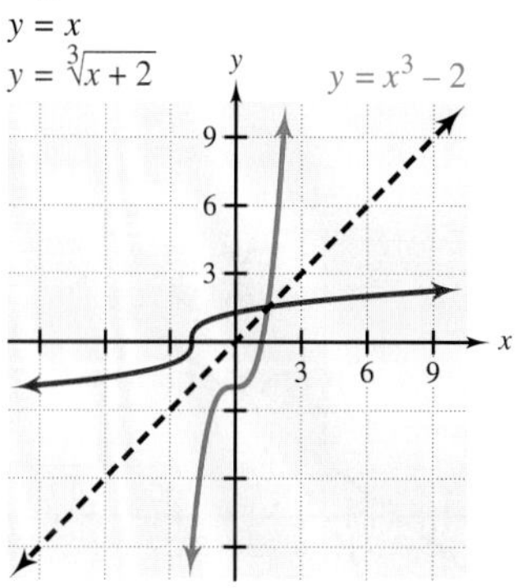

(b) $(f \circ g)(x) = f(x^3 - 2) = \sqrt[3]{(x^3 - 2) + 2} = \sqrt[3]{x^3} = x$

$(g \circ f)(x) = g\left(\sqrt[3]{x + 2}\right) = \left(\sqrt[3]{x + 2}\right)^3 - 2 = (x + 2) - 2 = x$

Because $(f \circ g)(x) = x$ and $(g \circ f)(x) = x$, f and g are inverse functions. Figure 11.8 shows their graphs and the line of symmetry.

In Example 2 the graphs of the inverse functions would coincide if they were folded along the diagonal line shown in the figures. In general, the graphs of f and f^{-1} are symmetric with respect to the line of symmetry $y = x$.

Determining the Inverse

If a one-to-one function is defined by a rule, we can often determine a rule for the inverse function by interchanging the roles of x and y and solving for y.

Determining the Inverse of a Function

To determine the inverse of a function f, perform the following steps:

1. Replace the function notation $f(x)$ with y.
2. Interchange x and y.
3. Solve for y, and replace y with $f^{-1}(x)$.

EXAMPLE 3 Determining an Inverse Function

Determine the inverse of the given function.

(a) $f(x) = 2x - 3$ (b) $g(x) = \sqrt[3]{x + 5}$

A more formal approach to Example 3 is based on the fact that $f(f^{-1}(x)) = x$.

$$f(f^{-1}(x)) = 2 \cdot f^{-1}(x) - 3$$
$$x = 2 \cdot f^{-1}(x) - 3$$
$$f^{-1}(x) = \frac{x+3}{2}$$

Solution

(a)

$f(x) = 2x - 3$

$y = 2x - 3$ Replace $f(x)$ with y.

$x = 2y - 3$ Interchange x and y.

$x + 3 = 2y$ Solve for y.

$\frac{x+3}{2} = y$

$f^{-1}(x) = \frac{x+3}{2}$ Replace y with $f^{-1}(x)$.

Figure 11.9 shows the graphs of f and f^{-1}. Notice that the graphs are symmetric with respect to the line $y = x$. Producing the graphs of f and f^{-1} gives you a visual check on your work. You can also use the definition of inverse functions to check that $(f \circ f^{-1})(x) = (f^{-1} \circ f)(x) = x$.

Figure 11.9

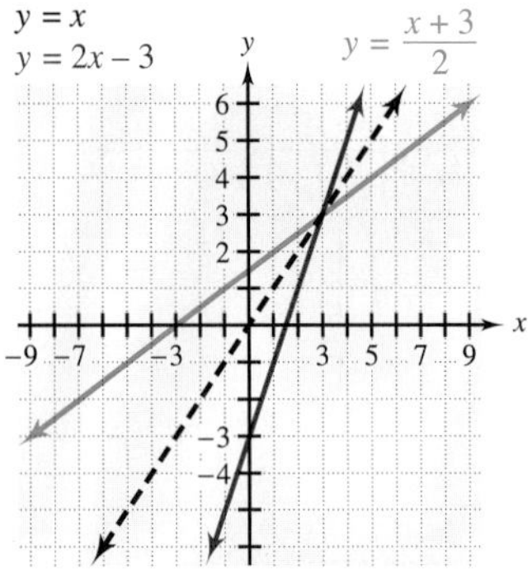

(b)

$g(x) = \sqrt[3]{x + 5}$

$y = \sqrt[3]{x + 5}$ Replace $g(x)$ with y.

$x = \sqrt[3]{y + 5}$ Interchange x and y.

$x^3 = y + 5$ Cube both sides of the equation.

$x^3 - 5 = y$ Solve for y.

$g^{-1}(x) = x^3 - 5$ Replace y with $g^{-1}(x)$.

Quick Reference 11.2

One-to-One Functions

- A function is a **one-to-one function** if no two different ordered pairs have the same second coordinate.
- The graph of a one-to-one function passes the Horizontal Line Test: Any horizontal line intersects the graph at most once.

Inverse Functions

- If the x- and y-coordinates of a one-to-one function f are interchanged, the resulting function is the **inverse function** f^{-1}.
- The domain of f is the same as the range of f^{-1}, and the range of f is the same as the domain of f^{-1}.
- The graphs of f and f^{-1} are symmetric with respect to the line $y = x$.
- An inverse function f^{-1} reverses the rule of function f.
- Functions f and g are inverse functions if $(f \circ g)(x) = x$ for each x in the domain of g and if $(g \circ f)(x) = x$ for each x in the domain of f.
- This definition can be used to determine whether two functions are inverse functions.

Determining the Inverse

- To determine the inverse of a function, perform the following steps.
  1. Replace the function notation $f(x)$ with y.
  2. Interchange x and y.
  3. Solve for y and replace y with $f^{-1}(x)$.

Speaking the Language 11.2

1. If no two different ordered pairs of a function have the same second coordinate, we say that the function is ________.
2. The symbol f^{-1} represents the ________ of function f.
3. We can determine whether a graph represents a function by using the ________ Line Test; to test for one-to-one functions, we can use the ________ Line Test.
4. We say that the graphs of functions f and f^{-1} are ________ with respect to the line $y = x$.

Exercises 11.2

Concepts and Skills

1. Suppose that no two ordered pairs of a set have the same second coordinate. Is the set a one-to-one function? Explain.

2. Describe how the Horizontal Line Test is used to determine whether a function is a one-to-one function.

In Exercises 3–6, determine whether the given function is a one-to-one function.

3. $\{(5, 2), (2, 5), (3, 2), (6, -4)\}$
4. $\{(5, 2), (3, 0), (1, -2), (7, 4)\}$
5. $\{(-3, -1), (-1, 1), (3, 5), (1, 3)\}$
6. $\{(-3, -1), (-1, -3), (3, -2), (2, -1)\}$

In Exercises 7–12, determine whether the given graph represents a one-to-one function.

7.
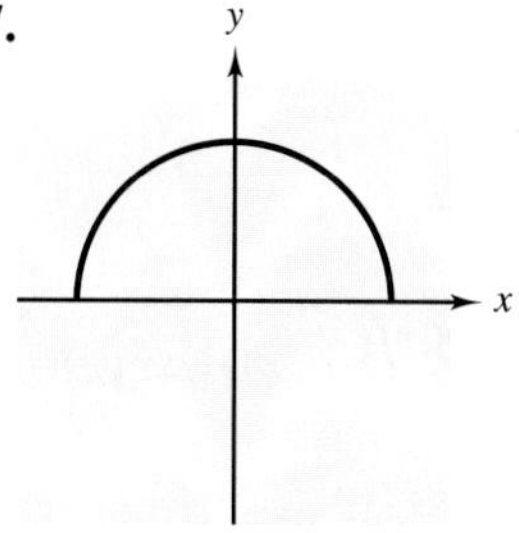

8.
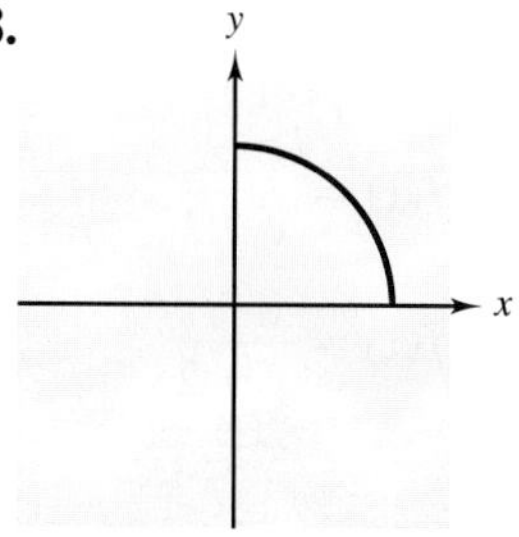

9.
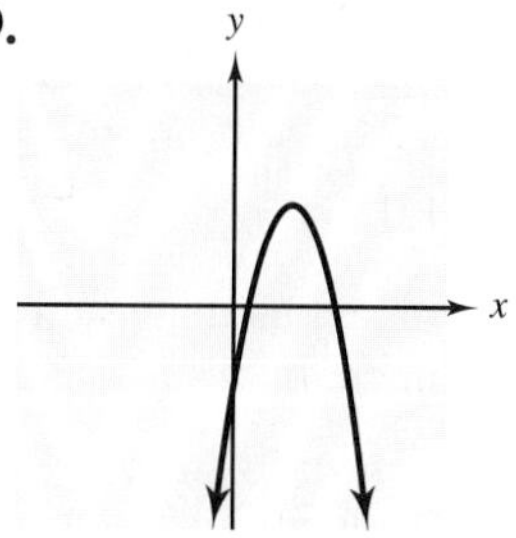

10.
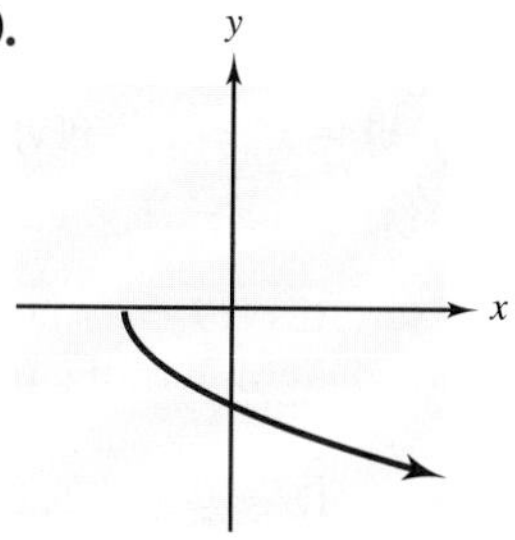

11.
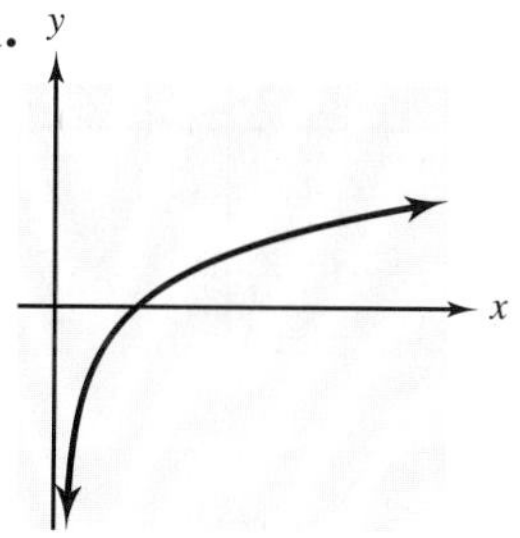

12.
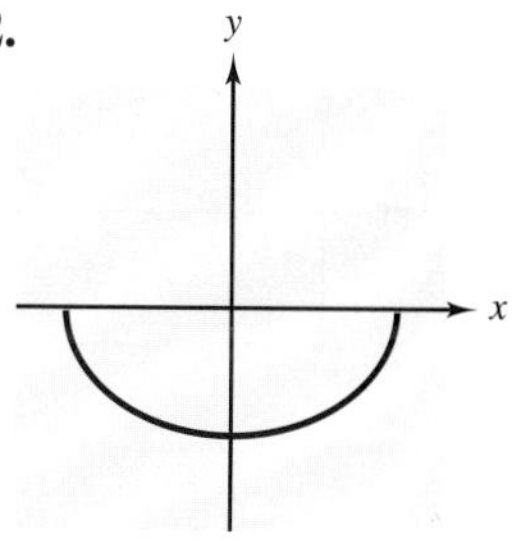

In Exercises 13–16, determine f^{-1}. State the domain and range of f^{-1}.

13. $f = \{(1, 2), (3, 4), (5, 6), (7, -1)\}$
14. $f = \{(-1, 2), (-3, 4), (-5, -7), (5, 8)\}$
15. $f = \{(3, 2), (4, 3), (-5, 1), (-7, -2)\}$
16. $f = \{(5, -6), (7, -8), (-8, 5), (-6, 7)\}$

CE In Exercises 17–20, f is a one-to-one function. Suppose that $f(3) = 4$, $f(5) = 7$, $f(-2) = -3$, and $f(-4) = -5$. Evaluate f^{-1} as indicated.

17. $f^{-1}(-3)$

18. $f^{-1}(-5)$

19. $f^{-1}(4)$

20. $f^{-1}(7)$

21. Consider the functions $f(x) = x$ and $g(x) = x^{-1}$. Is g the inverse of f? Explain.

22. Suppose that you sketch the graphs of functions f and g and you notice that the graphs are mirror images about the line $y = x$. What is your conjecture about the two functions?

In Exercises 23–28, sketch the graphs of each given pair of functions. Although the graphs are not conclusive, decide whether it is reasonable to believe that the functions may be inverses.

23. $f(x) = 2x + 1, \quad g(x) = \dfrac{x-1}{2}$

24. $f(x) = 3x + 1, \quad g(x) = 3x - 2$

25. $f(x) = x^2 + 1, \quad g(x) = 1 - x^2$

26. $f(x) = |x - 1|, \quad g(x) = x + 1$

27. $f(x) = x^3 - 2, \quad g(x) = \sqrt[3]{x+2}$

28. $f(x) = \sqrt[3]{x} - 1, \quad g(x) = x^3 + 1$

In Exercises 29–32, the graph of a function is given. Sketch the graph of the inverse function.

29.

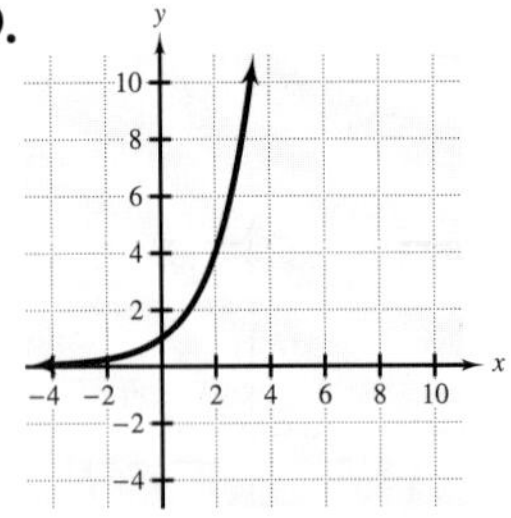

30.

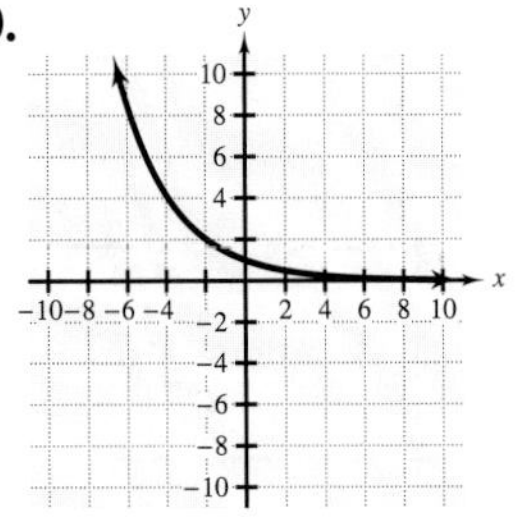

31.

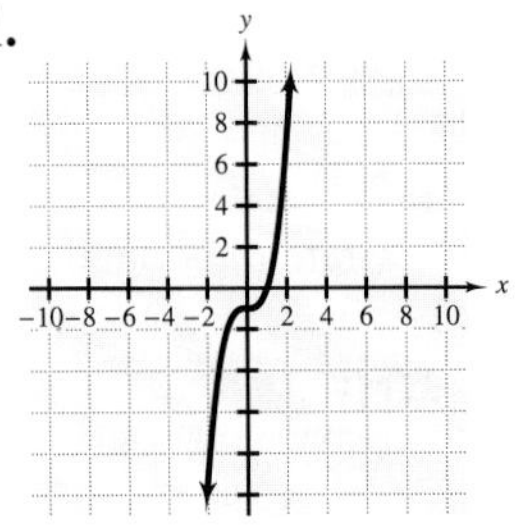

32.

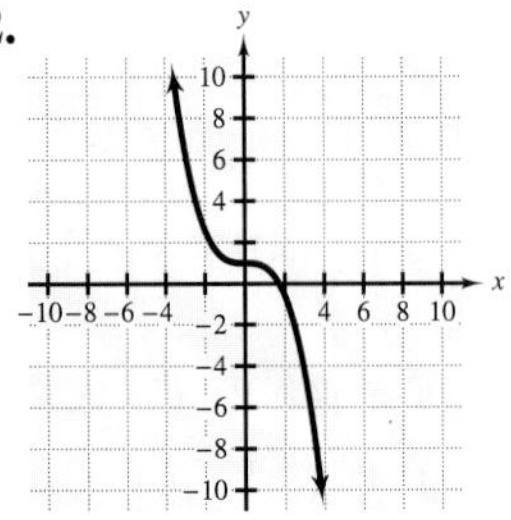

33. Explain why a linear function always has an inverse.

34. For what positive integers n does $f(x) = x^n$ have an inverse function? Explain. Determine the inverse of $f(x) = x^n$ for those values of n.

In Exercises 35–46, determine whether the pairs are inverses by showing that $(f \circ g)(x) = (g \circ f)(x) = x$.

35. $f(x) = 2x$, $g(x) = -2x$

36. $f(x) = x + 2$, $g(x) = x - 2$

37. $f(x) = 2 - 3x$, $g(x) = \dfrac{1}{3}(2 - x)$

38. $f(x) = \dfrac{3}{4}x$, $g(x) = \dfrac{4}{3}x$

39. $f(x) = 2 - x$, $g(x) = x - 2$

40. $f(x) = x - 1$, $g(x) = \dfrac{1}{x-1}$

41. $f(x) = x^{1/5}$, $g(x) = x^5$

42. $f(x) = -3(x - 1)$, $g(x) = 1 - \dfrac{x}{3}$

43. $f(x) = \dfrac{3}{x}$, $g(x) = \dfrac{3}{x}$

44. $f(x) = x^{1/3}$, $g(x) = x^{-1/3}$

45. $f(x) = \sqrt[3]{x+1}$, $g(x) = x^3 - 1$

46. $f(x) = \dfrac{1}{4}x^2, x \geq 0$, $g(x) = 2\sqrt{x}, x \geq 0$

In Exercises 47–68, determine f^{-1} and verify that $(f \circ f^{-1})(x) = x$ for all x in the domain of f^{-1} and that $(f^{-1} \circ f)(x) = x$ for all x in the domain of f.

47. $f(x) = 3x + 5$

48. $f(x) = 5x - 7$

49. $f(x) = -2x$

50. $f(x) = 2(x - 1)$

51. $f(x) = \dfrac{x}{3} + 5$

52. $f(x) = \dfrac{2x-1}{3}$

53. $f(x) = \dfrac{1}{x}$

54. $f(x) = \dfrac{1}{x} - 3$

55. $f(x) = \dfrac{1}{x+2}$

56. $f(x) = \dfrac{3}{1-x}$

57. $f(x) = \dfrac{x+1}{x-2}$

58. $f(x) = \dfrac{x+3}{2x+1}$

59. $f(x) = x^3 + 1$

60. $f(x) = -\frac{1}{8}x^3$

61. $f(x) = x^{-3}$

62. $f(x) = x^{-5} + 1$

63. $f(x) = \frac{\sqrt{x}}{2}$

64. $f(x) = \sqrt{3x}$

65. $f(x) = \sqrt[3]{2 - x}$

66. $f(x) = \sqrt[5]{3 - x}$

67. $f(x) = \sqrt[5]{x} + 3$

68. $f(x) = 1 - \sqrt[3]{x}$

CE In Exercises 69 72, a verbal description of function f is given. Write a verbal description of the inverse function f^{-1}.

69. Function f adds 2 to x and then divides the result by 3.

70. Function f multiplies x by -2 and then subtracts 1 from the result.

71. Function f cubes 1 more than x.

72. Function f adds 4 to x and then takes half of the result.

CE **73.** From the following list, identify the pairs of functions that are inverse functions.

(i) $f(x) = \frac{1}{3}$, $g(x) = 3$

(ii) $f(x) = 5$, $g(x) = -5$

(iii) $f(x) = \frac{5}{x}$, $g(x) = \frac{5}{x}$

CE **74.** Suppose that the graph of function f lies entirely in the given quadrants. In what quadrants does the graph of function f^{-1} lie?

(a) I and II

(b) II and IV

(c) I and III

(d) II and III

75. Explain why $y = \sqrt{9 - x^2}$, $x \geq 0$, has an inverse but $y = \sqrt{9 - x^2}$ does not.

CE **76.** Let $f(x) = 2x - 1$. Determine $f^{-1}(x)$ by evaluating f at $f^{-1}(x)$ and then solving for $f^{-1}(x)$.

Real-Life Applications

77. The function $C(t) = \frac{5}{9}(t - 32)$ converts Fahrenheit temperatures to Celsius. Determine the inverse function and describe what it does.

78. Write a function that converts hours to minutes. Determine the inverse function and describe what it does.

79. Suppose that a British pound is worth \$1.50. Write a function that gives the value d in dollars of p British pounds. Then write the inverse of the function and describe what it does.

80. One meter is 1.094 yards. Write a function that gives the number y of yards in m meters. Then write the inverse of the function and describe what it does.

Modeling with Real Data

81. The accompanying table shows that the number of cellular phone subscribers has risen along with the number of cell sites. (Source: Cellular Telecommunications Industry Association.)

Year	Cell sites (in thousands)	Subscribers (in millions)
1995	22.7	33.8
1996	30.0	44.0
1997	51.6	55.3
1998	65.9	69.2

The number of cell sites C (in thousands) can be written as a function of the number of subscribers s (in millions):

$$C(s) = 1.3s - 23$$

(a) What kind of function is C? Visualize its graph to help you decide whether C is a one-to-one function.

(b) Solve $C(s)$ for s. What does the resulting function represent?

82. Consider the relation R represented by the following mapping diagram.

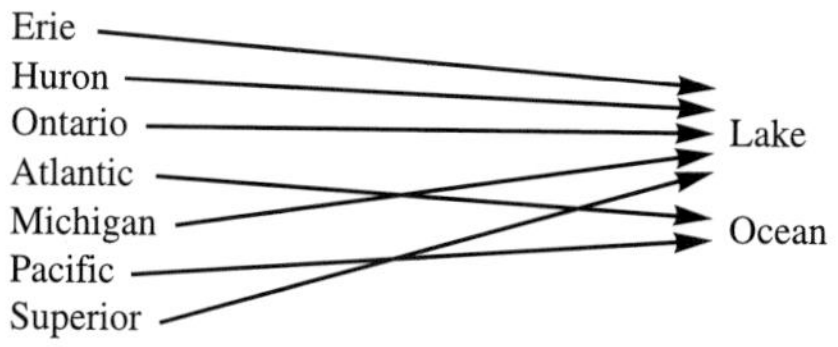

(a) Is R a function? Why or why not?

(b) If R is a function, does the function have an inverse? Why or why not?

Data Analysis: *World Population*

The bar graph shows the years in which the world population is predicted to reach certain milestones. (Source: United Nations Population Division.)

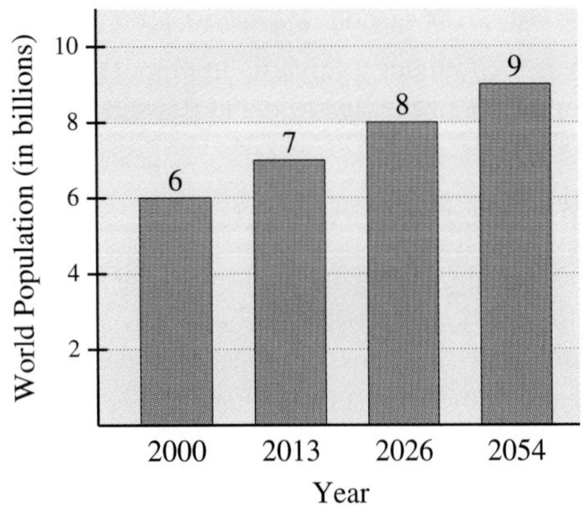

83. Use the data for 2000 and 2054 to write a linear function $y(p)$ to model the number of years y after 2000 as a function of the population p (in millions).

84. Use the data for 2000 and 2054 to write a linear function $p(y)$ to model the population p (in millions) as a function of the number of years after 2000.

85. On the basis of the definitions of y and p in Exercises 83 and 84, what appears to be the relationship between the two functions?

86. Determine $y \circ p$. Does $y \circ p$ lead to the same conclusion as in Exercise 85?

Challenge

In Exercises 87 and 88, sketch the graph of the function. Then sketch the graph of the inverse function.

87. $f(x) = x^2 - 4x;\ x \leq 2$

88. $f(x) = x^2 - 4x + 3;\ x \leq 2$

In Exercises 89–92, determine f^{-1} and verify that $(f \circ f^{-1})(x) = x$ for all x in the domain of f^{-1} and that $(f^{-1} \circ f)(x) = x$ for all x in the domain of f.

89. $f(x) = x^2 + 1;\ x \geq 0$

90. $f(x) = |x - 4|;\ x \leq 4$

91. $f(x) = -|x + 5|;\ x \leq -5$

92. $f(x) = x^2 - 6x + 5;\ x \leq 3$

11.3 Exponential Functions

Definition • Natural Base *e* • Graphs of Exponential Functions • Exponential Equations • Real-Life Applications

Definition

In the familiar function $g(x) = x^2$, the base is a variable, and the exponent is a constant. In this section, we consider functions such as $f(x) = 2^x$ in which the base is a constant and the exponent is a variable.

Informally, you can illustrate how to define 2^x for irrational numbers by using a calculator to evaluate 2^π by successively evaluating $2^3, 2^{3.1}, 2^{3.14}, 2^{3.142}$, etc.

Having defined rational exponents, we know that the expression 2^x is defined for all rational numbers. Although the details are beyond the scope of this book, the expression 2^x can also be defined for any irrational number. Thus the function $f(x) = 2^x$ is defined for all real numbers.

Definition of an Exponential Function

An **exponential function** can be written as $f(x) = b^x$, where $b > 0$, $b \neq 1$, and x is any real number.

The two restrictions on b in the definition are important. First, the definition excludes $b = 1$ because 1^x has a value of 1 for all values of x and the function would simply be the constant function $f(x) = 1$.

The definition also requires b to be positive so that the function can be defined for all real numbers. For example, if $b = -4$ and $x = \frac{1}{2}$, then evaluating the function would result in the expression $(-4)^{1/2}$, which is not a real number.

EXAMPLE 1

Evaluating an Exponential Expression

(a) If $f(x) = 3^x$, then $f(2) = 3^2 = 9$.

(b) If $f(x) = -5^x$, then $f(2) = -5^2 = -1 \cdot 5^2 = -25$.

(c) If $g(x) = 2^x$, then $g(-4) = 2^{-4} = \dfrac{1}{2^4} = \dfrac{1}{16}$.

(d) If $g(x) = \left(\dfrac{1}{4}\right)^x$, then $g(-2) = \left(\dfrac{1}{4}\right)^{-2} = 4^2 = 16$.

(e) If $h(x) = 9^x$, then $h\left(\dfrac{1}{2}\right) = 9^{1/2} = \sqrt{9} = 3$.

Note: In Example 1(b), $f(x) = -5^x$ is an exponential function. However, $(-5)^x$ cannot be used to define an exponential function because the base is negative.

Natural Base *e*

Stress that *e* is not a variable. Performing calculations, as in Example 2, should help students who have trouble thinking of *e* as representing a specific number.

An irrational number approximately equal to 2.718281828 occurs frequently in applications in science, engineering, business, and other areas. Because it occurs so frequently, the letter e is used to represent it, just as the Greek letter π is used to represent the irrational number approximately equal to 3.14159.

The number e is often the base in an exponential function.

$$f(x) = e^x$$

This function is called the **natural exponential function,** but more often it is called simply *the* exponential function.

Note: Most scientific calculators have a key for evaluating the natural exponential function.

EXAMPLE 2

Evaluating e^x

Let $f(x) = e^x$. Evaluate the function at the indicated values. Round results to three decimal places.

(a) $f(1)$ (b) $f(-1)$ (c) $f(2)$

Solution

(a) $f(1) = e^1 = 2.718$

(b) $f(-1) = e^{-1} = 0.368$

(c) $f(2) = e^2 = 7.389$

Graphs of Exponential Functions

As with other functions, we graph exponential functions by determining several pairs that belong to the function and plotting the corresponding points. Drawing a smooth curve through these points gives the graph.

EXAMPLE 3

Graphing an Exponential Function

Sketch the graph of the given function.

(a) $f(x) = 2^x$ (b) $g(x) = 2^{-x}$

Solution

(a) The table shows the result of choosing several values for x and evaluating the function. Figure 11.10 shows the plotted points and the graph of the function.

x	-3	-2	-1	0	1	2	3
2^x	$\frac{1}{8}$	$\frac{1}{4}$	$\frac{1}{2}$	1	2	4	8

Figure 11.10

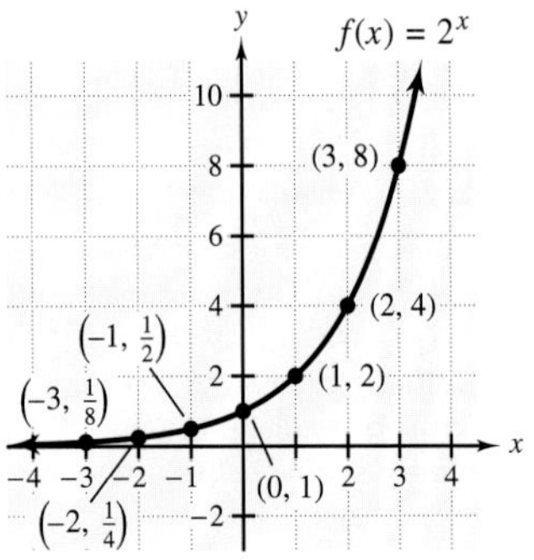

(b) The table shows several pairs that belong to the function. Figure 11.11 shows the graph of the function.

x	-3	-2	-1	0	1	2	3
2^{-x}	8	4	2	1	$\frac{1}{2}$	$\frac{1}{4}$	$\frac{1}{8}$

Figure 11.11

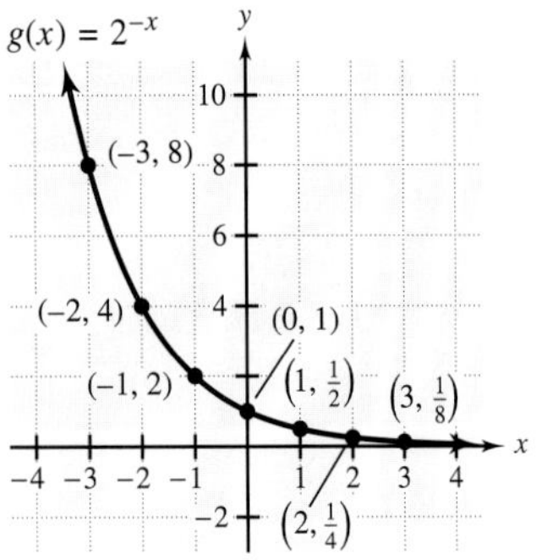

Exploring the Concept

Graphs of Exponential Functions

To investigate the behavior of exponential functions, we consider their graphs for $0 < b < 1$ and for $b > 1$.

Students may wonder about these choices of values for b. The definition of an exponential function states that $b > 0$ and $b \neq 1$. Therefore, $0 < b < 1$ or $b > 1$.

Figure 11.12

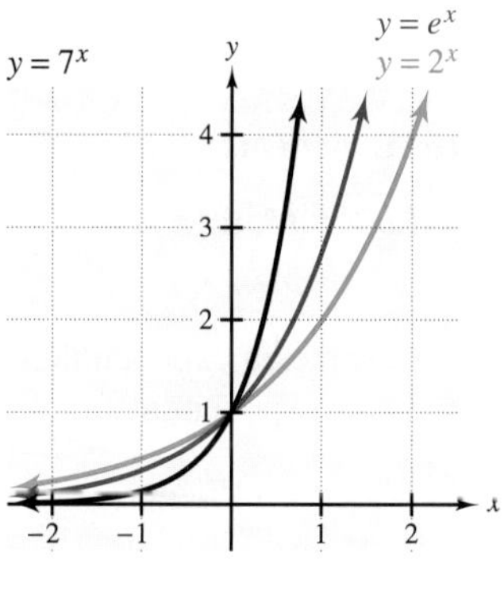

Figure 11.13

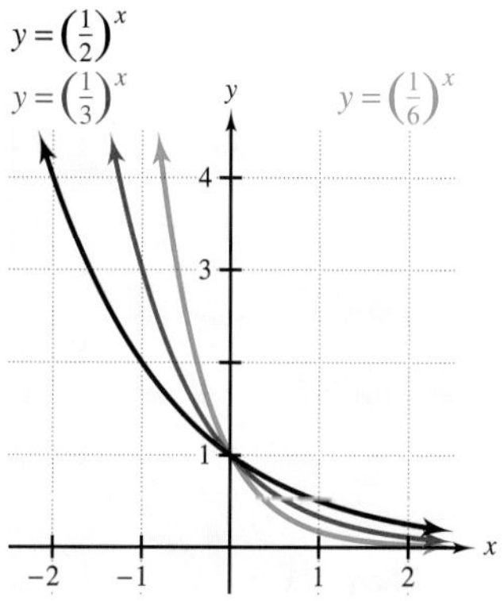

Note that $b > 1$ for each function in Figure 11.12 and that the graphs rise from left to right. As the base increases, the graphs are steeper. In Figure 11.13, $0 < b < 1$ for each function, and the graphs fall from left to right. As the base decreases, the graphs are steeper.

Although the graphs may be quite steep, they do extend forever to the left and right. Thus the domain of each of the functions is **R**. Because $b > 0$, b^x can never be 0. Therefore, the graphs approach but do not reach the x-axis. The range of each of the functions is $\{y \mid y > 0\}$.

Because $b^0 = 1$, the y-intercept of each graph is (0, 1). Applying the Horizontal Line Test, we see that all the functions are one-to-one functions.

LEARNING TIP

Being familiar with these properties will be helpful in graphing exponential functions. For instance, the graph of $y = 2^x + 3$ is simply the graph of $y = 2^x$ shifted 3 units upward.

Here is a summary of our findings about exponential functions.

Properties of Exponential Functions and Their Graphs

Let $f(x) = b^x$, $b > 0$ and $b \neq 1$.

1. The y-intercept is (0, 1); there is no x-intercept.
2. The function f is a one-to-one function.
3. The domain is the set of all real numbers; the range is $\{y \mid y > 0\}$.
4. The graph approaches but does not reach the x-axis.
5. For $b > 1$, the graph rises from left to right; for $0 < b < 1$, the graph falls from left to right.

Think About It

Describe the graph of the exponential function $f(x) = -b^x$? What are the domain, the range, and the intercepts?

Exponential Equations

Equations that involve a variable in the exponent are called **exponential equations.** The following are some examples of exponential equations.

$$2^x = 8 \qquad 2^{3x} = 16 \qquad 27^x = 81 \qquad \left(\frac{1}{3}\right)^x = 9$$

To solve exponential equations algebraically requires new techniques.

We observed that an exponential function is a one-to-one function. Therefore, if $b^m = b^n$, then m must equal n.

Sometimes students think they are "canceling" or dividing out the number b. Explain that we can equate the exponents because the functions are one-to-one.

Property of Equivalent Exponents

For $b > 0$ and $b \neq 1$, if $b^m = b^n$, then $m = n$.

We can use this property to solve an exponential equation. If we can write each side of an exponential equation as an exponential function with the same base, then we can equate the exponents.

EXAMPLE 4

Equating Exponents to Solve an Exponential Equation

Solve the equation $4^x = 8$.

Solution

$$4^x = 8$$

$$(2^2)^x = 2^3 \quad \text{Write each side with a base of 2.}$$

$$2^{2x} = 2^3 \quad \text{Power to a Power Rule}$$

$$2x = 3 \quad \text{Equate the exponents.}$$

$$x = \frac{3}{2}$$

To do this, we need logarithms. Equations such as $3^x = 2$ are discussed in a later section.

As we will see in the next section, it is always possible to write exponential expressions with the same base. However, the method of equating exponents is recommended only for those special cases in which changing the base is easy to do.

EXAMPLE 5

Solving Exponential Equations

Solve each exponential equation.

(a) $4^{2x+1} = 32$

(b) $\left(\frac{2}{3}\right)^{1-x} = \left(\frac{9}{4}\right)^{2x}$

(c) $e^{x^2-1} = 1$

Solution

(a)
$$4^{2x+1} = 32$$

$$(2^2)^{2x+1} = 2^5 \quad \text{Write both sides with base 2.}$$

$$2^{4x+2} = 2^5 \quad \text{Power to a Power Rule}$$

$$4x + 2 = 5 \quad \text{Equate exponents.}$$

$$4x = 3 \quad \text{Solve for } x.$$

$$x = \frac{3}{4}$$

(b) $\left(\frac{2}{3}\right)^{1-x} = \left(\frac{9}{4}\right)^{2x}$

$\left(\frac{2}{3}\right)^{1-x} = \left[\left(\frac{3}{2}\right)^{2}\right]^{2x}$ Write both sides with base $\frac{2}{3}$.

$\left(\frac{2}{3}\right)^{1-x} = \left[\left(\frac{2}{3}\right)^{-2}\right]^{2x}$ $\left(\frac{a}{b}\right)^{-n} = \left(\frac{b}{a}\right)^{n}$

$\left(\frac{2}{3}\right)^{1-x} = \left(\frac{2}{3}\right)^{-4x}$ Power to a Power Rule

$1 - x = -4x$ Equate exponents.

$1 = 3x$ Solve for x.

$x = -\frac{1}{3}$

(c) $e^{x^2 - 1} = 1$

$e^{x^2 - 1} = e^0$ Write both sides with base e.

$x^2 - 1 = 0$ Equate exponents.

$(x + 1)(x - 1) = 0$ Factor.

$x + 1 = 0$ or $x - 1 = 0$ Zero Factor Property

$x = -1$ or $x = 1$

Note that we would not be able to use this procedure to solve the equation $e^x = 2$ because we presently have no way of expressing 2 as a power of e. The techniques for solving this type of equation will be introduced later.

Real-Life Applications

In this section and in the remaining sections of the chapter, a number of formulas are used for application problems. You may want to decide which, if any, of these formulas you want students to know.

Exponential equations play a major role in describing the conditions of certain application problems. Sometimes the exponential equation is a formula that has been derived to relate the variables of the problem.

For example, the formula to compute the value A of an investment of P dollars at an annual interest rate r compounded n times per year for t years is

$$A = P\left(1 + \frac{r}{n}\right)^{nt}$$

EXAMPLE 6

Compound Interest

A certain credit card company charges 14% interest compounded monthly. If a person charges \$1500 on the card and plans to wait 1.5 years to pay the bill, how much interest will be due?

Solution

The interest is compounded 12 times per year during the 1.5 years.

$P = 1500 \qquad r = 0.14 \qquad n = 12 \qquad t = 1.5$

$$A = P\left(1 + \frac{r}{n}\right)^{nt}$$ Compound interest formula

$$A = 1500\left(1 + \frac{0.14}{12}\right)^{12(1.5)}$$ Substitute for $P, r, n,$ and t.

$$A \approx 1848.27$$

At the end of 1.5 years, the balance is approximately \$1848.27. Thus the amount of interest will be \$348.27.

Quick Reference 11.3

Definition

- An **exponential function** can be written $f(x) = b^x$ where $b > 0$, $b \neq 1$, and x is any real number.

Natural Base e

- The irrational number 2.71828 . . . is represented by the letter e.
- The function $f(x) = e^x$ is called the **natural exponential function.**

Graphs of Exponential Functions

- The following are properties of the graph of an exponential function $f(x) = b^x$, $b > 0$ and $b \neq 1$.
  1. The y-intercept is (0, 1); there is no x-intercept.
  2. The function f is a one-to-one function.
  3. The domain is the set of all real numbers; the range is $\{y \mid y > 0\}$.
  4. The graph approaches but does not reach the x-axis.
  5. For $b > 1$, the graph rises from left to right; for $0 < b < 1$, the graph falls from left to right.

Exponential Equations

- An equation in which the variable appears in the exponent is called an **exponential equation.**
- Because an exponential function is a one-to-one function, if $b^m = b^n$, then $m = n$. This rule can be used to solve an exponential equation as follows.
  1. Write each side of the equation as a single exponential expression with the same base.
  2. Use the rules of exponents to simplify both sides.
  3. Equate the exponents.
  4. Solve the resulting equation.

Speaking the Language 11.3

1. The irrational number 2.71828 ... is represented by the symbol ______.

2. The function $f(x) = b^x$, $b > 0$ and $b \neq 1$, is called a(n) ______ function.

3. The graph of $f(x) = b^x$, $b > 0$ and $b \neq 1$, has no ________ -intercept, and the domain of the function is ________ .

4. If $b^m = b^n$, where m and n are variable expressions, then we can solve the equation by ________ the exponents.

Exercises 11.3

Concepts and Skills

1. Explain the conditions under which the graph of $f(x) = b^x$
   (a) rises from left to right.
   (b) falls from left to right.

2. If $f(x) = (1/b)^x$ with $b > 1$, what is the effect on the graph of f as b increases?

In Exercises 3–10, let $f(x) = 2^{3x-1}$ and $g(x) = \left(\frac{1}{3}\right)^{2x}$. Evaluate each of the following.

For Exercises 3–18, use a calculator when needed.

3. $f(2)$
4. $g(-1)$
5. $g(1)$
6. $f(-1)$
7. $g(0)$
8. $f(0)$
9. $f\left(\frac{2}{3}\right)$
10. $g\left(-\frac{3}{2}\right)$

In Exercises 11–18, let $g(x) = e^{-2x}$ and $h(x) = \left(\frac{1}{5}\right)^x$. Determine each of the following values to the nearest hundredth.

11. $g(2)$
12. $h(-2)$
13. $g(-3)$
14. $h(0)$
15. $h\left(-\frac{3}{2}\right)$
16. $h\left(\frac{1}{2}\right)$
17. $g(-2)$
18. $g(1)$

In Exercises 19–26, evaluate the given function as indicated. Round results to the nearest hundredth.

19. $f(x) = 4e^{3x}$, $f(2)$
20. $g(x) = 100e^{-2x}$, $g(3)$
21. $g(x) = \left(\sqrt{3}\right)^{2x}$, $g(5)$
22. $f(x) = \left(\sqrt{5}\right)^{-3x}$, $f(-3)$
23. $f(x) = (0.23)^{3x-1}$, $f(-2)$
24. $g(x) = (1.45)^{1-2x}$, $g(3)$
25. $g(x) = -3e^{2x} + 4$, $g(-2)$
26. $f(x) = 5 + 3e^{-x+1}$, $f(5)$

27. From the following list, identify the functions that have the same graph and explain why.
    (i) $f(x) = 3^x$ (ii) $g(x) = 3^{-x}$
    (iii) $h(x) = \left(\frac{1}{3}\right)^x$ (iv) $s(x) = \frac{1}{3^x}$

28. If $f(x) = b^x$ with $b > 1$, what is the effect on the graph of f as b increases?

29. Sketch the graphs of the following three functions.
    $f(x) = 3^x$
    $g(x) = 3^x + 2$
    $h(x) = 3^x - 4$
    In general, compare the graphs of b^x and $b^x + c$ when $c > 0$ and when $c < 0$.

30. Sketch the graphs of the following three functions.
    $f(x) = 3^x$
    $g(x) = 3^{x+2}$
    $h(x) = 3^{x-4}$
    In general, compare the graphs of b^x and b^{x+k} when $k > 0$ and when $k < 0$.

In Exercises 31 and 32, compare the graphs of the given pair of functions to the graph of $h(x) = 2^x$.

31. (a) $f(x) = 2^x + 1$ (b) $g(x) = 2^x - 3$
32. (a) $f(x) = 2^{x+3}$ (b) $g(x) = 2^{x-4}$

Exercises 31–36 help students understand how graphs of exponential functions are shifted by the addition of constants and by changes in the value of the exponent.

In Exercises 33 and 34, compare the graphs of the given pair of functions to the graph of $h(x) = \left(\frac{1}{3}\right)^x$.

33. (a) $f(x) = \left(\frac{1}{3}\right)^x + 3$ (b) $g(x) = \left(\frac{1}{3}\right)^{x+3}$
34. (a) $f(x) = \left(\frac{1}{3}\right)^{-x}$ (b) $g(x) = 3^{-x}$

CE The following figure shows the graph of $f(x) = b^x$. In Exercises 35 and 36, use this graph to assist you in matching the given functions to graph (A), (B), (C), or (D).

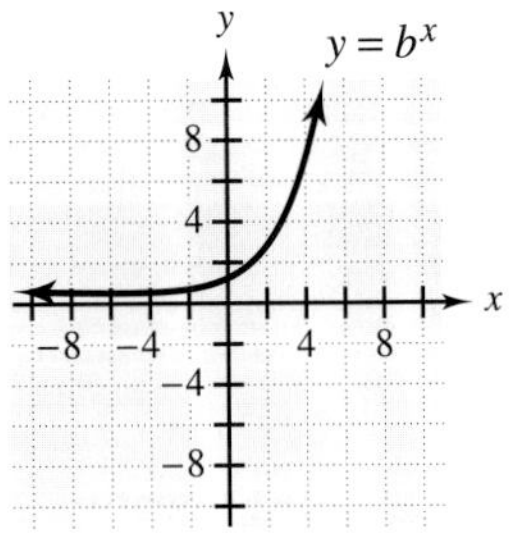

(A)

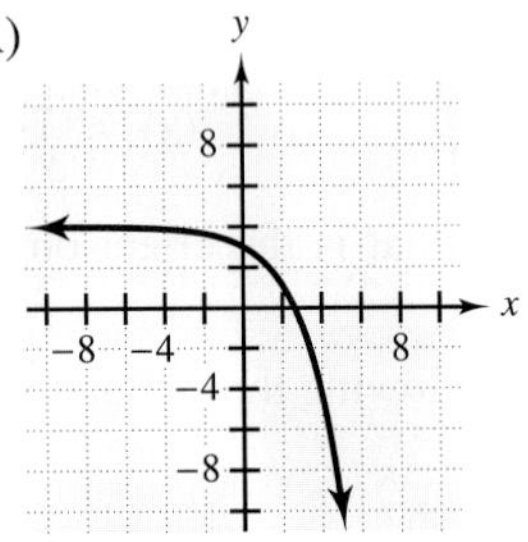

(B)

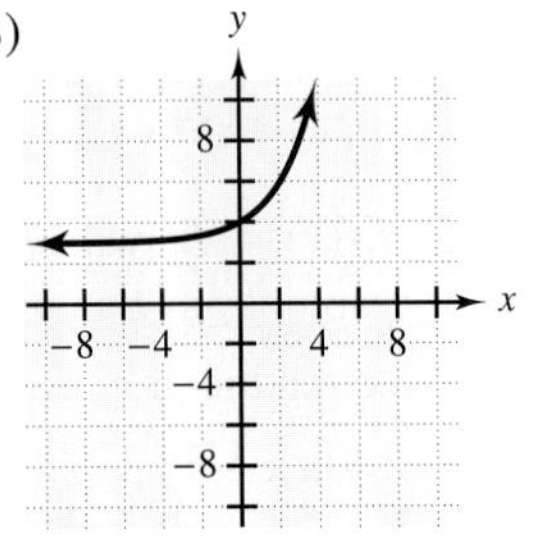

(C)

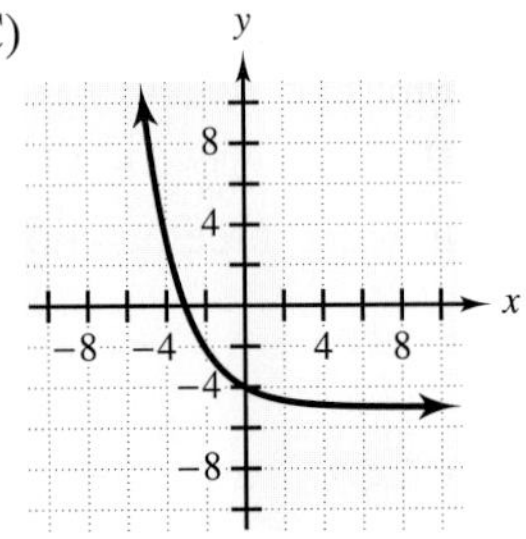

(D)

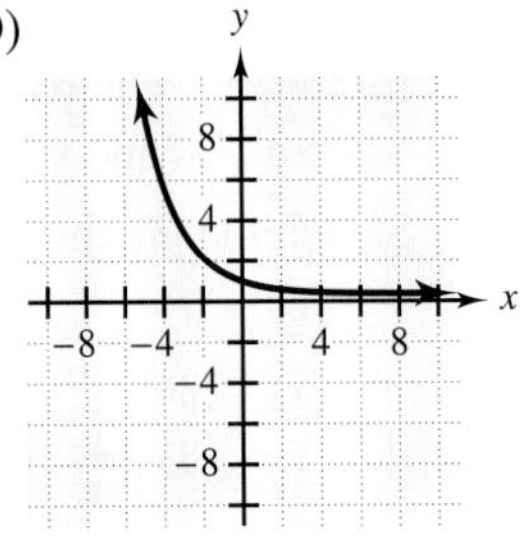

35. (a) $g(x) = b^x + 3$ (b) $h(x) = 4 - b^x$

36. (a) $gx) = b^{-x}$ (b) $h(x) = b^{-x} - 5$

In Exercises 37–48, sketch the graph of the function.

37. $f(x) = 5^x$

38. $g(x) = 4^{-x}$

39. $h(x) = \left(\frac{1}{6}\right)^x$

40. $g(x) = \left(\frac{2}{3}\right)^x$

41. $f(x) = -6^{-x}$

42. $h(x) = -2^x$

43. $f(x) = 2^{x-2}$

44. $h(x) = 3^x + 1$

45. $g(x) = -3^x - 2$

46. $f(x) = 1 - 4^{1-x}$

47. $h(x) = 7^{|x|}$

48. $g(x) = 3^{x^2}$

49. Explain why we cannot use the same method to solve $4^x = 12$ as we use to solve $4^x = 8$.

50. Explain why the graphs of $f(x) = 2^{2x}$ and $g(x) = 4^x$ are the same.

In Exercises 51–58, solve the exponential equation.

51. $5^{-x} = 125$

52. $2^{x-1} = 16$

53. $\left(\frac{2}{5}\right)^x = \frac{25}{4}$

54. $\left(\frac{3}{2}\right)^x = \frac{8}{27}$

55. $e^{3x+1} = e^{x-2}$

56. $e^{x+1} = e^{2x}$

57. $\frac{1}{9^x} = 27$

58. $25^x = \frac{1}{5}$

In Exercises 59–82, solve the exponential equation.

59. $49^x = 343$

60. $9^{-2x} = 27$

61. $16 = 8^{x-2}$

62. $e^{x+4} = \frac{1}{e^{3x}}$

63. $3 \cdot 8^x = 12$

64. $-3^x = -9$

65. $9 \cdot 3^x = \frac{1}{3}$

66. $4^x \cdot 16^{2-3x} = 8$

67. $2^{-x-3} = -4$

68. $-4^{2x} = 2$

69. $\frac{5^{x+1}}{5^{1-x}} = \frac{1}{25^x}$

70. $\frac{3^{x+1}}{27^x} = \frac{1}{9^x}$

71. $e^x + 4 = 3$

72. $7 - e^{3-2x} = 6$

All problems in this group can be solved by equating exponents.

73. $\left(\frac{1}{3}\right)^x 27^{x+1} = 9^{3-2x}$

74. $\left(\frac{1}{3}\right)^{x^2} = 27^{(2/3)x-1}$

75. $32^{1-x} = 8^{x^2-x}$

76. $125^{2-3x} = 25^{3x-5}$

77. $(0.25)^{2x+1} = 8^{2-x}$

78. $(0.2)^{x-3} = 25^{1-x}$

79. $e^{x^2-1} = 7^0$

80. $(e^{x+2})^x = (e^{x+2})^2$

81. $\left(\sqrt{3}\right)^{2-2x} = 9^{x^2}$

82. $\left(\sqrt{2}\right)^{4-2x} = \left(\sqrt[3]{2}\right)^{6x}$

For Exercises 83–86, remind students about the formula preceding Example 6.

Real-Life Applications

83. A couple plans to invest \$1000 each year to save for a vacation home. They find two banks that pay the same interest rate of 6%, but one bank pays only simple interest, whereas at the other bank, the interest is compounded daily (365 days). What would be the difference in the amount of interest they would earn at the two banks after the first year?

84. Each of two sisters received a gift of \$2000 from their aunt with the stipulation that they invest the money for at least 1 year. The older sister invested the money in an account that paid 7% compounded quarterly, and the younger sister invested her money in another bank that paid 6% compounded daily (365 days). What is the difference in the amount of interest that the two sisters earned?

85. When a company downsized 2 years ago, an employee received a lump-sum severance pay, which he invested in a growth fund that yielded an annual rate of 9% in dividends, compounded quarterly. The account is now worth \$5974.16. How much severance pay did the person receive?

86. At the beginning of the year, an employee invested in the company savings plan as well as in the company retirement plan. The savings plan paid 11% compounded daily (365 days), and the retirement fund paid 7% compounded quarterly. At the end of 1 year, the savings plan had a balance of \$3348.79 and the retirement account balance was \$3322.76. What was the total original investment in both plans?

Modeling with Real Data

87. In 1997, the average annual per-capita expenditure for books was \$98.69. (Source: Book Industry Study Group.) Suppose that the amount was expected to increase by 6% annually for the next 5 years.

(a) Complete the accompanying table by writing exponential expressions for the expenditures in each year through 2002.

Year	n	Expenditures
1997	0	98.69
1998	1	$98.69(1.06)^1$
1999	2	$98.69(1.06)^2$
2000	3	
2001	4	
2002	5	

(b) Write an exponential function that models the expenditure E in year n, where n is the number of years since 1997.

(c) What expenditure does the model project for 2002? For 2003?

88. The number of households using personal computers to conduct their banking is projected to increase exponentially. (Source: Home Banking Report.)

Year	Households
1995	754,000
1996	1,884,000
1997	3,297,000
1998	5,769,000
1999	8,653,000
2000	12,980,000

The exponential function $B(x) = 0.95e^{0.55x}$ models the number of households B (in millions) subscribing to online banking, where x is the number of years since 1995.

(a) According to the model, how many households are projected to use the online banking service in 2002?

(b) Estimate the year in which about 26 million households are expected to have online banking services. Evaluate function B to confirm your estimate.

Data Analysis: *Cost of Medicare*

Not until the mid-1990s was any serious effort made to curtail the cost of health care in America. The accompanying figure shows the rapid increase in federal spending for Medicare during the period 1980–2005. (Source: U.S. Health Care Financing Administration.)

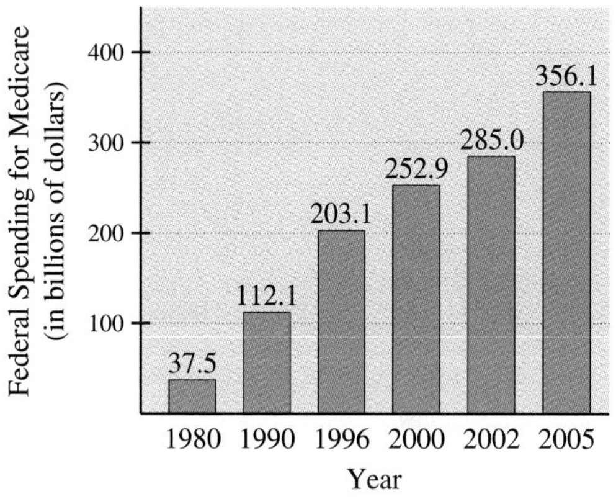

The data can be modeled by $M(t) = 41.7(1.09)^t$, where M is the federal spending (in billions) for Medicare, and t is the number of years since 1980.

89. By looking at the model function, how do you know that the model predicts a rise in costs?

90. Use the model function to estimate the cost of Medicare in the year 2007.

91. On the basis of the model function, what is the expected percentage increase in cost from 2007 to 2017?

92. People enrolled in Medicare must pay a monthly premium. Assuming a Medicare enrollment of 40,000,000 in 2005, what monthly premium would be required to cover the entire projected cost of the program in that year?

Challenge

In Exercises 93 and 94, sketch the graphs of the left and right sides of the given equation. Then use the graphs to estimate the solution(s) of the equation.

93. $x^2 = 2^x$

94. $x^3 = 3^x$

In Exercises 95–98, sketch the graphs of the left and right sides of the given inequality. Then use the graphs to estimate the solution(s) of the inequality.

95. $2^x < 3^x$

96. $e^x < e^{-x}$

97. $2^x > x$

98. $\left(\frac{1}{3}\right)^x \le -x$

In Exercises 99 and 100, use a graph of the given function to estimate the range of the function.

99. $f(x) = 2 - 3^{-x}$

100. $g(x) = e^{|x|}$

11.4 Logarithmic Functions

Inverses of Exponential Functions • Evaluation of Logarithms • Logarithmic Equations • Graphs of Logarithmic Functions

Inverses of Exponential Functions

In the preceding section, we learned that the exponential function $f(x) = b^x$ is a one-to-one function. This means that an exponential function has an inverse.

Because the graph of a function and its inverse are symmetric with respect to the line $y = x$, we can sketch the graph of the inverse of a function by drawing the mirror image of the function on the other side of the line of symmetry.

Exploring the Concept

In this illustration, we learn much about the graph of the logarithmic function even before we define what the logarithmic function is.

Inverses of Exponential Functions

Consider the one-to-one functions $f(x) = 2^x$ and $g(x) = \left(\frac{1}{2}\right)^x$.

Figure 11.14 shows the graphs of f and f^{-1}, and Figure 11.15 shows the graphs of g and g^{-1}.

Figure 11.14

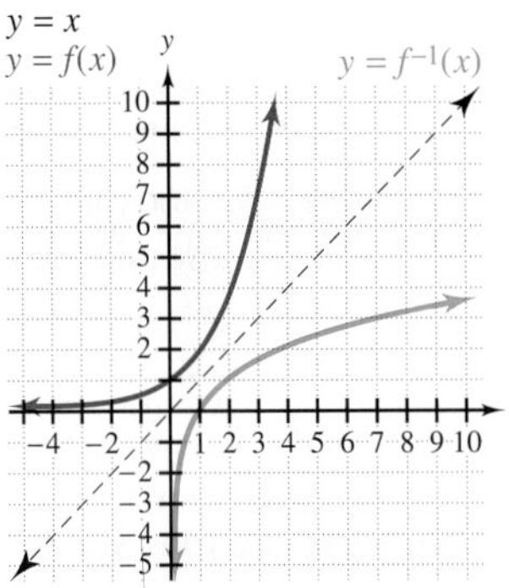

Figure 11.15

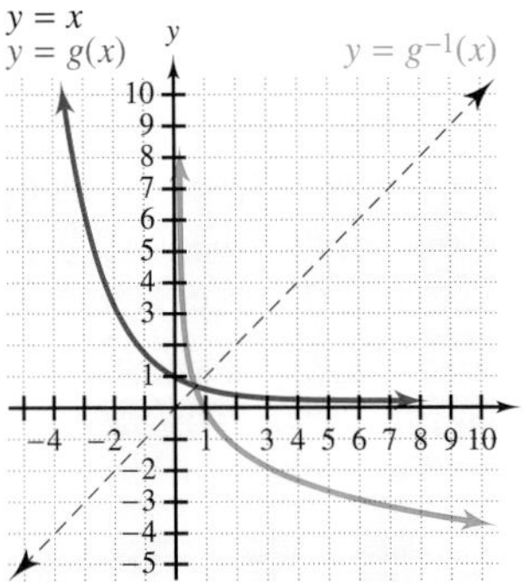

Observe that the x-intercept for the graphs of both inverse functions is (1, 0). The graphs approach but never reach the y-axis. Therefore, there is no y-intercept.

Because the graphs of the inverse functions lie entirely to the right of the y-axis, the domain of each is $\{x \mid x > 0\}$. The range of each inverse function is **R**.

These observations are true for any exponential function $y = b^x$. Each exponential function has an inverse. We can sketch the graph of the inverse, and we can discover properties of the inverse. However, unlike the functions that we studied in Section 11.2, we cannot interchange the variables x and y and solve for y to obtain an algebraic description of the function. We simply call the inverse function the **logarithmic function with base *b*.**

Definition of Logarithmic Function

For $x > 0$, $b > 0$, and $b \neq 1$, the **logarithm of *x* with base *b*,** represented by $\log_b x$, is defined by the following:

$$\log_b x = y \text{ if and only if } x = b^y.$$

The function $g(x) = \log_b x$ is called the **logarithmic function with base *b*** and is the inverse of the exponential function $f(x) = b^x$.

For $f(x) = b^x$, rather than use the name f^{-1} for the inverse of f, we use the name $\log_b$.

The notation $\log_b x$ is often read more simply as the "log base b of x." The number x is called the **argument** of the logarithm.

Natural and Common Logarithms

The logarithmic function with base e is called the **natural logarithmic function** and is written $y = \ln x$ rather than $y = \log_e x$.

The logarithmic function with base 10 is called the **common logarithmic function** and is written $y = \log x$ rather than $y = \log_{10} x$.

Evaluation of Logarithms

Expressions can be changed from exponential to logarithmic form or from logarithmic to exponential form.

Exponential Form	*Logarithmic Form*
$x = b^y$	$\log_b x = y$
base: b; exponent: y	base: b; exponent: y

Observe that if $y = \log_b x$, then, in the exponential form, $x = b^y = b^{\log_b x}$. Thus $\log_b x$ is an exponent.

EXAMPLE 1

Exponential Form to Logarithmic Form

Write in logarithmic form.

The exponential form and the logarithmic form are two different ways of expressing the same relationship among three numbers.

Exponential Form	*Logarithmic Form*
(a) $3^4 = 81$	$\log_3 81 = 4$
(b) $5^{-2} = 0.04$	$\log_5 0.04 = -2$
(c) $10^3 = 1000$	$\log 1000 = 3$

Recall that log 1000 means $\log_{10} 1000$.

EXAMPLE 2

Logarithmic Form to Exponential Form

Write in exponential form.

Logarithmic Form	*Exponential Form*	
(a) $\log_4 2 = \frac{1}{2}$	$4^{1/2} = 2$	
(b) $\ln e = 1$	$e^1 = e$	Recall that ln *e* means $\log_e e$.
(c) $\log_5 \frac{1}{5} = -1$	$5^{-1} = \frac{1}{5}$	

One method for evaluating a logarithmic expression begins with writing the expression in exponential form.

EXAMPLE 3

Evaluating Logarithmic Expressions

Evaluate the logarithms.

(a) $\log_6 6$ (b) $\log_{1/4} 2$ (c) $\log_5 125$

Solution

The method in each part of this example works because we can easily write both sides of the equation with the same base.

(a) Let $\log_6 6 = y$.

$6^y = 6$ Write in exponential form.

$6^y = 6^1$

$y = 1$ Equate exponents.

Therefore, $\log_6 6 = 1$.

(b) Let $\log_{1/4} 2 = y$.

$\left(\frac{1}{4}\right)^y = 2$ Write in exponential form.

$\left(\frac{1}{2^2}\right)^y = 2$ Write both sides with the same base.

$(2^{-2})^y = 2$ Definition of negative exponent

$2^{-2y} = 2^1$ Power to a Power Rule

$-2y = 1$ Equate exponents.

$y = -\frac{1}{2}$

Therefore, $\log_{1/4} 2 = -\frac{1}{2}$.

(c) Let $\log_5 125 = y$.

$5^y = 125$ Write in exponential form.

$5^y = 5^3$ Write both sides with the same base.

$y = 3$ Equate exponents.

Therefore, $\log_5 125 = 3$.

Note: Most scientific calculators have keys for both common logarithms and natural logarithms.

EXAMPLE 4

Evaluating Logarithms with a Calculator

Evaluate the following logarithms. Round results to two decimal places.

(a) $\ln 8$ (b) $\log 20$

Solution

(a) $\ln 8 \approx 2.08$

(b) $\log 20 \approx 1.30$

Example 4 shows how easy it is to use a calculator to evaluate common logarithms and natural logarithms. However, most calculators do not have keys for evaluating logarithms with bases other than 10 or e.

It can be shown that a logarithm with any base b can be evaluated in terms of logarithms with any other base a.

LEARNING TIP

Noting that x is above b in the expression $\log_b x$ might help you to remember that x is in the numerator and b is in the denominator in the Change of Base Formula.

Change of Base Formula

For $x > 0$ and for any positive bases a and b where $a \neq 1$ and $b \neq 1$,

$$\log_b x = \frac{\log_a x}{\log_a b}$$

One use of this formula is for evaluating a logarithm with any base in terms of natural or common logarithms.

EXAMPLE 5

Evaluating Logarithms with Bases Other Than 10 or *e*

We can use either natural logarithms or common logarithms to change the base.

(a) $\log_3 5 = \frac{\log 5}{\log 3} \approx 1.46$ Use $\log_b x = \frac{\log_a x}{\log_a b}$ with $a = 10$.

(b) $\log_{2.3} 4.72 = \frac{\ln 4.72}{\ln 2.3} \approx 1.86$ Use $a = e$.

Logarithmic Equations

We can often solve equations involving logarithms by converting logarithmic forms to exponential forms.

EXAMPLE 6

Solving Logarithmic Equations

Solve each equation for x.

(a) $\log_x 27 = 3$ (b) $\log_{2/3} x = 2$

(c) $\log x = -1$ (d) $\ln x = 0$

Solution

(a) $\log_x 27 = 3$

$x^3 = 27$ Write in exponential form.

$x = 3$ Odd Root Property

(b) $\log_{2/3} x = 2$

$\left(\frac{2}{3}\right)^2 = x$ Write in exponential form.

$x = \frac{4}{9}$

(c) $\log x = -1$

$10^{-1} = x$ Common logarithm base is 10.

$x = \frac{1}{10}$

(d) $\ln x = 0$

$e^0 = x$ Natural logarithm base is *e*.

$x = 1$

Graphs of Logarithmic Functions

We can learn more about the behavior of logarithmic functions by examining their graphs.

EXAMPLE 7

Graphs of Logarithmic Functions

Sketch the graphs of the following logarithmic functions.

(a) $f(x) = \log_3 x$ (b) $g(x) = \log_{1/3} x$

Solution

For each function, create a table of values and plot the points. Each part shows a typical table of values along with the graph of the function.

(a)

x	$\frac{1}{9}$	$\frac{1}{3}$	1	3	9	27
$\log_3 x$	−2	−1	0	1	2	3

Figure 11.16

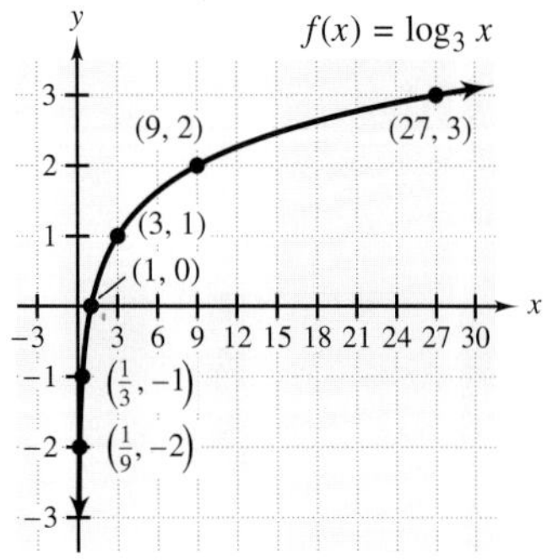

Think About It

Show how to evaluate the following expressions without a calculator.

(a) $(\log_2 3)(\log_3 4)(\log_4 5)(\log_5 6)(\log_6 7)(\log_7 8)$

(b) $(\log_2 4)(\log_4 6)(\log_6 8)$

Compare the functions in parts (a) and (b) by using the Change of Base Formula to write the functions in terms of $\log x$:

$f(x) = \dfrac{\log x}{\log 3}$ and $g(x) = \dfrac{\log x}{\log \frac{1}{3}}$.

Note that $\log 3$ is positive and $\log \frac{1}{3}$ is negative. That is why the graphs are flipped over.

(b)

x	$\frac{1}{9}$	$\frac{1}{3}$	1	3	9	27
$\log_{1/3} x$	2	1	0	−1	−2	−3

Figure 11.17

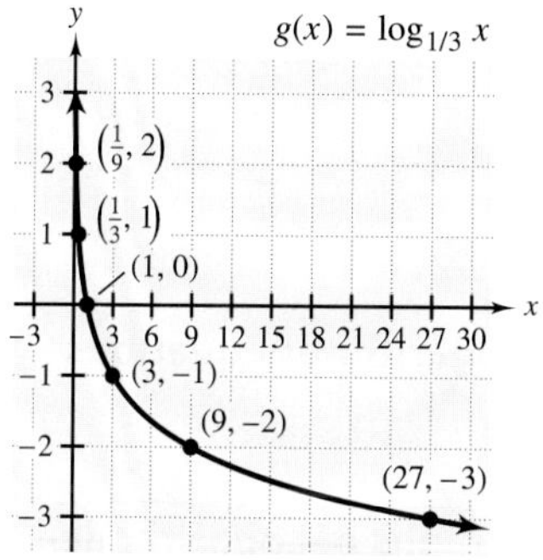

Observe that in part (a) of Example 7, $b > 1$ and the graph rises from left to right. In part (b) of Example 7, $0 < b < 1$ and the graph falls from left to right.

Here is a summary of properties of $f(x) = \log_b x$.

Properties of $f(x) = \log_b x$ and Its Graph

1. The domain of f is $\{x \mid x > 0\}$.
2. The range of f is the set of all real numbers.
3. The x-intercept is $(1, 0)$.
4. The graph approaches but does not reach the y-axis; there is no y-intercept.
5. For $b > 1$, the graph rises from left to right.
6. For $0 < b < 1$, the graph falls from left to right.

Quick Reference 11.4

Inverses of Exponential Functions

- For $x > 0$, $b > 0$, and $b \neq 1$, the **logarithm of x with base b,** represented by $\log_b x$, is defined by the following:

  $\log_b x = y$ if and only if $x = b^y$

  The function $f(x) = \log_b x$ is called the **logarithmic function with base b** and is the inverse of the exponential function $f(x) = b^x$.
- The logarithmic function with base e is called the **natural logarithmic function** and is written $y = \ln x$.
- The logarithmic function with base 10 is called the **common logarithmic function** and is written $y = \log x$.

Evaluation of Logarithms

- Exponential expressions can be changed to logarithmic form, and logarithmic expressions can be changed to exponential form.

  $x = b^y \longleftrightarrow \log_b x = y$
- Logarithmic expressions can sometimes be evaluated without a calculator if they are first converted to exponential forms.

- In order to use a calculator to evaluate a logarithm with a base other than 10 or e, the Change of Base Formula is needed.

  For $x > 0$, and for any positive bases a and b where $a \neq 1$ and $b \neq 1$,

$$\log_b x = \frac{\log_a x}{\log_a b}$$

Logarithmic Equations

- We can solve some (but not all) logarithmic equations by converting the logarithmic expressions to exponential expressions and equating exponents.

Graphs of Logarithmic Functions

- The following are properties of $f(x) = \log_b x$ and its graph.
  1. The domain of f is $\{x \mid x > 0\}$.
  2. The range of f is the set of all real numbers.
  3. The x-intercept is $(1, 0)$.
  4. The graph approaches but does not reach the y-axis; there is no y-intercept.
  5. For $b > 1$, the graph rises from left to right.
  6. For $0 < b < 1$, the graph falls from left to right.

Speaking the Language 11.4

1. The inverse of an exponential function is a(n) ________ function.

2. The expression $\log x$ is called a(n) ________ logarithm, whereas the expression $\ln x$ is called a(n) ________ logarithm.

3. The graph of $f(x) = \log_b x$, $x > 0$ and $b > 1$, has no ________-intercept and ________ from left to right.

4. To change $\log_5 x$ to a logarithm with base e, we can use the ________ Formula.

Exercises 11.4

Concepts and Skills

1. For the logarithmic function $y = \log_b x$, explain why b must be positive. Also explain why b cannot be 1.

2. Explain why $\log 100$ can be evaluated without a calculator but $\ln 100$ cannot.

In Exercises 3–10, write the given logarithmic equation in exponential form.

3. $\log_7 49 = 2$

4. $\log_4 8 = \frac{3}{2}$

5. $\log 0.001 = -3$

6. $\log_{1/2} 8 = -3$

7. $\ln \sqrt[3]{e} = \frac{1}{3}$

8. $\log_8 \frac{1}{4} = -\frac{2}{3}$

9. $\log_m n = 2$

10. $\ln 3 = t$

In Exercises 11–18, write the given exponential equation in logarithmic form.

11. $4^3 = 64$

12. $27^{2/3} = 9$

13. $10^{-5} = 0.00001$

14. $8^{-2/3} = \frac{1}{4}$

15. $\left(\frac{1}{2}\right)^{-3} = 8$

16. $5^{-3} = \frac{1}{125}$

17. $P = e^{rt}$

18. $2^x = y$

To do Exercises 19–26, students must understand that exponential functions and logarithmic functions are inverses.

In Exercises 19–26, for each function, determine f^{-1}.

19. $f(x) = e^x$

20. $f(x) = 3^x$

21. $f(x) = 4^{-x}$

22. $f(x) = \left(\frac{3}{4}\right)^x$

23. $f(x) = \log x$

24. $f(x) = \log_5 x$

25. $f(x) = \log_{1.5} x$

26. $f(x) = \ln x$

In Exercises 27–50, evaluate the given logarithms, if possible. Assume all variables represent positive numbers.

27. $\log_4 16$ **28.** $\log_3 81$

29. $\ln 1$ **30.** $\log_{11} 11$

31. $\log_9 243$ **32.** $\log_9 \frac{1}{27}$

33. $\log 1000$ **34.** $\ln e^5$

35. $\ln(-3)$ **36.** $\log_4(-1)$

37. $\log_{1/9} 3$ **38.** $\log_{2/3}\left(\frac{9}{4}\right)$

39. $\log_2 8^{1/3}$ **40.** $\log_5 25^{-1/2}$

41. $\log_3 \sqrt{3^5}$ **42.** $\log \sqrt[3]{10^2}$

43. $\log 0$ **44.** $\log_7(-7)$

45. $\log_{2/3}\left(\frac{3}{2}\right)$ **46.** $\log_{3/4}\left(\frac{4}{3}\right)$

47. $\log_b b^3$ **48.** $\log_b b$

49. $\log_a \sqrt{a}$ **50.** $\log_b 1$

Parentheses are important when entering these expressions in a calculator.

In Exercises 51–60, evaluate the given logarithm. Round the results to two decimal places.

Remind students about the Change of Base Formula for bases other than 10 and e.

51. $\ln 4$

52. $\log 40$

53. $\log_4 7$

54. $\log_2 5$

55. $\ln\left(\frac{\sqrt{3}}{2}\right)$

56. $\frac{\ln \sqrt{5}}{\ln 3}$

57. $\log\left(7 - \sqrt{2}\right)$

58. $\log_7(\log 2)$

59. $\log_5 7^3$ **60.** $\log \sqrt{5}$

61. What is the first step in solving a logarithmic equation $c = \log_b x$?

62. If $\log_b(x + 2) = c$, explain how you know that the solution of the equation is greater than -2.

In Exercises 63–82, solve the given equation.

63. $\log_3 x = 5$ **64.** $\log_3 x = -2$

65. $\ln x = 1$ **66.** $\ln t = 0$

67. $\log_8 t = -\frac{2}{3}$ **68.** $\log_{27} x = \frac{4}{3}$

69. $\log_x 81 = 2$ **70.** $\log_x e^{17} = 17$

71. $\log_x 25 = \frac{2}{3}$ **72.** $\log_x 0.001 = -\frac{3}{2}$

73. $\log_x 1 = 0$

74. $\log_x x = 1$

75. $\log_2 \sqrt{x} = \frac{1}{2}$

76. $\log_2 \sqrt[3]{x} = \frac{2}{3}$

77. $\log_6 6^x = 3$

78. $\log_3 9^x = 2$

79. $\log_4(x + 1) = 2$

80. $\log_5(2x - 1) = 1$

81. $\log_5 x^2 = 4$

82. $\log_3(x^2 - 1) = 1$

83. Under what conditions will the graph of $\log_b x$

(a) rise from left to right?

(b) fall from left to right?

84. If $f(x) = 7^x$ and $g(x) = \log_7 x$, what is $(f \circ g)(x)$? Explain how you know.

In Exercises 85–88, compare the graphs of each pair of functions to the graph of $f(x) = \ln x$.

85. (a) $g(x) = 3 + \ln x$

(b) $h(x) = \ln(3 + x)$

86. (a) $g(x) = \ln(x - 5)$

(b) $h(x) = -5 + \ln x$

87. (a) $g(x) = \ln(-x)$

(b) $h(x) = -\ln x$

88. (a) $g(x) = \log_2 x$

(b) $h(x) = \log_3 x$

CE The following figure shows the graph of a function $f(x) = \log_b x$. In Exercises 89 and 90, use this graph to assist you in matching each of the functions to graph (A), (B), (C), or (D).

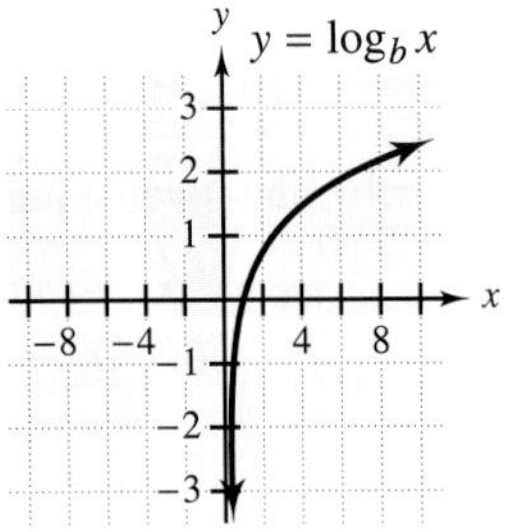

(A)

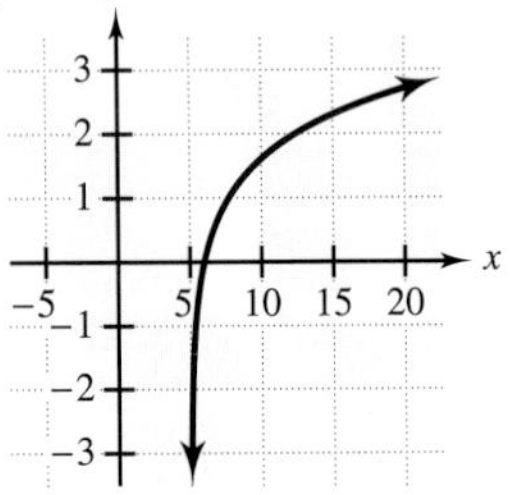

(B)

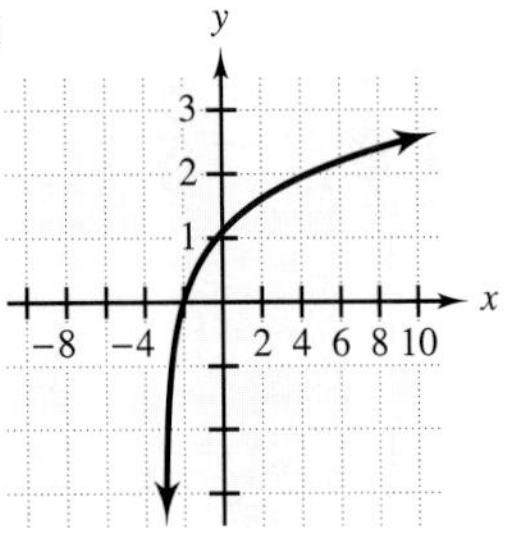

(C)

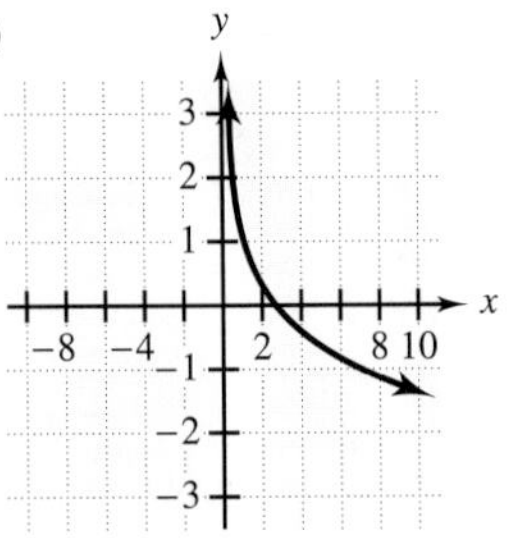

(D)

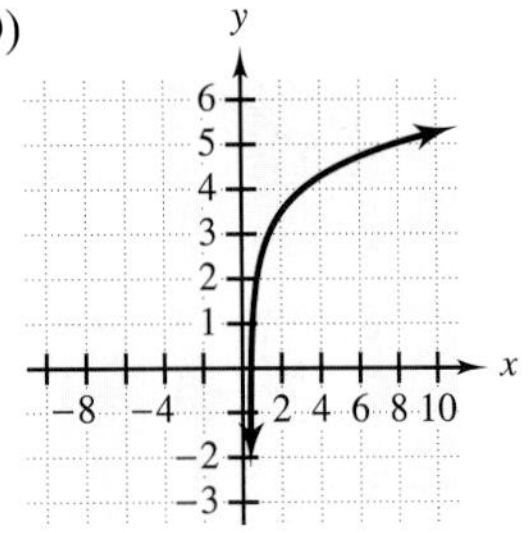

89. (a) $g(x) = \log_b (x + 3)$ (b) $h(x) = 3 + \log_b x$

90. (a) $h(x) = 1 - \log_b x$ (b) $g(x) = \log_b (x - 5)$

In Exercises 91–102, sketch the graph of the function.

91. $f(x) = \log_4 x$

92. $g(x) = \log_5 x$

93. $g(x) = \log_{1/2} x$

94. $h(x) = \log_{1/6} x$

95. $h(x) = -2 \log_7 x$

96. $f(x) = 3 + \log_{10} x$

97. $f(x) = 2 + \log_{1/5} x$

98. $f(x) = -\log_{1/8} x$

99. $h(x) = \log_3 (x - 2)$

100. $k(x) = \log_2 (3 + x)$

101. $f(x) = 4 - \log_2 (1 - x)$

102. $g(x) = 1 + \log_3 (x + 5)$

Real-Life Applications

103. An amplifier produces a power gain G (in decibels) given by $G = 10 \log (P_o/P_i)$, where P_o is the power output and P_i is the power input.

(a) To the nearest hundredth, what is the power gain when $P_o = 16$ and $P_i = 0.004$?

(b) What is the power output for an amplifier that produces a 50-decibel gain when the power input is 0.006?

104. The number of cell divisions d that occur in a given time is given by $d = \log_2 (n/n_0)$, where n_0 is the initial cell count and n is the final cell count.

(a) Approximately how many cell divisions occurred if the initial cell count was 50 and the final cell count was 100,000?

(b) Determine the final cell count if an initial cell count of 100 cells undergoes 12 cell divisions.

105. The acidity of a substance is measured by the pH value. To determine the acidity of a substance, chemists measure the concentration H^+ of hydrogen ions in the substance and then calculate the pH with the formula $\text{pH} = -\log H^+$.

(a) The hydrogen ion concentration of lemon juice is about 0.005. What is the pH value?

(b) The water in a pond is best for fish if the pH is 7. Determine the hydrogen ion concentration.

106. The air pressure P is related to the altitude y (in kilometers) by the formula $\ln (P/P_0) = -0.116y$, where P_0 is the air pressure at sea level.

(a) To the nearest tenth, at what altitude is the air pressure 75% of the air pressure at sea level?

(b) The air pressure at sea level is 760 millimeters. To the nearest whole number, what is the air pressure at an altitude of 5 kilometers?

Modeling with Real Data

107. The accompanying table shows the average number of home runs per game for the National and American Leagues for selected years in the period 1992–1999. (Source: Major League Baseball.)

	Home Runs	
Year	**National**	**American**
1992	1.30	1.57
1994	1.91	2.23
1996	2.11	2.45
1999	2.24	2.32

The following natural logarithmic functions model the number of home runs per game, where x is the number of years since 1990.

National: $N(x) = 0.93 + 0.63 \ln x$

American: $A(x) = 1.32 + 0.54 \ln x$

(a) According to the models, how many home runs per game will each league average in 2002?

(b) Sketch the graphs of the two functions in the same coordinate system. For the period 2000–2025, what two conclusions can be drawn from the graphs?

108. Recycling takes 95% less energy than making new metal from ore. The accompanying table shows the percentage of aluminum cans that were recycled for selected years in the period 1975–1998. (Source: The Aluminum Association.)

Year	**Percentage of Aluminum Cans Recycled**
1975	26.9
1985	51.0
1995	62.2
1996	63.5
1998	62.8

The percentage R of recycled cans x years after 1970 is given by $R(x) = -7.65 + 21.57 \ln x$.

(a) According to the model, what percentage of cans is projected to be recycled in 2003?

(b) In what year does the model predict that 75% of all aluminum cans will be recycled?

Data Analysis: *Telecommuting*

Telecommuting—using a computer or phone technology to work from home— is a developing trend for American workers. The accompanying figure shows the number (in millions) of employees involved in telecommuting for selected years from 1990 to 2002. (Source: Gil Gordon Associates.)

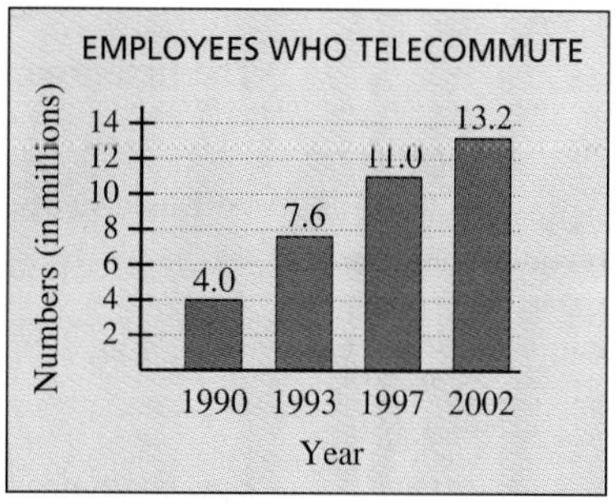

If x represents the number of years since 1985, the number (in millions) of employees who telecommute can be modeled by either of the following functions.

$N_1(x) = -8.2 + 7.6 \ln x$

$N_2(x) = 17.5 - 21.7e^{-0.095x}$

109. Create a table of values for N_1 and N_2 for the years given in the bar graph and the year 2050. (Round all results to the nearest tenth.)

110. How well do functions N_1 and N_2 model the actual data?

111. Using the year 2050 as an example, compare functions N_1 and N_2 in terms of long-range trends?

112. If the number of telecommuters had been 2 million greater for each year given in the table, how would the two model functions need to be modified to obtain new model functions?

Challenge

113. Suppose $a > b$. If you multiply both sides of the inequality by $\ln \frac{1}{2}$, what is the result? Why?

114. Show that $e^{x - \ln x} = x^{-1}e^x$.

In Exercises 115 and 116, evaluate the given expression.

115. $e^{\ln 5 - \ln 2}$

116. $\ln e^{\ln e}$

In Exercises 117 and 118, for what values of x is the function positive?

117. $\ln |x|$

118. $\ln (x^2 - 16)$

In Exercises 119 and 120, solve the given inequalities.

119. $\log_3 x < \log_4 x$

120. $\ln x > \log x$

11.5 Properties of Logarithms

Basic Properties • Product, Quotient, and Power Rules for Logarithms • Combining the Properties

Basic Properties

Because logarithms are defined in terms of exponents, the properties of logarithms are closely related to the properties of exponents.

The first group of properties follows directly from the definition of logarithms.

Think About It

To solve exponential equations, we wrote both sides with the same base and equated the exponents. For the equation $3^x = 6$, use Property 4 to write 6 with a base of 3. Then solve the equation.

Basic Properties of Logarithms

1. $\log_b 1 = 0$
2. $\log_b b = 1$
3. $\log_b b^x = x$
4. $b^{\log_b x} = x$

For each property an equivalent exponential form can be written. In the following table, the logarithmic properties in the left column are justified by the corresponding exponential forms in the center column.

Logarithmic Form	*Exponential Form*	*Example*
$\log_b 1 = 0$	$b^0 = 1$	$\log_7 1 = 0$
$\log_b b = 1$	$b^1 = b$	$\log_{3/4} \frac{3}{4} = 1$
$\log_b b^x = x$	$b^x = b^x$	$\log_8 8^5 = 5$
$\log_b x = \log_b x$	$b^{\log_b x} = x$	$5^{\log_5 3} = 3$

Product, Quotient, and Power Rules for Logarithms

The following properties of logarithms correspond to the Product, Quotient, and Power Rules for Exponents.

LEARNING TIP

To help remember the properties of logarithms, compare them with the corresponding properties of exponents. For example:

Product	*Sum*
$b^m b^n = b^{m+n}$	
$\log(MN) = \log M + \log N$	

Product, Quotient, and Power Rules for Logarithms

For $b > 0$, $b \neq 1$, positive real numbers M and N, and any real number c, the following are properties of logarithms.

1. $\log_b MN = \log_b M + \log_b N$ — Product Rule for Logarithms
2. $\log_b \frac{M}{N} = \log_b M - \log_b N$ — Quotient Rule for Logarithms
3. $\log_b M^c = c(\log_b M)$ — Power Rule for Logarithms

The following demonstrates why the Product Rule for Logarithms is true.

Let $\log_b M = r$ and $\log_b N = s$. Then $b^r = M$ and $b^s = N$. Therefore, the product $MN = b^r b^s = b^{r+s}$. But $b^{r+s} = MN$ implies that

$$\log_b MN = r + s = \log_b M + \log_b N$$

The following are some other common errors.

$$\frac{\log_b M}{\log_b N} = \log_b M - \log_b N$$

$$(\log_b M)(\log_b N) = \log_b M + \log_b N$$

Justifications for the Quotient and Power Rules for Logarithms are similar to that used for the Product Rule for Logarithms. The proofs are left as exercises.

Note: There is no property for the logarithm of a *sum* of two numbers or a *difference* of two numbers. In particular, $\log_b (x + y)$ is not equal to $\log_b x + \log_b y$, and $\log_b (x - y)$ is not equal to $\log_b x - \log_b y$.

In the remainder of this section, we assume that all logarithmic expressions are defined.

The following examples illustrate the properties.

Property of Logarithms	*Example*
Product Rule for Logarithms	$\ln (5 \cdot 2) = \ln 5 + \ln 2$
	$\log_2 7 + \log_2 y = \log_2 7y$
Quotient Rule for Logarithms	$\log \frac{9}{4} = \log 9 - \log 4$
	$\log_3 5 - \log_3 x = \log_3 \frac{5}{x}$
Power Rule for Logarithms	$\ln 2^3 = 3 \ln 2$
	$5 \log_6 t = \log_6 t^5$

EXAMPLE 1

Rewriting Logarithms Whose Arguments Are Products, Quotients, or Exponential Expressions

(a) $\log_4 5(t + 2) = \log_4 5 + \log_4 (t + 2)$ — Product Rule for Logarithms

(b) $\ln e(e - 1) = \ln e + \ln (e - 1)$ — Product Rule for Logarithms

$= 1 + \ln (e - 1)$ — $\ln e = 1$

(c) $\ln \frac{1}{y} = \ln 1 - \ln y$ — Quotient Rule for Logarithms

$= 0 - \ln y$ — $\ln 1 = 0$

$= -\ln y$

(d) $\log_6 \frac{y + 4}{y} = \log_6 (y + 4) - \log_6 y$ — Quotient Rule for Logarithms

(e) $\log \sqrt[3]{t} = \log t^{1/3} = \frac{1}{3} \log t$ — Power Rule for Logarithms

(f) $\ln \frac{1}{x^3} = \ln x^{-3} = -3 \ln x$ — Power Rule for Logarithms

EXAMPLE 2

Writing a Logarithmic Expression as a Single Logarithm

(a) $\ln 3 + \ln x = \ln (3x)$ — Product Rule for Logarithms

(b) $\log (x + 1) + \log (x - 1) = \log [(x + 1)(x - 1)]$

$= \log (x^2 - 1)$

(c) $\ln (2x) - \ln 3 = \ln \frac{2x}{3}$ — Quotient Rule for Logarithms

(d) $\log (x^2 + 4) - \log (x + 2) = \log \dfrac{x^2 + 4}{x + 2}$

(e) $\dfrac{1}{2} \ln t = \ln t^{1/2} = \ln \sqrt{t}$ Power Rule for Logarithms

(f) $-2 \log_7 t = \log_7 t^{-2} = \log_7 \dfrac{1}{t^2}$

Combining the Properties

When we need to expand a single logarithm into two or more logarithms or to write a logarithmic expression with a single logarithm, we will sometimes need to use more than one of the rules for logarithms.

EXAMPLE 3

Expanding a Logarithmic Expression

(a) $\log_5 3(x + 4)^2 = \log_5 3 + \log_5 (x + 4)^2$ Product Rule for Logarithms

$= \log_5 3 + 2 \log_5 (x + 4)$ Power Rule for Logarithms

(b) $\ln \left(\dfrac{x + 2}{x + 1}\right)^2 = 2 \ln \dfrac{x + 2}{x + 1}$ Power Rule for Logarithms

$= 2 [\ln (x + 2) - \ln (x + 1)]$ Quotient Rule for Logarithms

$= 2 \ln (x + 2) - 2 \ln (x + 1)$ Distributive Property

(c) $\log \sqrt{\dfrac{x^3}{x + 2}} = \log \left(\dfrac{x^3}{x + 2}\right)^{1/2}$ Definition of rational exponent

$= \dfrac{1}{2} \log \dfrac{x^3}{x + 2}$ Power Rule for Logarithms

$= \dfrac{1}{2} [\log x^3 - \log (x + 2)]$ Quotient Rule for Logarithms

$= \dfrac{1}{2} [3 \log x - \log (x + 2)]$ Power Rule for Logarithms

$= \dfrac{3}{2} \log x - \dfrac{1}{2} \log (x + 2)$ Distributive Property

EXAMPLE 4

Combining Logarithms

(a) $2 \ln t + 3 \ln (t + 2) = \ln t^2 + \ln (t + 2)^3$ Power Rule for Logarithms

$= \ln t^2(t + 2)^3$ Product Rule for Logarithms

(b) $4[\log_6 (y + 1) - \log_6 y] = 4 \log_6 \dfrac{y + 1}{y}$ Quotient Rule for Logarithms

$= \log_6 \left(\dfrac{y + 1}{y}\right)^4$ Power Rule for Logarithms

Note that we first write each logarithm with 1 as the coefficient. Then we use the Product or Quotient Rule for Logarithms.

(c) $2 \log (2x + 1) + \log x - 4 \log (x + 3)$

$= \log (2x + 1)^2 + \log x - \log (x + 3)^4$ Power Rule for Logarithms

$= \log x(2x + 1)^2 - \log (x + 3)^4$ Product Rule for Logarithms

$= \log \dfrac{x(2x + 1)^2}{(x + 3)^4}$ Quotient Rule for Logarithms

Quick Reference 11.5

Basic Properties

- Certain properties of logarithms follow directly from the definition of logarithm and can be used to evaluate a logarithmic expression.

  1. $\log_b 1 = 0$
  2. $\log_b b = 1$
  3. $\log_b b^x = x$
  4. $b^{\log_b x} = x$

Product, Quotient, and Power Rules for Logarithms

- For $b > 0$, $b \neq 1$, positive real numbers M and N, and any real number c, the following are properties of logarithms.

  1. $\log_b MN = \log_b M + \log_b N$ — Product Rule for Logarithms
  2. $\log_b \dfrac{M}{N} = \log_b M - \log_b N$ — Quotient Rule for Logarithms
  3. $\log_b M^c = c(\log_b M)$ — Power Rule for Logarithms

Combining the Properties

- Logarithmic expressions may involve combinations of products, quotients, and powers. The Product, Quotient, and Power Rules for Logarithms can be used to expand a single logarithm into two or more logarithmic expressions.
- These same rules can be used to combine two or more logarithms into a single logarithmic expression.

Speaking the Language 11.5

1. To convert $\log 5 + \log 9$ into a single logarithm, we use the ________.
2. The logarithm of a quotient is equal to the ________ of the logarithms of the numerator and the denominator.
3. The ________ Rule for Logarithms can be applied to write $\log x^2$ as $2 \log x$, provided $x > 0$.
4. We know immediately that $(\log 4)(\log 3)(\log 2)(\log 1)$ is 0 because ________.

Exercises 11.5

Concepts and Skills

1. Treating a logarithm as an exponent, we can evaluate $\log_4 16$ by asking, "What exponent on 4 results in 16?" Because the answer is 2, $\log_4 16 = 2$. What question would you ask in order to evaluate $\log_9 3$? What is the answer?

2. What rule of logarithms are we using if we write $\log x^2 = \log(x \cdot x) = \log x + \log x = 2 \log x$? What rule of logarithms would be more efficient in obtaining the same result?

In Exercises 3–18, evaluate the given expression.

This group uses the Basic Properties of Logarithms.

3. $\log_5 5^7$
4. $\ln e^2$
5. $\log_{1.2} 1$
6. $\log_7 7$
7. $7^{\log_7 2}$
8. $3.5^{\log_{3.5} 10}$
9. $\log_9 \sqrt{3}$
10. $\log_4 \sqrt{8}$
11. $9^{-\log_3 2}$
12. $125^{-\log_5 4}$
13. $e^{2 \ln 3}$
14. $10^{3 \log 2}$
15. $\log_3(\log_4 4)$
16. $\ln(\log 10)$
17. $\log_3(\log_2 8)$
18. $\log_2(\log_2 16)$

In Exercises 19–22, write the given logarithm as a sum or difference of logarithms.

19. $\log_7 3x$

20. $\log_8 5mn$

21. $\log_7 \frac{5}{y}$

22. $\log_6 \frac{y}{t}$

In Exercises 23–26, write the given logarithm with no exponents or radicals in the argument.

23. $\log x^2$

24. $\log_3 7^{-4}$

25. $\log_9 \sqrt{n}$

26. $\ln \sqrt[3]{x}$

In Exercises 27–34, use the rules of logarithms to expand the logarithm in column A. Then, letting $\log_3 2 = A$ and $\log_3 5 = B$, match the result with one of the forms in column B.

Column A	*Column B*
27. $\log_3 \left(\frac{5}{6}\right)^3$	(a) $A - \frac{1}{2}B + 1$
28. $\log_3 \frac{45}{8}$	(b) $-\frac{2}{3}A + \frac{1}{3}B - \frac{2}{3}$
29. $\log_3 \frac{6\sqrt{5}}{5}$	(c) $-3A + 3B - 3$
30. $\log_3 \frac{36\sqrt{2}}{125}$	(d) $-\frac{3}{4}A - \frac{3}{4}$
31. $\log_3 \frac{5\sqrt{10}}{9}$	(e) $\frac{1}{5}A - \frac{8}{5}B - 1$
32. $\log_3 \frac{\sqrt[3]{30}}{6}$	(f) $-3A + B + 2$
33. $\log_3 \frac{\sqrt[4]{96}}{12}$	(g) $\frac{1}{2}A + \frac{3}{2}B - 2$
34. $\log_3 \frac{\sqrt[5]{50}}{75}$	(h) $\frac{5}{2}A - 3B + 2$

35. To evaluate $\log (AB)^n$, should you first apply the Product Rule for Logarithms or the Power Rule for Logarithms? Why?

36. From the following list, identify two logarithms that cannot be expanded and explain why.

(i) $\log (AB)$ (ii) $\log (A - B)$

(iii) $\log (A + B)$ (iv) $\log \frac{A}{B}$

In Exercises 37–58, if possible, use the properties of logarithms to expand the given expression.

More than one property may be needed in this group.

37. $\log_9 x^2y^{-3}$

38. $\log_3 27(x + 5)^3$

39. $\log \frac{(x + 1)^2}{x + 2}$

40. $\log \frac{x^2y^3}{z^4}$

41. $\ln \left(\frac{2x + 5}{x + 7}\right)^5$

42. $\log_3 \frac{(x^2z)^3}{(y^2z)^5}$

43. $\log \frac{\sqrt{10}}{x^2}$

44. $\ln \sqrt[4]{\frac{x^3y}{z^2}}$

45. $\log_5 \sqrt[3]{xy^2}$

46. $\log_3 81x^2\sqrt{y}$

47. $\log_2 \frac{\sqrt{x + 1}}{32}$

48. $\log_8 \sqrt{x(x + 2)}$

49. $\log_2 \frac{x^2\sqrt{y}}{\sqrt[3]{x}}$

50. $\ln \frac{(x + 2)\sqrt{x}}{x^2}$

51. $\log (4x^2 + 4x + 1)$

52. $\log (x^2 + 3x + 2)$

53. $\ln 1000e^{0.08t}$

54. $\ln (e^2 - e)$

55. $\log_4 (x + 4)$

56. $\ln (e^2 + 1)$

57. $\ln Pe^{rt}$

58. $\ln P\left(1 + \frac{r}{t}\right)^{nt}$

59. Explain how to combine $2 \log 3 + \log 3$ into a single logarithm without the Product Rule for Logarithms. Can you combine $2 \log x + \log y$ into a single logarithm without the Product Rule for Logarithms if $x \neq y$? Explain.

60. Which rule of logarithms would you use to show that $-\log x = \log (1/x)$? Would you use the same rule to show that $\log (1/x) = -\log x$? Write the steps to show that both equations are true.

In Exercises 61–64, write the given logarithmic expression as a single logarithm.

61. $\log_4 3 + \log_4 5$

62. $\ln x^2 + \ln x^{-1}$

63. $\log_3 10 - \log_3 5$

64. $\log_7 u^5 - \log_7 u^2$

In Exercises 65–68, write the given logarithmic expression as a single logarithm with a coefficient of 1.

65. $3 \log_2 y$

66. $-\log_5 x$

67. $\frac{1}{2} \log x$

68. $\frac{2}{3} \ln t$

In Exercises 69–72, write the given expression as a single logarithm and evaluate the result.

69. $\log_4 60 - \log_4 15$

70. $\log_3 4 + \log_3 6 - \log_3 8$

71. $\log_5 50 + 2 \log_5 10 - \log_5 40$

72. $\frac{1}{3}\log_6 27 - \log_6 8 - \log_6 81$

In Exercises 73–88, use the rules of logarithms to combine the given expression into a single logarithmic expression with a coefficient of 1.

These exercises involve more than one property.

73. $\log_8 3 + \log_8 x - \log_8 y$

74. $\ln x - \ln 3 - \ln 2$

75. $2\log_7 x - \log_7 y$

76. $\log_8 5 + 2\log_8 t$

77. $\frac{1}{2}\log x - \frac{2}{3}\log y$

78. $\frac{3}{2}\log_6 r + \frac{1}{2}\log_6 s$

79. $3\log y - 2\log x - 4\log z$

80. $\log y + 2\log(x+2) - 5\log z$

81. $2\log(x+3) - 3\log(x+1)$

82. $\frac{1}{2}\log(t+1) - 3\log t$

83. $\frac{1}{2}[\log x - \log(x+1)]$

84. $\frac{1}{3}[2\log(x+2) - \log(x-5)]$

For Exercises 85–88, you may want to provide some additional guidance by showing how a number c can be written as a logarithm.
$c = c\cdot 1 = c\log_b b = \log_b b^c$

85. $5 + 2\log_3 x$

86. $1 + 4\log_2\sqrt{x}$

87. $rt + \ln P_0$

88. $0.06t + \ln 1000$

CE In Exercises 89–94, determine whether the equation is true.

89. $\frac{\log_8 3}{\log_8 5} = \log_8\frac{3}{5}$

90. $\frac{\log_3 25}{\log_3 5} = 2$

91. $\frac{\log_3 81}{\log_3 9} = \log_3 9$

92. $\frac{\log_2 12}{\log_2 4} = \log_2 3$

93. $(\log 3)(\log 2) = \log 6$

94. $-2 = (\log_7 9)(\log_{1/3} 7)$

CE In Exercises 95–100, determine whether the given equation is always true.

95. $\log_2 x^x = x\log_2 x$

96. $\log 10^{x+1} = x + 1$

97. $\log_5 x^2y = 2\log_5 x + \log_5 y$

98. $\ln\frac{e}{x} = 1 - \ln x$

99. $\log_3(x+10) = \log_3 x + \log_3 10$

100. $\log x^2 = (\log x)^2$

Real-Life Applications

101. A consultant measured the effect of travel fatigue on productivity of business travelers. The productivity was modeled by $P(t) = \frac{10}{\log 5 + \log(t+2)}$, where t is the number of hours that the person spent in transit and waiting for transit (such as sitting in airports).

(a) By what percentage did the productivity decline with 4 hours of travel time?

(b) Write the function with a single logarithm in the denominator.

102. Suppose that the sales (in thousands of units) of a certain new brand of cereal is given by the function $S(t) = 100\log_2(t+2)$, where t is the number of months since the brand was introduced. How many units of the product were sold six months after its introduction? Show that an equivalent formula can be written with natural logarithms:

$$S(t) = \frac{\ln(t+2)^{100}}{\ln 2}$$

Challenge

103. Let $\log_b M = r$ and $\log_b N = s$. Using steps similar to those used to justify the Product Rule for Logarithms, show that the Quotient Rule for Logarithms is true.

104. Let $\log_b M = r$. Show that the Power Rule for Logarithms is true.

105. Show that $2 + \log_3 x = \log_3 9x$.

106. Show that

$$\log_2 x + \log_3 x = \frac{(\log 2 + \log 3)\log x}{(\log 2)(\log 3)}$$

107. Show that $\log_2(xy) = \frac{\ln x + \ln y}{\ln 2}$.

108. Suppose $a^2 + b^2 = 7ab$. Show that the following is true.

$$\log(a+b) - \log 3 = \frac{1}{2}(\log a + \log b)$$

11.6 Exponential and Logarithmic Equations

Definitions and Rules • Exponential Equations • Logarithmic Equations • Real-Life Applications

Definitions and Rules

An equation in which a variable appears in the argument of a logarithm is called a **logarithmic equation.** An equation in which a variable appears as an exponent is called an **exponential equation.**

Because both exponential and logarithmic functions are one-to-one functions, the following properties can be used to solve exponential and logarithmic equations.

Properties of Exponential and Logarithmic Equations

Assuming that all expressions are defined, the following properties hold for exponential and logarithmic equations.

1. If $x = y$, then $b^x = b^y$. Equating exponential expressions
2. If $b^x = b^y$, then $x = y$. Equating exponents
3. If $x = y$, then $\log_b x = \log_b y$. Equating logarithmic expressions
4. If $\log_b x = \log_b y$, then $x = y$. Equating arguments of logarithms

Exponential Equations

Recall that when an exponential equation can be written with both sides having the same base, we can equate the exponents and solve the resulting equation. However, if both sides of an equation cannot be written with the same base, other algebraic techniques are needed to solve the equation.

Exploring the Concept

Solving Exponential Equations

Consider the equation $3^{2x} = 5$. To solve the equation, we choose a logarithm with a convenient base, usually ln or log, and equate the logarithms of both sides.

$$3^{2x} = 5$$

$$\ln 3^{2x} = \ln 5 \quad \text{Equate the logarithms of both sides.}$$

$$2x \cdot \ln 3 = \ln 5 \quad \text{Power Rule for Logarithms}$$

$$(2 \ln 3)x = \ln 5$$

$$x = \frac{\ln 5}{2 \ln 3} \quad \text{Divide both sides by 2 ln 3.}$$

$$x \approx 0.73$$

Note that this is a first-degree equation, where the coefficient of x is 2 ln 3.

EXAMPLE 1 Solving Exponential Equations

(a)

$$7^{x+1} = 14$$

$$\log 7^{x+1} = \log 14 \quad \text{Equate the logarithms of both sides.}$$

$$(x+1)\log 7 = \log 14 \quad \text{Power Rule for Logarithms}$$

$$x \log 7 + \log 7 = \log 14$$ Distributive Property

$$x \log 7 = \log 14 - \log 7$$ Isolate the variable term.

$$x = \frac{\log 14 - \log 7}{\log 7}$$

$$x \approx 0.36$$

To avoid rounding-off errors, evaluate the logarithmic expression as the last step.

(b) When an equation involves the number e, we often choose to use natural logarithms to solve.

$$e^x = 2$$

$$\ln e^x = \ln 2$$ Equate the logarithms of both sides.

$$x \ln e = \ln 2$$ Power Rule for Logarithms

$$x \cdot 1 = \ln 2$$ $\ln e = 1$

$$x = \ln 2$$

$$x \approx 0.69$$

(c)

$$8 - 3^{2x-1} = 1$$ Isolate the exponential term.

$$-3^{2x-1} = -7$$ Subtract 8 from both sides.

$$3^{2x-1} = 7$$ Multiply both sides by -1.

$$\log 3^{2x-1} = \log 7$$ Equate the logarithms of both sides.

$$(2x - 1) \log 3 = \log 7$$ Power Rule for Logarithms

$$2x - 1 = \frac{\log 7}{\log 3}$$ Divide both sides by log 3.

$$2x = \frac{\log 7}{\log 3} + 1$$ Solve for x.

$$x = \frac{1}{2}\left(\frac{\log 7}{\log 3} + 1\right)$$

$$x \approx 1.39$$

Logarithmic Equations

One method for solving logarithmic equations is to equate the arguments and then use Property 4.

Exploring the Concept

Solving Logarithmic Equations

Consider the equation $\ln(x - 3) + \ln(x + 1) = \ln 5$.

To solve the equation, we write the left side as a single logarithm and then equate the arguments of the logarithms.

$$\ln(x - 3) + \ln(x + 1) = \ln 5$$

$$\ln(x - 3)(x + 1) = \ln 5$$ Product Rule for Logarithms

$$(x - 3)(x + 1) = 5$$ Equate the arguments of the logarithms.

$$x^2 - 2x - 3 = 5$$ FOIL

$$x^2 - 2x - 8 = 0$$ Write the quadratic equation in standard form.

$$(x - 4)(x + 2) = 0$$ Factor.

You may need to correct the mistaken belief that we are "canceling" ln or dividing both sides by ln. We can equate arguments because the logarithmic function is a one-to-one function.

LEARNING TIP

For a logarithmic equation, use the properties of logarithms to write each side of the equation as a single term.

$$x - 4 = 0 \quad \text{or} \quad x + 2 = 0 \qquad \text{Zero Factor Property}$$
$$x = 4 \quad \text{or} \quad x = -2$$

Check the two values for x.

$$\ln (x - 3) + \ln (x + 1) = \ln 5$$
$$\ln 1 + \ln 5 = \ln 5 \qquad \text{Replace } x \text{ with } 4.$$
$$\ln 5 = \ln 5 \qquad \ln 1 = 0$$

The result is true, so 4 is a solution.

Replacing x with -2 gives $\ln (-5) + \ln (-1) = \ln 5$. But logarithmic functions are defined only for positive numbers. Therefore, -2 is an extraneous solution. The only solution of the equation is 4.

EXAMPLE 2 Solving Logarithmic Equations

Solve the given equation.

(a) $\ln (1 - 2x) = \ln 5$

(b) $\log_3 x + \log_3 (x - 2) = 1$

(c) $\ln (4x - 5) - \ln (x - 2) = \ln (2x + 1)$

Solution

(a) $\ln (1 - 2x) = \ln 5$

$$1 - 2x = 5 \qquad \text{Equate arguments.}$$
$$-2x = 4$$
$$x = -2$$

The solution is easily verified by substitution.

(b) $\log_3 x + \log_3 (x - 2) = 1$

$$\log_3 x(x - 2) = 1 \qquad \text{Product Rule for Logarithms}$$
$$x(x - 2) = 3^1 \qquad \text{Exponential form}$$
$$x^2 - 2x = 3 \qquad \text{Distributive Property}$$
$$x^2 - 2x - 3 = 0 \qquad \text{Quadratic equation in standard form}$$
$$(x - 3)(x + 1) = 0 \qquad \text{Factor.}$$
$$x - 3 = 0 \quad \text{or} \quad x + 1 = 0 \qquad \text{Zero Factor Property}$$
$$x = 3 \quad \text{or} \quad x = -1$$

The value -1 is not in the domain of $\log_3 (x - 2)$, so it is an extraneous solution. The only solution is $x = 3$.

In part (c), we equate arguments because all of the terms are logarithmic expressions. In part (b), this was not the case, so we converted to exponential form.

(c) $\ln (4x - 5) - \ln (x - 2) = \ln (2x + 1)$

$$\ln \frac{4x - 5}{x - 2} = \ln (2x + 1) \qquad \text{Quotient Rule for Logarithms}$$
$$\frac{4x - 5}{x - 2} = 2x + 1 \qquad \text{Equate arguments.}$$

$$4x - 5 = (2x + 1)(x - 2) \quad \text{Clear the fraction.}$$
$$4x - 5 = 2x^2 - 3x - 2 \quad \text{FOIL}$$
$$0 = 2x^2 - 7x + 3 \quad \text{Standard form}$$
$$0 = (2x - 1)(x - 3) \quad \text{Factor.}$$
$$2x - 1 = 0 \quad \text{or} \quad x - 3 = 0 \quad \text{Zero Factor Property}$$
$$2x = 1 \quad \text{or} \quad x = 3$$
$$x = \frac{1}{2} \quad \text{or} \quad x = 3$$

Think About It

Show that $\log 10^{\ln e} = \ln e^{\log 10}$.

Because $\frac{1}{2}$ is not in the domain of $\ln(4x - 5) - \ln(x - 2)$, it is an extraneous solution. Therefore, 3 is the only solution.

Real-Life Applications

When interest is *compounded continuously*, the amount A on deposit after t years is given by $A = Pe^{rt}$, where r is the interest rate and P is the initial investment.

EXAMPLE 3

Continuously Compounded Interest

At age 24, a person has dreams of becoming a millionaire. If this person's current net worth is \$150,000, what growth rate is required to reach the \$1,000,000 goal by age 50?

Solution

$A = 1{,}000{,}000$ — Desired net worth at age 50

$P = 150{,}000$ — Current net worth

$t = 26$ — Number of years until the age is 50

$$A = Pe^{rt} \quad \text{Model for continuous compounding}$$
$$1{,}000{,}000 = 150{,}000e^{26r} \quad \text{Substitute } A, P, \text{ and } t.$$
$$6.667 \approx e^{26r} \quad \text{Divide both sides by 150,000.}$$
$$\ln 6.667 \approx 26r \quad \text{Convert to logarithmic form.}$$
$$r \approx \frac{\ln 6.667}{26}$$
$$r \approx 0.073$$

For the person to be a millionaire by age 50, the current assets must be invested at an interest rate of 7.3%, compounded continuously.

Example 3 illustrates exponential growth.

If a quantity increases (growth) or decreases (decay) over a period of time, the formula $P = P_0e^{kt}$ can sometimes be used to describe the amount P of the quantity after a certain time t. The initial amount of the quantity (at time $t = 0$) is P_0. The constant k is called the *growth constant* or the *decay constant*. A positive value of k indicates growth, and a negative value of k indicates decay.

Note that the graph of $y = P_0e^{kt}$ increases when $k > 0$ and decreases when $k < 0$.

Growth or decay that can be approximated by the model $P = P_0e^{kt}$ is called *exponential growth* or *exponential decay*.

EXAMPLE 4

Exponential Decay of Toxic Waste

Containers of toxic waste totaling 500 pounds are buried in a landfill. After 30 years, 400 pounds of toxic waste still remain. If the decay is exponential, how much will remain after 100 years?

Solution

In growth or decay problems, we often have to determine the constant k.

$P_0 = 500$	Initial amount of toxic waste
$P = 500e^{kt}$	Exponential decay model with $P_0 = 500$
$400 = 500e^{30k}$	When $t = 30, P = 400$.
$0.8 = e^{30k}$	Divide both sides by 500.
$\ln 0.8 = 30k$	Convert to logarithmic form.
$k = \frac{\ln 0.8}{30}$	
$k \approx -0.0074$	Decay constant is negative.

Now we use the formula again, this time with the known decay constant.

$P \approx 500e^{-0.0074t}$	
$P \approx 500e^{-0.0074(100)}$	Time t is 100 years.
$P \approx 500e^{-0.74}$	
$P \approx 238.56$	

After 100 years, approximately 238.56 pounds of toxic waste will remain in the landfill.

Quick Reference 11.6

Definitions and Rules

- An equation in which a variable appears in the argument of a logarithm is a **logarithmic equation.** An equation in which a variable appears as an exponent is an **exponential equation.**
- Assuming that all expressions are defined, the following properties hold for exponential and logarithmic equations.
  1. If $x = y$, then $b^x = b^y$.
  2. If $b^x = b^y$, then $x = y$.
  3. If $x = y$, then $\log_b x = \log_b y$.
  4. If $\log_b x = \log_b y$, then $x = y$.

Exponential Equations

- To solve an exponential equation that can be written with both sides having the same base, equate the exponents and solve the resulting equation.
- To solve an exponential equation for which the bases are different, take the logarithm (with a convenient base) of both sides, simplify, and solve the resulting equation.

Logarithmic Equations

- To solve a logarithmic equation algebraically, follow these steps.

1. Write one side of the equation as a single logarithm and the other as a number or as a single logarithm. The logarithms on each side must have the same base.
2. If one side of the equation is a number, convert the equation to an exponential equation and solve.
3. If both sides involve a logarithmic expression with the same base, equate the arguments and solve the resulting equation.
4. Check all solutions in the original equation. Using the properties of logarithms can introduce extraneous solutions.

Speaking the Language 11.6

1. The equation $\log(x + 1) = 4$ is called a(n) ________ equation because a variable appears in the ________ of the logarithm.

2. A method for solving $e^{x-1} = 5$ is to equate the ________ of both sides.

3. A method for solving $\log(2x - 5) = 2$ is to write the equation in ________ form.

4. If properties of logarithms are used to solve an equation, checking solutions is important because ________ solutions might be introduced.

Exercises 11.6

Concepts and Skills

1. To solve the equation $b^x = 3$, under what conditions is it not necessary to write the equation in logarithmic form?

2. To solve the equation $e^x = 5$, why is it not possible to use the method of equating exponents?

In Exercises 3–6, solve the given equation.

In this group, both sides can be written with the same base.

3. $3^{5x} = \dfrac{1}{27}$ **4.** $5^{2x} = \dfrac{1}{25}$

5. $7^{1-x} = 49$ **6.** $\left(\dfrac{1}{9}\right)^{x-3} = 27$

In Exercises 7–10, solve for x. Express the result to the nearest hundredth.

In Exercises 7–10, logarithms must be used.

7. $3^x = 5$ **8.** $e^{3x} = 5$

9. $3^{x-3} = 2^{x+2}$ **10.** $4^{2x-1} = 10^{1-x}$

In Exercises 11–28, solve the exponential equation. Express the result to the nearest hundredth.

11. $e^{20k} = 0.6$ **12.** $e^{0.25t} = 10$

13. $4e^{2x-1} = 5$ **14.** $\dfrac{2}{3}e^{-x/2} = 10$

15. $\left(\dfrac{5}{3}\right)^{-x} = 20$ **16.** $5 \cdot 4^{1-3x} = 12$

17. $3 + e^x = 7$ **18.** $2e^x - 5 = 19$

19. $5^x + 7 = 2$ **20.** $15 - 2^x = 12$

21. $600(1 + 0.02)^{4t} = 1000$

22. $750\left(1 + \dfrac{0.06}{12}\right)^{12t} = 1200$

23. $5^{-x^2} = 0.2$ **24.** $3^{x^2} = 5$

25. $5^{|x+1|} = 2$ **26.** $3^{|x|} = 0.5$

27. $|2 - 3^x| = 1$ **28.** $|3e^x + 1| = 2$

29. To solve $\log_5 x = \log 8$, can you use the method of equating the arguments of the logarithms? Explain.

30. Solving $\log x + \log(x + 3) = \log 18$ leads to two values for x: 3 and -6. Assuming that all steps of the solving process are correct, what is the solution of the equation? Why?

In Exercises 31–44, solve for x.

31. $\log_5 x = \log_5(2x - 3)$

32. $\log(3 - x) = \log(2x)$

33. $\log_5(x - 1) = 2$

34. $7 + \log_2 (x + 1) = 10$

35. $\log_3 \sqrt[3]{3x - 5} = 1$

36. $1 + \log_2 \sqrt{x} = 5$

37. $\frac{1}{3} \log_8 (x - 7) = \log_8 3$

38. $\frac{1}{2} \ln (3x - 1) = \ln 5$

39. $\log_2 (x^2 - 7x) = 3$

40. $\log_6 x(x + 1) = 1$

41. $\log_4 (5 - 3x)^3 = 6$

42. $\log_2 (x^2 + 4x)^2 = 10$

43. $\frac{1}{2} \log_9 (6 - x) = \log_9 x$

44. $\frac{1}{2} \log_2 (x - 1) = \log_2 (x - 3)$

In Exercises 45–62, solve the given equation.

In this group, use properties of logarithms to combine logarithms into a single logarithmic expression.

45. $2 - \log_9 x^2 = 1$

46. $1 + \log_4 (x + 1)^2 = 4$

47. $\log_4 6x - \log_4 (x - 5) = 2$

48. $\log (3t - 1) = 1 + \log (5t - 2)$

49. $\log x + \log (7 - x) = 1$

50. $2 \log_3 x = \log_3 (5x - 4)$

51. $1 - \log (x - 5) = \log \left(\frac{x}{5}\right)$

52. $\log t = 2 - \log (t + 21)$

53. $\log_5 (2x - 3) - \log_5 x = 1$

54. $\log_2 (4x + 5) - \log_2 (x - 7) = 2$

55. $\log_5 x(x + 2) = 0$

56. $\log_5 x(x + 3) = 1$

57. $\log x + \log (4 + x) = \log 5$

58. $\ln (2x + 1) = \ln 14 - \ln (x + 2)$

59. $\log (x + 6) - 2 \log x = \log 12$

60. $\ln 3 = \ln (6 - 7x) - 2 \ln x$

61. $|2 - \log_2 (x - 1)| = 3$

62. $\log_6 |2x + 1| = 1$

63. The following is one of the basic exponent properties.

For $b > 0$ and $b \neq 1$, if $b^x = b^y$, then $x = y$.

Using an example, explain why $b \neq 1$ is a necessary condition.

64. Use what you know about the graphs of the functions $f(x) = e^x$ and $g(x) = \ln x$ to explain why the equation $e^x = \ln x$ has no solution.

CE In Exercises 65–72, each equation is quadratic in form. Use the substitution methods discussed in Section 10.4 to solve the equation.

65. $(\log x)^2 + \log x^3 - 4 = 0$

66. $(\ln x)^2 - \ln x - 6 = 0$

67. $\log_4 x - 3\sqrt{\log_4 x} + 2 = 0$

68. $\log_3 x - 4\sqrt{\log_3 x} + 3 = 0$

69. $e^{2x} + 6 = 5e^x$

70. $e^{2x} - 4e^x + 3 = 0$

71. $5^{2x} = 5^x + 42$

72. $7^{2x} + 7^x - 2 = 0$

CE In Exercises 73–80, solve for the indicated variable.

73. $R = \log I$ for I

74. $P = \log \left(\frac{10P_I}{P_0}\right)$ for P_0

75. $L = 10 \log \left(\frac{I}{I_0}\right)$ for I

76. $\ln M = \ln Q - \ln (1 - Q)$ for Q

77. $P = P_0 e^{rt}$ for r

78. $A = A_0 e^{-kt}$ for k

79. $2 = \left(1 + \frac{r}{n}\right)^{nt}$ for t

80. $A = P(1 + r)^t$ for t

In the following application problems, the required formula is given in each case. After substitution of the given values, the resulting equation can be solved with known methods.

Real-Life Applications

Compound Interest

If P dollars is invested at an interest rate r for a period of t years and if the interest is compounded continuously, then the value A of the investment is given by $A = Pe^{rt}$.

81. A certain bank advertises that customers can double their investment in 14 years because interest is compounded continuously. What is the annual interest rate offered by the bank?

82. At the time of their first child's birth, a couple invested \$5000 at 5.5% compounded continuously. Their plan was to give the money to the child when the value of the investment reached \$20,000. At what age will the child receive the money?

Exponential Growth

During certain periods of time, populations of people or other life forms may grow exponentially. Models describing such growth may be exponential equations.

83. A planning study for a small country predicted that the country's population P (in millions) could be modeled as a function of the number t of years since 1990: $P(t) = 14e^{0.03t}$.

(a) What was the population in 1990?

(b) What will the population be in 2010?

(c) What will the population be in 2020?

84. In an industrial city, the number of toxic particles in the atmosphere is found to be increasing over time according to the formula $P(t) = 400 \cdot 2^{0.4t}$ where P is measured in standard units and t is the number of years since 1994. Estimate the number of toxic particles in the year 2006.

Exponential Decay

Certain radioactive substances decay according to the formula $P(t) = P_0e^{rt}$, where P_0 is the original amount of the substance, $P(t)$ is the amount of substance remaining after time t, and r is the rate of decay.

The *half-life* of a substance is the time t_h it takes for half of the original amount of the substance to remain: $P(t_h) = \frac{1}{2}P_0$.

85. The half-life of carbon-14 is 5730 years. In how many years will only 30% of an original amount of carbon-14 remain?

86. What is the half-life of a substance when only 30% of the substance remains after 5 days?

Depreciation

Accountants refer to the decrease in the value of an item over time as *depreciation*. Depreciation is sometimes modeled with an exponential equation.

87. A bulldozer depreciates so that the value after t years is given by $V(t) = V_0e^{-0.19t}$ where V_0 is the initial cost. After 7 years the bulldozer is worth \$22,480.57. What was the original cost?

88. A front-end loader depreciates so that the value after t years is given by $V(t) = V_0e^{-0.27t}$ where V_0 is the initial cost. After 6 years the front-end loader is worth \$9894.94. What was the original cost?

Law of Cooling

The temperature T of an object after an elapsed period of time t is given by the formula $T = T_m + D_0e^{kt}$, where T_m is the constant temperature of the surrounding medium, D_0 is the difference between the initial temperature of the object and the temperature of the surrounding medium, and k is a cooling constant that depends on the composition of the object.

89. The coroner arrived at the murder scene at 8:00 A.M. The body was in a 60°F room and the body's temperature was 88°F. After 45 minutes, the body's temperature was 85°F. If you assume that a normal body temperature is 98.6°F, at what time did the murder occur?

90. A cake is removed from a 350°F oven and placed in a room whose temperature is 60°F. How long will it take for the cake to cool to 75°F? (If t is in minutes, $k = -0.1$.)

Richter Scale

A Richter scale reading is given by $R = \log I$, where I is the number of times more intense an earthquake is than a very small quake.

91. The 1988 earthquake in Armenia registered 6.9 on the Richter scale. What would be the Richter scale reading for an earthquake that is 10 times as intense?

92. The Richter scale reading for the 1964 Alaskan earthquake was 8.5. For the 1985 Mexico City earthquake, the Richter scale reading was 7.8. How much more intense was the Alaskan earthquake?

Modeling with Real Data

93. The number of wild horses on Cumberland Island has increased dramatically since 1960 to the point where they endanger island beaches by eating sea oats that hold the sand dunes together and keep the beach from eroding. (Source: National Park Service.)

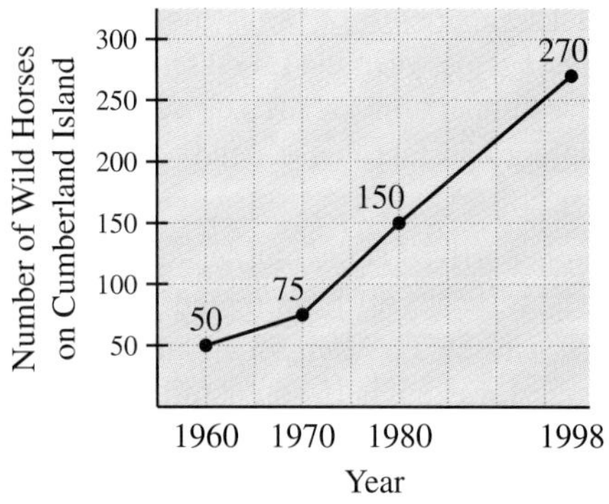

The population of horses can be modeled by $h(x) = 51.59(1.045)^x$, where x is the number of years since 1960.

(a) Write an equation to estimate the year in which the horse population is projected to reach 340.

(b) Solve the equation in part (a).

94. The bar graph shows the number (in millions) of instant lottery tickets sold for selected years from 1990 to 1998. (Source: 1999 World Lottery Almanac.)

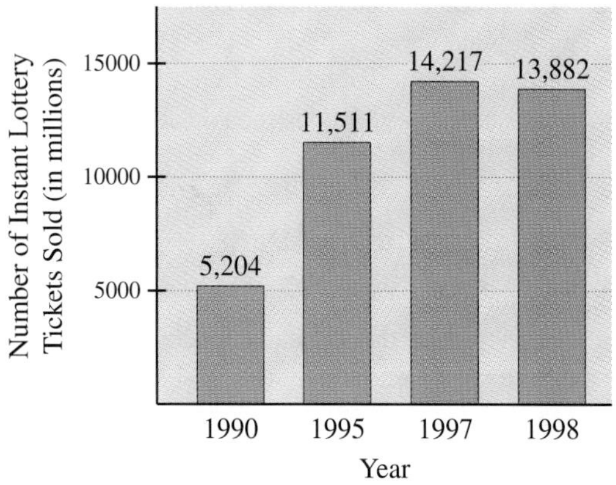

The function $T(x) = -30{,}761 + 15{,}637 \ln x$ models the data as a function of the number of years since 1980.

(a) Write an equation to determine the year in which sales will reach 17.5 billion.

(b) Solve the equation in part (a).

Data Analysis: *Debit Cards*

An increasing number of shoppers use bank cards that can be used at the point of sale (POS) to debit their checking accounts. The table shows that the number of such transactions has increased with the increasing number of terminals. (Source: Faulkner & Gray.)

Year	Point-of-Sale Transactions (in millions)	Point-of-Sale Terminals (in thousands)
1990	191	53
1993	430	155
1994	624	341
1995	775	529
1996	1146	875
1997	1442	1300
1998	1800	1700

The following functions model the data, where x is the number of years since 1990.

Transactions (millions): $P(x) = 190.6e^{0.26x}$

Terminals (thousands): $T(x) = 50.5e^{0.45x}$

95. Write and solve an equation to estimate the year in which the number of point-of-sale terminals will be 7 million.

96. Write and solve an equation to estimate the year in which the number of transactions will be approximately 12 billion.

97. Write a function $A(x)$ to model the number of transactions per terminal. (Remember that $P(x)$ is in millions whereas $T(x)$ is in thousands.) On the basis of the model $A(x)$, what is the trend in the number of transactions per terminal?

98. Write and solve an equation to determine the year in which the number of point-of-sale transactions per terminal will be about 320.

Challenge

In Exercises 99 and 100, $f(x) = \log x$ and $g(x) = \ln x$. Solve the given equation.

99. $(g \circ f)(x) = 1$

100. $(f \circ g)(x) = 1$

101. Solve the following equations and explain why the solution sets are different.

(a) $2 \log_3 x + \log_3 (10 - x^2) = 2$

(b) $\log_3 x^2 + \log_3 (10 - x^2) = 2$

102. The following is a proof of the Change of Base Formula. Fill in the blanks with a rule or definition that justifies each step.

$x = b^{\log_b x}$ (a) ______

$\log_a x = \log_a b^{\log_b x}$ (b) ______

$\log_a x = (\log_b x)(\log_a b)$ (c) ______

$\dfrac{\log_a x}{\log_a b} = \log_b x$ (d) ______

103. Consider the following equation.

$$\ln(3 - 4x) - \ln(x - 2) = \ln(2x + 3)$$

(a) Determine the domain of each logarithmic expression.

(b) On the basis of your results in part (a), what are the permissible replacements for x in the given equation?

(c) What is the solution set of the equation?

11 • Chapter Project: Telecommuting

In Section 11.4, we found that a logarithmic function is a suitable model for the given data about the trend in the number of telecommuters. Although a certain function may be a good global model for long time periods, it may not accurately reflect what went on during selected shorter periods.

The following bar graph expands on the data given in Section 11.4 and shows a little more history during the earlier years. (Source: Gil Gordon Associates.)

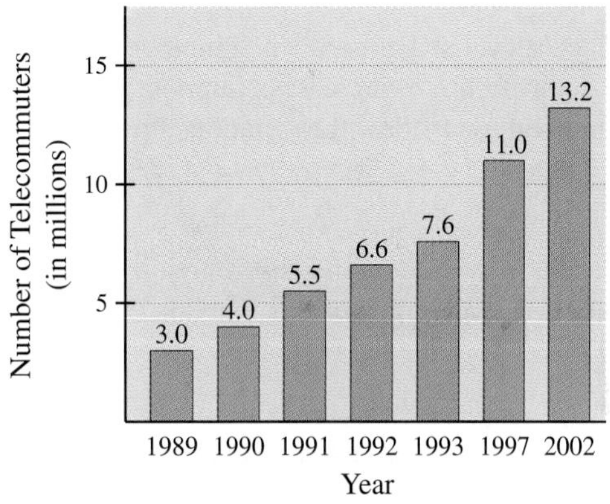

In the following exercises, let x represent the number of years since 1988.

1. Write ordered pairs with the first coordinate the number of years since 1988 and the second coordinate the number of telecommuters.
2. Using the ordered pairs in Exercise 1, plot the points associated with the first three years shown in the bar graph. For what reason might we choose an *exponential* function to model the data rather than a linear function?
3. Make a table of values for the exponential function $E(x) = 2.2(1.35)^x$ for the first three years. How well does the function appear to model the data for those years?
4. Plot the points associated with the data for 1989–1993. For what reason might we choose a *linear* function to model these data?
5. Plot the points associated with *all* the data. For what reason might we choose a *logarithmic* function to model these data?
6. Make a table of values for the logarithmic function $L(x) = 1.72 + 4.04 \ln x$ for the years in the bar graph. How well does the function appear to model the data for those years?

Chapter 11 Review Exercises

Section 11.1

In Exercises 1–8, let $f(x) = x^2 - 3x$, $g(x) = 3x + 4$, and $h(x) = \sqrt{x + 1}$. Determine each of the following.

1. $(f + g)(x)$

2. $(f - g)(2.4)$

3. $\left(\frac{f}{g}\right)(3.7)$

4. $(fg)(x)$

5. $(f \circ g)(x)$

6. $(g \circ h)(x)$

7. $(f \circ h)(3.56)$

8. $(h \circ f)(4.23)$

9. Let $f(x) = \sqrt{x - 1}$ and $g(x) = x + 1$. Because $g(-1)$ is defined, can we conclude that -1 is in the domain of $f \circ g$? Explain.

In Exercises 10 and 11, for the given functions f and g, determine $(f \circ g)(x)$ and $(g \circ f)(x)$.

10. $f(x) = 8, \quad g(x) = 2x^2 - 3x + 4$

11. $f(x) = \frac{1}{x + 2}, \quad g(x) = |2x + 5|$

12. For a 50-mile ride, a cyclist's time t (in minutes) increases as the temperature x increases. Suppose the following function describes the relationship between t and x, where $70 < x < 90$.

$$t(x) = 2x + 160$$

If the cyclist's rate is $r = 50/t$, write the function $(r \circ t)(x)$. What is your interpretation of this function?

Section 11.2

13. Explain the difference between the Vertical Line Test and the Horizontal Line Test.

In Exercises 14 and 15, determine whether the given graph represents a one-to-one function.

14.

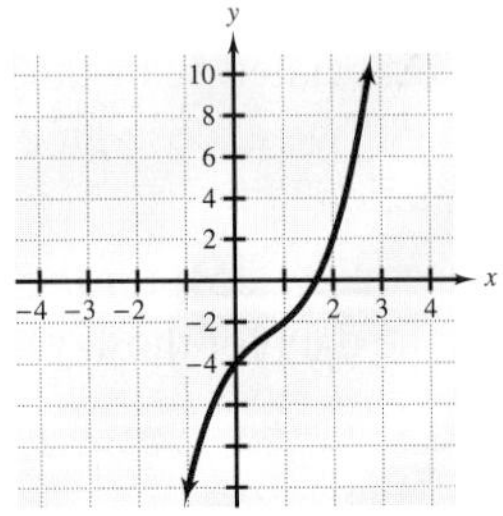

15.

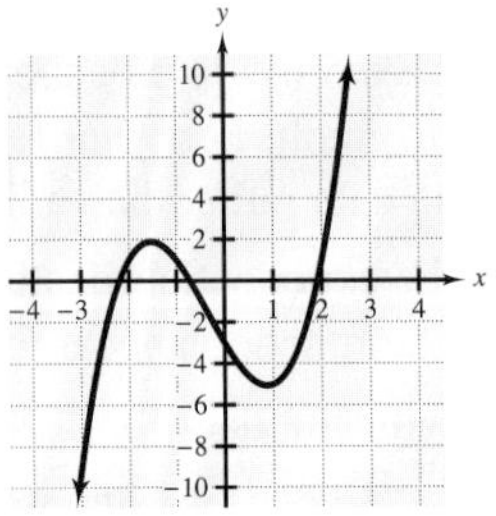

16. Determine whether the function is a one-to-one function.

(a) $f = \{(3, 6), (-1, 5), (5, 6), (1, 4)\}$

(b) $g = \{(7, 2), (-3, 6), (-2, -1), (0, 4)\}$

17. For each function in Exercise 16 that is one-to-one, determine the inverse function. State the domain and range of the inverse function.

In Exercises 18 and 19, determine whether the pairs of functions are inverses by comparing $f \circ g$ and $g \circ f$.

18. $f(x) = 1 - 2x, \quad g(x) = 2x + 1$

19. $f(x) = \frac{1}{x} + 1, \quad g(x) = \frac{1}{x - 1}$

In Exercises 20 and 21, determine f^{-1}. Then verify that $(f \circ f^{-1})(x) = x$ for all x in the domain of f^{-1}.

20. $f(x) = 5x + 8$

21. $f(x) = x^3 - 5$

22. What is the equation of the line of symmetry for the graphs of inverse functions? Describe the graphs of f and f^{-1} if the equation $f(x) = f^{-1}(x)$ has no solution.

Section 11.3

23. For the exponential function $f(x) = b^x$, what restrictions, if any, are placed on b and x?

24. For the graph of the exponential function $f(x) = b^x$, state each of the following.

(a) x-intercept

(b) y-intercept

(c) domain

(d) range

25. Evaluate the given function as indicated.

(a) $f(x) = -8e^{-2x}, \quad f(1)$

(b) $g(x) = \left(\frac{9}{16}\right)^{3x}, \quad g\left(\frac{1}{2}\right)$

26. Sketch the graph of $f(x) = 5 - 2^x$.

In Exercises 27–32, solve the given exponential equation.

27. $9^{2x} = 27$

28. $e^{2x-5} = 1$

29. $4^x = 2048$

30. $16^{2-3x} = 8^{5-6x}$

31. $\left(\frac{1}{27}\right)^{3x} = 9^{4x-5}$

32. $\left(\sqrt{5}\right)^{6-2x} = 25^{x^2}$

33. Suppose that you want to model the growth behavior of a certain quantity over time t, and you decide to use the function $f(t) = b^t$. Would the base b be greater than 1 or would it be between 0 and 1? Why?

34. Five years ago, some money was invested in an account at 7.5% interest compounded monthly. Today the account is worth \$7629.80. How much was originally invested?

Section 11.4

In Exercises 35 and 36, write the given exponential equation in logarithmic form.

35. $8^{2/3} = 4$ **36.** $49^{-1/2} = \frac{1}{7}$

In Exercises 37 and 38, write the given logarithmic equation in exponential form.

37. $\log_{25} \frac{1}{5} = -\frac{1}{2}$ **38.** $\log_{1/2} 16 = -4$

39. For the definition of the logarithmic function, explain why $b = 1$ is excluded.

40. What are the bases of the following logarithms?

(a) $\log_3 x$ (b) $\ln x$ (c) $\log x$

In Exercises 41–44, evaluate the logarithms.

41. $\ln e^{3.23}$ **42.** $\log_{27} \frac{1}{9}$

43. $\log_{1/4} 32$ **44.** $\log 100^{2.5}$

In Exercises 45 and 46, solve for x. Round the results to two decimal places.

45. $\ln x = 5.2$ **46.** $\log x = 5.2$

In Exercises 47–50, solve for x.

47. $\log_{343} x = \frac{2}{3}$ **48.** $\log_x 676 = 2$

49. $\log_4 32 = x$ **50.** $\log_x 128 = \frac{7}{3}$

51. For the graph of the function $f(x) = \log_b x$, state each of the following.

(a) x-intercept

(b) y-intercept

(c) domain

(d) range

52. Suppose $g(t) = \log_b t$, where $0 < b < 1$. In the 20th century, and for appropriate values of time t, would g be more likely to model the change in the cost of living or the incidence of polio? Why?

Section 11.5

In Exercises 53–56, use the basic properties of logarithms to simplify the given expression.

53. $\log_{21} 21^{41}$ **54.** $21^{\log_{21} 43}$

55. $\log_{21} 21$ **56.** $\log_b 1$

57. Describe how you can use the fact that $\log 7 \approx 0.845$ to estimate $\log 49$.

58. Let $\log_2 3 = C$ and $\log_2 5 = D$. Write

$$\log_2 \frac{5\sqrt[3]{60}}{24}$$

as an expression involving C and D.

In Exercises 59 and 60, use the properties of logarithms to expand the given expression. Write your answer in terms of logarithms of x, y, and z. Assume x, y, and z represent positive numbers.

59. $\log \frac{\sqrt[4]{x^2y^3}}{x^3z^4}$ **60.** $\ln \sqrt[6]{\frac{x^5y^4}{z^3}}$

In Exercises 61 and 62, use the properties of logarithms to combine the given expression into a single logarithmic expression. Assume x, y, and z represent positive numbers.

61. $\frac{1}{4}[3 \log (x + 4) - \log y - \log 2z]$

62. $3 \log (x + 2) - 5 \log (x + 1)$

63. Write $\log_2 24 + 2 \log_2 5 - \log_2 75$ as a single logarithm. Then evaluate it.

64. Write $\log_3 (x + 5)$ in terms of natural logarithms.

Section 11.6

65. Explain why $3^x = \frac{1}{9}$ can be solved by equating exponents, but $3^x = 4$ cannot. For the latter equation, what method should you use to solve it?

66. If you wished to solve $\log_2 x = 4$ by equating arguments of logarithms, you could replace 4 with $\log_2 16$. What is an easier method for solving the equation?

In Exercises 67–72, solve for x. Round your solutions to two decimal places.

67. $3^x = 7$

68. $e^{-x} = 0.78$

69. $\log_4 (x + 3) + \log_4 (x - 3) = 2$

70. $2^{x-1} = 3^{x+2}$

71. $\ln (x - 4) - \ln (3x - 10) = \ln \frac{1}{x}$

72. $\log_2 (x + 1) + \log_2 (3x - 5) = \log_2 (5x - 3) + 2$

73. If $a = 6.9e^{-2.3b}$, find b when $a = 0.45$.

74. Suppose \$7000 is invested in a savings fund in which interest is compounded continuously at the rate of 12% per year. How long will it take for the money to double in value?

75. Suppose the number N of bacteria present in a certain culture after t minutes is $N = ke^{0.04t}$ where k is a constant. If 5500 bacteria are present after 11 minutes, how many bacteria were present initially?

76. An amount of \$800 is deposited in a savings account and earns interest for 8 years at 5.25% compounded monthly. If there were no withdrawals or additional deposits, how much is in the account at the end of the 8 years?

Chapter 11 Test

For Questions 1–4, let $f(x) = x^2 - 4$, $g(x) = x + 2$, and $h(x) = \sqrt{x}$. Determine the combined function.

1. $(f + g)(x)$

2. $\left(\frac{f}{g}\right)(x)$

3. $(f \circ g)(x)$

4. $(h \circ g)(x)$

5. Verify that $f(x) = \frac{x - 5}{3}$ and $g(x) = 3x + 5$ are inverse functions.

6. For the function $f(x) = \sqrt[3]{x + 2}$, determine f^{-1}.

7. The graph of a function is given in the figure. Sketch the graph of the inverse function.

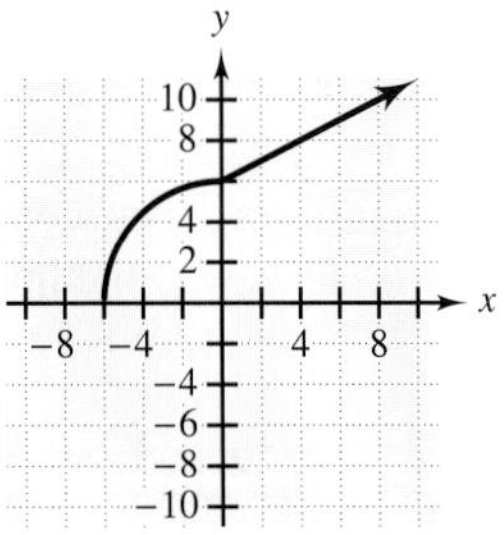

For Questions 8 and 9, evaluate the expression.

8. $\log_{25} 5$

9. $\log_2 4^{3/4}$

For Questions 10–13, solve the equation.

10. $5^{2-3x} = 25$

11. $\left(\frac{1}{16}\right)^{x-1} = 8$

12. $8^x = 3^{x+1}$

13. $e^{2x} = 10$

14. Suppose that you know the value of log 2. Explain how to evaluate $\log_2 10$.

15. Write the expression $\log\left(\frac{A^2}{\sqrt{B}}\right)$, where A and B represent positive numbers, in terms of $\log A$ and $\log B$.

For Questions 16 and 17, write the given expression as a single logarithmic expression.

16. $\log(x^2 - x - 2) - \log(x - 2)$

17. $2\log_6 x + \log_6(x - 1) - 1$

For Questions 18–21, solve the given equation.

18. $\log(2x + 1) = \log 5$

19. $\log x + \log(x + 3) = 1$

20. $\ln(2t + 1) - \ln t = \ln 3$

21. $2 + \ln x = 0$

22. At the birth of a child, \$5000 is invested in a college fund. What rate of interest, compounded continuously, is required for the investment to be worth \$23,000 eighteen years later?

23. The demand (in thousands) for a certain brand of soap t weeks after a major advertising campaign is given by $D(t) = 2 + 5(0.89)^t$.

(a) What was the demand when the ad campaign began?

(b) Using a graph of $y = D(t)$, state your opinion about the effectiveness of the ad campaign.

(c) After approximately how many weeks had the demand for this brand of soap fallen to half the initial demand?

24. The Richter scale rating of an earthquake is given by the formula $R = \log I$, where I is the number of times more intense the earthquake is than a very small quake. The San Francisco earthquake that occurred during the World Series in 1989 registered 7.1 on the Richter scale. How much more intense was the 1906 San Francisco earthquake, which registered 8.3 on the Richter scale?

25. The temperature T of an object after an elapsed period of time t is given by $T = T_m + D_0e^{kt}$, where T_m is the constant temperature of the surrounding medium, D_0 is the difference between the initial temperature of the object and the temperature of the surrounding medium, and k is a cooling constant.

A can of soda had been left in a hot car where the soda reached a temperature of 90°F. If the can was placed in a cooler of ice at 32°F, how long did the beverage take to chill to 40°F? (With time measured in hours, the cooling constant is $k = -1.32$.)

Cumulative Test, Chapters 9–11

1. Simplify. Assume that y represents any real number.

(a) $\sqrt{y^8}$ (b) $\sqrt{y^{10}}$

2. Determine the real number value, if any.

(a) $(-27)^{2/3}$ (b) $(-27)^{-2/3}$

(c) $(-16)^{3/4}$ (d) $-16^{-3/4}$

3. Simplify and write the result with positive exponents.

(a) $\dfrac{a^{-2}b^{1/3}}{a^{1/2}b}$ (b) $(x^{-1}y^2)^{-1/2}$

(c) $\left(\dfrac{x^{-1}}{y^2}\right)^{-3}$

4. Simplify.

(a) $\sqrt[3]{\dfrac{x^6}{8y^3}}$ (b) $\sqrt[5]{3x}\,\sqrt[5]{2x^2y}$ (c) $\dfrac{\sqrt{x^3y}}{\sqrt{xy^3}}$

5. Simplify.

(a) $\sqrt{20x^9}$ (b) $\sqrt[14]{y^{21}}$

6. Perform the indicated operations.

(a) $\sqrt{12x^3} - x\sqrt{3x}$ (b) $\left(\sqrt{a} + \sqrt{2}\right)^2$

7. What is the product of $3 - \sqrt{5}$ and its conjugate?

8. Rationalize the denominator.

(a) $\dfrac{5}{\sqrt[3]{2}}$ (b) $\dfrac{\sqrt{x}}{\sqrt{2x} + \sqrt{y}}$

9. Solve.

(a) $\sqrt{2x + 1} = x\sqrt{3}$ (b) $(2x - 3)^{2/3} = 9$

10. Simplify.

(a) i^{19} (b) $(3 - i)^2$ (c) $\dfrac{i}{2 + i}$

11. In addition to the method of completing the square, name three methods for solving a quadratic equation. Which of these three methods works for all quadratic equations?

12. Solve $(3x - 1)(2x + 3) = x(x - 2)$. Round your solution(s) to the nearest hundredth.

13. Six less than a number is 7 times the reciprocal of the number. What is the number?

14. If $ax^2 + bx + c = 0$ has two imaginary solutions, what is the relationship between b^2 and $4ac$? Why?

15. Solve $a + a^{1/2} = 12$.

16. A person jogged 2 miles and then walked back to the starting point. He jogged 3 mph faster than he walked. If the total time was 39 minutes, how fast did he jog?

17. Suppose $p(x) = x^2 + 2x - 24$. Solve each of the following.

(a) $p(x) > 0$

(b) $p(x) < -25$

(c) $p(x) > -26$

18. If $f(x) = ax^2 - 6x + c$, tell what you know about a or c for each of the following descriptions of the graph.

(a) The y-intercept is 3.

(b) The parabola opens downward.

(c) The x-coordinate of the vertex is -1.

19. Suppose that the daily profit P from producing x items is given by $P(x) = 8x - x^2$.

(a) How many items should be produced each day to maximize the profit?

(b) What will the maximum daily profit be?

20. For $f(x) = x^2$ and $g(x) = 2x$, evaluate each of the following.

(a) $(f + g)(0)$

(b) $(f - g)(2)$

(c) $(fg)(-2)$

21. For $f(x) = |x - 1|$ and $g(x) = 2x + 1$, evaluate each of the following, if possible.

(a) $(f \circ g)\left(-\frac{1}{2}\right)$ (b) $\left(\frac{f}{g}\right)\left(-\frac{1}{2}\right)$

22. If $f(x) = \frac{x}{x - 3}$, determine f^{-1}. Then show that $(f \circ f^{-1})(x) = x$ for all x in the domain of f^{-1} and that $(f^{-1} \circ f)(x) = x$ for all x in the domain of f.

23. What properties do the graphs of $f(x) = 5^x$ and $g(x) = \left(\frac{1}{5}\right)^x$ have in common? Describe how the graphs are different.

24. Solve each of the following equations.

(a) $\log_2 32 = x$ (b) $\log_3 (x - 1) = 2$

(c) $\log_x 7 = 2$

25. (a) Expand $\ln \frac{x^2 y}{\sqrt{z}}$, where $x, y, z > 0$.

(b) Write $3 \log (x - 1) - \log (x + 1)$, where $x > 1$, as a single logarithm.

26. Solve each of the following equations. Round your results to the nearest hundredth.

(a) $\log_3 x(x - 6) = 3$

(b) $3^{x+1} = 5$

(c) $4e^{2x} = 1$

27. For continuously compounded interest, the value A of an investment of P dollars after t years is given by $A = Pe^{rt}$, where r is the interest rate. Approximately what interest rate is needed for a \$1200 investment to be worth \$2187 after 10 years?

Appendix

Conic Sections

Circle • Parabola • Ellipse • Hyperbola

A **conic section** is obtained from the intersection of a plane with an **infinite right circular cone.** Figure A.1 shows that altering the angle of the plane produces four types of curves: the parabola, the circle, the ellipse, and the hyperbola.

Figure A.1

Conic sections can also be defined as the graph of second-degree equations of the form $Ax^2 + By^2 + Cx + Dy + E = 0$. We will use a third approach by defining conic sections geometrically in terms of distances.

Circle

Definition of a Circle

A **circle** is the set of points in a plane that are equidistant from a fixed point. The distance is called the **radius,** and the fixed point is called the **center.**

Figure A.2

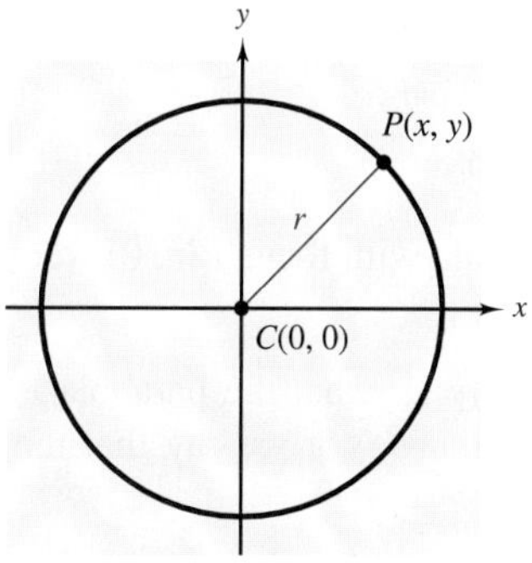

Figure A.2 shows a circle with center $C(0, 0)$ and radius r and a point $P(x, y)$ of the circle.

To write the equation of the circle, we apply the Distance Formula.

$$\sqrt{(x-0)^2 + (y-0)^2} = r \qquad \text{The distance between } C(0,0) \text{ and } P(x,y) \text{ is } r.$$

$$\sqrt{x^2 + y^2} = r \qquad \text{Simplify.}$$

$$x^2 + y^2 = r^2 \qquad \text{Square both sides of the equation.}$$

Equation of a Circle Whose Center Is the Origin

The standard form of the equation of a circle with center $C(0, 0)$ and radius $r > 0$ is

$$x^2 + y^2 = r^2$$

EXAMPLE 1

Writing an Equation of a Circle

Write an equation of a circle whose center is $C(0, 0)$ and that contains the point $P(3, -4)$.

Solution

We use the distance formula to determine the radius.

$$r = CP = \sqrt{(3 - 0)^2 + (-4 - 0)^2} = \sqrt{9 + 16} = \sqrt{25} = 5$$

The equation is $x^2 + y^2 = 25$.

Parabola

Definition of a Parabola

A **parabola** is the set of all points that are equidistant from a fixed point called the **focus** and a fixed line called the **directrix.**

Figure A.3

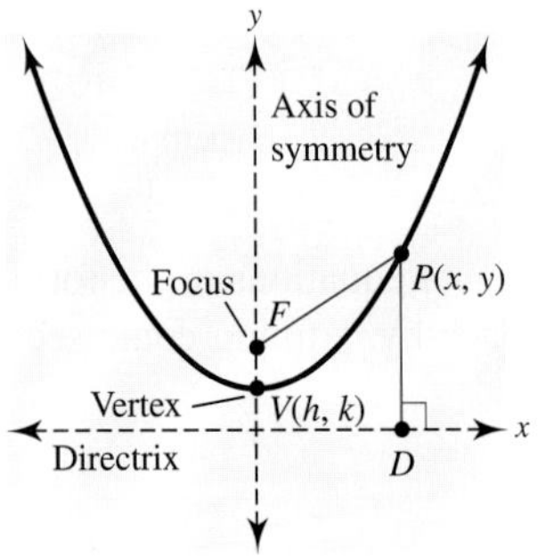

As shown in Figure A.3, the **vertex** is the midpoint between the focus and the directrix. The **axis of symmetry** is the line containing the vertex and the focus.

Figure A.4 shows a parabola with focus $F(0, c)$, vertex $V(0, 0)$, and directrix $y = -c$. To write the equation of this parabola, we use the fact that for any point $P(x, y)$ of the parabola, $FP = PQ$.

Figure A.4

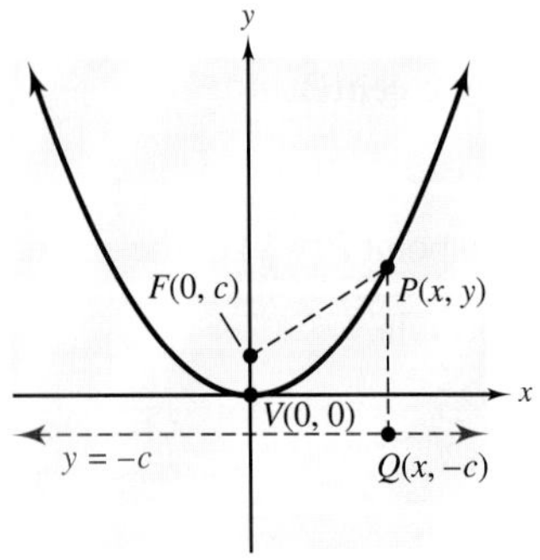

$$\sqrt{(x - 0)^2 + (y - c)^2} = \sqrt{(x - x)^2 + [y - (-c)]^2} \qquad FP = FQ$$

$$\sqrt{x^2 + (y - c)^2} = \sqrt{(y + c)^2} \qquad \text{Simplify.}$$

$$x^2 + (y - c)^2 = (y + c)^2 \qquad \text{Square both sides.}$$

$$x^2 + y^2 - 2cy + c^2 = y^2 + 2cy + c^2 \qquad \text{Simplify.}$$

$$x^2 = 4cy \qquad \text{Solve for } x^2.$$

A similar derivation gives the equation $y^2 = 4cx$ for a parabola with focus $F(c, 0)$, vertex $V(0, 0)$, and directrix $x = -c$. (See Fig. A.5.)

When the axis of symmetry is vertical, as in Figure A.4, we say that the parabola is *vertical;* when the axis of symmetry is horizontal, as in Figure A.5, we say that the parabola is *horizontal.*

Figure A.5

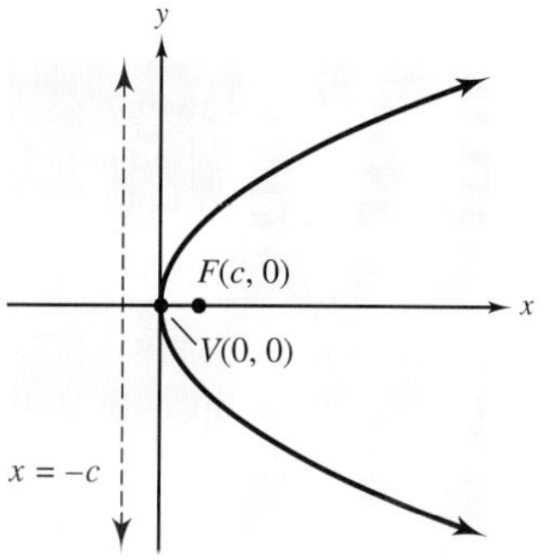

Equations and Properties of a Parabola Whose Vertex Is the Origin

	Vertical Parabola	*Horizontal Parabola*
Equation	$x^2 = 4cy$	$y^2 = 4cx$
Focus	$F(0, c)$	$F(c, 0)$
Directrix	$y = -c$	$x = -c$
Axis of symmetry	y-axis	x-axis
Opens	Upward if $c > 0$ Downward if $c < 0$	Right if $c > 0$ Left if $c < 0$

EXAMPLE 2

Figure A.6

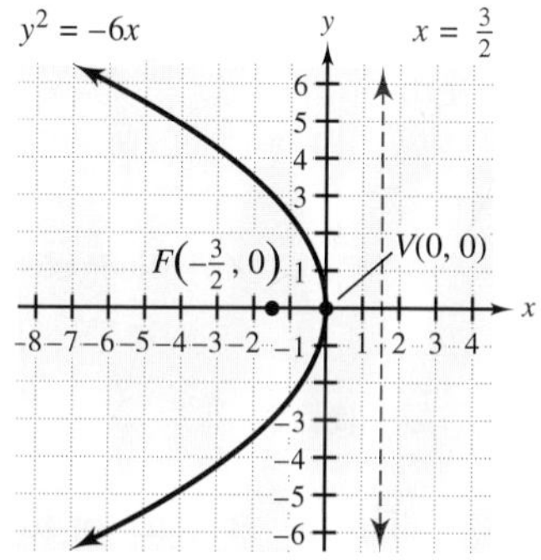

Determining the Focus and Directrix of a Parabola

Determine the focus and directrix of the parabola whose equation is $y^2 = -6x$.

Solution

The equation $y^2 = -6x$ describes a horizontal parabola with focus $F(c, 0)$ and directrix $x = -c$.

Comparing $y^2 = -6x$ to the model $y^2 = 4cx$, we see that $4c = -6$, and so $c = -\frac{3}{2}$. Therefore, the focus is $F\left(-\frac{3}{2}, 0\right)$ and the directrix is $x = -\left(-\frac{3}{2}\right) = \frac{3}{2}$. (See Fig. A.6.)

Note that because $c < 0$, the parabola opens to the left.

When we are given an equation of a parabola, as in Example 2, we can use it to determine the characteristics of the curve. Conversely, if we have enough information about a parabola, then we can use one of the models to write its equation.

EXAMPLE 3

Figure A.7

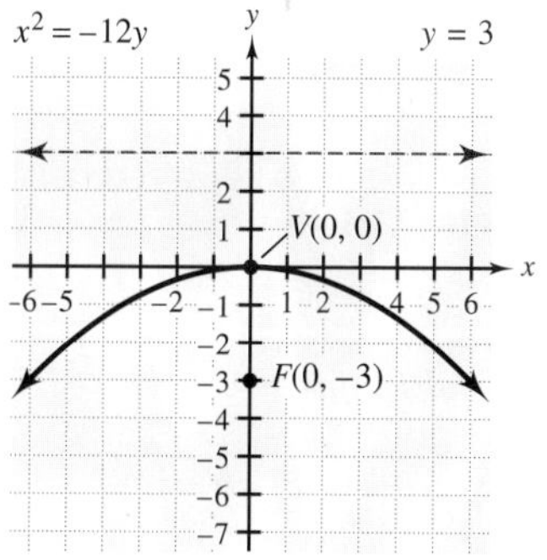

Writing an Equation of a Parabola

Write an equation of the parabola whose focus is $F(0, -3)$ and whose vertex is $V(0, 0)$.

Solution

The focus is $F(0, -3)$, so we know that $c = -3$. Because the focus is a point of the y-axis and c is negative, the parabola is vertical and opens downward.

$x^2 = 4cy$ Model for vertical parabola

$x^2 = -12y$ $c = -3$

Figure A.7 shows the parabola. The equation of the directrix is $y = -c = -(-3) = 3$.

Ellipse

A circle has just one center, and a parabola has just one focus. The shape of the next conic section that we will discuss, the ellipse, is governed by two points, each called a focus. (The plural of *focus* is *foci,* pronounced fō-sī.)

Definition of an Ellipse

An **ellipse** is the set of all points in a plane such that the sum of the distances from each point to two fixed points is a constant. Each of the two fixed points is called a **focus** of the ellipse, and the two together are called the **foci** of the ellipse.

Figure A.8(a) shows an ellipse with foci $F_1(-c, 0)$ and $F_2(c, 0)$. Figure A.8(b) shows an ellipse with foci $F_1(0, -c)$ and $F_2(0, c)$. For any point $P(x, y)$ of the ellipse, by definition, $PF_1 + PF_2 = k$, where k is a constant.

Figure A.8

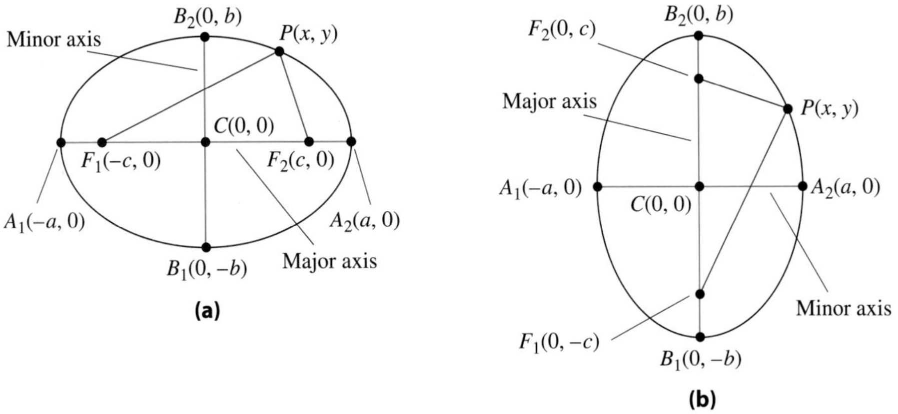

The line containing the foci intersects the ellipse at points called the **vertices** of the ellipse. The line segment connecting the vertices is the **major axis,** and its midpoint is the **center** of the ellipse. If the major axis is horizontal, we call the ellipse a *horizontal* ellipse; if the major axis is vertical, we call the ellipse a *vertical* ellipse.

The line perpendicular to the major axis at the center of the ellipse intersects the ellipse at points called **co-vertices.** The line segment connecting the co-vertices is the **minor axis.**

Using the definition of an ellipse and the Distance Formula, we can derive the equation for an ellipse whose center is the origin.

Equations and Properties of an Ellipse Whose Center Is the Origin

The following properties hold for $a > 0$ and $b > 0$.

	Horizontal Ellipse	*Vertical Ellipse*
Equation	$\frac{x^2}{a^2} + \frac{y^2}{b^2} = 1, a > b$	$\frac{x^2}{a^2} + \frac{y^2}{b^2} = 1, a < b$
Vertices	$A_1(-a, 0)$ and $A_2(a, 0)$	$B_1(0, -b)$ and $B_2(0, b)$
Co-vertices	$B_1(0, -b)$ and $B_2(0, b)$	$A_1(-a, 0)$ and $A_2(a, 0)$
Foci	$F_1(-c, 0)$ and $F_2(c, 0)$, where $c^2 = a^2 - b^2$	$F_1(0, -c)$ and $F_2(0, c)$, where $c^2 = b^2 - a^2$

To sketch an ellipse, the intercepts $A(\pm a, 0)$ and $B(0, \pm b)$ are the key points to plot.

EXAMPLE 4

Sketching an Ellipse

Sketch the ellipse whose equation is $9x^2 + 16y^2 = 144$.

Figure A.9

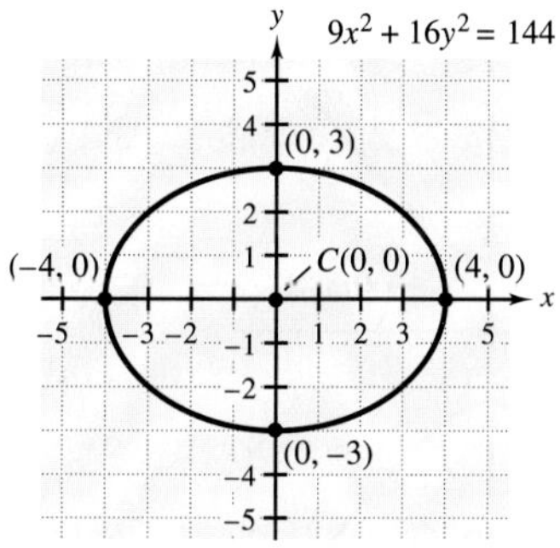

Solution

$$\frac{9x^2}{144} + \frac{16y^2}{144} = \frac{144}{144}$$ Divide both sides by 144 to write the equation in standard form.

$$\frac{x^2}{16} + \frac{y^2}{9} = 1$$ From this form, we can see that $a = 4$ and $b = 3$.

Because $a > b$, the ellipse is a horizontal ellipse. The center is $C(0, 0)$, and the intercepts are $(\pm 4, 0)$ and $(0, \pm 3)$. (See Fig. A.9.)

EXAMPLE 5

Writing an Equation of an Ellipse

Write an equation of the ellipse whose foci are $(0, \pm 4)$ and whose x-intercepts are $(\pm 3, 0)$.

Solution

Because the foci are points of the y-axis, the ellipse is a vertical ellipse. From the foci, we see that $c = 4$, and from the x-intercepts, we see that $a = 3$. We use those values to determine b.

$$c^2 = b^2 - a^2$$
$$16 = b^2 - 9$$
$$b^2 = 25$$
$$b = 5$$

We use $a = 3$ and $b = 5$ to write an equation of the ellipse.

$$\frac{x^2}{9} + \frac{y^2}{25} = 1$$

Hyperbola

Figure A.10

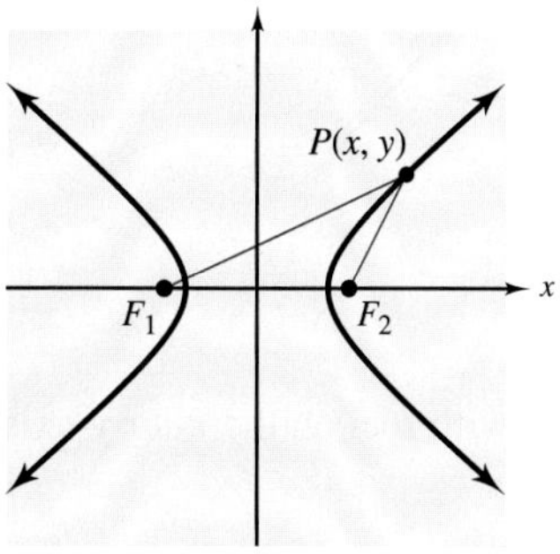

Definition of a Hyperbola

A **hyperbola** is the set of all points in a plane such that the absolute value of the difference of the distances from each point to two fixed points is a constant. The fixed points are called **foci,** and the line containing the foci is the **transverse axis.**

Figure A.10 shows a hyperbola whose center is the origin and whose foci F_1 and F_2 are points of the x-axis. From the definition of a hyperbola, $|PF_1 - PF_2| = k$, where k is a constant.

If the transverse axis is horizontal, we say that the hyperbola is *horizontal* [see Fig. A.11(a)]. If the transverse axis is vertical, we say that the hyperbola is *vertical* [see Fig. A.11(b)].

Figure A.11

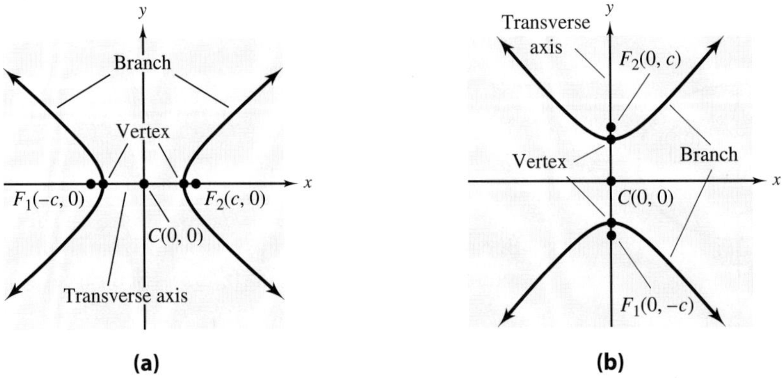

The two separate parts of the graph are called **branches.** Each branch intersects the transverse axis at a point called the **vertex.** The midpoint of the line segment connecting the vertices is the **center** of the hyperbola.

As shown in Figure A.12, as the two branches are extended, they become closer to (but never touch) two lines called **asymptotes.** The equations of the asymptotes are $y = \pm\frac{b}{a}x$. The asymptotes of the hyperbola are the extended diagonals of the **central rectangle.** Observe that the midpoints of the sides of the central rectangle are $(\pm a, 0)$ and $(0, \pm b)$.

Figure A.12

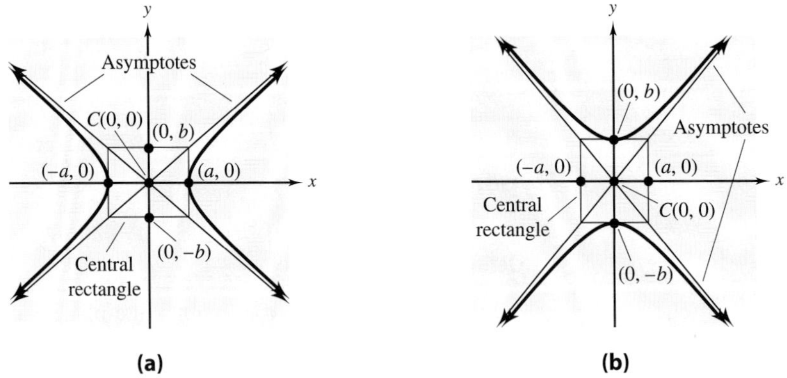

To write an equation of a hyperbola centered at the origin with foci on the x- or y-axis, we use the definition and the Distance Formula.

Equations and Properties of a Hyperbola Whose Center Is the Origin

The following properties hold for $a > 0$ and $b > 0$.

	Horizontal Hyperbola	*Vertical Hyperbola*
Equation	$\frac{x^2}{a^2} - \frac{y^2}{b^2} = 1$	$\frac{y^2}{b^2} - \frac{x^2}{a^2} = 1$
Vertices	$V_1(-a, 0)$ and $V_2(a, 0)$	$V_1(0, -b)$ and $V_2(0, b)$
Foci	$F_1(-c, 0)$ and $F_2(c, 0)$, where $c^2 = a^2 + b^2$	$F_1(0, -c)$ and $F_2(0, c)$, where $c^2 = a^2 + b^2$

To sketch a hyperbola, we begin by using the values of a and b to draw the central rectangle and asymptotes.

EXAMPLE 6

Sketching a Hyperbola

Describe the hyperbola whose equation is $9x^2 - 25y^2 = 225$.

Solution

$$\frac{9x^2}{225} - \frac{25y^2}{225} = \frac{225}{225} \quad \text{Divide both sides by 225 to write the equation in standard form.}$$

$$\frac{x^2}{25} - \frac{y^2}{9} = 1 \quad \text{Standard form of a horizontal hyperbola}$$

With the equation in standard form, we can see that a is 5, which means that the vertices are $V_1(-5, 0)$ and $V_2(5, 0)$. We can also see that b is 3, which means that the midpoints of the other sides of the central rectangle are $(0, -3)$ and $(0, 3)$. With this information, we can draw the central rectangle and the asymptotes. [See Fig. A.13(a).]

Using Figure A.13(a) as a template, we can sketch the hyperbola, as shown in Figure A.13(b).

Figure A.13

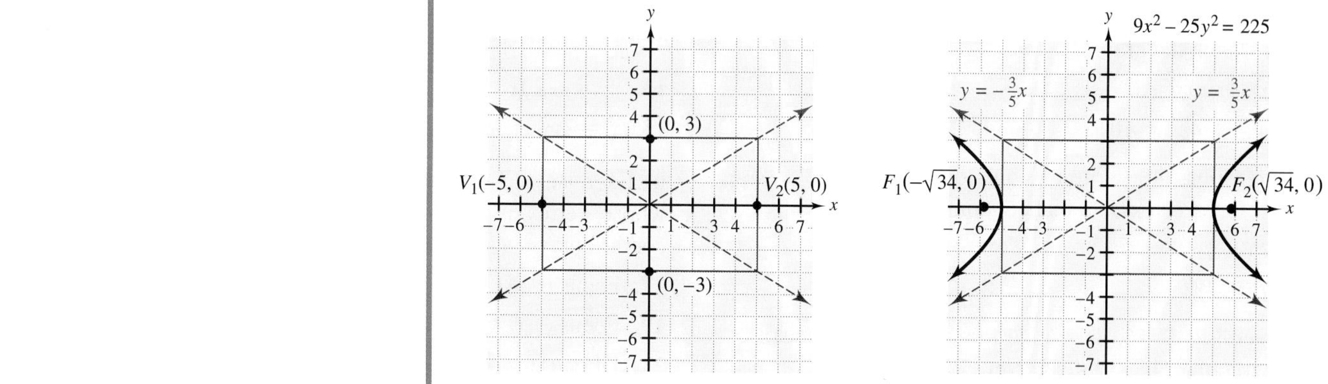

(a) (b)

To determine the foci, we use the relationship

$$c^2 = a^2 + b^2$$
$$c^2 = 5^2 + 3^2 = 25 + 9 = 34$$
$$c = \pm\sqrt{34} \approx \pm 5.83$$

Thus the foci are $F_1\left(-\sqrt{34}, 0\right)$ and $F_2\left(\sqrt{34}, 0\right)$.

The equations of the asymptotes are $y = \pm\frac{b}{a}x = \pm\frac{3}{5}x$.

EXAMPLE 7

Writing an Equation of a Hyperbola

Write an equation of a hyperbola with vertices $V(0, \pm 2)$ and foci $F(0, \pm 3)$.

Solution

Because the vertices are points of the y-axis, the hyperbola is vertical and the model equation is

$$\frac{y^2}{b^2} - \frac{x^2}{a^2} = 1$$

From the given vertices, we know that $b = 2$, and from the given foci, we know that $c = 3$. To determine a^2, we use the relationship $c^2 = a^2 + b^2$ or $a^2 = c^2 - b^2$.

$$a^2 = c^2 - b^2$$
$$a^2 = 3^2 - 2^2 = 9 - 4 = 5$$

Now we use the values of b^2 and a^2 to write the equation.

$$\frac{y^2}{4} - \frac{x^2}{5} = 1 \qquad b^2 = 4 \text{ and } a^2 = 5$$

Appendix Exercises

In Exercises 1–4, determine the radius of the circle. Then sketch the graph.

1. $x^2 + y^2 = 16$ **2.** $x^2 + y^2 = 25$

3. $4x^2 + 4y^2 = 40$ **4.** $2x^2 + 2y^2 = 16$

In Exercises 5–8, write an equation of the circle whose center is the origin and that contains the given point.

5. $P(3, 4)$ **6.** $P(3, -5)$

7. $P(0, 7)$ **8.** $P(-2, -6)$

In Exercises 9–16, determine the focus and directrix of the parabola. Then sketch the graph.

9. $x^2 = y$ **10.** $x^2 = -4y$

11. $y = -\frac{1}{2}x^2$ **12.** $y = 3x^2$

13. $y^2 = -20x$ **14.** $y^2 = 3x$

15. $x = 2y^2$ **16.** $x = -y^2$

In Exercises 17–24, write an equation of the parabola whose vertex is the origin and that satisfies the given conditions.

17. Focus $(0, 4)$ **18.** Focus $(0, -2)$

19. Focus $(-6, 0)$ **20.** Focus $(0.5, 0)$

21. Directrix $y = -2$ **22.** Directrix $x = 3$

23. Vertical parabola containing $P(6, -3)$

24. Horizontal parabola containing $P(10, 5)$

In Exercises 25–32, determine the vertices and co-vertices of the ellipse. Then sketch the graph.

25. $\frac{x^2}{4} + \frac{y^2}{9} = 1$

26. $\frac{x^2}{16} + \frac{y^2}{4} = 1$

27. $x^2 + \frac{y^2}{36} = 1$

28. $\frac{x^2}{10} + \frac{y^2}{25} = 1$

29. $25x^2 + y^2 = 25$

30. $3x^2 + 4y^2 = 12$

31. $4x^2 + y^2 = 16$

32. $x^2 + 9y^2 = 9$

In Exercises 33–40, write an equation of an ellipse whose center is the origin and that satisfies the given conditions.

33. Vertices $(0, \pm 3)$, length of minor axis is 2

34. Vertices $(\pm 4, 0)$, co-vertices $(0, \pm 3)$

35. Vertices $(\pm 4, 0)$, foci $(\pm 3, 0)$

36. Vertices $(0, \pm 5)$, foci $(0, \pm 3)$

37. Horizontal ellipse, length of major axis is 10, length of minor axis is 6

38. Vertical ellipse, length of major axis is 12, foci $(0, \pm 5)$

39. Vertices $(\pm 7, 0)$, contains $P(0, -6)$

40. Horizontal ellipse, length of major axis is 10, contains $P(-4, 2.4)$

In Exercises 41–48, determine the vertices of the hyperbola and the equations of the asymptotes. Then sketch the graph.

41. $\frac{x^2}{9} - y^2 = 1$

42. $\frac{x^2}{16} - \frac{y^2}{25} = 1$

43. $y^2 - x^2 = 36$

44. $4x^2 - y^2 = 16$

45. $9x^2 - 4y^2 = 4$

46. $y^2 - 8x^2 = 8$

47. $4y^2 - 9x^2 = 36$

48. $4x^2 - y^2 = 36$

In Exercises 49–54, write an equation of the hyperbola whose center is the origin and that satisfies the given conditions.

49. Vertices $(\pm 4, 0)$, foci $(\pm 5, 0)$

50. Vertices $(0, \pm 2)$, foci $(0, \pm 3)$

51. Vertices $(0, \pm 3)$, asymptotes $y = \pm\frac{3}{4}x$

52. Vertices $(\pm 5, 0)$, asymptotes $y = \pm 2x$

53. Horizontal hyperbola, midpoints of the central rectangle are $(\pm 6, 0)$ and $(0, \pm 4)$

54. Vertical hyperbola, midpoints of the central rectangle are $(\pm 5, 0)$ and $(0, \pm 2)$

Chapter 1

Think About It

Section 1.1 *(page 2)*

The number 0.121121112 . . . has a pattern, but it is not a repeating pattern. The number is not a rational number.

Section 1.2 *(page 11)*

Once the fractions have been written with a common denominator, the fraction with the larger absolute value is the fraction whose numerator has the larger absolute value.

Section 1.3 *(page 19)*

If the coordinates of the points are x and y, then the distance between the points is $|x - y|$, regardless of the values of x and y. Thus the positioning of the number line does not affect the result.

Section 1.4 *(page 25)*

Because $9.987 \approx 10$ and $-2.01 \approx -2$, the product in (i) is easier to estimate as -20. More rounding error would occur in (ii).

Section 1.5 *(page 32)*

To divide a number by a fraction, we multiply the number by the reciprocal of the fraction. Therefore, $n \div 2 = n \cdot \frac{1}{2}$, not $n \div \frac{1}{2}$. An equivalent way to write $n \div \frac{1}{2}$ is $n \cdot 2$.

Section 1.6 *(page 37)*

If $a = b$, then $(a - b)^2 = 0$. Otherwise, squaring $a - b$ means multiplying two numbers with the same sign, in which case the result is positive. Therefore, $(a - b)^2$ is nonnegative for all a and b.

Section 1.7 *(page 46)*

To evaluate $3(7 + 2)$, the Order of Operations can be used to add 7 and 2 and then multiply the result by 3. The Order of Operations cannot be used to simplify $3(x + 2)$, because we cannot add x and 2. The Distributive Property must be applied in this case.

Speaking the Language

Section 1.1 *(page 6)*

1. variable **2.** graph **3.** the opposite of x
4. absolute value

Section 1.2 *(page 13)*

1. addends **2.** like signs
3. approximately equal to **4.** grouping symbols

Section 1.3 *(page 21)*

1. difference **2.** subtrahend
3. distance between the points **4.** inequality

Section 1.4 *(page 27)*

1. factors **2.** have like signs **3.** odd
4. numerators and the denominators

Section 1.5 *(page 33)*

1. divisor **2.** undefined, zero **3.** reciprocal
4. opposites

Section 1.6 *(page 40)*

1. exponent **2.** base **3.** radicand
4. Order of Operations

Section 1.7 *(page 49)*

1. associative **2.** reciprocal **3.** Distributive
4. factoring

Chapter Project *(page 52)*

1,2.

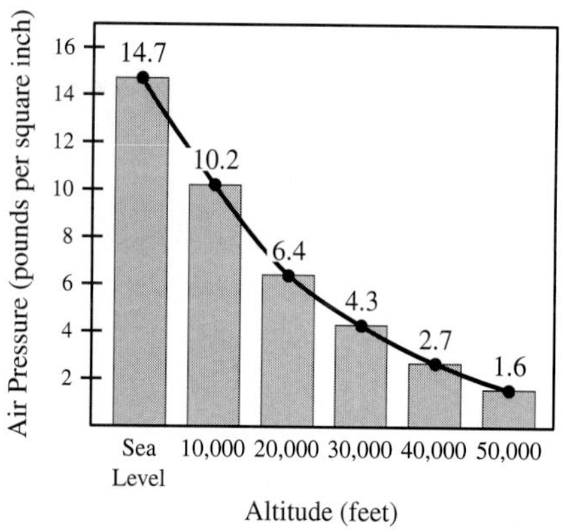

3.

Highest Point	*Estimated Air Pressure*
Mount Elbert	8.5
Iron Mountain	14.5
Charles Mound	14.1
Durfee Hill	14.3
Clingmans Dome	11.7
Mount Rainier	8.5

4. Answers will vary. **5.** 29,000 feet
6. Difference in elevation: 20,602 feet
Difference in air pressure: 8.5 pounds per square inch

Looking Back (page 55)

1. 5 **2.** $\frac{2}{5}$ **3.** 5 **4.** 23 **5.** \$315 **6.** 126 miles

7. $6x - 21$ **8.** $-3 + y$ **9.** $(a + 3)x$ **10.** $-4(a + b)$

11. 10 **12.** The squares of opposites are equal.

Chapter 2

Think About It

Section 2.1 (page 60)

The sum of the lengths of any two sides of a triangle must exceed the length of the third side. In this case, $2 + 3 < 7$, so 2, 3, and 7 cannot be the lengths of the sides of a triangle.

Section 2.2 (page 66)

3; 2; 1

Section 2.3 (page 74)

The system is a coordinate space consisting of three number lines (axes) that are mutually perpendicular.

Section 2.4 (page 85)

Assuming no aliases, the relation is a function. The relation formed by reversing the pairs is not a function, because a given name can be paired with more than one person.

Section 2.5 (page 96)

$$f(a + b) = a + b + 1$$
$$f(a) = a + 1$$
$$f(b) = b + 1$$
$$f(a) + f(b) = (a + 1) + (b + 1) = a + b + 2$$

Therefore, $f(a + b) \neq f(a) + f(b)$.

Speaking the Language

Section 2.1 (page 62)

1. hypotenuse **2.** algebraic, numerical **3.** solution **4.** Pythagorean Theorem

Section 2.2 (page 69)

1. coefficients **2.** Distributive **3.** equivalent **4.** like

Section 2.3 (page 78)

1. quadrants **2.** first (or x-) coordinate **3.** y-axis **4.** origin

Section 2.4 (page 90)

1. relation **2.** function **3.** Vertical Line Test **4.** domain

Section 2.5 (page 96)

1. variable **2.** value of y **3.** x-coordinate **4.** evaluating

Chapter Project (page 101)

1.

t	*Actual Percent A*	*Modeled Percent M*	$M - A$	$(M - A)^2$
0	18%	18.7%	0.7	0.49
2	30	29.4	−0.6	0.36
4	41	40.1	−0.9	0.81
6	50	50.8	0.8	0.64

2. $D \approx 1.52$

3. (i)

t	*Actual Percent A*	*Modeled Percent M*	$M - A$	$(M - A)^2$
0	18%	18%	0	0
2	30	30	0	0
4	41	42	1	1
6	50	54	4	16

(ii)

t	*Actual Percent A*	*Modeled Percent M*	$M - A$	$(M - A)^2$
0	18%	19%	1	1
2	30	30	0	0
4	41	41	0	0
6	50	52	2	4

(iii)

t	*Actual Percent A*	*Modeled Percent M*	$M - A$	$(M - A)^2$
0	18%	20%	2	4
2	30	30	0	0
4	41	40	−1	1
6	50	50	0	0

(iv)

t	*Actual Percent A*	*Modeled Percent M*	$M - A$	$(M - A)^2$
0	18%	18%	0	0
2	30	29.5	−0.5	0.25
4	41	41	0	0
6	50	52.5	2.5	6.25

4. (i) $D \approx 4.12$

(ii) $D \approx 2.24$

(iii) $D \approx 2.24$

(iv) $D \approx 2.55$

5. Answers will vary.

Looking Back (page 104)

1. −3 **2.** 30 **3.** 2 **4.** 7 **5.** $5x + 8$ **6.** $-6x + 5$

7. Yes **8.** No

9. **(a)** $4L - 18$ **(b)** $90 - 5n$ **(c)** $1.08P$

10. $\{x \mid x < 3\}, (-\infty, 3)$

11. For both values of x, the value of the expression is 5.

12. 3

Chapter 3

Think About It

Section 3.1 *(page 109)*

The value of $3x + 1$ is less than the value of $x + 2$ when $x = 0$, but the value of $3x + 1$ is greater than the value of $x + 2$ when $x = 1$. Therefore, $3x + 1$ must equal $x + 2$ for some x-value in the interval $(0, 1)$.

Section 3.2 *(page 117)*

Multiply both sides of (a) by 10 and of (b) by 1000. If the greatest number of digits to the right of the decimal point is n, multiply by 10^n.

Section 3.3 *(page 124)*

Solving $C = 2\pi r$ for r, we obtain $r = \dfrac{C}{2\pi}$. If $r_1 = \dfrac{C}{2\pi}$ and $r_2 = \dfrac{C+1}{2\pi}$, then $r_2 - r_1 = \dfrac{C+1}{2\pi} - \dfrac{C}{2\pi} = \dfrac{1}{2\pi}$. Thus increasing the circumference by 1 foot increases the radius by $\dfrac{1}{2\pi}$ feet. Note that the result is independent of the size of the circle.

Section 3.4 *(page 134)*

For the 28-foot wire, the lengths could be 1, 2, 3, 22, or 2, 3, 4, 18, and so on. Thus the lengths of the pieces cannot be determined. For the 7-foot wire, the lengths of the four pieces must be 1, 2, 3, and 1.

Section 3.5 *(page 143)*

The retail price is $1.30w$. The discounted price is $0.80(1.30w) = 1.04w$. Therefore, the seller will make 4%.

Section 3.6 *(page 157)*

If the weights are exchanged, the pans will reverse positions. If $a < b$, then it is not true that $b < a$.

Section 3.7 *(page 167)*

(a) $\emptyset$ **(b)** **R**

Section 3.8 *(page 174)*

The given set includes -2 but not 2. Thus the set could not be the solution set of either of the given forms.

Speaking the Language

Section 3.1 *(page 113)*

1. inconsistent equation (or contradiction) **2.** Addition
3. identity **4.** linear equations in one variable

Section 3.2 *(page 120)*

1. simplify **2.** inconsistent equation (or contradiction)
3. the LCD of all the fractions **4.** Multiplication

Section 3.3 *(page 127)*

1. equation **2.** constants **3.** trapezoid
4. solving the formula

Section 3.6 *(page 160)*

1. infinity **2.** positive, negative
3. interval notation, a number line graph **4.** equivalent

Section 3.7 *(page 168)*

1. double inequality **2.** disjunction **3.** and
4. union, intersection

Section 3.8 *(page 177)*

1. conjunction **2.** zero, negative **3.** positive
4. disjunction

Chapter Project *(page 179)*

1. 6,465,000

2. 16%

3. 18–24: 8,424,000
25–44: 16,524,000
45–64: 7,452,000

4. More uninsured persons are in the 25–44 age group than in the other two age groups combined.

5. 49% **6.** 19.8 million

Looking Back *(page 185)*

1. **(a)** 7 **(b)** 18

2. (i), (iii), (iv)

3. **(a)** -10

(b) $\frac{1}{2}x - 10 = 0;\ x = 20$

4. **(a)** Because the second coordinate is 0, the point belongs to the x-axis.

(b) Points of the y-axis have a first coordinate of 0.

5. $y = x - 5$ **6.** $y = -\frac{2}{3}x + 4$ **7.** $\frac{4}{7}$ **8.** $-\frac{3}{4}, 5$

9. **(a)** $1.5p + 2$ **(b)** $L = W - 4$

10. **(a)** $x \le -4$ **(b)** -12

11. **(a)** Yes **(b)** No

12. Use the Distance Formula and the Pythagorean Theorem.

Chapter 4

Think About It

Section 4.1 *(page 192)*

$(-b/a, 0)$

Section 4.2 *(page 205)*

According to the Property of Signs in Quotients (see Section 1.5), $-\frac{-a}{-b} = -\frac{a}{b} = \frac{-a}{b} = \frac{a}{-b}$. Thus all four lines have the same slope.

Section 4.3 *(page 212)*

The Distance Formula would be needed to show that the four sides have the same length.

Section 4.4 *(page 225)*

To write the equations of the lines, we must know the slopes of the lines. Although we know that the product of the slopes is -1, we do not know the slope of either line.

Section 4.5 *(page 238)*

The solution set of $|x - y| < 0$ is Ø, so there is no graph. The solution set of $|x - y| \ge 0$ is all pairs (x, y) of real numbers, so the graph is the entire coordinate plane.

Speaking the Language

Section 4.1 *(page 196)*

1. $Ax + By = C; f(x) = ax + b$ **2.** intercepts

3. horizontal; vertical **4.** solution

Section 4.2 *(page 207)*

1. run; rise **2.** negative

3. second coordinate of the y-intercept; slope

4. undefined; 0

Section 4.3 *(page 217)*

1. direct variation **2.** parallel **3.** slope

4. perpendicular

Section 4.4 *(page 227)*

1. point–slope **2.** vertical **3.** slope–intercept **4.** slope

Section 4.5 *(page 239)*

1. half-plane **2.** boundary line **3.** intersection

4. test point

Chapter Project *(page 243)*

1. 326.27 pairs

2. For 1963, $x = -21$ and $E(-21) = -5686.81$. The model is invalid for 1963 and should be considered unreliable for years prior to 1986.

3. 763 years

4. For 2006, $x = 22$ and $E(22) = 8342.8$. Therefore, the slope (average rate of increase) of function E is not great enough to reach the goal.

5. About \$196

6. In 2004, $x = 20$ and $E(20) = 7690.26$. At \$196 per pair, the level of funding would be about \$1.5 million.

Looking Back *(page 246)*

1. **(a)**

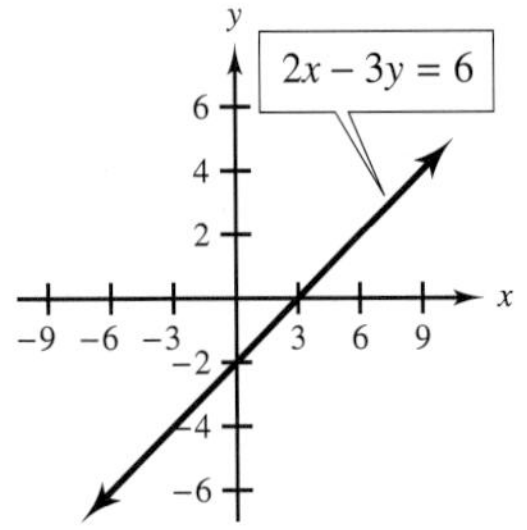

(b)

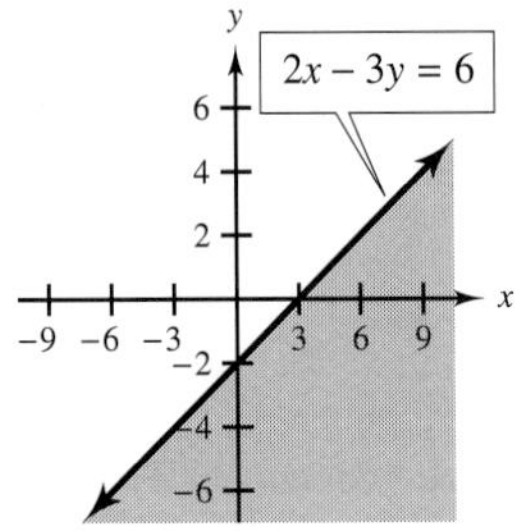

2. (i), (iii)

3. The graph of $x = 3$ is a vertical line containing (3, 0), and the graph of $y = 3$ is a horizontal line containing (0, 3).

4. 3 **5.** 5 **6.** -22 **7.** -1 **8.** No **9.** Yes

10. **(a)** The equation is an identity, and the solution set is **R**.

(b) The equation is a contradiction, and the solution set is Ø.

11. **(a)** -13 **(b)** -2

12. **(a)** $0.34x + 0.20y$ **(b)** $0.2x + 0.09y$ **(c)** $2x + 3y$

Chapter 5

Think About It

Section 5.1 *(page 253)*

Because the graphs of the two equations have the same y-intercept and different slopes, the solution is $(0, -7)$.

Section 5.2 *(page 264)*

Only pairs (x, y) that are represented by points of the coinciding lines are solutions.

Section 5.3 *(page 272)*

A linear equation in four variables is an equation that can be written $Ax + By + Cz + Dw = E$, where A, B, C, D, and E are real numbers and A, B, C, and D are not all zero. The graph would require a four-dimensional coordinate space.

Section 5.4 *(page 283)*

Because the augmented matrix has one additional column, its dimensions are $n \times (n + 1)$.

Section 5.5 *(page 291)*

The matrices are equivalent. The determinants are opposites.

Section 5.6 *(page 304)*

The revised condition produces equations that are dependent, and the new system has infinitely many solutions.

Section 5.7 *(page 312)*

The solution set of a system of linear inequalities is the intersection of half-planes. It is impossible for this intersection to be the entire coordinate plane.

Speaking the Language

Section 5.1 *(page 256)*

1. solution **2.** inconsistent **3.** dependent
4. substitution

Section 5.2 *(page 266)*

1. coefficients **2.** parallel, coincide **3.** one variable
4. elimination

Section 5.3 *(page 277)*

1. plane **2.** ordered triple **3.** y, (1) and (3), (2) and (3)
4. at exactly one point

Section 5.4 *(page 287)*

1. square **2.** augmented **3.** diagonal **4.** no

Section 5.5 *(page 297)*

1. determinant **2.** minor **3.** Cramer's
4. coefficient matrix

Section 5.7 *(page 315)*

1. system of inequalities **2.** intersection **3.** conjunction
4. first

Chapter Project *(page 318)*

1. 1994: 77%, 1997: 87%

2. Oral communication dropped from 23% in 1994 to 13% in 1997.

3. 1999 **4.** 100 reams

5. The percentages reflect the consumption of paper only for office communications. Paper is consumed for many other purposes, such as printing e-mail.

6. 101%. Because the percentage cannot exceed 100%, the prediction is not realistic.

Looking Back *(page 321)*

1. **(a)** 81 **(b)** -32 **(c)** 10,000

2. 1 **3.** 32 **4.** $-15xy$ **5.** $10x - 5$ **6.** $-4 - y$

7. $2x + 5$ **8.** $13x^3 - 7x^2$ **9.** $2t - 6$ **10.** $-a + 10$

11. **(a)** x^5 **(b)** $(a - 1)^2$

12. **(a)** $3 \cdot y \cdot y$ **(b)** $(t + 2)(t + 2)(t + 2)$

Chapter 6

Think About It

Section 6.1 *(page 326)*

Beginning with x, repeated squaring results in the pattern $x^1, x^2, x^4, x^8, x^{16}, x^{32}, x^{64}, \ldots$.

Section 6.2 *(page 334)*

$$\left(\frac{a}{b}\right)^{-2} = \frac{1}{\left(\frac{a}{b}\right)^2} = \frac{1}{\frac{a^2}{b^2}} = 1 \cdot \frac{b^2}{a^2} = \frac{b^2}{a^2} = \left(\frac{b}{a}\right)^2$$

Section 6.3 *(page 339)*

The scientific notation for a negative number is $n \cdot 10^p$, where $-10 < n \le -1$ and p is an integer.

Section 6.4 *(page 348)*

The product function is $(pq)(x)$, and the quotient function is $\left(\frac{p}{q}\right)(x)$.

Section 6.5 *(page 356)*

The columns for $(x + 3)^2$ and $x^2 + 6x + 9$ would be the same because those expressions are equivalent.

Speaking the Language

Section 6.1 *(page 330)*

1. base, exponent **2.** multiply **3.** Product, Quotient **4.** natural

Section 6.2 *(page 336)*

1. exponent, base **2.** reciprocal **3.** zero **4.** integer

Section 6.3 *(page 342)*

1. right, left **2.** scientific notation **3.** decimal **4.** exponent on 10

Section 6.4 *(page 349)*

1. monomial, trinomial **2.** coefficient, degree **3.** descending, missing **4.** polynomial

Section 6.5 *(page 358)*

1. FOIL **2.** Distributive **3.** squaring **4.** sum, difference

Chapter Project *(page 361)*

1. $6.37 \cdot 10^6$ kilometers
2. Nearly 18 times as far
3. 1.7 pounds
4. $1.764 \cdot 10^4$ mph; 17,640 mph
5. 84.7 minutes
6. $1.723 \cdot 10^4$ mph

Looking Back *(page 363)*

1. **(a)** $10 - 15y$ **(b)** $3(x + y)$
2. **(a)** $x^2 + 2x - 15$ **(b)** $-15x^2 - 8x + 7$
3. $y^2 - 25$ **4.** $1 - 16x^2$ **5.** $a^2 + 2a + 1$
6. $4x^2 - 12x + 9$ **7.** $3x^2 - 14x - 5$ **8.** $x^3 + 27$
9. **(a)** 4, 5 **(b)** $-6, 1$ **(c)** $-3, -4$ **(d)** $8, -2$
10. **(a)** $-\frac{1}{2}$ **(b)** 0
11. $-1, 15$
12. Replace y with 0 and solve the resulting equation for x.

Chapter 7

Think About It

Section 7.1 *(page 371)*

$$\begin{aligned} xy(x + 1) + x(x + 1) + y^2(x + 1) + y(x + 1) &= (x + 1)(xy + x + y^2 + y) \\ &= (x + 1)\,[x(y + 1) + y(y + 1)] \\ &= (x + 1)(y + 1)(x + y) \end{aligned}$$

Section 7.2 *(page 376)*

$x^3 + x^3 = (x + x)\,(x^2 - x^2 + x^2) = 2x(x^2) = 2x^3$

Section 7.3 *(page 384)*

$(3x + 9)(2x - 4) = 3 \cdot (x + 3) \cdot 2 \cdot (x - 2) = 6(x + 3)\,(x - 2)$

Section 7.4 *(page 394)*

$-1, 1, 2$

Section 7.5 *(page 403)*

Q: $\frac{1}{2}x + \frac{11}{4}$, R: $-\frac{5}{8}$

Speaking the Language

Section 7.1 *(page 372)*

1. greatest common factor (GCF) **2.** prime **3.** factoring **4.** grouping

Section 7.2 *(page 376)*

1. difference of two squares **2.** perfect square trinomial **3.** sum of two cubes **4.** multiplying

Section 7.3 *(page 387)*

1. quadratic, linear **2.** leading **3.** trial-and-check **4.** greatest common factor

Section 7.4 *(page 394)*

1. quadratic **2.** quadratic, linear **3.** Zero Factor **4.** double

Section 7.5 *(page 404)*

1. Remainder, Dividend **2.** missing **3.** synthetic division **4.** remainder, divisor

Chapter Project (page 407)

1.

Country	*Exchange for $1534*
England	997.10 pounds
Italy	2,327,078 lire
Germany	2270.32 marks
Norway	9832.94 crowns
Spain	191,750 pesetas

2. $c = 6.41d$, $p = 0.65d$
3. $d = 0.16c$, $d = 1.54p$, $c = 9.63p$

   The formula $c = 9.63p$ can be used to convert pounds to crowns.
4. 8667 crowns **5.** $416.07 **6.** Answers will vary.

Looking Back (page 409)

1. (a) $-\frac{1}{2}$ (b) $\frac{13}{40}$ (c) $-\frac{6}{35}$ (d) $\frac{3}{10}$ **2.** $-\frac{3}{4}$
3. (a) 0 (b) undefined (c) -8
4. (a) $\frac{3}{x}$ (b) $\frac{2y^3}{3x^2}$
5. $2x(x + 3)$ **6.** $(x + 9)(x - 9)$ **7.** $(3 - 2x)(2 + x)$
8. Factor out the common factor of 2 to obtain $2(1 - x)$ or the opposite of the common factor to obtain $-2(x - 1)$.
9. $-4, 2$ **10.** $\frac{7}{4}$ **11.** $\frac{3}{2}$ **12.** (a) $\frac{250}{r}$ (b) $\frac{84}{f}$

Chapter 8

Think About It

Section 8.1 (page 414)

$k = 3$ or $k = 5$

Section 8.2 (page 422)

After simplifying, $f(x) = 1$ provided that $x \neq -5$ and $x \neq 5$. Thus the domain of f is $\{x \mid x \neq -5 \text{ and } x \neq 5\}$, and the domain of g is **R**.

Section 8.3 (page 429)

GCF: $2^m \cdot 3^m$; LCM: $2^n \cdot 3^n$

Section 8.4 (page 439)

The complex fraction simplifies to $x^2 + \frac{a^2 + b^2}{ab}x + 1$. If $a = b$, the expression becomes $x^2 + 2x + 1 = (x + 1)^2$.

Section 8.5 (page 446)

$k = 2$

Section 8.6 (page 456)

The faster person's work rate is k times the slower person's work rate.

Speaking the Language

Section 8.1 (page 417)

1. polynomials **2.** restricted value
3. Equivalent Fractions **4.** factoring

Section 8.2 (page 423)

1. reciprocals **2.** divisor **3.** common factors
4. Q, R, and S

Section 8.3 (page 433)

1. numerators, denominator
2. least common multiple (LCM)
3. factor **4.** rational

Section 8.4 (page 440)

1. complex fraction **2.** adding a/b and c/d
3. multiplying the numerator and denominator by bd
4. zero

Section 8.5 (page 448)

1. denominators, clear **2.** extraneous **3.** equation
4. denominator

Chapter Project (page 462)

1. 228
2. The model's value is 238, whereas the actual number was 215. Therefore, the model overestimates the number.
3. For 2001 ($x = 11$), $N(x) \approx 654$. The result is not valid because it is greater than the total number of players.
4. $P(x) = 1.9x^3 - 25.9x^2 + 101.9x + 138.4$
5. The domain of P is **R**, whereas the domain of N excludes $x = -1$.
6. Because $P(x) = N(x)$ for the given values of x, a table for function P would be identical to the table for function N.

Looking Back (page 465)

1. (a) x^8 (b) $\frac{1}{a^{15}}$ **2.** (a) a^8 (b) $\frac{1}{x^2}$
3. (a) a^8b^{12} (b) $\frac{9}{x^{10}}$ **4.** (a) $\frac{3y^4}{x^2}$ (b) $\frac{8b^3}{a^5}$
5. (a) $\frac{1}{25}$ (b) -3 (c) 8 **6.** $(y + 3)^2$

7. $a^2b^6(a^2 + b^2)$ **8.** $2n + 5n = (2 + 5)n = 7n$

9. $-4, 4$ **10.** 3 **11.** $-1, 4$ **12.** $9^2 + 12^2 = 15^2$

Chapter 9

Think About It

Section 9.1 *(page 472)*

Other examples include addition and subtraction as well as multiplication and division. Forming a reciprocal is an operation that undoes itself.

Section 9.2 *(page 478)*

Writing $(-1)^{2/4}$ as $\sqrt[4]{(-1)^2} = \sqrt[4]{1} = 1$ but writing it as $\left(\sqrt[4]{-1}\right)^2$ results in an expression that is not defined.

Section 9.3 *(page 484)*

$x - 9 = (x^{1/2})^2 - 3^2 = (x^{1/2} + 3)(x^{1/2} - 3)$

$x^{2/3} - 4 = (x^{1/3})^2 - 2^2 = (x^{1/3} + 2)(x^{1/3} - 2)$

Section 9.4 *(page 488)*

(i) $n \geq 0$ **(ii)** 0 **(iii)** 0

Section 9.5 *(page 495)*

$$\frac{\sqrt{8}}{\sqrt[3]{2}} = \frac{2^{3/2}}{2^{1/3}} = 2^{7/6} = 2\sqrt[6]{2}$$

Section 9.6 *(page 506)*

$$\frac{9\sqrt{\sqrt{x}} + \sqrt{x\sqrt{x}} - 3\sqrt{x} - 27}{x - 81}$$

Section 9.7 *(page 516)*

The properties state that both sides can be raised to an integer exponent. Raising both sides to the $\frac{3}{2}$ power results in only one equation: $x - 2 = 8$.

Section 9.8 *(page 523)*

For even n, the result is a real number, and for odd n, the result is an imaginary number. The result is always a complex number.

Speaking the Language

Section 9.1 *(page 473)*

1. index, radicand **2.** fourth roots **3.** principal (positive)

4. simplifying

Section 9.2 *(page 479)*

1. rational **2.** index **3.** raise the result to the third power

4. exponent, base

Section 9.3 *(page 485)*

1. Power to a Power **2.** the bases are the same **3.** index

4. rational exponents

Section 9.4 *(page 491)*

1. indices **2.** perfect third power **3.** simplified

4. Product

Section 9.5 *(page 499)*

1. Quotient, Product **2.** fraction **3.** denominator

4. rationalizing the denominator

Section 9.6 *(page 507)*

1. like radicals **2.** conjugates **3.** Distributive

4. radical, rationalize

Section 9.7 *(page 517)*

1. Squaring Property of Equality **2.** extraneous solution

3. radicand **4.** Odd Root, Even Root

Section 9.8 *(page 526)*

1. complex **2.** imaginary **3.** complex conjugates

4. binomials

Chapter Project *(page 529)*

1. 2013 **2.** about 4% **3.** $\frac{27.3x^{1/5}}{15.2}$; 3

4. 8.03 million acres **5.** About 30 feet

6. Answers will vary.

Looking Back *(page 533)*

1. (a) 6 **(b)** $-3, 2$ **2.** $-1, \frac{5}{2}$ **3.** $\frac{2}{3}$ **4.** $0, \frac{1}{3}$

5. $-3, -2, 2$ **6.** $-\frac{3}{2}, \frac{5}{2}$ **7.** 3 **8.** 2.54 **9.** -1

10. 5

11. Graph the solid boundary line $y = x - 6$. Because the inequality is in the form $y \leq mx + b$, shade below the line.

12. 2:00 P.M.

Chapter 10

Think About It

Section 10.1 *(page 537)*

(i) $k > 0$ **(ii)** $k = 0$ **(iii)** $k < 0$

Section 10.2 *(page 542)*

The constant term is the real number $-\frac{1}{4}$.

Section 10.3 *(page 548)*

If a were 0, the equation would be a linear equation, not a quadratic equation

Section 10.4 *(page 559)*

Because the radical is nonnegative, u^2 and $5u$ are nonnegative. Thus $u^2 + 5u + 6$ must be positive, not 0.

Section 10.5 *(page 564)*

43

Section 10.6 *(page 579)*

Three noncollinear points determine a parabola. The given information is insufficient for writing the quadratic function.

Section 10.7 *(page 587)*

$x^2 + 2x - 15 \le 0, -x^2 - 2x + 15 \ge 0$

Speaking the Language

Section 10.1 *(page 539)*

1. quadratic **2.** double root **3.** Even Root Property
4. Square Root, imaginary

Section 10.2 *(page 545)*

1. half, square **2.** divide both sides
3. perfect square trinomial **4.** $3 + i, 3 - i$

Section 10.3 *(page 553)*

1. standard form **2.** discriminant **3.** imaginary numbers
4. factoring

Section 10.4 *(page 560)*

1. quadratic **2.** restricted **3.** extraneous **4.** $x^{1/5}$

Section 10.6 *(page 581)*

1. parabola **2.** vertex **3.** axis of symmetry
4. vertical, horizontal

Section 10.7 *(page 591)*

1. quadratic inequality **2.** solutions, intervals
3. test point, solution set **4.** $\le, \ge$

Chapter Project *(page 594)*

1.

Age	Target Heart Rate Women	Men
20	160	156
30	152	152
40	144	148
50	136	144
60	128	140
70	120	136

2. 30 **3. (a)** 158 pounds **(b)** 3 pounds
4. (a) 444 calories **(b)** 6.5 mph
5. At age 60, men and women have about the same cholesterol level.
6. Answers will vary.

Looking Back *(page 597)*

1. $\{x \mid x \ne 3\}$ **2.** $\{x \mid x \ge -5\}$
3. If every vertical line intersects the graph at most once, the graph represents a function.
4. (a) Yes **(b)** No **5. (a)** $x^2 + 5x$ **(b)** $9t - 3$
6. (a) $x^2 + x - 6$ **(b)** $x - 3$
7. 3 **8.** $\frac{1}{3}$ **9.** $3t + 5$ **10.** $3t^2 - 7$
11. (a) 2 **(b)** -5 **(c)** 8
12. (a) $\frac{1}{16}$ **(b)** 27 **(c)** 1 **(d)** 8 **(e)** $\frac{1}{25}$

Chapter 11

Think About It

Section 11.1 *(page 604)*

No. Consider starting at the origin and facing north. The rule for $f \circ g$ takes you to the point (3, 0), and the rule for $g \circ f$ takes you to the point (0, 3).

Section 11.2 *(page 610)*

No, two different people may have the same mother.

Section 11.3 *(page 621)*

The graph of $-b^x$ is the graph of b^x folded across the x-axis. The domain is **R**, the range is $(-\infty, 0)$, and the y-intercept is $(0, -1)$. There is no x-intercept.

Section 11.4 *(page 632)*

In each part, use the Change of Base Formula to write each factor with the same base. Then simplify the product.

(a) 3 **(b)** 3

Section 11.5 *(page 638)*

$3^x = 3^{\log_3 6}$

$x = \log_3 6$

Section 11.6 *(page 647)*

$\log 10^{\ln e} = (\ln e)(\log 10) = (\log 10)(\ln e) = \ln e^{\log 10}$

Speaking the Language

Section 11.1 *(page 606)*

1. first **2.** range **3.** composition **4.** intersection

Section 11.2 *(page 615)*

1. one-to-one **2.** inverse **3.** Vertical, Horizontal

4. symmetric

Section 11.3 *(page 624)*

1. e **2.** exponential **3.** x, **R** **4.** equating

Section 11.4 *(page 634)*

1. logarithmic **2.** common, natural **3.** y, rises

4. Change of Base

Section 11.5 *(page 641)*

1. Product Rule for Logarithms **2.** difference **3.** Power

4. $\log 1 = 0$

Section 11.6 *(page 649)*

1. logarithmic, argument **2.** logarithms **3.** exponential

4. extraneous

Chapter Project *(page 653)*

1. (1, 3.0), (2, 4.0), (3, 5.5), (4, 6.6), (5, 7.6), (9, 11.0), (14, 13.2)

2. The rate of increase is increasing.

3. The function models the data well.

4. The rate of increase is relatively constant.

5. In the long run, the rate of increase is decreasing.

6. The function models the data well.

Chapter 1

Section 1.1 *(page 6)*

1. The decimal name of a rational number is a terminating or a repeating decimal.
2. A decimal that neither terminates nor repeats is an irrational number.
3. Terminating **4.** Terminating **5.** Repeating
6. Repeating **7.** Negative integers
8. Numbers that cannot be written as $n/1$, where n is an integer.
9. Numbers that can be written as p/q such that $q = 1$.

10. Zero **11.** False **12.** True **13.** True **14.** True

15. False **16.** False

17. The number 1.75 is a terminating decimal, and the digit 5 is followed by zeros. The number $1.\overline{75}$ is a repeating decimal, and the block of digits, 75, repeats without end.

18. The symbol < means "is less than"; the symbol ≤ means "is less than or equal to."

19. Rational **20.** Rational **21.** Irrational **22.** Irrational

23. Rational **24.** Rational **25.** Irrational **26.** Irrational

27. **(a)** $\sqrt{16}$ **(b)** $0, \sqrt{16}$ **(c)** $-4, 0, -17, \sqrt{16}$
(d) $-4, 0, \frac{3}{5}, 0.25, -17, 0.\overline{63}, \sqrt{16}$ **(e)** $\sqrt{7}, \pi$

28. **(a)** None **(b)** None **(c)** -24
(d) $-24, -4\frac{7}{8}, 0.\overline{36}, 0.2863$ **(e)** $-\sqrt{5}, \frac{\sqrt{3}}{2}, \frac{\pi}{7}$

29.
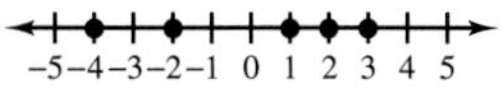

30.
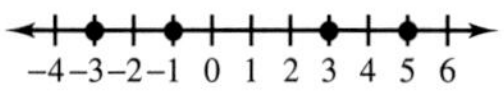

31.
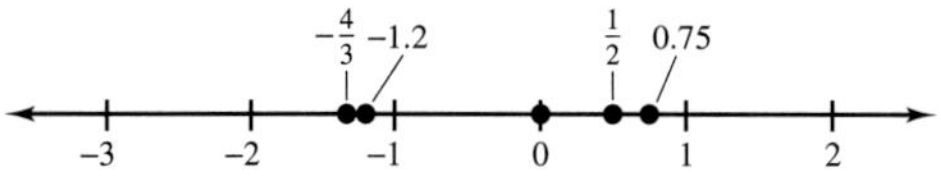

32.
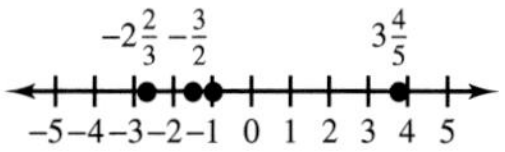

33.
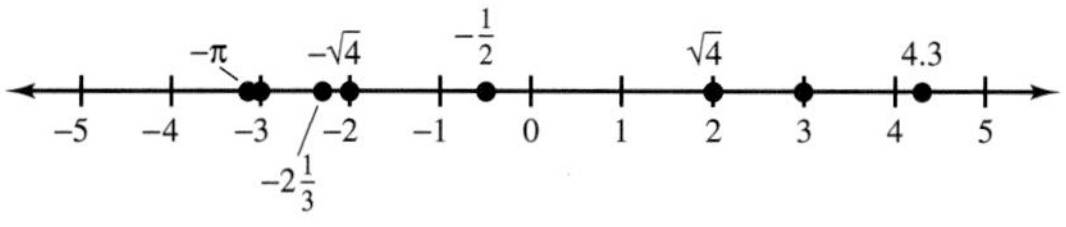

34.
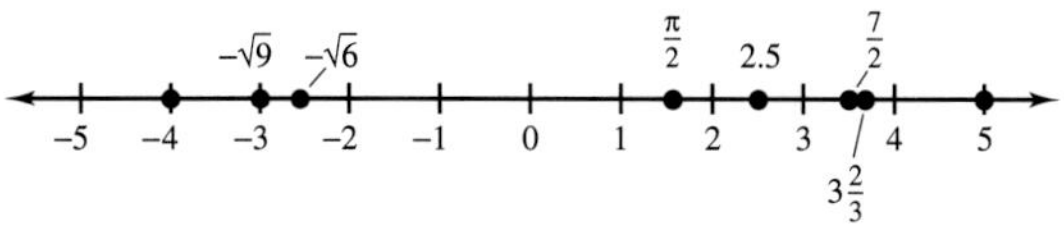

35. < **36.** > **37.** < **38.** < **39.** > **40.** <

41. > **42.** < **43.** > **44.** < **45.** 6 **46.** −13

47. 0 **48.** −4

49. The symbols $|7|$ and $|-7|$ represent the distances from 0 to 7 and from 0 to −7, respectively. Because these distances are the same, $|7| = |-7|$.

50. If $a = -2$, then $|-a| = |-(-2)| = |2| = 2$, which is not the value of a. Thus, in this case, $|-a| \neq a$.

51. −5, 5 **52.** −5, 1 **53.** −6, 2 **54.** −8, 8 **55.** 6

56. 4 **57.** −7 **58.** −10 **59.** 0 **60.** −12 **61.** 0

62. 9 **63.** 1 **64.** 2 **65.** \$450 **66.** \$1800

67. \$1.90 **68.** \$1000 **69.** 1867

70. **(a)** 65%
(b) The length of the bar for San Francisco is about half the length of the bar for San Jose. The percentage of homes that people with median incomes can afford in San Francisco is about half the percentage of homes that people with median incomes can afford in San Jose.

71. 3.15 gallons **72.** 10% **73.** 19.2 ounces

74. 21 gallons **75.** > **76.** > **77.** > **78.** <

79. $\frac{27}{10}$ **80.** $\frac{44}{125}$ **81.** $\frac{1}{3}$ **82.** $\frac{5}{3}$

Section 1.2 *(page 13)*

1. Add the absolute values and retain the common sign.
2. If the addends have unlike signs, find their absolute values, subtract the smaller absolute value from the larger, and use the sign of the addend with the larger absolute value as the sign of the sum.

3. 3 **4.** 5 **5.** −4 **6.** −3 **7.** −10 **8.** −15

9. −8 **10.** −14 **11.** 0 **12.** 0 **13.** 2 **14.** −2

15. 8 **16.** 6 **17.** 8 **18.** −6 **19.** −7 **20.** −9

21. −55 **22.** −22 **23.** 8 **24.** −6 **25.** 0 **26.** 7

27. −22 **28.** −21 **29.** −1 **30.** −55 **31.** −11

32. 27 **33.** −17 **34.** −15 **35.** −6 **36.** −20

37. 1 **38.** −15 **39.** −9 **40.** 10 **41.** 0 **42.** −1

43. −6.8 **44.** −1.8 **45.** 2.8 **46.** −1.2 **47.** −51.57

48. 77.95 **49.** −10.31 **50.** −1512.67 **51.** 5

52. −4 **53.** −7 **54.** −14 **55.** 9 **56.** 8 **57.** $\frac{4}{3}$

58. $-\frac{6}{5}$ **59.** $-\frac{1}{15}$ **60.** $\frac{5}{21}$ **61.** $-\frac{1}{16}$ **62.** $-\frac{1}{2}$

63. $-\frac{17}{3}$ **64.** $-\frac{31}{4}$ **65.** −7 **66.** 10 **67.** −12

68. 0 **69.** > **70.** < **71.** > **72.** > **73.** 9

74. −45 **75.** −4 **76.** −2 **77.** 2 **78.** 7 **79.** 0

80. −1.01 **81.** $-\frac{27}{5}$ **82.** $-\frac{139}{12}$ **83.** −45 **84.** −17

85. -18.3 **86.** -0.2 **87.** -7 **88.** 0 **89.** 0 **90.** 7
91. $-\frac{27}{8}$ **92.** $-\frac{31}{24}$ **93.** \$8.41 **94.** 8 printers

95. *Weight at the End of Each Week*

211
212
210
209
213
210
212

His weight was 212 pounds at the end of the period.

96. 1.86°F

97.

Fenway:	88
Wrigley:	86
Yankee:	77
Dodger:	38
Turner:	3

98. **(a)** 27,136 megawatts
(b) 30,733 megawatts
(c) 67%

99. 41% **100.** 20%

101. The number who use their cards 2 times per month is not necessarily equal to the number who use their cards 3 times per month.

102. The total number of ATM cards is not given.

103. Positive **104.** Positive **105.** Negative
106. Negative **107.** 81
108. **(a)** -4 **(b)** -8 **(c)** The sum is -48.

Section 1.3 *(page 21)*

1. Change the minus sign to a plus sign, and change 8 to -8: $-5 + (-8)$.

2. In the expression $7 - 5$, the 7 is the minuend and the 5 is the subtrahend. Because $7 + 5 = 5 + 7$, there is no ambiguity in calling both 7 and 5 addends. Because $7 - 5 \neq 5 - 7$, it is necessary to give 7 and 5 different names in each expression.

3. -11 **4.** -14 **5.** -10 **6.** -11 **7.** 13 **8.** 15
9. 6 **10.** 8 **11.** -13 **12.** -21 **13.** 52 **14.** -25
15. 18 **16.** 23 **17.** -24 **18.** 24 **19.** -3 **20.** -5
21. 48 **22.** 29 **23.** -20 **24.** -26 **25.** -3
26. -41 **27.** -10 **28.** -15 **29.** -5 **30.** 3
31. 13 **32.** -15 **33.** -8 **34.** 3 **35.** -14 **36.** 15
37. 0 **38.** -6 **39.** 1 **40.** -7 **41.** -10 **42.** 13
43. $-\frac{5}{3}$ **44.** $\frac{2}{5}$ **45.** $-\frac{1}{10}$ **46.** $-\frac{1}{12}$ **47.** $-\frac{14}{3}$
48. $-\frac{29}{4}$ **49.** $\frac{17}{16}$ **50.** $-\frac{19}{14}$

51. The opposite of the difference of a and b is the difference of b and a.

52. If the value of $x - y$ is -4, then the value of $y - x$ is 4 because $y - x$ and $x - y$ are opposites.

53. -34 **54.** -18 **55.** 74.23 **56.** 49.92 **57.** 5
58. -2 **59.** 6 **60.** 7 **61.** $AB = 6, BC = 10, AC = 16$
62. $AB = 8, BC = 10, AC = 18$ **63.** -5 **64.** 0
65. -41.13 **66.** -76.96 **67.** 43 **68.** 56 **69.** 5
70. 16 **71.** $-\frac{22}{15}$ **72.** $\frac{13}{12}$ **73.** -21 **74.** 6 **75.** -32
76. -59 **77.** $-\frac{19}{4}$ **78.** $-\frac{55}{8}$ **79.** 32 **80.** 0
81. 59.76 **82.** -295.91 **83.** -45 **84.** 46 **85.** 0
86. -10 **87.** -8 **88.** -6 **89.** -80 **90.** -12
91. -15 **92.** 4 **93.** 14,775 feet **94.** 27,855 feet

95. **(a)**

Category	*Difference*
Always	+12%
Sometimes	−2%
Seldom	−6%
Never	−4%

(b) The sum of the differences is 0. The total is 100% for both years. Thus the increase in the Always category is offset by decreases in the other categories.

96. 620 per 1000 homes

97. 11,931, −2175, −1259, −1695, 4142, 11,908

98. The mean difference of approximately 3809 feet is large because of the dominance of two large numbers.

99. 2.73 miles **100.** Answers will vary. **101.** Negative
102. Positive **103.** Negative **104.** Positive

105. The statement $|a - b| = |b - a|$ is true for all real numbers a and b because the distance between a and b on the number line is the same regardless of the starting point.

106. $1 - 2 + 3 - (4 + 5 - 6) + 7 - 8 + 9 = 7$

107. Of the nine numbers, five are odd and four are even. Regardless of the signs, the result will be an odd number.

Section 1.4 *(page 27)*

1. Determine the product of the absolute values. The answer is positive.

2. Determine the product of the absolute values. The answer is negative.

3. -28 **4.** -40 **5.** -10 **6.** -42 **7.** 12 **8.** 60
9. 12 **10.** 40 **11.** -100 **12.** -81 **13.** -44
14. -56 **15.** -7 **16.** 0 **17.** -24 **18.** -5 **19.** 0
20. 92 **21.** 6.21 **22.** 15.3 **23.** -4.83 **24.** -2.15

25. 1525.92 **26.** -1456.73 **27.** -3.24 **28.** -48
29. -24 **30.** -3.8 **31.** -30 **32.** -42 **33.** $-\frac{1}{5}$
34. $\frac{2}{5}$ **35.** $\frac{1}{4}$ **36.** $\frac{2}{3}$ **37.** $\frac{10}{3}$ **38.** -6 **39.** $\frac{85}{2}$ **40.** 28
41. The result is positive because the number of negative factors is even.
42. If any factor in a product is zero, then the product is zero.
43. -24 **44.** 40 **45.** 70 **46.** -36 **47.** 12
48. -24 **49.** 48 **50.** -120 **51.** -36 **52.** 30
53. -32 **54.** 25 **55.** 81 **56.** -64 **57.** $<$ **58.** $>$
59. $<$ **60.** $<$ **61.** -3 **62.** -4 **63.** 6 **64.** 0
65. $-\frac{11}{3}$ **66.** -2 **67.** 28.512 **68.** -8.602 **69.** -20
70. 90 **71.** -40 **72.** 12 **73.** -163.0209
74. 5467.361 **75.** 12 **76.** -12 **77.** 72 **78.** -42
79. $-\frac{1}{3}$ **80.** $\frac{3}{35}$ **81.** 96 **82.** 63 **83.** -140
84. -320 **85.** 2, 3 **86.** $-2, -4$ **87.** $4, -3$
88. $5, -3$ **89.** $-2, -6$ **90.** $-4, -6$ **91.** $-8, 1$
92. $-7, 3$
93. $7(-2) = -2 + (-2) + (-2) + (-2) + (-2) + (-2) + (-2) = -14$
94. Choice (i) is easier to calculate. Because all the factors are -1 and there is an even number of negative signs, the result is 1.
95. 75 **96.** 13
97. **(a)** Total annual consumption (billions of gallons):

Brazil	1.235
China	0.305
Germany	1.030
Mexico	1.892
Russia	0.055
United States	5.638
Zimbabwe	0.041

(b) United States
98. **(a)** \$2000 **(b)** 60%
99. $-\$5.7$ trillion **100.** $-\$399$ billion **101.** $-\$1456$
102. $-\$41{,}606$ **103.** Positive **104.** Negative
105. Negative **106.** Positive **107.** Positive
108. Negative

Section 1.5 *(page 33)*

1. Divide the absolute values. The result is positive.
2. Divide the absolute values. The result is negative.
3. -2 **4.** -7 **5.** -3 **6.** -4 **7.** 8 **8.** 6 **9.** 8
10. -1 **11.** 16 **12.** -1 **13.** -3 **14.** -1 **15.** -4
16. -8 **17.** -7 **18.** Undefined **19.** Undefined
20. 0 **21.** 0 **22.** 1 **23.** $\frac{1}{2}$ **24.** -3 **25.** 8
26. -6 **27.** -4 **28.** 12 **29.** -1 **30.** 2
31. A nonzero number divided by itself is 1.
32. When any number is divided by 1, the result is the number.
33. 2.19 **34.** -9.61 **35.** -0.51 **36.** -3.14 **37.** 4
38. $-\frac{1}{2}$ **39.** 0 **40.** Undefined **41.** $\frac{1}{2}$ **42.** 24
43. $-\frac{3}{5}$ **44.** -4 **45.** $\frac{4}{3}$ **46.** $-\frac{5}{6}$ **47.** $\frac{15}{2}$ **48.** $\frac{28}{3}$
49. $-\frac{3}{32}$ **50.** $-\frac{7}{128}$
51. **(a)** -5 **(b)** $\frac{1}{5}$ **52.** **(a)** 8 **(b)** $-\frac{1}{8}$
53. **(a)** $\frac{3}{4}$ **(b)** $-\frac{4}{3}$ **54.** **(a)** $-\frac{2}{5}$ **(b)** $\frac{5}{2}$
55. **(a)** 0 **(b)** No reciprocal **56.** **(a)** $\frac{4}{5}$ **(b)** $-\frac{5}{4}$
57. $>$ **58.** $<$ **59.** $<$ **60.** $>$ **61.** 14 **62.** -2
63. -40 **64.** -1 **65.** 0.46 **66.** 1.16 **67.** $-\frac{9}{4}$
68. -8 **69.** -1 **70.** -7 **71.** -28 **72.** -2
73. -3 **74.** 8 **75.** 3 **76.** 2 **77.** -2 **78.** 5
79. 0 **80.** -1 **81.** $-\frac{1}{3}$ **82.** $\frac{1}{4}$ **83.** -9 **84.** 3
85. -36 **86.** -20 **87.** Undefined **88.** -8
89. Division by 0 is not defined.
90. We could claim that $0 \div 0$ is any number c because $c \cdot 0 = 0$. Because the result is not unique, $0 \div 0$ is undefined.
91. $\frac{5}{8}$ **92.** $-\frac{6}{11}$ **93.** $-\frac{1}{6}$ **94.** $\frac{3}{5}$ **95.** $\frac{4}{7}$ **96.** $\frac{7}{9}$
97. **(a)** 4 **(b)** 80% **98.** 0.05
99. **(a)** 65–74: 30 million
75–84: 15 million
85 and older: 7 million
(b) 65–74; 75–84 **(c)** 85 and older
100. $-29,940$ species per year
101. -127°F **102.** -3.5°F **103.** -17°F
104. 8000 feet **105.** Negative **106.** Positive
107. Positive **108.** Negative **109.** -4 **110.** $\frac{2}{3}$

Section 1.6 *(page 40)*

1. In 3^4, 3 is the base. In $\sqrt{3}$, 3 is the radicand.
2. Parentheses: $4 \div (3-8)$
Brackets: $9 - [(3+2)^2 - 2]$
Braces: $4\{1 - [4 \cdot 3 - 6]\}$
Fraction bar: $\dfrac{4+3}{8}$
Radical symbol: $\sqrt{25-6}$
3. 7^6 **4.** $(-11)^4$ **5.** $-(-5)^3$ **6.** -2^5 **7.** a^3
8. x^3y^2 **9.** $(2b)^4$ **10.** $(xy)^3$ **11.** $\left(\frac{7}{9}\right) \cdot \left(\frac{7}{9}\right) \cdot \left(\frac{7}{9}\right) \cdot \left(\frac{7}{9}\right)$
12. $11 \cdot 11 \cdot 11 \cdot 11 \cdot 11$ **13.** $2 \cdot x \cdot x$
14. $-1 \cdot x \cdot x \cdot x \cdot x$ **15.** $5 \cdot 5 \cdot y \cdot y \cdot y$

16. $2 \cdot x \cdot x \cdot x \cdot y \cdot y$ **17.** 25 **18.** 81 **19.** -8

20. 16 **21.** -49 **22.** -36 **23.** 32 **24.** 100

25. $\frac{4}{9}$ **26.** $-\frac{27}{64}$ **27.** 1 **28.** -1 **29.** 248,832

30. 14,641 **31.** $-177{,}147$ **32.** $-262{,}144$

33. 6057.68 **34.** 1963.17

35. Because $6^2 = 36$ and $(-6)^2 = 36$, the two square roots of 36 are 6 and -6. The radical symbol refers to the principal (positive) square root. Thus $\sqrt{36} = 6$.

36. If $53^2 = 2809$, then $\sqrt{2809} = 53$. **37.** $2, -2$

38. $4, -4$ **39.** $6, -6$ **40.** $10, -10$

41. Not a real number **42.** 0 **43.** $23, -23$

44. $42, -42$ **45.** 2 **46.** 4 **47.** -6 **48.** -10

49. $\frac{2}{3}$ **50.** $-\frac{5}{6}$ **51.** 26 **52.** 43 **53.** 2.9

54. 3.1 **55.** 23.11 **56.** 27.33 **57.** 252.74 **58.** 0.09

59. For $(-6)^2$, multiply $(-6)(-6)$ to obtain 36. For -6^2, multiply $-1(-6)(-6)$ to obtain -36.

60. For $5 - 3^2$, we square 3 and subtract the result from 5. For $(5 - 3)^2$, we subtract 3 from 5 and square the result.

61. -6 **62.** -11 **63.** 10 **64.** 36 **65.** 10

66. -15 **67.** -3 **68.** -11 **69.** 2 **70.** -14

71. 16 **72.** -2 **73.** -1 **74.** 3 **75.** 14 **76.** 0

77. -6 **78.** 14 **79.** $-2 \cdot 7 - 4 = -18$

80. $\dfrac{(-8)^2 + 6}{10} = 7$ **81.** $\dfrac{-3 + 7}{2} - (-4) = 6$

82. $[3 - (-2)]^2 = 25$

83. **(a)** $-3(4 - 5)$ **(b)** $(-3 \cdot 4) - 5$ **(c)** $-(3 \cdot 4 - 5)$

84. **(a)** $(7 - 5)^2 + 1$ **(b)** $7 - (5^2 + 1)$ **(c)** $(7 - 5^2) + 1$

85. -36 **86.** 24 **87.** -4 **88.** 7 **89.** -11

90. 4 **91.** 25 **92.** 8.54 **93.** 17 **94.** 4.90

95. 0.22 **96.** -1

97.

x	-4	-1	0	1	4	9	16	25
$\sqrt{x}$	X	X	0	1	2	3	4	5
x^2	16	1	0	1	16	81	256	625

98.

x	-4	-1	0	1	4
$\sqrt{x}$	X	X	0	1	2
$\sqrt{-x}$	2	1	0	X	X

99. 109 **100.** Lose $0.75

101. $26 + 160 \cdot 0.12$; multiplication **102.** 172

103. 4.1 million

104. Griffey: 48
McGwire: 65

105. **(a)** No values of x **(b)** $x = 0$ **(c)** $x \neq 0$

106. **(a)** $x \neq 0$ **(b)** $x = 0$ **(c)** No values of x

107. **(a)** Positive **(b)** Either **(c)** Positive **(d)** Negative **(e)** Negative **(f)** Negative **(g)** Positive **(h)** Negative

108. $1 + 2 + 3 + 4 + 5 + 6 + 7 + 8 \cdot 9$ **109.** $3^3 + 3$

110. $3 = \dfrac{4 + 4 + 4}{4}$ $7 = 4 + 4 - \dfrac{4}{4}$

$4 = 4(4 - 4) + 4$ $8 = 4 + 4 + 4 - 4$

$5 = \dfrac{4 \cdot 4 + 4}{4}$ $9 = \dfrac{4}{4} + 4 + 4$

$6 = 4 + \dfrac{4 + 4}{4}$ $10 = \dfrac{44 - 4}{4}$

Section 1.7 *(page 49)*

1. When $-9 - 4x + x^2$ is written as $-9 + (-4x) + x^2$, we can apply the Commutative Property of Addition to write the expression as $x^2 + (-4x) + (-9) = x^2 - 4x - 9$.

2. **(a)** The number 0 is its own opposite. Because $0 + 0 = 0$, the Property of Additive Inverses is satisfied.

(b) The number 1 is its own reciprocal, as is the number -1. Because $1 \cdot \dfrac{1}{1} = 1$ and $-1 \cdot \dfrac{1}{-1} = 1$, the Property of Multiplicative Inverses is satisfied in both cases.

3. Associative Property of Addition

4. Commutative Property of Addition

5. Additive Identity Property

6. Property of Additive Inverses

7. Distributive Property

8. Property of Multiplicative Inverses

9. Distributive Property

10. Associative Property of Multiplication

11. Multiplicative Identity Property

12. Multiplication Property of 0

13. Multiplication Property of -1

14. Distributive Property **15.** Distributive Property

16. Commutative Property of Multiplication **17.** (i)

18. (a) **19.** (b) **20.** (g) **21.** (c) **22.** (h) **23.** (f)

24. (d) **25.** $x + (4 + 3)$ **26.** $3(x + 6)$ **27.** $(6 \cdot 2)x$

28. $-x + 8$ **29.** $x + 6$ **30.** 0 **31.** $1x$ **32.** $n - 4$

33. $3(4z + 5)$ **34.** $-3a + 18$

35. Suppose that 0 had a reciprocal r. Then, according to the Property of Multiplicative Inverses, $0 \cdot r = 1$. But the Multiplication Property of 0 states that $0 \cdot r = 0$. Therefore, 0 does not have a reciprocal.

36. The reciprocal of $\frac{p}{q}$ is $\frac{1}{\frac{p}{q}} = 1 \div \frac{p}{q} = 1 \cdot \frac{q}{p} = \frac{q}{p}$.

37. $\frac{1}{5}$; Property of Multiplicative Inverses

38. 7; Property of Additive Inverses

39. 1; Multiplicative Identity Property

40. 0; Property of Additive Inverses

41. -1; Distributive Property

42. $(x - 5)$; Distributive Property

43. **(a)** Commutative Property of Addition
(b) Associative Property of Addition
(c) Property of Additive Inverses
(d) Additive Identity Property

44. **(a)** Multiplication Property of -1
(b) Distributive Property
(c) Commutative Property of Addition

45. **(a)** Definition of Subtraction
(b) Associative Property of Addition
(c) Property of Additive Inverses
(d) Additive Identity Property

46. **(a)** Associative Property of Multiplication
(b) Property of Multiplicative Inverses
(c) Multiplicative Identity Property

47. $x + (3 + 2) = x + 5$ **48.** $(-4 + 8) + y = 4 + y$

49. $2z + (7 - 2) = 2z + 5$ **50.** $3x + (-5 + 7) = 3x + 2$

51. $(-5 \cdot 3)x = -15x$ **52.** $2(y \cdot y) = 2y^2$

53. $\left[-\frac{5}{6}\left(-\frac{9}{10}\right)\right]x = \frac{3}{4}x$ **54.** $\left(\frac{3}{4} \cdot 2\right)xy = \frac{3}{2}xy$

55. $20 + 15b$ **56.** $-6a + 14$ **57.** $-3x - 12$

58. $4x - 1$ **59.** $-15 + 6x$ **60.** $12x - 20$ **61.** $xy + xz$

62. $st + 3s$ **63.** $5(x + y)$ **64.** $11(a - b)$ **65.** $3(x + 4)$

66. $7(y - 2)$ **67.** $5(y + 1)$ **68.** $2(1 + b)$ **69.** $3(2 - x)$

70. $2(x - 1)$ **71.** $-x + 4$ **72.** $-3 - 5x$

73. $-2x - 5y + 3$ **74.** $4x + y - 7$ **75.** $12x$

76. $3y^2$ **77.** $-12x^2$ **78.** $12xy$ **79.** $x + 8$ **80.** 7

81. 3 **82.** $-10 + 2x$ **83.** $\left(\frac{3}{7} + \frac{4}{7}\right) + \left(\frac{2}{9} + \frac{5}{9} + \frac{2}{9}\right) = 2$

84. $(77 + 23) + (40 + 60) + (30 + 70) = 300$

85. $(2 \cdot 5)[-87(-1)] = 870$ **86.** $\left(5 \cdot \frac{1}{5}\right)\left(\frac{1}{3} \cdot 3\right) = 1$

87. $7(100 + 8) = 700 + 56 = 756$

88. $30(100 + 9) = 3000 + 270 = 3270$

89. $15(100 - 2) = 1500 - 30 = 1470$

90. $4(50 - 1) = 200 - 4 = 196$

91. If n is the number, $9n = (10 - 1)n = 10n - n$.

92. A nonzero number and its opposite have opposite signs, whereas a nonzero number and its reciprocal have the same sign.

93. $-\frac{5}{7}, \frac{7}{5}$ **94.** $8, -\frac{1}{8}$ **95.** $-\frac{2}{3}x, \frac{3}{2x}$ **96.** $-\frac{2}{y}, \frac{y}{2}$

97. $-x - 3, \frac{1}{x + 3}$ **98.** $5 - y, \frac{1}{y - 5}$

99. $3x - 4, \frac{1}{4 - 3x}$ **100.** $-2y, \frac{1}{2y}$

101. Property of Additive Inverses

102. Associative Property of Multiplication

103. Commutative Property of Addition

104. $2(20 + 35)$; $2 \cdot 20 + 2 \cdot 35$
Distributive Property

105 Sometimes: True if $a = b$, otherwise false.

106. Sometimes: True if $a = b$ or $a = -b$, otherwise false.

107. Sometimes: True if a and b have like signs, otherwise false.

108. Sometimes: True if, for example, $a = 5$ and $b = 2$; false if, for example, $a = 2$ and $b = 5$.

109. False **110.** False

111. $\frac{a + b}{3} = \frac{1}{3}(a + b) = \frac{1}{3}a + \frac{1}{3}b = \frac{a}{3} + \frac{b}{3}$

Chapter Review Exercises *(page 53)*

1. False **2.** False **3.** True **4.** True **5.** False

6. True **7.** False

8. **(a)** $-2, -\sqrt{9}, 5$
(b) $-2, \frac{5}{9}, -\sqrt{9}, 2.34, 5$
(c) $\frac{\pi}{2}, -\sqrt{10}$

9. Because $0.5 = \frac{1}{2}$, the statement $0.5 < \frac{1}{2}$ is false. However, $0.5 \le \frac{1}{2}$ is true because $\le$ means "less than or equal to."

10. If $n < 0$, then $-n$ is positive.

11. **(a)** -2 or 4 **(b)** -7 or 7

12. **(a)** $<$ **(b)** $>$ **(c)** $>$ **(d)** $<$

13. Addends are numbers that are added. In $4 + 6$, 4 and 6 are addends.

14. If the addends have unlike signs, then we subtract the absolute values of the addends to determine the absolute value of the sum.

15. -3 **16.** -20 **17.** $\frac{1}{12}$ **18.** $-\frac{49}{5}$ **19.** 3 **20.** -14

21. 11 **22.** -2 **23.** 11°F **24.** \$319.12

25. In $6 - 2$, 6 is the minuend and 2 is the subtrahend.

26. The numbers $a - b$ and $b - a$ are opposites. **27.** 20

28. -9 **29.** $-\frac{23}{20}$ **30.** $-\frac{19}{3}$ **31.** -10 **32.** -6

33. 4 **34.** 13 units **35.** 10,350 feet **36.** 27°F

37. The factors are the numbers that are multiplied.

38. The product is positive if there is an even number of negative factors. Otherwise, the product is negative.

39. -20 **40.** 35 **41.** $\frac{5}{2}$ **42.** $-\frac{7}{6}$ **43.** $-\frac{1}{6}$

44. -18 **45.** 0 **46.** 20 **47.** -4.5 **48.** 0

49. 3000 gallons less **50.** 1

51. For $10 \div 2$, 10 is the dividend and 2 is the divisor.

52. If $a \div 0 = c$, then $0 \cdot c = a$. If $a \neq 0$, then this is impossible because 0 multiplied by any real number is zero. If $a = 0$, then any value for c would satisfy the equality and the value for c would not be unique.

53. -5 **54.** 5 **55.** 0 **56.** $-\frac{18}{25}$ **57.** $\frac{4}{81}$ **58.** $-\frac{7}{32}$

59. $-\frac{1}{2}$ **60.** 24 hours

61. For -5^2, multiply $-1(5)(5)$ to obtain -25. For $(-5)^2$, multiply $(-5)(-5)$ to obtain 25.

62. The statement is not true. The left side is not a real number, whereas the right side is -5.

63. $(-3)^5 = -243$ **64.** $-5(-7 + 3^2) + (-21) = -31$

65. 16 **66.** 16 **67.** 0 **68.** -1 **69.** 1 **70.** $-\frac{8}{27}$

71. 7 **72.** -7 **73.** 25 **74.** 64 **75.** -24 **76.** 9

77. 5 **78.** 2

79. The Multiplication Property of -1 states that $-1x = -x$.

80. The task is not commutative because the order makes a difference in the result.

81. Commutative Property of Addition

82. Multiplicative Identity Property

83. Associative Property of Multiplication

84. Distributive Property **85.** Additive Identity Property

86. Multiplication Property of -1 **87.** $7 \cdot (c + 5)$

88. $[x + (-2)] + 2$ **89.** $5(x + 8)$ **90.** $x + 0$

91. $-5\left(-\frac{1}{5}\right)$ **92.** -7

Chapter Test *(page 55)*

1. [*1.1*] $-2, 0, 6, \sqrt{9}$ **2.** [*1.1*] $\pi, \sqrt{7}$

3. [*1.1*] $-3.7, -2, -0.\overline{14}, 0, \frac{2}{3}, 6, \sqrt{9}$

4. [*1.1*] **(a)** 5 **(b)** 5 **(c)** -5

5. [*1.3*] 23

6. [*1.1*]

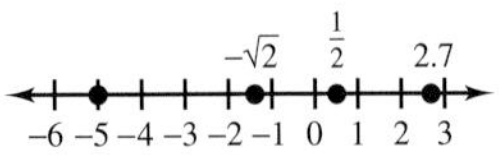

7. [*1.3*] -4 **8.** [*1.4*] $-\frac{1}{15}$ **9.** [*1.6*] $\frac{5}{37}$ **10.** [*1.6*] -7

11. [*1.6*] -1 **12.** [*1.3*] 4 **13.** [*1.2*] $-\frac{5}{4}$

14. [*1.3*] -9.2 **15.** [*1.4*] 0 **16.** [*1.5*] 8

17. [*1.5*] **(a)** 0 **(b)** Division by zero is undefined.

18. [*1.7*] $3x - 12$ **19.** [*1.7*] $-x + 4$

20. [*1.7*] $3 + (t - 3r)$ **21.** [*1.3*] 4 **22.** [*1.5*] $-\frac{1}{2}$

23. [*1.6*] 1.40 **24.** [*1.6*] -5 **25.** [*1.3*] -3.49°C

Chapter 2

Section 2.1 *(page 62)*

1. For $x = 2$, the denominator is 0. Division by 0 is undefined.

2. If $x \neq 7$, $\dfrac{x-7}{7-x} = \dfrac{x-7}{-(x-7)} = -\dfrac{x-7}{x-7} = -1$. If $x = 7$, the denominator is 0 and the expression is undefined.

3. 8 **4.** 14 **5.** 4 **6.** 10 **7.** -13 **8.** 5 **9.** -3

10. 14 **11.** 3 **12.** 0 **13.** Undefined **14.** Undefined

15. -1 **16.** $\frac{2}{5}$ **17.** 3 **18.** 5 **19.** -2 **20.** 7

21. -5 **22.** 9 **23.** 10 **24.** 4 **25.** Undefined

26. Undefined **27.** 0 **28.** 0 **29.** $\frac{5}{4}$ **30.** 17 **31.** 2

32. 1 **33.** $-\frac{2}{3}$ **34.** 2 **35.** 2 **36.** 0 **37.** $-\frac{2}{5}$

38. Undefined **39.** -9.41 **40.** 34.47 **41.** -13.2

42. -96.81 **43.** 19.02 **44.** -246.84 **45.** 0

46. Not a real number **47.** 7 **48.** 2.45

49. Replacing t with 5 results in $\sqrt{-1}$, which is not a real number.

50. The value of x could have been 3 or -3, because the result of squaring either number is 9.

51. 4 **52.** -4 **53.** $\frac{9}{5}$ **54.** 45 **55.** -4 or 14 **56.** 9

57. When we evaluate an expression, we perform all indicated operations to determine the value of the expression. To verify that a number is a solution of an equation, we substitute the number for the variable in the equation and then evaluate the two sides to determine whether they are equal.

58. Every member of A is a solution, because according to the Associative Property of Multiplication, the equation is true for all real numbers.

59. 2 **60.** $-\frac{1}{2}$ **61.** 4 **62.** -3 **63.** $\frac{5}{2}$ **64.** -5

65. Yes **66.** No **67.** No **68.** Yes **69.** Yes **70.** No

71. No **72.** Yes **73.** No **74.** No **75.** Yes **76.** Yes

77. Yes **78.** No **79.** No **80.** Yes **81.** No; 84.9 feet

82. 13.9 inches

83. No. The ladder, building, and ground form a triangle whose sides are 6, 22, and 24. Because $6^2 + 22^2 \neq 24^2$, the building is not perfectly vertical with respect to the ground.

84. Measure from pin A to pin C. If $80^2 + 150^2 = (AC)^2$, the triangle is a right triangle and the sides are perpendicular. The distance from A to C should be 170 feet.

85. 10°C **86.** 14°F **87.** \$1300

88. **(a)** 196 feet **(b)** 96 feet

89.

State	*High (in °C)*	*Low (in °F)*
Hawaii	37.8°	14.0°
Oklahoma	49.4°	−40.0°
Florida	42.8°	−2.2°
Wisconsin	45.6°	−54.4°

90. 2664 **91.** 0, 2, 4, 6

92. The value $t = 10$ corresponds to the year 2002. The model indicates that 72% of riders will wear helmets in 2002.

93. 45%

94. When $t = 16$ (2008), the value of the model expression is 104.3. Because the percent cannot exceed 100%, the model is invalid for that year, and it is probably not reliable beyond 1998.

95. Suppose that n is the line number in the table. Then for $a^2 + b^2 = c^2$, $a = 2n + 1$, $b = n(a + 1)$, and $c = b + 1$. The next two lines are: $9^2 + 40^2 = 41^2$, $11^2 + 60^2 = 61^2$.

96. By the Pythagorean Theorem, $(AC)^2 + (BC)^2 = (AB)^2$. Thus the area of square I equals the sum of the areas of squares II and III.

Section 2.2 *(page 69)*

1. Suppose that A is an algebraic expression with no explicitly written constant term. Because $A = A + 0$, the constant term of the expression is 0.

2. Suppose that A is an algebraic term with no explicitly written coefficient. Because $A = 1 \cdot A$, the coefficient of the term is 1.

3. Terms: $2x, -y, 5$ Coefficients: $2, -1, 5$

4. Terms: $6a, -5b, c, -2d$ Coefficients: $6, -5, 1, -2$

5. Terms: $3a^3, -4b^2, 5c, -6d$ Coefficients: $3, -4, 5, -6$

6. Terms: $3x^2, 5x, -6$ Coefficients: $3, 5, -6$

7. Terms: $7y^2, -2(3x - 4), \frac{2}{3}$ Coefficients: $7, -2, \frac{2}{3}$

8. Terms: $\frac{5x}{3}, -2(y - 3)$ Coefficients: $\frac{5}{3}, -2$

9. Terms: $\frac{x + 4}{5}, -5x, 2y^3$ Coefficients: $\frac{1}{5}, -5, 2$

10. Terms: $\frac{2x^2}{5}, -y^2, -8x, 2$ Coefficients: $\frac{2}{5}, -1, -8, 2$

11. The expression $2x + 3$ contains one plus sign that separates the terms $2x$ and 3. The plus sign in the expression $2(x + 3)$ is inside parentheses and is not considered when identifying terms. Thus this expression has only one term.

12. The terms do not have the same exponents on the variable factors.

13. $-x + 3$ **14.** $2x - 2$ **15.** $3x - 3y - 5$

16. $x - 4y + 4$ **17.** $6x - 7y$ **18.** $-3y + 3x$ **19.** 4

20. $-6 + 2x$ **21.** $-x + y$ **22.** $m + \frac{1}{4}n$ **23.** $17x^2 + 3x$

24. $22x^3 - 11x^2$ **25.** $10ac^2 - 13ac$ **26.** $2ax^2 + 8a^2x$

27. $2.75x^3 - 4.40x^2$ **28.** $12.22a^2 - 8.97b^2$

29. We can write $-(2x - 5)$ as $-1(2x - 5)$, and we can write $3 - (4a + 1)$ as $3 + (-1)(4a + 1)$. In both cases, distributing -1 changes the signs of the terms inside the grouping symbols.

30. Because $2x^2z = 2 \cdot x \cdot x \cdot z$ and $2xyz = 2 \cdot x \cdot y \cdot z$, both terms have four factors.

31. $-a + b$ **32.** $-3 - n$ **33.** $10x - 15$

34. $-12 + 3x$ **35.** $-2a + 3b - 7d$

36. $-3h + 7k + 8g$ **37.** $3x - 6y - z + 2$

38. $-6x - 12y - 7z + 14$ **39.** (d) **40.** (e) **41.** (f)

42. (c) **43.** (a) **44.** (b) **45.** $-x - 1$ **46.** $1 + 3x$

47. $x - 12$ **48.** $-10 + 2x$ **49.** $-7a - 2b$

50. $-2a - b$ **51.** -8 **52.** $17 - 5x$ **53.** $-3x + 17$

54. $-2 - x$ **55.** $x + 2$ **56.** $6x - 6$ **57.** $3x + 5$

58. $-x - 2y$ **59.** $8t + 13$ **60.** $-3t$ **61.** $2x - 4$

62. $17y - 3x$ **63.** $c - 125$ **64.** $10a - 63$

65. $-14x - 37$ **66.** $x - 27$ **67.** 4 **68.** $-22 + 4x$

69. $x^2y^3 + 3x^3y^2$ **70.** $2x^4y - x^2y^2$ **71.** 0

72. $2xy^2$ **73.** $\frac{2n}{12} = \frac{n}{6}$ **74.** $n(n + 4) = n^2 + 4n$

75. $3n + n = 4n$ **76.** $n + 8 - 2n = 8 - n$ **77.** (b)

78. (a) **79.** (b) **80.** (a) **81.** $x - 5 = 3$

82. $2x + 1 = 7$ **83.** $8x = 11$ **84.** $\frac{x}{4} = 3x$

85. $2x = x - 3$ **86.** $x^2 + 25 = 10x$ **87.** $x - 0.4x = 50$

88. $x + \frac{1}{2}y = 16$ **89.** $x - 2y = -5$ **90.** $\frac{x + 2}{3} = 7$

91. $3L$ **92.** h^2 **93.** $25q + 10d$ **94.** $55h$

95. $2.5x + 15$ **96.** $\frac{1}{h}$ **97.** $3n + 3$ **98.** $\frac{n}{n + 2}$

99. $0.08d$ **100.** $1.09d$ **101.** $\sqrt{a^2 + (a - 2)^2}$

102. $0.65c$ **103.** $1.05s$ **104.** $0.20L$

105. The phrase is ambiguous because we can read it as "(the product of 3 and a number) increased by 7" or as "the product of 3 and (a number increased by 7)." The corresponding translations are $3x + 7$ and $3(x + 7)$.

106. For $a = 1$ and $b = 0$, the expressions have the same value. However, the expressions are not equivalent because they do not have the same value for all replacements for a and b. For example, the expressions are not equal if $a = 4$ and $b = 1$.

107. **(a)** $1000x + 6000$ **(b)** 9000

(c) In 1998 ($x = 3$), the estimated number of complaints is 9000.

(d) 13,000

108. $0.00526P$ **109.** $\dfrac{d}{250{,}000{,}000}$ **110.** $\$0.25x$ billion

111. \$6.51 billion **112.** $\$250{,}000{,}000\,(T + L + F)$

113. (a) **114.** (b) **115.** (a) **116.** (a) **117.** (b)

118. (a) **119.** $2x^2$ **120.** 0

Section 2.3 *(page 78)*

1. Starting at the origin, move 2 units to the left along the x-axis and then 4 units vertically upward. The destination is the point $A(-2, 4)$.

2. The points $P(4, 0)$ and $Q(0, 3)$ are examples of points that are not in any quadrant. They are points of the axes.

3.

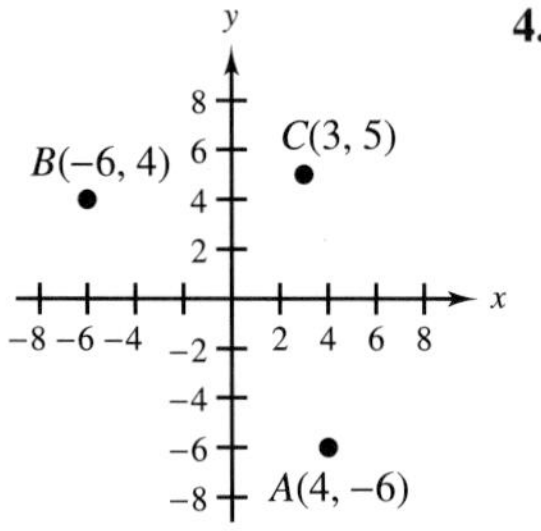

4.

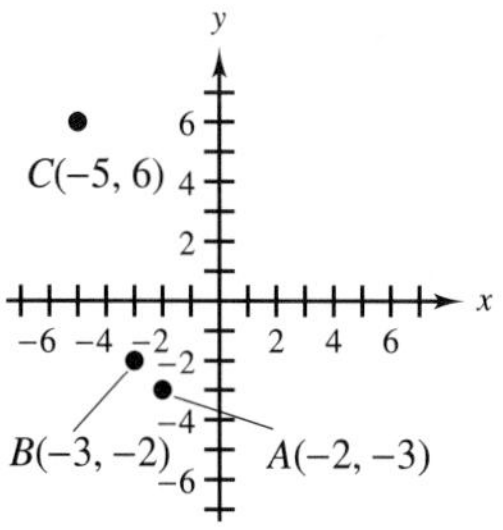

5.

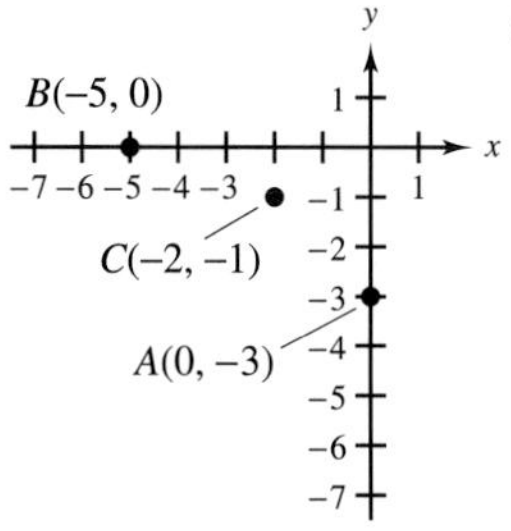

6.

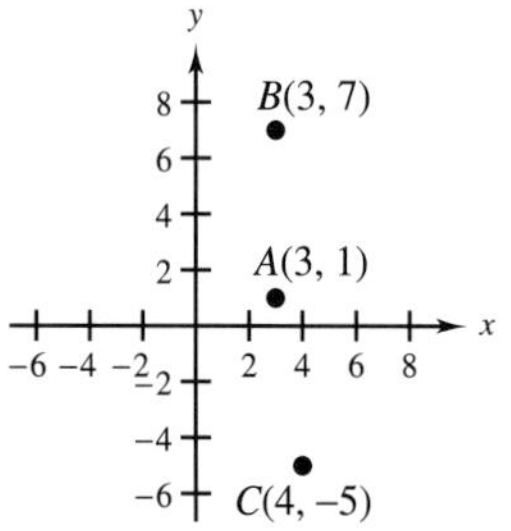

7. (8, 3)

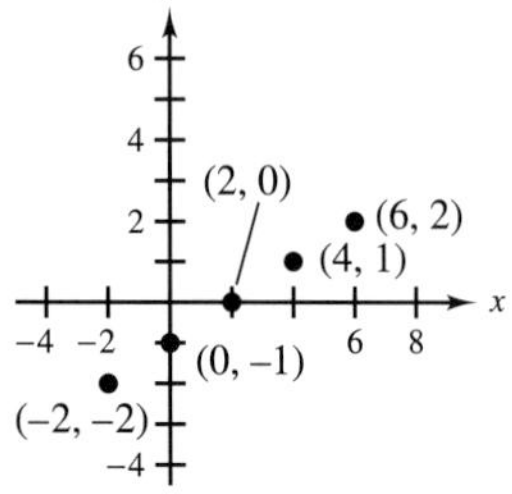

8. (10, −4)

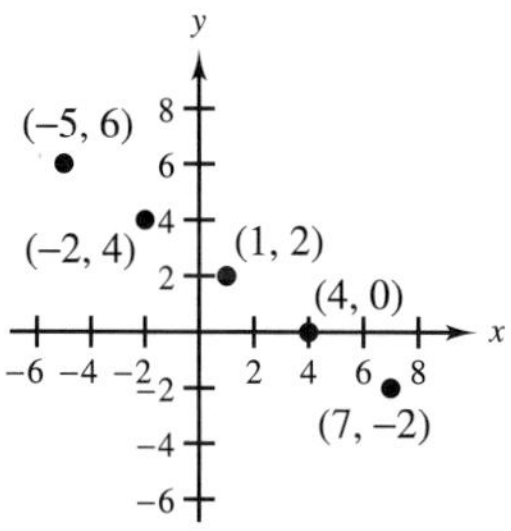

9. (2, 5)

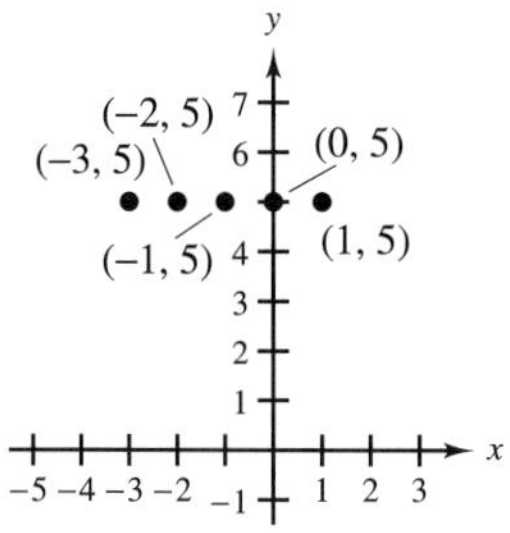

10. (4, −3)

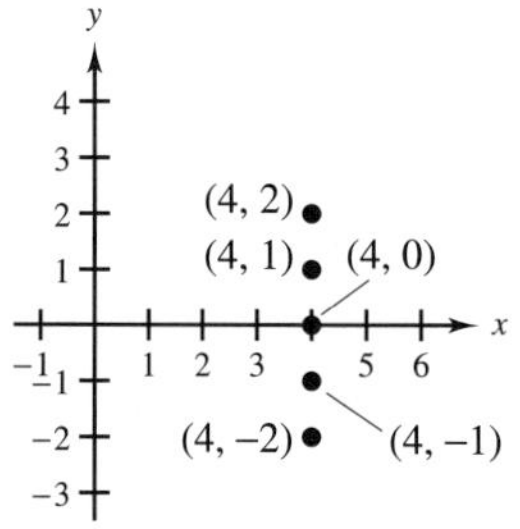

11. (6, 3)

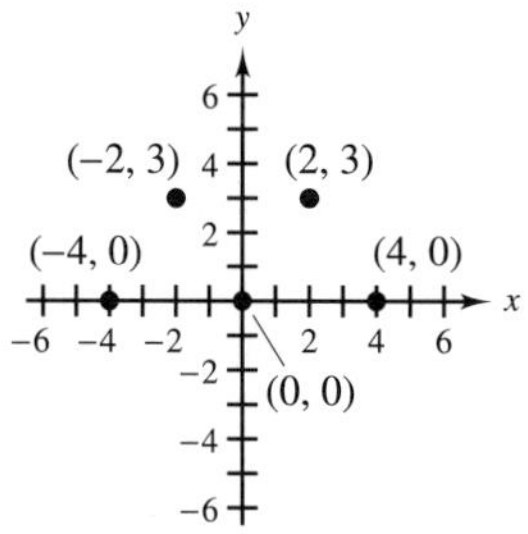

12. (4, 0)

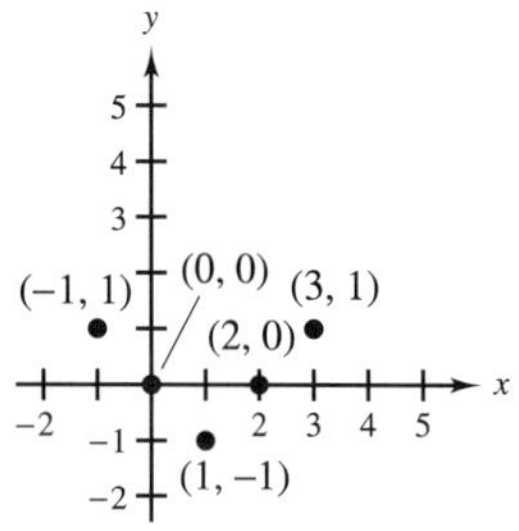

13. III **14.** IV **15.** I **16.** III **17.** IV **18.** II

19. $(-6, 4)$ **20.** $(4, -6)$ **21.** $(0, 0)$ **22.** $(-5, -3)$

23. $(5, 0)$ **24.** $(0, 7)$

25. Points whose x-coordinates are 0 belong to the y-axis.

26. Points whose y-coordinates are 0 belong to the x-axis.

27. I, IV, or x-axis **28.** II, III, or x-axis

29. III, IV, or y-axis **30.** I, II, or y-axis **31.** x-axis

32. y-axis **33.** II, III, or x-axis **34.** I, II, or y-axis

35. II, IV **36.** I, III **37.** $x < 0, y = 0$

38. $x = 0, y < 0$ **39.** $x = 0, y > 0$ **40.** $x > 0, y = 0$

41. **(a)** $(+, -)$ **(b)** $(+, +)$ **(c)** $(-, +)$ **(d)** $(-, -)$

42. **(a)** I **(b)** IV **(c)** III

43. **(a)** x-axis **(b)** y-axis
(c) Quadrant IV **(d)** Quadrant I or III

44. **(a)** Quadrant IV **(b)** Quadrant II
(c) Quadrant I **(d)** Quadrant III

45. $(0, 13)$ **46.** $(-14, 0)$ **47.** $(9, 9)$ **48.** $(-11, -11)$

49. $(-12, 6)$ **50.** $(-5, -20)$ **51.** $(-4, 6)$

52. $(19, -13)$ **53.** $(17, 0)$ **54.** $(6, -14)$

55. Find the square of the difference of the x-coordinates and the square of the difference of the y-coordinates. Add these quantities and take the square root of the result.

56. The x-coordinate of the midpoint is the average of the x-coordinates of the endpoints. The y-coordinate is the average of the y-coordinates of the endpoints.

57. 17 **58.** 24 **59.** 16 **60.** 27 **61.** 10

62. $\sqrt{162} \approx 12.73$ **63.** 5 **64.** $\sqrt{421} \approx 20.52$

65. $\sqrt{104} \approx 10.20$ **66.** $\sqrt{109} \approx 10.44$ **67.** Collinear

68. Collinear **69.** Noncollinear **70.** Collinear

71. $P = 16, A = 15$ **72.** $P = 12, A = 9$

73. $P = 12, A = 9$ **74.** $P = 18, A = 20$ **75.** $(4, -2)$

76. $(0, 0)$ **77.** $(1, 5)$ **78.** $(-1, 1.5)$ **79.** $(0, 1.75)$

80. $(1, -1.5)$ **85.** Yes **86.** Yes **87.** No **88.** Yes

89. No

90. You should feel the force of the wind at noon. The predicted rate is approximately 22.4 mph.

91. 21.6 inches **92.** 0.65 mile

93. **(a)** Employed full-time with one child
(b) Employed full-time with four or more children

94. 2 **95.** **(a)**

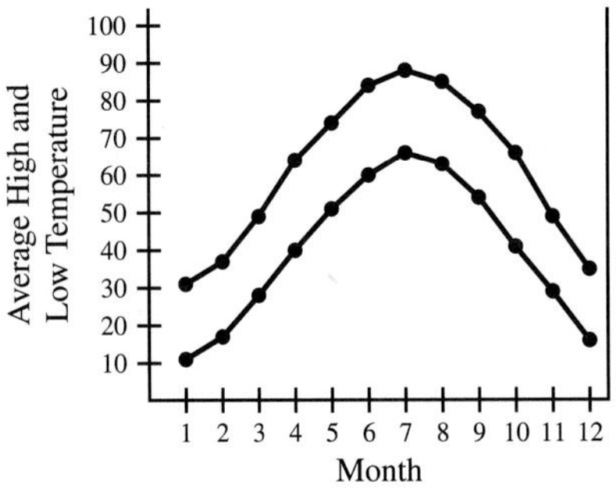

(b) Choose the two points that represent the average high and the average low for any month. If the y-coordinates of those points are y_H and y_L, then the distance $y_H - y_L$ is the difference between the average high and low temperatures for that month.

96. From October to November, the average high drops 17°.

97. (1970, 63), (1980, 75), (1990, 85), (2000, 96)

98. (1985, 79.5)

99. The estimated population in 1985 was 79.5 million.

100. 11 million; 107 million

101.

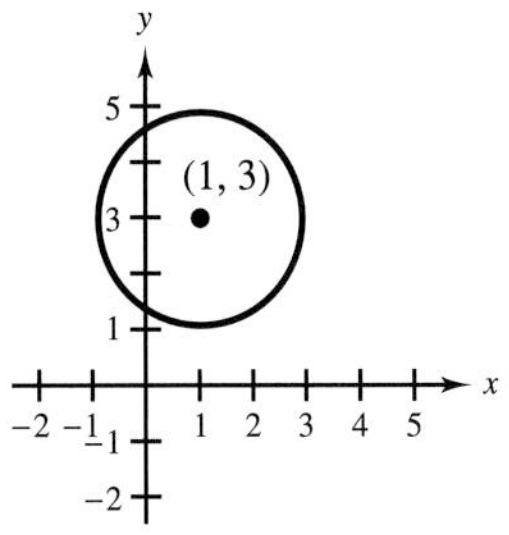

The graph is a circle with center $C(1, 3)$ and radius 2.

102.

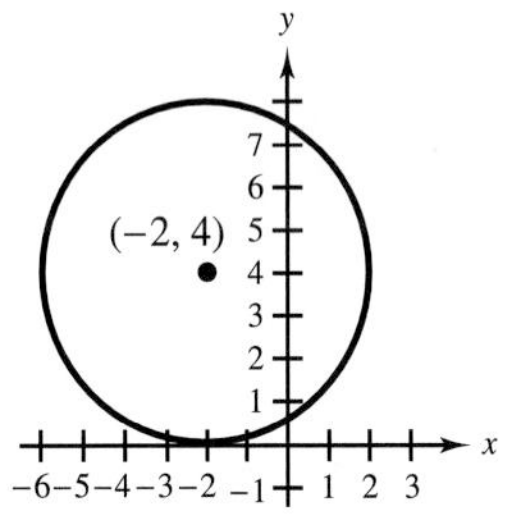

The graph is a circle with center $C(-2, 4)$ and radius 4.

103. $B(8, 8)$ **104.** $B(-4, -9)$

Section 2.4 *(page 90)*

1. A relation is a set of ordered pairs. A function is a relation in which each first coordinate is paired with exactly one second coordinate. Every function is a relation, but not every relation is a function.

2. The domain of a relation is the set of first coordinates of the ordered pairs. The range is the set of second coordinates.

3. $\{(-3, 9), (-2, 4), (-1, 1), (0, 0), (1, 1), (2, 4), (3, 9)\}$
Domain: $\{-3, -2, -1, 0, 1, 2, 3\}$
Range: $\{0, 1, 4, 9\}$

4. $\{(1, -3), (2, -2), (3, -1), (4, 0), (5, 1), (6, 2), (7, 3)\}$
Domain: $\{1, 2, 3, 4, 5, 6, 7\}$
Range: $\{-3, -2, -1, 0, 1, 2, 3\}$

5. $\{(1, 4), (2, 8), (3, 12), (4, 16), (5, 20)\}$
Domain: $\{1, 2, 3, 4, 5\}$
Range: $\{4, 8, 12, 16, 20\}$

6. $\{(1, \pi), (2, 4\pi), (3, 9\pi), (4, 16\pi)\}$
Domain: $\{1, 2, 3, 4\}$
Range: $\{\pi, 4\pi, 9\pi, 16\pi\}$

7. $\{(10, 70), (20, 140), (30, 210), (40, 280)\}$
Domain: $\{10, 20, 30, 40\}$
Range: $\{70, 140, 210, 280\}$

8. $\{(1, 8), (2, 9), (3, 10), (4, 11), (5, 12)\}$
Domain: $\{1, 2, 3, 4, 5\}$
Range: $\{8, 9, 10, 11, 12\}$

9.

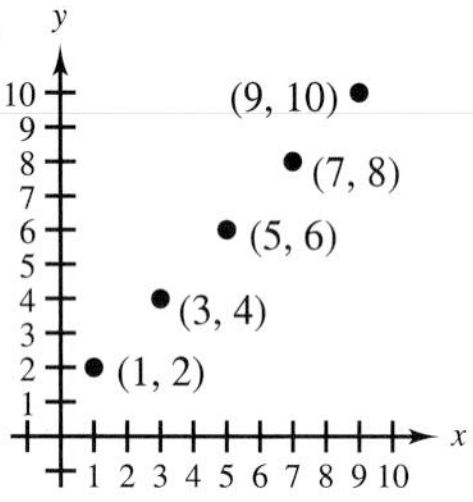

Domain: $\{1, 3, 5, 7, 9\}$
Range: $\{2, 4, 6, 8, 10\}$
Function

10.

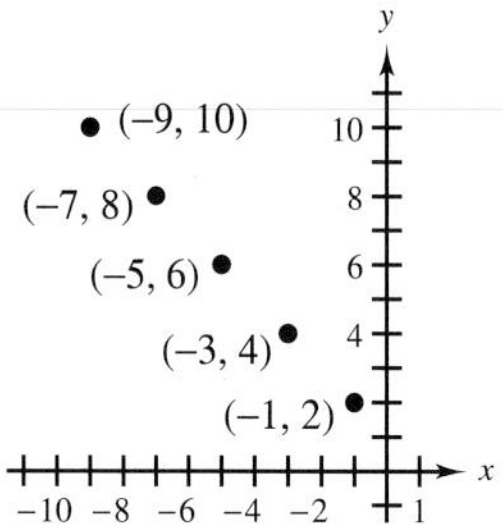

Domain: $\{-1, -3, -5, -7, -9\}$
Range: $\{2, 4, 6, 8, 10\}$
Function

11.

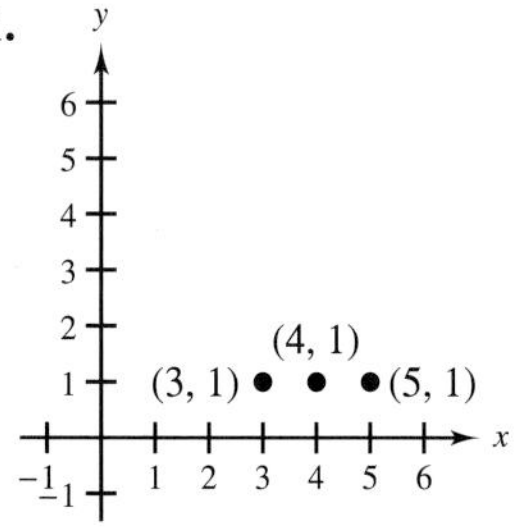

Domain: $\{3, 4, 5\}$
Range: $\{1\}$
Function

12.

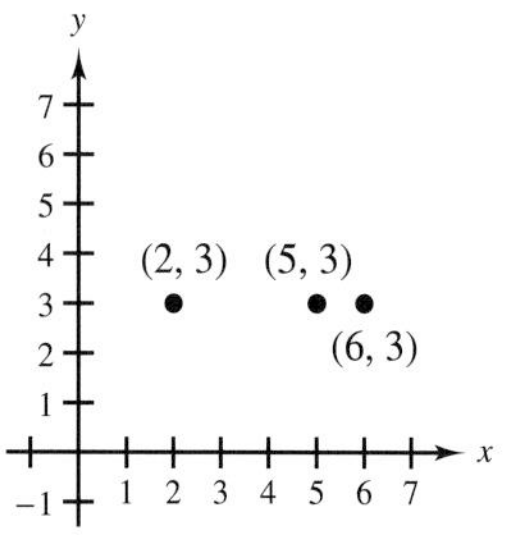

Domain: $\{2, 5, 6\}$
Range: $\{3\}$
Function

13.

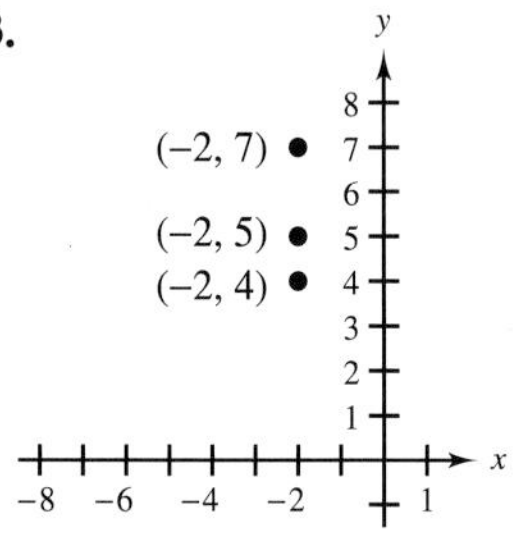

Domain: $\{-2\}$
Range: $\{4, 5, 7\}$
Not a function

14.

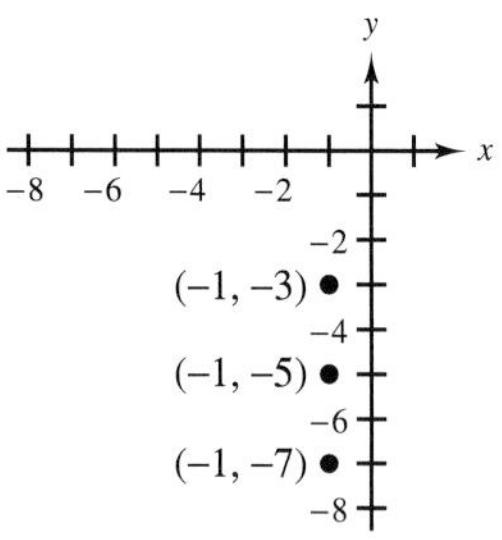

Domain: $\{-1\}$
Range: $\{-3, -5, -7\}$
Not a function

15.

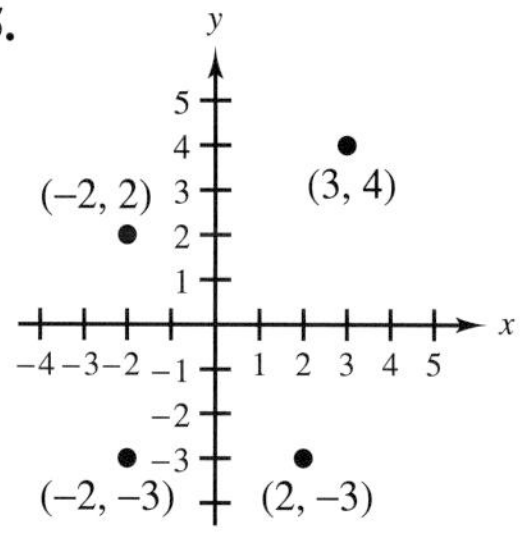

Domain: $\{-2, 2, 3\}$
Range: $\{-3, 2, 4\}$
Not a function

16.

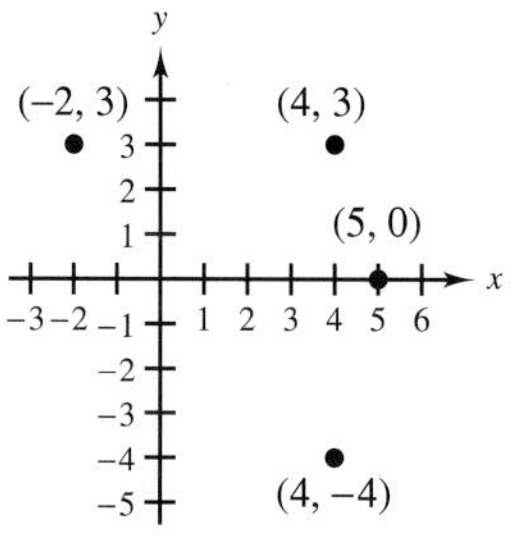

Domain: $\{-2, 4, 5\}$
Range: $\{-4, 0, 3\}$
Not a function

17.

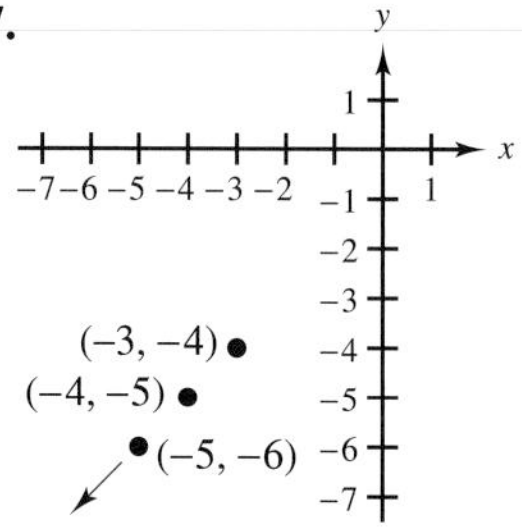

Domain: $\{\ldots, -5, -4, -3\}$
Range: $\{\ldots, -6, -5, -4\}$
Function

18.

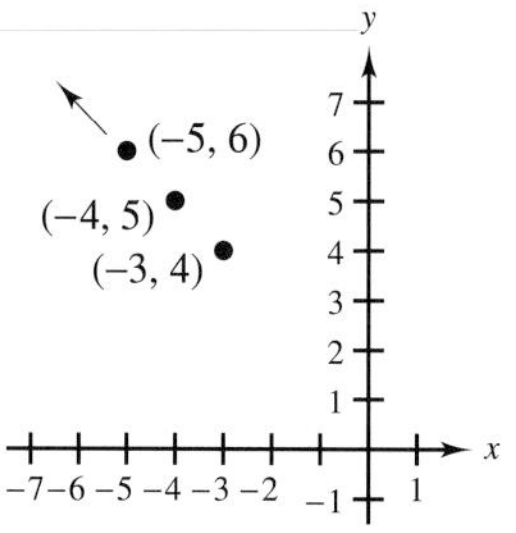

Domain: $\{\ldots, -5, -4, -3\}$
Range: $\{4, 5, 6, \ldots\}$

19. The domain of a relation can be estimated from a graph by determining the x-coordinates of the points of the graph.

20. The range of a relation can be estimated from a graph by determining the y-coordinates of the points of the graph.

21. For a graph to represent a function, no vertical line can intersect the graph in more than one point. If a vertical line intersects a graph in more than one point, there are two ordered pairs with the same first coordinate. This violates the definition of a function.

22. All functions are relations because a function is a set of ordered pairs. However, a relation with two ordered pairs having the same first coordinate is not a function.

23. $(-\infty, -4)$ **24.** $[-5, \infty)$ **25.** $(-3, 7]$ **26.** $[3, 8)$

27. $\{x \mid -3 \leq x < 7)$ **28.** $\{x \mid 0 \leq x \leq 5\}$

29. $\{x \mid x \leq 0\}$ **30.** $\{x \mid x > -1\}$

31. Domain: $[-7, 1]$; Range: $[-6, 2]$; Not a function

32. Domain: $[-5, 4]$; Range: $[-5, 5]$; Not a function

33. Domain: $\{-5, -4, -3, -2, -1, 0, 1, 2, 3\}$
Range: $\{3\}$; Function

34. Domain: $\{-2\}$
Range: $\{-3, -2, -1, 0, 1, 2, 3, 4\}$
Not a function

35. Domain: $(-\infty, 0]$; Range: **R**; Not a function
36. Domain: **R**; Range: $[0, \infty)$; Function
37. Domain: **R**; Range: $[3, \infty)$ Function
38. Domain: $(-\infty, 4]$; Range: $[0, \infty)$; Function
39. Domain: **R**; Range: **R**; Function
40. Domain: **R**; Range: $(-\infty, 3]$; Function
41. Domain: **R**; Range: $[-3, \infty)$; Function
42. Domain: **R**; Range: **R**; Function
43. Function **44.** Function **45.** Function
46. Not a function **47.** Not a function **48.** Function
49. Function **50.** Function **51.** Not a function
52. Function **53.** Function **54.** Function
55. Not a function **56.** Not a function **57.** Not a function
58. Not a function **59.** Function **60.** Function
61. Domain: **R**; Range: **R** **62.** Domain: **R**; Range: **R**
63. Domain: **R**; Range: $(-\infty, 9]$
64. Domain: **R**; Range: $[-5, \infty)$
65. Domain: $(-\infty, 9]$; Range: $[0, \infty)$
66. Domain: $[-7, \infty)$; Range: $[0, \infty)$
67. Domain: **R**; Range: $[-1, \infty)$
68. Domain: **R**; Range: $(-\infty, 4]$
69. Domain: **R**; Range: $[0, \infty)$
70. Domain: **R**; Range: $(-\infty, 3]$
71. (a) Domain: $\{1, 2, 3, 4, 5, 6\}$;
Range: $\{70, 84, 105, 140, 210, 420\}$
(b) $\{(1, 420), (2, 210), (3, 140), (4, 105), (5, 84), (6, 70)\}$
72. (a) Domain: $\{1, 2, 3, 4\}$; Range: $\{9.5, 11, 12.5, 14\}$
(b) $\{(1, 9.5), (2, 11), (3, 12.5), (4, 14)\}$
73. (a) $A = 20x - x^2$

(b)

x	0	4	8	10	12	16	20
Area	0	64	96	100	96	64	0

(c) 100 square inches **(d)** 10 inches by 10 inches
(e) 4 inches by 16 inches **(f)** (0, 20)
74. (a) $P = x(30 - x)$

(b)

x	0	5	10	15	20	25	30
Income	0	125	200	225	200	125	0

(c) [0, 225]
(d) To produce the greatest income, 15 cards should be sold.
(e) 10 or 20 **(f)** [0, 30]

75. (a) {(Birds, 12.6), (Dogs, 52.9), (Rabbits, 5.7), (Cats, 59)}
(b) Yes; no two ordered pairs have the same first coordinate.
76. No; Clinton and Reagan would appear as first coordinates in more than one pair.
77. No; red and silver would appear as first coordinates in more than one pair.
78. Yes; each year is unique, so no two ordered pairs would have the same first coordinate.
79. $A =$ {(New Orleans, 256), (Tampa, 246), (Jackson, 231), (Mobile, 217), (Casper, 217)}
Set A is a function.
80. Yes **81.** No **82.** \$6.40 **83.** No
84. No **85.** Yes **86.** No **87.** No **88.** No
89. Domain: $\{x \mid x \neq 0\}$; Range: $\{-1, 1\}$
90. Domain: **R**; Range: $[0, \infty)$
91. Domain: **R**; Range: $[0, \infty)$
92. Domain: **R**; Range: **R**

Section 2.5 *(page 96)*

1. To evaluate $f(3)$, replace x in the expression $2x + 1$ with 3 and perform the indicated operations. $f(3) = 2(3) + 1 = 7$.
2. To find $g(t + 4)$, replace x with $t + 4$ and perform the indicated operations. $g(t + 4) = 1 - 3(t + 4) = -3t - 11$.
3. 3 **4.** 1 **5.** 2 **6.** 0 **7.** $-2, -2$ **8.** $-1, -1$
9. 2 **10.** 1 **11.** -1 **12.** 1 **13.** 1 **14.** 4.5
15. $3 + t$ **16.** $b - 4$ **17.** a **18.** a **19.** 3, 8
20. $8, -1$ **21.** 3, 17 **22.** $-25, 29$ **23.** $-3, -3$
24. 7, 9
25. One point of the graph is $(4, -7)$. The x-coordinate of each point of the graph is the value of the variable, and the y-coordinate is the value of the function.
26. We know that $g(-3) = 5$ because the x-coordinate of each point of the graph is the value of the variable, and the y-coordinate is the value of the function.
27. (a) -13 **(b)** -23 **(c)** -11 **(d)** -5
28. (a) 9 **(b)** -3 **(c)** 9 **(d)** 11
29. (a) 2 **(b)** 2 **(c)** 18 **(d)** 0
30. (a) -3 **(b)** -2 **(c)** -5 **(d)** 21
31. (a) 1 **(b)** 4 **(c)** 7 **(d)** 1
32. (a) 1 **(b)** 3 **(c)** 1 **(d)** 5
33. (a) -32.9 **(b)** -2.75 **(c)** 2 **(d)** 397
34. (a) 17.8 **(b)** $4.3\overline{3}$ **(c)** 19.75 **(d)** 1.6
35. (a) -4.44 **(b)** -559.04 **(c)** -4165.54 **(d)** 4.02
36. (a) -30.134 **(b)** -21.12 **(c)** 0.60 **(d)** -1.44

37. **(a)** $h(5) = -2, k(5) = -\frac{3}{4}$
(b) The parentheses in the function k require the subtraction to be performed before the division.

38. **(a)** $B(1) = 7, C(1) = 12$
(b) The parentheses in the function C require the addition and subtraction to be performed before the division.

39. -2 **40.** -9 **41.** 5 **42.** -18 **43.** -59 **44.** 11

45. 26 **46.** 21

47. **(a)** $G(1) = G(-4) = G(2.3) = -3$
(b) The results in part (a) suggest that $G(x) = -3$ for all x.
(c) $G(x) = -3$

48. **(a)** $F(3) = F(1) = F(-3.7) = -2$
(b) The results in part (a) suggest that $F(x) = -2$ for all x.
(c) $F(x) = -2$

49. **(a)** $\sqrt{66.4} \approx 8.15$ **(b)** $\sqrt{6.7} \approx 2.59$

50. **(a)** $\sqrt{14.6} \approx 3.82$ **(b)** $\sqrt{24.6} \approx 4.96$

51. **(a)** $\sqrt{90.09} \approx 9.49$ **(b)** $\sqrt{14.25} \approx 3.77$

52. **(a)** $\sqrt{18.72} \approx 4.33$ **(b)** $\sqrt{69.68} \approx 8.35$

53. **(a)** 11,441 **(b)** 7736 **54.** **(a)** 3304 **(b)** 10,710

55. **(a)** 0 **(b)** 1.87 **56.** **(a)** 0.4 **(b)** -0.19

57. **(a)** $f(7) = -49, g(7) = 49, h(7) = 49$
(b) $f(-7) = -49, g(-7) = 49, h(-7) = 49$
(c) f: B, A; g: B; h: A, B
(d) Functions g and h have the same graph because $x^2 = (-x)^2$.

58. **(a)** $f(2) = -8, g(2) = 8, h(2) = -8$
(b) $f(-2) = 8, g(-2) = -8, h(-2) = 8$
(c) Choice (iii) because $-x^3 = (-x)^3$.

59. **(a)** $B(0) = C(0) = 5$
(b) $B(3) = \sqrt{34}, C(3) = 8$
(c) $B(-6) = \sqrt{61}, C(-6) = -1$
(d) The two functions do not have the same graph because $B(x) \neq C(x)$ for all x.

60. **(a)** $f(3) = g(3) = h(3) = 3$
(b) $f(-5) = 5, g(-5) = -5, h(-5) = 5$
(c) Functions f and h have the same graph because $\sqrt{x^2} = |x|$.

61. **(a)** Yes **(b)** No

62. **(a)** $A(14) = B(14) = 9$
(b) $B(1) = 4$, but $A(1)$ is undefined.
(c) No, because the domains are different.

63. **(a)** $10a - 7$ **(b)** $5a + 3$

64. **(a)** $6z + 20$ **(b)** $18z + 2$

65. **(a)** $t^2 - 5t + 6$ **(b)** $4t^2 - 10t + 6$

66. **(a)** $9t^2 - 12t - 8$ **(b)** $t^2 + 4t - 8$

67. **(a)** $\sqrt{a^2 + 5}$ **(b)** $\sqrt{4a^2 + 5}$

68. **(a)** $\sqrt{6 - 9b^2}$ **(b)** $\sqrt{6 - m^2}$

69. **(a)** $|-s^3 - 2|$ **(b)** $|27s^3 - 2|$

70. **(a)** $|8r^3 + 4|$ **(b)** $|-r^3 + 4|$

71. **(a)** 1 **(b)** $2h + 1$ **(c)** $2h$ **(d)** 2

72. **(a)** -25 **(b)** $4h - 25$ **(c)** $4h$ **(d)** 4

73. **(a)** 11 **(b)** $11 - 3h$ **(c)** $-3h$ **(d)** -3

74. **(a)** 27 **(b)** $27 + 5h$ **(c)** $5h$ **(d)** 5

75. 55 paces per minute **76.** 51 squares **77.** 1292

78. **(a)** \$30 **(b)** \$50 **(c)** \$40

79.

Year	1997	1999	2001	2002
x	2	4	6	7
Semiprivate	8577	9012	9447	9665
Private	4734	4708	4683	4670

80. **(a)** 229.5 milligrams

(b)

Weight (pounds)	6	8	10	12	14
Dosage (milligrams)	148.5	175.5	202.5	229.5	256.5

A cat receiving 202 milligrams weighs about 10 pounds.

81.

Year	$f(x)$	f Difference
1960	500	300
1970	1020	-120
1980	1540	-240
1990	2060	-360
2000	2580	420

82.

Year	$g(x)$	g Difference
1960	841	-41
1970	842	58
1980	1183	117
1990	1864	-164
2000	2885	115

83. g

84. Although both functions predict an increase in the population, g predicts a faster increase.

85. **(a)** 3, 30, 300, 3000, 30,000

(b) As x becomes smaller, $3/x$ becomes larger.

86. **(a)** 121; 12,321; 1,234,321; 123,454,321

(b) On the basis of the pattern in part (a), a reasonable prediction is that $g(11{,}111{,}111) = 123{,}456{,}787{,}654{,}321$.

Review Exercises *(page 102)*

1. Replace each occurrence of x with 3 and evaluate the resulting numerical expression.

2. A solution of an equation is a value of the variable that makes the equation true.

3. 1 **4.** 23 **5.** -12 **6.** 6.24 **7.** $-\frac{2}{3}$ **8.** 6

9. -1 **10.** 35°C **11.** **(a)** Yes **(b)** No

12. Yes **13.** The exponents on x and y are not the same.

14. The terms are the addends of the expression.

15. Terms: $3x$, $-2y$, $-z$; Coefficients: 3, -2, -1

16. Terms: $3x^2$, $-7(h-5)$; Coefficients: 3, -7 **17.** No

18. Yes **19.** $-2a - 3b + 4c$ **20.** $-6x + 8y + 6h - 15k$

21. $-9x^3 + 16x^2$ **22.** $17ab^2 - 20a^2b$

23. **(a)** $-15n - 20$ **(b)** $10 - 3n$

24. **(a)** $2x - 3 = 7$ **(b)** No

25. $-2x + 6$ **26.** $-8y + 60$ **27.** $(-4, 3)$ **28.** $(5, 0)$

29. $(-2, -3)$ **30.** $(1, 1)$ **31.** $(3, -4)$ **32.** $(0, -2)$

33. II **34.** IV **35.** II **36.** I **37.** y-axis **38.** II

39.

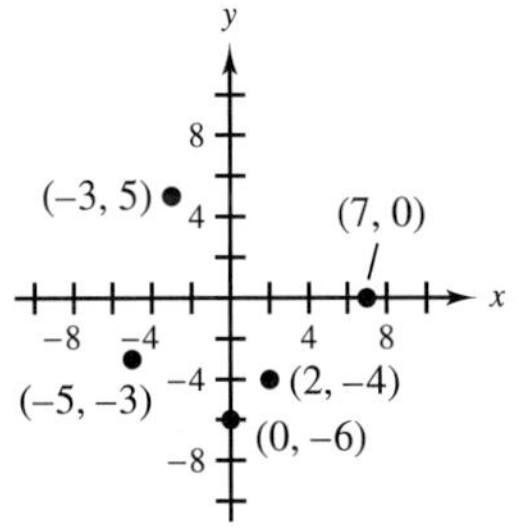

Domain: $\{\ldots, -5, -4, -3\}$
Range: $\{\ldots, -6, -5, -4\}$
Function

40. $(2, 4)$ **41.** **(a)** $\sqrt{90} \approx 9.49$ **(b)** $(-6.5, -0.5)$

42. **(a)** 10 **(b)** $(10, -5)$ **43.** 36 **44.** Yes

45. Domain: $\{-4, 0, 3, 5\}$; Range: $\{-6, 1, 5, 7\}$; Function

46. Domain: $[-7, 4]$; Range: $[-5, 6]$; Not a function

47. Function **48.** Function **49.** Not a function

50. Because a function is a set of ordered pairs, it is also a relation. Statement (ii) is true. A relation may have two different ordered pairs with the same first coordinate. Thus statement (i) is not true.

51. Domain: $[-2, \infty)$; Range: $[0, \infty)$

52. Domain: **R**; Range: $(-\infty, 16]$ **53.** (i)

54. **(a)** $A = b(24 - b)$, where b is the length of the base

(b)

b	2	4	8	12	16	20	22
Area	44	80	128	144	128	80	44

(c) 144 square inches **(d)** $(0, 24)$

55. 8 **56.** 5 **57.** 4.58 **58.** 13 **59.** 20 **60.** -74

61. **(a)** 3 **(b)** -7.5 **62.** $1 - 3h$

Chapter Test *(page 104)*

1. [*2.1*] 5 **2.** [*2.1*] 20.216 **3.** [*2.1*] -2 **4.** [*2.1*] Yes

5. [*2.2*] Terms are addends of an expression. Factors are quantities that are multiplied.

6. [*2.2*] $4s - t$ **7.** [*2.2*] $-4x + 2$

8. [*2.2*] **(a)** $n + 3n = 2n - 2$ **(b)** Yes

9. [*2.3*] **(a)** $(-3, 6)$ **(b)** $(0, 3)$ **(c)** $(3, 0)$ **(d)** $(6, -3)$

10. [*2.3*] **(a)** I, IV, or x-axis **(b)** II

11. [*2.3*] **(a)** $(2.5, 1)$ **(b)** $\sqrt{34} \approx 5.83$

12. [*2.3*] **(a)** 76.2 miles; No

(b) No. The distance from the tornado to the home plus the distance from the home to the radar site is not equal to the distance from the tornado to the radar site.

13. [*2.4*] When x is replaced with -4, the value of the expression is -14. Thus the y-coordinate is -14.

14. [*2.4*] 5 **15.** [*2.5*] 3 **16.** [*2.5*] 11

17. [*2.5*] $3t - 8$ **18.** [*2.5*] $3t$

19. [*2.4*] **(a)** Domain: $[-4, 4]$; Range: $[-5, 5]$

(b) Domain: $\{1, 2, 3, 5\}$; Range: $\{-1, 3, 7\}$

20. [*2.4*] **(a)** The graph represents a function because it passes the Vertical Line Test.

(b) The graph does not represent a function because it does not pass the Vertical Line Test.

Cumulative Test, Chapters 1–2 *(page 106)*

1. [*1.1*] $(-2, 3]$

2. [*1.1*]

(a) False

(b) True

(c) True

(d) False

3. [*1.3*] 3 **4.** [*1.2*] $-\frac{5}{21}$ **5.** [*1.3*] -10 **6.** [*1.3*] $-\frac{16}{15}$

7. [*1.4*] -12 **8.** [*1.4*] $-\frac{9}{4}$ **9.** [*1.5*] Undefined

10. [*1.5*] 2 **11.** [*1.6*] -25 **12.** [*1.6*] 5 **13.** [*1.6*] 4

14. [*1.6*] Not a real number **15.** [*1.6*] 5 **16.** [*1.6*] 2 **17.** [*1.7*] $3(3x + 4)$ **18.** [*1.7*] $-x$ **19.** [*1.7*] $(2 + 5)x$ **20.** [*1.7*] $1y$ **21.** [*2.1*] 0.5 **22.** [*2.1*] $-\frac{50}{49}$ **23.** [*2.1*] 6

24. [2.1] **(a)** No **(b)** Yes

25. [2.2] $-x + 6$ **26.** [2.2] 0

27. [2.2] **(a)** $\frac{x}{7} + 3$ **(b)** $2(6 - x) = 2x$

28. [2.3] **(a)** II **(b)** x-axis **(c)** y-axis **(d)** IV

29. [2.3] **(a)** $\sqrt{296} \approx 17.20$ **(b)** $(1, -2)$

30. [2.4] When x is replaced with 4, the value of the expression is 5.

31. [2.5] **(a)** -19 **(b)** -76

32. [2.5] Replacing x with 1 would result in $\frac{1}{0}$, which is undefined.

33. [2.4] (i) and (iii)

34. [2.4] The graph of a relation is the graph of a function if no vertical line intersects the graph in more than one point.

Chapter 3

Section 3.1 *(page 113)*

1. An equation is a statement that two algebraic expressions have the same value.

2. A linear equation in one variable is an equation that can be written in the form $Ax + B = 0$, where $A \neq 0$. Examples of equations that do not satisfy this definition are $x^2 + 4 = -3x$ and $\frac{3}{x} + 2 = 0$.

3. Yes **4.** Yes **5.** No **6.** No **7.** Yes **8.** Yes **9.** No **10.** Yes **11.** Yes **12.** No **13.** Yes **14.** Yes **15.** No **16.** No **17.** Yes **18.** Yes **19.** No **20.** Yes

21. Replace each x with 7. Because the resulting values of both sides of the equation are equal, the solution is verified.

22. The solution of the equation is the value of x for which the value of the expression $2x - 3$ is 5.

33. 5 **34.** 2 **35.** -5 **36.** 5 **37.** $-3, 0, 2$ **38.** $-5, -1, 4$ **39.** -11 **40.** 0 **41.** -1 **42.** 2 **43.** No solution **44.** No solution **45.** -3 **46.** 0 **47.** All real numbers **48.** All real numbers

49. For $x + 2 = 3$, 2 is added to x, so we add -2 to both sides to isolate the variable. For $-2x = 6$, x is multiplied by -2, so we divide both sides by -2 to isolate the variable.

50. Equivalent equations are equations that have exactly the same solution sets.

51. 4 **52.** -14 **53.** -8 **54.** 13 **55.** 0 **56.** 8 **57.** 3 **58.** 5 **59.** -17 **60.** 0 **61.** 6 **62.** 0 **63.** 0 **64.** 4 **65.** -9 **66.** -6 **67.** $-\frac{2}{15}$ **68.** $\frac{9}{4}$ **69.** 36 **70.** -20 **71.** 6 **72.** -9 **73.** 3 **74.** -7 **75.** 2 **76.** 2 **77.** $-\frac{9}{2}$ **78.** 2 **79.** -12 **80.** -5 **81.** -2 **82.** 2 **83.** $\frac{3}{2}$ **84.** $-\frac{4}{3}$

85. (a) $C_1(h) = 20 + 2h$; $C_2(h) = 10 + 2.25h$

(b)

h	10	20	30	40	50	60
$C_1(h)$	40.00	60.00	80.00	100.00	120.00	140.00
$C_2(h)$	32.50	55.00	77.50	100.00	122.50	145.00

(c) $20 + 2h = 10 + 2.25h$ **(d)** 40 hours

86. (a) $-5x + 500 = 100$ **(b)** $-5x + 500 = 50$

(c)

x	60	70	80	90	100
$-5x + 500$	200	150	100	50	0

The solutions are (a) 80 and (b) 90.

(d) A price increase of \$10 caused sales to be cut in half from 100 to 50.

(e) \$100

87. The simplified equation is $0.002x + 0.65 = 0.7$. The solution is 25, which corresponds to the year 2015. In that year, there are projected to be 0.7 million more elementary school teachers than secondary school teachers.

88. The simplified equation is $0.038x + 2.83 = 3.5$. The approximate solution is 18, which corresponds to the year 2008. In that year, there will be a total of 3.5 million teachers.

89. $1.16a + 9.27 = 40$; $a \approx 26.5$ **90.** 78 **91.** -8.51 **92.** 4.69 **93.** $-1, 3$ **94.** $0, -3$ **95.** $x = \frac{b - c}{a}$

96. $x = \frac{d - b}{a - c}$

97. The equation $0 \cdot x + B = C$ is not a linear equation in one variable because it cannot be written as $Ax + B = 0$, where $A \neq 0$. The equation is an identity if $B = C$.

98. Answers will vary.

Section 3.2 *(page 120)*

1. (i) Because $\frac{3}{2}$ and $\frac{2}{3}$ are reciprocals, their product is 1 and the left side becomes $1x$, or x.

(ii) Dividing a nonzero number by itself results in 1, so the left side becomes $1x$, or x.

(iii) Multiplying the left side by 3 results in $2x$. Then, when we divide by 2, we obtain x.

2. The algebraic method produces exact solutions.

3. -6 **4.** -5 **5.** $\mathbf{R}$ **6.** $\mathbf{R}$ **7.** -8 **8.** 2 **9.** 17
10. 4 **11.** $\emptyset$ **12.** $\emptyset$ **13.** 6 **14.** $\frac{11}{9}$

15. The Multiplication Property of Equations permits us to multiply both *sides* by 15: $15\left(\frac{1}{3} - 2x\right) = 15\left(\frac{2}{5}\right)$. By the Distributive Property, we obtain $15 \cdot \frac{1}{3} - 15 \cdot 2x = 15 \cdot \frac{2}{5}$.

16. The solution of $3x = 2x$ is 0. Dividing both sides by x results in $3 = 2$, which has no solution. Because the solutions of the two equations are different, the equations are not equivalent.

17. $-\frac{1}{24}$ **18.** $\frac{7}{4}$ **19.** 1 **20.** 19 **21.** $-\frac{10}{3}$ **22.** -36
23. $\frac{40}{3}$ **24.** $\frac{5}{8}$ **25.** 5 **26.** $\frac{1}{2}$ **27.** 4 **28.** 5
29. -100 **30.** $-\frac{1}{2}$ **31.** Yes **32.** Yes **33.** Yes
34. Yes **35.** No **36.** No

37. Because an inconsistent equation has no solution, the solution set is empty.

38. Because an identity is true for every real number replacement of the variable, the solution set contains all real numbers.

39. 6; Conditional **40.** 11; Conditional **41.** $\mathbf{R}$; Identity
42. $\emptyset$; Inconsistent **43.** $\emptyset$; Inconsistent **44.** $\mathbf{R}$; Identity
45. 15; Conditional **46.** 24; Conditional

47. **(a)** Use the Distributive Property to write the left side of the equation as $-x - 1$, which is never equal to $-x + 1$.
(b) Applying the Distributive Property to the left side of the equation results in $-3 + x + 5$, which simplifies to the right side of the equation, $x + 2$.

48. The equation is true for some values of x ($x \geq 0$), but it is false for other values of x ($x < 0$).

49. -15 **50.** -19 **51.** 8 **52.** 5 **53.** 3 **54.** -7
55. 1 **56.** 5 **57.** 3 **58.** 4 **59.** 5 **60.** 5 **61.** 16
62. -10 **63.** -22 **64.** $\frac{20}{7}$ **65.** $\frac{5}{2}$ **66.** -2 **67.** -5
68. $\emptyset$ **69.** $\mathbf{R}$ **70.** $\mathbf{R}$ **71.** $\frac{1}{3}$ **72.** 3 **73.** $\emptyset$ **74.** $\frac{15}{19}$
75. $\frac{21}{2}$ **76.** $-\frac{1}{9}$ **77.** $\frac{23}{5}$ **78.** $-\frac{24}{11}$ **79.** 3 **80.** 3
81. 5 **82.** -4

83. **(a)** $30 + 8(s - 1)$ **(b)** $8s + 22 = 62$; $s = 5$

84. **(a)** $\frac{1}{3}x + \frac{1}{2}x + 35$
(b) $\frac{1}{3}x + \frac{1}{2}x + 35 = x$
(c) $x = 210$; 70

85. $1.475x + 13.4 = 50$; $x \approx 25$ (2015)

86. **(a)** 1.25 million **(b)** $0.01x + 1.20 = 1.3$; $x = 10$ (2010)

87. **(a)** $0.28x - 1.07 = 20$
(b) $x \approx 75$; The tax rate will reach 20% by the year 2010.

88. 32.5% **89.** $h(t) = 1.42t + 52.55$; 73.85 million

90. $t \approx 21$; The African-American and Hispanic populations will be equal in 2011.

91. The equation describes the Hispanic population exceeding the African-American population by 2 million people. $t \approx 26$; In the year 2016, the Hispanic population will be 2 million greater than the African-American population.

92. $0.9t + 22.3 = 1.10(0.52t + 30.25)$
$t \approx 33.46$; The Hispanic population will be 10% greater than the African-American population in 2023.

93. $\frac{c + ab}{a}$, $a \neq 0$ **94.** $3b - a$ **95.** $-\frac{26}{77}$ **96.** $\frac{43}{19}$

97. No **98.** No

Section 3.3 *(page 127)*

1. To solve a formula for a given variable, treat all other variables as constants and use normal equation-solving procedures to isolate the given variable.

2. We regard r as the variable and treat all other letters as representing constants.

3. $m = \frac{F}{a}$ **4.** $d = \frac{C}{\pi}$ **5.** $r = \frac{I}{Pt}$ **6.** $W = \frac{V}{LH}$

7. $t = \frac{s}{v}$ **8.** $t = \frac{d}{r}$ **9.** $b = \frac{2A}{h}$ **10.** $h = \frac{3V}{B}$

11. $a = \frac{2A - bh}{h}$ **12.** $h = \frac{2A}{a + b}$ **13.** $\pi = \frac{A}{r^2}$

14. $y = \frac{x^2}{4p}$ **15.** $m = \frac{E}{c^2}$ **16.** $\pi = \frac{3V}{4r^3}$ **17.** $v = at + w$

18. $u = x - sz$ **19.** $r = \frac{A - P}{Pt}$ **20.** $r = \frac{E - IR}{I}$

21. $R = \frac{PV}{nT}$ **22.** $V = \frac{nRT}{P}$ **23.** $R_1 = 2R_T - R_2$

24. $\theta = \frac{2A}{r^2}$ **25.** $y = -\frac{ax + c}{b}$ **26.** $m = \frac{y - b}{x}$

27. $v = \frac{2s - gt^2}{2t}$ **28.** $g = \frac{2s - 2vt}{t^2}$ **29.** $n = \frac{A - a + d}{d}$

30. $d = \frac{A - a}{n - 1}$ **31.** $a = \frac{2S - n(n - 1)d}{2n}$

32. $d = \frac{2S - 2an}{n(n - 1)}$

33. To isolate y, we can divide both sides by 3, or we can divide each term of both sides by 3.

34. Creating a table of values is easier when the equation is solved for y.

35. $y = -\frac{2}{5}x + 2$ **36.** $y = 3x - 9$ **37.** $y = x + 5$
38. $y = \frac{1}{5}x + 2$ **39.** $y = -\frac{2}{3}x + 3$ **40.** $y = \frac{5}{6}x + 2$
41. $y = \frac{3}{4}x - \frac{11}{4}$ **42.** $y = \frac{5}{7}x + \frac{15}{7}$ **43.** $y = -\frac{1}{2}x$
44. $y = x + 1$ **45.** $y = \frac{4}{3}x + 40$ **46.** $y = \frac{5}{4}x + 25$
47. $y = \frac{1}{2}x + 7$ **48.** $y = -\frac{3}{4}x + 2$ **49.** $y = \frac{6}{5}x + 12$
50. $y = 2x$ **51.** $y = \frac{3}{4}x + \frac{17}{4}$ **52.** $y = -\frac{3}{2}x + \frac{37}{2}$

53. **(a)** 144.71 square inches

(b) $h = \frac{A - 2\pi r^2}{2\pi r}$ **(c)** 22 centimeters

54. **(a)** 86.26 cubic meters **(b)** $h = \frac{3V}{\pi r^2}$ **(c)** 12 feet

55. All except (v)

56. The total travel time was $\frac{1}{2} + \frac{2}{5} + \frac{1}{3} = \frac{37}{30}$ hours. The total distance traveled was 60 miles. Therefore, her average speed r was $r = \frac{d}{t} = \frac{60}{\frac{37}{30}} \approx 48.6$ mph.

57. 672.27 feet **58.** Yes **59.** 67 feet

60. Because the diameter d is 6 feet, the circumference C of the tablecloth is $C = \pi d = 6\pi \approx 18.85$ feet. The 6 yards of fringe on hand is 18 feet, which is not quite enough.

61. The depth H of the water in the second tank is calculated by $H = \frac{V}{LW} = \frac{15}{6(1.5)} = \frac{15}{9} \approx 1.7$ feet. Thus the water in the second tank will be deeper.

62. 1.43 and 1.47 inches **63.** 555 feet **64.** 28 feet

65. $S = 2WH + 2LH + 2WL$ **66.** 63.66 meters

67. **(a)** \$102.96 **(b)** 15%

68. **(a)** \$67.20 **(b)** 45%

69. **(a)** 127 miles **(b)** 159 miles **(c)** 239 miles

70. **(ii)** **71.** 20.56 meters

72. Beulah: 94.7; Carla: 145.0; Donna: 113.0; Juan: 63.3

73. Function S

74. **(i)** 0, **(ii)** 0; The two models appear to be equally good.

75. **(i)** 1826, **(ii)** 194

Function S

76. To measure the extent to which actual data and modeled data differ, the sum of $|M - A|$ is better because it reflects only positive differences in the data. The sum of $M - A$ is not a valid measure because large positive and negative differences tend to cancel each other out.

77. $x = \frac{ac + bc}{b - a}$ **78.** $x = \frac{3y + 1}{2 - y}$ **79.** $r = \frac{S - a}{S}$

80. $r = \frac{mv^2}{F}$ **81.** $V_2 = \frac{P_1V_1T_2}{P_2T_1}$ **82.** $I = \frac{E}{R + r}$

83. $L = \frac{A - 2WH}{2W + 2H}$ **84.** $P = \frac{A}{1 + rt}$

Section 3.4 *(page 137)*

1. The other number is $T - n$. The sum of the numbers is $n + (T - n) = T$.

2. If your equation is incorrect, the solution may satisfy the equation but not the conditions of the problem.

3. 13 **4.** 19 **5.** 36 **6.** 6

7. Both consecutive even integers and consecutive odd integers are 2 units apart on the number line. Whether the integers are even or odd depends on whether x represents an even integer or an odd integer. The representations are the same in both cases.

8. The equation is $x + (x + 2) = 2(x + 1)$, which is true for all integers x. The conditions are satisfied by any three consecutive integers.

9. 47, 48 **10.** 0, 1, 2 **11.** -1, 1 **12.** 86, 88

13. 4 feet, 8 feet, 33 feet **14.** 16 feet, 18 feet, 20 feet

15. 30 yards, 25 yards, 50 yards **16.** 8 feet, 9 feet, 15 feet

17. 40°, 60°, 80° **18.** 22°, 73°, 85°

19. Yes; 30°, 60°, 90° **20.** Yes; 50°, 50°, 80°

21. 20 feet, 12 feet **22.** 30 inches, 14 inches

23. 13 yards, 17 yards **24.** 30 meters, 37 meters

25. 21 **26.** 16 **27.** 19 **28.** 22

29. Answers will vary. **30.** Answers will vary.

31. Because the sum of even integers is always even, the sum cannot be 19.

32. The solution of $x + (x + 2) + (x + 4) = 19$ is not an integer. The conditions cannot be met.

33. -3 **34.** 17 **35.** 50°, 40°, 90° **36.** 16 feet

37. 80° **38.** 30° **39.** No; the book is 14 inches high.

40. First poster: 22 inches; Second poster: 21 inches

41. 9 inches **42.** 8.5 inches

43. No; the window is 45 inches high.

44. Yes; the height of the computer is 16 inches.

45. Yes; the angles are 26°, 50°, and 104°.

46. Yes; the angles are 50°, 60°, and 70°.

47. 56.4° **48.** 70° **49.** 12°

50. No; the frame is only 8 inches wide. **51.** 26

52. Younger niece: \$500; Older niece: \$1000; Nephew: \$1500

53. \$400, \$800 **54.** 4 **55.** 4 **56.** 5

57. 2 points **58.** 37.5 acres

59. The sum of an odd integer and an even integer cannot be an even integer.

60. 79 **61.** 49 **62.** 96 or 97

63. 8 meters, 10 meters, 22 meters

64. 18 yards, 22 yards, 66 yards **65.** 12 **66.** 388

67. 58 **68.** 360 **69.** \$40,000 **70.** 8 acres

71. 38 children, 152 men, 76 women **72.** 34

73. She cannot know for sure. The solution of $x + (x + 2) + (x + 4) = 597$ is 197. Therefore, the highest score was $x + 4 = 201$, but it was not necessarily the husband's score.

74. 240

75. **(a)** $N(t) = 0.175t + 5.60$

(b) $0.175t + 5.60 = 8.4$; $t = 16$ (2012)

76. $p + 1.25p = 78.9$; $p \approx 35$

77. 1.77 metric tons

78. 2007

79. $-1.77x + 99.37 = 78.24$

80. $x \approx 12$ (2002)

81. Answers will vary.

82. Answers will vary.

Section 3.5 *(page 148)*

1. In mathematics, *of* means *times:* 0.15(30).

2. In equation (i), x is the cost including tax. In Equation (ii), x is the cost before tax.

3. \$205.33 **4.** \$16,427.88 **5.** \$350, lost money

6. 18 **7.** \$2000 **8.** \$357.50

9. \$182,000 **10.** \$3500 **11.** \$25 **12.** \$36 **13.** \$299

14. No; original price at the Sweater Shop: \$85; current price at Cardigans Galore: \$80.

15. The costs in (a), (c), and (d) are variable costs; the costs in (b) and (e) are fixed costs.

16. The term b represents the fixed cost, and the variable cost is represented by the term ax.

17. 7.6 hours **18.** 482 miles **19.** 200 **20.** \$8

21. The value is the principal P plus 7% of the principal: (iii).

22. The 18% interest rate applies to one card, and the 20% interest rate applies to the other card. There is no connection between the two cards, so using both will not result in a 38% interest rate.

23. \$571.43 **24.** \$947.87 **25.** \$1100

26. \$1250 **27.** \$20,000 **28.** \$34,355.56

29. \$2500 at 9%, \$4000 at 6%

30. \$11,000 at 10%, \$15,000 at 6% **31.** \$4000

32. \$2300 at 8%, \$2700 at 8.3%

33. 45 mph **34.** 3.2 hours **35.** bus: 56 mph; car: 62 mph

36. 140 miles **37.** 2:24 P.M., 168 miles

38. 2.25 hours, 11.25 miles

39. 10 pounds

40. Fescue: 62.5 pounds; bluegrass: 37.5 pounds

41. 21 pounds

42. Cinnamon tea: 12 kilograms; black tea: 8 kilograms

43. 3 pounds **44.** \$9.33

45. 50 gallons of 3%, 100 gallons of 4.5%

46. 50% silver: 64 pounds; 25% silver: 96 pounds

47. 2 ounces of each **48.** Pure: 9 gallons; 50%: 6 gallons

49. 64 ounces **50.** 0.8 ounce **51.** \$1475 **52.** \$1325

53. English: \$63; Biology: \$67; Math: \$79

54. 37 **55.** 4 pounds

56. Cheaper: 60 pounds; more expensive: 40 pounds

57. 20° **58.** 40°, 50°

59. 1.54 cups of 12%, 0.46 cup of 25% **60.** 4.5 gallons

61. 16.4 mph **62.** 12:00 noon **63.** \$82,000

64. \$120,000

65. Rabbit: 5.84 seconds; Cat: 6.82 seconds; Elephant: 8.18 seconds; Coyote: 4.76 seconds; Tortoise: 20.1 minutes; Snail: 1.9 hours

66. \$569

67. Before trial: 131,528, After trial: 20,275, Dismissed: 11,707

68. 4492 **69.** 79.7%

70. Although the percentage is higher in 1998–1999, the total number of cases is much smaller. The number of cases settled before trial actually decreased by 42,010.

71. If x = wholesale cost, then $1.25x$ = retail value. The buyer's offer is $0.75(1.25x) = 0.9375x$. Because this is less than the amount you paid for the inventory, you would lose money.

72. **(a)** Big Ed's: \$65, Pop's: \$63

(b) $f(m) = 25 + 0.4m$; $g(m) = 33 + 0.3m$

(c) The first coordinate of the point of intersection is the number of miles for which the cost would be the same. The second coordinate is the cost when the cost is the same.

(d) $m = 80$

(e) $f(80) = g(80) = 57$. The cost of an 80-mile trip is \$57 using either rental agency.

(f) (80, 57)

(g) For trips under 80 miles, use Big Ed's; for trips over 80 miles, use Pop's.

Section 3.6 *(page 160)*

1. The interval $[a, b]$ contains a, but the interval $(a, b]$ does not.

2. The notation $[2, \infty)$ is correct. The symbol ∞ is not a real number and cannot be the endpoint of an interval.

3. $(-\infty, 2)$, $\{x \mid x < 2\}$

4. $(-5, \infty)$, $\{x \mid x > -5\}$

5. $[3, \infty)$, $\{x \mid x \geq 3\}$

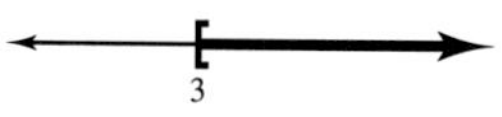

6. $(-\infty, -4]$, $\{x \mid x \le -4\}$

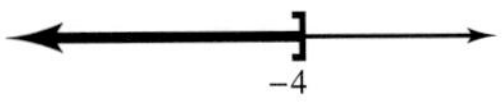

7. $(-3, 2]$, $\{x \mid -3 < x \le 2\}$

8. $[-2, 4)$, $\{x \mid -2 \le x < 4\}$

9. $(-\infty, 5)$ **10.** $(5, \infty)$ **11.** $(-\infty, 0]$ **12.** $[-15, \infty)$
13. $(-\infty, -7)$ **14.** $[2, \infty)$ **15.** $(-\infty, -1]$ **16.** $[0, \infty)$

17. **(a)** Because $-2 < 3$ is true, the solution set is **R**.
(b) Because $3 < -5$ is false, the solution set is Ø.
(c) Because $4 > 4$ is false, the solution set is Ø.
(d) Because $-7 \le -7$ is true, the solution set is **R**.
(e) Because $8 \ge -8$ is true, the solution set is **R**.

18. If $-x < -z$, then $x > z$. Multiplying both sides by -1 reverses the inequality symbol.

19. $<$ **20.** $<$ **21.** $>$ **22.** $>$ **23.** $>$ **24.** $<$
25. $<$ **26.** $>$ **27.** $c < 0$ **28.** $c > 0$
29. Any real number **30.** Any real number **31.** $c > 0$
32. $c < 0$

33. The methods are the same except that we must reverse the inequality symbol if we multiply or divide both sides of an inequality by a negative number.

34. The solution for $x < a$ excludes a, whereas the solution set for $x \le a$ includes a.

35. $(-\infty, 4]$ **36.** $(7, \infty)$ **37.** $(-5, \infty)$ **38.** $(-\infty, 8]$
39. $(-\infty, -8]$ **40.** $(-\infty, -9)$ **41.** $(-\infty, 15)$
42. $\left(-\infty, \frac{1}{9}\right)$ **43.** $(-\infty, 6)$ **44.** $[2, \infty)$ **45.** $[5, \infty)$
46. $(-\infty, 5)$ **47.** **R** **48.** **R** **49.** $[3, \infty)$ **50.** $(0, \infty)$
51. $(-\infty, 8)$ **52.** $(0, \infty)$ **53.** $\left(-\infty, \frac{2}{5}\right]$ **54.** $\left(-\infty, \frac{5}{6}\right)$
55. Ø **56.** **R** **57.** $[0, \infty)$ **58.** $(-15, \infty)$ **59.** Ø
60. Ø **61.** $\left(-\infty, \frac{5}{2}\right)$ **62.** $(-\infty, -3)$ **63.** $(-\infty, -8]$
64. $\left[-\frac{15}{2}, \infty\right)$ **65.** $\left(-\infty, \frac{35}{4}\right)$ **66.** $[-3, \infty)$ **67.** $\left(-\frac{19}{2}, \infty\right)$
68. $\left(-\infty, -\frac{1}{10}\right)$ **69.** $\left(-\infty, \frac{5}{29}\right]$ **70.** $[-3, \infty)$
71. $\left(-\infty, -\frac{5}{11}\right)$ **72.** $\left(\frac{2}{3}, \infty\right)$ **73.** $[0.23, \infty)$ **74.** $[7, \infty)$
75. $(-\infty, 0.27]$ **76.** $\left(-\infty, \frac{164}{35}\right]$

77. The sum of the lengths of any two sides must be greater than the length of the third side: $a + b > c$, $a + c > b$, $b + c > a$.

78. 9 inches **79.** 4 feet **80.** 61.5 feet

81. Third place, need at least 4 points

82. No, you would need at least 102 on the last test to average at least 70.

83. 790 miles **84.** 6 **85.** \$6000 **86.** 46 minutes

87. **(a)** $0.25x + 4.7 \ge 7.2$
(b) $[10, \infty)$; At least \$7.2 million will be spent by the winning candidate from 2006 onward.
(c) In the bar graph, the bar of $W(x)$ starts higher than the bar for $L(x)$, and the values of $W(x)$ are increasing at a faster rate.

88. **(a)** Approximately $[7, \infty)$
(b) From 1997 onward, the number of travelers exceeded 40 million.

89. **(a)** $[76, \infty)$
(b) At the present rate, the average number of runs scored per game in the National League will be at least as many as in the American League in the year 2069 and beyond.

90. $(0.16x + 9.80) - (0.17x + 9.04) \le 0.6$; 2009 and later

91. $\left(-\infty, \dfrac{b-a}{5}\right]$ **92.** $\left(\dfrac{10}{c}, \infty\right)$ **93.** $\left[\dfrac{8}{a^2}, \infty\right)$

94. $(-\infty, b - a)$ **95.** The first number is at most -2.

96. **(a)** The inequality symbol is either $<$ or $>$.
(b) The inequality symbol is either $\le$ or $\ge$.

Section 3.7 *(page 168)*

1. The solution set is the intersection of the solution sets of the individual inequalities.

2. The solution set is the union of the solution sets of the individual inequalities.

3. Yes, No **4.** No, Yes **5.** Yes, Yes **6.** Yes, No
7. Yes, Yes **8.** No, No **9.** Yes, No **10.** No, Yes, Yes

11. A conjunction is a compound inequality in which the connective is *and*. A disjunction is a compound inequality in which the connective is *or*.

12. **(a)** For $x < -7$ or $x \ge 1$, all numbers that satisfy either condition belong to the solution set. For $x < -7$ and $x \ge 1$, there are no numbers that satisfy both conditions, so the solution set is Ø.
(b) For $x < 1$ or $x \ge -7$, every real number satisfies at least one of the conditions, so the solution set is **R**. For $x < 1$ and $x \ge -7$, only numbers between -7 and 1, not including 1, satisfy both conditions.

13.

14.

15.

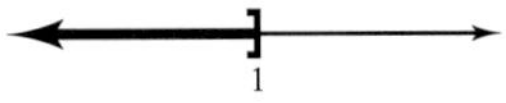

16.

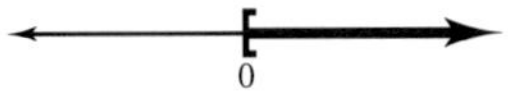

17. Ø **18.** Ø

19. 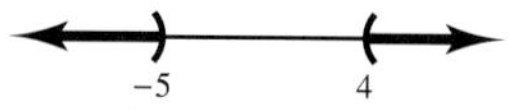

20. 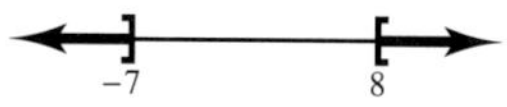

21. 

22. 

23. 

24. 

25. **(a)** The only solution is the number c: $\{c\}$.
(b) No number can be less than c and also greater than c: Ø.
(c) Every real number is either at most c or at least c: **R**.
(d) The only number that does not satisfy at least one of the conditions is c: $\{x \mid x \neq c\}$.

26. **(a)** The double inequality is another way to write the conjunction. They are equivalent.
(b) A disjunction cannot be written as a double inequality. They are not equivalent.
(c) In $1 \le x \ge 5$, the inequality symbols are not in the same direction. They are not equivalent.

27. $-7 < x < 4$ **28.** $-1 < x < 7$ **29.** $-6 \le t \le -\frac{2}{3}$

30. $3 < y \le 7$ **31.** $-1.5 \le x \le 4$ **32.** $-1 \le x < 3$

33. $-4.5 < x \le -1$ **34.** $1 < x < 8$ **35.** $-1 \le x < 4$

36. $-2 \le t < 4$ **37.** $-4 < x \le 5.75$ **38.** $-3.5 < x < 4$

39. $-4 < x \le 8$ **40.** $-5 < x < 5$

41. The solution set of a disjunction is the union of intervals, whereas the solution set of a conjunction is the intersection of intervals. Thus, for example, the union of $(-\infty, 1)$ and $[1, \infty)$ is **R**, but the intersection of those intervals is Ø.

42. The solution set of a linear inequality such as $Ax + B < 0$ is not empty. Therefore, the solution set of a disjunction of linear inequalities cannot be empty.

43. $[-5, 7]$ **44.** $[-10, 12]$ **45.** $[4, 14)$ **46.** $(5, 9)$

47. $(-11, 6]$ **48.** $[-2, 5)$ **49.** $(-\infty, -4] \cup [3, \infty)$

50. $(-\infty, -8] \cup [2, \infty)$ **51.** $(-\infty, -7) \cup (-1, \infty)$

52. $(-\infty, 1) \cup (9, \infty)$ **53.** $(-\infty, -1.25) \cup [1, \infty)$

54. $(-\infty, -3) \cup \left[\frac{1}{2}, \infty\right)$ **55.** $(-9, -6]$ **56.** $[-3, 4]$

57. $(-\infty, -1) \cup (5, \infty)$ **58.** $(-\infty, -1] \cup (4, \infty)$

59. $(-\infty, -3] \cup (4, \infty)$ **60.** $(-\infty, -4) \cup (2, \infty)$ **61.** Ø

62. Ø **63.** $[4, \infty)$ **64.** $(-\infty, 3)$ **65.** $[0, 2]$

66. $(-2, -1)$ **67.** $(-\infty, 0) \cup (6, \infty)$ **68.** $(3, \infty)$

69. $(-5, 1)$ **70.** $\left(-\frac{7}{3}, \frac{1}{3}\right]$ **71.** Ø **72.** Ø **73.** $[3, \infty)$

74. $(-\infty, -3]$ **75.** $(2, \infty)$ **76.** $[-2, \infty)$ **77.** 4 **78.** 0

79. $(-2, \infty)$ **80.** $[-4, \infty)$ **81.** Ø **82.** Ø **83.** 10

84. −3 **85.** **R** **86.** **R**

87. Because $3 < x > 8$ is incorrect notation, it has no valid interpretation.

88. Because there are infinitely many real numbers between 0.99 and 1, the inequality has infinitely many solutions.

89. 45, 40 **90.** 20 hours

91. Between 2.5 hours and 3.75 hours, inclusive

92. Between 2.6 miles and 2.8 miles, inclusive

93. Between 29 and 35, inclusive

94. Between 20 feet and 32 feet, inclusive

95. Between 3 and 6, inclusive

96. Between 200 and 225, inclusive

97. Between 80 and 100, inclusive

98. Between 15 and 20, inclusive

99. **(a)** $155x + 1835 \ge 2300$ **(b)** $[3, \infty)$
(c) For 1998 and beyond, the number of road rage incidents was 2300 or more.

100. **(a)** $[2, 4]$
(b) During the period 1997–1999, the number of incidents was between 2145 and 2455, inclusive.

101. **(a)** 65%

(b)

x	$V_1(x)$	$V_2(x)$
1	8.12	8
2	11.27	15
3	14.42	22
4	17.57	29

(c) 1996

102. **(a)** V_2 **(b)** V_1

103. $[2, 4]$

104. In the years 2000–2002, the number of ATM terminals was between 240,000 and 280,000, inclusive.

105. $f(x) = (12)(3491)(20x + 200)$ **106.** 2560

107. $k \ge 3$ **108.** $k > -1.5$ **109.** $[-7, 3]$ **110.** $(-\infty, 2)$

111. **R** **112.** Ø

Section 3.8 *(page 177)*

1. The number represented by x is 5 units from 2.

2. The equation $|x + 3| = 5$ has two solutions, 2 and −8. The equation $x + 3 = 5$ has only one solution, 2.

3. −12, 12 **4.** −1, 17 **5.** −2.5, 7.5 **6.** −2.25, 3.75

7. 0.4 **8.** 5 **9.** 5 **10.** −8, 20 **11.** −14, 6

12. 2, 4 **13.** −25, 5 **14.** ∅

15. The inequality $|x| \geq 5$ is equivalent to the disjunction $x \leq -5$ or $x \geq 5$ and has the solution set $(-\infty, -5] \cup [5, \infty)$. The inequality $|x| \leq 5$ is equivalent to the conjunction $-5 \leq x \leq 5$ and has the solution set $[-5, 5]$.

16. The symbol $|x|$ represents a nonnegative number. Therefore, $|x|$ cannot be less than a negative number.

17. $(-4, 5)$ **18.** $\left(-\frac{16}{3}, 4\right)$ **19.** −7 **20.** 1

21. $[-3, 11]$ **22.** $[-6, 7]$ **23.** ∅ **24.** ∅

25. $(-\infty, -3] \cup \left[\frac{5}{3}, \infty\right)$ **26.** $(-\infty, -3] \cup [8, \infty)$

27. $(-\infty, 5) \cup (5, \infty)$ **28.** $(-\infty, -3) \cup (-3, \infty)$

29. **R** **30.** **R**

31. $\left(-\infty, -\frac{11}{4}\right) \cup \left(\frac{5}{4}, \infty\right)$ **32.** $(-\infty, -5) \cup \left(\frac{19}{3}, \infty\right)$

33. $|x - 5| < 7$ **34.** $|x + 5| \leq 4$ **35.** $|x + 7| \geq 10$

36. $|x - 7| > 12$ **37.** $(-12, 12)$ **38.** $(-\infty, -1] \cup [17, \infty)$

39. $(-3, 15)$ **40.** $(-\infty, -25] \cup [5, \infty)$ **41.** **R** **42.** **R**

43. $|x + 1| < 3$ **44.** $|x - 2| \leq 3$ **45.** $|x| \geq 2$

46. $|x - 1| > 3$ **47.** −2.5, 5.5 **48.** $-\frac{11}{3}, \frac{19}{3}$ **49.** −11, 9

50. 7, −10 **51.** ∅ **52.** ∅ **53.** −1, 11 **54.** 8, −10

55. ∅ **56.** ∅ **57.** $\frac{4}{3}$ **58.** $\frac{5}{2}$ **59.** −3, 2 **60.** $-3, \frac{11}{3}$

61. $-\frac{7}{3}, \frac{17}{3}$ **62.** $-\frac{1}{12}, \frac{17}{12}$ **63.** 10 **64.** 0 **65.** $\frac{5}{3}$, 5

66. $\frac{3}{5}$, 1 **67.** **R** **68.** **R**

69. Divide both sides by −3 to obtain $|x - 2| \leq -4$.

70. **(a)** The inequality $|x| > 5$ is satisfied by −7, for example, but −7 is not a solution of $x > 5$. The inequalities are not equivalent.

(b) The conjunction $|x - 1| \leq 3$ can be written as the double inequality $-3 \leq x - 1 \leq 3$, so they are equivalent.

71. $(-\infty, -5) \cup (2, \infty)$ **72.** $(-1, 15)$ **73.** $\left(\frac{1}{3}, \frac{7}{3}\right)$

74. $\left(-\infty, \frac{3}{5}\right) \cup \left(\frac{9}{5}, \infty\right)$ **75.** **R** **76.** ∅

77. $\left(-\infty, -\frac{1}{3}\right] \cup \left[\frac{11}{3}, \infty\right)$ **78.** $[-2, 9]$

79. $\frac{5}{3}$ **80.** **R** **81.** $(-\infty, 1) \cup \left(\frac{7}{3}, \infty\right)$ **82.** $\left(-1, \frac{11}{3}\right)$

83. ∅ **84.** **R** **85.** $(-1, 2)$ **86.** $(-\infty, -9) \cup (-1, \infty)$

87. $\left(-\infty, -\frac{5}{2}\right) \cup \left(\frac{3}{2}, \infty\right)$ **88.** $\left[-\frac{1}{4}, \frac{7}{4}\right]$

89. By taking the absolute value, we guarantee that the distance is not negative.

90. The expression $|-x|$ is not necessarily equal to x. If x is negative, then $|-x| = -x$. The inequality $|-x| < 3$ implies that $-3 < x < 3$.

91. $|L - 12| \leq 0.24$; $-0.24 \leq L - 12 \leq 0.24$

92. $|w - 135| \leq 3$; $-3 \leq w - 135 \leq 3$

93. $|d - 3| \leq \frac{1}{16}$; $-\frac{1}{16} \leq d - 3 \leq \frac{1}{16}$

94. $|t - 5| \leq 6$; $-6 \leq t - 5 \leq 6$

95. $|50 - s| \leq 6$; between 44 and 56, inclusive

96. $|72 - h| \leq 2$; between 70 inches and 74 inches, inclusive

97. $|98.6 - t| \leq 0.5$; between 98.1°F and 99.1°F, inclusive

98. $|220 - v| < 5$; between 215 volts and 225 volts

99. 1989: 33.7 million; 1991: 35.6 million; 1993: 37.4 million; 1995: 39.3 million; 1997: 41.2 million

100. 16% **101.** Increased from 13% in 1987

102. 17 million people

103. **(a)** By the definition of absolute value, $|x - 2| = x - 2$ only if $x - 2 \geq 0$ or $x \geq 2$.

(b) By the definition of absolute value, $|x - 2| = -(x - 2)$ only if $x - 2 \leq 0$ or $x \leq 2$.

(c) The absolute value of a number and that of its opposite are equal. Therefore, the equation is true for all real numbers.

104. 1 **105.** 0 **106.** 4.5 **107.** −2.5 **108.** $x \geq 0$

109. $x \leq 0$

110. The derived equations are as follows.

(i) $x - 2 = 2x + 1$

(ii) $x - 2 = -(2x + 1)$

(iii) $-(x - 2) = 2x + 1$

(iv) $-(x - 2) = -(2x + 1)$

Multiplying both sides of (iv) by −1 produces (i), and multiplying both sides of (iii) by −1 produces (ii).

Review Exercises *(page 181)*

1. No **2.** Yes

5. The goal is to find an equivalent equation in which the variable is isolated.

6. We apply the Addition Property to isolate the variable term and the Multiplication Property to isolate the variable.

7. 11 **8.** 0 **9.** −8 **10.** 4 **11.** 6 **12.** 8

13. −30 **14.** 8 **15.** 55 **16.** 1 **17.** 0 **18.** −2

19. −7 **20.** 6 **21.** 1.4; Conditional

22. ∅; Inconsistent **23.** **R**; Identity

24. $\frac{19}{9}$; Conditional **25.** −5 **26.** −3 **27.** $H = \dfrac{V}{LW}$

28. $R_2 = 2R_T - R_1$ **29.** $d = \dfrac{6xy - acy}{cx}$

30. $b = \dfrac{2A - ah}{h}$ **31.** $I = \dfrac{E}{R}$ **32.** $v = \dfrac{h + 16t^2}{t}$

33. $y = \frac{3}{4}x - \frac{25}{4}$ **34.** $y = \frac{6}{5}x + \frac{3}{5}$ **35.** $r = \dfrac{A - P}{Pt}$; 6.5%

36. A paved area of 4875 square feet meets the requirement.

37. 2 feet **38.** 12 **39.** 32 feet, 33 feet, 34 feet

40. 25°, 70°, 85° **41.** 225 **42.** 53 and 55 **43.** \$155
44. \$1560 **45.** 9 hours
46. 12 gallons of 10%, 3 gallons of 20% **47.** 70 mph, 64 mph
48. \$2700 at 7.5%, \$5300 at 9.3% **49.** 15°, 75°
50. Smaller: 12 gallons per hour; Larger: 18 gallons per hour
51. $[2, \infty)$ **52.** $(2, 7]$ **53.** $(-\infty, 4)$

54. 

55. 

56. 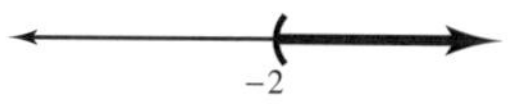

57. $(-\infty, 6)$ **58.** $[5, \infty)$ **59.** $[3, \infty)$ **60.** $(-\infty, -8]$
61. $(-2.5, -0.25]$ **62.** $[-4, \infty)$ **63.** $[11.5, \infty)$
64. $(-2, 2)$ **65.** $(-7, 4)$ **66.** $(7, \infty)$ **67.** $(-5, 2]$
68. $(-\infty, 27]$ **69.** 60 meters
70. Width cannot exceed 34 inches, length cannot exceed 68 inches, height cannot exceed 51 inches
71. 0 to 600 miles **72.** Answers will vary. **73.** **R**
74. $[2, 12)$ **75.** $[-6, 2]$ **76.** $[4, \infty)$ **77.** **R**
78. $(-\infty, -1)$ **79.** $(-\infty, -2] \cup [7, \infty)$ **80.** **R**
81. $(-\infty, -2) \cup (3, \infty)$ **82.** $[2, 3]$ **83.** $(-\infty, -1)$
84. Ø **85.** Between \$1875 and \$2500, inclusive
86. Between \$1700 and \$2800, inclusive **87.** $-8, 8$
88. $7.5, -10.5$ **89.** $-4.5, 7.5$ **90.** Ø **91.** 1.5, 2.5
92. Ø **93.** $\left(-\frac{19}{3}, 3\right)$ **94.** $(-\infty, -1) \cup (9, \infty)$
95. $(-\infty, 2] \cup [8, \infty)$ **96.** **R** **97.** $(-\infty, -7] \cup [7, \infty)$
98. Ø **99.** $-5 \le x - 3 \le 5$ **100.** $|x - 2| > 6$
101. 6 or 12
102. Between 131 pounds and 141 pounds, inclusive
103. Answers will vary.

Chapter Test *(page 185)*

1. [*3.1*] **(a)** Addition Property of Equations
(b) Property of Additive Inverses
(c) Multiplication Property of Equations
2. [*3.2*] 3; Conditional **3.** [*3.2*] **R**; Identity
4. [*3.2*] Ø; Inconsistent **5.** [*3.3*] $t = \dfrac{A - P}{Pr}$
6. [*3.3*] $y = -\frac{2}{3}x + 4$ **7.** [*3.5*] \$2300
8. [*3.4*] 62.5 meters by 37.5 meters **9.** [*3.5*] 1.2 gallons
10. [*3.5*] 42 mph, 54 mph **11.** [*3.6*] $[-5, 2)$
12. [*3.6*] $(4, \infty)$ **13.** [*3.6*] $(-\infty, -3)$
14. [*3.6*] $(-\infty, -7.5]$ **15.** [*3.6*] $(-\infty, 1.5)$
16. [*3.7*] $[-2, 2]$ **17.** [*3.7*] $(-2, 3)$ **18.** [*3.7*] $[-3, \infty)$
19. [*3.7*] **R** **20.** [*3.8*] $-4, 0$
21. [*3.8*] $(-\infty, -1) \cup (3, \infty)$ **22.** [*3.8*] $[-2, 3]$
23. [*3.8*] Ø
24. [*3.7*] Between 12 minutes and 42 minutes, inclusive
25. [*3.6*] At most 1.35 inches

Chapter 4

Section 4.1 *(page 196)*

1. The equation $x = -5$ can be written as $1x + 0y = -5$, which is the standard form of a linear equation in two variables.
2. There is a one-to-one correspondence between the points of the graph and the solutions of the equation.
3. Yes **4.** Yes **5.** No **6.** No **7.** No **8.** Yes
9. Yes **10.** No **11.** No **12.** Yes **13.** No **14.** Yes
15. 5 **16.** 1 **17.** 5 **18.** 3 **19.** -0.5 **20.** 2
21. $a = 11, b = -3, c = 6$ **22.** $a = 2, b = 0, c = 9$
23. $a = 3, b = 3, c = 3$ **24.** $a = 3, b = 3, c = 3$
25. $a = -5, b = -1, c = 16$ **26.** $a = 0, b = -5, c = 4$
27. For $|x| = 2$, there are two solutions: -2 and 2. For $|x| < 2$, the solutions are all numbers in the open interval from -2 to 2. For $y = x + 4$, one solution is the ordered pair $(-2, 2)$.
28. The solution set for the linear equation $x = 1$ in one variable is $\{1\}$. The solution set for the linear equation $x = 1$ in two variables is a set of ordered pairs of the form $(1, y)$.

29.

x	10	-6	7
y	-6	10	-3

30.

x	5	-13	-9
y	1	-8	-6

31.

x	0	3	-5
$f(x)$	1	10	-14
(x, y)	$(0, 1)$	$(3, 10)$	$(-5, -14)$

32.

x	-2	-7	3
$g(x)$	-1	9	-11
(x, y)	$(-2, -1)$	$(-7, 9)$	$(3, -11)$

33. **(a)** $2x - 3y - 6 = 0$ **(b)** $c = -1.5, d = -\frac{2}{3}$

34. **(a)** $3x + 4y + 9 = 0$ **(b)** $c = 2, d = -1.5$

35.

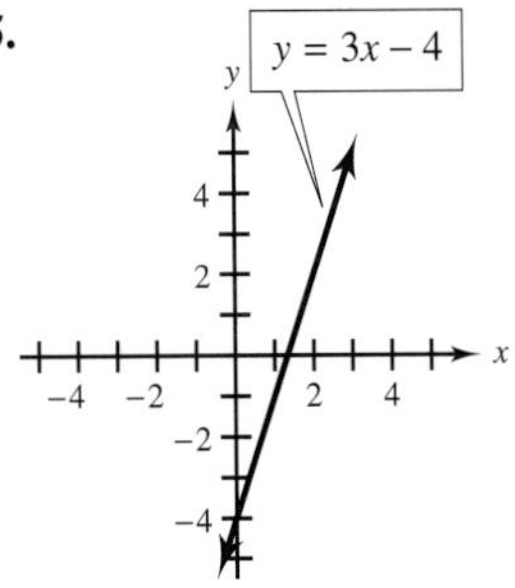

36.

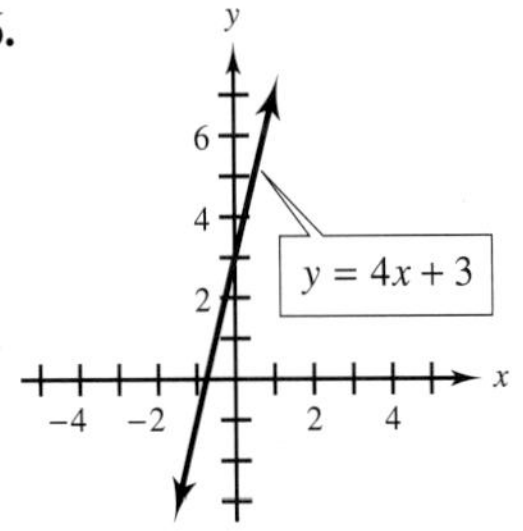

37.

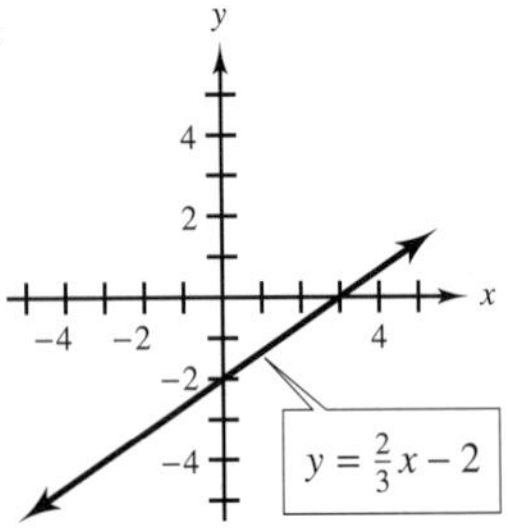

38.

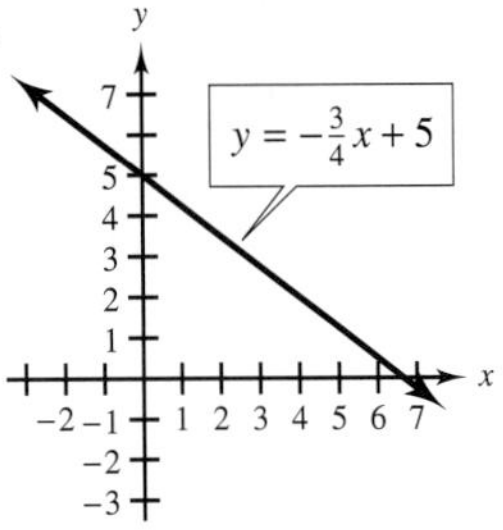

39.

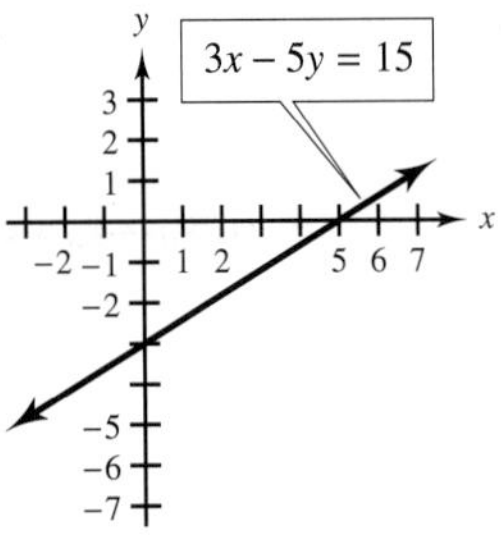

40.

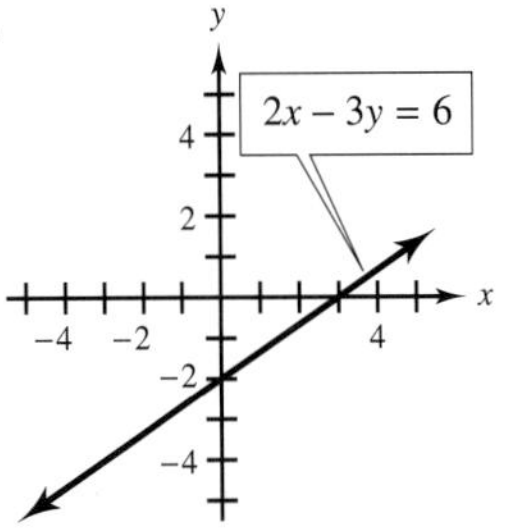

41.

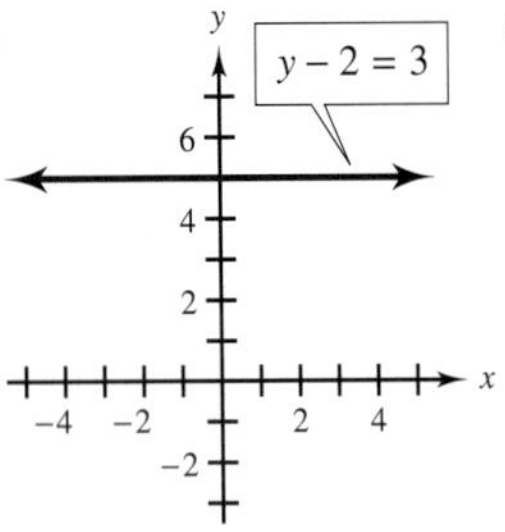

42.

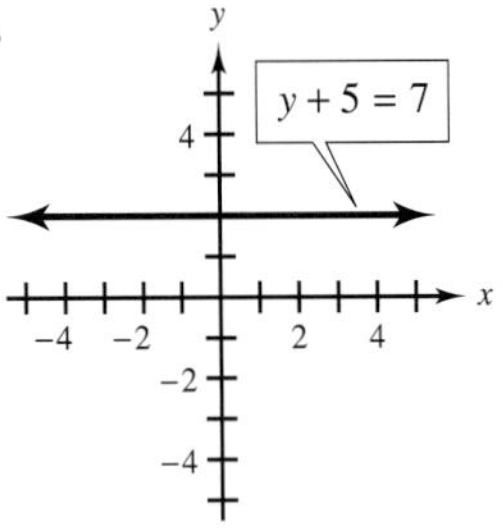

43.

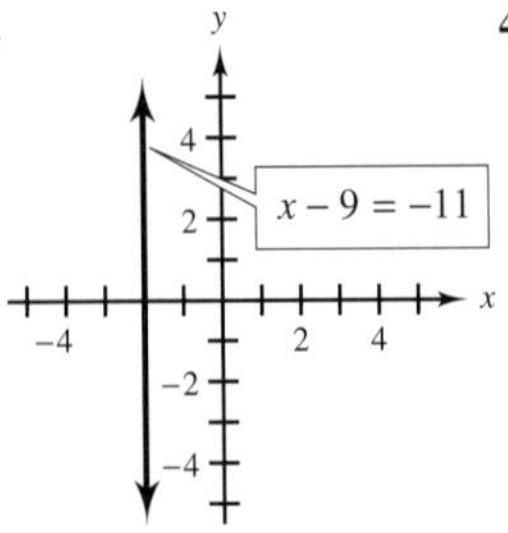

44.

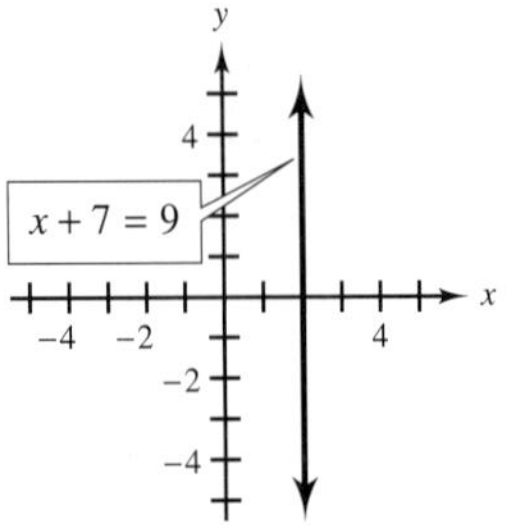

45.

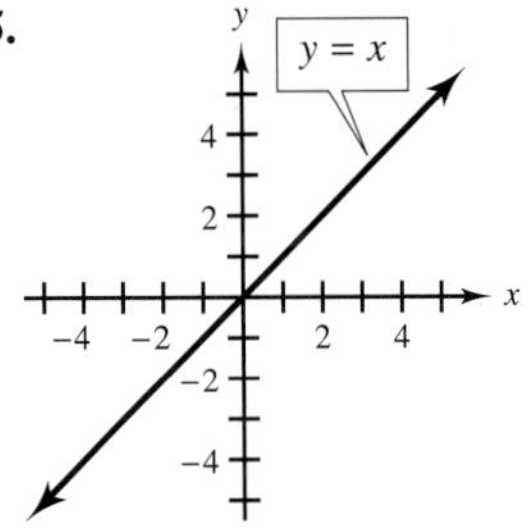

46.

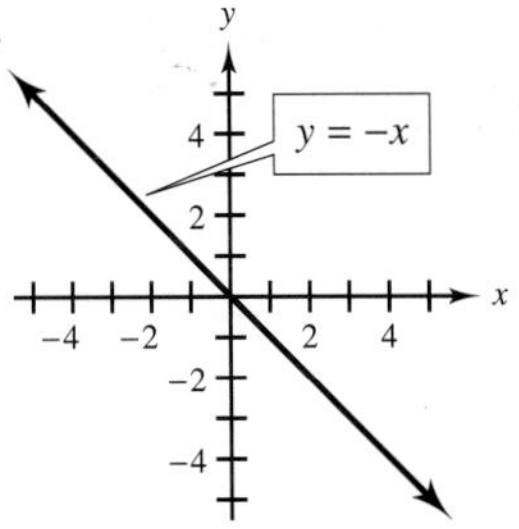

47.

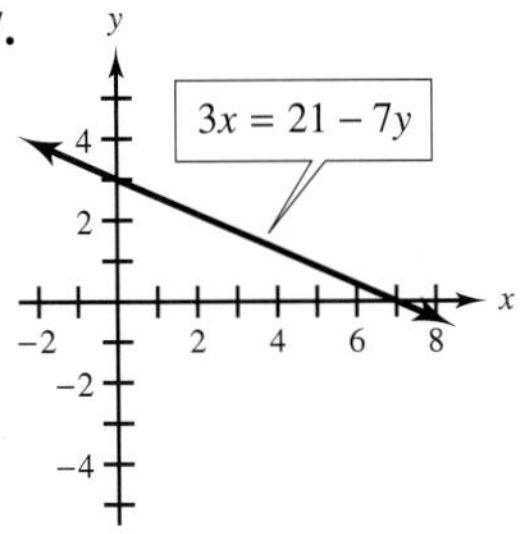

48.

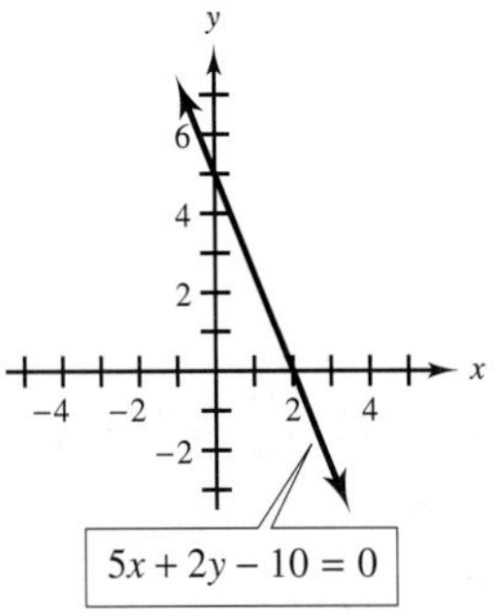

49. (0, 12), (4, 0)

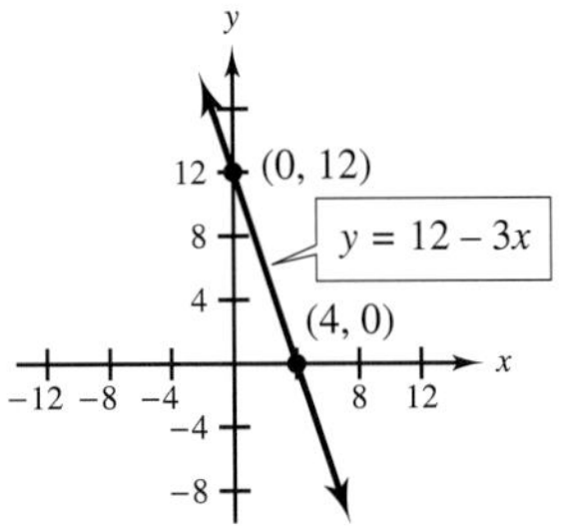

50. (12, 0), (0, −12)

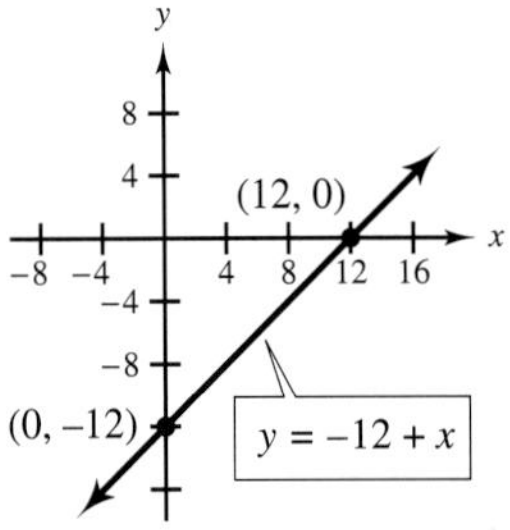

51. (0, −10), (30, 0)

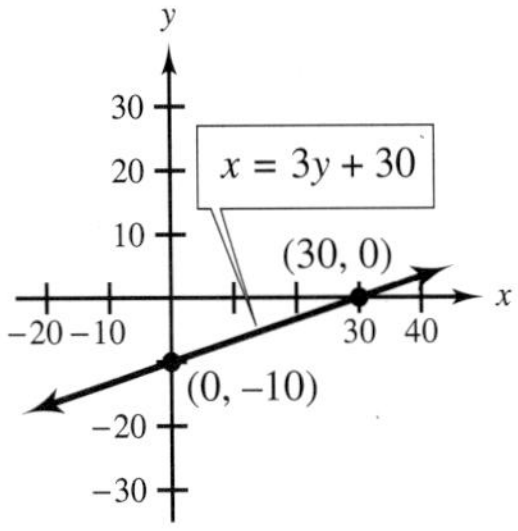

52. (15, 0), (0, 15)

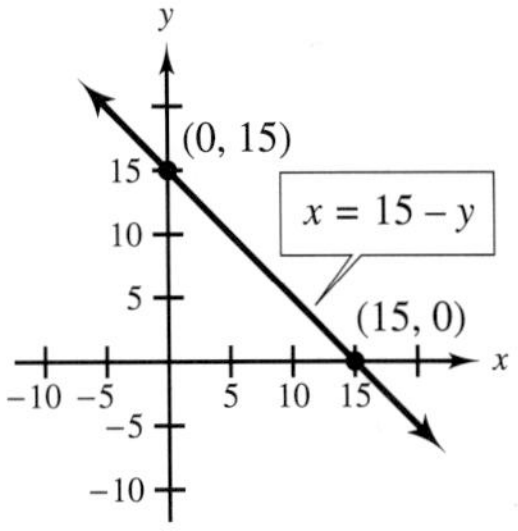

53. (0, −9), (12, 0)

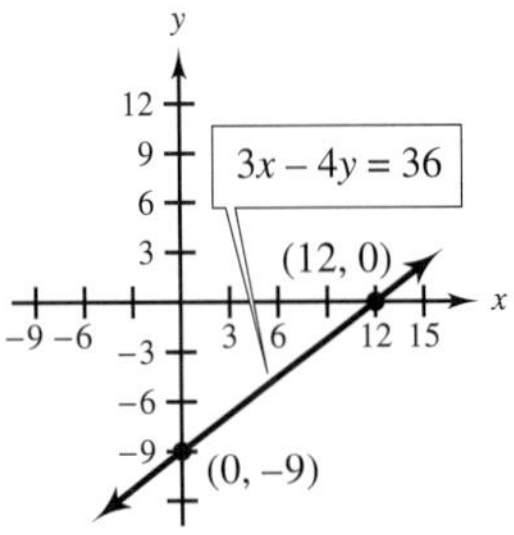

54. (15, 0), (0, 10)

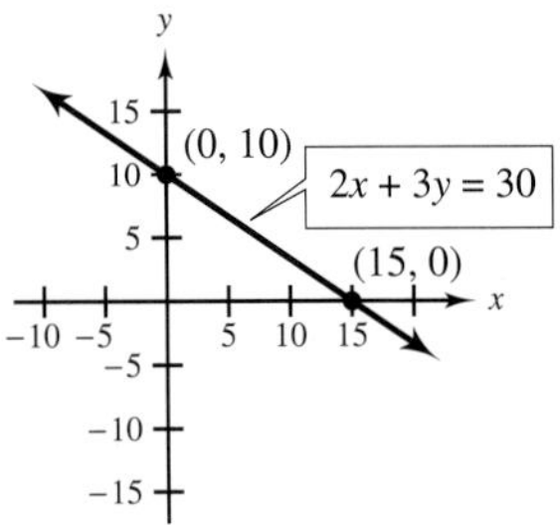

55. (i) and (ii) **56.** (ii) and (iii)

57. $y = -\frac{2}{5}x + \frac{14}{5}$; $\left(0, \frac{14}{5}\right)$ **58.** $y = \frac{4}{3}x - 5$; $(0, -5)$

59. $y = -4x - 10$; $(0, -10)$ **60.** $y = -\frac{4}{9}x + 12$; $(0, 12)$

61. $(0, 0)$ **62.** $(0, 0)$ **63.** $(0, -3)$, $(4, 0)$

64. $(6, 0)$, $(0, -4)$ **65.** $(0, 4)$, $(3, 0)$ **66.** $(5, 0)$, $(0, -3)$

67. $(0, 15)$ **68.** $(-1, 0)$

69. If the x- and y-intercepts are the same, the line contains the origin. Therefore, $b = 0$.

70. **(a)** The form of the equation is $x = c$, where c is a constant.

(b) The form of the equation is $y = b$, where b is a constant.

71. **(a)** $y = x + 4$

(b) $y = x - 3$

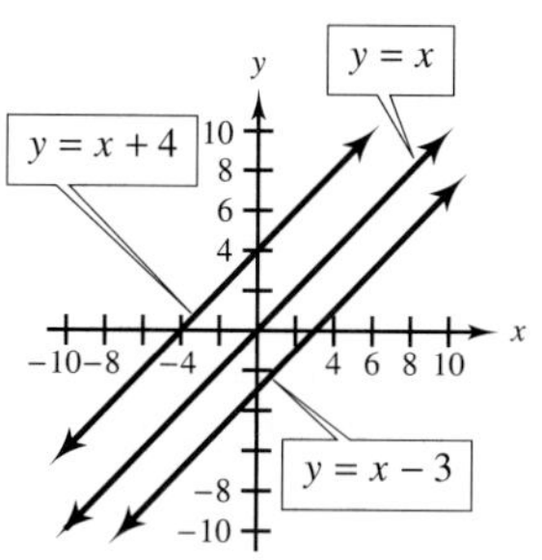

72. **(a)** $y = 4$

(b) $y = -5$

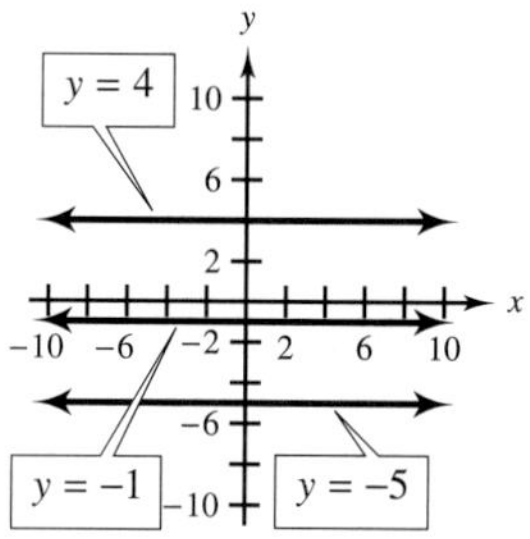

73. $a = -2, c = 8$ **74.** $b = 0.1, c = -1$

75. $a = 5, b = -3$ **76.** $a = 1, b = 1$ **77.** Vertical

78. Vertical **79.** Neither **80.** Neither **81.** Horizontal

82. Horizontal **83.** $x = -y$ **84.** $y = x$ **85.** $y = \frac{1}{2}x$

86. $y = 2x - 3$ **87.** $x + y = 5$ **88.** $x - y = 2$

89. **(a)** $y = 22 - x$

(b) The y-intercept $(0, 22)$ corresponds to "no windows washed," and the x-intercept $(22, 0)$ corresponds to "all windows washed."

90. **(a)** $y = 40 - x$

(b) The y-intercept $(0, 40)$ represents the situation of all drivers being women; the x-intercept $(40, 0)$ represents the situation of all drivers being men.

91. \$7.00, \$7.15, \$7.30, \$7.45, \$7.60, \$7.75

92. Between 80°F and 83°F, inclusive

93. **(a)** The graph of a linear function is a line, and a line can be drawn relatively close to the tops of the bars.

(b) (ii)

(c) 2004

94. **(a)** (i) $(0, 152)$ (ii) $(0, 125.5)$ (iii) $(0, 168)$

(b) The y-intercepts represent the expenditures in 1990. Both (i) and (iii) are too high.

95. **(a)** $(0, 73.7)$; When $x = 0$ (1980), the life expectancy was 73.7 years.

(b) 0.15 year

(c) About $(-491, 0)$; When $x = -491$ (the year 1489), the life expectancy was 0, which is invalid.

96. **(a)** \$1.29 billion

(b) The annual increase is the same as the coefficient of x.

97. 1100

98. $(0, 967)$; When $x = 0$ (1995), the estimated production was 967 million gallons.

99. $k = 1267$; The graph of $y = 1267$ is a horizontal line, which implies that the annual production was constant throughout the 3-year period.

100. Declining production would cause the model line to become more horizontal or even fall from left to right. Therefore, the y-intercept of the line would be greater than the y-intercept of the graph of P.

101. The x-intercept is $(g, 0)$, and the y-intercept is $(0, h)$.

102.

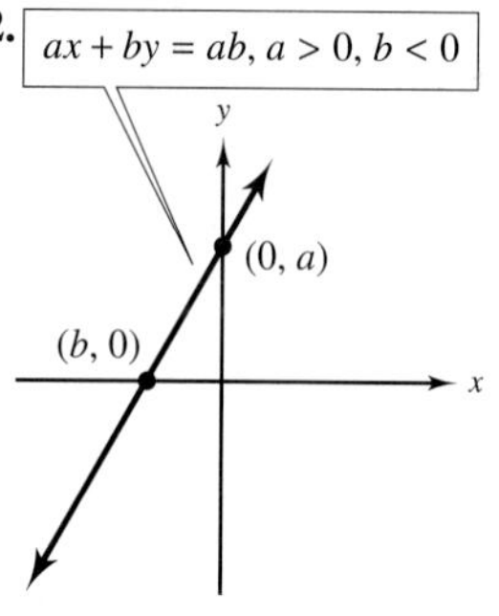

103.

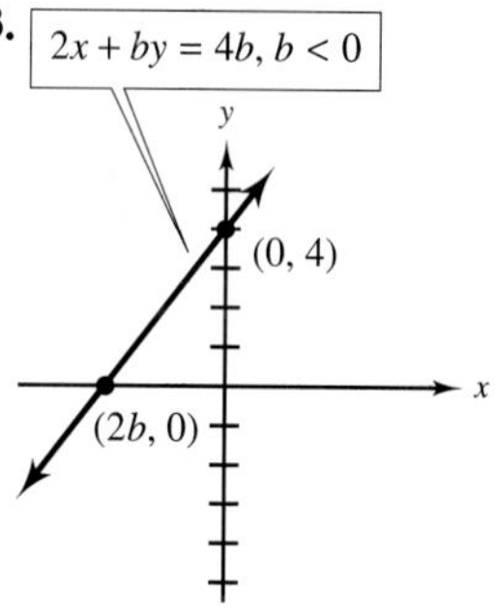

104.

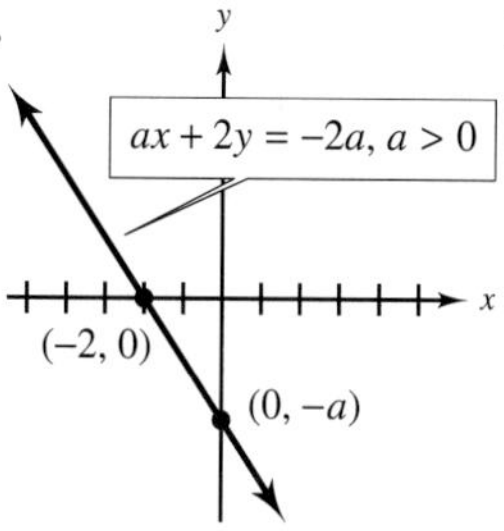

105.

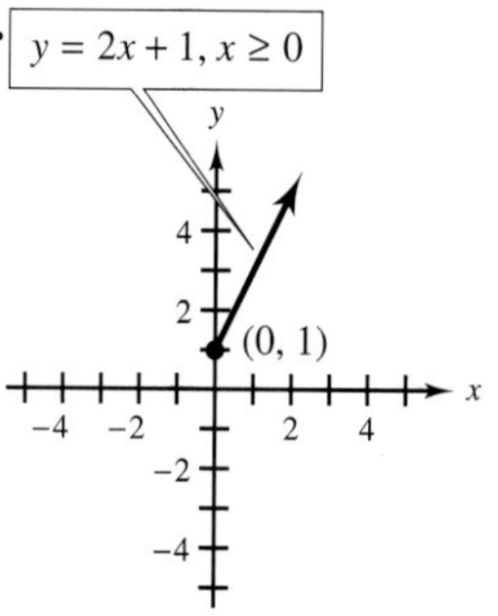

106.

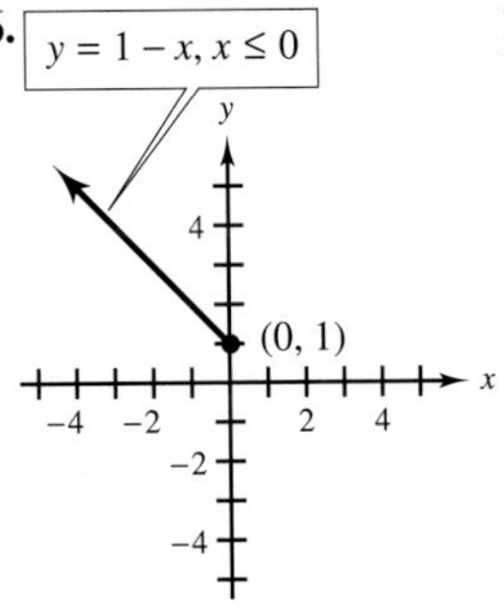

107.

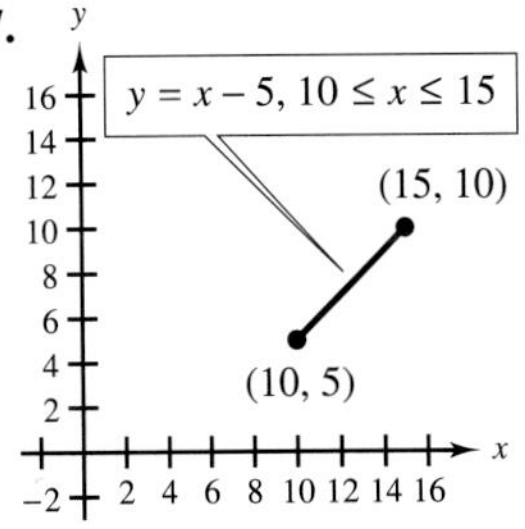

108.

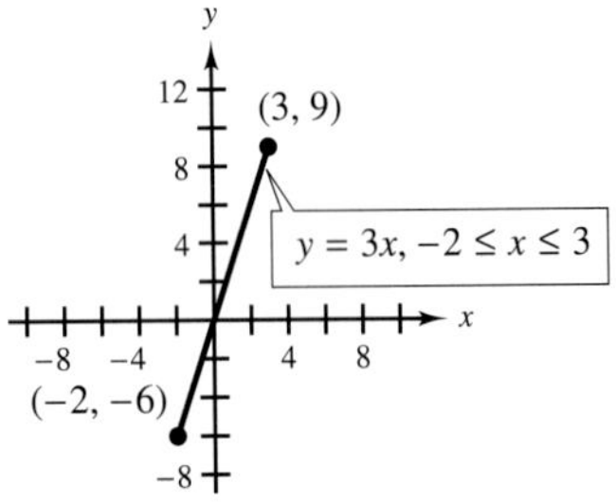

Section 4.2 *(page 207)*

1. The geometric interpretation is the rise divided by the run.

2. (a) A line with a positive slope rises from left to right.
(b) A line with a negative slope falls from left to right.

3. 2 **4.** $\frac{5}{2}$ **5.** -1 **6.** $-\frac{2}{3}$ **7.** 0 **8.** 0

9. Undefined **10.** Undefined **11.** 5 **12.** 2 **13.** $\frac{7}{2}$

14. $\frac{4}{7}$ **15.** Undefined **16.** 0 **17.** -2.825

18. -4.26 **19.** $-\frac{4}{5}$ **20.** $-\frac{1}{3}$ **21.** 0 **22.** Undefined

23. $\frac{5}{2}$ **24.** $-\frac{2}{5}$

25. Procedures (i) and (iii) are correct. Procedure (ii) refers to a rise and a run that are both negative. Thus the slope is $+3$.

26. Slope is 0 when $y_1 = y_2$. This occurs when the line is horizontal. Slope is undefined when $x_1 = x_2$. This occurs when the line is vertical.

27. $y = 3$ **28.** $y = 5.5$ **29.** $x = 2$ **30.** $x = -5$

31. $y = 4$ **32.** $y = 5$

33. Method (ii) is the correct use of the slope formula. In method (i), the coordinates are not subtracted in the same order in the numerator and the denominator.

34. No. For example, if $y_2 - y_1 = 8$ and $x_2 - x_1 = 10$, then the slope is $\frac{4}{5}$. We know only that the ratio of the rise to the run is $\frac{4}{5}$.

35. $-\frac{3}{5}$ **36.** $\frac{1}{2}$ **37.** 0 **38.** Undefined **39.** $\frac{3}{4}$ **40.** $\frac{2}{3}$

41. $y = -2x + 17$; $m = -2$ **42.** $y = 3x + 10$; $m = 3$

43. $y = 2x$; $m = 2$ **44.** $y = -3x$; $m = -3$

45. $y = 15$; $m = 0$ **46.** $y = 8$; $m = 0$

47. $y = -\frac{1}{3}x + 2$; $m = -\frac{1}{3}$ **48.** $y = -\frac{3}{4}x + \frac{5}{2}$; $m = -\frac{3}{4}$

49. Undefined **50.** Undefined **51.** $y = \frac{5}{3}x - \frac{7}{3}$; $m = \frac{5}{3}$

52. $y = \frac{7}{5}x - \frac{8}{5}$; $m = \frac{7}{5}$ **53.** $y = \frac{9}{8}x - 18$; $m = \frac{9}{8}$

54. $y = -\frac{5}{8}x + 25$; $m = -\frac{5}{8}$ **57.** (i), (iii) **58.** (i)

59. 0 **60.** 0

61.

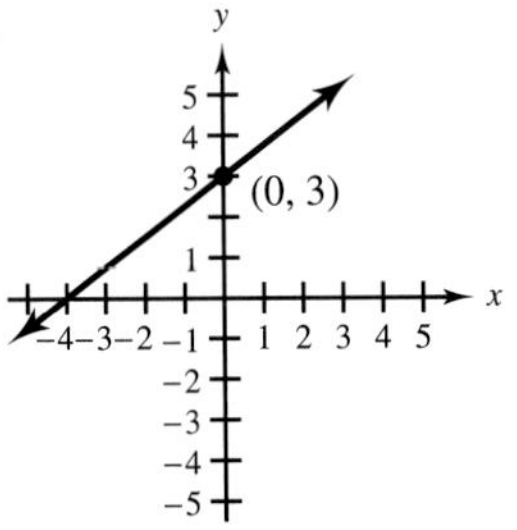

62.

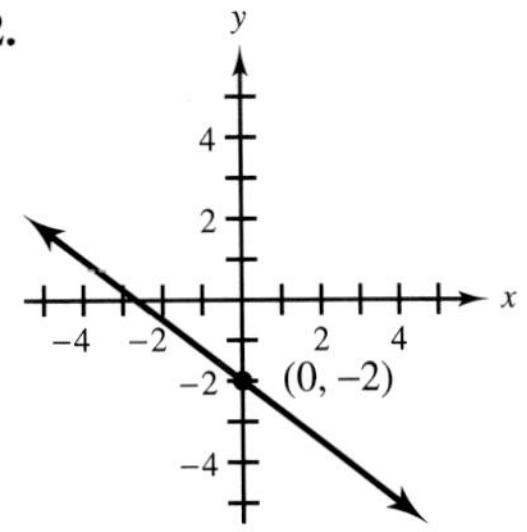

63.

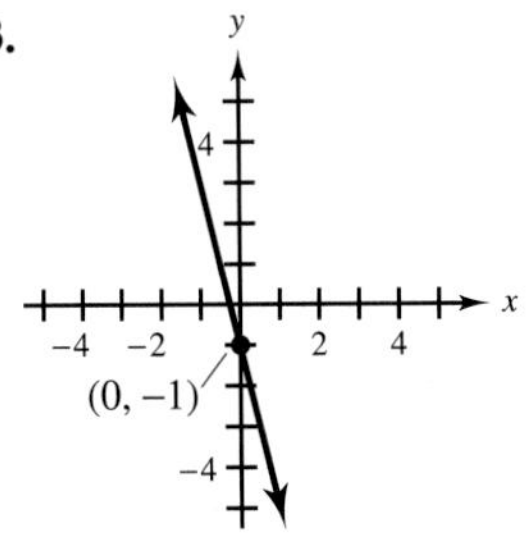

64.

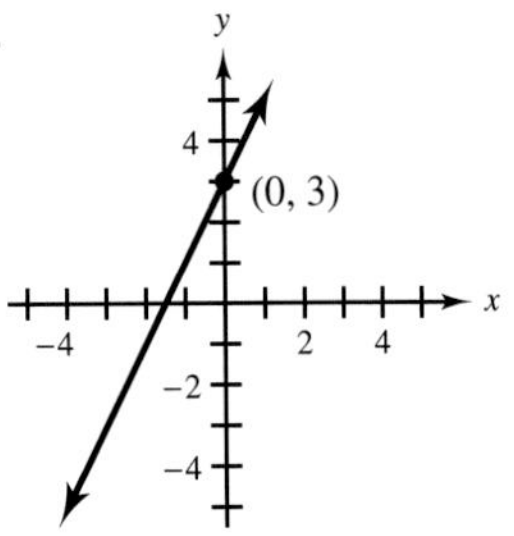

65.

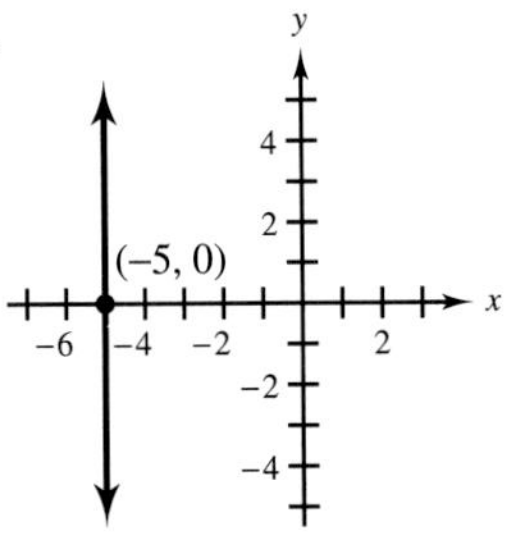

66.

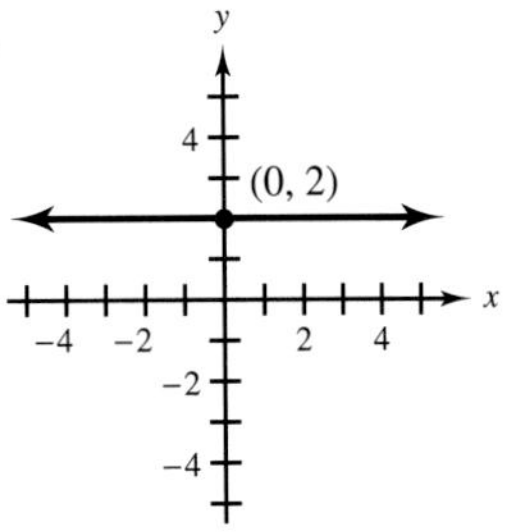

67.

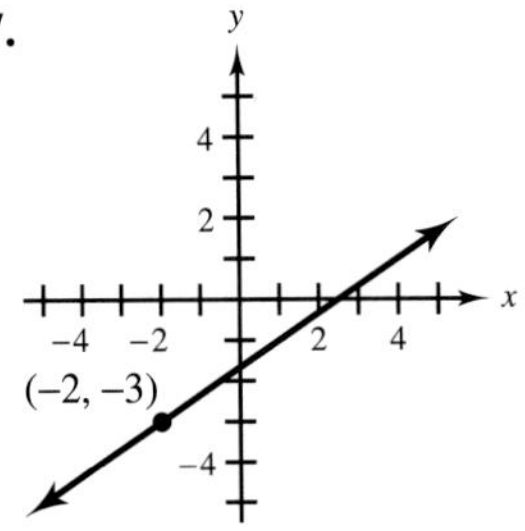

68.

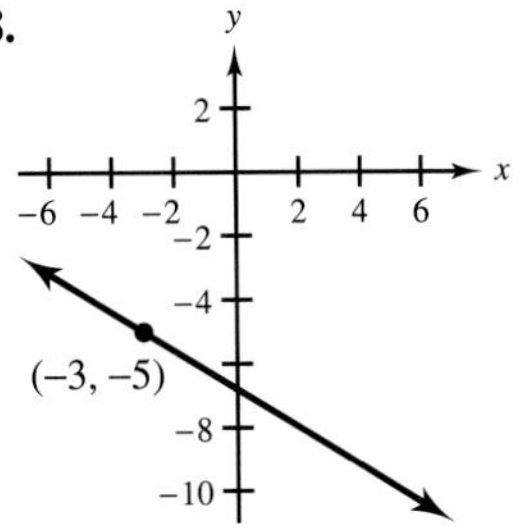

69.

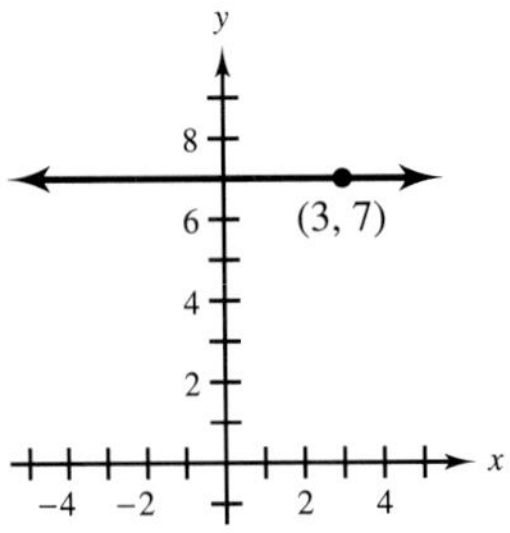

70.

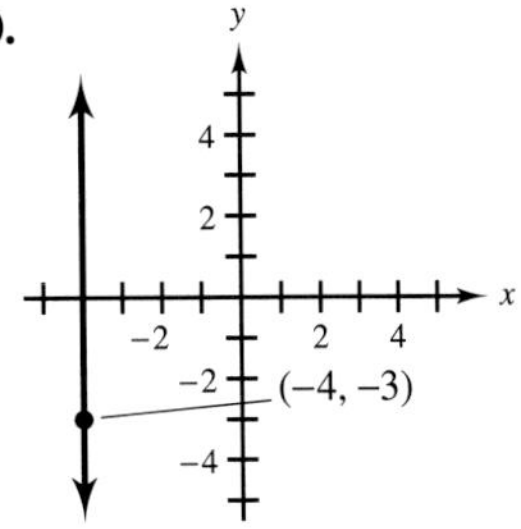

71.

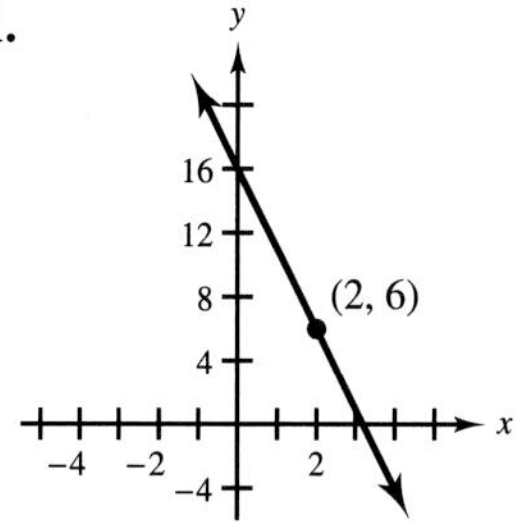

72.

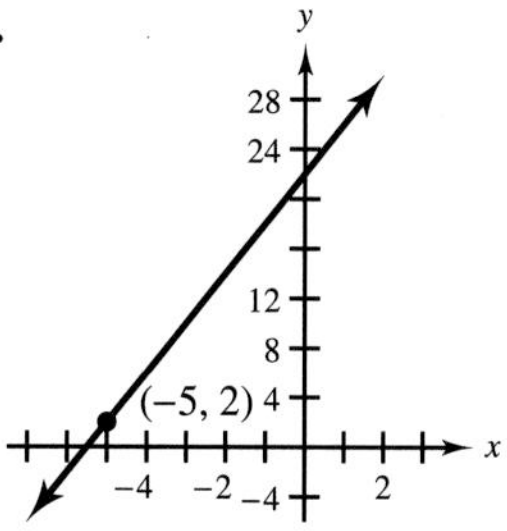

73.

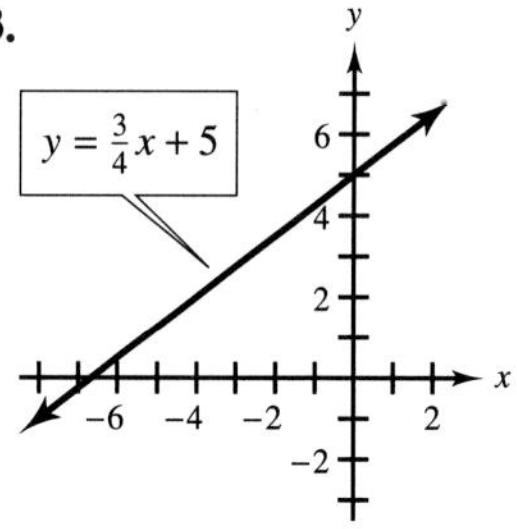

74.

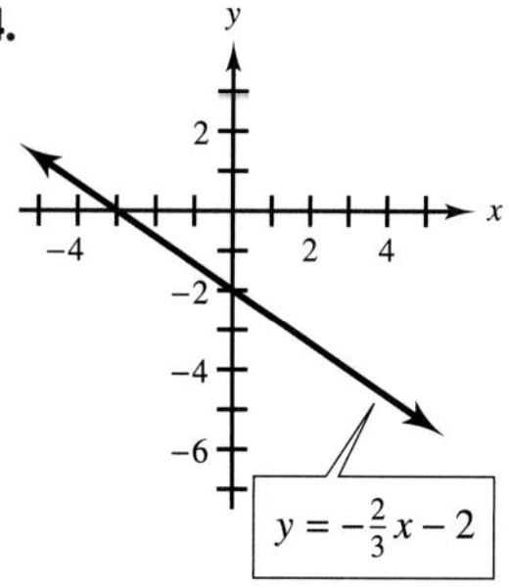

75.

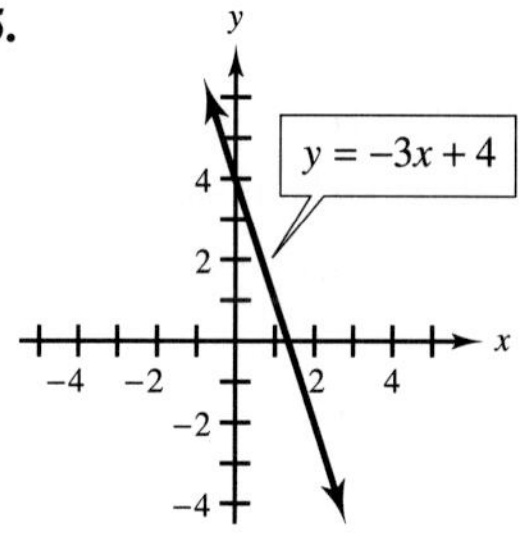

76.

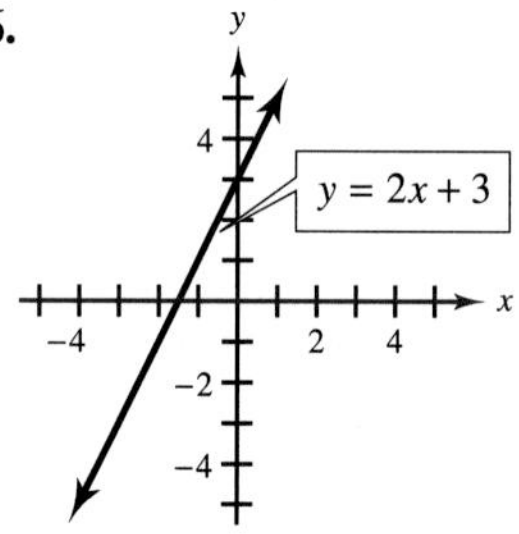

77.

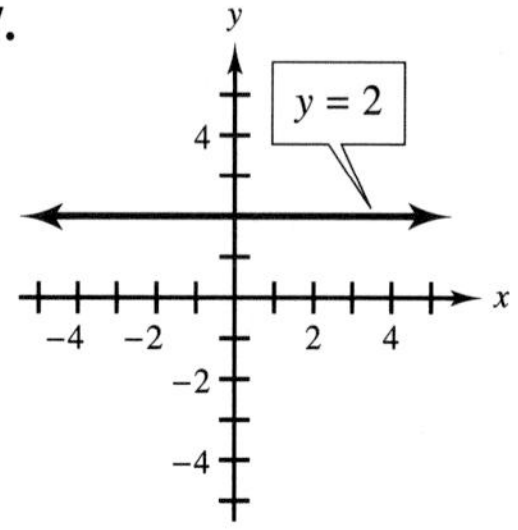

78.

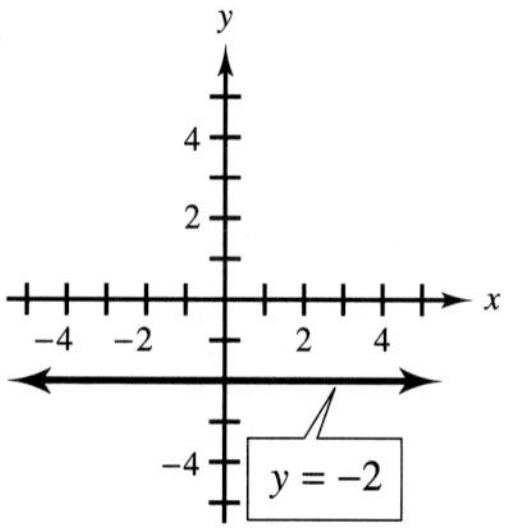

79.

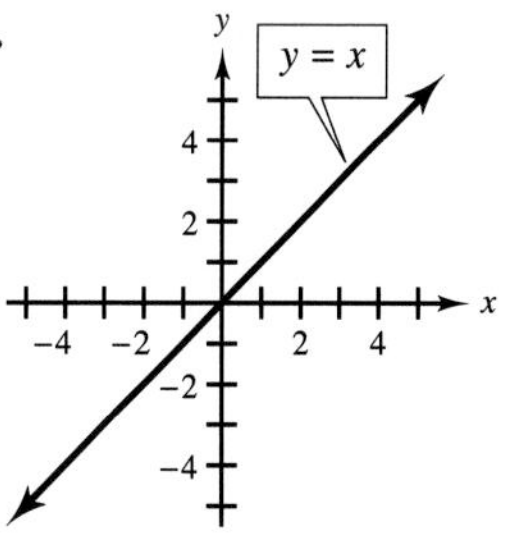

80.

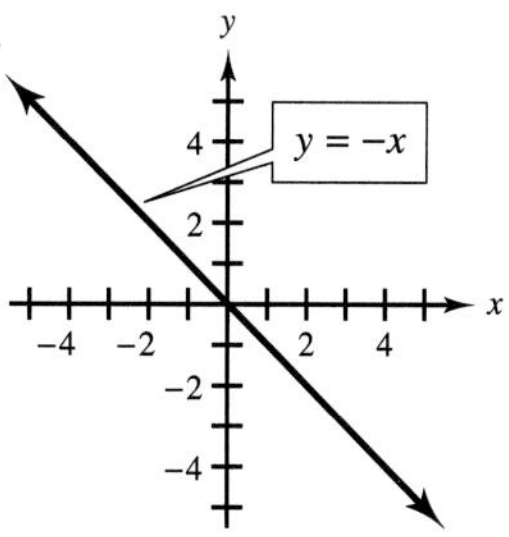

81.

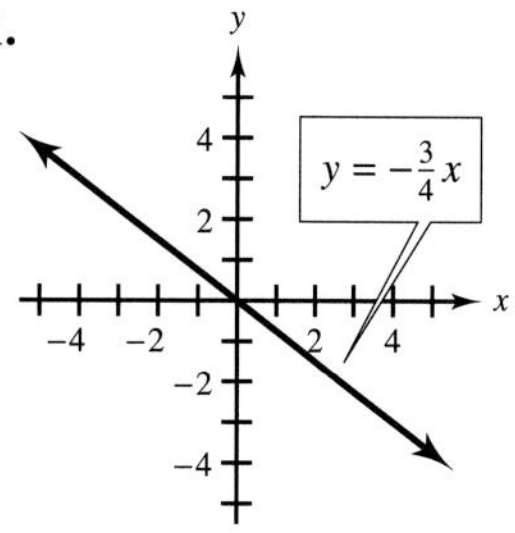

82.

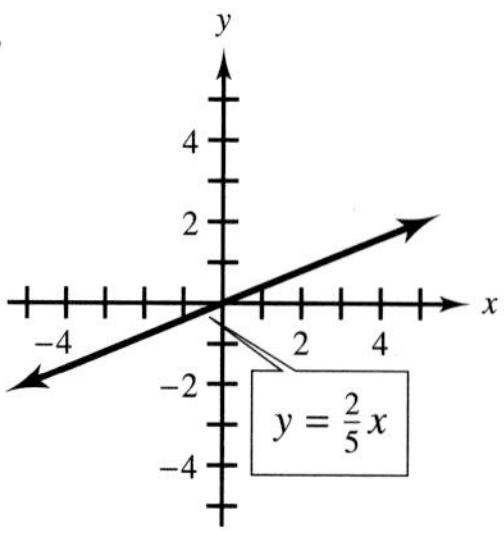

83.

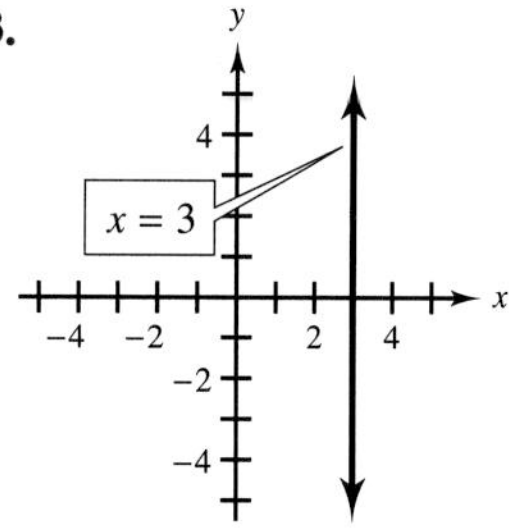

84.

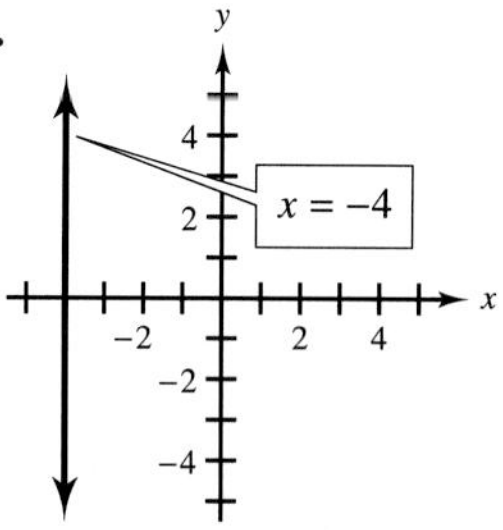

85. The graph retains the same slope, but it is shifted upward.

86. The graph retains the same y-intercept, but it is steeper.

87. $y = -\frac{2}{5}x + 2$

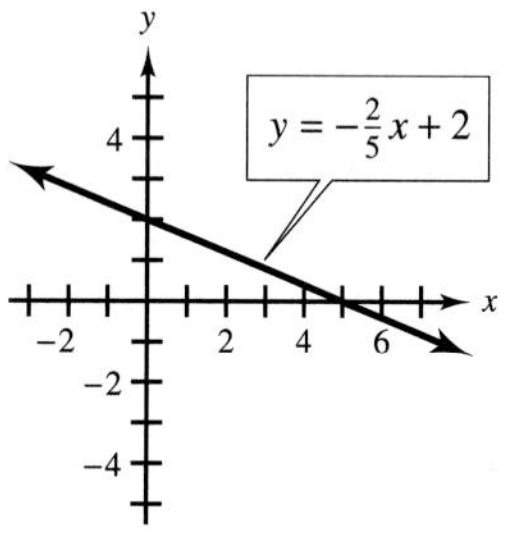

88. $y = \frac{4}{5}x - 4$

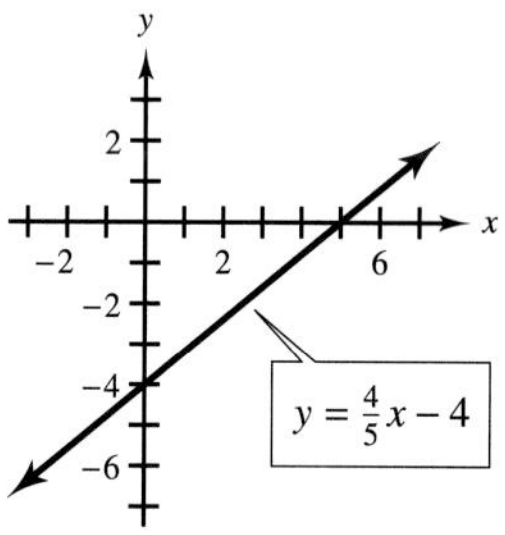

89. $x = 7$

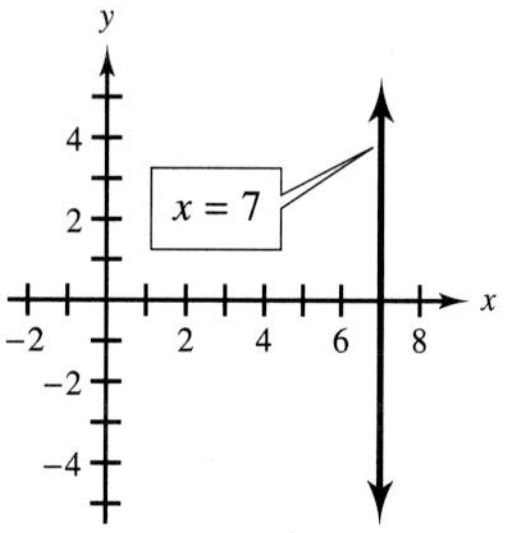

90. $x = -4$

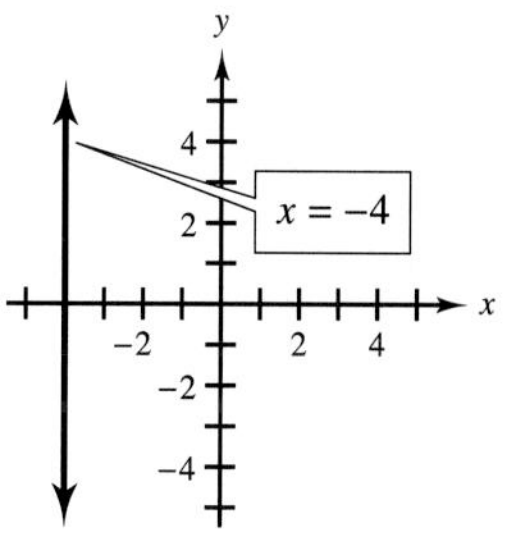

91. $y = 2x - 8$

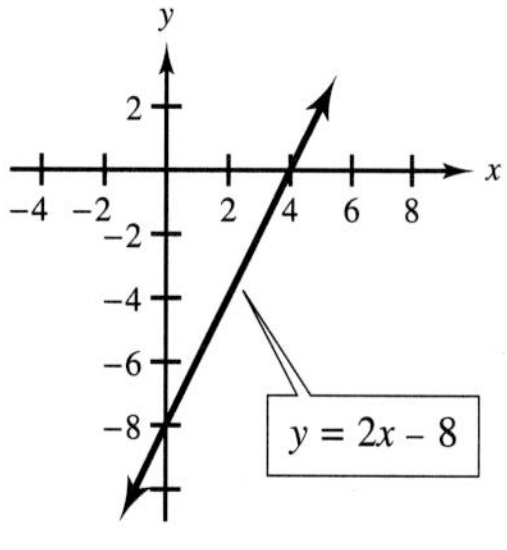

92. $y = -\frac{1}{3}x + 2$

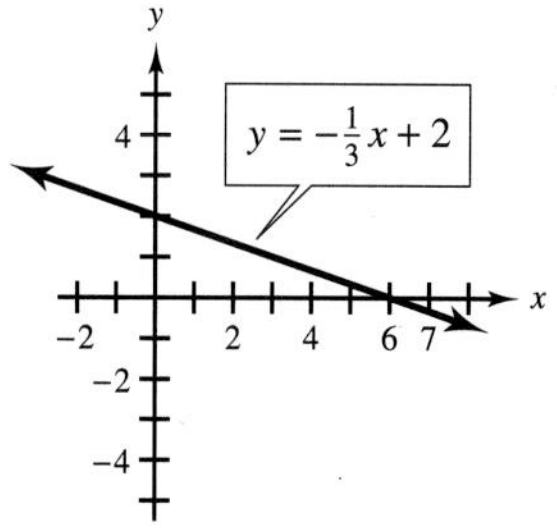

93. $y = 7$

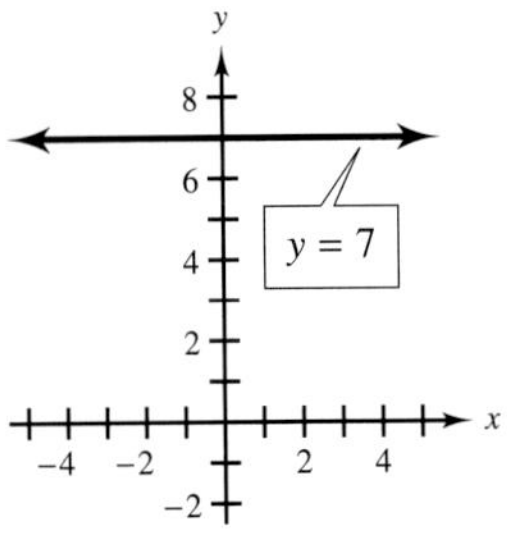

94. $y = 6$

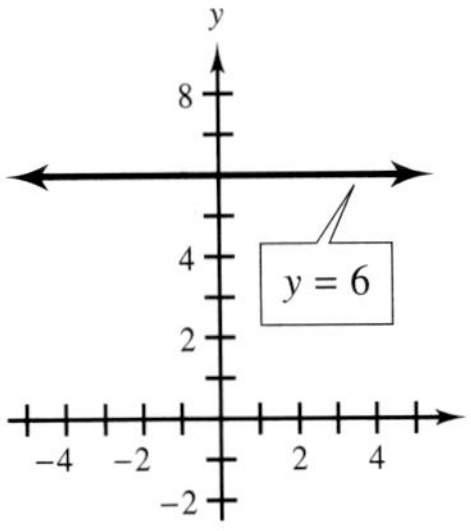

95. $y = 2x$

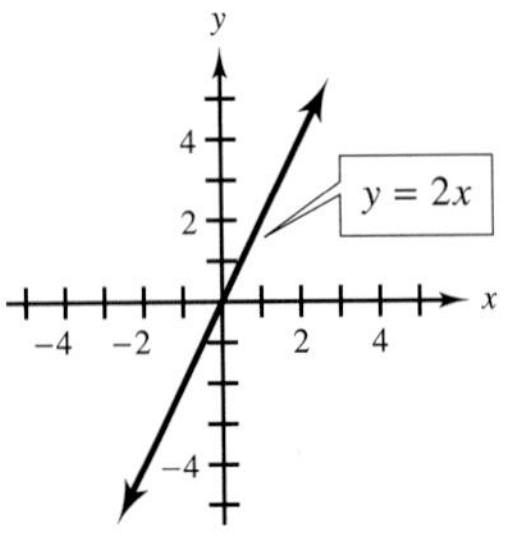

96. $y = -3x$

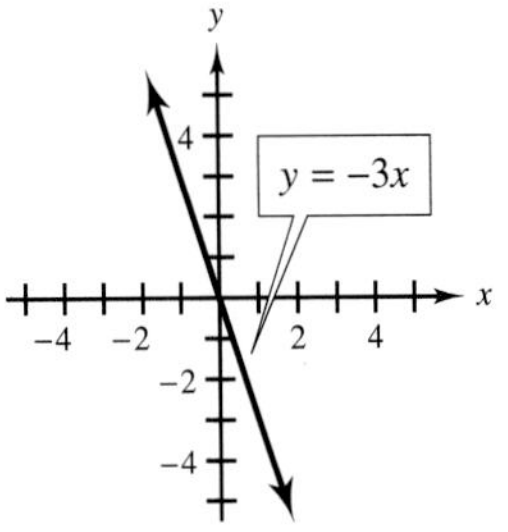

97. **(a)** 5 feet 1 inch **(b)** 2
(c) The slope of the graph is the same as the average annual growth rate.
(d) The age is 10 and the height is 4 feet 5 inches.

98. The slope would be close to 0.

99. **(a)** -75 **(b)** 1200 **(c)** Rise: -525; Run: 7
(d) 2006

100. 7.2 inches

101. **(a)** 5.6 **(b)** $5.\overline{3}$

102. The slope represents the average annual increase in the percentage of cars with air bags.

103. **(a)** Semiprivate: $+217.50$, Private: -12.75
(b) The positive slope indicates an increase in the number of semiprivate courses, whereas the negative slope indicates a decrease in the number of private courses.

104. **(a)** 1995: 3383, 1999: 4304 **(b)** 230.25
(c) Because the slope is positive, the difference between the number of semiprivate and private courses is increasing.

105. The graph is rising from left to right, which is consistent with the bar graph.

106. 2058

107. Because the percentage of women is increasing, the percentage of men is decreasing. Therefore, the coefficient of x would be negative.

108. About (50, 2058); In the year 2058, 50% of the legislators will be women and 50% will be men.

109. $m = \frac{3}{4}$

110. **(a)** $y_1 = ax_1 + b$
(b) $y_2 = ax_2 + b$
(c) $y_1 - y_2 = ax_1 - ax_2 = a(x_1 - x_2)$; $a = \dfrac{y_1 - y_2}{x_1 - x_2}$
(d) Slope is defined as $\dfrac{y_1 - y_2}{x_1 - x_2}$. Thus the number a in $y = ax + b$ is the slope of the line.

111. **(a)** $m = 3$ **(b)** $m = 2$ **(c)** $m = 1$

112. $m = 0.1$; As we select points closer and closer to P, the value of the slope becomes closer to 0.

Section 4.3 *(page 217)*

1. Perpendicular lines are lines that intersect to form a right angle.

2. Parallel lines are lines in the same plane that do not intersect.

3. Neither **4.** Neither **5.** Parallel **6.** Perpendicular

7. Perpendicular **8.** Neither

9. **(a)** -3 **(b)** $\frac{1}{3}$ **10.** **(a)** $\frac{5}{8}$ **(b)** $-\frac{8}{5}$

11. **(a)** $-\frac{3}{4}$ **(b)** $\frac{4}{3}$ **12.** **(a)** 2.5 **(b)** -0.4

13. **(a)** 0 **(b)** Undefined **14.** **(a)** Undefined **(b)** 0

15. Parallel **16.** Parallel **17.** Neither **18.** Neither

19. Perpendicular **20.** Perpendicular

21. The slope of a vertical line is undefined.

22. Because $m_1 = -\dfrac{1}{m_2}$ implies that $m_1m_2 = -1$, the lines are perpendicular.

23. Parallel **24.** Perpendicular **25.** Perpendicular

26. Parallel **27.** Neither **28.** Parallel

29. Perpendicular **30.** Parallel **31.** Parallel

32. Perpendicular **33.** Neither **34.** Perpendicular

35. Undefined **36.** Undefined **37.** 0

38. 0 **39.** **(a)** 2 **(b)** $-\frac{1}{2}$ **40.** **(a)** -3 **(b)** $\frac{1}{3}$

41. **(a)** $-\frac{4}{3}$ **(b)** $\frac{3}{4}$ **42.** **(a)** $\frac{2}{5}$ **(b)** $-\frac{5}{2}$

43. **(a)** 0 **(b)** Undefined **44.** **(a)** Undefined **(b)** 0

45. $-\frac{2}{3}$

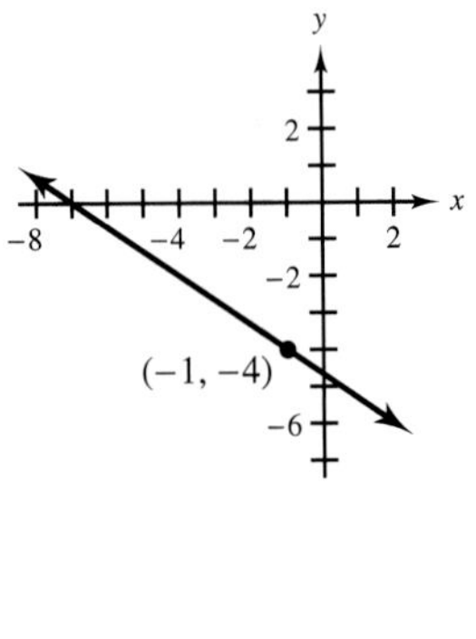

46. $-\frac{1}{2}$

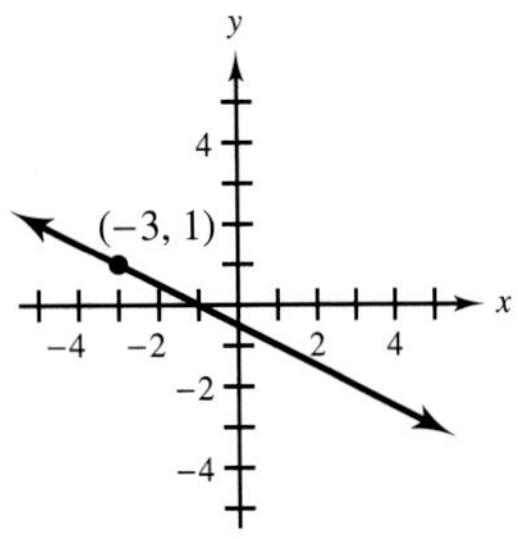

47. Undefined

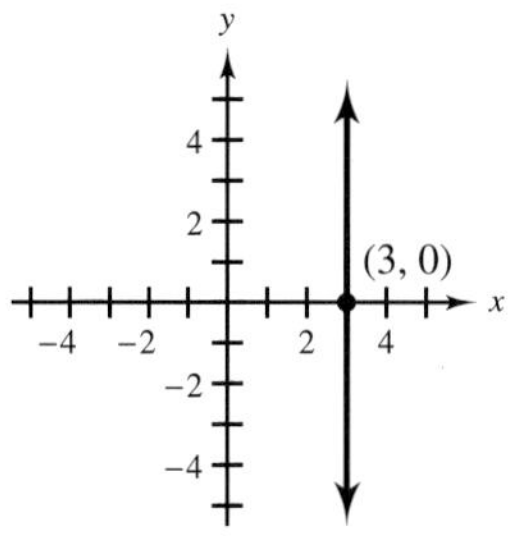

48. 0

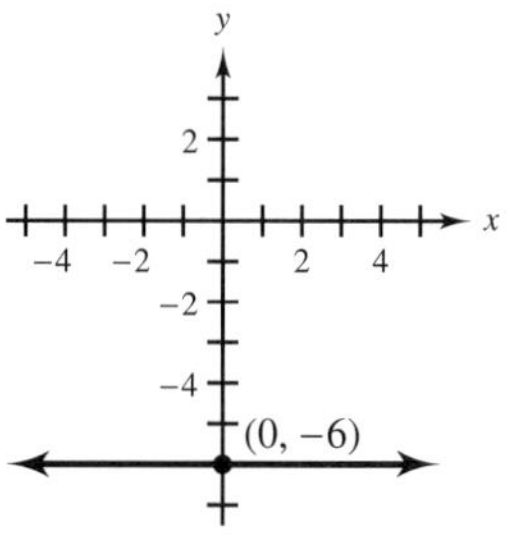

49. **(a)** 10.5 **(b)** $-\frac{14}{3}$ **50.** **(a)** $-\frac{1}{4}$ **(b)** 4

53. Collinear **54.** Not collinear

59. The slope of a line represents the rate at which y changes with respect to a change in x.

60. The rate of change of a linear function is the slope of the line, and the slope is the same for all points of the line.

61. $\frac{1}{2}$ **62.** $-\frac{2}{3}$ **63.** -4 **64.** 3 **65.** $\frac{3}{2}$ **66.** $-\frac{2}{5}$

67. $-\frac{7}{3}$ **68.** $\frac{4}{5}$ **69.** y increased by 6

70. x decreased by 3 **71.** x decreased by 4

72. y increased by 3 **73.** y decreased by 10

74. x increased by 3 **75.** x increased by 4

76. y decreased by 12

77. There exists a constant k such that $A = kB$.

78. Because y varies directly with x, $y = kx$ for some constant k. Thus $b = ka$ and $d = kc$. Therefore, $k = \frac{b}{a} = \frac{d}{c}$, which implies that $\frac{a}{b} = \frac{c}{d}$.

79. Yes **80.** No **81.** Yes **82.** No **83.** No

84. No **85.** increases, decreases **86.** $kx, \frac{y}{x}$

87. **(a)** $c = 0.1m + 29$ **(b)** 0.1

88. **(a)** $I = 0.3s + 100$ **(b)** 0.3

89. 1 cent **90.** 4.5°F **91.** 2820 pounds

92. 1590 pounds **93.** 247.5 square feet **94.** $6\frac{1}{3}$ inches

95. $156.25 **96.** $16

97. **(a)** The cost C is $C(d) = 233d$, where d is the number of days of vacation travel.

(b) The function is in the form $y = kx$.

(c) 233

98. $1.35 billion per year

99. **(a)** $m = 9.8$; The average two-year increase in nuclear-based electricity generation for the period was 9.8 billion kilowatt-hours.

(b) 9.8, which is the same as the slope

100. **(a)** The y-intercept of a direct variation is the origin. Because one data point is (0, 1.38), a linear function describing the data would not be a direct variation.

(b) 0.17

(c) During this period, per-share dividends have increased by 17 cents per year.

101.

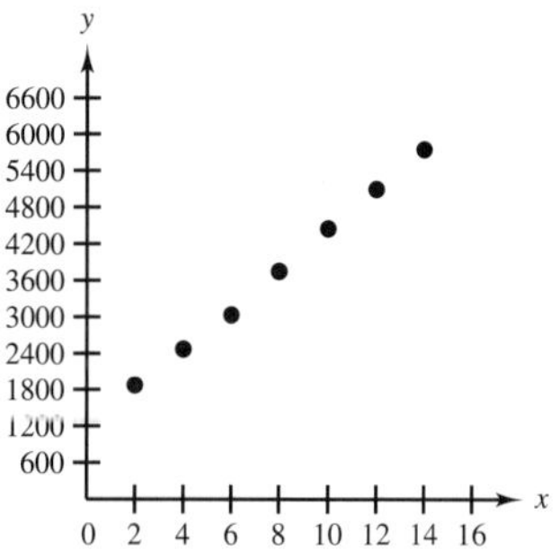

102. 322.75; The average 2-year increase in the number of bald eagles during the period 1986–1998 was 323 pairs.

103. $p(x) = 322.75x + 1229.5$ **104.** 2008 **105.** 6

106. 4 **107.** 1 **108.** 0 **109.** 0 **110.** Any $k \neq 0$

Section 4.4 *(page 228)*

1. We must know the slope and the y-intercept.

2. A horizontal line has a slope of 0, but a vertical line has no defined slope.

3. $y = -4x - 3$ **4.** $y = 3x + 4$ **5.** $y = \frac{1}{2}x + 3$

6. $y = -\frac{2}{3}x - 2$ **7.** $y = -5x + 1$ **8.** $y = 6x - 6$

9. $y = -\frac{3}{4}x - 5$ **10.** $y = \frac{5}{8}x + 8$ **11.** $y = 7x - \frac{2}{3}$

12. $y = -4x + \frac{1}{2}$ **13.** $y = \frac{3}{4}x + 2$ **14.** $y = \frac{3}{5}x - 1$

15. $x = 4$ **16.** $y = -2$ **17.** $y = -\frac{5}{2}x - 3$

18. $y = -\frac{1}{4}x + 3$

19. A horizontal line has a slope of 0, but a vertical line has no defined slope.

20. We must know the slope and the coordinates of a point of the line.

21. $y = 2x + 1$ **22.** $y = -3x + 11$ **23.** $y = \frac{2}{5}x - 8$

24. $y = -\frac{1}{3}x + 1$ **25.** $y = -\frac{2}{3}x - \frac{19}{3}$ **26.** $y = \frac{4}{5}x - \frac{6}{5}$

27. $y = -5$ **28.** $y = -4$ **29.** $x = -3$ **30.** $x = -4$

31. $x - y = 3$ **32.** $4x - 2y = 7$ **33.** $5x - 3y = 10$

34. $6x + 3y = -4$ **35.** $2x + 8y = -7$ **36.** $5x + 6y = 3$

37. Use the slope formula to determine the slope.

38. The slope of the line is undefined. **39.** $y = -2x + 10$

40. $y = 2x - 1$ **41.** $y = -2x + 3$ **42.** $y = -3x + 6$

43. $y = x + 1$ **44.** $y = \frac{1}{4}x - 2$ **45.** $y = -\frac{2}{3}x - \frac{5}{3}$

46. $y = -\frac{3}{4}x + \frac{1}{4}$ **47.** $x = -4$ **48.** $x = 2$

49. $y = \frac{1}{2}x - \frac{5}{2}$ **50.** $y = -\frac{2}{3}x - \frac{2}{3}$ **51.** $y = 5$

52. $y = -3$

53. **(a)** A line that is perpendicular to a vertical line is a horizontal line. The equation is in the form $y = b$.

(b) A line that is parallel to the y-axis is a vertical line. The equation is in the form $x = c, c \neq 0$.

54. The coordinates of a point of L_1 are also needed.

55. $y = -4$ **56.** $y = 6$ **57.** $y = 5$ **58.** $y = -3$

59. $y = -\frac{2}{3}x - \frac{13}{3}$ **60.** $y = \frac{3}{4}x + \frac{11}{4}$ **61.** $x = -2$

62. $x = 7$ **63.** $y = 3x + 2$ **64.** $y = \frac{7}{2}x - 4$

65. $x = -2$ **66.** $x = 3$ **67.** $y = 4x + 2$

68. $y = -\frac{3}{2}x - 4$ **69.** $x = -2$ **70.** $x = 3$

71. L_1: $y = -\frac{4}{3}x + \frac{11}{3}$ L_2: $y = \frac{3}{4}x - \frac{1}{2}$

72. L_1: $y = -\frac{1}{4}x + \frac{3}{2}$ L_2: $y = 4x + 10$

73. $y = -x - 7$ **74.** $y = -\frac{1}{3}x + 3$

75. **(a)** $y = -x + 3$ **(b)** $y = -\frac{3}{4}x - 3$

76. **(a)** $y = \frac{1}{2}x + 4$ **(b)** $y = -2x + 8$

77. 6 **78.** $-\frac{3}{2}$ **79.** 4 **80.** $\frac{2}{9}$ **81.** $-\frac{1}{6}$

82. -10 **83.** 1 **84.** -5

85. **(a)** $v = -200t + 1500$ **(b)** \$700

86. $P = 2.7x + 35$, where $x =$ the number of years since 1992; The population increased by 2.7 persons per year.

87. \$153.3 million; Net loans increased by \$6.8 million per year.

88. 2003

89. **(a)** $N(x) = 0.014x + 1.134$

(b) The value of b would be 1.161.

(c) The graph of N comes as close as possible to all the data points, but it does not necessarily contain any actual point.

90. $P(x) = \frac{5}{3}x + 10$

91. **(a)** $x =$ number of years since 1995

(b) $E(x) = 1.1x + 87.2$

(c) 2003

92. **(a)** Because the circulation decreased, the graph of the function would fall from left to right. Therefore, m would be negative.

(b) $C = -0.32x + 16$; Circulation decreased at an average rate of 320,000 units per year.

93. $F = 0.06x + 6.56$; The model is the same for any two data points because all the data points are points of the same line.

94. The slope is unchanged, but the y-intercept is different.

95. The slope of the line connecting the data points for 1997 and 2010 will be less than that of the line connecting the data points for 1997 and 2000. Therefore, the coefficient of x would be smaller.

96. $F = 0.045x + 6.605$

99. $(3, 0), (0, 2)$ **100.** $(-4, 0), (0, -2)$

101. $\left(\frac{5}{2}, 0\right), \left(0, -\frac{4}{3}\right)$ **102.** $(3, 0), (0, 4)$

103. $bx - ay = 0$ **104.** $ax + by = 2ab$

Section 4.5 *(page 239)*

1. The solution set of $x + 2y \leq 6$ contains ordered pairs that satisfy $x + 2y = 6$, but the solution set of $x + 2y < 6$ does not.

2. The inequality $y > x + 2$ does not define a function. For example, (5, 9) and (5, 8) are both solutions, which violates the definition of a function.

3. No, yes, no **4.** Yes, yes, no **5.** Yes, yes, no

6. Yes, no, yes **7.** No, yes, no **8.** Yes, yes, no

9. In the first case, we are assuming that $x + 3 > 7$ is an inequality in one variable. In the second case, we are assuming that $x + 3 > 7$ is an inequality in two variables.

10. The reasoning is faulty because the inequality must be solved for y before using the inequality symbol to determine the half-plane to shade. In this instance, the half-plane above the line is shaded.

11. Below the line **12.** Above the line

13. Above the line **14.** Below the line

15. Below the line **16.** Above the line

17.

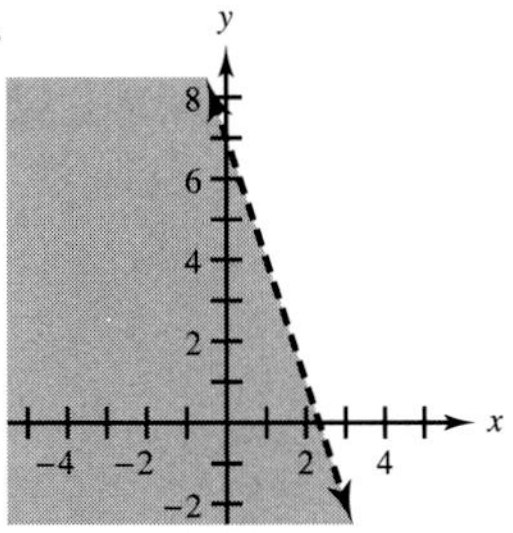

18.

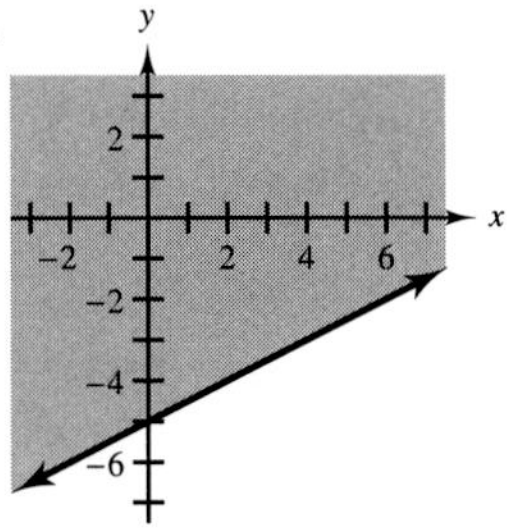

19.

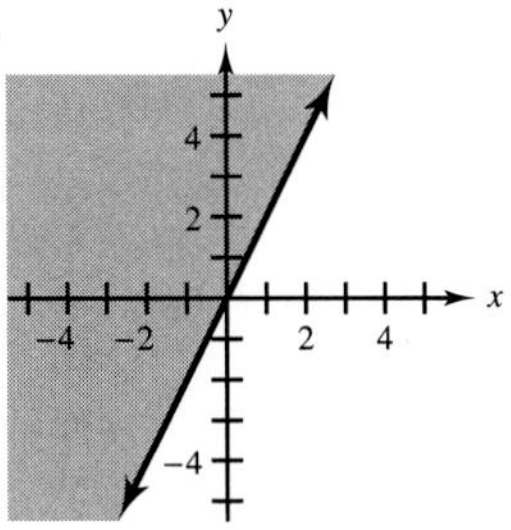

20.

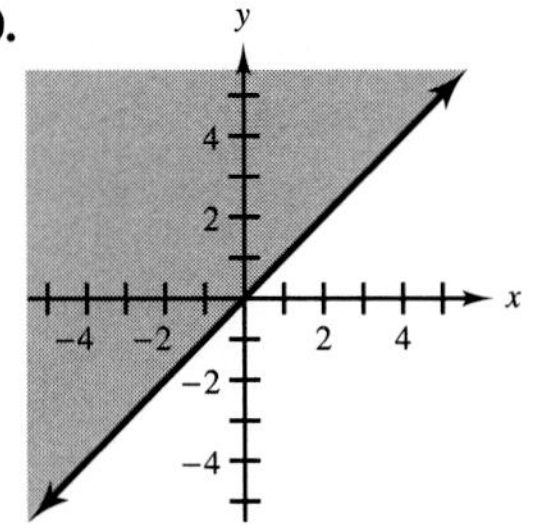

21.

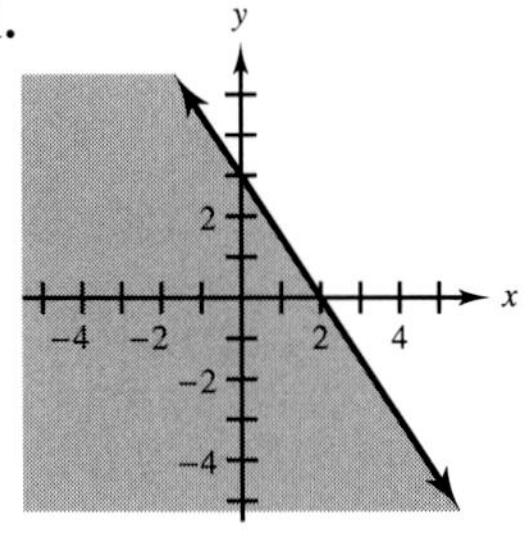

22.

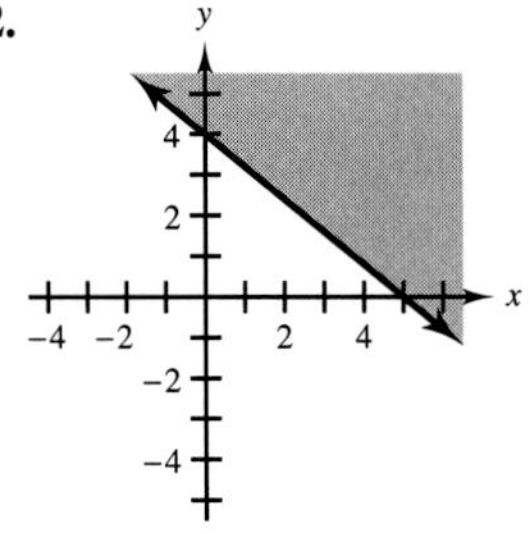

23.
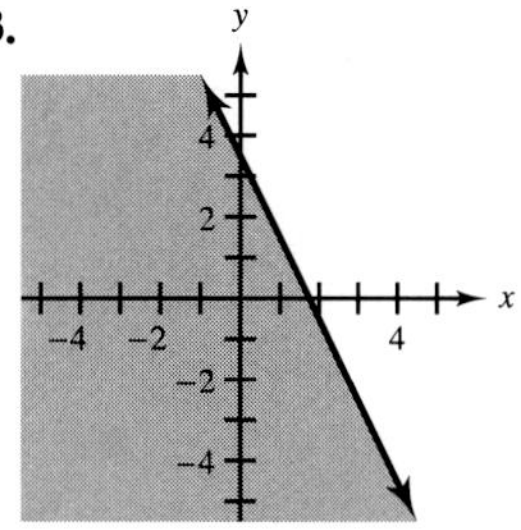

24.
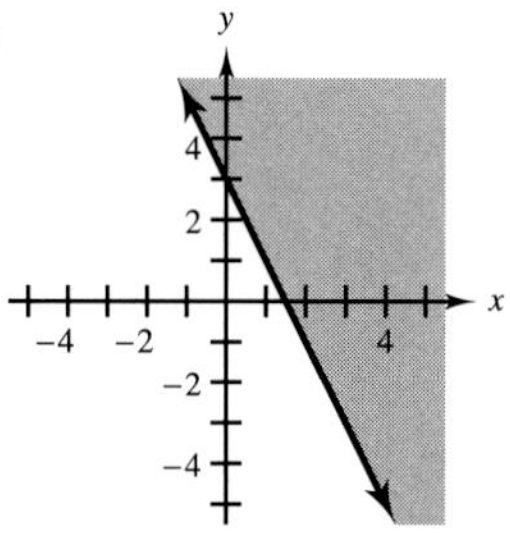

25.
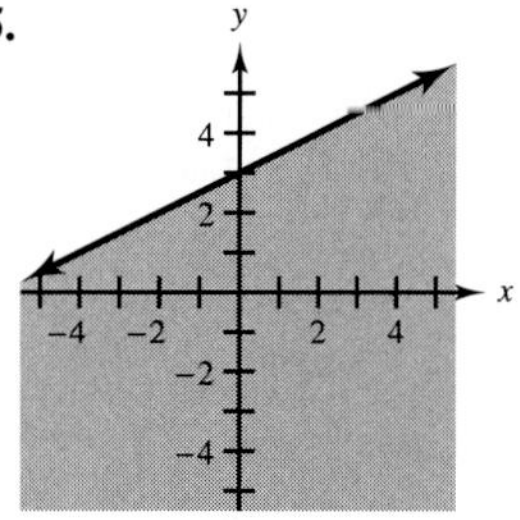

26.
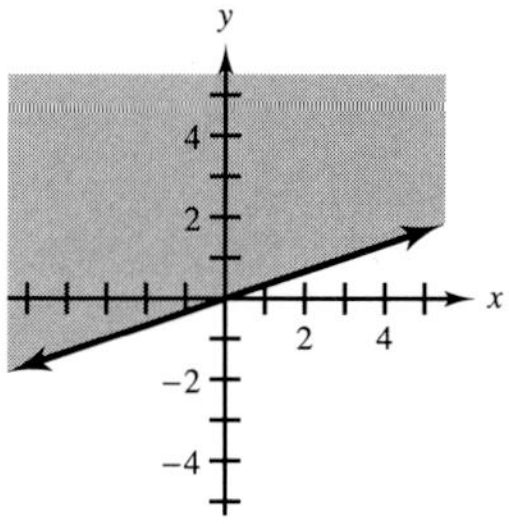

27.
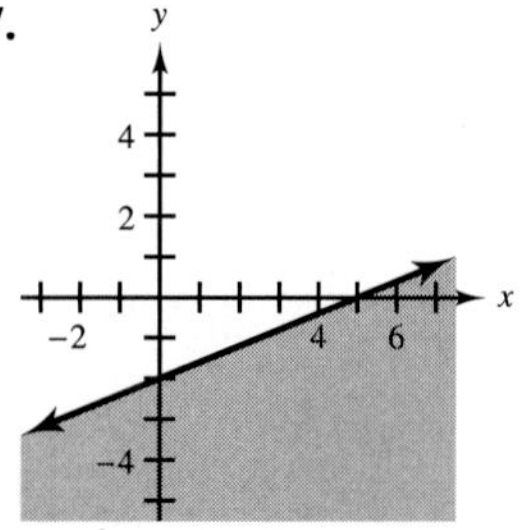

28.
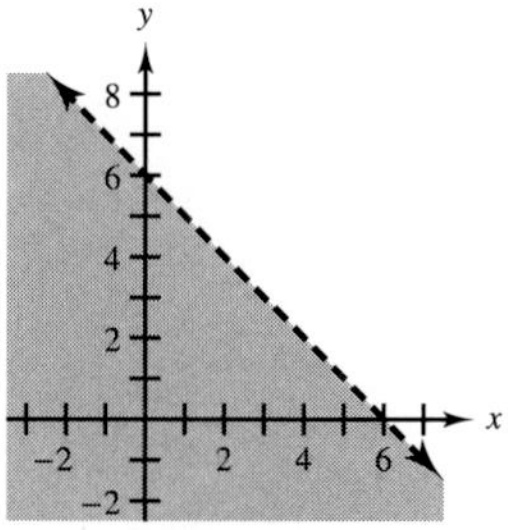

29.
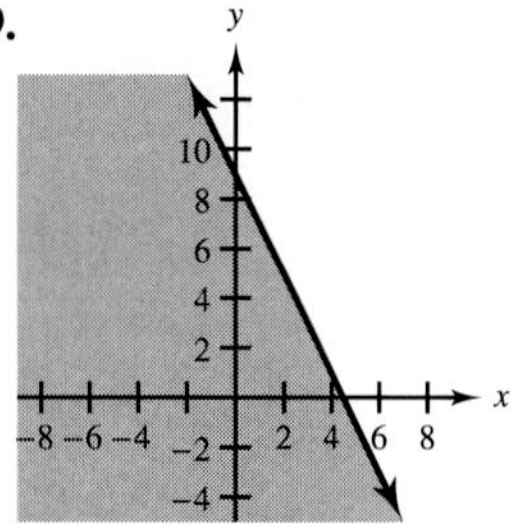

30.
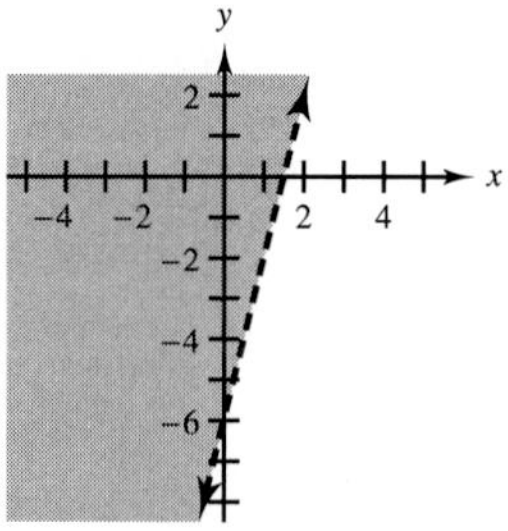

31.
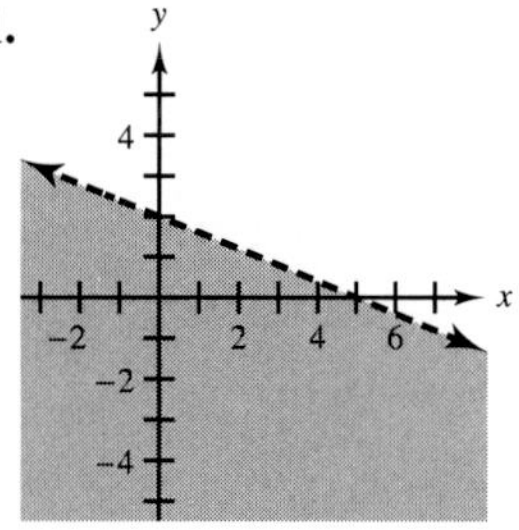

32.
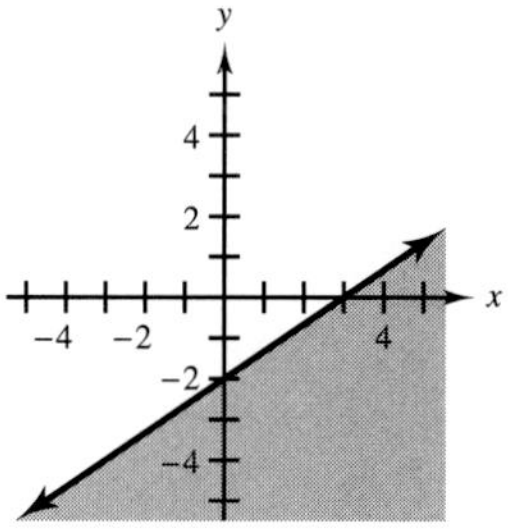

33.
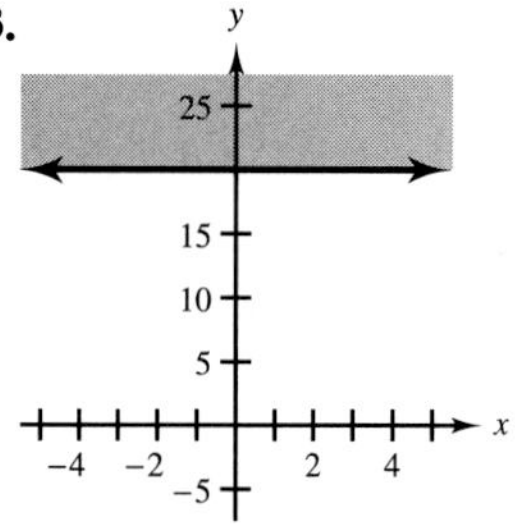

34.
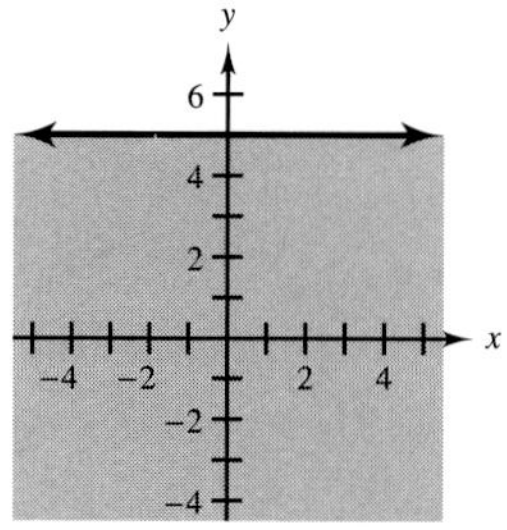

35.
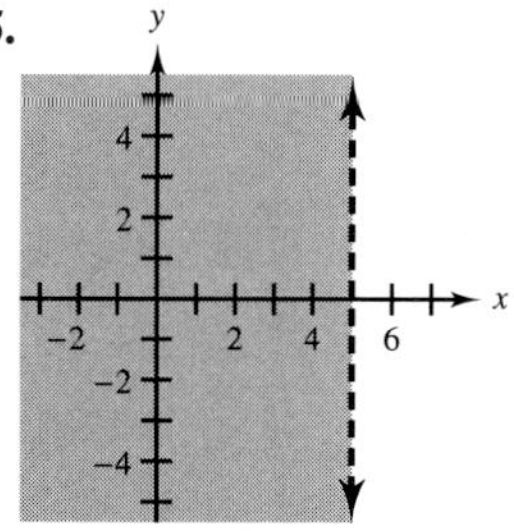

36.
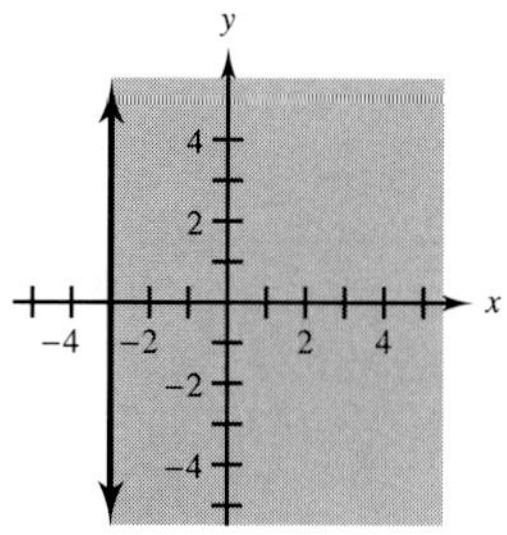

37. $y \le 3$ **38.** $y > -4$ **39.** $y < 0.5x + 1$
40. $y \le -x - 2$ **41.** $x \le 2$ **42.** $x > -2$

43. $x \ge y + 1$
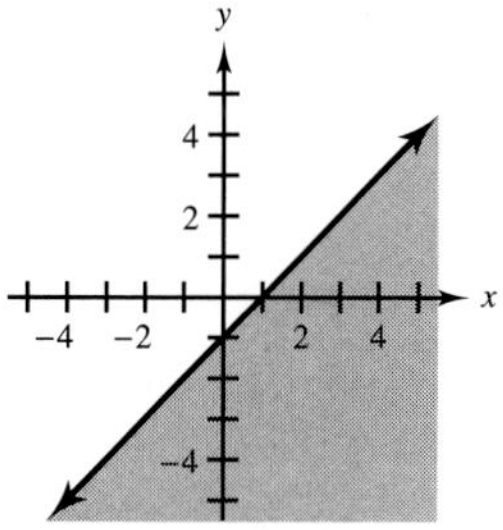

44. $y \le 2x$
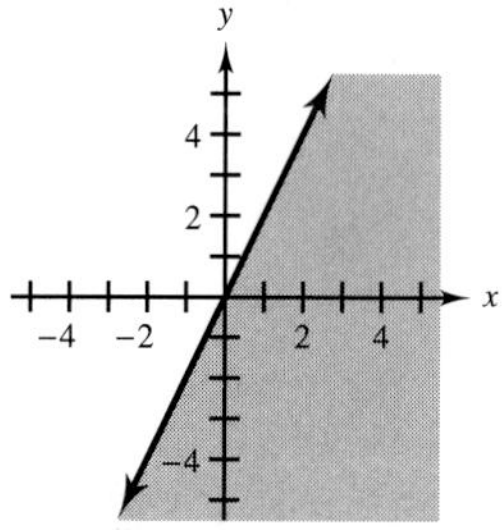

45. $y \ge 3x - 4$
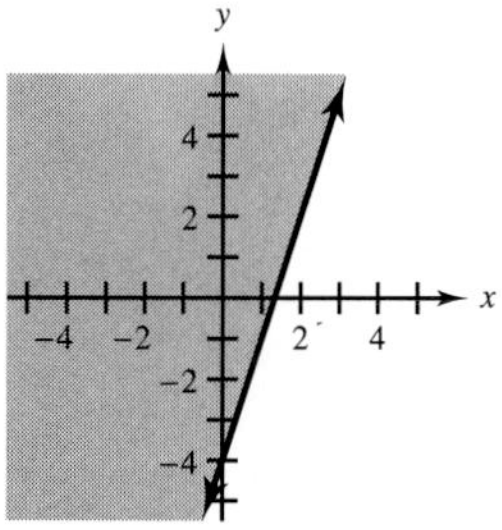

46. $x \le 0.5y$
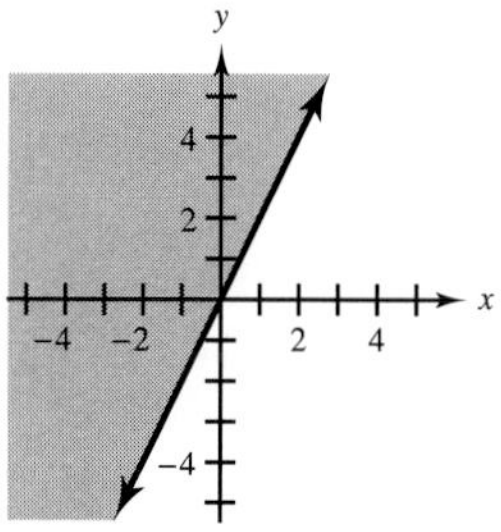

47. $x + 2y > 10$
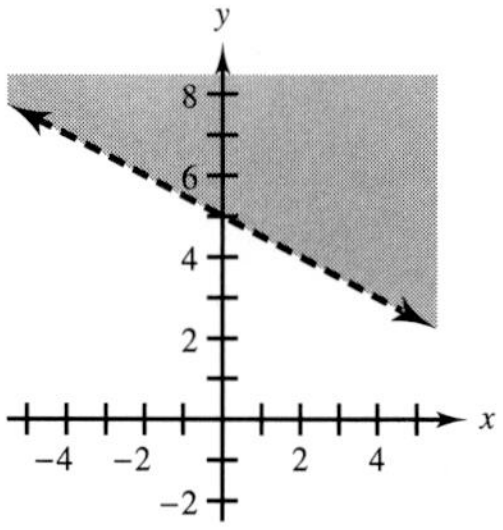

48. $x - \frac{2}{3}y > 7$
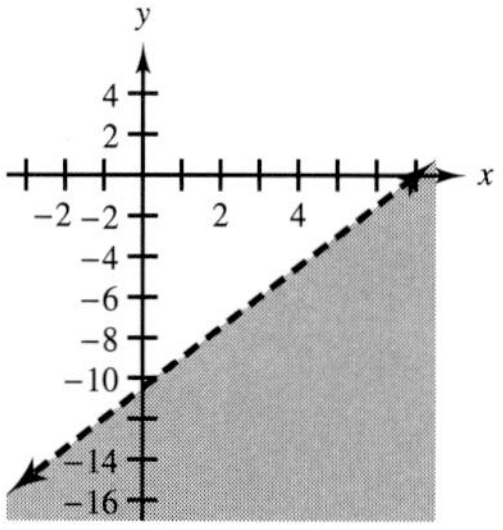

49.

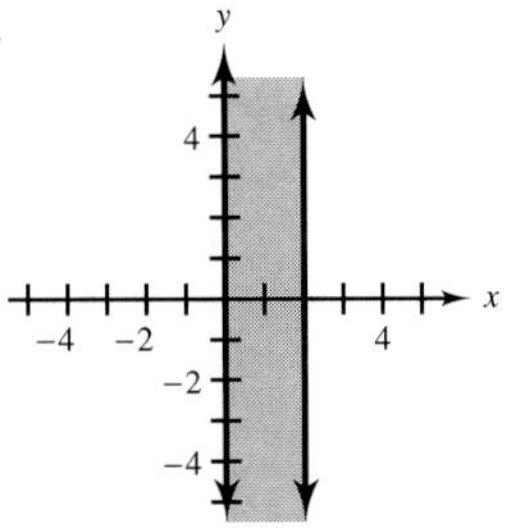

50.

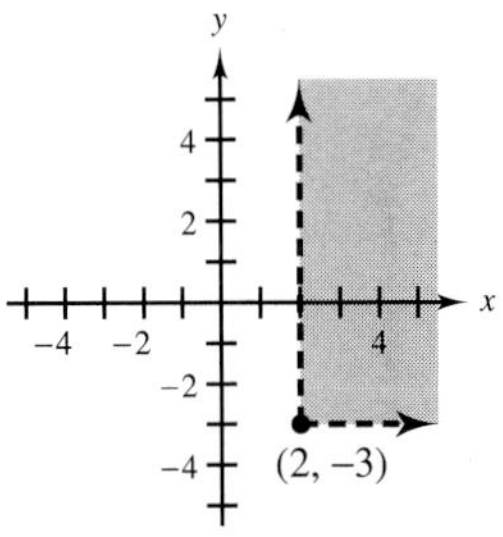

51.

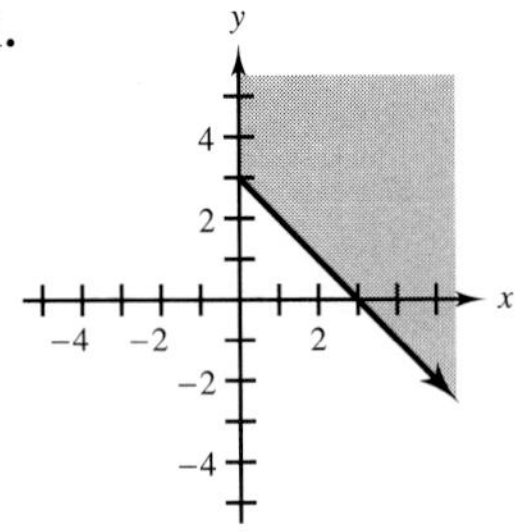

52.

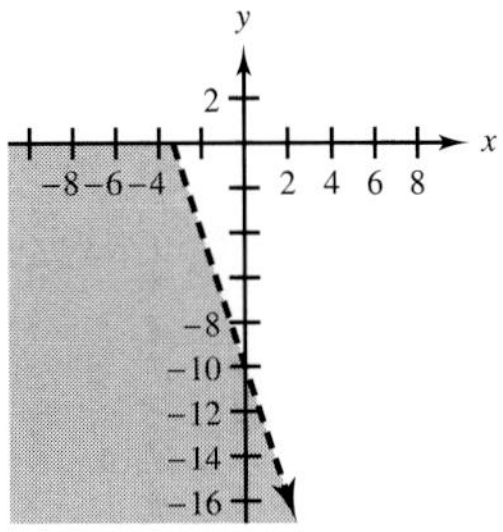

53.

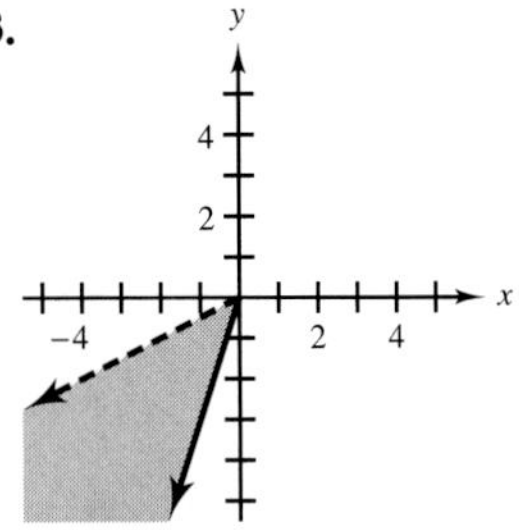

54.

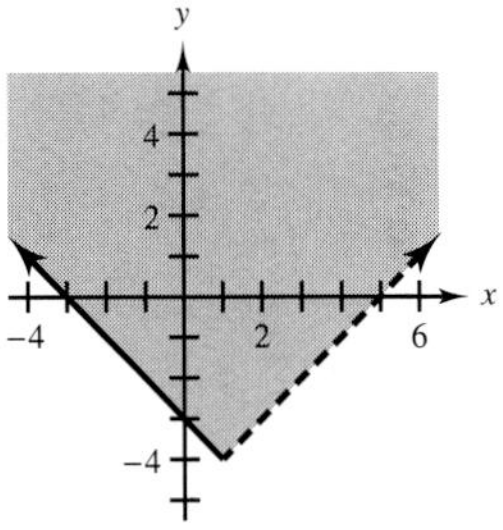

55.

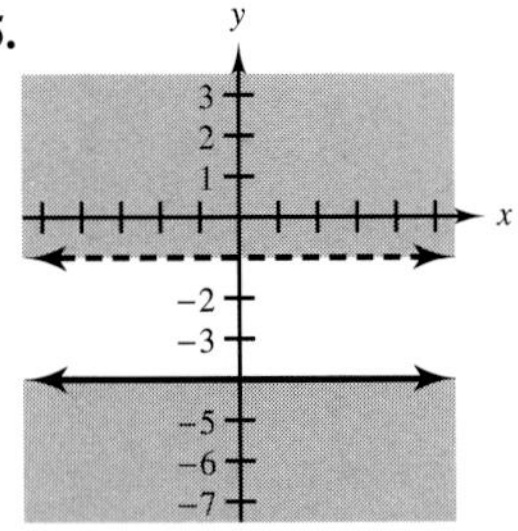

56.

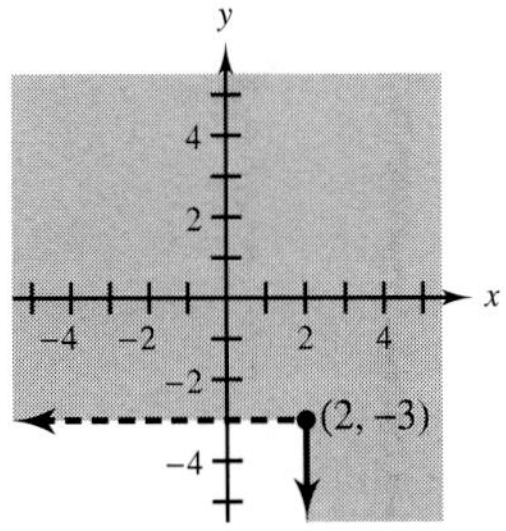

57.

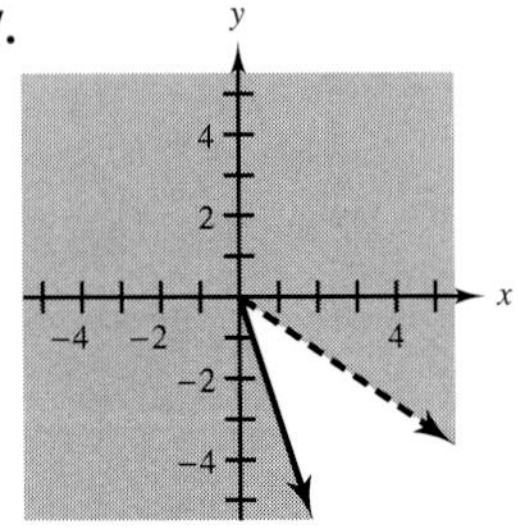

58.

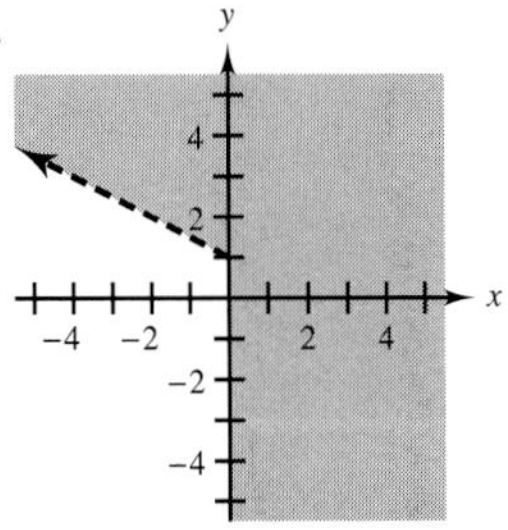

59.

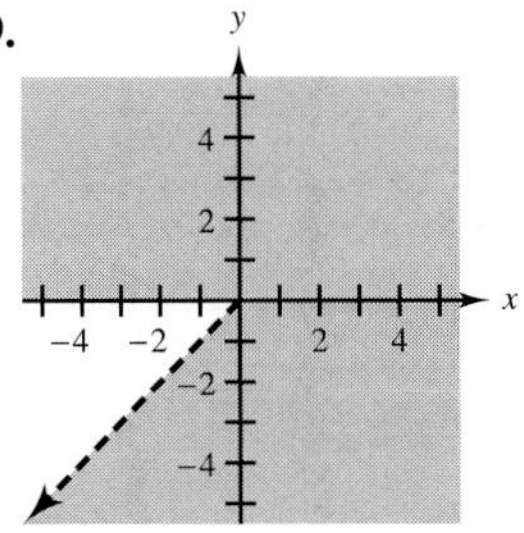

60.

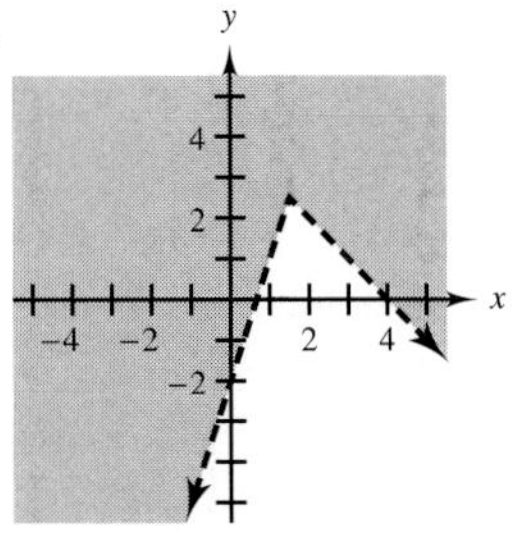

61.

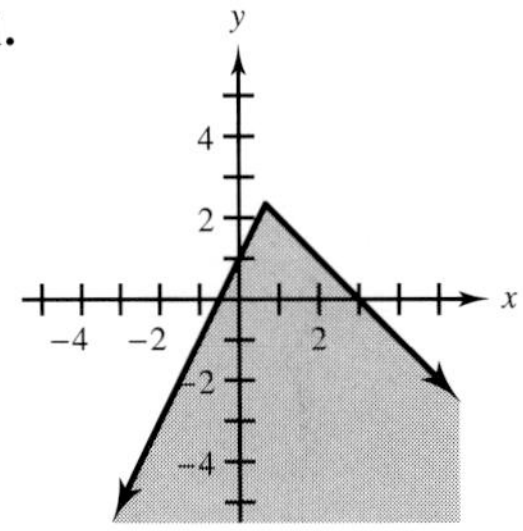

62.

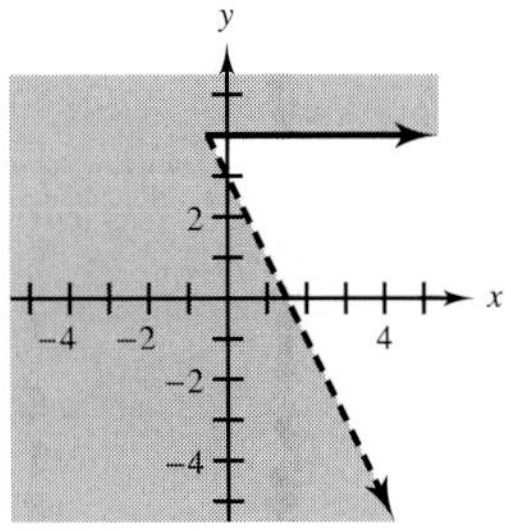

63.

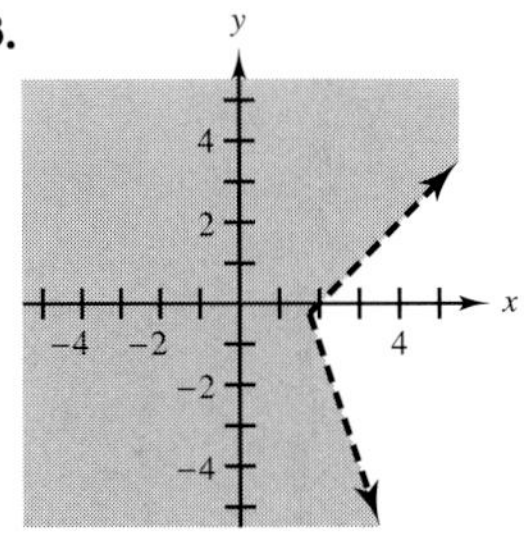

64.

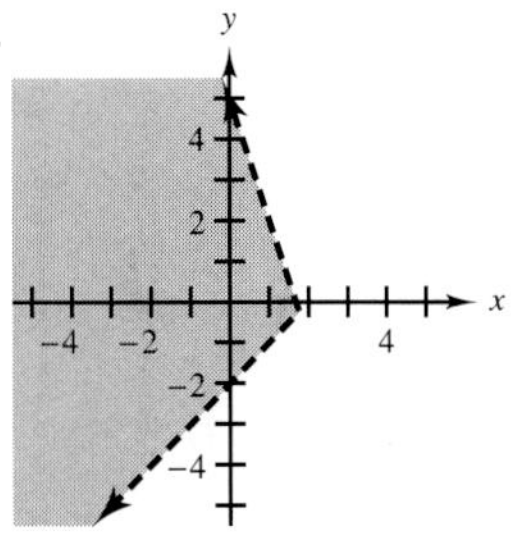

65.

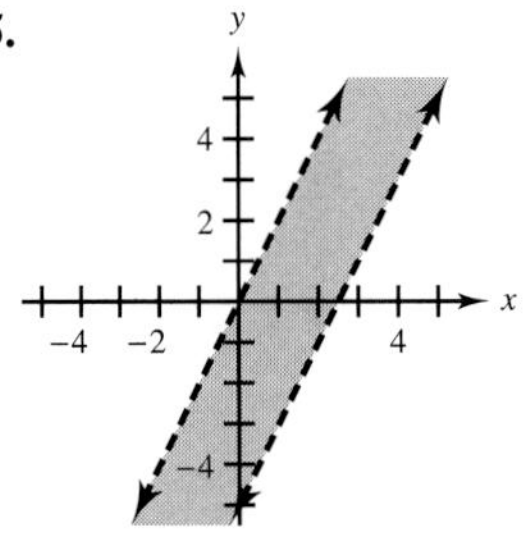

66.

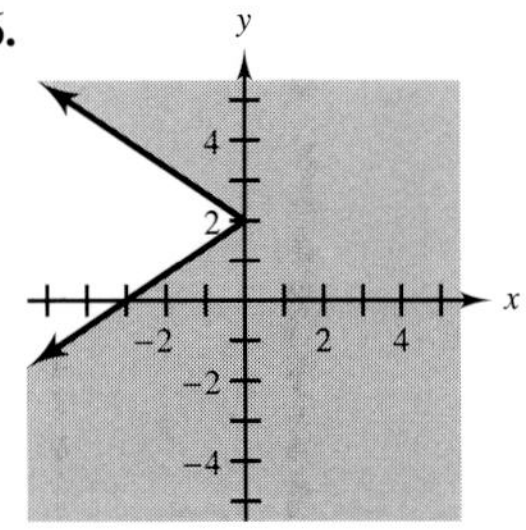

67.

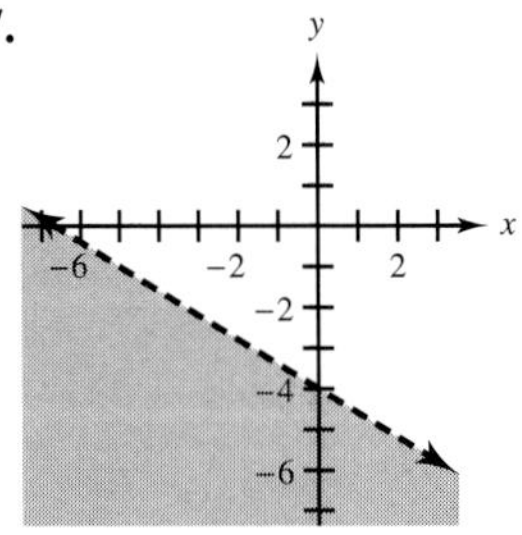

68.

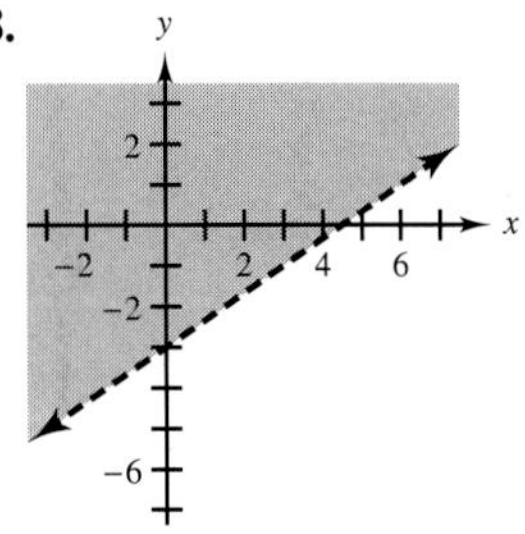

69.

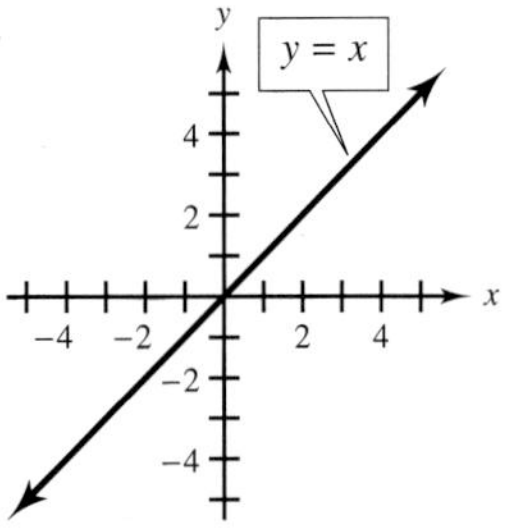

70.

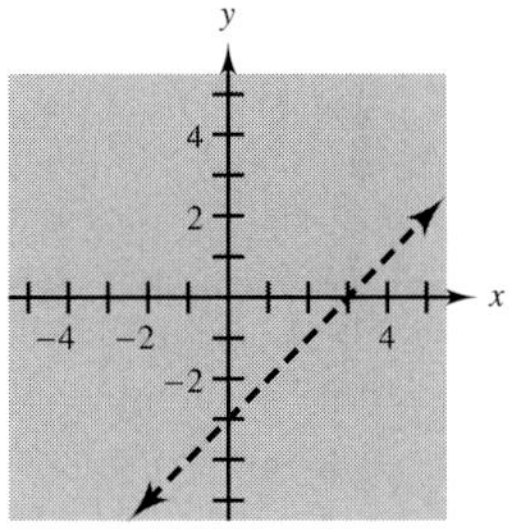

71.

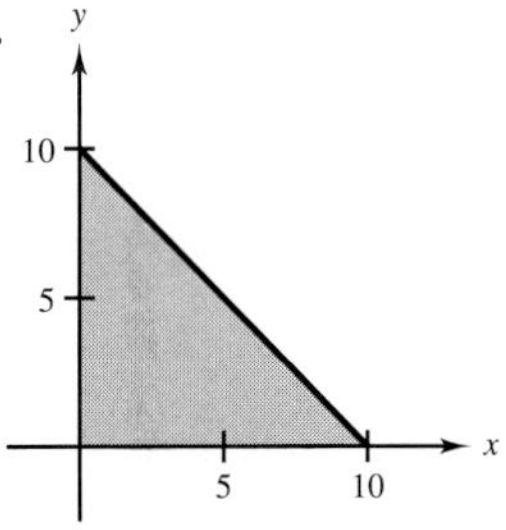

72.

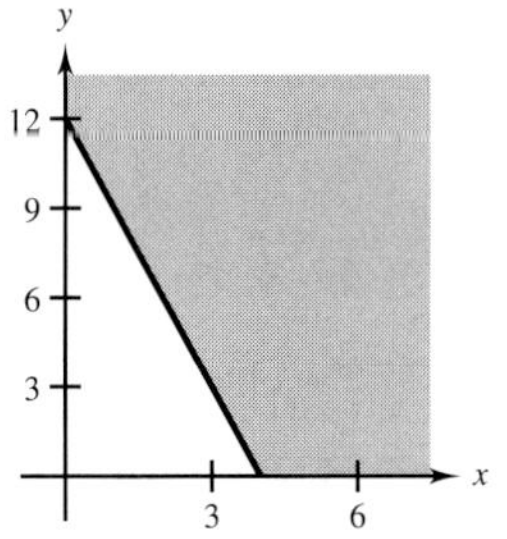

73. $y < 2$ and $y \geq -3$ **74.** $y \geq 4$ or $y < -1$

75. $x \leq -2$ or $x > 1$ **76.** $x > -6$ and $x \leq -2$

77. $y \geq x + 3$ or $y \leq x - 2$ **78.** $y < x + 5$ and $y > x + 1$

79.

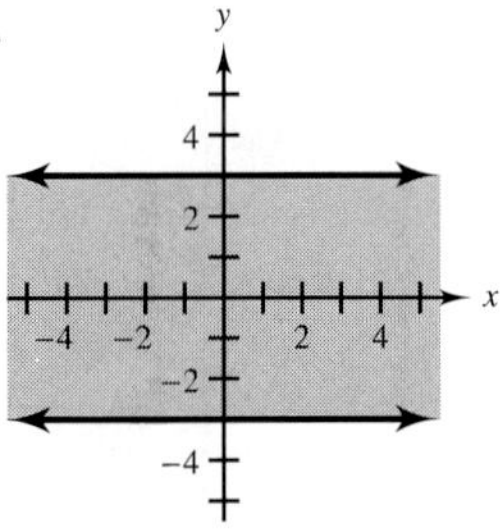

80.

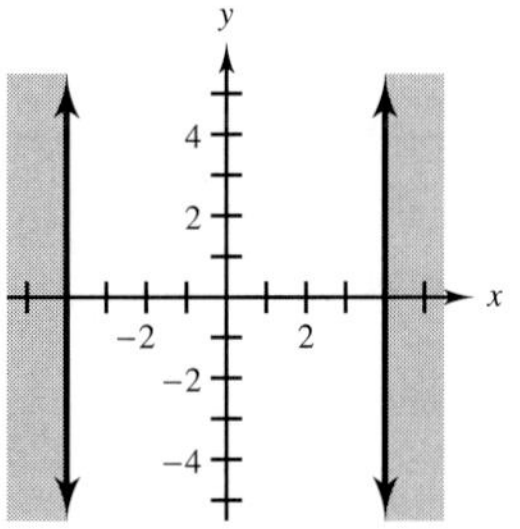

81.

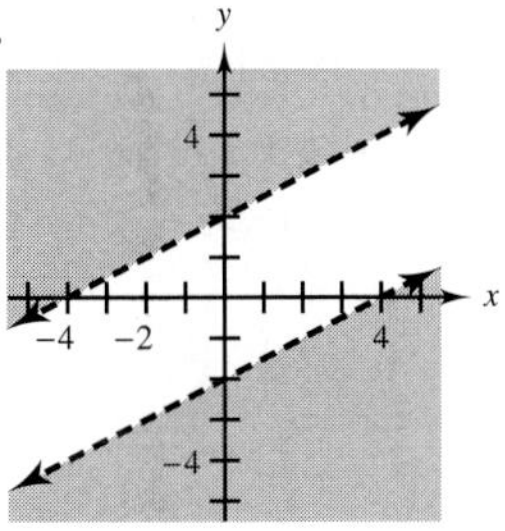

82.

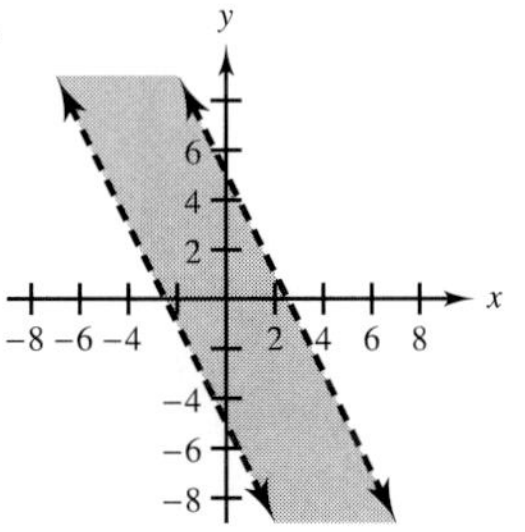

83.

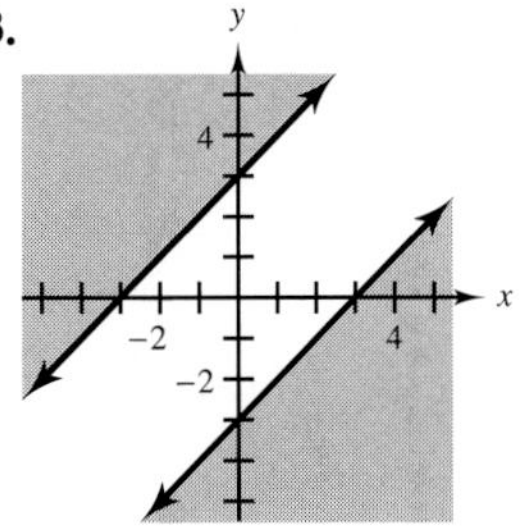

84.

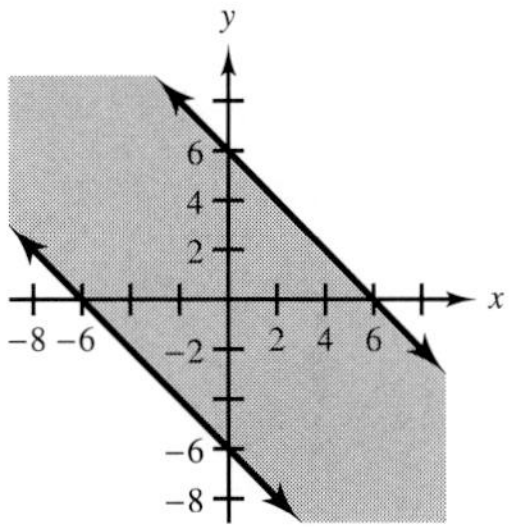

85. $\varnothing$ **86.** Entire plane

87.

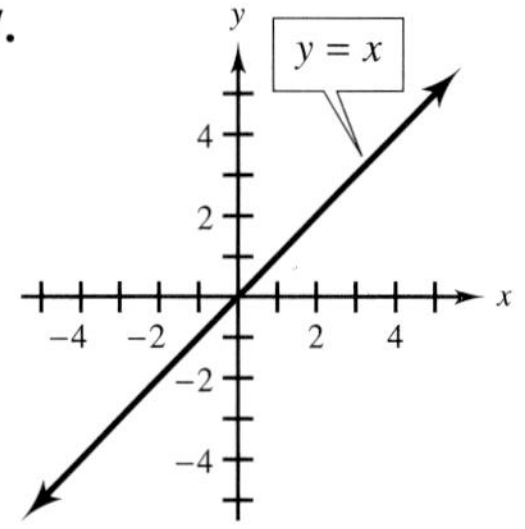

88.

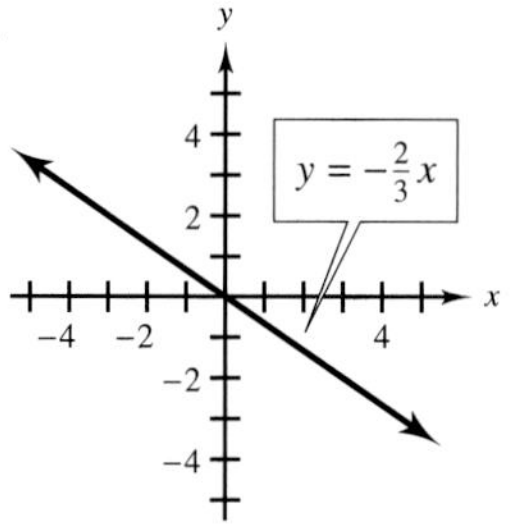

89.

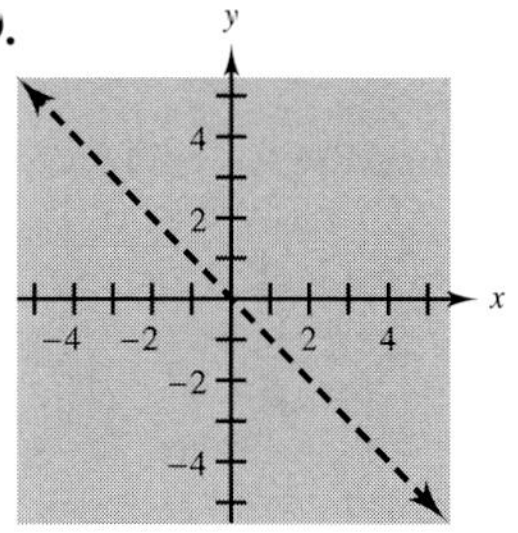

90.

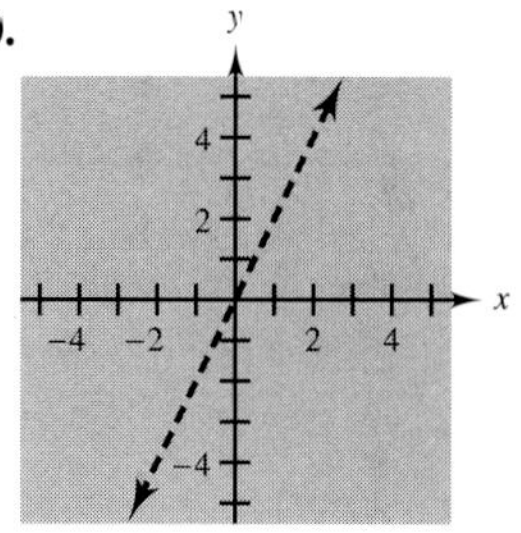

91. x = number of 4×4 pallets; y = number of 3×5 pallets
$16x + 15y \leq 6000$, $x \geq 0$, $y \geq 0$

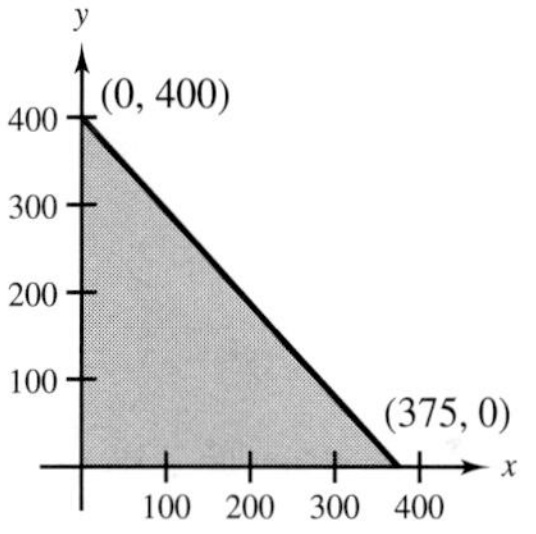

92. x = number of patients receiving 3 meals per day
y = number of patients receiving 4 meals per day
$3x + 4y \leq 1000$, $x \geq 0$, $y \geq 0$

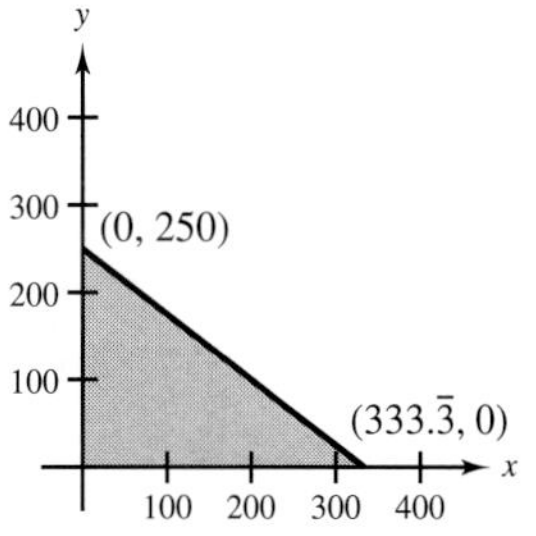

93. x = score of one team, y = score of the other team
$|x - y| \leq 5, x \geq 0, y \geq 0$

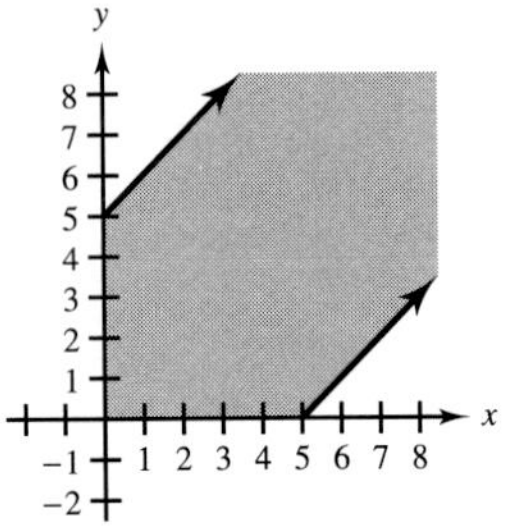

94. f = number of female members, m = number of male members
$|f - m| < 10, f \geq 0, m \geq 0$

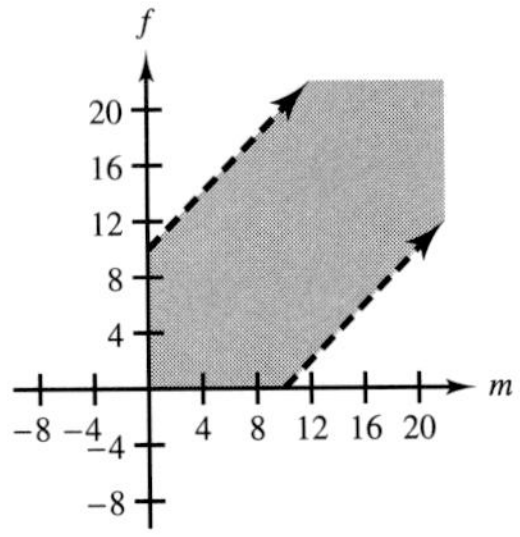

95. c = number of cats; d = number of dogs
$20 \leq c + d \leq 100, c \geq 0, d \geq 0$

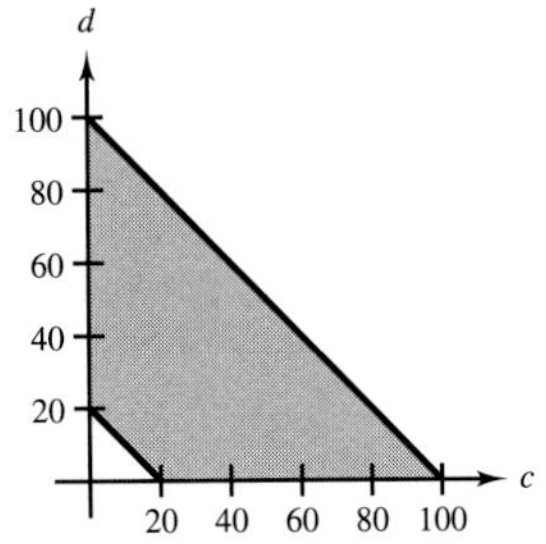

96. $m + f \geq 50$ and $-10 \leq m - f \leq 10$

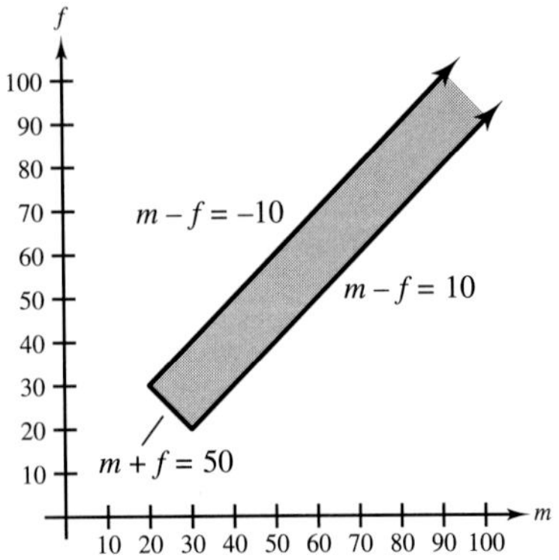

97. (a) $R + D \leq 101.84$

(b) The coordinates represent the number of votes for the two candidates.

98. (a) Use $<$ rather than $\leq$.

(b) The boundary line would be dashed rather than solid.

99. (a) $y \leq 6457.50 + 0.28(x - 43{,}050)$ **(b)** Below

100. The slope of the boundary line would increase. **101.** 3

102. Combinations of weights and numbers of drinks that would produce a blood alcohol level in excess of 0.05

103. Below **104.** Above

105.

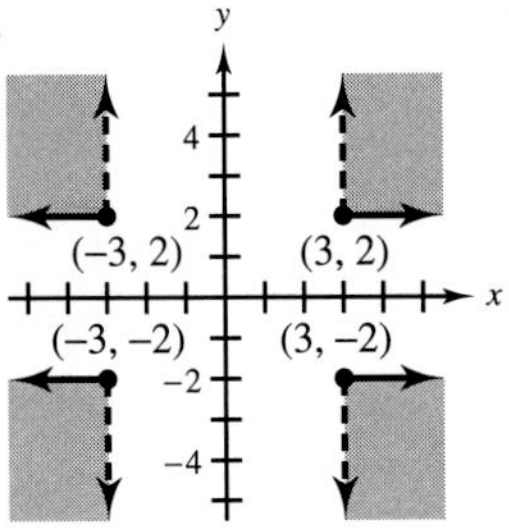

106.

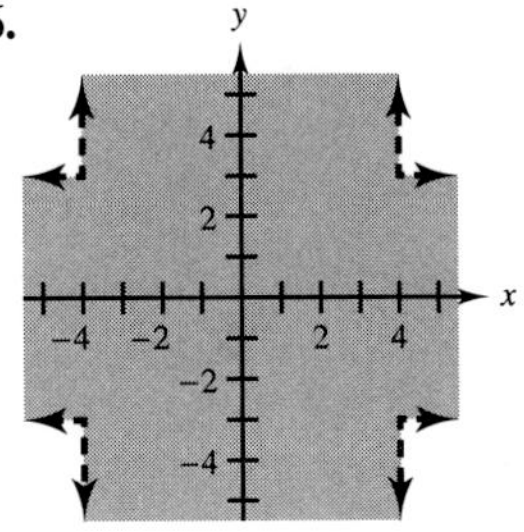

107. Entire plane **108.** Ø

109.

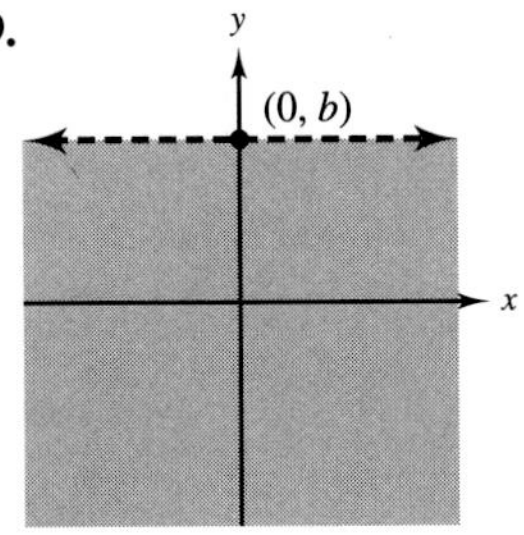

110.

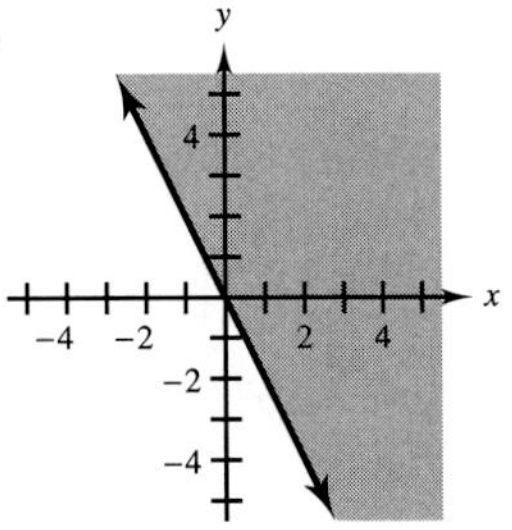

111.

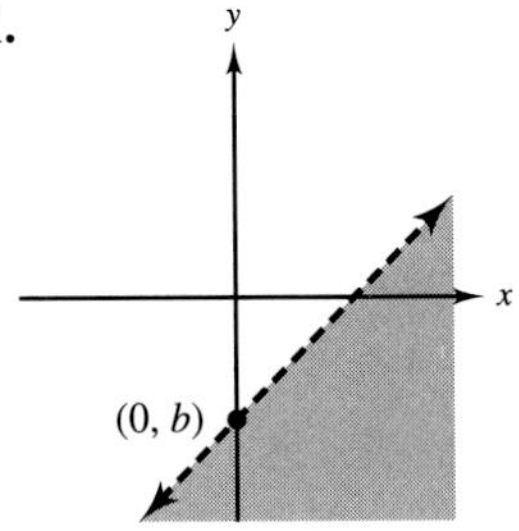

112.

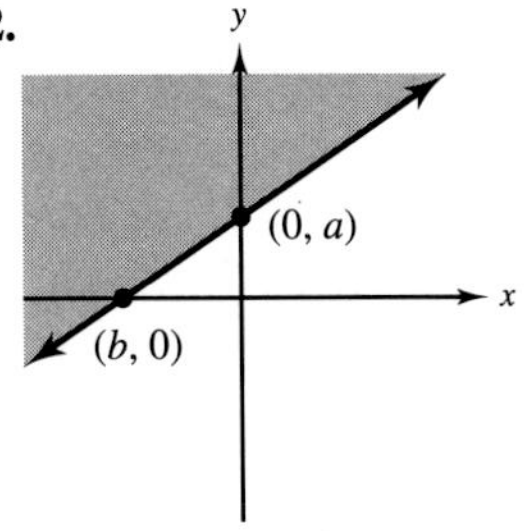

Chapter Review Exercises *(page 244)*

1. No **2.** Yes **3.** Yes **4.** No

7. $a = 9, b = -7, c = 12$ **8.** $a = 2.5, b = 3, c = -5$

9.

x	3	7	-7
y	9	15	-6

10.

x	2	11	−4
y	0	6	−4

11. (0, −7), (3, 0) **12.** (1.5, 0), (0, 3)
13. (0, 20); No x-intercept **14.** (6, 0); No y-intercept
15. $y = \frac{1}{2}x - \frac{5}{2}$; $\left(0, -\frac{5}{2}\right)$ **16.** $y = \frac{4}{5}x - \frac{19}{5}$; $\left(0, -\frac{19}{5}\right)$
17. $y = 2x - 8$; (0, −8)
18. $y = \frac{5}{4}x - 15$; (0, −15)

19.
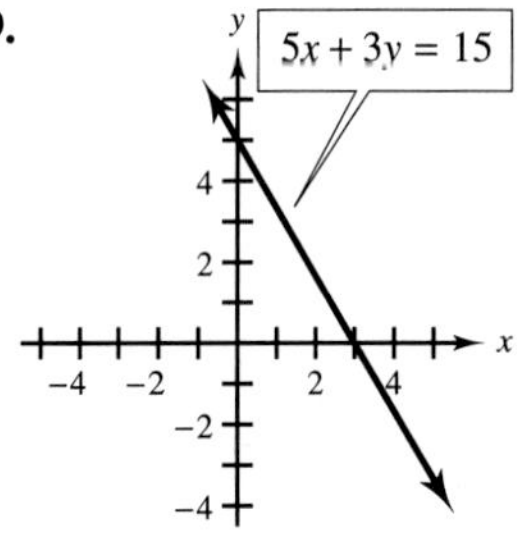

20.
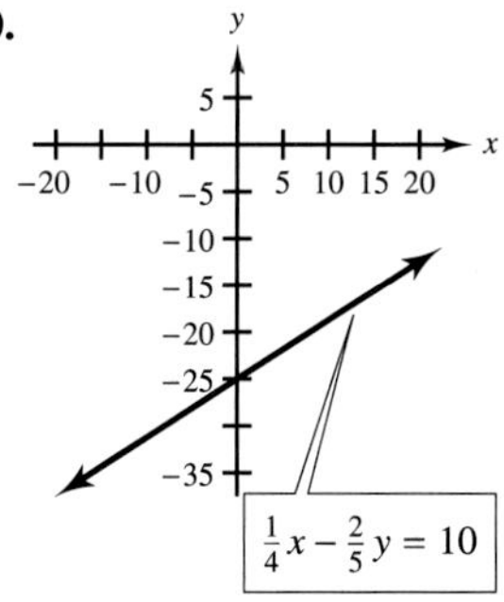

21.
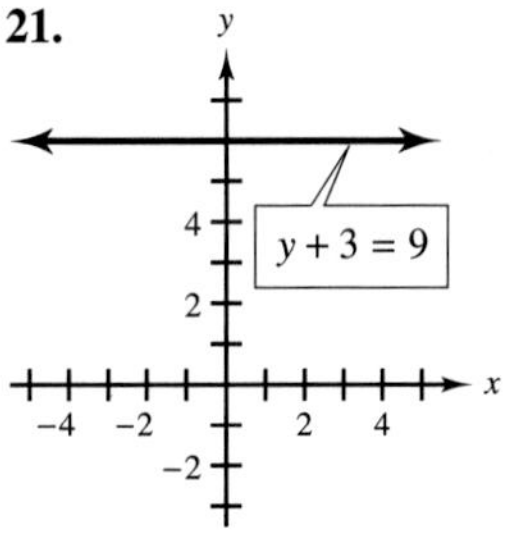

22.
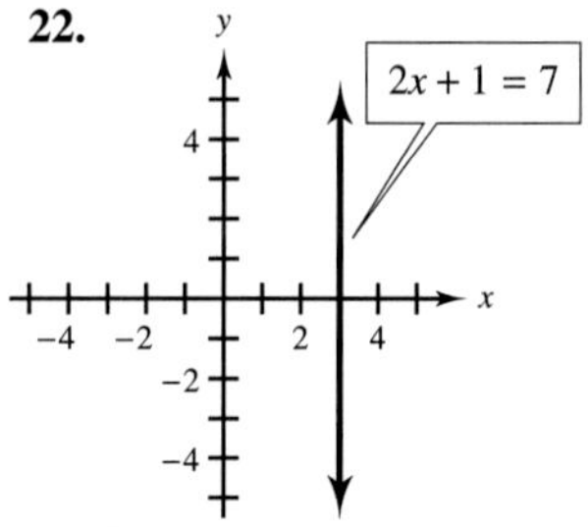

23. (ii) **24. (a)** L_4 **(b)** L_2 **(c)** L_1 **(d)** L_3
25. $\frac{5}{7}$ **26.** Undefined **27.** $-\frac{10}{9}$ **28.** 0
29. $y = \frac{3}{4}x - \frac{15}{4}$; $\frac{3}{4}$ **30.** $y = -\frac{1}{2}x + 2$; $-\frac{1}{2}$

31.
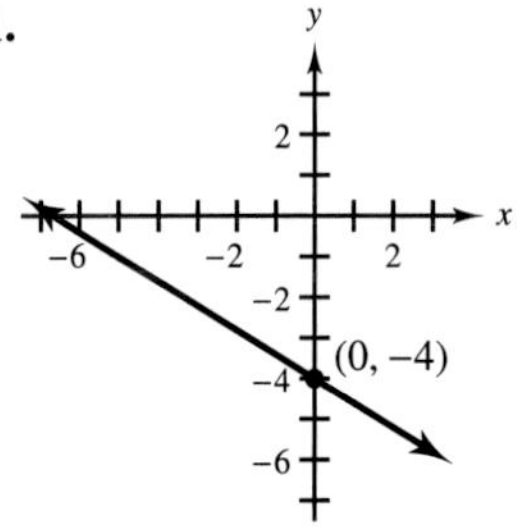

32.
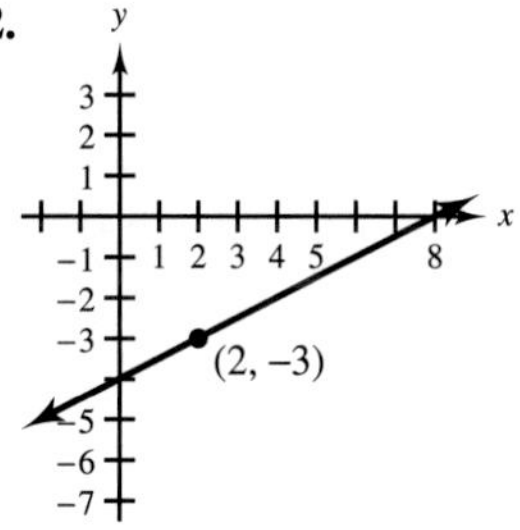

33.
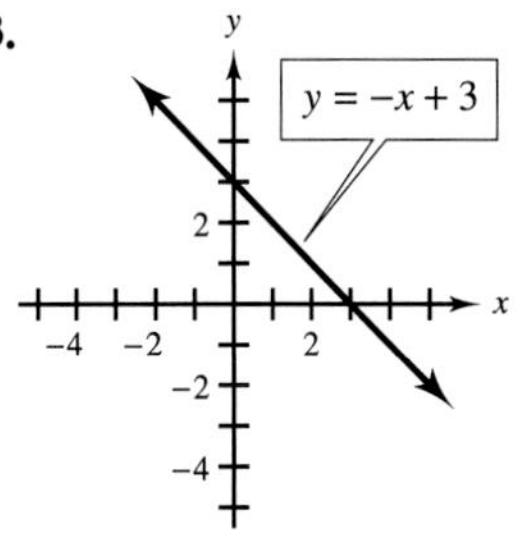

34.
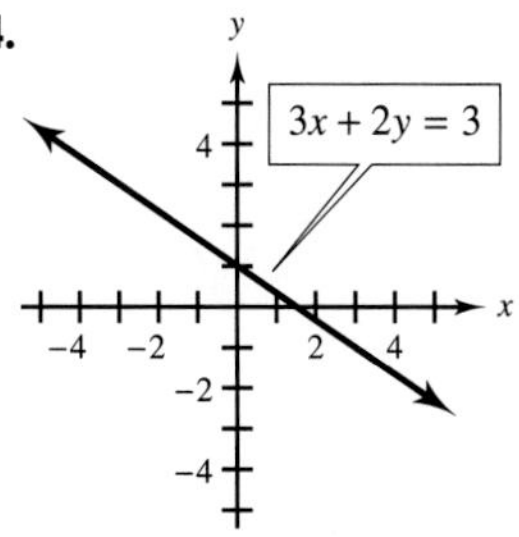

35.
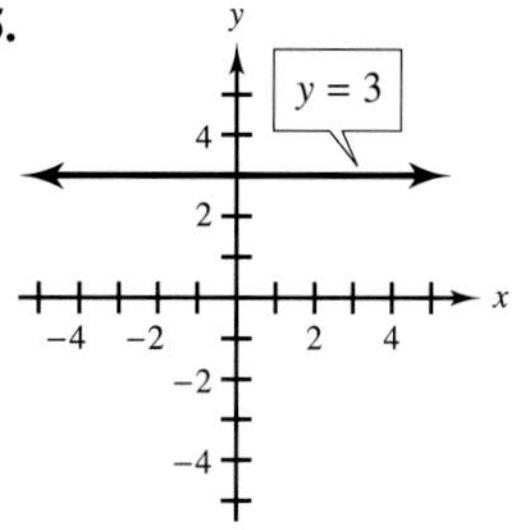

36.
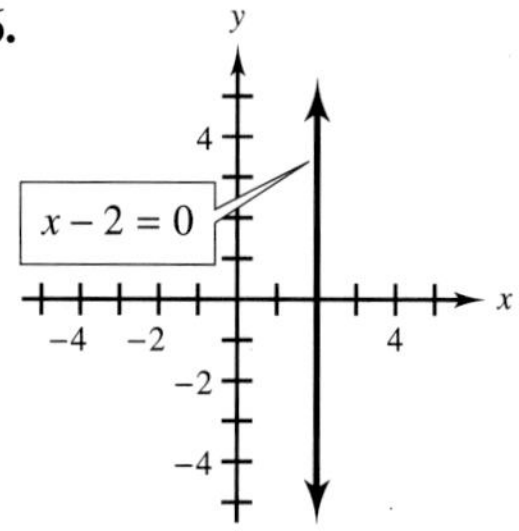

37. Parallel **38.** Neither **39.** Perpendicular
40. Perpendicular **41.** Perpendicular **42.** Neither
43. Yes **44.** Not collinear **45.** −3 **46.** $\frac{2}{3}$
47. y is decreased by 15. **48.** x is increased by 6.
49. \$1.25 **50.** −2°F **51.** No **52.** Yes **53.** Yes
54. No **55.** 112,500 **56.** \$5.33 **57.** $y = -3x - 1$
58. $y = \frac{4}{3}x + \frac{2}{3}$ **59.** $y = -x - 2$ **60.** $x = 2$
61. $y = \frac{5}{4}x - \frac{9}{4}$ **62.** $y = -5$ **63.** $y = 6$ **64.** $x = 6$
65. $y = -\frac{2}{3}x + \frac{17}{3}$ **66.** $y = -\frac{3}{2}x$
67. $y = -0.4x + 2$ **68.** $y = -\frac{1}{3}x - 5$
69. 2003 **70.** No **71.** Yes, no, yes **72.** No, yes, no
73. No, yes, no **74.** No, no, yes

75.
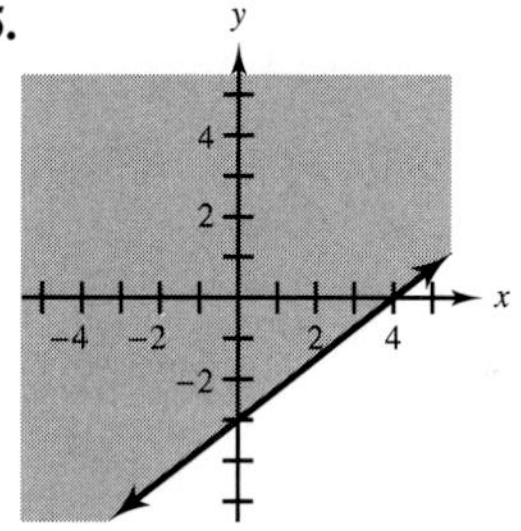

76.
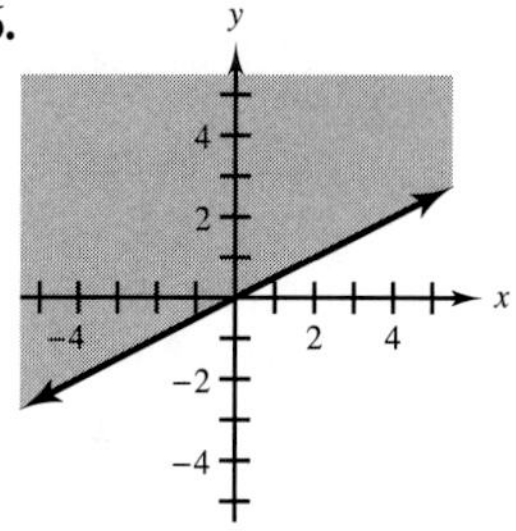

77.
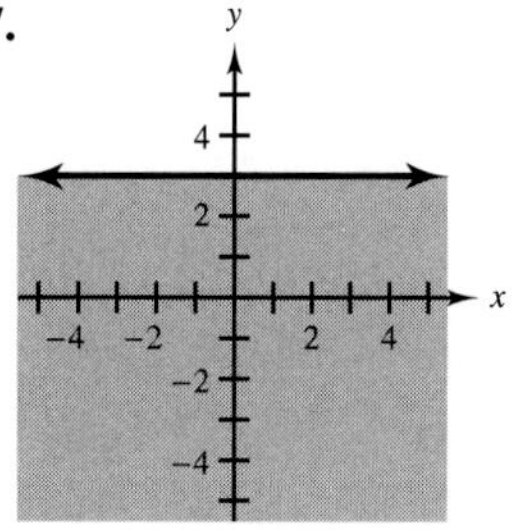

78.
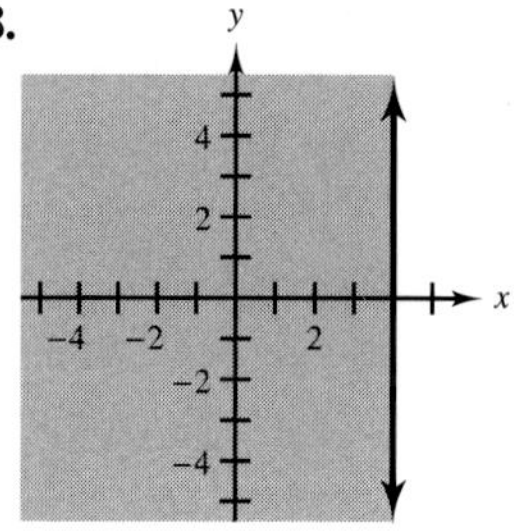

79.
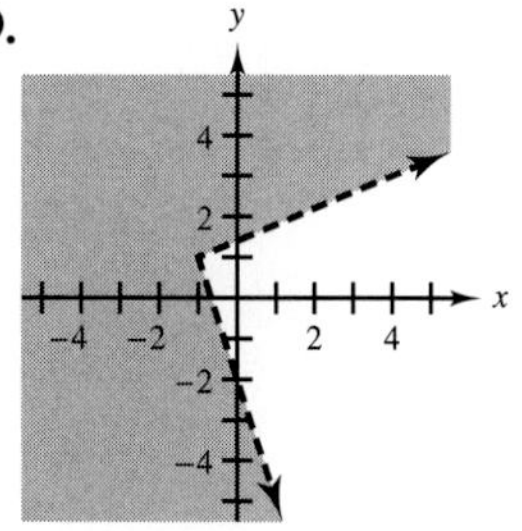

80.
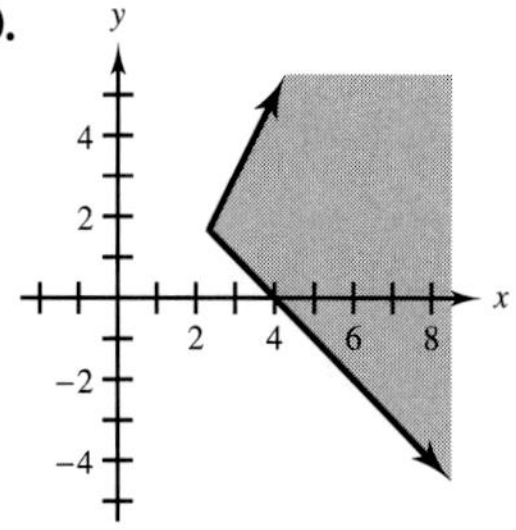

81.

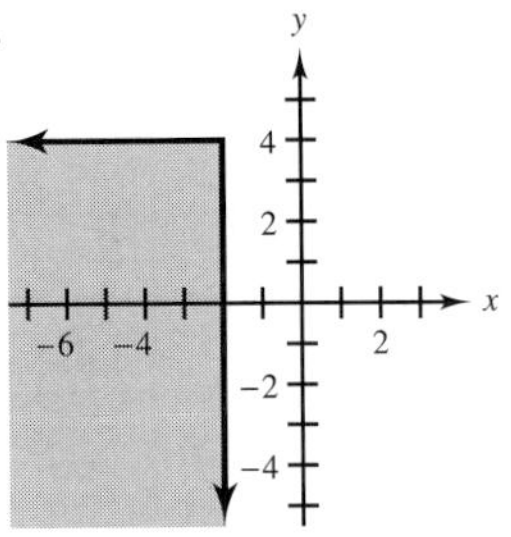

82.

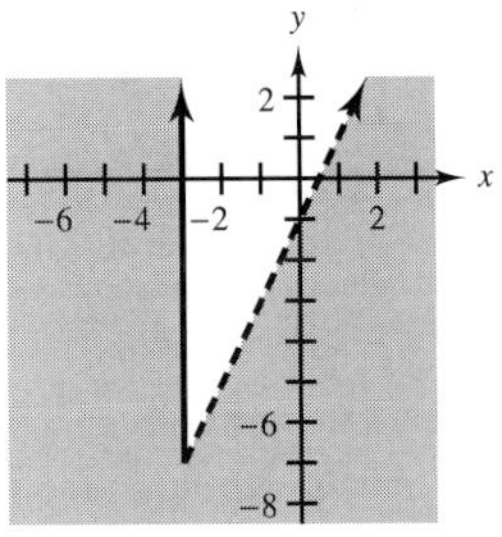

83.

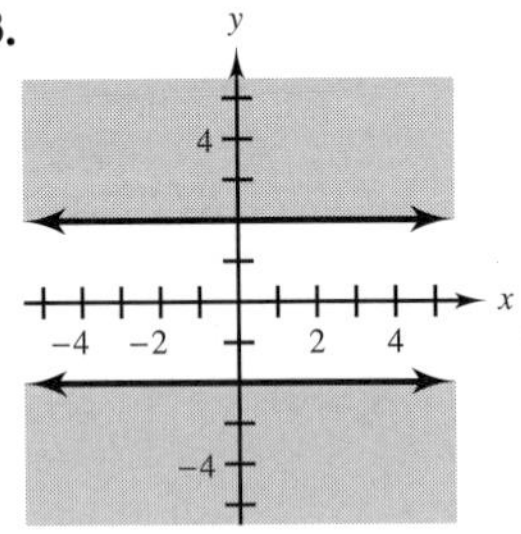

84.

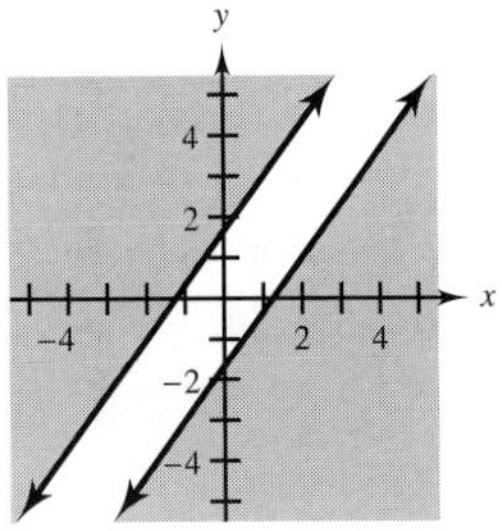

85.

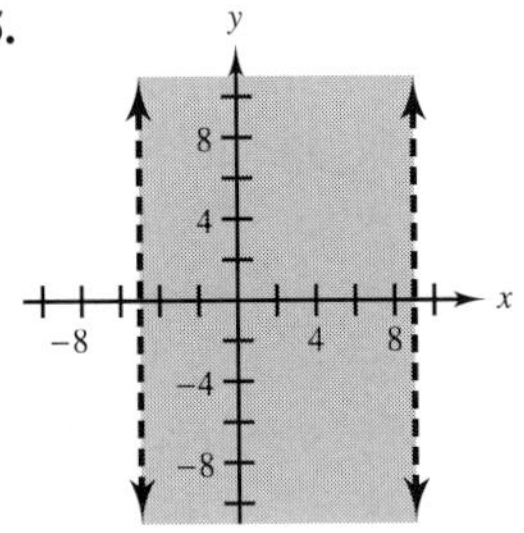

86. ∅

87. x = number of grandstand tickets sold
y = number of bleacher tickets sold
$6x + 4y < 11{,}400$, $x \geq 0$, $y \geq 0$

88. x = length of each of the two equal sides
y = length of the third side
$2x + y \leq 42$, $x > 0$, $y > 0$

89. x = number of girls, y = number of boys
$x + y \leq 400$ and $x \geq 100$, $y \geq 0$

90. x = actual percentage of those in favor
y = percentage of sample polled who were in favor
$|x - y| \leq 3$, $x \geq 0$, $y \geq 0$

Chapter Test *(page 247)*

1. [*4.1*] **(a)** $(-1, 0)$ **(b)** $(0, 2)$ **(c)** $(-2, -2)$ **(d)** $(0.5, 3)$

2. [*4.2*] $(0, 0)$; $m = 1$

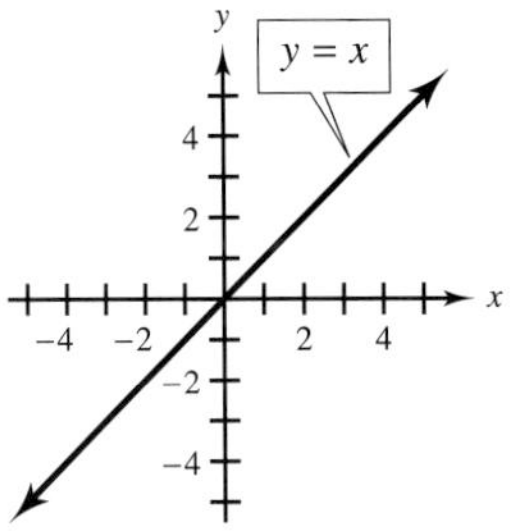

3. [*4.2*] $(0, -2)$, $(-2, 0)$; $m = -1$

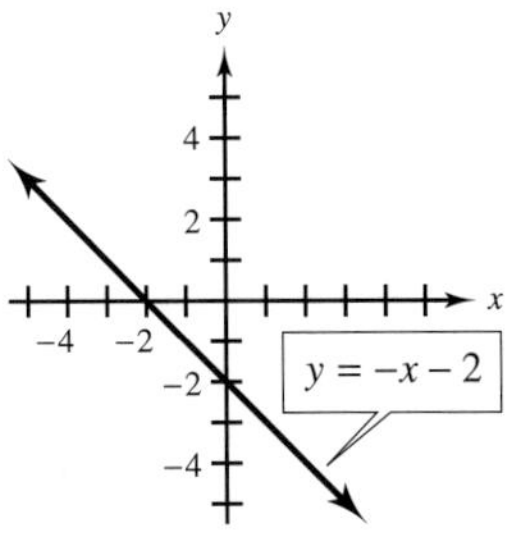

4. [*4.2*] $(0, -3)$, $(4.5, 0)$; $m = \frac{2}{3}$

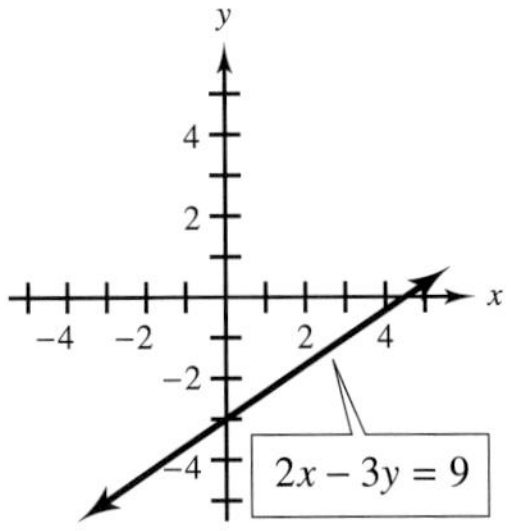

5. [*4.2*] $(5, 0)$; slope undefined

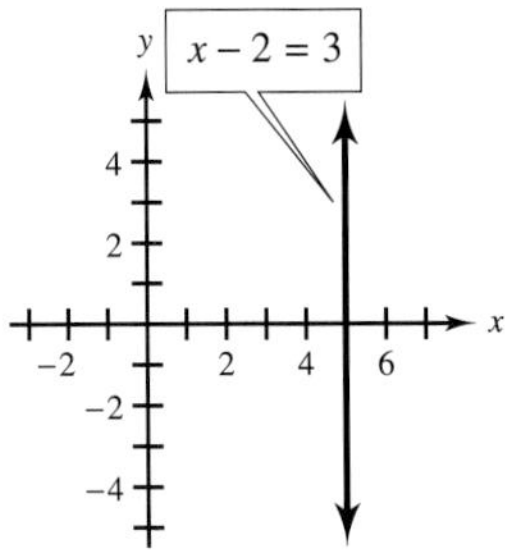

6. [*4.2*] $m < 0, b < 0$ **7.** [*4.4*] $y = \frac{2}{3}x - \frac{13}{3}$
8. [*4.4*] $y = 2x + 5$ **9.** [*4.4*] $x = 2$ **10.** [*4.4*] $y = 3$
11. [*4.4*] $y = 2x + 2$ **12.** [*4.4*] $y = 0.75x + 3$

13. [*4.5*]

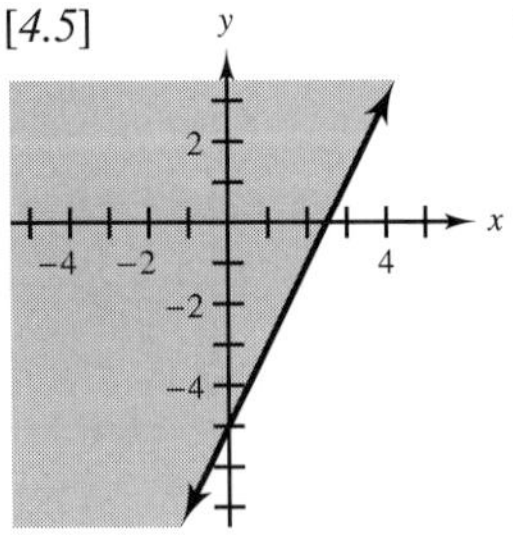

14. [*4.5*]

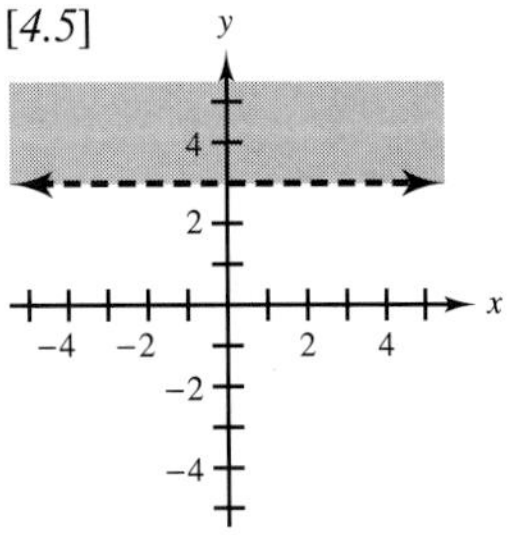

15. [*4.5*]

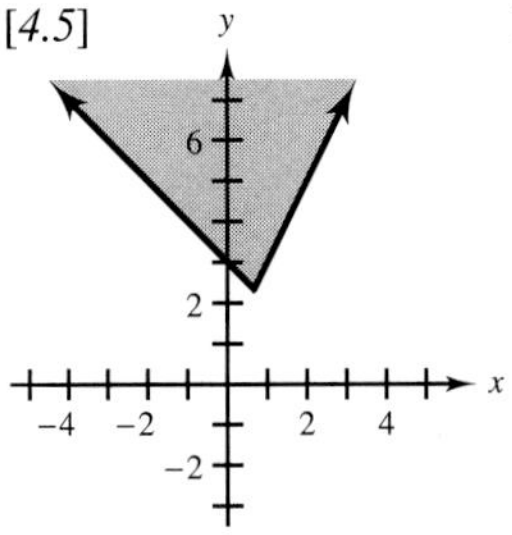

16. [*4.5*]

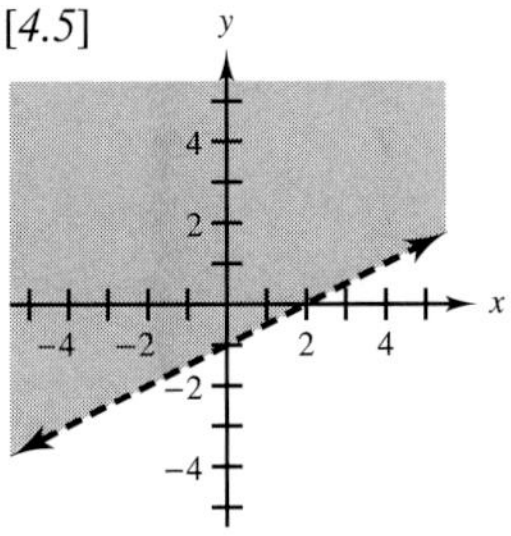

17. [*4.5*]

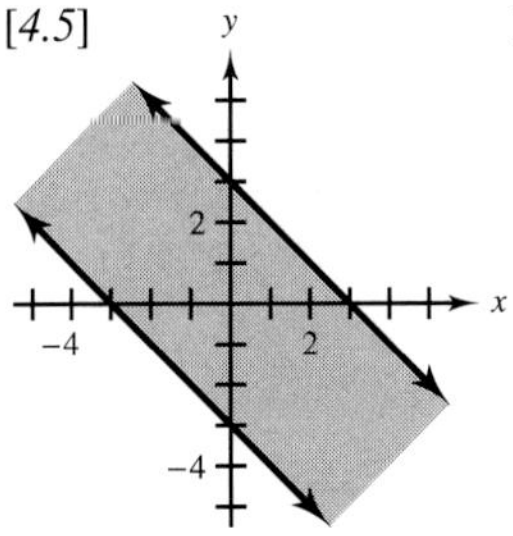

18. [*4.5*]

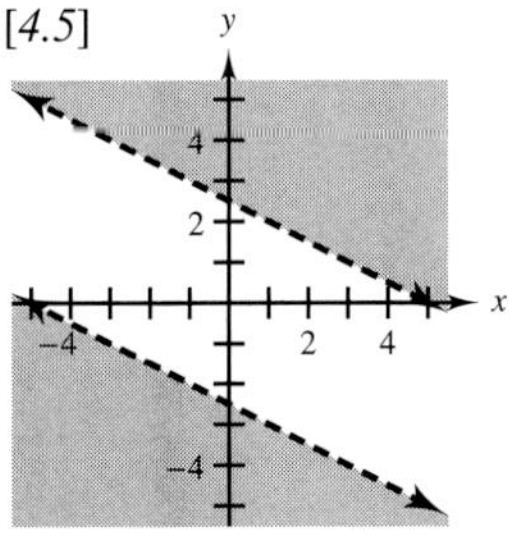

19. Answers will vary. **20.** [*4.2*]

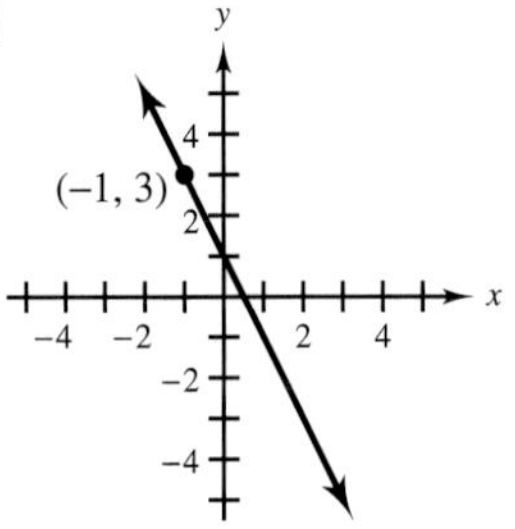

21. [*4.3*] Perpendicular **22.** [*4.3*] Neither

23. [*4.3*] $v = 32t$; 64 feet per second **24.** [*4.3*] −\$0.45

25. [*4.3*] t = number of hours since 6:00 A.M.

$T = 2.5t + 88$

At 3:00 P.M., $t = 9$ and $T = 110.5$°F.

Chapter 5

Section 5.1 *(page 256)*

1. Substitute the numbers into each equation. If both equations are satisfied, then the ordered pair is a solution.

2. To be a solution, the ordered pair must satisfy both equations. Because (1, 2) does not satisfy the first equation, it is not a solution of the system.

3. No, yes, no **4.** No, no, yes **5.** No, yes, yes

6. Yes, no, yes **7.** $a = 1, b = 2$ **8.** $a = 3, b = 3$

9. $a = 4, c = 7$ **10.** $b = -3, c = 22$

11. Two lines cannot intersect at exactly two points.

12. **(a)** The graphs are parallel lines.

(b) The graphs intersect at a point.

(c) The graphs coincide.

13. None **14.** None **15.** 1 **16.** 1

17. Infinitely many **18.** Infinitely many

19. **(a)** (i) The graphs do not coincide.

(ii) The graphs coincide.

(b) Yes, the lines could be parallel.

20. **(a)** (i) The graphs are not parallel.

(ii) The graphs are parallel.

(b) Yes, the graphs could coincide.

21. (0, 6) **22.** (3, 3) **23.** (0, 0) **24.** (−2, −3)

25. (3, 4) **26.** (5, 7) **27.** No solution **28.** No solution

29. (5, −2) **30.** (2, 4) **31.** (−3, 4) **32.** (2, 1)

33. (1, 3) **34.** (−4, 3) **35.** (−4, 7) **36.** (3, −5)

37. Infinitely many solutions **38.** Infinitely many solutions

39. (4, −5) **40.** (−6, −2)

41. **(a)** Yes, either equation can be solved for either variable.

(b) Solving the second equation for y is easier.

42. **(a)** The equations are dependent, and the system has infinitely many solutions.

(b) The system is inconsistent and has no solution.

43. (−22, −27) **44.** (24, −21) **45.** Dependent

46. Dependent **47.** Inconsistent **48.** Inconsistent

49. Dependent **50.** Dependent **51.** (2, −3) **52.** (3, 4)

53. Inconsistent **54.** Inconsistent **55.** (0, 5) **56.** (0, 9)

57. (80, 80) **58.** (−720, −252) **59.** −2 **60.** 3

61. 5 **62.** 3 **63.** −2 **64.** −2 **65.** −4 **66.** $-\frac{1}{3}$

67. 10, 6 **68.** 18, 24 **69.** 20, 12 **70.** 18, 12

71. 67°, 23° **72.** 20°, 70° **73.** 112°, 68° **74.** 54°, 126°

75. 27 feet, 10 feet **76.** 13 meters, 9 meters

77. 10 inches, 30 inches **78.** 5 meters, 11 meters

79. 3500 **80.** 14 **81.** 6 **82.** 48

83. **(a)** Because the lines have different slopes, the graphs have one point of intersection.

(b) (52.125, 8.57875)

(c) In 2002 ($x \approx 52$), the percentages of men and of women over age 65 in the work force will be equal (about 8.6%).

84. **(a)** The solution is approximately (−14, 827), which means that 14 years before 1995 (1981), the tuitions were equal (about \$827). However, this result is too far removed from the given data to be considered reliable.

(b) The graphs are different straight lines that intersect at about (−14, 827). Therefore, the lines cannot intersect again at some other point.

85. $W(x) = 0.93x + 11.13$

86. **(a)** The function represents the percentage of Ph.D.s awarded to men.

(b) $M(x) = -0.93x + 88.87$

(c) The functions are the same.

87. At the point of intersection, the percentages of Ph.D.s awarded to men and to women are the same. Therefore, the y-coordinate would be 50 (50%).

88. About 2007 **89.** $(8, -4)$ **90.** $(-5, 8)$

91. No solution **92.** No solution

93. $(-2, 5), (4, -1), (-4.5, 2.5), (1.5, -3.5)$
The figure is a rectangle.

94. $(-3, 6), (3, 6), (-3, -1)$
The triangle is not an isosceles triangle.

95. No, either infinitely many solutions or no solution.

96. **(a)** If $a = 0$, there is no solution.

(b) If $a \neq 0$, the lines have different slopes and cannot coincide.

Section 5.2 *(page 266)*

1. **(a)** Multiply the second equation by -3 and add the equations.

(b) Multiply the first equation by 4 and add the equations.

2. Eliminating y requires only one step. Multiply the first equation by -4 and add the equations.

3. $(6, 4)$ **4.** $(5, 3)$ **5.** $(-10, 7)$ **6.** $\left(2, -\frac{8}{3}\right)$

7. $(-2, 5)$ **8.** $(-3, -4)$ **9.** $\left(2, \frac{2}{3}\right)$ **10.** $(2, -5)$

11. No, the result means that there are infinitely many solutions.

12. No, a solution of a system is an ordered pair. Replace x with 4 in either equation and solve for y.

13. $(10, -7)$ **14.** $(2, 5)$ **15.** $(5, -3)$ **16.** $(-1, 4)$

17. $\left(\frac{1}{2}, \frac{2}{3}\right)$ **18.** $\left(-\frac{4}{3}, \frac{3}{4}\right)$ **19.** $(5, 4)$ **20.** $(-1, 2)$

21. The lines intersect at $(3, 2)$. Many other pairs of lines intersect at this point.

22. Neither variable can be eliminated by adding the equations. The solution is $(-3, 7)$.

23. $(1, 1)$ **24.** $(7, 11)$ **25.** $(0, 2)$ **26.** $\left(-\frac{22}{29}, \frac{32}{29}\right)$

27. $(-2, 1)$ **28.** $(-1, -2)$ **29.** $(0, 1)$ **30.** $(2, 0)$

31. Both variables are eliminated and the resulting equation is true.

32. Both variables are eliminated and the resulting equation is false.

33. $(-3, -4)$ **34.** $\left(\frac{21}{11}, \frac{8}{11}\right)$ **35.** No solution

36. No solution **37.** Infinitely many solutions

38. Infinitely many solutions **39.** No solution

40. No solution **41.** $(4, -2)$ **42.** $(-1, -2)$

43. Infinitely many solutions **44.** Infinitely many solutions

45. $\left(\frac{41}{7}, \frac{23}{7}\right)$ **46.** $(5, 0)$ **47.** No solution

48. No solution **49.** Infinitely many solutions

50. Infinitely many solutions **51.** $(3, -3)$ **52.** $(2, 3)$

53. $(1, 2)$ **54.** $(2, -2)$ **55.** Infinitely many solutions

56. $\left(\frac{31}{3}, \frac{1}{3}\right)$ **57.** $a = 3, b = 2$ **58.** $a = 3, c = -2$

59. $a = 1, b = 3$ **60.** $a = 1, c = 0$ **61.** -4 **62.** -9

63. -6 **64.** -8 **65.** $c = 4$ **66.** 4 **67.** $a = 3, b = -3$

68. $a = -4, c = 8$ **69.** $\left(\frac{b+c}{2}, \frac{b-c}{2}\right)$

70. $(3 - c, 2c - 3)$ **71.** $\left(\frac{c+3}{a+1}, \frac{3-ac}{a+1}\right), a \neq -1$

72. $(c, 0), b \neq 0$ **73.** $\left(\frac{1}{a}, 1\right), a \neq 0$

74. $\left(\frac{c}{a+b}, \frac{c}{a+b}\right), a \neq \pm b$ **75.** $(1, 0), a \neq -b$

76. $\left(\frac{6+c}{3}, \frac{c}{3}\right)$ **77.** Hamburger: \$1.90; Milk shake: \$1.20

78. Root beer: \$0.75; Fries: \$0.50 **79.** 27 **80.** 21

81. 43 nickels, 21 dimes **82.** 42 nickels, 19 quarters

83. 6 **84.** 78 T-shirts, 43 sweatshirts **85.** 15%

86. Corn: \$2.50; Wheat: \$3.40

87. **(a)** About (23.5, 107)

(b) In 2018, adults will spend 107 hours per year reading and 107 hours per year on the Internet.

88. **(a)** The rates of increase are the same for male and female M.D.s.

(b) The system has no solution. If the model functions remain valid, the number of male and female M.D.s will never be equal.

89. When we use $y_1 \approx 6x - 1$ and $y_2 \approx -3x + 69$, an estimated solution is (8, 46).

90. (7.9, 44.2)

91. There are other forms of communication, such as telephone, voice mail, and personal conversations.

92. In about 1998, both percentages were about 46%.

93. $(2, 4)$ **94.** $(3, -2)$ **95.** $(-5, 2.5)$ **96.** $\left(-\frac{7}{3}, -\frac{7}{4}\right)$

97. The system has no solution because $\frac{1}{x} = 0$ has no solution.

98. $a \neq 1$ **99.** $b \neq -4$ **100.** $c = 5$ **101.** $c = 5$

102. $b \neq 14$ **103.** $a \neq 1$ **104.** $c \neq -1$ **105.** $c \neq 10$

Section 5.3 *(page 278)*

1. No, not all coefficients can be zero.

2. A solution of a system must satisfy each equation of the system. Only (1, 0, 0) satisfies all three equations.

3. Yes, no **4.** No, yes **5.** No, yes **6.** Yes, no

7. $a = 3, b = 0, c = -2$ **8.** $a = b = c = 0$

9. $b = 3, c = 0, d = 1$ **10.** $a = 1, c = -3, d = -4$

11. The graph of a linear equation in two variables is a line in a plane. The graph of a linear equation in three variables is a plane in three-dimensional space.

12. The planes intersect at exactly one point.

13. **(a)** At least two planes are parallel or a plane is parallel to the line of intersection of the other two planes.

(b) The three planes coincide or intersect in a common line.

14. No, you must eliminate the same variable in each pair.

15. $(-1, 3, -2)$ **16.** $(2, -3, 1)$ **17.** $(7, 5, 4)$

18. $(-2, -1, 1)$ **19.** $(3, -2, 1)$ **20.** $(5, -4, -1)$

21. $(4, -3, -1)$ **22.** $(1, 2, 3)$ **23.** $(1, 3, -2)$

24. $\left(-\frac{33}{4}, -8, -\frac{1}{4}\right)$ **25.** $\left(3, \frac{1}{2}, -\frac{1}{3}\right)$ **26.** $(2, -1, 1)$

27. $\left(1, -\frac{1}{3}, \frac{1}{2}\right)$ **28.** $(-2, 0, -1)$ **29.** $\left(\frac{1}{2}, \frac{1}{3}, -\frac{2}{3}\right)$

30. $\left(\frac{3}{2}, -\frac{3}{4}, \frac{1}{3}\right)$

31. **(a)** After elimination of one variable, the resulting system of two equations in two variables is inconsistent.

(b) After elimination of one variable, the two equations in two variables in the resulting system are dependent.

32. Replace z with 3 in the second equation to obtain $y = -1$. Replace y with -1 and z with 3 in the first equation to obtain $x = 12$. The solution is $(12, -1, 3)$.

33. Ø **34.** Ø

35. The set of all solutions to $x + 2y - z = 3$

36. The set of all solutions to $2x - y + 4z = 6$

37. $(3, -2, 1)$ **38.** $(-2, 4, -5)$ **39.** $(1, 2, 3)$

40. $(3, 1, 2)$ **41.** $(5, -4, -1)$ **42.** $(1, 2, -3)$

43. $(5, -4, -3)$ **44.** $(-2, 3, -1)$ **45.** $(-7, -4, 3)$

46. $(-1, 2, -2)$ **47.** Ø **48.** Ø

49. Set of all solutions to $4x - 3y + 6z = 12$

50. Set of all solutions to $x - 0.4y + 3z = 4$

51. $(-3, -6, 4)$ **52.** $\left(1, \frac{1}{8}, -\frac{1}{4}\right)$ **53.** $(-3, 2, 5)$

54. $(3, 3, 2)$ **55.** $(-1, -3, 2)$ **56.** $(3, 2, -2)$

57. 23°, 64°, 93° **58.** 35°, 78°, 67°

59. 27 yards, 52 yards, 21 yards

60. 3.5 feet, 5 feet, 5.5 feet

61. 27 nickels, 56 dimes, 72 quarters

62. 3 pennies, 31 nickels, 42 dimes, 14 quarters

63. **(a)**
$$\begin{aligned} R + D &= 0.08(1200) \\ R + D + I &= 1200 \\ R &= D + 4 \end{aligned}$$

(b) Republican: 50; Democrat: 46; Independent: 1104

64. T-shirt: \$9; Sweatshirt: \$15; Tank top: \$4

65. **(a)**
$$\begin{aligned} b + w + p &= 100 \\ b &= 2w \\ w &= 3p \end{aligned}$$

(b) Brass: 60; Woodwind: 30; Percussion: 10

66. Growth fund: 6%; Income fund: 5%; Money market fund: 3%

67. (30, 11.6), (70, 4.8), (100, 9.5)

68.
$$\begin{aligned} 11.6 &= 900a + 30b + c \\ 4.8 &= 4900a + 70b + c \\ 9.5 &= 10{,}000a + 100b + c \end{aligned}$$

69. $p(t) = 0.005t^2 - 0.67t + 27.2$ **70.** 14%

71. $(1, 2, -1, 3)$ **72.** $(-1, 1, 2, -2)$ **73.** $(1, -1, 0.5)$

74. $(2, 3, 1)$

75. $\left(\dfrac{a - b + c}{2}, a - \dfrac{a - b + c}{2}, c - \dfrac{a - b + c}{2}\right)$

76. $a \neq -1$

Section 5.4 *(page 287)*

1. A 2×4 matrix is an array of numbers with two rows and four columns.

2. The elements of the second column are 5 and 9. Columns are read vertically.

3. 2×4 **4.** 3×2 **5.** 2×2 **6.** 3×3

7. 3×4 **8.** 4×1

9. **(a)** A coefficient matrix is a matrix whose elements are the coefficients of the variables of the equations in a system of equations.

(b) A constant matrix is a matrix whose elements are the constants of the equations in a system of equations.

(c) An augmented matrix is a matrix whose elements are the coefficients and the constants of the equations in a system of equations.

10. The number of equations determines the number of rows, and the number of variables determines the number of columns. Thus a system of n linear equations in n variables gives an $n \times n$ matrix, which is a square matrix.

11. $\left[\begin{array}{cc|c} 3 & 2 & 6 \\ 1 & -4 & 9 \end{array}\right]$ **12.** $\left[\begin{array}{cc|c} 3 & -2 & 8 \\ 6 & -4 & 17 \end{array}\right]$

13. $\left[\begin{array}{ccc|c} 1 & 2 & -3 & 5 \\ 1 & 0 & 2 & 15 \\ 0 & 2 & -1 & 6 \end{array}\right]$ **14.** $\left[\begin{array}{ccc|c} 1 & 1 & -1 & 0 \\ 0 & 2 & 3 & 1 \\ 1 & 0 & 2 & 0 \end{array}\right]$

15.
$$\begin{aligned} 2x + y &= 1 \\ 3x - 2y &= 12 \end{aligned}$$

16.
$$\begin{aligned} 2x + y &= 8 \\ y &= 2 \end{aligned}$$

17.
$$\begin{aligned} x + y - z &= 2 \\ 2x - 3y + z &= 5 \\ 3x + 2y - 4z &= 3 \end{aligned}$$

18.
$$\begin{aligned} x + y - z &= 3 \\ y - 3z &= 8 \\ x - 2z &= 3 \end{aligned}$$

19. The goal is to have 1's along the main diagonal of the matrix and 0's elsewhere.

20. The n equations lead to n rows, but a constant matrix has only one column. Thus the dimensions are $n \times 1$.

21. **(a)** $R_1 \div 3$ **(b)** $-2R_1 + R_2 \to R_2$ **(c)** $R_2 \div (-7)$ **(d)** $-2R_2 + R_1 \to R_1$ **(e)** (2, 1)

22. **(a)** $R_1 \div 2$ **(b)** $-3R_1 + R_2 \to R_2$ **(c)** $R_2 \div 13$ **(d)** $3R_2 + R_1 \to R_1$ **(e)** (−3, 2)

23. **(a)** $-2R_1 + R_2 \to R_2$ **(b)** $-1R_1 + R_3 \to R_3$ **(c)** $R_2 \div (-3)$ **(d)** $-1R_3$ **(e)** $R_3 + R_1 \to R_1$ **(f)** $R_3 + R_2 \to R_2$ **(g)** $-1R_2 + R_1 \to R_1$ **(h)** (1, 2, −1)

24. **(a)** $R_3 \div 7$ **(b)** $R_3 + R_1 \to R_1$ **(c)** $5R_3 + R_2 \to R_2$ **(d)** $R_2 \div 2$ **(e)** $-R_2 + R_1 \to R_1$ **(f)** (3, 1, 0)

25. **(a)** 2 **(b)** 2 **(c)** −56 **(d)** 2 **(e)** 1 **(f)** −3 **(g)** 1 **(h)** (−3, 8)

26. **(a)** 1 **(b)** 1 **(c)** 1 **(d)** 1 **(e)** 8 **(f)** 1 **(g)** 1 **(h)** −1 **(i)** 1 **(j)** 0 **(k)** −1 **(l)** (0, −1)

27. **(a)** 3 **(b)** −3 **(c)** 3 **(d)** −2 **(e)** 1 **(f)** −2 **(g)** −6 **(h)** −2 **(i)** 1 **(j)** 0 **(k)** 1 **(l)** 0 **(m)** 0 **(n)** 0 **(o)** (1, 0, −2)

28. **(a)** 1 **(b)** 1 **(c)** −4 **(d)** −5 **(e)** 0 **(f)** 12 **(g)** 1 **(h)** −6 **(i)** 3 **(j)** 4 **(k)** (3, 2, 4)

29. The equations must be written in standard form.

30. **(a)** The system is inconsistent and the solution set is Ø.
(b) The equations are dependent and there are infinitely many solutions.
(c) If the other row is $1\ 0 \mid a$, then the solution is of the form $(a, 0)$. Otherwise, we cannot describe the solution set with the given information.

31. (3, −4) **32.** (−4, 3) **33.** $\left(\frac{1}{2}, -\frac{1}{4}\right)$

34. $\left(-\frac{1}{3}, \frac{2}{3}\right)$ **35.** Inconsistent **36.** Inconsistent

37. Dependent **38.** Dependent **39.** (4, 2, −3)

40. (5, −3, 2) **41.** (1, −3, 2) **42.** (3, −2, 0)

43. $\left[\begin{array}{rr|r} 2 & -5 & -5 \\ 5 & 2 & 2 \end{array}\right]$ **44.** $\left[\begin{array}{rr|r} 1 & -3 & 9 \\ -1 & 5 & -17 \end{array}\right]$

45. $\left[\begin{array}{rrr|r} 3 & -4 & 1 & 2 \\ 1 & 1 & 3 & 1 \\ 2 & 3 & -1 & 0 \end{array}\right]$ **46.** $\left[\begin{array}{rrr|r} 1 & 1 & -1 & 0 \\ 2 & -1 & 3 & -2 \\ 3 & 2 & -1 & -1 \end{array}\right]$

47. (−3, 1) **48.** (2, 6) **49.** (3, −2, −2)

50. (−1, 2, −3) **51.** (0, −1, 2) **52.** (−1, −1, 2)

53. (1, −2, −1) **54.** (1, 2, −3)

55. **(a)** $\begin{bmatrix} 428 & 453 \\ 408 & 430 \\ 483 & 544 \end{bmatrix}$ The average number of minutes in a woman's work day in developing countries

(b) $\begin{bmatrix} 428 & 408 & 483 \\ 453 & 430 & 544 \end{bmatrix}$; column 1

56. The average number of hours

57. (1, −0.5, 3, −5.5) **58.** (−2, −3, 1, 2)

59. (2, 3, −4, −5) **60.** (1, −1, −1, 2)

61. If $c \neq 1$, then the solution is unique; if $c = 1$, then there is no solution.

62. If $a \neq \frac{1}{3}$, then the solution is unique; if $a = \frac{1}{3}$, then there is no solution.

63. **(a)** There are no values of p, q, and r for which the solution is unique.
(b) If $p + q + r \neq 0$, then there is no solution.
(c) If $p + q + r = 0$, then there are infinitely many solutions.

Section 5.5 *(page 297)*

1. A matrix is an array of numbers. A determinant is a number associated with a matrix.

2. No, the determinant of A is $ps - qr$, whereas the determinant of B is $qr - ps$. The two determinants are opposites.

3. 7 **4.** 14 **5.** 0 **6.** −17 **7.** 0 **8.** 1 **9.** 1

10. 0 **11.** −1 **12.** 34 **13.** 0 **14.** −8 **15.** −51

16. 8 **17.** −8 **18.** 2 **19.** 0 **20.** 0 **21.** 16 **22.** 7

23. 0 **24.** 0 **25.** $x^2 - 9$ **26.** $16 - y^2$ **27.** $x - 5y$

28. −5 **29.** 7 **30.** 9 **31.** 1 **32.** −8 **33.** 4

34. −11 **35.** Division by zero is not defined.

36. If $D = 0$, Cramer's Rule does not apply. **37.** No

38. Yes **39.** No **40.** No **41.** Yes **42.** Yes

43. (2, 5) **44.** (−2, −3) **45.** (−1, 2) **46.** (5, −7)

47. (5, 3) **48.** $\left(-\frac{7}{6}, \frac{8}{3}\right)$ **49.** (3, 2) **50.** (0, 10)

51. (5, 2) **52.** (3, −2) **53.** (3, 0.5) **54.** (−1, 3)

55. Cramer's Rule does not apply.

56. Cramer's Rule does not apply.

57. $\left(-\frac{1}{4}, \frac{2}{3}\right)$ **58.** $\left(\frac{5}{3}, -4\right)$ **59.** (1, 3, 2) **60.** (2, −4, 5)

61. (−2, −3, 4) **62.** (3, 2, −1)

63. Cramer's Rule does not apply.

64. Cramer's Rule does not apply. **65.** (2, −1, −2)

66. (−1.6, 4.2, −1.4) **67.** (3, −1, 0) **68.** (4, 0, 2)

69. $\left(-1, 7, \frac{1}{3}\right)$ **70.** $(15, 8, -8)$ **73.** -9 **74.** -9

75. 16 **76.** 12 **77.** -2.5 **78.** -2 **79.** -4 **80.** 5

81. Grocery store: 14 years; Pharmacy: 10 years

82. Home: 17 years; Next door: 8 years

83. $1600 in checking, $800 in savings

84. 8 centimeters, 11 centimeters

85. 27 nickels, 9 dimes, 5 quarters **86.** 32°, 64°, 84°

87. $\left(2\sqrt{3}, -\sqrt{2}\right)$ **88.** $\left(-3\sqrt{5}, 2\sqrt{6}\right)$

89. $\begin{vmatrix} 2 & y \\ 3 & x \end{vmatrix}$ **90.** $\begin{vmatrix} 1 & y \\ -4 & x \end{vmatrix}$ **91.** $\begin{vmatrix} a & y \\ -b & x \end{vmatrix}$ **92.** $\begin{vmatrix} m & -1 \\ b & x \end{vmatrix}$

93. Answers will vary.

94. The system that describes the conditions is as follows.

$$\begin{aligned} n + d + q &= 82 \\ 5n + 10d + 25q &= 820 \\ n - 3q &= 0 \end{aligned}$$

The determinant of the coefficient matrix is zero.

Section 5.6 *(page 306)*

1. 12 pounds **2.** Cashews: 18 pounds; Pecans: 27 pounds

3. 5 pounds **4.** Azaleas: 70; Rhododendrons: 30

5. 42 cakes, 48 pies, 50 dozen cookies

6. Peanuts: 10 pounds; Cashews: 8 pounds; Chocolates: 12 pounds

7. 237 patrons, 219 others **8.** 821 students, 422 nonstudents

9. 14,261 **10.** 298

11. 1450 orchestra, 1550 loge, 2500 balcony **12.** 3000

13. $3490 at 8.5%; $6510 at 6.9%

14. $6366 at 6.5%; $4634 at 9%

15. $3800 at 6%; $3200 at 7%

16. $4100 at 9.5%; $4400 at 8%

17. Stock fund: $5400; Bond fund: $4100; Mutual fund: $7000

18. Growth fund: $10,000; GNMA: $28,000; Money market fund: $22,000

19. The number of ounces of acid is found by multiplying the total volume of solution (20 ounces) by the concentration (0.10).

20. Because two of the three solutions have concentrations that are less than 20%, the concentration of the resulting solution should be less than 20%.

21. 34%: 50 liters; 55%: 20 liters

22. A: 11 liters; B: 15 liters **23.** 200 gallons

24. 20%: 9 liters; 70%: 6 liters

25. 10%: 6 liters; 20%: 4 liters; 50%: 8 liters

26. Solution 1: 40%; Solution 2: 25%; Solution 3: 55%

27. Because the boat is pushed by the current, the effective rate is the sum of the rates of the boat and the current.

28. Because the times and distances are the same, the effective rates must also be the same. Therefore, the upstream rate of the boat must be greater than the downstream rate.

29. Canoe: 5 mph; Current: 2 mph

30. Boat: 18 mph; Current: 7 mph

31. Airplane: 450 mph; Wind: 30 mph

32. Airplane: 480 mph; Wind: 40 mph

33. Canoe: 4 mph; Current: 2 mph

34. Jogging: 12 mph; Biking: 30 mph; Swimming: 3.4 mph

35. Father: $34; Son: $17 **36.** 4.5 cups

37. 39 five-dollar bills, 31 ten-dollar bills

38. Biology: 16; French: 11

39. Heavy duty: 12; Homeowner: 19 **40.** 400

41. Gloss: 32 gallons; Flat: 43 gallons

42. 20%: 26.175 ounces; 28%: 43.625 ounces

43. Orchid: 37; Carnation: 58

44. Candy bars: 207; Popcorn: 158 **45.** 140 **46.** 550

47. 8-mile: 22; 6-mile: 10; 5.5 mile: 8

48. Organist: 2 hours; Volunteer: 2.5 hours; Secretary: 1.5 hours

49. A: 5; B: 8; C: 4 **50.** 6426

51. First set: 15 volumes; Second set: 14 volumes; Third set: 6 volumes

52. The longest side is 18 inches. The sum of the other two sides is 22 inches.

53. Hamburger: $1.95; Shake: $1.50; Fries: $0.95

54. 20 pounds

55. **(a)** $C(x) = -6x + 28$; $H(x) = 6x + 5$

(b) Approximately $(1.9, 16.5)$

(c) In 2000 ($x \approx 2$), the percentages of news stories involving crime and human interest were approximately equal (about 16.5%).

56. **(a)** Public: $y = 0.53x + 6.46$; Private: $y = 0.92x + 16.10$

(b) The graph for private colleges has a greater y-intercept and a greater slope than the graph for public colleges. Therefore, the graphs cannot intersect for $x > 0$.

(c) 2011

57.
$$\begin{aligned} 343 &= 4a + 2b + c \\ 1033 &= 36a + 6b + c \\ 2124 &= 100a + 10b + c \end{aligned}$$

58. $a \approx 12.53$, $b \approx 72.25$, $c \approx 148.38$

$y = 12.53x^2 + 72.25x + 148.38$

59.

x	y
2	343
4	638
6	1033
8	1528
10	2124

60. The equation appears to model the data well.

61. 3.94 ounces **62.** 2

63. Crew chief: 5 mph; inspector: 7 mph

64. \$250,000; greater than \$250,000; less than \$250,000

Section 5.7 *(page 315)*

1. The graph of each linear inequality is a half-plane. The graph of the solution set of the system is the intersection of the half-planes.

2. Because all the inequalities must be satisfied, the inequalities of the system are implicitly connected with *and,* and the system is therefore a conjunction.

3.

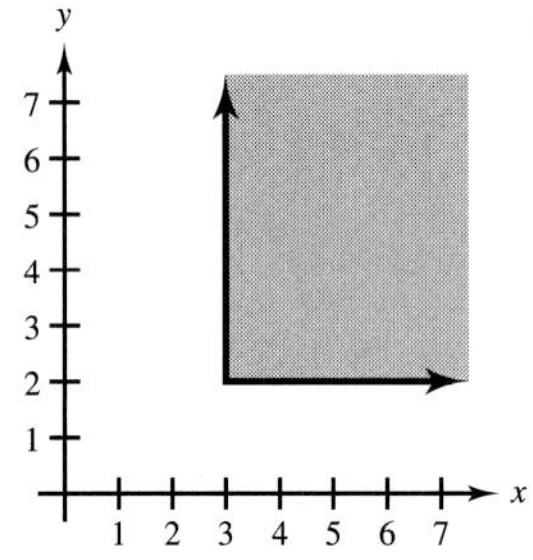

4.

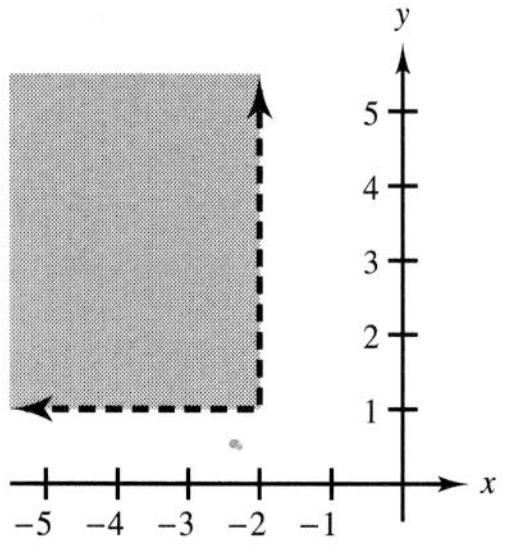

5.

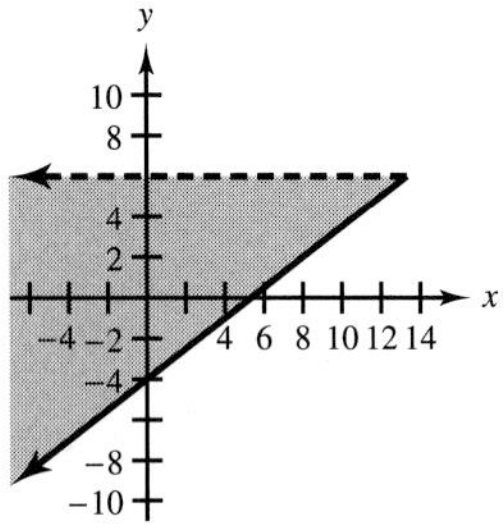

6.

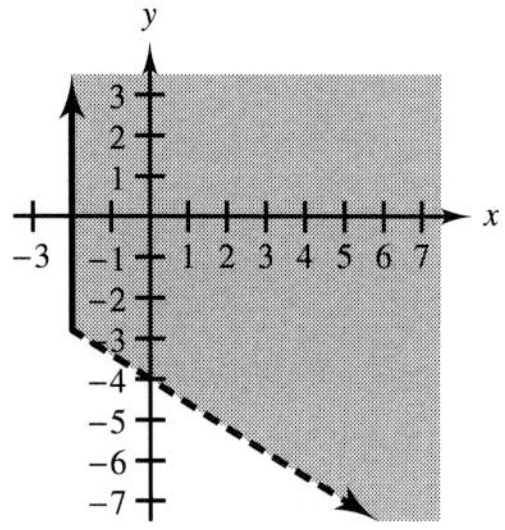

7.

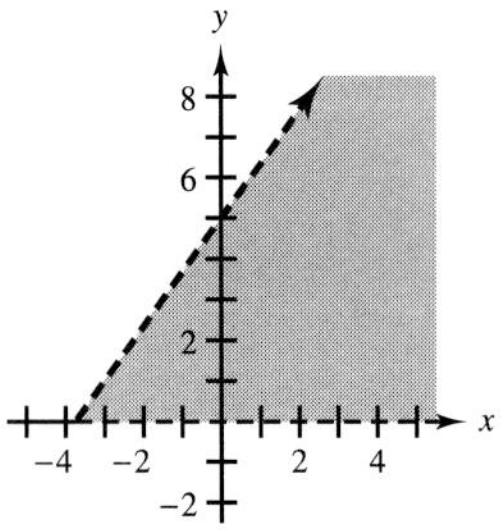

8.

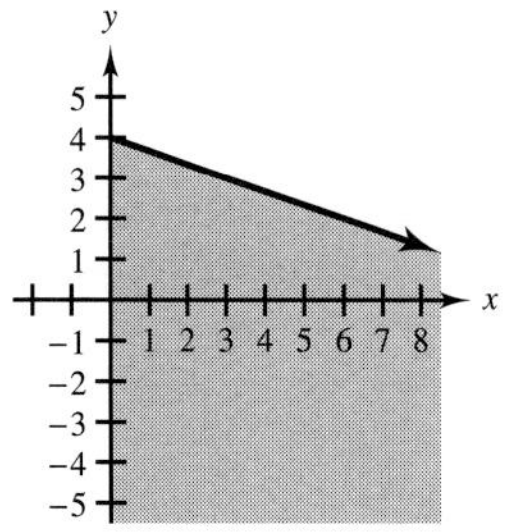

9.

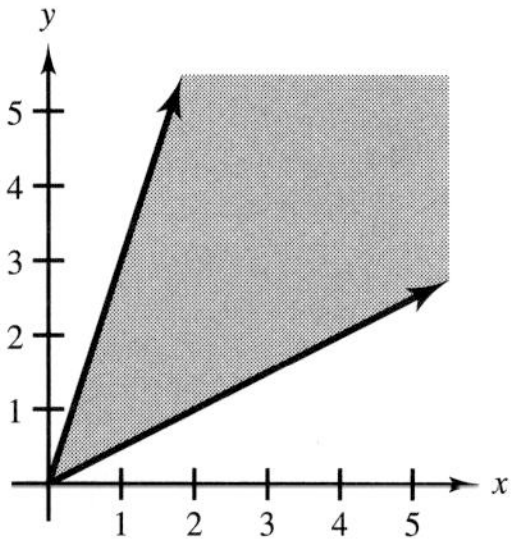

10.

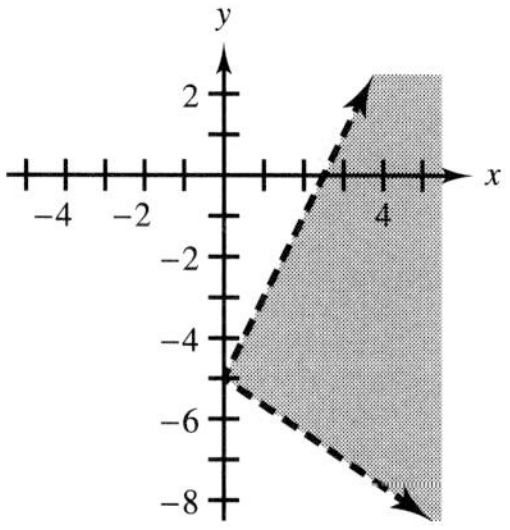

11.

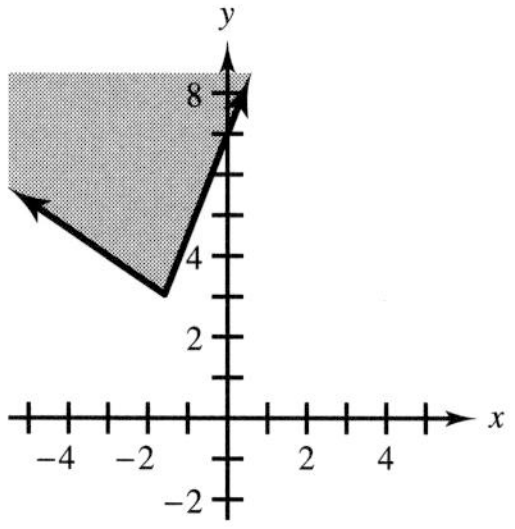

12.

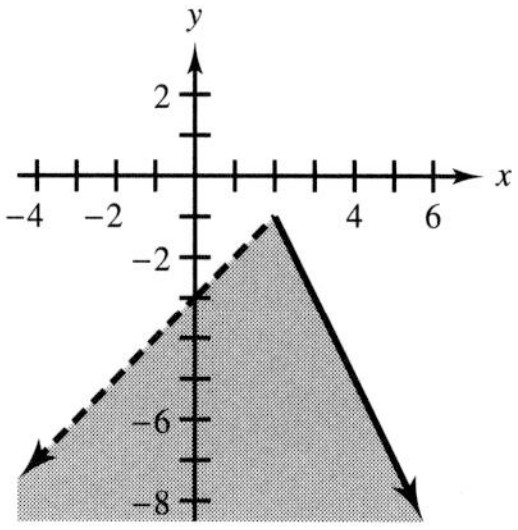

13. Ø **14.** Ø **15.** $y \leq 4, x \leq 5, x \geq 0, y \geq 0$

16. $y \leq -x + 6, x \geq 0, y \geq 0$

17. $y \leq -8x + 8, y > -\frac{1}{3}x - 2$ **18.** $y \geq x - 3, x > -3$

19. The graph is a square, centered at the origin, and its interior. The sides are parallel to the axes, and the length of each side is 10.

20. The pair (a, b) must satisfy every inequality in order to be a solution. If (a, b) fails to satisfy the first inequality, it is not a solution.

21.

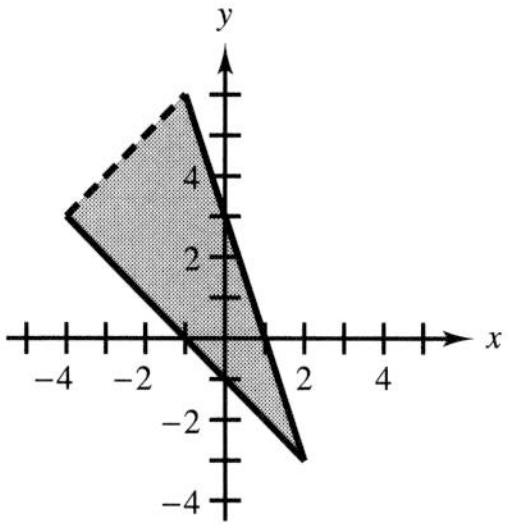

22.

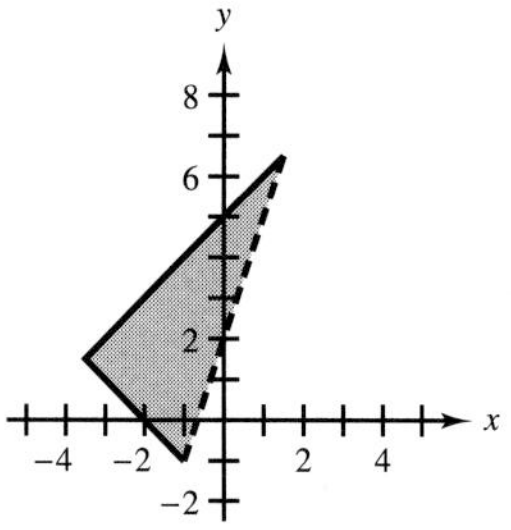

23.

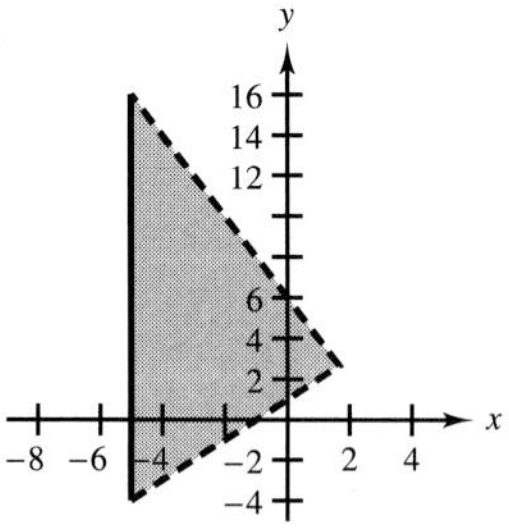

24.

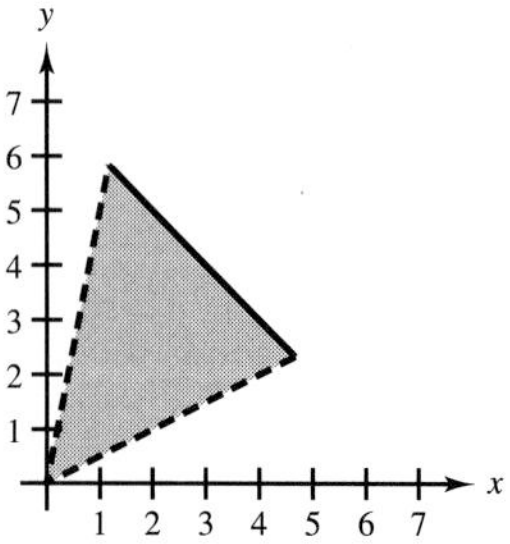

25.

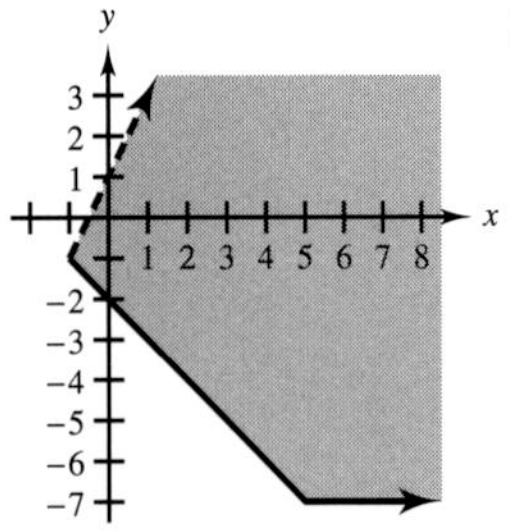

26.

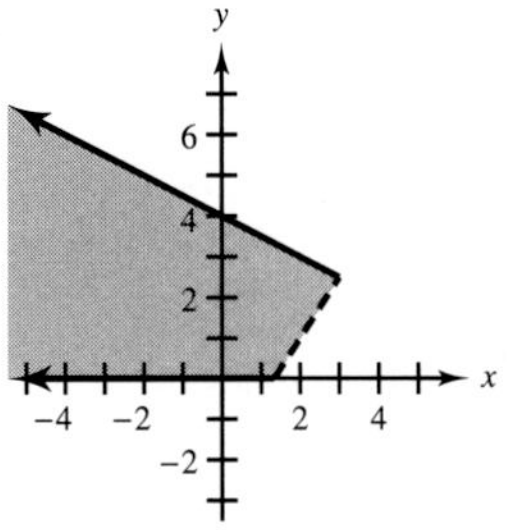

27. The number of people cannot be negative: $x \geq 0$ and $y \geq 0$.

28. Because the variables represent numbers of people, only solutions that involve whole numbers are valid.

29.

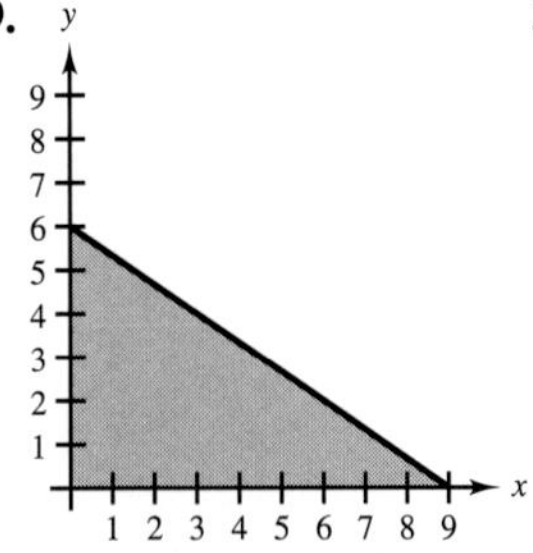

30.

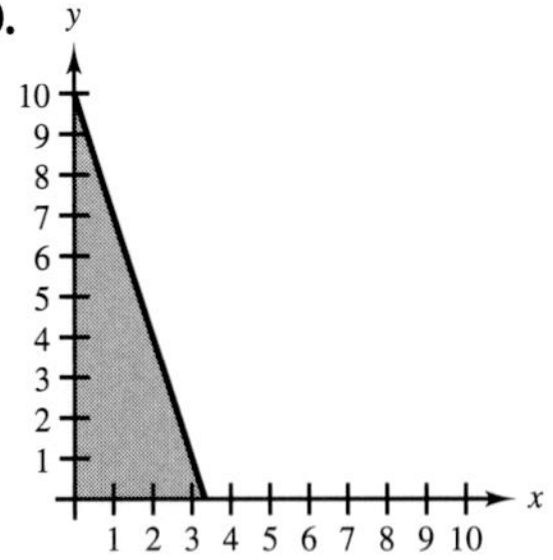

31.

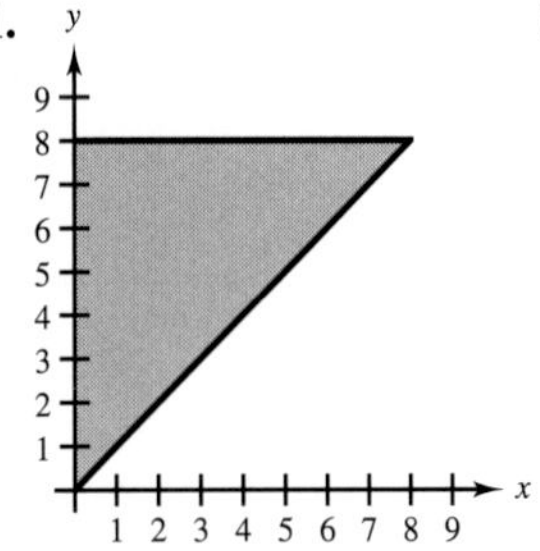

32.

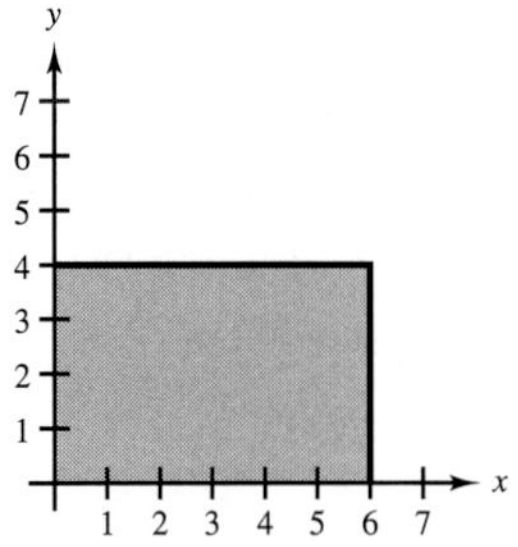

33.

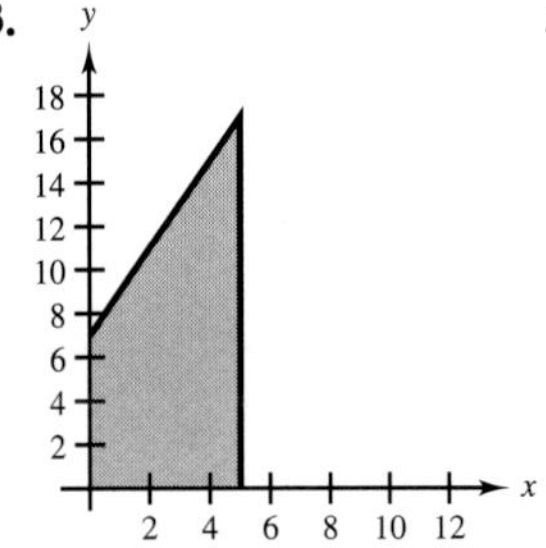

34.

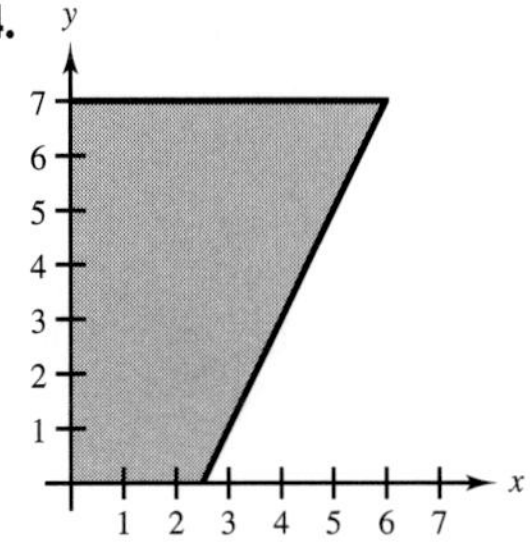

35.

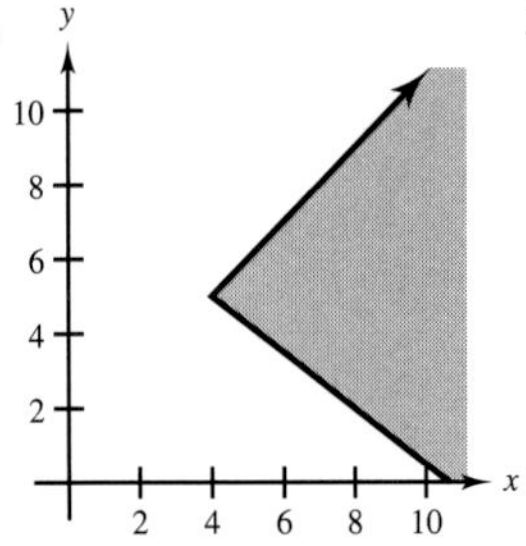

36.

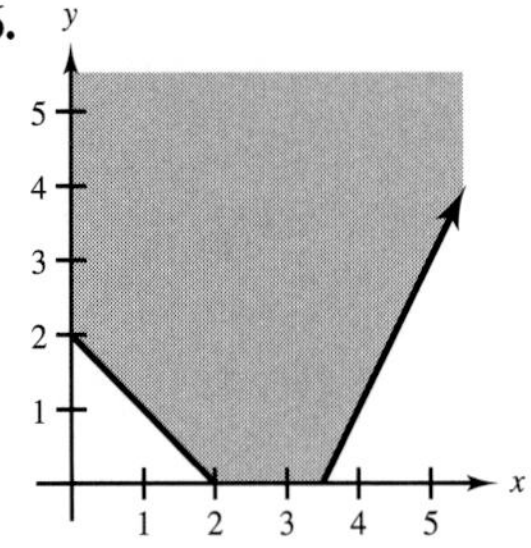

37.

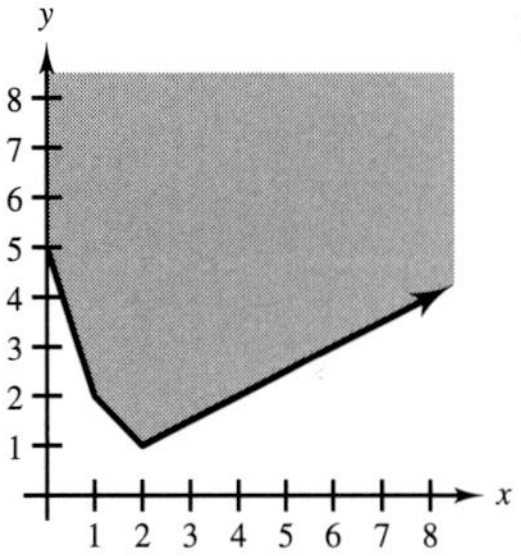

38.

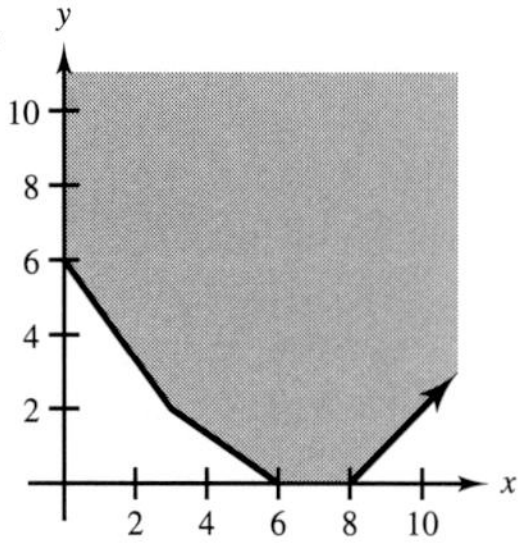

39. Ø **40.** Ø

41. x = number of millions of dollars in short-term notes
y = number of millions of dollars in mortgages
$x + y \leq 10, y \leq 7, x \geq 0, y \geq 0$
\$7 million, \$4 million

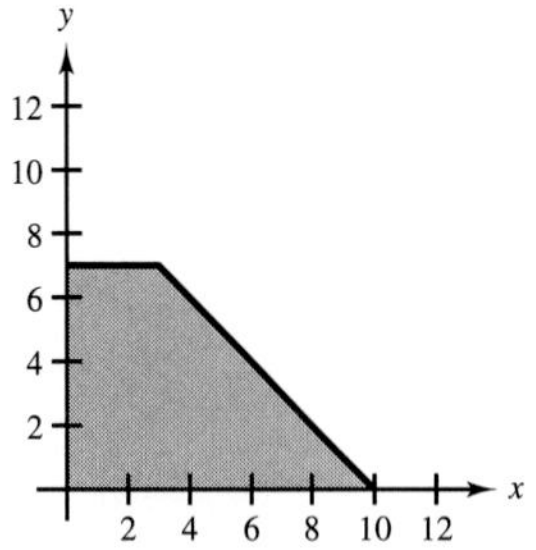

42. x = number of acres of corn
y = number of acres of soybeans
$x + y \leq 2000, y \leq 800, x \leq y + 700, x \geq 0, y \geq 0$
Greatest: 800 acres; Least: 400 acres

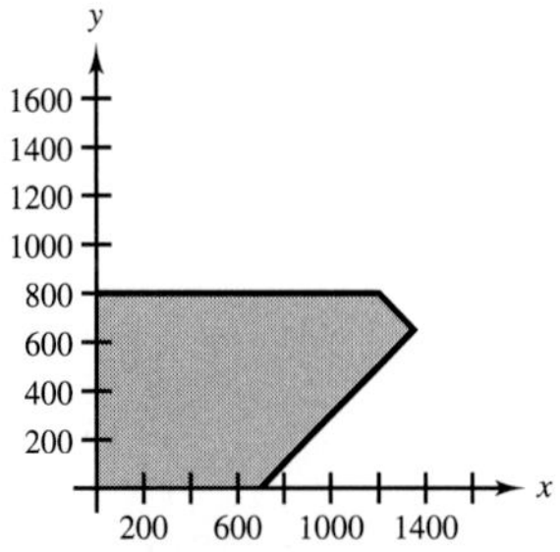

43. x = number of tuna salads, y = number of chicken salads
$x \leq \frac{2}{3}y, x + y \geq 50, x \geq 0, y \geq 0$
From 0 to 40 tuna salads can be prepared.

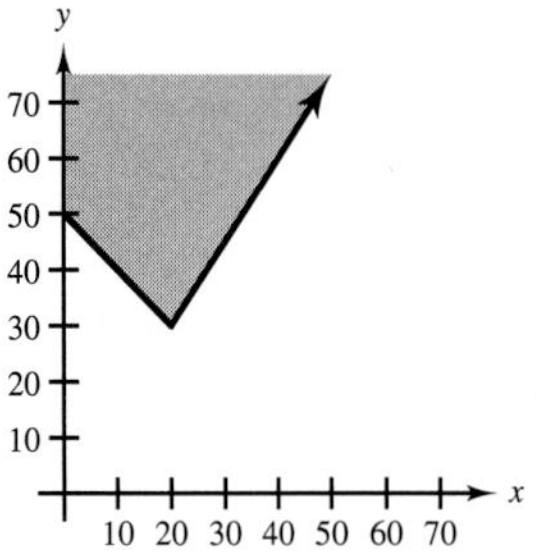

44. x = number of pounds of Irish cream coffee
y = number of pounds of macadamia coffee
$x + y \geq 100, x \leq 2y, x \geq 0, y \geq 0$
Least number of pounds of macadamia nuts: $33\frac{1}{3}$ pounds

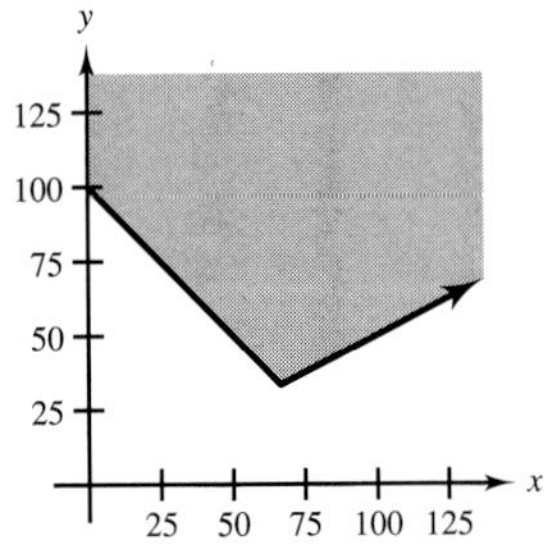

45. **(a)** x = number of barns, y = number of tool sheds
$15x + 5y \leq 90, 7x + 6y \leq 84, x \geq 0, y \geq 0$
(b) Yes **(c)** No

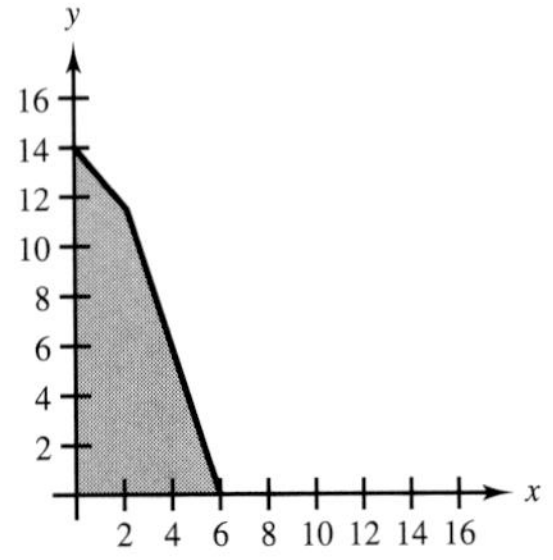

46. **(a)** x = number of banquets, y = number of receptions
$13x + 3y \leq 39, 8x + 5y \leq 40, x \geq 0, y \geq 0$
(b) No **(c)** Yes

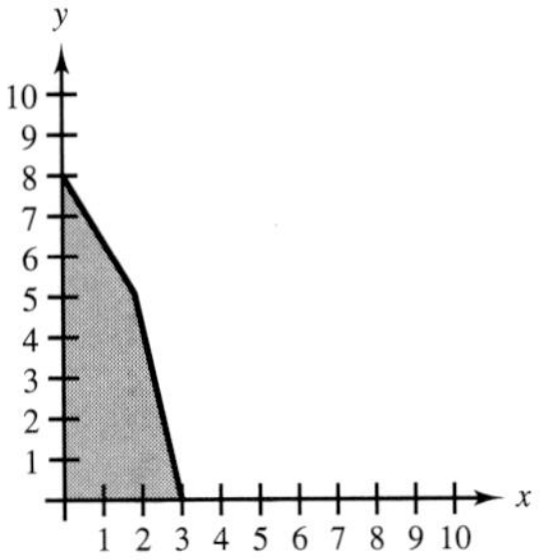

47. Short-term notes: \$3 million; Mortgages: \$7 million

48. Corn: 1350 acres; Soybeans: 650 acres

Chapter Review Exercises *(page 318)*

1. The point represents a solution of both equations.

2. Because the second equation is already solved for y, substitute the expression for y into the first equation: $2x + 3(7x + 1) = 3$.

3. Yes **4.** No **5.** (3, 1) **6.** No solution

7. Infinitely many solutions **8.** (−5, 8) **9.** 1

10. Infinitely many **11.** 1 **12.** None **13.** (0.6, 1.6)

14. No solution **15.** Infinitely many solutions

16. (2.8, −1.6) **17.** (0.75, −0.25) **18.** $\left(\frac{6}{5}, -\frac{2}{3}\right)$

19. Return to the original system and use the addition method again, this time to eliminate x.

20. If the resulting equation is true, the system has infinitely many solutions. If the resulting equation is false, the system has no solution.

21. (2, −4) **22.** (−2, 1) **23.** Infinitely many solutions

24. No solution **25.** (8, −6) **26.** Infinitely many solutions

27. No solution **28.** (−10, 12) **29.** −12 **30.** 3

31. −3 **32.** 2

33. The solution set is represented by a plane drawn in a three-dimensional coordinate space.

34. The method does not work. The same variable must be eliminated from both pairs of equations.

35. No **36.** No **37.** (−2, 3, 1) **38.** (3, 5, −1)

39. (2, −3, −5) **40.** (1, −1, −3) **41.** No solution

42. (0, 2, 1) **43.** (−8, 1, 3) **44.** No solution

45. (1, 2, −5) **46.** No solution

47. A coefficient matrix consists of the coefficients of the variables of a system of equations. An augmented matrix is the combination of a coefficient matrix and a constant matrix.

48. Multiply row 2 by 2, and add the result to row 1.

49. $\left[\begin{array}{cc|c} 4 & -3 & 12 \\ 1 & -2 & -3 \end{array}\right]$ **50.** $\left[\begin{array}{ccc|c} 0 & 1 & -2 & -4 \\ -2 & 0 & 2 & 5 \\ -6 & 2 & 0 & 1 \end{array}\right]$

51. $\begin{aligned} x + 2y - 3z &= 4 \\ 3x + y - 2z &= 5 \\ x + y &= 0 \end{aligned}$ **52.** $\begin{aligned} 5x - 2y &= -2 \\ 6x &= 7 \end{aligned}$

53. (2, −1) **54.** (15, 9) **55.** (−22, −27)

56. $\left(-1, -\frac{5}{3}, -\frac{5}{6}\right)$ **57.** (0, 0, 1) **58.** (1, 2, 3)

59. (−10, 9) **60.** (−0.91, 2.78) **61.** (−1, 4, 0)

62. (3, 2, 1)

63. Because the value of the determinant is 0, the expression is undefined. Thus the system does not have a unique solution.

64. The number 13 is called the determinant, and it is calculated as $2(5) - 3(-1)$.

65. 10 **66.** 0 **67.** −6 **68.** 0 **69.** (−1, 1)

70. (1, 0, −3) **71.** 33 fives, 27 twenties

72. Any two numbers that satisfy one condition also satisfy the other condition.

73. Sledge: 15; Claw: 27 **74.** 4 hours; 3.5 miles **75.** 435

76. Oak: 3; Sourwood: 2; Pine: 12

77.

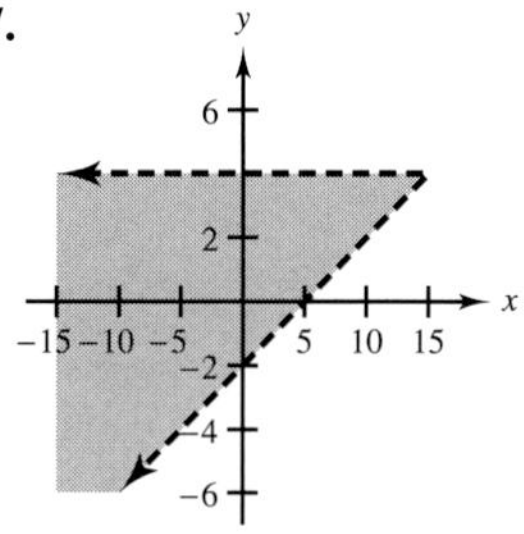

78.

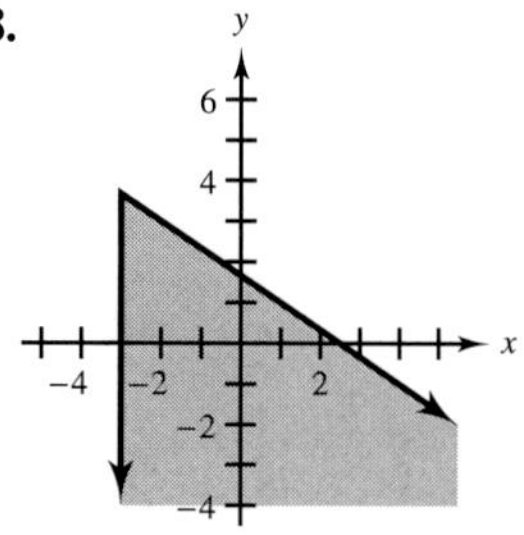

79.

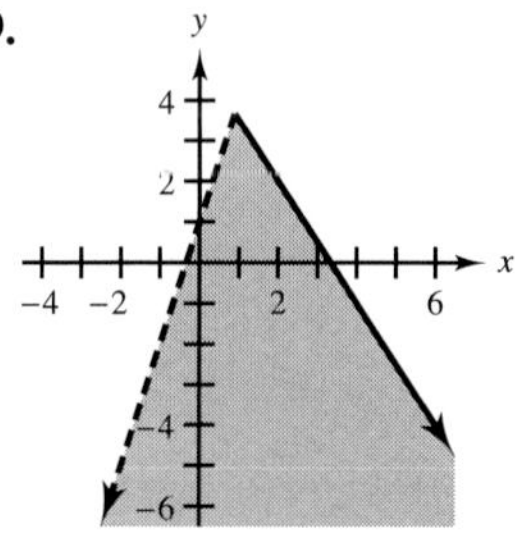

80.

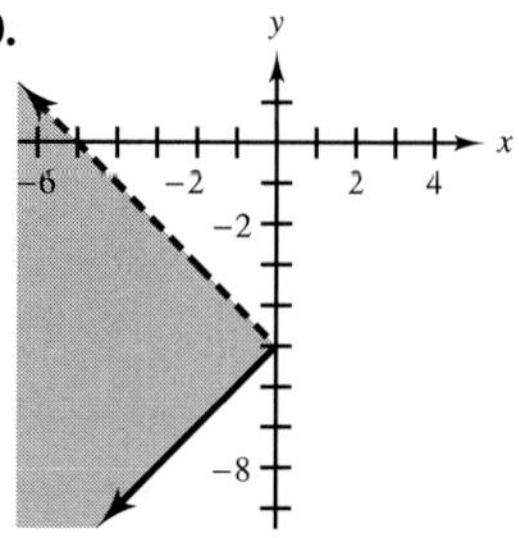

81.

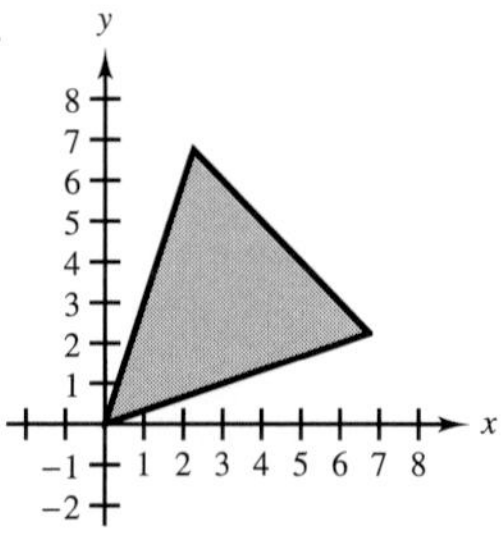

82.

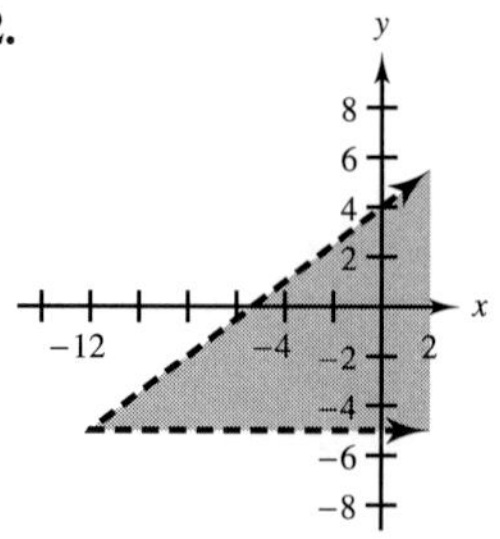

83.

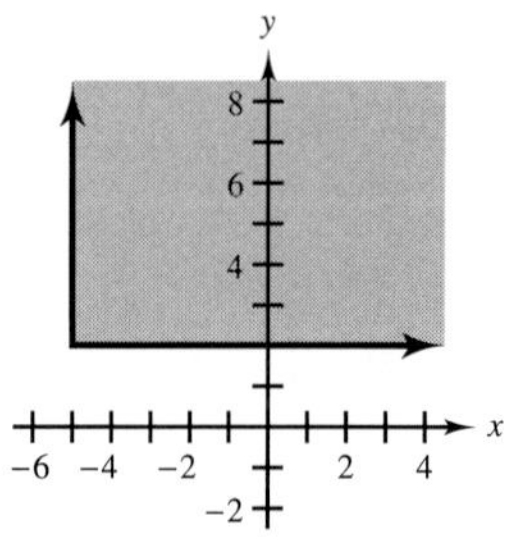

84. ∅

85.

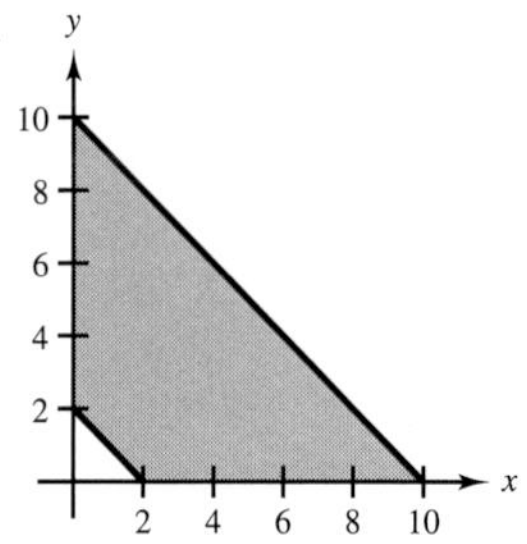

86.

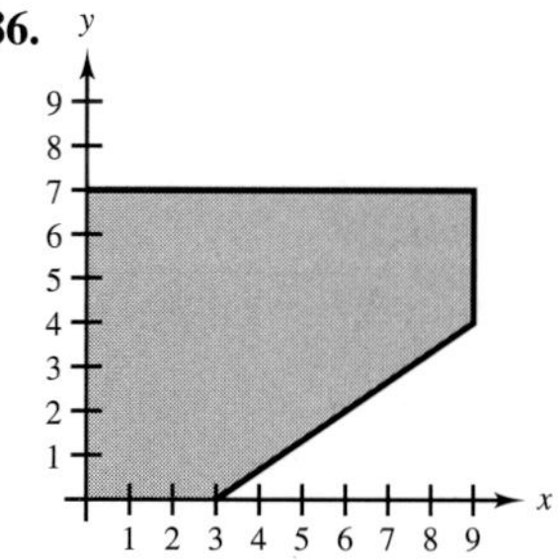

87. x = number of bags of mortar, y = number of bags of lime
$x + y \leq 20$, $80x + 40y \leq 1000$, $x \geq 0$, $y \geq 0$

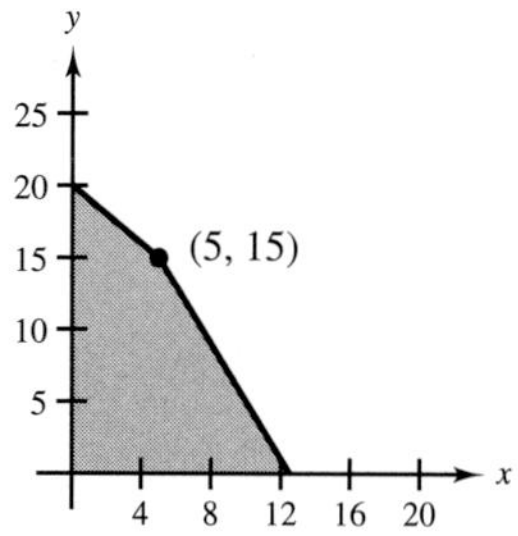

Chapter Test *(page 322)*

1. [*5.1*] (2, −1) **2.** [*5.1*] (−12, 11) **3.** [*5.1*] (5, −3)
4. [*5.2*] Infinitely many solutions **5.** [*5.2*] (10, −2)
6. [*5.3*] (2, −1, 1) **7.** [*5.4*] (−3, 0.5, 1) **8.** [*5.4*] (4, −2)
9. [*5.4*] (1, 3, −2) **10.** [*5.1*] Dependent
11. [*5.1*] Inconsistent **12.** [*5.1*] Unique solution
13. [*5.1*] **(a)** The graphs coincide.
(b) The graphs intersect at one point.
(c) The graphs are parallel.
14. [*5.5*] 1 **15.** [*5.5*] (−5, 1)
16. [*5.7*]

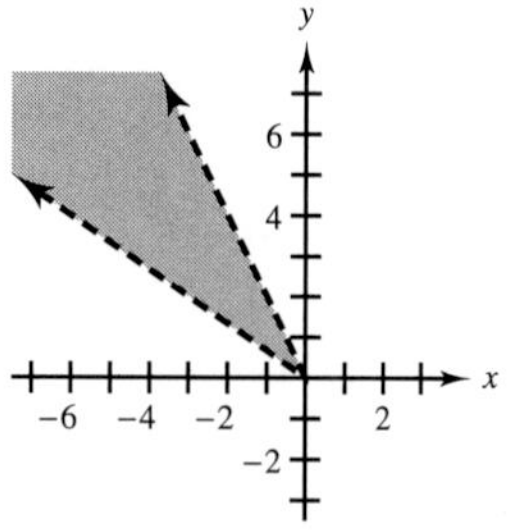

17. [*5.7*]

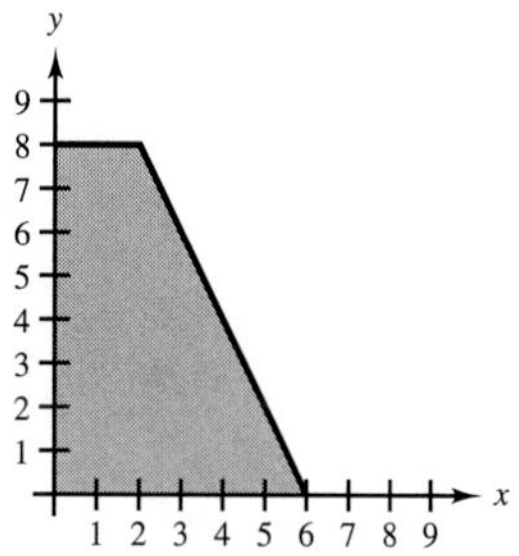

18. [*5.6*] Wind speed: 25 mph; Airspeed of plane: 175 mph
19. [*5.6*] $31\frac{2}{3}^\circ$, $58\frac{1}{3}^\circ$
20. [*5.6*] 40 liters of the 15% solution, 60 liters of the 40% solution
21. [*5.6*] Shrimp: 45.5 pounds; Crab: 43 pounds; Lobster: 48 pounds

22. [*5.7*] p = number of pines, d = number of dogwoods
$p + d \le 20,\ p \ge 2d,\ p \ge 0,\ d \ge 0$

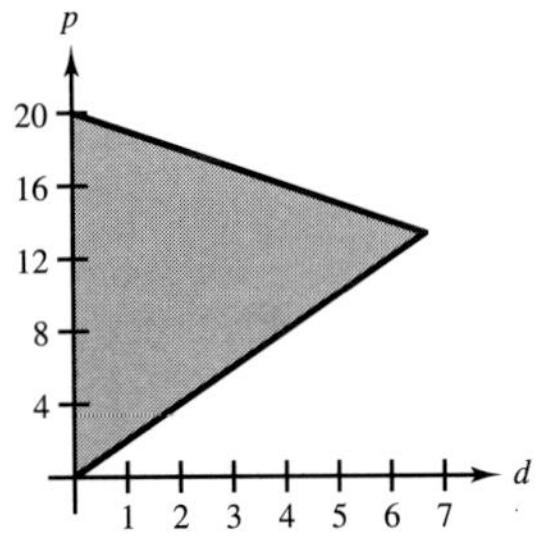

Cumulative Test: Chapters 3–5 *(page 323)*

1. [*3.1*] The solution is 4. When x is replaced with 4, the value of each side of the equation is 17.

2. [*3.2*] **(a)** $\frac{6}{5}$ **(b)** $\frac{49}{26}$ **3.** [*3.3*] $b = \frac{2A}{h}$

4. [*3.3*] 72 **5.** [*3.4*] 1:45 P.M.

6. [*3.5*] \$3000 at 5%, \$5000 at 6.5%

7. [*3.6*] **(a)** $\left(-\frac{2}{3}, \infty\right)$

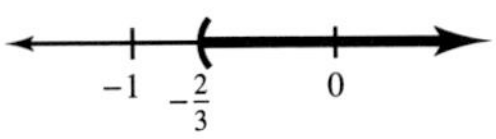

(b) $(-\infty, -1]$

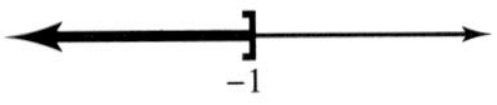

8. [*3.7*] **(a)** $(-\infty, -1) \cup [3, \infty)$

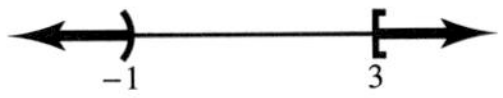

(b) $(3, 6)$

9. [*3.8*] **(a)** $(-5, 11)$ **(b)** $3, -5$
(c) $(-\infty, -5] \cup [-1, \infty)$

10. [*3.6*] 15 inches **11.** [*4.1*] $a = \frac{1}{3}, b = 2$

12. [*4.1*] $(0, -10), (-2, 0)$

13. [*4.2*] **(a)** Negative **(b)** Undefined **(c)** Zero **(d)** Positive

14. [*4.2*] $y = \frac{1}{2}x + 2$

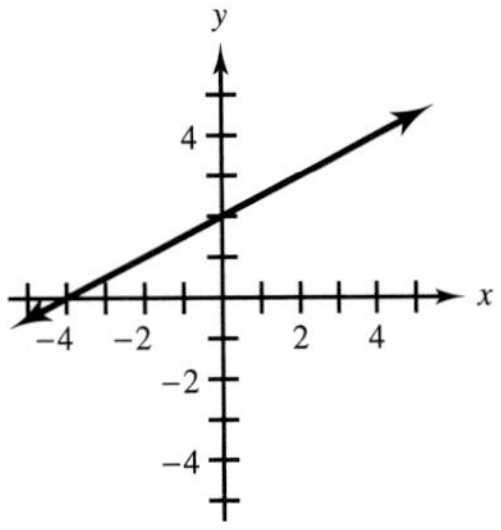

15. [*4.3*] The rate of change with respect to x is the slope of the line: $\frac{2}{3}$.

16. [*4.3*] $\frac{7}{5}$ **17.** [*4.4*] $x - y = 3$ **18.** [*4.4*] $y = -\frac{2}{3}x - \frac{10}{3}$

19. [*4.5*] **(a)** The boundary line is the dashed line $y = x - 7$.
(b) Either test a point or write the inequality as $y > x - 7$ and shade above the line.

20. [*4.5*]

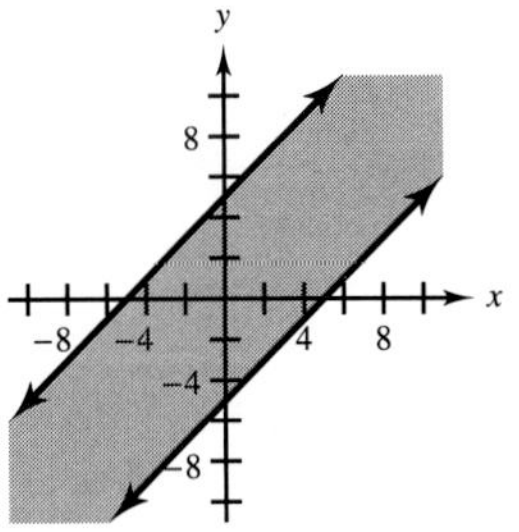

21. [*5.1*] **(a)** The solution set is empty.
(b) The solution set contains exactly one ordered pair.
(c) The solution set contains infinitely many ordered pairs.

22. [*5.1*] $(-2, 6)$

23. [*5.2*] **(a)** Dependent
(b) $(-1, 5)$
(c) Inconsistent

24. [*5.2*] Math: 200; English: 220

25. [*5.3*] $(-1, 0, 4)$ **26.** [*5.3*] $(3, -2, 1)$

27. [*5.4*] $(5, -3, 2)$ **28.** [*5.5*] -3

29. [*5.5*] The determinant of the coefficient matrix is zero.

30. [*5.6*] 4 mph **31.** [*5.6*] 150

32. [*5.7*]

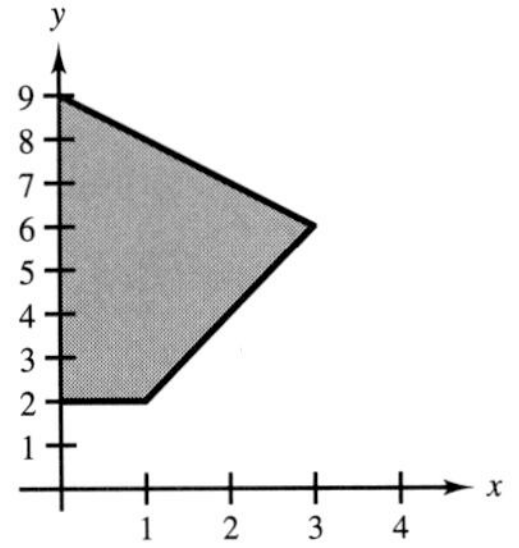

33. [*5.7*] $2L + 2W \le 40,\ W \le \frac{1}{2}L,\ L \ge 0,\ W \ge 0$

Chapter 6

Section 6.1 *(page 330)*

1. Any two monomials can be multiplied, but monomials must be like terms in order for us to add them.

2. The expression -5^3 means $-1 \cdot 5 \cdot 5 \cdot 5 = -125$, and the expression $(-5)^3$ means $(-5)(-5)(-5) = -125$. The expression -5^2 means $-1 \cdot 5 \cdot 5 = -25$, whereas $(-5)^2$ means $(-5)(-5) = 25$.

3. **(a)** b^3 **(b)** $-b^3$ **(c)** $-b^3$ **(d)** b^3 **(e)** $-b^3$

Parts (a) and (d) are equivalent, and parts (b), (c), and (e) are equivalent.

4. **(a)** $-b^2$ **(b)** $-b^2$ **(c)** b^2 **(d)** b^2

Parts (a) and (b) are equivalent, and parts (c) and (d) are equivalent.

5. 2^{11} **6.** t^5 **7.** $-5x^6$ **8.** $2y^8$ **9.** $18t^{10}$ **10.** $-2y^{10}$

11. $(6+t)^9$ **12.** $(x-y)^5$ **13.** $3y^7$ **14.** $6x^3$

15. $-6x^8$ **16.** $-40t^{16}$ **17.** $-6x^4y^3$ **18.** $2x^7y^9$

19. **(a)** Not possible **(b)** $64y^9$

20. **(a)** $-t^5$ **(b)** $-42t^{10}$ **21.** **(a)** $-2n$ **(b)** $-18n^3$

22. **(a)** 0 **(b)** $-6m^6$ **23.** **(a)** $2y^3$ **(b)** $-20y^9$

24. **(a)** Not possible **(b)** x^8 **25.** **(a)** $8x^5$ **(b)** $8x^5$

26. **(a)** $12y^3$ **(b)** $12y^3$

27. When we multiply, we add exponents: $x^3 \cdot x^3 = x^6$. When we combine terms, we do nothing with the exponents: $x^3 + x^3 = 2x^3$. When we raise a power to a power, we multiply the exponents: $(x^3)^3 = x^9$.

28. Expressions (i) and (iii) are equivalent to $(2x^3)^5$. Because $(2x^3)^5$ is a product to a power, the coefficient 2 must be raised to the fifth power, which is not the case in expression (ii).

29. x^{12} **30.** t^{10} **31.** $(x-3)^{12}$ **32.** $(2-5x)^{12}$

33. x^5y^5 **34.** $-8t^3$ **35.** $-x^6y^3$ **36.** $a^{30}b^{54}$

37. $-8x^6y^{12}$ **38.** $81x^{12}y^{20}$ **39.** $\frac{a^6}{b^6}$ **40.** $-\frac{243}{v^5}$

41. $\frac{16x^{12}}{y^{10}}$ **42.** $\frac{x^3y^6}{z^3}$ **43.** $\frac{49x^6}{25y^{10}}$ **44.** $\frac{625a^{20}}{16b^{24}}$ **45.** -7^6

46. z^3 **47.** x^{11} **48.** y **49.** $(x+1)^2$ **50.** $2x-1$

51. $-\frac{x}{2}$ **52.** $3y^3$ **53.** $\frac{3z^3}{4}$ **54.** $-\frac{3}{5}x^2$ **55.** a^2b

56. x^3y^7 **57.** a^4b^3 **58.** x^3y^8 **59.** $\frac{x^6y^4}{2}$ **60.** $8a^{13}b^8$

61. $(7-2)^2$ **62.** $(9 \div 3)^2$ **63.** $(8 \cdot 5)^2$ **64.** $(5+3)^2 \div 4$

65. False **66.** False **67.** True **68.** True **69.** True

70. False **71.** $-20x^8$ **72.** $x^{12}y^{15}$ **73.** $27s^9t^{24}$

74. $-32a^5b^{35}c^{20}$ **75.** $-t^{11}$ **76.** x^8 **77.** $-2a^{10}$

78. $48x^{17}$ **79.** x^4y^6 **80.** $24x^{14}$ **81.** $-\frac{x^8}{4}$ **82.** $\frac{3}{2}s^3t^5$

83. $2x^6$ **84.** $\frac{1}{3}y^5$ **85.** $\frac{2x^4}{5}$ **86.** y **87.** a^4b^2 **88.** $25xy^2$

89. We can write $x^6 = (x^2)^3 = (5)^3 = 125$. Thus $x^6 = 125$.

90. For example, let $c = 1$ and $d = 1$. Then $(1+1)^2 = 4$, but $1^2 + 1^2 = 2$.

91. 4 **92.** 9 **93.** 6 **94.** 4

95.

Year	F(t)	S(t)
2002	8,490	3641
2004	8,999	3786
2006	9,528	3933
2008	10,078	4082
2010	10,648	4235

Per capita federal taxes rose much more rapidly than per capita taxes for state and local governments.

96. $R(t) = 0.023t$

97.

Year	R(t)
2002	2.35
2004	2.39
2006	2.44
2008	2.48
2010	2.53

The increasing values of $R(t)$ show that the ratio of federal tax to state/local taxes increased.

98. $r(t) = 2$; According to this model, the ratio of federal to state/local taxes remained constant, and federal taxes were twice state/local taxes throughout the century.

99. 3^{13} **100.** 2^{19} **101.** 2^5 **102.** 3^2 **107.** y^{3n+2}

108. y^{3m+1} **109.** a^{12n^2} **110.** $\frac{b^{2n^2}}{a^{mn}}$

Section 6.2 *(page 336)*

1. If 0^0 were defined, then according to the Quotient Rule for Exponents, $\frac{0^5}{0^5} = 0^{5-5} = 0^0$. But $0^5 = 0$, and division by 0 is undefined.

2. The expression $2x^0$ means $2 \cdot x^0 = 2 \cdot 1 = 2$. But the expression $(2x)^0 = 1$. Unless grouping symbols indicate otherwise, exponentiation takes place first.

3. 1 **4.** 1 **5.** 3 **6.** -7 **7.** -1 **8.** 1 **9.** $\frac{1}{25}$

10. $\frac{1}{7}$ **11.** $-\frac{1}{8}$ **12.** $\frac{1}{16}$ **13.** 15 **14.** -9 **15.** 4

16. $-\frac{27}{8}$ **17.** $\frac{1}{8}$ **18.** $\frac{27}{2}$ **19.** $\frac{7}{12}$ **20.** $\frac{3}{10}$

21. If b were 0, b^{-n} would be $0^{-n} = \frac{1}{0^n} = \frac{1}{0}$, which is undefined.

22. Write the reciprocal of the number with a positive exponent.

23. $\frac{1}{x^7}$ **24.** y^3 **25.** $\frac{4}{x^4}$ **26.** $-\frac{5}{y}$ **27.** $\frac{x^3}{32}$ **28.** $\frac{4v^2}{u^3}$

29. $-\frac{1}{x^3y^4}$ **30.** $\frac{y}{z^3}$ **31.** $-\frac{k^6}{5}$ **32.** $\frac{t^5}{3}$ **33.** $\frac{1}{(2+y)^3}$

34. $3(x+1)^5$ **35.** $-\frac{3b}{2a}$ **36.** $-\frac{2b^3}{a^4}$ **37.** $<$ **38.** $=$

39. $<$ **40.** $=$ **41.** $-1-4^0, -|8|^0, 0, (-10)^0, 4^0+x^0$

42. $(-2)(-x)^0, -x^0+1, |-6|^0, (-x)^0+1$ **43.** 1 **44.** $\frac{1}{y}$

45. $-\frac{12}{x^3}$ **46.** $\frac{3}{x^3}$ **47.** $\frac{14y}{x^7}$ **48.** $\frac{x^5}{y}$ **49.** $\frac{1}{z^7}$ **50.** x^{13}

51. $\frac{1}{y^5}$ **52.** $-2t^{12}$ **53.** $\frac{3}{2z^3}$ **54.** $\frac{3}{t^6}$ **55.** $\frac{y^{10}}{x^5}$ **56.** $\frac{x}{y^2}$

57. $-\frac{1}{64x}$ **58.** $\frac{y^2}{3}$ **59.** $\frac{y^2}{36x^{10}}$ **60.** $\frac{2x^2}{y^4}$ **61.** $\frac{x^{39}}{y^{48}}$

62. $\frac{y^5}{3x^5}$ **63.** x^{-8} **64.** $3x^{-2}$ **65.** x^4 **66.** x^{-4} **67.** $\frac{1}{x^{30}}$

68. $\frac{1}{y^8}$ **69.** $\frac{a^{20}}{16}$ **70.** $\frac{x^2}{y^6}$ **71.** $-\frac{8y^{12}}{x^6}$ **72.** $\frac{y^{16}}{4x^{12}}$

73. $-\frac{k^5}{32}$ **74.** $\frac{1}{x^9y^{12}}$ **75.** $\frac{-8000}{x^9y^{12}}$ **76.** $\frac{9x^6y^4}{25}$

77. $(4-3)^{-9}$ **78.** $(5\cdot 2)^{-1}$ **79.** $(2^5+1)^0$

80. $(5\div 3)^{-1}$ **81.** $2(xy)^{-2}$ **82.** $x^5(y^{-3}z)^0$ **83.** $\frac{x^{12}}{16}$

84. $-\frac{x^2}{5y}$ **85.** $\frac{-4}{x^5}$ **86.** $-\frac{1}{t^5}$ **87.** $-\frac{12}{y^2}$ **88.** $-\frac{6x^2}{y^4}$

89. $\frac{1}{a^{17}b^{50}}$ **90.** $-\frac{16t^5}{s^{13}}$ **91.** $-\frac{x^2}{2y^7}$ **92.** $-\frac{3y^7}{x^5}$

93. $-\frac{x^6y^{15}}{8}$ **94.** $\frac{x^{24}}{27}$ **95.** $\frac{1}{x^8}$ **96.** $\frac{1}{t^4}$ **97.** $\frac{b^2}{a^2}$

98. $\frac{27}{b}$ **99.** $\frac{4}{3}$ or $-\frac{4}{3}$ **100.** x^4 or $-x^4$

105. $m=2, n=-6$ **106.** $n=2$

Section 6.3 *(page 342)*

1. The number n is such that $1 \le n < 10$, and p is an integer.

2. Expression (ii) is in scientific notation. In expressions (i) and (iii), the numbers 35 and 10 are not in the required interval $1 \le n < 10$.

3. $1.25\cdot 10^5$ **4.** $1.5\cdot 10^4$ **5.** $2.5704\cdot 10^3$

6. $1.1025\cdot 10^2$ **7.** $-5.376\cdot 10^8$ **8.** $1.0\cdot 10^9$

9. $4.567\cdot 10^{11}$ **10.** $-3.5795\cdot 10^7$ **11.** $-4\cdot 10^{-1}$

12. $2.5\cdot 10^{-2}$ **13.** $6.45\cdot 10^{-6}$ **14.** $-1.0\cdot 10^{-8}$

15. $3.749\cdot 10^{-7}$ **16.** $2.46\cdot 10^{-7}$ **17.** 1,340,000

18. 0.000007358 **19.** −0.0000004214 **20.** 68,700

21. 0.000001 **22.** 1,000,000,000 **23.** 30,000

24. −0.05 **25.** −5,720,000,000 **26.** 0.0000000063

27. 0.00089 **28.** −48,300,000 **29.** $4.3\cdot 10^{52}$

30. $-8.8\cdot 10^{35}$ **31.** $-6.7\cdot 10^{-24}$ **32.** $3.3\cdot 10^{-61}$

33. $1.2\cdot 10^{-11}$ **34.** $1.1\cdot 10^{-11}$ **35.** $5.7\cdot 10^{12}$

36. $-5.7\cdot 10^{12}$ **37.** $>$ **38.** $=$ **39.** $>$ **40.** $=$

41. 7153 **42.** 5491 **43.** 2.34 **44.** 9.28 **45.** 6000

46. 0.05 **47.** 200 **48.** 50 **49.** 5 **50.** 25

51. $3.65\cdot 10^{11}$ **52.** $3.95\cdot 10^4$ **53.** $8.13\cdot 10^{12}$

54. $1.76\cdot 10^{-10}$ **55.** $1.06\cdot 10^{15}$ **56.** $3.27\cdot 10^{-13}$

57. $1.20\cdot 10^{-6}$ **58.** $2.22\cdot 10^4$ **59.** $3.88\cdot 10^{-10}$

60. $8.48\cdot 10^{-14}$ **61.** $3.456\cdot 10^9$ **62.** $4.23\cdot 10^{11}$

63. $1.234\cdot 10^8$ **64.** $2.58\cdot 10^{-13}$ **65.** $9.75\cdot 10^{-12}$

66. $7.5394\cdot 10^{-5}$ **67.** 37,700,000 acres

68. 0.000000000053 meter **69.** $7\cdot 10^{-12}$ millimeter

70. $1.63\cdot 10^9$ **71.** 5,000,000

72. \$2,279,000; $\$2.279\cdot 10^6$ **73.** $\$1.06\cdot 10^{10}$

74. $1.17\cdot 10^9$ **75.** $2.5\cdot 10^{-4}$ inch **76.** $1.14\cdot 10^{-4}$ %

77. $7.2\cdot 10^6$ deaths **78.** $1.5\cdot 10^{11}$ dollars

79. $1.1\cdot 10^8$ person-hours **80.** $2.7\cdot 10^{11}$ family-hours

81. $8.0\cdot 10^{-4}$ pound **82.** $6.25\cdot 10^{-7}$ second

83. $1.3653\cdot 10^{10}$; $2.986\cdot 10^9$ **84.** About 22%

85. About \$11 billion **86.** About \$138 million **87.** 6

88. −6 **89.** 13 **90.** −5 **91.** $1.2\cdot 10^3$ **92.** $50\cdot 10^3$

93. $43.7\cdot 10^{-15}$ **94.** $1\cdot 10^{-6}$ **95.** $931.3\cdot 10^{18}$

96. $83.8\cdot 10^6$

Section 6.4 *(page 350)*

1. **(a)** The variable is in the denominator.
(b) The exponent is negative.

2. Statements (i) and (iv) are true. Statement (ii) is false because −8 is also a coefficient. Statement (iii) is false because 2^3 is a constant with degree 0.

3. Yes **4.** No **5.** Yes **6.** Yes **7.** 5, 0 **8.** −1, 1

9. −4, 3 **10.** 2π, 2 **11.** 1, 2 **12.** 4, 3

13. The degree of a term is the value of the exponent on the variable (or the sum of the exponents if more than one variable). The degree of a polynomial is the degree of the term with the highest degree.

14. Because x^2+0x+5 has only two nonzero terms, the expression is a binomial.

15. 1 **16.** 2 **17.** 5 **18.** 4 **19.** 5, trinomial

20. 6, monomial **21.** 3, none of these **22.** 1, binomial

23. 5, binomial **24.** 0, monomial

25. $2, -3y^2+6y+6, 6+6y-3y^2$

26. $3, -2y^3+7, 7-2y^3$

27. $9, 4x^9-7x^8-5x^6+x^3, x^3-5x^6-7x^8+4x^9$

28. $11, -7x^{11}+5x^7+3x^4-6x^2+4x,$
$4x-6x^2+3x^4+5x^7-7x^{11}$

29. $12, n^{12}+4n^8-2n^6-3n^4, -3n^4-2n^6+4n^8+n^{12}$

30. $9, -p^9 + 8p^5 + 22p^4 - 5^7, -5^7 + 22p^4 + 8p^5 - p^9$

31. -4 **32.** -7 **33.** 15 **34.** 0 **35.** -24 **36.** 0

37. $4a$ **38.** $4x^2 - 9x - 2$ **39.** $3xy^2$ **40.** $m^2n + 2mn + 1$

41. $-7x + 6$ **42.** $9x - 14$ **43.** $4x^2 - x + 14$

44. $2x^2 + 2x + 3$ **45.** $-2t^2 - 3t + 1$ **46.** $m^2 + 3m - 1$

47. $n^2 - m^2 + 7mn$ **48.** $2x^2y + 1$ **49.** $-2xy + 5y^2$

50. $4ab^2 - 5ab$ **51.** $-2a^2 + 5a^2b$ **52.** $7y^3 - 5xy^2 + 6$

53. $-2x^3 - 8x$ **54.** $8y - 8$ **55.** $-4x^2 + 5x - 8$

56. $4x^3 - 3x^2 - 5x + 13$ **57.** $x^2 - 12$

58. $-x^4 + 3x^2 + 2x - 11$ **59.** $2x^2 + 2x + 2$

60. $-3y^3 + 8y^2 - y - 8$

61. (ii) The minuend is $3x + 4$ and the subtrahend is $x - 2$.

62. Multiplying both sides of an equation results in an equivalent equation, but multiplying the polynomial by a nonzero constant changes the value of the polynomial.

63. $3x^2 - 2x + 7$ **64.** $y^3 + y^2 + 5$ **65.** $-9x^2 - 3x - 9$

66. $-x^3$ **67.** $-x + 7$ **68.** $-2x + 4$ **69.** $x - 3$

70. $2x - 11$ **71.** $x^2 - 5x + 2$ **72.** $4x^2 + 3$ **73.** $-x - 1$

74. $-4y$ **75.** -2 **76.** $2 - 8x$ **77.** $-3y + 3$ **78.** $4a$

79. $-2x^2 - 6xy + 2y^2 + 4xy^2$ **80.** $a^2 + 17ab - 3a^2b - 4b^2$

81. 5, 0 **82.** 3, 5 **83.** $-1, -11$ **84.** $-113, 11$

85. $-130.05, 68.33$ **86.** $-60.32, 1.83$ **87.** $x^2 + 4x - 2$

88. $-x^2 + 7x + 3$ **89.** $3x^2 - 11x - 16$ **90.** $x^2 - 11x - 6$

91. **(a)** 1600 **(b)** 1000 **(c)** 232 **(d)** 0 **(e)** 0 **(f)** $40 \leq t \leq 100$

92. **(a)** 300 **(b)** 375 **(c)** 144 **(d)** 0 **(e)** 0 **(f)** $50 \leq t \leq 90$

93. **(a)** \$2012 **(b)** \$9710 **(c)** $T(x) = x^2 + 17x + 22$ **(d)** \$11,722 **(e)** Answers will vary. **(f)** \$20,800 **(g)** $P(x) = x^2 - 9x - 22$ **(h)** \$9078 **(i)** Answers will vary.

94. **(a)** \$2450 **(b)** \$4775 **(c)** $T(x) = x^2 + 23x + 715$ **(d)** \$7225 **(e)** Answers will vary. **(f)** \$19,810 **(g)** $P(x) = 3x^2 - 20x - 715$ **(h)** \$12,585 **(i)** Answers will vary.

95. $a = -1, b = 1, c = 4$ **96.** $a = -1, b = -2, c = 4$

97. $a = 0, b = 6, c = -9$ **98.** $a = 1, b = 0, c = -7$

99. When we simplify the left side, we obtain $(a + b)x^2 - 2 = 4x^2 - 2$, which means that $a + b = 4$. There are infinitely many values of a and b such that $a + b = 4$.

100. The quantity $p(3)$ is the value of $p(x)$ when $x = 3$. Thus $p(3)$ is a constant with degree 0 if $p(3) \neq 0$ or with no degree if $p(3) = 0$.

101. 2, 3, 5

Section 6.5 *(page 358)*

1. Statement (ii) is false. For example, $(x + 2)(x - 2) = x^2 - 4$, which is a binomial.

2. **(a)** $(x^2 + 3)(x^2 + x) = x^4 + x^3 + 3x^2 + 3x$ **(b)** $(x + 3)(x - 4) = x^2 - x - 12$ **(c)** $(x + 3)(x - 3) = x^2 - 9$

3. $6x^3 - 9x^2$ **4.** $6x^4 + 8x^3$ **5.** $-8x^5 - 20x^3$

6. $a^6 - 4a^5 + 3a^4$ **7.** $-x^3y^2 + x^2y^3 + 2xy^2$

8. $-6x^4y + 12x^3y^2 - 42x^3y^3$ **9.** $x^2 - 2x - 24$

10. $x^2 + 4x - 21$ **11.** $y^2 - 9y + 14$ **12.** $x^2 - 11x + 30$

13. $-y^2 - 2y + 15$ **14.** $-x^2 + 6x + 16$

15. $20x^2 - 9x - 18$ **16.** $8x^2 + 2x - 15$

17. $9x^2 + 29x + 6$ **18.** $-16y^2 + 56y - 49$

19. $-10y^2 + 3y + 4$ **20.** $-28x^2 - x + 2$

21. $x^2 - 13xy + 36y^2$ **22.** $x^2 - 8xy + 15y^2$

23. $a^2 + 3ab - 10b^2$ **24.** $m^2 + 3mn - 4n^2$

25. $12x^2 - 28xy + 15y^2$ **26.** $12x^2 + 11xy + 2y^2$

27. $6x^4 - 13x^2 + 6$ **28.** $12x^4 - 43x^2 + 35$

29. $x^3 - 2x^2 - 3x + 10$ **30.** $x^3 + 2x^2 - 18x + 9$

31. $6x^3 - 7x^2 - 5x + 3$ **32.** $6x^3 - 2x^2 - x + 2$

33. $x^3 + y^3$ **34.** $x^3 - 8y^3$ **35.** $4x^4 - 4x^3 + 7x^2 - 11x + 4$

36. $9x^4 + 11x^2 + 4x - 4$ **37.** $x^4 - 2x^2 + 1$

38. $x^4 - 8x^2 + 16$ **39.** $x^3 - 8$ **40.** $x^3 + 27$

41. $2x^4 - 8x^3 + 19x^2 - 22x + 15$

42. $2x^4 - x^3 - 6x^2 + 7x - 2$

43. Yes, but recognizing the special pattern saves the step of determining the inner and outer products.

44. No, the expression is not a product of two binomials.

45. $x^2 + 6x + 9$ **46.** $x^2 - 10x + 25$ **47.** $4 - 4y + y^2$

48. $16 + 8x + x^2$ **49.** $49x^2 - 42x + 9$

50. $81 + 36a + 4a^2$ **51.** $9a^2 - 6ab + b^2$

52. $m^2 - 12mn + 36n^2$ **53.** $4x^2 + 20xy + 25y^2$

54. $9x^2 - 24xy + 16y^2$ **55.** $x^2 - 16$ **56.** $x^2 - 9$

57. $49 - y^2$ **58.** $9 - n^2$ **59.** $4x^2 - 9$ **60.** $36 - 25a^2$

61. $16x^2 - 25y^2$ **62.** $4x^2 - 49y^2$ **63.** $81n^2 - 16m^2$

64. $a^2 - 4b^2$

65. The Associative Property of Multiplication guarantees that either grouping is correct. Method (i) is easier because it can be done in one step.

66. The Associative Property of Multiplication guarantees that the two expressions are equivalent. The grouping in (ii) is easier because it leads to the product of two binomials.

67. $a^6 - b^6$ **68.** $a^4 - b^4$ **69.** $16 + 4p^2q - 4pq^2 - p^3q^3$

70. $a^4b^4 - 4a^2b^2 + 3$ **71.** $x^4y^4 + 4x^2y^2 + 4$

72. $x^6 + 8x^3yz^2 + 16y^2z^4$ **73.** $12x^5 - 26x^4 - 10x^3$

74. $-3y^5 + 3y$ **75.** $-19x - 18$ **76.** $14x - 15$

77. $2b^3 - 5b^2 - 13b + 30$ **78.** $3a^3 - 2a^2 - 51a - 70$

79. $16x^4 - 1$ **80.** $81a^4 - 18a^2 + 1$ **81.** $-20x - 50$

82. -40 **83.** $-x^2 + x - 18$ **84.** $2x^2 + 51x - 10$

85. $12x$ **86.** $-x^2 - x + 9$ **87.** $32x^5 - 48x^4 + 18x^3$

88. $9y^3 + 21y^2 + 16y + 4$ **89.** $5y^2 + 20y - 29$

90. $12x^2 + 22x - 6$ **91.** $a = 4, b = -2$

92. $a = 2, b = -3$ **93.** $a = 3, b = 2$

94. $a = 1, b = -7$ or $a = -1, b = 7$

95. $2rt - r + 5t - 5$ **96.** $x^3 + 3x^2 + 2x$ **97.** $300w - 2w^2$

98. $10x^2 + 2x + 29$

99. Because the rate of change is not approximately constant, a linear model does not appear to be appropriate.

100. $P(x) = 0.48x + 16.85$

101. $E(x) = 0.39x^3 + 11.90x^2 + 5.60x + 2355.46$

102. $E(x)$ represents the total sales (in dollars).

103. $x^{2n} + 6x^n + 8$ **104.** $x^{2n} - 4x^n + 4$ **105.** $x^{2n} - 16$

106. $x^{2n+1} + x^{2n} - x^{n+1} - x^n + x + 1$

109. Write the expression as $[(x + 3y) - 4]\,[(x + 3y) + 4]$ and use the special product form $(A - B)(A + B)$. The result is $x^2 + 6xy + 9y^2 - 16$.

110. $$[(x + y) + 2]^2 = (x + y)^2 + 4(x + y) + 4 = x^2 + 2xy + y^2 + 4x + 4y + 4$$

111. $$[(x^2 + 5x) + 3]\,[(x^2 + 5x) - 3] = (x^2 + 5x)^2 - 9 = x^4 + 10x^3 + 25x^2 - 9$$

112. $4y^2 - 4xy + 12y + x^2 - 6x + 9$

113. $x^2 - 6x + 9 - 25y^2$

Chapter Review Exercises *(page 361)*

1. In the expression $x^2 \cdot x^3$, there is a total of 5 factors of x. This total can be found by adding the exponents according to the Product Rule for Exponents.

2. After using the Product to a Power Rule on $(x^2y^3)^4$, we have $(x^2)^4(y^3)^4$, to which the Power to a Power Rule applies.

3. -3^{12} **4.** k^{12} **5.** 4^4 **6.** x^4y^3 **7.** $\frac{-8}{125}$ **8.** c^{42}

9. $\frac{8h^9}{k^6}$ **10.** $-2x^{16}$ **11.** $-35x^{12}y^9$ **12.** $5c^{12}$

13. Any nonzero number divided by itself is 1.

14. If $b^{-n} = a$, then $a = \frac{1}{b^n}$, which implies that $ab^n = 1$. Therefore, neither a nor b can be 0.

15. 1 **16.** $-\frac{1}{4}$ **17.** $-\frac{1}{27}$ **18.** $\frac{8}{15}$ **19.** $\frac{1}{8}$ **20.** $\frac{1}{16}$

21. $-\frac{k}{h}$ **22.** $-\frac{4k^3}{3h^2}$ **23.** $\frac{1}{r^9}$ **24.** $-\frac{4}{x^3}$ **25.** $\frac{1}{y^2}$

26. $\frac{x}{32}$ **27.** $\frac{d^7}{c^{14}}$ **28.** $\frac{y^{15}}{8x^{12}}$ **29.** $\frac{h^{20}k^{12}}{16}$ **30.** $\frac{1}{y^4}$

31. The number p is the number of places to move the decimal point in 2.3 to obtain 23,000.

32. The calculator display for 5^{25} is 2.980232239E17, where E17 means 10^{17}. Thus $5^{25} \approx 2.98 \cdot 10^{17}$.

33. $2.45 \cdot 10^{10}$ **34.** $5.63 \cdot 10^{-6}$ **35.** 0.000000683

36. 849,000 **37.** $-1.6 \cdot 10^{87}$ **38.** $1.7 \cdot 10^{-16}$

39. **(a)** $6.9 \cdot 10^4$, $5.1 \cdot 10^6$ **(b)** 74 people per square mile

40. 0.49 cent

41. **(a)** Binomial **(b)** 2 **(c)** $-2x^2 + 3x$ **(d)** -2

42. **(a)** Monomial **(b)** 4 **(c)** x^4 **(d)** 0

43. **(a)** Trinomial **(b)** 5 **(c)** $9x^5 + x^2 - 3$ **(d)** 1

44. **(a)** None **(b)** 6 **(c)** $x^6 + 5x^5 - 7x^2 - 4x$ **(d)** -7

45. 6 **46.** 4 **47.** 2 **48.** -3 **49.** -55 **50.** 4

51. $2x^2 - 8x + 13$ **52.** $2x - 9$ **53.** $3x^2 - 2x + 7$

54. $7x^2 - 4x + 7$ **55.** $-8x^8$ **56.** $6x^5 - 8x^3$

57. $10a^2b - 2a^2b^3 + 6a^2b^2$ **58.** $x^2 - x - 12$

59. $-6x^2 + 7x + 24$ **60.** $2x^3 + x^2 - 5x + 2$

61. $-x^2 - 12x - 19$ **62.** $3x^3 + 3x^2 - 6x$

63. $8x^2 - 22x + 15$

64. $7xy - 12x + 4y^2 - 12y - x^2y + 3x^2 - xy^2$

65. A difference of two squares consists of two perfect square terms separated by a minus sign.

66. The template is used for squaring a binomial.

67. $9x^2 - 4$ **68.** $1 - 9y^2$ **69.** $9y^2 - 12y + 4$

70. $a^2b^2 + 8ab + 16$ **71.** $x^2 - x - 6$

72. $x^2 + 4xy + 4y^2 - 25$ **73.** $x^3 - x$ **74.** $2x^2 + 3x + 1$

Chapter Test *(page 364)*

1. [6.1] x^4y^6 **2.** [6.2] t^2 **3.** [6.2] $\frac{4b^8}{a^{10}}$ **4.** [6.2] $\frac{1}{a^4b^6}$

5. [6.2] $\frac{4b^2}{5a^3}$ **6.** [6.2] c^2d^6 **7.** [6.2] $-\frac{1}{32}$ **8.** [6.2] 3

9. [6.2] $-\frac{1}{8}$ **10.** [6.3] 27,000,000

11. [6.3] 0.0000456 **12.** [6.3] $5.7 \cdot 10^{-7}$

13. [6.3] $8 \cdot 10^{-4}$, $2.6 \cdot 10^8$

The product $\$2.08 \cdot 10^5$ represents the President's salary.

14. [6.4] -16

15. [6.4] **(a)** $(p + q)(x) = 3x^4 + x^3 - 5x^2 - 2x - 5$
$(p + q)(1) = -8$
$p(1) + q(1) = -7 + (-1) = -8$

(b) $(p - q)(x) = 3x^4 - 9x^3 - 5x^2 + 14x - 9$
$(p - q)(-1) = -16$
$p(-1) - q(-1) = -11 - 5 = -16$

16. [6.5] $-2x^5 + 10x^4 - 6x^3$ **17.** [6.5] $6t^2 + 11st + 4s^2$

18. [6.5] $x^3 - 11x + 6$ **19.** [6.5] $x^4 + 6x^2y + 9y^2$

20. [6.5] $25x^2 - 49z^2$ **21.** [6.5] $-6a^5b^6$

22. [6.5] $P(x) = x^5 + x^3 - x^2 - 1$

Chapter 7

Section 7.1 *(page 372)*

1. Choose the smallest exponent with each variable.

2. The GCF of an expression is not always a monomial. For example, in $(x + y)z + (x + y)^2$, the GCF is $x + y$.

3. 12 **4.** 7 **5.** 4 **6.** 5 **7.** x^3 **8.** y^4 **9.** mn^3

10. pq **11.** $2x - 1$ **12.** $4x - 3$ **13.** 5 **14.** 3

15. $3(x + 4)$ **16.** $2(x + 3)$ **17.** $5(x + 2y - 6)$

18. $6(1 - 4a - 3b)$ **19.** $4x(3x - 1)$ **20.** $2x(3x + 1)$

21. Prime **22.** Prime **23.** $x^2(x - 1)$ **24.** $c^3(c + 3)$

25. $3x(x^2 - 3x + 4)$ **26.** $5b^4(3b + b^3 - 2)$

27. $xy^2(x^4y^2 - x^3y + 1)$ **28.** $xy(x^6y^7 + y^4 - x^4)$

29. $(a + by)(3x - 2y)$ **30.** $(x - 3)(2x - 5y)$

31. $(2x + 3)(4x^2 - 7y^2)$ **32.** $(x - 2)(5x^5 - 9y^3)$

33. All are correct because the product in each case is $4x - 12$.

34. By the Multiplication Property of -1, $x - 2y = -1(2y - x)$.

35. $4(x - 3)$, $-4(3 - x)$ **36.** $2(5 - 2x)$, $-2(2x - 5)$

37. $y(6 + 3x - x^2)$, $-y(x^2 - 3x - 6)$

38. $b(-a^2 + 5)$, $-b(a^2 - 5)$ **39.** $(y - 3)(y - 2)$

40. $(x - 2)(x + 3)$ **41.** $(3 - z)(5 + z)$ **42.** $(6 - a)(4 - a)$

43. $(x - 4)(2x + 5)$ **44.** $(3 - a)(1 + a)$

45. The answer is obtained from multiplying the factors.

$$(2x - 5)(x + 3) = 2x^2 + 6x - 5x - 15$$

46. The correct method is (ii). Removing the parentheses in (i) will not result in the original expression.

47. $(b + y)(a + x)$ **48.** $(x + 3)(a + 2)$ **49.** $(d - 4)(c + 3)$

50. $(c + 5)(a - 2)$ **51.** Prime **52.** Prime

53. $(x^2 + y)(a + 3)$ **54.** $(x^2 + 2)(a - y)$

55. $(x^2 + 3)(x + 2)$ **56.** $(x + 4)(x^2 - 3)$

57. $3(y^2 + 4)(y + 3)$ **58.** $4(y - 3)(y^2 + 5)$

59. $(a^2 + 1)(b + y)$ **60.** $(b - x)(ab - 1)$

61. $(x^2 + 3)(x - 5)$ **62.** $(x - 4)(2x^2 - 3)$

63. $(3x^2 + 4)(2x - 3)$ **64.** $(3x + 2)(2x^2 + 3)$

65. $3(2x^2 - 5)(3x + 1)$ **66.** $2(a + b)(a^2 + 3b)$

67. $x(3y^2 + 7)(y - 2)$ **68.** $rt(t + 1)(r + 1)$ **69.** $4y - 5$

70. $4 - 3x$ **71.** $3x^2 + 2x + 6$ **72.** $4x^2 + x - 2$

73. $\frac{1}{3}(x + 15)$ **74.** $\frac{1}{4}(a - 6b)$ **75.** $\frac{1}{12}(2x^2 + 3x + 36)$

76. $\frac{1}{10}(10x^2 - 5x + 2)$ **77.** $d^{15}(d^5 - 1)$ **78.** $y^{12}(1 + y^8)$

79. $11x^{48}(8 + x^3)$ **80.** $10a^{25}(10a^{25} + 1)$

81. $6ab^2(4a^4b - b^2 + 2a^2)$ **82.** $4c^2d(4c^4d^3 + 2d^2 - 3)$

83. $4a^2b^2(x - 4)^2(b + 3ax - 12a)$

84. $5c^4d^4(x + 3)^3[3d(x + 3)^2 + c]$ **85.** $A = P(1 + r)$

86. $h = 16t(35 - t)$ **87.** $A = 2\pi r(h + r)$

88. $A = \frac{1}{2}h(b_1 + b_2)$

89. Use the Distributive Property to multiply $x^{-6}(3x^2 + 2)$.

90. The GCF is x^{-4} and the other factor is $2x + 5$.

91. $2x - 5$ **92.** $1 + x^{10}$ **93.** $2x^{-4}(x^7 - 5)$

94. $y^{-4}(y^2 + 3)$ **95.** $5x^{-4}y^{-2}(y^9 - 3x^7)$

96. $3x^{-1}y^{-3}(y + 3x^2)$ **97.** $(x + 3)(a^2 + b - 5)$

98. Factor $4n^2 + 2n = 2n(2n + 1)$. Then $2n$ is an even integer and $2n + 1$ is an odd integer.

Section 7.2 *(page 377)*

1. A difference of two squares consists of two perfect square terms separated by a minus sign.

2. The expression $9x^2 + 36$ has a GCF of 9. The expression $9x^2 + 16$ has no GCF other than 1, and this sum of two squares cannot be factored.

3. $(x + 3)(x - 3)$ **4.** $(3x + 4)(3x - 4)$

5. $(1 + 2x)(1 - 2x)$ **6.** $(5 + a)(5 - a)$

7. $(4x + 5y)(4x - 5y)$ **8.** $(7x + 8y)(7x - 8y)$

9. $\left(\frac{1}{2}x + 1\right)\left(\frac{1}{2}x - 1\right)$ **10.** $\left(\frac{1}{3}y + \frac{1}{4}\right)\left(\frac{1}{3}y - \frac{1}{4}\right)$

11. $(a^3 + b^8)(a^3 - b^8)$ **12.** $(x^2 + y)(x^2 - y)$

13. $(a + b + 2)(a + b - 2)$ **14.** $(x - 3 + y)(x - 3 - y)$

15. $(4 + x - y)(4 - x + y)$ **16.** $(9 + x)(1 - x)$

17. A perfect square trinomial is a trinomial with two terms being perfect squares and the third term being twice the product of the square roots of the other two terms.

18. Square the binomial to obtain the original expression.

19. $(a - 2)^2$ **20.** $(x + 4)^2$ **21.** $(2x + 3)^2$ **22.** $(3x - 2)^2$

23. $(x - 7y)^2$ **24.** $(x + 6y)^2$ **25.** $\left(x + \frac{1}{2}\right)^2$

26. $\left(y - \frac{2}{3}\right)^2$ **27.** $(x^2 - 5)^2$ **28.** $(y^2 + 3)^2$

29. $(x + 1)(x^2 - x + 1)$ **30.** $(2 - a)(4 + 2a + a^2)$

31. $(3y - 2)(9y^2 + 6y + 4)$ **32.** $(4y + 3)(16y^2 - 12y + 9)$

33. $(5 - 2x)(25 + 10x + 4x^2)$

34. $(4 - 3z)(16 + 12z + 9z^2)$

35. $(3x + y)(9x^2 - 3xy + y^2)$

36. $(2a - bc)(4a^2 + 2abc + b^2c^2)$

37. $(x^2 + 1)(x^4 - x^2 + 1)$ **38.** $(b^2 - 5)(b^4 + 5b^2 + 25)$

39. $(x + 1)^2$ **40.** $(x - 5)^2$ **41.** $(0.5x + 0.7)(0.5x - 0.7)$

42. $(1.2a + 0.04)(1.2a - 0.04)$ **43.** $(x - 1)(x^2 + x + 1)$

44. $(3 + b)(9 - 3b + b^2)$ **45.** $(c - 10)^2$ **46.** $(d + 9)^2$

47. $2(x + 3)(x^2 - 3x + 9)$ **48.** $3(x - 2)(x^2 + 2x + 4)$

49. Prime **50.** $2(16a^2 + 25)$ **51.** $(6 - 7y)^2$

52. $(5 + 6p)^2$ **53.** $4(x + 4)(x - 4)$ **54.** $4(2 + y)(2 - y)$
55. $(5x - 4y)^2$ **56.** $(4x + 5y)^2$
57. $(x^2 + 4)(x + 2)(x - 2)$
58. $(9a^2 + b^2)(3a + b)(3a - b)$
59. $(2 + 5x^2)(4 - 10x^2 + 25x^4)$
60. $(5 - 3x^2)(25 + 15x^2 + 9x^4)$ **61.** $(y^4 - 3)^2$
62. $(x^4 + 2)^2$ **63.** $(6x - y)(36x^2 + 6xy + y^2)$
64. $(ab + 2)(a^2b^2 - 2ab + 4)$ **65.** $(x - 2)^2(x^2 + 2x + 4)^2$
66. $(x + 2)^2(x^2 - 2x + 4)^2$ **67.** $(x^6 + 2y^{10})(x^6 - 2y^{10})$
68. $(7 + 9x^2)(7 - 9x^2)$ **69.** $(x^3 - 2)^2$ **70.** $(y^3 + 3)^2$
71. $2(7 + y^3)(7 - y^3)$ **72.** $5x^2(5 + x^3)(5 - x^3)$
73. $x^2(5xy - 1)(25x^2y^2 + 5xy + 1)$
74. $y^4(3x^3y - 2)(9x^6y^2 + 6x^3y + 4)$ **75.** $x^3(7x^2 - 5)^2$
76. $x^2(3x^2 - 7)^2$ **77.** Prime **78.** Prime
79. $(4 + 3xy^3)(16 - 12xy^3 + 9x^2y^6)$
80. $(1 - 3x^2y)(1 + 3x^2y + 9x^4y^2)$ **81.** $(x + 2)^2(x - 2)^2$
82. $(x + 3)^2(x - 3)^2$ **83.** $(4x^2 + 5y)(4x^2 - 5y)$
84. $(c^4 + 1)(c^2 + 1)(c + 1)(c - 1)$ **85.** $(8x^2 + 5y)^2$
86. $(7x^3 + 9y^2)^2$ **87.** $(4a^2 + 9d^2)(2a + 3d)(2a - 3d)$
88. $(x^2 + c^2)(x + c)(x - c)$ **89.** $2x(x + 2)(x^2 - 2x + 4)$
90. $3y^2(y - 3)(y^2 + 3y + 9)$
91. $(5x^2 - 3y^4)(25x^4 + 15x^2y^4 + 9y^8)$
92. $(3x^3 + 2y^2)(9x^6 - 6x^3y^2 + 4y^4)$
93. $4(a + 2b)(x + y)(x - y)$ **94.** $5(a + b)(a - b)(2x + y)$
95. $(a + b)(x + y)(x - y)$ **96.** $(x - y)(2a + b)(2a - b)$
97. $4(x - 2)^2$ **98.** $9(2x - 1)^2$ **99.** $(x + 2)^2(x - 2)^2$
100. $(x + 1)^2(x - 1)^2$
101. Because $x^6 - y^6 = (x^3)^2 - (y^3)^2$, the expression is a difference of two squares. Because $x^6 - y^6 = (x^2)^3 - (y^2)^3$, the expression is a difference of two cubes.
102. Because a sum of two squares does not factor, the expression should be regarded as a sum of two cubes.
103. $-6, 6$ **104.** $-20, 20$ **105.** 1 **106.** 1 **107.** 16
108. 25 **109.** $A = \pi R^2 - \pi r^2, A = \pi(R + r)(R - r)$
110. Answers will vary.
111. $\frac{1}{2}mv_1^2 - \frac{1}{2}mv_2^2, \frac{1}{2}m(v_1 + v_2)(v_1 - v_2)$
112. $x^4 - x, x(x - 1)(x^2 + x + 1)$
113. $V(x) = 80(7x^2 + 555)$ **114.** $E(x) = 20(4x + 71)$
115. $F(x) = 0.08(7x^2 + 555)$
116. $C(x) = 1.6(7x^2 + 555)(4x + 71)$
117. $(x + 2)(x - 2)(x^2 - 2x + 4)(x^2 + 2x + 4)$
118. $(c + 1)(c - 1)(c^2 - c + 1)(c^2 + c + 1)$
119. $(a + b)(a - b)(a^2 - ab + b^2)(a^2 + ab + b^2)$
120. $(x + 2y)(x - 2y)(x^2 - 2xy + 4y^2)(x^2 + 2xy + 4y^2)$
121. $(2x + 1)(2x - 1)(x - 2)(x^2 + 2x + 4)$
122. $(x - 2)(x + 2)^2(x^2 - 2x + 4)$
123. $(x + 1)(x - 1)(y - z)(y^2 + yz + z^2)$
124. $(y + 3x - 2)(y - 3x + 2)$

Section 7.3 *(page 387)*

1. The Commutative Property of Multiplication allows us to write the factors in either order.
2. Multiply the factors to determine whether you obtain the original expression.
3. $(x + 1)(x + 2)$ **4.** $(x - 1)(x - 3)$ **5.** $(x - 3)(x - 6)$
6. $(x + 6)(x - 3)$ **7.** $(x + 2)(x + 7)$ **8.** $(x + 7)(x + 3)$
9. Prime **10.** Prime **11.** $(x - 8y)(x + 2y)$
12. $(x - 9y)(x + 5y)$ **13.** $(2x + 1)(x + 2)$
14. $(x + 2)(2x + 3)$ **15.** $(2 + x)(6 - x)$
16. $(1 + x)(4 - x)$ **17.** $(3x + 1)(x - 3)$
18. $(x - 3)(3x + 5)$ **19.** $(x - 5)(4x + 3)$
20. $(3 - 2x)(4 + 5x)$ **21.** Prime **22.** Prime
23. $(c + 2)(c - 1)$ **24.** $(x - 3)(x + 1)$
25. $(6x - 5)(x + 3)$ **26.** $(x + 1)(15x - 7)$
27. $(x + 8)(x - 3)$ **28.** $(x + 8)(x - 5)$
29. $(4 + x)(6 - x)$ **30.** $(3 + x)(5 - x)$
31. $(b - 5)(b + 3)$ **32.** $(c - 6)(c + 2)$
33. $(3x + 2)(x + 1)$ **34.** $(x - 3)(3x + 2)$
35. $(3x - 5)(2x - 3)$ **36.** $(2 + 3x)(3 - 4x)$
37. $(x + 5)^2$ **38.** $(y + 4)^2$ **39.** $(2 - 5x)(1 + 3x)$
40. $(2 - 5x)(1 - 3x)$ **41.** $(9 - 10x)(2 + x)$
42. $(3x - 2)(x + 4)$ **43.** $(a - 6)(a - 4)$
44. $(b - 6)(b + 4)$ **45.** $(3x + 2)(3x - 5)$
46. $(2x - 3)(3x + 5)$ **47.** $(x - 7)(x - 9)$
48. $(x - 9)(x + 7)$ **49.** $9(x - 2)(x + 1)$
50. $9(x - 3)(x + 2)$ **51.** $(2x + 3)(6x - 1)$
52. $(5x + 4)(7x - 3)$ **53.** $(4x - 7y)(2x + 3y)$
54. $(2x - 3y)(4x + 5y)$ **55.** $2(x + 3)(x - 2)$
56. $3(x + 4)(x - 1)$ **57.** $4(x + 3)^2$ **58.** $8(x + 5)(x + 1)$
59. There are fewer combinations to try because 17 and 23 have fewer factors than 24 and -18.
60. Both factorizations are correct.
61. $8xy(x - 5y)(x + 2y)$ **62.** $6x^4(x - 8)(x - 2)$
63. $-3x(2x - 5)(2x + 3)$ **64.** $5y^4(2y + 5)(y + 3)$
65. $2y(2x - 3)(7x + 3)$ **66.** $3y(y - 6)(5y + 2)$
67. $(ab - 5)(ab + 2)$ **68.** $(cd - 7)(cd + 3)$
69. $xy(x + 6)^2$ **70.** $2x(x + 7)^2$ **71.** $(xy - 8)(xy + 6)$
72. $(3 - ab)(12 + ab)$ **73.** $2(3x - y)^2$

74. $(2xy + 5)(4xy - 1)$ **75.** $(6m - n)(m + n)$

76. $2n^4(4m + n)(2m + n)$ **77.** $2xy^2(4x - y)(2x - y)$

78. $6ab^2(5b - a)(b + a)$ **79.** $x(x - 1)(x - 3)(x + 14)$

80. $x(2x - 1)(x + 7)(x - 3)$

81. $(x + 5)(x - 5)(x + 3)(x - 3)(x + 1)(x - 1)$

82. $(y + 2)(y - 2)(x^2 + 3)(3x^2 - 2)$

83. $(xy + 1)(2xy - 3)(xy + 3)$

84. $(2 - y)(2x + 5y)(x - 2y)$

85. No, the factor $x^2 - 4$ can be factored as $(x + 2)(x - 2)$.

86. The trinomial is not a quadratic trinomial.

87. $(x^2 + 30)(x^2 - 2)$ **88.** $(x^2 - 20)(x^2 + 3)$

89. $(x^3 + 5)(x^3 + 12)$ **90.** $(x^3 - 10)(x^3 - 6)$

91. $(x + 3)(x - 3)(x + 2)(x - 2)$

92. $(x + 2)(x - 2)(x + 1)(x - 1)$ **93.** $(5x^4 - 7)(x^4 + 2)$

94. $(x^4 + 9)(3x^2 + 2)(3x^2 - 2)$

95. $(2x + 1)(2x - 1)(x + 3)(x - 3)$

96. $(x^2 + 3)(2x + 1)(2x - 1)$

97. $(3x + 2)(3x - 2)(x + 1)(x - 1)$

98. $(x + 2)(x - 2)(9x^2 + 5)$ **99.** $(x + 3)(x + 6)$

100. $(x + 10)(x + 5)$ **101.** $2(x - 3)(2x - 7)$

102. $(4 - 9y)(1 - 12y)$ **103.** $(2x^2 + 4x - 3)(x^2 + 2x + 2)$

104. $(2x + xy - 3)(4x + 2xy + 5)$ **105.** $(x + 8)(x - 3)$

106. $\frac{1}{2}(x + 4)(x + 10)$; The difference is 6.

107. $f(t) = 52(1.5t + 75)$

108. The other factor represents the average weekly food cost.

109. The coefficient of t indicates that food costs were increasing at an annual rate of \$78 during this period. In the factored form, the coefficient of t indicates that food costs were increasing at a weekly rate of \$1.50.

110. The constant term indicates that annual food costs were \$3900 in 1990. In the factored form, the constant term indicates that weekly food costs were \$75.

111. $-7, 7, -11, 11$ **112.** $-1, 1, -4, 4, -11, 11$

113. $-8, 8, -16, 16$ **114.** $-5, 5, -13, 13$ **115.** 2

116. 1 **117.** $(x^n + 8)(x^n + 3)$ **118.** $(y^n + 5)(y^n + 6)$

119. $(y^{2n} - 11)(y^n + 3)(y^n - 3)$

120. $(x^{2n} - 8)(x^n + 3)(x^n - 3)$ **121.** $(5x^n - 3)(x^n + 1)$

122. $(x^n + 3)(7x^n - 5)$ **123.** $x^{2n}(3x^n + 2)(x^n + 1)$

124. Answers will vary.

Section 7.3 Supplementary Exercises *(page 389)*

1. $(5x + 3)(x - 2)$ **2.** $(x - 6)(5x - 1)$

3. $(5 + 9y)(5 - 9y)$ **4.** $9(x + 3)(x - 3)$

5. $(x^2 - 3)(2x + 1)$ **6.** $(x + 3)(3x^2 + 1)$

7. $(x - 2)(3x + 5y)$ **8.** $(y - 3)(y^2 - 5)$

9. $(5x + 1)(25x^2 - 5x + 1)$

10. $(4 - 3a)(16 + 12a + 9a^2)$ **11.** $(x + 2)(x + 4)$

12. $(x - 4)(x + 3)$ **13.** Prime **14.** Prime **15.** $(2x + 5)^2$

16. $5(3a - 1)^2$ **17.** $(a - b)(x - 3)$ **18.** $(x + 3)(y - 5)$

19. $25(y - 1)(y^2 + y + 1)$ **20.** $(ab + 2)(a^2b^2 - 2ab + 4)$

21. $(y^2 + 4x^2)(y + 2x)(y - 2x)$ **22.** $5(x + 2)(x - 2)$

23. $a^2b^4(b^3 + 3ab - 1)$ **24.** $3x^2(x^2 + 5x - 1)$

25. $(3x + 1)(7x - 2)$ **26.** $(3 - y)(1 - 7y)$ **27.** $(4 - a)^2$

28. $(x + 3y)^2$ **29.** $2(5 + 3x^2)(5 - 3x^2)$

30. $(3x^2 + 2)(x^2 + 4)$ **31.** $(y + 5)(2 - 3x)$

32. $(2x - 3)(x + 3)(x - 3)$ **33.** $4x^2(3x - 2)(2x + 3)$

34. $b(8a - 3)(a - 3)$ **35.** $(c^{10} + d^4)(c^5 + d^2)(c^5 - d^2)$

36. $(9 + t^2)(3 + t)(3 - t)$ **37.** Prime **38.** Prime

39. $(a + 2)(a - 2)(x - y)(x^2 + xy + y^2)$

40. $(3a + 2)(3a - 2)(2a + 5)$ **41.** $a^5b^3(b - a^2b^5 + a)$

42. $(a + b)(3d - c)$ **43.** $(m - 8)(m - 2)$

44. $(1 + y)(8 - y)$ **45.** $4(x^2 + 4y^2)(x + 2y)(x - 2y)$

46. $2(2x + 7y)(2x - 7y)$

47. $(a + b - 2)(a^2 + 2ab + b^2 + 2a + 2b + 4)$

48. $(c + d + 4)(c + d - 4)$ **49.** $5(2x - 3)^2$

50. $x^2(y + 11)(y - 8)$ **51.** $2a^4(3a - 5)(5a - 9)$

52. $2(a - 3)(6a + 7)$

53. $(c - 1)(c^2 + c + 1)(c + 2)(c - 2)$

54. $2x^2(2x - 3)(4x^2 + 6x + 9)$

55. $(x^2 + 11y^2)(x + 3y)(x - 3y)$ **56.** $(a - 8)(a - 5)$

Section 7.4 *(page 395)*

1. The first person is setting all three factors equal to 0. The second person regards the left side as $(3x)(x + 5)$ and is setting the two factors equal to 0.

2. The property applies only if one side of the equation is 0.

3. $-3, 4$ **4.** $-2.5, -1$ **5.** $0, 6$ **6.** $0, 0.5$ **7.** 5

8. -4 **9.** $-5, -1, 3$ **10.** $-1.5, 0, 1$

11. Assuming that the equation has only one variable, the largest exponent on the variable is 1 for a first-degree equation and 2 for a second-degree equation.

12. The factors must be identical. **13.** $-7, 0$ **14.** $0, 2.5$

15. $-3, 3$ **16.** $-7, 7$ **17.** $2, 3$ **18.** $-3, \frac{2}{3}$

19. $-1, 2.5$ **20.** $-\frac{1}{3}, \frac{3}{2}$

21. The fact that the product of two numbers is 1 does not imply that the two numbers are 1. There are infinitely many pairs of reciprocals, each with a product of 1.

22. The two solutions of $x^2 = 7x$ are 0 and 7. If we divide both sides by x, we are assuming that x is not 0. Thus 0 is lost as a solution.

23. 0, 5 **24.** $-5, 5$ **25.** $-3, 7$ **26.** $-11, 5$

27. $-6, 1$ **28.** $-2, 4$ **29.** 5, 7 **30.** 3, 4 **31.** 1.5, 4

32. $\frac{5}{7}, 3$ **33.** $-2, \frac{4}{9}$ **34.** $-\frac{2}{3}, \frac{3}{4}$ **35.** $-6, 5$ **36.** $-6, 7$

37. 2, 30 **38.** $-20, -3$ **39.** $-4, 2$ **40.** $\frac{1}{3}$ **41.** 2.5, 2

42. $-3, \frac{2}{3}$ **43.** $-3, 4$ **44.** 3, 5 **45.** $-4, 1.5$

46. $-3, 6$ **47.** 0, 1.25 **48.** $-\frac{13}{5}, 2$ **49.** $-6, 3, 5$

50. $-2.5, 0.75, 2.5$ **51.** $-3, 0, 3$ **52.** $-\frac{2}{5}, 0, \frac{2}{5}$

53. $-1, 0, \frac{2}{3}$ **54.** $-1, 0, \frac{2}{3}$ **55.** $-1, 1, 5$

56. $-3, -1, 1$ **57.** $-2, 1.5, 2$ **58.** $-3, -\frac{4}{3}, 3$

59. $-3, -2, 2, 3$ **60.** $-5, -1, 1, 5$ **61.** $2, \frac{5}{2}, \frac{5}{3}$

62. $-1.5, 1.5, 4, 12$ **63.** $-3, 2, 3$ **64.** $\frac{1}{5}, \frac{3}{2}, \frac{5}{2}$

65. $b = 5; -0.5$ **66.** $a = -2; -\frac{5}{3}$ **67.** $c = 6; -3.5$

68. $c = 2; 2$ **69.** $x^2 + x - 12 = 0$

70. $x^2 - 3x - 10 = 0$ **71.** $x^2 - 8x + 16 = 0$

72. $x^2 + 4x + 4 = 0$ **73.** $2x^2 - x = 0$

74. $3x^2 + 4x - 4 = 0$ **75.** $-c, 2c$ **76.** $-2y, y$

77. $-\frac{3y}{2}, \frac{y}{2}$ **78.** $\frac{2c}{3}, -\frac{c}{2}$ **79.** $-3a, a$ **80.** $-\frac{4a}{5}, \frac{a}{2}$

81. $-6, -3, -2, 1$ **82.** $-\frac{1}{3}, \frac{2}{3}, 1, 2$ **83.** $-9, -7, -3, -1$

84. $-6, -5, -2, -1$ **85.** $-1, 1, 2, 4$ **86.** 1, 2, 3, 4

87. -9 or 8 **88.** -9 or 7 **89.** 3, 4, 5 or $-5, -4, -3$

90. $-6, -5, -4$ or 4, 5, 6 **91.** 0, 2 or 3 **92.** 0, 2, or 6

93. -2 and 5 **94.** 5 and 7 **95.** 22 meters, 27 meters

96. 14 feet, 21 feet **97.** 8 inches, 15 inches, 17 inches

98. 15 feet, 20 feet, 25 feet **99.** 2 feet

100. 100 feet, 150 feet **101.** 15 inches, 20 inches **102.** 10

103. **(a)** $C(x) = -0.05(x^2 - 88x + 640)$
(b) $C(x) = -0.05(x - 80)(x - 8)$
(c) 8, 80
(d) No person of age 8 or 80 becomes upset about not having coffee at a regular time.

104. **(a)** $-1.4(x - 10)(x - 92)$
(b) 10, 92
(c) The average annual cost of utilities is 0 for persons age 10 and 92. The results are not very meaningful because the cost is 0 for children older than age 10, and people age 92 will still have utilities costs.

105. $x = 1$, y is any number **106.** $-\frac{1}{6}, \frac{3}{4}$ **107.** $-2, 1$

108. \$1356.80

Section 7.5 *(page 404)*

1. Dividend: $x + 7$, Divisor: x, Quotient: 1, Remainder: 7

2. Multiply $x + 4$ by $x + 1$ and add 5 to the result. You should obtain the dividend $x^2 + 5x + 9$.

3. $2x^2 + 4x$ **4.** $2x - 1$ **5.** $2x^3 + 3x + \frac{1}{2}$

6. $2x^3 - x^2 + \frac{2}{3}$ **7.** $8a^3 + 6a^2 - 3a + 4$

8. $6a^3 - 4a^2 - 5a + 3$ **9.** $5x^3 - 3x^2 + \frac{3}{2} - \frac{2}{x}$

10. $4x^3 + 2x^2 - \frac{3}{5} + \frac{3}{x}$ **11.** $5c^3d^5 + 2c^2d^3 - \frac{3d}{c}$

12. $9c^3d^4 - 5cd^2 + \frac{4}{d}$ **13.** $9x^2 + 7x - 6$

14. $6x^2 - 4x + 2$ **15.** 8 **16.** -4

17. Stop dividing when the remainder is 0 or when the degree of the remainder is less than the degree of the divisor.

18. The remainder could be 0 (which has no degree) or it could be of degree 0 or 1.

19. $Q(x) = 2, R(x) = 7$ **20.** $Q(x) = 4, R(x) = -5$

21. $Q(x) = x + 2, R(x) = -1$ **22.** $Q(x) = x - 4, R(x) = 6$

23. $Q(x) = 3x + 3, R(x) = 5$

24. $Q(x) = x - 1, R(x) = -1$

25. $Q(a) = a^2 - a + 2, R(a) = -2$

26. $Q(x) = x^3 + x^2 - x + 2, R(x) = -3$

27. $Q(x) = 3x^2 + 4x + 1, R(x) = 0$

28. $Q(x) = 2x^2 - x + 3, R(x) = 0$ **29.** $Q(x) = 1, R(x) = 4$

30. $Q(x) = 1, R(x) = -5$ **31.** $Q(x) = 1, R(x) = x - 4$

32. $Q(x) = 3x, R(x) = -4x - 1$

33. $Q(a) = a^2 + 2a, R(a) = a - 3$

34. $Q(c) = 2c^2 + 6c + 16, R(c) = 34c - 13$

35. Q: $x + 6y$, R: $15y^2$ **36.** Q: $2x - 2y$, R: y^2

37. Q: $4x^2 - 4xy + y^2$, R: $-3y^3$

38. Q: $x^2 + xy + 2y^2$, R: $3y^3$

39. Write the divisor as $x - (-3)$. Then the divider is -3.

40. When we use 0 as the coefficient of each missing term, the row of dividend coefficients is $-1 \quad 0 \quad 0 \quad 0 \quad 1 \quad 0$.

41. $Q(x) = 3x + 2, R(x) = 0$ **42.** $Q(x) = x - 5, R(x) = 0$

43. $Q(x) = 6x + 4, R(x) = 3$

44. $Q(x) = 2x - 8, R(x) = 10$

45. $Q(x) = 2x^3 + x^2 + 3x + 3, R(x) = 10$

46. $Q(x) = 3x^2 + 5x + 13, R(x) = 34$

47. $Q(x) = x^3 - 2x^2 + 5x - 10, R(x) = 6$

48. $Q(x) = 2x^2 - 7x + 10, R(x) = 2$

49. $Q(x) = x^2 + 4x + 5, R(x) = 4$

50. $Q(x) = x^2 + 3x + 13, R(x) = 26$

51. No, the divisor cannot be second-degree. It must be in the form $x - c$.

52. The first expression is the dividend and the second expression is the divisor, so we mean divide $x + 2$ by $x^2 + 5x - 3$.

53. $a = 1, b = -2, c = 0, d = 1, e = -2$

54. $a = 2, b = 1, c = -16, d = 1, e = 6$

55. Dividend: $x^3 + 3x^2 - 2x - 4$, divisor: $x + 1$

56. Dividend: $x^3 - 4x^2 + 3x - 11$, divisor: $x - 4$

57. $x + 4 + \frac{15}{x - 6}$ **58.** $3x + 5 - \frac{10}{x + 3}$

59. $3x + 14 + \frac{44}{x - 3}$ **60.** $2x - 8 + \frac{15}{x + 2}$

61. $x + 5$ **62.** $x - 4$ **63.** $2a^2 + 3a + 2 + \frac{7}{3a - 2}$

64. $2x^3 - x^2 + 3x - 5 + \frac{5}{2x + 1}$

65. $2x^2 + x - 5 + \frac{12}{x + 2}$

66. $3x^3 - 2x^2 - 3x + 5 - \frac{8}{x + 1}$

67. $x^4 - 2x^3 + 2x^2 - 4x + 9 - \frac{13}{x + 2}$

68. $x^4 - x^3 + 3x^2 - 9x + 2 - \frac{11}{x + 3}$

69. $(x - 1)(x + 4)(x - 3)$ **70.** $(x + 8)(x - 5)(x - 3)$

71. $(x + 3)(x - 4)(x - 2)$ **72.** $(x - 2)(x - 3)(x - 1)$

73. $(2x + 1)(x - 1)^2$ **74.** $(3x + 2)(x - 2)(x + 1)$

75. $2x - 13 + \frac{21}{x + 2}$ **76.** $1 + \frac{6}{x - 2}$

77. $9x + 15 + \frac{16}{x - 1}$ **78.** $x^2 + 3 + \frac{1}{x^2 + 1}$

79. $2x^2 + 5x - 1$ **80.** $x^3 + x^2 + x - 2$ **81.** $x + 2$

82. $x^2 + 1$ **83.** $Q(x) = x^4 + 3x^3 + 6x^2 + x + 4, R(x) = 0$

84. $Q(x) = x^3 - x^2 + 2x - 2, R(x) = 4$

85. $Q(x) = x + 3, R(x) = 11x - 3$ **86.** $-x + 8$

87. $9x^5 + 12x^4 + 3x^3 - 12x^2$ **88.** $P(x) = x - 4$

89. $x + 8$ **90.** $w + 2$ **91.** **(a)** $x^2 - 8x - 1$ **(b)** 2 feet

92. **(a)** $n + 52$ **(b)** Decrease, \$50 **93.** -8 **94.** 2

95. 4 **96.** -3 **97.** $4x^{2n} - 16$ **98.** $2x^{4n} + 4x^{3n} - 3x^{2n}$

99. $Q(x) = x^{3n} - x^n + 5, R(x) = -14$

100. $Q(x) = x^n + 1, R(x) = -5$ **101.** 2 **102.** $a = 2, c = 1$

Chapter Review Exercises *(page 408)*

1. Factor out the common factor $-x^2y^2$.

2. The expression is equivalent to $5x + 13y$, which has no common factor and cannot be factored by any other means.

3. $x^{25}(1 + x^{10})$ **4.** $6x^2(x - 6)$ **5.** $x^3y^3(x^3 - x^4y - y^2)$

6. $(x - a)(3x - 4y)$ **7.** $(x^2 - 2)(x + 3)$

8. $(c - f)(a - b)$ **9.** $7x^3(2x^4 - 5x^2 + 6)$

10. $3x(3x - 2)^3$ **11.** Answers will vary. **12.** $xy^2 - 4$

13. The constant term is not a perfect square.

14. Multiplying the factors, we obtain $x^3 + 1$.

15. $(7x + 4y)(7x - 4y)$ **16.** $(a^{10} + b^5)(a^{10} - b^5)$

17. $(4 + a^2)(2 + a)(2 - a)$ **18.** $(2a - 1)(4a^2 + 2a + 1)$

19. $(3b^2 + 2a)(9b^4 - 6ab^2 + 4a^2)$ **20.** $(2x + 3y)^2$

21. $(c + 8)^2$ **22.** $(c^2 + 3)^2$ **23.** Answers will vary.

24. $(2x - 3y)^2$ **25.** $(x - 6)(x + 1)$ **26.** $(x + 4)(x + 1)$

27. $2(x + 7)(x - 1)$ **28.** $(x^2 + 12)(x^2 - 5)$

29. Because the numbers 6 and 24 have several factors, they lead to a large number of possible factorizations to try.

30. The approach is correct, but factoring out the common factor $(2x - 1)$ is easier.

31. $(3x + 5)(x + 3)$ **32.** $(x - 1)(8x + 5)$

33. $(6x - 5)(2x + 3)$ **34.** $3(2x - 3)(2x + 1)$

35. $(x + 4)(x - 2)$ **36.** $5(5x + 4y)(5x - 4y)$

37. $(5x - 4)(25x^2 + 20x + 16)$ **38.** $3(x - 5)(x + 4)$

39. No, but you can factor out the 2 to obtain $[2(x + 2y)]^2 = 4(x + 2y)^2$.

40. The designer decreased the length by 5 feet and increased the width by 2 feet.

41. $(3x - 5)(x + 3)$ **42.** $(2x + 5)(x + 3)(x - 3)$

43. $(x^3 + 3y^2)(x^6 - 3x^3y^2 + 9y^4)$

44. $(y^2 + x^2)(y + x)(y - x)$ **45.** $x^3y^3(y - x)$

46. $25x^3(2 - x)$ **47.** $x(x + 3)(6x + 7)$

48. $3(8x - 9)(2x - 5)$

49. In $a + b = 0$, we know that a and b are opposites, but we do not know what the numbers are. In $ab = 0$, either a or b must be 0.

50. One of the factors must be 0, but the factor 5 is not 0.

51. 0, 8 **52.** $-9, 9$ **53.** $-7, 6$ **54.** $5, \frac{3}{4}$ **55.** $-6, 4$

56. $-4, 3$ **57.** $-3, -2, 2$ **58.** $-2, 6$ **59.** $-3, -2, -\frac{1}{3}, \frac{2}{3}$

60. $-1, \frac{3}{5}$ **61.** -9 or 7 **62.** 5 seconds

63. Multiplying the quotient times the divisor and adding the remainder should result in the dividend.

64. The 2, -1, and 1 are the coefficients of the quotient, and the 3 is the remainder.

65. $3x^3 - 4x$ **66.** $3a^3b - 2a + \frac{4}{b^2}$

67. $Q(x) = 3x^2 - 4x - 5, R(x) = -5$

68. $Q(x) = 6x^2 - 2x - 9, R(x) = 8x + 10$

69. Q: $x^2 + 3y^2$, R: $8y^3$

70. $Q(x) = x^2 + 2x + 3, R(x) = 4x - 4$

71. $P(x) = 2x^3 + x^2 + 4$ **72.** $a = 1, b = -2, c = -1, d = 7$

73. $Q(x) = 2x^2 - x - 1, R(x) = 4$
74. $Q(x) = x^4 - 2x^3 + 3, R(x) = -4$
75. $Q(x) = x^4 + 2x^3 + 4x^2 + 8x + 16, R(x) = 0$
76. $Q(x) = 2x^2 + 4x + 2, R(x) = -4$

Chapter Test *(page 410)*

1. [7.1] $6r^3s^4(3s - 2r^2)$ **2.** [7.1] $(y - 1)(x + 3)$
3. [7.2] $3(3 + 2x)(3 - 2x)$ **4.** [7.2] $(2y - 5x)^2$
5. [7.2] $(1 - 2x)(1 + 2x + 4x^2)$ **6.** [7.3] $(3x + 4)(2x + 1)$
7. [7.4] $0, \frac{4}{3}$ **8.** [7.4] $-3, 0.4$ **9.** [7.4] $-2, \frac{1}{2}, 2$
10. [7.4] $-1, 0.5, 1, 2.5$
11. [7.5] $P(x) = 2x^3 + 5x^2 - 1$
$2x^3 + 5x^2 - 1 = (2x^2 - x + 3)(x + 3) - 10$
12. [7.5] $Q(x) = 3x^2 - 8x + 21, R(x) = -49$
13. [7.5] $Q(x) = -5x^5y^2 + 7x^4y^6, R(x) = 0$
14. [7.5] $Q(x) = x + 3, R(x) = -2x + 4$
15. [7.5] $Q(x) = x^4 - 2x^3 + 4x^2 - 8x + 16, R(x) = -33$
16. [7.5] The number k is the remainder after dividing $P(x) = ax^3 + bx^2 + cx + d$ by $x + 2$.
17. [7.5] Answers will vary.
18. [7.4] 7 feet, 10 feet **19.** [7.4] 100 feet, 120 feet

Chapter 8

Section 8.1 *(page 417)*

1. Any number that makes the denominator zero is excluded from the domain.
2. Solve the equation $x^2 - 2x - 15 = 0$. The solutions are excluded from the domain.
3. 0.5, 0, undefined **4.** undefined, -0.5, 0
5. -0.875, 0, undefined **6.** -0.75, undefined, 0
7. $\{x \mid x \neq 4\}$ **8.** $\{x \mid x \neq -3\}$ **9.** $\{x \mid x \neq 5\}$
10. $\{x \mid x \neq 2\}$ **11.** $\{x \mid x \neq -2\}$ **12.** $\{x \mid x \neq 0\}$
13. $\{x \mid x \neq -3, 3\}$ **14.** $\{x \mid x \neq -3, 0\}$
15. $\{x \mid x \neq -4, 2\}$ **16.** $\{x \mid x \neq -0.2, 2\}$
17. **R** **18.** **R** **19.** + **20.** − **21.** -25 **22.** -9
23. The expression in (i) has a common factor $(x + 3)$ that can be divided out. In (ii), there is no common factor in the numerator and the denominator.
24. No; if $x = 4$, for example, the value of the expression is 9, not 3. The expression $(x - 3)$ is not a common factor that can be divided out.
25. $\frac{2x - 1}{3}$ **26.** $\frac{a(1 + 2a^2)}{3}$ **27.** 1 **28.** 1
29. $x - 9$ **30.** $\frac{1}{x - 8}$ **31.** Cannot be simplified
32. Cannot be simplified
33. $\frac{1}{1 + 5x}$ **34.** $\frac{1}{4 - x}$ **35.** $\frac{x + 4}{x + 3}$ **36.** $\frac{5x + 2}{3x + 2}$
37. $\frac{a(a + 1)}{6}$ **38.** $\frac{5}{a(a - 2)}$ **39.** $\frac{1}{2x + 1}$ **40.** $\frac{1}{3 - 2x}$
41. $\frac{x - 5}{x - 8}$ **42.** $\frac{4a + 3}{a + 3}$ **43.** 1 **44.** $\frac{1}{a + 3}$ **45.** $\frac{a}{b + 3}$
46. $\frac{5b}{a - 5}$ **47.** $\frac{1}{x - y}$ **48.** $\frac{1}{a - b}$ **49.** $\frac{c - 2d}{2c + d}$
50. $\frac{x + 4y}{x + 6y}$ **51.** -2 **52.** $3 - 2x$ **53.** $3 - x$
54. $x - 2$ **55.** **(a)** Neither **(b)** 1 **(c)** -1
56. **(a)** -1 **(b)** Neither **(c)** 1
57. **(a)** -1 **(b)** 1 **(c)** Neither
58. **(a)** 1 **(b)** -1 **(c)** Neither
59. -1 **60.** -1 **61.** $-\frac{1}{3}$ **62.** $-\frac{3}{4}$ **63.** $-(a + 7)$
64. $-(x + 2)$ **65.** $\frac{-1}{5w + z}$ **66.** $\frac{x + 2}{x - 2}$ **67.** $-\frac{x + 4}{x + 3}$
68. $-\frac{x + 4}{2x - 1}$ **69.** $x - 3$ **70.** $\frac{4 + 6y + 9y^2}{1 + x}$
71. $\frac{x^2 - 2x + 4}{x - 2}$ **72.** $\frac{1}{1 - x}$ **73.** $\frac{3x(3x - 2)}{2(2x + 3)}$
74. $2 - x$ **75.** $12x^2y$ **76.** $-9ab^3$ **77.** 25 **78.** -35
79. $(x + 5)^2$ **80.** $6x^3$ **81.** $5(x - 3)$ **82.** $9x - 9$
83. $-3(x + 4)$ **84.** $5x - 20$ **85.** $9(x - 5)$
86. $8x + 32$ **87.** $-12t$ **88.** $15x^5y$
89. $-1(1 + x)(1 - x)$ **90.** $-(3x - 2)^2$
91. **(a)** \$2 million **(b)** \$10 million
(c) The cost doubles.
92. **(a)** 8.0 **(b)** 28 **(c)** 0.66 **93.** $-0.293x + 28.873$
94. The expression $x - 28$ is a factor of $N(x)$.
95. $N(x) = (x - 28)(-0.293x + 28.873)$
96. $x = 28, 98.5$; The result indicates that people age 28 and 98.5 have a net worth of zero.
97. $-\frac{1}{x^3}$ **98.** $\frac{1}{x^3}$ **99.** x^5 **100.** x^4
101. For even n, $\{x \mid x \neq -1, 1\}$; for odd n, $\{x \mid x \neq 1\}$
102. For even n, **R**; for odd n, $\{x \mid x \neq -1\}$ **103.** **R**
104. $\{x \mid x \neq -1, 1\}$ **105.** $\frac{(x + 1)^2 - (x - 2)^2}{x(x + 4) - (x + 1)^2} = 3$

Section 8.2 *(page 424)*

1. An easier method is to factor first and then divide out common factors.
2. $R(x) = x^{-1} \cdot x = \frac{1}{x} \cdot x$; The domain of $y = 1$ is **R**, whereas the domain of $y = R(x)$ is $\{x \mid x \neq 0\}$.

3. $\frac{1}{5a}$ **4.** $\frac{2y^5z}{x^2}$ **5.** $\frac{a-4}{3}$ **6.** $-3(2x+1)$ **7.** 1

8. $\frac{15}{p}$ **9.** $\frac{c-6}{20c}$ **10.** $\frac{v+4}{4v}$ **11.** $\frac{1}{3}$ **12.** $\frac{z-9}{6z^2}$

13. $\frac{y+1}{y+8}$ **14.** $\frac{z+7}{z+2}$ **15.** $3-x$ **16.** x^2-4

17. $2x^2-x-1$ **18.** $x^2+7x+10$

19. The opposite of $x+3$ is $-(x+3)=-x-3$. The reciprocal of $x+3$ is $\frac{1}{x+3}$.

20. When we write $\frac{5}{x-1} \div \frac{x-2}{3} = \frac{5}{x-1} \cdot \frac{3}{x-2}$, we see that replacing x with 2 makes the second denominator 0.

21. $\frac{x^3}{6y^2}$ **22.** $5x^3$ **23.** $\frac{6}{x}$ **24.** $\frac{x}{7}$ **25.** $\frac{-x(x-2)}{x+2}$

26. $\frac{x-8}{3x(x+3)}$ **27.** $\frac{x-1}{8x}$ **28.** $\frac{1}{3}$ **29.** $\frac{x-6}{x-2}$

30. $\frac{x+1}{6x}$ **31.** $\frac{1}{2x(3x-2)}$ **32.** $\frac{1}{(x-3)(x+3)}$

33. $6x^2-11x+4$ **34.** $7x+49$ **35.** x^3-8 **36.** x^4

37. $\frac{-20x^3}{3y}$ **38.** $\frac{7y}{6x^4}$ **39.** $\frac{2(x+3)}{x-4}$ **40.** $\frac{1-3x}{-3x(3-x)}$

41. $\frac{x+2}{3x(x-1)}$ **42.** $\frac{b-4}{2b(b-3)}$ **43.** $\frac{1}{2}$ **44.** $2b$

45. $\frac{a+9}{a-9}$ **46.** $\frac{3}{x+8}$ **47.** $\frac{x+1}{x+4}$ **48.** $\frac{t+5}{t+4}$

49. $\frac{x}{(2x-3)(2x-5)}$ **50.** $\frac{3(w-4)}{w-3}$ **51.** 1 **52.** $\frac{a-c}{a-2c}$

53. $x+5$ **54.** x^2-xy+y^2 **55.** $-\frac{9}{10}$ **56.** $\frac{49x^3y^3}{2(x+1)}$

57. 4 **58.** $x-1$ **59.** $3x^2(x-1)$ **60.** x^2

61. $\frac{x-3}{x(x+1)}$ **62.** $\frac{1}{(x+4)(x+2)}$

63. **(a)** $C(x)=\frac{5}{4}x+\frac{11}{6}$

(b) The graph of C for the combined data lies between the graphs of B and G.

(c) Boys: 7; Girls: 9; Combined: 8

64. **(a)** $A(x)=\frac{1313x+16.241}{x+19}$ **(b)** 1993

65. $\frac{x-2}{x+3}$ **66.** $\frac{1}{x^2+x-1}$ **67.** $\frac{2x-3}{x+1}$ **68.** $\frac{x^2-16}{x+1}$

69. As x approaches 0, the y-values become larger. By selecting x-values closer and closer to 0, we can make the y-value as large as we want.

70. For larger and larger x-values, the corresponding y-values become smaller and smaller. By selecting a sufficiently large x-value, we can make the y-value as close to 0 as we want.

Section 8.3 *(page 433)*

1. Add the numerators and retain the common denominator.

2. When factored, the numerator has $x-3$ as a factor that can be divided out to simplify the expression.

3. $\frac{8}{a}$ **4.** $\frac{4}{t}$ **5.** $\frac{-3x-9}{x}$ **6.** $\frac{12a-7b}{b}$ **7.** 6

8. 11 **9.** -6 **10.** -3 **11.** $\frac{3}{x-4}$ **12.** $\frac{1}{t+5}$

13. $x+7$ **14.** $\frac{x-6}{2x+3}$ **15.** $\frac{1}{x+2}$ **16.** $\frac{1}{x-1}$

17. $\frac{9-2x}{2x+3}$ **18.** $\frac{x+4}{x-4}$ **19.** $\frac{1}{x-2}$ **20.** -2

21. $x+6$ **22.** $x-6$ **23.** $3t-3$ **24.** $2x^2+6x-6$

25. $2x^2-x$ **26.** $ab-b$

27. Multiply the numerator and the denominator of the second fraction by -1.

28. Multiply the numerator and the denominator of the second fraction by -1.

29. 5; $a+5$ **30.** -2; 5 **31.** $8x$; $11x+9$

32. $-7x$; $-5x-1$ **33.** $\frac{9x+3}{7x-8}$ **34.** $\frac{-x+9}{9x-2}$

35. $x+4$ **36.** $x+2y$ **37.** $\frac{a^2+b^2}{a-b}$ **38.** $\frac{2}{x+5}$

39. $30x^3y^2$ **40.** $18x^5y^7$ **41.** $6x(x-4)$ **42.** $4t(t+2)$

43. $-30t(t+5)$ **44.** $12x(x-7)$

45. $(3x-2)(x+1)(4x+1)$ **46.** $(3x-5)(x+3)(x+2)$

47. $3(x+1)(x+2)(x-2)$ **48.** $-3y(y+3)(y-1)(y+2)$

49. Multiply the numerator and the denominator by x.

50. The LCD is the product of the two denominators.

51. $\frac{5b+7a^2}{a^3b^2}$ **52.** $\frac{9d^2-4c}{15c^3d^4}$ **53.** $\frac{x^2-3}{x}$ **54.** $\frac{8x-6}{x-2}$

55. $\frac{11x-12}{x(x-4)}$ **56.** $\frac{t-6}{t(t+3)}$ **57.** $\frac{12}{(x-5)(x+7)}$

58. $\frac{x^2-5x-9}{(x-8)(x-3)}$ **59.** $\frac{t^2+2t+3}{t+1}$ **60.** $\frac{-2t^2-7t+3}{t+3}$

61. $\frac{2t+5}{18(t-5)}$ **62.** $\frac{13x-4}{6(2x-1)}$ **63.** $\frac{2t-1}{t(t+1)^2}$

64. $\frac{5y-1}{y^2(y-2)}$ **65.** $\frac{13t^2+2t-2}{3t^2(2t-1)(t+1)}$

66. $\frac{4x^2+3x-18}{2x(2x^2-x-6)}$ **67.** $\frac{1}{2x+1}$ **68.** $\frac{10x-27}{(x-3)^2}$

69. $\frac{x+5}{4(x-9)}$ **70.** $\frac{x-5}{2(x+3)}$ **71.** $\frac{8x-49}{(x+7)(x+8)(x-8)}$

72. $\frac{14x-27}{(x-6)(x+2)(x-9)}$ **73.** $\frac{x-12}{x-6}$ **74.** $\frac{x-6}{x-3}$

75. $\frac{x-6}{(x+3)(x-3)}$ **76.** $\frac{x+12}{(x+4)(x+10)}$

77. $\frac{x+9}{(x+3)(x-1)}$ **78.** $\frac{x-9}{(x+3)(x-5)}$

79. $(x+y)^{-1} = \frac{1}{x+y}$ and $x^{-1}+y^{-1} = \frac{1}{x}+\frac{1}{y}$. The expressions are not equivalent.

80. $(xy)^{-1} = \frac{1}{xy}$ and $x^{-1}y^{-1} = \frac{1}{x}\cdot\frac{1}{y} = \frac{1}{xy}$. The expressions are equivalent.

81. $\frac{y-x}{xy}$ **82.** $\frac{3y+4x}{xy}$ **83.** $\frac{x^2+x+9}{(x+3)(x-2)}$

84. $\frac{5x-12}{(x-3)^2}$

85. **(a)** The table suggests that increasing the hours of training will reduce errors, but errors can never be reduced to zero.

(b) $E_1(x) = \frac{5}{x(x+1)}$; The table suggests that the incentive plan will reduce errors even further.

86. **(a)** 8 **(b)** 2

87. $V_1(t) = \frac{100(28t+157)}{t+25}$

88. $100\left(\frac{28t+157}{t+25} + \frac{10(9t+73)}{t+25}\right)$

89. $100\left(\frac{118t+887}{t+25}\right)$; The expression represents the total annual fixed and variable costs (in dollars) for 10,000 miles of driving.

90. The total cost in 1997 was $4228 + 10{,}000(0.108) = \$5308$. The IRS allowed $10{,}000(0.315) = \$3150$. Thus the tax deduction was considerably less than the actual cost.

91. $\frac{-1}{x(x+h)}$ **92.** $-\frac{2x+h}{x^2(x+h)^2}$ **93.** $\frac{1-x^2}{1+x^2}$

94. $\frac{x^2+x+1}{1-x^2}$

Section 8.4 *(page 440)*

1. The expressions do not mean the same thing. The first expression means $\frac{2}{3}$ divided by 4, and its value is $\frac{1}{6}$. The second expression means 2 divided by $\frac{3}{4}$, and its value is $\frac{8}{3}$.

2. The expression can be written as $\dfrac{\frac{a+2}{a-1}}{\frac{a-3}{a+5}}$.

3. $\frac{x^2}{3}$ **4.** $-\frac{6}{5y^3}$ **5.** $\frac{1}{3x}$ **6.** $\frac{3(c+d)}{4(c-d)}$ **7.** $\frac{x-5}{2}$

8. $\frac{1}{x+1}$ **9.** $\frac{x-9}{48x}$ **10.** $\frac{x+8}{2x}$ **11.** $\frac{x-4}{2x}$ **12.** $\frac{7}{x^2}$

13. $\frac{x-3}{7x(x-2)}$ **14.** $\frac{x-9}{4x(x-2)}$ **15.** $-\frac{x+2}{x-3}$ **16.** $\frac{x}{x+2}$

17. Add or subtract the fractions in the numerator and the denominator. Then multiply the numerator by the reciprocal of the denominator.

18. The fraction bar serves as a grouping symbol.

19. $\frac{4(t+6)}{3(t+2)}$ **20.** $\frac{27-5x}{12x-36}$ **21.** $\frac{1}{y}$ **22.** $2x+1$

23. $\frac{2x-1}{x}$ **24.** $\frac{1}{x-3}$ **25.** -1 **26.** $\frac{y+x}{y+1}$ **27.** $\frac{1}{x}$

28. x **29.** xy **30.** $\frac{3b-2a^2}{a^2b(a+2b)}$ **31.** $\frac{-x-15}{2x^2-17}$

32. $-\frac{x}{2}$ **33.** $-\frac{2x^2+3x-2}{2x^2-5x-5}$ **34.** $\frac{2x-3}{x-1}$ **35.** $\frac{1}{2}$

36. $\frac{x^2+16}{x}$ **37.** $\frac{x-2}{2x-1}$ **38.** $\frac{3x+7}{x-3}$ **39.** $\frac{x+2y}{x-2y}$

40. $-\frac{3x^2+7y}{x^2+2y}$ **41.** $x+2$ **42.** $\frac{4+6y+9y^2}{2-3y}$

43. Exclude values for which Q, R, or S is 0.

44. If $x = 3$, the fraction in the denominator of the complex fraction is 0. Thus the complex fraction is undefined.

45. $\frac{x+y}{x-3y}$ **46.** $\frac{2x+1}{2x+3}$ **47.** $\frac{-1}{x(x+h)}$ **48.** $\frac{-1}{3(x+3)}$

49. $a-b$ **50.** $\frac{18}{a-3}$ **51.** $\frac{x-1}{(x-2)(x+1)}$

52. $\frac{a-c}{a+c}$ **53.** $\frac{a+b}{a^3b^3}$ **54.** $\frac{a-3}{27a^3}$

55. Multiplying the given expression by the LCD would change the value of the expression.

56. The methods in parts (a) and (b) both result in $\frac{1-x}{2+x}$. The method in part (a) requires fewer steps.

57. $\frac{1}{a+b}$ **58.** $\frac{ab}{a+b}$ **59.** $\frac{x^3+x}{x^4+1}$ **60.** $\frac{b}{a}$ **61.** $\frac{x-1}{3x-4}$

62. $\frac{x-2}{2x+1}$ **63.** $\frac{x^3+y^4}{x^2y^3(x-y)}$ **64.** $\frac{-2a^3}{1-a}$

65. $\frac{x+1}{x^2+x+5}$ **66.** $\frac{2a-1}{2a(a-1)}$

67. **(a)** $\dfrac{\frac{x}{x+1}}{\frac{x+1}{x+2}}$ **(b)** $\frac{x^2+2x}{x^2+2x+1}$ **68.** 9

69. $p(t)\cdot r(t)/100 = \frac{-0.0026t^3+0.1755t^2-0.312t}{0.7t+10}$

70. $p(t)\cdot v(t)/100 = \frac{-0.00232t^3+0.1566t^2-0.2784t}{0.7t+10}$

71. $\frac{13(34t^3-390t^2+2857t+7)}{20(102t-75)}$ **72.** \$13.43

73. Attendance increased by about 6.4%, whereas salaries increased by about 13%.

74. 253% **75.** $\dfrac{-2}{(2x + 2h - 1)(2x - 1)}$

76. $\dfrac{2}{(x + h + 2)(x + 2)}$ **77.** $\dfrac{-3}{(x + h - 2)(x - 2)}$

78. $\dfrac{-2}{x(x + h)}$ **79.** $\dfrac{14x^2 - 9x - 3}{7x - 2}$ **80.** $\dfrac{15x^2 - 5x + 2}{5x - 3}$

81. $\frac{8}{5}$ **82.** $-\frac{2}{7}$

Section 8.5 *(page 449)*

1. We call -3 and 0 restricted values because each one makes a denominator 0.

2. The resulting equation is $x^2 + 9 = 0$. Because $x \neq 0$, x^2 is positive, and so $x^2 + 9$ is positive, not 0.

3. $-3, 0, 4$ **4.** 0, 2 **5.** $-3, 0, 3$ **6.** $-5, 3, 5$

7. When we simplify, we obtain $x + 2$ only if $x \neq 2$.

8. The lack of x^2 terms in the original equation does not imply that the resulting equation will have no x^2 terms. In this case, we obtain $x^2 = 2x + 2$.

9. 20 **10.** $\frac{84}{31}$ **11.** 4 **12.** $\frac{13}{3}$ **13.** 7 **14.** $\frac{8}{9}$ **15.** $-\frac{3}{2}$

16. 3 **17.** $\frac{2}{3}$ **18.** $\frac{2}{5}$ **19.** 8 **20.** 6 **21.** $-1.5, 5$

22. $\frac{8}{3}, -\frac{9}{4}$ **23.** $-4, 5$ **24.** $-2, 2$ **25.** 3 **26.** $\frac{2}{3}, 4$

27. $-4, 1$ **28.** $-2, 1$

29. In (i), 9 is an extraneous solution. In (ii), clearing fractions leads to $-4 = 4$, which is false, and the equation has no solution.

30. No, because -1 is a restricted value. The solution set is **R**, $x \neq -1$.

31. 4 **32.** 3 **33.** -5 **34.** $-\frac{8}{5}$ **35.** No solution

36. No solution **37.** $\frac{9}{8}, -\frac{1}{2}$ **38.** $\frac{1}{2}, -6$ **39.** $-6, \frac{1}{2}$

40. -6 **41.** $-8, \frac{1}{2}$ **42.** $-4, \frac{1}{3}$ **43.** -2 **44.** -4

45. $-\frac{121}{32}$ **46.** $\frac{49}{6}$ **47.** $-\frac{10}{7}$ **48.** -4 **49.** 2

50. $-\frac{3}{2}, -1$ **51.** No solution **52.** No solution

53. No solution **54.** No solution

55. To solve the equation, clear the fractions by multiplying both sides by the LCD. To perform the addition, rewrite each expression with the LCD as the denominator.

56. The LCD is not needed in (iii). To perform the division, multiply by the reciprocal of the divisor.

57. $\dfrac{x^2 - 5x - 9}{(x - 8)(x - 3)}$ **58.** $\dfrac{9}{x}$ **59.** -1 **60.** $\dfrac{9}{2}$ **61.** $\dfrac{y}{8}$

62. $\dfrac{t + 2}{2t(t + 5)}$ **63.** $\dfrac{-2t^2 - 7t + 3}{t + 3}$ **64.** $\dfrac{5 + 2t}{18(t - 5)}$

65. 0, 6 **66.** $-\dfrac{2}{3}$ **67.** $\dfrac{4}{x + 2}$ **68.** $\dfrac{x}{(2x - 3)(2x - 5)}$

69. $R = \dfrac{E}{I}$ **70.** $q_1 = \dfrac{kr^2F}{q_2}$ **71.** $T_1 = \dfrac{P_1V_1T_2}{P_2V_2}$

72. $k = \dfrac{L_t - L_0}{L_0t}$ **73.** $r = \dfrac{I}{Pt}$ **74.** $b = \dfrac{h^2}{a}$

75. $m_2 = \dfrac{F\mu r^2}{m_1}$ **76.** $r = \dfrac{2T}{p}$ **77.** $a = \dfrac{S(1 - r)}{1 - r^n}$

78. $f = \dfrac{pq}{(k - 1)(q + p)}$ **79.** $a = \dfrac{bx}{b - y}$

80. $c = \dfrac{abz}{ab - bx - ay}$ **81.** $y = -2x - 2$

82. $y = \frac{2}{3}x - \frac{16}{3}$ **83.** $y = -\frac{4}{5}(x + 4)$ **84.** $y = -\frac{3}{4}x + 1$

85. The greatest frequency of DUI arrests per 100,000 drivers occurs among younger drivers.

86. About 25

87. The highest frequency of DUI arrests per 100,000 drivers is 356, which occurs at about age 25.

88. Because the equation has no solution, the number of arrests per 100,000 never reaches 400.

89. 7 **90.** $-2, \frac{3}{2}$ **91.** $-3, -2, 3$ **92.** $-2, -1, 1$

93. $y = m(x - x_1) + y_1$ **94.** $y = mx + b$ **95.** $-\frac{5}{4}, 0$

96. $-5, -13$

Section 8.6 *(page 458)*

1. $\frac{1}{2}, 3$ **2.** $-\frac{2}{3}, 4$ **3.** 6 and 10 or $-\frac{10}{3}$ and $\frac{2}{3}$

4. $-\frac{4}{3}$ and $\frac{5}{3}$ or 5 and 8

5. In (i) we are asked to find the sum of two fractions. Write the fractions with the LCD and add. In (ii) we are asked to solve an equation. Clear the fractions by multiplying both sides by the LCD.

6. If we multiply both sides by the LCD, which is bd, we obtain $ad = bc$.

7. 4 inches **8.** $AC = 9.375$, $DE \approx 5.87$ **9.** 540

10. 21 **11.** 9 **12.** 34 **13.** 1200 **14.** 90 **15.** 15

16. $\frac{12}{5}$ **17.** 6 **18.** 6 **19.** 7.2 hours **20.** 7.5 hours

21. $19.50 **22.** 267 **23.** 15 minutes **24.** 18 minutes

25. 6 hours **26.** Larger: 36 minutes; smaller: 45 minutes

27. Plane: 150 mph; train: 60 mph

28. Car: 55 mph; plane: 99 mph **29.** 12 mph

30. 0.5 second

31. Choices (c) and (e) illustrate inverse relations.

32. In 90 minutes, $\frac{5}{3}$ of the vat is filled and $\frac{1}{2}$ of the vat is drained. Because $\frac{5}{3} - \frac{1}{2} = \frac{7}{6}$, the vat would be overflowing.

33. 60 feet **34.** 140 **35.** 24 minutes **36.** 66 minutes

37. 30° **38.** 60 mph **39.** 6 seconds

40. $30,400, $45,600 **41.** 200 mph **42.** 6 hours

43. 5.5 hours **44.** 10 mph **45.** 28 days

46. 1190 parkas; 6.25°F

47. (a) $\frac{100(0.04x + 0.4)}{0.07x + 1.9}$ **(b)** $\frac{100(0.12x + 1.02)}{-0.16x + 14.9}$

48. (a) Aluminum: $\frac{0.04x + 0.4}{0.07x + 1.9} = 0.4$; glass: $\frac{0.12x + 1.02}{-0.16x + 14.9}$

(b) Aluminum: 2010; glass: 2007

49. 47

50. 12 pounds of pecans and 15 pounds of cashews or 15 pounds of pecans and 18 pounds of cashews

51. Small: 8 pounds; Medium: 12 pounds; Large: 5 pounds

52. 6, 8, 10

53. Square: 0.6 inch by 0.6 inch
Rectangle: 0.6 inch by 5.6 inches

Chapter Review Exercises *(page 463)*

1. The numbers are 5 and −3, which are called restricted values.

2. (a) 1, 2, 100, 10,000, undefined

(b) As x approaches 5, $g(x)$ becomes larger and larger.

(c) Replacing x with 5 causes the denominator of g to be 0.

3. $\{x \mid x \neq 2.5\}$ **4.** $\{x \mid x \neq -5, \frac{4}{3}\}$

5. In the first expression, 3 is a factor in both the numerator and the denominator. There is no common factor in the numerator and denominator of the second expression.

6. $\frac{1}{2x + 3}$ **7.** $\frac{x + 4}{2x - 3}$ **8.** $-\frac{x + 3}{x + 4}$ **9.** $\frac{b + 4}{b + 2}$

10. $x - 7$ (thousand dollars) **11.** $5(x - 4)$ **12.** $3(x - 5)$

13. $\frac{8}{x^2}$ **14.** $\frac{8}{x - 7}$ **15.** $\frac{x + 1}{x - 2}$ **16.** $\frac{x(x - 5)}{2(x - 3)}$

17. (a) $\frac{1}{xy}$ **(b)** $\frac{1}{x} \cdot \frac{1}{y}$

(c) The expressions are equivalent because $\frac{1}{x} \cdot \frac{1}{y} = \frac{1}{xy}$.

18. $\frac{x - 2}{5x}$ **19.** $\frac{6}{x + 2}$ **20.** $\frac{3}{x^2}$ **21.** $\frac{x - 1}{x - 4}$

22. 6, −4, 4, 1 **23.** $\frac{2}{x - 2}$

24. Dividing by $\frac{a^{-1}}{b^{-1}}$ is the same as multiplying by $\frac{b^{-1}}{a^{-1}} = \frac{a}{b}$.

25. $9(x + 7)(x - 7)$ **26.** $(x + 1)(x + 2)(x - 2)$

27. If the denominators were not the same, the terms would not have a common factor, and the Distributive Property would not apply.

28. −4 **29.** $\frac{-4x + 5}{5x - 4}$ **30.** $\frac{3x + 16}{10x^2}$

31. $\frac{10x - 37}{(x + 4)(x - 4)(x - 3)}$ **32.** $x(x - 7)$ **33.** $\frac{5x + 4}{x(x + 2)}$

34. $\frac{x + 9}{(x + 3)(x - 1)}$ **35.** $\frac{6x + 27}{(x + 8)(x - 6)(x + 5)}$

36. $\frac{2(x + 1)}{x(x + 2)}$

37. Multiply the numerator by the reciprocal of the denominator or multiply the numerator and the denominator by the LCD of all the fractions.

38. $\frac{3x - 1}{x}$ **39.** $\frac{a - 3}{12}$ **40.** $\frac{x - 1}{x - 3}$

41. Answers will vary.

42. $\frac{x - 3}{2x - 3}$ **43.** $\frac{x^2 + 16}{x}$ **44.** $-\frac{3x^2 + 7y}{x^2 + 2y}$

45. An apparent solution does not check if it makes a denominator zero. Such a solution is called an extraneous solution.

46. If $a \neq b$, the equation has no solution. If $a = b$, then the solution set consists of all real numbers except restricted values.

47. $\frac{1}{2}, -\frac{2}{5}$ **48.** $\frac{5}{4}, 5$ **49.** No solution **50.** $-\frac{4}{13}$

51. 2 **52.** **R**, $x \neq -3, 3$ **53.** No solution **54.** $\frac{15}{4}$

55. $x = \frac{3y}{y - 3}$ **56.** 24 hours **57.** $4\frac{2}{3}$ hours

58. 15 hours **59.** 8 **60.** 54 mph **61.** 45

Chapter Test *(page 465)*

1. [*8.1*] When $x = 3$, the denominator is zero.

2. [*8.1*] $\{x \mid x \neq 0, -3\}$ **3.** [*8.1*] $\frac{2(x + 2)}{x(x - 2)}$

4. [*8.1*] $\frac{x + 5}{x - 9}$ **5.** [*8.1*] $-\frac{x + 2y}{x + y}$ **6.** [*8.2*] $\frac{t + 2}{t + 3}$

7. [*8.2*] $\frac{1}{2(y - 5)}$ **8.** [*8.3*] $3x(x - 2)$

9. [*8.3*] The expression can be written $\frac{r}{m + n} - \frac{r - 1}{m + n}$. Because the denominators are the same, the subtraction is easy to perform. The result is $\frac{1}{m + n}$.

10. [*8.3*] 1 **11.** [*8.3*] $\frac{3 - 5t}{t^2}$

12. [*8.3*] $\frac{2x^2 - 16x - 25}{(2x - 1)(x + 5)(4x + 3)}$

13. [*8.4*] All are complex fractions. The last two are equivalent.

14. [*8.4*] $\frac{6}{x - 3y}$ **15.** [*8.4*] $\frac{x - 3}{x + 2}$ **16.** [*8.4*] $\frac{b^2}{b^2 - a^2}$

17. [*8.5*] 12 **18.** [*8.5*] −2, 5

19. [*8.5*] No solution **20.** [*8.5*] −3.2

21. [*8.5*] In (a) the apparent solution is 1, but it is an extraneous solution. In (b) the equation leads to $-2 = 3$, so the equation has no solution.

22. [*8.5*] $B = \dfrac{A}{2A - 1}$ **23.** [*8.6*] 40%

24. [*8.6*] 2.4 hours **25.** [*8.6*] 5 mph

Cumulative Test: Chapters 6–8 *(page 466)*

1. [*6.2*] 1 **2.** [*6.2*] $\dfrac{1}{x}$ **3.** [*6.2*] $\dfrac{1}{8x^6}$

4. [*6.2*] x^2y^2 **5.** [*6.2*] $\dfrac{2z^2}{3x^3}$ **6.** [*6.2*] $108x^4$

7. [*6.3*] **(a)** 0.000027 **(b)** $8.3 \cdot 10^{-9}$

8. [*6.3*] $2.5 \cdot 10^4$ **9.** [*6.4*] $x^3 + x^2 + x - 1$

10. [*6.5*] $-6x + 18$

11. [*7.1*] **(a)** $x^3y^3(x - y)$ **(b)** $(x + 3y)(a - 7b)$

12. [*7.2*] **(a)** $(2x + 5)(2x - 3)$ **(b)** $2(x - 3)^2$
(c) $(y + 2)(y^2 - 2y + 4)$

13. [*7.3*] **(a)** $(z + 9)(z - 7)$ **(b)** $(a - 9b)(a + 5b)$
(c) $3(4 - x)(1 + x)$

14. [*7.4*] After factoring the left side, set each factor equal to 0. Then solve each case for x.

15. [*7.4*] **(a)** 0, 7 **(b)** $-1, 1, -5, 5$ **16.** [*7.4*] 8 or -3

17. [*7.5*] $Q(c) = 2c^3 + c^2 + 3c - 2$, $R(c) = -2$

18. [*7.5*] $Q(x) = x^4 + 3x^3 + 6x^2 + x + 4$, $R(x) = -1$

19. [*8.1*] $\dfrac{x + 3}{4x + 3}$ **20.** [*8.2*] $\dfrac{a + 4}{a - 4}$

21. [*8.3*] $\dfrac{y + 12}{(y + 10)(y + 4)}$ **22.** [*8.4*] $\dfrac{x}{3}$ **23.** [*8.5*] -4

24. [*8.6*] $3\frac{3}{7}$ hours **25.** [*8.6*] 55 mph

Chapter 9

Section 9.1 *(page 474)*

1. The number 9 has two square roots, -3 and 3, because squaring either results in 9. The expression $\sqrt{9}$ means the principal or positive square root of 9, which is 3.

2. We call n the index and b the radicand.

3. 6, -6 **4.** $\frac{4}{5}, -\frac{4}{5}$ **5.** Not a real number

6. Not a real number **7.** 3 **8.** -4 **9.** 5, -5

10. 2, -2 **11.** -3 **12.** -2 **13.** Not a real number

14. 2, -2 **15.** 4 **16.** 8 **17.** -4 **18.** 2

19. $\frac{2}{3}$ **20.** $-\frac{5}{12}$ **21.** Not a real number

22. Not a real number **23.** -4 **24.** -6 **25.** 20

26. -6 **27.** 4 **28.** 2 **29.** 2 **30.** 5 **31.** 7

32. -1.5 **33.** 2.69 **34.** 4.21 **35.** -0.92 **36.** 2.47

37. The expression x^3 could be positive or negative depending on the value of x. The expression x^2 is nonnegative for all x.

38. The statement is true only if $\sqrt[n]{x}$ is defined—that is, if n is odd or if n is even and $x \geq 0$.

39. 4 **40.** 3 **41.** 7 **42.** -3 **43.** 10 **44.** -5

45. 2 **46.** 3 **47.** 21 **48.** 23 **49.** $2\sqrt{x}$ **50.** $\sqrt[3]{y} - 5$

51. $\sqrt[5]{a} - 9$ **52.** $\sqrt{c} + b$ **53.** $\sqrt{c + b}$ **54.** $\sqrt[4]{xy}$

55. x^5 **56.** x^8 **57.** $4y^3$ **58.** $-3z^4$ **59.** x^7 **60.** y^4

61. $-y^5$ **62.** $4a^2$ **63.** x^7 **64.** x^5 **65.** $-x^2$ **66.** $5y^3$

67. $(3x)^3$ **68.** $(5ac)^2$ **69.** $(x + 3)^5$ **70.** $(x - 2)^8$

71. **(a)** Because $\sqrt{x^2} = x$ only if $x \geq 0$, the functions are not equivalent.
(b) Because $\sqrt[3]{x^3} = x$ for all real numbers x, the functions are equivalent.

72. The radical in (iv) cannot be simplified with present methods because the exponent on x is not a multiple of the index.

73. $|x^3|$ **74.** $|t^5|$ **75.** $-x^4$ **76.** $5|y^7|$ **77.** x^3

78. y^3 **79.** $|a|$ **80.** $|y|$ **81.** $|x + 2|$ **82.** $|x^2 - 9|$

83. $-2|x - 1|$ **84.** $|3x^3 - 5x|$ **85.** $1 - x$ **86.** $|x - 7|$

87. $>$ **88.** $<$ **89.** $>$ **90.** $>$ **91.** 4 **92.** 12

93. 8 **94.** 4 **95.** -3 **96.** -1

97. **(a)** \$251 million **(b)** 15%

98. **(a)** Over 65: 999,000; over 85: 119,000
(b) Over 65: 26%; over 85: 41%

99. Cheyenne: -10.1°F; Boston: -3.6°F; Seattle: 21.8°F; Norfolk: 24.1°F

100. About 15 mph **101.** 37.5°F

102. The wind-chill temperature cannot be greater than the actual temperature. We restrict x to values for which $Y \leq T$.

103. -1 **104.** Not a real number **105.** 2

106. -2 **107.** 2 **108.** Not a real number

109. **(a)** If we did not exclude $n = 1$, then $\sqrt[1]{b} = a$ would imply that $a^1 = b$, which would mean that $\sqrt[1]{b} = b$. Thus nothing is gained by including $n = 1$ in the definition.
(b) For $b = 1$, $\sqrt[0]{b} = a$ implies that $a^0 = b$ or $a^0 = 1$. But this statement is true for all $a \neq 0$. Thus $\sqrt[0]{b}$ would not have a unique value if $b = 1$.
For $b \neq 1$, $\sqrt[0]{b} = a$ implies that $a^0 = b$, which implies that $a^0 \neq 1$. Because this statement is false, $\sqrt[0]{b}$ has no value.

Section 9.2 *(page 480)*

1. Choice (iii) is false because $16^{1/2}$ is the principal square root 4.

2. The expression in (iii) is the real number $\frac{1}{2}$. Because the other expressions are equivalent to radicals with negative radicands and even indices, they are not real numbers.

3. 7 **4.** -2 **5.** -4 **6.** 3 **7.** Not a real number

8. Not a real number **9.** $\frac{2}{3}$ **10.** $\frac{1}{2}$ **11.** -3

12. 2 **13.** -2 **14.** -4 **15.** Not a real number

16. Not a real number **17.** -5 **18.** -6 **19.** 1.97

20. 2.84 **21.** Not a real number **22.** Not a real number

23. -1.41 **24.** -4.47

25. The numerator is the exponent and the denominator is the index.

26. Choice (i) is correct because 2 is the base, 3 is the index, and 5 is the power.

27. $3\sqrt[3]{x^2}$ **28.** $-4\sqrt[6]{y^5}$ **29.** $\sqrt[3]{(3x)^2}$ **30.** $\sqrt[4]{(2ab)^3}$

31. $\sqrt[3]{(2x+3)^2}$ **32.** $\sqrt[4]{(t+2)^5}$ **33.** $3\sqrt{x} + \sqrt{2y}$

34. $\sqrt[3]{-2a} - 2\sqrt[3]{b}$ **35.** $y^{2/3}$ **36.** $z^{3/5}$ **37.** $(3x^3)^{1/4}$

38. $(7x^2y)^{1/5}$ **39.** $(x^2+4)^{1/2}$ **40.** $(x+3)^{1/2}$

41. $t^{1/2} - 5^{1/2}$ **42.** $(2y)^{1/2} + 1$ **43.** $(2x-1)^{2/3}$

44. $(x+1)^{3/4}$ **45.** -2.46 **46.** 2.11 **47.** -3.01

48. 13.09 **49.** 29.93 **50.** 5.30

51. Use the $\sqrt[x]{\ }$ key with $x = 5$ or evaluate $470^{1/5}$.

52. Because $8^{2/(-3)} = 8^{(-2)/3}$, we can write the expression with 3 as the index and -2 as the power.

53. $\frac{1}{3}$ **54.** $-\frac{1}{3}$ **55.** 3 **56.** 3 **57.** 9 **58.** 216

59. $\frac{1}{9}$ **60.** $-\frac{1}{64}$ **61.** 8 **62.** 125

63. Not a real number **64.** Not a real number

65. $\frac{1}{4}$ **66.** $\frac{1}{16}$ **67.** $\frac{9}{4}$ **68.** $\frac{81}{16}$ **69.** $-\frac{1}{16}$ **70.** -32

71. Not a real number **72.** -32 **73.** $-\frac{1}{32}$ **74.** $\frac{1}{4}$

75. 11 **76.** 56 **77.** $\frac{7}{2}$ **78.** $\frac{28}{729}$ **79.** 8 **80.** $-\frac{1}{9}$

81. 2 **82.** -3 **83.** $\frac{1}{3}$ **84.** 343 **85.** 6.35 **86.** 0.16

87. 0.72 **88.** 6.31 **89.** 0.53 **90.** 4.33 **91.** 16

92. 16 **93.** -8 **94.** 27 or -27 **95.** $\frac{2}{3}$ **96.** $-\frac{4}{3}$

97. **(a)** The graphs intersect at (0, 0) and (1, 1). If $x = 0$, then $x^{1/n} = 0^{1/n} = 0$ for all $n > 1$. If $x = 1$, then $x^{1/n} = 1^{1/n} = 1$ for all $n > 1$.

(b) $0 < x < 1$ **(c)** $x > 1$

98. Answers will vary.

99. **(a)** 2 hours **(b)** 5 additional hours

(c) 0; Yes. Even if a student did not study for the final, it is unlikely that the exam grade would be 0. The model indicates that a perfect score can be earned with about 10 hours of study. This seems realistic.

100. **(a)** \$5.85 billion **(b)** \$6.8 billion

(c) If F were 0, then $R^{-1/3}$ would be 0, which is impossible.

101. **(a)** 81% **(b)** 4.8%

102. **(a)** 1042 million **(b)** 189 million

103. $\frac{1}{2}$ **104.** $\frac{1}{3}$ **105.** $\frac{1}{2}$ **106.** $\frac{1}{3}$

107. Only odd integer values of x are in the domain.

108. The functions are (a) equal if $x \ge 0$ and (b) not equal if $x < 0$.

Section 9.3 *(page 485)*

1. Choice (i) is true according to the Power of a Product Rule and the Power to a Power Rule. Choice (ii) is untrue for $a = 3$ and $b = 4$, for example.

2. In the first case, we multiply fractions, which is easier than the addition of fractions that is required in the second case.

3. 36 **4.** $\frac{1}{9}$ **5.** 16 **6.** 81 **7.** 6 **8.** 4

9. $\frac{16}{49}$ **10.** $\frac{25}{27}$ **11.** 65 **12.** 1 **13.** 7 **14.** $\frac{1}{4}$ **15.** 2

16. 2 **17.** 6 **18.** 25 **19.** $\frac{1}{3}$ **20.** $\frac{1}{2}$ **21.** $\frac{1}{2}$ **22.** $\frac{1}{5}$

23. Choice (i) is false because the exponents were multiplied when they should have been added. Choice (ii) is true because the Product Rule for Exponents was applied properly.

24. **(a)** The statement is true when n is odd.

(b) The statement is true when n is even.

25. x **26.** $a^{2/5}$ **27.** $t^{12/7}$ **28.** $b^{3/4}$ **29.** $a^{1/4}b^{3/2}$

30. $\dfrac{x}{y}$ **31.** $z^{1/6}$ **32.** $\dfrac{1}{m^{3/2}}$ **33.** $a^{1/3}b^{4/3}$ **34.** $\dfrac{1}{x^{1/2}y^{5/2}}$

35. $\dfrac{1}{a}$ **36.** y^3 **37.** y **38.** $\dfrac{1}{z}$ **39.** a^9b^6 **40.** $\dfrac{b^4}{a^3}$

41. $\frac{1}{2}$ **42.** 1 **43.** $\frac{3}{2}$ **44.** -1

45. In the first expression, the index and the exponent have a common factor. Thus, when the expression is written with a rational exponent, the exponent can be reduced. None of this is true for the second expression.

46. If n is even, then x must be nonnegative. If n is odd, then x can be any real number.

47. 8 **48.** -3 **49.** a^3 **50.** $|x|$ **51.** $a^2|b|$ **52.** xy^2

53. $|x+y|$ **54.** $|2x-1|$ **55.** $\dfrac{b^{1/6}}{a^2}$ **56.** $\dfrac{y^{1/2}}{x}$ **57.** $\dfrac{a^4}{b^6}$

58. $5a^2b^3$ **59.** $\dfrac{-9x^{3/2}}{y}$ **60.** $\dfrac{x^6}{y^3z^2}$ **61.** $\dfrac{b^{10/3}}{a^3}$ **62.** $\dfrac{1}{x^4y}$

63. $\dfrac{x^2}{y^3}$ **64.** $\dfrac{a^{10}}{b^5}$ **65.** $\dfrac{2y^2}{x^3z}$ **66.** $\dfrac{3y^3z^2}{x^4}$ **67.** ab^4

68. $-12x^2$ **69.** $\dfrac{a^4c^{7/2}}{b^7}$ **70.** $\dfrac{x^2}{y^{10}z^3}$ **71.** $\dfrac{y^2}{3x^4}$ **72.** $\dfrac{2y}{xz}$

73. $a^2 + 1$ **74.** $y + 1$ **75.** $x^2 - x^{5/12}$ **76.** $b - b^{5/3}$

77. $x - y$ **78.** 1 **79.** $x + 3 - 2(3x)^{1/2}$

80. $x + 2(xy)^{1/2} + y$ **81.** $\sqrt{x}$ **82.** $\sqrt{x}$ **83.** $\sqrt[3]{2x}$

84. $\sqrt{3x}$ **85.** $\sqrt[5]{x^2y^3}$ **86.** $\sqrt[4]{xy^2}$ **87.** $\sqrt{a}$ **88.** $\sqrt[3]{a}$

89. $\sqrt[12]{x^{11}}$ **90.** $\sqrt[15]{x^8}$ **91.** $\dfrac{1}{\sqrt[12]{x^5}}$ **92.** $\sqrt[3]{x^2}$ **93.** $\sqrt[12]{7}$

94. $\sqrt[20]{x^3}$ **95.** $\sqrt[6]{200}$ **96.** $\sqrt[6]{1323}$

97. **(a)** 49 **(b)** $16p^{5/4}$ **(c)** 89,975

98. **(a)** $\dfrac{400}{w^{7/15}}$ **(b)** \$2.09

99. \$2094 million; 305 million **100.** $\dfrac{23t^{1/6}}{5}$

101. The function $C(t)$ models the average expenditure per visitor.

102. The trend in the expenditure per visitor shows slight increases.

103. $x^{-1/2}(3 + 5x^2)$ **104.** $a^{-5/2}(a^2 + 1)$ **105.** $5a^{-1/2}(a - 2)$

106. $3x^{1/3}(2 - x)$ **107.** $(t^{1/5} + 5)(t^{1/5} - 1)$

108. $(3x^{2/3} + 5)(x^{1/3} + 1)(x^{1/3} - 1)$ **109.** $(2x^{1/6} + 3)(x^{1/6} - 5)$

110. $(3y^{1/3} - 1)(2y^{1/3} - 1)$ **111.** $\sqrt[8]{x}$ **112.** $\sqrt[24]{x}$

Section 9.4 *(page 492)*

1. The Product Rule for Radicals applies to products of real numbers. The factors $\sqrt{-3}$ and $\sqrt{-12}$ are not real numbers.

2. The rule does not apply because the indices are not the same.

3. $\sqrt{6}$ **4.** $\sqrt{7x}$ **5.** $\sqrt{15xy}$ **6.** $\sqrt{2abc}$ **7.** $-12\sqrt{14}$

8. $2\sqrt[5]{18}$ **9.** $\sqrt[3]{x^2 + x - 6}$ **10.** $\sqrt[4]{x^2 + 2x}$

11. Factor the trinomial: $\sqrt{(x + 5)^2} = x + 5$. This step is necessary because there is no rule for simplifying a radical whose radicand is a sum. Although $x^2 + 10x + 21$ can be factored, the result is not a perfect square. Therefore, the radical cannot be simplified.

12. Both factors of $8x^6$ can be written as perfect cubes: $(2)^3(x^2)^3$. Thus $8x^6$ is a perfect cube.

13. $5y$ **14.** $4x^3y^5$ **15.** $7w^4t^2$ **16.** $6x^4y^6$ **17.** $2t^2$

18. $2x^2y^4$ **19.** $3x^3y^5$ **20.** $3x^9y^4$ **21.** $x(y - 3)^2$

22. $y^2(2x + 7)$ **23.** $x + 6$ **24.** $y + 3$ **25.** $2y + 1$

26. $3t + 2$

27. The radicand 32 has a perfect square factor, 16. The radicand 30 does not have a perfect square factor.

28. We take the square root of the coefficient, but we divide the exponent by 2.

29. $2\sqrt{3}$ **30.** $2\sqrt{5}$ **31.** $5\sqrt{2}$ **32.** $3\sqrt{2}$ **33.** $3\sqrt[3]{2}$

34. $4\sqrt[3]{3}$ **35.** $-2\sqrt[3]{6}$ **36.** $-3\sqrt[3]{2}$ **37.** $2\sqrt[4]{2}$

38. $3\sqrt[4]{2}$ **39.** $2\sqrt[5]{8}$ **40.** $2\sqrt[5]{4}$ **41.** $a^2b\sqrt{ab}$

42. $x^4y^2z^3\sqrt{xz}$ **43.** $a^3b^2\sqrt[3]{b^2}$ **44.** $x^5y^5\sqrt[3]{x^2}$

45. $x^4y^2\sqrt[4]{y}$ **46.** $ab\sqrt[4]{ab^3}$ **47.** $2\sqrt{15t}$ **48.** $2x^5\sqrt{5}$

49. $5x^{12}\sqrt{x}$ **50.** $3a^4\sqrt{2a}$ **51.** $10a^3b^7\sqrt{5a}$

52. $10a\sqrt{ab}$ **53.** $-7x^3y^3\sqrt{y}$ **54.** $-2r^2t^4\sqrt{2rt}$

55. $2w^3t^4\sqrt{7t}$ **56.** $5x^2y^3z^4\sqrt{3xyz}$ **57.** $3x^3y^3\sqrt[3]{3y^2}$

58. $3x^2\sqrt[3]{2x^2}$ **59.** $-2xy\sqrt[5]{x^2y^4}$ **60.** $2xy^2\sqrt[4]{2x^2}$

61. Factor out the common factor 4 from the radicand. Then apply the Product Rule for Radicals.

62. **(a)** $\sqrt{8x}\sqrt{2x^3} = \sqrt{4}\sqrt{2x}\sqrt{x^2}\sqrt{2x} = 2\sqrt{2x}\cdot x\sqrt{2x}$
$= (2x)(2x) = 4x^2$

(b) $\sqrt{8x}\sqrt{2x^3} = \sqrt{16x^4} = 4x^2$

63. $2\sqrt{3x + 2y}$ **64.** $b\sqrt{a + b}$ **65.** $2\sqrt{x^2 + 4}$

66. $x\sqrt{x^2 + 1}$ **67.** $3x^3y^3\sqrt{x + y}$ **68.** $5a\sqrt{2a + 3a^2b}$

69. $3|y|\sqrt{1 + y^2}$ **70.** $t^2\sqrt{t^2 + 1}$ **71.** $r^2|t|\sqrt{1 + r^2t^2}$

72. $|ab^3|\sqrt{a^2 + b^2}$ **73.** $|y - 5|$ **74.** $\sqrt{2}\cdot|t - 7|$

75. -20 **76.** $-16x\sqrt{3}$ **77.** $3\sqrt[3]{6}$ **78.** $2\sqrt[3]{5}$

79. $7x\sqrt{2}$ **80.** $3y\sqrt{5y}$ **81.** $x^4\sqrt{6x}$ **82.** $3x^2y$

83. 20 **84.** $3a^2$ **85.** $6x^2y^2\sqrt{2y}$ **86.** $3c^{11}d^{13}\sqrt{5cd}$

87. $5xy^2\sqrt[3]{x^2}$ **88.** $3xy^2\sqrt[3]{3xy^2}$ **89.** $3xy\sqrt[4]{2xy}$

90. $3xy^2\sqrt[4]{x}$ **91.** $\sqrt{16a^4b^3}$ **92.** $-3\sqrt{4x^7y^5}$

93. $\sqrt{3x^2y^3}\sqrt{3x^2y^3}$ **94.** $\sqrt{6a^4b^3}\sqrt{3ab^5}$ **95.** $\sqrt[3]{16x^4y^3}$

96. $\sqrt[3]{250x^6y^5}$

97. **(a)** Design **(b)** $5y^3\sqrt{2}$ **(c)** \$23,864.85

98. **(a)** 8 **(b)** $\dfrac{h^2\sqrt{5h}}{500}$

99. **(a)** $N_1(x) = 0.0483x\sqrt[5]{x^2}$ **(b)** $N_2(x) = 0.00651x^2$

100. **(a)** 14 **(b)** $N(x) = \dfrac{823\sqrt{1150x^2 + 4150x}}{x}$

101. The radical is not simplified because the radicand has a perfect cube factor.

102. $x\sqrt[3]{x}$ **103.** $xy^2\sqrt{2xy}$ **104.** $\sqrt[4]{3x^2y^3}$ **105.** $\sqrt[7]{4x^6y^5}$

106. $x^2y\sqrt{2xy}$

107. If n were divisible by a number larger than $\sqrt{n}$, then the quotient would be a number less than $\sqrt{n}$, and that number would have already been tested. Thus only numbers that do not exceed $\sqrt{n}$ need to be tested.

Section 9.5 *(page 500)*

1. The rule does not apply because the indices are different.

2. The denominator can be simplified to a rational number.

3. $\sqrt{6}$ **4.** $\sqrt{6t}$ **5.** 4 **6.** 3 **7.** x^3 **8.** $3a^5b$

9. $\frac{7a^3b^2}{2}$ **10.** $\frac{9ab}{4}$ **11.** x **12.** x **13.** $x^2y^4\sqrt{y}$

14. $2a^3\sqrt{5a}$ **15.** $3w^3t^2\sqrt{5wt}$ **16.** $2a^2b\sqrt{2ab}$

17. $3x^2d\sqrt{xd}$ **18.** $6r^3s^6\sqrt{2r}$ **19.** $3x^3\sqrt[3]{2x^2}$

20. $2ab^2\sqrt[4]{ab^2}$ **21.** $\frac{1}{6}$ **22.** $\frac{1}{2}$ **23.** $\frac{1}{t^3}$ **24.** $\frac{2b}{a^3}$

25. $\frac{1}{5a}$ **26.** $\frac{3}{5t^3}$ **27.** $\frac{4}{3xy^2}$ **28.** $\frac{1}{a^3b}$ **29.** $\frac{x}{2}$ **30.** $\frac{2}{x^5}$

31. The radical could be simplified if Q is a perfect square or if Q divides P.

32. The factor x^3 contains a perfect square factor.

33. $\frac{\sqrt{5}}{7}$ **34.** $\frac{\sqrt{17x}}{10}$ **35.** $\frac{2}{x^4}$ **36.** $\frac{y^2}{3}$ **37.** $\frac{2\sqrt{7}}{3}$

38. $\frac{5\sqrt{5a}}{12}$ **39.** $-\frac{\sqrt[3]{2a}}{5}$ **40.** $\frac{3\sqrt[4]{3w}}{2}$ **41.** $2x^5y\sqrt{5x}$

42. $2x^5y^2\sqrt{3y}$ **43.** $5x^2y^4\sqrt{3}$ **44.** $5x^7y^2\sqrt{3x}$

45. $7x^2y^3z^3\sqrt{yz}$ **46.** $3r^3s^2w^4\sqrt{2s}$ **47.** $2a^2b^3\sqrt[3]{3a}$

48. $3ab^2c^2\sqrt[3]{2ac^2}$ **49.** $\frac{y^2}{x}$ **50.** $\frac{\sqrt{b}}{a^2}$ **51.** $\frac{a\sqrt{a}}{3b^3}$

52. $\frac{2\sqrt{c}}{b}$ **53.** $\frac{x\sqrt[4]{x}}{2}$ **54.** $\frac{2x^3}{y^2}$

55. Either method is correct, but multiplying by $\sqrt{3}$ requires fewer steps.

56. Because the resulting exponent must be a multiple of the index, $k + m$ must be a multiple of n.

57. $\sqrt{5}$ **58.** $2\sqrt{3}$ **59.** $-\frac{\sqrt{15}}{3}$ **60.** $\frac{\sqrt{2}}{6}$ **61.** $\frac{3\sqrt{10}}{2}$

62. $-\frac{\sqrt{2}}{6}$ **63.** $\frac{7\sqrt[3]{9}}{3}$ **64.** $\frac{9\sqrt[3]{5}}{5}$ **65.** $\frac{9\sqrt[4]{2}}{2}$ **66.** $\sqrt[5]{4}$

67. $\frac{2\sqrt{3x}}{x}$ **68.** $\frac{3\sqrt{t}}{5t}$ **69.** $\frac{2\sqrt{x+2}}{x+2}$ **70.** $\frac{4\sqrt{4+y^2}}{4+y^2}$

71. $\frac{a\sqrt{2a}}{4}$ **72.** $\frac{x^2\sqrt{5}}{10}$ **73.** $\frac{\sqrt{5y}}{y^3}$ **74.** $x\sqrt{x}$ **75.** $\frac{\sqrt{15x}}{6x^3}$

76. $\frac{\sqrt{2a}}{6a}$ **77.** $6x\sqrt{3y}$ **78.** $\frac{a\sqrt{b}}{2b}$ **79.** $\frac{3\sqrt{3x}}{2x^2}$

80. $\frac{2y^2\sqrt{3x}}{3x}$ **81.** $\frac{x^2\sqrt{5xy}}{10y}$ **82.** $\frac{\sqrt{y}}{2xy}$ **83.** 5 **84.** 4

85. $\sqrt{14}$ **86.** $3\sqrt{35}$ **87.** $\sqrt{3}$ **88.** $\sqrt{5x}$ **89.** $2\sqrt[3]{b}$

90. $\frac{7\sqrt[4]{3}}{3}$ **91.** $\frac{\sqrt[3]{10}}{2}$ **92.** $\sqrt[4]{5}$ **93.** $\frac{\sqrt[4]{x^3y}}{y}$ **94.** $5\sqrt[5]{x^3}$

95. $\frac{x\sqrt[4]{y}}{2y}$ **96.** $\frac{2\sqrt[5]{2a^3b}}{ab}$ **97.** 16 **98.** 100 **99.** 36

100. 4 **101.** 4 **102.** 6 **103.** (a) $3000\sqrt{d}$ (b) \$6000

104. (a) 246 (b) $40\sqrt[5]{m}$
(c) The requirement per bird decreases, but the total requirement increases.

105. The graph suggests that the percentage of taxpayers who expect a refund decreases with the age of the taxpayer.

106. About 32 years old **107.** 75 **108.** About 28%

109. 1.414, 1.848, 1.962, 1.990, 1.998
The values appear to be approaching 2.
The pattern does not hold for $n = 3$.

110. $\frac{\sqrt[4]{x^3}}{x}$ **111.** $\frac{\sqrt[6]{x^5}}{x}$

112. (a) All three numbers must be positive.
(b) The expression is defined if exactly one of the numbers is positive or if all three numbers are positive.

113. Answers will vary.

Section 9.6 *(page 507)*

1. The *product* $\sqrt{7} \cdot \sqrt{7} = 7$, whereas the *sum* $\sqrt{7} + \sqrt{7} = 2\sqrt{7}$.

2. In the second expression, the terms in parentheses cannot be combined, so the parentheses are needed to indicate that both terms are to be multiplied by $\sqrt{5}$.

3. $6\sqrt{13}$ **4.** $-\sqrt{7}$ **5.** $16\sqrt{x}$ **6.** $-\sqrt{ax}$ **7.** $-3\sqrt[3]{y}$

8. $10\sqrt[3]{x^2}$ **9.** $-4\sqrt{2x} + \sqrt{3}$ **10.** $-ab\sqrt{a} + 3\sqrt{b}$

11. The terms must have a common factor for the Distributive Property to apply. The radicals should be simplified before we can decide whether the terms can be combined.

12. The Commutative Property allows us to reorder the factors, and the Associative Property allows us to group the factors.

$$(3\sqrt{2})(5\sqrt{7}) = (3 \cdot 5)(\sqrt{2} \cdot \sqrt{7}) = 15\sqrt{14}$$

13. $\sqrt{3}$ **14.** $7\sqrt{5}$ **15.** $8\sqrt[3]{2}$ **16.** $-2\sqrt[3]{3}$ **17.** $\sqrt[4]{2}$

18. $5\sqrt[4]{3}$ **19.** $62\sqrt{2}$ **20.** $11\sqrt{5}$ **21.** $2\sqrt{3} + 10\sqrt{2}$

22. $33\sqrt{2} - 4\sqrt{5}$ **23.** $\sqrt{2x}$ **24.** $6\sqrt{5c}$ **25.** $13x\sqrt{3}$

26. $2x\sqrt{3y}$ **27.** $11a^2\sqrt{5b}$ **28.** $19a^2\sqrt{a}$

29. $6x\sqrt{2x} - 6x\sqrt{2}$ **30.** $3x^2 - x\sqrt{3x}$

31. $4x\sqrt{xy} - 7y\sqrt{xy}$ **32.** $3x$ **33.** 0 **34.** $3\sqrt[4]{2x^3}$

35. $-x\sqrt[3]{2}$ **36.** 0 **37.** $24 - 4\sqrt{21}$ **38.** $24 + 2\sqrt{10}$

39. $20\sqrt{21} + 35\sqrt{7}$ **40.** $30\sqrt{55} - 48\sqrt{11}$

41. $5\sqrt{7} - 5\sqrt{3}$ **42.** $14\sqrt{x} + 21x$ **43.** $\sqrt{3y} - 3\sqrt{y}$

44. $5\sqrt{2y} - 2y\sqrt{5}$ **45.** $\sqrt{7}$ **46.** $-2\sqrt{y}$ **47.** $x + 2\sqrt{3}$

48. $\sqrt{15} - 5$

49. Simplifying each radical, we obtain $8 + 4 + 2 = 14$.

50. The quantity in (ii) is not a rational number because the square of $2 + \sqrt{5}$ is $9 + 4\sqrt{5}$, which is an irrational number.

51. $\sqrt{6} + 3 - 2\sqrt{2} - 2\sqrt{3}$

52. $3\sqrt{15} - 15\sqrt{5} - 6 + 10\sqrt{3}$

53. $\sqrt{21x} + 7x - 3\sqrt{3} - 3\sqrt{7x}$

54. $3\sqrt{t} + 4t\sqrt{3t} - 2\sqrt{3} - 8t$

55. $3\sqrt{5} - 8\sqrt{10}$ **56.** $2x - 7\sqrt{x} - 15$ **57.** $8 - 4\sqrt{3}$

58. $18 + 6\sqrt{5}$ **59.** $15 + 10\sqrt{2}$ **60.** $12 - 4\sqrt{5}$

61. -1 **62.** 14 **63.** -78 **64.** 3 **65.** $100 - 50t^2$

66. $49w^2 - 27$ **67.** $-7 - 4\sqrt{3}$

68. $\sqrt{15} - \sqrt{10} - 3 + \sqrt{6}$ **69.** $\dfrac{7 - \sqrt{21}}{4}$

70. $\dfrac{5\sqrt{3} + 5}{2}$ **71.** $3 + \sqrt{6}$ **72.** $4(\sqrt{5} + \sqrt{2})$

73. $4 - \sqrt{15}$ **74.** $\dfrac{\sqrt{35} - \sqrt{15} - \sqrt{14} + \sqrt{6}}{4}$

75. $\dfrac{y + 9\sqrt{y} + 20}{y - 16}$ **76.** $\dfrac{5(\sqrt{z} - z)}{1 - z}$ **77.** $\dfrac{x - 2\sqrt{x}}{x - 4}$

78. $\dfrac{3 + \sqrt{3x}}{3 - x}$ **79.** $\dfrac{\sqrt{x+1} + 2}{x - 3}$ **80.** $\dfrac{2(3 - \sqrt{x+2})}{7 - x}$

81. $\dfrac{x - 2\sqrt{xy} + y}{x - y}$ **82.** $\dfrac{x\sqrt{y} - y\sqrt{x}}{x - y}$

87. The operation cannot be performed because the denominators are not the same. It is not necessary to rationalize the denominators in order to add. It is only necessary that the fractions have the same denominator.

88. The results are the same, but the first approach requires fewer steps.

89. $\dfrac{x - 3\sqrt{x}}{x}$ **90.** $\dfrac{5\sqrt{t} + t\sqrt{2}}{2t}$ **91.** $\dfrac{16\sqrt{5}}{5}$ **92.** $\dfrac{\sqrt{6}}{2}$

93. $\dfrac{2 - 3\sqrt{2}}{2}$ **94.** $-\dfrac{7\sqrt{2}}{2}$ **95.** $\dfrac{2\sqrt{x+1}}{x+1}$

96. $\dfrac{-3\sqrt{x+3}}{x+3}$ **97.** $6 - \sqrt{3}$ **98.** $\dfrac{3x + 5\sqrt{x}}{x - 25}$

99. $\dfrac{-1}{2(\sqrt{5} - \sqrt{7})}$ **100.** $\dfrac{1}{\sqrt{11} + \sqrt{3}}$ **101.** $\dfrac{x - 9}{x - 6\sqrt{x} + 9}$

102. $\dfrac{25 - a}{25 + 10\sqrt{a} + a}$ **103.** $\dfrac{1}{\sqrt{x+2} + \sqrt{x}}$

104. $\dfrac{1}{\sqrt{x+h} + \sqrt{x}}$ **105.** 0 and 1 **106.** Answers will vary.

107. **(a)** $d = \sqrt{x^2 - 2500}$ **(b)** $h = \sqrt{2600 - 20x}$

108. Answers will vary.

109. **(a)** 92% **(b)** $\dfrac{T + 100\sqrt{T}}{10{,}000 - T}$

110. **(a)** $\dfrac{0.0001D - 0.01\sqrt{D}}{0.0001D - 1}$ **(b)** 0.05

111. $F(t) = 1252\sqrt[5]{t^2}$ **112.** $B(t) = 15.3t\sqrt[5]{t^2} + 135\sqrt[5]{t^2}$

113. $T(t) = 15.3t\sqrt[5]{t^2} + 1387\sqrt[5]{t^2}$ **114.** 6254 **115.** 1.63

116. 2.62 **117.** $2\sqrt{x}$ **118.** $3\sqrt{3}$ **119.** $-\sqrt{2}$ **120.** 0

121. $\dfrac{-\sqrt{6} - 2\sqrt{15} + 3\sqrt{10}}{12}$

122. $A^3 - B^3 = (A - B)(A^2 + AB + B^2)$
Multiply the numerator and the denominator by $\sqrt[3]{x^2} + \sqrt[3]{xy} + \sqrt[3]{y^2}$.

123. $\dfrac{3(\sqrt[3]{x^2} + \sqrt[3]{2x} + \sqrt[3]{4})}{x - 2}$ **124.** $\dfrac{5(\sqrt[3]{x^2} + \sqrt[3]{3x} + \sqrt[3]{9})}{x - 3}$

125. $\dfrac{\sqrt{x} + 3}{x + 3}$ **126.** 3

Section 9.7 *(page 517)*

1. To determine the domain of the function, solve the inequality $R \geq 0$.

2. On the left side, $x \geq 3$. On the right side, $x \leq 2$. Because both conditions cannot be satisfied, the equation has no solution.

3. $\{x \mid x \geq 2\}$ **4.** $\{x \mid x \geq -7\}$ **5.** $\{x \mid x \geq 1.5\}$

6. $\{x \mid x \geq 0\}$ **7.** $\{x \mid x \leq 2\}$ **8.** $\{x \mid x \leq 0.5\}$

9. **R** **10.** **R** **11.** **R** **12.** **R** **13.** $\{x \mid x \geq -3\}$

14. $\{x \mid x \leq 5\}$

15. The equations are not equivalent; the first has no solution, whereas the second has the solution 25. The methods used to solve a radical equation may introduce an extraneous solution.

16. Squaring both sides results in $x + 2\sqrt{x} + 1 = 25$, which still contains a radical. The first step should be to isolate the radical.

17. 3 **18.** $-\frac{3}{8}$ **19.** 3 **20.** -7 **21.** No solution

22. $-3, 3$ **23.** 15 **24.** 5 **25.** $4, -1$ **26.** $-3, \frac{2}{3}$

27. 9 **28.** 8 **29.** $\frac{1}{2}, 2$ **30.** 5 **31.** 3 **32.** 2

33. $-3, 2$ **34.** $-4, -2$ **35.** No solution **36.** 2, 5

37. No solution **38.** No solution **39.** No solution

40. No solution **41.** 4 **42.** $\frac{1}{4}, 4$

43. The radical in each equation can be eliminated by squaring once. The first two resulting equations are second-degree.

44. The first equation must be squared twice. In the second equation, isolate the radical and square once.

45. 4 **46.** 4 **47.** 3, 7 **48.** 7, 3 **49.** No solution
50. No solution **51.** 3 **52.** $\frac{3}{2}$ **53.** 6 **54.** 2 **55.** 2, 3
56. 5, 1 **57.** 2, 6 **58.** 11, 3 **59.** 4 **60.** 5

61. Raise both sides to the $\frac{b}{a}$ power in order to isolate x.

62. The value of the radical is at least 0, and adding 1 makes the value of the left side at least 1.

63. $\pm\sqrt{6}$ **64.** $\pm 3\sqrt{5}$ **65.** No solution **66.** ± 3

67. 4.5, 0.5 **68.** $-\frac{6}{5}, -\frac{14}{5}$ **69.** $\frac{1 \pm \sqrt{3}}{3}$

70. $-2 \pm 3\sqrt{2}$ **71.** 5 **72.** -2 **73.** $\frac{1}{2}$

74. No solution **75.** ± 27 **76.** No solution **77.** $\frac{1}{81}$

78. $\pm\frac{1}{125}$ **79.** $\frac{25}{8}, \frac{23}{8}$ **80.** $-\frac{159}{32}$ **81.** 6 **82.** $\frac{67}{2}, -\frac{61}{2}$

83. 4 **84.** 16,777,216 **85.** $\sqrt{57} \approx 7.55$ **86.** 16

87. $\sqrt{5} \approx 2.24$ **88.** $\sqrt{11} \approx 3.32$ **89.** 25.04 feet

90. 41.71 feet **91.** 31.62 feet **92.** 31.43 feet

93. 82.92 feet **94.** 106.4 miles **95.** Yes, $r = 5.499$ feet

96. 24.8 centimeters **97.** 4 weeks **98.** 6 years

99. 2004 **100.** $x \approx 20$ (2010)

101. $9.75\sqrt{x} + 3.45 - \sqrt{3.2x + 37.4} = 35$

102. In 2008, the amounts differ by approximately $35 billion.

103. $-\frac{6}{5}, -1$ **104.** 8 **105.** No solution **106.** 1

107. **(a)** $\{x \mid x \geq 0\}$ **(b)** $\{x \mid x \leq 0\}$ **(c)** **R**

Section 9.8 *(page 527)*

1. The number is i and the properties are $i^2 = -1$ and $i = \sqrt{-1}$.

2. The Associative and Commutative Properties of Multiplication are used to regroup and reorder factors.

3. $3i$ **4.** $i\sqrt{5}$ **5.** $3i\sqrt{2}$ **6.** $\frac{4}{5}i$ **7.** $5 + 6i$

8. $4\sqrt{5} + 3i\sqrt{3}$ **9.** $-2 - i\sqrt{3}$ **10.** $1 - i\sqrt{3}$

11. Write i^{34} as a power of i^2: $i^{34} = (i^2)^{17} = (-1)^{17} = -1$.

12. Add the real parts and add the imaginary parts.

13. $a = 2, b = -1$ **14.** $a = 3, b = 2$

15. $a = -1, b = 5$ **16.** $a = 0, b = 3$ **17.** (b)

18. **(a)** Real **(b)** Imaginary **(c)** Imaginary **(d)** Imaginary **(e)** Real **(f)** Real

19. $1 + i$ **20.** $-17 + 8i$ **21.** $15 + 10i$ **22.** $-i$

23. $6 + 21i$ **24.** $-7 - 6i$ **25.** 14 **26.** $6 - 24i$

27. $13i$ **28.** $-2i\sqrt{2}$ **29.** $1 - 4i$ **30.** $-8 + 29i\sqrt{2}$

31. $a = -2, b = -2$ **32.** $a = 3, b = 6$

33. $a = -1, b = -2$ **34.** $a = -3, b = -2$ **35.** -10

36. 12 **37.** $28 + 21i$ **38.** $-10 + 6i$ **39.** $-6 - 17i$

40. $-8 + 2i$ **41.** $8 + 6i$ **42.** $-5 + 10i$ **43.** -4

44. $2i\sqrt{5}$ **45.** $-9 - 3\sqrt{2}$ **46.** $-20 + 21i$ **47.** $-i$

48. i **49.** 1 **50.** -1 **51.** -1 **52.** $-i$

53. i **54.** 1

55. In part (a) the result is $a^2 - b^2$, but in part (b) the result is $a^2 + b^2$.

56. Multiply the numerator and the denominator by $c - di$.

57. $-7i$, 49 **58.** $4i$, 16 **59.** $2 - 3i$, 13 **60.** $3 + 4i$, 25

61. $\sqrt{5} + 4i$, 21 **62.** $7 + i\sqrt{2}$, 51 **63.** $\frac{3}{2}i$ **64.** $-\frac{4}{5}i$

65. $2 - 3i$ **66.** $-5 - 2i$ **67.** $\frac{1 - 7i}{5}$ **68.** $\frac{2 + 23i}{13}$

69. $\frac{-1 + 43i}{74}$ **70.** $\frac{5 - 31i}{34}$ **71.** $\frac{6 - 3i\sqrt{3}}{7}$

72. $\frac{-4 - 4i\sqrt{2}}{3}$ **73.** $\frac{\sqrt{15} + 3\sqrt{3} + 3i - 3i\sqrt{5}}{12}$

74. $\frac{2\sqrt{6} + 5i\sqrt{3}}{11}$ **75.** $8 - 3i$ **76.** $-4 + 10i$ **77.** 28

78. -3 **79.** $\frac{2 - i}{3}$ **80.** $\frac{-1 - 4i}{2}$ **81.** $9 + 6i$

82. $1 + 5i$ **83.** $11 - 60i$ **84.** $-4 - 7i$ **85.** $\frac{-7 + 24i}{25}$

86. $\frac{-1 + 5i}{13}$ **87.** $7i$ **88.** $16i\sqrt{7}$ **89.** $5 - 3i\sqrt{3}$

90. $4 + 6i$ **91.** 5 **92.** $\sqrt{30}$ **97.** $(x + 3i)(x - 3i)$

98. $(x + 5i)(x - 5i)$ **99.** $(2x + i)(2x - i)$

100. $(7x + 4i)(7x - 4i)$

101. **(a)** -1 **(b)** i **(c)** $\frac{2 - 3i}{13}$ **(d)** $\frac{-3 + 4i}{25}$

104. $\frac{27 - 6i}{17}$ **105.** $\frac{-23 + 37i}{26}$

Chapter Review Exercises *(page 530)*

1. The number x^2 has two square roots, x and $-x$, but $\sqrt{x^2}$ is the principal (nonnegative) square root. Thus we use absolute value symbols to ensure that the result is nonnegative. The number x^3 has only one cube root, and absolute value symbols are not used.

2. The radical symbol is used to indicate the principal (nonnegative) square root, 3.

3. -8 **4.** 4 **5.** -4 **6.** Not a real number

7. x^2 **8.** $9x^2$ **9.** $x + 4$ **10.** $7x^3$

11. **(a)** $\sqrt{10} - a$ **(b)** $\sqrt{10 - a}$ **12.** 1.6 inches

13. The expression $(-4)^{1/2}$ means $\sqrt{-4}$, but there is no real number whose square is -4.

14. The number n is the index, and m is the power to which the radical is raised.

15. 64 **16.** $-\frac{2}{3}$ **17.** $\frac{1}{9}$ **18.** $-\frac{1}{3}$

19. Not a real number **20.** $4\sqrt[7]{x^6}$, $\sqrt[7]{(4x)^6}$

21. $25^{1.5} = 25^{3/2} = (\sqrt{25})^3 = 5^3 = 125$ **22.** 16.55 feet

23. When n is even, the initial result is always $|x|$. However, if x is nonnegative, then $|x| = x$.

24. Both methods are correct. In (i) we use the rule $\left(\frac{a}{b}\right)^{-n} = \left(\frac{b}{a}\right)^n$, and in (ii) we use the Quotient to a Power Rule.

25. **(a)** 6 **(b)** 4

26. $f^{2/7}$ **27.** $\frac{c^{18}}{b^{12}}$ **28.** $a^{1/2}$ **29.** $\frac{3x^2y}{z^3}$ **30.** $\frac{x^{3/2}}{2}$

31. $\frac{x^{3/2}}{9}$ **32.** $a = 8, b = 3$ **33.** $\frac{1}{x} - x^3$

34. **(a)** $\sqrt[15]{x}$ **(b)** $\sqrt[5]{x^4y^3}$

35. The Product Rule cannot be used because the indices are different. We can use rational exponents to write $\sqrt[3]{5} \cdot \sqrt{5} = 5^{1/3} \cdot 5^{1/2} = 5^{2/6} \cdot 5^{3/6} = 5^{5/6} = \sqrt[6]{5^5}$.

36. The radicand 24 has a perfect cube factor, 8, but the radicand 25 does not.

37. $45x$ **38.** $3a^2b^3$ **39.** $\sqrt{42x}$ **40.** $\sqrt{a^2 + a}$

41. $x^2\sqrt{x}$ **42.** $3\sqrt{5}$ **43.** $c^8d^{10}\sqrt{c}$

44. Cannot be simplified **45.** Answers will vary.

46. $10\sqrt{x^2 + 3x}$

47. The radicand must contain no perfect nth power factors. The radicand must not contain a fraction. There must be no radical in the denominator of a fraction.

48. Multiplying the numerator and the denominator by $\sqrt[3]{x}$ results in $\frac{1}{\sqrt[3]{x^2}}$, and the denominator cannot be simplified.

49. x **50.** $\frac{3xy}{2}$ **51.** $\frac{a^3\sqrt{5a}}{3b^2}$ **52.** $\frac{x^5}{8}$ **53.** $\sqrt{6}$

54. $\frac{\sqrt{39}}{3}$ **55.** $\frac{3\sqrt{x}}{x^2}$ **56.** $\frac{\sqrt[4]{2}}{2}$ **57.** $\frac{3y\sqrt{x}}{x}$ **58.** $\sqrt[6]{x}$

59. Answers will vary.

60. **(a)** $\sqrt{5}$ **(b)** Requires more than one operation

(c) $\frac{\sqrt{15}}{3}$

61. **(a)** The indices are the same, but the radicands are different.

(b) The radicands are the same, but the indices are different.

62. Multiplying by $\sqrt{2}$ results in $\frac{\sqrt{2}}{3\sqrt{2} + 2}$, and the denominator is still not rationalized. Multiply by $3 - \sqrt{2}$.

63. $13\sqrt{5x}$ **64.** $\sqrt[3]{2}$ **65.** $8 + 2\sqrt{15}$ **66.** $x - 2$

67. $\sqrt{2}$

68. **(a)** 45 **(b)** 4 **(c)** $14 + 6\sqrt{5}$

Parts (a) and (b) are rational numbers.

69. $\frac{5(5 + \sqrt{3})}{22}$ **70.** $\frac{x + 4\sqrt{x} - 21}{x - 9}$ **71.** $2\sqrt{3}$

72. $\frac{2a}{a^2 - 5}$ **73.** $13\sqrt{5}$ **74.** $\sqrt{x}$

75. When an algebraic method is correctly used to solve an equation, an apparent solution may not satisfy the equation. Such a number is an extraneous solution.

76. Raise both sides to the nth power.

77. 3 **78.** 4 **79.** 1 **80.** $-29, 25$ **81.** 15.20 feet

82. 5 **83.** 4 and 9 **84.** $30\sqrt{2} \approx 42.43$ miles

85. The square root of a negative number is not a real number.
$\sqrt{-4}\sqrt{-9} = (2i)(3i) = 6i^2 = -6$

86. The first product is $a^2 - b^2$, and the second product is $a^2 + b^2$.

87. $-2 + 2i$ **88.** $4 - 2i$ **89.** $33 + 4i$ **90.** $24 + 10i$

91. $4 - 2i$ **92.** 23 **93.** i **94.** $-1 + 2i$

95. $a = 0, b = -6$ **96.** (d)

Chapter Test *(page 533)*

1. *[9.1]* $\{x \mid x \leq 3\}$ **2.** *[9.2]* $\frac{-2y^4}{x^2}$

3. *[9.2]* $-25^{1/2} = -\sqrt{25} = -5$
$(-25)^{1/2} = \sqrt{-25}$, which is not a real number

4. *[9.5]* 20 **5.** *[9.3]* $\sqrt[12]{x^7}$ **6.** *[9.3]* $\sqrt[6]{x^5}$

7. *[9.6]* $\sqrt{3}$ **8.** *[9.6]* $26 - 5\sqrt{5}$

9. *[9.5]* The conditions are (i) b is divisible by a, or (ii) a is a perfect square.

10. *[9.6]* $\frac{10\sqrt{2} - 9}{7}$ **11.** *[9.4]* Answers will vary.

12. *[9.4]* $7x^5y^8\sqrt{2x}$ **13.** *[9.3]* $\frac{125}{64}$ **14.** *[9.7]* $-\frac{7}{3}, 3$

15. *[9.7]* Extraneous solutions can occur.

16. *[9.7]* 9 **17.** *[9.7]* 1 **18.** *[9.8]* $-1 + 6i$

19. *[9.8]* $5 + 12i$ **20.** *[9.8]* $\frac{1 - i}{2}$

21. *[9.8]* $i\sqrt{7}, i\sqrt{3}$ **22.** *[9.8]* (i) and (iv)

23. *[9.2]* $s = V^{1/3}$, 5.85 inches

24. *[9.8]* $\sqrt{3} - 3 + i\sqrt{3} + 3i$

25. *[9.7]* **(a)** $B < 0$ *[9.1]* **(b)** $a > 0$ or n is odd

Chapter 10

Section 10.1 *(page 539)*

1. The Zero Factor Property can be used only if the quadratic expression can be factored.

2. The product must be zero in order for us to apply the Zero Factor Property.

3. $-6, -1$ **4.** $\frac{1}{2}, \frac{3}{2}$ **5.** $0, 1$ **6.** 5 **7.** $-\frac{4}{3}, 2$

8. $-7, 4$ **9.** $1, 3$ **10.** $-\frac{1}{2}, \frac{1}{8}$ **11.** $-1, \frac{2}{5}$ **12.** $-\frac{5}{2}, -3$

13. If $A^2 = c$ and $c > 0$, then $A = \pm\sqrt{c}$. Thus the equation $x^2 = 6$ has two solutions, $\pm\sqrt{6}$. If $A^2 = c$ and $c < 0$, then there are no real number solutions. Thus $x^2 = -6$ has no real number solution.

14. Because the Even Root Property deals with real numbers, $x^2 = c$, where $c < 0$, has no real number solution. The Square Root Property deals with complex numbers, so $x^2 = c$, where $c < 0$, has two imaginary number solutions.

15. $\frac{1}{5}, \frac{7}{5}$ **16.** $\dfrac{2 \pm \sqrt{15}}{3} \approx 1.96$ or -0.62

17. $-5 \pm \dfrac{\sqrt{5}}{4} \approx -4.44, -5.56$ **18.** $-\frac{7}{3}, -\frac{11}{3}$ **19.** $0, 3$

20. $\frac{26}{5}, \frac{14}{5}$ **21.** $-3, 3$ **22.** $-6, 6$ **23.** $\pm 4i$ **24.** $\pm 5i$

25. $\pm 2i\sqrt{6} \approx \pm 4.90i$ **26.** $\pm 3i\sqrt{2} \approx \pm 4.24i$ **27.** $3 \pm 2i$

28. $-4 \pm 3i$ **29.** $1 \pm 5i$ **30.** $-\frac{1}{2} \pm \frac{3}{2}i$

31. $\dfrac{-5 \pm i\sqrt{6}}{2} \approx -2.5 \pm 1.22i$

32. $\dfrac{-7 \pm i\sqrt{5}}{3} \approx -2.33 \pm 0.75i$ **33.** $-1 \pm \frac{3}{2}i$

34. $-2 \pm \dfrac{\sqrt{11}}{5}i \approx -2 \pm 0.66i$ **35.** $\pm\frac{2}{3}i$ **36.** $\pm\frac{5}{6}i$

37. $-\frac{1}{2}, \frac{5}{2}$ **38.** $\frac{18}{7}, \frac{10}{7}$ **39.** $\pm 2i$ **40.** $\pm 4i$

41. **(a)** $ax^2 + bx + c = 0, a \neq 0$ **(b)** $(x - h)^2 = k$

42. After writing the equation in standard form, use the factoring method if the expression can be factored. Use the Square Root Property if the equation can be written in the form $A^2 = c$.

43. $0, 3$ **44.** $-\frac{7}{6}, \frac{3}{2}$ **45.** $\pm 3i\sqrt{5} \approx \pm 6.71i$

46. $\pm 2i\sqrt{5} \approx \pm 4.47i$ **47.** $-\frac{1}{6}$ **48.** 3 **49.** $\pm 1.2i$

50. ± 1.1 **51.** $\pm\sqrt{10} \approx \pm 3.16$ **52.** $\pm 3\sqrt{2} \approx \pm 4.24$

53. $-5, 7$ **54.** $-3, \frac{5}{4}$ **55.** $-\frac{5}{4} \pm \frac{7}{4}i$ **56.** $\frac{4}{3} \pm 2i$

57. $\dfrac{-2 \pm 8\sqrt{3}}{3} \approx 3.95$ or -5.29

58. $1 \pm 2\sqrt{6} \approx -3.90, 5.90$ **59.** $\pm i\sqrt{6} \approx \pm 2.45i$

60. $\pm 5i$ **61.** $1, 4$ **62.** $-\frac{1}{2}, -\frac{2}{5}$ **63.** $\frac{3}{2}$ **64.** $-\frac{1}{5}$

65. $x^2 - 8x + 15 = 0$ **66.** $x^2 + 3x + 2 = 0$

67. $x^2 - 6x + 9 = 0$ **68.** $x^2 = 0$

69. $3x^2 - 13x + 12 = 0$ **70.** $12x^2 - 17x + 6 = 0$

71. $x^2 + 16 = 0$ **72.** $x^2 + 3 = 0$ **73.** $-8, -9$

74. 42 and 43 **75.** 18, 20 **76.** 6 seconds **77.** 5 minutes

78. Northbound: 30 miles; Eastbound: 40 miles

79. **(a)** $0.82t^2 + 1.5t + 5 = 212$ **(b)** $t = 15(2005)$

80. **(a)** $-0.028x^2 + 3x - 6.7 = 66.7$

81. $-2 \pm i\sqrt{y}$ **82.** $-a \pm 3$ **83.** $-c \pm 4i$ **84.** $\dfrac{-1 \pm 4i}{a}$

85. $1 \pm i$ **86.** $2 \pm 3i$ **87.** $-6, -\frac{4}{3}$ **88.** ± 1

Section 10.2 *(page 545)*

1. Expression A is a first-degree binomial.

2. The value of k is the square of half of b.

3. 9 **4.** 16 **5.** $\frac{49}{4}$ **6.** $\frac{81}{4}$ **7.** $\frac{4}{9}$ **8.** $\frac{25}{16}$

9. Both methods are correct. The factoring method can be used because $x^2 - 8x + 12$ is factorable. The method of completing the square can be used for any quadratic equation.

10. The expression $x^2 + x + 1$ is not factorable, so the factoring method will not work. The method of completing the square can be used for any quadratic equation.

11. $-3, -1$ **12.** $-2, -4$ **13.** $-5, 3$ **14.** $-5, 11$

15. $-3, 4$ **16.** $-6, 1$ **17.** $-3, 2$ **18.** $-2, 7$

19. $3 \pm \sqrt{5} \approx 0.76$ or 5.24 **20.** $-1 \pm \sqrt{6} \approx -3.45, 1.45$

21. $-6, 4$ **22.** $-1, 5$ **23.** $2 \pm 2\sqrt{2} \approx -0.83$ or 4.83

24. $2 \pm 2\sqrt{3} \approx -1.46$ or 5.46

25. Divide both sides by 3 to make the leading coefficient 1.

26. After creating a perfect square trinomial, we factor it.

27. $-1, -\frac{1}{3}$ **28.** $-2, \frac{4}{5}$ **29.** $-5, \frac{3}{2}$ **30.** $\frac{1}{3}, \frac{3}{2}$

31. $-\frac{3}{2}, \frac{7}{2}$ **32.** $-\frac{3}{2}, \frac{9}{2}$ **33.** $\frac{5}{2}$ **34.** $\frac{3}{4}$

35. $\dfrac{3 \pm \sqrt{17}}{4} \approx -0.28$ or 1.78

36. $\dfrac{3 \pm \sqrt{57}}{8} \approx -0.57$ or 1.32

37. $\dfrac{-7 \pm \sqrt{57}}{4} \approx -3.64$ or 0.14

38. $\dfrac{-5 \pm \sqrt{73}}{8} \approx -1.69$ or 0.44 **39.** 9 **40.** ± 20

41. $2 \pm 2i$ **42.** $-3 \pm i$ **43.** $2 \pm i\sqrt{2} \approx 2 \pm 1.41i$

44. $1 \pm i\sqrt{6} \approx 1 \pm 2.45i$ **45.** $\dfrac{-3 \pm i\sqrt{11}}{2} \approx -1.5 \pm 1.66i$

46. $\dfrac{-5 \pm i\sqrt{7}}{2} \approx -2.5 \pm 1.32i$

47. $\dfrac{-1 \pm i\sqrt{23}}{4} \approx -0.25 \pm 1.20i$

48. $\frac{-1 \pm 2i\sqrt{5}}{3} \approx -0.33 \pm 1.49i$ **49.** Answers will vary.

50. $k = \frac{-b^2}{4}$ **51.** 2, 8 **52.** $-2, \frac{3}{2}$ **53.** $\pm\frac{7}{6}$

54. $\pm 7\sqrt{7} \approx \pm 18.52$ **55.** $-2, 5$ **56.** $-9, 4$

57. $-3 \pm \frac{\sqrt{7}}{2} \approx -4.32$ or -1.68 **58.** $-\frac{2}{3}, 2$

59. $\frac{-1 \pm i\sqrt{35}}{2} \approx -0.5 \pm 2.96i$

60. $\frac{-2 \pm \sqrt{10}}{2} \approx -2.58, 0.58$ **61.** $\frac{9 \pm 3\sqrt{5}}{2} \approx 1.15, 7.85$

62. $\frac{1}{4} \pm \frac{1}{4}i$ **63.** $\frac{-3 \pm \sqrt{21}}{4} \approx -1.90, 0.40$ **64.** $-\frac{5}{3}, \frac{1}{3}$

65. $\frac{1 \pm i\sqrt{35}}{2} \approx 0.5 \pm 2.96i$

66. $\frac{1 \pm \sqrt{17}}{2} \approx -1.56$ or 2.56

67. $\frac{-7 \pm \sqrt{97}}{8} \approx -2.11$ or 0.36

68. $\frac{1 \pm i\sqrt{95}}{6} \approx 0.17 \pm 1.62i$ **69.** $-\frac{2}{3}, \frac{1}{3}$

70. $\frac{4 \pm \sqrt{30}}{2} \approx -0.74$ or 4.74 **71.** $-1, 2$

72. $\frac{1 \pm \sqrt{21}}{2} \approx -1.79$ or 2.79

73. $\frac{1 \pm i\sqrt{35}}{4} \approx 0.25 \pm 1.48i$

74. $\frac{5 \pm i\sqrt{47}}{4} \approx 1.25 \pm 1.71i$ **75.** $a = 4; -\frac{5}{4}$

76. $a = 18; -\frac{5}{6}$ **77.** $b = 7; -\frac{3}{2}$ **78.** $b = -2; -\frac{3}{5}$

79. $c = -48; -8$ **80.** $c = 3; -3$ **81.** 5 **82.** $-2, 10$

83. 5.54 **84.** -0.14 **85.** 31 feet **86.** \$95

87. 2 hours and 17 minutes **88.** 11 hours **89.** 13

90. The polynomial is prime. **91.** $-7.3, 12.7$

92. The solutions indicate that in 1982 and 2002, the average household spent \$3000 on eating out. We might expect the amount spent to continue to increase in the future but to be lower for years prior to 1994. The prediction for 2002 is more likely to be valid.

93. $c \pm ci\sqrt{3}$ **94.** $\frac{-1 \pm \sqrt{1 - a^2}}{a}$

95. $-\sqrt{5} \pm 3 \approx -5.24$ or 0.76

96. $3\sqrt{3} \pm 2 \approx 3.20$ or 7.20

97. **(a)** $k \geq 2$ or $k \leq -2$ **(b)** $-2 < k < 2$

98. **(a)** $k \leq 1$ **(b)** $k > 1$

99. $-2i \pm 2$ **100.** $4i \pm 3$

101. **(a)** $x^2 + 8x$ **(b)** 16 **(c)** $x^2 + 8x + 16$
(d) $(x + 4)^2$ **(e)** $(x + 4)^2 = x^2 + 8x + 16$

Section 10.3 *(page 553)*

1. The equation must be in standard form.

2. Although $x^2 - 2x - 1088 = (x - 36)(x + 34)$, the factorization may not be apparent, in which case using the Quadratic Formula is the better method.

3. $3, -9, -6$ **4.** $1, 0, 4$ **5.** $1, -1, 2$ **6.** $8, -20, -1$

7. $-1, 2$ **8.** 2, 1 **9.** 4 **10.** 1 **11.** $-4, \frac{1}{2}$ **12.** $3, \frac{1}{2}$

13. 0, 5 **14.** $0, \frac{3}{2}$ **15.** $\frac{9 \pm \sqrt{129}}{8} \approx -0.29$ or 2.54

16. $\frac{-7 \pm \sqrt{97}}{8} \approx 0.36$ or -2.11

17. $1 \pm 2i\sqrt{2} \approx 1 \pm 2.83i$

18. $-2 \pm i\sqrt{6} \approx -2 \pm 2.45i$

19. $-1 \pm \sqrt{6} \approx -3.45$ or 1.45

20. $-1 \pm \sqrt{5} \approx 1.24$ or -3.24

21. $\frac{-1 \pm i\sqrt{31}}{4} \approx -0.25 \pm 1.39i$

22. $\frac{-5 \pm i\sqrt{83}}{6} \approx -0.83 \pm 1.52i$ **23.** $0, \frac{2}{3}$ **24.** $\frac{1}{3}, -\frac{5}{2}$

25. $\pm\frac{3}{7}$ **26.** $\pm\frac{\sqrt{3}}{2} \approx \pm 0.87$ **27.** 7 **28.** $\frac{1}{3}$

29. $\frac{4 \pm \sqrt{10}}{5} \approx 0.17$ or 1.43 **30.** $\frac{1}{2}, -3$ **31.** $-2, 4$

32. $\frac{-2 \pm \sqrt{11}}{2} \approx 0.66$ or -2.66 **33.** $-5, 1$

34. $3 \pm 2\sqrt{3} \approx 6.46$ or -0.46

35. $\frac{-1 \pm i\sqrt{34}}{5} \approx -0.2 \pm 1.17i$

36. $\frac{2 \pm 3\sqrt{5}}{2} \approx 4.35$ or -2.35

37. $\frac{3 \pm i\sqrt{23}}{2} \approx 1.5 \pm 2.40i$

38. $\frac{5 \pm 3\sqrt{5}}{2} \approx 5.85$ or -0.85 **39.** $3 \pm \sqrt{5} \approx 0.76$ or 5.24

40. $-1 \pm \sqrt{5} \approx 1.24$ or -3.24

41. $\frac{-5 \pm i\sqrt{11}}{2} \approx -2.5 \pm 1.66i$

42. $\frac{1 \pm i\sqrt{29}}{6} \approx 0.17 \pm 0.90i$

43. $\frac{2 \pm 2i\sqrt{5}}{3} \approx 0.67 \pm 1.49i$

44. $\dfrac{3 \pm 3i\sqrt{3}}{2} \approx 1.5 \pm 2.60i$

45. $\dfrac{5 \pm \sqrt{33}}{4} \approx -0.19$ or 2.69

46. $\dfrac{-9 \pm \sqrt{111}}{6} \approx 0.26$ or -3.26

47. $\dfrac{1 \pm \sqrt{13}}{2} \approx -1.30$ or 2.30

48. $\dfrac{1 \pm \sqrt{5}}{4} \approx 0.81$ or -0.31

49. 2 rational **50.** 2 rational **51.** 2 imaginary

52. 2 imaginary **53.** 1 rational (double root)

54. 1 rational (double root) **55.** 2 irrational

56. 2 irrational **57.** $k < 16$ **58.** $k < \frac{1}{3}$ **59.** $k < 4$

60. $k < \frac{9}{8}$ **61.** $k > 9$ **62.** $k > \frac{4}{5}$ **63.** $k > \frac{9}{4}$

64. $k > \frac{25}{16}$ **65.** ± 6 **66.** $\pm 2\sqrt{3}$ **67.** 0, 8 **68.** 4, 1

69. **(a)** The discriminant is $a^2 + 20$, which is positive for all values of a. Therefore, the solutions are always real numbers.

(b) The discriminant is $a^2 + 4k$. Because k is positive, the discriminant is always positive, and the solutions are always real numbers.

70. **(a)** The discriminant is 81, which is a perfect square. Thus the solutions are rational numbers and the factoring method can be used.

(b) The discriminant is 73, which is not a perfect square. Thus the solutions are irrational numbers and the factoring method cannot be used.

71. $(x - 20)(x + 24)$ **72.** $(x + 12)(x + 20)$

73. $(x - 18)(x + 25)$ **74.** $(x - 24)(x + 28)$

75. Answers will vary.

76. The sum of the roots is $\frac{3}{2}$. Because $5 + (-4) \neq \frac{3}{2}$, the solutions are not 5 and -4.

77. Answers will vary.

78. The product of the roots is -10. Because $5(-4) \neq -10$, the solutions are not 5 and -4.

79. $\frac{2}{3}, -\frac{1}{3}$ **80.** $1, \frac{1}{2}$ **81.** $-1, 1$ **82.** $4, -1$ **85.** Yes

86. 19.92 feet **87.** \$50 or \$80 **88.** \$75

89. 85% **90.** Between 1984 and 1994 **91.** 1990

92. The sign is negative, which indicates that the graph of $p(x)$ has no x-intercept. The model indicates that there is no year in which no GED recipient planned additional schooling.

93. $x = \dfrac{-a \pm \sqrt{a^2 + 8a}}{2}$ **94.** $x = \dfrac{y \pm \sqrt{y^2 + 8}}{2}$

95. $-1 \pm \sqrt{9 + x}$ **96.** $1 \pm \sqrt{-x - 4}$

97. $x^2 + 10x + 5 = 0$ **98.** $x^2 + 4x + 13 = 0$

99. $\dfrac{-\sqrt{3} \pm \sqrt{7}}{2} \approx 0.46$ or -2.19

100. $\dfrac{-\sqrt{10} \pm \sqrt{42}}{4} \approx 0.83$ or -2.41

101. $i \pm \sqrt{2} \approx i \pm 1.41$ **102.** $i, 2i$

Section 10.4 *(page 560)*

1. The quadratic equation has two solutions, -4 and 1. However, 1 is a restricted value in the original equation and must be rejected. In the second method, simplification requires dividing out the factor $x - 1$, which is possible only if x is not 1. Thus in both cases, 1 is rejected as a possible solution.

2. The methods used to solve equations with rational expressions may introduce extraneous solutions.

3. $\dfrac{3 \pm i\sqrt{23}}{4} \approx 0.75 \pm 1.20i$ **4.** $\dfrac{3 \pm i\sqrt{11}}{2} \approx 1.5 \pm 1.66i$

5. $4 \pm \sqrt{7} \approx 6.65$ or 1.35 **6.** $\dfrac{1 \pm i\sqrt{3}}{2} \approx 0.5 \pm 0.87i$

7. $\dfrac{-1 \pm i\sqrt{15}}{2} \approx -0.5 \pm 1.94i$

8. $\dfrac{-5 \pm \sqrt{17}}{4} \approx -0.22$ or -2.28

9. $\dfrac{-3 \pm \sqrt{21}}{2} \approx -3.79$ or 0.79

10. $\dfrac{5 \pm \sqrt{41}}{2} \approx 5.70$ or -0.70

11. $\dfrac{3 \pm \sqrt{17}}{2} \approx -0.56$ or 3.56

12. $-2 \pm \sqrt{15} \approx -5.87$ or 1.87 **13.** -2

14. No solution

15. Both $\sqrt{x^2 - 5}$ and $2\sqrt{x}$ are nonnegative, and not both are 0 for the same value of x. Thus the sum cannot be 0.

16. The methods used to solve equations with radical expressions may introduce extraneous solutions.

17. $1, \frac{4}{9}$ **18.** 3 **19.** $\dfrac{7 \pm \sqrt{17}}{8} \approx 1.39$ or 0.36 **20.** -1

21. 5 **22.** $\dfrac{-4 + \sqrt{5}}{2} \approx -0.88$ **23.** $\dfrac{6 - \sqrt{19}}{2} \approx 0.82$

24. No solution **25.** $\dfrac{-3 + \sqrt{13}}{2} \approx 0.30$ **26.** 6

27. By making an appropriate substitution, we write the given equation in the form $au^2 + bu + c = 0$, where u represents an expression. After solving this equation for u, we replace u with the expression that it represents and solve for the original variable.

28. Let $u = x^{1/6}$ so that $u^2 = (x^{1/6})^2 = x^{1/3}$.

29. $\pm 2, \pm\sqrt{3} \approx \pm 1.73$ **30.** $\pm 3, \pm\sqrt{5} \approx \pm 2.24$

31. $\pm 3i, \pm\sqrt{7} \approx \pm 2.65$ **32.** $\pm 4i, \pm i$ **33.** 4, 16

34. 49, 9 **35.** $-1, \frac{1}{2}$ **36.** 3 **37.** 64, 27 **38.** 1

39. 7, -2, 3, 2 **40.** 5, -4, 3, -2

41. $-2 \pm \sqrt{3} \approx -0.27$ or -3.73

42. $3 \pm \sqrt{5} \approx 5.24$ or 0.76 **43.** $\pm 3, \pm\sqrt{2} \approx \pm 1.41$

44. $\pm 2, \pm 2i$ **45.** $16, \frac{49}{9}$ **46.** 9, 16 **47.** 9 **48.** 4

49. $\pm 2, \pm\sqrt{3} \approx \pm 1.73$

50. $\pm 2\sqrt{2} \approx \pm 2.83, \pm\sqrt{2} \approx \pm 1.41$

51. $\dfrac{-3 \pm \sqrt{41}}{4} \approx 0.85$ or -2.35

52. $\dfrac{1 \pm i\sqrt{35}}{3} \approx 0.33 \pm 1.97i$ **53.** $\dfrac{-4 + \sqrt{3}}{2} \approx -1.13$

54. 2 **55.** $\pm 2, \pm i\sqrt{5} \approx \pm 2.24i$

56. $\pm i\sqrt{3} \approx \pm 1.73i, \pm i\sqrt{2} \approx \pm 1.41i$ **57.** 49, 36 **58.** 49

59. $1 \pm 2i$ **60.** $\dfrac{-1 \pm i\sqrt{15}}{2} \approx -0.50 \pm 1.94i$ **61.** 8

62. 6 **63.** -27, 125 **64.** 64 **65.** $\dfrac{-7 + \sqrt{73}}{6} \approx 0.26$

66. $\dfrac{3 + \sqrt{21}}{4} \approx 1.90$ **67.** 1

68. $\dfrac{-5 \pm \sqrt{29}}{2} \approx 0.19$ or -5.19

69. $\dfrac{3 \pm \sqrt{33}}{2} \approx -1.37$ or 4.37

70. $\dfrac{-3 \pm \sqrt{13}}{2} \approx 0.30$ or -3.30 **71.** $53 - 8\sqrt{39} \approx 3.04$

72. $7 - 2\sqrt{7} \approx 1.71$ **73.** $\frac{11}{4}, 4$ **74.** $-\frac{8}{7}, -\frac{2}{3}$

75. No solution **76.** -6

77. $\dfrac{1 \pm i\sqrt{7}}{2} \approx 0.5 \pm 1.32i, \dfrac{1 \pm i\sqrt{3}}{2} \approx 0.5 \pm 0.87i$

78. $-\dfrac{1}{2}, 4, \dfrac{7 \pm \sqrt{33}}{4} \approx 3.19$ or 0.31 **79.** 900 motors

80. 700 brochures **81.** 300 students, \$650 per person

82. 50 minutes

83. **(a)** $-0.3(x + 5)^2 + 5.14(x + 5) + 2.34 = 15.5$
(b) $x = 9(2004)$

84. **(a)** $0.037(x + 40)^2 - 3.01(x + 40) + 109.14 = 50$
(b) Letting $u = x + 40$, $0.037u^2 - 3.01u + 109.14 = 50$
$x \approx 33$ or $x \approx 48$

The model indicates that 50% of the respondents age 33 and 50% of the respondents age 48 do not believe that a neat house is very important.

85. \$27.94

86. $t \approx 20.7$; The model projects the cost of the school lunch program to be \$10 billion in the year 2010.

87. $C(-10) = 2246$, which is the cost (in millions of dollars) of the school lunch program in 1980.

88. \$243.63 **89.** $\dfrac{1 + \sqrt{13}}{2} \approx 2.30$ **90.** -9, 3

91. $2, -1, -1 \pm i\sqrt{3} \approx -1 \pm 1.73i, \dfrac{1 \pm i\sqrt{3}}{2} \approx 0.5 \pm 0.87i$

92. $-3, \dfrac{3 \pm 3i\sqrt{3}}{2} \approx 1.50 \pm 2.60i,$
$1, \dfrac{-1 \pm i\sqrt{3}}{2} \approx -0.50 \pm 0.87i$

93. $1, -2 - \sqrt{3} \approx -3.73$ **94.** 3, -1

95. $\dfrac{1 \pm i\sqrt{3}}{2} \approx 0.5 \pm 0.87i$ **96.** $\dfrac{2 + \sqrt{3}}{2} \approx 1.87$

Section 10.5 *(page 568)*

1. 16.97 feet **2.** 15.51 yards **3.** 97.14 feet, 127.14 feet

4. 20.5 feet **5.** 17.3 inches, 10.70 inches

6. 5.9 inches, 9.6 inches **7.** $AC = 12.26$, $DE = 3.26$

8. $AB = 5$, $DF = 6$ **9.** 26.88 mph **10.** 5 mph

11. 57.16 mph **12.** 6.63 meters/second, 7.18 meters/second

13. 11.08 hours, 13.08 hours **14.** 39 hours **15.** 9 hours

16. 4 hours **17.** 370 shares **18.** 500 shares

19. Last year: 459 boxes at \$18.67; This year: 561 boxes at \$20.67

20. 452 **21.** 60 **22.** 40, \$2.50

23. 11 and 13 or -13 and -11 **24.** 7 and 8 or -8 and -7

25. 90.38 feet **26.** 0.46 foot **27.** 8 feet

28. 4.85 feet, 7.85 feet **29.** 9 **30.** 4.8 seconds

31. 5.41 feet **32.** 12 inches **33.** 8 yards **34.** 4 feet

35. 25.08 feet **36.** Car: 123.6 miles; Truck: 93.6 miles

37. 3.12 feet **38.** 4.96 feet **39.** 3 **40.** 2 **41.** 4 miles

42. 60 mph **43.** About 12:49 **44.** 9.4 hours **45.** 12

46. 55.4 mph

47. The x-intercepts are about (1, 0) and (108, 0). The interpretation is that no one age 1 or 108 has a credit card.

48. The model indicates that about 23% of 10-year-olds have credit cards.

49. Because the percent who always pay the balance due increases with age, the linear model $y = ax + b$ is the better model. A model is $y = 0.88x + 13.13$.

50. Because the slope is positive, the graph rises from left to right. A function that models the percent who do not pay their balances would have a graph that falls from left to right.

51. \$15.63 **52.** 1.52 feet **53.** 4.31 inches

54. 1108.32 feet

Section 10.6 *(page 581)*

1. The graph is a parabola. For positive a the graph opens upward, and for negative a the graph opens downward. The graph becomes narrower as $|a|$ increases.

2. **(a)** The graph of f is the same as the graph of $y = x^2$ shifted k units upward if $k > 0$ or shifted $|k|$ units downward if $k < 0$.

(b) The graph of g is the same as the graph of $y = x^2$ shifted h units to the right if $h > 0$ or shifted $|h|$ units to the left if $h < 0$.

3. (e) **4.** (f) **5.** (c) **6.** (a) **7.** (b) **8.** (d)

9. **(a)** Same vertex and width but opens downward.

(b) Same vertex but opens downward and is narrower.

10. **(a)** Shifted up 4 units **(b)** Shifted down 4 units

11. **(a)** Shifted left 3 units **(b)** Shifted right 3 units

12. **(a)** Shifted down 1 unit **(b)** Shifted right 1 unit

13. **(a)** Shifted right 3 units and down 2 units

(b) Shifted left 1 unit and up 4 units

14. **(a)** Opens downward, is narrower, and vertex shifted left one unit and down one unit.

(b) Narrower, vertex shifted right 1 unit and up 2 units

15. The coordinate is $-\dfrac{b}{2a}$. Determine the y-coordinate by evaluating the function at $-\dfrac{b}{2a}$.

16. If the vertex is $V(h, k)$, then the minimum is k if the parabola opens upward, and the maximum is k if the parabola opens downward.

17. (3, 4); Minimum: 4

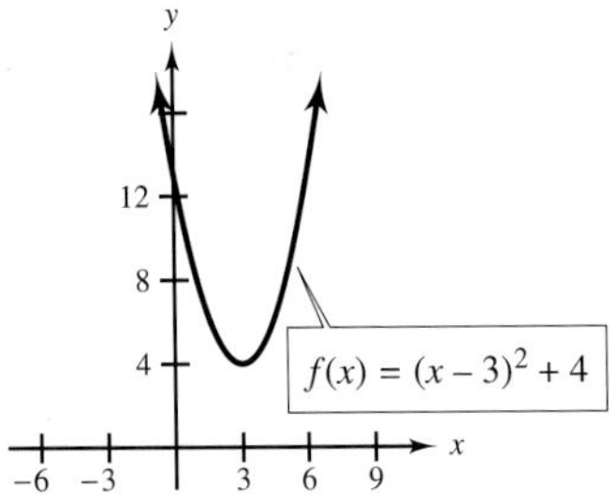

18. $(-3, -4)$: Minimum: -4

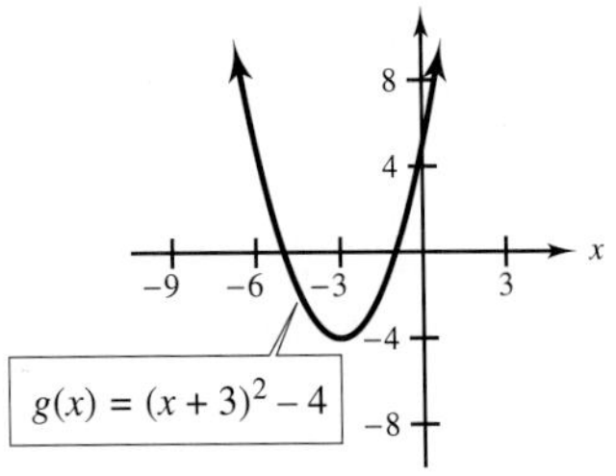

19. $(-1, -4)$; Minimum: -4

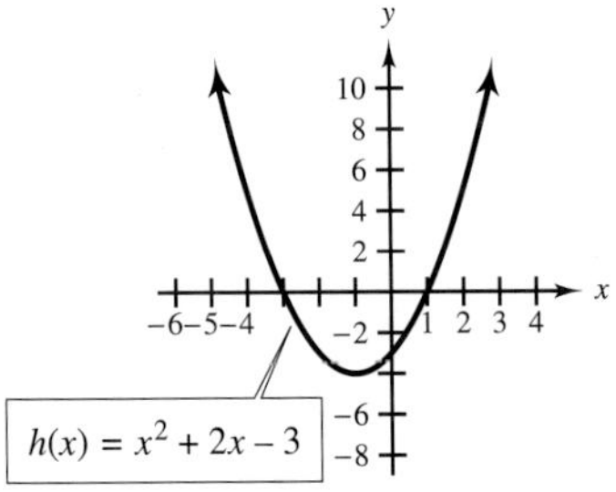

20. $(3, -4)$; Minimum: -4

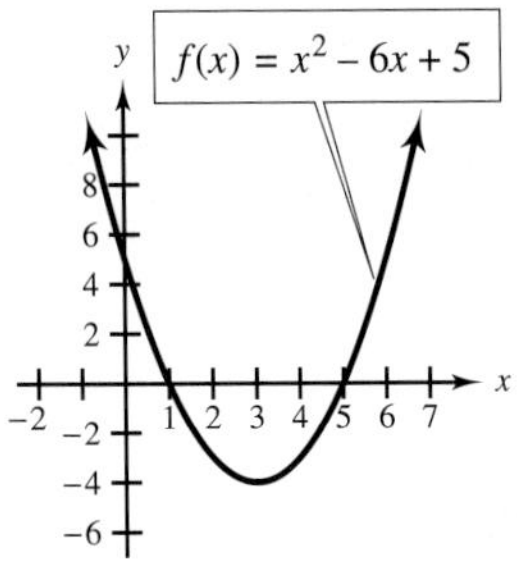

21. $\left(\frac{3}{2}, 0\right)$; Maximum: 0

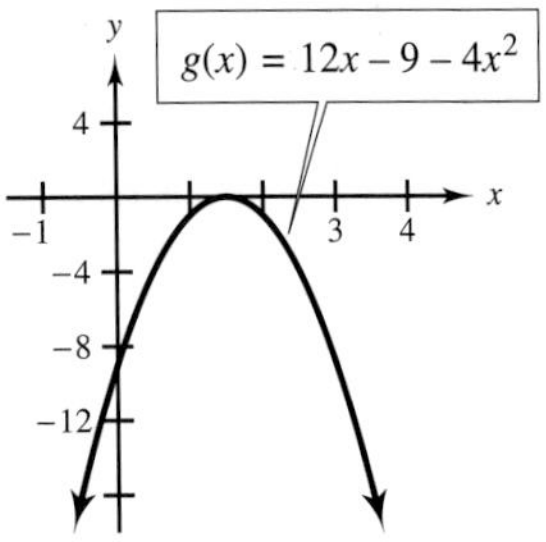

22. $\left(\frac{5}{6}, -\frac{47}{12}\right)$; Maximum: $-\frac{47}{12}$

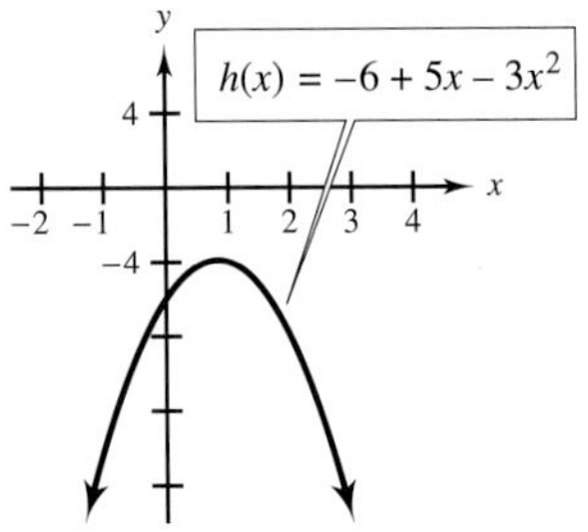

23. $(3, 0)$, $x = 3$ **24.** $(-2, 5)$, $x = -2$

25. $\left(-\frac{5}{4}, -\frac{81}{8}\right)$, $x = -\frac{5}{4}$ **26.** $(-1, 5)$, $x = -1$

27. $(-2, 9)$, $x = -2$ **28.** $\left(\frac{5}{4}, \frac{25}{8}\right)$, $x = \frac{5}{4}$

29. Because the domain is **R**, the function is defined for $x = 0$. The y-intercept is $(0, c)$, where c is the constant term of the function.

30. The x-intercepts are $(m, 0)$ and $(n, 0)$.

31. $(0, -28)$, $(7, 0)$, $(-4, 0)$ **32.** $(0, 12)$, $(1.5, 0)$, $(4, 0)$

33. $(0, 9)$, no x-intercept **34.** $(0, -3)$, no x-intercept

35. $(0, -4)$, $\left(\frac{2}{3}, 0\right)$ **36.** $(0, 1)$, $\left(-\frac{1}{2}, 0\right)$

37. $(0, 0)$, $\left(-\frac{5}{3}, 0\right)$ **38.** $(0, 0)$, $\left(\frac{1}{2}, 0\right)$

39. **(a)** (0, 8), (4, 0), (2, 0) **(b)** (3, −1), $x = 3$
(c) Domain: **R**; Range: $\{y \mid y \geq -1\}$

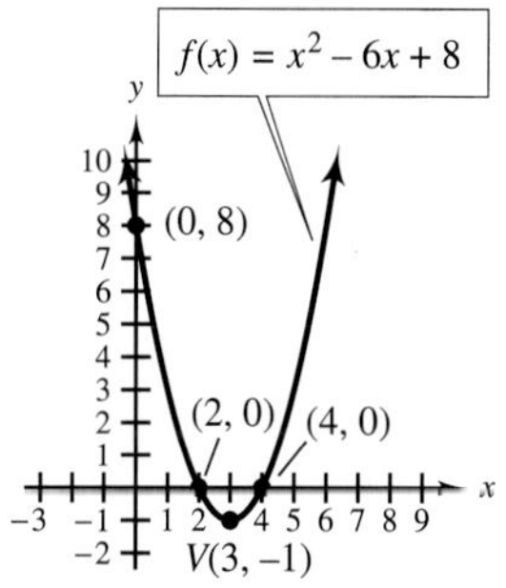

40. **(a)** (0, 12), (−6, 0), (−2, 0) **(b)** (−4, −4), $x = -4$
(c) Domain: **R**; Range: $\{y \mid y \geq -4\}$

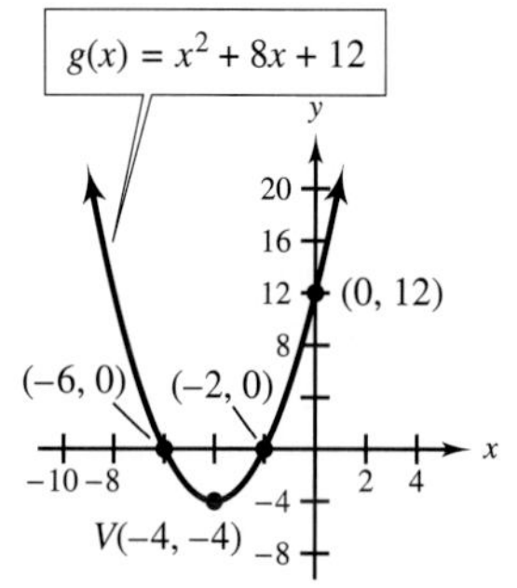

41. **(a)** (0, 5), (−5, 0), (1, 0) **(b)** (−2, 9), $x = -2$
(c) Domain: **R**; Range: $\{y \mid y \leq 9\}$

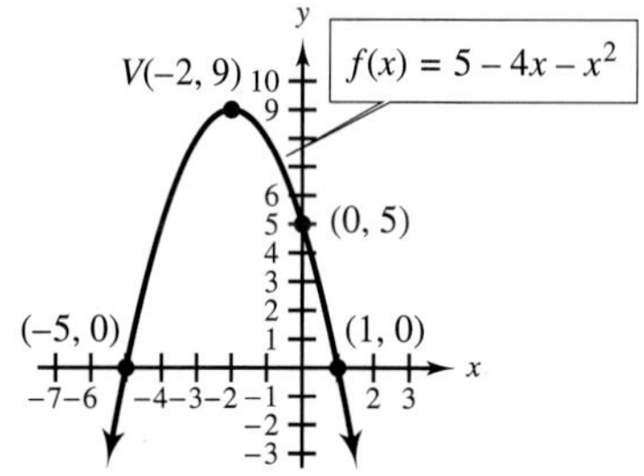

42. **(a)** (0, 12), (−6, 0), (2, 0) **(b)** (−2, 16), $x = -2$
(c) Domain: **R**; Range: $\{y \mid y \leq 16\}$

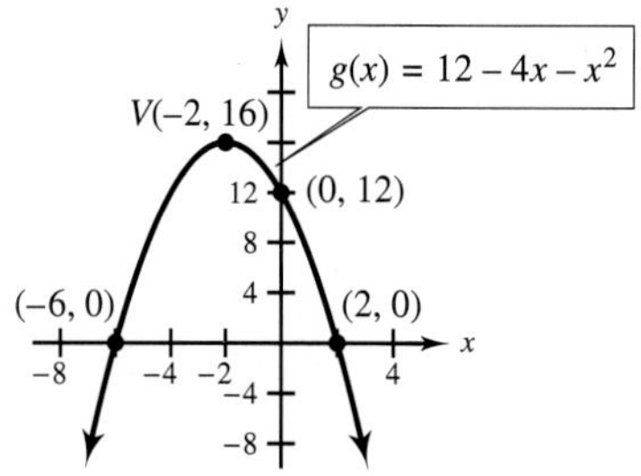

43. **(a)** (0, 9) **(b)** (−1, 7), $x = -1$
(c) Domain: **R**; Range: $\{y \mid y \geq 7\}$

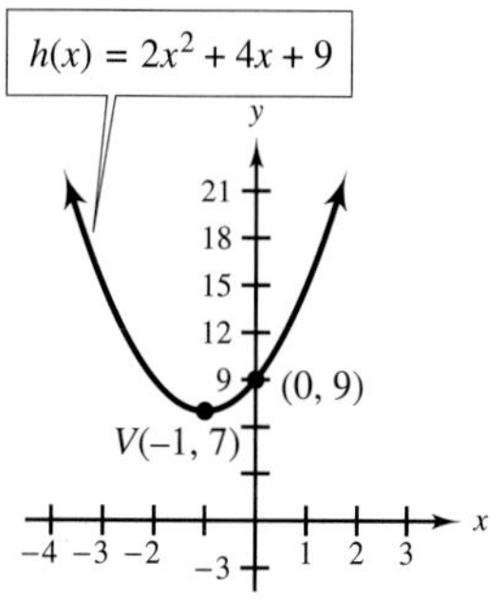

44. **(a)** (0, −8) **(b)** (1, −7), $x = 1$
(c) Domain: **R**; Range: $\{y \mid y \leq -7\}$

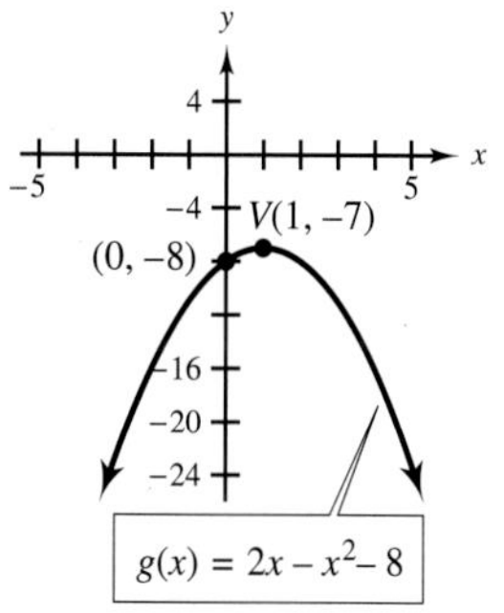

45. The domain of any quadratic function is **R**. Knowing only the vertex is not enough to specify the range. We must also know whether the graph opens upward or downward.

46. **(a)** Because the line of symmetry contains the vertex, the equation is $x = h$.
(b) The first coordinate of the vertex is a.

47. $\{y \mid y \leq 3\}$ **48.** $\{y \mid y \geq 1\}$ **49.** $\{y \mid y \geq 2\}$
50. $\{y \mid y \leq 0\}$ **51.** $x = 0$ **52.** $x = -1$ **53.** $x = 3$
54. $x = 0$ **55.** (0, 3) **56.** (−3, −2) **57.** (5, −1)
58. (−2, 0) **59.** None **60.** One, rational
61. Two, rational **62.** Two, irrational **63.** None
64. 2 **65.** 2 **66.** None **67.** 1 **68.** 2 **69.** 2
70. None **71.** 1 **72.** None **73.** 2 **74.** 1
75. $h = 0, k = 3$ **76.** $h = 0, k = -\frac{1}{2}$
77. $h = -\frac{3}{5}, k = 0$ **78.** $h = 1, k = 0$
79. $h = -4, k = 2$ **80.** $h = -1, k = -4$
81. $h = 3, k = 2$ **82.** $h = 2, k = -3$
83. $b = -2, c = -15$ **84.** $b = 7, c = 10$
85. $b = 2, c = 0$ **86.** $b = 0, c = -16$
87. $b = -8, c = 16$ **88.** $b = 2, c = 1$

89. $f(x) = x^2 - 4x + 3$ **90.** $f(x) = 2x^2 - 8x + 5$

91. $f(x) = 3x^2 + x - 2$ **92.** $f(x) = 4x^2 - 8x - 5$

93. 15, 15 **94.** 18, -18 **95.** $\frac{15}{16}$ second

96. 187.5 square feet **97.** \$230

98. $y = -0.024x^2 + 2.4x$ **99.** Downward **100.** Upward

101. For the graph of I_1, the vertex is about (8.5, 158.2), which means that the highest number of accounts, 158.2 million, occurred in 1998. Because the graph is falling from left to right for years after 1998, the model predicts that the number of accounts will decrease.

102. For the graph of I_2, the vertex is about (1.8, 89.4), which means that the lowest number of accounts, 89.4 million, occurred in 1991. Because the graph is rising from left to right for years after 1991, the model predicts that the number of accounts will increase.

103. $\left(\frac{3}{4}, -\frac{49}{8}\right)$ **104.** $\left(\frac{5}{2}, \frac{33}{4}\right)$ **105.** $\left(-\frac{1}{2}, \frac{3}{4}\right)$ **106.** $\left(-\frac{1}{2}, 0\right)$

107. **(a)** \$21,000

(b) $V(1, 21025)$; The income is maximized if one store is vacant.

(c) 30

108. 18

109. $f(x) = ax^2 + bx + c$

$$f(x) - c = ax^2 + bx$$

$$\frac{f(x)}{a} - \frac{c}{a} = x^2 + \frac{b}{a}x$$

$$\frac{f(x)}{a} - \frac{c}{a} + \frac{b^2}{4a^2} = x^2 + \frac{b}{a}x + \frac{b^2}{4a^2}$$

$$\frac{f(x)}{a} - \frac{4ac - b^2}{4a^2} = \left(x + \frac{b}{2a}\right)^2$$

$$\frac{f(x)}{a} = \left(x + \frac{b}{2a}\right)^2 + \frac{4ac - b^2}{4a^2}$$

$$f(x) = a\left(x + \frac{b}{2a}\right)^2 + \frac{4ac - b^2}{4a}$$

With the function written in this form, we see that

$$h = -\frac{b}{2a} \text{ and } k = \frac{4ac - b^2}{4a}.$$

Section 10.7 *(page 591)*

1. Plot the solutions to divide the number line into intervals. Then test a value in each interval to determine whether the inequality is true or false. The solution set is the union of all "true" intervals.

2. The Quadratic Formula will work for any quadratic equation. Other methods include factoring and the Square Root Property.

3. $(-1, 3)$ **4.** $(-\infty, -3) \cup \left(\frac{1}{2}, \infty\right)$ **5.** $(-\infty, 0) \cup \left(\frac{8}{5}, \infty\right)$

6. $(-1.37, 4.37)$ **7.** $(2, 5)$ **8.** $(-\infty, -2) \cup (6, \infty)$

9. $(-\infty, -1) \cup \left(\frac{3}{2}, \infty\right)$ **10.** $(-1.37, 4.37)$

11. Solutions of the equation are solutions of a $\le$ or $\ge$ inequality.

12. If the solutions are a and b, where $a < b$, then the possible solution intervals are as follows.

(i) (a, b) or $(-\infty, a) \cup (b, \infty)$

(ii) $[a, b]$ or $(-\infty, a] \cup [b, \infty)$

13. $(-\infty, -3) \cup (3, \infty)$ **14.** $(-\infty, -3] \cup [0, \infty)$

15. $\left[-\frac{5}{2}, \frac{1}{2}\right]$ **16.** $(-\infty, -2) \cup (4, \infty)$ **17.** $[-7, 1]$

18. $[-1, 3]$ **19.** $(-\infty, -6) \cup (3, \infty)$ **20.** $[-1, 5]$

21. $[0, 5]$ **22.** $(-1.32, 1.32)$ **23.** $\left(-\infty, -\frac{2}{3}\right] \cup \left[\frac{1}{4}, \infty\right)$

24. $(-\infty, -3) \cup \left(\frac{3}{4}, \infty\right)$ **25.** $(-1.12, 3.12)$

26. $(-\infty, -3.83] \cup [1.83, \infty)$ **27.** (b) **28.** (d) **29.** (a)

30. (c) **31.** $(-\infty, -3] \cup [2, \infty)$ **32.** $(-5, 3)$

33. $[-2, 3]$ **34.** $\left(-2, \frac{1}{3}\right)$ **35.** $\left(-\frac{2}{3}, \frac{3}{2}\right)$ **36.** $\left(-2, \frac{3}{5}\right)$

37. $(-\infty, -1.12] \cup [3.12, \infty)$ **38.** $[-1.12, 3.12]$

39. $[-2, 4]$ **40.** $(-\infty, 1] \cup [3, \infty)$ **41.** $\left(-\infty, 0\right] \cup \left[\frac{7}{4}, \infty\right)$

42. $[-1.26, 1.26]$ **43.** $(-\infty, -4] \cup [4, \infty)$

44. $\left(-\infty, -3\right] \cup \left[\frac{1}{3}, \infty\right)$ **45.** $[-0.58, 8.58]$ **46.** $\left[-\frac{4}{3}, \frac{2}{3}\right]$

47. The solution set could be **R** or Ø.

48. The solution set could be **R**, Ø, $(-\infty, 3) \cup (3, \infty)$, or $\{3\}$.

49. Because $x - 2$ is a factor of $ax^2 + bx + c$, (2, 0) is an x-intercept of the graph of $f(x) = ax^2 + bx + c$. But the solutions of $ax^2 + bx + c > 0$ are represented by points of the graph that are above the x-axis. Thus 2 is not a solution.

50. We are given that the graph of $f(x) = x^2 - x - 6$ is below the x-axis in the interval $(-2, 3)$. Therefore, the graph of f is on or above the x-axis in $(-\infty, -2] \cup [3, \infty)$, which is the solution set of the inequality.

51. $\left(-\infty, -5\right] \cup \left[-\frac{3}{4}, \infty\right)$ **52.** $(-6, -1)$

53. $(-\infty, 0) \cup (5, \infty)$ **54.** $[-5, 5]$ **55.** Ø **56.** Ø

57. $\{-2.5\}$ **58.** $\left[\frac{10}{3}, \frac{7}{2}\right]$ **59.** $(-\infty, -0.83] \cup [4.83, \infty)$

60. Ø **61.** $\{3\}$ **62.** $\{4\}$ **63.** $(-0.87, 2.87)$

64. $(-\infty, -0.64] \cup [3.14, \infty)$ **65.** $(-\infty, 1] \cup [5, \infty)$

66. **R** **67.** **R** **68.** Ø **69.** $\left[-\frac{7}{3}, -\frac{1}{2}\right]$ **70.** $\left\{\frac{1}{3}\right\}$

71. $(-\infty, -2) \cup \left(\frac{5}{4}, \infty\right)$ **72.** $(-\infty, -3.65] \cup [1.65, \infty)$

73. $(-2.53, 0.53)$ **74.** **R** **75.** $[-4, -2]$

76. $\left(-\infty, -\frac{3}{2}\right) \cup \left(\frac{1}{3}, \infty\right)$ **77.** $k > \frac{9}{8}$ **78.** $k > \frac{9}{4}$

79. Not possible **80.** Not possible **81.** $k \ge \frac{25}{4}$

82. $k \ge 1$ **83.** $k \le -3.2$ **84.** $k \ge \frac{1}{5}$ **85.** 36 **86.** $\frac{9}{4}$

87. ± 6 **88.** ± 4 **89.** 1 **90.** Between 0 and 2

91. If x and $x + 35$ represent the width and length of the home, then $x(x + 35) > 2556$, from which $x > 35$. Thus the length of the home is more than 70 feet, which violates provision (b).

92. At least 6.69 feet

93. Between 10.74 inches and 29.26 inches

94. Between 1.3 and 6.7 seconds

95. $-1.51t^2 + 15.3t + 696 = 680$

96. The solutions, approximately -1 and 11, indicate that in 1989 and 2001, the number of deaths was approximately 680,000.

97. The years in which deaths were no more than 680,000.

98. The solutions, approximately $-1 < t < 11$, indicate that from about 1989 to 2001, the number of deaths exceeded 680,000.

101. $-4 < k < 4$ **102.** $-6 \le k \le 6$ **103.** None

104. $-4 \le k \le 4$ **105.** About -1.08 and 2.08

106. About 1.11 and -0.11

Chapter Review Exercises *(page 595)*

1. Write the equation as $4x^2 - 25 = 0$ and use the factoring method. Write the equation as $x^2 = \frac{25}{4}$ and use the Even Root Property.

2. $\frac{1}{14}, \frac{1}{10}$ **3.** $-6, 7$ **4.** $\frac{4}{3}$ **5.** $-5, 3$ **6.** $2, 5$ **7.** ± 5

8. $-\frac{24}{5}, -\frac{16}{5}$ **9.** $\pm\frac{\sqrt{39}}{3} \approx \pm 2.08$ **10.** $\pm 6i$

11. $2 \pm i\sqrt{37} \approx 2 \pm 6.08i$ **12.** -9 and -7 or 7 and 9

13. After dividing both sides by 2, we obtain an x-coefficient of $\frac{5}{2}$. To complete the square, add the square of half the x-coefficient to both sides. This number is $\frac{25}{16}$.

14. $\frac{49}{4}$ **15.** $-5 \pm \sqrt{26} \approx 0.10$ or -10.10

16. $\frac{5 \pm \sqrt{41}}{2} \approx -0.70$ or 5.70 **17.** $3 \pm i$

18. $\frac{1 \pm i\sqrt{11}}{3} \approx 0.33 \pm 1.11i$ **19.** $\frac{5 \pm \sqrt{13}}{2} \approx 4.30$ or 0.70

20. $\frac{3 \pm i\sqrt{19}}{2} \approx 1.5 \pm 2.18i$ **21.** $-\frac{3}{2}, 1$

22. $\frac{-1 \pm i\sqrt{7}}{4} \approx -0.25 \pm 0.66i$

23. $\frac{-3 \pm \sqrt{11}}{2} \approx 0.16$ or -3.16 **24.** 3 and 4

25. All can be solved with the Quadratic Formula or by completing the square. Parts (i) and (iv) can be solved with the Square Root Property. Parts (i) and (ii) can be solved by factoring.

26. $\frac{-3 \pm \sqrt{37}}{2} \approx 1.54$ or -4.54 **27.** $\pm\frac{4\sqrt{5}}{5} \approx \pm 1.79$

28. $\frac{4}{3}, 1$ **29.** $\frac{5 \pm \sqrt{73}}{8} \approx -0.44$ or 1.69

30. $-2 \pm i\sqrt{3} \approx -2 \pm 1.73i$ **31.** $\frac{1 \pm i\sqrt{71}}{4} \approx 0.25 \pm 2.11i$

32. 1, rational **33.** 2, rational **34.** 2, irrational

35. 2, imaginary **36.** $\frac{-5 \pm \sqrt{53}}{2} \approx 1.14$ or -6.14

37. All are quadratic in form.

(a) $u = \frac{1}{x}$; $2u^2 + 3u - 5 = 0$

(b) $u = \sqrt{2x}$; $u^2 + u - 3 = 0$

(c) $u = x + 1$; $u^2 + 2u + 1 = 0$

(d) $u = x^{1/4}$; $u - u^2 + 3 = 0$

38. $18, -2$ **39.** $-1 \pm \sqrt{3} \approx 0.73$ or -2.73

40. No solution **41.** 0

42. $\pm\sqrt{3} \approx \pm 1.73, \pm\sqrt{2} \approx \pm 1.41$ **43.** 25 **44.** 1

45. $-4, -3, 2, 3$ **46.** $\pm 1, \pm 3i$

47. $\frac{-3 \pm i\sqrt{7}}{2} \approx -1.5 \pm 1.32i$ **48.** $-2, -\frac{4}{3}$

49. 10.95 square yards **50.** 52 **51.** 5 **52.** 9.34 mph

53. 15 hours **54.** 27.7 inches

55. To determine the range of $f(x) = ax^2 + bx + c$, find the vertex (h, k) using $h = -\frac{b}{2a}$. If $a > 0$, the range is $\{y \mid y \ge k\}$. If $a < 0$, the range is $\{y \mid y \le k\}$.

56. To determine the x-intercepts of the graph of $f(x) = ax^2 + bx + c$, solve the equation $ax^2 + bx + c = 0$.

57. $V(3, 0)$; $(3, 0)$, $(0, 9)$

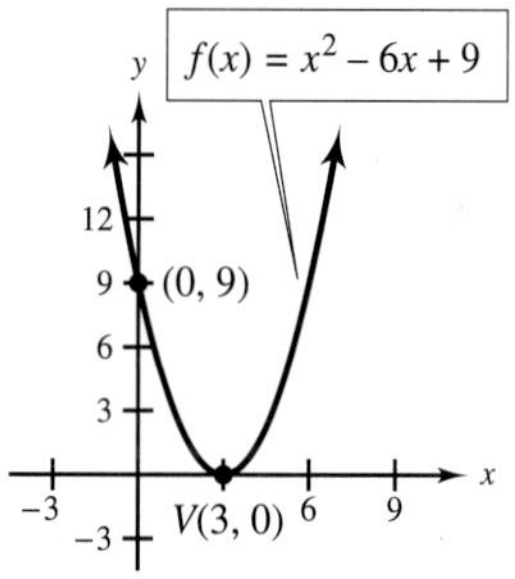

58. $V\left(\frac{2}{3}, -\frac{25}{3}\right)$; $(0, -7)$, $\left(\frac{7}{3}, 0\right)$, $(-1, 0)$

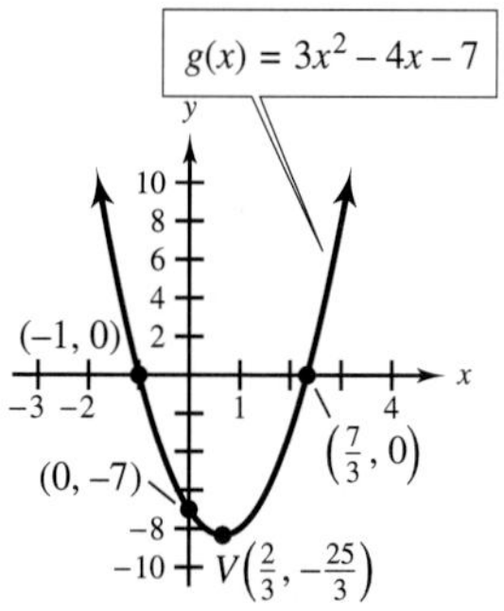

59. Downward; $\{y \mid y \le 16\}$ **60.** Downward; $\left\{y \mid y \le \frac{49}{12}\right\}$

61. Upward; $\left\{y \mid y \ge -\frac{25}{8}\right\}$ **62.** Upward; $\left\{y \mid y \ge \frac{19}{4}\right\}$

63. 62.5 feet

64. **(a)** 2 seconds **(b)** 64 feet

65. The graph of $f(x) = x^2 + 5x + 6$ is below the x-axis in the x-interval $(-3, -2)$. Therefore, the graph of f is on or above the x-axis for $x \leq -3$ and for $x \geq -2$. Thus the solution set is $(-\infty, -3] \cup [-2, \infty)$.

66. $\left(-\infty, \frac{3}{2}\right] \cup \left[2, \infty\right)$ **67.** $\{3\}$ **68.** $\left[-\frac{9}{2}, -\frac{1}{2}\right]$ **69.** $\mathbf{R}$

70. $(-0.73, 2.73)$ **71.** $\emptyset$

72. Between 42.34 feet and 82.66 feet **73.** $(-1.39, 2.89)$

74. $\mathbf{R}$ **75.** $k > \frac{1}{3}$ **76.** $k \geq \frac{9}{8}$

Chapter Test *(page 598)*

1. [*10.2*] $\dfrac{-2 \pm \sqrt{7}}{3} \approx -1.55$ or 0.22 **2.** [*10.1*] $-\frac{2}{3}$ or $\frac{3}{2}$

3. [*10.1*] $\dfrac{1 \pm \sqrt{5}}{2} \approx -0.62$ or 1.62

4. [*10.3*] $\dfrac{1 \pm \sqrt{3}}{2} \approx -0.37$ or 1.37

5. [*10.3*] $\dfrac{-7 \pm \sqrt{41}}{4} \approx -0.15$ or -3.35

6. [*10.3*] $\dfrac{1 \pm i\sqrt{23}}{6} \approx 0.17 \pm 0.80i$

7. [*10.3*] $\pm i, \pm\dfrac{\sqrt{2}}{3} \approx \pm 0.47$ **8.** [*10.1*] -2

9. [*10.5*] $h = 5$ feet, $b = 8$ feet **10.** [*10.5*] 5.12 hours

11. [*10.5*] 3 inches

12. [*10.6*] If $f(x) = 0$ has no solution, then the graph of f might lie entirely above the x-axis. In that case, $f(x) > 0$ would be true for all real numbers x.

13. [*10.7*] $(-\infty, -2] \cup [3, \infty)$

14. [*10.6*] $f(x) = (x - 3)^2 + 2$; $V(3, 2)$; $x = 3$

15. [*10.6*] **(a)** Domain: $\mathbf{R}$; Range: $\{y \mid y \geq -4\}$ **(b)** $a > 0$ **(c)** 2, 6

16. [*10.6*] $(2.39, 0)$, $(-8.39, 0)$

Chapter 11

Section 11.1 *(page 606)*

1. Evaluate f and g at 1 and add the results, or algebraically add f and g and then evaluate the resulting function at 1.

2. Because $g(-1) = 0$, the domain of $(f/g)(x)$ is $\mathbf{R}$, $x \neq -1$.

3. $\mathbf{R}$ **4.** $\mathbf{R}$ **5.** $(-\infty, 0.5]$ **6.** $(-5, \infty)$

7. $\mathbf{R}$, $x \neq -4, 0, 1$ **8.** $\mathbf{R}$ **9.** $(-\infty, -4] \cup [4, \infty)$

10. $(-\infty, -4] \cup [2, \infty)$

11. **(a)** -5 **(b)** 3 **(c)** 0 **(d)** 0

12. **(a)** 3 **(b)** $\sqrt{2} - 3$ **(c)** 3 **(d)** $\frac{1}{3}$

13. **(a)** 2 **(b)** 2 **(c)** 0 **(d)** Undefined

14. **(a)** 2 **(b)** $\frac{2}{9}$ **(c)** $\frac{1}{8}$ **(d)** 2

15. **(a)** 2.5 **(b)** 1.5 **(c)** 1 **(d)** 1

16. **(a)** $\frac{82}{9}$ **(b)** 0 **(c)** 1 **(d)** $\frac{1}{9}$

17. **(a)** 21 **(b)** -11 **(c)** 0 **(d)** -4

18. **(a)** 4.6 **(b)** -7 **(c)** $-\frac{17}{7}$ **(d)** -16

19. **(a)** 2 **(b)** 12 **(c)** -9 **(d)** -0.8

20. **(a)** 0 **(b)** 12 **(c)** 0 **(d)** $-\frac{1}{2}$

21. **(a)** -2 **(b)** -2 **(c)** $\frac{727}{3}$ **(d)** 124

22. **(a)** 35 **(b)** 3 **(c)** $3\sqrt{2}$ **(d)** 0

23. No, 0 is in the domain of h, but 0 is not in the domain of f/g because $g(0) = 0$.

24. The domain of all three functions is $(-3, 4)$.

25. **(a)** $x^2 - x - 12$; $\mathbf{R}$ **(b)** $x^2 + x - 6$; $\mathbf{R}$ **(c)** $x - 3$; $\mathbf{R}$, $x \neq -3$ **(d)** $x^3 + 3x^2 - 9x - 27$; $\mathbf{R}$

26. **(a)** $x^2 + 5x + 6$; $\mathbf{R}$ **(b)** $-x^2 - 3x$; $\mathbf{R}$ **(c)** $x^3 + 7x^2 + 15x + 9$; $\mathbf{R}$ **(d)** $\dfrac{1}{x + 1}$; $\mathbf{R}$, $x \neq -1, -3$

27. **(a)** $\dfrac{x^2 + 4}{x^2 - 4}$; $\mathbf{R}$, $x \neq -2, 2$ **(b)** $\dfrac{x^2 - 4}{x^2 + 4}$; $\mathbf{R}$ **(c)** 1; $\mathbf{R}$ **(d)** 1; $\mathbf{R}$, $x \neq -2, 2$

28. **(a)** $\dfrac{1}{x - 3}$; $\mathbf{R}$, $x \neq 2, 3$ **(b)** $x - 3$; $\mathbf{R}$, $x \neq 2$ **(c)** 1; $\mathbf{R}$, $x \neq 2$ **(d)** 1; $\mathbf{R}$, $x \neq 2, 3$

29. **(a)** $\dfrac{-5}{(x + 2)(x - 3)}$; $\mathbf{R}$, $x \neq -2, 3$
(b) $\dfrac{2x - 1}{(x + 2)(x - 3)}$; $\mathbf{R}$, $x \neq -2, 3$
(c) $\dfrac{x - 3}{x + 2}$; $\mathbf{R}$, $x \neq -2, 3$
(d) $\dfrac{1}{(x + 2)(x - 3)}$; $\mathbf{R}$, $x \neq -2, 3$

30. **(a)** $\dfrac{7x + 11}{(x + 3)(2 - x)}$; $\mathbf{R}$, $x \neq -3, 2$
(b) $\dfrac{2x^2 + 5x + 7}{(x + 3)(x - 2)}$; $\mathbf{R}$, $x \neq -3, 2$
(c) $\dfrac{x^2 + 6x + 9}{-x^2 + x + 2}$; $\mathbf{R}$, $x \neq -3, -1, 2$
(d) $\dfrac{-x^2 + x + 2}{x^2 + 6x + 9}$; $\mathbf{R}$, $x \neq -3, 2$

31. $4t^2 - 2t - 3$ **32.** $t^2 - t - 5$ **33.** $\dfrac{t^2 - 16}{2t - 4}$

34. $t^3 + 5t^2 + 4t$ **35.** $9t^2 + 3t + 2$ **36.** $-t^2 - t + 8$

37. $-8t^3 + 20t^2 + 6t - 15$ **38.** $\dfrac{3t + 45}{t^2 - 27}$

39. Produce the graphs of f and g and note the intersection of the x-values for which f and g are defined.

40. The domain of f is $\{x \mid x \leq 1\}$, and the domain of g is $\{x \mid x \geq 5\}$. Because the two domains do not have an element in common, f and g cannot be added.

41. **(a)** $[1.5, 5]$ **(b)** $[1.5, 5]$

42. **(a)** $(-\infty, 2]$ **(b)** $(-\infty, 2]$

43. **(a)** $(1, 8]$ **(b)** $[1, 8]$ **44.** **(a)** $\varnothing$ **(b)** $\varnothing$

45. First evaluate $g(5)$ and then evaluate $f(g(5))$.

46. Because $f(1) = 2$ and $g(2) = 5$, $(g \circ f)(1)$ is defined. Although $g(1) = 3$, $f(3)$ is undefined. Thus $(f \circ g)(1)$ is undefined.

47. **(a)** -2 **(b)** -1 **(c)** 2 **(d)** -14

48. **(a)** -1 **(b)** 17 **(c)** 1 **(d)** -1

49. **(a)** 15 **(b)** 16 **(c)** -3 **(d)** 1

50. **(a)** 6 **(b)** 10 **(c)** 6 **(d)** 10

51. **(a)** 5 **(b)** -2 **(c)** -4 **52.** **(a)** 6 **(b)** 3 **(c)** 1

53. $-4x^2 - 4x + 2; -2x^2 + 7$

54. $-3x^2 + 12x - 2; -9x^2 + 24x - 12$

55. $9x^2 - 21x + 15; 3x^2 - 9x + 13$

56. $2x^2 - 8x - 15; 4x^2 - 28x + 40$

57. $x - 2; \sqrt[3]{x^3 - 2}$ **58.** $-\sqrt[3]{2x^3 + 3}; -2x + 3$

59. $5; 7$ **60.** $46; 7$ **61.** $9x - 16; x + 10$

62. $x - 4; 9x - 8$ **63.** $\left||x - 3| - 3\right|; x^4 + 6x^2 + 12$

64. $x^4 - 4x^2 + 2; \left||x + 3| + 3\right|$

65. $8x^4 - 24x^3 + 12x^2 + 9x; x$

66. $x - 6; 27x^4 - 36x^3 + 6x^2 + 4x$ **67.** $3; 2$

68. $-5; -2$ **73.** $(g \circ f)(x)$ **74.** $(h \circ g)(x)$

75. $(g \circ g)(x)$ **76.** $(f \circ f)(x)$ **77.** $(f \circ h)(x)$

78. $(h \circ f)(x)$ **79.** $(g \circ h)(x)$ **80.** $(f \circ g)(x)$

81. **(a)** $g(d(t)) = \dfrac{54t}{25}$ **(b)** $c(g(d(t))) = 1.03\left(\dfrac{54t}{25}\right)$

(c) \$26.70

82. $(A \circ r)(t) = 200\pi t^2$; $(A \circ r)(5) \approx 15{,}700$ square feet

83. **(a)** $f(x) = x$ **(b)** $c(x) = 10 - 0.05x$

(c) $T(x) = 10x - 0.05x^2$ **(d)** \$487.20

84. **(a)** $f(x) = \sqrt{x^2 + 90^2}$ **(b)** $x(t) = 60t$

(c) The function $(f \circ x)(t) = \sqrt{(60t)^2 + 90^2}$ represents the distance from the ball to first base at any time t.

85. The composition of F and G is the number of transactions as a function of the number of years since 1990.

86. $G(8) = 182.2$. The actual entry is 187.0, which is the number (in thousands) of ATM terminals in 1998.

87. $F(187) = 11{,}166$. The actual entry is 11,160, which is the number (in millions) of ATM transactions when the number of terminals was 187.

88. $(F \circ G)(8) = 11{,}210$. The actual entry is 11,160, which is the number (in millions) of ATM transactions in 1998.

89. $f = g \circ h$, where $g(x) = \sqrt{x}$ and $h(x) = x^2 - 5x + 6$

90. $f = g \circ h$, where $g(x) = \dfrac{1}{x}$ and $h(x) = x^2 + 4x$

91. $f = g \circ h$, where $g(x) = x^4$ and $h(x) = 3x + 2$

92. $f = g \circ h$, where $g(x) = 3x - 5$ and $h(x) = \sqrt{x}$

93. $f = g \circ h$, where $g(x) = x + 8$ and $h(x) = |2x + 5|$

94. 3; Not defined

Section 11.2 *(page 615)*

1. The set is not necessarily a one-to-one function. If the set has any ordered pairs with the same first coordinate, then the set is not even a function.

2. If any horizontal line intersects the graph of the function in at most one point, then the function is a one-to-one function.

3. No **4.** Yes **5.** Yes **6.** No **7.** No **8.** Yes

9. No **10.** Yes **11.** Yes **12.** No

13. $f^{-1} = \{(2, 1), (4, 3), (6, 5), (-1, 7)\}$
Domain: $\{2, 4, 6, -1\}$; Range: $\{1, 3, 5, 7\}$

14. $f^{-1} = \{(2, -1), (4, -3), (-7, -5), (8, 5)\}$
Domain: $\{2, 4, -7, 8\}$; Range: $\{-1, -3, -5, 5\}$

15. $f^{-1} = \{(2, 3), (3, 4), (1, -5), (-2, -7)\}$
Domain: $\{2, 3, 1, -2\}$; Range: $\{3, 4, -5, -7\}$

16. $f^{-1} = \{(-6, 5), (-8, 7), (5, -8), (7, -6)\}$
Domain: $\{-6, -8, 5, 7\}$; Range: $\{5, 7, -8, -6\}$

17. -2 **18.** -4 **19.** 3 **20.** 5

21. No, $f^{-1}(x) = f(x) = x$. The symbol x^{-1} means $1/x$, not the inverse function.

22. The functions f and g appear to be inverse functions.

23. Yes **24.** No **25.** No **26.** No **27.** Yes **28.** No

29.

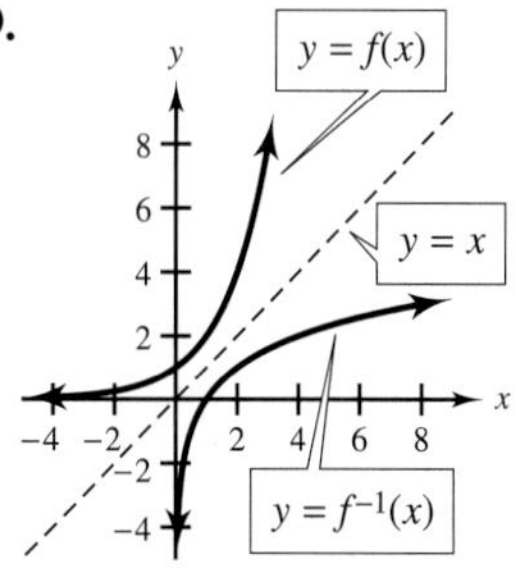

30.

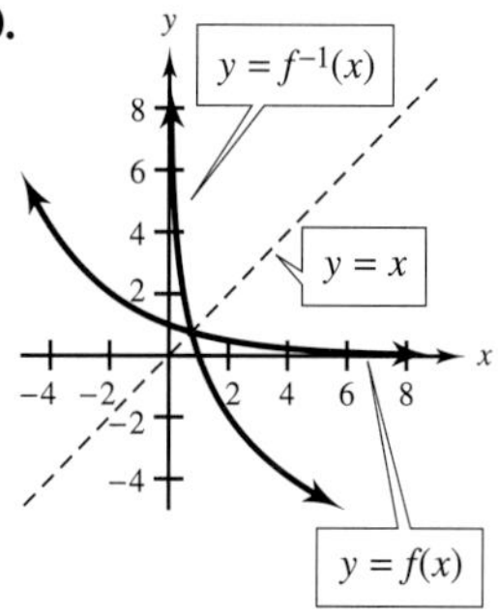

31.

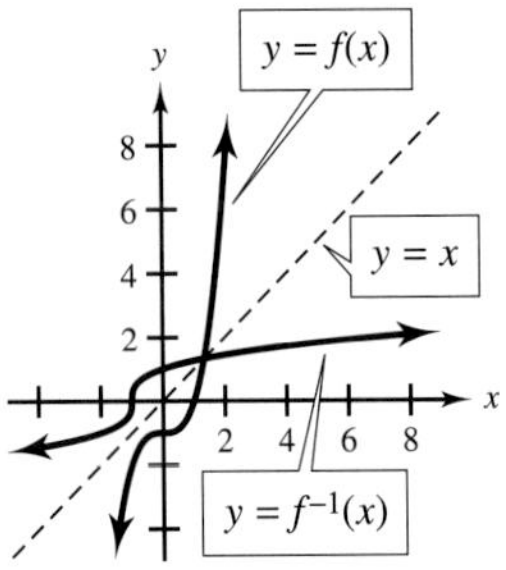

32.

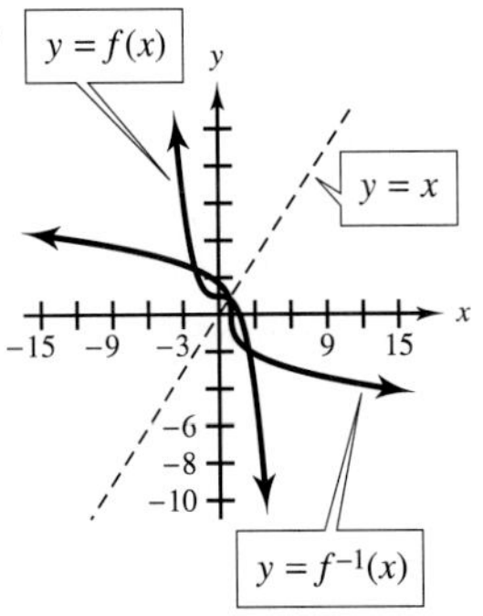

33. The graph of a linear function is a nonvertical line that passes the Horizontal Line Test. Therefore, the function has an inverse.

34. The function $f(x) = x^n$ provided n is odd because any value of $f(x)$ corresponds to exactly one x-value. If n were even, then a positive value of $f(x)$ would correspond to more than one value of x. For example, if $f(x) = x^2$, then two pairs of f are $(2, 4)$ and $(-2, 4)$, which means that the graph of f would not pass the Horizontal Line Test. If n is odd, then $f^{-1}(x) = x^{1/n}$.

35. No **36.** Yes **37.** Yes **38.** Yes **39.** No **40.** No

41. Yes **42.** Yes **43.** Yes **44.** No **45.** Yes

46. Yes **47.** $\frac{x-5}{3}$ **48.** $\frac{x+7}{5}$ **49.** $-0.5x$ **50.** $\frac{x+2}{2}$

51. $3(x-5)$ **52.** $\frac{3x+1}{2}$ **53.** $\frac{1}{x}$ **54.** $\frac{1}{x+3}$

55. $\frac{1}{x} - 2$ **56.** $\frac{x-3}{x}$ **57.** $\frac{2x+1}{x-1}$ **58.** $\frac{3-x}{2x-1}$

59. $\sqrt[3]{x-1}$ **60.** $-2\sqrt[3]{x}$ **61.** $x^{-1/3}$ **62.** $\sqrt[5]{\frac{1}{x-1}}$

63. $4x^2, x \geq 0$ **64.** $\frac{x^2}{3}, x \geq 0$ **65.** $2 - x^3$ **66.** $3 - x^5$

67. $(x-3)^5$ **68.** $(1-x)^3$

69. Multiply x by 3 and then subtract 2.

70. Add 1 to x and then divide the result by -2.

71. Take the cube root of x and then subtract 1.

72. Double the value of x and then subtract 4. **73.** (iii)

74. **(a)** I and IV **(b)** II and IV
(c) I and III **(d)** III and IV

75. The first is a one-to-one function, but the second is not.

76. $\frac{x+1}{2}$

77. $\frac{9}{5}t + 32$; Converts Celsius temperatures to Fahrenheit

78. If m is the number of minutes and h is the number of hours, then $m(h) = 60h$ converts hours to minutes. The inverse function is $h(m) = \frac{m}{60}$, which converts minutes to hours.

79. $d = 1.50p$; The inverse function is $p = \frac{d}{1.50}$, which gives the value p in pounds of d dollars.

80. $y = 1.094m$; The inverse function is $m = \frac{y}{1.094}$, which gives the number of meters in y yards.

81. **(a)** Function C is a linear function. Its graph is neither horizontal nor vertical, so C is a one-to-one function.
(b) $s = \frac{C + 23}{1.3}$. The inverse function represents the number of subscribers (in millions) as a function of the number of cell sites (in thousands).

82. **(a)** Relation R is a function because each named body of water is paired with exactly one of lake or ocean.
(b) The function does not have an inverse because the function is not one-to-one.

83. $y(p) = 18p - 108$ **84.** $p(y) = \frac{1}{18}y + 6$

85. The functions appear to be inverse functions. **86.** Yes

87.

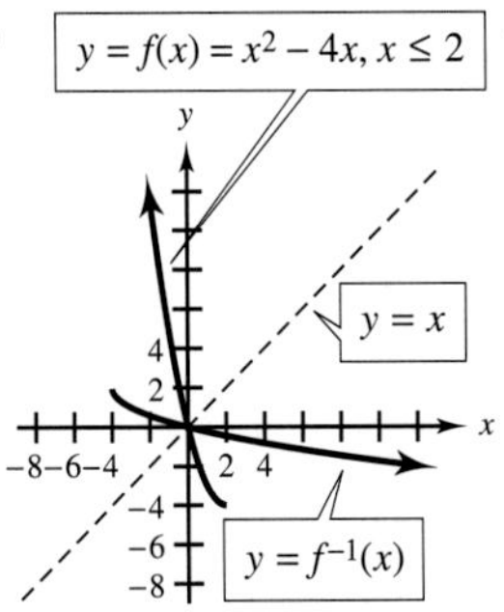

88.

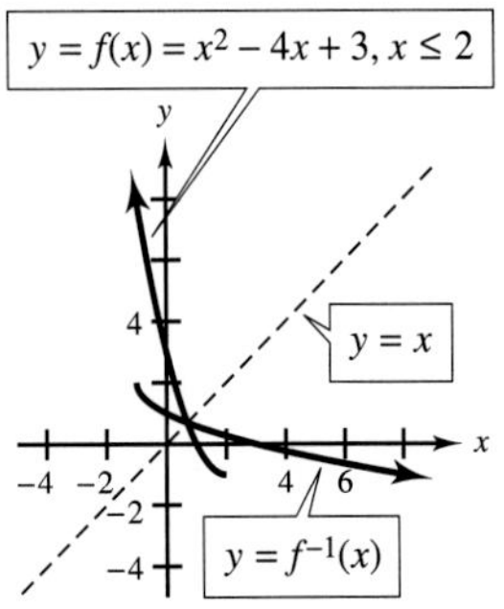

89. $\sqrt{x-1}, x \geq 1$ **90.** $4 - x, x \geq 0$ **91.** $x - 5, x \leq 0$

92. $3 - \sqrt{4 + x}, x \geq -4$

Section 11.3 *(page 625)*

1. **(a)** The graph rises from left to right if $b > 1$.
(b) The graph falls from left to right if $0 < b < 1$.

2. As b increases, the graph becomes steeper.

3. 32 **4.** 9 **5.** $\frac{1}{9}$ **6.** $\frac{1}{16}$ **7.** 1 **8.** $\frac{1}{2}$ **9.** 2

10. 27 **11.** 0.02 **12.** 25 **13.** 403.43 **14.** 1

15. 11.18 **16.** 0.45 **17.** 54.60 **18.** 0.14 **19.** 1613.72

20. 0.25 **21.** 243 **22.** 1397.54 **23.** 29,370.08

24. 0.16 **25.** 3.95 **26.** 5.05

27. Because $3^{-x} = \left(\frac{1}{3}\right)^x = \frac{1}{3^x}$, functions (ii), (iii), and (iv) all have the same graph.

28. As b increases, the graph becomes steeper.

29. For $c > 0$ the graph is shifted up c units, and for $c < 0$ the graph is shifted down $|c|$ units.

30. For $k > 0$ the graph is shifted left k units, and for $k < 0$ the graph is shifted right $|k|$ units.

31. **(a)** Up 1 unit **(b)** Down 3 units

32. **(a)** Left 3 units **(b)** Right 4 units

33. **(a)** Up 3 units **(b)** Left 3 units

34. **(a)** Reflection of the graph of h in the y-axis

(b) The graphs are the same.

35. **(a)** (B) **(b)** (A) **36.** **(a)** (D) **(b)** (C)

37.

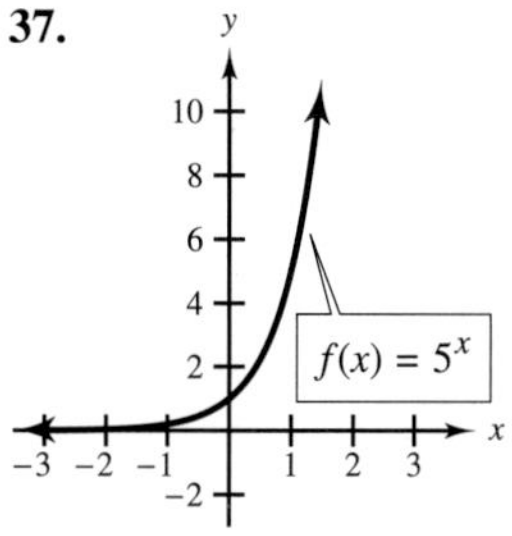

38.

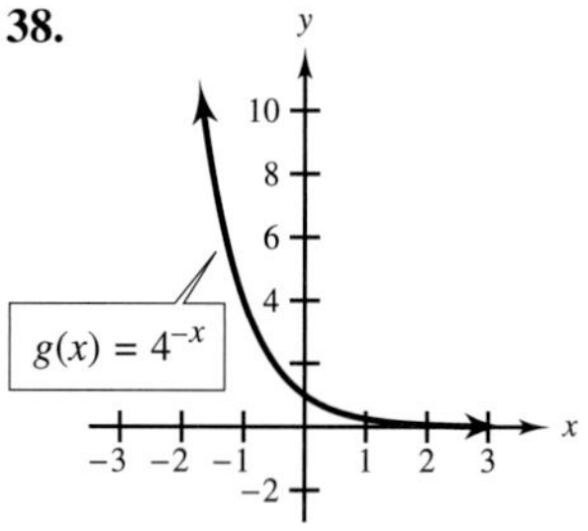

39.

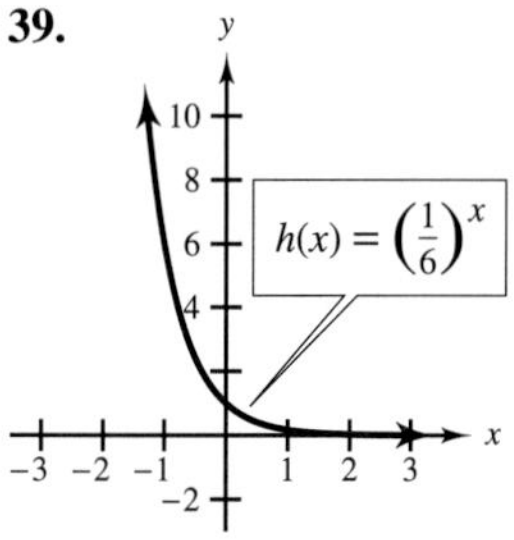

40.

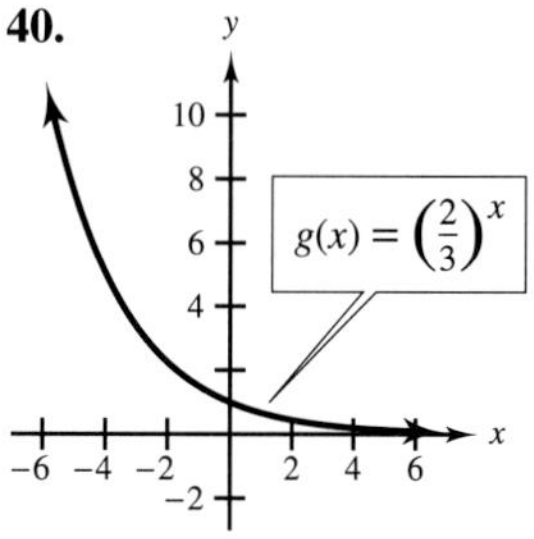

41.

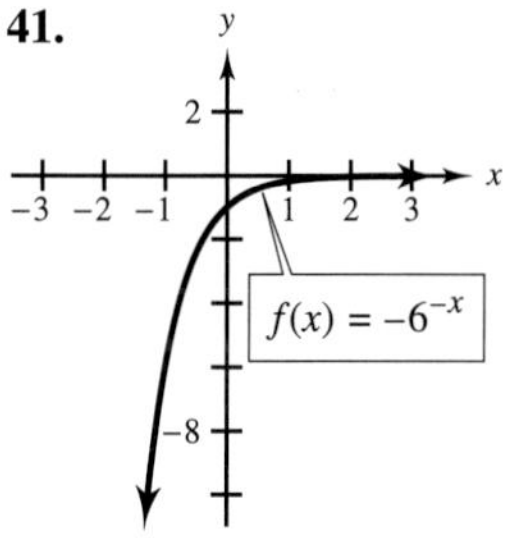

42.

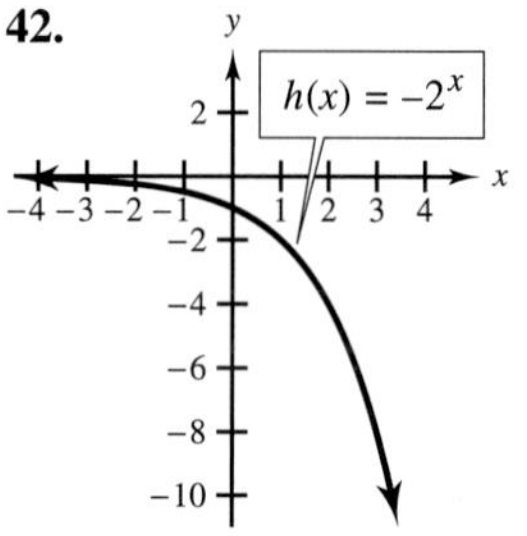

43.

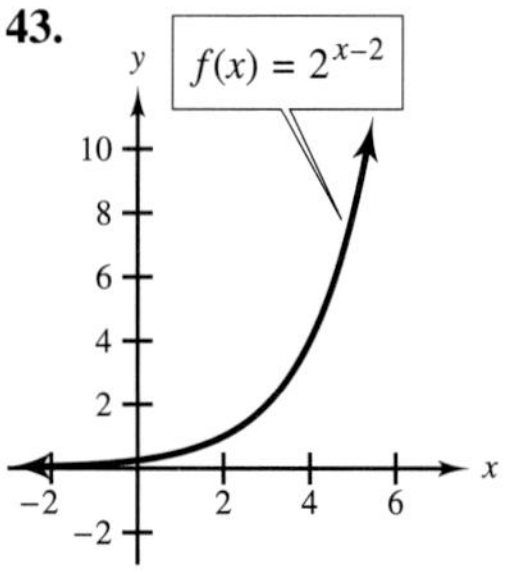

44.

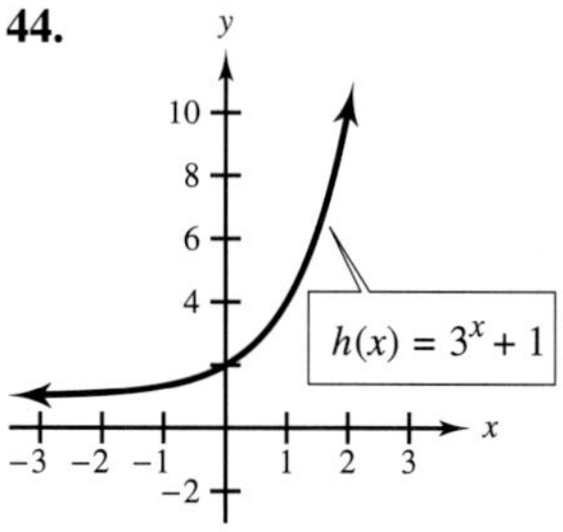

45.

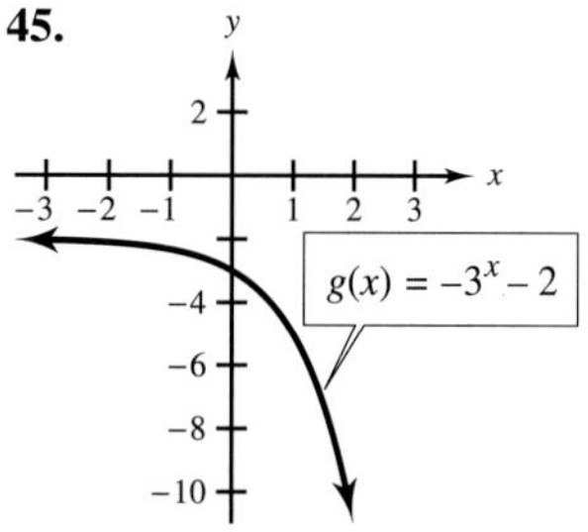

46.

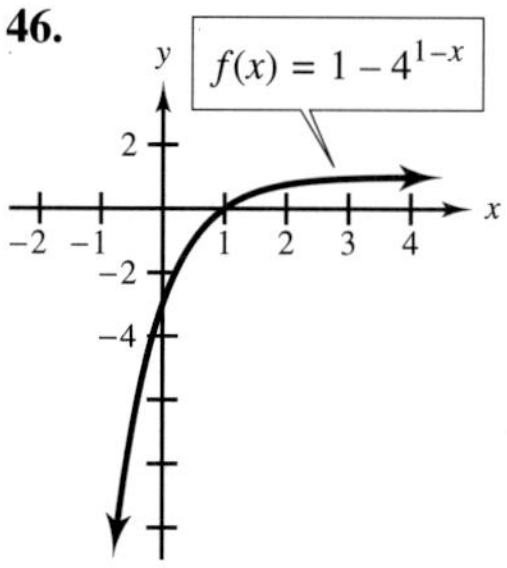

47.

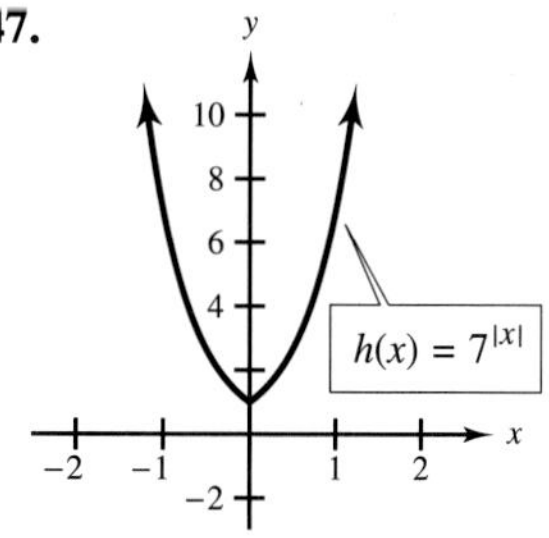

48.

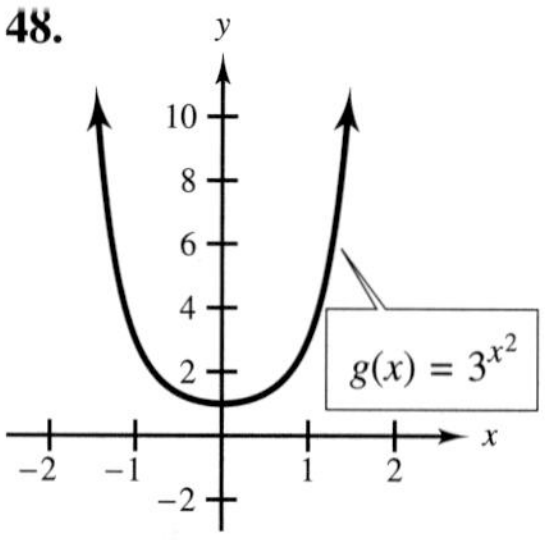

49. Both sides of the equation $4^x = 8$ can be written with the same base, and then the exponents can be equated. The two sides of $4^x = 12$ cannot be written with the same integer base.

50. Because $f(x) = 2^{2x} = (2^2)^x = 4^x = g(x)$, the graphs of f and g are the same.

51. -3 **52.** 5 **53.** -2 **54.** -3 **55.** -1.5 **56.** 1

57. -1.5 **58.** -0.5 **59.** 1.5 **60.** -0.75 **61.** $\frac{10}{3}$

62. -1 **63.** $\frac{2}{3}$ **64.** 2 **65.** -3 **66.** 0.5

67. No solution **68.** No solution **69.** 0

70. No solution **71.** No solution **72.** 1.5 **73.** 0.5

74. $-3, 1$ **75.** $1, -\frac{5}{3}$ **76.** $\frac{16}{15}$ **77.** -8 **78.** -1

79. ± 1 **80.** ± 2 **81.** $-1, 0.5$ **82.** $\frac{2}{3}$ **83.** \$1.83

84. \$20.06 **85.** \$5000 **86.** \$6100

87. **(a)** $98.69(1.06)^3, 98.69(1.06)^4, 98.69(1.06)^5$

(b) $E(n) = 98.69(1.06)^n$ **(c)** \$132.07, \$139.99

88. **(a)** 44.6 million **(b)** 2001

89. The value of b, 1.09, is greater than 1.

90. About \$427.2 billion **91.** 137% **92.** \$741.88

93. $2, -0.77$ **94.** 3 **95.** $(0, \infty)$ **96.** $(-\infty, 0)$ **97.** **R**

98. $\varnothing$ **99.** $\{y \mid y < 2\}$ **100.** $\{y \mid y \geq 1\}$

Section 11.4 *(page 634)*

1. The logarithmic function is the inverse of the exponential function, which is not defined for $b \leq 0$ or for $b = 1$.

2. If $x = \log 100$, then $10^x = 100 = 10^2$. Thus $x = 2$. If $x = \ln 100$, then $e^x = 100$, but 100 cannot easily be written as a power of e.

3. $7^2 = 49$ **4.** $4^{3/2} = 8$ **5.** $10^{-3} = 0.001$

6. $\left(\frac{1}{2}\right)^{-3} = 8$ **7.** $e^{1/3} = \sqrt[3]{e}$ **8.** $8^{-2/3} = \frac{1}{4}$ **9.** $m^2 = n$

10. $e^t = 3$ **11.** $\log_4 64 = 3$ **12.** $\log_{27} 9 = \frac{2}{3}$

13. $\log 0.00001 = -5$ **14.** $\log_8 \frac{1}{4} = -\frac{2}{3}$

15. $\log_{1/2} 8 = -3$ **16.** $\log_5 \frac{1}{125} = -3$

17. $\ln P = rt$ **18.** $\log_2 y = x$ **19.** $\ln x$ **20.** $\log_3 x$

21. $\log_{1/4} x$ **22.** $\log_{3/4} x$ **23.** 10^x **24.** 5^x **25.** 1.5^x

26. e^x **27.** 2 **28.** 4 **29.** 0 **30.** 1 **31.** 2.5

32. −1.5 **33.** 3 **34.** 5 **35.** Not defined

36. Not defined **37.** −0.5 **38.** −2 **39.** 1 **40.** −1

41. 2.5 **42.** $\frac{2}{3}$ **43.** Not defined **44.** Not defined

45. −1 **46.** −1 **47.** 3 **48.** 1 **49.** 0.5 **50.** 0

51. 1.39 **52.** 1.60 **53.** 1.40 **54.** 2.32 **55.** −0.14

56. 0.73 **57.** 0.75 **58.** −0.62 **59.** 3.63 **60.** 0.35

61. Write the equation in exponential form.

62. Because the domain of logarithmic functions is $(0, \infty)$, $x + 2 > 0$. Thus the solution must be greater than -2.

63. 243 **64.** $\frac{1}{9}$ **65.** e **66.** 1 **67.** 0.25 **68.** 81

69. 9 **70.** e **71.** 125 **72.** 100 **73.** $x > 0, x \neq 1$

74. $x > 0, x \neq 1$ **75.** 2 **76.** 4 **77.** 3 **78.** 1

79. 15 **80.** 3 **81.** ± 25 **82.** ± 2

83. **(a)** The graph rises from left to right if $b > 1$.
(b) The graph falls from left to right if $0 < b < 1$.

84. Because f and g are inverse functions, $(f \circ g)(x) = x$.

85. **(a)** Up 3 units **(b)** Left 3 units

86. **(a)** Right 5 units **(b)** Down 5 units

87. **(a)** Folded across the y-axis
(b) Folded across the x-axis

88. **(a)** Steeper **(b)** Flatter

89. **(a)** (B) **(b)** (D) **90.** **(a)** (C) **(b)** (A)

91.

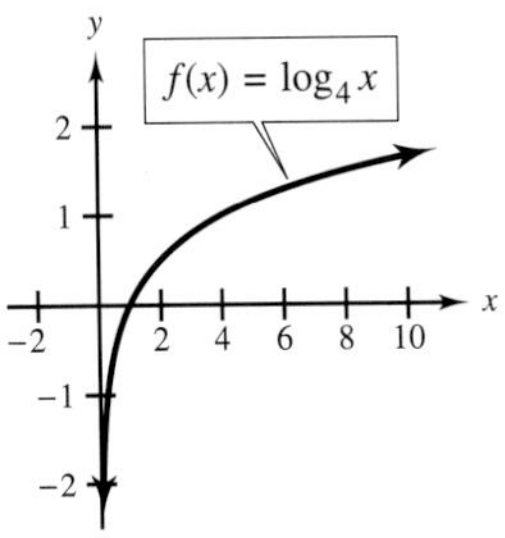

92.

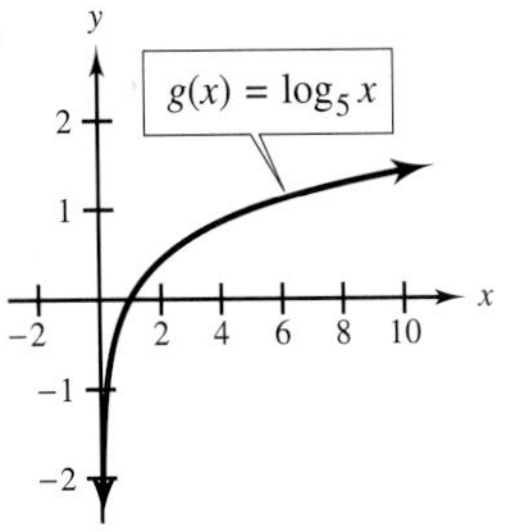

93.

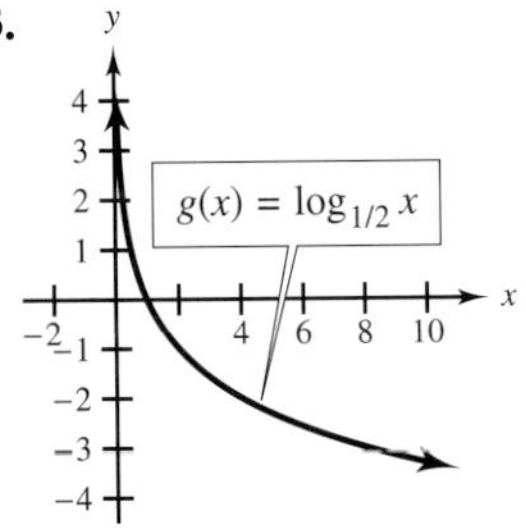

94.

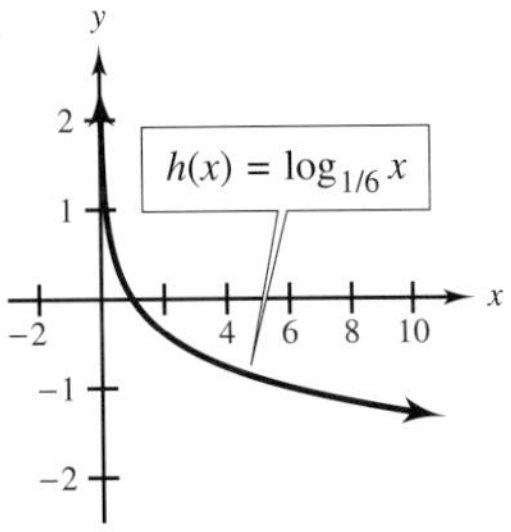

95.

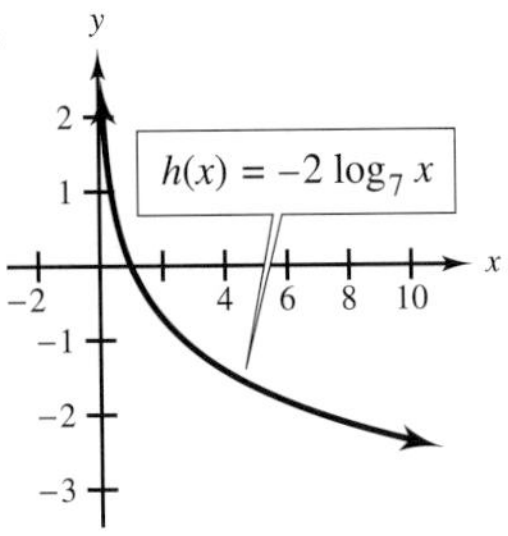

96.

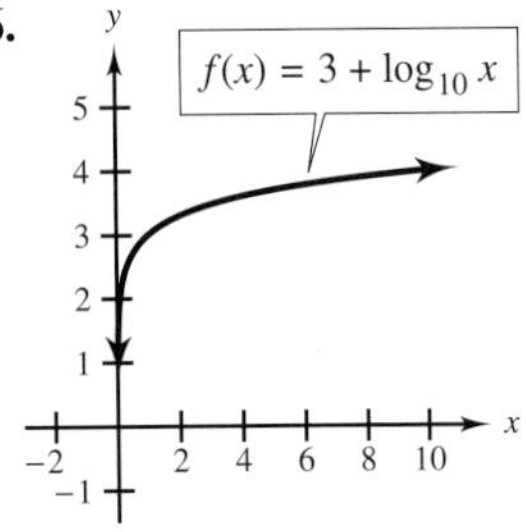

97.

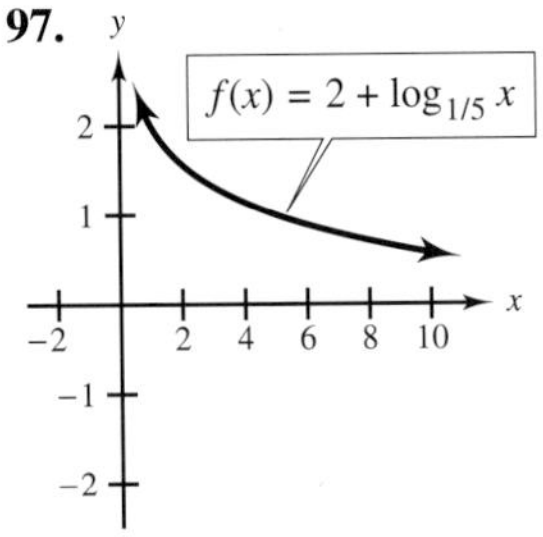

98.

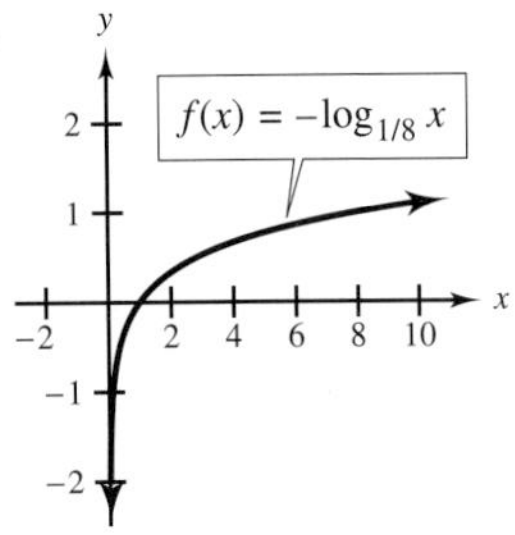

99.

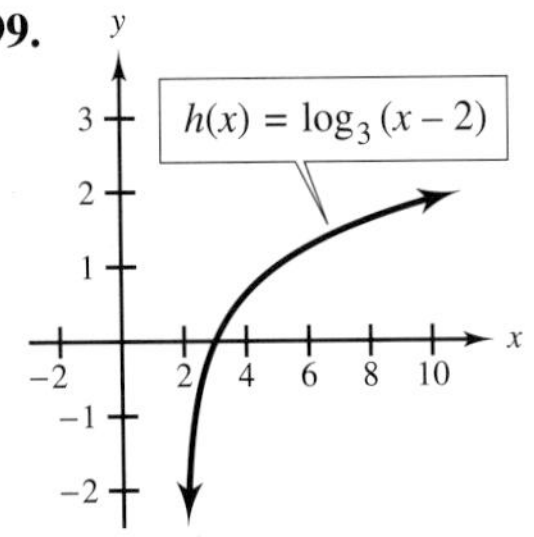

100.

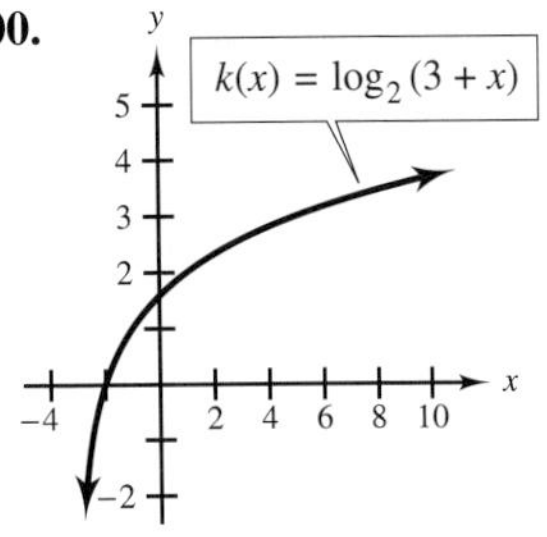

101.

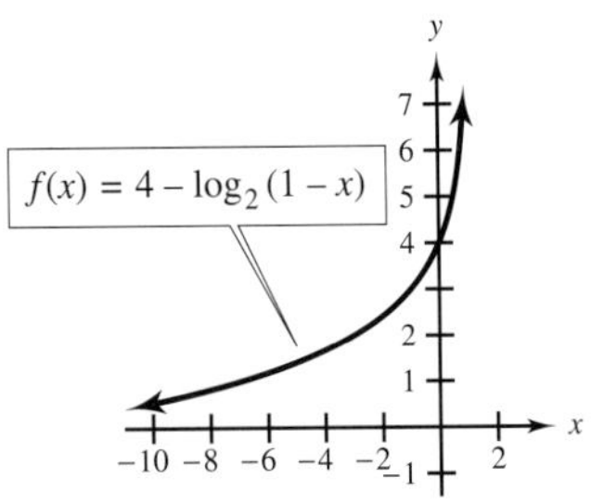

102.

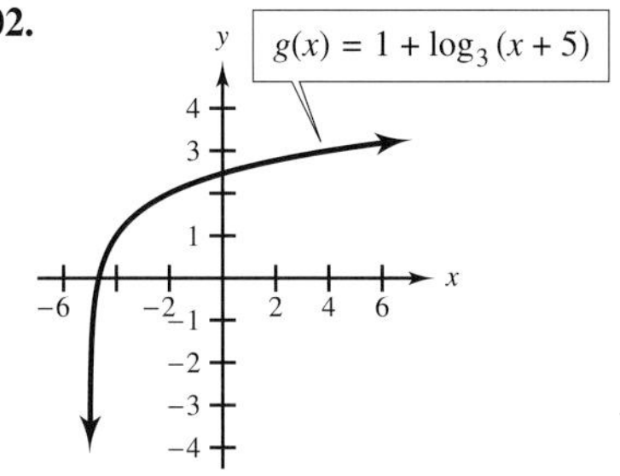

103. **(a)** 36.02 decibels **(b)** 600

104. **(a)** 11 **(b)** 409,600

105. **(a)** 2.3 **(b)** 10^{-7}

106. **(a)** 2.5 kilometers **(b)** 426 millimeters

107. **(a)** National: 2.50; American: 2.66

(b) The American League average will be higher than the National League average, but the difference in the averages will decrease.

108. **(a)** 67.8%

(b) 2017

109.

Year	x	N_1	N_2
1990	5	4.0	4.0
1993	8	7.6	7.4
1997	12	10.7	10.6
2002	17	13.3	13.2
2050	65	23.6	17.5

110. Both functions model the data well.

111. Function N_1 predicts a greater long-range increase in the number of telecommuters.

112. Add 2 to the function.

113. Because $\ln \frac{1}{2} < 0$, reverse the inequality symbol to obtain $a\left(\ln \frac{1}{2}\right) < b\left(\ln \frac{1}{2}\right)$.

114. Answers will vary. **115.** 2.5 **116.** 1

117. $(-\infty, -1) \cup (1, \infty)$ **118.** $\left(-\infty, -\sqrt{17}\right) \cup \left(\sqrt{17}, \infty\right)$

119. $(0, 1)$ **120.** $(1, \infty)$

Section 11.5 *(page 641)*

1. What exponent on 9 results in 3? The answer is $\frac{1}{2}$.

2. In the first instance, we are using the Product Rule for Logarithms. Using the Power Rule for Logarithms would be more efficient.

3. 7 **4.** 2 **5.** 0 **6.** 1 **7.** 2 **8.** 10 **9.** 0.25 **10.** 0.75 **11.** 0.25 **12.** $\frac{1}{64}$ **13.** 9 **14.** 8 **15.** 0

16. 0 **17.** 1 **18.** 2 **19.** $\log_7 3 + \log_7 x$

20. $\log_8 5 + \log_8 m + \log_8 n$ **21.** $\log_7 5 - \log_7 y$

22. $\log_6 y - \log_6 t$ **23.** $2 \log x$ **24.** $-4 \log_3 7$

25. $\frac{1}{2} \log_9 n$ **26.** $\frac{1}{3} \ln x$ **27.** (c) **28.** (f) **29.** (a)

30. (h) **31.** (g) **32.** (b) **33.** (d) **34.** (e)

35. Begin with the Power Rule for Logarithms to obtain $n \log (AB)$. In the Order of Operations, exponents have priority over products.

36. The logarithms in (ii) and (iii) cannot be expanded because there are no rules for expanding the logarithm of a sum or difference.

37. $2 \log_9 x - 3 \log_9 y$ **38.** $3 + 3 \log_3 (x + 5)$

39. $2 \log (x + 1) - \log (x + 2)$

40. $2 \log x + 3 \log y - 4 \log z$

41. $5 \ln (2x + 5) - 5 \ln (x + 7)$

42. $6 \log_3 x - 2 \log_3 z - 10 \log_3 y$ **43.** $\frac{1}{2} - 2 \log x$

44. $\frac{3}{4} \ln x + \frac{1}{4} \ln y - \frac{1}{2} \ln z$ **45.** $\frac{1}{3} \log_5 x + \frac{2}{3} \log_5 y$

46. $4 + 2 \log_3 x + \frac{1}{2} \log_3 y$ **47.** $-5 + \frac{1}{2} \log_2 (x + 1)$

48. $\frac{1}{2} \log_8 x + \frac{1}{2} \log_8 (x + 2)$ **49.** $\frac{5}{3} \log_2 x + \frac{1}{2} \log_2 y$

50. $\ln (x + 2) - \frac{3}{2} \ln x$ **51.** $2 \log (2x + 1)$

52. $\log (x + 2) + \log (x + 1)$ **53.** $0.08t + 3 \ln 10$

54. $1 + \ln (e - 1)$ **55.** Cannot be expanded

56. Cannot be expanded **57.** $rt + \ln P$

58. $\ln P + nt \ln \left(1 + \dfrac{r}{t}\right)$

59. Use the Distributive Property: $(2 + 1) \log 3 = 3 \log 3$. For $2 \log x + \log y$, $x \neq y$, the Distributive Property does not apply.

60. In the first case, the Power Rule gives us

$$-\log x = \log x^{-1} = \log \frac{1}{x}.$$

In the second case, the Quotient Rule gives us

$$\log \frac{1}{x} = \log 1 - \log x = 0 - \log x = -\log x.$$

61. $\log_4 15$ **62.** $\ln x$ **63.** $\log_3 2$ **64.** $\log_7 u^3$

65. $\log_2 y^3$ **66.** $\log_5 \dfrac{1}{x}$ **67.** $\log \sqrt{x}$ **68.** $\ln \sqrt[3]{t^2}$

69. 1 **70.** 1 **71.** 3 **72.** −3 **73.** $\log_8 \dfrac{3x}{y}$

74. $\ln \dfrac{x}{6}$ **75.** $\log_7 \dfrac{x^2}{y}$ **76.** $\log_8 5t^2$ **77.** $\log \dfrac{\sqrt{x}}{\sqrt[3]{y^2}}$

78. $\log_6 r\sqrt{rs}$ **79.** $\log \dfrac{y^3}{x^2 z^4}$ **80.** $\log \dfrac{y(x + 2)^2}{z^5}$

81. $\log \dfrac{(x + 3)^2}{(x + 1)^3}$ **82.** $\log \dfrac{\sqrt{t + 1}}{t^3}$ **83.** $\log \sqrt{\dfrac{x}{x + 1}}$

84. $\log \sqrt[3]{\dfrac{(x+2)^2}{x-5}}$ **85.** $\log_3 243x^2$ **86.** $\log_2 2x^2$

87. $\ln P_0e^{rt}$ **88.** $\ln 1000e^{0.06t}$ **89.** False **90.** True

91. True **92.** False **93.** False **94.** True **95.** True

96. True **97.** True **98.** True **99.** False **100.** False

101. **(a)** 32.3% **(b)** $P(t) = \dfrac{10}{\log(5t+10)}$

102. 300,000 units

Section 11.6 *(page 649)*

1. If b can be written as a power of 3, then we can equate exponents instead of writing the equation in logarithmic form.

2. Because 5 cannot easily be written as a power of e, equating exponents is not a useful method to solve the equation.

3. -0.6 **4.** -1 **5.** -1 **6.** 1.5 **7.** 1.46 **8.** 0.54

9. 11.55 **10.** 0.73 **11.** -0.03 **12.** 9.21 **13.** 0.61

14. -5.42 **15.** -5.86 **16.** 0.12 **17.** 1.39 **18.** 2.48

19. No solution **20.** 1.58 **21.** 6.45 **22.** 7.85

23. ± 1 **24.** ± 1.21 **25.** $-0.57, -1.43$

26. No solution **27.** 0, 1 **28.** -1.10

29. No; the bases must be the same for us to equate arguments.

30. The only solution is 3 because -6 is not in the domain of the logarithmic function.

31. 3 **32.** 1 **33.** 26 **34.** 7 **35.** $\frac{32}{3}$ **36.** 256

37. 34 **38.** $\frac{26}{3}$ **39.** $8, -1$ **40.** $-3, 2$ **41.** $-\frac{11}{3}$

42. $-8, 4$ **43.** 2 **44.** 5 **45.** $-3, 3$ **46.** $-9, 7$

47. 8 **48.** $\frac{19}{47}$ **49.** 2, 5 **50.** 4, 1 **51.** 10 **52.** 4

53. No solution **54.** No solution **55.** $-2.41, 0.41$

56. $1.19, -4.19$ **57.** 1 **58.** 1.5 **59.** 0.75 **60.** $\frac{2}{3}$

61. 33, 1.5 **62.** $2.5, -3.5$

63. If b were permitted to be 1, then $1^3 = 1^8$, but $3 \neq 8$.

64. Because f and g are inverse functions, their graphs are symmetric about the line $y = x$. Moreover, neither graph intersects $y = x$. Therefore, the graphs of f and g do not intersect, so the equation has no solution.

65. $10, 10^{-4}$ **66.** e^3, e^{-2} **67.** 4, 256 **68.** $3^9, 3$

69. $\ln 3, \ln 2$ **70.** $\ln 3, 0$ **71.** $\dfrac{\ln 7}{\ln 5}$ **72.** 0

73. $I = 10^R$ **74.** $P_0 = 10^{1-P}P_I$ **75.** $I = I_0\sqrt[10]{10^L}$

76. $Q = \dfrac{M}{M+1}$ **77.** $r = \dfrac{1}{t}\ln\left(\dfrac{P}{P_0}\right)$ **78.** $k = -\dfrac{1}{t}\ln\left(\dfrac{A}{A_0}\right)$

79. $t = \dfrac{\ln 2}{n\ln\left(1+\dfrac{r}{n}\right)}$ **80.** $t = \dfrac{\ln\left(\dfrac{A}{P}\right)}{\ln(1+r)}$ **81.** 4.95%

82. 25.21 years

83. **(a)** 14 million **(b)** 25.51 million **(c)** 34.43 million

84. 11,143 units **85.** 9953 years **86.** 2.88 days

87. \$85,000 **88.** \$50,000 **89.** 5:53 A.M.

90. 30 minutes **91.** 7.9 **92.** 5 times as intense

93. **(a)** $51.59(1.045)^x = 340$

(b) $x \approx 43(2003)$

94. **(a)** $-30{,}761 + 15{,}637 \ln x = 17{,}500$

(b) $x \approx 22(2002)$

95. 2001 **96.** About 2006

97. $A(x) = 3774e^{-0.19x}$; The number of transactions per terminal is declining.

98. 2003 **99.** 10^e **100.** e^{10}

101. **(a)** 1, 3 **(b)** $\pm 1, \pm 3$

In (a) negative numbers are not in the domain of $\log_3 x$, but in (b) negative numbers are in the domain of both logarithmic functions.

102. **(a)** Basic Property of Logarithms

(b) Equate Logarithmic Expressions

(c) Power Rule for Logarithms

(d) Divide both sides, $\log_a b \neq 0$

103. **(a)** The domain of $\ln(3-4x)$ is $(-\infty, 0.75)$, whereas the domain of $\ln(x-2)$ is $(2, \infty)$.

(b) Because the domains have no x-values in common, the equation has no permissible replacements for x.

(c) $\emptyset$

Chapter Review Exercises *(page 654)*

1. $x^2 + 4$ **2.** -12.64 **3.** 0.17 **4.** $3x^3 - 5x^2 - 12x$

5. $9x^2 + 15x + 4$ **6.** $3\sqrt{x+1} + 4$ **7.** -1.85

8. 2.49 **9.** No, $g(-1) = 0$, but $f(0)$ is not defined.

10. 8, 108 **11.** $\dfrac{1}{|2x+5|+2}, \left|\dfrac{2}{x+2}+5\right|$

12. $\dfrac{50}{2x+160}$; This function is the cyclist's rate as a function of the temperature x.

13. Use the Vertical Line Test to determine whether a relation is a function. Use the Horizontal Line Test to determine whether a function is one-to-one.

14. One-to-one **15.** Not one-to-one

16. **(a)** Not one-to-one **(b)** One-to-one

17. $g^{-1} = \{(2, 7), (6, -3), (-1, -2), (4, 0)\}$
Domain: $\{2, 6, -1, 4\}$; Range: $\{7, -3, -2, 0\}$

18. Not inverses **19.** Inverses **20.** $\frac{x-8}{5}$ **21.** $\sqrt[3]{x+5}$

22. The line of symmetry is $y = x$. If $f(x) = f^{-1}(x)$ has no solution, then neither graph intersects the line of symmetry.

23. The exponent x can be any real number, but $b > 0$ and $b \neq 1$.

24. **(a)** No x-intercept **(b)** $(0, 1)$ **(c)** **R** **(d)** $(0, \infty)$

25. **(a)** $-\frac{8}{e^2}$ **(b)** $\frac{27}{64}$ **26.**

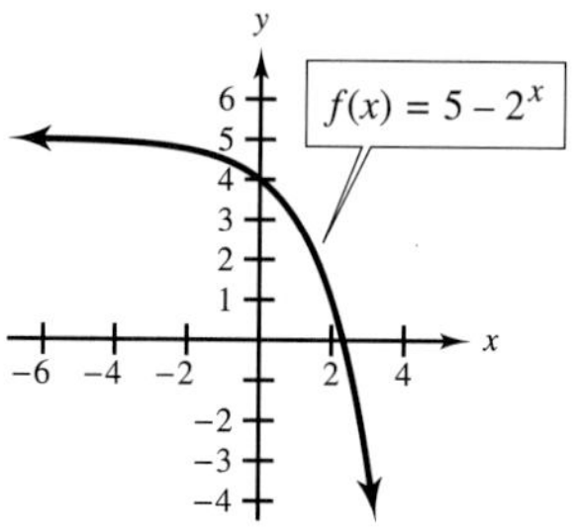

27. $\frac{3}{4}$ **28.** $\frac{5}{2}$ **29.** $\frac{11}{2}$ **30.** $\frac{7}{6}$ **31.** $\frac{10}{17}$ **32.** $-1.5, 1$

33. For growth, the function should be increasing: $b > 1$.

34. \$5250 **35.** $\log_8 4 = \frac{2}{3}$ **36.** $\log_{49} \frac{1}{7} = -\frac{1}{2}$

37. $25^{-1/2} = \frac{1}{5}$ **38.** $\left(\frac{1}{2}\right)^{-4} = 16$

39. The function $y = 1^x$ is not one-to-one. Thus it does not have an inverse.

40. **(a)** 3 **(b)** e **(c)** 10

41. 3.23 **42.** $-\frac{2}{3}$ **43.** -2.5 **44.** 5 **45.** 181.27

46. 158,489.32 **47.** 49 **48.** 26 **49.** 2.5 **50.** 8

51. **(a)** $(1, 0)$ **(b)** None **(c)** $(0, \infty)$ **(d)** **R**

52. Because the described function is a decreasing function, it is more likely to model the incidence of polio.

53. 41 **54.** 43 **55.** 1 **56.** 0

57. Because $\log 49 = 2 \log 7$, we estimate $\log 49$ as $2(0.845) = 1.69$.

58. $\frac{4}{3}D - \frac{7}{3} - \frac{2}{3}C$ **59.** $\frac{3}{4} \log y - \frac{5}{2} \log x - 4 \log z$

60. $\frac{5}{6} \ln x + \frac{2}{3} \ln y - \frac{1}{2} \ln z$ **61.** $\log \sqrt[4]{\frac{(x+4)^3}{2yz}}$

62. $\log \frac{(x+2)^3}{(x+1)^5}$ **63.** 3 **64.** $\frac{\ln(x+5)}{\ln 3}$

65. In the first equation, both sides can be written as powers of 3. In the second equation, 4 cannot easily be written as a power of 3. The equation can be solved with logarithms.

66. Change the equation into exponential form: $x = 2^4$.

67. 1.77 **68.** 0.25 **69.** 5 **70.** -7.13 **71.** 5 **72.** 7

73. 1.19 **74.** 5.78 years **75.** 3542 **76.** \$1216.45

Chapter Test *(page 656)*

1. *[11.1]* $x^2 + x - 2$ **2.** *[11.1]* $x - 2, x \neq -2$

3. *[11.1]* $x^2 + 4x$ **4.** *[11.1]* $\sqrt{x+2}$

5. *[11.2]* $(f \circ g)(x) = (g \circ f)(x) = x$ **6.** *[11.2]* $x^3 - 2$

7. *[11.2]*

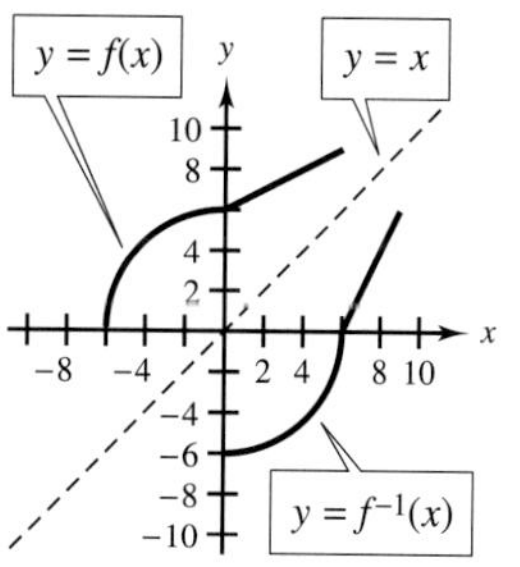

8. *[11.4]* 0.5 **9.** *[11.4]* 1.5 **10.** *[11.3]* 0

11. *[11.3]* 0.25 **12.** *[11.6]* 1.12 **13.** *[11.6]* 1.15

14. *[11.4]* Use the Change of Base Formula to obtain $\log_2 10 = \frac{\log 10}{\log 2} = \frac{1}{\log 2}$. Because you know the value of log 2, you can determine the result.

15. *[11.5]* $2 \log A - \frac{1}{2} \log B$

16. *[11.5]* $\log(x + 1)$ **17.** *[11.5]* $\log_6\left[\frac{x^2(x-1)}{6}\right]$

18. *[11.6]* 2 **19.** *[11.6]* 2 **20.** *[11.6]* 1

21. *[11.6]* $e^{-2} \approx 0.14$ **22.** *[11.6]* 8.48%

23. *[11.5]* **(a)** 7000

(b) As time t increases, demand D decreases. The ad campaign is ineffective.

(c) 10.3 weeks

24. *[11.6]* 15.85 times as intense **25.** *[11.6]* 1.5 hours

Cumulative Test: Chapters 9–11 *(page 657)*

1. [9.1] **(a)** y^4 **(b)** $|y^5|$

2. [9.2] **(a)** 9 **(b)** $\frac{1}{9}$ **(c)** Not a real number **(d)** $-\frac{1}{8}$

3. [9.3] **(a)** $\frac{1}{a^{5/2}b^{2/3}}$ **(b)** $\frac{x^{1/2}}{y}$ **(c)** x^3y^6

4. [9.4] **(a)** $\frac{x^2}{2y}$ **(b)** $\sqrt[5]{6x^3y}$ **(c)** $\frac{x}{y}$

5. [9.5] **(a)** $2x^4\sqrt{5x}$ **(b)** $y\sqrt{y}$

6. [9.6] **(a)** $x\sqrt{3x}$ **(b)** $a + 2\sqrt{2a} + 2$ **7.** [9.6] 4

8. [9.7] **(a)** $\frac{5\sqrt[3]{4}}{2}$ **(b)** $\frac{x\sqrt{2} - \sqrt{xy}}{2x - y}$

9. [9.8] **(a)** 1 **(b)** $-12, 15$

10. [9.9] **(a)** $-i$ **(b)** $8 - 6i$ **(c)** $\frac{1+2i}{5}$

11. [*10.3*] Methods for solving a quadratic equation include factoring, using the Square Root Property, and using the Quadratic Formula. The Quadratic Formula always works.

12. [*10.3*] $-2.09, 0.29$ **13.** [*10.5*] $-1, 7$

14. [*10.3*] If the equation has imaginary number solutions, then the discriminant is negative. Thus $b^2 - 4ac < 0$, and so $b^2 < 4ac$.

15. [*10.4*] 9 **16.** [*10.5*] 8 mph

17. [*10.7*] **(a)** $(-\infty, -6) \cup (4, \infty)$ **(b)** $\emptyset$ **(c)** **R**

18. [*10.6*] **(a)** $c = 3$ **(b)** $a < 0$ **(c)** $a = -3$

19. [*10.6*] **(a)** 4 **(b)** 16

20. [*11.1*] **(a)** 0 **(b)** 0 **(c)** -16

21. [*11.1*] **(a)** 1 **(b)** Not defined

22. [*11.2*] $f^{-1}(x) = \dfrac{3x}{x - 1}$

23. [*11.3*] Both have a y-intercept of $(0, 1)$ and no x-intercept. The graph of f rises and the graph of g falls from left to right.

24. [*11.4*] **(a)** 5 **(b)** 10 **(c)** $\sqrt{7}$

25. [*11.5*] **(a)** $2 \ln x + \ln y - \frac{1}{2} \ln z$ **(b)** $\log \dfrac{(x-1)^3}{x+1}$

26. [*11.6*] **(a)** $9, -3$ **(b)** 0.46 **(c)** -0.69

27. [*11.6*] 6%

Appendix *(page 666)*

1. $r = 4$

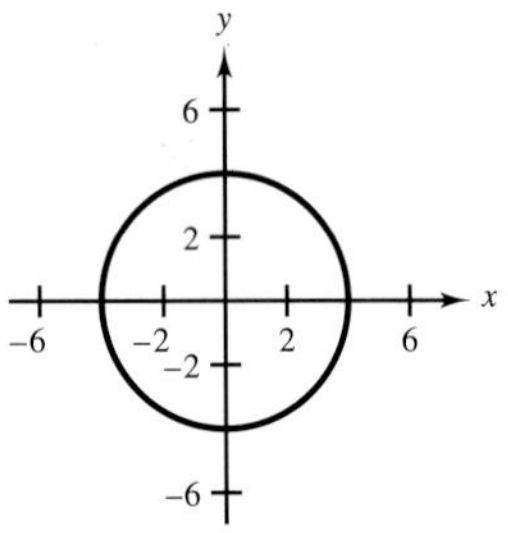

2. $r = 5$

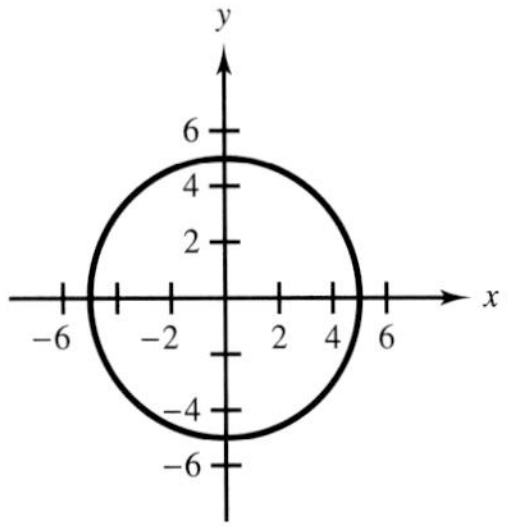

3. $r = \sqrt{10} \approx 3.16$

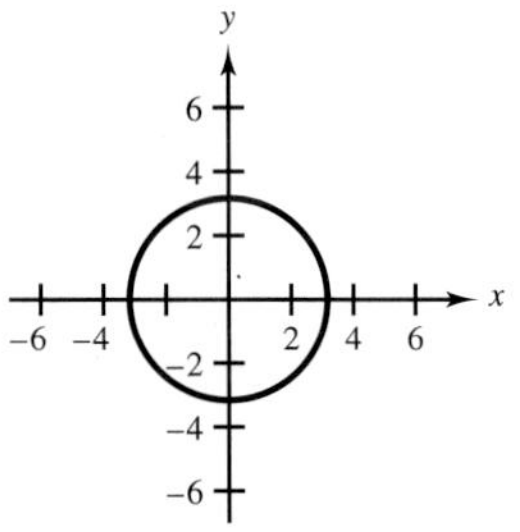

4. $r = 2\sqrt{2} \approx 2.83$

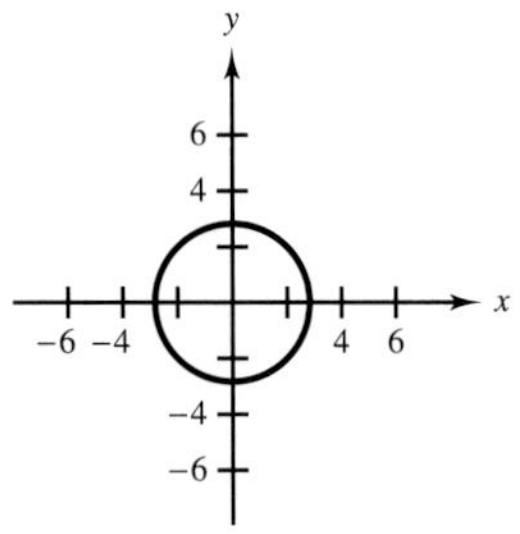

5. $x^2 + y^2 = 25$ **6.** $x^2 + y^2 = 34$ **7.** $x^2 + y^2 = 49$

8. $x^2 + y^2 = 40$

9. $F\left(0, \frac{1}{4}\right)$; $y = -\frac{1}{4}$

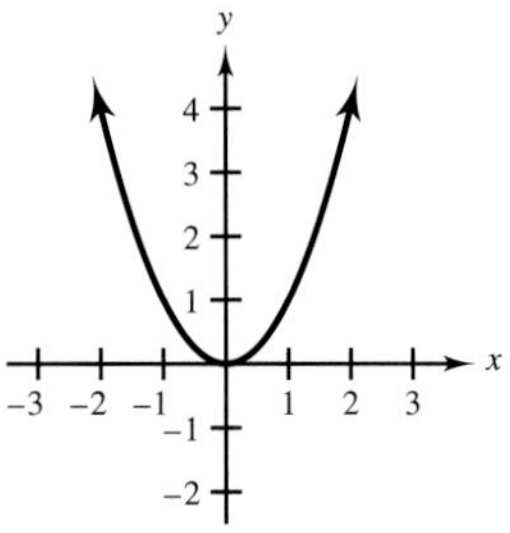

10. $F(0, -1)$; $y = 1$

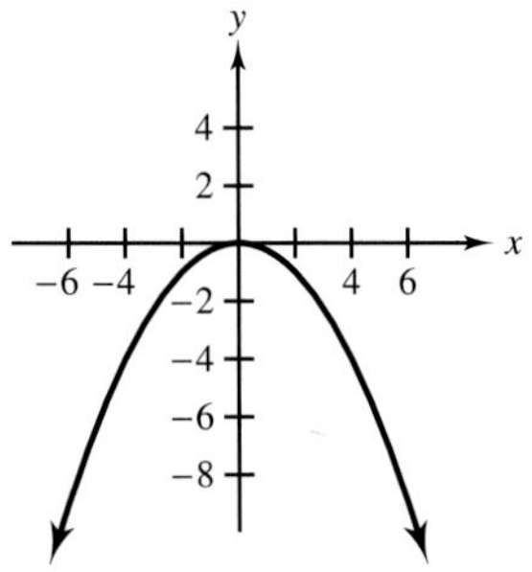

11. $F\left(0, -\frac{1}{2}\right)$; $y = \frac{1}{2}$

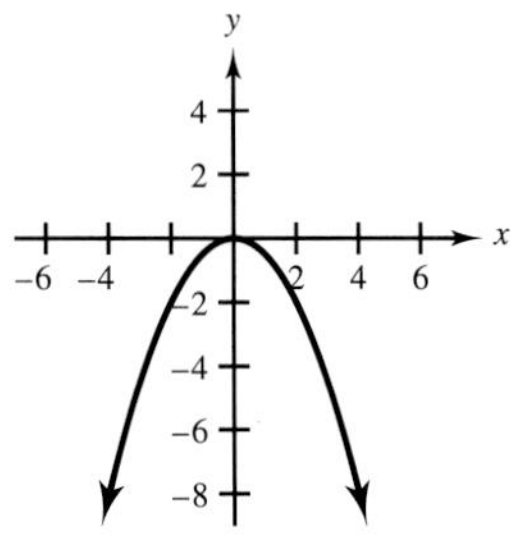

12. $F\left(0, \frac{1}{12}\right); y = -\frac{1}{12}$

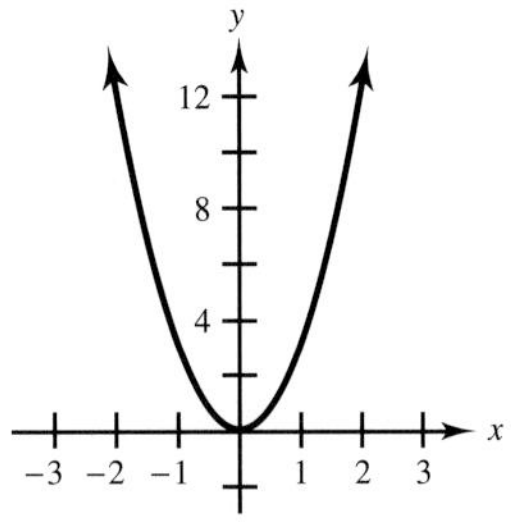

13. $F(-5, 0); x = 5$

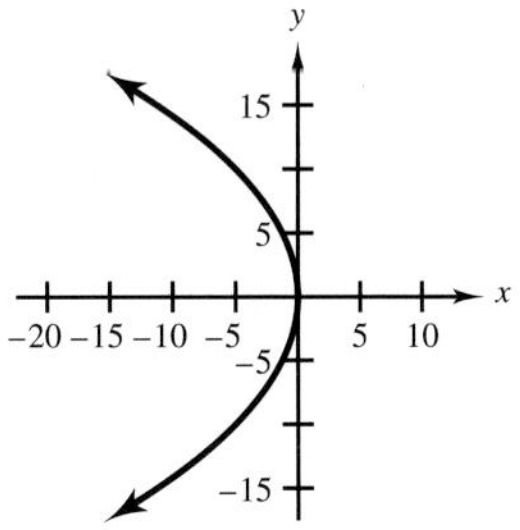

14. $F\left(\frac{3}{4}, 0\right); x = -\frac{3}{4}$

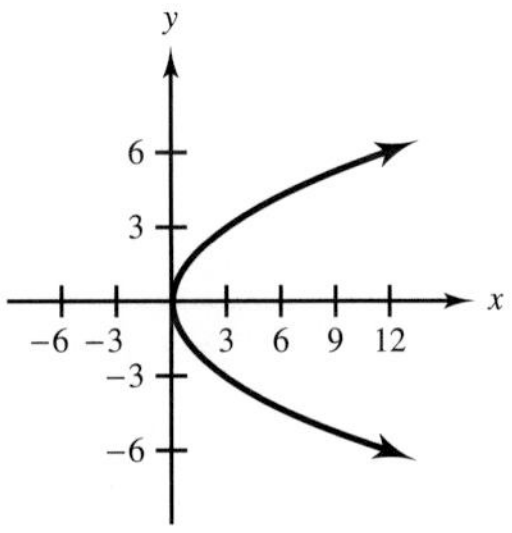

15. $F\left(\frac{1}{8}, 0\right); x = -\frac{1}{8}$

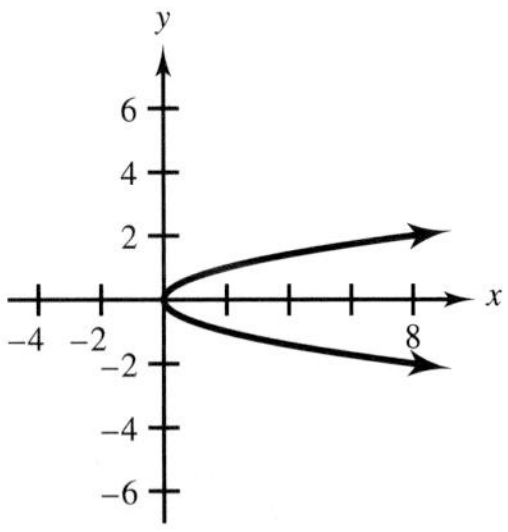

16. $F\left(-\frac{1}{4}, 0\right); x = \frac{1}{4}$

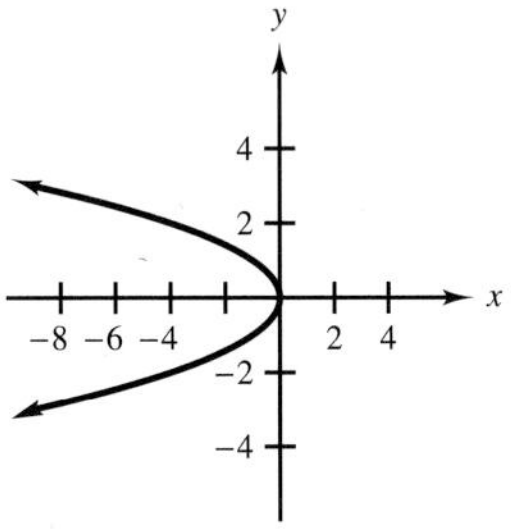

17. $x^2 = 16y$ **18.** $x^2 = -8y$ **19.** $y^2 = -24x$
20. $y^2 = 2x$ **21.** $x^2 = 8y$ **22.** $y^2 = -12x$

23. $x^2 = -12y$ **24.** $y^2 = \frac{5}{2}x$

25. $(\pm 2, 0), (0, \pm 3)$

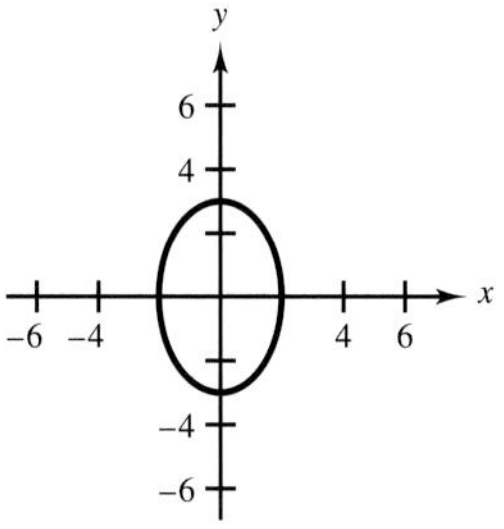

26. $(\pm 4, 0), (0, \pm 2)$

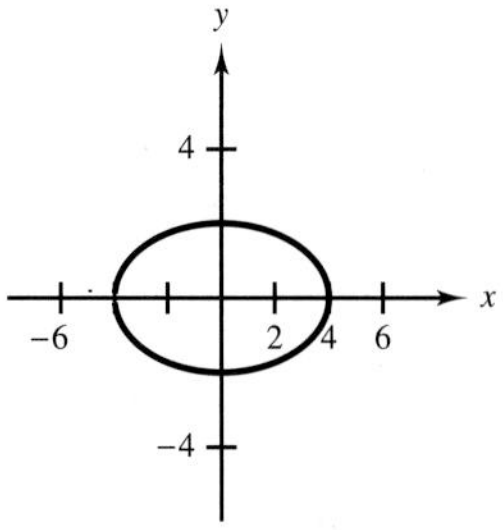

27. $(\pm 1, 0), (0, \pm 6)$

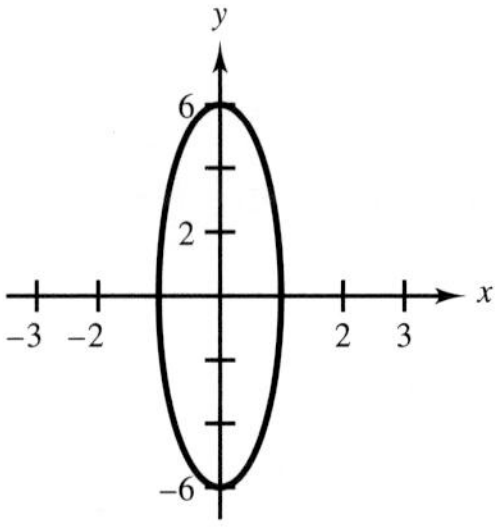

28. $\left(\pm\sqrt{10}, 0\right), (0, \pm 5)$

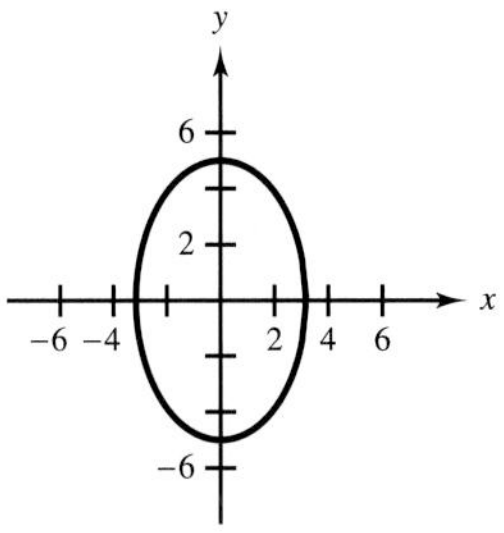

29. $(\pm 1, 0), (0, \pm 5)$

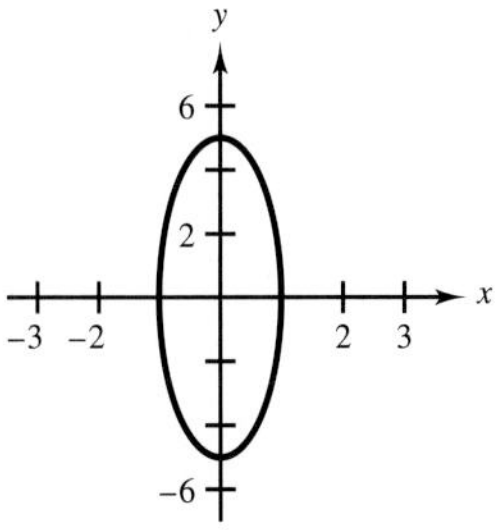

30. $(\pm 2, 0), \left(0, \pm\sqrt{3}\right)$

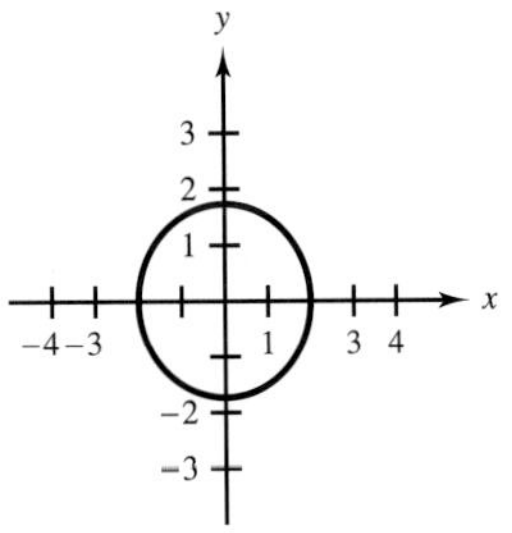

31. $(\pm 2, 0), (0, \pm 4)$

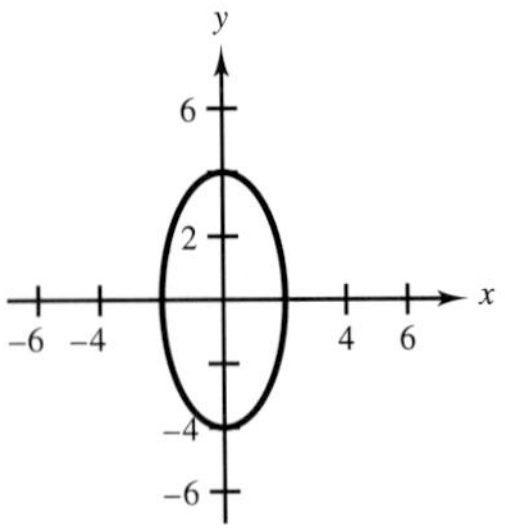

32. $(\pm 3, 0), (0, \pm 1)$

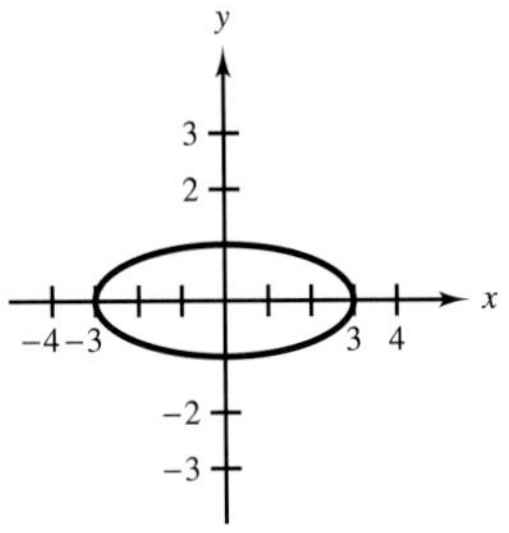

33. $x^2 + \dfrac{y^2}{9} = 1$ **34.** $\dfrac{x^2}{16} + \dfrac{y^2}{9} = 1$ **35.** $\dfrac{x^2}{16} + \dfrac{y^2}{7} = 1$

36. $\dfrac{x^2}{16} + \dfrac{y^2}{25} = 1$ **37.** $\dfrac{x^2}{25} + \dfrac{y^2}{9} = 1$ **38.** $\dfrac{x^2}{11} + \dfrac{y^2}{36} = 1$

39. $\dfrac{x^2}{49} + \dfrac{y^2}{36} = 1$ **40.** $\dfrac{x^2}{25} + \dfrac{y^2}{16} = 1$

41. $(\pm 3, 0); y = \pm\frac{1}{3}x$

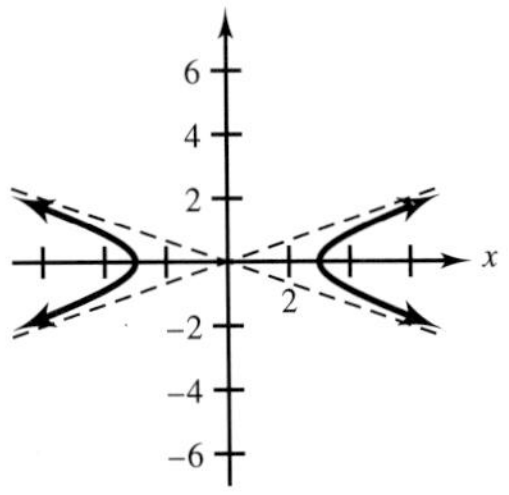

42. $(\pm 4, 0); y = \pm\frac{5}{4}x$

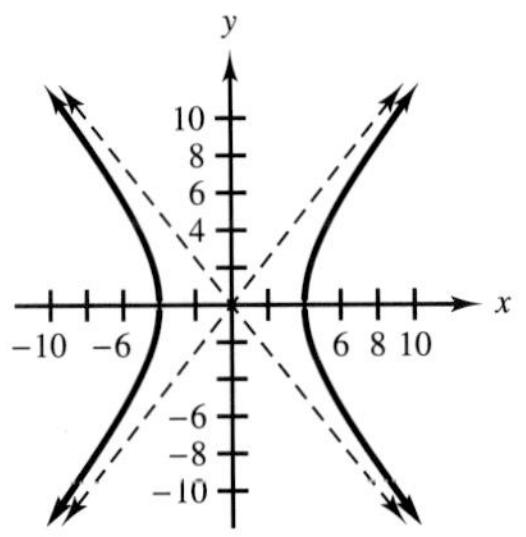

43. $(0, \pm 6); y = \pm x$

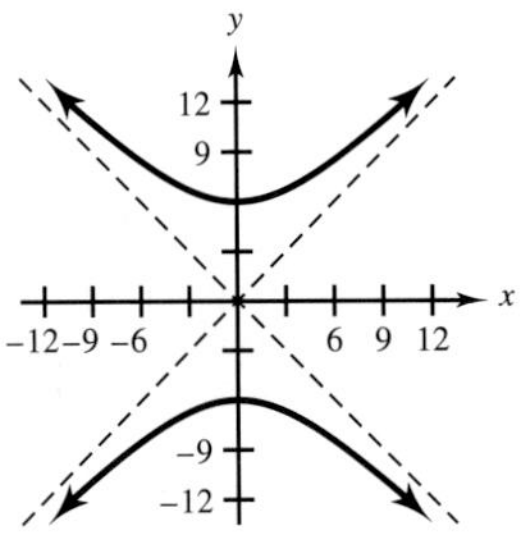

44. $(\pm 2, 0); y = \pm 2x$

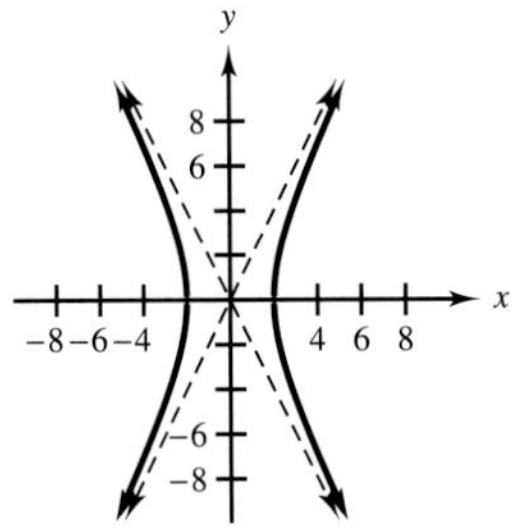

45. $\left(\pm\frac{2}{3}, 0\right); y = \pm\frac{3}{2}x$

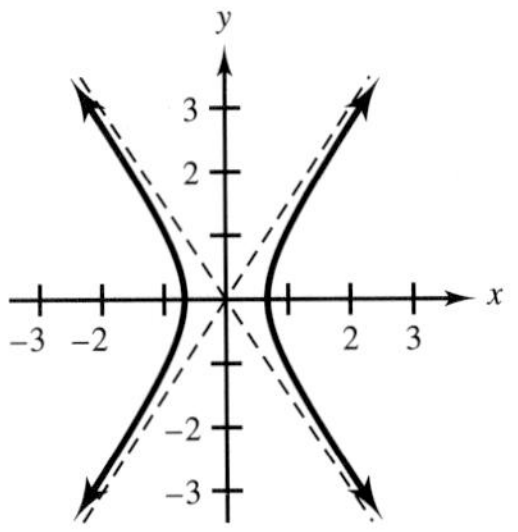

46. $\left(0, \pm 2\sqrt{2}\right); y = \pm 2\sqrt{2}x$

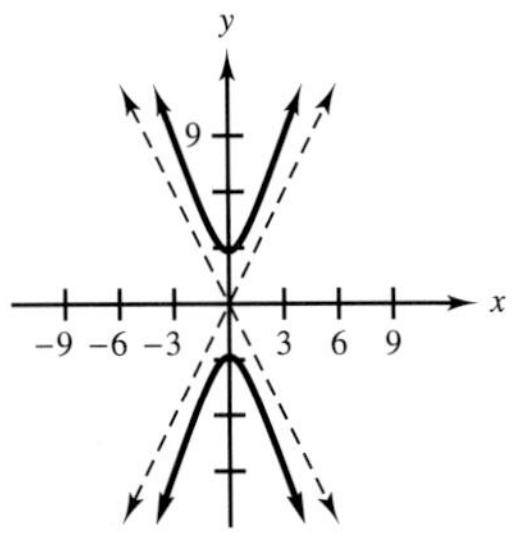

47. $(0, \pm 3)$; $y = \pm\frac{3}{2}x$

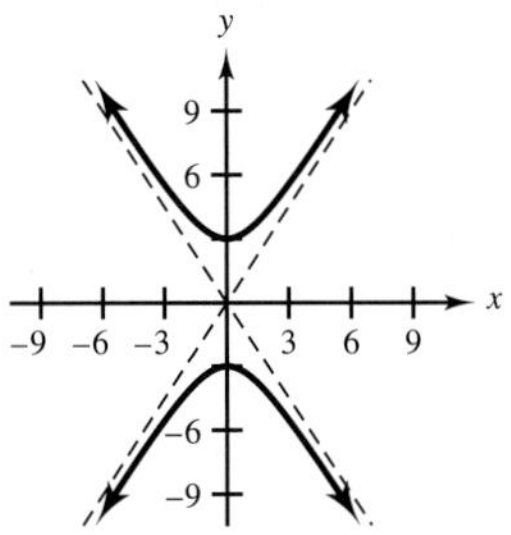

48. $(\pm 3, 0)$; $y = \pm 2x$

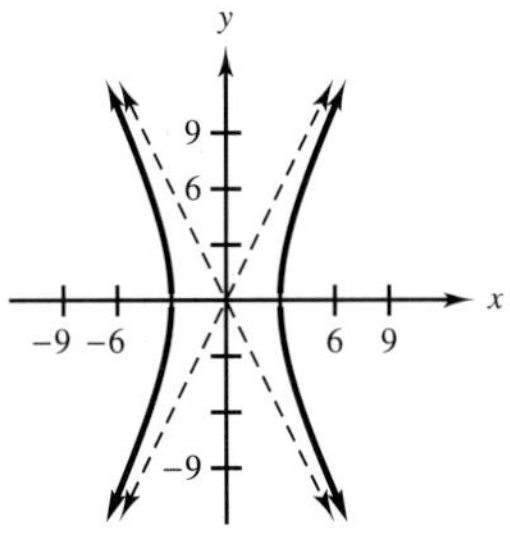

49. $\dfrac{x^2}{16} - \dfrac{y^2}{9} = 1$ **50.** $\dfrac{y^2}{4} - \dfrac{x^2}{5} = 1$ **51.** $\dfrac{y^2}{9} - \dfrac{x^2}{16} = 1$

52. $\dfrac{x^2}{25} - \dfrac{y^2}{100} = 1$ **53.** $\dfrac{x^2}{36} - \dfrac{y^2}{16} = 1$ **54.** $\dfrac{y^2}{4} - \dfrac{x^2}{25} = 1$

Index of Chapter Projects

Index of Data Analysis

Index of Modeling with Real Data

Index of Real-Life Applications

General Index